C

ı

Cytokines

Cytokines

edited by

Anthony R. Mire-Sluis *and* Robin Thorpe

Division of Immunobiology,
NIBSC, Blanche Lane,
South Mimms, Herts, EN6 3QG, UK

ACADEMIC PRESS

San Diego   London   Boston
New York   Sydney   Tokyo   Toronto

Academic Press Inc.
525 B Street, Suite 1900, San Diego, California 92101-4495, USA
http://www.apnet.com

Academic Press Limited
24–28 Oval Road, London NW1 7DX, UK
http://www.hbuk.co.uk/ap/

ISBN 0-12-498340-5

A catalogue record for this book is available from the British Library

Typeset by Phoenix Photosetting, Chatham, Kent
Printed in Great Britain by The Bath Press, Bath, Avon
98 99 00 01 02 03 BP 9 8 7 6 5 4 3 2 1

Contents

3. Interleukin-3   35

Hanna Harant *and* Ivan J.D. Lindley

4. Interleukin-4   53

Anthony R. Mire-Sluis

5. Interleukin-5   69

Colin J. Sanderson, Stephane Karlen, Sigrid Cornelis, Geert Plaetinck, Jan Tavernier *and* Rene Devos

10. Interleukin-10 151

René de Waal Malefyt

11. Interleukin-11 169

Paul F. Schendel *and* Katherine J. Turner

12. Interleukin-12 183

Richard Chizzonite, Ueli Gubler, Jeanne Magram *and* Alvin S. Stern

25.   Interferons Alpha, Beta, and Omega   361

Anthony Meager

26.   Interferon-Gamma   391

Edward De Maeyer *and* Jaqueline De Maeyer-Guignard

27.   Oncostatin M   401

Mohammed Shoyab, Najma Malik *and* Philip M. Wallace

28. *Transforming Growth Factor β₁* 415

Francis W. Ruscetti, Maria C. Birchenall-Roberts, John M. McPherson *and* Robert H. Wiltrout

29. *RANTES* 433

Peter J. Nelson, James M. Pattison *and* Alan M. Krensky

30. NAP-2/ENA-78 449

Alfred Walz

31. Macrophage Inflammatory Protein 1-α 467

Robert J.B. Nibbs, Gerard J. Graham *and* Ian B. Pragnell

32. Monocyte Chemotactic Proteins 1, 2 and 3 489

Paul Proost, Anja Wuyts, Ghislain Opdenakker *and* Jo Van Damme

33. GRO/MGSA 507

Stephen Haskill *and* Susanne Becker

Summary Tables

Contributors

Claudia Ballaun
Institute of Applied Microbiology,
Nussdorfer Laende 11,
A-1190 Vienna,
Austria

Susanne Becker
US Environmental Protection Agency,
National Health and Environmental Effects
Research Laboratory,
Research Triangle Park,
NC 27211,
USA

Rudi Beyaert
Laboratory of Molecular Biology,
Flanders Interuniversity Institute for Biotechnology and
University of Ghent,
K.L. Ledeganckstraat 35,
B-9000 Ghent,
Belgium

Maria C. Birchenall-Roberts
Biological Carcinogenisis Development Program SAIC,
Frederick, MD,
USA

Robin E. Callard
Immunobiology Unit,
Institute of Child Health,
30 Guildford Street,
London WC1N 1EH,
UK

Robin Campbell
Clinical Development,
Amgen Inc.,
Thousand Oaks, CA 91320,
USA

Richard Chizzonite
Dept of Inflammation/Autoimmune Diseases,
Hoffmann La Roche Inc.,
Nutley, NJ 07110,
USA

Francesco Colotta
Department of Immunology and Microbiology,
Preclinical Research Immunology/Oncology,
Pharmacia Research Center,
Nerviano (MI),
Italy

Sigrid Cornelis
Laboratory Molecular Biology,
K. L. Ledeganckstraat 35,
B-9000 Ghent,
Belgium

Edward De Maeyer
CNRS – UMR 177,
Institut Curie,
Bât. 110,
Centre Universitaire,
91405 Orsay,
France

Jaqueline De Maeyer-Guignard
CNRS – UMR 177,
Institut Curie,
Bât. 110,
Centre Universitaire,
91405 Orsay,
France

Stephen Devereux
Department of Haematology,
University College London Medical School,
98 Chenies Mews,
London WC1E 6HX,
UK

Rene Devos
Roche Research Ghent,
Jozef Plateaustraat 22,
B-9000 Ghent,
Belgium

T. Michael Dexter
Paterson Institute for Cancer Research,
Christie Hospital/HS Trust,
Wilmslow Road,
Manchester M20 48X,
UK

Andrew J. Dorner
Genetics Institute Inc.,
87 Cambridgepark Drive,
Cambridge, MA 02140,
USA

Walter Fiers
Laboratory of Molecular Biology,
Flanders Interuniversity Institute for Biotechnology and
University of Ghent,
K.L. Ledeganckstraat 35,
B-9000 Ghent,
Belgium

MaryAnn Foote
Clinical Development,
Amgen Inc.,
Thousand Oaks, CA 91320,
USA

Marc B. Garnick
Pharmaceutical Peptides Inc,
Cambridge, MA 02139,
USA

Pietro Ghezzi
Istituto di Ricerche Farmacologiche "Mario Negri",
via Eritrea 62,
Milano,
Italy

Brian Gliniak
Immunex Research and Development Corporation,
51 University Street,
Seattle, WA 98101,
USA

Nicholas M. Gough
AMRAD Operations Pty Ltd,
17–27 Cotham Road,
Kew, 3101 Victoria,
Australia
and The Cooperative Research Centre for Cellular
Growth Factors

Gerard J. Graham
Cancer Research Campaign Laboratories,
Beatson Institute for Cancer Research,
Garscube Estate,
Bearsden,
Glasgow G61 1BD,
UK

Ueli Gubler
Dept of Inflammation/Autoimmune Diseases,
Hoffmann La Roche Inc.,
Nutley, NJ 07110,
USA

Hanna Harant
Department of Cellular and Molecular Biology,
Novartis Forschungsinstitut,
PO Box 80,
Brunner Strasse 59,
A-1235 Vienna,
Austria

Helmut Hasibeder
Professional Services,
Amgen Inc.,
Thousand Oaks, CA 91320,
USA

Stephen Haskill
Department of Obstetrics and Gynecology,
Microbiology and Immunology,
Lineberger Cancer Research Center,
University of North Carolina at Chapel Hill,
Chapel Hill, NC 27599,
USA

Douglas J. Hilton
The Walter and Eliza Hall Institute for Medical Research,
Parkville,
3050 Victoria,
Australia
and The Cooperative Research Centre for Cellular
Growth Factors

Stephane Karlen
Institute for Child Health Research,
PO Box 855,
West Perth,
Western Australia 6872

James C. Keith Jr
Genetics Institute, Inc.,
87 Cambridgepark Drive,
Cambridge, MA 02140,
USA

Alan M. Krensky
Division of Immunology and Transplantation Biology,
Department of Pediatrics,
Stanford University,
Stanford, CA,
USA

David C. Linch
Department of Haematology,
University College London Medical School,
98 Chenies Mews,
London WC1E 6HX,
UK

Ivan J. D. Lindley
Department of Cellular and Molecular Biology,
Novartis Research Institute,
PO Box 80,
Brunner Strasse 59,
A-1235 Vienna,
Austria

Stewart Lyman
Immunex Research and Development Corporation,
51 University Street,
Seattle, WA 98101,
USA

Jeanne Magram
Dept of Inflammation/Autoimmune Diseases,
Hoffmann La Roche Inc.,
Nutley, NJ 07110,
USA

Najma Malik
Bristol-Myers Squibb Pharmaceutical Research Institute,
3005 First Avenue,
Seattle, WA 98121,
USA

Alberto Mantovani
Istituto di Ricerche Farmacologiche "Mario Negri",
via Eritrea 62,
Milano,
Italy

David J. Matthews
Immunobiology Unit,
Institute of Child Health,
30 Guildford Street,
London WC1N 1EH,
UK

John McPherson
Genzyme Tissue Repair,
Framingham, MA,
USA

Anthony Meager
Division of Immunobiology,
NIBSC,
Blanche Lane,
South Mimms,
Herts EN6 3QG,
UK

Anthony Mire-Sluis
Division of Immunobiology,
NIBSC,
Blanche Lane,
South Mimms,
Herts EN6 3QG,
UK

George Morstyn
Clinical Development,
Amgen Inc.,
Thousand Oaks, CA 91320,
USA

David H. Munn
Medical College of Georgia,
Augusta, GA 30912,
USA

Anthony E. Namen
Immunex Research and Development Corporation,
51 University Street,
Seattle, WA 98101,
USA

Peter J. Nelson
Clinical Biochemistry Group,
Department of Internal Medicine,
Klinikum Innenstadt,
Ludwig-Maximilian-University,
Munich,
Germany

Robert J.B. Nibbs
Cancer Research Campaign Laboratories,
Beatson Institute for Cancer Research,
Garscube Estate,
Bearsden,
Glasgow G61 1BD,
UK

Ghislain Opdenakker
Rega Institute For Medical Research,
Laboratory of Molecular Immunology,
University of Leuven,
Minderbroedersstraat 10,
B-3000 Leuven,
Belgium

James M. Pattison
Guy's Hospital,
St. Thomas's Street,
London SE1 9RT,
UK

Geert Plaetinck
Roche Research Ghent,
Jozef Plateaustraat 22,
B-9000 Ghent,
Belgium

Ian B. Pragnell
Cancer Research Campaign Laboratories,
Beatson Institute for Cancer Research,
Garscube Estate,
Bearsden,
Glasgow G61 1BD,
UK

Paul Proost
Rega Institute For Medical Research,
Laboratory of Molecular Immunology,
University of Leuven,
Minderbroedersstraat 10,
B-3000 Leuven,
Belgium

Jean-Christophe Renauld
Ludwig Institute for Cancer Research,
Brussels Branch and Experimental Medicine Unit,
Catholic University of Louvain,
74 Avenue Hippocrate,
B-1200 Brussels,
Belgium

Carl D. Richards
Department of Pathology,
McMaster University,
1200 Main Street West,
Hamilton,
Ontario,
Canada L8N 3Z5

Francis W. Ruscetti
Laboratory of Leukocyte Biology,
NCI Frederick Cancer Research and Development Center,
Frederick, MD,
USA

Colin J. Sanderson
Institute for Child Health Research,
PO Box 855,
West Perth, WA 6872,
Australia

Robert G. Schaub
Genetics Institute, Inc.,
87 Cambridgepark Drive,
Cambridge, MA 02140,
USA

Paul F. Schendel
Genetics Institute, Inc.,
87 Cambridgepark Drive,
Cambridge, MA 02140,
USA

Matthew L. Sherman
Genetics Institute, Inc.,
87 Cambridgepark Drive,
Cambridge, MA 02140,
USA

Mohammed Shoyab
Department of Pathology,
F-167 Health Sciences Building,
SC-38,
University of Washington,
Seattle, WA 98195,
USA

Alvin S. Stern
Dept of Inflammation/Autoimmune Diseases,
Hoffmann La Roche Inc.,
Nutley, NJ 07110,
USA

Joseph P. Sypek
Genetics Institute, Inc.,
87 Cambridgepark Drive,
Cambridge, MA 02140,
USA

Jan Tavernier
Flanders Interuniversity Institute for Biotechnology,
University of Ghent,
K. L. Ledeganckstraat 35,
B-9000 Ghent,
Belgium

Robin Thorpe
Division of Immunobiology,
NIBSC,
Blanche Lane,
South Mimms,
Herts EN6 3QG,
UK

Katherine J. Turner
Genetics Institute, Inc.,
87 Cambridgepark Drive,
Cambridge, MA 02140,
USA

Jo Van Damme
Rega Institute For Medical Research,
Laboratory of Molecular Immunology,
University of Leuven,
Minderbroedersstraat 10,
B-3000 Leuven,
Belgium

René de Waal Malefyt
Department of Human Immunology,
DNAX Research Institute,
901 California Ave,
Palo Alto, CA 94304,
USA

Philip M. Wallace
Xcyte Therapies,
2203 Airport Way South,
Seattle, WA 98134,
USA

Alfred Walz
Theodor Kocher Institute,
University of Bern,
Freiestrasse 1,
CH-3012 Bern,
Switzerland

Robert H. Wiltrout
Experimental Immunology,
NCI-Frederick Cancer Research and Development
Center,
Frederick, MD,
USA

Anja Wuyts
Rega Institute For Medical Research,
Laboratory of Molecular Immunology,
University of Leuven,
Minderbroedersstraat 10,
B-3000 Leuven,
Belgium

Preface

The cytokines are probably the most important biologically active group of molecules to be identified since the discovery of the classical endocrine hormones. Most of the cytokines are small to medium sized proteins or glycoproteins which mediate potent biological affects on most cell types. Originally identified as being important in inflammatory processes, the development and maintenance of immune respones and for haematopoiesis, it is now becoming evident that cytokines are involved, at least to some extent, in most if not all physiological processes. Progress with the identification with new cytokine molecules is particularly fast moving and new molecules are discovered with alarming frequency. Most cytokine molecules show more than one property and many seem able to mediate a vast array of different biological functions. More than one hundred different cytokine molecules have now been identified and when this is considered with the pleiotropic functions mediated by many of these substances, their importance is difficult to overestimate. Cytokine involvement in pathological processes seems likely if the delicate balance of their production and control is disturbed and therapeutic use of cytokines for a broad range of clinical indications is now established.

This book contains a concise description of all the important aspects of cytokine structure, biochemistry and biology. In most cases individual chapters are devoted to single cytokines. However, insufficient accurate information is at present available for interleukin 17 and 18 to allow complete chapters to be written for these molecules. Their properties (as far as these are known to date) are included in the summary tables which are located at the end of the book.

We would like to thank all the authors for their contributions to this book, in addition to Sarah Stafford, Heather Burroughs and Andrew Davies, whose invaluable help with the development of the book is very gratefully received.

Tony and Robin

1. Interleukin 1

Francesco Colotta, Pietro Ghezzi *and* Alberto Mantovani

1. Introduction

Interleukin-1 (IL-1) is the term for two polypeptide mediators (IL-1α and IL-1β) that are among the most potent and multifunctional cell activators described in immunology and cell biology. The spectrum of action of IL-1 encompasses cells of hematopoietic origin, from immature precursors to differentiated leukocytes, vessel wall elements, and cells of mesenchymal, nervous, and epithelial origin (Dinarello, 1991, 1994, 1996). Occupancy of few receptors, perhaps one, per cell is sufficient to elicit cellular responses.

The activity of IL-1 overlaps largely with that of tumor necrosis factor (TNF) and other cytokines. This overlap renders it difficult to trace unequivocally the history of the discovery of IL-1 in the "premolecular" era. Many of the early descriptions of IL-1 activities, for instance that of endogeneous pyrogen, can as well pertain to other cytokines. The identification of lymphocyte-activating factor (LAF) was a landmark because it provided a reliable, easy, but still considerably specific *in vitro* assay for purification and, eventually, cloning of IL-1 (for review see Oppenheim and Gery, 1993). It is ironic that the importance of IL-1 in the generation of T cell immunity is still not clearly and unequivocally defined despite its role in the discovery.

The production and action of IL-1 is regulated by multiple control pathways, some of which are unique to this cytokine (Figure 1.1). This complexity and uniqueness is best represented by the term "IL-1 system" (Colotta *et al.*, 1994a). The IL-1 system consists of the two agonists IL-1α and IL-1β, a specific activation system (IL-1 converting enzyme, ICE), a receptor antagonist (IL-1Ra) produced in different isoforms (Arend, 1993), and two high-affinity surface binding molecules (Colotta *et al.*, 1994a). This chapter will focus largely on IL-1 and its receptors, with a concise summary of the properties of other elements in the system (IL-1Ra, ICE).

Cytokines
ISBN 0–12–498340–5

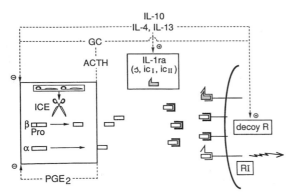

Figure 1.1   An overview of the IL-1 system. ICE, IL-1-converting enzyme; GC, glucocorticoid hormones; Ra, receptor antagonist. The + and – signs indicate stimulation or inhibition of production.

2.   *The Cytokine Gene*

2.1   The IL-1 Gene

In 1984, the human IL-1β (Auron *et al.*, 1984) cDNA was identified and sequenced. Next, additional sequences were obtained from human IL-1 (March *et al.*, 1985). Earlier biochemical studies had identified two forms of IL-1 on the basis of their distinct isoelectric points (Dinarello *et al.*, 1974). That indeed two forms of IL-1 exist was confirmed by the identification of two different cDNAs.

Human IL-1α cDNA is shown in Figure 1.2. The 5′ untranslated region is 59 bp long, followed by a single open reading frame of 813 bp which codes for a precursor protein of 271 amino acids. The sequence does not contain any N-terminal signal peptide of hydrophobic residues, nor does it have an internal long hydrophobic stretch. The coding sequence is followed by a 3′ untranslated region of 1141 bp, followed by the poly(A) tail. The 3′ untranslated sequence of IL-1α, and of IL-1β also, contains several copies of the ATTTA motif which is thought to be involved in shortening transcript stability. Human IL-1α mRNA comprises 2000–2200 nucleotides. Human IL-1β cDNA has a 5′ untranslated region of approximately 70 bp. The coding region is 807 bp long, thus coding for a 269-amino-acid precursor protein. The protein encoded by this sequence does not contain classical leader peptide. IL-1β transcripts are 1800 nucleotides in length. Human IL-1α and IL-1β share only 26% homology at the protein level, but significantly more occurs at the nucleotide level (45%), thus raising the possibility that these genes arise from a duplication event.

2.2   IL-1 Gene Structure

The IL-1α gene is approximately 10 kb in length (Furutani *et al.*, 1986), whereas the IL-1β gene is 7 kb

(Clark *et al.*, 1986). The human (Lafage *et al.*, 1989; Webb *et al.*, 1986) IL-1 genes are located on chromosome 2, in position 2q13, in the same region as IL-1Ra and the type I and type II receptors. Although IL-1α and IL-1β exon sequences are considerably different, the two genes share a highly conserved intron–exon structure (Clark *et al.*, 1986). Moreover, IL-1α and IL-1β genes show considerable homology in the intron sequences, suggesting a regulatory role for these regions in IL-1 expression (see below).

Both genes have 7 exons (Figure 1.3). The first intron of IL-1β contains a highly conserved homopurine tract. Other sequences in the first intron of the human IL-1β gene may exert either positive or negative regulatory activity in transfection experiments. The fourth intron and third intron of, respectively, the human IL-1α and IL-1β genes contain Alu sequences. There is a 46 bp tandem repeat in intron 6 of human IL-1α that could act as an enhancer or a suppressor. Each repeat contains a binding sequence for the transcriptional factor SP-1, an imperfect copy of a viral enhancer element and an inverse complementary copy of the glucocorticoid (GC)-responsive elements. A polymorphism within this intron is generated by a variability in the number of repeats. Also intron 5 of the IL-1β gene contains a sequence which resembles the GC-responsive consensus element (TGTYCT). A role, if any, for these sequences in mediating the GC-induced suppression of IL-1 gene transcription has not been determined. Only one variant IL-1β allele has been described.

2.3   IL-1 Promoter

The IL-1 genes are not expressed in unstimulated blood monocytes, vascular cells (smooth-muscle and endothelial cells), and fibroblasts. The IL-1α promoter, in contrast to the IL-1β promoter, lacks a CAT box and has a very poor TATA box. Transient transfection experiments using upstream sequences of the human IL-1β gene demonstrated that CAT expression is detectable in the human promonocytic cell line THP-1 but not in HeLa cells (Clark *et al.*, 1988). In contrast, a fusion gene containing also the first intron of IL-1β gene was expressed in HeLa cells (Bensi *et al.*, 1990). Moreover, when the first intron is present in the expression construct, only 132 bp of IL-1β promoter is sufficient to promote transcriptional activity.

A phorbol myristate acetate (PMA)-responsive enhancer within the human IL-1β gene was identified between positions –2983 and –2795 upstream from the transcriptional start site (Bensi *et al.*, 1990). This enhancer sequence contains a DNA motif similar to the AP-1-responsive elements. A lypopolysaccharide (LPS)-responsive element within the human IL-1β gene is located between –3757 and –2729. The LPS-inducible element also appeared to mediate PMA and IL-1 responsiveness in monocytes and fibroblasts.

A   Human IL-1α

```
                                    *********
  1401        GCCTACTTAAGACAATTACAAAAGGCGAAGAAGACTGACTCAGGCTT           1447


  1448        AAGCTGCCAGCCAGAGAGGGAGTCATTTCATTGGCGTTTGAGTCAGCA           1498        Exon 1
              AAG


  2163        AAGTCAAGATGGCCAAAGTTCCAGACATGTTTGAAGACCTGAAGA                          Exon 2
              ACTGTTACAG                                                 2217


  3176        TGAAAATGAAGAAGACAGTTCCTCCATTGATCATCTGTCTCTG                            Exon 3
              AATCAG                                                     3224


  4113        AAATCCTTCTATCATGTAAGCTATGGCCCACTCCATGAAGGCTGCA
              TGGATCAATCTGTGTCTCTGAGTATCTCTGAAACCTCTAAAACATC
              CAAGCTTACCTTCAAGGAGAGCATGGTGGTAGTAGCAACCAACGGG             Exon 4
              AAGGTTCTGAAGAAGAGACGGTTGAGTTTAAGCCAATCCATCACT
              GATGATGACCTGGAGGCCATCGCCAATGACTCAGAGGAAG                   4335


  6272        AAATCATCAAGCCTAGGTCATCACCTTTTAGCTTCCTGAGCAATGT
              GAAATACAACTTTATGAGGATCATCAAATACGAATTCATCCTGA               Exon 5
              ATGACGCCCTCAATCAAAGTATAATTCGAGCCAATGATCAGTACC
              TCACGGCTGCTGCATTACATAATCTGGATGAAGCAG                       6442


  7825        TGAAATTTGACATGGGTGCTTATAAGTCATCAAAGGATGATGCT
              AAAATTACCGTGATTCTAAGAATCTCAAAAACTCAATTGTATG                Exon 6
              TGACTGCCCAAGATGAAGACCAACCAGTGCTGCTGAAG                     7949


 10300        GAGATGCCTGAGATACCCAAAACCATCACAGGTAGTGAGACCAA
              CCTCCTCTTCTTCTGGGAAACTCACGGCACTAAGAACTATTTCACA
              TCAGTTGCCCATCCAAACTTGTTTATTGCCACAAAGCAAGACTAC
              TGGGTGTGCTTGGCAGGGGGGGCCACCCTCTATCACTGACTTTCAGAT
              ACTGGAAAACCAGGCGTAGGTCTGGAGTCTCACTTGTCTCACTTGTGCAG
              TGTTGACAGTTCATATGTACCATGTACATGAAGAAGCTAAATCCTTTACT
              GTTAGTCATTTGCTGAGCATGTACTGAGCCTTGTAATTCTAAATGAATGT
              TTACACTCTTTGTAAGAGTGGAACCAACACTAACATATAATGTTGTTATT
              TAAAGAACACCCTATATTTTGCATAGTACCAATCATTTTAATTATTATTC
              TTCATAACAATTTTAGGAGGACCAGAGCTACTGACTATGGCTACCAAAAA
              GACTCTACCCATATTACAGATGGGCAAATTAAGGCATAAGAAAACTAAG
              AAATATGCACAATAGCAGTCGAAACAAGAAGCCACAGACCTAGGATTT
              CATGATTTCATTTCAACTGTTTGCCTTCTGCTTTTAAGTTGCTGATGAAC
              TCTTAATCAAATAGCATAAGTTTCTGGGACCTCAGTTTTATCATTTTCAA
              AATGGAGGGA ATAATACCTAAGCCTTCCTGCCGCAACAGTTTTTTATGCT       Exon 7
              AATCAGGGAGGTCATTTTGGTAAAATACTTCTCGAAGCCGAGCCTCAAGA
              TGAAGGCAAAGCACGAAATGTTATTTTTTAATTATTATTTATATATGTAT
              TTATAAATATATTTAAGATAATTATAATATACTATATTTATGGGAACCC
              CTTCATCCTCT GAGTGTGACCAGGCATCCTCCACAATAGCAGACAGTGTTTT
              CTGGGATAAGTAAGTTTGATTTCATTAATACAGGGCATTTTGGTCCAAGTT
              GTGCTTATCCCATAGCCAGGAAACTCTGCATTCTAGTACTTGGGAGACCTGT
              AATCATATAATAAATGTACATTAATTACCTTGAGCCAGTAATTGGTCC
              GATCTTTGACTCTTTTGCCATTAAACTTACCTGGGCATTCTTGTTTCATTCA
              ATTCCACCTGCAATCAAGTCCTACAAGCTAAAATTAGATGAACTCAACTT
              TGACAACCATGAGACCACTGTTATCAAAACTTTCTTTTCTGGAATGTAAT
              CAATGTTTCTTCTAGGTTCTAAAAAATTGTGATCAGACCATAATGTTACAT
              TATTATCAACAATAGTGATTGATAGAGTGTTATCAGTCATAACTAAATA
              AAGCTTGCAACAAAATTCTCTG                                     11654
```

Figure 1.2   Gene sequences of IL-1α and IL-1β. Sequences of (a) IL-1α and (b) IL-1β are reported. TATA box is indicated by asterisks. Untranslated regions are underlined.

B   HUMAN IL-1β


```
                        ***********
1891    GCTTTTGAAAGCTATAAAAACAGCGAGGGAGAAACTGGCAGAT        1933


1934    ACCAACCTCTTCGAGGCACAAGGCACAACAGGCTGCTCTGGGATTCT              Exon 1
        CTTCAGCCAATCTTCATTGCTCAA                          2004


2465    GTGTCTGAAGCAGCCATGGCAGAAGTACCTGAGCTCGCCAGTG                 Exon 2
        AAATGATGGCTTATTACAG                               2526


3091    TGGCAATGAGGATGACTTGTTCTTTGAAGCTGATGGCCCTAA                  Exon 3
        ACAGATGAAG                                        3142


5124    TGCTCCTTCCAGGACCTGGACCTCTGCCCTCTGGATGGCGGCATC
        CAGCTACGAATCTCCGACCACCACTACAGCAAGGGCTTCAGGC
        AGGCCGCGTCAGTTGTTGTGGCCATGGACAAGCTGAGGAAGAT                 Exon 4
        GCTGGTTCCCTGCCCACAGACCTTCCAGGAGAATGACCTGAGCA
        CCTTCTTTCCCTTCATCTTTGAAGAAG                       5325


5873    AACCTATCTTCTTCGACACATGGGATAACGAGGCTTATGTGCA
        CGATGCACCTGTACGATCACTGAACTGCACGCTCCGGGACTCAC               Exon 5
        AGCAAAAAAGCTTGGTGATGTCTGGTCCATATGAACTGAAAG
        CTCTCCACCTCCAGGGACAGGATATGGAGCAACAAG              6037


7274    TGGTGTTCTCCATGTCCTTTGTACAAGGAGAAGAAAGTAATGA
        CAAAATACCTGTGGCCTTGGGCCTCAAGGAAAAGAATCTGTAC                Exon 6
        CTGTCCTGCGTGTTGAAAGATGATAAGCCCACTCTACAGCTGG
        AG                                                7404


8126    AGTGTAGATCCCAAAAATTACCCAAAGAAGAAGATGGAAA
        AGCGATTTGTCTTCAACAAGATAGAAATCAATAACAAGCTGG
        AATTTGAGTCTGCCCAGTTCCCCAACTGGTACATCAGCACCTCT
        CAAGCAGAAAACATGCCCGTCTTCCTGGGAGGGACCAAAGGCG
        GCCAGGATATAACTGACTTCACCATGCAATTTGTGTCTTCCTAA
        AGAGAGCTGTACCCAGAGAGTCCTGTGCTGAATGTGGACTCAATCCCTA
        GGGCTGGCAGAAAGGGAACAGAAAGGTTTTTTGAGTACGGCTATAGCCT
        GGACTTTCCTGTTGTCTACACCAATGCCCAACTGCCTGCCTTAGGGTAGT           Exon 7
        GCTAAGAGGATCTCCTGTCCATCAGCCAGGACAGTCAGCTCTCTCCTTTC
        AGGGCCAATCCCCAGCCCTTTTGTTGAGCCAGGCCTCTCTCACCTCTCCTA
        CTCACTTAAAGCCCGCCTGACAGAAACCACGGCCACATTTGGTTCTAAG
        AAACCCTCTGTCATTCGCTCCCACATTCTGATGAGCAACCGCTTCCCTAT

        TTATTTATTTATTTGTTTGTTTGTTTTATTCATTGGTCTAATTTATTCA
        AAGGGGGCAAGAAGTAGCAGTGTCTGTAAAAGAGCCTAGTTTTTAAT
        AGCTATGGAATCAATTCAATTTGGACTGGTGTGCTCTC TTAAATCAA
        GTCCTTTAATTAAGACTGAAAATATATAAGCTCAGATTATTTAAAT
        GGGAATATTTATAAATGAGCAAATATCATACTGTTCAATGGTTCTGA
        AATAAACTTCTCTGAAG                                 8947
```

Figure 1.2   Gene sequences of IL-1α and IL-1β. Sequences of (a) IL-1α and (b) IL-1β are reported. TATA box is indicated by asterisks. Untranslated regions are underlined.

2.4   REGULATION OF IL-1 GENE EXPRESSION

A variety of external stimuli can activate transcription of IL-1 genes. Among these are endotoxins from Gram-negative and exotoxins from Gram-positive bacteria, phorbol esters, calcium ionophores (Yamato *et al.*, 1989), UV light (Kupper *et al.*, 1987), T cells (Wasik *et al.*, 1988), complement components (Schindler *et al.*, 1990a), and adhesion to extracellular matrix molecules (Haskill *et al.*, 1988). IL-1β induction by LPS does not require intact protein synthesis, thus suggesting the activation of pre-existing transcriptional factors (Turner *et al.*, 1989). Among cytokines, IL-1 gene transcription

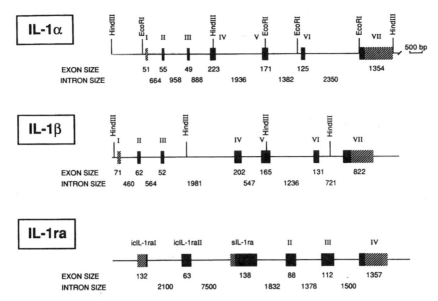

Figure 1.3   Gene structure of IL-1α, IL-1β, and IL-1Ra. Exon and intron sizes are indicated. Selected restriction sites are also reported.

is induced by IL-1 itself, TNF (Dinarello *et al.*, 1987; Warner *et al.*, 1987a,b), and IL-2 (Kovacs *et al.*, 1989; Numerof *et al.*, 1990). Certain cytokines synergize with LPS in inducing IL-1 gene transcription, as demonstrated for IFN-γ (Haq *et al.*, 1985; Ucla *et al.*, 1990) and GM-CSF (Frendl *et al.*, 1990). Histamine alone did not induce protein synthesis or IL-1β transcripts in human peripheral blood mononuclear cells (PBMC), but enhanced (2- to 3-fold) IL-1α-induced IL-1β transcripts and protein via H_2 receptors (Vannier and Dinarello, 1993). In contrast, histamine reduced LPS-induced IL-1 expression in monocytes isolated by adherence to plastic (Dohlsten *et al.*, 1988). Early gene products from cytomegalovirus are able to induce CAT activity of a fusion gene including IL-1 gene sequence spanning from −1097 to +14 of human IL-1β promoter (Iwamoto *et al.*, 1990). LPS-induced transcripts in mononuclear phagocytes peak at 4 to 6 h after stimulation.

IL-1 gene transcription is also under negative control. The T_H2-derived cytokines IL-4 (Essner *et al.*, 1989; Hart *et al.*, 1989), IL-10 (Moore *et al.*, 1990), and IL-13 (Minty *et al.*, 1993; de Waal Malefyt *et al.*, 1993) suppress IL-1 expression in LPS-treated monocytes. Also IL-6 downregulates IL-1 expression induced by LPS (Schindler *et al.*, 1990b). IL-4 reduces IL-1 expression also when IL-1 is induced synergistically by IFN-γ and LPS (Donnelly *et al.*, 1990). In contrast, IFN-γ downregulates IL-1β and IL-1α induction by IL-1 (Ghezzi and Dinarello, 1988). Apart from cytokines, the best inhibitors of IL-1 transcription are glucocorticoids (GC). GC reduce IL-1 production at both transcriptional (Nishida *et al.*, 1988) and post-

transcriptional levels (Knudsen *et al.*, 1987; Lee *et al.*, 1988). GC block IL-1β transcript expression in IL-1-stimulated astrocytoma cells (Nishida *et al.*, 1989) and in LPS-activated PBMC (Lew *et al.*, 1988) and U937 cells (Knudsen *et al.*, 1987). One study failed to report GC-induced suppression of IL-1 transcription, possibly because of the high LPS concentration (10 μg/ml), which could have overcome GC activity (Kern *et al.*, 1988). Prostaglandins (PG) do not affect IL-1 gene expression induced by IL-1 in vascular cells (Warner *et al.*, 1987a,b), but block IL-1 release at the post-transcriptional level (Knudsen *et al.*, 1986). The same was found in LPS-stimulated PBMC (Sung and Walters, 1991) and murine macrophages (Ohmori *et al.*, 1990). In contrast with these studies, PGE_2 and cAMP agonists were shown to augment IL-1β mRNA in LPS-activated murine macrophages, without any effect on IL-1α transcripts (Ohmori *et al.*, 1990). Also, IL-1-induced expression of both IL-1α and IL-1β production was found to be enhanced by PGE_2 in human PBMC (Vannier and Dinarello, 1993). These discrepancies have yet to be explained. Heat shock response inducers downregulate IL-1 expression (Schmidt and Abdulla, 1988). Thus, IL-1 gene transcription is modulated by positive and negative signals and can be modulated by different cytokines, either alone or in combination.

IL-1 expression is also differentially regulated during the maturation of blood monocytes into macrophages. Gene expression studies revealed that 4 h after LPS stimulation, IL-1β transcripts in macrophages were 3-fold lower than in monocytes (Herzyk *et al.*, 1992). However, total (i.e., intracellular and secreted) IL-1β protein production was higher in macrophages than in monocytes, thus suggesting a higher translation efficiency in macrophages. Nevertheless, macrophages secrete only 1–5% of intracellular IL-1β whereas monocytes release 5–20%, thus explaining the finding that monocytes secrete more IL-1β than macrophages (Herzyk *et al.*, 1992).

IL-1α and IL-1β promoters share low homology; it is thus not surprising that these two genes are differentially expressed in different cell types. Keratinocytes (Kupper *et al.*, 1986) and T cell clones (Acres *et al.*, 1987) express more IL-1α (2- to 4-fold) than IL-1β transcripts. By contrast, LPS-stimulated mononuclear cells (PBMC) express predominantly IL-1β mRNA (Burchett *et al.*, 1988). In U937 cells PMA is able to induce IL-1β, but not IL-1α, transcripts. IL-1α induced both transcripts in PBMC (Dinarello *et al.*, 1987) but only IL-1β transcripts in vascular cells (smooth-muscle and endothelial cells) (Warner *et al.*, 1987a,b). In these latter cell types, IL-1α transcripts were observed when cells were treated with the protein synthesis inhibitor cycloheximide (CH). CH is also known to superinduce LPS-induced IL-1 transcripts, via a stabilization of IL-1 transcript stability (Turner *et al.*, 1989). By contrast, PMA-induced IL-1 mRNA is more stable and is not affected by the presence of CH (Fenton *et al.*, 1988). Thus, LPS and PMA induce IL-1 gene expression by different mechanisms.

Sequences involved in binding of transcriptional regulatory proteins were identified in the IL-1β gene promoter (Fenton, 1990). Using electrophoretic mobility shift assay, a DNA binding activity was found in the promoter region between –58 and +11 using nuclear extracts from resting and activated PBMC. This region is highly conserved between human and murine IL-1β promoters, and contains the TATA box sequence. A more detailed analysis revealed that a protein termed NFIL-1βA binds upstream to the TATA box, at –49 to –38, suggesting that this nuclear factor may interact with TATA box-binding factors (Fenton, 1990).

2.5 POST-TRANSCRIPTIONAL REGULATION

The regulation of IL-1 mRNA stability represents a further level of control of IL-1 expression. As mentioned above, the IL-1 3′ untranslated regions contain multiple AU-rich regions, which are known to be involved in transcript decay. LPS-activated THP-1 cells transcribe IL-1α and IL-1β genes at similar rates, but IL-1α transcripts are much more unstable than those coding for IL-1β (Turner *et al.*, 1988). The ability of IFN-γ to increase LPS-induced IL-1 production is mediated by both enhanced IL-1β transcription and increased mRNA stability (Arend *et al.*, 1989). Identical results were obtained with PMA-induced expression of IL-1β in fibroblasts (Yamato *et al.*, 1989) and THP-1 cells (Fenton *et al.*, 1988). Since a transfected mutated H-*ras*

gene increases both IL-1β transcription and mRNA stability, a role for G-proteins in these processes is likely (Demetri *et al.*, 1990). A different IL-1β mRNA stability was described in monocytes ($t_{1/2} \sim 2$–3 h) compared to macrophages, ($t_{1/2} \sim 10$ h) (Herzyk *et al.*, 1992). Nevertheless, since IL-1β transcripts are more abundant in monocytes than in macrophages, this implies that the transcriptional rate of this gene is higher in monocytes. Histamine decreases IL-1β mRNA stability, but augments IL-1β transcripts in IL-1α-treated PBMC, presumably by an augmented transcriptional rate (Vannier and Dinarello, 1993). TNF augmented IL-1β transcripts in fibrosarcoma cells by augmenting transcript stability, while leaving gene transcription unchanged (Gorospe *et al.*, 1993). This effect of TNF was mediated by activating protein kinase C.

Among negative regulators of IL-1 expression, both IL-4 (Donnelly *et al.*, 1991) and the GC (Lee *et al.*, 1988) hormone dexamethasone (Dex), in addition to inhibiting IL-1 gene transcription, also increase IL-1 mRNA decay. The destabilizing effect of IL-4 and Dex on IL-1 mRNAs is blocked by the presence of a protein synthesis inhibitor, thus indicating that this effect requires *de novo* protein synthesis.

3. Proteins

Mature human IL-1α (pl = 5) and IL-1β (pl = 7) are polypeptides which share 26% amino acid (aa) identity. Comparison of mature IL-1β from different animal species indicates that the amino acid sequence is conserved by 75–78%. IL-1α sequences are less conserved among species (60–70%). The primary translation products of IL-1α and IL-1β are 271 and 269 aa long, corresponding to molecular masses of 30 606 and 30 749 Da, respectively (Figure 1.4). The IL-1α propeptide is biologically active, whereas the IL-1β precursor is not. Although the sequence contains glycosylation sites, sugars are not important in the biological activity of IL-1. IL-1α may be glycosylated and mannose sites may be important for association to the cell membrane (Brody and Durum, 1989).

IL-1α and IL-1β lack a signal peptide. IL-1α remains mostly in the cytosol and associated with the plasma membrane. The pathway of secretion of mature IL-1β (aa 117–269 of the precursor) is not defined and may be common with that of other leaderless proteins (Rubartelli *et al.*, 1990).

Cleavage of pro-IL-1β to mature IL-1β is mediated by ICE; ICE is a cysteine protease representative of a novel class of proteolytic enzymes (Cerretti *et al.*, 1992; Miura *et al.*, 1993; Thornberry *et al.*, 1992; Walker *et al.*, 1994; Wilson *et al.*, 1994). ICE and related molecules are involved in the regulation of apoptosis (Miura *et al.*, 1993). It remains unclear whether and how regulation of

A   IL-1α

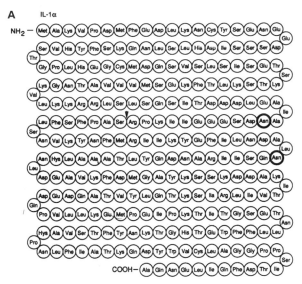

The arrow before Ser¹¹³ indicates the position of the mature NH₂ terminal.

Asn glycosylation sites are marked in bold circles.

B   IL-1β

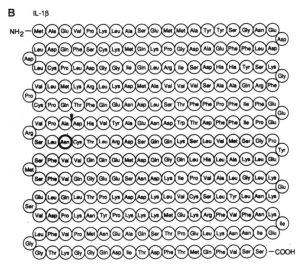

The arrow before Ala¹¹⁷ indicates the position of the mature NH₂ terminal after cleavage by IL-1-converting enzyme (ICE).

Asn glycosylation sites are marked in bold circles.

C   IL-1 Ra

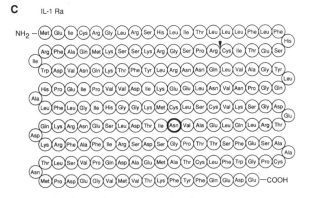

The arrow before Arg²⁶ indicates the position of the mature NH₂ terminal after removal of the leader peptide.

Asn glycosylation sites are marked in bold circles.

A

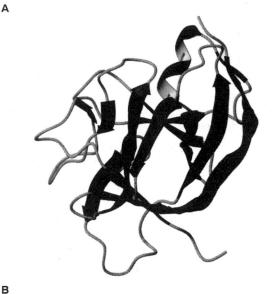

B

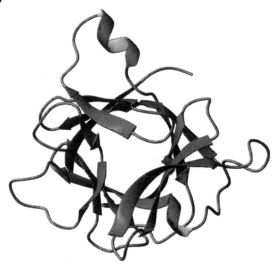

Figure 1.5   Three-dimensional structure of (a) IL-1α and (b) IL-1β (Priestle *et al.*, 1988, 1989).

cell death and IL-1β processing are related. It is noteworthy that induction of apoptosis induces IL-1 in monocytes (Dinarello, 1994).

IL-1α and IL-1β have been crystallized (Priestle *et al.*, 1988, 1989) (Figure 1.5). In spite of limited sequence similarity, IL-1α and IL-1β have similar three-dimensional structure, consisting of a β-barrel with four triangular faces forming a tetrahedron. A number of site-derived mutagenesis studies have been attempted to identify critical residues for receptor binding and

Figure 1.4   Amino acid sequence of (a) human IL-1α, (b) human IL-1β, and (c) human IL-1Ra. Glycosylation sites and the ICE cleavage site of IL-1β are indicated.

biological activity. The two cysteine residues of IL-1β are essential for activity, whereas exposed lysines are not critical. Mutation studies indicated that arginine 4, leucine 6, threonine 9, arginine 11, histidine 30 and aspartic acid 146 are important for biological activity. Arginine 120 is also important either in stabilizing the tertiary structure or in the interaction with the receptor. Substitution of arginine 127 to glycine generates a mutein with no biological activity, which binds to IL-1R (Yamayoshi *et al.*, 1990).

Multiple forms of IL-1Ra have been identified (Arend, 1993; Carter *et al.*, 1990; Eisenberg *et al.*, 1990). Soluble IL-1Ra (sIL-1Ra) is a 152-aa protein (pl = 5.2), secreted both in unglycosylated (18 kDa) and glycosylated (22 kDa) forms. Intracellular IL-1Ra type I (icIL-1Ra I) was cloned in keratinocytes and consists of 159 aa (Haskill *et al.*, 1991). It lacks a signal sequence and remains intracellular. We recently cloned a new isoform of icIL-1Ra, icIL-1Ra II, which contains extra 21 amino acids at the N-terminus (Muzio *et al.*, 1995).

Stimuli which induce IL-1 also cause production of IL-1Ra, which may counterbalance the action of IL-1. IL-1Ra and icIL-1Ra I and II are expressed in mononuclear phagocytes and PMN. icIL-1Ra I and II are also expressed by fibroblasts and epithelial cells (Haskill *et al.*, 1991 and unpublished). By and large, signals which induce IL-1 also cause production of IL-1Ra, which may counteract the action of agonist molecules. However, expression of IL-1 and IL-1Ra can be dissociated. Immune complexes and glycans preferentially trigger production of IL-1Ra versus IL-1 (Arend, 1991, 1993; Poutsiaka *et al.*, 1993; Re *et al.*, 1993; Roux Lombard *et al.*, 1989). Monocytes cultured *in vitro* to resemble mature tissue macrophages, and alveolar macrophages, express IL-1Ra constitutively with stimulation in response to GM-CSF. Finally, IL-4, IL-13, and IL-10 inhibit IL-1 expression but amplify IL-1Ra production (Cassatella *et al.*, 1994; Muzio *et al.*, 1994a; Re *et al.*, 1993).

4.   Cellular Sources and Production

Cells of the monocyte-macrophage lineage are the main cellular source of IL-1, although most cell types have the potential to express this cytokine. In the absence of *in vitro* or *in vivo* stimulation, the IL-1 genes are not expressed. Diverse inducers including bacterial products (e.g., LPS), complement components, cytokines (TNF, IFN-γ, GM-CSF, IL-1 itself), cause transcription (see above), which does not necessarily result in translation. For instance, adhesion causes accumulation of IL-1 mRNA, which requires a triggering stimulus (minute amounts of LPS) for translation into protein.

The synthesis of IL-1 is inhibited by endogenous agents, particularly PG and GC. In monocytes or monocytic cell lines *in vitro*, IL-1 production is inhibited by the addition of PGE_2. This inhibition is also to be considered a negative feedback mechanism, since when cells are stimulated (with LPS, or phorbol esters) to produce IL-1 they also produce PGE_2, which will limit IL-1 production (Knudsen *et al.*, 1987; Kunkel *et al.*, 1986). Accordingly, addition of inhibitors of PG upregulates IL-1 production (Knudsen *et al.*, 1987; Kunkel *et al.*, 1986).

More importantly, GC inhibit the synthesis of IL-1 (Snyder and Unanue, 1982) both at the transcriptional and the translational level (Knudsen *et al.*, 1987), as they do the synthesis of most proinflammatory cytokines (including TNF, IL-8, IL-6, IL-2, MCP-1). Also in this case, GC act as endogenous inhibitors of IL-1 production as demonstrated by the increase in IL-1 production observed in adrenalectomized animals (Perretti *et al.*, 1989) (for a role of endogenous GC, see Section 5.1.3 on the endocrine system).

5.   Biological Activities

5.1   SPECTRUM OF ACTION OF IL-1

IL-1 affects a wide range of cells and organs. Its spectrum of action is similar to that of TNF and, to a lesser degree, of IL-6 (Dinarello, 1996). Induction of secondary cytokines, including IL-6, colony-stimulating factors (CSF), and chemokines, is involved in many of the *in vitro* and *in vivo* activities of IL-1. The vast phenomenological literature on the activities of IL-1 is concisely summarized here based on target organs/tissues and the reader is referred to previous reviews for more detailed analysis (Dinarello, 1991, 1994; Oppenheim and Gery, 1993).

5.1.1   Hematopoietic Cells

IL-1 affects the hematopoietic system at various levels, from immature precursors to mature myelomonocytic and lymphoid elements. "Hemopoietin-1" activity was found to be mediated by IL-1α (Bagby, 1989). IL-1 induces production of colony-stimulating factors (CSFs) in a variety of cell types, including elements of the bone marrow stroma. It acts synergistically with hematopoietic growth factors on various stages of hematopoietic differentiation. IL-1 is also active as a hematopoietic growth factor *in vivo*: it stimulates production of CSFs, accelerates bone marrow recovery after cytotoxic chemotherapy or irradiation, and has radioprotective activity (Fibbe *et al.*, 1989; Neta *et al.*, 1988; Oppenheim *et al.*, 1989).

IL-1 acts on T and B lymphocytes (Dinarello, 1991). In particular, it costimulates T cell proliferation in the classic costimulator assay (Oppenheim and Gery, 1993). The AP-1 transcription complex in the promoter of IL-2 represents one molecular target for the costimulatory activity of IL-1 (Muegge *et al.*, 1989). The LAF assay has been invaluable for the identification of IL-1, but the actual role of IL-1 costimulation in T cell physiology has not been fully established. Recent data showing that IL-1Ra favors the development of T_H1-type responses may suggest a role for IL-1 in the generation of T_H2-type responses (Manetti *et al.*, 1994).

IL-1 affects mature myelomonocytic elements. It induces cytokine production in monocytes, while it is a poor inducer in polymorphonuclear leukocytes (PMN) (see above). It prolongs the *in vitro* survival of PMN by blocking apoptosis (Colotta *et al.*, 1992).

5.1.2   Vascular Cells

IL-1 profoundly affects the function of vessel wall elements, endothelial cells in particular (Bevilacqua *et al.*, 1984; Dejana *et al.*, 1984; Rossi *et al.*, 1985). IL-1 activates endothelial cells in a proinflammatory, prothrombotic sense (Mantovani *et al.*, 1992). IL-1 induces production of tissue factor and platelet-activating factor, downmodulates the protein C-dependent anti-coagulation pathway, and induces production of an inhibitor of thrombus dissolution (PAI-1). IL-1 induces gene expression-dependent production of vasodilatory mediators (NO, PGI_2), expression of adhesion molecules, and production chemokines in cultured endothelial cells. The concerted action of changes in rheology, adhesion, and chemotactic factors underlies leukocyte recruitment at sites of IL-1 production or injection (Mantovani *et al.*, 1992).

5.1.3   Neuroendocrine System

Infection and inflammation induce an elevation of blood corticosteroids (CS) through an activation of the so-called hypothalamic–pituitary–adrenal axis (HPAA) very similar to that observed with stress. This is the result of a central action whereby IL-1 stimulates the release of corticotropin-releasing hormone (CRH) by the hypothalamus that will induce adrenocorticotropin production by the pituitary, ultimately causing a release of CS in the bloodstream by the adrenals (Basedovsky *et al.*, 1986; Sapolsky *et al.*, 1987; Woloski *et al.*, 1985). This release is inhibited by anti-CRH antibodies. The increase in CS resulting from HPAA activation by IL-1 may have several consequences in view of the wide range of immunosuppressive, anti-inflammatory and metabolic actions of GC (Fantuzzi and Ghezzi, 1993). We stress here that CS are (as previously mentioned in Section 4) potent inhibitors of the synthesis of IL-1 (and of other cytokines). They also prevent the hemodynamic shock associated with injection of IL-1, TNF, or LPS, and protect against IL-1 or LPS toxicity. Therefore, activation of HPAA may be considered a feedback mechanism for control of IL-1 production and toxicity. The fact that adrenalectomized animals are extremely susceptible to IL-1 toxicity strongly supports this hypothesis (Bertini *et al.*, 1988).

5.1.4   The Acute-phase Response

IL-1 is a key mediator of the series of host responses to infection and inflammation known as acute-phase response (reviewed in Dinarello, 1984). One of the early acronyms of IL-1 before it was named under the current system was LEM (leukocytic endogenous mediator). This was originally identified as a major mediator of the acute-phase response, particularly hypoferremia and induction of acute-phase proteins (Kampschmidt *et al.*, 1980; Kampschmidt and Pulliam, 1978).

Hypoferremia seems to be mediated by an effect on neutrophils, which would be stimulated to release lactoferrin to sequester iron in the tissues (Van Snick *et al.*, 1974). This could constitute a "nutritional immunity" against infection since iron is essential for the growth of many bacteria.

Acute-phase proteins (APP) are proteins whose synthesis is increased during inflammation (Kushner, 1982). They include C-reactive protein, serum amyloids, fibrinogen, hemopexin, and various proteinase inhibitors; these molecules may have protective, antitoxic, and other functions yet to be defined. The synthesis of some of these proteins (e.g., serum amyloid A) can be directly induced in hepatocytes by IL-1 (Ramadori *et al.*, 1985). Others, such as fibrinogen, are induced indirectly through IL-6 (Heinrich *et al.*, 1990). Therefore, both IL-1 and IL-6 (like other cytokines) behave like HSFs (hepatocyte-stimulating factors, another acronym used to name cytokines that stimulate liver APP synthesis).

The increased synthesis of APP is part of a rearrangement of liver metabolism where the synthesis of "normal liver proteins" is decreased: one such negative acute-phase reactant is albumin, whose gene expression is decreased by IL-1 (Ramadori *et al.*, 1985).

5.1.5   Central Nervous System

IL-1 is the main endogenous pyrogen. In 1943 Menkins suggested that leukocytes release a pyrogenic substance, "pyrexin", that was subsequently detected in the circulation of febrile rabbits (Atkins and Woods, 1955). Human leukocytic pyrogen was purified in 1977 by Dinarello and colleagues (Dinarello *et al.*, 1977) and an immunoassay for it was developed. It is now clear that the main endogenous pyrogen was IL-1, and recombinant IL-1 induces fever in experimental animals (Dinarello *et al.*, 1986a), an activity shared with other cytokines including TNF and IL-6 (although these are much less potent than IL-1) (Dinarello *et al.*, 1986b, 1991). The pyrogenic action of IL-1 is due to the increased production of PG. In fact, IL-1 fever is abolished by PG synthesis inhibitors and IL-1 directly stimulate PGE_2 release by hypothalamic tissue (Dinarello *et al.*, 1986b; Rettori *et al.*, 1991).

In addition to fever, IL-1 has other effects on the central nervous system. These include induction of slow-wave sleep (Krueger *et al.*, 1994) and anorexia

(Hellerstein *et al.*, 1989; Uehara *et al.*, 1990), typically associated with infections. It also activates the hypothalamus to produce CRH as described in Section 5.1.3.

5.1.6   Other Effects

It is almost impossible to list all the activities of IL-1. Since it has been available as a recombinant protein, a large number of papers have been published reporting the most disparate effects of this cytokine. IL-1 has a number of local effects that have been termed "catabolic" and play a role in destructive joint and bone diseases. In particular, IL-1 induces production of collagenase by synovial cells (Dayer *et al.*, 1986) and of metalloproteinases by chondrocytes (Saklatvala *et al.*, 1985).

IL-1 also stimulates fibroblast proliferation (Schmidt *et al.*, 1989) and collagen synthesis (Postlethwaite *et al.*, 1983) and thus play a role in fibrosis. It induces a profound hypotension (Okusawa *et al.*, 1988), an effect which is inhibited by a cyclooxygenase inhibitor and in which IL-1 and TNF act synergistically.

Another important action of IL-1 is its toxicity for insulin-producing β cells in Langerhans islets (Bendtzen *et al.*, 1986), supporting a role of IL-1 in the pathogenesis of insulin-dependent type I diabetes.

6.   Receptors and Signal Transduction

The first IL-1R was cloned from murine (Sims *et al.*, 1988) and then human T cells (Sims *et al.*, 1989). Soon after identification and cloning of this T cell IL-1R, thereafter named type I R (IL-1RI), it was evident that a second receptor exists for IL-1, expressed in B lymphocytes and myelomonocytic cells, referred to as type II (IL-1RII) (McMahan *et al.*, 1991) (Figure 1.6).

Transcripts of human and mouse IL-1RI are approximately 5 kb in length. A single open reading frame encodes for a protein of 552 amino acids, with a molecular mass, when fully glycosylated, of 80–85 kDa. The molecular mass of the unglycosylated protein is 62 kDa. On the basis of their structures, IL-1RI and IL-1RII have been assigned to the IgG-like superfamily of receptors, with the extracellular portion containing three IgG-like domains. The extracellular region (319 aa long), which contains seven potential sites of *N*-glycosylation, of IL-1RI is followed by a 20 aa long transmembrane region, and then by a 213 aa cytoplasmic portion. The cytosolic region has no homology with any kinase described so far, and the only protein which shows some homology is the *Drosophila* TOLL protein.

IL-1RII transcripts are approximately 1803 bp long. The human transcript encodes for a 386-aa protein of 68 kDa. Treatment with *N*-glycosidases reduced the molecular mass to 55 kDa. Five potential sites of *N*-glycosylation have been identified. The extracellular region of 332 amino acids shares only 28% homology

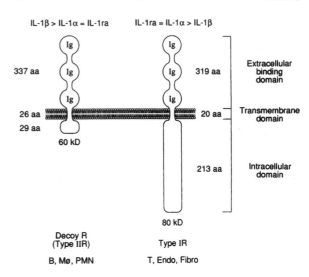

Figure 1.6   IL-1 receptors: a true receptor and a decoy. Numbers refer to amino acids' relative affinity for different ligands and predominant expression is shown.

with the corresponding region of IL-1RI in humans. A 26-aa transmembrane domain is then followed by a very short cytoplasmic domain of 29 residues (McMahan *et al.*, 1991).

As already mentioned, the genes of IL-1Rs are located on chromosome 2 (band q12.22) in humans and in the centromeric region of chromosome 1 in mice (McMahan *et al.*, 1991). The promoter region of IL-1RI has been recently described (Ye *et al.*, 1993).

IL-1RI and IL-1RII are usually coexpressed. However IL-1RI is expressed as the predominant form in fibroblasts and T cells. By contrast B cells, monocytes, and PMN express preferentially the IL-1RII (McMahan *et al.*, 1991).

IL-1RI and IL-1RII have different affinities for the three ligands of the IL-1 family. Although some differences are evident among different studies, IL-1RI binds IL-1α with higher affinity than IL-1β ($k_d = 10^{-10}$ M and 10^{-9} M, respectively). By contrast, IL-1RII binds IL-1β more avidly than IL-1α ($k_d = 10^{-9}$–10^{-10} M and 10^{-8} M, respectively). IL-1Ra binds to IL-1RI with an affinity similar to that of IL-1α, whereas IL-1RII binds IL-1Ra 100-fold less efficiently than IL-1RI (McMahan *et al.*, 1991).

Given the existence of two distinct IL-1Rs, a number of studies have investigated the actual role played by each of them in IL-1 signaling. As summarized briefly hereafter, all available evidence indicates that IL-1-induced activities are mediated exclusively via the IL-1RI, whereas IL-1RII has no signaling activity and inhibits IL-1 activities by acting as a decoy for IL-1 (for a recent review, see Colotta *et al.*, 1994a).

In a number of different cell types, circumstantial evidence has been obtained that different IL-1 activities are mediated by the IL-1RI. Human endothelial cells, in which IL-1 regulates functions related to inflammation and thrombosis, express exclusively IL-1RI, indicating that IL-1RII is at least dispensable for IL-1 signaling (Boraschi et al., 1991; Colotta et al., 1993b).

Blocking monoclonal antibodies directed against IL-1RI inhibited IL-1 activities in the hepatoma cell line HEPG2 (Sims et al., 1993), which expresses almost equal amounts of IL-1RI and IL-1RII. IL-1α-induced costimulatory activity in CD4$^+$ murine T cell clones, which express both receptors, was mediated solely by IL-1RI (McKean et al., 1993).

IL-1RI appears to be the only signaling receptor also in cell types expressing predominantly the IL-1RII and only minute amounts of IL-1RI. Blocking monoclonal antibodies against IL-1RI totally blocked IL-1-induced expression of cytokines and adhesion molecules in the human monocytic cell line THP-1, human circulating monocytes, and PMN (Sims et al., 1993; Colotta et al., 1993a). Also IL-1-induced survival of PMN was blocked by anti-type I blocking antibodies (Colotta et al., 1993a).

Whereas the signaling activity of IL-1RI is well established, unequivocal evidence supporting a signaling function of IL-1RII is still lacking. Blocking monoclonal antibodies against IL-1RII did not inhibit the biological activities of IL-1 in a number of different cell types, including lymphocytes, monocytes, PMN, and hepatoma cells (Sims et al., 1993; Colotta et al., 1993a). In monocytes, anti-type I antibodies blocked IL-1 activities, whereas anti-type II antibodies did not block, but rather augmented, the responsiveness of cells to IL-1, consistent with a model in which the IL-1RII is an inhibitor of IL-1 (see below) (Sims et al., 1993; Colotta et al., 1993a).

In addition to lacking any signaling function, the IL-1RII is shed in a soluble (sIL-1RII) form. sIL-1RII was found in the supernatants of the B lymphoblastoid cell line Raji and of mitogen-activated mononuclear cells (reviewed in Colotta et al., 1994a). IL-1RII is also released by cytokine- and dexamethasone-treated PMN and monocytes (Colotta et al., 1993a, 1994b; Re et al., 1994). sIL-1RII is shed rapidly, within minutes after treatment of PMN and monocytes with chemotactic stimuli and oxygen radicals, indicating that release of this receptor represents an aspect of the complex reprogramming of myelomonocytic cells in response to these mediators (Colotta et al., 1995, 1996).

The findings that the IL-1RII has no signaling function and that it is shed in a soluble form suggested that this molecule could act as an inhibitor of IL-1. We examined this hypothesis in human PMN, in which we found that IL-1 is a potent inducer of PMN survival in culture (Colotta et al., 1992). Since IL-4 inhibited IL-1-mediated survival, and IL-4 upregulated IL-1RII expression and release in these cells (Colotta et al., 1993a), we reasoned that the inhibitory activity of IL-4 on IL-1 activity could be mediated by an upregulation of

IL-1RII. The inhibitory activity of IL-4 was totally abrogated by the presence of blocking antibodies directed against IL-1RII, thus demonstrating that IL-1RII inhibits IL-1 activity (Colotta et al., 1993a). We proposed that the mechanism of inhibition is to act as a decoy target for IL-1, consisting in IL-1RII binding IL-1 without any signaling function, thus sequestering it and preventing the cytokine from binding to the IL-1RI, the only IL-1R with a cell signaling function.

Consistently with the decoy model of action of the IL-1RII, blocking antibodies to IL-1RII augmented the activity of suboptimal concentrations of IL-1 on IL-1-induced expression of cytokines and adhesion molecules in human circulating monocytes (Sims et al., 1993; Colotta et al., 1993a).

To validate the decoy model of action of IL-1RII, we overexpressed this receptor in type I-expressing human fibroblasts. As expected, IL-1 activity was reduced in fibroblasts expressing high levels of IL-1RII. The inhibitory effect of transfected type II receptors was evident at suboptimal concentrations of IL-1, whereas saturating amounts of IL-1 overcame the IL-1RII-mediated inhibition of IL-1 activity (Re et al., 1996).

The finding that GC and T$_H$2-derived cytokines (IL-4 and IL-13) upregulate IL-1RII expression and release (Colotta et al., 1993a, 1994b; Re et al., 1994; Shieh et al., 1993) is in keeping with the concept that the IL-1RII may represent a physiological pathway of inhibition of IL-1. Induction of expression and release of the IL-1RII may contribute to the anti-inflammatory properties of T$_H$2-derived cytokines and GC.

7. Transduction Pathways

The mechanism of transduction of the IL-1 signal is a highly controversial and confused area (Mizel, 1990; O'Neill et al., 1990). In certain cellular systems, but not in others, elevations of cAMP have been reported after exposure to IL-1. Among lipid mediators, diacylglycerol (DAG) was found increased in the absence of phosphoinositide hydrolysis (Rosoff et al., 1988). The source of DAG was phosphatidylcholine or phosphatidylethanolamine in different cell types (Rossi, 1993). Recently, in analogy with TNF, ceramide originating from hydrolysis of sphingomyelin was involved in IL-1 signaling (Mathias et al., 1993). However, cells from patients with Nieman–Pick disease, who have a profound deficiency of acid sphyngomyelinase, have full responsiveness to IL-1 and TNF (Kuno et al., 1994).

IL-1 causes rapid Ser/Thr phosphorylation of diverse proteins, including cytoskeletal components, a membrane receptor and HSP 27 (Bird and Saklatvala, 1990; Guesdon and Saklatvala, 1991; Stylianou et al., 1992). Tyr phosphorylation has been detected after exposure to IL-1 (Rosoff et al., 1988). It was recently found that IL-1 activates a novel cascade of protein kinases which eventually results in the phosphorylation of HSP 27 (Freshney et al., 1994).

8. Mouse IL-1

The murine IL-1α (Lomedico et al., 1984) cDNA was identified and sequenced in 1984. Other sequences for mouse (Gray et al., 1986), rat, and other species were then identified. As in humans, mouse IL-1α mRNA is 2000–2200 nucleotides long. Murine IL-1α cDNA codes for a protein 62% homologous to the human counterpart. The amino acid sequences of human and murine IL-1β are 67% homologous.

The murine IL-1 genes are also located on chromosome 2 (D'Eustachio et al., 1987; Telford et al., 1986), and, as in humans, the mouse IL-1β gene is 7 kb long (Telford et al., 1986) and its first intron contains a highly conserved homopurine tract. An unusual restriction fragment length polymorphism has been identified within the sixth intron of the mouse IL-1α gene (Haugen et al., 1989), generating at least six different alleles.

9. IL-1 Receptor Antagonist (IL-1Ra)

The molecule now termed IL-1Ra was discovered as a 22 to 25 kDa inhibitory activity against IL-1 in supernatants from stimulated human monocytes (see Arend, 1993, for a recent review). The cDNA for this molecule was cloned from a library obtained from IgG-treated monocytes (Eisenberg et al., 1990) and from the myelomonocytic cell line U937 (Carter et al., 1990). The cDNA contains a single open reading frame coding for a 177-aa protein, preceeded by a short 15 bp 5′ untranslated region and followed by a long 1133 bp 3′ untranslated region that does not contain any AUUUA motif involved in mRNA stability. The total length of IL-1Ra cDNA is 1.8 kb. Structural analysis of IL-1Ra sequence revealed the presence of a 25-aa hydrophobic stretch at the N-terminus resembling a signal peptide. Removal of this leader peptide generates a 152-aa mature protein, with a predicted molecular mass of 17 775 Da. The protein sequence contains a potential site of N-glycosylation.

A structural variant of IL-1Ra has been cloned from adherent human monocytes (Haskill et al., 1991). This new cDNA is identical to the sequence described earlier (see above) except at 5′ end, in which the first 85 base pairs are substituted by a new sequence of 130 bp. This variant (hereafter referred to as intracellular IL-1Ra (icIL-1Ra) as opposed to soluble IL-1Ra (sIL-1Ra)) is generated by an alternative splicing event, in which a new first exon is spliced into an internal acceptor site located in the first exon of sIL-1Ra. As a consequence, icIL-1Ra

is composed of 159 amino acids, of which 152 are identical to sIL-1Ra, lacking the signal peptide found in sIL-1Ra. Thus this new form of IL-1Ra remains intracellular. Although icIL-1Ra is also referred to as "keratinocyte or epithelial type IL-1Ra", it is now evident that cells of different origin can also express icIL-1Ra, including monocytes (Muzio et al., 1994a), polymorphonuclear (PMN) cells (Muzio et al., 1994a,b), and fibroblasts (Krzesicki et al., 1993). A substantial proportion of total IL-1Ra produced by monocytes and PMN remains cell-associated, even after stimulation (Muzio et al., 1994a,b). The biological function played by the cell-associated fraction of IL-1Ra remains largely obscure.

The gene for IL-1Ra (Figure 1.3) has been localized to the long arm of chromosome 2, in the same region as IL-1α and IL-1β (Lennard et al., 1992; Steinkasserer et al., 1992). In humans this chromosomal region also contains IL-1RI and IL-1RII genes, whereas in mice IL-1R genes are localized in chromosome 1.

The IL-1Ra gene contains two alternative first exons followed by three exons. Compared to IL-1 sequences, the first alternative exons appear unique as the N-terminal region of IL-1Ra proteins have little homology with the corresponding regions of IL-1α and IL-1β. The gene is 6.4 kb long, and the first icIL-1Ra-specific exon is a further 9.6 kb upstream. Thus, the sIL-1Ra promoter is within the first intron of icIL-1Ra. This promoter (Smith et al., 1992b) contains a TATAA box and sequences for binding of transcriptional factors, including NFκB, NFIL-1bA, AP-1, and CRE. In transfection experiments, sIL-1Ra promoter was active only in cell types which naturally can be induced to express IL-1Ra (Arend, 1993). The putative promoter region of icIL-Ra (Arend, 1993) apparently lacks any TATAA or CAAT motif, and thus could utilize alternative mechanisms to start transcription.

The genomic structure of IL-1Ra is probably more complex than detailed above, as a new exon has been localized between the exons coding for soluble and intracellular forms of IL-1Ra. Usage of this new exon generates a new isoform of icIL-1Ra that we propose to term icIL-1Ra type II (icIL-1RaII). By RT-PCR we found that icIL-1RaII is exposed in keratinocytes and, at lower levels, in monocytes and PMN (Muzio M. et al., in preparation).

IL-1Ra is produced by different cell types, including monocyte-macrophages, PMN cells, and fibroblasts (reviewed in Arend, 1993). Keratinocytes and other cells of epithelial origin produce almost exclusively icIL-1Ra (Haskill et al., 1991).

Adherent human monocytes stimulated with LPS produce near equivalent amounts of IL-1β and IL-1Ra. By contrast, IgG-induced monocytes produce little, if any, IL-1, whereas transcription and translation of IL-1Ra is sustained. IL-1Ra transcripts induced by LPS have the same half-life as IL-1β ($t_{1/2}$ = 2–4 h), whereas those induced by adherent IgG have extraordinarily high stability ($t_{1/2}$ > 15 h) (Arend *et al.*, 1991).

Among cytokines, GM-CSF augments IL-1Ra production in monocytes (Shields *et al.*, 1990). IL-1α and IL-1β are weak inducers of IL-1Ra in monocytes, whereas IL-1β enhances the induction by IgG (Vannier *et al.*, 1992). IL-4, while inhibiting IL-1 expression and production (see above), induces IL-1Ra in monocytes and PMN (Fenton *et al.*, 1992; Re *et al.*, 1993; Vannier *et al.*, 1992). In the same vein, IL-10 enhances LPS-induced expression of IL-1Ra in both monocytes (Arend, 1993) and PMN (Cassatella *et al.*, 1994). Also IL-13 has been found to induce IL-1Ra in human myelomonocytic cells (Muzio *et al.*, 1994a). IL-13 was found to induce both ic- and sIL-1Ra transcripts (Muzio *et al.*, 1994a). We found that TGF-β induces both s- and icIL-1Ra transcripts in human polymorphonuclear cells (Muzio *et al.*, 1994b).

Levels of IL-1Ra produced by cells of the mononuclear phagocyte lineage depend upon the differentiation state. Differentiation *in vitro* of monocytes into macrophages augments levels of IL-1Ra production, which in macrophages appears to be constitutive whereas monocytes require an appropriate activation. IL-1Ra production in macrophages is further augmented by GM-CSF but not by LPS and IgG which, as mentioned above, are good inducers of IL-1Ra in monocytes (Janson *et al.*, 1991).

Transcripts for ic- and sIL-1Ra are differentially regulated. Although both transcripts are induced in myelomonocytic cells by IL-13 (Muzio *et al.*, 1994a) and TGF-β (Muzio *et al.*, 1994b), in human fibroblasts LPS preferentially induced sIL-1Ra, whereas PMA induced icIL-1Ra transcripts (Krzesicki *et al.*, 1993).

10. Clinical Implications

IL-1 is central mediator of local and systemic inflammatory reactions. Blood IL-1 levels reach relatively modest levels in response to septic conditions, but IL-1Ra increases to levels orders of magnitude higher (Dinarello, 1994). This may represent a feedback control mechanism. Most of the interest of IL-1 is in its pathogenetic role in septic shock syndrome and rheumatoid arthritis. However, *in vitro* studies, animal models, and results of studies reporting levels of IL-1 in human diseases have indicated other pathologies where blockade of IL-1 might be beneficial. These include vasculitis, disseminated intravascular coagulation, osteoporosis, neurodegenerative disorders such as Alzheimer disease, diabetes, lupus nephritis, immune complex glomerulonephritis and autoimmune diseases in general (Dinarello, 1991). Inhibition of IL-1 by IL-1Ra, anti-IL-1, or anti-IL-1R antibodies are protective in various animal models including endotoxin-induced

hemodynamic shock and lethality, arthritis, inflammatory bowel disease, spontaneous diabetes in BB rats, graft-versus-host disease in mice, heart allograft rejection, and experimental autoimmune encephalomyelitis.

Since TNF is produced concomitantly and IL-1 synergizes with TNF (but not with IL-6) in many systems, it is likely that these two cytokines act in concert in the pathogenesis of these disorders.

IL-1 is an autocrine/paracrine growth factor for acute and chronic myeloid leukemia cells (Rambaldi *et al.*, 1991), plasmacytoma (via IL-6; Klein *et al.*, 1995), and possibly some solid tumors such as ovarian carcinoma (Mantovani, 1994). Interestingly AML blasts usually express IL-1β, but not IL-1Ra. Blockage of IL-1 using IL-1Ra inhibits AML proliferation *in vitro* (Rambaldi *et al.*, 1991).

The exploration of the therapeutic potential of IL-1 has provided an opportunity to examine the *in vivo* activity of systemic IL-1 administration in humans. Phase I trials have been conducted with IL-1α and IL-1β in studies ultimately aimed at exploiting the hematopoietic/radioprotective action of these molecules (Smith *et al.*, 1992a; Tewari *et al.*, 1990). By and large, results obtained confirm data from animal experimentation. Systemic (intravenous) IL-1 (1–10 ng/ml) causes fever, sleepiness, anorexia, myalgia, arthralgia, and headache. At doses of 100 ng/ml or higher, a rapid fall in blood pressure occurs.

Therapeutic strategies aimed at blocking IL-1 have received considerable attention in experimental models and in humans (Dinarello, 1994; Fisher *et al.*, 1994a,b; Pribble *et al.*, 1994). To date, IL-1Ra and engineered soluble Type 1 receptor have been tested in the clinic. While initial results with IL-1Ra in septic shock syndrome encouraged optimism, subsequent phase III study failed to substantiate the phase II data (Pribble *et al.*, 1994). *A posteriori*, IL-1Ra may have been beneficial in patients with more serious disease and organ failure (Fisher *et al.*, 1994a). Initial results with IL-1Ra in graft-versus-host disease are encouraging (Antin *et al.*, 1994).

Another possible site for pharmacological action could be the processing of IL-1β by ICE. Inhibitors of IL-1β secretion acting by inhibiting ICE are under study, on the basis that IL-1β is quantitatively more important than IL-1α. Unfortunately, to date only peptides have been identified, with all the limitation of this type of molecule, particularly the very low bioavailability.

11. References

Acres, R.B., Larsen, A. and Conlon, P.J. (1987). IL 1 expression in a clone of human T cells. J. Immunol. 138, 2132.

Antin, J.H., Weinstein, H.J., Guinan, E.C., *et al.* (1994). Recombinant human interleukin-1 receptor antagonist in the treatment of steroid-resistant graft-versus-host disease. Blood 84, 1342.

Arend, W.P. (1991). Interleukin 1 receptor antagonist. A new member of the interleukin 1 family. J. Clin. Invest. 88, 1445.

Arend, W.P. (1993). Interleukin-1 receptor antagonist. Adv. Immunol. 54, 167.

Arend, W.P., Gordon, D.F., Wood, W.M., Janson, R.W., Joslin, F.G. and Jameel, S. (1989). IL-1 beta production in cultured human monocytes is regulated at multiple levels. J. Immunol. 143, 118.

Arend, W.P., Smith, M.F.J., Janson, R.W. and Joslin, F.G. (1991). IL-1 receptor antagonist and IL-1 beta production in human monocytes are regulated differently. J. Immunol. 147, 1530.

Atkins, E. and Wood, W.B.J. (1955). Studies on the pathogenesis of fever. II. Identification of an endogenous pyrogen in the bloodstream following the injection of typhoid vaccine. J. Exp. Med. 102, 499.

Auron, P.E., Webb, A.C., Rosenwasser, L.J., et al. (1984). Nucleotide sequence of human monocyte interleukin 1 precursor cDNA. Proc. Natl Acad. Sci. USA 81, 7907.

Bagby, G.C.J. (1989). Interleukin-1 and hematopoiesis. Blood Rev. 3, 152.

Basedovsky, H., Del Rey, A. and Dinarello, C.A. (1986). Immunoregulatory feedback between interleukin-1 and glucocorticoid hormones. Science 233, 652.

Bendtzen, K., Mandrup-Poulsen, T., Nerup, J., Nielsen, J.H., Dinarello, C.A. and Svenson, M. (1986). Cytotoxicity of human pI 7 interleukin-1 for pancreatic islets of Langerhans. Science 232, 1545.

Bensi, G., Mora, M., Raugei, G., Buonamassa, D.T., Rossini, M. and Melli, M. (1990). An inducible enhancer controls the expression of the human interleukin 1 beta gene. Cell Growth Diff. 1, 491.

Bertini, R., Bianchi, M. and Ghezzi, P. (1988). Adrenalectomy sensitizes mice to the lethal effects of interleukin 1 and tumor necrosis factor. J. Exp. Med. 167, 1708.

Bevilacqua, M.P., Pober, J.S., Majeau, G.R., Cotran, R.S. and Gimbrone, M.A. Jr (1984). Interleukin-1 (IL-1) induces biosynthesis and cell surface expression of procoagulant activity in human vascular endothelial cells. J. Exp. Med. 160, 618.

Bird, T.A. and Saklatvala, J. (1990). Down-modulation of epidermal growth factor receptor affinity in fibroblasts treated with interleukin 1 or tumor necrosis factor is associated with phosphorylation at a site other than threonine 654. J. Biol. Chem. 265, 235.

Boraschi, D., Rambaldi, A., Sica, A., et al. (1991). Endothelial cells express the interleukin-1 receptor type I. Blood 78, 1262.

Brody, D.T. and Durum, S.K. (1989). Membrane IL-1: IL-1 alpha precursor binds to the plasma membrane via a lectin-like interaction. J. Immunol. 143, 1183.

Burchett, S.K., Weaver, W.M., Westall, J.A., Larsen, A., Kronheim, S. and Wilson, C.B. (1988). Regulation of tumor necrosis factor/cachectin and IL-1 secretion in human mononuclear phagocytes. J. Immunol. 140, 3473.

Carter, D.B., Deibel, M.R.J., Dunn, C.J., et al. (1990). Purification, cloning, expression and biological characterization of an interleukin-1 receptor antagonist protein. Nature 344, 633.

Cassatella, M.A., Meda, L., Gasperini, S., Calzetti, F. and Bonora, S. (1994). Interleukin 10 (IL-10) upregulates IL-1 receptor antagonist production from lipopolysaccharide-stimulated human polymorphonuclear leukocytes by delaying mRNA degradation. J. Exp. Med. 179, 1695.

Cerretti, D.P., Kozlosky, C.J., Mosley, B., et al. (1992). Molecular cloning of the interleukin-1 beta converting enzyme. Science 256, 97.

Clark, B.D., Collins, K.L., Gandy, M.S., Webb, A.C. and Auron, P.E. (1986). Genomic sequence for human prointerleukin 1 beta: possible evolution from a reverse transcribed prointerleukin 1 alpha gene. Nucleic Acids Res. 14, 7897.

Clark, B.D., Fenton, M.J., Rey, H., Webb, A.C. and Auron, P.E. (1988). Characterization of cis and trans acting elements involved in human pro IL-1 beta gene expression. In "Monokines and Other Non-lymphocytic Cytokines", pp. 47–53. Alan R. Liss, New York.

Colotta, F., Re, F., Polentarutti, N., Sozzani, S. and Mantovani, A. (1992). Modulation of granulocyte survival and programmed cell death by cytokines and bacterial products. Blood 80, 2012.

Colotta, F., Re, F., Muzio, M., et al. (1993a). Interleukin-1 type II receptor: a decoy target for IL-1 that is regulated by IL-4. Science 261, 472.

Colotta, F., Sironi, M., Borre, A., et al. (1993b). Type II interleukin-1 receptor is not expressed in cultured endothelial cells and is not involved in endothelial cell activation. Blood 81, 1347.

Colotta, F., Dower, S.K., Sims, J.E. and Mantovani, A. (1994a). The type II "decoy" receptor: novel regulatory pathway for interleukin-1. Immunol. Today, 15, 562.

Colotta, F., Re, F., Muzio, M., et al. (1994b). IL-13 induces expression and release of the IL-1 decoy receptor in human polymorphonuclear cells. J. Biol. Chem. 269, 12403.

Colotta, F., Orlando, S., Fadlon, E.J., Sozzani, S., Matteucci, C. and Mantovani, A. (1995). Chemoattractants induce rapid release of the interleukin 1 type II decoy receptor in human polymorphonuclear cells. J. Exp. Med. 181, 2181.

Colotta, F., Saccani, S., Giri, J.G., et al. (1996). Regulated expression and release of the interleukin-1 decoy receptor in human mononuclear phagocytes. J. Immunol. 156, 2534.

D'Eustachio, P., Jadidi, S., Fuhlbrigge, R.C., Gray, P.W. and Chaplin, D.D. (1987). Interleukin-1 alpha and beta genes: linkage on chromosome 2 in the mouse. Immunogenetics 26, 339.

Dayer, J.M., de Rochemonteix, B., Burrus, B., Demczuk, S. and Dinarello, C.A. (1986). Human recombinant interleukin 1 stimulates collagenase and prostaglandin E2 production by human synovial cells. J. Clin. Invest. 77, 645.

de Waal Malefyt, R., Figdor, C.G., Huijbens, R., et al. (1993). Effects of IL-13 on phenotype, cytokine production, and cytotoxic function of human monocytes. Comparison with IL-4 and modulation by IFN-gamma or IL-10. J. Immunol. 151, 6370.

Dejana, E., Breviario, F., Balconi, G., et al. (1984). Stimulation of prostacyclin synthesis in vascular cells by mononuclear cell products. Blood 64, 1280.

Demetri, G.D., Ernst, T.J., Pratt, E.S., Zenzie, B.W., Rheinwald, J.G. and Griffin, J.D. (1990). Expression of ras oncogenes in cultured human cells alters the transcriptional and posttranscriptional regulation of cytokine genes. J. Clin. Invest. 86, 1261.

Dinarello, C.A. (1984). Interleukin-1 and the pathogenesis of the acute-phase response. N. Engl. J. Med. 311, 1413.

Dinarello, C.A. (1991). Interleukin-1 and interleukin-1 antagonism. Blood 77, 1627.

Dinarello, C.A. (1994). Blocking interleukin-1 receptors. Int. J. Clin. Lab. Res. 24, 61.

Dinarello, C.A. (1996). Biological basis for IL-1 in disease. Blood 87, 2095.

Dinarello, C., Golden, N. and Wolff, S. (1974). Demonstration and characterization of two distinct human leukocytic pyrogens. J. Exp. Med. 139, 1369.

Dinarello, C.A., Renfer, L. and Wolff, S.M. (1977). Human leukocytic pyrogen: purification and development of a radioimmunoassay. Proc. Natl Acad. Sci. USA 74, 4624.

Dinarello, C.A., Cannon, J.G., Mier, J.W., et al. (1986a). Multiple biological activities of human recombinant interleukin 1. J. Clin. Invest. 77, 1734.

Dinarello, C.A., Cannon, J.G., Wolff, S.M., Bernheim, H.A., Beutler, B. and Cerami, A. (1986b). Tumor necrosis factor (cachectin) is an endogenous pyrogen and induces production of interleukin 1. J. Exp. Med. 163, 1433.

Dinarello, C.A., Ikejima, T., Warner, S.J., et al. (1987). Interleukin 1 induces interleukin 1. I. Induction of circulating interleukin 1 in rabbits in vivo and in human mononuclear cells in vitro. J. Immunol. 139, 1902.

Dinarello, C.A., Canno, J.G., Mancilla, J., Bishai, I., Lees, J. and Coceani, F. (1991). Interleukin-6 as an endogenous pyrogen: induction of prostaglandin E_2 in brain but not in peripheral blood mononuclear cells. Brain Res. 562, 199.

Dohlsten, M., Kalland, T., Sjogren, H.O. and Carlsson, R. (1988). Histamine inhibits interleukin 1 production by lipopolysaccharide-stimulated human peripheral blood monocytes. Scand. J. Immunol. 27, 527.

Donnelly, R.P., Fenton, M.J., Finbloom, D.S. and Gerrard, T.L. (1990). Differential regulation of IL-1 production in human monocytes by IFN-gamma and IL-4. J. Immunol. 145, 569.

Donnelly, R.P., Fenton, M.J., Kaufman, J.D. and Gerrard, T.L. (1991). IL-1 expression in human monocytes is transcriptionally and posttranscriptionally regulated by IL-4. J. Immunol. 146, 3431.

Eisenberg, S.P., Evans, R.J., Arend, W.P., et al. (1990). Primary structure and functional expression from complementary DNA of a human interleukin-1 receptor antagonist. Nature 343, 341.

Essner, R., Rhoades, K., McBride, W.H., Morton, D.L. and Economou, J.S. (1989). IL-4 down-regulates IL-1 and TNF gene expression in human monocytes. J. Immunol. 142, 3857.

Fantuzzi, G. and Ghezzi, P. (1993). Glucocorticoids as cytokine inhibitors: role in neuroendocrine control and therapy of inflammatory diseases. Mediat. Inflamm. 2, 263.

Fenton, M.J. (1990). Transcriptional factors that regulate human IL-1/hemopoietin 1 gene expression. Hematopoiesis 120, 67.

Fenton, M.J., Vermeulen, M.W., Clark, B.D., Webb, A.C. and Auron, P.E. (1988). Human pro-IL-1 beta gene expression in monocytic cells is regulated by two distinct pathways. J. Immunol. 140, 2267.

Fenton, M.J., Buras, J.A. and Donnelly, R.P. (1992). IL-4 reciprocally regulates IL-1 and IL-1 receptor antagonist expression in human monocytes. J. Immunol. 149, 1283.

Fibbe, W.E., van der Meer, J.W., Falkenburg, J.H., Hamilton, M.S., Kluin, P.M. and Dinarello, C.A. (1989). A single low dose of human recombinant interleukin 1 accelerates the recovery of neutrophils in mice with cyclophosphamide-induced neutropenia. Exp. Hematol. 17, 805.

Fisher, C.J.J., Dhainaut, J.F., Opal, S.M., et al. (1994a). Recombinant human interleukin 1 receptor antagonist in the treatment of patients with sepsis syndrome. Results from a randomized, double-blind, placebo-controlled trial. Phase III rhIL-1ra Sepsis Syndrome Study Group. J. Am. Med. Assoc. 271, 1836.

Fisher, C.J.J., Slotman, G.J., Opal, S.M., et al. (1994b). Initial evaluation of human recombinant interleukin-1 receptor antagonist in the treatment of sepsis syndrome: a randomized, open-label, placebo-controlled multicenter trial. The IL-1RA Sepsis Syndrome Study Group. Crit. Care Med. 22, 12.

Frendl, G., Fenton, M.J. and Beller, D.I. (1990). Regulation of macrophage activation by IL-3. II. IL-3 and lipopolysaccharide act synergistically in the regulation of IL-1 expression. J. Immunol. 144, 3400.

Freshney, N.W., Rawlinson, L., Guesdon, F., et al. (1994). Interleukin-1 activates a novel protein kinase cascade that results in phosphorylation of Hsp27. Cell 78, 1039.

Furutani, Y., Notake, M., Fukui, T., et al. (1986). Complete nucleotide sequence of the gene for human interleukin 1 alpha. Nucleic Acids Res. 14, 3167.

Ghezzi, P. and Dinarello, C.A. (1988). IL-1 induces IL-1. III. Specific inhibition of IL-1 production by IFN-gamma. J. Immunol. 140, 4238.

Gorospe, M., Kumar, S. and Baglioni, C. (1993). Tumor necrosis factor increases stability of interleukin-1 mRNA by activating protein kinase C. J. Biol. Chem. 268, 6214.

Gray, P.W., Glaister, D., Chen, E., Goeddel, D.V. and Pennica, D. (1986). Two interleukin 1 genes in the mouse: cloning and expression of the cDNA for murine interleukin 1 beta. J. Immunol. 137, 3644.

Guesdon, F. and Saklatvala, J. (1991). Identification of a cytoplasmic protein kinase regulated by IL-1 that phosphorylates the small heat shock protein, hsp27. J. Immunol. 147, 3402.

Haq, A.U., Rinehart, J.J. and Maca, R.D. (1985). The effect of gamma interferon on IL-1 secretion of in vitro differentiated human macrophages. J. Leukoc. Biol. 38, 735.

Hart, P.H., Vitti, G.F., Burgess, D.R., Whitty, G.A., Piccoli, D.S. and Hamilton, J.A. (1989). Potential antiinflammatory effects of interleukin 4: suppression of human monocyte tumor necrosis factor alpha, interleukin 1, and prostaglandin E2. Proc. Natl Acad. Sci. USA 86, 3803.

Haskill, S., Johnson, C., Eierman, D., Becker, S. and Warren, K. (1988). Adherence induces selective mRNA expression of monocyte mediators and proto-oncogenes. J. Immunol. 140, 1690.

Haskill, S., Martin, G., Van Le, L., et al. (1991). cDNA cloning of an intracellular form of the human interleukin 1 receptor antagonist associated with epithelium. Proc. Natl Acad. Sci. USA 88, 3681.

Haugen, A., Mann, D., Murray, C., Weston, A. and Willey, J.C. (1989). Interleukin-1 alpha gene intron containing variable repeat region coding for the SP1 transcription factor recognition sequence is polymorphic. Mol. Carcinogen. 2, 68.

Heinrich, P.C., Castell, J.V. and Andus, T. (1990). Interleukin 6 and the acute-phase response. Biochem. J. 265, 621.

Hellerstein, M.K., Meydani, S.N., Meydani, M. and Dinarello, C.A. (1989). Interleukin-1-induced anorexia in rat. J. Clin. Invest. 84, 228.

Herzyk, D.J., Allen, J.N., Marsh, C.B. and Wewers, M.D. (1992). Macrophage and monocyte IL-1 beta regulation differs at multiple sites. Messenger RNA expression, translation, and post-translational processing. J. Immunol. 149, 3052.

Iwamoto, G.K., Monick, M.M., Clark, B.D., Auron, P.E., Stinski, M.F. and Hunninghake, G.W. (1990). Modulation of interleukin 1 beta gene expression by the immediate early genes of human cytomegalovirus. J. Clin. Invest. 85, 1853.

Janson, R.W., Hance, K.R. and Arend, W.P. (1991). Production of IL-1 receptor antagonist by human in vitro-derived macrophages. Effects of lipopolysaccharide and granulocyte-macrophage colony-stimulating factor. J. Immunol. 147, 4218.

Kampschmidt, R.F. and Pulliam, L.A. (1978). Effect of human monocyte pyrogen on plasma iron, plasma zinc, and blood neutrophils in rabbits and rats. Proc. Soc. Exp. Biol. Med. 158, 32.

Kampschmidt, R.F., Pulliam, L.A. and Upchurch, H.F. (1980). The activity of partially purified leukocytic endogenous mediator in endotoxin-resistant C3H/HeJ mice. J. Lab. Clin. Med. 95, 616.

Kern, J.A., Lamb, R.J., Reed, J.C., Daniele, R.P. and Nowell, P.C. (1988). Dexamethasone inhibition of interleukin 1 beta production by human monocytes. Posttranscriptional mechanisms. J. Clin. Invest. 81, 237.

Klein, B., Zhang, X.G., Lu, Z.Y. and Bataille, R. (1995). IL-6 in multiple myeloma. Blood 85, 863.

Knudsen, P.J., Dinarello, C.A. and Strom, T.B. (1986). Prostaglandins posttranscriptionally inhibit monocyte expression of interleukin 1 activity by increasing intracellular cyclic adenosine monophosphate. J. Immunol. 137, 3189.

Knudsen, P.J., Dinarello, C.A. and Strom, T.B. (1987). Glucocorticoids inhibit transcriptional and post-transcriptional expression of interleukin 1 in U937 cells. J. Immunol. 139, 4129.

Kovacs, E.J., Brock, B., Varesio, L. and Young, H.A. (1989). IL-2 induction of IL-1 beta mRNA expression in monocytes. Regulation by agents that block second messenger pathways. J. Immunol. 143, 3532.

Krueger, J.M., Toth, L.A., Floyd, R., et al. (1994). Sleep, microbes and cytokines. Neuroimmunomodulation 1, 100.

Krzesicki, R.F., Hatfield, C.A., Bienkowski, M.J., et al. (1993). Regulation of expression of IL-1 receptor antagonist protein in human synovial and dermal fibroblasts. J. Immunol. 150, 4008.

Kunkel, S.L., Chensue, S.W. and Phan, S.H. (1986). Prostaglandins as endogenous mediators of interleukin 1 production. J. Immunol. 136, 186.

Kuno, K., Sukegawa, K., Ishikawa, Y., Orii, T. and Matsushima, K. (1994). Acid shingomyelinase is not essential for the IL-1 and tumor necrosis factor receptor signaling pathway leading to NFkB activation. Int. Immunol. 6, 1269.

Kupper, T.S., Ballard, D.W., Chua, A.O., et al. (1986). Human keratinocytes contain mRNA indistinguishable from monocyte interleukin 1 alpha and beta mRNA. Keratinocyte epidermal cell-derived thymocyte-activating factor is identical to interleukin 1. J. Exp. Med. 164, 2095.

Kupper, T.S., Chua, A.O., Flood, P., McGuire, J. and Gubler, U. (1987). Interleukin 1 gene expression in cultured human keratinocytes is augmented by ultraviolet irradiation. J. Clin. Invest. 80, 430.

Kushner, I. (1982). The phenomenon of the acute-phase response. Ann. N.Y. Acad. Sci. 389, 39.

Lafage, M., Maroc, N., Dubreuil, P., et al. (1989). The human interleukin-1 alpha gene is located on the long arm of chromosome 2 at band q13. Blood 73, 104.

Lee, S.W., Tsou, A.P., Chan, H., et al. (1988). Glucocorticoids selectively inhibit the transcription of the interleukin 1 beta gene and decrease the stability of interleukin 1 beta mRNA. Proc. Natl Acad. Sci. USA 85, 1204.

Lennard, A., Gorman, P., Carrier, M., et al. (1992). Cloning and chromosome mapping of the human interleukin-1 receptor antagonist gene. Cytokine 4, 83.

Lew, W., Oppenheim, J.J. and Matsushima, K. (1988). Analysis of the suppression of IL-1 alpha and IL-1 beta production in human peripheral blood mononuclear adherent cells by a glucocorticoid hormone. J. Immunol. 140, 1895.

Lomedico, P.T., Gubler, U., Hellmann, C.P., et al. (1984). Cloning and expression of murine interleukin-1 cDNA in Escherichia coli. Nature 312, 458.

Manetti, R., Barak, V., Piccinni, M.P., et al. (1994). Interleukin-1 favours the in vitro development of type 2 T helper (Th2) human T-cell clones. Res. Immunol. 145, 93.

Mantovani, A. (1994). Tumor-assisted macrophages in neoplastic progression: A paradigm for the in vivo function of chemokines. Lab. Invest. 71, 5.

Mantovani, A., Bussolino, F. and Dejana, E. (1992). Cytokine regulation of endothelial cell function. FASEB J. 6, 2591.

March, C.J., Mosley, B., Larsen, A., et al. (1985). Cloning, sequence and expression of two distinct human interleukin-1 complementary DNAs. Nature 315, 641.

Mathias, S., Younes, A., Kan, C.C., Orlow, I., Joseph, C. and Kolesnick, R.N. (1993). Activation of the sphingomyelin signaling pathway in intact EL4 cells and in a cell-free system by IL-1 beta. Science 259, 519.

McKean, D.J., Podzorski, R.P., Bell, M.P., et al. (1993). Murine T helper cell-2 lymphocytes express type I and type II IL-1 receptors, but only the type I receptor mediates costimulatory activity. J. Immunol. 151, 3500.

McMahan, C.J., Slack, J.L., Mosley, B., et al. (1991). A novel IL-1 receptor, cloned from B cells by mammalian expression, is expressed in many cell types. EMBO J. 10, 2821.

Minty, A., Chalon, P., Derocq, J.M., et al. (1993). Interleukin-13 is a new human lymphokine regulating inflammatory and immune responses. Nature 362, 248.

Miura, M., Zhu, H., Rotello, R., Hartwieg, E.A. and Yuan, J. (1993). Induction of apoptosis in fibroblasts by IL-1 beta-converting enzyme, a mammalian homolog of the C. elegans cell death gene ced-3. Cell 75, 653.

Mizel, S.B. (1990). How does interleukin 1 activate cells? Cyclic AMP and interleukin 1 signal transduction. Immunol. Today 11, 390.

Moore, K.W., Vieira, P., Fiorentino, D.F., Trounstine, M.L., Khan, T.A. and Mosmann, T.R. (1990). Homology of cytokine synthesis inhibitory factor (IL-10) to the Epstein–Barr virus gene BCRFI. Science 248, 1230.

Muegge, K., Williams, T.M., Kant, J., et al. (1989). Interleukin-1 costimulatory activity on the interleukin-2 promoter via AP-1. Science 246, 249.

Muzio, M., Re, F., Sironi, M., et al. (1994a). Interleukin-13 induces the production of interleukin-1 receptor antagonist (IL-1ra) and the expression of the mRNA for the intracellular (keratinocyte) form of IL-1ra in human myelomonocytic cells. Blood 83, 1738.

Muzio, M., Sironi, M., Polentarutti, N., Mantovani, A. and Colotta, F. (1994b). Induction by transforming growth factor-beta 1 of the interleukin-1 receptor antagonist and of its intracellular form in human polymorphonuclear cells. Eur. J. Immunol. 24, 3194.

Muzio, M., Polentarutti, N., Sironi, M., *et al.* (1995). Cloning and characterization of a new isoform of the interleukin-1 receptor antagonist. J. Exp. Med. 182, 623.

Neta, R., Oppenheim, J.J. and Douches, S.D. (1988). Interdependence of the radioprotective effects of human recombinant interleukin 1 alpha, tumor necrosis factor alpha, granulocyte colony-stimulating factor, and murine recombinant granulocyte-macrophage colony-stimulating factor. J. Immunol. 140, 108.

Nishida, T., Takano, M., Kawakami, T., Nishino, N., Nakai, S. and Hirai, Y. (1988). The transcription of the interleukin 1 beta gene is induced with PMA and inhibited with dexamethasone in U937 cells. Biochem. Biophys. Res. Commun. 156, 269.

Nishida, T., Nakai, S., Kawakami, T., Aihara, K., Nishino, N. and Hirai, Y. (1989). Dexamethasone regulation of the expression of cytokine mRNAs induced by interleukin-1 in the astrocytoma cell line U373MG. FEBS Lett. 243, 25.

Numerof, R.P., Kotik, A.N., Dinarello, C.A. and Mier, J.W. (1990). Pro-interleukin-1 beta production by a subpopulation of human T cells, but not NK cells, in response to interleukin-2. Cell. Immunol. 130, 118.

O'Neill, L.A., Bird, T.A. and Saklatvala, J. (1990). How does interleukin 1 activate cells? Interleukin 1 signal transduction. Immunol. Today 11, 392.

Ohmori, Y., Strassman, G. and Hamilton, T.A. (1990). cAMP differentially regulates expression of mRNA encoding IL-1 alpha and IL-1 beta in murine peritoneal macrophages. J. Immunol. 145, 3333.

Okusawa, S., Gelfand, J.A., Ikejima, T., Connolly, R.J. and Dinarello, C.A. (1988). Interleukin 1 induces a shock-like state in rabbits. Synergism with tumor necrosis factor and the effect of cyclooxygenase inhibition. J. Clin. Invest. 81, 1162.

Oppenheim, J.J. and Gery, I. (1993). From lymphodrek to interleukin 1 (IL-1). Immunol. Today 14, 232.

Oppenheim, J.J., Neta, R., Tiberghien, P., Gress, R., Kenny, J.J. and Longo, D.L. (1989). Interleukin-1 enhances survival of lethally irradiated mice treated with allogeneic bone marrow cells. Blood 74, 2257.

Perretti, M., Becherucci, C., Scapigliati, G. and Parente, L. (1989). The effect of adrenalectomy on interleukin-1 release in vitro and in vivo. Br. J. Pharmacol. 98, 1137.

Postlethwaite, A.E., Lachman, L.B., Mainardi, C.L. and Kang, A.H. (1983). Interleukin 1 stimulation of collagenase production by cultured fibroblasts. J. Exp. Med. 157, 801.

Poutsiaka, D.D., Mengozzi, M., Vannier, E., Sinha, B. and Dinarello, C.A. (1993). Cross-linking of the beta-glucan receptor on human monocytes results in interleukin-1 receptor antagonist but not interleukin-1 production. Blood 82, 3695.

Pribble, J., Fisher, C., Opal, S., *et al.* (1994). Human recombinant receptor antagonist increases survival time in patients with sepsis syndrome and end organ dysfunction. Crit. Care Med. 22, A192.

Priestle, J.P., Schar, H.P. and Grutter, M.G. (1988). Crystal structure of the cytokine interleukin-1 beta. EMBO J. 7, 339.

Priestle, J.P., Schar, H.P. and Grutter, M.G. (1989). Crystallographic refinement of interleukin 1 beta at 2.0 Å resolution. Proc. Natl Acad. Sci. USA 86, 9667.

Ramadori, G., Sipe, J.D., Dinarello, C.A., Mizel, S.B. and Colten, H.R. (1985). Pretranslational modulation of acute-phase hepatic protein synthesis by murine recombinant interleukin 1 (IL-1) and purified human IL-1. J. Exp. Med. 162, 930.

Rambaldi, A., Torcia, M., Bettoni, S., *et al.* (1991). Modulation of cell proliferation and cytokine production in acute myeloblastic leukemia by interleukin-1 receptor antagonist and lack of its expression by leukemic cells. Blood 78, 3248.

Re, F., Mengozzi, M., Muzio, M., Dinarello, C.A., Mantovani, A. and Colotta, F. (1993). Expression of interleukin-1 receptor antagonist (IL-1ra) by human circulating polymorphonuclear cells. Eur. J. Immunol. 23, 570.

Re, F., Muzio, M., De Rossi, M., *et al.* (1994). The type II "receptor" as a decoy target for IL-1 in polymorphonuclear leukocytes: characterization of induction by dexamethasone and ligand binding properties of the released decoy receptor. J. Exp. Med. 179, 739.

Re, F., Sironi, M., Muzio, M., *et al.* (1996). Inhibition of interleukin-1 responsiveness by type II receptor gene transfer: a surface "receptor" with anti-interleukin-1 function. J. Exp. Med. 183, 1841.

Rettori, V., Gimeno, M.F., Karara, A., Gonzalez, M.C. and McCann, S.M. (1991). Interleukin 1α inhibits prostaglandin E_2 release to suppress pulsatile release of luteinizing hormone but not follicle-stimulating hormone. Proc. Natl Acad. Sci. USA 88, 2763.

Rosoff, P.M., Savage, N. and Dinarello, C.A. (1988). Interleukin-1 stimulates diacylglycerol production in T lymphocytes by a novel mechanism. Cell 54, 73.

Rossi, B. (1993). IL-1 transduction signals. Eur. Cytokine Netw. 4, 181.

Rossi, E., Breviario, F., Dejana, E. and Mantovani, A. (1985). Prostacyclin synthesis induced in vascular cells by interleukin-1. Science 229, 174.

Roux Lombard, P., Modoux, C. and Dayer, J.M. (1989). Production of interleukin-1 (IL-1) and a specific IL-1 inhibitor during human monocyte-macrophage differentiation: influence of GM-CSF. Cytokine 1, 45.

Rubartelli, A., Cozzolino, F., Talio, M. and Sitia, R. (1990). A novel secretory pathway for interleukin-1 beta, a protein lacking a signal sequence. EMBO J. 9, 1503.

Saklatvala, J., Sarsfield, S.J. and Townsend, Y. (1985). Pig Interleukin-1. J. Exp. Med. 162, 1208.

Sapolsky, R., Rivier, C., Yamamoto, G., Plotsky, P. and Vale, W. (1987). Interleukin-1 stimulates the secretion of hypothalamic corticotropin-releasing factor. Science 238, 522.

Schindler, R., Gelfand, J.A. and Dinarello, C.A. (1990a). Recombinant C5a stimulates transcription rather than translation of interleukin-1 (IL-1) and tumor necrosis factor: translational signal provided by lipopolysaccharide or IL-1 itself. Blood 76, 1631.

Schindler, R., Mancilla, J., Endres, S., Ghorbani, R., Clark, S.C. and Dinarello, C.A. (1990b). Correlations and interactions in the production of interleukin-6 (IL-6), IL-1, and tumor necrosis factor (TNF) in human blood mononuclear cells: IL-6 suppresses IL-1 and TNF. Blood 75, 40.

Schmidt, J.A. and Abdulla, E. (1988). Down-regulation of IL-1 beta biosynthesis by inducers of the heat-shock response. J. Immunol. 141, 2027.

Schmidt, J.A., Mizel, S.B., Cohen, D. and Green, I. (1989). Interleukin 1, a potential regulator of fibroblast proliferation. J. Immunol. 128, 2177.

Shieh, J.H., Peterson, R.H. and Moore, M.A. (1993). Cytokines and dexamethasone modulation of IL-1 receptors on human neutrophils in vitro. J. Immunol. 150, 3515.

Shields, J., Bernasconi, L.M., Benotto, W., Shaw, A.R. and Mazzei, G.J. (1990). Production of a 26,000-dalton interleukin 1 inhibitor by human monocytes is regulated by granulocyte-macrophage colony-stimulating factor. Cytokine 2, 122.

Sims, J.E., March, C.J., Cosman, D., *et al.* (1988). cDNA espression cloning of the IL-1 receptor, a member of the immunoglobulin superfamily. Science 241, 585.

Sims, J.E., Acres, R.B., Grubin, C.E., *et al.* (1989). Cloning of the interleukin 1 receptor from human T cells. Proc. Natl Acad. Sci. USA 86, 8946.

Sims, J.E., Gayle, M.A., Slack, J.L., *et al.* (1993). Interleukin 1 signaling occurs exclusively via the type I receptor. Proc. Natl Acad. Sci. USA 90, 6155.

Smith, J.W., Urba, W.J., Curti, B.D., *et al.* (1992a). The toxic and hematologic effects of interleukin-1 alpha administered in a phase I trial to patients with advanced malignancies. J. Clin. Oncol. 10, 1141.

Smith, M.F.J., Eidlen, D., Brewer, M.T., Eisenberg, S.P., Arend, W.P. and Gutierrez Hartmann, A. (1992b). Human IL-1 receptor antagonist promoter. Cell type-specific activity and identification of regulatory regions. J. Immunol. 149, 2000.

Snyder, D.S. and Unanue, E.R. (1982). Corticosteroids inhibit murine macrophage Ia expression and interleukin 1 production. J. Immunol. 129, 1803.

Steinkasserer, A., Spurr, N.K., Cox, S., Jeggo, P. and Sim, R.B. (1992). The human IL-1 receptor antagonist gene (IL1RN) maps to chromosome 2q14-q21, in the region of the IL-1 alpha and IL-1 beta loci. Genomics 13, 654.

Stylianou, E., O'Neill, L.A., Rawlinson, L., Edbrooke, M.R., Woo, P. and Saklatvala, J. (1992). Interleukin 1 induces NF-kappa B through its type I but not its type II receptor in lymphocytes. J. Biol. Chem. 267, 15836.

Sung, S.S. and Walters, J.A. (1991). Increased cyclic AMP levels enhance IL-1 alpha and IL-1 beta mRNA expression and protein production in human myelomonocytic cell lines and monocytes. J. Clin. Invest. 88, 1915.

Telford, J.L., Macchia, G., Massone, A., Carinci, V., Palla, E. and Melli, M. (1986). The murine interleukin 1 beta gene: structure and evolution. Nucleic Acids Res. 14, 9955.

Tewari, A., Buhles, W.C.J. and Starnes, H.F.J. (1990). Preliminary report: effects of interleukin-1 on platelet counts. Lancet 336, 712. [Retracted by Buhles W.C., Starnes H.F. (1992). Lancet 340(8817), 496]

Thornberry, N.A., Bull, H.G., Calaycay, J.R., *et al.* (1992). A novel heterodimeric cysteine protease is required for interleukin-1 beta processing in monocytes. Nature 356, 768.

Turner, M., Chantry, D. and Feldmann, M. (1988). Post-transcriptional control of IL-1 gene expression in the acute monocytic leukemia line THP-1. Biochem. Biophys. Res. Commun. 156, 830.

Turner, M., Chantry, D., Buchan, G., Barrett, K. and Feldmann, M. (1989). Regulation of expression of human IL-1 alpha and IL-1 beta genes. J. Immunol. 143, 3556.

Ucla, C., Roux Lombard, P., Fey, S., Dayer, J.M. and Mach, B. (1990). Interferon gamma drastically modifies the regulation of interleukin 1 genes by endotoxin in U937 cells. J. Clin. Invest. 85, 185.

Uehara, A., Sekiya, C., Takasugi, Y., Namiki, M. and Arimura, A. (1990). Anorexia induced by interleukin-1: involvement of corticotropin-releasing factor. Am. J. Physiol. 137, 1173.

Vannier, E. and Dinarello, C.A. (1993). Histamine enhances interleukin (IL)-1-induced IL-1 gene expression and protein synthesis via H2 receptors in peripheral blood mononuclear cells. Comparison with IL-1 receptor antagonist. J. Clin. Invest. 92, 281.

Vannier, E., Miller, L.C. and Dinarello, C.A. (1992). Coordinated antiinflammatory effects of interleukin 4: interleukin 4 suppresses interleukin 1 production but up-regulates gene expression and synthesis of interleukin 1 receptor antagonist. Proc. Natl Acad. Sci. USA 89, 4076.

Van Snick, J., Masson, P.L. and Heremans, J.F. (1974). The involvement of lactoferrin in the hyposideremia of acute inflammation. J. Exp. Med. 140, 1068.

Walker, N.P., Talanian, R.V., Brady, K.D., *et al.* (1994). Crystal structure of the cysteine protease interleukin-1 beta-converting enzyme: a (p20/p10)2 homodimer. Cell 78, 343.

Warner, S.J., Auger, K.R. and Libby, P. (1987a). Interleukin 1 induces interleukin 1. II. Recombinant human interleukin 1 induces interleukin 1 production by adult human vascular endothelial cells. J. Immunol. 139, 1911.

Warner, S.J., Auger, K.R. and Libby, P. (1987b). Human interleukin 1 induces interleukin 1 gene expression in human vascular smooth muscle cells. J. Exp. Med. 165, 1316.

Wasik, M.A., Donnelly, R.P. and Beller, D.I. (1988). Lymphokine-independent induction of macrophage membrane IL-1 by autoreactive T cells recognizing either class I or class II MHC determinants. J. Immunol. 141, 3456.

Webb, A.C., Collins, K.L., Auron, P.E., *et al.* (1986). Interleukin-1 gene (IL1) assigned to long arm of human chromosome 2. Lymphokine Res. 5, 77.

Wilson, K.P., Black, J.A., Thomson, J.A., *et al.* (1994). Structure and mechanism of interleukin-1 beta converting enzyme. Nature 370, 270.

Woloski, B.M.R.N.J., Smith, E.M., Meyer, W.J.I., Fuller, G.M. and Blalock, J.E. (1985). Corticotropin-releasing activity of monokines. Science 230, 1035.

Yamato, K., el Hajjaoui, Z. and Koeffler, H.P. (1989). Regulation of levels of IL-1 mRNA in human fibroblasts. J. Cell. Physiol. 139, 610.

Yamayoshi, M., Ohue, M., Kawashima, H., *et al.* (1990). A human IL-1 alpha derivative which lacks prostaglandin E2 inducing activity and inhibits the activity of IL-1 through receptor competition. Lymphokine Res. 9, 405.

Ye, K., Dinarello, C.A. and Clark, B.D. (1993). Identification of the promoter region of human interleukin 1 type I receptor gene: multiple initiation sites, high G+C content, and constitutive expression. Proc. Natl Acad. Sci. USA 90, 2295.

2. Interleukin-2

Robin Thorpe

1. Introduction

Interleukin-2 was originally described as "T cell growth factor" present in lymphocyte conditioned medium which was able to maintain the growth of cytotoxic T cells for relatively long periods (Morgan *et al.*, 1976). Subsequent studies using rDNA-derived IL-2 have revealed that the cytokine is a potent growth and differentiation factor for T cells as well as stimulating natural killer (NK) cells, and lymphocyte-activated killer (LAK) cells. IL-2 also shows strong B cell growth factor activity and can stimulate monocytic lineage cells. As with many cytokines, some of the activities of IL-2 are in part shared with other molecules, e.g., IL-4, IL-7, and IL-15.

Interleukin-2 has been the subject of a very large literature, and is often described in early reports by a variety of alternative names, such as blastogenic factor, costimulator activity, or thymocyte differentiation factor. In some cases such activities relate to the use of crude preparations which probably contain cytokines and other biologically active molecules as well as IL-2, and so the precise "factor" responsible for the action described is not always clear.

Cytokines
ISBN 0–12–498340–5

2. *The IL-2 Gene*

The human IL-2 gene is present as a single copy on chromosome 4q26-28.

2.1 Genomic Organization

The complete sequence and structure of the human IL-2 gene were derived using cloned DNA from peripheral blood lymphocytes (Fujita *et al.*, 1983) or cDNA from the Jurkat T cell line (Holbrook *et al.*, 1984). A Jurkat-derived clone has been expressed in monkey cells (Taniguchi *et al.*, 1983). The gene contains three introns, two of which are large (see Figures 2.1 and 2.2). The first exon contains the

5′ untranslated region of 47 bp and the fourth exon contains the 3′ untranslated region of 279 bp. The gene spans 5737 bp. The promoter region contains a "TATA" sequence 75 bp upstream from the translation unit site and also contains a palindromic sequence. There are sequences in the promoter which show some homology to the γ-interferon gene, and the second intron contains a motif which closely resembles the core sequence for the viral enhancer elements (Fujita *et al.*, 1983).

2.2 Regulation of Gene Expression

Stimulation of T cells via the antigen receptor (either directly with antigen or indirectly using plant lectins like

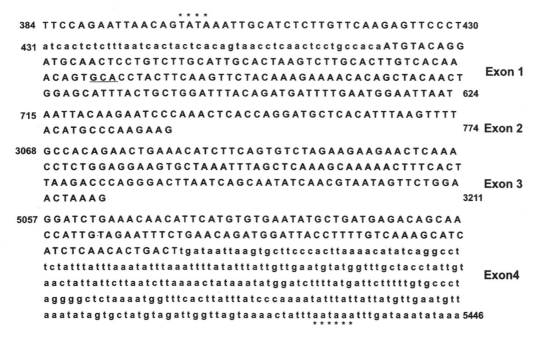

Figure 2.1   The sequence of the IL-2 gene. The four exons of the IL-2 gene are indicated. The TATA box and the AATAAA sequence are marked with asterisks. The underlined GCA is the coding sequence for the first amino acid residue of the mature protein. The lower-case bases in exons 1 and 4 are nontranslated regions.

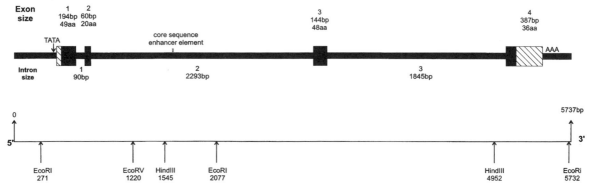

Figure 2.2   Organization of the IL-2 gene. The four exons are shown, with the positions of regulatory sequences indicated. The shaded portions in exons 1 and 4 show the untranslated regions.

concanavalin A (Con A) or phytohaemagglutinin (PHA)) results in production and secretion of IL-2. Expression is controlled by an enhancer element upstream of the 5' end of the TATA box (between bp −319 and −52). This includes binding sites for NFAT-1, NFκB, AP-1, and Oct-1/NF-IL-2-A proteins. Evidence suggests that all the binding sites for the above proteins must be occupied if efficient activation of the gene is to follow. This may explain the requirement for a dual stimulus for maximal IL-2 secretion, i.e., signaling via the antigen receptor and activation of protein kinase C (often achieved *in vitro* with phorbol esters like phorbol myristate acetate) – the latter is associated with full activation of production of AP-1 and NFκB proteins (Crabtree, 1989; Hoyos *et al.*, 1989; McGuire and Iacobelli, 1997; Muegge and Durum 1989; Ullman *et al.*, 1990). IL-2 gene expression is also regulated by stabilization of IL-2 mRNA mediated by motifs in the 3' untranslated region of the message; UA-rich regions are associated with instability and negative translational control (Shaw and Kamen, 1986; Kruys *et al.*, 1989). A negative response element which suppresses IL-2 production has been identified in the IL-2 gene promoter/enhancer region. A labile repressor which acts to reduce processing of IL-2 mRNA precursors has also been described (Kaempfer *et al.*, 1987).

3.   The IL-2 Protein

The mature human IL-2 protein consists of 133 amino acids (see Figure 2.3) (Robb *et al.*, 1984). The gene sequence suggests that it is synthesized as a precursor containing 153 amino acids and that the 20-residue hydrophobic leader sequence (signal peptide) is cleaved to produce the mature protein prior to or during secretion (Clark *et al.*, 1984; Smith *et al.*, 1983; Stern *et al.*, 1984).

The molecule contains a single *N*-linked glycosylation site at position 3 and differences in glycosylation cause size and charge heterogeneity in both "natural" and cell line-derived IL-2 (Robb, 1984; Robb and Smith, 1981). Glycosylation is not necessary for biological activity, but is necessary for the binding of at least one monoclonal antibody. Several rDNA-derived IL-2 preparations have been produced; nonglycosylated material prepared using prokaryotic organisms is both stable and fully biologically active.

The natural molecule contains three cysteine residues at positions 58, 105, and 125, two of which (58 and 105) form a disulfide bridge that is essential for the biological activity of the molecule (Wang *et al.*, 1984). The cysteine at position 125 can form intermolecular disulfide bridges, leading to the production of dimers and higher aggregates. Molecules folded with an internal disulfide bridge involving the cysteine at position 125 (i.e., between residues 58–125 and 105–125) show minimal, if any, biological activity. Although these do not seem to occur in "natural" IL-2 preparations, they can be present or even predominate in rDNA-derived products. Refolding of rDNA IL-2 to produce bioactive cytokine is a critical step in production of such preparations; controlling the conditions so that the "correct" disulfide bridge is produced has been found problematical. For this reason many rDNA preparations (especially those produced using prokaryotic systems) are produced as muteins with the cysteihe at position 125 replaced by another amino acid, e.g., serine or alanine (Liang *et al.*, 1986; Wang *et al.*, 1984). Such muteins show similar biological activity to the native molecule. Genetically engineered IL-2 preparations produced using bacterial systems may contain an additional methionine residue at the amino terminus which does not seem to influence biological activity.

Natural IL-2 occurs as a group of glycoproteins with apparent M_r values between 13 000 and 17 500 (as defined by SDS-PAGE) and pI values between 6.6 and

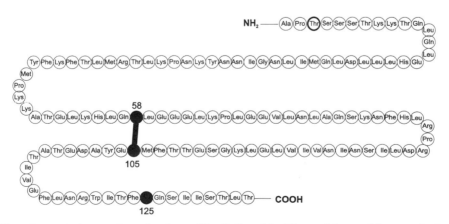

Figure 2.3   The primary and secondary structure of the IL-2 protein. The cysteine residues are indicated by solid circles and the glycosylation site by a bold circle.

8.2 (Robb and Smith, 1981). It is a hydrophobic molecule which is stable to moderate heat and stable at low pH (Robb et al., 1983).

Crystallography has shown IL-2 to consist of six α-helical domains (A–F) with no apparent β structural regions. An antiparallel helical bundle is formed from helices C–E. Helices A and B are joined through a bent loop to the rest of the molecule (see Figure 2.4) (Brandhuber et al., 1987; Mott et al., 1995). An alternative structure has also been proposed (Bazan, 1992). A region of the N-terminus of the A helix has been proposed as a binding site for the β (p70) chain of the IL-2 receptor (Collins et al., 1988) and part of the loop connecting the A and B helices is involved in binding the α (p55) receptor chain (Grant et al., 1992; Sauve et al., 1991). The binding site for the γ chain of the receptor possibly involves the C terminal of helix D.

4.   Cellular Sources and Production

IL-2 appears to be produced exclusively by T lymphocytes and both CD4$^+$ and CD8$^+$ T cells can secrete the cytokine. T_H1 and T_H2 subpopulations of helper T cells can produce IL-2, although T_H1 cells may produce higher levels than T_H2 cells (this is not yet clear, especially for human cells). Some T cell clones, lines and

hybridomas can also produce IL-2. Relatively few continuously growing lymphoblastoid T cell lines can produce significant levels of IL-2.

T cells require stimulation with antigen or mitogenic lectins (like PHA or Con A) if they are to secrete detectable IL-2. This involves activation via the T cell receptor. A second signal such as that mediated by IL-1 or phorbol esters is also required if significant production of IL-2 is to be achieved (see Section 2.2 above). T cell clones, lines, and hybridomas normally also need some stimulus if they are to produce the cytokine.

Requirements for stimulation of lymphoblastoid T cell lines for production of IL-2 vary. Some clones of the Jurkat human T cell line will secrete relatively substantial amounts of IL-2 following stimulation with PHA alone, although this can be dramatically increased by addition of phorbol myristate acetate (PMA). The mouse EL-4 thymoma line produces maximal amounts of IL-2 following stimulation with PMA alone, and the gibbon line MLA144 constitutively secretes appreciable quantities of IL-2.

The time course of IL-2 production by T cell preparations, lines, clones, hybridomas, and lympho-blastoid lines may vary. Most "normal" human activated T cells produce optimal IL-2 preparations at 40–48 h following activation with PHA or PHA/PMA. The Jurkat human lymphoblastoid line produces potent IL-2 preparations 18 h after stimulation with PHA or PHA/PMA.

IL-2 from most other species studied seem similar to the human molecule, but the mouse protein contains an unusual insertion comprising 12 repeat glutamine residues which are not found in IL-2 from other species (Yokota et al., 1985). The mouse molecule occurs mainly as a dimer, whereas the human protein is largely monomeric.

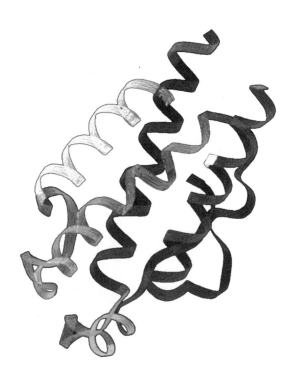

Figure 2.4   The three-dimensional structure of the IL-2 protein. Generated using the INSIGHT II graphic modeling program (Biosym Technology, San Diego, USA). Residues 98–104 are not included.

5.   Biological Activity

The most significant effects of IL-2 are exerted on leukocytes, although less pronounced action on other cell types has been reported. As with many cytokines, IL-2-stimulated cells can secrete other cytokines or other biologically active substances which can mediate "secondary" effects.

5.1   EFFECTS ON T LYMPHOCYTES

IL-2 stimulates dramatic proliferation of activated (antigen or lectin) T lymphocytes. It acts on all subpopulations of T cells and promotes progression through the G_1 phase of the cell cycle, resulting in growth of cells and increase in cell numbers (Smith, 1980). It can also cause proliferation of "resting" T cells, but as these cells do not express significant amounts of the IL-2R α chain, this requires a much higher amount of IL-2 than is required for activated cells (see Section 6 below); the effect of "physiological" amounts of IL-2 on nonactivated lymphocytes is questionable.

IL-2 also stimulates cytolytic activity of subsets of T lymphocytes, enhances T cell motility, and induces secretion of other cytokines such as IFN-γ, IL-4, and TNF. It is, therefore, a T cell differentiation factor (Farrar et al., 1982; Howard et al., 1983).

Proliferative effects of IL-2 on thymocytes have been demonstrated (Zuniga Pflucker et al., 1990) and the cytokine may play a role in thymic development (Carding et al., 1989; Tentori et al., 1988). This is supported by studies involving neutralizing antibodies specific for IL-2 (Jenkinson et al., 1987), but perhaps questioned by knockout mouse studies (Schorle et al., 1991). It seems most likely that the in vivo role of IL-2 in development of thymic function is complex and involves the presence of and interaction with many other cytokines and biological substances.

The effects of the cytokine generally enhance and potentiate immune responses. Activated T cells show increased responses to IL-2 if IL-4 is also present, but nonactivated T cells seem refractory to this (Martinez et al., 1990; Mitchell et al., 1989). In vitro, IL-2 is capable of maintenance of relatively long-term growth of activated T cell lines and clones (Morgan et al., 1976; Schreier and Tees, 1981).

5.2   EFFECTS ON B LYMPHOCYTES

IL-2 stimulates the proliferation of activated (antigen or anti-IgM) B lymphocytes (Gearing et al., 1985; Tsudo et al., 1984). It also promotes the induction of immunoglobulin secretion (Ralph et al., 1984) and J chain synthesis (Blackman et al., 1986) by B cells. As with T cells, IL-2 acts to enhance immune effects mediated by activated B cells (Mingari et al., 1984).

5.3   EFFECTS ON LARGE GRANULAR LYMPHOCYTES

IL-2 causes the proliferation of large granular lymphocytes. It enhances natural killer (NK) cell activity and induces the so-called "lymphokine activated killer" (LAK) cell activity (see Section 9) which seem at least partly to be mediated by large granular lymphocytes. IL-2 stimulates the cytolytic activity of these cell types and induces their secretion of other cytokines including IFN-γ (Ortaldo et al., 1984; Trinchieri et al., 1984).

5.4   EFFECTS ON MONOCYTIC CELLS

IL-2 promotes the cytolytic activity of blood monocytes and some reports suggest that it promotes proliferation and differentiation of these cells (Baccarini et al., 1989).

The cytokine enhances macrophage antibody-dependent tumoricidal activity (Ralph et al., 1988). Other cells of monocytic lineage, e.g., oligodendrocytes, show enhanced growth and proliferation in response to IL-2 (Benveniste and Merrill, 1986; Saneto et al., 1986).

5.5   EFFECTS ON OTHER CELL TYPES

Effects of IL-2 on nonhemopoietic cells are not entirely clear, although IL-2 receptors have been detected on a variety of cell types including fibroblasts (Plaisance et al., 1992), squamous cell carcinoma lines (Weidmann et al., 1992), and rat epithelial cells (Ciacci et al., 1993). Reported evidence suggests that IL-2 can enhance the proliferation of at least some of these cell types.

6.   The IL-2 Receptor

The IL-2 receptor consists of three chains (known as α, β and γ) which interact with each other and IL-2 to effectively signal IL-2-mediated events to the cell (Minami et al., 1993). It has been suggested that there may be other "receptor" or at least receptor-associated chains present at least on some cell types, but this is still controversial.

6.1   STRUCTURE OF THE IL-2 RECEPTOR CHAINS

6.1.1   The α Chain

The α chain of the IL-2R is an approximately 55 kDa glycoprotein originally described as occurring on activated T cells as p55 or the Tac antigen (Leonard et al., 1985a). It has subsequently been classified as the CD25 surface marker. The mature protein (see Figure 2.5) consists of 251 amino acids (aa) most of which make up the extracellular portion of the receptor chain. There is a 19 aa membrane spanning region and a very short (13-aa) intracellular "tail". The protein is synthesized as a 272-aa precursor which contains a 21-aa signal peptide which is cleaved to yield the mature functional form. The molecule contains five internal disulfide bonds which seem essential for function. Amino acid residues 1–6 and 35–43 appear critical for binding to IL-2.

The IL-2R α chain is encoded by a single gene on chromosome 10p14-15 (Leonard et al., 1985b). It is not a member of the cytokine receptor superfamily, but has regions with some homology with complement protein Ba fragment. A soluble form of the IL-2R α chain (sCD25) which is able to bind IL-2 can be generated by proteolysis and released from the cell surface (Robb and Kutny, 1987; Rubin et al., 1985).

6.1.2 The β Chain

The β chain of the IL-2R (p70 or p75; CD122) is a 70–75 kDa glycoprotein coded by a single gene on chromosome 22q11.2-12 (Robb and Kutny, 1987). The protein consists of 525 amino acids (see Figure 2.4) of which 214 comprise the extracellular portion and 286 make up the intracellular domain; there is a 25-aa transmembrane portion between these two regions (Hatakeyama *et al.*, 1989). The protein is synthesized as a 551-aa precursor.

The extracellular part of the molecule shows it to a member of the cytokine receptor superfamily (Cosman *et al.*, 1990), with the four characteristic conserved cysteine residues and WSXWS motif (see Figure 2.5). The large intracellular domain contains a relatively large number (42) of proline residues and also a serine-rich region, which are like regions linked with signal transduction function in other receptors (Hatakeyama *et al.*, 1989). It also possesses an acidic region in the center of the domain (residues 345–390) containing high numbers of glutamic and aspartic acid residues. This latter structure is associated with interaction with members of the *src* family of tyrosine kinases. However, no intrinsic kinase activity is predicted from the sequence of the intracellular domain.

6.1.3 The γ Chain

The γ chain of the IL-2R is a 64 kDA glycoprotein (p64; CD132) containing 347 amino acids (see Figure 2.5). It comprises a 232-aa extracellular domain, a 29-aa transmembrane region and a 64-aa intracellular portion (Takeshita *et al.*, 1992). It is encoded by a single gene on chromosome Xq13 (Noguchi *et al.*, 1993) and is synthesized as a 369-aa precursor.

The extracellular domain contains the WSXWS element and the four conserved cysteine residues confirming the molecule as a member of the cytokine receptor superfamily. It also has a leucine zipper motif (see Figure 2.5). The cytoplasmic domain shows no apparent kinase-like sequences but does contain a region like the *src* homology domain (SH2). The γ chain of the IL-2R is also a component of the IL-4, IL-7, and IL-15 receptors, and may be part of the receptors for IL-9 and IL-13. It is thus sometimes known as the cytokine receptor common γ chain (γ_c).

6.2 IL-2–IL-2 RECEPTOR BINDING

The α chain of the IL-2R binds IL-2 with relatively low affinity ($k_d \sim 10^{-8}$ M) and the isolated β chains show even lower affinity for the cytokine ($k_d \sim 10^{-7}$ M). The γ chain

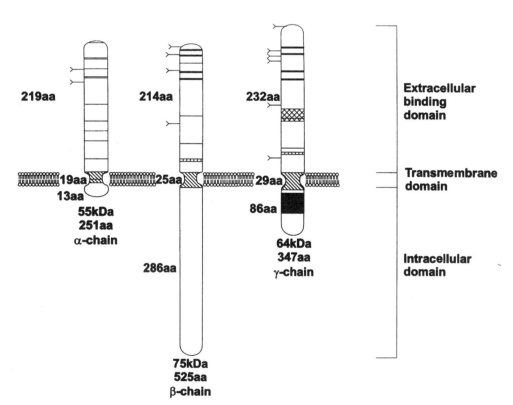

Figure 2.5 The structure of the IL-2 receptor. The thick lines indicate the four conserved cysteine residues characteristic of members of the cytokine (hemopoietic) receptor superfamily. Thin lines show nonconserved cysteines. The transmembrane regions are represented by diagonal hatching and the SH2 domain of the K_c chain by solid shading. The areas shaded with vertical lines represent the WSXWS motif and the leucine zipper in the K_c chain is shown by cross-hatching.

shows little significant binding capacity in isolation for IL-2. However, combination of the various chains results in increased affinity for IL-2 (Wu *et al.*, 1995). Heterodimers of βγ or αγ chains show intermediate affinity for IL-2 ($k_d \sim 10^{-9}$ M). However, the affinity of heterotrimers of αβγ chains for IL-2 is dramatically increased compared with all other combinations ($k_d \sim 10^{-11}$ M) and such receptor complexes are normally referred to as "high affinity". It seems that in complexes containing β and γ chains (including αβγ trimers) the interaction with IL-2 causes molecular reorganization which allows direct contact/binding between the γ chain and the bound IL-2 molecule (Voss *et al.*, 1993).

7. Receptor Expression and Signal Transduction

Several cell types including resting T and B lymphocytes, NK cells, and monocytic cells constitutively express the β and γ chains of the IL-2R; these are probably present as βγ heterodimers. The structure of both chains suggests that they are capable of signal transduction, but the relatively low affinity of the complexes for ligand limits the potential for signaling IL-2 mediated effects. As would be predicted from its very short intracellular "tail", the IL-2R α chain does not mediate any signal transduction functions. The α chain is not significantly expressed by nonactivated cells, but activation of T and B cells dramatically enhances expression of the α chain, which leads to the formation of αβγ chain high-affinity complexes that are able both to bind IL-2 at relatively low concentrations and also to transduce signals which mediate IL-2 effector functions (Taniguchi *et al.*, 1983). IL-2 is able to upregulate expression of IL-2R α chains (Reem and Yeh, 1984), which potentiates the observed potency of IL-2 mediated proliferation, etc., with activated T and B lymphocytes. NK and monocytic cells seem to respond primarily to IL-2 through intermediate-affinity (βγ) receptors, although subpopulations may express the high-affinity complexes. Nonhemopoietic cells are unable to produce IL-2R α chains, although "functional" intermediate-affinity receptors have been reported on some apparently responding cells (Plaisance *et al.*, 1992). The presence of the γ chain in receptor complexes seems essential for receptor internalization, which is necessary for signal transduction (Takeshita *et al.*, 1992).

The biochemical events associated with IL-2 signal transduction are complex and often unclear. Several different possible pathways associated with this have been proposed (Karnitz and Abraham, 1996; Thèze *et al.*, 1996). Phosphorylation of a number of proteins, including the β and γ chains of the receptor themselves, occurs following binding of IL-2, and phosphorylation of Raf-1 kinase occurs early in this process (Turner *et al.*, 1991). As the receptor chains possess no intrinsic kinase

activity, the phosphorylations must be catalyzed by nonreceptor kinases as "secondary" phenomena. A rapid increase in intracellular pH occurs following IL 2 binding to receptor as a result of activation of an Na^+/H^+ antiport (Mills and May, 1987).

The serine-rich region in the cytoplasmic domain of the receptor β chain activates *syc* tyrosine kinase and seems responsible for induction of c-*myc*, which is associated with cell proliferation (Shibuya *et al.*, 1992). The acidic domain on the β chain binds the *src* kinase-like molecule p56^{lck} and causes activation of p21^{ras} (probably involving the *shc* adaptor protein) and induction of *fos/jun* expression (Minami *et al.*, 1993). It is possible that the receptor β chain may bind other *src* family kinases such as p59^{fyn}, p55^{blk} and p56^{lyn} which are known to be present in B lymphocytes and other cell types. It has been shown that Ras is activated following IL-2–IL-2R binding and that this requires effects mediated by two domains on the IL-2R (Satoh *et al.*, 1992).

Activation of Ras may trigger the MAP kinase cascade, although the precise role of this pathway in IL-2 signaling is not clear. The IL-2R β chain binds the janus kinase JAK1 on activation, which may play a role in signal transduction. It has also been proposed that phosphoinositide-3-kinase may be involved in IL-2 signaling (Reif *et al.*, 1997; Remillard *et al.*, 1991). The p70 S6 kinase (p70^{s6k}) is a critical component in regulating IL-2 responses and this process seems to involve the lipid kinase mTOR (Kuo *et al.*, 1992; Price *et al.*, 1992).

The γ chain of the activated receptor binds JAK3 followed by phosphorylation of the transcription-activating proteins STAT3 and STAT5. JAK3 signaling involving the γ_c chain has been shown to be essential for lymphoid development in mice (Nosaka *et al.*, 1995). The SH2 domain of the γ chain may also be involved in binding phosphotyrosine residues on a variety of signaling proteins involved in IL-2 signal transduction.

8. IL-2–IL-2R Involvement in Disease

The potent effects of IL-2 on the immune system and other physiological pathways suggest that abnormalities associated with IL-2 action in these biological processes could cause pathology. Production of and response to IL2 have been shown to decline with age in rats and mice (Rosenberg *et al.*, 1994; Thoman and Weigle, 1982) and suggested as causing diminished immune responses; however, precise data to demonstrate the importance of this effect and whether it occurs in humans is generally lacking. Knockout mouse studies may provide information relevant to potential effects of abnormal IL-2–IL-2R expression in humans (Thèze *et al.*, 1996). However, there are two clinical disorders that appear to be caused or at least compounded by abnormalities in IL-2R or IL-2R–IL-2 expression and activity.

8.1   X-LINKED SEVERE COMBINED IMMUNODEFICIENCY (X-LINKED SCID)

X-linked SCID is associated with severe impairment of humoral and cellular immune functions. It has been shown that the gene defect responsible for the disorder maps to the same locus as the IL-2Rγ (γ_c) chain. Such mutations limit the production of IL-2Rγ chains to truncated variants which presumably lack function (Noguchi et al., 1993). It therefore seems probable that inability to respond to IL-2 may cause or at least play a part in the underlying basis of SCID. However, the γ_c protein is an essential component of receptors for other cytokines which are clearly necessary for development and/or maintenance of immune functions (e.g., IL-4, IL-7, IL-15). It is therefore possible that inability to respond to IL-2 only partially explains the severe immunodeficiency shown by SCID patients.

8.2   ADULT T CELL LEUKEMIA/LYMPHOMA

Infection with HTLV I causes abnormal expression of numerous proteins, including the IL-2R α chain, and most of these effects appear to be mediated by the viral Tax gene product. Although the contribution of IL-2–IL-2R abnormalities to development of the fatal clinical disorder adult T cell leukemia is difficult to interpret owing to the multiple effects on expression of several proteins/genes and the inconsistent IL-2 dependence of leukemic T cells derived from leukemia patients, a role for disregulated IL-2 response seems likely for at least some of the pathology in this clinical condition (Yodoi and Uchiyama, 1992).

9.   IL-2 Therapy

The observed stimulatory effects of IL-2 on immune processes, and especially cellular immune mechanisms, have prompted a large number of trials to assess the therapeutic potential of the cytokine in a range of clinical indications (Lotze and Rosenberg, 1988).

IL-2 therapy has involved direct injection of IL-2 itself and/or expansion and stimulation of patient's lymphoid cells with IL-2 in vitro followed by introducing such cells back into the patient with concurrent IL-2 administration to maintain the active cell fraction. The latter procedure is known as "lymphokine activated killer cell" (LAK) therapy (Rosenberg et al., 1985; Rosenstein et al., 1984).

9.1   TREATMENT OF CANCER

Most therapeutic applications of IL-2 have involved treatment of neoplastic conditions (Kolitz and Mertelsmann, 1991). Such therapeutic approaches have been used with variable success for treatment of metastatic melanoma (Dillman et al., 1991; Goedegebuure et al., 1995; Leong et al., 1995; Lotze et al., 1992; Rosenberg et al., 1994; Schwartzentruber et al., 1994), ovarian cancer (Canevari et al., 1995), breast cancer (Addison et al., 1995; Ziegler et al., 1992), colorectal carcinoma (Hjelm et al., 1995; Okuno et al., 1993), hemangioendothelioma (Bhutto et al., 1995), neuroblastoma (Bauer et al., 1995; Frost et al., 1997; Marti et al., 1995), myeloblastoid leukemia (Foa, 1993; Stevenson et al., 1995), endocrine tumors (Lissoni et al., 1995), non-small-cell (Ardizzoni et al., 1994; Scudeletti et al., 1993; Yang et al., 1991) and small-cell lung carcinoma (Clamon et al., 1993), hairy-cell leukemia (Liberati et al., 1994), squamous cell carcinoma (Mattijssen et al., 1994), bladder carcinoma (Gomella et al., 1993; Nouri et al., 1994), glioma (Pedersen et al., 1994), Hodgkin's lymphoma (Margolin et al., 1991), pulmonary metastases (Barna et al., 1994), sarcoma (Kawakami et al., 1993), myelogenous leukemia (Benyunes et al., 1993; Hamon et al., 1993; Meloni et al., 1996), hepatocellular carcinoma (Yamamoto et al., 1993), metastatic pleural effusions (Viallat et al., 1993), adenocarcinoma (Bjermer et al., 1993), B cell lymphocytosis (Tiberghien et al., 1992), metastatic renal carcinoma (Gambacorti Passerini et al., 1993; Schoof et al., 1993; Sleijfer et al., 1992; Thompson et al., 1992; Weiss et al., 1992), and malignancies associated with AIDS (Bernstein et al., 1995).

To date, the only indication for which an IL-2 product is licensed for therapeutic use in Europe and the United States is renal carcinoma.

9.2   TREATMENT OF INFECTIOUS DISEASES

The use of IL-2 to treat infectious diseases such as hepatitis B (Tilg et al., 1993) and herpes simplex virus genital infection (Weinberg et al., 1986) has been described. A recent trial has shown that infusion of IL-2 in selected AIDS patients can induce substantial sustained increases in CD4$^+$ lymphocytes without associated increases in plasma HIV RNA levels (Kovacs et al., 1996). Low-dose IL-2 therapy seems promising in this clinical condition (Jacobsen et al., 1996). IL-2 therapy for HTLV-1 infection has been described (Davey et al., 1997). Oral IL-2 treatment for Campylobacter infection in mice has been reported (Rollwagen and Baqar, 1996).

9.3   GENE THERAPY

Therapy using cells transfected with IL-2 genes has been described for treatment of glioblastoma, melanoma, and other conditions (Flemming et al., 1997; Sobol et al., 1995; Tohmatsu et al., 1993; Uchiyama et al., 1993).

9.4 BONE MARROW TRANSPLANTATION

A small number of studies have been carried out to assess the usefulness of IL-2 in bone marrow and stem cell transplantation. Mouse studies have suggested that such approaches may be potentially valuable (Charak et al., 1992; Riccardi et al., 1986) and human trials have shown some success (Beaujean et al., 1995; Charak et al., 1992; De-Laurenzi et al., 1995; Lopez-Jimenez et al., 1997).

9.5 ADVERSE REACTIONS RELATING TO IL-2 THERAPY

Administration of IL-2 to patients is often associated with adverse effects. These range from relatively mild discomfort to severe effects which requires the halting of therapy. A particularly frequent complication with IL-2 therapy is the development of "capillary" or "vascular leak syndrome" (Lotze et al., 1986). This adverse effect disappears if treatment is stopped, but the very severe vascular permeability which results in interstitial pulmonary edema and eventual multiorgan failure often necessitates reduction or discontinuance of therapeutic administration of IL-2 (Funke et al., 1994; Pockaj et al., 1994). The precise cause of the syndrome is not clear, but a mouse study has characterized the condition (Rosenstein et al., 1986). IL-2 induction of inducible nitric oxide synthetase and production of nitric oxide may be involved in the process (Orucevic et al., 1997).

Other, less frequently observed, adverse effects noted during or following IL-2 administration include cardiomyopathy (Goel et al., 1992), scleroderma (Puett and Fuchs, 1994), myelodysplasia (Toze et al., 1993), hypothyroidism (Sauter et al., 1992; Weijl et al., 1993), diabetes (Whitehead et al., 1995), renal disease (Guleria et al., 1994), colonic ischemia (Sparano et al., 1991), inflammatory arthritis (Massarotti et al., 1992), hypoprothrombinemia (Birchfield et al., 1992), fever, diarrhea and asthenia (Ridolfi et al., 1992). Occasionally patients may develop antibodies against IL-2, which may (Kirchner et al., 1991), or may not (Scharenberg et al., 1994) compromise therapy.

Therapeutic strategies, including dosing and route of administration, are largely influenced by attempts to limit adverse effects associated with IL-2 therapy (Banks et al., 1997).

9.6 IL-2 COMBINATION THERAPY

Several therapeutic regimes involving combination of IL-2 with other cytokines or substances have been devised. The primary aim is either to increase therapeutic effect or to allow adjustment of dosing to minimize adverse effects. Combination therapy of IL-2 with IFN-α, IFN-β, TNF-α, GM-CSF, and IL-4 (Fossa et al., 1993; Nagler et al., 1997; Keilholz et al., 1997; Krigel et al., 1995; Liberati et al., 1994; Shaffer et al., 1995; Sosman et al., 1994, 1995), has been described, as has IL-2 with 5-fluorouracil (Atzpodien et al., 1993; Ridolfi et al., 1994) or zidovudine (Wood et al., 1993). Therapeutic use of IL-2 modified with polyethylene glycol has also been reported (Menzel et al., 1993). Monoclonal antibody–IL-2 fusion proteins have been devised for tumor therapy (Hornick et al., 1997).

9.7 ANTI-IL-2R THERAPY

Anti-IL-2R therapy has been tried to induce immunosuppression or as antitumor treatment.

Monoclonal antibodies specific for CD25 have been used to immunosuppress patients to prevent allograft rejection of various tissues (Kirkman et al., 1991; Nashan et al., 1995; Tinubu et al., 1994). Such antibodies, either unmodified or conjugated to toxins, have also been used for treatment of Hodgkin lymphoma (Engert et al., 1991), adult T cell leukemia (Hakimi et al., 1993; Waldmann et al., 1993), and chronic lymphocytic leukemia (Kreitman et al., 1992). Fusion proteins consisting of IL-2 with toxins have been produced for therapy of rheumatoid arthritis (Sewell et al., 1993), Hodgkin lymphoma (Tepler et al., 1994), T cell lymphoma (Hesketh et al., 1993), and chronic lymphocytic leukemia (LeMaistre et al., 1991).

10. Acknowledgments

I thank Barbara Bolgiano for Figure 2.4, Andrew Davies for drawing the figures, and Jenni Haynes for processing the manuscript.

11. References

Addison, C.L., Braciak, T., Ralston, R., Muller, W.J., Gauldie, J. and Graham, F.L. (1995). Intratumoral injection of an adenovirus expressing interleukin 2 induces regression and immunity in a murine breast cancer model. Proc. Natl Acad. Sci. USA 92, 8522–8526.

Ardizzoni, A., Bonavia, M., et al. (1994). Biologic and clinical effects of continuous infusion interleukin-2 in patients with non-small cell lung cancer. Cancer 73, 1353–1360.

Atzpodien, J., Kirchner, H., Hanninen, E.L., Deckert, M., Fenner, M. and Poliwoda, H. (1993). Interleukin-2 in combination with interferon-alpha and 5-fluorouracil for metastatic renal cell cancer. Eur. J. Cancer 5, S6–S8.

Baccarini, M., Schwinzer, R. and Lohmann Matthes, M.L. (1989). Effect of human recombinant IL-2 on murine macrophage precursors. Involvement of a receptor distinct from the p55 (Tac) protein. J. Immunol. 142, 118–125.

Banks, R.E., Forbes, M.A., Hallan, S., et al. (1997). Treatment of metastatic renal cell carcinoma with subcutaneous interleukin 2: evidence for non-renal clearance of cytokines. Brit. J. Cancer 75, 1842–1848.

Barna, B.P., Thomassen, M.J., Maier, M., *et al.* (1994). Combination therapy with a synthetic peptide of C-reactive protein and interleukin 2: augmented survival and eradication of pulmonary metastases. Cancer Immunol. Immunother. 38, 38–42.

Bauer, M., Reaman, G.H., Hank, J.A., *et al.* (1995). A phase II trial of human recombinant interleukin-2 administered as a 4-day continuous infusion for children with refractory neuroblastoma, non-Hodgkin's lymphoma, sarcoma, renal cell carcinoma, and malignant melanoma. A Childrens Cancer Group study. Cancer 75, 2959–2965.

Bazan, J. (1992). Unravelling the structure of IL-2. Science 257, 410–412.

Beaujean, F., Bernaudin, F., Kuentz, M., *et al.* (1995). Successful engraftment after autologous transplantation of 10-day cultured bone marrow activated by interleukin 2 in patients with acute lymphoblastic leukemia. Bone Marrow Transplant. 15, 691–696.

Benveniste, E.N. and Merrill, J.E. (1986). Stimulation of oligodendroglial proliferation and maturation by interleukin-2. Nature 321, 610–613.

Benyunes, M.C., Massumoto, C., York, A., *et al.* (1993). Interleukin-2 with or without lymphokine-activated killer cells as consolidative immunotherapy after autologous bone marrow transplantation for acute myelogenous leukemia. Bone Marrow Transplant. 12, 159–163.

Bernstein, Z.P., Porter, M.M., Gould, M., *et al.* (1995). Prolonged administration of low-dose interleukin-2 in human immunodeficiency virus-associated malignancy results in selective expansion of innate immune effectors without significant clinical toxicity. Blood 86, 3287–3294.

Bhutto, A.M., Uehara, K., Takamiyagi, A., Hagiwara, K. and Nonaka, S. (1995). Cutaneous malignant hemangioendothelioma: clinical and histopathological observations of nine patients and a review of the literature. J. Dermatol. 22, 253–261.

Birchfield, G.R., Rodgers, G.M., Girodias, K.W., Ward, J.H. and Samlowski, W.E. (1992). Hypoprothrombinemia associated with interleukin-2 therapy: correction with vitamin K. J. Immunother. 11, 71–75.

Bjermer, L., Gronberg, H., Roos, G. and Henriksson, R. (1993). Interleukin-2-administration intravenously and intrapleurally in a patient with primary pulmonary adenocarcinoma. Cellular responses in peripheral blood, intrapleural fluid and bronchoalveolar lavage. Biotherapy 6, 1–7.

Blackman, M.A., Tigges, M.A., Minie, M.E. and Koshland, M.E. (1986). A model system for peptide hormone action in differentiation: interleukin 2 induces a B lymphoma to transcribe the J chain gene. Cell 47, 609–617.

Brandhuber, B.J., Boone, T., Kenney, W.C. and McKay, D.B. (1987). Three-dimensional structure of interleukin-2. Science 238, 1707–1709.

Canevari, S., Stoter, G., Arienti, F., *et al.* (1995). Regression of advanced ovarian carcinoma by intraperitoneal treatment with autologous T lymphocytes retargeted by a bispecific monoclonal antibody. J. Natl Cancer Inst. 87, 1463–1469.

Carding, S.R., Jenkinson, E.J., Kingston, R., Hayday, A.C., Bottomly, K. and Owen, J.J. (1989). Developmental control of lymphokine gene expression in fetal thymocytes during T-cell ontogeny. Proc. Natl Acad. Sci. USA 86, 3342–3345.

Charak, B.S., Choudhary, G.D., Tefft, M. and Mazumder, A.

(1992). Interleukin-2 in bone marrow transplantation: preclinical studies. Bone Marrow Transplant. 10, 103–111.

Ciacci, C., Mahida, Y.R., Dignass, A., Koizumi, M. and Podolsky, D.K. (1993). Functional interleukin-2 receptors on intestinal epithelial cells. J. Clin. Invest. 92, 527–532.

Clamon, G., Herndon, J., Perry, M.C., *et al.* (1993). Interleukin-2 activity in patients with extensive small-cell lung cancer: a phase II trial of Cancer and Leukemia Group B. J. Natl Cancer Inst. 85, 316–320.

Clark, S.C., Arya, S.K., Wong Staal, F., *et al.* (1984). Human T-cell growth factor: partial amino acid sequence, cDNA cloning, and organization and expression in normal and leukemic cells. Proc. Natl Acad. Sci. USA 81, 2543–2547.

Collins, L., Tsien, W.H., Seals, C., *et al.* (1988). Identification of specific residues of human interleukin 2 that affect binding to the 70-kDa subunit (p70) of the interleukin 2 receptor. Proc. Natl Acad. Sci. USA 85, 7709–7713.

Cosman, D., Lyman, S.D., Idzerda, R.L., *et al.* (1990). A new cytokine receptor superfamily. Trends Biochem. Sci. 15, 265–270.

Crabtree, G.R. (1989). Contingent genetic regulatory events in T lymphocyte activation. Science 243, 355–361.

Davey, R.T. Jr., Chaitt, D.G., Piscitelli, S.C., *et al.* (1997). Subcutaneous administration of interleukin-2 in human immunodeficiency virus type 1-infected persons. J. Infect. Dis. 175, 781–789.

De-Laurenzi, A., Iudicone, P., Zoli, V., *et al.* (1995). Recombinant interleukin-2 treatment before and after autologous stem cell transplantation in hematologic malignancies: clinical and immunologic effects. J. Hematother. 4, 113–120.

Dillman, R.O., Oldham, R.K., Barth, N.M., *et al.* (1991). Continuous interleukin-2 and tumor-infiltrating lymphocytes as treatment of advanced melanoma. A national biotherapy study group trial. Cancer 68, 1–8.

Engert, A., Martin, G., Amlot, P., Wijdenes, J., Diehl, V. and Thorpe, P. (1991). Immunotoxins constructed with anti-CD25 monoclonal antibodies and deglycosylated ricin A-chain have potent anti-tumour effects against human Hodgkin cells in vitro and solid Hodgkin tumours in mice. Int. J. Cancer 49, 450–456.

Farrar, J.J., Benjamin, W.R., Hilfiker, M.L., Howard, M., Farrar, W.L. and Fuller-Farrar, J. (1982). The biochemistry, biology, and role of interleukin 2 in the induction of cytotoxic T cell and antibody forming B cell responses. Immunol. Rev. 63, 129–166.

Flemming, C.L., Patel, P.M., Box, G., *et al.* (1997). Sarcoma cells engineered to secrete IFN-gamma or IL-2 acquire sensitization to immune cell killing via different mechanisms. Cytokine 9, 328–332.

Foa, R. (1993). Does interleukin-2 have a role in the management of acute leukemia? J. Clin. Oncol. 11, 1817–1825.

Fossa, S.D., Aune, H., Baggerud, E., Granerud, T., Heilo, A. and Theodorsen, L. (1993). Continuous intravenous interleukin-2 infusion and subcutaneous interferon-alpha in metastatic renal cell carcinoma. Eur. J. Cancer 9, 1313–1315.

Frost, J.D., Hank, J.A., Reaman, G.H., *et al.* (1997). A phase I/IB trial of murine monoclonal anti-GD2 antibody 14.G2a plus interleukin-2 in children with refractory neuroblastoma: a report of the Children's Cancer Group. Cancer 80, 317–333.

Fujita, T., Takaoka, C., Matsui, H. and Taniguchi, T. (1983).

Structure of the human interleukin-2 gene. Proc. Natl Acad. Sci. USA 80, 7437–7441.

Funke, I., Prummer, O., Schrezenmeier, H., *et al.* (1994). Capillary leak syndrome associated with elevated IL-2 serum levels after allogeneic bone marrow transplantation. Ann. Hematol. 68, 49–52.

Gambacorti Passerini, C., Hank, J.A., Albertini, M.R., *et al.* (1993). A pilot phase II trial of continuous-infusion interleukin-2 followed by lymphokine-activated killer cell therapy and bolus-infusion interleukin-2 in renal cancer. J. Immunother. 13, 43–48.

Gearing, A., Thorpe, R., Bird, C. and Spitz, M. (1985). Human B cell proliferation is stimulated by interleukin 2. Immunol. Lett. 9, 105–108.

Goedegebuure, P.S., Douville, L.M., Li, H., *et al.* (1995). Adoptive immunotherapy with tumor-infiltrating lymphocytes and interleukin-2 in patients with metastatic malignant melanoma and renal cell carcinoma: a pilot study. J. Clin. Oncol. 13, 1939–1949.

Goel, M., Flaherty, L., Lavine, S. and Redman, B.G. (1992). Reversible cardiomyopathy after high-dose interleukin-2 therapy. J. Immunother. 11, 225–229.

Gomella, L.G., McGinnis, D.E., Lattime, E.C., *et al.* (1993). Treatment of transitional cell carcinoma of the bladder with intravesical interleukin-2: a pilot study. Cancer Biother. 8, 223–227.

Grant, A.J., Roessler, E., Ju, G., Tsudo, M., Sugamura, K. and Waldmann, T.A. (1992). The interleukin 2 receptor (IL-2R): the IL-2R alpha subunit alters the function of the IL-2R beta subunit to enhance IL-2 binding and signaling by mechanisms that do not require binding of IL-2 to IL-2R alpha subunit. Proc. Natl Acad. Sci. USA 89, 2165–2169.

Guleria, A.S., Yang, J.C., Topalian, S.L., *et al.* (1994). Renal dysfunction associated with the administration of high-dose interleukin-2 in 199 consecutive patients with metastatic melanoma or renal carcinoma. J. Clin. Oncol. 12, 2714–2722.

Hakimi, J., Ha, V.C., Lin, P., *et al.* (1993). Humanized Mik beta 1, a humanized antibody to the IL-2 receptor beta-chain that acts synergistically with humanized anti-TAC. J. Immunol. 151, 1075–1085.

Hamon, M.D., Prentice, H.G., Gottlieb, D.J., *et al.* (1993). Immunotherapy with interleukin 2 after ABMT in AML. Bone Marrow Transplant. 11, 399–401.

Hatakeyama, M., Mori, H., Doi, T. and Taniguchi, T. (1989). A restricted cytoplasmic region of IL-2 receptor beta chain is essential for growth signal transduction but not for ligand binding and internalization. Cell 59, 837–845.

Hesketh, P., Caguioa, P., Koh, H., *et al.* (1993). Clinical activity of a cytotoxic fusion protein in the treatment of cutaneous T-cell lymphoma. J. Clin. Oncol. 11, 1682–1690.

Hjelm, A.L., Ragnhammar, P., Fagerberg, J., *et al.* (1995). Subcutaneous interleukin-2 and alpha-interferon in advanced colorectal carcinoma. A phase II study. Cancer Biother. 10, 5–12.

Holbrook, N.J., Smith, K.A., Fornace, A.J., Jr., Comeau, C.M., Wiskocil, R.L. and Crabtree, G.R. (1984). T-cell growth factor: complete nucleotide sequence and organization of the gene in normal and malignant cells. Proc. Natl Acad. Sci. USA 81, 1634–1638.

Hornick, J.L., Khawli, L.A., Hu, P., *et al.* (1997). Chimeric CLL-1 antibody fusion proteins containing granulocyte-macrophage colony-stimulating factor or interleukin-2 with specificity for B-cell malignancies exhibit enhanced effector functions while retaining tumor targeting properties. Blood 89, 4437–4447.

Howard, M., Matis, L., Malek, T.R., *et al.* (1983). Interleukin 2 induces antigen-reactive T cell lines to secrete BCGF-I. J. Exp. Med. 158, 2024–2039.

Hoyos, B., Ballard, D.W., Bohnlein, E., Siekevitz, M. and Greene, W.C. (1989). Kappa B-specific DNA binding proteins: role in the regulation of human interleukin-2 gene expression. Science 244, 457–460.

Jacobsen, E., Pilaro, F. and Smith, K.A. (1996). Rational interleukin 2 therapy for HIV positive individuals: Daily low doses enhance immune function without toxicity. Proc. Natl Acad. Sci. USA 93, 10405–10410.

Jenkinson, E.J., Kingston, R. and Owen, J.J. (1987). Importance of IL-2 receptors in intra-thymic generation of cells expressing T-cell receptors. Nature 329, 160–162.

Kaempfer, R., Efrat, S. and Marsh, S. (1987). Regulation of human interleukin 2 gene expression. In "Lymphokines, Vol. 13, Molecular Cloning and Analysis of Lymphokines", (eds. D.R. Webb and D.V. Goeddel), pp. 59–72. Academic Press, San Diego.

Karnitz, L.M. and Abraham, R.T. (1996). Interleukin-2 receptor signaling mechanisms. Adv. Immun. 61, 147–191.

Kawakami, Y., Haas, G.P. and Lotze, M.T. (1993). Expansion of tumor-infiltrating lymphocytes from human tumors using the T-cell growth factors interleukin-2 and interleukin-4. J. Immunother. 14, 336–347.

Keilholz, U., Goey, S.H., Punt, C.J., *et al.* (1997). Interferon alfa-2a and interleukin-2 with or without cisplatin in metastatic melanoma: a randomized trial of the European Organization for Research and Treatment of Cancer Melanoma Cooperative Group. J. Clin. Oncol. 15, 2579–2588.

Kirchner, H., Korfer, A., Evers, P., *et al.* (1991). The development of neutralizing antibodies in a patient receiving subcutaneous recombinant and natural interleukin-2. Cancer 67, 1862–1864.

Kirkman, R.L., Shapiro, M.E., Carpenter, C.B., *et al.* (1991). A randomized prospective trial of anti-Tac monoclonal antibody in human renal transplantation. Transplantation 51, 107–113.

Kolitz, J.E. and Mertelsmann, R. (1991). The immunotherapy of human cancer with interleukin 2: present status and future directions. Cancer Invest. 9, 529–542.

Kovacs, J.A., Vogel, S., Jeffrey, B.S., *et al.* (1996). Controlled trial of interleukin-2 infusions in patients infected with the human immunodeficiency virus. N. Engl. J. Med. 335, 1350–1356.

Kreitman, R.J., Chaudhary, V.K., Kozak, R.W., FitzGerald, D.J., Waldman, T.A. and Pastan, I. (1992). Recombinant toxins containing the variable domains of the anti-Tac monoclonal antibody to the interleukin-2 receptor kill malignant cells from patients with chronic lymphocytic leukemia. Blood 80, 2344–2352.

Krigel, R.L., Padavic Shaller, K., Toomey, C., Comis, R.L. and Weiner, L.M. (1995). Phase I study of sequentially administered recombinant tumor necrosis factor and recombinant interleukin-2. J. Immunother. Emphasis Tumor Immunol. 17, 161–170.

Kruys, V., Marinx, O., Shaw, G., Deschamps, J. and Huez, G. (1989). Translational blockade imposed by cytokine-derived UA-rich sequences. Science 245, 852–855.

Kuo, C.J., Chung, J., Fiorentino, D.F., Flanagan, W.M., Blenis, J. and Crabtree, G.R. (1992). Rapamycin selectively inhibits interleukin-2 activation of p70 S6 kinase. Nature 358, 70–73.

LeMaistre, C.F., Rosenblum, M.G., Reuben, J.M., et al. (1991). Therapeutic effects of genetically engineered toxin (DAB486IL-2) in patient with chronic lymphocytic leukaemia. Lancet 337, 1124–1125.

Leonard, W.J., Depper, J.M., Kanehisa, M., et al. (1985a). Structure of the human interleukin-2 receptor gene. Science 230, 633–639.

Leonard, W.J., Donlon, T.A., Lebo, R.V. and Greene, W.C. (1985b). Localization of the gene encoding the human interleukin-2 receptor on chromosome 10. Science 228, 1547–1549.

Leong, S.P., Zhou, Y.M., Granberry, M.E., et al. (1995). Generation of cytotoxic effector cells against human melanoma. Cancer Immunol. Immunother. 40, 397–409.

Liang, S.M., Thatcher, D.R., Liang, C.M. and Allet, B. (1986). Studies of structure–activity relationships of human interleukin-2. J. Biol. Chem. 261, 334–337.

Liberati, A.M., De Angelis, V., Fizzotti, M., et al. (1994). Natural-killer-stimulatory effect of combined low-dose interleukin-2 and interferon beta in hairy-cell leukemia patients. Cancer Immunol. Immunother. 38, 323–331.

Lissoni, P., Barni, S., Tancini, G., et al. (1995). Immunoendocrine therapy with low-dose subcutaneous interleukin-2 plus melatonin of locally advanced or metastatic endocrine tumors. Oncology 52, 163–166.

Lopez-Jimenez, J., Perez-Oteyza, J., Munoz, A., et al. (1997). Subcutaneous versus intravenous low-dose IL-2 therapy after autologous transplantation: results of a prospective, non-randomized study. Bone Marrow Transplant 19, 429–434.

Lotze, M.T., Matory, Y.L., Rayner, A.A., et al. (1986). Clinical effects and toxicity of interleukin-2 in patients with cancer. Cancer 58, 2764–2772.

Lotze, M. and Rosenberg, S.A. (1988). Interleukin 2 as a pharmacologic reagent. In "Interleukin", Vol. 2 (ed. K.A. Smith), pp. 237–294. Academic Press, San Diego.

Lotze, M.T., Zeh, H.J.D., Elder, E.M., et al. (1992). Use of T-cell growth factors (interleukins 2, 4, 7, 10, and 12) in the evaluation of T-cell reactivity to melanoma. J. Immunother. 12, 212–217.

Margolin, K.A., Aronson, F.R., Sznol, M., et al. (1991). Phase II trial of high-dose interleukin-2 and lymphokine-activated killer cells in Hodgkin's disease and non-Hodgkin's lymphoma. J. Immunother. 10, 214–220.

Marti, F., Pardo, N., Peiro, M., et al. (1995). Progression of natural immunity during one-year treatment of residual disease in neuroblastoma patients with high doses of interleukin-2 after autologous bone marrow transplantation. Exp. Hematol. 23, 1445–1452.

Martinez, O.M., Gibbons, R.S., Garovoy, M.R. and Aronson, F.R. (1990). IL-4 inhibits IL-2 receptor expression and IL-2-dependent proliferation of human T cells. J. Immunol. 144, 2211–2215.

Massarotti, E.M., Liu, N.Y., Mier, J. and Atkins, M.B. (1992). Chronic inflammatory arthritis after treatment with high-dose interleukin-2 for malignancy. Am. J. Med. 92, 693–697.

Mattijssen, V., De Mulder, P.H., De Graeff, A., et al. (1994). Intratumoral PEG-interleukin-2 therapy in patients with locoregionally recurrent head and neck squamous-cell carcinoma. Ann. Oncol. 5, 957–960.

McGuire, K.L. and Iacobelli, M. (1997). Involvement of Rel, Fos, and Jun proteins in binding activity to the IL-2 promoter CD28 response element/AP-1 sequence in human T cells. J. Immunol. 159, 1319–1327.

Meloni, G., Vignetti, M., Andrizzi, C., et al. (1996). Interleukin-2 for the treatment of advanced acute myelogenous leukemia patients with limited disease: updated experience with 20 cases. Leuk. Lymphoma 21, 429–435.

Menzel, T., Schomburg, A., Korfer, A., et al. (1993). Clinical and preclinical evaluation of recombinant PEG-IL-2 in human. Cancer Biother. 8, 199–212.

Mills, G.B. and May, C. (1987). Binding of interleukin 2 to its 75-kDa intermediate affinity receptor is sufficient to activate Na^+/H^+ exchange. J. Immunol. 139, 4083–4087.

Minami, Y., Kono, T., Miyazaki, T. and Taniguchi, T. (1993). The IL-2 receptor complex: its structure, function, and target genes. Annu. Rev. Immunol. 11, 245–268.

Mingari, M.C., Gerosa, F., Carra, G., et al. (1984). Human interleukin-2 promotes proliferation of activated B cells via surface receptors similar to those of activated T cells. Nature 312, 641–643.

Mitchell, L.C., Davis, L.S. and Lipsky, P.E. (1989). Promotion of human T lymphocyte proliferation by IL-4. J. Immunol. 142, 1548–1557.

Morgan, D.A., Ruscetti, F.W. and Gallo, R. (1976). Selective in vitro growth of T lymphocytes from normal human bone marrows. Science 193, 1007–1008.

Mott, H.R., Baines, B.S., Hall, R.M., et al. (1995). The solution structure of the F42A mutant of human interleukin 2. J. Mol. Biol. 247, 979–994.

Muegge, K. and Durum, S.K. (1989). From cell code to gene code: cytokines and transcription factors. New Biol. 1, 239–246.

Nagler, A., Ackerstein, A., Or, R., et al. (1997). Immunotherapy with recombinant human interleukin-2 and recombinant interferon-alpha in lymphoma patients postautologous marrow or stem cell transplantation. Blood 89, 3951–3959.

Nashan, B., Schwinzer, R., Schlitt, H.J., Wonigeit, K. and Pichlmayr, R. (1995). Immunological effects of the anti-IL-2 receptor monoclonal antibody BT 563 in liver allografted patients. Transplant. Immunol. 3, 203–211.

Noguchi, M., Yi, H., Rosenblatt, H.M., et al. (1993). Interleukin-2 receptor gamma chain mutation results in X-linked severe combined immunodeficiency in humans. Cell 73, 147–157.

Nosaka, T., van Deursen, J.M., Tripp, R.A., et al. (1995). Defective lymphoid development in mice lacking Jak3. Science 270, 800–802.

Nouri, A.M., Hyde, R. and Oliver, R.T. (1994). Clinical and immunological effect of intravesical interleukin-2 on superficial bladder cancer. Cancer Immunol. Immunother. 39, 68–70.

Okuno, K., Hirohata, T., Nakamura, K., et al. (1993). Hepatic arterial infusions of interleukin-2-based immunochemo-therapy in the treatment of unresectable liver metastases from colorectal cancer. Clin. Ther. 15, 672–683.

Ortaldo, J.R., Mason, A.T., Gerard, J.P., et al. (1984). Effects of natural and recombinant IL 2 on regulation of IFN gamma production and natural killer activity: lack of involvement of the Tac antigen for these immunoregulatory effects. J. Immunol. 133, 779–783.

Orucevic, A., Hearn, S. and Lala, P.K. (1997). The role of active inducible nitric oxide synthase expression in the pathogenesis of capillary leak syndrome resulting from interleukin-2 therapy in mice. Lab. Invest. 76, 53–65.

Pedersen, P.H., Ness, G.O., Engebraaten, O., Bjerkvig, R., Lillehaug, J.R. and Laerum, O.D. (1994). Heterogeneous response to the growth factors [EGF, PDGF (bb), TGF-alpha, bFGF, IL-2] on glioma spheroid growth, migration and invasion. Int. J. Cancer 56, 255–261.

Plaisance, S., Rubinstein, E., Alileche, A., et al. (1992). Expression of the interleukin-2 receptor on human fibroblasts and its biological significance. Int. Immunol. 4, 739–746.

Pockaj, B.A., Yang, J.C., Lotze, M.T., et al. (1994). A prospective randomized trial evaluating colloid versus crystalloid resuscitation in the treatment of the vascular leak syndrome associated with interleukin-2 therapy. J. Immunother. 15, 22–28.

Price, D.J., Grove, J.R., Calvo, V., Avruch, J. and Bierer, B.E. (1992). Rapamycin-induced inhibition of the 70-kilodalton S6 protein kinase. Science 257, 973–977.

Puett, D.W. and Fuchs, H.A. (1994). Rapid exacerbation of scleroderma in a patient treated with interleukin 2 and lymphokine activated killer cells for renal cell carcinoma. J. Rheumatol. 21, 752–753.

Ralph, P., Jeong, G., Welte, K., et al. (1984). Stimulation of immunoglobulin secretion in human B lymphocytes as a direct effect of high concentrations of IL 2. J. Immunol. 133, 2442–2445.

Ralph, P., Nakoinz, I. and Rennick, D. (1988). Role of interleukin 2, interleukin 4, and alpha, beta, and gamma interferon in stimulating macrophage antibody-dependent tumoricidal activity. J. Exp. Med. 167, 712–717.

Reem, G.H. and Yeh, N.H. (1984). Interleukin 2 regulates expression of its receptor and synthesis of gamma interferon by human T lymphocytes. Science 225, 429–430.

Reif, K., Burgering, B.M. and Cantrell, D.A. (1997). Phosphatidylinositol 3-kinase links the interleukin-2 receptor to protein kinase B and p70 S6 kinase. J. Biol. Chem. 272, 14426–14433.

Remillard, B., Petrillo, R., Maslinski, W., et al. (1991). Interleukin-2 receptor regulates activation of phosphatidylinositol 3-kinase. J. Biol. Chem. 266, 14167–14170.

Riccardi, C., Giampietri, A., Migliorati, G., Cannarile, L. and Herberman, R.B. (1986). Generation of mouse natural killer (NK) cell activity: effect of interleukin-2 (IL-2) and interferon (IFN) on the in vivo development of natural killer cells from bone marrow (BM) progenitor cells. Int. J. Cancer 38, 553–562.

Ridolfi, R., Maltoni, R., Riccobon, A., Flamini, E. and Amadori, D. (1992). Evaluation of toxicity in 22 patients treated with subcutaneous interleukin-2, alpha-interferon with and without chemotherapy. J. Chemother. 4, 394–398.

Ridolfi, R., Maltoni, R., Riccobon, A., et al. (1994). A phase II study of advanced colorectal cancer patients treated with combination 5-fluorouracil plus leucovorin and subcutaneous interleukin-2 plus alpha interferon. J. Chemother. 6, 265–271.

Robb, R. (1984). Interleukin-2: the molecule and its function. Immunol. Today, 5, 203.

Robb, R.J. and Kutny, R.M. (1987). Structure–function relationships for the IL 2-receptor system. IV. Analysis of the sequence and ligand-binding properties of soluble Tac protein. J. Immunol. 139, 855–862.

Robb, R.J. and Smith, K.A. (1981). Heterogeneity of human T-cell growth factor(s) due to variable glycosylation. Mol. Immunol. 18, 1087–1094.

Robb, R.J., Kutny, R.M. and Chowdhry, V. (1983). Purification and partial sequence analysis of human T-cell growth factor. Proc. Natl Acad. Sci. USA 80, 5990–5994.

Robb, R.J., Kutny, R.M., Panico, M., Morris, H.R. and Chowdhry, V. (1984). Amino acid sequence and post-translational modification of human interleukin 2. Proc. Natl Acad. Sci. USA 81, 6486–6490.

Rollwagen, F.M. and Baqar, S. (1996). Oral cytokine administration. Trends Immunol. Today, 17, 548–550.

Rosenberg, S.A., Lotze, M.T., Muul, L.M., et al. (1985). Observations on the systemic administration of autologous lymphokine-activated killer cells and recombinant interleukin-2 to patients with metastatic cancer. N. Engl. J. Med. 313, 1485–1492.

Rosenberg, S.A., Yannelli, J.R., Yang, J.C., et al. (1994). Treatment of patients with metastatic melanoma with autologous tumor-infiltrating lymphocytes and interleukin 2. J. Natl Cancer Inst. 86, 1159–1166.

Rosenstein, M., Yron, I., Kaufmann, Y. and Rosenberg, S.A. (1984). Lymphokine-activated killer cells: lysis of fresh syngeneic natural killer-resistant murine tumor cells by lymphocytes cultured in interleukin 2. Cancer Res. 44, 1946–1953.

Rosenstein, M., Ettinghausen, S.E. and Rosenberg, S.A. (1986). Extravasation of intravascular fluid mediated by the systemic administration of recombinant interleukin 2. J. Immunol. 137, 1735–1742.

Rubin, L.A., Kurman, C.C., Fritz, M.E., et al. (1985). Soluble interleukin 2 receptors are released from activated human lymphoid cells in vitro. J. Immunol. 135, 3172–3177.

Saneto, R.P., Altman, A., Knobler, R.L., Johnson, H.M. and de Vellis, J. (1986). Interleukin 2 mediates the inhibition of oligodendrocyte progenitor cell proliferation in vitro. Proc. Natl Acad. Sci. USA 83, 9221–9225.

Satoh, T., Minami, Y., Kono, T., et al. (1992). Interleukin 2-induced activation of Ras requires two domains of interleukin 2 receptor beta subunit, the essential region for growth stimulation and Lck-binding domain. J. Biol. Chem. 267, 25423–25427.

Sauter, N.P., Atkins, M.B., Mier, J.W. and Lechan, R.M. (1992). Transient thyrotoxicosis and persistent hypothyroidism due to acute autoimmune thyroiditis after interleukin-2 and interferon-alpha therapy for metastatic carcinoma: a case report. Am. J. Med. 92, 441–444.

Sauve, K., Nachman, M., Spence, C., et al. (1991). Localization in human interleukin 2 of the binding site to the alpha chain (p55) of the interleukin 2 receptor. Proc. Natl Acad. Sci. USA 88, 4636–4640.

Scharenberg, J.G., Stam, A.G., von Blomberg, B.M., et al. (1994). The development of anti-interleukin-2 (IL-2) antibodies in cancer patients treated with recombinant IL-2. Eur. J. Cancer 12, 1804–1809.

Schoof, D.D., Terashima, Y., Batter, S., Douville, L., Richie, J.P. and Eberlein, T.J. (1993). Survival characteristics of metastatic renal cell carcinoma patients treated with lymphokine-activated killer cells plus interleukin-2. Urology 41, 534–539.

Schorle, H., Holtschke, T., Hunig, T., Schimpl, A. and Horak, I. (1991). Development and function of T cells in mice rendered interleukin-2 deficient by gene targeting. Nature 352, 621–624.

Schreier, M.H. and Tees, R. (1981). Long-term culture and cloning of specific helper T cells. Immunol. Methods 2, 263.

Schwartzentruber, D.J., Hom, S.S., Dadmarz, R., et al. (1994). In vitro predictors of therapeutic response in melanoma patients receiving tumor-infiltrating lymphocytes and interleukin-2. J. Clin. Oncol. 12, 1475–1483.

Scudeletti, M., Filaci, G., Imro, M.A., et al. (1993). Immunotherapy with intralesional and systemic interleukin-2 of patients with non-small-cell lung cancer. Cancer Immunol. Immunother. 37, 119–124.

Sewell, K.L., Parker, K.C., Woodworth, T.G., Reuben, J., Swartz, W. and Trentham, D.E. (1993). DAB486IL-2 fusion toxin in refractory rheumatoid arthritis. Arthritis Rheum. 36, 1223–1233.

Shaffer, L., Giralt, S., Champlin, R. and Chan, K.W. (1995). Treatment of leukemia relapse after bone marrow transplantation with interferon-alpha and interleukin 2. Bone Marrow Transplant. 15, 317–319.

Shaw, G. and Kamen, R. (1986). A conserved AU sequence from the 3′ untranslated region of GM-CSF mRNA mediates selective mRNA degradation. Cell 46, 659–667.

Shibuya, H., Yoneyama, M., Ninomiya Tsuji, J., Matsumoto, K. and Taniguchi, T. (1992). IL-2 and EGF receptors stimulate the hematopoietic cell cycle via different signaling pathways: demonstration of a novel role for c-myc. Cell 70, 57–67.

Sleijfer, D.T., Janssen, R.A., Buter, J., de Vries, E.G., Willemse, P.H. and Mulder, N.H. (1992). Phase II study of subcutaneous interleukin-2 in unselected patients with advanced renal cell cancer on an outpatient basis. J. Clin. Oncol. 10, 1119–1123.

Smith, K.A. (1980). T-cell growth factor. Immunol. Rev. 51, 337–357.

Smith, K.A., Favata, M.F. and Oroszlan, S. (1983). Production and characterization of monoclonal antibodies to human interleukin 2: strategy and tactics. J. Immunol. 131, 1808–1815.

Sobol, R.E., Fakhrai, H., Shawler, D., et al. (1995). Interleukin-2 gene therapy in a patient with glioblastoma. Gene Ther. 2, 164–167.

Sosman, J.A., Fisher, S.G., Kefer, C., Fisher, R.I. and Ellis, T.M. (1994). A phase I trial of continuous infusion interleukin-4 (IL-4) alone and following interleukin-2 (IL-2) in cancer patients. Ann. Oncol. 5, 447–452.

Sosman, J.A., Kefer, C., Fisher, R.I., Jacobs, C.D., Pumfery, P. and Ellis, T.M. (1995). A phase IA/IB trial of anti-CD3 murine monoclonal antibody plus low-dose continuous-infusion interleukin-2 in advanced cancer patients. J. Immunother. Emphasis Tumor Immunol. 17, 171–180.

Sparano, J.A., Dutcher, J.P., Kaleya, R., et al. (1991). Colonic ischemia complicating immunotherapy with interleukin-2 and interferon-alpha. Cancer 68, 1538–1544.

Stern, A.S., Pan, Y.C., Urdal, D.L., et al. (1984). Purification to homogeneity and partial characterization of interleukin 2 from a human T-cell leukemia. Proc. Natl Acad. Sci. USA 81, 871–875.

Stevenson, F.K., Zhu, D., King, C.A., Ashworth, L.J., Kumar, S. and Hawkins, R.E. (1995). Idiotypic DNA vaccines against B-cell lymphoma. Immunol. Rev. 145, 211–228.

Takeshita, T., Asao, H., Ohtani, K., et al. (1992). Cloning of the gamma chain of the human IL-2 receptor. Science 257, 379–382.

Taniguchi, T., Matsui, H., Fujita, T., et al. (1983). Structure and expression of a cloned cDNA for human interleukin-2. Nature 302, 305–310.

Tentori, L., Longo, D.L., Zuniga Pflucker, J.C., Wing, C. and Kruisbeek, A.M. (1988). Essential role of the interleukin 2–interleukin 2 receptor pathway in thymocyte maturation in vivo. J. Exp. Med. 168, 1741–1747.

Tepler, I., Schwartz, G., Parker, K., et al. (1994). Phase I trial of an interleukin-2 fusion toxin (DAB486IL-2) in hematologic malignancies: complete response in a patient with Hodgkin's disease refractory to chemotherapy. Cancer 73, 1276–1285.

Thèze, J., Alzari, P.M. and Bertoglio, J. (1996). Interleukin 2 and its receptors: recent advances and new immunological functions. Immunol. Today, 17, 481–486.

Thoman, M.L. and Weigle, W.O. (1982). Cell-mediated immunity in aged mice: an underlying lesion in IL 2 synthesis. J. Immunol. 128, 2358–2361.

Thompson, J.A., Shulman, K.L., Benyunes, M.C., et al. (1992). Prolonged continuous intravenous infusion interleukin-2 and lymphokine-activated killer-cell therapy for metastatic renal cell carcinoma. J. Clin. Oncol. 10, 960–968.

Tiberghien, P., Racadot, E., Deschaseaux, M.L., et al. (1992). Interleukin-2-induced increase of a monoclonal B-cell lymphocytosis. A novel in vivo interleukin-2 effect? Cancer 69, 2583–2588.

Tilg, H., Vogel, W., Tratkiewicz, J., et al. (1993). Pilot study of natural human interleukin-2 in patients with chronic hepatitis B. Immunomodulatory and antiviral effects. J. Hepatol. 19, 259–267.

Tinubu, S.A., Hakimi, J., Kondas, J.A., et al. (1994). Humanized antibody directed to the IL-2 receptor beta-chain prolongs primate cardiac allograft survival. J. Immunol. 153, 4330–4338.

Tohmatsu, A., Okino, T., Stabach, P., Padula, S.J., Ergin, M.T. and Mukherji, B. (1993). Analysis of cytolytic effector cell response in vitro against autologous human tumor cells genetically altered to synthesize interleukin-2. Immunol. Lett. 35, 51–57.

Toze, C.L., Barnett, M.J. and Klingemann, H.G. (1993). Response of therapy-related myelodysplasia to low-dose interleukin-2. Leukemia 7, 463–465.

Trinchieri, G., Matsumoto Kobayashi, M., Clark, S.C., Seehra, J., London, L. and Perussia, B. (1984). Response of resting human peripheral blood natural killer cells to interleukin 2. J. Exp. Med. 160, 1147–1169.

Tsudo, M., Uchiyama, T. and Uchino, H. (1984). Expression of Tac antigen on activated normal human B cells. J. Exp. Med. 160, 612–617.

Turner, B., Rapp, U., App, H., Greene, M., Dobashi, K. and Reed, J. (1991). Interleukin 2 induces tyrosine phosphorylation and activation of p72-74 Raf-1 kinase in a T-cell line. Proc. Natl Acad. Sci. USA 88, 1227–1231.

Uchiyama, A., Hoon, D.S., Morisaki, T., Kaneda, Y., Yuzuki, D.H. and Morton, D.L. (1993). Transfection of interleukin 2 gene into human melanoma cells augments cellular immune response. Cancer Res. 53, 949–952.

Ullman, K.S., Northrop, J.P., Verweij, C.L. and Crabtree, G.R. (1990). Transmission of signals from the T lymphocyte antigen receptor to the genes responsible for cell proliferation and immune function: the missing link. Annu. Rev. Immunol. 8, 421–452.

Viallat, J.R., Boutin, C., Rey, F., Astoul, P., Farisse, P. and Brandely, M. (1993). Intrapleural immunotherapy with escalating doses of interleukin-2 in metastatic pleural effusions. Cancer 71, 4067–4071.

Voss, S.D., Leary, T.P., Sondel, P.M. and Robb, R.J. (1993). Identification of a direct interaction between interleukin 2 and the p64 interleukin 2 receptor gamma chain. Proc. Natl Acad. Sci. USA 90, 2428–2432.

Waldmann, T.A., White, J.D., Goldman, C.K., *et al.* (1993). The interleukin-2 receptor: a target for monoclonal antibody treatment of human T-cell lymphotrophic virus I-induced adult T-cell leukemia. Blood 82, 1701–1712.

Wang, A., Lu, S.D. and Mark, D.F. (1984). Site-specific mutagenesis of the human interleukin-2 gene: structure–function analysis of the cysteine residues. Science 224, 1431–1433.

Weidmann, E., Sacchi, M., Plaisance, S., *et al.* (1992). Receptors for interleukin 2 on human squamous cell carcinoma cell lines and tumor in situ. Cancer Res. 52, 5963–5970.

Weijl, N.I., Van der Harst, D., Brand, A., *et al.* (1993). Hypothyroidism during immunotherapy with interleukin-2 is associated with antithyroid antibodies and response to treatment. J. Clin. Oncol. 11, 1376–1383.

Weinberg, A., Basham, T.Y. and Merigan, T.C. (1986). Regulation of guinea-pig immune functions by interleukin 2: critical role of natural killer activity in acute HSV-2 genital infection. J. Immunol. 137, 3310–3317.

Weiss, G.R., Margolin, K.A., Aronson, F.R., *et al.* (1992). A randomized phase II trial of continuous infusion interleukin-2 or bolus injection interleukin-2 plus lymphokine-activated killer cells for advanced renal cell carcinoma. J. Clin. Oncol. 10, 275–281.

Whitehead, R.P., Hauschild, A., Christophers, E. and Figlin, R. (1995). Diabetes mellitus in cancer patients treated with combination interleukin 2 and alpha-interferon. Cancer Biother. 10, 45–51.

Wood, R., Montoya, J.G., Kundu, S.K., Schwartz, D.H. and Merigan, T.C. (1993). Safety and efficacy of polyethylene glycol-modified interleukin-2 and zidovudine in human immunodeficiency virus type 1 infection: a phase I/II study. J. Infect. Dis. 167, 519–525.

Wu, Z., Johnson, K.W. and Ciardelli, T.L. (1995). Ligand binding analysis of interleukin-2 receptor complexes using surface plasmon resonance. J. Immunol. Methods 183, 127–130.

Yamamoto, M., Iizuka, H., Fujii, H., Matsuda, M. and Miura, K. (1993). Hepatic arterial infusion of interleukin-2 in advanced hepatocellular carcinoma. Acta Oncol. 32, 43–51.

Yang, S.C., Grimm, E.A., Parkinson, D.R., *et al.* (1991). Clinical and immunomodulatory effects of combination immunotherapy with low-dose interleukin 2 and tumor necrosis factor alpha in patients with advanced non-small cell lung cancer: a phase I trial. Cancer Res. 51, 3669–3676.

Yodoi, J. and Uchiyama, T. (1992). Diseases associated with HTLV-1: virus, IL-2 receptor dysregulation and redox regulation. Immunol. Today, 13, 405–411.

Yokota, T., Arai, N., Lee, F., Rennick, D., Mosmann, T. and Arai, K. (1985). Use of a cDNA expression vector for isolation of mouse interleukin 2 cDNA clones: expression of T-cell growth-factor activity after transfection of monkey cells. Proc. Natl Acad. Sci. USA 82, 68–72.

Ziegler, L.D., Palazzolo, P., Cunningham, J., *et al.* (1992). Phase I trial of murine monoclonal antibody L6 in combination with subcutaneous interleukin-2 in patients with advanced carcinoma of the breast, colorectum, and lung. J. Clin. Oncol. 10, 1470–1478.

Zuniga Pflucker, J.C., Smith, K.A., Tentori, L., Pardoll, D.M., Longo, D.L. and Kruisbeek, A.M. (1990). Are the IL-2 receptors expressed in the murine fetal thymus functional? Dev. Immunol. 1, 59–66.

3. Interleukin-3

Hanna Harant *and* Ivan J.D. Lindley

1. Introduction

Interleukin-3 was orginally identified as a multilineage hemopoietic growth factor, which stimulates colony formation of erythroid, megakaryocytic, granulocytic, and monocytic lineages (Clark and Kamen, 1987; Clark-Lewis and Schrader, 1988; Metcalf, 1991; Ihle, 1992). A variety of synonyms exists for IL-3, such as multicolony stimulating factor (multi-CSF), hemopoietic cell growth factor (HCGF), and colony-forming unit-simulating activity (CFU-SA) (Moore, 1988). IL-3 supports proliferation of myeloid precursors and also plays an essential role in supporting survival of these cells by prevention of apoptosis. The production of IL-3 is, in contrast to other hemopoietic growth factors, restricted to a few cell types, such as activated T lymphocytes and mast cells, underlying its role in inducible but not constitutive hemopoiesis. Beside its effects on immature myeloid progenitor cells, IL-3 acts on mature end-stage myeloid cells, such as basophils and eosinophils, by priming histamine release and generation of leukotrienes by IgE-dependent and -independent mechanisms, suggesting an essential role in allergic responses. It further acts on monocytes and endothelial cells, and thus also plays an important role in inflammatory processes. The biological effects of IL-3 on hemopoiesis led to its use in clinical trials, to support and accelerate hemopoiesis in myeloid disorders and after high-dose chemotherapy or autologous bone marrow transplantation. Further development of IL-3 agonists, with specific effects on hemopoiesis and lower activity on mature leukocytes, is currently in progress (Clark and Kamen, 1987; Clark-Lewis and Schrader, 1988; Metcalf, 1991; Ihle, 1992; Lopez *et al.*, 1992; Cosman, 1993).

2. *The Cytokine Gene*

Human IL-3 is encoded by a single mRNA transcript of approximately 1 kb length. The human IL-3 gene is located on chromosome 5 at 5q23-31, together with the genes encoding granulocyte-macrophage colony-stimulating factor (GM-CSF), IL-4, IL-5, IL-9, macrophage colony-stimulating factor (M-CSF), the M-CSF receptor and various other genes encoding growth factors or their receptors. This small segment is termed the cytokine gene cluster. The IL-3 and GM-CSF genes are tightly linked and are only 9 kb apart (Yang *et al.*, 1988; van Leeuwen *et al.*, 1989).

2.1   Genomic Organization of the IL-3 Gene

The highly homologous gibbon IL-3 cDNA was used to screen for the human IL-3 gene. The human IL-3 gene is approximately 3 kb in length and consists of five exons (coding for 54, 14, 30, 14, and 40 amino acids) interrupted by four introns. The sequence of the exons of the human IL-3 gene is more than 99.5% homologous with the sequence of the gibbon cDNA clone. The human DNA sequence has relatively high homology to that of the mouse gene in the 3′ noncoding region (73%), while only 45% homology was found between the respective coding regions (Yang *et al.*, 1986). The human IL-3 gene contains a single consensus signal (AATAAA) for polyadenylation in the 3′ noncoding region (nucleotides 2871–2876). This sequence is found at the 3′-end of most RNA polymerase II transcripts and typically signals the addition of a polyadenylate sequence 10–25 nucleotides 3′ to the AAUAAA (Yang and Clark, 1988) (Figures 3.1, 3.2).

The 3′ noncoding sequence of both the gibbon and the human IL-3 cDNA contains an AT-rich segment characterized by several repeats of the sequence ATTTA, which has been identified in cDNA clones of numerous lymphokines and other regulatory proteins that are transiently expressed, and has been shown to affect stability of mRNA (Shaw and Kamen, 1986).

1
GATCCAAAC ATG AGC CGC CTG CCC GTC CTG CTC CTG   Exon 1
CTC CAA CTC CTG GTC CGC CCC GGA CTC CAA GCT CCC
ATG ACC CAG ACA ACG TCC TTG AAG ACA AGC TGG GTT
AAC TGC TCT AAC ATG ATC GAT G AA ATT ATA ACA CAC
TTA AAG CAG CCA CCT TTG CCT TTG CTG
171
GAC TTC AAC AAC CTC AAT GGG GAA GAC CAA GAC ATT   Exon 2
CTG ATG
214
GAA AAT AAC CTT CGA AGG CCA AAC CTG GAG GCA TTC   Exon 3
AAC AGG GCT GTC AAG AGT TTA CAG AAC GCA TCA GCA
ATT GAG AGC ATT CTT AAA
304
AAT CTC CTG CCA TGT CTG CCC CTG GCC ACG GCC GCA   Exon 4
CCC ACG
345
CGA CAT CCA ATC CAT ATC AAG GAC GGT GAC TGG AAT   Exon 5
GAA TTC CGG AGG AAA CTG ACG TTC TAT CTG AAA ACC
CTT GAG AAT GCG CAG GCT CAA CAG ACG ACT TTG AGC
CTC GCG ATC TTT TAG TCCAACGTCCAGCTCGTTCTCT
GGGCCTTCTCACCACAGCGCCTCGGGACATCAAAAACAGCA
GAACTTCTGAAACCTCTGGGTCATCTCTCACACATTCCAGGA
CCAGAAGCATTTCACCTTTTCCTGCGGCATCAGATGAATTGT
TAATTATCTAATTTCTGAAATGTGCAGCTCCCATTTGGCCTTG
TGCGGTTGTGTTCTCA
674

Figure 3.1   Nucleotide sequence of the human IL-3 gene (accession code GenEMBL M14743). The five exons of the human IL-3 gene are shown. The underlined bases in exon 1 and exon 5 are nontranslated regions (Yang *et al.*, 1986).

2.2   Transcriptional Control

The 5′ upstream regulatory sequences of the IL-3 gene contain binding sites for several nuclear transcription factors. The AP-1 binding site (nucleotides (nt) −301 to −295 relative to the transcription start) is required for transcriptional activation by phorbol myristate acetate (PMA), phytohemagglutinin (PHA), and anti-CD2 in human primary T lymphocytes (Park *et al.*, 1993). Six base pairs downstream of the AP-1 site, a binding site for the transcription factor Elf-1, a member of the Ets family, is present. Cooperation of AP-1 with Elf-1 or NF-IL-3A was shown to be required for maximal promoter activity in the stimulated MLA-144 gibbon T cell line (Gottschalk *et al.*, 1993). The NF-IL-3A/ACT-1 binding site, located between nt −156 and −147 relative to the transcription start, has been shown to be a stimulation-responsive element in Jurkat, gibbon MLA-144, and primary human T cells (Shoemaker *et al.*, 1990). The transcription factor NF-IL-3A has recently been cloned and is nearly identical to the E4BP4 transcriptional repressor protein, and also binds to regulatory sequences in the adenovirus E4 promoter and interferon-γ (IFN-γ) promoter (Zhang *et al.*, 1995). Over the 5′ end of the ACT-1 site binds an inducible, T cell-specific factor that shares functional properties with the OAP[40] complex of the ARRE-1 element of the IL-2 promoter. Over its 3′ end, the ACT-1 site binds the octamer factor Oct-1 (Davies *et al.*, 1993; Kaushansky *et al.*, 1994). Transcription of the IL-3 gene is negatively regulated by a strong repressor element, termed nuclear inhibitory protein (NIP) (nt −271 to −250 relative to the transcription start), which inhibits IL-3 gene expression in the absence of the AP-1/Elf-1 site in MLA-144 cells (Mathey-Prevot *et al.*, 1990). Three protein complexes bind to the NIP region of the IL-3 promoter in MLA-144 cells, and one complex (NIP-C3) was identified as the repressive factor. The presence of NIP-C3 was demonstrated in a variety of cell types (Engeland *et al.*, 1995). The GT/GC-rich region (nt −76 to −47 relative to the transcription start) is required for basal promoter activity and serves as a response element for the human T cell leukemia virus type-I-encoded Tax. This region binds the transcription factors EGR1, EGR2 and DB1 (Koyano-Nakagawa *et al.*, 1994).

2.3   POST-TRANSCRIPTIONAL CONTROL

The 3′ untranslated region (UTR) of the IL-3 mRNA contains the repetitive AUUUA motif, found in various lymphokine and cytokine genes, which is responsible for mRNA destabilization (Shaw and Kamen, 1986; Caput *et al.*, 1986; Brewer and Ross, 1988; Malter, 1989; Brewer, 1991). In human Jurkat T cells, upregulation of IL-3 mRNA stimulated by treatment with PMA and PHA occurs both at the transcriptional level and post-transcriptionally, by increasing the half-life of the IL-3 mRNA (Ryan *et al.*, 1991). In human primary T lymphocytes, PMA has been shown to augment the concanavalin A (Con A)-induced IL-3 mRNA accumulation in a similar manner, by increasing the transcription rate and mRNA stabilization (Dokter *et al.*, 1993a). The stromal-derived growth factor IL-7 was shown to enhance Con A-induced IL-3 mRNA accumulation in human primary T lymphocytes, by an effect on post-transcriptional mechanisms (Dokter *et al.*, 1993b). In mast cells, expression of IL-3 is induced following IgE receptor activation via calcium-dependent mRNA stabilization (Wodnar-Filipowicz and Moroni, 1990). Analysis of the 8 AUUUA motifs located in the 3′ UTR of the IL-3 mRNA demonstrated that within a cluster of six AUUUA motifs, three adjacent motifs are required for binding of a factor which mediates rapid IL-3 mRNA decay (Stoecklin *et al.*, 1994). The murine cell

line FL-*IL3*-R, which was derived from the IL-3 dependent cell line FL5.12 by transformation with a retrovirus (MLV-*src*) encoding the v-*src* oncogene, contains a rearrangement at the IL-3 locus and produces IL-3 constitutively. The rearrangement resulted from the transposition of an intracisternal A particle (IAP) provirus into the 3′ untranslated region of the IL-3 gene, within the ATTTA sequence motifs, resulting in an increase in the half-life of the IL-3 mRNA (Algate and McCubrey, 1993; Mayo *et al.*, 1995).

3.   *The Protein*

The human IL-3 cDNA was derived from a library from an activated T cell clone (Otsuka *et al.*, 1988). The cDNA is composed of 922 bases and has a single open reading frame of 152 amino acids. The encoded protein consists of a hydrophobic leader peptide of 19 amino acids (aa), and the mature protein is 133-aa long, with a predicted molecular mass of 15.1 kDa. SDS-PAGE determinations showed that IL-3 expressed in COS cells has an apparent molecular mass between 15 and 25 kDa, while expression in yeast gives an apparent molecular mass between 18 and 23 kDa. Human IL-3 contains two potential *N*-linked glycosylation sites (aa 15–17 and 70–72), but glycosylation is not required for biological activity. Two conserved cysteine residues are present in the human protein (aa 16 and 84). Human IL-3 has 29% homology with the murine IL-3 at the amino acid level (Dorrsers *et al.*, 1987) (Figure 3.3 on Page 43).

Human IL-3 protein, expressed in *E. coli*, was found to be insoluble at the concentration and pH required for NMR studies, and no x-ray studies have yet been reported. However, a 112-residue truncated variant of recombinant hIL-3, SC-65369, has been used for NMR studies. This variant lacks the N-terminal 13 residues and C-terminal 8 residues, and has 14 amino acid substitutions relative to the wild-type protein, and is in fact slightly more active than the native protein. Data derived using this mutant molecule indicates that IL-3 contains four helical bundles (helices A–D) and three reverse turns, similar to GM-CSF. A fifth short helix (A′) is located in the loop connecting the first and second helices. This helix A′ is absent in GM-CSF and IL-5, suggesting an essential role in the recognition of the α chain of the IL-3 receptor (Feng *et al.*, 1995).

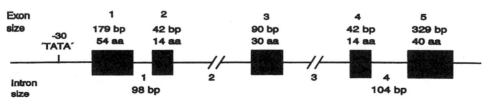

Figure 3.2   Gene structure of IL-3 gene.

4. Cellular Sources and Production

IL-3 is produced by human T cells stimulated with PHA and PMA or a combination of IL-1 with anti-CD3 monoclonal antibody. IL-3 is also produced by human T cell lines, such as the Jurkat cell line, in response to treatment with PHA. The gibbon cell line MLA-144 is dependent on stimulation with PMA for expression of IL-3 (Otsuka *et al.*, 1988; Niemeyer *et al.*, 1989).

Human natural killer (NK) cells have been shown to produce IL-3 upon stimulation with phorbol dibutyrate and calcium ionophore A23187 (Cuturi *et al.*, 1989).

Cross-linkage of high-affinity Fcε receptors (FcεRI) with IgE on mast cells leads to induction of allergic responses and secretion of histamine and newly synthesized arachidonic acid metabolites. Cross-linkage of FcεRI or treatment with calcium ionophore stimulates expression of various lymphokines by murine mast cell lines, lymphokines which were originally thought to be produced only by a subset of murine T cell lines (T_H2 cells). The factors expressed by cross-linked mast cells include IL-3, IL-4, IL-5, and IL-6 (Plaut *et al.*, 1989; Wodnar-Filipowicz *et al.*, 1989). The human mast cell line HMC-1, which lacks the high-affinity receptor for IgE, can express transcripts for IL-3, amongst other cytokines, after stimulation with TPA or ionomycin (Nilsson *et al.*, 1995).

Another source of IL-3 is the murine WEHI-3 myelomonocytic leukemia cell line, which produces IL-3 constitutively, due to insertion of an endogenous retroviral element with its 5′ long terminal repeat in the proximity of the promoter region of the IL-3 gene (Ymer *et al.*, 1985). IL-3-dependent murine PB-3c mast cells express IL-3 and can form IL-3 autocrine tumors after v-H-*ras* expression. In one tumor, an insertion of an endogenous retroviral element (IAP) into the promoter of the IL-3 gene was shown (Hirsch *et al.*, 1993).

Other sources of IL-3 include astrocytes and neuronal cells of the mouse brain, which have been shown to express IL-3 mRNA (Farrar *et al.*, 1989). Human thymic epithelial cells also express IL-3 (Dalloul *et al.*, 1991).

There are some conflicting reports concerning the expression of IL-3 by epidermal keratinocytes. Expression of an interleukin-3-like activity was described for murine keratinocytes, and the presence of IL-3 mRNA was demonstrated in murine cell lines and also in normal murine neonatal primary cultures of keratinocytes, indicating that murine keratinocytes express IL-3 (Luger *et al.*, 1985). Human keratinocytes also produce an IL-3-like activity (Danner and Luger, 1987; Dalloul *et al.*, 1992). However, in a recent report it is proposed that the observed IL-3-like bioactivity described for human keratinocytes can be attributed to GM-CSF and IL-6 (Kondo *et al.*, 1995).

5. Biological Activity

5.1 THE ROLE OF IL-3 IN HEMOPOIESIS

Homeostasis of the hemopoietic system is maintained by continuous growth and differentiation of blood cells. Mast cells, megakaryocytes, erythrocytes, neutrophils, eosinophils, basophils, and monocytes are derived from myeloid precursor cells, while T and B lymphocytes are derived from the lymphoid precursors. Both lineages develop from pluripotent stem cells, which have the capacity for both differentiation and self-renewal. Stem cells in contact with stromal cells are thought to be in the G_0 resting phase and enter into the cell cycle at a constant rate, termed *constitutive hemopoiesis*, which is regulated by humoral factors and cell-to-cell interaction. *Inducible hemopoiesis* is a response to acute situations such as infection and bleeding, and is driven by humoral factors that promote rapid expansion and maturation of specific sets of hemopoietic cells at the affected sites. The factors that control both constitutive and inducible hemopoiesis are the colony-stimulating factors (GM-CSF, G-CSF, M-CSF, IL-3, and other factors such as SCF (stem cell factor, steel factor, c-*kit* ligand), erythropoietin (EPO), IL-6, IL-1, and tumor necrosis factor-α (TNF-α). In contrast to other factors, many of which are produced by a variety of cells, production of IL-3 is restricted to only a few cell types, such as activated T cells and mast cells. This observation led to the concept that IL-3 is involved only in *inducible hemopoiesis* and does not play a role in *constitutive hemopoiesis*. This hypothesis is underlined by the fact that IL-3 is not produced by fetal tissues (Azoulay *et al.*, 1987) or by bone marrow stromal cells (Naperstek *et al.*, 1986; Gualtieri *et al.*, 1987). IL-3 has three effects on committed progenitor cells: the induction of proliferation and differentiation, and the support of cell survival by prevention of apoptosis (Clark and Kamen, 1987; Metcalf, 1989, 1991).

5.1.1 Role of IL-3 in the Proliferation and Differentiation of Pluripotent Myeloid Progenitor Cells

IL-3 is a multilineage hemopoietic growth factor that acts on early myeloid progenitor cells and supports their proliferation and terminal differentiation into the granulocytic, monocytic, megakaryocytic, and erythroid lineage, in both the murine and human systems (Ikebuchi *et al.*, 1987, 1988). IL-3 acts on early stages of hemopoiesis, rather than on cells in the terminal process of maturation (Sonoda *et al.*, 1988). The myeloid progenitors are characterized by expression of the CD34 cell surface molecule (CD34+), and represent 0.5–2.5% of the cells in the bone marrow (BM) or umbilical cord blood. CD34+ cells proliferate in the presence of a variety of hemopoietic growth factors, including IL-3, GM-CSF, and G-CSF (Litzow *et al.*, 1991; Terstappen *et al.*,

1991). The development of CD34$^+$ cells into the erythroid lineage by IL-3 is supported by EPO (Saeland et al., 1988). SCF can synergize with IL-3, GM-CSF or G-CSF to induce unseparated human bone marrow (BM) cells to form colony-forming units-granulocyte-monocyte (CFU-GM), and to form erythroid burst-forming units (BFU-E) in the presence of IL-3 and EPO (Bernstein et al., 1991; McNiece et al., 1991; Brandt et al., 1992).

CD34$^+$ cells can be further subdivided into cells at differing stages of differentiation, as assayed by the expression of specific surface molecules. For instance, CD34$^+$ cells expressing low levels of HLA-DR, but high levels of c-kit (CD34$^+$ HLA-DR$^-$ c-kit$^+$) represent more primitive progenitors, compared to those expressing HLA-DR (CD34$^+$ HLA-DR$^+$). The latter are enriched for more differentiated human progenitors (HPCs), including CFU-GM, BFU-E, CFU-granulocyte-erythroid-monocyte-megakaryoctye (CFU-GEMM), and CFU-megakaryocyte (CFU-Mk). The response of CD34$^+$ cells to IL-3 or other hemopoietic factors depends upon their stage of differentiation. SCF has been shown to support the survival of the more primitive progenitors (CD34$^+$ HLA-DR$^-$ c-kit$^+$), whereas IL-3 supports the survival of the more differentiated progenitors (CD34$^+$ HLA-DR$^+$) (Brandt et al., 1994). It has further been demonstrated that CD34$^+$ cells from cord blood, expressing high levels of c-kit coupled with low levels of IL-3 receptors, represent erythroid progenitors, whereas cells expressing low levels of c-kit represent granulocytic, monocytic and primitive progenitors (Sato et al., 1993b).

IL-3 interacts with other cytokines, and the lineage development can be influenced by the combination of factors supplied. IL-3, GM-CSF, and IL-5 can support eosinophilic differentiation, and it has been suggested that IL-3 and GM-CSF support the early and intermediate stages of this process (Lopez et al., 1986; Saito et al., 1988), while IL-5 primarily supports the late and terminal proliferation and maturation (Yamaguchi et al., 1988). Among the CSFs, IL-3 and GM-CSF have been reported to support development of CD34$^+$ toward the megakaryocytic lineage (Egeland et al., 1991). IL-3 induces early stages of megakaryocytopoiesis, whereas the hemopoietic factor thrombopoietin (c-mpl ligand) is required for full megakaryocytic maturation (Banu et al., 1995; Guerriro et al., 1995). TNF-α potentiates the IL-3 and GM-CSF-induced short-term proliferation of CD34$^+$ cells, whereas long-term treatment results in inhibition of the granulocytic differentiation and support of the maturation into the monocytic lineage (Caux et al., 1990, 1991). IL-3 also cooperates with TNF-α for the development of dendritic/Langerhans cells from cord blood CD34$^+$ hemopoietic progenitor cells (Caux et al., 1996). TNF-α has also been shown to increase expression of β_c of the IL-3, IL-5, and GM-CSF receptors on CD34$^+$ cells, which then become more

sensitive to IL-3 (Sato et al., 1993b). IL-1 has been shown to upregulate the β_c in the human factor-dependent cell line TF-1 (Watanabe et al., 1992). IL-3 has further been reported to support the growth-promoting effect of SCF on murine mast cell precursors (Rennick et al., 1995).

The in vivo data are mainly consistent with those obtained from in vitro studies. Thus, infection of mouse bone marrow cells with retroviral vectors expressing IL-3 cDNA and transplantation into irradiated mice induced a nonneoplastic myeloproliferative disorder. Mice over-expressing IL-3 exhibited increased spleen, peritoneal, and peripheral blood cellularity, and extensive neutro-philic infiltration of the spleen, lung, liver, and muscle, whereas bone marrow cellularity decreased (Wong et al., 1989; Chang et al., 1989).

5.1.2   Role of IL-3 in Proliferation and Maturation of Lymphoid Progenitors

For long-term proliferation, B cell progenitors require a variety of hemopoietic growth factors and are dependent on factors derived from stroma cells, such as IL-7, and on direct cell-to-cell contact. These activities involve CD44 and hyaluronate, very late activation antigen-4 (VLA-4) and fibronectin, and SCF–c-kit and IL-7–IL-7 receptor interactions (Whitlock and Witte, 1982; Takeda et al., 1989). IL-3 has been proposed to be a growth factor for mouse B cell precursors, since it has been shown to support the growth of freshly isolated fetal liver murine pre-B cells and the long-term culture of IL-3-dependent pre-B cell clones which can be converted into mature antibody-secreting cells in vitro (Palacios et al., 1984). Murine B cell precursors die in the absence of IL-3, underlining its role in the development of early B cell precursors (Palacios and Steinmetz, 1985). Optimal growth of pre-B cells was obtained by combination of IL-3 with the stroma-derived cytokine IL-7 (Rennick et al., 1989). IL-3 has also been reported to be an alternative growth factor for mouse CD45RO (B-220$^+$) and c-kit$^+$ pre-B cells, and can replace IL-7 (Winkler et al., 1995). The cell line SPGM-1, which was derived from a mouse progranulocyte/ promacrophage tumor and exhibits a pre-B cell phenotype, can be induced to switch from pre-B cells to macrophages in the presence of IL-3 (Martin et al., 1993).

Treatment of mouse lymphoprogenitors with two-factor combinations, including SCF plus IL-6, IL-11, or G-CSF, supported the lymphomyeloid potential of primary colonies. IL-3 could not replace or synergize with SCF in maintaining the B-lymphoid potential of the primary colonies, although the frequency of colony formation was similar with IL-3 or SCF alone. IL-3 and IL-1α have further been shown to suppress the B-lymphoid potential of primitive progenitors, suggesting a role of IL-3 and IL-1α as negative regulators in early stages of B-lymphopoiesis (Hirayama et al., 1992, 1994). The possible negative regulatory role of IL-3 on B-cell development is also evident from the development of a nonmalignant B cell lymphoproliferative disorder in transgenic mice expressing antisense IL-3 RNA (Cockayne et al., 1994).

There are few reports on the effect of IL-3 on human B cell proliferation. Xia and colleagues reported that IL-3 is a late-stage B cell growth factor on *Staphylococcus aureus* Cowan 1 (SAC)-activated B cells. IL-3 synergizes with IL-2 to enhance both B cell proliferation and differentiation by enhancing IL-2 receptor expression (Xia et al., 1992). IL-3 has further been shown to support the growth of human B cell lymphoma cells (Clayberger et al., 1992). Plasma cell precursors isolated from peripheral blood mononuclear cells (PBMC) of patients suffering from multiple myeloma (MM), a human B cell malignancy characterized by bone marrow accumulation of plasma cells, have been shown to be controlled in their growth by IL-3 and synergistically by IL-3 and IL-6 (Bergui et al., 1989). Human leukemic B-cell precursors (BCPs) from patients with acute lymphoblastic leukemia (ALL) express functional receptors for IL-3, and proliferate in response to IL-3 (Uckun et al., 1989; Eder et al., 1992).

Expansion and maturation of early thymic progenitor cells and pro-T cells are still poorly understood. The thymic environment expresses a variety of cytokines, including IL-3. The earliest intrathymic progenitors express low levels of CD4 (CD4^{lo}, CD3$^-$, CD8$^-$, CD44$^+$, CD25$^-$, c-kit^+), contain T, B, and NK and dendritic cell precursor activities, and are not yet committed to the T-cell lineage. CD4^{lo} were shown to proliferate best in the presence of combinations of cytokines (IL-3 + IL-6 + SCF; IL-1 + IL-3 + IL-6 + SCF; IL-1 + IL-3 + IL-6 + IL-7 + SCF) (Moore and Zlotnik, 1995).

5.1.3   Role of IL-3 in Prevention of Apoptosis

Expansion and maturation of the immature progenitor cells of bone marrow appear to be controlled by a balance between proliferation and apoptosis. Proliferation of these cells is stimulated by hemopoietic cytokines, such as IL-3 and GM-CSF, which act on early precursor cells. Upon withdrawal of IL-3 or GM-CSF from progenitor cells or primary IL-3-dependent cells from the bone marrow, they undergo programmed cell death, or apoptosis, although the enhancement of cell survival is distinct from the stimulation of proliferation (Williams et al., 1990).

Apoptosis induced by IL-3 withdrawal from the IL-3-dependent hematopoietic cell line 32D was shown to be associated with repartitioning of intracellular calcium and is blocked by overexpression of *bcl*-2 (Baffy et al., 1993). In the human IL-3-dependent erythroleukemia cell line TF-1, *bcl*-2 mRNA and protein levels decrease upon withdrawal of IL-3 and it was suggested that IL-3 plays a role in maintaining *bcl*-2 transcription through activation of protein kinase C in this cell line (Rinaudo et al., 1995). Two different signaling pathways have been demonstrated

for induction of proliferation and prevention of apoptosis by IL-3, but both signals are required for long-term proliferation. The signaling pathway for DNA synthesis can be clearly dissociated from the anti-apoptotic function induced by ligand binding. The distal region of the IL-3/GM-CSF receptor β chain, which activates the Ras/Raf-1/MAPK signaling pathway, has been shown to be implicated in the prevention of apoptosis. Thus the Ras signaling pathway appears to provide another level of growth control, presumably by regulation of *bcl*-2 expression (Kinoshita et al., 1995).

IL-3 has also been shown to prevent other types of apoptosis, such as apoptotic cell death induced by DNA-damaging irradiation or treatment with the topoisomerase II inhibitor etoposide in the murine IL-3-dependent pre-B cell line BAF3. Apoptosis induced by DNA damage can also be prevented by overexpression of *bcl*-2, and IL-3 and *bcl*-2 can act in a cooperative manner (Collins et al., 1992; Ascaso et al., 1994).

5.2   EFFECTS OF IL-3 ON MATURE LEUKOCYTE FUNCTION

5.2.1   Monocytes/Macrophages

In addition to its effects on hemopoietic precursors, IL-3 modulates leukocyte function *in vitro* and *in vivo*. IL-3 does not induce proliferation in mature macrophages, but exerts macrophage-activating properties, regulating the expression of various surface antigens and cytokines in murine peritoneal macrophages, human peripheral blood monocytes, and cells of the human THP-1 line. IL-3 induces class II MHC antigen (Ia) and LFA-1 (CD11a/CD18) expression in murine peritoneal macrophages. IL-3 is a less potent stimulus than IFN-γ on class II MHC (Ia) antigen expression, but can induce LFA-1 with a time-course similar to the induction of LFA-1 by either IFN-γ or LPS. IL-3 can interact synergistically with IFN-γ in inducing class II MHC (Ia) antigen expression, while IL-3 and IFN-γ have only an additive effect on LFA-1 induction (Frendl and Beller, 1990). IL-3 and GM-CSF can increase monocyte adhesiveness to endothelial cells via induction of CD18, the common β chain of the LFA-1 family, on the surface of monocytes (Elliott et al., 1990). In human monocytes, GM-CSF or IL-3 increase the expression of HLA-DR, CD14 (LPS receptor), and IL-1α (Dimri et al., 1994). CD23 (FcεRII) expression by B cells and monocytes can be regulated by IL-4, IFN-γ, and IFN-α, as well as by GM-CSF and IL-3. GM-CSF is able to enhance CD23 expression by monocytes and IL-3 enhances CD23 expression by both monocytes and B cells. IL-4 acts synergistically with GM-CSF or IL-3 to induce monocyte CD23 expression (Alderson et al., 1994).

In murine peritoneal macrophages, IL-3 alone does not induce IL-1 bioactivity, whereas IL-1 is induced synergistically in the presence of IL-3 and suboptimal concentrations of LPS (Frendl et al., 1990). IL-3 and LPS are also able to induce IL-6 and TNF-α in murine macrophages (Frendl, 1992).

IL-3, like GM-CSF, can regulate macrophage tumoricidal activity in vitro. In response to IL-3 and GM-CSF, human peripheral monocytes express TNF-α, which itself may be involved in monocyte cytotoxicity (Cannistra et al., 1988). Using the malignant melanoma cell line A375 as target, tumor cell killing by activated macrophages was demonstrated by treatment with IFN-γ and suboptimal concentrations of microbial products such as muramyl peptide, whereas IL-3 or GM-CSF were able to induce peripheral blood monocytes to kill A375 tumor cells in vitro without any requirement for a microbial cofactor. The mechanisms by which activated monocytes kill A375 tumor cells are the production of reactive oxygen and nitrogen intermediates, release of soluble cytotoxic molecules and the engagement of cell surface cytotoxic molecules (Grabstein and Alderson, 1993). CD40, a member of the NF receptor family of cell surface proteins, is expressed on B cells, dendritic cells, some carcinoma cell lines, and human thymic epithelium. Treatment of human primary monocytes with GM-CSF, IL-3, or IFN-γ resulted in induction of CD40 mRNA and enhancement of CD40 cell surface expression. CD40 was found to mediate monocyte adhesion to cells expressing recombinant CD40 ligand. Coincubation of monocytes with CV-1/EBNA cells transfected with CD40 ligand induced them to become tumoricidal against A375 melanoma cells (Alderson et al., 1993).

A further effect of human IL-3 is to enhance the Candida albicans killing capacity of human monocytes in vitro (Wang et al., 1989). In vivo treatment with human IL-3 as a single infusion in hematologically normal individuals did not produce significant changes in expression of the adhesion molecules CD11b or L-selectin or of monocyte respiratory burst activity, but demonstrated a significant increase in monocyte phagocytosis and killing of Candida albicans. These data suggest that IL-3 has significant effects also on monocyte function in vivo (Khwaja et al., 1994).

5.2.2 Basophils

Mature basophils express receptors for IL-3, and IL-3 has been recognized as a differentiation factor for human basophils (Valent et al., 1989a,b). IL-3 has the capacity to enhance histamine release from human basophils stimulated by IgE-dependent and IgE-independent mechanisms (Hirai et al., 1988). Beside its priming effect on histamine release by basophils, IL-3 was shown to induce direct histamine release in basophils from some humans, with cells from atopics responding to a greater extent than those from nonatopic donors (Haak-Frendscho et al., 1988; MacDonald et al., 1989). Removal of surface IgE from basophils rendered them

unresponsive to IL-3, and the response could be restored by passive sensitization of basophils with IgE since IgE is known to bind histamine-releasing factors (MacDonald et al., 1989; Miadonna et al., 1993). IL-3- and IgE-dependent histamine release was shown to be inhibited by Na^+ (Tedeschi et al., 1995). According to other reports IL-3 can induce basophil adhesion to endothelial cells and expression of CD11b on the basophil surface without causing histamine release (Bochner et al., 1990), and can induce homotypic aggregation of basophils without induction of degranulation (Knol et al., 1993).

IgE-independent induction of histamine release and generation of leukotriene C_4 (LTC_4) can be observed upon stimulation of basophils with calcium ionophore A23187, the formylated peptide formylmethionylleucyl-phenylalanine (FMLP), anaphylatoxins C5a and C3a, or platelet-activating factor (PAF), and is enhanced by prestimulation of the basophil with IL-3 (Kurimoto et al., 1989; Bischoff et al., 1990; Brunner et al., 1991). The chemokine cytokine IL-8 was shown to induce histamine release and generation of LTC_4 in an IgE-independent fashion in basophils which were preexposed to IL-3 (Dahinden et al., 1989). Another report describes an inhibitory effect of IL-8 on basophils stimulated with histamine-releasing factors or IL-3 (Kuna et al., 1991).

5.2.3 Eosinophils

Eosinophilia is a response of the mammalian host in helminthic infection. Eosinophil production appears to be regulated by IL-3, IL-5, and GM-CSF, and eosinophils express receptors for all three cytokines (Lopez et al., 1991). Long-term treatment of eosinophils with IL-3 results in increased viability, induces normodense cells to become hypodense, and enhances their cytotoxicity against antibody-coated Schistosoma mansoni larvae. Moreover, IL-3 has been shown to augment calcium ionophore-induced LTC_4 generation (Rothenberg et al., 1988). IL-5 was identified as the most potent enhancer of Ig-induced degranulation and increased eosinophil-derived neurotoxin (EDN) release, but IL-3 and GM-CSF also enhanced Ig-induced EDN release, although they were less potent than IL-5 (Fujisawa et al., 1990). Pretreatment of human eosinophils with IL-3 primes them for enhanced generation of LTC_4 after stimulation with FMLP, C5a and PAF (Takafuji et al., 1991, 1995).

5.2.4 Neutrophils

IL-3, like GM-CSF and G-CSF, stimulates the production of neutrophils from bone marrow precursors, but unlike the other two factors, it is unable to affect the function of mature neutrophils, consistent with the loss of IL-3 receptors during differentiation into the neutrophilic series (Lopez et al., 1988). In a recent report, neutrophils cultured in the presence of GM-CSF were shown to express IL-3R α chain mRNA and both high- and low-affinity IL-3 receptors. Moreover, coincubation of neutrophils with IL-3 and GM-CSF resulted in synergistic upregulation of HLA-DR expression (Smith *et al.*, 1995).

5.2.5 Endothelial Cells

In an inflamed tissue, proliferation and activation of endothelial cells are regulated by humoral factors released by T lymphocytes and monocytes infiltrating the perivascular space. Human umbilical vein endothelial cells (HUVECs) express low levels of IL-3 receptors and can be stimulated by IL-3 to proliferate. Moreover, in these cells, IL-3 induces transcription of endothelial-leukocyte adhesion molecule-1 (ELAM-1), which is associated with their activation. IL-3-induced ELAM-1 transcription is followed by enhanced adhesion of neutrophils and CD4$^+$ T cells to HUVECs (Brizzi *et al.*, 1993). HUVECs express the common β chain of the GM-CSF, IL-3, and IL-5 receptors and the α chain of the IL-3 receptor, but expression of the α chains of GM-CSF and IL-5 receptors is minimal or absent (Collotta *et al.*, 1993). The expression of the IL-3 receptor α chain was shown to be upregulated by TNF-α, IL-1β, and lipopolysaccharide (LPS), whereas the α chains for GM-CSF and IL-5 remained unaffected. Addition of IL-3 to TNF-α-activated HUVECs enhanced the expression of E-selectin and the release of IL-8. E-Selectin is a surface molecule required for the initial interaction of neutrophils and memory T cells with endothelium, whereas IL-8 directs neutrophil firm adhesion and transmigration (Korpelainen *et al.*, 1993). IFN-γ is also a stimulator of IL-3R expression in HUVECs. IL-3 and IFN-γ exert a synergistic effect on major histocompatibility complex (MHC) class II gene expression and on the production of the early-acting hemopoietic cytokines IL-6 and G-CSF, while the expression of GM-CSF and IL-8 remained unaffected (Korpelainen *et al.*, 1995).

6.   *The Receptor*

The IL-3 receptor (IL-3R) belongs to the hemopoietin receptor (HR) superfamily. Members of this family include the receptors for IL-2 (β and γ chains), IL-3, IL-4, IL-5, IL-6, IL-7, IL-9, IL-11, GM-CSF, granulocyte colony-stimulating factor (G-CSF), growth hormone (GH), prolactin (PRL), erythropoietin (EPO), leukemia inhibitory factor (LIF), and ciliary neurotrophic factor (CNTF) (Cosman, 1993).

IL-3, GM-CSF, and IL-5 have specific unique low-affinity receptors (α chains) with similar structures, characterized by a highly conserved N-terminus containing cysteine residues, a C-terminal WSXWS motif within a less conserved 200-amino-acid stretch, and a short cytoplasmic domain. These receptors can bind the

ligand, but are unable to transduce a signal. The IL-3 receptor α chain (IL-3Rα) has a predicted size of 41 kDa and is expressed as a 70 kDa glycosylated protein. The cDNA of IL-3Rα contains an open reading frame of 378 amino acids. The N-terminal 18 amino acid residues constitute a hydrophobic sequence typical of signal peptides. The transmembrane segment contains 20 amino acids and the intracellular domain 53 amino acids. The predicted extracellular domain contains four conserved cysteine residues and the "WSXWS" motif, although in the case of the IL-3Rα there is a leucine instead of tryptophan (LSXWS). The IL-3Rα chain binds IL-3 with low affinity (K_d = 120 ± 60 nM). The high-affinity IL-3 receptor is formed by association of the low-affinity α-chain with the β chain (β$_c$, originally termed KH97). This protein is unable to bind the ligand on its own, but converts to a high-affinity binding (K_d = 140 ± 30 pM) in the presence of the α chain and ligand (Kitamura *et al.*, 1991a,b). β$_c$ has a longer cytoplasmic domain and is able to transduce a biological signal. β$_c$ has a predicted molecular mass of 96 kDa and is expressed as a 120 kDa glycosylated protein. The cDNA for β$_c$ contains an open reading frame of 897 amino acids, of which the N-terminal 16 amino acids constitute a hydrophobic domain. After cleavage of the signal peptide, the mature receptor consists of 881 amino acids. The extracellular domain contains the "WSXWS" motif and conserved cysteine residues, and the transmembrane domain spans 27 amino acids (Fig. 3.4). β$_c$ is shared with the α chains of the GM-CSF and IL-5 receptors, and the specificity of cytokine action is then determined by the distribution of the different α and β chains (Hayashida *et al.*, 1990) (Fig. 3.5). Comparison of the β$_c$ cDNA with that of the mouse β chains, AIC2A and AIC2B, showed 56% and 55% identity, respectively.

The human IL-3 receptor α chain gene is located in the pseudoautosomal region of the X and Y chromosomes at bands Xp22.3 and Yp13.3, tightly linked to the GM-CSF receptor α chain gene (Kremer *et al.*, 1993; Milatovich *et al.*, 1993). β$_c$ is mapped to chromosome 22, at band 22q13.1 (Takai *et al.*, 1994).

6.1   Receptor Expression

Ligand-binding studies demonstrated that there is cross-competition between GM-CSF, IL-3, and IL-5 on mature leukocytes, where binding of one of the cytokines can be competed by excess of one of the others, explained by the existence of the common β subunit of the high-affinity receptors for GM-CSF, IL-3, and IL-5 (Hayashida *et al.*, 1990). Cross-competition exists between GM-CSF and IL-3 on monocytes, and between GM-CSF, IL-3, and IL-5 on eosinophils and basophils (Elliott *et al.*, 1992). While GM-CSF associates more rapidly with its receptors on monocytes and shows faster stimulation of cell functions than does IL-3, the latter dissociates from its receptor complex more slowly than

Figure 3.3   Amino acid sequence of mature human IL-3 (accession code Swissprot P08700). ▨, **conflicting amino acid: Ser (Yang *et al*., 1986). Pro (Otsuka *et al*., 1988). Disulfide bond between Cys[16] and Cys[84].** <u>Asn-X-Ser</u>, **two potential glycosylation sites.**

GM-CSF (Elliott *et al.*, 1992; Lopez *et al.*, 1992). Eosinophils and basophils express all three receptors, and cross-competition occurs between GM-CSF, IL-3, and IL-5. For eosinophils the binding hierarchy was defined as GM-CSF>IL-3>IL-5, indicating that bound IL-5 can be competed by excess amounts of GM-CSF or IL-3 (Lopez *et al.*, 1991). On human basophils IL-3 was shown to compete for GM-CSF and IL-5 binding, while IL-5 and GM-CSF did not compete for IL-3 binding. The hierarchy of basophil stimulation was defined as IL-3>GM-CSF>IL-5 (Lopez *et al.*, 1990).

7. Signal Transduction

Signal transduction following IL-3 binding to the high-affinity α/β receptor is mediated by the β$_c$ subunit of the IL-3R (Sakamaki *et al.*, 1992). β$_c$ itself does not possess intrinsic tyrosine kinase activity, but one of the earliest events that occur after IL-3 binding is induction of protein tyrosine phosphorylation of many cellular proteins and of β$_c$ itself (Isfort *et al.*, 1988; Sorensen *et al.*, 1989; Isfort and Ihle, 1990; Duronio *et al.*, 1992; Linnekin *et al.*, 1992). Two Src family kinases, Lyn and Fyn, have been implicated in the action of IL-3 in certain cell types (Torigoe *et al.*, 1992). A non-Src type kinase, janus kinase (JAK2), which belongs to the family that includes JAK1, JAK3, and TYK2, has also been shown to be tyrosine-phosphorylated and activated in response to stimulation with IL-3 (Silvennoinen *et al.*, 1993). JAK2 associates with β$_c$ of the IL-3R, and JAK2 activation following ligand binding requires the membrane-proximal domain. Activation of JAKs has been reported

for the receptors of many cytokines, including GM-CSF, IL-6, and GH. In the interferon system, JAKs activate STAT (signal transducers and activators of transcription) factors, which bind directly to DNA after activation and translocation to the nucleus (Darnell *et al.*, 1994; Ihle *et al.*, 1994). IL-3, GM-CSF, and IL-5 have been shown to transduce signals through two STAT5 homologs, presumably via activation of JAK2 (Mui *et al.*, 1995a,b). The membrane-proximal region of the β$_c$ has also been reported to be involved in the induction of c-*myc* by IL-3 or GM-CSF (Sato *et al.*, 1993a). The distal region of β$_c$ is required for activation of another proposed signaling pathway, which involves receptor phosphorylation after ligand binding, providing a binding site for proteins, such as Shc, which contain SH2 domains. Binding and phosphorylation of Shc leads to association with Grb2, an increase in GTP-bound Ras, activation of Raf-1, and activation of the MAP kinase pathway and thus induction of c-*jun* and c-*fos* (Satoh *et al.*, 1991; Sato *et al.*, 1993a; Ihle *et al.*, 1994). Intracellular signaling mediated by IL-3, IL-5, and GM-CSF receptors is, for the most part, indistinguishable and the biological differences arise from differential receptor expression.

8. Murine Cytokine

The mouse IL-3 gene exists as a single copy on chromosome 11, 14 kb away from the GM-CSF gene (Barlow *et al.*, 1987; Lee and Young, 1989). The murine IL-3 gene is approximately 3 kb length, and contains five exons, encoding 55, 14, 32, 14, and 51 amino acids, interrupted by four introns of 96, 993, 135, and 122 bp, located within the coding region. In the 5′ flanking region, a TATA-like sequence is present 31 nucleotides upstream from the transcription start. Several potential CAAT box sequences are located upstream from the TATA box and are separated by a GC-rich region. At the 3′ flanking region of the second intron are nine repeats of a closely related 14-bp sequence, which has homology with a 20 bp repeated sequence in the human genome shown to have enhancer activity (Miyatake et al., 1985). A single AATAAA sequence is present in the 1123 bp region beyond the translational termination codon. This conserved sequence, which is present near the 3′ end of eukaryotic mRNAs, is involved in endonucleolytic cleavage of the precursor mRNA to yield the 3′ poly(A) attachment site of the mature mRNA (Campbell et al., 1985).

Murine IL-3 is encoded by a single mRNA species approximately 0.9 kb in length. The murine IL-3 cDNA was cloned from a cDNA library derived from the myelomonocytic leukemia cell line WEHI-3 (Fung et al., 1984). Another cDNA clone derived from the Con A-stimulated T cell clone Cl.Ly1$^+$2$^-$/9, encoding the mast cell growth factor (MCGF) was identical with that of IL-3 except for a single nucleotide difference (Yokota et al., 1984). Both cDNA clones code for a precursor polypeptide of 166 amino acids including a signal peptide of 26 amino acids, cleavage of which yields a mature protein of 140 amino acids. The predicted molecular mass of mature IL-3 is 15.7 kDa. Murine IL-3 was purified from WEHI-3 cells and also chemically synthesized (Ihle et al., 1982, 1983; Bazill et al., 1983; Clark-Lewis et al., 1984, 1986). Analysis of the molecular mass showed heterogeneity in the size of secreted IL-3 (20–30 kDa according to gel filtration and SDS-PAGE). The isoelectric point is heterogeneous (pI = 4–8), possibly owing to post-translational modifications, such as glycosylation, and still remained heterogeneous (pI = 6–8) after neuraminidase treatment or treatment of IL-3 producing cells with tunicamycin. The murine IL-3 has four potential N-linked glycosylation sites (Asn-X-Ser at positions 16–18, 44–46, 51–53, and 86–88), but glycosylation is not required for biological activity. The murine IL-3 has four cysteines, at amino acids 17, 79, 80, and 140, and the disulfide bridge between cysteines 17 and 80 is important for biological activity. Mouse IL-3 is trypsin sensitive, S-S reduction-sensitive, and heat stable (65°C, 30 min; 50% loss of activity at 80°C, 30 min) (Clark-Lewis and Schrader, 1988).

The cDNA of the mouse IL-3 receptor α chain contains an open reading frame of 396 amino acids, including a 16-aa signal peptide. It has a predicted molecular mass of 41.2 kDa, but is expressed as a 60–70 kDa glycosylated protein. The membrane-spanning region is predicted to consist of 24 amino acids (Hara and Miyajima, 1992). In the murine system, there are two β subunits, named β$_{IL3}$ (originally termed AIC2A) and β$_c$ (originally termed AIC2B). β$_{IL3}$ is expressed as a 110–120 kDa protein (predicted size 94.7 kDa) and β$_c$ as a 120–140 kDa protein (predicted size 96.6 kDa). β$_{IL3}$ and β$_c$ have 91% identity at the amino acid level but are encoded by two distinct genes. The cDNAs of β$_{IL3}$ and β$_c$ contain an open reading frame of 878 and 896 amino acids, respectively. Both receptors contain a 26-aa transmembrane domain and a 22-aa signal peptide. β$_c$ has one additional N-linked glycosylation site and an extra 18 amino acids in the cytoplasmic domain. The β$_c$ receptor subunit appears to play a similar role as human β$_c$ (Gorman et al., 1990; Kitamura et al., 1991a), whereas β$_{IL3}$ itself binds IL-3 with low affinity (K_d = 17.9 ± 3.6 nM) and can contribute to high-affinity binding in a complex with the murine α chain. However, unlike human and mouse β$_c$, β$_{IL3}$ does not interact with the GM-CSF or IL-5 receptor α chains (Itoh et al., 1990). Both mouse β chains are downregulated at the post-transcriptional level in an IL-3-nonresponsive variant of the mouse mast cell line MC/9 (Hara and Miyajima, 1994). The biological significance of these two parallel IL-3 receptor systems is not yet understood, as β$_{IL3}$ and β$_c$ appear to be coordinately expressed (Hara and Miyajima, 1992).

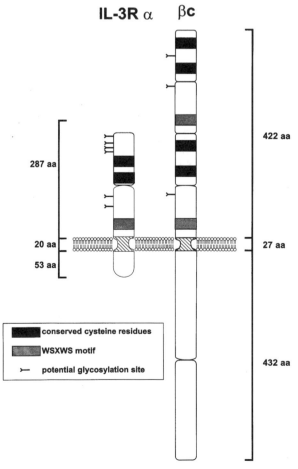

IL-3R α   βc

287 aa

20 aa

53 aa

422 aa

27 aa

432 aa

■ conserved cysteine residues
▨ WSXWS motif
⤙ potential glycosylation site

Figure 3.4   Schematic representation of the IL-3 receptor α and β chains.

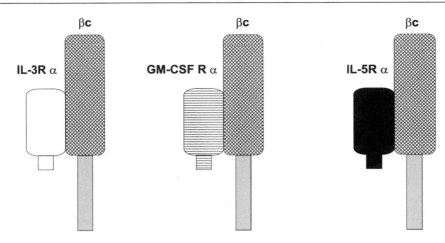

Figure 3.5   A model of IL-3R/GM-CSF R/IL-5R chain interactions.

The gene for murine α chain is located in the proximal region of the mouse chromosome 14, separated from the gene for the mouse GM-CSF receptor α chain, which is located on mouse chromosome 19 (Miyajima *et al.*, 1995). The mouse β chains (β$_{IL3}$ and β$_c$) are located on mouse chromosome 15 (Gorman *et al.*, 1992).

9.   IL-3 in Disease and Therapy

9.1   IL-3 ASSOCIATED PATHOLOGY

The chromosomal region representing the cytokine gene cluster is frequently deleted in patients with myelodysplastic syndrome (MDS), or acute myeloid leukemia (AML), occurring after cytotoxic therapy for the treatment of malignant or nonmalignant diseases (t-MDS/t-AML). t-MDS/t-AML is often associated with the loss of an entire chromosome 5 or 7 or a deletion of the long arm of these chromosomes [del(5q)/del(7q)]. The band 5q31 was identified as a critically and most frequently deleted band in patients with malignant myeloid diseases (Le Beau *et al.*, 1987, 1993).

9.2   IL-3 IN THERAPY *IN VIVO*

The effects of IL-3 *in vivo* have been consistent with most of the *in vitro* data obtained from hemopoietic cells. Overexpression of IL-3 in mice resulted in a nonneoplastic myeloproliferative disorder (Chang *et al.*, 1989; Wong *et al.*, 1989). Intravenous administration of IL-3 to nonhuman primates followed by GM-CSF resulted in enhanced leukocyte counts, mainly neutrophils, megakaryocytes, basophils, and eosinophils, as well as increased levels of histamine (Donahue *et al.*, 1988; Mayer *et al.*, 1989; Stahl *et al.*, 1992).

The efficacy of hemopoietic growth factors to support hemopoiesis after high-dose chemotherapy or for autologous bone marrow support has been tested in several clinical trials. IL-3 was administered in clinical trials to patients with bone marrow failure and myelodysplastic syndromes and showed clear effects in stimulating hemopoiesis (Ganser *et al.*, 1990a,b), although it was less effective in the treatment of aplastic anemia (Ganser *et al.*, 1990c). IL-3 or combinations of IL-3 and GM-CSF have been used to shorten hemopoietic recovery (Biesma *et al.*, 1992; Brugger *et al.*, 1992; Orazi *et al.*, 1992; Postmus *et al.*, 1992; Fay and Bernstein, 1996; Hovgaard *et al.*, 1997).

9.3   FUSION PROTEINS AND IL-3 AGONISTS

PIXY321 is a recombinant fusion protein containing IL-3 and GM-CSF joined by a synthetic linker sequence. Ligand binding studies showed an equivalent binding capacity compared to GM-CSF, but there was a 10-fold greater binding affinity than that of IL-3 (Curtis *et al.*, 1991). PIXY321 sustained megakaryocytopoiesis *in vitro*, comparable to a combination of GM-CSF and IL-3 (Bruno *et al.*, 1992). *In vivo* administration of PIXY321 to sarcoma patients increased monocyte numbers (Hamilton *et al.*, 1993). Another clinical study describes significant effects of PIXY321 on neutrophil, platelet, and reticulocyte counts before chemotherapy of patients with sarcoma (Vadhan-Raj *et al.*, 1995).

IL-3/EPO fusion proteins have been constructed, and tested for biological activity *in vitro*. The effects of three IL-3/EPO fusion proteins on enhancement of proliferation of BFU-E and CFU-GEMM were similar to the combination of IL-3 with EPO (Lu *et al.*, 1995).

A deletion mutant of IL-3 (hIL-3$_{15-125}$) revealed full activity in an AML193.1.3 cell proliferation assay. Further single-site mutations on hIL-3$_{15-125}$ were screened for bioactivity, resulting in mutants with 5-fold to 26-fold greater activity than that of native IL-3, showing that amino acids can be substituted without loss of activity (Olins *et al.*, 1995). Another genetically engineered IL-3 agonist (SC-55494) exhibits 10-fold to 20-fold greater biological activity on human hemopoietic cell proliferation and a 20-fold greater affinity for α/β receptor binding on intact cells, but is only twice as active as IL-3 in priming synthesis of LTC_4 and triggering histamine release from peripheral blood leukocytes (Thomas *et al.*, 1995). Thus the development of novel IL-3 agonists may result in enhanced activity on hemopoiesis without inducing undesirable side-effects.

10. References

Alderson, M.R., Armitage, R.J., Tough, T.W., Strockbine, L., Fanslow, W.C. and Spriggs, M.K. (1993). CD40 expression by human monocytes: regulation by cytokines and activation of monocytes by the ligand for CD40. J. Exp. Med. 178, 669–674.

Alderson, M.R., Armitage, R.J., Tough, T.W. and Ziegler, S.F. (1994). Synergistic effects of IL-4 and either GM-CSF or IL-3 on the induction of CD23 expression by human monocytes: regulatory effects of IFN-α and IFN-γ. Cytokine 6, 407–413.

Algate, P.A. and McCubrey, J.A. (1993). Autocrine transformation of hemopoietic cells resulting from cytokine message stabilization after intracisternal A particle transposition. Oncogene 8, 1221–1232.

Ascaso, R., Marvel, J., Collins, M.K.L. and Lopez-Rivas, A. (1994). Interleukin-3 and Bcl-2 cooperatively inhibit etoposide-induced apoptosis in a murine pre-B cell line. Eur. J. Immunol. 24, 537–541.

Azoulay, M., Webb, C.G. and Sachs, L. (1987). Control of hematopoietic cell growth regulators during mouse fetal development. Mol. Cell. Biol. 7, 3361–3364.

Baffy, G., Miyashita, T., Williamson, J.R. and Reed, J.C. (1993). Apoptosis induced by withdrawal of interleukin-3 (IL-3) from an IL-3 dependent hematopoietic cell line is associated with repartitioning of intracellular calcium and is blocked by enforced bcl-2 oncoprotein production. J. Biol. Chem. 268, 6511–6519.

Banu, N., Wong, J., Deng, B., Groopman, J. and Avrahm, H. (1995). Modulation of megakaryocytopoiesis by thrombopoietin: the c-Mpl ligand. Blood 86, 1331–1338.

Barlow, D.P., Bucan, M., Lehrach, H., Hogan, B.L.M. and Gough, N.M. (1987). Close genetic and physical linkage between the murine haemopoietic growth factor genes GM-CSF and multi-CSF (IL3). EMBO J. 6, 617–623.

Bazill, G.W., Haynes, M., Garland, J. and Dexter, T.M. (1983). Characterization and partial purification of a haemopoietic cell growth factor in WEHI-3 cell conditioned medium. Biochem. J. 210, 747–759.

Bergui, L., Schena, M., Gaidano, H., Riva, M. and Caligaris-Cappio, F. (1989). Interleukin 3 and interleukin 6 synergistically promote the proliferation and differentiation of malignant plasma cell precursors in multiple myeloma. J. Exp. Med. 170, 613–618.

Bernstein, I.D., Andrews, R.G. and Zsebo, K.M. (1991). Recombinant human stem cell factor enhances the formation of colonies by CD34+ and CD34+lin− cells, and the generation of colony-forming cell progeny from CD34+lin− cells cultured with interleukin-3, granulocyte colony-stimulating factor, or granulocyte-macrophage colony-stimulating factor. Blood 77, 3216–3221.

Biesma, B., Willemse, P.H.B., Mulder, N.H., et al. (1992). Effects of interleukin-3 after chemotherapy for advanced ovarian cancer. Blood 80, 1141–1148.

Bischoff, S.C., De Weck, A.L. and Dahinden, C.A. (1990). Interleukin 3 and granulocyte/macrophage-colony-stimulating factor render human basophils responsive to low concentrations of complement component C3a. Proc. Natl Acad. Sci. USA 87, 6813–6817.

Bochner, B.S., McKelvey, A.A., Sterbinsky, S.A., et al. (1990). IL-3 augments adhesiveness for endothelium and CD11b expression in human basophils but not neutrophils. J. Immunol. 145, 1832–1837.

Brandt, J., Briddell, R.A., Srour, E.F., Leemhuis, T.B. and Hoffman, R. (1992). Role of c-*kit* ligand in the expansion of human hematopoietic progenitor cells. Blood 79, 634–641.

Brandt, J.E., Bhalla, K. and Hoffman, R. (1994). Effects of interleukin-3 and c-*kit* ligand on the survival of various classes of human hematopoietic progenitor cells. Blood 83, 1507–1514.

Brewer, G. (1991). An A+U-rich element RNA-binding factor regulates c-myc mRNA stability in vitro. Mol. Cell. Biol. 11, 2460–2466.

Brewer, G. and Ross, J. (1988). Poly (A) shortening and degradation of the 3′ A+U rich sequences of human c-myc mRNA in a cell-free system. Mol. Cell. Biol. 8, 1697–1708.

Brizzi, M.F., Garbarino, G., Rossi, P.R., et al. (1993). Interleukin 3 stimulates proliferation and triggers endothelial-leukocyte adhesion molecule 1 gene activation of human endothelial cells. J. Clin. Invest. 91, 2887–2892.

Brugger, W., Bross, K., Frisch, J., et al. (1992). Mobilization of peripheral blood progenitor cells by sequential administration of interleukin-3 and granulocyte-macrophage colony-stimulating factor following polychemotherapy with etoposide, ifosfamide, and cisplatin. Blood 79, 1193–1200.

Brunner, T., De Weck, A.L. and Dahinden, C.A. (1991). Platelet-activating factor induces mediator release by human basophils primed with IL-3, granulocyte-macrophage colony-stimulating factor, or IL-5. J. Immunol. 147, 237–242.

Bruno, E., Briddell, R.A., Cooper, R.J., Brandt, J.E. and Hoffman, R. (1992). Recombinant GM-CSF/IL-3 fusion protein: its effect on in vitro human megakaryocytopoiesis. Exp. Hematol. 20, 494–499.

Campbell, H.D., Ymer, S., Fung, M. and Young, I.G. (1985). Cloning and nucleotide sequence of the murine interleukin-3 gene. Eur. J. Biochem. 150, 297–304.

Cannistra, S.A., Vellenga, E., Groshek, P., Rambaldi, A. and Griffin, J.D. (1988). Human granulocyte-monocyte colony-stimulating factor and interleukin 3 stimulate monocyte cytotoxicity through a tumor necrosis factor-dependent mechanism. Blood 71, 672–676.

Caput, D., Beutler, B., Hartog, K., Thayer, R., Brown-Shirmen, S. and Cerami, A. (1986). Identification of a common nucleotide sequence in the 3′ untranslated region of mRNA molecules specifying inflammatory mediators. Proc. Natl Acad. Sci. USA 83, 1670–1676.

Caux, C., Saeland, S., Favre, C., Duvert, V., Mannoni, P. and Bancherou, J. (1990). Tumor necrosis factor-alpha strongly potentiates interleukin-3 and granulocyte-macrophage colony-stimulating factor-induced proliferation of human CD34+ hematopoietic progenitor cells. Blood 75, 2292–2298.

Caux, C, Favre, C., Saeland, S., *et al.* (1991). Potentiation of early hematopoiesis by tumor necrosis factor-α is followed by inhibition of granulopoietic differentiation and proliferation. Blood 78, 635–644.

Caux, C., Vanbervliet, B., Massacrier, C., *et al.* (1996). Interleukin-3 cooperates with tumor necrosis factor α for the development of human dendritic/ Langerhans cells from cord blood CD34$^+$ hematopoietic progenitor cells. Blood 87, 2376–2385.

Chang, J.M., Metcalf, D., Lang, R.A., Gonda, T.J. and Johnson, G.R. (1989). Nonneoplastic hematopoietic myeloproliferative syndrome induced by dysregulated multi-CSF (IL-3) expression. Blood 73, 1487–1497.

Clark, S.C. and Kamen, R. (1987). The human hematopoietic colony-stimulating factors. Science 236, 1229–1237.

Clark-Lewis, I. and Schrader, J.W. (1988). In "Interleukin 3: The Panspecific Hemopoietin" (ed. J.W. Schrader), pp. 1–37. Academic Press, San Diego.

Clark-Lewis, I., Kent, S.B.H. and Schrader, J.W. (1984). Purification to apparent homogeneity of a factor stimulating the growth of multiple lineages of hemopoietic cells. J. Biol. Chem. 259, 7488–7494.

Clark-Lewis, I., Aebersold, R., Ziltener, H., Schrader, J.W., Hood, L.E. and Kent, S.B.H. (1986). Automated chemical synthesis of a protein growth factor for hemopoietic cells, interleukin-3. Science 231, 134–139.

Clayberger, C., Luna-Fineman, S., Lee, J.E., *et al.* (1992). Interleukin 3 is a growth factor for human follicular B cell lymphoma. J. Exp. Med. 175, 371–376.

Cockayne, D.A., Bodine, D.M., Cline, A., Nienhuis, A.W. and Dunbar, C.E. (1994). Transgenic mice expressing antisense interleukin-3 RNA develop a B-cell lymphoproliferative syndrome or neurological dysfunction. Blood 84, 2699–2710.

Cohen, D.R., Hapel, A.J. and Young, I.G. (1986). Cloning and expression of the rat interleukin-3 gene. Nucleic Acids Res. 14, 3641–3658.

Collins, M.K.L., Marvel, J., Malde, P. and Lopez-Rivas, A. (1992). Interleukin 3 protects murine bone marrow cells from apoptosis induced by DNA damaging agents. J. Exp. Med. 176, 1043–1051.

Collotta, F., Bussolino, F., Polentarutti, N., *et al.* (1993). Differential expression of the common β subunit and specific α chains of the receptors for GM-CSF, IL-3 and IL-5 in endothelial cells. Exp. Cell Res. 206, 311–317.

Cosman, D. (1993). The hematopoietin receptor superfamily. Cytokine 5, 95–106.

Curtis, B.M., Williams, D.E., Broxmeyer, H.E., *et al.* (1991). Enhanced hematopoietic activity of a human granulocyte/macrophage colony-stimulating factor-interleukin 3 fusion protein. Proc. Natl Acad. Sci. USA 88, 5809–5813.

Cuturi, M.C., Anegon, I., Sherman, F., *et al.* (1989). Production of hematopoietic colony-stimulating factors by human natural killer cells. J. Exp. Med. 169, 569–583.

Dahinden, C.A., Kurimoto, Y., De Weck, A.L., Lindley, I., Dewald, B. and Baggiolini, M. (1989). The neutrophil-activating peptide NAF/NAP-1 induces histamine and leukotriene release by interleukin-3 primed basophils. J. Exp. Med. 170, 1787–1792.

Dalloul, A.H., Arock, M., Fourcade, C., *et al.* (1991). Human thymic epithelial cells produce interleukin-3. Blood 77, 69–74.

Dalloul, A.H., Arock, M., Fourcade, C., *et al.* (1992).

Epidermal keratinocyte-derived basophil promoting activity. J. Clin. Invest. 90, 1242–1247.

Danner, M. and Luger, T.A. (1987). Human keratinocytes and epidermoid carcinoma cell lines produce a cytokine with interleukin-3 activity. J. Invest. Dermatol. 88, 353–361.

Darnell, J.E., Jr, Kerr, I.M. and Stark, G.R. (1994). Jak-STAT pathways and transcriptional activation in response to IFNs and other extracellular signalling proteins. Science 264, 1415–1421.

Davies, K., TePas, E.C., Nathan, D.G. and Mathey-Prevot, B. (1993). Interleukin-3 expression by activated T cells involves an inducible, T-cell-specific factor and an octamer binding protein. Blood 81, 928–934.

Dimri, R., Nissimov, L. and Keisari, Y. (1994). Effect of human recombinant granulocyte-macrophage colony-stimulating factor and IL-3 on the expression of surface markers of human monocyte-derived macrophages in long-term cultures. Lymphokine Cytokine Res. 13, 239–245.

Dokter, W.H.A., Esselink, M.T., Sierdsema, S.J., Halie, M.R. and Vellenga, E. (1993a). Transcriptional and posttranscriptional regulation of the interleukin-4 and interleukin-3 genes in human T cells. Blood 81, 35–40.

Dokter, W.H.A., Sierdsema, S.J., Esselink, M.T., Halie, M.R. and Vellenga, E. (1993b). IL-7 enhances the expression of IL-3 and granulocyte-macrophage-CSF mRNA in activated human T cells by post-transcriptional mechanisms. J. Immunol. 150, 2584–2590.

Donahue, R.E., Seehra, J., Metzger, M., *et al.* (1988). Human IL-3 and GM-CSF act synergistically in stimulating hematopoiesis in primates. Science 241, 1820–1823.

Dorssers, L., Burger, H., Bot, F., *et al.* (1987). Characterization of a human multilineage-colony-stimulating factor cDNA clone identified by a conserved noncoding sequence in mouse interleukin-3. Gene 55, 115–124.

Duronio, V., Clark-Lewis, I., Federsppiel, B., Wieler, J.S. and Schrader, J.W. (1992). Tyrosine phosphorylation of receptor β subunits and common substrates in response to interleukin-3 and granulocyte-macrophage colony-stimulating factor. J. Biol. Chem. 267, 21856–21863.

Eder, M., Ottmann, O.G., Hansen-Hagge, T.E., *et al.* (1992). In vitro culture of common acute lymphoblastic leukemia blasts: effects of interleukin-3, interleukin-7, and accessory cells. Blood 79, 3274–3284.

Egeland, T., Steen, R., Quarsten, H., Gaudernack, G., Yang, Y.-C. and Thorsby, E. (1991). Myeloid differentiation of purified CD34$^+$ cells after stimulation with recombinant human granulocyte-monocyte colony-stimulating factor (CSF), granulocyte-CSF, monocyte-CSF, and interleukin-3. Blood 78, 3192–3199.

Elliott, M.J., Vadas, M.A., Cleland, L.G., Gamble, J.R. and Lopez, A.F. (1990). IL-3 and granulocyte-macrophage colony-stimulating factor stimulate two distinct phases of adhesion in human monocytes. J. Immunol. 145, 167–176.

Elliott, M.J., Moss, J., Dottore, M., Park, L.S., Vadas, M.A. and Lopez, A.F. (1992). Differential binding of IL-3 and GM-CSF to human monocytes. Growth Factors 6, 15–29.

Engeland, K., Andrews, N.C. and Mathey-Prevot, B. (1995). Multiple proteins interact with the nuclear inhibitory protein repressor element in the human interleukin-3 promoter. J. Biol. Chem. 270, 24572–24579.

Farrar, W.L., Vinocour, M. and Hill, J.M. (1989). In situ hybridization histochemistry localization of interleukin-3 mRNA in mouse brain. Blood 73, 137–140.

Fay, J.W. and Bernstein, S.H. (1996). Recombinant human interleukin-3 and granulocyte-macrophage colony-stimulating factor after autologous bone marrow transplantation for malignant lymphoma. Semin. Oncol. 23 (2 Suppl. 4), 22–27.

Feng, Y., Klein, B.K., Vu, L., Aykent, S. and McWherter, C.A. (1995). 1H, 13C, and 15N NMR resonance assignments, secondary structure, and backbone topology of a variant of human interleukin-3. Biochemistry 34, 6540–6551.

Frendl, G. (1992). Interleukin 3: from colony-stimulating factor to pluripotent immunoregulatory cytokine. Int. J. Immunopharmacol. 14, 421–430.

Frendl, G. and Beller, D.I. (1990). Regulation of macrophage activation by IL-3. I. IL-3 functions as a macrophage-activating factor with unique properties, inducing Ia and lymphocyte function-associated antigen-1 but not cytotoxicity. J. Immunol. 144, 3392–3399.

Frendl, G., Fenton, M.J. and Beller, D.I. (1990). Regulation of macrophage activation by IL-3. II. IL-3 and lipopolysaccharide act synergistically in the regulation of IL-1 expression. J. Immunol. 144, 3400–3410.

Fujisawa, T., Abu-Ghazaleh, R., Kita, H., Sanderson, C.J. and Gleich, G.J. (1990). Regulatory effect of cytokines on eosinophil degranulation. J. Immunol. 144, 642–646.

Fung, M.C., Hapel, A.J., Ymer, S., et al. (1984). Molecular cloning of cDNA for murine interleukin-3. Nature 307, 233–237.

Ganser, A., Seipelt, G., Lindemann, A., et al. (1990a). Effects of recombinant human interleukin-3 in patients with myelodysplastic syndromes. Blood 76, 455–462.

Ganser, A., Lindemann, A., Seipelt, G., et al. (1990b). Effects of recombinant human interleukin-3 in patients with normal hematopoiesis and in patients with bone marrow failure. Blood 76, 666–676.

Ganser, A., Lindemann, A., Seipelt, G., et al. (1990c). Effects of recombinant human interleukin-3 in aplastic anemia. Blood 76, 1287–1292.

Gorman, D.M., Itoh, N., Kitamura, T., et al. (1990). Cloning and expression of a gene encoding an interleukin 3 receptor-like protein: identification of another member of the cytokine receptor gene family. Proc. Natl Acad. Sci. USA 87, 5459–5463.

Gorman, D.M., Itoh, N., Jenkins, N.A., Gilbert, D.J., Copeland, N.G. and Miyajima, A. (1992). Chromosomal localization and organization of the murine genes encoding the β subunits (AIC2A and AIC2B) of the interleukin 3, granulocyte/macrophage colony-stimulating factor, and interleukin 5 receptors. J. Biol. Chem. 267, 15842–15848.

Gottschalk, L.R., Giannola, D.M. and Emerson, S.G. (1993). Molecular regulation of the human IL-3 gene: inducible T cell-restricted expression requires intact AP-1 and Elf-1 nuclear protein binding sites. J. Exp. Med. 178, 1681–1692.

Grabstein, H. and Alderson, M.R. (1993). Regulation of macrophage tumoricidal activity by granulocyte-macrophage colony-stimulating factor. In "Hemopoietic Growth Factor and Mononuclear Phagocytes" (ed. van Furth), pp. 140–147. Karger, Basel.

Gualtieri, R.J., Liang, C.-M., Shadduck, R.K., Waheed, A. and Banks, J. (1987). Identification of the hematopoietic growth factors elaborated by bone marrow stromal cells using antibody neutralization analysis. Exp. Hematol. 15, 883–889.

Guerriro, R., Testa, U., Gabbianelli, M., et al. (1995). Unilineage megakaryocytic proliferation and differentiation

of purified hematopoietic progenitors in serum-free liquid culture. Blood 86, 3725–3736.

Haak-Frendscho, M., Arai, N., Arai, K., Baeza, M.L., Finn, A. and Kaplan, A.P. (1988). Human recombinant granulocyte-macrophage colony-stimulating factor and interleukin 3 cause basophil histamine release. J. Clin. Invest. 82, 17–20.

Hamilton, R.F., Jr., Vadhan-Raj, S., Uthman, M., Grey, M. and Holian, A. (1993). The in vivo effects of pIXY321 therapy on human monocyte activity. J. Leukocyte Biol. 53, 640–650.

Hara, T. and Miyajima, A. (1992). Two distinct functional high affinity receptors for mouse interleukin-3 (IL-3). EMBO J. 11, 1875–1884.

Hara, T. and Miyajima, A. (1994). Regulation of IL-3 receptor expression: evidence for a post-transcriptional mechanism that dominantly suppresses the expression of β subunits. Int. Immunol. 6, 1525–1533.

Hayashida, K., Kitamura, T., Gorman, D.M., Arai, K., Yokota, T. and Miyajima, A. (1990). Molecular cloning of a second subunit of the receptor for human granulocyte-macrophage colony-stimulating factor (GM-CSF): reconstitution of a high-affinity GM-CSF receptor. Proc. Natl Acad. Sci. USA 87, 9655–9659.

Hirai, K., Morita, Y., Misaki, Y., et al. (1988). Modulation of human basophil histamine release by hemopoietic growth factors. J. Immunol. 141, 3958–3964.

Hirayama, F., Shih, J.-P., Awgulewitsch, A., Warr, G.W., Clark, S.C. and Ogawa, M. (1992). Clonal proliferation of murine lymphohemopoietic progenitors in culture. Proc. Natl Acad. Sci. USA 89, 5907–5911.

Hirayama, F., Clark, S.C. and Ogawa, M. (1994). Negative regulation of early B lymphopoiesis by interleukin 3 and interleukin 1α. Proc. Natl Acad. Sci. USA 91, 469–473.

Hirsch, H.H., Nair, A.P.K. and Moroni, C. (1993). Suppressible and nonsuppressible autocrine mast cell tumors are distinguished by insertion of an endogenous retroviral element (IAP) into the interleukin 3 gene. J. Exp. Med. 178, 403–411.

Hovgaard, D.J., Stahl-Skov, P. and Nissen, N.I. (1997). The in vivo effects of interleukin-3 on histamine levels in non-Hodgkin's lymphoma patients. Pharmacol. Toxicol. 80, 290–294.

Ihle, J.N. (1992). Interleukin-3 and hematopoiesis. Chem. Immunol. 51, 65–106.

Ihle, J.N., Keller, J., Henderson, L., Klein, F. and Palaszynski, E. (1982). Procedures for the purification of interleukin-3 to homogeneity. J. Immunol. 129, 2431–2436.

Ihle, J.N., Keller, J., Oroszlan, S., et al. (1983). Biological properties of homogenous interleukin 3. J. Immunol. 131, 282–287.

Ihle, J.N., Witthuhn, B.A., Quelle, F.W., et al. (1994). Signaling by the cytokine receptor superfamily: JAKs and STATs. TIBS 19, 222–227.

Ikebuchi, K., Wong, G.G., Clark, S.C., Ihle, J.N., Hirai, Y. and Ogawa, M. (1987). Interleukin 6 enhancement of interleukin 3-dependent proliferation of multipotential hemopoietic progenitors. Proc. Natl Acad. Sci. USA 84, 9035–9039.

Ikebuchi, K., Clark, S.C., Ihle, J.N., Souza, L.M. and Ogawa, M. (1988). Granulocyte colony-stimulating factor enhances interleukin 3-dependent proliferation of multipotential hemopoietic progenitors. Proc. Natl Acad. Sci. USA 85, 3445–3449.

Isfort, R.J. and Ihle, J.N. (1990). Multiple hematopoietic growth factors signal through tyrosine phosphorylation. Growth Factors 2, 213–220.

Isfort, R.J., Stevens, D., May, W.S. and Ihle, J.N. (1988). Interleukin 3 binds to a 140-kDa phosphotyrosine-containing cell surface protein. Proc. Natl Acad. Sci. USA 85, 7982–7986.

Itoh, N., Yonehara, S., Schreurs, J., et al. (1990). Cloning of an interleukin-3 receptor gene: a member of a distinct receptor gene family. Science 247, 324–327.

Kaushansky, K., Shoemaker, S.G., O'Rork, C.A. and McCarthy, J.M. (1994). Coordinate regulation of multiple human lymphokine genes by Oct-1 and potentially novel 45 and 43 kDa polypeptides. J. Immunol. 152, 1812–1820.

Khwaja, A., Addison, I.E., Yong, K. and Linch, D.C. (1994). Interleukin-3 administration enhances human monocyte function in vivo. Br. J. Haematol. 88, 515–519.

Kinoshita, T., Yokota, T., Arai, K. and Miyajima, A. (1995). Suppression of apoptotic death in hematopoietic cells by signalling through the IL-3/GM-CSF receptors. EMBO J. 14, 266–275.

Kitamura, T., Hayashida, K., Sakamaki, K., Yokota, T., Arai, K. and Miyajima, A. (1991a). Reconstitution of functional receptors for human granulocyte/macrophage colony-stimulating factor (GM-CSF). Evidence that the protein encoded by AIC2B cDNA is a subunit of the murine GM-CSF receptor. Proc. Natl Acad. Sci. USA 88, 5082–5086.

Kitamura, T., Sato, N., Arai, K. and Miyajima, A. (1991b). Expression cloning of the human IL-3 receptor cDNA reveals a shared β subunit for the human IL-3 and GM-CSF receptors. Cell 66, 1165–1174.

Knol, E.F., Kuijpers, T.W., Mul, F.P.J. and Roos, D. (1993). Stimulation of human basophils results in homotypic aggregation. J. Immunol. 151, 4926–4933.

Kondo, S., Ciarletta, A., Turner, K.I., Sauder, D.N. and McKenzie, R.C. (1995). Failure to detect interleukin (IL)-3 mRNA or protein in human keratinocytes: antibodies to granulocyte macrophage-colony-stimulating factor or IL-6 (but not IL-3) neutralize "IL-3" bioactivity. J. Invest. Dermatol. 104, 335–339.

Korpelainen, E.I., Gamble, J.R., Smith, W.B., et al. (1993). The receptor for interleukin 3 is selectively induced in human endothelial cells by tumor necrosis factor α and potentiates interleukin 8 secretion and neutrophil transmigration. Proc. Natl Acad. Sci. USA 90, 11137–11141.

Korpelainen, E.I., Gamble, J.R., Smith, W.B., Dottore, M., Vadas, M.A. and Lopez, A.F. (1995). Interferon-γ upregulates interleukin-3 (IL-3) receptor expression in human endothelial cells and synergizes with IL-3 in stimulating major histocompatibility complex class II expression and cytokine production. Blood 86, 176–182.

Koyano-Nakagawa, N., Nishida, J., Baldwin, D., Arai, K. and Yokota, T. (1994). Molecular cloning of a novel human cDNA encoding a zinc finger protein that binds to the interleukin-3 promoter. Mol. Cell. Biol. 14, 5099–5107.

Kremer, E., Baker, E., D'Andrea, R.J., et al. (1993). A cytokine receptor gene cluster in the X-Y pseudoautosomal region? Blood 82, 22–28.

Kuna, P., Reddigari, S.R., Kornfeld, D. and Kaplan, A.P. (1991). IL-8 inhibits histamine release from human basophils induced by histamine-releasing factors, connective tissue activating peptide III, and IL-3. J. Immunol. 147, 1920–1924.

Kurimoto, Y., de Weck, A.L. and Dahinden, C.A. (1989). Interleukin-3 dependent mediator release in basophils triggered by C5a. J. Exp. Med. 170, 467–479.

Le Beau, M.M., Epstein, N.D., O'Brien, J.S., et al. (1987). The interleukin 3 gene is located on human chromosome 5 and is deleted in myeloid leukemias with a deletion of 5q. Proc. Natl Acad. Sci. USA 84, 5913–5917.

Le Beau, M.M., Espinosa III, R., Neuman, W.L., et al. (1993). Cytogenetic and molecular delineation of the smallest commonly deleted region of chromosome 5 in malignant myeloid diseases. Proc. Natl Acad. Sci. USA 90, 5484–5488.

Lee, J.S. and Young, I.G. (1989). Fine-structure mapping of the murine IL-3, GM-CSF genes by pulsed-filed electrophoresis and molecular cloning. Genomics 5, 359–362.

Linnekin, D., Evans, G., Michiel, D. and Farrar, W.L. (1992). Characterization of a 97-kDa phosphotyrosylprotein regulated by multiple cytokines. J. Biol. Chem. 267, 23993–23998.

Litzow, M.R., Brashem-Stein, C., Andrews, R.G. and Bernstein, I.D. (1991). Proliferative responses to interleukin-3 and granulocyte colony-stimulating factor distinguish a minor subpopulation of CD34-positive marrow progenitors that do not express CD33 and a novel antigen, 7B9. Blood 77, 2354–2359.

Lopez, A.F., Williamson, D.J., Gamble, J.R., et al. (1986). Recombinant human granulocyte-macrophage colony-stimulating factor stimulates in vitro mature human neutrophil and eosinophil function, surface receptor expression, and survival. J. Clin. Invest. 78, 1220–1228.

Lopez, A.F., Dyson, P.G., Bik, L., et al. (1988). Recombinant human interleukin-3 stimulation of hematopoiesis in humans: loss of responsiveness with differentation in the neutrophilic myeloid series. Blood 72, 1797–1804.

Lopez, A.F., Eglinton, J.M., Lyons, B., et al. (1990). Human interleukin-3 inhibits the binding of granulocyte-macrophage colony-stimulating factor and interleukin-5 to basophils and strongly enhances their functional activity. J. Cell. Physiol. 145, 69–77.

Lopez, A.F., Vadas, M.A., Woodcock, J.M., et al. (1991). Interleukin-5, interleukin-3, and granulocyte-macrophage colony-stimulating factor cross-compete for binding to cell surface receptors on human eosinophils. J. Biol. Chem. 266, 24741–24747.

Lopez, A.F., Elliott, M.J., Woodcock, J. and Vadas, M.A. (1992). GM-CSF, IL-3 and IL-5: cross-competition on human haemopoietic cells. Immunol. Today 13, 495–500.

Lu, L., Xiao, M., Li, Z.-H., et al. (1995). Influence in vitro of IL-3/Epo fusion proteins compared with the combination of IL-3 plus Epo in enhancing the proliferation of single isolated erythroid and multipotential progenitor cells from human umbilical cord blood and adult bone marrow. Exp. Hematol. 23, 1130–1134.

Luger, T.A., Wirth, U. and Köck, A. (1985). Epidermal cells synthesize a cytokine with interleukin 3-like properties. J. Immunol. 134, 915–919.

MacDonald, S.M., Schleimer, R.P., Kagey-Sobotka, A., Gillis, S. and Lichtenstein, L.M. (1989). Recombinant IL-3 induces histamine release from human basophils. J. Immunol. 142, 3527–3532.

Malter, J.S. (1989). Identification of an AUUUA-specific messenger RNA binding protein. Science 246, 664–666.

Martin, M., Strasser, A., Baumgarth, N., et al. (1993). A novel cellular model (SPGM 1) of switching between the pre-B cell and myelomonocytic lineages. J. Immunol. 150, 4395–4406.

Mathey-Prevot, B., Andrews, N.C., Murphy, H.S., Kreissman, S.G. and Nathan, D.G. (1990). Positive and negative elements regulate human interleukin 3 expression. Proc. Natl Acad. Sci. USA 87, 5046–5050.

Mayer, P., Valent, P., Schmidt, G., Liehl, E. and Bettelheim, P. (1989). The in vivo effects of recombinant human interleukin-3: demonstration of basophil differentiation factor, histamine-producing activity, and priming of GM-CSF-responsive progenitors in nonhuman primates. Blood 74, 613–621.

Mayo, M.W., Wang, X.Y., Algate, P.A., et al. (1995). Synergy between AUUUA motif disruption and enhancer insertion results in autocrine transformation of interleukin-3-dependent hematopoietic cells. Blood 86, 3139–3150.

McNiece, I.K., Langley, K.E. and Zsebo, K.M. (1991). Recombinant human stem cell factor synergizes with GM-CSF, G-CSF, IL-3 and Epo to stimulate human progenitor cells of the myeloid and erythroid lineages. Exp. Hematol. 19, 226–231.

Metcalf, D. (1989). The molecular control of cell division, differentiation commitment and maturation in haemopoietic cells. Nature 339, 27–30.

Metcalf, D. (1991). Control of granulocytes and macrophages: molecular, cellular, and clinical aspects. Science 254, 529–533.

Miadonna, A., Roncarolo, M.G., Lorini, M. and Tedeschi, A. (1993). Inducing and enhancing effects of IL-3, -5 and -6 and GM-CSF on histamine release from human basophils. Clin. Immunol. Immunopathol. 67, 210–215.

Milatovich, A., Kitamura, T., Miyajima, A. and Francke, U. (1993). Gene for the alpha-subunit of the human interleukin-3 receptor (IL3A) localized to the X-Y pseudoautosomal region. Am. J. Hum. Genet. 53, 1146–1153.

Miyajima, I., Levitt, L., Hara, T., et al. (1995). The murine interleukin-3 receptor α subunit gene: chromomosomal localization, genomic structure, and promoter function. Blood 85, 1246–1253.

Miyatake, S., Yokota, T., Lee, F. and Arai, K. (1985). Structure of the chromosomal gene for murine interleukin 3. Proc. Natl Acad. Sci. USA 82, 316–320.

Moore, M.A.S. (1988). In "Lymphokines, Interleukin 3: the Panspecific Hemopoietin" (ed. J.W. Schrader), pp. 219–280. Academic Press, San Diego.

Moore, T.A. and Zlotnik, A. (1995). T-cell lineage commitment and cytokine responses of thymic progenitors. Blood 86, 1850–1860.

Mui, A.-L., Wakao, H., O'Farrell, A.-M., Harada, N. and Miyajima, A. (1995a). Interleukin-3, granulocyte-macrophage colony stimulating factor and interleukin-5 transduce signals through two STAT5 homologs. EMBO J. 14, 1166–1175.

Mui, A.-L., Wakao, H., Harada, N., O'Farrell, A.-M. and Miyajima, A. (1995b). Interleukin-3, granulocyte-macrophage colony-stimulating factor, and interleukin-5 transduce signals through two forms of STAT5. J. Leukocyte Biol. 57, 799–803.

Naparstek, E., Pierce, J., Metcalf, D., et al. (1986). Induction of growth alterations in factor-dependent hematopoietic progenitor cell lines by cocultivation with irradiated bone marrow stromal cell lines. Blood 67, 1395–1403.

Niemeyer, C.M., Sieff, C.A., Mathey-Prevot, B., et al. (1989). Expression of human interleukin-3 (multi-CSF) is restricted to human lymphocytes and T-cell tumor lines. Blood 73, 945–951.

Nilsson, G., Svensson, V. and Nilsson, K. (1995). Constitutive and inducible cytokine mRNA expression in the human mast cell line HMC-1. Scand. J. Immunol. 42, 76–81.

Olins, P.O., Bauer, S.C., Braford-Goldberg, S., et al. (1995). Saturation mutagenesis of human interleukin-3. J. Biol. Chem. 277, 23754–23760.

Orazi, A., Cattoretti, G., Schiro, R., et al. (1992). Recombinant human interleukin-3 and recombinant human granulocyte-macrophage colony-stimulating factor administered in vivo after high-dose cyclophosphamide cancer chemotherapy: effect on hematopoiesis and microenvironment in human bone marrow. Blood 79, 2610–2619.

Otsuka, T., Miyajima, A., Brown, N., et al. (1988). Isolation and characterization of an expressible cDNA encoding human IL-3. J. Immunol. 140, 2288–2295.

Palacios, R. and Steinmetz, M. (1985). IL3-dependent mouse clones that express B-220 surface antigen, contain Ig genes in germ-line configuration, and generate B lymphocytes in vivo. Cell 41, 727–734.

Palacios, R., Henson, G., Steinmetz, M. and McKearn, J.P. (1984). Interleukin-3 supports growth of mouse pre-B-cell clones in vitro. Nature 309, 126–131.

Park, J.-H., Kaushansky, K. and Levitt, L. (1993). Transcriptional regulation of interleukin 3 (IL3) in primary human T lymphocytes. J. Biol. Chem. 268, 6299–6308.

Plaut, M., Pierce, J.H., Watson, C.J., Hanley-Hyde, J., Nordan, R.P. and Paul, W.E. (1989). Mast cell lines produce lymphokines in response to cross-linkage of Fcε RI or to calcium ionophores. Nature 339, 64–67.

Postmus, P.E., Gietema, J.A., Damsma, O., et al. (1992). Effects of recombinant human interleukin-3 in patients with relapsed small-cell lung cancer treated with chemotherapy: a dose-finding study. J. Clin. Oncol. 10, 1131–1140.

Rennick, D., Jackson, J., Moulds, C., Lee, F. and Yang, G. (1989). IL-3 and stromal cell-derived factor synergistically stimulate the growth of pre-B cell lines cloned from long-term lymphoid bone marrow cultures. J. Immunol. 142, 161–166.

Rennick, D., Hunte, B., Holland, G. and Thompson-Snipes, L. (1995). Cofactors are essential for stem cell factor-dependent growth and maturation of mast cell progenitors: comparative effects of interleukin-3 (IL-3), IL-4, IL-10, and fibroblasts. Blood 85, 57–65.

Rinaudo, M.S., Su, K., Falk, L.A., Haldar, S. and Mufson, R.A. (1995). Human interleukin-3 receptor modulates bcl-2 mRNA and protein levels through protein kinase C in TF-1 cells. Blood 86, 80–88.

Rothenberg, M.E., Owen, W.F., Jr., Silberstein, D.S., et al. (1988). Human eosinophils have prolonged survival, enhanced functional properties, and become hypodense when exposed to human interleukin 3. J. Clin. Invest. 81, 1986–1992.

Ryan, G.R., Milton, S.E., Lopez, A.F., Bardy, P.G., Vadas, M.A. and Shannon, M.F. (1991). Human interleukin-3 mRNA accumulation is controlled at both the transcriptional and posttranscriptional level. Blood 77, 1195–1202.

Saeland, S., Caux, C., Favre, C., et al. (1988). Effects of recombinant human interleukin-3 on CD34-enriched normal hematopoietic progenitors and on myeloblastic leukemia cells. Blood 72, 1580–1588.

Saito, H., Hatake, K., Dvorak, A.M., et al. (1988). Selective differentiation and proliferation of hematopoietic cells induced by recombinant human interleukins. Proc. Natl Acad. Sci. USA 85, 2288–2292.

Sakamaki, K., Miyajima, I., Kitamura, T. and Miyajima, A. (1992). Critical cytoplasmic domains of the common β subunit of the human GM-CSF, IL-3 and IL-5 receptors for growth signal transduction and tyrosine phosphorylation. EMBO J. 11, 3541–3549.

Sato, N., Sakamaki, K., Terada, N., Arai, K. and Miyajima, A. (1993a). Signal transduction by the high affinity GM-CSF receptor: two distinct cytoplasmic regions of the common β subunit responsible for different signaling. EMBO J. 12, 4181–4189.

Sato, N., Caux, C., Kitamura, T., et al. (1993b). Expression and factor-dependent modulation of the interleukin-3 receptor subunits on human hematopoietic cells. Blood 82, 752–761.

Satoh, T., Nakafuku, M., Miyajima, A. and Kaziro, Y. (1991). Involvement of ras p21 protein in signal-transduction pathways from interleukin 2, interleukin 3, and granulocyte/macrophage colony-stimulating factor, but not from interleukin 4. Proc. Natl Acad. Sci. USA 88, 3314–3318.

Shaw, G. and Kamen, R. (1986). A conserved AU sequence from the 3′ untranslated region of GM-CSF mRNA mediated selective mRNA degradation. Cell 46, 659–667.

Shoemaker, S.G., Hromas, R. and Kaushansky, K. (1990). Transcriptional regulation of interleukin 3 gene expression in T lymphocytes. Proc. Natl Acad. Sci. USA 87, 9650–9654.

Silvennoinen, O., Witthuhn, B.A., Quelle, F.W., Cleveland, J.L., Yi, T. and Ihle, J.N. (1993). Structure of the murine Jak2 protein-tyrosine kinase and its role in interleukin 3 signal transduction. Proc. Natl Acad. Sci. USA 90, 8429–8433.

Smith, W.B., Guida, L., Sun, Q., et al. (1995). Neutrophils activated by granulocyte-macrophage colony-stimulating factor express receptors for interleukin-3 which mediate class II expression. Blood 86, 3938–3944.

Sonoda, Y., Yang, Y.-C., Wong, G.G., Clark, S.C. and Ogawa, M. (1988). Analysis in serum-free culture of the targets of recombinant human hemopoietic growth factors: interleukin 3 and granulocyte/macrophage-colony-stimulating factor are specific for early developmental stages. Proc. Natl Acad. Sci. USA 85, 4360–4364.

Sorensen, P.H.B., Mui, A.L.-F., Murthy, S.C. and Krystal, G. (1989). Interleukin-3, GM-CSF, and TPA induce distinct phosphorylation events in an interleukin-3 dependent multipotential cell line. Blood 73, 406–418.

Stahl, C.P., Winton, E.F., Monroe, M.C., et al. (1992). Differential effects of sequential, simultaneous, and single agent interleukin-3 and granulocyte-macrophage colony-stimulating factor on megakaryocyte maturation and platelet response in primates. Blood 80, 2479–2485.

Stoecklin, G., Hahn, S. and Moroni, C. (1994). Functional hierarchy of AUUUA motifs in mediating rapid interleukin-3 mRNA decay. J. Biol. Chem. 269, 28591–28597.

Takafuji, S., Bischoff, S.C., De Weck, A.L. and Dahinden, C.A. (1991). IL-3 and IL-5 prime normal human eosinophils to produce leukotriene C_4 in response to soluble agonists. J. Immunol. 147, 3855–3861.

Takafuji, S., Tadokoro, K., Ito, K. and Dahinden, C.A. (1995). Effects of physiologic soluble agonists on leukotriene C_4 production and degranulation by human eosinophils. Int. Arch. Allergy Immunol. 108, 36–38.

Takai, S., Yamada, K., Hirayama, N., Miyajima, A. and Taniyama, T. (1994). Mapping of the human gene encoding the mutual signal-transducing subunit (beta-chain) of granulocyte-macrophage colony-stimulating factor (GM-CSF), interleukin-3 (IL-3), and interleukin-5 (IL-5) receptor complexes to chromosome 22q13.1. Hum. Genet. 93, 198–200.

Takeda, S., Gillis, S. and Palacios, R. (1989). In vitro effects of recombinant interleukin 7 on growth and differentiation of bone marrow pro-B- and pro-T-lymphocyte clones and fetal thymocyte clones. Proc. Natl Acad. Sci. USA 86, 1634–1638.

Tedeschi, A., Palella, M., Milazzo, N., Lorini, M. and Miadonna, A. (1995). IL-3 induced histamine release from human basophils. J. Immunol. 155, 2652–2660.

Terstappen, L.W.M.M., Huang, S., Safford, M., Lansdorp, P.M. and Loken, M.R. (1991). Sequential generations of hematopoietic colonies derived from single nonlineage-committed $CD34^+CD38^-$ progenitor cells. Blood 77, 1218–1227.

Thomas, J.W., Baum, C.M., Hood, W.F., et al. (1995). Potent interleukin 3 receptor agonist with selectively enhanced hematopoietic activity relative to recombinant human interleukin 3. Proc. Natl Acad. Sci. USA 92, 3779–3783.

Torigoe, T., O'Connor, R., Santoli, D. and Reed, J.C. (1992). Interleukin-3 regulates the activity of the LYN protein-tyrosine kinase in myeloid-committed leukemic cell lines. Blood 80, 617–624.

Uckun, F.M., Gesner, T.G., Song, C.W., Myers, D.E. and Mufson, A. (1989). Leukemic B-cell precursors express functional receptors for human interleukin-3. Blood 73, 533–542.

Vadhan-Raj, S., Broxmeyer, H.E., Andreeff, M., et al. (1995). In vivo biological effects of PIXY321, a synthetic hybrid protein of recombinant human granulocyte-macrophage colony-stimulating factor and interleukin-3 in cancer patients with normal hematopoiesis: a phase I study. Blood 86, 2098–2105.

Valent, P., Schmidt, G., Besemer, J., et al. (1989a). Interleukin-3 is a differentiation factor for human basophils. Blood 73, 1763–1769.

Valent, P., Besemer, J., Muhm, M., Majdic, O., Lechner, K. and Bettelheim, P. (1989b). Interleukin 3 activates human blood basophils via high-affinity binding sites. Proc. Natl Acad. Sci. USA 86, 5542–5546.

van Leeuwen, B.H., Matrinson, M.E., Webb, G.C. and Young, I.G. (1989). Molecular organization of the cytokine gene cluster, involving the human IL-3, IL-4, IL-5, and GM-CSF genes, on human chromosome 5. Blood 73, 1142–1148.

Wang, M., Friedman, H. and Djeu, J.Y. (1989). Enhancement of human monocyte function against Candida albicans by the colony-stimulating factors (CSF): IL-3, and granulocyte-macrophage-CSF. J. Immunol. 143, 671–677.

Watanabe, Y., Kitamura, T., Hayashida, K. and Miyajima, A. (1992). Monoclonal antibody against the common β subunit ($β_c$) of the human interleukin-3 (IL-3), IL-5, and granulocyte-macrophage colony-stimulating factor receptors shows upregulation of $β_c$ by IL-1 and tumor necrosis factor-β. Blood 80, 2215–2220.

Whitlock, C.A. and Witte, O.N. (1982). Long-term culture of B lymphocytes and their precursors from murine bone marrow. Proc. Natl Acad. Sci. USA 79, 3608–3612.

Williams, G.T., Smith, C.A., Spooncer, E., Dexter, T.M. and Taylor, D.R. (1990). Haemopoietic colony stimulating factors promote cell survival by suppressing apoptosis. Nature 343, 76–79.

Winkler, T.H., Melchers, F. and Rolink, A.G. (1995). Interleukin-3 and interleukin-7 are alternative growth factors for the same B-cell precursors in the mouse. Blood 85, 2045–2051.

Wodnar-Filipowicz, A., Heusser, C.H. and Moroni, C. (1989). Production of the haemopoietic growth factors GM-CSF and interleukin-3 by mast cells in response to IgE receptor-mediated activation. Nature 339, 150–152.

Wodnar-Filipowicz, A. and Moroni, C. (1990). Regulation of interleukin-3 mRNA expression in mast cells occurs at the post-transcriptional level and is mediated by calcium ions. Proc. Natl Acad, Sci. USA 87, 777–781.

Wong, P.M.C., Chung, S.-W., Dunbar, C.E., Bodine, D.M., Ruscetti, S. and Nienhuis, A.W. (1989). Retrovirus-mediated transfer and expression of the interleukin-3 gene in mouse hematopoietic cells result in a myeloproliferative disorder. Mol. Cell. Biol. 9, 798–808.

Xia, X., Li, L. and Choi, Y.S. (1992). Human recombinant IL-3 is a growth factor for normal B cells. J. Immunol. 148, 491–497.

Yamaguchi, Y., Suda, T., Suda, J., et al. (1988). Purified interleukin 5 supports the terminal differentiation and proliferation of murine eosinophilic precursors. J. Exp. Med. 167, 43–56.

Yang, Y.-C. and Clark, S.C. (1988). In "Interleukin 3: The Panspecific Hemopoietin" (ed. J.W. Schrader), pp. 375–391. Academic Press, San Diego.

Yang, Y.-C., Ciarletta, A.B., Temple, P.A., et al. (1986). Human IL-3 (multi-CSF): Identification by expression cloning of a novel hematopoietic growth factor related to murine IL-3. Cell 47, 3–10.

Yang, Y.C., Kovacic, S., Kriz, R., et al. (1988). The human genes for GM-CSF and IL-3 are closely linked in tandem on chromosome 5. Blood 71, 958–961.

Ymer, S., Tucker, W.Q.J., Sanderson, C.J., Hapel, A.J., Campbell, H.D. and Young, I.G. (1985). Constitutive synthesis of interleukin-3 by leukaemia cell line WEHI-3B is due to retroviral insertion near the gene. Nature 317, 255–258.

Yokota, T., Lee, F., Rennick, D., et al. (1984). Isolation and characterization of a mouse cDNA clone that expresses mast-cell growth-factor activity in monkey cells. Proc. Natl Acad. Sci. USA 81, 1070–1074.

Zhang, W., Zhang, J., Kornuc, M., Kwan, K., Frank, R. and Nimer, S.D. (1995). Molecular cloning and characterization of NF-IL3A, a transcriptional activator of the human interleukin-3 promoter. Mol. Cell. Biol. 15, 6055–6063.

4. Interleukin-4

Anthony R. Mire-Sluis

1. Introduction

Interleukin-4 (IL-4) was originally identified as a molecule that was able to stimulate DNA synthesis of anti-IgM-stimulated murine B lymphocytes. This activity lead to the molecule being named initially as B cell growth factor (BCGF) (Howard *et al.*, 1982). Concomitantly the protein was shown to induce lipopolysaccharide (LPS)-activated B cells to produce IgG1 and this activity was named BCDF (Isakson *et al.*, 1982). Further studies showed that this molecule also increased the expression of class II MHC molecules on resting B cells. Therefore, the name of the protein was proposed as B cell stimulatory factor-1 (BSF-1) (Noelle *et al.*, 1984; Paul, 1984; Roehm *et al.*, 1984). Following the cloning of the cDNA, the protein was given the name IL-4 (Noma *et al.*, 1986; Yokota *et al.*, 1986). Subsequent work has described the diversity and complex biological activities of IL-4, including growth of T lymphocytes and mast cells as well as a wide range of effects on many hematopoietic and endothelial cell types (Ohara, 1989; Callard, 1991; Banchereau *et al.*, 1994).

2. The Cytokine Gene

Human IL-4 is encoded by a single mRNA transcript of approximately 0.9 kb in length. The IL-4 gene consists of four exons arranged over 10 kb of DNA and exists as a single copy on chromosome 5 at 5q23.3–31.2 (Le Beau *et al.*, 1988; Arai *et al.*, 1989).

Cytokines
ISBN 0–12–498340–5

2.1   Genomic Organization of the IL-4 Gene

Following the isolation of the human cDNA clone for IL-4 from concanavalin A-stimulated T lymphocytes, the complete nucleotide sequence and structure of the chromosomal gene was reported (Yokota *et al.*, 1986; Arai *et al.*, 1989). The gene spans 10 kb of DNA contains four exons (Figures 4.1 and 4.2). The first exon contains the 5′ untranslated region of 65 bp and the fourth exon contains a 3′ untranslated region of 91 bp. A conventional "TATA" sequence exists 27 bp upstream from the transcription initiation site, starting at base pair position 1078. The gene contains the octamer enhancer motif sequence (ATGCAAAG) of the sv40 72 bp repeat sequence 234 bp upstream from the initiation site (Davidson *et al.*, 1986). Within intron 2 exists the feature of a TG element which may be involved in the modulation of gene expression.

The gene also contains three tandem repeats at position 7694 to 7902 within intron 3.

2.2   Transcriptional Control

Initial studies following the complete analysis of the nucleotide sequence for the IL-4 gene showed that the gene, when transfected, can be induced by phorbol ester and the calcium ionophore A23187, but only weakly using A23187 alone (Arai *et al.*, 1989). An 11 bp region of the gene termed the "P sequence" (CGAAAATTTCC) is responsible for the responsiveness of the gene to phorbol ester/A23187, occurring at positions –79 to –69 from the cap site (Abe *et al.*, 1992). It appears from data derived from work in the murine gene that this site binds a novel protein termed the NF-AT$_p$ that may control the expression of the IL-4 gene

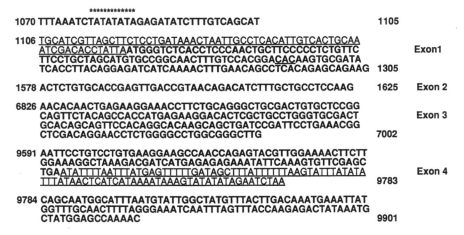

Figure 4.1   Gene sequence of the IL-4 gene. The four exons of the interleukin gene are indicated. Feint, underlined bases in exon 1 and exon 4 are nontranslated regions. The bold-underlined CAC is the coding sequence for the first mature amino acid of the protein. The TATA box is marked by asterisks.

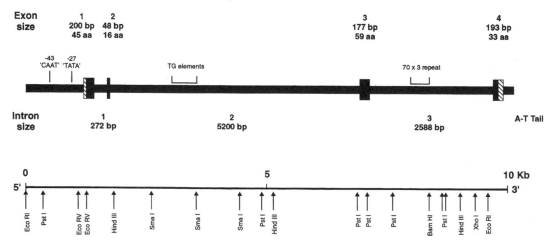

Figure 4.2   Gene structure of the IL-4 gene. The four exons are indicated by hatched sections indicating the untranslated regions.

(Kubo *et al.*, 1994). A negative regulatory element (NRE) also exists in the form of two protein binding sites: NRE 1 ($^{-311}$CTCCCTTCT^{-303}) and NRE 2 ($^{-288}$CTTTTTGCTTTGC^{-303}). The proteins Neg-1 and Neg-2 bind to these sites respectively. Neg-1 has only been detected in T cell lines and may be responsible for T cell specific inhibition of IL-4 gene transcription. The binding of both Neg-1 and Neg-2 are required for functional suppression of the gene (Li-Weber *et al.*, 1992). A positive regulatory element (PRE-1) has also been detected and shown to lie between positions −240 and −223 (Li-Weber *et al.*, 1992). Two additional regulatory elements, a CCAAT element and a 15-nucleotide IFN- and virus-stimulation response element (ISRE) have been shown to interact with IRF-2 (a transcription repressor of IFN genes) and a NF-1 like factor. NF-Y, a transcription factor involved in MHC class II expression, binds the CCAAT element as well as the "P sequence" (Li-Weber *et al.*, 1994; Brown and Hural, 1997). T-cell subset expression is regulated by a silencer element and STAT6 (Kubo *et al.*, 1997).

3.   *The Protein*

The cDNA for human IL-4 encodes for a 153-amino-acid (aa) protein containing a 24-aa signal sequence. The mature IL-4 protein consists of 129 amino acids, 6 of which are cysteine residues (Figure 4.3). The protein exists as a monomer and contains two *N*-linked glycosylation sites at positions 38 and 105. The protein is *N*-glycosylated, but the exact site of glycosylation has yet to be determined, although it is known that position 38 is occupied. No *O*-linked glycosylation occurs (Yokota *et*

al., 1986; Le *et al.*, 1988; Carr *et al.*, 1990). The deglycosylated protein has a molecular mass of approximately 15 kDa and the glycosylated material can have a molecular mass of 18–19 kDa up to a 60 kDa hyperglycosylated protein produced in yeast (Park *et al.*, 1987). Glycosylation of IL-4 does not seem to be important for biological activity (Le *et al.*, 1988).

The cysteine residues form three disulfide bonds (C^3–C^{127}, C^{24}–C^{65} and C^{46}–C^{99}) (Windsor *et al.*, 1990; Carr *et al.*, 1990). All three disulfide bonds are required for full biological activity, but of the three, C^{46}–C^{99} appears the most important for the structural and biological integrity of the protein (Kruse *et al.*, 1991). Human IL-4 has an isoelectric point of 10.5 and no sites of phosphorylation (Trotta, 1992).

Three-dimensional studies of IL-4 show it to exist as a left-handed four-helix bundle with an up–up–down–down arrangement of the four helices. Two long loops, each consisting of two helical sections, run the length of the molecule (AB and CD) (Figure 4.4). A short region of β-sheet connects the two loops. The helices are held together by a core of hydrophobic side-chains whilst the exterior of the protein is highly hydrophilic (Walter *et al.*, 1992; Redfield *et al.*, 1994b; Smith *et al.*, 1994). The molecule is stable at extreme values of pH and is stable at 4°C for more than 3 months (Windsor *et al.*, 1991; Redfield *et al.*, 1994b).

4.   *Cellular Sources and Production*

IL-4 is mainly produced by subsets of the T lymphocyte lineage, mast cells and basophils. In humans, CD4$^+$ CD45RA$^-$ (memory) cells produce the majority of IL-4

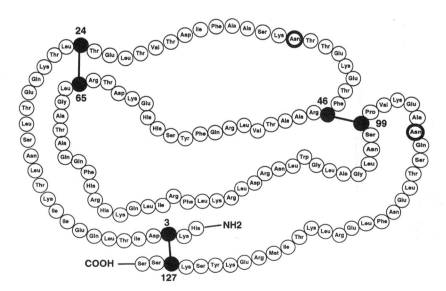

Figure 4.3   The primary and secondary structure of the IL-4 protein. Cysteine residues are indicated by solid black circles and possible glycosylation sites are indicated by thick circles.

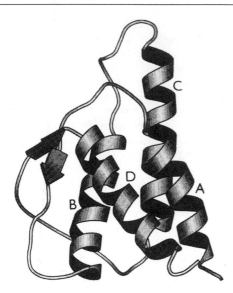

Figure 4.4   The three-dimensional structure of IL-4. A schematic diagram showing the four-helix bundle folding of human IL-4 generated using MOLSCRIPT (Kraulis, 1991). This figure is kindly provided by C. Redfield and C.M. Dobson (Redfield, *et al.*, 1994b).

(DeKruyff *et al.*, 1995), whereas only a proportion of CD8[+] T cell clones secrete detectable amounts of IL-4 (Lewis *et al.*, 1988; Yokota *et al.*, 1988; Seder *et al.*, 1992). IL-4 is induced in these cells via stimulation of the T cell receptor with lectins or antigen, CD3 or CD2 via addition of antibodies, or on addition of phorbol esters with or without the presence of calcium ionophore (Yokota *et al.*, 1988). Concanavalin A (Con A) induces mRNA expression peaking at 6–12 h but with a low transcription rate. mRNA decays after 90 min post stimulation. However, phorbol myristate acetate (PMA) increases Con A IL-4 expression by stabilizing mRNA (Dokter *et al.*, 1993). Signaling via the cAMP pathway in T cells appears to downregulate IL-4 production by Con A but not other stimulators (Borger *et al.*, 1996). Mast cells can be induced to produce IL-4 via cross-linkage of high-affinity Fc receptors (Brown *et al.*, 1987; Plaut *et al.*, 1989, Tunon de Lara *et al.*, 1994). There have been reports in mice that there is a population of splenic non-T non-B cells that can be induced to produce large quantities of IL-4 via activation of Fc, R or FcγRII (Ben-Sasson *et al.*, 1990). IL-4 production can be induced in basophils after challenge with anti-IgE antibodies, FMLP or C5a (MacGlashan *et al.*, 1994). Eosinophils produce IL-4 on addition of anti-IgE antibodies (Nonaka *et al.*, 1995).

5.   Biological Activity

The most striking features of the biological activity of IL-4 are the number of cell types on which IL-4 acts, the numerous activities it induces, and the plieotropic manner in which the activities occur. The function of IL-4 on any particular cell type appears to be heavily influenced by the state of activation of that cell and the different components of the surrounding environment (Callard, 1991; Spits, 1992).

5.1   B Cells

5.1.1   Resting B Cells

IL-4 alone does not induce resting B cells to proliferate, but does induce phenotypic changes and a small increase in cell volume (Yokota *et al.*, 1988; Valle *et al.*, 1989). IL-4 induces expression of surface IgM, CD23, soluble CD23, CD40 and small increases of MHC class II (Noelle *et al.*, 1984; Conrad *et al.*, 1987; Gordon *et al.*, 1988; Shields *et al.*, 1989). The increase in surface IgM does not cause an IL-4 induced increase in responsiveness to anti-IgM (Shields *et al.*, 1989). Therefore, although IL-4 induces a certain degree of B cell activation, it is not providing a competent signal since it is unable to induce resting B cells to enter G1 in the cell cycle (Clark *et al.*, 1989). IL-4 also plays a role in B cell chemotaxis (Komai-Koma *et al.*, 1995).

5.1.2   B Cell Proliferation

IL-4 acts as a costimulator of B cell proliferation with anti-IgM (Howard *et al.*, 1982). IL-4 can also induce proliferation in conjunction with *Staphylococcus aureus* (SAC) or phorbol ester and calcium ionophore, or immobilized anti-CD40 antibodies (Defrance *et al.*, 1992; Gordon *et al.*, 1988; Valle *et al.*, 1989). Antigen-stimulated B cells also proliferate in response to IL-4 (Llorente *et al.*, 1990).

5.1.3   B Cell Differentiation and Ig Chain Switching

IL-4 alone does not induce the secretion of immunoglobulin by B cells. B cells preincubated with SAC or with phorbol ester and ionophore secrete IgG and IgM and small levels of IgA (Banchereau *et al.*, 1994). IL-4 is able to induce B cells to produce IgE, but only in the presence of mononuclear cells or CD4[+] T cells (Pène *et al.*, 1988a,b). IgG4 production is also induced in this manner (Lundgren *et al.*, 1989). IL-6 has been identified as being involved in this process (Vercelli *et al.*, 1989). IL-4 is able to induce the switching of expression of cells from producing IgM to IgG1 and/or IgE (Callard, 1991; Schultz and Coffman, 1992). Recently, it has been shown that highly purified human pre-B cells differentiate into IgG-secreting cells induced by IL-4 and CD4[+] T cells (Punnonen *et al.*, 1993). In IL-4-deficient mice levels of IgG1 and IgE are strongly reduced, although T and B cell development is normal (Kuhn *et al.*, 1991).

5.2   T Cells

IL-4 has been shown to promote the growth of T cells that have been preactivated by phytohemagglutinin (Yokota *et al.*, 1986; Spits *et al.*, 1987). Both activated CD8[+] and CD4[+] T cells proliferate in response to IL-4 and have an increased response to IL-2 (Mitchell *et al.*, 1989). It is interesting that resting T cells treated with IL-2 are inhibited from proliferating in the presence of IL-4, although IL-4 has to be present throughout the culture period (Martinez *et al.*, 1990). IL-4 can induce the expression of CD8 on CD4[+] T cells (Paliard *et al.*, 1988; Hori *et al.*, 1991). IL-4 regulates the induction of cell-mediated cytolytic activity by T cells and is inhibited by IL-12 p 40 (Abdi and Mermann, 1997). IL-4 added to mixed lymphocyte cultures early on and then removed results in inhibition of the generation of cytotoxic activity (Widmer *et al.*, 1987; Horohov *et al.*, 1988). However, if IL-4 remains in the cultures throughout, an increased level of generation of cytotoxic activity occurs (Spits *et al.*, 1988). IL-4 does, however, block the generation of antigen nonspecific T cell cytotoxicity (Spits *et al.*, 1989). IL-4 induces proliferation of postnatal thymocytes, resulting in the growth of CD3[+] thymocytes and the differentiation of pro-T cells into mature T cells (Bàrcena *et al.*, 1990). In addition, IL-4 promotes T cell chemotaxis in association with other cytokines (Tan *et al.*, 1995).

5.3   Monocytes and Macrophages

IL-4 has several effects on the surface antigen expression of monocytes. Addition of IL-4 to monocytes increases expression of class II MHC antigens, CD13, CD23 and CD18, CD11b and CD11c (Van Hal *et al.*, 1992; Vercilli *et al.*, 1988; Littman *et al.*, 1989; Te Velde *et al.*, 1988). IL-4 also down regulates expression of CD14, CD64, CD32 and CD16 (Lauener *et al.*, 1990; Te Velde *et al.*, 1990a,b). Although IL-4 induces the differentiation of monocytes to macrophages (Te Velde *et al.*, 1988) it inhibits macrophage growth (Jansen *et al.*, 1989). IL-4 inhibits the production of a wide range of cytokines from stimulated monocytes (Hart *et al.*, 1989; Essner *et al.*, 1990; Standiford *et al.*, 1990; Cluitmans *et al.*, 1994) whilst inducing the production of IL-1 receptor antagonist, G-CSF and GM-CSF (Weiser *et al.*, 1989; Fenton *et al.*, 1992). The effects of IL-4 on the functional status of monocytes remains enigmatic. IL-4 enhances antitumor macrophage activity, yet it downregulates production of superoxide, metalo-proteinases and PGE$_2$ (Crawford *et al.*, 1987; Abramson and Gallin, 1990; Corcoran *et al.*, 1992). IL-4 also abrogates the monocytic activation activity induced by IFN-γ (Te Velde *et al.*, 1990a,b).

5.4   Hematopoiesis

IL-4 alone does not affect the proliferation or differentiation of myeloid progenitors *in vitro* (Peschel *et al.*, 1987; Broxmeyer *et al.*, 1988; Sonada *et al.*, 1990). IL-4 enhances the growth of progenitors in the presence of G-CSF, IL-6 or IL-11, but inhibits colony formation in the presence of either GM-CSF, M-CSF or IL-3 (Rennick *et al.*, 1987, 1989; Vellenga *et al.*, 1990; Jacobsen *et al.*, 1995), although the inhibition appears to be restricted to the monocytic lineage (Jansen *et al.*, 1989). IL-4 synergizes with EPO to produce modest growth of erythroid colony-forming units and dramatically increases the growth of erythroid burst-forming units (Rennick *et al.*, 1987; Broxmeyer *et al.*, 1988). However, the combination of IL-3, or GM-CSF and EPO in the presence of IL-4 results in the inhibition of erythroid colony formation (de Wolf *et al.*, 1990; Vellenga *et al.*, 1993). IL-4 in combination with IL-3 induces the proliferation and differentiation of precursors of basophils, mast cells and eosinophils into mature cells (Favre *et al.*, 1990; Tsuji *et al.*, 1990).

5.5   LAK and NK Cells

IL-4 alone is unable to induce lymphocyte-activated killer (LAK) cell activity in mononuclear cells and strongly inhibits IL-2 induced production of LAK cells, and particularly natural killer cells, when incubated throughout the culture (Spits *et al.*, 1988; Banchereau, 1990; Phillips *et al.*, 1992). The inhibition of IL-2 induced nonspecific cytotoxicity occurs at the induction stage, but does not affect the cytotoxic machinery of resting or preactivated NK cells (Nagler *et al.*, 1988; Phillips *et al.*, 1992). IL-4 inhibits IL-2-induced expression of CD69 and production of IFN-γ by NK cells (Phillips *et al.*, 1992). Preincubation of NK cells with IL-4 inhibits the adhesion of the cells to vascular endothelial cells (Paganin *et al.*, 1994).

5.6   Other Cell Types

IL-4 upregulates the expression of CD23 on eosinophils and decreases their expression of Fcγ receptors. This results in a reduced capacity to respond to IgG binding (Tanaka *et al.*, 1989; Baskar *et al.*, 1990). IL-4 also acts as a neutrophil activator since it stimulates the respiratory burst mediated by F-met-leu-phe (Boey *et al.*, 1989; Bober *et al.*, 1995). Mast cells and astrocytes grow in response to IL-4 (Banchereau *et al.*, 1994; Lui *et al.*, 1997) whilst osteoblast growth is inhibited by IL-4 (Riancho *et al.*, 1993, 1995). IL-4 induces adhesion molecule expression on endothelial cells (Palmer-Crocker and Pober, 1995).

6.   *IL-4 Receptor*

The human IL-4 receptor consists of two, possibly three, chains (Duschl and Sebald, 1996). An IL-4-specific binding α chain of 130–140 kDa has been cloned and shown to confer biological responsiveness to IL-4 (Galizzi *et al.*, 1990; Backmann, 1990). The second chain of the IL-4 receptor is the 64 kDa chain of the IL-2 receptor (Takashita *et al.*, 1992; Kondo *et al.*, 1993; Russel *et al.*, 1993). There have been several reports of IL-4 binding proteins in the range of 60–70 kDa, although the significance of these proteins (some possibly degradation products of the 130 kDa receptor) remains unclear (Foxwell *et al.*, 1989; Galizzi *et al.*, 1989; Fanslow *et al.*, 1993; Gauchat *et al.*, 1997; Dawson *et al.*, 1997). IL-4 binds directly to the 130 kDa receptor with a high affinity and the presence of the IL-2 γ chain increases this affinity by 2–3-fold, although IL-4 does not bind this molecule

directly (Russell *et al.*, 1993). Since the discovery of the role of the IL-2 γ chain in IL-4 binding cells, it is difficult to interpret data provided previously concerning actual affinities and receptor numbers. However, direct binding of iodinated IL-4 to a wide variety of cell types reveals that most have between 100 and 400 IL-4 binding sites per cell (Urdal and Park, 1988; Ohara *et al.*, 1987). The gene for the 130 kDa receptor exists on chromosome 16p11-12 and encodes for a 800-aa protein containing a 24-aa transmembrane domain, a 207-aa cytokine binding domain, and a 569-aa intracellular domain (Figure 4.5) (Pritchard *et al.*, 1991). The extracellular domain contains a WSXWS motif and four conserved cysteines from a superfamily of receptors of hematopoietic cytokines

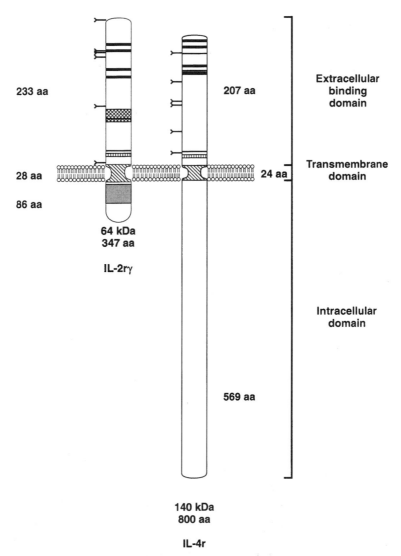

Figure 4.5   The structure of the receptors for human IL-4. Thick lines indicate the four conserved cysteines of the hemopoietic receptor family. Thin horizontal lines represent nonconserved cysteines. Vertical bars donate the WSXWS motif. The transmembrane domain is indicated by hatching. The criss-crossed section of the IL-2 receptor γ-chain extracellular domain contains the leucine zipper sequence and intracellularly, the dotted section indicates the SH3 binding domain.

(Bazan, 1990; D'Andrea *et al.*, 1990). The α chain is also glycosylated with bi-, tri-, and tetraantenny structures (Rajan *et al.*, 1995). The 64 kDa IL-2 receptor γ chain exists as a 347-amino-acid protein containing a 28-aa transmembrane domain, an 86-aa cytoplasmic domain, and a 233-amino-acid extracellular domain. The extracellular domain also contains the four conserved cysteines and the WS motif, together with a leucine zipper sequence (Takashita *et al.*, 1992). T cells stimulated with low concentrations of IL-4 or with phorbol ester, calcium ionophore, anti-CD23 or Con A increase levels of IL-4 binding sites (Ohara and Paul, 1988; Armitage *et al.*, 1990). Activation of B cells with anti-IgM or *Staphylococcus aureus* appears to increase IL-4 receptor levels, but this may be due to increases in the size of the cells (Banchereau, 1990). Monocytes do not increase levels of IL-4 receptor in response to IL-4, but do respond to phorbol ester or Protein kinase A activators (de Wit *et al.*, 1994). The IL-2 γ chain is not present on nonlymphoid cells and could account for differences in the affinities and regulation of IL-4 receptors on different cell types (Takashita *et al.*, 1992). Models of the mechanism of binding of IL-4 to its receptor suggest that a 1:1 complex is formed, leading to a dimerization of the α chain and the IL-2 γ chain (Hoffman *et al.*, 1994; Gustchina *et al.*, 1995; Duschl and Sebold, 1996). IL-4 function can be transferred to the IL-2 receptor by transfection of tyrosine-containing sequences of the IL-4 receptor α chain (Wang *et al.*, 1996). Structural studies illustrate that helices A and C of IL-4 provide α chain binding sites and helix D possibly IL-2 γ chain binding (Müller *et al.*, 1994; Reusch *et al.*, 1994).

7.   *Signal Transduction*

The biochemical events occurring on binding of human IL-4 to its receptor are only recently being reported to any extent and remain poorly understood, having been confused by a vast amount of data concerning the signal transduction of murine IL-4. It must be stressed that all available data pertaining to human IL-4 binding to normal human cells clearly shows that murine IL-4 uses distinctly different pathways for signal transduction. Addition of human IL-4 to resting human B cells has been reported to cause a transient increase of inositol 1,4,5-trisphosphate and calcium levels followed by a sustained increase in cellular cyclic adenosine monophosphate (Finney *et al.*, 1990). However, there are also reports that these cells do not induce calcium ion release nor activate protein kinase C in response to IL-4 (Clark *et al.*, 1989; Galizzi *et al.*, 1989; Banchereau, 1990; Lee *et al.*, 1993). IL-4 appears to utilize a tyrosine protein kinase and phosphatase system as part of its signaling mechanisms in a variety of human cell lines (Mire-Sluis and Thorpe, 1991; Kotanides and Reich,

1993; Mire-Sluis *et al.*, 1994). IL-4 induces the tyrosine phosphorylation of a nuclear factor termed IL-4-NAF which subsequently binds to the genes of IL-4-induced proteins (Kotanides and Reich, 1993). IL-4 induces serine/threonine phosphorylation of a set of substrates in the human B cell line SKW6.4 (Goldstein and Kim, 1993). Protein kinase C translocates to the nucleus of monocytes treated with IL-4 (Arruda and Ho, 1992). IL-4 activates the nuclear factor STAT 6 and the kinase JAK 1 in T cells (Kotanides *et al.*, 1995; Kaplan *et al.*, 1996; Takeda *et al.*, 1996; Reichel *et al.*, 1997), the kinases JAK2 and JAK3 in B cells and carcinoma lines (Fenghao *et al.*, 1995; Malabarba *et al.*, 1995; Murata *et al.*, 1996), and MAP kinase in keratinocytes (Wery *et al.*, 1996). The insulin receptor substrate-1 (IRS-1) is phosphorylated on growth inhibition by IL-4 (Schnyder *et al.*, 1996), whereas IRS-2 is phosphorylated in hematopoietic cells (Sun *et al.*, 1995). The data suggest that signaling mechanisms of IL-4 vary between cell types and are also dependent upon the specific biological responses elicited (Callard, 1991). Models have been proposed that may shed light on such a complex signaling network (Duschl and Sebold, 1996).

8.   *Murine IL-4*

The gene for murine IL-4 has been cloned and exists on chromosome 11.5q23 (Lee *et al.*, 1986; Noma *et al.*, 1986; D'Eustachio *et al.*, 1988). The murine IL-4 gene spans 6 kb and consists of four exons coding for 44, 16, 51, and 29 amino acids, respectively. The gene includes three introns of 254, 4000 and 1300 base pairs (Arai, 1987; Yokota *et al.*, 1988). The murine gene promoter contains five cis-acting P sequences which mediate gene control by the nuclear factor NF(P). The gene also contains a Y box $^{-114}$CTGATTGG^{-107} with homology to those of the MHC class II gene promoters (Szabo *et al.*, 1993). The gene can be induced through the activity of calcineurin (Kubo *et al.*, 1993).

The gene encodes for a 140-aa protein that is clipped to produce the 120-aa mature IL-4 protein. The protein has a molecular mass of 13 558 Da, an isoelectric point of 9.7, three possible glycosylation sites at 41, 71, 97, and three internal disulfide bridges (Grabstein *et al.*, 1986; Ohara *et al.*, 1987). The disulfide bridges are between C^5–C^{87}, C^{27}–C^{67} and C^{49}–C^{94} (Carr *et al.*, 1990). Production of murine IL-4 is similar to that of human IL-4 except that there is a more clear-cut subtyping of T helper cells (the T_H2 subtype) that produce IL-4 (Spits, 1992). The biological activities of murine IL-4 are also similar to those for human IL-4. The murine IL-4 receptor system is similar to the human receptor system described except for slight variations in the structure of the proteins involved. The 130–140 kDa murine receptor has been cloned and codes for an 810-aa precursor protein of the mature 785-aa mature receptor. The receptor contains a 22-aa transmembrane domain, a 553-aa cytoplasmic domain and a 210-aa ligand-binding extracellular domain (Mosley *et al.*, 1989; Harada *et al.*, 1990). The murine receptor also includes the four conserved cysteines and the WS motif. In addition to the 130 kDa protein, in the murine system an mRNA transcript exists that codes for a soluble IL-4 receptor (Mosley *et al.*, 1989). The mRNA codes for a 37–40 kDa protein and has been shown to be produced by T cells, B cells, and macrophages and can be induced by IL-4 itself (Chilton and Fernandez-Botran, 1993). The soluble receptor has been found in different biological fluids in mice and has been proposed as a carrier for IL-4 (Fanslow *et al.*, 1990; Fernandez-Botran and Vitetta, 1990).

There have been several reports covering the signal transduction pathways of murine IL-4, and reviews tend to mix data from human and murine sources (Duschl and Sebald, 1996). Although neither the 130 kDa receptor or the IL-2 receptor γ chain contains a kinase domain, addition of IL-4 to murine cells induces tyrosine phosphorylation of several substrates. In particular, a 170 kDa substrate termed insulin receptor substrate-1 (IRS-1) is phosphorylated together with a wide range of substrates in myeloid cells (Morla *et al.*, 1988; Isfort and Ihle, 1990; Wang *et al.*, 1992; Keegan *et al.*, 1994). IRS-1 has also been found to be phosphorylated in a wide range of other cell types and normal B cells and T cells (Keegan and Pierce, 1994). The 130 kDa IL-4 receptor is phosphorylated on tyrosine residues on IL-4 addition in the HT-2 cell line (Izuhara and Harada, 1993). IL-4 does not induce inositol lipid metabolism or calcium changes (Justement *et al.*, 1986; Mizuguchi *et al.*, 1986). The c-*fes* protooncogene product, a non-receptor tyrosine kinase, is phosphorylated on tyrosine residues in T cell lines in addition of IL-4 and may mediate murine IL-4-induced tyrosine phosphorylation events (Izuhara *et al.*, 1994).

9. IL-4 in Disease and Therapy

As an immunomodulator with diverse biological activities, it is not surprising that IL-4 is present in the lesions associated with a variety of disorders. The presence or absence of IL-4 in disease has resulted in different strategies to increase IL-4 levels or inhibit its activity depending on the disorder.

9.1 IL-4 ASSOCIATED PATHOLOGY

Increased levels of IL-4 have been detected in biological samples from several disease states. Levels of IL-4 are increased in scleroderma, bronchial asthma, allergy, multiple sclerosis, autoimmune thyroid disease, atopic dermatitis, inflammatory bowel disease, endometriosis and systemic sclerosis (Needleman *et al.*, 1992; Renz *et al.*, 1992; Bradding *et al.*, 1993; Hirooka *et al.*, 1993, Link *et al.*, 1994; Matsumoto *et al.*, 1994; Corrigan, 1995; Imada *et al.*, 1995; Nonaka *et al.*, 1995; Salerno *et al.*, 1995; Nielsen *et al.*, 1996; Hsu *et al.*, 1997; Hasegawa *et al.*, 1997).

9.2 IL-4 AS AN ANTITUMOR AGENT *IN VITRO*

IL-4 appears to have antitumor activity on several different types of tumor cells *in vitro*. IL-4 inhibits the growth of tumorigenic hemopoietic cells, such as Philadelphia positive acute lymphocytic leukemia, non-Hodgkin B-lymphoma, chronic lymphocytic leukemia, and chronic myogenous leukemia (Luo *et al.*, 1991; Defrance *et al.*, 1992; Okabe *et al.*, 1992; Estrov *et al.*, 1993; Puri and Siegel, 1993; Reittie and Hoffbrand, 1994). IL-4 can also inhibit the growth of solid tumor cells *in vitro*, such as those from melanoma, renal cell carcinoma, colon cancer, breast cancer, and lung cancer (Hoon *et al.*, 1991; Toi *et al.*, 1992; Obiri *et al.*, 1993; Puri and Siegel, 1993; Topp *et al.*, 1993).

9.3 IL-4 IN ANTITUMOR THERAPY *IN VIVO*

Several preclinical and phase I trials of IL-4 have been undertaken (Leach *et al.*, 1997a,b). IL-4 has a dose limiting toxicity of approximately 5–10 μg/kg per day. Side-effects include flu-like symptoms, fatigue, diarrhea, vascular leak syndrome, elevated liver enzymes, and possibly myocardial toxicity (Atkins *et al.*, 1992; Cornacoff *et al.*, 1992; Dean *et al.*, 1992; Gilleece *et al.*, 1992; Ghosh *et al.*, 1993; Prendiville *et al.*, 1993; Sosman *et al.*, 1994; Trehu *et al.*, 1993). No significant solid tumor responses to IL-4 have been reported to date. A phase II trial for renal cell carcinoma and malignant melanoma produced no responses, whilst an in-depth study of the monocytic responses of patients undergoing IL-4 treatment showed that monocyte responses (as *in vitro*) are clearly depressed by IL-4 administration (Wong *et al.*, 1992; Margolin *et al.*, 1994). A phase II study of IL-4 treatment in metastatic renal cell carcinoma at 1 μg/kg three times weekly showed no responses (Stadler *et al.*, 1995). Attempts are underway to increase success, through the use of IL-4, to stimulate the growth of tumor infiltrating lymphocytes in melanoma (Casanelli *et al.*, 1994). In hematopoietic malignancies there have been a few partial responses using IL-4, although a great deal of work is required before such data becomes meaningful and IL-4 a useful antitumor therapeutic (Puri and Siegel, 1993). Intralesional injection of IL-4 in a murine renal tumor model led to a localized effect but no impact on subsequent metastasis (Younes *et al.*, 1995). IL-4 has also been tested in the treatment of Kaposi's sarcoma (Tulpule *et al.*, 1997).

9.4   IL-4 AND DIABETES

Studies of mouse models of diabetes and some human studies illustrate the decreased responsiveness of T cells to stimulation through the T cell receptor (Zipris *et al.*, 1991a,b; Zier *et al.*, 1984). Treatment of mice susceptible to developing diabetes with IL-4 not only reversed the T cell unresponsiveness but protected the mice from developing diabetes (Rapoport *et al.*, 1993). IL-4 has, therefore, been suggested as a possible therapeutic for preventing diabetes (Robinson, 1994).

9.5   THE THERAPEUTIC POTENTIAL OF THE SOLUBLE IL-4 RECEPTOR

The production of recombinant murine soluble IL-4 receptor has allowed the investigation of its effects *in vivo* (Jacobs *et al.*, 1991). The soluble IL-4 receptor can regulate alloreactivity responses and delay allograft rejection as well as inhibiting IgE responses (Fanslow *et al.*, 1991; Maliszewski *et al.*, 1992; Sato *et al.*, 1993, Renz *et al.*, 1995). Although these responses may have clinical relevance, it has also been noted that the soluble receptor can enhance the activity of IL-4 in inducing IgE responses if added as a complex. The therapeutic potential of the soluble IL-4 receptor therefore needs further investigation.

10. *References*

Abdi, K. and Herrmann, S.H. (1997). CTL generation in the presence of IL-4 is inhibited by free p40: evidence for early and late IL-12 function. J. Immunol. 159(7), 3148–3155.

Abe, E., De Waal Malefyt, R., Matsuda, I., Arai, K. and Arai, N. (1992). An 11-base-pair DNA sequence motif apparently unique to the human interleukin-4 gene confers responsiveness to T-cell activation signals. Proc. Natl Acad. Sci. USA 89, 2864–2868.

Abramson, S.L. and Gallin, J.I. (1990). IL-4 inhibits superoxide production by human mononuclear phagocytes. J. Immunol. 144, 625–630.

Arai, K. (1987). Structural analysis of the mouse chromosomal gene encoding interleukin 4 which expresses B cell, T cell and mast cell stimulating activities. Nucleic Acids Res. 15, 333–339.

Arai, N., Nomura, D., Villaret, D., *et al.* (1989). Complete nucleotide sequence of the chromosomal gene for human IL-4 and its expression. J. Immunol. 142(1), 274–282.

Armitage, R.J., Beckmann, M.P., Idzerda, R.L., Alpert, A. and Fanslow, W.C. (1990). Regulation of interleukin-4 receptors on human T cells. Int. Immunol. 2, 1039–1045.

Arruda, S. and Ho, J.L. (1992). IL-4 receptor signal transduction in human monocytes is associated with protein kinase C translocation. J. Immunol. 149, 1258–1264.

Atkins, M.B., Vachino, G., Tilg, H.J., *et al.* (1992). Phase I evaluation of thrice-daily intravenous bolus interleukin-4 in patients with refractory malignancy. J. Clin. Oncol. 10, 1802–1811.

Backmann, M.P. (1990). Human interleukin-4 receptor confers biological responsiveness and defines a novel receptor superfamily. J. Exp. Med. 171, 861–869.

Banchereau, J. (1990). Human interleukin-4 and its receptor. In "Hematological Growth Factors in Clinical Applications" (eds. R. Mertelsmann and F. Herrman), pp. 433–469. New York, Marcel Dekker.

Banchereau, J., Brière, F., Galizzi, J.P., Miossec, P. and Rousset, F. (1994). Human interleukin-4. J. Lipid Mediators Cell Signal. 9, 43–53.

Bàrcena, A., Toribio, M.-L., Pezzi, L. and Martinez, A.-C. (1990). A role for interleukin-4 in the differentiation of mature T cell receptor γ/δ^+ cells from human intrathymic T cell precursors. J. Exp. Med. 172, 439–464.

Baskar, P., Silberstein, D.S. and Pincus, S.H. (1990). Inhibition of IgG-triggered human eosinophil function by IL-4. J. Immunol. 144, 2321–2326.

Bazan, J.F. (1990). Structural design and molecular evolution of a cytokine receptor superfamily. Proc. Natl Acad. Sci. USA 87, 6934–6938.

Ben-Sasson, S., Le Gros, G., Conrad, D.H., Finkelman, F.D. and Paul, W.E. (1990). Cross-linking Fc receptors stimulate splenic non-B, non-T cells to secrete interleukin-4 and other lymphokines. Proc. Natl Acad. Sci. USA 87, 1421–1425.

Bober, L.A., Waters, T.A., Pugliese-Sivo, C.C., Sullivan, L.M., Narula, S.K. and Grace, M.J. (1995). IL-4 induces neutrophilic maturation of HL-60 cells and activation of human peripheral blood neutrophils. Clin. Exp. Immunol. 99(1), 129–136.

Boey, H., Rosenbaum, R., Castracane, J. and Borish, L. (1989). Interleukin-4 is a neutrophil activator. J. Allergy Clin. Immunol. 83, 978–984.

Borger, P., Kauffman, H.F., Postma, D.S. and Vellenga, E. (1996). Interleukin-4 gene expression in activated human T lymphocytes is regulated by the cyclic adenosine monophosphate-dependent signaling pathway. Blood 87(2), 691–698.

Bradding, P., Feather, I.H., Wilson, S., *et al.* (1993). Immunolocalization of cytokines in the nasal mucosa of normal and perennial rhinitic subjects. J. Immunol. 151, 3853–3865.

Brown, M.A. and Hural, J. (1997). Functions of IL-4 and control of its expression. Crit. Rev. Immunol. 17(1), 1–32.

Brown, M.A., Pierce, J.H., Watson, C.J., Falco, J., Ihle, J.N. and Paul, W.E. (1987). B cell stimulatory factor-1/interleukin-4 mRNA is expressed by normal and transformed mast cells. Cell 50, 809–818.

Broxmeyer, H.E., Lu, L., Cooper, S., Rubin, B.Y., Gillis, S. and Williams, D.E. (1988). Synergistic effects of purified recombinant human and murine B-cell growth factor/IL-4 on colony formation *in vitro* by hematopoietic progenitor cells. J. Immunol. 141, 3852–3862.

Callard, R.E. (1989). Increased expression of surface IgM but no IgD or IgG on human B cells in response to interleukin-4. Immunology 66, 224–227.

Callard, R.E. (1991). Immunoregulation by interleukin-4 in man. Br. J. Haematol. 78, 293–299.

Carr, C., Ayken, S., Kimack, N.M. and Levine, A.D. (1990). Disulfide assignments in recombinant mouse and human interleukin-4. Biochemistry 30, 1515–1523.

Casanelli, N., Foà, R. and Pormiani, G. (1994). Active immunization of metastatic melanoma patients with interleukin-4 transduced, allogeneic melanoma cells. A phase I–II study. Hum. Gene. Ther. 5, 1059–1064.

Chilton, P.M. and Fernandez-Botran, R. (1993). The production of soluble interleukin-4 receptors by murine spleen cells is regulated by T cell activation and interleukin-4. J. Immunol. 151, 5907–5913.

Clark, E.A., Shu, G.L., Lüscher, B., et al. (1989). Activation of human B cells. Comparison of the signal transduced by interleukin-4 (IL-4) to four different competence signals. J. Immunol. 143, 3873–3879.

Cluitmans, F.H.M., Esendam, B.H.J., Landegent, J.E., Willemze, R. and Falkenburg, J.H.F. (1994). IL-4 down-regulates IL-2-, IL-3-, and GM-CSF-induced cytokine gene expression in peripheral blood monocytes. Ann. Hematol. 68, 293–298.

Conrad, D.H., Waldschmidt, T.J., Lee, W.T., et al. (1987). Effect of B cell stimulatory factor-1 (interleukin-4) on Fcε, and Fcγ receptor expression on murine B cells and B cell lines. J. Immunol. 139, 2290–2296.

Corcoran, M.L., Stetler-Stevenson, W.G., Brown, P.D. and Wahl, L.M. (1992). Interleukin-4 inhibition of prostaglandin E2 synthesis blocks interstitial collagenase and 92-kDa type IV collagenase/gelatinase production by human monocytes. J. Biol. Chem. 267, 515–519.

Cornacoff, J.B., Gossett, K.A., Barbolt, T.A. and Dean, J.H. (1992). Preclinical evaluation of recombinant human interleukin-4. Toxicol. Lett. 64/65, 299–309.

Corrigan, C.J. (1995). Elevated interleukin-4 secretion by T lymphocytes: a feature of atopy or of asthma? Clin. Exp. Allergy 25(6), 485–487. [Editorial]

Crawford, R.M., Finbloom, D.S., Ohara, J., Paul, W.E. and Meltzer, M.S. (1987). BSF-1: a new macrophage activation factor: B cell stimulatory factor-1 (interleukin-4) activates macrophages for increased tumoricidal activity and expression of Ia antigens. J. Immunol. 139, 135–141.

D'Andrea, A.D., Fasman, G.D. and Lodish, H.F. (1990). A new hematopoietic growth factor receptor superfamily: structural features and implications for signal transduction. Curr. Opin. Cell Biol. 2, 648–651.

Davidson, I., Fromental, C., Augereau, P., Wildeman, A., Zenke, M. and Chambon, P. (1986). Cell-type specific protein binding to the enhancer of simian virus 40 in nuclear extracts. Nature 323, 544–548.

Dawson, C.H., Brown, B.L. and Dobson, P.R. (1997). A 70-kDa protein facilitates interleukin-4 signal transduction in the absence of the common gamma receptor chain. Biochem. Biophys. Res. Commun. 233(1), 279–282.

Dean, J.H., Cornacoff, J.B., Barbolt, T.A., Gossett, K.A. and LaBrie, T. (1992). Pre-clinical toxicity of IL-4: a model for studying protein therapeutics. Int. J. Immunopharmacol. 14, 391–399.

D'Eustachio, P., Brown, M., Watson, C. and Paul, W.E. (1988). The IL-4 gene maps to chromosome 11, near the gene encoding IL-3. J. Immunol. 141, 3067–3073.

Defrance, T., Fluckiger, A.C., Rossi, J.F., Magaud, J.P., Sotto, J.J. and Banchereau, J. (1992). Antiproliferative effects of interleukin-4 on freshly isolated non-Hodgkin malignant B-lymphoma cells. Blood 79, 990–1002.

DeKruyff, R.H., Fang, Y., Secrist, H. and Umetsu, D.T. (1995). IL-4 synthesis by in vivo-primed memory CD4$^+$ T cells: II. Presence of IL-4 is not required for IL-4 synthesis in primed CD4$^+$ T cells. J. Clin. Immunol. 15(2), 105–115.

De Wit, H., Hendriks, D.W., Halie, M.R. and Vellenga, E. (1994). Interleukin-4 receptor regulation in human monocytic cells. Blood 84, 608–615.

De Wolf, J.Th.M., Beentjes, J.A.M., Esselink, M.T., Smith, J.W., Halie, M.R. and Vellenga, E. (1990). IL-4 suppresses the IL-3 dependent erythroid colony formation from normal human bone marrow cells. Br. J. Hematol. 74, 246–250.

Dokter, W.H.A., Esselink, M.T., Sierdsema, S.J., Halie, M.R. and Vellenga, E. (1993). Transcriptional and posttranscriptional regulation of the interleukin-4 and interleukin-3 genes in human T cells. Blood 81, 35–40.

Duschl, A. and Sebald, W. (1996). Transmembrane and intracellular signalling by interleukin-4: receptor dimerization and beyond. Eur. Cytokine. Netw. 7(1), 37–49.

Essner, R., Rhoades, K., McBride, W.H., Morton, D.L. and Economou, J.S. (1990). IL-4 down-regulates IL-1 and TNF gene expression in human monocytes. J. Immunol. 142, 3857–3861.

Estrov, Z., Markowitz, A.B., Kurzrock, R., et al. (1993). Suppression of chronic myelogenous leukemia colony growth by interleukin-4. Leukemia 7, 214–226.

Fanslow, W.C., Clifford, K., VandenBos, T., Teel, A., Armitage, R.J. and Beckmann, M.P. (1990). A soluble form of the interleukin-4 receptor in biological fluids. Cytokine 2, 398–405.

Fanslow, W.C., Clifford, K.N., Park, L.S., et al. (1991). Regulation of alloreactivity in vivo by IL-4 and the soluble IL-4 receptor. J. Immunol. 147, 535–542.

Fanslow, W.C., Spriggs, M.K., Rauch, C.T., et al. (1993). Identification of a distinct low-affinity receptor for human interleukin-4 on pre-B cells. Blood 81, 2998–3005.

Favre, C., Saeland, S., Caux, C., Duvert, V. and De Vries, J.E. (1990). Interleukin-4 has basophilic and eosinophilic cell growth-promoting activity on cord blood cells. Blood 75, 67–-73.

Fenghao, X., Saxon, A., Nguyen, A., Ke, Z., Diaz-Sanchez, D. and Nel, A. (1995). Interleukin 4 activates a signal transducer and activator of transcription (Stat) protein which interacts with an interferon-gamma activation site-like sequence upstream of the I epsilon exon in a human B cell line. Evidence for the involvement of Janus kinase 3 and interleukin-4 Stat. J. Clin. Invest. 96(2), 907–914.

Fenton, M.J., Buras, J.A. and Donnelly, R.P. (1992). IL-4 reciprocally regulates IL-1 and IL-1 receptor antagonist expression in human monocytes. J. Immunol. 149, 1283–1288.

Fernandez-Botran, R. and Vitetta, E.S. (1990). A soluble, high-affinity, interleukin-4-binding protein is present in the biological fluids of mice. Proc. Natl Acad. Sci., USA 87, 4202–4213.

Finney, M., Guy, G.R., Michell, R.H., et al. (1990). Interleukin-4 activates human B lymphocytes via transient inositol lipid hydrolysis and delayed cyclic adenosine monophosphate generation. Eur. J. Immunol. 20, 151–158.

Foxwell, B.M.J., Woerly, G. and Ryffel, B. (1989). Identification of interleukin-4 receptor-associated proteins and expression of both high- and low-affinity binding on human lymphoid cells. Eur. J. Immunol. 13, 1637–1641.

Galizzi, J.P., Cabrillat, H., Rousset, F., Ménétrier, C., de Vries, J.E. and Banchereau, J. (1988). IFN-γ and prostaglandin E2 inhibit IL-4-induced expression of FcεR2/CD23 on B lymphocytes through different mechanisms without altering binding of IL-4 to its receptor. J. Immunol. 141, 1982–1989.

Galizzi, J.P., Zuber, C.E., Cabrillat, H., Djossou, O. and Banchereau, J. (1989). Internalization of human interleukin-4 and transient down-regulation of its receptor in the CD23-inducible Jiyoye cells. J. Biol. Chem. 264, 6984–6990.

Galizzi, J.P., Zuber, C.E., Harada, N., et al. (1990). Molecular cloning of a cDNA encoding the human interleukin 4 receptor. Int. Immunol. 2, 669–675.

Gauchat, J.F., Schlagenhauf, E., Feng, N.P., et al. (1997). A novel 4-kb interleukin-13 receptor alpha mRNA expressed in human B, T, and endothelial cells encoding an alternate type-II interleukin-4/interleukin-13 receptor. Eur. J. Immunol. 27(4), 971–978.

Ghosh, A.K., Smith, N.K., Prendiville, J., Thatcher, N., Crowther, D. and Stern, P.L. (1993). A phase I study of recombinant human interleukin-4 administered by the intravenous and subcutaneous route in patients with advanced cancer: immunological studies. Eur. Cytokine Netw. 4(3), 205–211.

Gilleece, M.H., Scarffe, J.H., Ghosh, A., et al. (1992). Recombinant human interleukin-4 (IL-4) given as daily subcutaneous injections – a phase I dose toxicity trial. Br. J. Cancer 66, 204–216.

Goldstein, H. and Kim, A. (1993). Immunoglobulin secretion and phosphorylation of common proteins are induced by IL-2, IL-4, and IL-6 in the factor responsive human B cell line, SKW6.4. J. Immunol. 151, 6701–6711.

Gordon, J., Misslum, M.H., Guy, G.R. and Ledbetter, J.A. (1988). Resting B lymphocytes can be triggered directly through the CD240 (Bp50) antigen: a comparison with IL-4 mediated signalling. J. Immunol. 140, 1425–1430.

Grabstein, K., Eisenman, J., Mochizuki, D., et al. (1986). Purification to homogeneity of B cell stimulating factor. J. Exp. Med. 163, 1405–1411.

Gustchina, A., Zdanov, A., Schalk-Hihi, C. and Wlodawer, A. (1995). A model of the complex between interleukin-4 and its receptors. Proteins 21(2), 140–148.

Harada, N., Castle, B.E., Gorman, D.M., et al. (1990). Expression cloning of a cDNA encoding the murine interleukin-4 receptor based on ligand binding. Proc. Natl Acad. Sci. USA 84, 3365–3371.

Hart, P.H., Vitti, D.R., Burgess, G.A., Whitty, G.A., Picolli, D.S. and Hamilton, J.A. (1989). Potential anti-inflammatory effects of IL-4. Suppression of human monocyte tumor necrosis factor, interleukin-1 and prostaglandin E2. Proc. Natl Acad. Sci. USA 68, 3803–3807.

Hasegawa, M., Fujimoto, M., Kikuchi, K. and Takehara, K. (1997). Elevated serum levels of interleukin 4 (IL-4), IL-10, and IL-13 in patients with systemic sclerosis. J. Rheumatol. 24(2), 328–332.

Hirooka, Y., Kayama, M., Ohga, S., et al. (1993). Deregulated production of interleukin-4 (IL-4) in autoimmune thyroid disease assayed with a new radioimmunoassay. Clin. Chim. Acta 216, 1–11.

Hoffman, R.C., Schalk-Hihi, C., Castner, B.J., et al. (1994). Stoichiometry of the complex of human interleukin-4 with its receptor. FEBS Lett. 347, 17–21.

Hoon, D.S.B., Banex, M. and Okun, E. (1991). Modulation of human melanoma cells by interleukin-4 and in combination with γ–interferon or α–tumor necrosis factor. Cancer Res. 51, 2002–2013.

Hori, T., Paliard, X., de Waal Malefyt, R., Ranes, M. and Spits, H. (1991). Comparative analysis of CD8 expressed on mature CD4+ CD8+ T cell clones cultured with IL-4 and that on CD8+ T cell clones: implications for functional significance of CD8b. Int. Immunol. 3, 737–742.

Horohov, D.W., Crim, J.A., Smith, P.L.S. and Siegel, J.P. (1988). IL-4 (B cell stimulatory factor 1) regulates multiple aspects of influenza virus-specific cell-mediated immunity. J. Immunol. 141, 4217–4223.

Howard, M., Farrar, J., Hilfiker, M., et al. (1982). Identification of a T-cell derived B-cell growth factor distinct from interleukin-2. J. Exp. Med. 155, 914–923.

Hsu, C.C., Yang, B.C., Wu, M.H. and Huang, K.E. (1997). Enhanced interleukin-4 expression in patients with endometriosis. Fertil. Steril. 67(6), 1059–1064.

Imada, M., Estelle, F., Simons, R., Jay, F.T. and Hayglass, K.T. (1995). Allergen-stimulated interleukin-4 and interferon-gamma production in primary culture: responses of subjects with allergic rhinitis and normal controls. Immunology 85(3), 373–380.

Isakson, P.C., Pure, E., Vitetta, S. and Krammer, P.H. (1982). T cell-derived B cell differentiation factor(s): effects on the isotype switch to murine B cells. J. Exp. Med. 155, 734–748.

Isfort, R.J. and Ihle, J.N. (1990). Multiple hematopoietic growth factors signal through tyrosine phosphorylation. Growth Factors 2, 213–218.

Izuhara, K. and Harada, N. (1993). Interleukin-4 (IL-4) induces protein tyrosine phosphorylation of the IL-4 receptor and association of phosphatidylinositol 3-kinase to the IL-4 receptor in a mouse T cell line, HT2. J. Biol. Chem. 268, 13097–13102.

Izuhara, K., Feldmann, R.A., Greer, P. and Harada, N. (1994). Interaction of the c-fes proto-oncogene product with the interleukin-4 receptor. J. Biol. Chem. 269, 18623–18629.

Jacobs, C.A., Lynch, D.H., Roux, E.R., et al. (1991). Characterization and pharmacokinetic parameters of recombinant soluble interleukin-4 receptor. Blood 77, 2396–2406.

Jacobsen, F.W., Keller, J.R., Ruscetti, F.W., Veiby, O.P. and Jacobsen, S.E. (1995). Direct synergistic effects of IL-4 and IL-11 on proliferation of primitive hematopoietic progenitor cells. Exp. Hematol. 23(9), 990–995.

Jansen, J.H., Wientjens, G.J., Fibbe, W.E., Willemze, R. and Kluin-Nelemans, H.C. (1989). Inhibition of human macrophage colony formation by interleukin-4. J. Exp. Med. 170, 577–582.

Justement, L., Chen, Z., Harris, L., et al. (1986). BSF-1 induces membrane protein phosphorylation but not phosphoinositide metabolism, Ca2+ mobilization, protein kinase C translocation, or membrane depolarization in resting murine B lymphocytes. J. Immunol. 137, 3664–3670.

Kaplan, M.H., Schindler, U., Smiley, S.T. and Grusby, M.J. (1996). Stat6 is required for mediating responses to IL-4 and for development of Th2 cells. Immunity 4(3), 313–319.

Keegan, A.D. and Pierce, J.H. (1994). The interleukin-4 receptor: signal transduction by a hematopoietin receptor. J. Leukocyte Biol. 55, 272–279.

Keegan, A.D., Nelms, K., White, M., Wang, L.-M., Pierce, J.H. and Paul, W.E. (1994). An IL-4 receptor region containing an insulin receptor motif is important for IL-4-mediated IRS-1 phosphorylation and cell growth. Cell 76, 811–820.

Komai-Koma, M., Liew, F.Y. and Wilkinson, P.C. (1995). Interactions between IL-4, anti-CD40, and anti-immunoglobulin as activators of iocomotion of human B cells. J. Immunol. 155(3), 1110–1116.

Kondo, M., Takeshita, T., Ishii, N., et al. (1993). Sharing of the interleukin-2 (IL-2) receptor γ chain between receptors for IL-2 and IL-4. Science 262, 1874–1878.

Kotanides, H. and Reich, N.C. (1993). Requirement of tyrosine phosphorylation for rapid activation of a DNA binding factor by IL-4. Science 262, 1265–1268.

Kotanides, H., Moczygemba, M., White, M.F. and Reich, N.C. (1995). Characterization of the interleukin-4 nuclear activated factor/STAT and its activation independent of the insulin receptor substrate proteins. J. Biol. Chem. 270(33), 19481–19486.

Kraulis, P. (1991). MOLSCRIPT: a programme to produce both detailed and schematic plots of protein structures. J. Appl. Crystallog. 24, 946–950.

Kruse, N., Lehrnbecher, T. and Sebald, W. (1991). Site-directed mutagenesis reveals the importance of disulfide bridges and aromatic residues for structure and proliferative activity of human interleukin-4. FEBS Lett. 286, 58–60.

Kubo, M., Kincald, R.L., Webb, D.R. and Ransom, J.T. (1993). The Ca^{2+}/calmodulin-activated, phosphoprotein phosphatase calcineurin is sufficient for positive transcriptional regulation of the mouse IL-4 gene. Int. Immunol. 6, 179–183.

Kubo, M., Kincaid, R.L. and Ransom, J.T. (1994). Activation of the interleukin-4 gene is controlled by the unique calcineurin-dependent transcriptional factor NF(P). J. Biol. Chem. 269(30), 19441–19446.

Kubo, M., Ransom, J., Webb, D., et al. (1997). T-cell subset-specific expression of the IL-4 gene is regulated by a silencer element and STAT6. EMBO. J. 16(13), 4007–4020.

Kuhn, R., Rajewsky, K. and Muller, W. (1991). Generation and analysis of interleukin-4 deficient mice. Science 254(5032), 707–710.

Lauener, R.P., Goyert, S.M., Geha, R.S. and Vercelli, D. (1990). Interleukin-4 down-regulates the expression of CD14 in normal human monocytes. Eur. J. Immunol. 20, 2375–2381.

Le, H.V., Ramanathan, L., Labdon, J.E., et al. (1988). Isolation and characterization of multiple variants of recombinant human interleukin-4 expressed in mammalian cells. J. Biol. Chem. 263(22), 10817–10823.

Le Beau, M.M., Lemons, E.R.S., Espinosa, R., III., Larson, R.A., Arai, N. and Rowley, J.D. (1988). IL-4 and IL-5 map to human chromosomes 5 in a region encoding growth factors and receptors and are deleted in myeloid leukemias with a del(5q). Blood 73, 647–650.

Leach, M.W., Snyder, E.A., Sinha, D.P. and Rosenblum, I.Y. (1997a). Safety evaluation of recombinant human interleukin-4. I. Preclinical studies. Clin. Immunol. Immunopathol. 83(1), 8–11.

Leach, M.W., Rybak, M.E. and Rosenblum, I.Y. (1997b). Safety evaluation of recombinant human interleukin-4. II. Clinical studies. Clin. Immunol. Immunopathol. 83(1), 12–14.

Lee, C.-E., Yoon, S.-R. and Pyun, K.-H. (1993). Interleukin-4 signals regulating CD23 gene expression in human B cells: protein kinase C-independent signalling pathways. Cell. Immunol. 146, 171–185.

Lee, F., Yokota, T., Otsuka, T., et al. (1986). Isolation and characterization of a mouse interleukin cDNA clone that expresses B-cell stimulatory factor-1 activities and T cell- and mast-cell-stimulating activities. Proc. Natl Acad. Sci. USA 83, 2061–2066.

Lewis, D.B., Prickett, K.S., Larsen, A., Grabstein, K., Weaver, M. and Wilson, C.B. (1988). Restricted production of interleukin-4 by activated human T cells. Proc. Natl Acad. Sci. USA 85, 9743–9747.

Link, J., Soderstrom, M., Olsson, I., Hojeberg, R., Ljungdahl, A. and Link, H. (1994). Increased transforming growth factor-beta, interleukin-4 and interferon-gamma in multiple sclerosis. Ann. Neurol. 36, 379–386.

Li-Weber, M., Eder, A., Krafft-Czepa, H. and Krammer, P.H. (1992). T cell-specific negative regulation of transcription of the human cytokine IL-4. J. Immunol. 148(6), 1913–1918.

Li-Weber, M., Davydov, I.V., Krafft, H. and Krammer, P.H. (1994). The role of NF-Y and IRF-2 in the regulation of human IL-4 gene expression. J. Immunol. 153, 4122–4133.

Littman, B.H., Dastvan, F.F., Carlson, P.L. and Sanders, K.M. (1989). Regulation of monocyte/macrophage C2 production and HLA-DR expression by IL-4 (BSF-1) and IFN-γ. J. Immunol. 142, 520–525.

Liu, J., Flanagan, W.M., Drazba, J.A., et al. (1997). The CDK inhibitor, p27Kip1, is required for IL-4 regulation of astrocyte proliferation. J. Immunol. 159(2), 812–819.

Llorente, L., Mitjavila, F., Crevon, M.-C. and Galanaud, P. (1990). Dual effects of interleukin-4 on antigen-activated human B cells: induction of proliferation and inhibition of interleukin-2-dependent differentiation. Eur. J. Immunol. 20, 1887–1892.

Lundgren, M., Persson, U., Larsson, P., et al. (1989). Interleukin-4 induces synthesis of IgE and IgG4 in human B cells. Eur. J. Immunol. 19, 1311–1315.

Luo, H.Y., Rubio, M., Biron, G., Delespesse, G. and Sarfati, M. (1991). Antiproliferative effect of interleukin-4 in B chronic lymphocytic leukemia. J. Immunother. 10, 418–426.

MacGlashan, D.M., Jr., White, J.M., Huang, S.K., Ono, S.J., Schroeder, J.T., and Lichtenstein, L.M. (1994). Secretion of IL-4 from human basophils. The relationship between IL-4 mRNA and protein in resting and stimulated basophils. J. Immunol. 152, 3006–3016.

Malabarba, M.G., Kirken, R.A., Rui, H., et al. (1995). Activation of JAK3, but not JAK1, is critical to interleukin-4 (IL4) stimulated proliferation and requires a membrane-proximal region of IL4 receptor alpha. J. Biol. Chem. 270(16), 9630–9637.

Maliszewski, C.R., Morrissey, P.J., Fanslow, W.C., Sato, T.A., Willis, C. and Davidson, B. (1992). Delayed allograft rejection in mice transgenic for a soluble form of the IL-4 receptor. Cell Immunol. 143, 434–441.

Maliszewski, C.R., Sato, T.A., Davison, B., Jacobs, C.A., Finkelman, F.D. and Fanslow, W.C. (1994). In vivo biological effects of recombinant soluble interleukin-4 receptor (43750). Journal, 233.

Margolin, K., Aronson, F.R., Sznol, M., et al. (1994). Phase II studies of recombinant human interleukin-4 in advanced renal cancer and malignant melanoma. J. Immunol. 147–159.

Martinez, O.M., Gibbons, R.S., Garovoy, M.R. and Aronson, F.R. (1990). IL-4 inhibits IL-2 receptor expression and IL-2-dependent proliferation of human T cells. J. Immunol. 144, 2211–2215.

Matsumoto, K., Taki, F., Miura, M., Matsuzaki, M. and Takagi, K. (1994). Serum levels of soluble IL-2R, IL-4, and soluble FcεRII in adult bronchial asthma. Chest 105, 681–686.

Mire-Sluis, A.R. and Thorpe, R. (1991). Interleukin-4 proliferative signal transduction involves the activation of a tyrosine-specific phosphatase and the dephosphorylation of an 80-kDa protein. J. Biol. Chem. 266, 18113–18118.

Mire-Sluis, A.R., Page, L., Wadhwa, M. and Thorpe, R. (1994). Transforming growth factor-β1 blocks interleukin-4 induced cell proliferation by inhibiting a protein tyrosine phosphatase essential for signal transduction. Cytokine 6, 389–398.

Mitchell, L.C., Davis, L.S. and Lipsky, P.E. (1989). Promotion of human T lymphocyte proliferation by IL-4. J. Immunol. 142, 1548–1557.

Mizuguchi, J., Beaven, M.A., Ohara, J. and Paul, W.E. (1986). BSF-1 action on resting B cells does not require elevation of inositol phospholipid metabolism or increased $[Ca^{2+}]_i$. J. Immunol. 137, 2215–2220.

Morla, A.O., Schreurs, J., Miyajima, A. and Wang, J.Y.J. (1988). Hematopoietic growth factors activate the tyrosine phosphorylation of distinct sets of proteins in interleukin-3-dependent murine cell lines. Mol. Cell. Biol. 88, 2214–2221.

Mosley, B., Beckmann, P., March, C.J., et al. (1989). The murine interleukin-4 receptor: molecular cloning and characterization of secreted and membrane bound forms. Cell 59, 335–342.

Müller, T., Dieckmann, T., Sebald, W. and Oschkinat, H. (1994). Aspects of receptor binding and signalling of interleukin-4 investigated by site-directed mutagenesis and NMR spectroscopy. J. Mol. Biol. 237, 423–436.

Murata, T., Noguchi, P.D. and Puri, R.K. (1996). IL-13 induces phosphorylation and activation of JAK2 Janus kinase in human colon carcinoma cell lines: similarities between IL-4 and IL 13 signaling. J. Immunol. 156(8), 2972–2978.

Nagler, A., Lanier, L.L. and Phillips, J.H. (1988). The effects of IL-4 on human natural killer cells. A potent regulator of IL-2 activation and proliferation. J. Immunol. 141, 2349–2356.

Needleman, B.M., Wigley, F.M. and Stair, R.W. (1992). Interleukin-1, interleukin-2, interleukin-4, interleukin-6, tumour necrosis factor α, and interferon-γ levels in sera from patients with scleroderma. Arthritis Rheum. 35, 35–43.

Nielsen, O.H., Koppen, T., Rudiger, N., Horn, T., Eriksen, J. and Kirman, I. (1996). Involvement of interleukin-4 and -10 in inflammatory bowel disease. Dig. Dis. Sci. 41(9), 1786–1793.

Noelle, R., Krammer, P.H., Ohara, J., Uhr, J.W. and Vitetta, E.S. (1984). Increased expression of Ia antigens on resting B cells: an additional role for B-cell growth factor. Proc. Natl Acad. Sci. USA 81, 6149–6154.

Noma, Y., Sideras, T., Naito, T., et al. (1986). Cloning of cDNA encoding the murine IgG1 induction factor by a novel strategy using SP6 promotor. Nature 319, 640–644.

Nonaka, M., Nonaka, R., Woolley, K., et al. (1995). Distinct immunohistochemical localization of IL-4 in human inflamed airway tissues. IL-4 is localized to eosinophils in vivo and is released by peripheral blood eosinophils. J. Immunol. 155(6), 3234–3244.

Obiri, N.I., Hillman, G., Haas, G., Sud, S. and Puri, R.K. (1993). Expression of high affinity interleukin-4 receptors on human renal carcinoma cells and inhibition of tumor cell growth in vitro by IL-4. J. Clin. Invest. 91, 88–98.

Ohara, J. (1989). Interleukin-4: molecular structure and biochemical characteristics, biological function and receptor expression. In "The Year in Immunology. Immunoregulatory Cytokines and Cell Growth" (eds. J.M. Cruse and R.E. Lewis, Jr.), pp.126–159. Karger, Basel.

Ohara, J. and Paul, W.E. (1987). Receptors for B cell stimulatory factor-1 expressed on cells of hematopoietic lineage. Nature 325, 537–541.

Ohara, J. and Paul, W.E. (1988). Up-regulation of interleukin-4/B-cell stimulatory factor 1 receptor expression. Proc. Natl Acad. Sci. USA 85, 8221–8226.

Ohara, J., Coligan, J.E., Zoon, K., Maloy, W.L. and Paul, W.E. (1987). High-efficiency purification and chemical characterization of B cell stimulatory factor-1/interleukin 4. J. Immunol. 139, 1127–1132.

Okabe, M., Saiki, I. and Miyazaki, T. (1992). Inhibitory anti-tumor effects of interleukin-4 on Philadelphia chromosome-positive acute lymphocytic leukemia and other hematopoietic malignancies. Leuk. Lymphoma 8, 57–69.

Paganin, C., Matteucci, C., Cenzuales, S., Mantovani, A. and Allavena, P. (1994). IL-4 inhibits binding and cytotoxicity of NK cells to vascular endothelium. Cytokine 6, 135–140.

Paliard, X., de Waal Malefyt, R., de Vries, J.E. and Spits, H. (1988). Interleukin-4 mediates CD8 induction on human CD4+ T clones. Nature 335, 642–644.

Palmer-Crocker, R.L. and Pober, J.S. (1995). IL-4 induction of VCAM-1 on endothelial cells involves activation of a protein tyrosine kinase. J. Immunol. 154(6), 2838–2845.

Park, L.S., Friend, D., Sassenfeld, H.M. and Urdal, D.L. (1987). Characterization of the human B cell stimulatory factor 1 receptor. J. Exp. Med. 166, 476–488.

Paul, W.E. (1984). Nomenclature of lymphokines which regulate B-lymphocytes. Mol. Immunol. 21, 343–348.

Paul, W.E. (1991). Interleukin-4: a prototypic immunoregulatory lymphokine. J. Am. Soc. Haematol. 77(9), 1859–1870.

Pène, J., Rousset, F., Bière, F., et al. (1988a). IgE production by normal human lymphocytes is induced by interleukin-4 and suppressed by interferons γ and α and prostaglandin E2. Proc. Natl Acad. Sci. USA 85, 6880–6884.

Pène, J., Crétien, I., Rousset, F., Brière, F., Bonnefoy, J.Y. and de Vries, J.E. (1988b). Modulation of IL-4 induced human IgE production in vitro by IFN-γ and IL-5: the role of soluble CD23 (s-CD23). Eur. J. Immunol. 142, 1558–1564.

Peschel, C., Paul, W.E., Ohara, J. and Green, I. (1987). Effects of B cell stimulatory factor-1/interleukin-4 on hematopoietic progenitor cells. Blood 70, 254–260.

Phillips, J.H., Nagler, A., Spits, H. and Lanier, L.L. (1992). Immunomodulating effects of IL-4 on human natural killer cells. In "IL-4: Structure and Function" (ed. H. Spits), pp.169–185. CRC Press, Ann Arbor.

Plaut, M., Pierce, J.H., Watson, C.J., Hanley-Hyde, J., Nordan, R.P. and Paul, W.E. (1989). Mast cell lines produce lymphokines in response to cross linkage of FcεRI or to calcium ionophores. Nature 339, 64–67.

Prendiville, J., Thatcher, N., Lind, M., et al. (1993). Recombinant human interleukin-4 (rhuIL-4) administered by the intravenous and subcutaneous routes in patients with advanced cancer – a phase I toxicity study and pharmacokinetic analysis. Eur. J. Cancer 29A, 1700–1711.

Pritchard, M.A., Baker, E., Whitmore, S.A., et al. (1991). The interleukin-4 receptor gene (IL-4R) maps to 16p11.2-16p12.1 in human and to the distal region of mouse chromosome 7. Genomics 10, 801–806.

Punnonen, J., Aversa, G. and de Vries, J.E. (1993). Human pre-B cells differentiate into Ig-secreting plasma cells in the presence of interleukin-4 and activated CD4⁺ T cells or their membranes. Blood 82, 2781–2789.

Puri, R.K. and Siegel, J.P. (1993). Interleukin-4 and cancer therapy. Cancer Invest. 11, 473–481.

Rajan, N., Tsarbopoulos, A., Kumarasamy, R., et al. (1995). Characterization of recombinant human interleukin 4 receptor from CHO cells: role of N-linked oligosaccharides. Biochem. Biophys. Res. Commun. 206, 694–702.

Rapoport, M.J., Jaramillo, A., Zipris, D., et al. (1993). Interleukin-4 reverses T cell proliferative unresponsiveness and prevents the onset of diabetes in nonobese diabetic mice. J. Exp. Med. 178, 87–95.

Redfield, C., Smith, R.A.G. and Dobson, C.M. (1994a). Structural characterization of a highly-ordered "molten globule" at low pH. Struct. Biol. 1(1), 23–29.

Redfield, C., Smith, L.J., Boyd, J., et al. (1994b). Analysis of the solution structure of human interleukin-4 determined by heteronuclear three-dimensional nuclear magnetic resonance techniques. J. Mol. Biol. 238, 23–41.

Reichel, M., Nelson, B.H., Greenberg, P.D. and Rothman, P.B. (1997). The IL-4 receptor alpha-chain cytoplasmic domain is sufficient for activation of JAK-1 and STAT6 and the induction of IL-4-specific gene expression. J. Immunol. 158(12), 5860–5867.

Reittie, J.E. and Hoffbrand, A.V. (1994). Interleukin-4 (IL-4) inhibits proliferation and spontaneous cytokine release by chronic lymphocytic leukaemia cells. Leukemia Res. 18, 55–62.

Rennick, D., Yang, G., Muller, S.C., et al. (1987). Interleukin-4 (B-cell stimulatory factor 1) can enhance or antagonize the factor-dependent growth of hemopoietic progenitor cells. Proc. Natl Acad. Sci. USA 84, 6889–6893.

Rennick, D., Jackson, J., Yang, G., Wideman, J., Lee, F. and Hudak, S. (1989). Interleukin-6 interacts with interleukin-4 and other hematopoietic growth factors to selectively enhance the growth of megakaryocytic, erythroid, myeloid, and multipotential progenitor cells. Blood 73, 1828–1835.

Renz, H., Jujo, K., Bradley, K.L., Domenico, J., Gelfand, F.W. and Leung, D.Y. (1992). Enhanced IL-4 production and IL-4 receptor expression in atopic dermatitis and their modulation by interferon-gamma. J. Invest. Dermatol. 99, 403–408.

Renz, H., Enssle, K., Lauffer, L., Kurrle, R. and Gelfand, E.W. (1995). Inhibition of allergen-induced IgE and IgG1 production by soluble IL-4 receptor. Int. Arch. Allergy Immunol. 106(1), 46–54.

Reusch, A., Arnold, S., Heusser, C., Wagner, K., Weston, B. and Sebald, W. (1994). Neutralizing monoclonal antibodies define two different functional sites of human interleukin-4. J. Biochem. 222, 491–499.

Riancho, J.A., Zarrabeitia, M.T., Olmos, J.M., Amado, J.A. and Gonzalez-Macias, J. (1993). Effects of interleukin-4 on human osteoblast-like cells. Bone Miner. 21, 53–61.

Riancho, J.A., Gonzalez-Marcias, J., Amado, J.A., Olmos, J.M. and Fernandez-Luna, J.L. (1995). Interleukin-4 as a bone regulatory factor: effects on murine osteoblast-like cells. J. Endocrinol. Invest. 18(3), 174–179.

Robinson, A. (1994). Interleukin-4: new diabetes champion? Can. Med. Assoc. J. 150, 219–226.

Roehm, N.W., Liebson, J., Zlotnik, A., Kappler, J., Marrack, P. and Cambier, C. (1984). Interleukin-induced increase in Ia expression by normal mouse B cells. J. Exp. Med. 160, 679–694.

Russell, S.M., Keegan, A.D., Harada, N., et al. (1993). Interleukin-2 receptor γ chain: A functional component of the interleukin-4 receptor. Science 262, 1880.

Salerno, A., Dieli, F., Sireci, G., Bellavia, A. and Asherson, G.L. (1995). Interleukin-4 is a critical cytokine in contact sensitivity. Immunology 84(3), 404–409.

Sato, T.A., Widmer, M.B., Finkelman, F.D., et al. (1993). Recombinant soluble murine IL-4 receptor can inhibit or enhance IgE responses in vivo. J. Immunol. 150, 2717–2726.

Schnyder, B., Lahm, H., Woerly, G., Odartchenko, N., Ryffel, B. and Car, B.D. (1996). Growth inhibition signalled through the interleukin-4/interleukin-13 receptor complex is associated with tyrosine phosphorylation of insulin receptor substrate-1. Biochem. J. 315(3), 767–774.

Schultz, C.L. and Coffman, R.L. (1992). Mechanisms of murine isotype regulation by IL-4. In "IL-4: Structure and Function" (ed. H. Spits), pp.15–35. CRC Press, Ann Arbor.

Seder, R.A., Boulay, J.-L., Finkelman, F., et al. (1992). CD8⁺ T cells can be primed in vitro to produce IL-4. J. Immunol. 148(6), 1652–1656.

Shields, J.G., Armitage, R.J., Jamieson, B.N., Beverley, P.C. and Callard, R.E. (1989). Increased expression of surface IgM but not IgD or IgG on human B cells in response to IL-4. Immunology 66, 224–228.

Smith, L.J., Redfield, C., Smith, R.A.G., et al. (1994). Comparison of four independently determined structures of human recombinant interleukin-4. Struct. Biol. 1(5), 301–310.

Sonada, Y., Okuda, T., Yokota, S., et al. (1990). Actions of human interleukin-4/B-cell stimulatory factor-1 on proliferation and differentiation of enriched hematopoietic progenitor cells in culture. Blood 75, 1615–1621.

Sosman, J.A., Fisher, S.G., Kefer, C., Fisher, R.I. and Ellis, T.M. (1994). A phase I trial of continuous infusion interleukin-4 (IL-4) alone and following interleukin-2 (IL-2) in cancer patients. Ann. Oncol. 5, 447–452.

Spits, H. (ed.) (1992). In "IL-4: Structure and Function". CRC Press, Ann Arbor.

Spits, H., Yssel, H., Takebe, Y., et al. (1987). Recombinant interleukin-4 promotes the growth of human T cells. J. Immunol. 139, 1142–1147.

Spits, H., Yssel, H., Paliard, X., Kastelein, R., Figdor, C. and de Vries, J.E. (1988). IL-4 inhibits IL-2 mediated induction of human lymphokine activated killer cells, but not the generation of antigen specific cytotoxic T lymphocytes in mixed leukocyte cultures. J. Immunol. 141, 29–35.

Spits, H., Paliard, X. and de Vries, J.E. (1989). Antigen-specific, but not natural killer, activity of T cell receptor-γδ cytotoxic T lymphocyte clones involves secretion of Nα-benzyloxycarbonyl-L-lysine thiobenzyl ester serine esterase and influx of Ca²⁺ ions. J. Immunol. 143, 1506–1511.

Stadler, W.M., Rybak, M.E. and Vogelzang, N.J. (1995). A phase II study of subcutaneous recombinant human interleukin-4 in metastatic renal cell carcinoma. Cancer 76(9), 1629–1633.

Standiford, T.J., Strieter, R.M., Chensue, S.E., Westwick, J., Kasahara, K. and Kunkel, S.L. (1990). IL-4 inhibits the expression of IL-8 from stimulated human monocytes. J. Immunol. 145, 1435–1439.

Sun, X.J., Wang, L.M., Zhang, Y., et al. (1995). Role of IRS-2 in insulin and cytokine signalling. Nature 377(6545), 173–177.

Szabo, S.J., Gold, J.S., Murphy, T.L. and Murphy, K.M. (1993). Identification of cis-acting regulatory elements controlling interleukin-4 gene expression in T cells: roles for NF-Y and NF-AT^c. Mol. Cell. Biol. 13, 4793–4799.

Takashita, T., Asao, H., Ohtani, K., *et al.* (1992). Cloning of the γ chain of the human IL-2 receptor. Science 257, 379–384.

Takeda, K., Tanaka, T. Shi, W., *et al.* (1996). Essential role of Stat6 in IL-4 signalling. Nature 380(6575), 627–630.

Tan, J., Deleuran, B., Gesser, B., *et al.* (1995). Regulation of human T lymphocyte chemotaxis in vitro by T cell-derived cytokines IL-2, IFN-gamma, IL-4, IL-10, and IL-13. J. Immunol. 154(8), 3742–3752.

Tanaka, M., Lee, K., Yodoi, J., *et al.* (1989). Regulation of Fcε receptor 2 (CD23) expression on a human eosinophilic cell line EoL3 and a human monocytic cell line U937 by transforming growth factor β. Cell. Immunol. 122, 96–107.

Te Velde, A.A., Klomp, J.P.G., Yard, B.A., de Vries, J.E. and Figdor, C. (1988). Modulation of phenotypic and functional properties of human peripheral blood monocytes by IL-4. J. Immunol. 140, 1548–1553.

Te Velde, A.A., Rousset, F., Peronne, C., de Vries, J.E. and Figdor, C.G. (1990a). IFN-α and IFN-γ have different regulatory effects on IL-4-induced membrane expression of FcεRIIb and release of soluble FcεRIIb by human monocytes. J. Immunol. 144, 3052–3059.

Te Velde, A.A., Huijbens, R.J.F., de Vries, J.E. and Figdor, C.G. (1990b). IL-4 decreases FcγR membrane expression and FcγR-mediated cytotoxic activity of human monocytes. J. Immunol. 144, 3046–3051.

Toi, M., Bicknell, R. and Harris, A.L. (1992). Inhibition of colon and breast carcinoma cell growth by interleukin-4. Cancer Res. 52, 275–283.

Topp, M.S., Koenigsmann, M., Mire-Sluis, A., *et al.* (1993). Recombinant human interleukin-4 inhibits growth of some human lung tumor cell lines *in vitro* and *in vivo*. Blood 82, 2837–2845.

Trehu, E.G., Isner, J.M., Mier, J.W., Karp, D.D. and Atkins, M.B. (1993). Possible myocardial toxicity associated with interleukin-4 therapy. J. Immunother. 14, 348–356.

Trotta, P.P. (1992). Physicochemical and structural characteristics of interleukin-4. In "IL-4: Structure and Function" (ed. H. Spits), pp.15–35. CRC Press, Ann Arbor.

Tulpule, A., Joshi, B., DeGuzman, N., *et al.* (1997). Interleukin-4 in the treatment of AIDS-related Kaposi's sarcoma. Ann. Oncol. 8(1), 79–83.

Tunon-de-Lara, J.M., Okayama, Y., McEuen, A.R., Heusser, C.H., Church, M.K. and Walls, A.F. (1994). Release and inactivation of interleukin-4 by mast cells. Ann. N.Y. Acad. Sci. 725, 50–58.

Tsuji, K., Nakahata, T., Takagi, M., *et al.* (1990). Effects of interleukin-3 and interleukin-4 on the development of 'connective tissue-type' mast cells: interleukin-3 supports their survival and interleukin-4 triggers and supports their proliferation synergistically with interleukin-3. Blood 75, 421–427.

Urdal, D.L. and Park, L.S. (1988). Studies on hematopoietic growth factor receptors using human recombinant IL-3, GM-CSF, G-CSF, M-CSF, IL-1 and IL-4. Behring Inst. Mitt. 83, 27–39.

Valle, A., Zuber, C.E., Defrance, T., Djossou, O., De Rie, M. and Banchereau, J. (1989). Activation of human B lymphocytes through CD40 and interleukin-4. Eur. J. Immunol. 19, 1463–1467.

Van Hal, P.Th.W., Hopstaken-Broos, J.P.M., Wijkhuijs, J.M., Te Velde, A.A., Figdor, C.G. and Hoogsteden, H.C. (1992). Regulation of aminopeptidase-N (CD13) and Fc εRIIb (CD23) expression by IL-4 depends on the stage of maturation of monocytes/macrophages. J. Immunol. 149, 1395–1401.

Vellenga, E., De Wolf, J.Th.M., Beentjes, J.A.M., Esselink, M.T., Smit, J.W. and Halie, M.R. (1990). Divergent effects of IL-4 on the G-CSF and IL-3 supported myeloid colony formation from normal and leukemic bone marrow cells. Blood 75, 633–637.

Vellenga, E., Dokter, W. and Halie, R.M. (1993). Interleukin-4 and its receptor: modulating effects on immature and mature hematopoietic cells. Leukemia 7, 1131–1141.

Vercilli, D., Jabara, H.H., Lee, B.-W., Woodland, N., Geha, R.S. and Leung, D.Y.M. (1988). Human recombinant interleukin-4 induces Fc εR2/CD23 on normal human monocytes. J. Exp. Med. 167, 1406–1416.

Vercelli, D., Jabara, H.H., Arai, K.-I., Yokota, T. and Geha, R.S. (1989). Endogenous interleukin-6 plays an obligatory role in interleukin-4 dependent human IgE synthesis. Eur. J. Immunol. 19, 1419–1424.

Walter, M.R., Cook, W.J., Zhao, B.G., *et al.* (1992). Crystal structure of recombinant human interleukin-4. J. Biol. Chem. 267(28), 20371–20376.

Wang, H.Y., Paul, W.E. and Keegan, A.D. (1996). IL-4 function can be transferred to the IL-2 receptor by tyrosine containing sequences found in the IL-4 receptor alpha chain. Immunity 4(2), 113–121.

Wang, L.-M., Keegan, A.D., Paul, W.E., Heidaran, M.A., Gukind, J.S. and Pierce, J.H. (1992). IL-4 activates a distinct signal transduction cascade from IL-3 in factor-dependent myeloid cells. EMBO J. 11, 4899–4903.

Wery, S., Letourneur, M., Bertoglio, J. and Pierre, J. (1996). Interleukin-4 induces activation of mitogen-activated protein kinase and phosphorylation of shc in human keratinocytes. J. Biol. Chem. 271(15), 8529–8532.

Widmer, M.B., Acres, R.B., Sassenfeld, N.M. and Grabstein, K.H. (1987). Regulation of cytolytic cell populations from human peripheral blood by B cell stimulatory factor 1 (interleukin-4). J. Exp. Med. 166, 1447–1455.

Wieser, M., Bonifer, R., Oster, W., Lindemann, A., Mertelsmann, R. and Herrmann, F. (1989). Interleukin-4 induces secretion of CSF for granulocytes and CSF for macrophages by peripheral blood monocytes. Blood 73, 1105–1108.

Windsor, W.T., Syto, R., Durkin, J., *et al.* (1990). Disulfide bond assignment of mammalian cell-derived recombinant human interleukin-4. Biophys. J. 57, 423–429.

Windsor, W., Syto, R., Le, H.V. and Trotta, P.P. (1991). Analysis of the conformation and stability of *E. coli*-derived recombinant human interleukin-4 by circular dichroism. Biochemistry 30, 1259–1264.

Wong, H.L., Lotze, M.T., Wahl, L.M. and Wahl, S.M. (1992). Administration of recombinant IL-4 to humans regulates gene expression, phenotype, and function in circulating monocytes. J. Immunol. 148, 2118–2128.

Yokota, T., Otsuka, T., Mosmann, T., *et al.* (1986). Isolation and characterization of a human interleukin cDNA clone, homologous to mouse B-cell stimulatory factor 1, that expresses B-cell- and T-cell-stimulating activities. Proc. Natl Acad. Sci. USA 83, 5894–5898.

Yokota, T., Arai, N., De Vries, J., *et al.* (1988). Molecular biology of interleukin-4 and interleukin-5 genes and biology of their products that stimulate B cells, T cells and hemopoietic cells. Immunol. Rev. 102, 137–185.

Younes, E., Haas, G.P., Visscher, D., Pontes, J.E., Puri, R.K. and Hillman, G.G. (1995). Intralesional treatment of established murine primary renal tumor with interleukin-4: localized effect on primary tumor with no impact on metastases. J. Urol. 153, 490–493.

Zier, K.S., Leo, M.M., Spielman, R.S. and Baker, L. (1984). Decreased synthesis of interleukin-2 (IL-2) in insulin-dependent diabetes mellitus. Diabetes 33, 552–561.

Zipris, D., Crow, A.R. and Delovitch, T.L. (1991a). Altered thymic and peripheral T-lymphocyte repertoire precedes the onset of diabetes in NOD mice. Diabetes 40, 429–438.

Zipris, D., Lazarus, A.H., Crow, A.R., Hadzija, M. and Delovitch, T.L. (1991b). Defective thymic T cell activation by concanavalin A and anti-CD3 in autoimmune nonobese diabetic mice. Evidence for thymic T cell energy that correlates with the onset of insulitis. J. Immunol. 146(11), 3761–3763.

5. Interleukin-5

Colin J. Sanderson, Stepane Karlen, Sigrid Cornelis, Geert Plaetinck,
Jan Tavernier *and* Rene Devos

1. Introduction

IL-5 is produced by T lymphocytes as a glycoprotein with a molecular mass of 40–45 kDa and is unusual among the T cell-produced cytokines in being a disulfide-linked homodimer. It is a highly conserved member of a group of cytokines which are closely linked on human chromosome 5 (5q31). This gene cluster includes the genes for IL-3, IL-4, IL-9, IL-13, and granulocyte-macrophage colony stimulating factor (GM-CSF).

Two lines of research converged when it was demonstrated that two very different biological activities were properties of this molecule. Three main groups were working with the mouse B cell activity: Takatsu in Japan who considered the activity to be various forms of T cell-replacing factor (TRF) (Takatsu *et al.*, 1988), Swain and Dutton in California, who used the term B cell growth factor II (BCGFII) (Swain *et al.*, 1988), and Vitetta in Texas, who called it B cell differentiation factor (BCDF) (Vitetta *et al.*, 1984). In 1985 Takatsu's group purified TRF and showed that it was identical to BCGFII (Harada *et al.*, 1985). On the other hand, work on the production of eosinophils *in vitro* culminated in the identification of murine eosinophil differentiation factor (EDF) (Sanderson *et al.*, 1985; Warren and Sanderson, 1985). These two lines of research came together in 1986, when EDF was purified and shown to be identical to BCGFII (Sanderson *et al.*, 1986). These data were confirmed when the mouse (Kinashi *et al.*, 1986; Campbell *et al.*, 1988) and human (Azuma *et al.*, 1986; Campbell *et al.*, 1987) molecules were cloned and expressed and designated IL-5.

Thus IL-5 stimulates the production activation and survival of eosinophils in all species studied. It has activity on mouse B cells *in vitro*; however, there is little strong evidence for an activity on mouse B cells *in vivo*. In man there is not yet any convincing evidence for an activity on B cells either *in vitro* or *in vivo*. Thus in man it is highly specific for the eosinophil/basophil lineage, and, while other cytokines have similar activities on eosinophils and basophils *in vitro*, these other cytokines are also active on other cell lineages. The extensive literature on mouse B cells has been reviewed elsewhere (Takatsu, 1992; Takatsu *et al.*, 1992; Sanderson, 1994; Sanderson *et al.*, 1994) and will not be discussed in this review.

2. Gene Structure and Expression

The coding sequence of the IL-5 gene forms four exons (Figures 5.1 and 5.2). The introns show areas of similarity between the mouse and human sequences, although the mouse has a considerable amount of sequence (including repeat sequences) which are not present in the human gene. The mouse gene includes a 738-base-pair segment in the 3′-untranslated region which is not present in the human gene and thus the mouse mRNA is 1.6 kb while the human is 0.9 kb (Campbell *et al.*, 1988). Each of the exons contains the codons for an exact number of amino acids. The gene structure is also shared by other members of the cytokine gene cluster on chromosome 5 in man (Campbell *et al.*,

1987; Sutherland *et al.*, 1988; van Leeuwen *et al.*, 1989; Chandrasekharappa *et al.*, 1990) and chromosome 11 in the mouse (J.S. Lee *et al.*, 1989). Although there is no overall sequence homology, at either the nucleotide or amino acid level, between any of these cytokines, the localization and gene structural similarities suggest a common evolutionary origin (Sanderson *et al.*, 1988), and thus they may be regarded as members of a gene family.

Activation of T cells requires the interaction of the T cell receptor (TCR) complex with antigen (Ag) in association with the major histocompatibility complex, which leads to an increase in intracellular calcium concentration and protein kinase C activation (Perlmutter *et al.*, 1993). These processes can be bypassed by stimulating T cells with phorbol esters (e.g., phorbol myristate acetate, PMA) and ionomycin. However, engagement of accessory molecules on the T cell with additional ligands may deliver additional signals that affect lymphokine gene expression. Such a costimulatory signal can be provided by activating the CD28 ligand expressed on the surface of T lymphocytes (Jenkins and Johnson, 1993). Efficient production of IL-5 requires activation of both the TCR and a second signaling pathway. Both anti-CD28 (Wierenga *et al.*, 1991; Kuiper *et al.*, 1994; Schandené *et al.*, 1994) or cyclic AMP (cAMP) (H.J. Lee *et al.*, 1993) in combination with PMA were shown to be necessary for induction of IL-5 synthesis.

Studies with T cell clones *in vitro* suggest that IL-4

```
   0  ggatcctaatcaagaccccagtgaacagaactcgaccctgccaaggcttggcagtttccatttcaatcactgtcttcccaccagtatttt
      caatttcttttaagacagattaatctagccacagtcatagtagaacatagccgatctgaaaaaaacattcccaatatttatgtattttag
      cataaaattctgtttagtggtctaccttatactttgttttgcacacatcttttaagagggaagttaattttctgattttaagaaatgcaaa
      tgtggggcaatgatgtattaacccaaagattcttcgtaatagaaatgttttttaaaggggggggaaacagggattttttattattaaaagata
      aaagtaaatttatttttaagatataaggcattggaaacatttagtttcacgatatgccattattaggcattctctatctgattgttaga
      aattattcatttcctcaaagacagacaataaattgactggggacgcagtcttgtactatgcactttctttgccaaaggcaaacgcagaac
      gtttcagagccATGAGGATGCTTCTGCATTTGAGTTTGCTAGCTCTTGGAGCTGCCTACGTGTATGCCATCCCCACAGAAATTCCCACAA
      GTGCATTGGTGAAAGAGACCTTGGCACTGCTTTCTACTCATCGAACTCTGCTGATAGCCAATGAG                                     695
```

EXON 1

```
 902  ACTCTGAGGATTCCTGTTCCTGTACATAAAAAT  935
```

EXON 2

```
1887  CACCAACTGTGCACTGAAGAAATCTTTCAGGGAAtAGGCACACTGGAGAGTCAAACTGTGCAAGGGGGTACTGTGGAAAGACTATTCAAA
      AACTtGTCCTTAATAAAGAAATACATTGACGGCCAAAAA  2015
```

EXON 3

```
2120  AAAAGTGTGGAGAAGAAAGACGGAGAGTAAACCAATTCCtAGACTACCTGCAAGAGTTTCTTGGTGTAATGAACACCGAGTGGATAATAGAAAGTTGAga
      ctaaactggtttgttgcagccaaagattttggaggagaaggacattttactgcagtgagaatgagggccaagaaagagtcaggccttaat
      tttcagtataatttaacttcagagggaaagtaaatatttcaggcatactgacactttgccagaaagcataaaattcttaaaatatatttc
      agatatcagaatcattgaagtattttcctccaggcaaaattgatatacttttttcttatttaacttaacattctgtaaaatgtctgttaa
      cttaatagtatttatgaaagatataaggcattggtaaattagtatttatttattttaatgttatgttgtgttctaataaaacaaaaatagacaac
      tgttcaatttgctgctggcctctgtcttagcaattgaagttagcacagtccattgagtacatgcccagtttggaggaagggtctgagcac
      atgtggctgagcatcccccatttctctggagaagtctcaaggttgcaaggcacaccagaggtggaagtgatctagcaggacttagtgggga
      tgtgggggagcagggacacaggcaggaggtgaacctggttttctctctacagtatatccagaacctgggatggtgcagggtaaatggtagg
      gaataaatgaatgaatgtgctttccaagactgattgtagaactaaaatgagttgtaaggcgtccccctggaagaagggcagtgtggggaacc
      tgtaactaggttcctgcccagcctgtgagaagaatttggcagatcaatctcattgccagtatagagaggaagccagaaaccctctctgcc
      aaggcctgcaggggttcttaccccacctgaccctgcaccataacaaaaggaacagagagacactggtagggcagtcccattagaaagact
      gagttccgtattcccggggcagggcagcaccaggccgcacaacactccattctgcctgcttatggctatcagtagcatcactagagatt
      cttctgtttgagaaaacttctcaaggatc  3240
```

EXON 4

Figure 5.1   Sequence of the human IL-5 genomic gene. Coding sequences are shown in uppercase letters. The polyadenylation signal is underlined. Genbank accession no.: J02971.

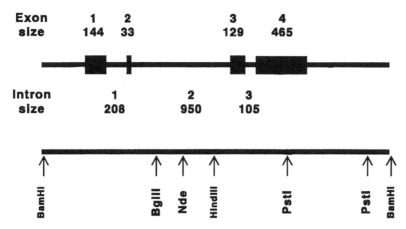

Figure 5.2   Map of the human IL-5 gene. Exons are shown as boxes. Exon and intron lengths are indicated in base pairs, and some restriction enzyme sites are shown in the lower panel.

and IL-5 are often coexpressed in clones designated T_H2 (Coffman *et al.*, 1989), and this provides a possible explanation for the frequent association of eosinophilia with high IgE levels. However, anti-CD3 induced the expression of IL-4, IL-5, and GM-CSF mRNA in mouse T cells, whereas treatment with IL-2 induced IL-5 mRNA expression but did not induce detectable IL-4 or GM-CSF (Bohjanen *et al.*, 1990), suggesting that independent control mechanisms exist for the regulation of IL-4 and IL-5. These differences are emphasized by the observation that cAMP and PMA have different effects on the induction of different cytokines in the mouse lymphoma EL4. PMA induces IL-2 and IL-4 but has only a low effect on IL-5 induction. While cAMP markedly enhances the effect of PMA on the induction of IL-5, it has an inhibitory effect on IL-2, IL-3, and IL-10 (H.J. Lee *et al.*, 1993) and no effect on IL-4 (Chen and Rothenberg, 1994). In addition, T-cell hybrids expressing IL-5 and no other known lymphokine have been produced (Warren and Sanderson, 1985). This observation, together with the fact that eosinophilia can occur without increase in the expression of other lymphokine than IL-5, strongly supports the hypothesis that a unique control exists for this lymphokine.

Corticosteroids inhibit IL-5 production both *in vivo* and *in vitro* (Rolfe *et al.*, 1992; Corrigan *et al.*, 1993; Wang *et al.*, 1993). This may be an important mechanism of corticosteroid activity in asthma. Both progesterone and testosterone induce IL-5 transcription, while dexamethasone is an inhibitor (Wang *et al.*, 1993). FK506 and cyclosporin A (CSA), two powerful inhibitors of T cell activation (Schreiber and Crabtree, 1992), were found to give conflicting results regarding their effect on IL-5 expression. They were found to inhibit the transcription of IL-5 in human peripheral blood mononuclear cells (Mori *et al.*, 1994), while CSA appeared to have no effect in mouse lymphoma cells (Naora and Young, 1994; Naora *et al.*, 1994a,b). On the

other hand, a strong effect of CSA on IL-5 transcription in mouse EL4 lymphoma cells has been observed (Karlen *et al.*, 1995).

Little is known of the mechanisms that regulate expression of IL-5 in activated T cells. Expression of most cytokine genes appear to be predominantly regulated at the transcriptional level (Crabtree, 1989). Analysis of the mouse IL-5 promoter region revealed that a sequence spanning the region −37 to −51 upstream of the RNA initiation site is also present in the promoters of the IL-3, IL-4, and GM-CSF genes (Miyatake *et al.*, 1991). This conserved lymphokine element (CLE0) was found to be essential for GM-CSF (Masuda *et al.*, 1993) and IL-5 (Naora *et al.*, 1994a,b) promoter activity. We recently observed that the mIL-5 CLE0 element mediates the induction signals generated by PMA/cAMP or PMA/anti-CD28 and is recognized by factors related to AP1 (Karlen *et al.*, 1996). The complex binding to the CLE0/AP1 motif was found to contain c-*fos* and *jun*B. As the regulation of *jun*B differs from the regulation of other members of the Jun family (de Groot *et al.*, 1991; Kobierski *et al.*, 1991), it has been suggested that different AP1 complex may be involved in cytokine gene expression and that the specificity for AP1 DNA binding may be provided by different signaling pathways resulting in differential cytokine gene expression (Su *et al.*, 1994). The mIL-5 CLE0/AP1 complex appears therefore to be a key factor for induction of IL-5 transcription.

3.   *Protein Structure*

The human IL-5 precursor is 134 residues long, including a signal peptide of 19 amino acids (Figure 5.3). The proteins show 70% sequence similarity, which is most pronounced in the C-terminal part. This high degree of

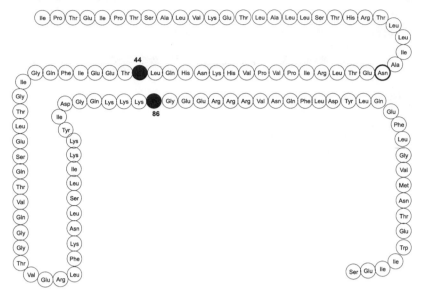

Figure 5.3   Amino acid sequence of human IL-5. Disulfide bonds to the other monomer in the dimer are indicated.

similarity explains the cross-species reactivity and immunological cross-reactivity observed between human and mouse IL-5 (Campbell *et al.*, 1987).

The secreted core polypeptide of hIL-5 corresponds to a calculated molecular mass of 13 149 kDa. Accordingly, *E. coli*-derived protein migrates as a single 13 kDa band on SDS-PAA gels under reducing conditions. However, in the absence of reducing agent, a molecular mass of 27 kDa is found, which represents the dimer (Proudfoot *et al.*, 1991). Likewise, the various recombinant IL-5s expressed by eukaryotic cells have native molecular masses between 35 and 60 kDa, changing to 16–32 kDa after treatment with reducing agents (Tominaga *et al.*, 1990; Ingley and Young, 1991; Kunimoto *et al.*, 1991; Guisez *et al.*, 1993). It has been shown that IL-5 exists as a cross-linked homodimer in an antiparallel (head-to-tail) arrangement (Minamitake *et al.*, 1990; McKenzie *et al.*, 1991a,b) by S-S bridging between cysteine residues 44 and 86. Reduction and alkylation of the two cysteines in recombinant hIL-5 or mutation to threonine residues lead to a biologically inactive monomer (Minamitake *et al.*, 1990; McKenzie *et al.*, 1991a,b), illustrating that dimer formation via disulfide bonds is essential for biological activity.

The variable, high molecular mass range observed for secreted IL-5 is explained by the heterogeneous addition of carbohydrate (*O*-linked glycosylation on Thr-3 and *N*-linked glycosylation on Asn-28) (Minamitake *et al.*, 1990; Tominaga *et al.*, 1990). Mouse IL-5 has an additional *N*-linked glycosylation at position Asn-55, which is absent in hIL-5 (Kodama *et al.*, 1993). Removal of the sugar moieties demonstrates that glycosylation is not necessary for receptor binding and biological activity *in vitro* (Tominaga *et al.*, 1990; Ingley and Young, 1991;

Kunimoto *et al.*, 1991). Deglycosylated CHO-derived hIL-5 even shows a marked increase in activity, with *O*-linked glycosylation identified as the suppressor of bioactivity, while *N*-linked glycosylation appears to contribute to thermostabilization of the protein (Tominaga *et al.*, 1990).

Crystals have been obtained from recombinant hIL-5 purified either from renatured *E. coli* inclusion bodies (Graber *et al.*, 1993; Hassell *et al.*, 1993) or from the supernatant of baculovirus-infected Sf9 cells (Guisez *et al.*, 1993) and the three-dimensional structure has been determined by x-ray diffraction at 2.4 Å resolution (Milburn *et al.*, 1993; Tavernier *et al.*, 1995) (Figure 5.4).

The diffraction data reveal a two-domain structure in which each domain (i.e., monomer) adopts the typical cytokine fold, as described for other members of the interleukin/haematopoietin family (Figure 5.4). Nevertheless, no significant sequence similarity exists between these proteins (Bazan, 1990). Typically, this fold consists of a four-helical bundle in an up–up, down–down array. However, IL-5 is unique among these structurally related cytokines in that one bundle is built up of three helices of one monomer, with the fourth helix (D′-helix swapping in Figure 5.4) being contributed by the second monomer. In addition, IL-5 also contains two short β-strands, located between helices A and B, and C and D, respectively, forming a small antiparallel β-sheet (β₁ with β′₂ in Figure 5.4). In contrast to the noncrystallographic 2-fold axis found for the hIL-5 structure determined from *E. coli*-derived protein (Milburn *et al.*, 1993), hIL-5 from Sf9 cells crystallizes as an exact crystallographic dimer (Tavernier *et al.*, 1995). In the latter case, hIL-5 recovered from the crystals is fully biologically active (Guisez *et al.*, 1993). One explanation for the

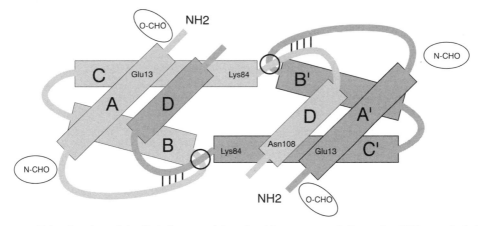

Figure 5.4   Tube drawing of the IL-5 dimer as determined by x-ray crystallography (Milburn *et al.*, 1993).

observed difference might be found in a differential folding behavior, as the *E. coli*-derived hIL-5 requires a denaturing/refolding procedure.

3.1 STRUCTURE–FUNCTION ANALYSIS OF IL-5

Using a series of human/mouse chimeric molecules, the species-specific interaction to the C-terminal one-third of human IL-5 has been investigated (McKenzie *et al.*, 1991a,b). CNBr-cleavage studies confirmed this conclusion (Kodama *et al.*, 1991). By mutating each of the amino acids on hIL-5 to the corresponding mouse residue, the interaction point has been localized to Asp-108, which lies at the C-terminal end of helix D, and to a lesser extent to Lys-84, which lies at the C-terminal end of helix B. Mutating only these two residues of hIL-5 to the mouse equivalent produces a molecule with the properties of mouse IL-5 (Cornelis *et al.*, 1995).

Further analysis by alanine scan has identified additional regions which contribute to the interaction with the α subunit: the loop between β-strand 1 and α-helix B, part of the β-strand 2, and the C-terminal region (Tavernier *et al.*, 1995). The areas identified are located close to the central axis of the hIL-5 dimer.

Although the interaction with the α subunit is a prerequisite for receptor triggering, signaling occurs only upon association with the β_c chain. Mutations at position 13 of hIL-5 are deficient in their interaction with the β_c chain and have antagonistic activities (Tavernier *et al.*, 1995), indicating that this residue is important for interaction with the β_c chain. This observation correlates well with previous reports where homologous residues were identified in GM-CSF (residue 21) (Scanafelt *et al.*, 1991; Lopez *et al.*, 1992c) and IL-3 (residue 22) (Lopez *et al.*, 1992b).

Using a solid-phase-based screening system, a class of isothiazolone derivatives have been identified which are potent inhibitors of the hIL-5/hIL-5Rα interaction. These compounds act through a covalent modification of the sulfhydryl group of the free cysteine at position 66, again pointing to the important role of the N-terminal domain in ligand binding (Devos *et al.*, 1994). Using a panel of neutralizing antibodies, it has been predicted that the helix A–loop 2 region engages the IL-5Rβ_c chain and the loop 3–helix D region engages the IL-5Rα chain (Dickason *et al.*, 1996).

4. *Cellular Sources*

Native mouse IL-5 has been isolated from T cell supernatants (Harada *et al.*, 1985; Sanderson *et al.*, 1985), while human IL-5 has not been purified from natural sources. The recombinant proteins have been expressed using either prokaryotic or eukaryotic expression systems: biologically active recombinant human and mouse IL-5 have been produced in *E. coli* (Proudfoot *et al.*, 1991), *Saccharomyces cerevisiae* (Ingley and Young, 1991; Tavernier *et al.*, 1989), baculovirus-infected Sf9 insect cells (Tavernier *et al.*, 1989; Ingley and Young, 1991; Kunimoto *et al.*, 1991; Guisez *et al.*, 1993), and CHO cells (Minamitake *et al.*, 1990).

Eosinophilia is T cell-dependent and it is therefore not surprising that the controlling factor is a T cell-derived cytokine (Sanderson *et al.*, 1985). It is characteristic of a limited number of disease states, most notably parasitic infections and allergy. Clearly, as eosinophilia is not characteristic of all immune responses, the factors controlling eosinophilia must not be produced by all T cells. Similarly, as it is now clear that IL-5 is the main controlling cytokine for eosinophilia (see below), then if IL-5 has other biological activities it is likely that these will coincide with the production of eosinophils.

All of the original reports on the characterization, purification, and cloning of murine IL-5 utilized T cell lines or lymphomas as the source of material, suggesting that T cells are an important source of the cytokine. The fact that eosinophilia is a T cell-dependent phenomenon illustrates the central role of the T cell in producing IL-5. However, the demonstration that IL-5 and other cytokine mRNAs are produced by mast cell lines opens the possibility that these cells may serve to amplify the development of eosinophilia (Plaut *et al.*, 1989; Burd *et al.*, 1989). Similarly, the observation that human Epstein–Barr virus-transformed B cells produce IL-5 raises the possibility that B cells may be an additional source of this cytokine (Paul *et al.*, 1990). Furthermore, eosinophils themselves have been demonstrated to produce IL-5 (Broide *et al.*, 1992). It is not clear whether the non-T cell-derived IL-5 plays a significant biological role in the development of eosinophilia.

Eosinophilia has been observed in a significant proportion of a wide range of human tumors. In many cases the presence of eosinophils has been found to be of positive prognostic significance (reviewed in Sanderson, 1992). Clearly, it is important to understand the mechanism of production of these eosinophils. In a study of Hodgkin disease with associated eosinophilia, all 16 cases gave a positive signal for IL-5 mRNA by in-situ hybridization (Samoszuk and Nansen, 1990). This suggests that IL-5 may be responsible for the production of eosinophils in these cases, and raises the possibility that eosinophilia in other tumors may also be due to the production of IL-5 by the tumor cells.

5. Biological Activities

One of the features of eosinophilia which has attracted the curiosity of hematologists for several decades is the apparent independence of eosinophil numbers of the numbers of other leukocytes. Thus, eosinophils are present in low numbers in normal individuals but can increase dramatically and independently of the number of neutrophils. Such changes are common during the summer months in individuals with allergic rhinitis (hay fever), or in certain parasitic infections. Clearly, such conditions will result in more broadly based leukocytosis when complicated by other infections. Although this specificity has been known for many years, somewhat surprisingly, it is not easy to find clear examples in the early literature. More recently, in experimental infection of volunteers with hookworms (*Necator americanus*), it was noted that an increase in eosinophils was the only significant change (Maxwell *et al.*, 1987), and our own work with *Mesocestoides corti* in the mouse demonstrated massive increases in eosinophils, independent of changes in neutrophils (Strath and Sanderson, 1986). This biological specificity suggests a mechanism of control that is independent of the control of other leukocytes. This, coupled with the

normally low numbers of eosinophils, provides a useful model for the study of the control of hematopoiesis by the immune system.

Liquid cultures of murine bone marrow produce neutrophils for extended periods without exogenous factors (Dexter *et al.*, 1977). It appears that the microenvironment of these cultures maintains the production of neutrophil precursors. No eosinophils are seen in the absence of exogenous factors, but IL-5 induces their production. This is lineage-specific, as only eosinophil numbers are increased in these cultures (Sanderson *et al.*, 1985). In contrast, both IL-3 and GM-CSF induce eosinophils as well as other cell types, most notably neutrophils and macrophages in bone marrow cultures (Campbell *et al.*, 1988). The production of eosinophils is considerably higher when the bone marrow is taken from mice infected with *M. corti* than it is from normal marrow. This suggests that marrow from infected mice contains more eosinophil precursors than marrow from normal mice (Sanderson *et al.*, 1985). Essentially similar results are obtained with human bone marrow cultures, where IL-5 induces the production of eosinophils in liquid culture (Clutterbuck and Sanderson, 1988; Clutterbuck *et al.*, 1989) and eosinophil colonies in semisolid medium (Clutterbuck and Sanderson, 1990).

5.1 Eosinophil Production *IN VIVO*

As IL-5 is normally a T cell product and the gene is transcribed for only a relatively short period of time after antigen stimulation, transgenic mice in which IL-5 is constitutively expressed by all T cells have been produced (Dent *et al.*, 1990). These mice have detectable levels of IL-5 in the serum. They show a profound and lifelong eosinophilia, with large numbers of eosinophils in the blood, spleen, and bone marrow. This indicates that the expression of IL-5 is sufficient to induce the full pathway of eosinophil differentiation. If other cytokines are required for the development of eosinophilia, then either they must be expressed constitutively or their expression is secondary to the expression of the IL-5 gene. This clear demonstration that the expression of the IL-5 gene in transgenic animals is sufficient for the production of eosinophilia provides an explanation for the biological specificity of eosinophilia. It therefore seems likely that because eosinophilia can occur without a concomitant neutrophilia or monocytosis, a mechanism must exist by which IL-5 is the dominant hematopoietic cytokine produced by the T cell system in natural eosinophilia. An interesting observation with these transgenic animals was that, despite their massive, long-lasting eosinophilia, the mice remained normal. This illustrates that an increased number of eosinophils is not of itself harmful, and that the tissue damage seen in allergic reactions and other diseases must be due to agents which trigger the eosinophils to degranulate.

5.2 MECHANISM OF CONTROL OF EOSINOPHILIA

Although it has become clear that IL-5 is the controlling factor in eosinophilia (Sanderson, 1992), the mechanisms which allow a selective production of IL-5 and thus provide a selective increase in eosinophils remain unknown. Furthermore, in other diseases eosinophils are not observed in significant numbers, suggesting that the expression of IL-5 is not induced in these cases. The cellular pathology typical of particular diseases may reflect the induction of different cytokines by the immune system, in response to different antigenic exposure.

It has been suggested that murine helper T cells fall into two groups (T helper-1 and T helper-2), the former producing predominantly interferon (IFN-γ) and IL-2, and the latter producing IL-4 and IL-5. This hypothesis provides an explanation for the frequently observed association between high levels of IgE antibody and eosinophilia, but does not provide an explanation for many of the complexities seen in the pathology of different diseases. For example, in the mouse IL-4 and IL-5 are thought to be the two major lymphokines in the control of antibody production; thus this classification leaves open the question of how antibody responses are controlled in the absence of eosinophilia. While the variety of cellular and antibody responses in different infections suggests some form of selection for T cells producing different patterns of cytokines, no clear mechanism has emerged to explain this selective response. The association between eosinophilia and helminth infections or allergic reactions could result either from features in common between the antigens involved in these reactions or from features in common between the tissues in which the immune system encounters these antigens. Clearly, understanding of these processes will be an important step forward in our understanding of the immune system, and these preliminary attempts to understand functional subsets in T cells are a basis for future studies.

An important approach to the understanding of the biological role of IL-5, was the administration of neutralizing antibody which blocked the production of eosinophils in mice infected with *Trichinella spiralis* (Coffman *et al.*, 1989). More recently the ablation of the IL-5 gene in mice was shown to block the production of eosinophils in both a lung sensitisation with ovalbumin and in parasite infection. Furthermore, and contrary to previous studies, no obligatory role for IL-5 in antibody responses could be demonstrated (Kopf *et al.*, 1996; Foster *et al.*, 1996). These experiments illustrate the unique role of IL-5 in the control of eosinophilia in this parasite infection. They also show that the apparent redundancy seen *in vitro*, where both IL3 and GM-CSF are also able to induce eosinophil production, does not operate *in vivo*. Furthermore, IL-5 plays no role in the development of IgE antibody (this activity is controlled by

IL-4), or in the development of the granuloma seen surrounding schistosomes in the tissues (Sher *et al.*, 1990).

5.3 ACTIVATION OF EOSINOPHILS

The ability of eosinophils to perform in functional assays can be increased markedly by incubation with a number of different agents, including IL-5. The phenomenon of activation is apparently independent of differentiation. It appears to have a counterpart *in vivo*, as eosinophils from different individuals vary in functional activity. It has been demonstrated that the ability of eosinophils to kill schistosomula increases in proportion to the degree of eosinophilia (Hagan *et al.*, 1985). This is consistent with a common control mechanism for both the production and activation of eosinophils in these cases.

The first observations on selective activation of human eosinophils by IL-5 showed that the ability of purified peripheral blood eosinophils to lyse antibody-coated tumor cells was increased when IL-5 was included in the assay medium (Lopez *et al.*, 1991). Similarly, the phagocytic ability of these eosinophils toward serum-opsonized yeast particles was increased in the presence of IL-5. There was a 90% increase in surface C3bi complement receptors, as well as an approximately 50% increase in the granulocyte functional antigens GFA1 and GFA2. Later studies demonstrated that IL-5 increases "polarization", including membrane ruffling and pseudopod formation, which appear to reflect changes in the cytoskeletal system. IL-5 also induces a rapid increase in superoxide anion production by eosinophils (Lopez *et al.*, 1991). In addition, IL-5 increases the survival of peripheral blood eosinophils.

A further interesting observation in this context was the demonstration that IL-5 is a potent inducer of Ig-induced eosinophil degranulation, as measured by the release of eosinophil-derived neurotoxin (EDN). IL-5 increased EDN release by 48% for secretory IgA and 136% for IgG. This enhancing effect appeared by 15 min and reached a maximum by 4 h (Fujisawa *et al.*, 1990). The finding that secretory IgA can induce eosinophil degranulation is particularly important because eosinophils are frequently found at mucosal surfaces where IgA is the most abundant immunoglobulin.

5.4 BASOPHILS

The most pronounced effect of IL-5 on cells other than eosinophils in man, is the effect on basophils. While early studies suggested that IL-5 induced only eosinophils (Clutterbuck and Sanderson, 1990; Clutterbuck and Sanderson, 1988; Clutterbuck *et al.*, 1989), a detailed study by electron microscopy of cells produced in human cord blood cultures revealed a small number of basophils (Dvorak *et al.*, 1989). Other studies have shown that IL-5 primes basophils for increased histamine production and leukotriene generation (Bischoff *et al.*, 1990; Hirai *et al.*, 1990), and basophils in the blood clearly express the IL-5 receptor (Lopez *et al.*, 1990). Thus while the effect of IL-5 on the production of basophils may be minor, the priming effect on mature basophils appears to be of significance in the allergic response.

In view of the well-characterized activity of IL-5 on mouse B cells, it was surprising when no activity could be demonstrated in a wide range of human B cell assay systems (Clutterbuck *et al.*, 1987). This lack of activity of human IL-5 has been confirmed in many different systems (Hagan *et al.*, 1985; Bende *et al.*, 1992). In contrast to the demonstration of IL-5 receptors on human eosinophils and basophils, discussed above, there are no reports of IL-5 receptors on human B cells. Until the true biological role of IL-5 in the mouse B cell system is understood, it is unlikely that the human activity will be clarified.

6.   IL-5 Receptor

6.1   The IL-5 Receptor Complex

The interleukin-5 receptor (IL-5R) consists of α and β subunits (Figure 5.5). The α subunit is ligand-specific, which it binds with intermediate affinity in man ($K_d \sim$

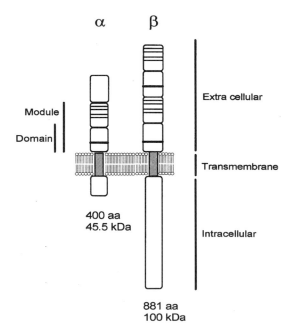

α

β

Module

Domain

Extra cellular

Transmembrane

400 aa
45.5 kDa

Intracellular

881 aa
100 kDa

Figure 5.5   Diagram of the IL-5 receptor, consisting of a ligand specific α chain and the β_c chain. These proteins are made up of ~100 residue domains, and cytokine receptor modules consisting of two domains. The β chain consists of two modules while the α chain has one module and an additional N-terminal domain. WS-WS motif shown as heavy bar, disulfide bonds are indicated by fine lines.

500 pM to 1 nM) and low affinity in the mouse ($K_d \sim$ 5–10 nM). Upon association with the α subunit, a high-affinity binding of approximately 150 pM is observed for both species. The β subunit does not have any detectable affinity for IL-5 by itself (Devos *et al.*, 1991a,b; Takaki *et al.*, 1991; Tavernier *et al.*, 1991; Murata *et al.*, 1992). The dissociation rate of mIL-5 from the α/β-complex is considerably slower ($t_{1/2} > 1$ h) than from the low-affinity binding site ($t_{1/2} < 2$ min) (Devos *et al.*, 1991a,b). IL-5 binds to its receptor with unidirectional species specificity: mIL-5 binds with comparable affinities to both murine and human α subunits, but hIL-5 displays a 100-fold lower binding affinity for the murine α chain, compared to its human counterpart.

The use of the β subunit is shared with IL-3 and GM-CSF (Kitamura *et al.*, 1991; Tavernier *et al.*, 1991), which both also have their own specific α subunits. For this reason, the β subunit is often referred to as β-common or β_c. This observation provides the molecular basis for the overlapping biological activities of these cytokines (Lopez *et al.*, 1992a,b; Nicola and Metcalf, 1991). When the IL-5R α and β_c subunits are coexpressed with the α subunits of IL-3 and GM-CSF, all three ligands will cross-compete for high-affinity binding. Interestingly, this cross-competition follows a hierarchical pattern: IL-3 = GM-CSF > IL-5.

The biological repertoire of IL-5, and also IL-3 and GM-CSF, is dependent on the expression pattern of their receptor α subunits. In the case of human IL-5, this expression is most prominent on eosinophils and basophils. In the mouse, in addition to these cell types, IL-5-specific binding sites are also found on activated B cells and some cell lines (Mita *et al.*, 1988; Rolink *et al.*, 1989). No comparable expression on B cells has been described so far in man. Depending on the relative expression levels of the two receptor subunits, only high- or both intermediate- (low in the mouse) and high-affinity receptors can be observed. In general, on human eosinophils or eosinophilic cell lines, only one class of high-affinity receptor is found (Plaetinck *et al.*, 1990; Ingley and Young, 1991; Lopez *et al.*, 1991; Migita *et al.*, 1991). In contrast, on both murine eosinophils and B cells, two affinities are often detected (Mita *et al.*, 1989; Barry *et al.*, 1991).

6.2   Receptor Subunit Genes

The human IL-5Rα gene is located on chromosome 3 in the region 3p26, which is syntenic with the murine chromosome 6 location (Isobe *et al.*, 1992; Tuypens *et al.*, 1992). The human and mouse β_c subunit gene loci are at chromosomes 22q12.3-13.1 and at chromosome 15, respectively (Gorman *et al.*, 1992; Shen *et al.*, 1992). AIC2A and AIC2B are closely linked. The gene organization of both receptor subunits is highly conserved and reflects their relationship with the cytokine/hematopoietin receptor superfamily (Tuypens *et al.*,

1992; Imamura *et al.*, 1994), although the β_c subunit gene is highly polymorphic (Freeburn *et al.*, 1996).

There is a unique *cis*-element of the IL-5Rα gene that appears to regulate the activity of its promoter activity (Sun *et al.*, 1996). Using a luciferase reporter construct, the promoter was found to be fairly myeloid and eosinophil lineage specific. Furthermore, the region between −432 and −398 was implicated for the promoter activity in the eosinophilic HL-60-Cl15 cell line. This region does not contain consensus sequences for known transcription factors, suggesting myeloid or even eosinophil-specific regulatory elements. Further insight into the transcriptional regulation of this highly eosinophil/basophil-specific gene might help to understand the critical steps involved in the commitment of the multipotential myeloid progenitors.

6.3   ALTERNATIVELY SPLICED FORMS OF IL-5Rα

Human eosinophils express through differential splicing three different transcripts from the same IL-5Rα locus (Tuypens *et al.*, 1992; Tavernier *et al.*, 1992). As a result, in addition to the membrane-anchored receptor, two soluble isoforms can be produced. Intriguingly, one of these soluble variants is the predominant (>90%) transcript detected in mature, circulating eosinophils or in eosinophils obtained from cord blood cultures. This major soluble isoform arises from splicing to a soluble-specific exon, which preceeds the exon encoding the transmembrane anchor (Tavernier *et al.*, 1992).

This soluble hIL-5Rα isoform can be produced in heterologous systems such as Cos1 cells or Sf9 cells, and has antagonistic properties *in vitro*. It inhibits the proliferation of IL-5-dependent cell lines, and also blocks the IL-5-induced differentiation from human cord blood cultures (Tavernier *et al.*, 1991). So far, however, translation of the message encoding this soluble variant in eosinophils *in vitro* or circulating soluble hIL-5Rα *in vivo* has been found. cDNAs encoding soluble receptor variants have also been detected in mouse B cells (Takaki *et al.*, 1990). They originate from a different splicing switch (Tavernier *et al.*, 1992; Imamura *et al.*, 1994).

6.4   RECEPTOR PROTEINS

Chemical cross-linking of hIL-5 on its receptor complex allows the detection of the α and β_c subunits as approximately 100 kDa and 160–170 kDa bands. Subtraction of the cross-linked IL-5 indicates sizes of 60 kDa and 120–130 kDa. The nucleotide sequence of hIL-5Rα cDNA predicts a polypeptide of 420 residues (Tavernier *et al.*, 1991, 1992; Murata *et al.*, 1992). It is characterized by a 20-residue N-terminal signal peptide, followed by a 322-aa long extracellular domain, a membrane anchor spanning 20 residues and a 58-aa long cytoplasmic tail. The predicted molecular mass for the α chain is 45.5 kDa, suggesting *N*-linked gylcosylation of one or more of the 6 potential *N*-glycosylation sites (and perhaps *O*-glycosylation). The β_c subunit totals 897 amino acids, including a predicted signal sequence of 16 residues, a 27-aa long membrane spanning domain and a cytoplasmic domain of 430 amino acids. This subunit is also likely to be glycosylated.

Both receptor chains belong to the cytokine/ hemopoietin receptor superfamily (Bazan, 1990), which is characterized by a modular structure, each module having a seven β-sheet scaffold (de Vos *et al.*, 1992). The hIL-5R α chain has three such domains: a juxtamembrane module containing a canonical Trp-Ser-Xxx-Trp-Ser motif (often referred to as WS-WS box); a central module containing four conserved cysteines involved in two disulfide bridges; and a third N-terminal module which is related to the WS-WS containing module (Tuypens *et al.*, 1992). The β_c subunit has a tandem array of the WS-WS and cysteine-containing modules.

The exact stoichiometry of α and β_c subunits within the receptor complex remains unknown. Despite the 2-fold symmetrical structure of IL-5, which opens the possibility for two receptor binding sites, association with only one α subunit is seen in solution (Devos *et al.*, 1993). A full high-affinity complex including α and β_c subunits cannot be reconstituted in solution (Devos, unpublished data). Perhaps steric hindrance or induced conformational changes may explain this observation. Consequently, the soluble hIL-5Rα subunit can act as an antagonist (see below). Whilst the data from solution binding experiments suggest the presence of only one α subunit in the complex, the stoichiometry of β_c subunits remains more elusive. It is also still unclear whether preformed $\alpha\beta_c$ complexes exist on the cell membrane.

The interaction with the β_c subunit causes a 2–3-fold increase in hIL-5 binding affinity, but gives a much more pronounced affinity conversion in the case of hGM-CSF (20–200-fold) (Hayashida *et al.*, 1990) or hIL-3 (500–1000-fold) (Kitamura *et al.*, 1991). This indicates that differences may exist in the way these three cytokines interact with the same β_c subunit, and correlates with observations on their nonreciprocal cross-competition (see above). In a mutagenesis study, the importance of the region Y^{365}–I^{368} was shown for the GM-CSF and IL-5, but not for IL-3, binding. These residues are located in a loop region, and interact with the Glu-21 of hGM-CSF, and therefore possibly with the homologous Glu-13 of hIL-5 (Woodcock *et al.*, 1994).

6.5   STRUCTURE–FUNCTION ANALYSIS OF IL-5Rα

A cross-species approach was used to localize IL-5 binding regions on the hIL-5R α subunit. Cos1 cells transfected with a series of chimeric receptors were tested for binding of both human IL-5 and mouse IL-5 as a control. It was concluded that the first half of the N-terminal module contains most critical residues involved in species-specific ligand binding. By subsequently substituting all residues differing between human and mouse within this region, the critical role of residues Asp-56 and Glu-58 could be demonstrated (Cornelis *et al.*, 1995).

6.5.1   Identification of Receptor Domains

The β_c chain has a relatively long cytoplasmic domain (~430 amino acid residues) (Gorman *et al.*, 1990; Hayashida *et al.*, 1990; Itoh *et al.*, 1990) compared to the α chains, which have rather short intracellular tails (~55 amino acid residues) (Gearing *et al.*, 1989; Takaki *et al.*, 1990; Kitamura *et al.*, 1991; Tavernier *et al.*, 1991). It has been demonstrated previously that the β_c chain is indispensable for signal transduction. Studies using a sequence of deletions in the human β_c receptor cytoplasmic domain have shown that at least two distinct regions are important for GM-CSF growth signaling (Sakamaki *et al.*, 1992). A membrane-proximal domain upstream of residue 517 containing a proline-rich element that is conserved among many cytokine receptors is essential for growth stimulation and for induction of c-*myc* and *pim*-1. A distal region of ~140 amino acid residues (between positions 626 and 763) is required for activation of *ras*, Raf-1, MAP kinase and p70 S6 kinase, as well as induction of c-*fos* and c-*jun* (Sato *et al.*, 1993). Upon ligand binding, the β_c receptor molecule is tyrosine-phosphorylated although the significance of this event is unclear. Deletion analysis of the β_c receptor cytoplasmic domain suggests that the phosphorylation site is dispensable for mitogenic signaling (Sakamaki *et al.*, 1992). However, it might be important for transducing alternative signals involved in differentiation or activation. Since these cytokines induce tyrosine phosphorylation of SH2-containing proteins (Wang *et al.*, 1992; Corey *et al.*, 1993), phosphorylation of the receptor can generate binding sites for signaling molecules such as p85 PI3 kinase, PLCg, Vav and Shc (Koch *et al.*, 1991). The latter is thought to link receptor triggering to *ras* signaling (Pelicci *et al.*, 1992). Tyrosine phosphorylation may also be involved in downregulation of the receptor as phosphorylation of the mIL-3R increases its susceptibility to protease degradation (Mui *et al.*, 1992).

It has been reported that the cytoplasmic domains of the α subunits of GM-CSF/IL-3/IL-5 receptors are required for signal transduction (Sakamaki *et al.*, 1992; Takaki *et al.*, 1993; Weiss *et al.*, 1993). Deletion of the cytoplasmic domain of the α subunits does not affect the binding affinity but abrogates biological response, suggesting that the α subunits are not only required for cytokine binding but are also important for signal transduction. Moreover, it also opens the possibility that α receptor chains may transmit cytokine-specific signals. In order to map critical residues required for signal transduction in the cytoplasmic domain of the human IL-5R α chain, a series of mutant receptors have been constructed. These were transfected in mouse BaF/3 cells expressing the resident mouse β_c chain (also known under the acronym AIC2B). All receptors were expressed and tested for IL-5 binding. As for the mouse receptor, the cytoplasmic domain of the hIL-5Ra subunit was found to be dispensable for ligand binding and association with the β_c chain. BaF/3 cells transfected with wild-type receptor can proliferate with hIL-5. In contrast, when the hIL-5Rα lacking the entire cytoplasmic tail was used, transfected cells were unable to grow with IL-5. The cytoplasmic domain is encoded by two separate exons. The carboxy-terminal part, which corresponds to the second exon, shows no significant sequence conservation between man and mouse. Deletion of this segment did not affect growth signaling. Within the region encoded by the first exon of the α chain of IL-5, a short fragment of homology with IL-3 and GM-CSF receptors can be observed (Kitamura *et al.*, 1991). This fragment contains a proline-rich element called "box1" that was first identified in the cytoplasmic region of the IL-6 signal transducer gp130 and that is conserved in almost all members of the cytokine receptor family (Murakami *et al.*, 1991), including the β_c subunit. Substitutions of conserved prolines or hydrophobic amino acid residues within this element abrogated growth signaling. Mutations in conserved residues outside box1 did not affect signal transduction.

By analogy with other cytokine receptor systems, it is likely that IL-5 induces heterodimers between α and β chains and that interaction of cytoplasmic regions of the dimerized receptor components is important for signaling. However, receptor complexes involving more than two components cannot be excluded. Despite its 2-fold symmetrical axis, human IL-5 dimer binds only one soluble receptor α subunit (Devos *et al.*, 1993). Therefore, the IL-5R complex probably contains only one α chain. The stoichiometry of binding of the β_c subunit is more elusive at present, also because association with β_c can only be observed with membrane-anchored receptors and not with soluble variants (our own observations). Although no physicochemical evidence for a β-dimer has been found so far, the formation of a GM-CSF receptor complex consisting of one GM-CSF α chain and two β_c receptors has been suggested by Budel *et al.* (1993). A chimeric receptor consisting of the extracellular domain of the murine erythropoietin (EPO) receptor and the cytoplasmic region of AIC2A (a murine β_c subunit homolog) can transmit an EPO-dependent growth response indicating that EPO-induced homodimerization of cytoplasmic β_c elements is sufficient for signaling (Zon *et al.*, 1992). On the other hand, the cytoplasmic region of IL-5Rα is absolutely required for signal transduction (Takaki *et al.*, 1993). It remains to be established whether the proline-rich sequence of IL-5Rα is essential for this dimerization (multimerization) of β_c or whether it interacts directly with a downstream signaling molecule.

In the latter model, cross-linking of two box1-containing receptors, as is the case for both αβ heterodimerization or ββ homodimerization should be sufficient to trigger signaling.

The studies described above all evaluate the role of various receptor components in growth stimulation. However, IL-5 induces not only proliferation but also differentiation in eosinophils. Further studies are required to determine whether IL-5-specific signal transduction pathways exist and, if so, which regions of the cytoplasmic domain of IL-5Rα are required.

7. *Signal Transduction Pathway*

Many tyrosine-phosphorylated proteins have been described in human and mouse cells stimulated with IL-5, GM-CSF, and IL-3 (Morla *et al.*, 1988; Isfort and Ihle, 1990; Murata *et al.*, 1990). The β_c-chain, in contrast to the α-chains, becomes tyrosine phosphorylated upon receptor triggering (Sorensen *et al.*, 1989; Sakamaki *et al.*, 1992). The importance of tyrosine phosphorylation as a major event in signal transduction is illustrated by the observation that specific inhibitors of tyrosine kinases such as herbimycin A inhibit the growth response to these cytokines (Satoh *et al.*, 1991). Activation of tyrosine kinases leads to the activation of Ras and phosphorylation of Raf, a Ser/Thr kinase (Satoh *et al.*, 1991; Kanakura *et al.*, 1991). However, activation of Ras or Raf alone is not sufficient to abrogate the requirements for GM-CSF and IL-3 (Rein *et al.*, 1985), suggesting that these cytokines activate multiple signal transduction pathways. It has been shown that IL-5 induces activation of STAT3 as well as a 45 kDa MAP kinase and Jak-2 kinase (JAK2) in human eosinophils, suggesting a role for tyrosine phosphorylation in IL-5 signaling (Caldenhoven *et al.*, 1995; Bates *et al.*, 1996).

Although most attention has been focused on tyrosine kinases, the action of tyrosine phosphatases is also essential.

The common receptor β_c chain shared by these three cytokines provides a molecular basis for the observation that IL-5, IL-3, and GM-CSF induce similar phosphorylation patterns and have several overlapping biological activities, especially on eosinophils and basophils. However, the fact that these cytokines do exhibit some different activities on these cells suggests that the α chain may have some controlling activity (Seelig *et al.*, 1994).

7.1 IDENTIFICATION OF CYTOPLASMIC COMPONENTS

Despite the absence of kinase domains in their receptors, cytokines rapidly induce tyrosine phosphorylation of cellular substrate proteins as well as of the receptors. The rapid induction of tyrosine phosphorylation, the phosphorylation of receptor chains, and the detection of protein kinase activity in receptor immunoprecipitates have led to the hypothesis that a protein tyrosine kinase physically associates with the receptor and becomes activated upon ligand binding. Recently, a number of studies have shown that cytokine receptors associate with and activate members of the JAK family of protein tyrosine kinases including JAK1, TYK2, JAK3 (previously called L-JAK) and probably more (Ihle *et al.*, 1994). Members of this family have tandem nonidentical catalytic domains and a large extracatalytic segment (Darnell *et al.*, 1994). Involvement of JAK2 tyrosine kinase in IL-3-mediated signal transduction has been reported (Silvennoinen *et al.*, 1993). In Ba/F3 cells transfected with hIL-5Rα, JAK2 also becomes phosphorylated upon IL-5 stimulation. Interestingly, phosphorylated β_c receptor could be coprecipitated with anti-JAK2 antibodies, suggesting that both molecules are physically associated. Mutations in the cytoplasmic domain of the hIL-5Rα that knock out biological activity also abrogate activation of JAK2. This shows that phosphorylation of JAK2 also requires an intact proline-rich box1 sequence. Moreover, it is an indication that JAK2 is a crucial step in IL-5 signaling.

Activated JAKs phosphorylate the receptors as well as cytoplasmic proteins belonging to the family of DNA binding proteins called STATs (signal transducers and activators of transcription) (Ihle *et al.*, 1994). These factors form a complex with homologous or heterologous subunits and migrate to the nucleus, where they bind to specific DNA regions. Importantly, tyrosine phosphorylation is required for complex formation and nuclear translocation. IL-5 as well as IL-3 and GM-CSF induce phosphorylation of a 90 kDa protein in transfected TF1 cells. This size corresponds to what has been reported for STAT proteins. It has been reported that certain STAT proteins recognize the IFN-γ response region (GRR) located in the promoter of the FcγRI gene (Larner *et al.*, 1993). Electrophoretic mobility shift assays with a ^{32}P-labeled GRR probe show that IL-5, IL-3, and GM-CSF activate DNA-binding complexes in cytoplasmic extracts of TF1 cells. Remarkably, stimulation with cytokines and subsequent lysis were carried out at 4°C, suggesting that receptor, kinase, and DNA-binding proteins must be in intimate contact with each other at the moment of receptor activation.

8. *Clinical Implications*

8.1 ASSOCIATION BETWEEN IL-5 AND EOSINOPHILIA

Evidence consistent with a key role for IL-5 in the control of eosinophilia comes from studies on eosinophilia in different diseases. If IL-5 were not the primary controlling factor in eosinophilia, it could be expected that cases of eosinophilia would exist without IL-5. As the following examples illustrate, eosinophilia in a wide variety of diseases is associated with IL-5 expression.

The development of eosinophilia in mice infected with *T. canis* is accompanied by the appearance of IL-5 mRNA in the spleen (Yamaguchi *et al.*, 1990). A comparison of the production of cytokines by normal individuals and by eosinophilic patients infected with the filarial parasite *Loa loa* gave interesting results. It was found that both groups produced similar levels of IL-3 and GM-CSF, but, whereas the normals gave relatively little IL-5, the cells from infected patients produced high levels of IL-5 (Limaye *et al.*, 1990). Elevations in IL-5 precede eosinophilia in onchocerciasis (Hagan *et al.*, 1996). Significant levels of IL-5 were also detected in the serum of patients with idiopathic hypereosinophilia (Owen *et al.*, 1989), and in patients with the eosinophilic-myalgia syndrome resulting from the ingestion of L-tryptophan (Owen *et al.*, 1990). Patients with eosinophilia associated with Hodgkin's disease were found to have IL-5 mRNA in the tumor cells, and a patient with angiolymphoid hyperplasia with eosinophilia (Kimura disease) was found to have constitutive expression of IL-5 mRNA in lymph node tissue (Inoue *et al.*, 1990).

Asthma is associated with an inflammatory reaction involving infiltration of eosinophils, and in-situ hybridization on mucosal bronchial biopsies indicated local expression of IL-5 which correlated with the number of infiltrating eosinophils (Hamid *et al.*, 1991). IL-5 has been detected in the bronchoalveolar lavage fluid of patients suffering pneumonia with distinct eosinophil involvement (Allen *et al.*, 1996). In addition, IL-5 is detected in several allergic disorders (Egan *et al.*, 1996).

8.2 TISSUE LOCALIZATION

Another aspect of the pathology of diseases characterized by eosinophilia is the preferential accumulation of eosinophils in tissues. As the blood contains both eosinophils and neutrophils, there must exist a specific mechanism that allows the eosinophils to pass preferentially from the blood vessels to the tissues. It is likely that IL-5 plays a role in this, for the following reasons.

The different tissue distribution of eosinophils in the two transgenic mouse systems probably results from the different tissue expression of IL-5. Using the metallothionein promoter, transgene expression was demonstrated in the liver and skeletal muscle, and eosinophils were observed in these tissues (Tominaga *et al.*, 1990). In contrast, the CD2–IL-5 mice with IL-5 expression in T cells did not have eosinophils in the liver or skeletal muscle (Dent *et al.*, 1990). This suggests that

eosinophils migrate into tissues where IL-5 is expressed. There are two mechanisms which might explain this. First, IL-5 is chemotactic for eosinophils (Yamaguchi *et al.*, 1988; Wang *et al.*, 1989). Second, IL-5 has been shown to upregulate adhesion molecules on eosinophils. Thus it was demonstrated that IL-5 increased the expression of the integrin CD11b on human eosinophils (Lopez *et al.*, 1986), and this increased expression was accompanied by an increased adhesion to endothelial cells. Adhesion was inhibited by antibody to CD11b or CD18, suggesting that the integrins are involved in eosinophil adhesion to endothelial cells (Walsh *et al.*, 1990). More recently it has been shown that eosinophils can use the integrin VLA4 (CD49d/CD29) in adherence to endothelial cells. In this case the ligand is VCAM-1. In contrast, neutrophils do not express VLA4 and do not use this adherence mechanism (Walsh *et al.*, 1991). IL-5 has also recently been identified as the major eosinophil chemoattractant in the asthmatic lung (Venge *et al.*, 1996) and is present in lesions of severe atopic dermatitis (Ishii *et al.*, 1996). Elevations of IL-5 have been detected in the autoimmune disease primary biliary cirrhosis (Krams *et al.*, 1996).

Thus, the critical role of IL-5 in the production of eosinophilia, coupled with a better understanding of the part played by eosinophils in the development of tissue damage in chronic allergy, suggests that IL-5 will be a major target for a new generation of anti-allergy drugs.

9. *References*

Allen, J.N., Liao, Z., Wewers, M.D., Altenberger, E.A., Moore, S.A. and Allen, E.D. (1996). Detection of IL-5 and IL-1 receptor antagonist in bronchoalveolar lavage fluid in acute eosinophilic pneumonia. J. Allergy Clin. Immunol. 97(6), 1366–1374.

Azuma, C., Tanabe, T., Konishi, M., *et al.* (1986). Cloning of human T-cell replacing factor (interleukin-5) and comparison with the murine homologue. Nucleic Acids Res. 14, 9149–9158.

Bates, M.E., Bertics, P.J. and Busse, W.W. (1996). IL-5 activates a 45-kilodalton mitogen-activated protein (MAP) kinase and Jak-2 tyrosine kinase in human eosinophils. J. Immunol. 156(2), 711–718.

Barry, S.C., McKenzie, A.N., Strath, M. and Sanderson, C.J. (1991). Analysis of interleukin 5 receptors on murine eosinophils: a comparison with receptors on B13 cells. Cytokine 3, 339–344.

Bazan, J.F. (1990). Structural design and molecular evolution of a cytokine receptor superfamily. Proc. Natl Acad. Sci. USA 87, 6934–6938.

Bende, R.J., Jochems, G.J., Frame, T.H., *et al.* (1992). Effects of IL-4, IL-5, and IL-6 on growth and immunoglobulin production of Epstein–Barr virus-infected human B cells. Cell. Immunol. 143, 310–323.

Bischoff, S.C., Brunner, T., De Weck, A.L. and Dahinden, C.A. (1990). Interleukin 5 modifies histamine release and leukotriene generation by human basophils in response to diverse agonists. J. Exp. Med. 172, 1577–1582.

Bohjanen, P.R., Okajima, M., and Hodes, R.J. (1990). Differential regulation of interleukin 4 and interleukin 5 gene expression: a comparison of T-cell gene induction by anti-CD3 antibody or by exogenous lymphokines. Proc. Natl Acad. Sci. USA 87, 5283–5287.

Broide, D.H., Paine, M.M. and Firestein, G.S. (1992). Eosinophils express interleukin 5 and granulocyte macrophage-colony-stimulating factor mRNA at sites of allergic inflammation in asthmatics. J. Clin. Invest. 90, 1414–1424.

Budel, L.M., Hoogerbrugge, H., Pouwels, K., van Buitenen, C., Delwel, R., Löwenberg, B. and Touw, I.P. (1993). Granulocyte-macrophage colony-stimulating factor receptors alter their binding characteristics during myeloid maturation through upregulation of the affinity converting β-subunit (KH97). J. Biol. Chem. 268, 10154–10159.

Burd, P.R., Rogers, H.W., Gordon, J.R., Martin, C.A., Jayaraman, S., Wilson, S.D., Dvorak, A.M., Galli, S.J. and Dorf, M.E. (1989). Interleukin 3-dependent and -independent mast cells stimulated with IgE and antigen express multiple cytokines. J. Exp. Med. 170, 245–257.

Caldenhoven, E., van Dijk, T., Raaijmakers, J.A., Lammers, J.W., Koenderman, L. and De Groot, R.P. (1995). Activation of the STAT3/acute phase response factor transcription factor by interleukin-5. J. Biol. Chem. 270(43), 25778–25784.

Campbell, H.D., Tucker, W.Q., Hort, Y., et al. (1987). Molecular cloning, nucleotide sequence, and expression of the gene encoding human eosinophil differentiation factor (interleukin 5). Proc. Natl Acad. Sci. USA 84, 6629–6633.

Campbell, H.D., Sanderson, C.J., Wang, Y., et al. (1988). Isolation, structure and expression of cDNA and genomic clones for murine eosinophil differentiation factor. Comparison with other eosinophilopoietic lymphokines and identity with interleukin-5. Eur. J. Biochem. 174, 345–352.

Chandrasekharappa, S.C., Rebelsky, M.S., Firak, T.A., Le Beau, M.M. and Westbrook, C.A. (1990). A long-range restriction map of the interleukin-4 and interleukin-5 linkage group on chromosome 5. Genomics 6, 94–99.

Chen, D. and Rothenberg, E.V. (1994). Interleukin 2 transcription factors as molecular targets of cAMP inhibition: delayed inhibition kinetics and combinatorial transcription roles. J. Exp. Med. 179, 931–942.

Clutterbuck, E.J. and Sanderson, C.J. (1988). Human eosinophil hematopoiesis studied in vitro by means of murine eosinophil differentiation factor (IL5): production of functionally active eosinophils from normal human bone marrow. Blood 71, 646–651.

Clutterbuck, E.J. and Sanderson, C.J. (1990). The regulation of human eosinophil precursor production by cytokines: a comparison of rhIL1, rhIL3, rhIL4, rhIL6 and GM-CSF. Blood 75, 1774–1779.

Clutterbuck, E., Shields, J.G., Gordon, J., et al. (1987). Recombinant human interleukin-5 is an eosinophil differentiation factor but has no activity in standard human B cell growth factor assays. Eur. J. Immunol. 17, 1743–1750.

Clutterbuck, E.J., Hirst, E.M. and Sanderson, C.J. (1989). Human interleukin-5 (IL-5) regulates the production of eosinophils in human bone marrow cultures: comparison and interaction with IL-1, IL-3, IL-6, and GM-CSF. Blood 73, 1504–1512.

Coffman, R.L., Seymour, B.W., Hudak, S., Jackson, J. and Rennick, D. (1989). Antibody to interleukin-5 inhibits helminth-induced eosinophilia in mice. Science 245, 308–310.

Corey, S., Eguinoa, A., Puyana-Theall, K., et al. (1993). Granulocyte macrophage-colony stimulating factor stimulates both association and activation of phosphoinositide 3OH-kinase and src-related tyrosine kinase(s) in human myeloid derived cells. EMBO J. 12, 2681–2690.

Cornelis, S., Plaetinck, G., Devos, R., et al. (1995). Detailed analysis of the IL5/IL5Rα interaction: characterisation of crucial residues on the ligand and the receptor. Embo. J. 14, 3395–3402.

Corrigan, C.J., Haczku, A., Gemou-Engesaeth, V., et al. (1993). CD4 T-lymphocyte activation in asthma is accompanied by increased serum concentrations of interleukin-5: Effect of glucocorticoid therapy. Am. Rev. Respir. Dis. 147, 540–547.

Crabtree, G.R. (1989). Contingent genetic regulatory events in T lymphocyte activation. Science 243, 355–361.

de Groot, R.P., Auwerx, J., Karperien, M., Staels, B. and Kruijer, W. (1991). Activation of junB by PKC and PKA signal transduction through a novel cis-acting element. Nucleic Acids Res. 19, 775–781.

Darnell, J.E. Jr., Kerr, I.M. and Stark, G.R. (1994). Jak-STAT pathways and transcriptional activation in response to IFNs and other extracellular signaling proteins. Science 264, 1415–1421.

Dent, L.A., Strath, M., Mellor, A.L. and Sanderson, C.J. (1990). Eosinophilia in transgenic mice expressing interleukin 5. J. Exp. Med. 172, 1425–1431.

de Vos, A.M., Ultsch, M. and Kossiakoff, A.A. (1992). Human growth hormone and extracellular domain of its receptor: crystal structure of the complex. Science 255, 306–312.

Devos, R., Plaetinck, G., Van der Heyden, J., et al. (1991a). Molecular basis of a high affinity murine interleukin-5 receptor. EMBO J. 10, 2133–2137.

Devos, R., Vandekerckhove, J., Rolink, A., et al. (1991b). Amino acid sequence analysis of a mouse interleukin 5 receptor protein reveals homology with a mouse interleukin 3 receptor protein. Eur. J. Immunol. 21, 1315–1317.

Devos, R., Guisez, Y., Cornelis, S., et al. (1993). Recombinant soluble human interleukin-5 (hIL-5) receptor molecules. Cross-linking and stoichiometry of binding to IL-5. J. Biol. Chem. 268, 6581–6587.

Devos, R., Guisez, Y., Plaetinck, G., et al. (1994). Covalent modification of the interleukin-5 receptor by isothiazolones leads to inhibition of the binding of interleukin-5. Eur. J. Biochem. 225, 635–640.

Devos, R., Plaetinck, G., Cornelis, S., Guisez, Y., Van der Heyden, J. and Tavernier, J. (1995). Interleukin-5 and its receptor: a drug target for eosinophilia associated with chronic allergic disease. J. Leukocyte Biol. 57, 813–819.

Dexter, T.M., Allen, T.D. and Lajtha, L.G. (1977). Conditions controlling the proliferation of haemopoietic stem cells in culture. J. Cell. Physiol. 91, 335–344.

Dickason, R.R., Huston, M.M. and Huston, D.P. (1996). Delineation of IL-5 domains predicted to engage the IL-5 receptor complex. J. Immunol. 156(3), 1030–1037.

Dvorak, A.M., Saito, H., Estrella, P., Kissell, S., Arai, N. and Ishizaka, T. (1989). Ultrastructure of eosinophils and basophils stimulated to develop in human cord blood mononuclear cell cultures containing recombinant human interleukin-5 or interleukin-3. Lab. Invest. 61, 116–132.

Egan, R.W., Umland, S.P., Cuss, F.M. and Chapman, R.W. (1996). Biology of interleukin-5 and its relevance to allergic disease. Allergy 51(2), 71–81.

Foster, P.S., Hogan, S.P., Ramsay, A.J., Matthaei, K.I. and Young, I.G. (1996). Interleukin 5 deficiency abolishes eosinophilia, airways hyperreactivity, and lung damage in a mouse asthma model [see comments]. J. Exp. Med. 183, 195–201.

Freeburn, R.W., Gale, R.E., Wagner, H.M. and Linch, D.C. (1996). The beta subunit common to the GM-CSF, IL-3 and IL-5 receptors is highly polymorphic but pathogenic point mutations in patients with acute myeloid leukaemia (AML) are rare. Leukemia 10(1), 123–129.

Fujisawa, T., Abu-Ghazaleh, R., Kita, H., *et al.* (1990). J. Immunol. 144, 642–646.

Gearing, D.P., King, J.A., Gough, N.M. and Nicola, N.A. (1989). Expression cloning or a receptor for human granulocyte-macrophage colony stimulating factor. EMBO J. 8, 3667–3676.

Goodall, G.J., Bagley, C.J., Vadas, M.A. and Lopez, A.F. (1993). A model for the interaction of the GM-CSF, IL-3 and IL-5 receptors with their ligands. Growth Factors 8, 87–97.

Gorman, D.M., Itoh, N., Kitamura, T., *et al.* (1990). Cloning and expression of a gene encoding an interleukin 3 receptor-like protein: identification of another member of the cytokine receptor gene family. Proc. Natl Acad. Sci. USA 87, 5459–5463.

Gorman, D.M., Itoh, N., Jenkins, N.A., Gilbert, D.J., Copeland, N.G. and Miyajima, A. (1992). Chromosomal localization and organization of the murine genes encoding the beta subunits (AIC2A and AIC2B) of the interleukin 3, granulocyte/macrophage colony-stimulating factor, and interleukin 5 receptors. J. Biol. Chem. 267, 15842–15848.

Graber, P., Bernard, A.R., Hassell, A.M., *et al.* (1993). Purification, characterisation and crystallisation of selenomethionyl recombinant human interleukin-5 from *Escherichia coli*. Eur. J. Biochem. 212, 751–755.

Guisez, Y., Oefner, C., Winkler, F.K., *et al.* (1993). Expression, purification and crystallization of fully active, glycosylated human interleukin-5. FEBS Lett. 331, 49–52.

Hagan, J.B., Bartemes, K.R., Kita, H., *et al.* (1996). Elevations in granulocyte-macrophage colony-stimulating factor and interleukin-5 levels precede posttreatment eosinophilia in onchocerciasis. J. Infect. Dis. 173(5), 1277–1280.

Hagan, P., Wilkins, H.A., Blumenthal, U.J., Hayes, R.J. and Greenwood, B.M. (1985). Eosinophilia and resistance to *Shistosoma haematobium* in man. Parasite Immunol. 7. 625–632.

Hamid, Q., Azzawi, M., Ying, S., *et al.* (1991). Expression of mRNA for interleukin-5 in mucosal bronchial biopsies from asthma. J. Clin. Invest. 87, 1541–1546.

Harada, N., Kikuchi, Y., Tominaga, S., Takaki, S. and Takatsu, K. (1985). BCGFII activity on activated B cells or purified T cell replacing factor (TRF) from a T cell hybridoma (B151K12). J. Immunol. 134, 3944–3951.

Hassell, A.M., Wells, T.N., Graber, P., *et al.* (1993). Crystallization and preliminary X-ray diffraction studies of recombinant human interleukin-5. J. Mol. Biol. 229, 1150–1152.

Hayashida, K., Kitamura, T., Gorman, D.M., Arai, K., Yokota, T. and Miyajima, A. (1990). Molecular cloning of a second subunit of the receptor for human granulocyte-macrophage colony-stimulating factor (GM-CSF): reconstitution of a high-affinity GM-CSF receptor. Proc. Natl Acad. Sci. USA 87, 9655–9659.

Hirai, K., Yamaguchi, M., Misaki, Y., *et al.* (1990). Enhancement of human basophil histamine release by interleukin 5. J Exp. Med. 172, 1525–1528.

Ihle, J.N. (1994). The Janus kinase family and signaling through members of the cytokine receptor superfamily. Pro. Soc. Exp. Bio. Med. 206, 268–272.

Imamura, F., Takaki, S., Akagi, K., *et al.* (1994). The murine interleukin-5 receptor α-subunit gene: characterization of the gene structure and chromosome mapping. DNA Cell Biol. 13, 283–292.

Ingley, E. and Young, I.G. (1991). Characterization of a receptor for interleukin-5 on human eosinophils and the myeloid leukemia line HL-60. Blood 78, 339–344.

Inoue, C., Ichikawa, A., Hotta, T. and Saito, H. (1990). Constitutive gene expression of interleukin-5 in Kimura's disease. Br. J. Haematol. 76, 554–555.

Isfort, R.J. and Ihle, J.N. (1990). Multiple hematopoietic growth factors signal through tyrosine phosphorylation. Growth Factors. 2, 213–220.

Ishii, E., Yamamoto, S., Sakai, R., Hamasaki, Y. and Miyazaki, S. (1996). Production of interleukin-5 and the suppressive effect of cyclosporin A in childhood severe atopic dermatitis. J. Pediatr. 128(1), 152–155.

Isobe, M., Kumura, Y., Murata, Y., *et al.* (1992). Localization of the gene encoding the alpha subunit of human interleukin-5 receptor (IL5RA) to chromosome region 3p24-3p26. Genomics 14, 755–758.

Itoh, N., Yonehara, S., Schreurs, J., *et al.* (1990). Cloning of an interleukin-3 receptor gene: a member of a distinct receptor gene family. Science 247, 324–327.

Jenkins, M.K. and Johnson, J.G. (1993). Molecules involved in T-cell costimulation. Curr. Opin. Immunol. 5, 361–367.

Kanakura, Y., Druker, B., Wood, K.W., *et al.* (1991). Granulocyte-macrophage colony-stimulating factor and interleukin-3 induce rapid phosphorylation and activation of the proto-oncogene Raf-1 in a human factor-dependent myeloid cell line. Blood 77, 243–248.

Karlen, S., D'Ercole, M. and Sanderson, C.J. (1996). Two pathways can activate the interleukin-5 gene and induce binding to the conserved lymphokine element 0. Blood 88, 211–221.

Kinashi, T., Harada, N., Severinson, E., *et al.* (1986). Cloning of a complementary DNA encoding T cell replacing factor and identity with B cell growth factor II. Nature 324, 70–73.

Kitamura, T., Sato, N., Arai, K. and Miyajima, A. (1991). Expression cloning of the human IL-3 receptor cDNA reveals a shared beta subunit for the human IL-3 and GM-CSF receptors. Cell 66, 1165–1174.

Kobierski, L.A., Chu, H.-M., Tan, Y. and Comb, M.J. (1991). cAMP-dependent regulation of proenkephalin by JunD and JunB: positive and negative effects of AP-1 proteins. Proc. Natl Acad. Sci. USA 88, 10222–10226.

Koch, C.A., Anderson, D., Moran, M.F., Ellis, C. and Pawson, T. (1991). SH2 and SH3 domains: elements that control interactions of cytoplasmic signaling proteins. Science 252, 668–674.

Kodama, S., Tsuruoka, N. and Tsujimoto, M. (1991). Role of the C-terminus in the biological activity of human interleukin 5. Biochem. Biophys. Res. Commun. 178, 514–519.

Kodama, S., Tsujimoto, M., Tsuruoka, N., Sugo, T., Endo, T.

and Kobata, A. (1993). Role of sugar chains in the in-vitro activity of recombinant human interleukin 5. Eur. J. Biochem. 211, 903–908.

Kopf, M., Brombacher, F., Hodgkin, P.D., *et al.* (1996). IL-5-deficient mice have a developmental defect in CD5+ B-1 cells and lack eosinophilia but have normal antibody and cytotoxic T cell responses. Immunity 4, 15–24.

Krams, S.M., Cao, S., Hayashi, M., Villanueva, J.C. and Martinez, O.M. (1996). Elevations in IFN-gamma, IL-5, and IL-10 in patients with the autoimmune disease primary biliary cirrhosis: association with autoantibodies and soluble CD30. Clin. Immunol. Immunopathol. 80(3 Pt 1), 311–320.

Kuiper, H.M., de Jong, R., Brouwer, M., Lammers, K., Wijdenes, J. and Van Lier, R.A.W. (1994). Influence of CD28 co-stimulation on cytokine production is mainly regulated via interleukin-2. Immunology 83, 38–44.

Kunimoto, D.Y., Allison, K.C., Watson, C., *et al.* (1991). High-level production of murine interleukin-5 (IL-5) utilizing recombinant baculovirus expression. Purification of the rIL-5 and its use in assessing the biologic role of IL-5 glycosylation. Cytokine 3, 224–230.

Larner, A.C., David, M., Feldman, G.M., *et al.* (1993). Tyrosine phosphorylation of DNA binding proteins by multiple cytokines. Science 261, 1730–1733.

Lee, H.J., Koyona-Nagagawa, N., Naito, Y., *et al.* (1993). cAMP activates the IL5 promoter synergistically with phorbol ester through the signally pathway involving protein kinase A in mouse thymoma EL-4. J. Immunol. 151, 6135–6142.

Lee, J.S., Campbell, H.D., Kozak, C.A. and Young, I.G. (1989). The IL-4 and IL-5 genes are closely linked and are part of a cytokine gene cluster on mouse chromosome 11. Somat. Cell. Genet. 15, 143–152.

Limaye, A.P., Abrams, J.S., Silver, J.E., Ottesen, E.A. and Nutman, T.B. (1990). Regulation of parasite-induced eosinophilia: selectively increased interleukin 5 production in helminth-infected patients. J. Exp. Med. 172, 399–402.

Lopez, A.F., Begley, C.G., Williamson, D.J., *et al.* (1986). Murine eosinophil differentiation factor. An eosinophil-specific colony-stimulating factor with activity for human cells. J. Exp. Med. 163, 1085–1099.

Lopez, A.F., Eglinton, J.M., Lyons, A.B., *et al.* (1990). Human interleukin-3 inhibits the binding of granulocyte-macrophage colony-stimulating factor and interleukin-5 to basophils and strongly enhances their functional activity. J. Cell Physiol. 145, 69–77.

Lopez, A.F., Vadas, M.A., Woodcock, J.M., *et al.* (1991). Interleukin-5, interleukin-3, and granulocyte-macrophage colony-stimulating factor cross-compete for binding to cell surface receptors on human eosinophils. J. Biol. Chem. 267, 24741–24747.

Lopez, A.F., Elliott, M.J., Woodcock, J. and Vadas, M.A. (1992a). GM-CSF, IL-3 and IL-5: cross-competition on human haemopoietic cells. Immunol. Today 13, 495–500.

Lopez, A.F., Shannon, M.F., Barry, S., *et al.* (1992b). A human interleukin 3 analog with increased biological and binding activities. Proc. Natl Acad. Sci. USA 89, 11842–11846.

Lopez, A.F., Shannon, M.F., Hercus, T., *et al.* (1992c). Residue 21 of human granulocyte-macrophage colony-stimulating factor is critical for biological activity and for high but not low affinity binding. EMBO J. 11. [In press]

Masuda, E.S., Tokumitsu, H., Tsuboi, A., *et al.* (1993). The granulocyte-macrophage colony-stimulating factor promoter

cis-acting element CLE0 mediates induction signals in T cells and is recognised by factors related to AP1 and NFAT. Mol. Cell. Biol. 13, 7399–7407.

Maxwell, C., Hussian, R., Nutman, T.B., *et al.* (1987). The clinical and immunologic responses of normal human volunteers to low dose hookworm (*Necator americanus*) infection. Am. J. Trop. Med. Hyg. 37, 126–134.

McKenzie, A.N.J., Barry, S.C., Strath, M. and Sanderson, C.J. (1991a). Structure–function analysis of interleukin-5 utilizing mouse/human chimeric molecules. EMBO J. 10, 1193–1199.

McKenzie, A.N.J., Ely, B. and Sanderson, C.J. (1991b). Mutated interleukin-5 monomers are biologically inactive. Mol. Immunol. 28, 155–158.

Migita, M., Yamaguchi, N., Mita, S., *et al.* (1991). Characterization of the human IL-5 receptors on eosinophils. Cell. Immunol. 133, 484–497.

Milburn, M., Hassell, A.M., Lambert, M.H., *et al.* (1993). A novel dimer configuration revealed by the crystal structure at 2.4 Å resolution of human interleukin-5. Nature 363, 172–176.

Minamitake, Y., Kodama, S., Katayama, T., Adachi, H., Tanaka, S. and Tsujimoto, M. (1990). Structure of recombinant human interleukin 5 produced by Chinese hamster ovary cells. J. Biochem. Tokyo. 107, 292–297.

Mita, S., Harada, N., Naomi, S., *et al.* (1988). Receptors for T cell-replacing factor/interleukin 5. Specificity, quantitation, and its implication. J. Exp. Med. 168, 863–878.

Mita, S., Tominaga, A., Hitoshi, Y., *et al.* (1989). Characterization of high-affinity receptors for interleukin 5 on interleukin 5-dependent cell lines. Proc. Natl Acad. Sci. USA 86, 2311–2315.

Miyatake, S., Shlomai, J., Arai, K. and Arai, N. (1991). Characterization of the mouse granulocyte-macrophage colony-stimulating factor (GM-CSF) gene promoter: nuclear factors that interact with an element shared by three lymphokine genes—those for GM-CSF, interleukin-4 (IL-4), and IL-5. Mol. Cell Biol. 11, 5894–5901.

Mori, A., Suko, M., Nishizaki, Y., *et al.* (1994). Regulation of interleukin-5 production by peripheral blood mononuclear cells from atopic patients with FK506, cyclosporin A and glucocorticoid. Int. Arch. Allergy Immunol. 104, 32–35.

Morla, A.O., Schreurs, J., Miyajima, A. and Wang, Y.J. (1988). Hematopoietic growth factors activate the tyrosine phosphorylation of distinct sets of proteins in interleukin-3-dependent murine cell lines. Mol. Cell. Biol. 8, 2214–2218.

Mui, A.L., Kay, R.J., Humphries, R.K. and Krystal, G. (1992). Ligand-induced phosphorylation of the murine interleukin 3 receptor signals its cleavage. Proc. Natl Acad. Sci. USA 89, 10812–10816.

Murakami, M., Narazaki, M., Hibi, M., *et al.* (1991). Critical cytoplasmic region of the interleukin 6 signal transducer gp130 is conserved in the cytokine receptor family. Proc. Natl Acad. Sci. USA 88, 11349–11353.

Murata, Y., Yamaguchi, N., Hitoshi, Y., Tominaga, A. and Takatsu, K. (1990). Interleukin 5 and interleukin 3 induce serine and tyrosine phosphorylations of several cellular proteins in an interleukin 5-dependent cell line. Biochem. Biophys. Res. Commun. 173, 1102–1108.

Murata, Y., Takaki, S., Migita, M., Kikuchi, Y., Tominaga, A. and Takatsu, K. (1992). Molecular cloning and expression of the human interleukin 5 receptor. J. Exp. Med. 175, 341–351.

Naora, H. and Young, I.G. (1994). Mechanisms regulating the mRNA levels of interleukin-5 and two other coordinately expressed lymphokines in the murine T lymphoma EL4.23. Blood 83, 3620–3628.

Naora, H., Altin, J.G. and Young, I.G. (1994a). TCR-dependent and -independent signaling mechanisms differentially regulate lymphokine gene expression in the murine T helper clone D10.G4.1. J. Immunol. 152, 5691–5702.

Naora, H., Van Leeuwen, B.H., Bourke, P.F. and Young, I.G. (1994b). Functional role and signal-induced modulation of proteins recognizing the conserved TCATTT-containing promoter elements in the murine IL-5 and GM-CSF genes in T lymphocytes. J. Immunol. 153, 3466–3475.

Nicola, N.A. and Metcalf, D. (1991). Subunit promiscuity among hemopoietic growth factor receptors. Cell 67, 1–4.

Owen, W.F., Rothenberg, M.E., Petersen, J., *et al.* (1989). Interleukin 5 and phenotypically altered eosinophils in the blood of patients with the idiopathic hypereosinophilic syndrome. J. Exp. Med. 170, 343–348.

Owen, W.F., Jr., Petersen, J., Sheff, D.M., *et al.* (1990). Hypodense eosinophils and interleukin 5 activity in the blood of patients with the eosinophilia-myalgia syndrome. Proc. Natl Acad. Sci. USA 87, 8647–8651.

Paul, C.C., Keller, J.R., Armpriester, J.M. and Baumann, M.A. (1990). Epstein–Barr virus transformed B lymphocytes produce interleukin-5. Blood 75, 1400–1403.

Pelicci, G., Lanfrancome, L., Grignani, F., *et al.* (1992). A novel transforming protein (SHC) with an SH2 domain is implicated in mitogenic signal transduction. Cell 70, 93–104.

Perlmutter, R.M., Levin, S.D., Appleby, M.W., Anderson, S.J. and Alberola-Ila, J. (1993). Regulation of lymphocyte function by protein phosphorylation. Annu. Rev. Immunol. 11, 451–499.

Plaetinck, G., der Heyden, J.V., Tavernier, J., *et al.* (1990). Characterization of interleukin 5 receptors on eosinophilic sublines from human promyelocytic leukemia (HL-60) Cells. J. Exp. Med. 172, 683–691.

Plaut, M., Pierce, J.H., Watson, C.J., Hanley Hyde, J., Nordan, R.P. and Paul, W.E. (1989). Mast cell lines produce lymphokines in response to cross-linkage of Fc epsilon RI or to calcium ionophores. Nature 339, 64–67.

Proudfoot, A.E., Davies, J.G., Turcatti, G. and Wingfield, P.T. (1991). Human interleukin-5 expressed in *Escherichia coli*: assignment of the disulfide bridges of the purified unglycosylated protein. FEBS Lett. 283, 61–64.

Rein, A., Keller, J., Schultz, A.M., Holmes, K.L., Medicus, R. and Ihle, J.N. (1985). Infection of immune mast cells by Harvey sarcoma virus: immortalization without loss of requirement for interleukin-3. Mol. Cell. Biol. 5, 2257–2264.

Rolfe, F.G., Hughes, J.M., Armour, C.L. and Sewell, W.A. (1992). Inhibition of interleukin-5 gene expression by dexamethasone. Immunology 77, 494–499.

Rolink, A.G., Melchers, F. and Palacios, R. (1989). Monoclonal antibodies reactive with the mouse interleukin 5 receptor. J. Exp. Med. 169, 1693–1701.

Sakamaki, K., Miyajima, I., Kitamura, T. and Miyajima, A. (1992). Critical cytoplasmic domains of the common beta subunit of the human GM-CSF, IL-3 and IL-5 receptors for growth signal transduction and tyrosine phosphorylation. EMBO J. 11, 3541–3549.

Samoszuk, M. and Nansen, L. (1990). Detection of interleukin-5 messenger RNA in Reed–Sternberg cells of Hodgkin's disease with eosinophilia. Blood 75, 13–16.

Sanderson, C.J. (1992). IL5, eosinophils and disease. Blood 79, 3101–3109.

Sanderson, C.J. (1994). In "The Cytokine Handbook" (ed. A.W. Thomson), pp. 127–143. Academic Press, London.

Sanderson, C.J., Warren, D.J. and Strath, M. (1985). Identification of a lymphokine that stimulates eosinophil differentiation in vitro. Its relationship to IL3, and functional properties of eosinophils produced in cultures. J. Exp. Med. 162, 60–74.

Sanderson, C.J., O'Garra, A., Warren, D.J. and Klaus, G.G. (1986). Eosinophil differentiation factor also has B-cell growth factor activity: proposed name interleukin 4. Proc. Natl Acad. Sci. USA 83, 437–440.

Sanderson, C.J., Campbell, H.D. and Young, I.G. (1988). Molecular and cellular biology of eosinophil differentiation factor (interleukin-5) and its effects on human and mouse B cells. Immunol. Rev. 102, 29–50.

Sanderson, C.J., Strath, M., Mudway, I. and Dent, L.A. (1994). In "Eosinophils: Immunological and Clinical Aspects" (eds. G.J. Gleich and A.B. Kay), pp. 335–351. Marcel Dekker, New York.

Sato, N., Sakamaki, K., Terada, N., Arai, K. and Miyajima, A. (1993). Signal transduction by the high-affinity GM-CSF receptor: two distinct cytoplasmic regions of the common beta subunit responsible for different signaling. EMBO J. 12, 4181–4189.

Satoh, T., Nakafuku, M., Miyajima, A. and Kaziro, Y. (1991). Involvement of ras p21 protein in signal-transduction pathways from interleukin 2, interleukin 3, and granulocyte/macrophage colony-stimulating factor, but not from interleukin 4. Proc. Natl Acad. Sci. USA 88, 3314–3318.

Scanafelt, A.B., Miyajima, A., Kitamura, T. and Kastelelein, R.A. (1991). The amino-terminal helix of GM-CSF and IL-5 governs high affinity binding to their receptors. EMBO J. 10, 4105–4112.

Schandené, L., Alonso-Vega, C., Willems, F., *et al.* (1994). B7/CD28-dependent IL-5 production by human resting T cells is inhibited by IL-10. J. Immunol. 152, 4368–4374.

Schreiber, S.L. and Crabtree, G.R. (1992). The mechanism of action of cyclosporin A and FK506. Immunol. Today 13, 136–142.

Seelig, G.F., Prosise, W.W. and Scheffler, J.E. (1994). A role for the carboxyl terminus of human granulocyte-macrophage colony-stimulating factor in the binding of ligand to the alpha-subunit of the high affinity receptor. J. Biol. Chem. 269, 5548–5553.

Shen, Y., Baker, E., Callen, D.F., *et al.* (1992). Localization of the human GM-CSF receptor beta chain gene (CSF2RB) to chromosome 22q12.2→q13.1. Cytogenet. Cell Genet. 61, 175–177.

Sher, A., Coffman, R.L., Hieny, S., Scott, P. and Cheever, A.W. (1990). Interleukin 5 is required for the blood and tissue eosinophilia but not granuloma formation induced by infection with *Schistosoma mansoni*. Proc. Natl Acad. Sci. USA 87, 61–65.

Silvennoinen, O., Wilthuhn, B.A., Quelle, F.W., *et al.* (1993). Structure of the murine Jak2 protein-tyrosine kinase and its role in interleukin 3 signal transduction. Proc. Natl Acad. Sci. USA 90, 8429–8433.

Sorensen, P., Mui, A.L. and Krystal, G. (1989). Interleukin-3 stimulates the tyrosine phosphorylation of the 140-kilodalton interleukin-3 receptor. J. Biol. Chem. 264, 19253–19259.

Strath, M. and Sanderson, C.J. (1986). Detection of eosinophil differentiation factor and its relationship to eosinophilia in *Mesocestoides corti*-infected mice. Exp. Hematol. 14, 16–20.

Su, B., Jacinto, E., Hibi, M., Kallunki, T., Karin, M. and Ben-Neriah, Y. (1994). JNK is involved in signal integration during costimulation of T lymphocytes. Cell 77, 727–736.

Sun, Z., Yergeau, D.A., Wong, I.C., *et al*. (1996). Interleukin-5 receptor alpha subunit gene regulation in human eosinophil development: identification of a unique cis-element that acts like an enhancer in regulating activity of the IL-5R alpha promoter. Curr. Top. Microbiol. Immunol. 211, 173–187.

Sutherland, G.R., Baker, E., Callen, D.F., *et al*. (1988). Interleukin-5 is at 5q31 and is deleted in the 5q- syndrome. Blood 71, 1150–1152.

Swain, S.L., McKenzie, D.T., Dutton, R.W., Tonkonogy, S.L. and English, M. (1988). The role of IL4 and IL5: characterization of a distinct helper T cell subset that makes IL4 and IL5 (Th2) and requires priming before induction of lymphokine secretion. Immunol. Rev. 102, 77–105.

Takaki, S., Tominaga, A., Hitoshi, Y., *et al*. (1990). Molecular cloning and expression of the murine interleukin-5 receptor. EMBO J. 9, 4367–4374.

Takaki, S., Mita, S., Kitamura, T., *et al*. (1991). Identification of the second subunit of the murine interleukin-5 receptor: interleukin-3 receptor-like protein, AIC2B is a component of the high affinity interleukin-5 receptor. EMBO J. 10, 2883.

Takaki, S., Murata, Y., Kitamura, T., Miyajima, A., Tominaga, A. and Takatsu, K. (1993). Reconstitution of the functional receptors for murine and human interleukin 5. J. Exp. Med. 177, 1523–1529.

Takatsu, K. (1992). Interleukin-5. Curr. Opin. Immunol. 4, 299–306. [Review]

Takatsu, K., Tominaga, A., Harada, N., *et al*. (1988). T cell-replacing factor (TRF)/interleukin 5 (IL-5): molecular and functional properties. Immunol. Rev. 102, 107–135.

Takatsu, K., Takaki, S., Hitoshi, Y., *et al*. (1992). Cytokine receptors on Ly-1 B cells. IL-5 and its receptor system. Ann. N.Y. Acad. Sci. 651, 241–258. [Review]

Tavernier, J., Devos, R., Van der Heyden, J., *et al*. (1989). Expression of human and murine interleukin-5 in eukaryotic systems. DNA 8, 481–501.

Tavernier, J., Devos, R., Cornelis, S., *et al*. (1991). A human high affinity interleukin-5 receptor (IL5R) is composed of an IL5-specific α chain and a β chain shared with the receptor for GM-CSF. Cell 66, 1175–1184.

Tavernier, J., Tuypens, T., Plaetinck, G., Verhee, A., Fiers, W. and Devos, R. (1992). Molecular basis of the membrane-anchored and two soluble isoforms of the human interleukin 5 receptor alpha subunit. Proc. Natl Acad. Sci. USA 89, 7041–7045.

Tavernier, J., Tuypens, T., Veehee, T., *et al*. (1995). Identification of receptor binding domains on human interleukin 5 (IL5) and design of an IL5-derived receptor antagonist. Proc. Natl Acad. Sci. USA 92, 5194–5198.

Tominaga, A., Takahashi, T., Kikuchi, Y., *et al*. (1990). Role of carbohydrate moiety of IL5: effect of tunicamycin on the glycosylation of IL5 and the biologic activity of deglycosylated IL5. J. Immunol. 144, 1345–1352.

Tuypens, T., Plaetinck, G., Baker, E., *et al*. (1992). Organization and chromosomal localization of the human interleukin 5 receptor alpha-chain gene. Eur. Cytokine. Netw. 3, 451–459.

van Leeuwen, B.H., Martinson, M.E., Webb, G.C. and Young, I.G. (1989). Molecular organization of the cytokine gene cluster, involving the human IL-3, IL-4, IL-5, and GM-CSF genes, on human chromosome. Blood 73, 1142–1148.

Venge, J., Lampinen, M., Hakansson, L., Rak, S. and Venge, P. (1996). Identification of IL-5 and RANTES as the major eosinophil chemoattractants in the asthmatic lung. J. Allergy. Clin. Immunol. 97(5), 1110–1115.

Vitetta, E.S., Brooks, K., Chen, Y., *et al*. (1984). T cell derived lymphokines that induce IgM and IgG secretion in activated murine B cells. Immunol. Rev. 78, 137–184.

Walsh, G.M., Hartnell, A., Wardlaw, A.J., *et al*. (1990). IL-5 enhances the in vitro adhesion of human eosinophils, but not neutrophils, in a leucocyte integrin (CD11/18)-dependent manner. Immunology 71, 258–265.

Walsh, G.M., Mermod, J.J., Hartnell, A., *et al*. (1991). Human eosinophil, but not neutrophil, adherence to IL-1-stimulated human umbilical vascular endothelial cells is alpha 4 beta 1 (very late antigen-4) dependent. J. Immunol. 146, 3419–3423.

Wang, J.M., Rambaldi, A., Biondi, A., *et al*. (1989). Recombinant human interleukin 5 is a selective eosinophil chemoattractant. Eur. J. Immunol. 19, 701–705.

Wang, L.M., Keegan, A.D., Paul, W.E., Heidaran, M.A., Gutkind, J.S. and Pierce, J.H. (1992). IL-4 activates a distinct signal transduction cascade from IL-3 in factor-dependent myeloid cells. EMBO J. 11, 4899–4908.

Wang, Y., Campbell, H.D. and Young, I.G. (1993). Sex hormones and dexamethasone modulate interleukin-5 gene expression in T lymphocytes. J. Steroid Biochem. Mol. Biol. 44, 203–210.

Warren, D.J. and Sanderson, C.J. (1985). Production of a T cell hybrid producing a lymphokine stimulating eosinophil differentiation. Immunology 54, 615–623.

Weiss, M., Yokoyama, C., Shikama, Y., Naugle, C., Druker, B. and Sieff, C.A. (1993). Human granulocyte-macrophage colony-stimulating factor receptor signal transduction requires the proximal cytoplasmic domains of the alpha and beta subunits. Blood 82, 3298–3306.

Wierenga, E.A., Snoek, M., Jansen, H.M., Bos, J.D., van Lier, R.A. and Kapsenberg, M.L. (1991). Human atopen-specific types 1 and 2 T helper cell clones. J. Immunol. 147, 2942–2949.

Woodcock, J.M., Zacharakis, B., Plaetinck, G., *et al*. (1994). Three residues in the common β chain of the human GM-CSF, IL-3 and IL-5 receptors are essential for GM-CSF and IL-5 but not IL-3 high affinity binding and interact with Glu21 of GM-CSF. EMBO J. 13, 5176–5185.

Yamaguchi, Y., Hayashi, Y., Sugama, Y., *et al*. (1988). Highly purified murine interleukin 5 (IL-5) stimulates eosinophil function and prolongs in vitro survival. IL-5 as an eosinophil chemotactic factor. J. Exp. Med. 167, 1737–1742.

Yamaguchi, Y., Matsui, T., Kasahara, T., *et al*. (1990). In vivo changes of hemopoietic progenitors and the expression of the interleukin 5 gene in eosinophilic mice infected with *Toxocara canis*. Exp. Hematol. 18, 1152-1157.

Zon, L.I., Moreau, J.F., Koo, J.W., Mathey-Prevot, B. and D'Andrea, A.D. (1992). The erythropoietin receptor transmembrane region is necessary for activation by the Friend spleen focus-forming virus gp55 glycoprotein. Mol. Cell. Biol. 12, 2949–2957.

6. Interleukin-6

Carl D. Richards

1. Introduction

Fahreas in 1921 noted that the erythrocyte sedimentation rate (ESR) was higher in serum from patients with inflammation compared to serum of those without. Not recognized until much later, this may have been the first observation of a manifestation of specific activity of IL-6, the induction of liver acute-phase proteins. In 1945 Homburger described a factor from purile exudate that induced fibrinogen responses (Homburger, 1945). The ESR and levels of another liver-derived acute-phase protein, C-reactive protein, and indeed more recently IL-6 itself, are currently used as measures of inflammation or clinical infection. Mixtures of cytokines that probably included IL-6, described as endogenous pyrogens or leukocyte endogenous mediator, were being characterized in the 1960s and 1970s. These products caused fever and marked increases in serum acute-phase proteins upon injection into rodents. Refinement of biochemical purification techniques in the 1970s and molecular cloning (1980s) showed that IL-6 was indeed among the impure preparations of previous studies that had such dramatic effects *in vivo*.

As has been the case for several other cytokines, the nature of individual *in vitro* activities of IL-6 had been pursued by several different investigators before the realization that one pleiotrophic molecule (IL-6) was responsible for what apparently seemed quite unique biological actions. In the mid 1980s it was realized that molecules termed B cell stimulatory factor, interferon-β_2, 26-kDa protein, macrophage granulocyte inducing factor 2 (MGI-2), hybridoma growth factor, and hepatocyte stimulating factor were due to a single polypeptide of molecular masses ranging from 19 kDa to 28 kDa, later designated interleukin-6 (Poupart *et al.*, 1987; Sehgal *et al.*, 1987). Other functions were rapidly attributed to IL-6 as reviewed previously (Wong and

Cytokines
ISBN 0–12–498340–5

Clark, 1988; Kishimoto, 1989; Revel, 1989; Akira *et al.*, 1990a), and more current studies are still identifying novel functions (see biological activities). IL-6 has important roles in inflammation, immunity, and hematopoiesis, and is found in serum and body fluids at levels that are active in bioassays *in vitro*. Elevated levels are significantly associated with many different disease states, and data from administration of exogenous IL-6 or antibodies to IL-6, and from transgenic and knockout mice, have yielded much information on the *in vivo* biological significance of this molecule. A number of excellent reviews examining different aspects of IL-6 actions have been published (Kishimoto and Hirano, 1988; Kishimoto, 1989; Revel, 1989; Heinrich *et al.*, 1990; Van Snick, 1990; Baumann and Gauldie, 1994). In addition, proceedings of meetings with an entire focus on IL-6 and IL-6-type cytokines are available, including two volumes of *Annals of the New York Academy of Sciences* (volume 55, 1989 and volume 62, 1995) as well as volume 88 of *Serono Symposia* publications entitled "IL-6: Physiopathology and Clinical Potentials", edited by M. Pevel, 1992. In addition, it is now clear that a group of cytokines share various bioactivities with IL-6 and have been described as IL-6-type cytokines or gp130 cytokines, on the basis of common utilization of the gp130 signal transducing molecule. IL-6 is the first-characterized and most-studied of this group, which also includes leukemia inhibitory factor (LIF), interleukin-11 (IL-11), oncostatin M (OM), ciliary neurotropic factor (CNTF), and more recently cardiotrophin-1.

2.   *The IL-6 Gene*

Human IL-6 is transcribed from a ~5 kb gene that contains four exons and six introns, and is spliced to a mature mRNA that is 1.2 to 1.3 kb (Yasukawa *et al.*, 1987). The human gene has been localized to the short arm of chromosome 7 (p15–21) (Sehgal *et al.*, 1986) and the mouse IL-6 gene to chromosome 5 (Billiau, 1987).

2.1   GENOMIC ORGANIZATION

The gene structure (see Figure 6.1) is similar to that of granulocyte colony-stimulating factor G-CSF (Tanabe *et al.*, 1989; Hirano *et al.*, 1989a) and other members of the IL-6 family (Bruce *et al.*, 1992). The major RNA start site at +1 is accompanied by a minor second RNA start site at approximately −21 (Zilberstein *et al.*, 1986). Three polyadenylation sites have been identified in human and mouse genes, but these do not result in substantial differences in mRNA size in these species (see Section 8). The cDNA sequence has been published for human (Haegeman *et al.*, 1986; Hirano *et al.*, 1986; Zilberstein *et al.*, 1986), bovine (Nakajima *et al.*, 1993), porcine (Richards and Saklatvala, 1991), rat (Northemann *et al.*,

1989) and mouse (Van Snick *et al.*, 1988; Fuller and Grenett, 1989). The predicted coding region (~636 bp in human) encodes 212 amino acids and the 3′ noncoding region (500 bp followed by poly(A) tail) encodes multiple copies of the AUUUA motif that is important in mRNA stability (Caput *et al.*, 1986; Shaw and Kamen, 1986; Cosman, 1987) (see Figures 6.1 and 6.2).

2.2   TRANSCRIPTION CONTROL

IL-6 expression can be induced by a broad range of cell stimuli, and different mechanisms may act individually or coordinately to enhance transcription. Phorbol ester and cytokines such as IL-1 and TNF are very strong inducers of IL-6 production (Walther *et al.*, 1988). Various groups have examined targets of signal transduction within the promoter region of human IL-6. These targets include the DNA-binding regions that are specific for nuclear factors referred to as nuclear factor for IL-6 (NFIL-6, both α and β), also termed CCAAT element binding protein (C/EBPβ and γ); nuclear factor for kappa light chain in B cells (NFκB); activator protein-1 (AP-1); cAMP response element binding protein (CREB); glucocorticoid receptor (GR) (Ray *et al.*, 1989; Isshiki *et al.*, 1990; Poli *et al.*, 1990; Chen-Kiang *et al.*, 1993). These sequences can be found within 200 bp of the transcriptional start site (see Fig. 6.2). Within cells stimulated by IL-1, NFIL-6α and -β as well as NFκB bind to their respective sites within the IL-6 promoter and activate transcription of downstream reporter genes (Akira *et al.*, 1990b; Libermann and Baltimore, 1990; Isshiki *et al.*, 1990; Zhang *et al.*, 1990). NFIL-6α and -β appear to act as heterodimers to enhance transcription more effectively (Kinoshita *et al.*, 1992) and NFIL-6 and NFκB can heterodimerize, suggesting interactive control of the IL-6 promoter (LeClair *et al.*, 1992). Both sites appear critical for maximum expression of IL-6. Cyclic AMP-inducing agents including PGE_2 have been found to enhance transcription (Zhang *et al.*, 1988) and more recently it has been suggested that elevated cAMP enhances efficiency of multiple regulatory sites including NFIL-6 and NFκB (Dendorfer *et al.*, 1994). Glucocorticoids inhibit IL-6 expression (Ray *et al.*, 1990; Waage *et al.*, 1990) and this appears to act at the transcriptional level through the ability of glucocorticoid receptor to bind AP-1 (Ray *et al.*, 1990) or NFκB p65 (Ray and Prefontaine, 1994) and inhibit binding of these factors to DNA.

Other regulatory factors have recently been implicated in IL-6 transcription. The tumor suppressor gene p53 inhibited IL-6 promoter-driven transcription upon overexpression (Santhanam *et al.*, 1991) while expression of p53 mutants did not. Wild-type retinoblastoma susceptibility gene (Rb) was also capable of repressing serum-induced IL-6 promoter-driven indicator expression in Hela cells (Santhanam *et al.*, 1991). Furthermore, expression vectors encoding adenovirus E1A proteins E1A289R and E1A243R caused a marked

-1223
                                                                GGATCCTCCTGCAGAGACACC
ATCCTGAGGGAAGAGGGCTTCTGAACCAGCTTGACCCAATAAGAAATTCTTGGGTGCCGACCGCGGACAGAGATTCAGAGCCGGTGCCTGCGTCCGTACTTTCCTTCCTTCTAGCTTCT
TTTGATTTCAAATCAAGACTTAGAGGAGGAGGGAGCGATAAACACAAACTCTGCAAGATGCCAACAAACTCTCCTTTGACATCCCAACAAAGAGTGAGTAGTAATCTCCCCTTTCTG
CCCTGAACCAAGTGGGCTTCAGTAATTTCAGGGCTCCAGAGACTGGTAGAAACAGTGGTGAAGAGACTCAGTGGCAGTCAGGGAGAGCACTGGCACCAGGCAAA
CCTCTGGCACAAGAGCAAAGTCCTACTGGAGATTCCAAGGGTCACTTGGGAGAGGGCAGGCAGCCAACCTCCTCTAAGTGGCTGAAGCAGTCAGTGAAGAAATGGCAGCAAGCGCGGT
GATGACTGGTAGTATTACCTTCTTCATAATCCAGGCTTGGGGGCTGCGATGAGTCAGGGAAACTCAGTCAGAGAACTCAGTTCTTTTTCTCTTTGTAAAACTTCGTGCATGACTTCAGCTTTACTCTTTGTCAA
TCTAGCCTGTTAATCTGGTCACTGAAAAAAAATTTTTTTTTTCAAAAACATAGCTTTAGCTTATTTTTTTCTCTTTGTAAAACTTAACTGAACGCTAAAT
GACATGCCCAAAGTGCTGAGTCACTAATAAAAGAAAAAAAGAAAGTAAGAGAAGTGGTTCTGCTCTTTAGCGCTAGCCTAAGCTGCACTTTCCCCGTAGTTGTGTC
TTGGCGATGCTAAAGGACGTCACATTGCACAATCTTAATAAGGTTTCAATCAGCCCACCTGGCCCCTCTGGCCCACCTCCTCCACCATGCTCCCTCGAGCGTCTATCTCCCTCCAGGAGCCCAGCT
GTCTCAATATTAGAGTCTCAACCCCAATAAATATAGGACTGGAGATGTCTGAGGCTCATTCTGAGCTCTATCTCCCTCCAGGAGCCCAGCT
ATGAACTCCTTCTCCACAA   19
      Exon 1

174
GCGCCTTCGGTCCAGTTGCCTTCTCCCTGGGGCTGCTCCTGGTGTTGCCTGCTCCTGCCTTCCCTGCCCCAGTACCCCCAGGAGAAGATTCCAAAGATGTAGCCGCCCCACACAGACAGCCAC
TCACCTTCTTCAGAACGAATTGCAAACAAATTCGGTACATCCTCGACGGCATCCTCAGCCCTGAGAAACGAG   364
                                                                Exon 2

1411
ACATGTAACAAGAGTAACATGTGTGAAAGCAGCAGCAAAGAGAGCCACTGCGAGAAACAACCTGACCTTCCAAAGATGGCTGAAAAAGATGGATGCTTCCAATCTGGATTCA
                                                                                    1526
                                                                Exon 3

2222
GAGACTTGCCTGGTGAAAATCATCACTGGTCTTTTGGAGTTTGAGGTATACCTAGAGTACCTCCAGAACAGATTTGAGAGTAGTGAGGAACAAGCCAGAGCTGTGCAGATGAGTACAAAA
GTCCTGATCCAGTTCCTGCAGAAAAAG   2377
                                                                Exon 4

4120 Exon 5
GCAAAGAATCTAGATGCAATAACCACCCCTGACCCAACCACCACAAATGCCAGCCTGCTGACGAAGCTGCAGGACATGACAACTCATCTCATTCTGCGCAGC
TTTAAGGAGTTCCTGCAGTCCAGCTCCTCGGCAAATGAGCATGGCACCTCAGATTGTTGTTGTTAATGGCATTCCTTCTTCTGTCAGAAACCTGTCCACTGGGCACAG
AACTTATGTTGTTCTCTATGGAGAACTATGAGCGTTAGGACACTATTTTAATTATTTTTTAATTATTAGTAAATATTTAAATATGTAAGTGCTATATTTA
TATTTTAAGGAGTACCACTTGAAACATTTATGTATTAGTTTTGAAATAATAATGGCTATGCAGTTGAAAGTGCATGCCAGCACATTTCTTGAAAGTGTAGG
CTTACCTCAAATAAATGGCTAACTTATACATATTTTTAAGAAATATTTATATTGTATTTATACCAATAAAATGTTTATAAAAATTCAGCACACTT
TGAGTGTGTCACGTGAAGCTT   4739

Figure 6.1   IL-6 gene DNA sequence. The nucleotide sequence of genomic DNA encoding BSF-2/IL-6. Nucleotides are numbered starting at A of the translation start codon. The "TATA"-like sequences are boxed. The sequence homologous to that found in the 5' flanking region of IL-2 gene is indicated by a dashed box.
(From Yasukawa et al. (1987).)

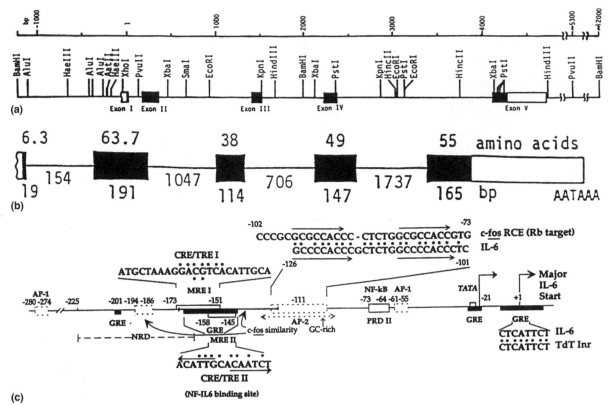

Figure 6.2   IL-6 gene structure: (a) restriction map; (b) exon and intron size; (c) promoter region. Modified from Hirano *et al.*, 1986; Ray *et al.*, 1990.

repression of IL-6 promoter-driven activity upon transfection into Hela cells or HepG2 cells (Janaswami *et al.*, 1992). A cell line stably transfected with E1A was incapable of responding to IL-1 or TNF in induction of IL-6. The mechanisms may involve interfering with binding of nuclear factors that normally participate in IL-6 gene regulation (Janaswami *et al.*, 1992). Interestingly, on the other side of the coin, IL-6 was able to inhibit p53-induced apoptosis of myeloid leukemia cell lines (Yonish-Rouach *et al.*, 1991) and suppress phosphorylation of Rb in hematopoietic cells (Resnitzky *et al.*, 1992). In addition, NF-IL-6 could substitute for adenovirus E1A in IL-6-stimulated HepG2 cells and bind to and stimulate transfected E1A promoters (Spergel and Chen-Kiang, 1991; Spergel *et al.*, 1992). Thus IL-6 and its mechanisms of action may interact with those of adenovirus expressed proteins.

IL-1-induced signal transduction has been studied in detail; however, the entire intracellular pathway by which a target gene such as IL-6 is regulated is not yet clear. IL-1 receptor activation activates the phosphorylation of inhibitor of κB (IκB), and thus its disassociation from NFκB which allows nuclear localization of NFκB. IL-1 receptor-ligand interaction also activates a cascade in which Raf and MAP kinase/ERK are phosphorylated (Bird *et al.*, 1992; Kishimoto *et al.*, 1994) apparently through a different pathway from IκB phosphorylation.

Activation of the membrane-associated GTP-binding protein p21 Ras may be an early event in this process, although it is not clear whether this is the only pathway in activation of MAP kinase. MAP kinase has been shown to phosphorylate NFIL-6 at threionine (residue 235) which results in transcriptional activation of the IL-6 gene (Nakajima *et al.*, 1993).

IL-1 also acts to modulate the mRNA stability of IL-6. TNF-stimulated cells were found to possess IL-6 mRNA with a short half-life (<1 h); however, IL-1-stimulated cells possessed IL-6 mRNA with a half-life of greater than 6 h (Ng *et al.*, 1994). The mechanism by which this occurs is not known. The AUUUA sequences in 3' noncoding region of IL-6 mRNA may play a role (Caput *et al.*, 1986; Cosman, 1987). This sequence binds particular protein species (Malter, 1989) which may act as modulators of RNA degradation or as targets for other mechanisms. Thus, multiple stimuli can activate IL-6 expression and this can occur at transcriptional and mRNA stability levels. This presumably underlies the importance of the role of this cytokine in bodily functions.

3.   *The IL-6 Protein*

Heterogeneity of preparations of IL-6 has been described by different laboratories, molecular masses ranging from

19 kDa to 30 kDa. Murine IL-6 lacks glycosylation, and human IL-6 shows microheterogeneity in size due to extent of glycosylation (Van Snick *et al.*, 1988; Fuller and Grenett, 1989; May *et al.*, 1988a, 1989). As a product of monocytes or macrophages, IL-6 has *N*- and *O*-linked glycosylation and is also phosphorylated at various serine residues (May *et al.*, 1988b). The IL-6 precursor peptide is 212 amino acids in size (211 in rat and mouse; 210 in bovine) and a typical signal sequence is cleaved off to make approximately 186-amino-acid mature forms, with an isoelectric point of ~5 (see Figure 6.3). For all species for which the IL-6 sequence is known, it shows four cysteine residues in conserved locations (rat IL-6 encodes an additional cysteine residue) and a number of consensus glycosylation sites (Asn-Y-Ser/Thr). Disulfide bonds play an important role in the correct folding of tertiary structure where linkages have been assigned between cysteine residues at C^{44}–C^{50}, and between C^{73}–C^{83} (Savino *et al.*, 1993). The homology of rat and mouse IL-6 is quite close at the amino acid level (93%), whereas human IL-6 is less homologous to these sequences and more to porcine (60%) and bovine (65%) IL-6. Recombinant derived IL-6 from *E. coli* or mammalian expression systems is available from many sources and products are stable at –20°C when stored in appropriate buffered conditions.

Data analysis suggests that the structure of IL-6 protein shows significant homology to IL-11, LIF, oncostatin M, and GM-CSF (Bazan, 1991; Rose and Bruce, 1991). The tertiary structure of IL-6 has not yet been solved by crystallography, but homology of hydrophobic and hydrophilic amino acids in α-helices is well conserved in comparison to α-helices of G-CSF (Bazan, 1990, 1991), as well as the two disulfide bridges that tether the molecule's structure. Models based on the x-ray crystallographic structure of bovine G-CSF have been altered to accommodate insertions and deletions in the loop regions (Savino *et al.*, 1993) to predict the IL-6 tertiary structure (see Figure 6.4). Thus IL-6 appears to be composed of a bundle of four antiparallel α-helices, A and B helices connected by a long AB loop, C and D helices connected by a CD loop (Savino *et al.*, , 1993). Since antibodies to peptides at either end of the IL-6 protein inhibit bioactivity (Brakenhoff *et al.*, 1990), the interaction with its receptor complex appears to involve multiple sites on the three-dimensional structure (see Section 6).

IL-6 acts as a monomeric entity on cells; however, in serum/plasma it can be found bound to various proteins that may or may not modify its bioactivity. α_2-Macroglobulin can act as a carrier for IL-6 (Matsuda *et al.*, 1989) as can complement components (Borth *et al.*, 1989). Analysis of IL-6 bioactivity from serum clearly shows multiple-size complexes suggesting that various serum proteins can bind but still allow IL-6 to interact with its receptor (May *et al.*, 1992). *In vitro* studies suggest that IL-6 complexed to the soluble form of IL-6R can activate signal transduction if added to HepG2 cells that possess membrane gp130 (Taga *et al.*, 1989; Hibi *et al.*, 1990; Mackiewicz *et al.*, 1992). The role of

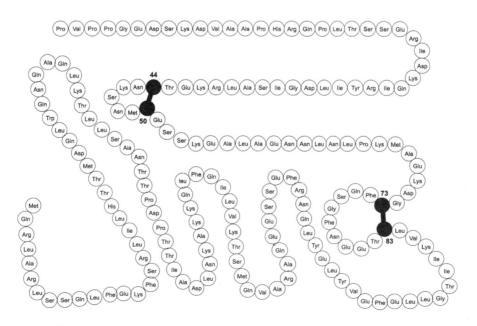

Figure 6.3   IL-6 amino acid sequence. (Taken from Hirano *et al.* (1986).)

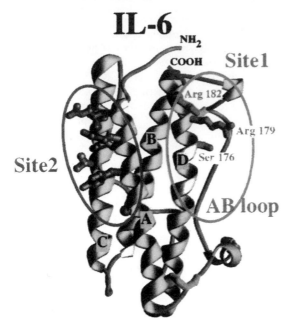

Figure 6.4   IL-6 tertiary structure. Depicted is a ribbon model of IL-6 tertiary structure. (Taken from Savino *et al*. (1993, 1994).)

soluble IL-6 receptor and modulation of function in other systems or *in vivo* remains to be fully understood (see Section 6).

4.   *Cellular Sources and Production*

A large number of cell types can express IL-6 *in vitro*. Various cytokines and other agents induce IL-6 expression, while glucocorticoids, for example, are potent inhibitors of IL-6 production. IL-6 is not stored preformed in cells to any great degree. Upon cell stimulation and increase in IL-6 mRNA levels, rapid synthesis of protein and secretion occur. Monocytes/macrophages may be the first to release IL-6 upon inflammation. The monocyte/macrophage products IL-1 and TNF are strong inducers of IL-6 release from stromal cell populations such as fibroblasts, endothelial cells, and keratinocytes among others. These represent a significant number of cells which potentially add to the rapid rise in local synthesis of IL-6, and thus to systemic levels. Intravenous injection of TNF can result in a very rapid increase in circulating IL-6, presumably derived from endothelial cells (Jablons *et al.*, 1989). Assessment of IL-6 levels in biological fluids has shown its presence in a wide variety of disease states (see Section 9), many of which are associated with inflammation. Levels may be seen to increase 10^3- to 10^4-fold in serum in certain conditions. Since many cells can express IL-6 *in vitro*, the same may be reflected *in vivo*. However,

assessment of which cells are expressing IL-6 in each of these conditions *in vivo* is not as clear. In-situ hybridization studies have not been productive in pinpointing IL-6-expressing tissue in disease states. Examination for specific positive cells in the context of many cells expressing small amounts of IL-6 may be fraught with technical difficulties.

IL-6 is produced by monocyte/macrophages in response to stimuli such as LPS (May *et al.*, 1988a; Gauldie *et al.*, 1987); IL-1α (Bauer *et al.*, 1984, 1988); TNF (Bauer *et al.*, 1988; May *et al.*, 1988a); IFN-γ and GM-CSF (Navarro *et al.*, 1989); fibrin fragments D,E (Ritchie and Fuller, 1983); protease complexes (Kurdowska and Travis, 1990), and some drugs (Stepien *et al.*, 1993) as well as possibly metal cations (Scuderi, 1990). IL-6 production in monocytes is inhibited by IL-4 (Donnelly *et al.*, 1993); IL-10 (de Waal Malefyt *et al.*, 1992; Howard and O'Garra, 1992) and glucocorticosteroid (Woloski *et al.*, 1985; Waage *et al.*, 1990; Amano *et al.*, 1992).

Also a source of marked amounts of IL-6 *in vitro*, fibroblasts from various tissues can be stimulated by IL-1α, IL-1β (Poupart *et al.*, 1987; Guerne *et al.*, 1989; Sehgal *et al.*, 1987; Haegeman *et al.*, 1986); TNF-α, TNF-β (May *et al.*, 1988a, 1989; Richards and Saklatvala, 1991); PDGF (Kohase *et al.*, 1987); LPS (Helfgott *et al.*, 1987); oncostatin M (Richards and Agro, 1994); PGE (Zhang *et al.*, 1988); and viruses (Sehgal *et al.*, 1988); but not IL-4 (Donnelly *et al.*, 1993).

In addition, other cells have been reported to produce IL-6 upon stimulation, such as endothelial cells with IL-1/TNF (Jirik *et al.*, 1989; Podor *et al.*, 1989) or oncostatin M (Brown *et al.*, 1991); chondrocytes with IL-1 (Bunning *et al.*, 1990; Malejczyk *et al.*, 1992); epithelial cells with IL-1/TNF (Luger *et al.*, 1989b); keratinocytes with IL-1 (Kupper *et al.*, 1989); bone marrow stroma with IL-1 (Chiu *et al.*, 1988) in which IL-6 is inhibited by estradiol (Jilka *et al.*, 1992); astrocytes and microglia with IL-1/TNF (Frei *et al.*, 1989; Rose-John and Heinrich, 1994); mesangial cells (Horii *et al.*, 1988); T cells and B cells (Hirano *et al.*, 1990; Horii *et al.*, 1988); and mast cells (Leal-Berumen *et al.*, 1994).

5.   *Biological Activities of IL-6*

A large array of effects of IL-6 include those which primarily act in a local environment, such as at sites of inflammatory cell and lymphocyte interaction and activation. These activities may be autocrine or paracrine in nature, depending on responding cells. IL-6 also acts in an endocrine fashion, such as regulation of systemic effects of distant target organs in acute inflammation. As with many other cytokines, the half-life of injected IL-6 protein *in vivo* is relatively short. This has made exogenously administered IL-6 effects more difficult to analyze; however, some interesting results are available. IL-6 transgenes and IL-6 knockout mice also provide valuable information on the *in vivo* biological functions of this molecule (see Section 8).

5.1   FUNCTIONS IN IMMUNE CELLS

IL-6 induces responses in B cells, T-helper cells, cytotoxic T lymphocytes (CTL), and natural killer (NK) cells *in vitro*, implying an importance in many aspects of immunity (reviewed in Kishimoto, 1989; Van Snick, 1990). IL-6 is well established as a late-stage differentiation factor for B cell to plasma cell transition (Muraguchi *et al.*, 1988; Kishimoto and Hirano, 1988), enhancing immunoglobulin production, and augmenting secondary antibody responses to antigen *in vivo* (Takatsuki *et al.*, 1988). IL-6 does not act on resting B cells, but acts on activated B cells that express IL-6 receptor. Stimulation then enhances IgM, IgG, and IgA production without an effect on B cell proliferation (Taga *et al.*, 1987). IL-6 also regulates T cell proliferation of peripheral and thymic T cells (Garman *et al.*, 1987; Takatsuki *et al.*, 1988). IL-6 was a possible contaminant of IL-1 preparations in the old lymphocyte-activating factor (LAF) assay so often used as a "specific" measure of IL-1, since investigators showed that IL-1 and IL-6 were synergistic in stimulation of PHA-costimulated mouse thymocytes (Houssiau *et al.*, 1988a). This appeared to depend on the presence of IL-2 and may indicate that IL-6 increases IL-2 responsiveness. In addition, IL-6 is also important in enhancement of CTL differentiation from precursors (Okada *et al.*, 1988; Takai *et al.*, 1988; Wong and Clark, 1988). Various studies have shown that IL-1 synergizes with IL-6 in proliferation of T cells and maturation of CTL (Houssiau and Van Snick, 1992).

Although IL-6 is a major product of monocytes and macrophages upon stimulation, there is evidence that IL-6 inhibits the production of IL-1 and TNF (Aderka *et al.*, 1989; Schindler *et al.*, 1990). IL-6 has also been shown to induce circulating levels of the IL-1 receptor antagonist and the soluble TNF receptor p55 *in vivo*, and this was reflected in its activity on GM-CSF-stimulated monocytes (Tilg *et al.*, 1994). IL-6 primes monocytes and neutrophils for enhanced oxidative respiratory burst responses (Kharazmi *et al.*, 1989). IL-6 can also augment natural killer cell activity *in vitro*, through the induction of IL-2 expression in peripheral blood mononuclear cell cultures (Luger *et al.*, 1989a). Although at first characterized by some investigators as having interferon-like antiviral activity, it is now presumed this action is rather weak if present at all (Poupart *et al.*, 1984; Hirano *et al.*, 1988a; Reis *et al.*, 1988).

5.2   HEMATOPOIESIS

The activities of IL-6 on hematopoietic cells and stem cells are several. Initially, Leary and colleagues (1988) found that IL-6 acted synergistically in combination with IL-3 to enhance multilineage colonies from murine spleen, and others showed that this action was primarily on multilineage progenitors (Koike *et al.*, 1988). IL-6 was also found to be identical to macrophage-granulocyte inducing factor-2 (MGI-2) (Sachs *et al.*, 1989), and to synergize with other cytokines in hematopoietic systems (Bot *et al.*, 1989; Rennick *et al.*, 1989). Similar to G-CSF, IL-11 and Steel factor, IL-6 appears to be involved in triggering the entry into the cell cycle, whereas IL-3 and GM-CSF support continued proliferation of multipotential progenitors. *In vivo* administration of IL-6 to mice with bone marrow transplantation was effective in aiding recovery of multilineage hematopoiesis (Okano *et al.*, 1989), as was IL-6 administration to irradiated mice (Patchen *et al.*, 1991). IL-6 has strong activity in enhancing megakaryocyte colony formation (Ishibashi *et al.*, 1989b; Kishimoto, 1989; Hirano *et al.*, 1990), an activity with which it shares with LIF, IL-11, and oncostatin M. These are synergistic with IL-3 in enhancing CFU-megakaryocyte derived colonies *in vitro* and can enhance platelet numbers in mice *in vivo* (Ishibashi *et al.*, 1989a). This action has been the base of clinical trials for the purpose of enhancing platelet levels in patients post chemotherapy or post bone marrow transplantation (see below).

5.3   IL-6 REGULATES CELL GROWTH

Another earlier-described activity of IL-6 was that as a positive growth factor for EBV-transformed B cells (Tosato and Pike, 1989), mouse plasmacytomas *in vitro* and *in vivo* (Hirano *et al.*, 1989b), as well as human myeloma cells (Kawano *et al.*, 1988) from patients with multiple myeloma. Myeloma cell growth in culture appears to be regulated in autocrine fashion by IL-6, since antibodies to IL-6 inhibit growth *in vitro* of various myeloma lines (Kawano *et al.*, 1988). Its action as a potent growth factor for mouse hybridomas is now used as a sensitive bioassay for IL-6 activity (Aarden *et al.*, 1987). IL-6 has also been shown to enhance growth of Kaposi sarcoma cell lines, and may do so in an autocrine fashion since transfection of antisense IL-6 RNA suppressed proliferation (Miles *et al.*, 1990) and antibodies specific for IL-6 decrease proliferation of these cells *in vitro* when stimulated with oncostatin M (Miles *et al.*, 1992). IL-6 has also shown to possess growth inhibitory activity, including suppression of the melanoma cell line A375 and breast carcinoma cell lines (Chen *et al.*, 1988; Novick *et al.*, 1992; Katz *et al.*, 1993) as well as inhibition and differentiation of a murine myeloid leukemic cell line M1 (Miyaura *et al.*, 1988). Lu and Kerbel (1993) have shown that early-stage melanoma cells are inhibited by IL-6 in proliferation, but that later-stage (advanced) melanoma cells are resistant to exogenous IL-6. Transfection of antisense IL-6 RNA inhibited proliferation (Lu and Kerbel, 1993), suggesting that IL-6 was acting as a positive growth factor in later-stage melanoma. Thus the state of tumor progression may play an important role in proliferative responses of melanoma cells to IL-6.

5.4 THE ACUTE-PHASE RESPONSE AND INFLAMMATION

One of the primary roles of IL-6 *in vivo* is as the major cytokine that initiates the hepatic acute-phase response. This is typified by a striking elevation of plasma concentrations of liver-derived (acute-phase) proteins and is a hallmark of the acute inflammatory response (Koj, 1974; Kushner, 1982; Koj and Gordon, 1985). Mediators responsible for this action are released at sites of inflammation and then circulate to act specifically on liver cells (Koj, 1974). IL-6 induces all positive acute-phase proteins, and reduces albumin and transferrin output (typical negative acute-phase proteins) by hepatocytes *in vitro* (Andus *et al.*, 1987, 1988; Morrone *et al.*, 1988; Castell *et al.*, 1989), and clearly induces liver responses in rodents upon administration *in vivo* similar to those induced by inflammation (Geiger *et al.*, 1988). Levels of IL-6 in serum after inflammatory stimulus show peak levels 6 to 12 h before the peak levels of serum acute-phase proteins (Nijsten *et al.*, 1987; Schreiber, 1987; Jablons *et al.*, 1989; Schreiber *et al.*, 1989; Gauldie *et al.*, 1992). The delay in serum increases in acute-phase proteins after initiation of inflammation reflects the time required to synthesize and release sufficient amounts of cytokines, their interaction with hepatocyte receptors, protein production by hepatocytes, and accumulation in the serum compartment. The expression of acute-phase proteins in liver is primarily at the level of transcription (Birch and Schreiber, 1986; May *et al.*, 1986; Schreiber *et al.*, 1989). The liver response subsides over a period of hours to days depending on the stimulus, and serum levels wane in 3–5 days depending on the acute-phase protein examined. IL-6 has also been shown to modulate the extent of glycosylation of liver-derived secreted acute-phase proteins.

The IL-6 mediated induction of acute-phase proteins is thought to represent a homeostatic response to inflammation (Gauldie, 1991; Baumann and Gauldie, 1990; Travis and Salvesen, 1983). Various acute-phase proteins (such as α_1-antitrypsin and α_1-antichymotrypsin) are inhibitory of protease action and thus could modulate enzyme action at sites of inflammation and a number of acute phase proteins show anti-inflammatory properties *in vivo* (reviewed in Tilg *et*

al., 1997). Other cytokines have a similar action on hepatocytes, including LIF (Baumann and Wong, 1989), IL-11 (Baumann and Schendel, 1991), oncostatin-M (Richards *et al.*, 1992; Richards and Shoyab, 1992) and cardiotrophin-1 (Richards *et al.*, 1996); however, the roles that these molecules play in liver responses *in vivo* are still to be clarified.

In examining the effects of IL-6 on local structural cells, it has more recently become clear that IL-6 is capable of modulating gene expression in connective-tissue cells including fibroblasts derived from various tissues, and chondrocytes, and is a modulator of growth in keratinocytes (Grossman *et al.*, 1989) and endothelial cells (May *et al.*, 1986). IL-6 induces the tissue inhibitor of metalloproteinases (TIMP-1) in fibroblasts and chondrocytes (Lotz and Guerne, 1991; Sato *et al.*, 1990), though some workers suggest not in skin fibroblasts (Emonard *et al.*, 1992). Since enzymatic breakdown of collagen in connective tissue is ultimately controlled by the balance of active metalloproteinases (MMP) and their inhibitors (such as TIMP-1), control of this balance by IL-6 and other cytokines will contribute to the regulation of extracellular matrix breakdown in normal and disease processes. Related cytokines LIF, IL-11, and oncostatin M also show this activity (Maier *et al.*, 1993; Richards *et al.*, 1993). IL-6 has further been shown to enhance production of α_1-antitrypsin (α_1-protease inhibitor, a major inhibitor of neutrophil elastase) by alveolar macrophages and mucosal enterocytes (Barbey-Morel *et al.*, 1987; Perlmutter, 1989). Thus, through upregulation of both metalloproteinase inhibitors and serine proteinase inhibitors by cells at sites of inflammation, IL-6 may further act to modulate enzymatic degradation of tissue. However, this may not be the case in all situations, since IL-6 can increase MMP expression *in vitro* when added in combination with the proinflammatory cytokine IL-1; IL-6 caused a further upregulation of collagenase and stromelysin (Ito *et al.*, 1992). Furthermore, in ovarectomized mice, subsequent bone loss due to decrease estrogen was mediated by IL-6, since antibodies to IL-6 could ameliorate the effects (Jilka *et al.*, 1992). Since estradiol can suppress IL-6 production by bone and marrow stromal cells, this suggests that post-menopausal bone loss is mediated by enhanced IL-6 production in these tissues.

In an *in vivo* model of acute inflammation, Ulich and colleagues found that IL-6 administration inhibits LPS-induced neutrophil accumulation in rat lungs (Ulich *et al.*, 1991, 1992). This may be due to IL-6-mediated inhibition of production of IL-1 and TNF (Aderka *et al.*, 1989; Schindler *et al.*, 1990), but this has not been established *in vivo*. Tilg and colleagues (1994) found that IL-6 augmented circulating levels of IL-1 receptor antagonist and soluble TNF receptor, both of which could have important anti-inflammatory effects by suppressing action of IL-1 or TNF. Furthermore, IL-6 administered by a miniosmotic pump was found to modify joint inflammation in collagen-induced arthritis in rats (Seckinger *et al.*, 1990). In the mouse model of endotoxic shock, treatment with monoclonal antibody to murine IL-6 resulted in protection from lethal challenge from *E. coli* or TNF (Starnes *et al.*, 1990). However, it is not clear how IL-6 might have this effect, since it has also been suggested that the serum half-life of IL-6 can be increased with antibody treatment, which could alter the responses invoked.

5.5   OTHER EFFECTS

IL-6 has activity in neurological systems which are integrated into the acute inflammatory response. IL-6 has been shown to cause fever upon injection into rabbits (Helle *et al.*, 1988), an activity also shown by IL-1 and TNF. Pituitary cells (At-20 cell line) will increase ACTH production in response to IL-6 or IL-1 *in vitro* (Woloski *et al.*, 1985) and IL-6 treatment was shown to enhance serum ACTH in rats (Naitoh *et al.*, 1988; Perlstein *et al.*, 1993). ACTH stimulates the adrenal gland and thus serum glucocorticoid levels, which in turn may modulate numerous cellular functions. Since many cytokines are repressed by glucocorticoids, sufficient concentrations may serve as a feedback system to turn off expression of cytokines such as IL-1, TNF, chemokines and IL-6 itself. Thus, effects of IL-6 in the central nervous system control multiple aspects of the systemic acute-phase response. IL-6 also induces c-*fos* expression in a chromaffin cell line PC12 and differentiation into neural cells (Satoh *et*

al., 1988) and may have further activities in neuronal tissues.

In summary, IL-6 appears to be a ubiquitous and necessary part of mammalian homeostatic response to body injury or tissue damage. Its effects in immunological, hematopoietic, and inflammatory systems reflect its importance in this respect. Studies on its interaction with the gp130-containing receptor complex have defined the IL-6-type cytokine family, all of which utilize gp130 as a signal transducing molecule. This provides a system where redundancy of cytokine action occurs in cells that possess the α receptor components for IL-6, LIF, IL-11, CNTF, and oncostatin M, possibly allowing for insurance of the response of that cell or tissue.

6.   *Receptors*

6.1   RECEPTOR SUBUNITS AND FUNCTIONAL COMPLEX

The nature of the cell surface receptor complex that binds the IL-6 ligand and resulting signal transduction has been examined in detail over the last several years (reviewed in Taga and Kishimoto, 1992; Kishimoto *et al.*, 1994). An 80 kDa IL-6 binding protein (IL-6R) from B cells was first cloned by Kishimoto and colleagues (Yamasaki *et al.*, 1988; Hirano *et al.*, 1989a). This contains a typical "cytokine receptor family" domain, characterized by four conserved cysteine residues and the WSXWS motif, as well as an immunoglobulin-like

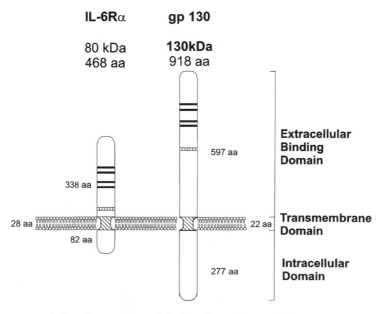

Figure 6.5   IL-6 receptor chains. Structural model of IL-6Rα and gp130. Thick bars represent four conserved cysteine residues and a WSXWS motif that feature the cytokine receptor family domain. The fibronectin type III module comprises ~100 amino acid residues. (From Taga *et al.* (1992).)

domain at the amino terminus (see Figure 6.5). A second component was characterized and cloned later as gp130 (Taga *et al.*, 1989; Hibi *et al.*, 1990 and mouse gp130 Saito *et al.*, 1992). The gp130 sequence also shows a "cytokine receptor family domain" as well as four fibronectin type III modules in the extracellular region. IL-6 ligand does not bind to gp130 itself but binds with higher affinity to complexes of IL-6R and gp130. Signal transduction was mediated by complexes containing both but not either of these two components alone (Hirano *et al.*, 1989a; Taga *et al.*, 1989). IL-6R displays very little intracytoplasmic domain in its structure, whereas the gp130 molecule displays a large cytoplasmic domain which is responsible for intracellular signaling. The sequences shows homology to box 1 and 2 of several other members of the cytokine receptor family (Murakami *et al.*, 1991; Kishimoto *et al.*, 1994) (see Figure 6.5) as well as a box 3 region also seen in receptors for LIF and oncostatin M. The gp130 molecule is found expressed in many tissues and cells, whereas the IL-6R is more restricted in its level of expression (Hibi *et al.*, 1990). Upon IL-6 ligand interaction, the IL-6/IL-6R interacts with a homodimer of gp130 molecules (Murakami *et al.*, 1993) which then initiates signal transduction (Figure 6.6). The bound IL-6 is rapidly internalized and degraded in rat hepatocytes (Nesbitt and Fuller, 1992a) and human HepG2 cells (Zohlnhoefer *et al.*, 1992) resulting in receptor downregulation. Both IL-6R and gp130 subunits are required, and a 10-amino-acid region of gp130 (intracellular) that contains a dileucine motif appears to be necessary for efficient internalization (Dittrich *et al.*, 1994).

Detailed structural analysis of IL-6 and its receptor awaits crystallographic data. However, on the basis of tertiary structural similarities to growth hormone and its receptor, models of IL-6 and its receptor complex have been predicted (Bazan, 1990). The bundle of α-helices of IL-6 putatively form two sites important in IL-6 activation of cells that are located on opposite sides (see Figure 6.4). Site 1 would be important for binding IL-6R and site 2 for binding and triggering dimerization of gp130 (Ehlers *et al.*, 1994; Brakenhoff *et al.*, 1994). Double mutations of Tyr-31 and Glu-35 (site 2) resulted in an IL-6 mutein with >80% reduction in bioactivity and a reduction in ability to associate with gp130 *in vitro*

(Savino *et al.*, 1994). Binding to IL-6Rα was not impaired, suggesting this region was critical for gp130 interaction but not IL-6Rα. This double mutant acts as an antagonist for IL-6 activity since it associates with the receptor but does not activate signals. Van Dam and colleagues (1993) have shown that in human/murine chimeric IL-6 molecules, the region Lys-41 – Glu-95 (between the first and second α-helices, AB loop containing "site 1") is important for IL-6 binding to the receptor. Thus site 1 appears critical for IL-6R binding and site 2 for association and activation of gp130 dimerization.

6.2 REGULATION OF MEMBRANE IL-6 RECEPTOR AND SOLUBLE RECEPTOR

The 80 kDa IL-6R mRNA levels can be enhanced *in vitro* in hepatoma cells by PMA, IL-1, IL-6, and glucocorticoid (Bauer *et al.*, 1989; Rose-John *et al.*, 1990; Gauldie *et al.*, 1992; Pietzko *et al.*, 1993); however, it is not clear whether this imparts significant differences in biological responses in these cells. Nesbitt and Fuller have shown that in primary rat hepatocytes, dexamethasone enhanced IL-6R by 2.7-fold and induced receptor number from 600 to >6000 per cell (Nesbitt and Fuller, 1992b). Alterations of IL-6R mRNA were also evident during acute inflammatory responses in liver of rats *in vivo* (Geisterfer *et al.*, 1993), and administration of IL-6 or corticosteroid can also increase IL-6R expression in liver cells. These transient modulations would modulate numbers of IL-6 receptor subunits on cells and thus presumably modify sensitivity to cytokine levels. The expression of gp130 mRNA is inherently much higher in various cell types, and can also be regulated by cytokines *in vitro* (Schooltink *et al.*, 1992; Snyers and Content, 1992). However, most cells or tissues examined express gp130, and thus specificity of IL-6 responsiveness may depend more on the expression and regulation of the IL-6R 80 kDa subunit.

The detection of soluble receptors for several cytokines has been demonstrated in body fluids (Novick *et al.*, 1989; Rose-John and Heinrich, 1994). Soluble IL-6R has been found in serum of patients with juvenile rheumatoid arthritis (Debenedetti *et al.*, 1994) or with HIV infection (Honda *et al.*, 1992). The production of the soluble IL-6R can occur due to proteolytic cleavage of the membrane bound 80 kDa receptor from COS cells transiently transfected with gp80 (Mullberg *et al.*, 1993, 1994; Rose-John and Heinrich, 1994). This shed receptor can bind IL-6 and the complex is active *in vitro* in stimulating HepG2 cells (Gottschall *et al.*, 1992). Other studies suggest that a differently spliced mRNA can be found for

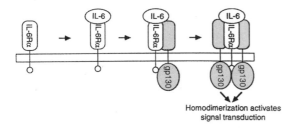

Homodimerization activates
signal transduction

Figure 6.6   Model of IL-6 receptor/ligand interaction. (Taken from Murakami *et al.* (1993).)

IL-6R that does not encode a transmembrane region (Lust *et al.*, 1992) and thus may produce a significant amount of secreted soluble receptor. In contrast to several other cytokine receptors which inhibit ligand function, soluble IL-6/IL-6R complexes can activate signal transduction and biological responses in cells that possess gp130 (Hibi *et al.*, 1990; Taga *et al.*, 1992; Mackiewicz *et al.*, 1992). This suggests that the soluble complex can activate cells that do not express 80 kDa IL-6R; but how this fits in to biological responses *in vivo* is not clear. Soluble gp130 has also been found in significant levels in serum, and may act as a competitor with cell-associated gp130 for soluble IL-6/IL-6R (Narazaki *et al.*, 1993).

7. Signal Transduction and Gene Regulation

The lack of a significant intracytoplasmic region of IL-6Rα, and identifiable sequences that could be implicated in signaling processes, is consistent with studies showing a lack of functional activity of this chain alone in gene regulation. Mutational analysis of the intracytoplasmic region of the gp130 molecule has led to identification of multiple regions implicated in signal transduction processes (Fukada *et al.*, 1996). This reflects the pleiotropic nature of IL-6, where separate sets of signals appear to be needed to regulate such a diverse array of genes and cellular effects. Multiple sites are phosphorylated on tyrosine residues of gp130; however, gp130 does not appear to have any inherent tyrosine kinase activity. Instead, gp130 associates with tyrosine kinases such as Janus kinase family members JAK/TYK (Stahl *et al.*, 1994) and possibly others such as the *src*-related tyrosine kinase Hck (Ernst *et al.*, 1994, 1996). Multiple defects in mouse embryos with targeted disruption of gp130 (Yoshida *et al.*, 1996) emphasize the importance of this molecule.

Deletion analysis of the cytoplasmic domains of gp130 has shown that the first 60–65 amino acids from the membrane are sufficient to mediate proliferative signals (Murakami *et al.*, 1991) (see Figure 6.5). This region contains homology to box 1 (10–18 amino acids from membrane) and box 2 (52–62 amino acids from membrane) of other hematopoietin receptors such as G-CSF, erythropoietin, growth hormone, and IL-2. More distant is a segment designated box 3 (123–130 amino acids from transmembrane region) that contains a tyrosine residue (protein 125 from membrane). Using chimeric receptor studies, Baumann and colleagues (1994) have suggested that box 3 domains are essential for signals that regulate IL-6R response element (IL-6RE)-transcription such as in the rat fibrinogen or haptoglobin genes. It appears that boxes 1 and 2 are necessary for STAT activation (signal transducers and activators of transcription) by associating with the JAK1 and JAK2 kinases that are responsible for phosphorylation of STAT (see below).

Two signal pathways induced by IL-6 have been described that may act simultaneously in certain cells or preferentially in different cell types. Certain genes are targets of the tyrosine phosphorylation pathway involving Janus kinases and the STAT family of nuclear factors. IL-6 ligand causes homodimerization of gp130 (Murakami *et al.*, 1993). There appears to be constitutive association between gp130 and JAK1, JAK2, and TYK2, and tyrosine kinases activity is activated with IL-6 binding (Guschin *et al.*, 1995), probably through conformational change of the receptor complex. Specific sites on the gp130 homodimer (box 1) are necessary for the JAK/TYK kinase interaction, since mutations in these sites abolish subsequent events (Stahl *et al.*, 1994). STAT family members then associate with gp130 dimers and act as substrates for JAK/TYK. Upon tyrosine phosphorylation, STATS translocate to the nucleus as homodimers or heterodimers to then bind to specific target DNA and activate transcription (Sadowski *et al.*, 1993; Yuan *et al.*, 1994; Akira *et al.*, 1994; Zhong *et al.*, 1994b). Mutation of box 3 motif resulted in inhibiting STAT activation, and transferring the motif to different receptors can confer a previously unrecognized STAT activation. Immediate events following IL-6 ligand and receptor interaction thus involve rapid tyrosine phosphorylation of gp130, JAK-kinase activation, phosphorylation, and translocation of STAT proteins. Fourcin and colleagues (1996) have found that cross-linking of gp130 itself with an antibody can activate gp130 dimerization, JAK kinase tyrosine phosphorylation of gp130, and results in biological effects. More recently, IL-6 has been shown to activate the adaptor molecule SHC that can bind to JAK2 and gp130 (Giordano *et al.*, 1997), as well as the phosphatase SHP-2 (Fukada *et al.*, 1996).

A second pathway that involves the GTP-binding protein Ras which may also be involved in many cytokine systems is reviewed in Satoh *et al.*, (1992). GTP-binding motifs are present in the gp130 intracytoplasmic region (Hibi *et al.*, 1990); however, their precise role if any is unclear. IL-6-induced differentiation of PC-12 cells results in elevated levels of GTP-binding Ras (Nakafuku *et al.*, 1992), and in a different system a Ras transgene showed enhanced trans-activating action of NFIL-6 (Nakajima *et al.*, 1993). This Ras-dependent pathway is dependent on intermediate steps involving Raf, MEK (MAP kinase kinase), and MAP kinase (Kishimoto *et al.*, 1994) and thus may overlap with pathways induced by IL-1 as described above (for IL-1 induction of the IL-6 gene). Other serine threonine protein kinases can also be activated by IL-6 (Yin and Yang, 1994).

IL-6 regulation of acute-phase protein genes in liver cells has provided a system for studying gene regulation by IL-6. Studies of IL-6-responsive genes initially characterized a consensus sequence [(T/A)T(C/G) TGGGA(A/C)] that confers transcriptional enhancement of downstream genes, variably referred to as the IL-6 response element (IL-6RE), acute-phase response element (APRE), or LIF response element (LIFRE). A consensus sequence was found at various locations and distances away from the transcriptional start site of acute-phase proteins including rat α-macroglobulin, haptoglobin, hemopexin, α_1-cysteine proteinase inhibitor, and α_1-acid glycoprotein (Prowse and Baumann, 1988; Baumann et al., 1989, 1990; Hattori et al., 1990; Won and Baumann, 1990) as well as human haptoglobin and CRP (Fowlkes et al., 1984; Tsuchiya et al., 1987; Castell et al., 1989; Fey et al., 1989). This sequence enhances transcription of downstream indicator genes in cells stimulated with IL-6 as well as HepG2 cells stimulated with other members of the IL-6-type cytokine family, including oncostatin M and LIF (Richards et al., 1992; Baumann et al., 1994). The factor(s) that recognize this sequence [termed acute-phase response factor (APRF) and IL-6 response element binding protein (IL-6REBP)] increase binding activity dramatically shortly after ligand signaling (Wegenka et al., 1993). The characterization and cloning of APRF revealed identity with STAT3, a novel member of the STAT family (Akira et al., 1994; Zhong et al., 1994b). STAT-3 and STAT-1 (or p91) are capable of interacting as homodimers (STAT1/1, STAT3/3) or heterodimers (STAT1/STAT3) (Sadowski et al., 1993; Akira et al., 1994; Yuan et al., 1994; Zhong et al., 1994a,b). STAT-1 has been shown to activate genes downstream of the interferon-γ-activated sequence (GAS) but not of mutated GAS elements (Shuai et al., 1994). Studies of M1 cells show that dominant negative forms of STAT3 block IL-6- and LIF-mediated growth arrest and apoptosis (Masashi et al., 1996), suggesting STAT3 is a critical step in gene regulation in these cells. In contrast, STAT3 mediates anti-apoptosis in B cells (Fukada et al., 1996). In addition to STAT3, STAT5β (recently cloned by Ripperger et al., 1995) can also bind IL-6 response elements, although Box 3 of gp130 does not appear necessary for activation of STAT5b (Lai et al., 1995). It is also possible that IL-6-induced gene regulation involves interaction between transcription factor families for maximal induction. For example, STAT3β cooperates with Jun in transcriptional activation (Schaefer et al., 1995); however, this has not yet been shown to occur in IL-6 signaling. In addition, STAT3 may be further phosphorylated at serine/threonine residues which could participate in regulation (Boulton et al., 1995).

As well as direct activation of nuclear factors by post-transcriptional modification, IL-6 also activates expression of "early" genes which can in turn regulate other targets. These include junB and EGR-1 in rat hepatoma cells (Baumann et al., 1991; Won et al., 1993), junB and TIS11 in B cells (Nakajima and Wall, 1991), CEBPδ or NFIL-6β (Akira et al., 1990b) in hepatocytes, and junB, c-jun, and jun-D in M1 cells (Lord et al., 1991). These may be linked to proliferation activities of IL-6.

8. Murine IL-6, Transgenics and Knockouts

The cDNA for mouse has been published (Van Snick et al., 1988; Fuller and Grenett, 1989) and the mouse IL-6 gene has been located on chromosome 5 (Billiau, 1987). As in human cells, the gene for mouse IL-6 is transcribed from a ~5 kb gene that contains four exons and six introns, and is spliced to 1.2 to 1.3 kb mature mRNA (Yasukawa et al., 1987). High homology to the human IL-6 sequence is evident in the promoter and regions 5′ of the transcriptional start site (Tanabe et al., 1989), suggesting that similar transcriptional control mechanisms are present in murine cells as previously discussed. The primary transcripts of rat IL-6 also shows a 2.5 kb mRNA species as well as the 1.3 kb message (Northemann et al., 1989). The murine mRNA encodes a 211-amino-acid peptide that is cleaved to form a 186-amino-acid mature protein; however, unlike human IL-6, murine IL-6 appears not to be glycosylated (Hirano et al., 1986). Many activities of IL-6 have been found upon characterization of the murine analog, as noted in previous sections.

On the basis of the phenotype of IL-6 knockout mice, the role of endogenous IL-6 in specific immune responses has become more clear. IL-6-deficient mice with a disrupted IL-6 gene (by homologous recombination) develop without any marked gross abnormalities. When challenged with vaccinia virus or vesicular stomatitis virus, they are unable to efficiently control infection and possess impaired antibody responses to vesicular stomatitis virus antigen (Kopf et al., 1994). Control of Listeria monocytogenes infection is also impaired.

In examining responses to inflammatory stimuli, the same authors found that subcutaneous turpentine-induced liver responses were largely inhibited, but that intravenous LPS-induced liver responses were not (Kopf et al., 1994). As described earlier, other cytokines such as LIF, IL-11, or oncostatin M can substitute for IL-6 action on liver cells in vitro, and thus may be involved in the LPS model in the knockout mouse. This suggests that the sites of stimuli play a large part in determining the IL-6 dependence or cytokines that are elicited and which in turn control hepatic acute-phase protein production.

Effects of IL-6 in transgenic mice are interesting. Under control of the human Ig heavy-chain enhancer, IL-6 overexpression caused increases in polyclonal serum Ig, splenomegaly, lymphoma, and thymoma (Suematsu et al., 1989). These animals also developed plasmacytosis and appeared to succumb primarily to proliferative glomerulonephritis. This is interesting in the light of studies which show that IL-6 is a growth factor for mesangial cells, and that mesangial cells from patients with proliferative glomerulonephritis constitutively produce IL-6 (Horii *et al.*, 1989). Thus IL-6 overproduction may play a significant role in this human disease.

Other vector systems have been used to examine the effects of IL-6 *in vivo*. Hawley and colleagues (1992) have transferred IL-6 to mouse bone marrow cells using a retrovirus vector. Upon transplantation of these cells into irradiated mice, a strong prolonged overexpression of IL-6 led to pronounced increases in serum acute-phase proteins and serum immunoglobulin levels. In addition, a lethal myeloproliferative disease characterized by large amounts of neutrophilic granulocytes in blood and spleen was evident (Hawley *et al.*, 1992). Using adenovirus 5 as a gene transfer method, Gauldie and associates have shown that recombinant adenovirus encoding IL-6 causes significant expression of IL-6 upon infection in rodents (Braciak *et al.*, 1993; Xing *et al.*, 1994). Intraperitoneal infection caused an increase in serum IL-6 and typical serum acute-phase protein response (Braciak *et al.*, 1993). Infection intratracheally caused marked local expression of virus-encoded IL-6 in lung tissue (Xing *et al.*, 1994) and this was associated with a transient lymphocytic expansion in lungs at day 7 that subsided by day 12. Thus the localization of IL-6 overexpression has an important influence on the nature of IL-6 effects seen *in vivo*.

9. Clinical Implications

The presence of elevated IL-6 in many different diseases implies that this cytokine is consistently an important part of either the disease process or the body's response to disease. Enhanced IL-6 in serum has been found in a wide variety of trauma or inflammatory conditions, such as in serum of patients in trauma/surgery (Nijsten *et al.*, 1987; Van Oers *et al.*, 1988) and in cerebral spinal fluid of patients with CNS infection (Honssiau *et al.*, 1988) or in vasculitis with CNS involvement (Hirohata *et al.*, 1993). IL-6 levels are enhanced in serum of patients with Crohn disease (Gross *et al.*, 1992), with systemic lupus erythematosus (Linker-Israeli *et al.*, 1991), and with alcoholic liver cirrhosis (Deviere *et al.*, 1989), and in patients with Castleman disease (Yoshizaka *et al.*, 1989). IL-6 is significantly enhanced in synovial fluid in rheumatoid arthritis (Hirano *et al.*, 1988b; Honssiau *et al.*, 1988) as well as in mesangial cells of patients with mesangial proliferative glomerulonephritis (Horii *et al.*, 1989). IL-6 has also been detected in multiple myeloma where it is expressed by tumor cells (Kawano *et al.*, 1988) or stromal cells (Klein *et al.*, 1989); in renal cell

carcinoma, expressed by tumor cells (Koo *et al.*, 1992); in cardiac myxoma patients, expressed by tumor cells (Jourdan *et al.*, 1990) and also found in serum (Hirano *et al.*, 1987).

Analysis of IL-6 as a clinical marker of inflammation or infection has shown it to correlate with other indices of disease activity, and it is now used for this purpose in combination with other clinical tests (Madhok *et al.*, 1993). Whether these levels could be determined as "overexpressed" (thus leading to IL-6-dependent pathology) is not clear. As is the case with most other regulatory molecules, the role of cytokines as stimulators or inhibitors of disease processes may depend on the concentration and the time over which the cytokines exert relevant biological actions. Overexpression of IL-6 during development may lead to quite different effects from those of overexpression in an adult mammal. IL-6 transgenic mice exhibit profound disease in many aspects; however, the relevance to human conditions of overexpression of such magnitude is not clear. The transgenics do provide valuable information in terms of what disease may be associated with IL-6, and a number of disease conditions have been examined in more detail.

IL-6-transgenic animals show a profound polyclonal plasmacytosis (Suematsu *et al.*, 1989). The ability of IL-6 to act as a growth factor for plasma cells and for human myeloma cells (Kawano *et al.*, 1988), which may or may not be autocrine in nature, suggests a role in neoplasia. Bone marrow stromal cells may be the most important source of IL-6 in myeloma patients (Klein *et al.*, 1989). Increased serum levels of IL-6 correlate well with disease severity in multiple myelomas and plasmacytomas (Bataille *et al.*, 1989). Interestingly, Suematsu and colleagues (1992) have shown a translocation in IL-6 transgenic mice involving c-*myc*, implying a role for overexpressed IL-6 in genetic mutations.

Others have examined the effects of exogenously administered IL-6 in mouse models of tumorigenicity. *E. coli*-derived recombinant IL-6 was able to substantially reduce metastases to lung and liver in syngeneic tumors, apparently indirectly through a host component (not directly) since irradiation abrogated the effect (Mule *et al.*, 1990). More recently this has been defined as the host cytotoxic T lymphocyte (CTL) response which is modulated by IL-6. Transfection of the IL-6 gene into 3LL tumor cells suppresses malignant growth and confers protection against parental metastatic cells (Porgador *et al.*, 1993) upon injection into mice. Long-term IL-6 therapy (1–10 μg/day) using human IL-6 in mice markedly inhibited lung metastasis of B16 melanoma cells after intravenous inoculation (Katz *et al.*, 1993). These effects suggest that IL-6 therapy has potential in certain tumors. However, since different states of melanoma progression have opposite responses to IL-6 *in vitro*, caution must prevail in predicting IL-6 actions in human disease.

One of the more pronounced effects of IL-6

recognized is that as a maturation factor for megakaryocytes *in vitro* and in IL-6 transgenic animals (Kawano *et al.*, 1988). In preclinical studies IL-6 has since been established (along with IL-11, LIF, and oncostatin M) as potentially useful in supporting thrombopoiesis and platelet numbers in patients with high-dose chemotherapy and bone marrow transplants. Along with its potential as an enhancer of CTL activity against certain tumors, this activity has encouraged phase I/II clinical trials in the United States (Weber *et al.*, 1993). The results of one set of studies have suggested that significant elevation of platelet counts, CRP, fibrinogen, ACTH, and cortisol occur with subcutaneous IL-6 therapy (Mule *et al.*, 1992; Weber *et al.*, 1993). However, no antitumor effects were seen against advanced metastatic melanoma, and a rapid (reversible) anemia was observed. High incidence of side-effects (fevers, chills, nausea, fatigue) and unexpected toxicity (hyperglycemia, atrial fibrillation, and reversible neurological symptoms) was also observed. In most cases the treatment was terminated. Using a separate product (glycosylated, mammalian expressed sigosix), another set of trials revealed fewer side-effects, albeit at 10-fold less IL-6 injected. Further analysis of efficacy must be completed in this study; however, it does appear that the usefulness of IL-6 as a therapeutic agent will be hampered by various side-effects, some of which are common to other cytokine treatments.

Since IL-6 has been strongly implicated as driving proliferation of multiple myelomas, inhibition of IL-6 activity would theoretically be useful in this disease. This could be accomplished using antibody to IL-6 or IL-6 muteins, both areas having received attention recently. Klein and associates have reported results of phase 1 trials using 20 mg/day for 6 days of monoclonal antibody to IL-6 in terminal multiple myeloma patients (Klein *et al.*, 1992). A significant antitumor effect was noted in 5 of 9 patients, with low toxicity observed. There was a partial drop in platelets and in neutrophil counts in 2 of 5 patients. These results have encouraged further trials.

IL-6 overexpression has also been implicated in a variety of autoimmune conditions (reviewed in Hirano *et al.*, 1990). Cardiac myxoma patients yielded myxoma cells that produced IL-6 (Jourdan *et al.*, 1990). Upon resection of the myxoma tumor, bone marrow plasmacytosis, hypergammaglobinemia, autoantibodies, serum IL-6, and serum acute phase proteins decreased. Owing to its activities as a growth factor for mesangial cells, its production by mesangial cells from patients, and its presence in patient urine, IL-6 has been implicated in a causative role in proliferative glomerulonephritis (Horii *et al.*, 1989). Castleman disease, characterized by hyperplastic lymphadenopathy, hypergammaglobinemia, and increases in platelets, is also associated with high IL-6 levels and polyclonal plasmacytosis (Yoshizaka *et al.*, 1989). Synovial fluid from patients with rheumatoid arthritis shows enhanced IL-6 levels (Hirano *et al.*,

1988b) and it has been suggested that this contributes to synovial B cell activation and Ig production (Nawata *et al.*, 1989). However, the abnormal expression of IL-6 is not yet linked to this particular disease as a primary cause, and may be a secondary response to other processes. Furthermore, high IL-6 alone may not be sufficient to cause organ-specific autoimmune disease.

10. References

Aarden, L.A., De Groot, E.R., Schaap, O.L. and Lansdorp, P.M. (1987). Production of hybridoma growth factor by human monocytes. Eur. J. Immunol. 17, 1411–1416.

Aderka, D., Le, J. and Vilcek, J. (1989). IL-6 inhibits lipopolysaccharide-induced tumor necrosis factor production in cultured human monocytes, U937 cells, and in mice. J. Immunol. 143, 3517–3523.

Akira, S., Hirano, T., Taga, T. and Kishimoto, T. (1990a). Biology of multifunctional cytokines: IL-6 and related molecules (IL 1 and TNF). FASEB J. 4, 2860–2867.

Akira, S., Isshiki, H., Sugita, T., *et al.* (1990b). A nuclear factor for IL-6 expression (NF-IL6) is a member of a C/EBP family. EMBO J. 9, 1897–2005.

Akira, S., Nishio, Y., Inoue, M., *et al.* (1994). Molecular cloning of APFR, a novel IFN-stimulated gene factor 3 p91-related transcription factor involved in the gp130-mediated signaling pathway. Cell 77, 63–71.

Amano, Y., Lee, S. and Allison, A. (1992). Inhibition of glucocorticoids of the formation of interleukin-1a, interleukin-1B and interleukin-6: mediation by decreased mRNA stability. Mol. Pharmacol. 43, 176–182.

Andus, T., Geiger, T., Hirano, T., *et al.* (1987). Recombinant human B cell stimulatory factor 2 (BSF-2/IFN-beta2) regulates beta-fibrinogen and albumin mRNA levels in Fao-9 cells. FEBS Lett. 221, 18–22.

Andus, T., Geiger, T., Hirano, T., *et al.* (1988). Regulation of synthesis and secretion of major rat acute-phase proteins by recombinant human interleukin-6 (BSF-2/IL-6) in hepatocyte primary cultures. Eur. J. Biochem. 173, 287–293.

Barbey-Morel, C., Pierce, J.A., Campbell, E.J. and Perlmutter, D.H. (1987). Lipopolysaccharide modulates the expression of α-1-proteinase inhibitor and other serine proteinase inhibitors in human monocytes and macrophages. J. Exp. Med. 166, 1041–1047.

Bataille, R., Jourdan, M., Zhang, X.G. and Klein, B. (1989). Serum levels of interleukin-6, a potent myeloma cell growth factor, as a reflect of disease severity in plasma cell dyscrasias. J. Clin. Invest. 84, 2008–2011.

Bauer, J., Birmelin, M., Northoff, G.H., *et al.* (1984). Induction of rat alpha-2-macroglobulin in vivo and in hepatocyte primary cultures: synergistic action of glucocorticoids and a Kupffer cell derived factor. FEBS Lett. 177, 89–94.

Bauer, J., Ganter, U., Geiger, T., *et al.* (1988). Regulation of interleukin-6 expression in cultured human blood monocytes and monocyte-derived macrophages. Blood 72, 1134–1140.

Bauer, J., Lengyel, G., Bauer, T.M., Acs, G. and Gerok, W. (1989). Regulation of interleukin-6 receptor expression in human monocytes and hepatocytes. FEBS Lett. 249, 27–30.

Baumann, H. and Gauldie, J. (1990). Regulation of hepatic acute phase plasma protein genes by hepatocyte stimulating factors and other mediators of inflammation. Mol. Biol. Med. 7, 147–159.

Baumann, H. and Gauldie, J. (1994). The acute phase response. Immunol. Today 15, 74–80.

Baumann, H. and Schendel, P. (1991). Interleukin-11 regulates the hepatic expression of the same plasma protein genes as interleukin-6. J. Biol. Chem. 266, 1–4.

Baumann, H. and Wong, G.G. (1989). Hepatocyte-stimulating Factor III shares structural and function identity with leukemia inhibitory factor. J. Immunol. 143, 1163–1167.

Baumann, H., Prowse, K.R., Marinkovic, S., Won, K.-A. and Jahreis, G.P. (1989). Stimulation of hepatic acute phase response by cytokines and glucocorticoids. Ann. N.Y. Acad. Sci. 557, 280–296.

Baumann, H., Morella, K.K., Jahreis, G.P. and Marinkovic, S. (1990). Distinct regulation of the interleukin-1 and interleukin-6 response elements of the rat haptoglobin gene in rat and human hepatoma cells. Mol. Cell. Biol. 10, 5967–5976.

Baumann, H., Jahreis, G.P., Morella, K.K., et al. (1991). Transcriptional regulation through cytokine and glucocorticoid response elements of rat acute phase plasma protein genes by C/EBP and JunB. J. Biol. Chem. 266, 20390–20399.

Baumann, H., Symes, A.J., Comeau, M.R., et al. (1994). Multiple regions within the cytoplasmic domains of the leukemia inhibitory factor receptor and gp130 cooperate in signal transduction in hepatic and neuronal cells. Mol. Cell. Biol. 14, 138–146.

Bazan, F. (1990). Structural design and molecular evolution of a cytokine receptor superfamily. Proc. Natl Acad. Sci. USA 87, 6934–6938.

Bazan, J.F. (1991). Neuropoietic cytokines in the hematopoietic fold. Neuron 7, 197–208.

Billiau, A. (1987). Interferon B2 as a promoter of growth and differentiation of B cells. Immunol.Today 8, 84–87.

Birch, H. and Schreiber, G. (1986). Transcriptional regulation of plasma protein synthesis during inflammation. J. Biol. Chem. 261, 8077–8080.

Bird, T.A., Schule, H.D., Delaney, P.B., Sims, J.E., Thoma, B. and Dower, S.K. (1992). Evidence that MAP (mitogegen-activated protein) kinase activation may be a necessary but not sufficient signal for a restricted subset of responses in IL-1-treated epidermoid cells. Cytokine 4, 429–440.

Borth, W., Urbanski, A. and Luger, T.A. (1989). Binding of cytokines to plasma proteins. Ann. N.Y. Acad. Sci. 557, 512–514.

Bot, F.J., Van Eijk, L., Broeders, L., Aarden, L.A. and Lowenberg, B. (1989). Interleuken-6 synergizes wth M-CSF in the formation of macrophage colonies from purified human marrow progenitor cells. Blood 73, 435–437.

Boulton, T.G., Zhong, Z., Wen, Z., et al. (1995). STAT3 activation by cytokines utilizing gp130 and related transducers involves a secondary modification requiring an H7-sensitive kinase. Proc. Natl Acad. Sci. 92, 6915–6919.

Braciak, T.A., Mittal, S.K., Graham, F.L., Richards, C.D. and Gauldie, J. (1993). Construction of recombinant human type 5 adenoviruses expressing rodent IL-6 genes. J. Immunol. 151, 5145–5153.

Brakenhoff, J.P., Hart, M., De Groot, E.R., De Padova, F. and Aarden, L.A. (1990). Structure–function analysis of human IL-6 epitope mapping of neutralizing monoclonal antibodies with amino- and carboxyl-terminal deletions. J. Immunol. 145, 561–568.

Brakenhoff, J.P., de Hon, F.D., Fontaine, V., et al. (1994). Development of a human interleukin-6 receptor antagonist. J. Biol. Chem. 269, 86–93.

Brown, T.J., Rowe, J.M., Lui, J. and Shoyab, M. (1991). Regulation of interleukin-6 expression by oncostatin M. J. Immunol. 147, 2175–2180.

Bruce, A.G., Linsley, P.S. and Rose, T.M. (1992). Oncostatin M. Prog. Growth Factor Res. 4, 157–170.

Bunning, R.A.D., Russell, R.G.G. and Van Damme, J. (1990). Independent induction of interleukin 6 and prostaglandin E2 by interleukin 1 in human articular chondrocytes. Biochem. Biophys. Res. Commun. 166, 1163–1170.

Caput, D., Beutler, B., Hartog, K., Thayer, R., Brown-Shimer, S. and Cerami, A. (1986). Identification of a common nucleotide sequence in the 3'-untranslated region of mRNA molecules specifying inflammatory mediators. Proc. Natl Acad. Sci. USA 83, 1670–1674.

Castell, J.V., Andus, T., Kunz, D. and Heinrich, P.C. (1989). Interleukin-6: the major regulator of acute-phase protein synthesis in man and rat. Ann. N.Y. Acad. Sci. USA 557, 86–101.

Chen, L., Mory, Y., Zilberstein, A. and Revel, M. (1988). Growth inhibition of human breast carcinoma and leukemia/lymphoma cell lines by recombinant interferon-B2. Proc. Natl Acad. Sci. USA 85, 8037–8041.

Chen-Kiang, S., Hsu, W., Natkunam, Y. and Zhang, X. (1993). Nuclear signaling by interleukin-6. Curr. Opin. Immunol. 5, 124–128.

Chiu, C.-P., Moulds, C., Coffman, R.L., Rennik, D. and Lee, F. (1988). Proc. Natl Acad. Sci. USA 85, 7099–7103.

Cosman, D. (1987). Control of messenger RNA stability. Immunol. Today 8, 16–17.

Debenedetti, F., Massa, M., Pignatti, P., Albani, S., Novick, D. and Martini, A. (1994). Serum soluble interleukin 6 (IL-6) receptor and IL-6/soluble IL-6 receptor complex in systemic juvenile rheumatoid arthritis. J. Clin. Invest. 93, 2114–2119.

Dendorfer, U., Oettgen, P. and Libermann, T.A. (1994). Multiple regulatory elements in the interleukin-6 gene mediate induction by prostaglandins, cyclic AMP, and lipopolysaccharide. Mol. Cell. Biol. 14, 4443–4454.

Deviere, J., Content, J., Denys, C., et al. (1989). High interleukin-6 serum levels and increased production by leucocytes in alcoholic liver cirrhosis. Correlation with IgA serum levels and lymphokines production. Clin. Exp. Immunol. 77, 221–225.

de Waal Malefyt, R., Yssel, H., Roncarolo, M.-G., Spits, H. and de Vries, J.E. (1992). Interleukin-10. Curr. Opin. Immunol. 4, 314–320.

Dittrich, E., Rose-John, S., Gerhartz, C., et al. (1994). Identification of a region within the cytoplasmic domain of the interleukin-6 (IL-6) signal transducer gp130 important for ligand-induced endocytosis of the IL-6 receptor. J. Biol. Chem. 269, 19014–19020.

Donnelly, R.P., Crofford, L.J., Freeman, S.L., et al. (1993). Tissue-specific regulation of IL-6 production by IL-4. J. Immunol. 151, 5603–5612.

Ehlers, M., Grotzinger, J., de Hon, F.D., et al. (1994). Identification of two novel regions of human IL-6 responsible for receptor binding and signal transduction. J. Immunol. 153, 1744–1753.

Emonard, H., Munaut, C., Melin, M., Lortat-Jacob, H. and Grimaud, J. (1992). Interleukin-6 does not regulate interstitial collagenase, stromelysin and tissue inhibitor of metalloproteinases synthesis by cultured human fibroblasts. Matrix 12, 471–474.

Ernst, M., Gearing, D.P. and Dunn, A.R. (1994). Functional and biochemical association of Hck with the LIF/IL-6 receptor signal transducing subunit gp130 in embryonic stem cells. EMBO J. 13, 1574–1584.

Ernst, M., Oates, A. and Dunn, A.R. (1996). Gp130-mediated signal transduction in embryonic stem cells involves activation of Jak and Ras/mitrogen-activated protein kinase pathways. J. Biol. Chem. 271, 30136–30143.

Fey, G.H., Hattori, M., Northemann, W., *et al.* (1989). Regulation of rat liver acute phase genes by interleukin-6 and production of hepatocyte stimulating factors by rat hepatoma cells. Ann. N.Y. Acad. Sci. 557, 317–331.

Fourcin, M., Chevalier, S., Guillet, C., Robledo, O., Froger, J., Pouplard-Barthelaix, A. and Gascan, H. (1996). gp130 Transducing receptor cross-linking is sufficient to induce interleukin-6 type responses. J. Biol. Chem. 271, 11756–11760.

Fowlkes, D.M., Mullis, N.T., Comeau, C.M. and Crabtree, G.R. (1984). Potential basis for regulation of the coordinately expressed fibrinogen genes: homology in the 5′ flanking regions. Proc. Natl Acad. Sci. USA 81, 2313–2316.

Frei, K., Malipiero, U.V., Leist, T.P., Zinkernagel, R.M., Schwab, M.E. and Fontana, A. (1989). On the cellular source and function of interleukin-6 produced in the central nervous system in viral diseases. Eur. J. Immunol. 19, 689–694.

Fukada, T., Hibi, M., Yamanaka, Y., *et al.* (1996). Two signals are necessary for cell proliferation induced by a cytokine receptor gp130: involvement of STAT3 in anti-apoptosis. Immunity 5, 449–460.

Fuller, G. and Grenett, H. (1989). The structure and function of the mouse hepatocyte stimulating factor. Ann. N.Y. Acad. Sci. 557, 31–45.

Garman, R.D., Jacobs, K.A., Clark, S.C. and Raulet, D.H. (1987). B-cell-stimulatory factor 2 (beta2 interferon) functions as a second signal for interleukin 2 production by mature murine T cells. Proc. Natl Acad. Sci. USA 84, 7629–7633.

Gauldie, J. (1991). Acute phase response. In "The Encyclopedia of Human Biology", Vol. 1, pp. 25–35. Academic Press, New York.

Gauldie, J., Richards, C., Harnish, D., Lansdorp, P. and Baumann, H. (1987). Interferon-beta2/B-cell stimulatory factor type 2 shares identity with monocyte hepatocyte-stimulating factor and regulates the major acute phase protein response in liver cells. Proc. Natl Acad. Sci. USA 84, 7251–7255.

Gauldie, J., Geisterfer, M., Richards, C. and Baumann, H. (1992). IL-6 regulation of the hepatic acute phase response. In "IL-6: Physiopathology and Clinical Potentials". (ed. M. Revel). Serono Symposia Publications, Vol. 88, pp. 151–162. Raven Press, New York.

Geiger, T., Andus, T., Klapproth, J., Hirano, T., Kishimoto, T. and Heinrich, P.C. (1988). Induction of rat acute-phase proteins by interleukin 6 in vivo. Eur. J. Immunol. 18, 717–721.

Geisterfer, M., Richards, C.D., Gwynne, D., Baumann, M.

and Gauldie, J. (1993). Regulation of IL-6 and hepatic IL-6 receptor in acute inflammation in vivo. Cytokine 5, 1–7.

Giordano, V., De Falco, G., Chiari, R., *et al.* (1997). Shc mediates IL-6 signaling by interacting with gp130 and Jak2 kinase. J. Immunol. 158, 4097–4103.

Gottschall, P., Komaki, G. and Arimura, A. (1992). Increased circulating interleukin-1 and interleukin-6 after intracerebroventricular injection of lipopolysaccharide. Neuroendocrinology 56, 935–938.

Gross, V., Andus, T., Caesar, I., Roth, M. and Scholmerich, J. (1992). Evidence for continuous stimulation of interleukin-6 production in Crohn's disease. Gastroenterology 102, 514–519.

Grossman, R.M., Krueger, J. and Yourish, D. *et al.* (1989). Interleukin-6 is expressed in high levels in psoriatic skin and stimulated proliferation of cultured human keratinocytes. Proc. Natl Acad. Sci. USA 86, 6367–6371.

Guerne, P.-A., Zuraw, B.L., Vaughan, J.H., Carson, D.A. and Lotz, M. (1989). Synovium as a source of interleukin 6 in vitro. J. Clin. Invest. 83, 585–592.

Guschin, D., Rogers, N., Briscoe, J., *et al.* (1995). A major role for the protein tyrosine kinase JAK1 in the JAK/STAT signal transduction pathway in response to interleukin-6. EMBO J. 14, 1421–1429.

Haegeman, G., Content, J., Volckaert, G., Derynck, R., Tavernier, J. and Fiers, W. (1986). Structural analysis of the sequence coding for an inducible 26-kDa protein in human fibroblasts. Eur. J. Biochem. 159, 625–632.

Hattori, M., Abraham, L.J., Northemann, W. and Fey, G.H. (1990). Acute-phase reaction induces a specific complex between hepatic nuclear proteins and the interleukin 6 response element of the rat alpha2-macroglobulin gene. Proc. Natl Acad. Sci. USA 87, 2364–2368.

Hawley, R.G., Fong, A.Z.C., Burns, B.F. and Hawley, T.S. (1992). Transplantable myeloproliferative disease induced in mice by an interleukin 6 retrovirus. J. Exp. Med. 176, 1149–1163.

Heinrich, P.C., Castell, J.V. and Andus, T. (1990). Interleukin-6 and the acute phase protein. Biochem. J. 265, 621–636.

Helfgott, D.C., May, L.T., Sthoeger, Z., Tamm, I. and Seghal, P.B. (1987). Bacterial lipopolysaccharide (endotoxin) entrances expression and secretion of B2 interferon by human fibroblasts. J. Exp. Med. 166, 1300.

Helle, M., Brakenhoff, J.P.J., De Groot, E.R. and Aarden, L.A. (1988). Interleukin-6 is involved in interleukin-1-induced activities. Eur. J. Immunol. 18, 957–959.

Hibi, M., Murakami, M., Saito, M., Hirano, T., Taga, T. and Kishimoto, T. (1990). Molecular cloning and expression of an IL-6 signal transducer, gp130. Cell 63, 1149–1157.

Hirano, T., Yasukawa, K., Harada, H., *et al.* (1986). Complementary DNA for a novel human interleukin (BSF-2) that induces B lymphocytes to produce immunoglobulin. Nature 324, 73–76.

Hirano, T., Taga, T., Yasukawa, K., *et al.* (1987). Human B-cell differentiation factor defined by an anti-peptide antibody and its possible role in autoantibody production. Proc. Natl Acad. Sci. USA 84, 228–231.

Hirano, T., Matsuda, T., Hosoi, K., Okana, A., Matsui, H. and Kishimoto, T. (1988a). Absence of antiviral activity in recombinant B cell stimulatory factor 2 (BSF-2) Immunol. Lett. 17, 41–45.

Hirano, T., Matsuda, T., Turner, M., *et al.* (1988b). Excessive production of interleukin-6/B cell stimulatory factor-2 in rheumatoid arthritis. Eur. J. Immunol. 18, 1797–1801.

Hirano, T., Taga, T., Yamasaki, F., *et al.* (1989a). Molecular cloning of the cDNAs for IL-6 and its receptor. Ann. N.Y. Acad. Sci. 557, 167–180.

Hirano, T., Taga, T., Yamasaki, K., *et al.* (1989b). Molecular cloning of the cDNAs for interleukin-6/B cell stimulatory factor 2 and its receptor. Ann. N.Y. Acad. Sci. 557, 167–178.

Hirano, T., Akira, S., Taga, T. and Kishimoto, T. (1990). Biological and clinical aspects of interleukin 6. Immunol. Today 11, 443–449.

Hirohata, S., Tanimoto, K. and Ito, K. (1993). Elevation of cerebrospinal fluid interleukin-6 activity in patients with vasculitides and central nervous system involvement. Clin. Immunol. Immunopathol. 66, 225–229.

Homburger, F. (1945). A plasma fibrinogen-increasing factor obtained from sterile abscesses in dogs. J. Clin. Invest. 24, 43–45.

Honda, M., Yamamoto, S., Cheng, M., *et al.* (1992). Human soluble IL-6 receptor: Its detection and enhanced release by HIV infection. J. Immunol. 148, 2175–2180.

Honssiau, F., Bukasa, K., Sindic, C., Van Damme, J. and Van Snick, J. (1988). Interleukin-6 in synovial fluid and serum of patients with rheumatoid arthritis and other inflammatory arthritides. Arthritis Rheum. 31, 784–788.

Horii, Y., Muraguchi, A., Suematsu, S., *et al.* (1988). Regulation of BSF-2/IL-6 production by human mononuclear cells: macrophage-dependent synthesis of BSF-2/IL-6 by T-cells. J. Immunol. 141, 1529–1535.

Horii, Y., Muraguchi, A. and Iwano, M. *et al.* (1989). Involvement of IL-6 in mesangial proliferative glomerulonephritis. J. Immunol. 143, 3949–3955.

Houssiau, F. and Van Snick, J. (1992). IL6 and the T-cell response. Res. Immunol. 143, 740–743.

Houssiau, F., Coulie, P.G., Olive, D. and Van Snick, J. (1988a). Synergistic activation of human T cells by interleukin 1 and interleukin 6. Eur. J. Immunol. 18, 653–656.

Houssiau, F.A., Bukasa, K., Sindic, C.J.M., Van Damme, J. and Van Snick, J. (1988b). Elevated levels of the 26K human hybridoma growth factor (interleukin 6) in cerebrospinal fluid of patients with acute infection of the central nervous system. Clin. Exp. Immunol. 71, 320–323.

Howard, M. and O'Garra, A. (1992). Biological properties of interleukin 10. Immunol. Today. 13, 198–200.

Ishibashi, T., Kimura, H., Shikama, Y., *et al.* (1989a). Interleukin-6 is a potent thrombopoietic factor in vivo in mice. Blood 74, 1241–1244.

Ishibashi, T., Kimura, H., Uchida, T., Kariyone, S., Friese, P. and Burstein, S.A. (1989b). Human interleukin 6 is a direct promoter of maturation of megakaryocytes in vitro. Proc. Natl Acad. Sci. USA 86, 5953–5957.

Isshiki, H., Akira, S., Tanabe, O., *et al.* (1990). Constitutive and interleukin-1 (IL-1)-inducible factors interact with the IL-1-responsive element in the IL-6 gene. Mol. Cell. Biol. 10, 2757–2764.

Ito, A., Itoh, Y., Sasaguri, Y., Morimatsu, M. and Mori, Y. (1992). Effects of interleukin-6 on the metabolism of connective tissue components in rheumatoid synovial fibroblasts. Arthritis Rheum. 35, 1197–1201.

Jablons, D.M., Mule, J.J., McIntosh, J.K., *et al.* (1989). IL-6/IFN-beta2 as a circulating hormone. Induction by cytokine administration in humans. J. Immunol. 142, 1542–1547.

Janaswami, P., Kalvakolanu, D., Zhang, Y. and Sen, G. (1992). Transcriptional repression of interleukin-6 gene by adenoviral E1A proteins. J. Biol. Chem. 267, 24886–24891.

Jilka *et al.*, (1992). Increased osteoclast development after estrogen loss: mediation by interleukin-6. Science. 257, 88–91.

Jirik, F.R., Podor, T.J., Hirano, T., *et al.* (1989). Bacterial lipopolysaccharides and inflammatory mediators augment IL-6 secretion by human endothelial cells. J. Immunol. 142, 144–147.

Jourdan, M., Bataille, R. and Seguin, J. *et al.* (1990). Constitutive production of interleukin-6 and immunological features in cardiac myxomas. Arthritis Rheum. 33, 398–402.

Katz, A., Shulman, L., Porgador, A., Revel, M., Feldman, M. and Eisenbach, L. (1993). Abrogation of B16 melanoma metastases by long-term low-dose interleukin-6 therapy. J. Immunother. 13, 98–109.

Kawano, M., Hirano, T., Matsuda, T., *et al.* (1988). Autocrine generation and essential requirement of BSF-2/IL-6 for human multiple myelomas. Nature 332, 83–85.

Kharazmi, A., Nielsen, H., Rechnitzer, C. and Bendtzen, K. (1989). Immunol. Lett. 21, 177–184.

Kinoshita, S., Akira, S. and Kishimoto, T. (1992). A member of the C/EBP family, NF-IL6B, forms a heterdimer and transcriptionally synergizes with NF-IL6. Proc. Natl Acad. Sci. USA 89, 1473–1476.

Kishimoto, T. (1989). The biology of interleukin-6. Blood 74, 1–10.

Kishimoto, T. and Hirano, T. (1988). Molecular regulation of B lymphocyte response. Annu. Rev. Immunol. 6, 485–512.

Kishimoto, T., Taga, T. and Akira, S. (1994). Cytokine signal transduction. Cell 76, 253–262.

Klein, B., Zhang, X.-G. and Jourdan, M., *et al.* (1989). Paracrine rather than autocrine regulation of myeloma-cell growth and differentiation by interleukin-6. Blood 73, 517–526.

Klein, B., Lu, Z.Y. and Bataille, R. (1992). Clinical applications of IL6 inhibitors. Res. Immunol. 143, 774–776.

Kohase, M., May, L.T., Tamm, I., Vilcek, J. and Sehgal, P.B. (1987). A cytokine network in human diploid fibroblasts: interactions of β-interferons, tumor necrosis factor, platelet-derived growth factor, and interleukin-1. Mol. Cell. Biol. 7, 273–280.

Koike, K., Nakahata, T., Takagi, M., *et al.* (1988). Synergism of BSF2/interleukin 6 and interleukin 3 on development of multipotential hemopoietic progenitors in serum free culture. J. Exp. Med. 168, 879–890.

Koj, A. (1974). Acute phase reactants – their synthesis, turnover and biological significance. Struct. Funct. Plasma Proteins 1, 73.

Koj, A. and Gordon, A.H. (1985). The acute-phase response to injury and infection, Introduction. In "Research Monographs in Cell and Tissue Physiology", vol. 10. (eds. A.H. Gordon and A. Koj), pp xxi–xxix. Elsevier, Amsterdam.

Koo, A., Armstrong, C., Bochner, B., *et al.* (1992). Interleukin-6 and renal cell cancer: production, regulation, and growth effects. Cancer Immunol. Immunother. 35, 97–105.

Kopf, M., Baumann, H., Freer, G., *et al.* (1994). Impaired immune and acute-phase responses in interleukin-6 deficient mice. Nature 368, 339–342.

Kupper, T.S., Min, K., Sehgal, P., *et al.* (1989). Production of IL-6 by keratinocytes. Ann. N.Y. Acad. Sci. 557, 454–464.

Kurdowska, A. and Travis, J. (1990). Acute phase protein stimulation by alpha1-antichymotrypsin-cathepsin G complexes. J. Biol. Chem. 265, 21023–21026.

Kushner, I. (1982). The phenomenon of the acute phase response. Ann. N.Y. Acad. Sci. 389, 39–98.

Lai, C.-F., Ripperger, J., Morella, K.K., et al. (1995). STAT3 and STAT5B are targets of two different signal pathways activated by hematopoietin receptors and control transcription via separate cytokine response elements. J. Biol. Chem. 270, 23254–23257.

Leal-Berumen, I., Conlon, P. and Marshall, J.S. (1994). IL-6 production by rat peritoneal mast cells is not necessarily preceded by histamine release and can be induced by bacterial lipopolysaccharide. J. Immunol. 152, 5468–5476.

Leary, A. G., Ikebuchi, K., Hirai, Y., et al. (1988). Synergism between interleukin-6 and interleukin-3 in supporting proliferation of human hematopoietic stem cells: comparison with interleukin-1 alpha. Blood 71, 1759–1763.

LeClair, K., Blanar, M. and Sharp, P. (1992). The p50 subunit of NF-kB associates with the NF-IL6 transcription factor. Proc. Natl Acad. Sci. USA 89, 8145–8149.

Libermann, T.A. and Baltimore, D. (1990). Activation of interleukin-6 gene expression through the NF-kB transcription factor. Mol. Cell. Biol. 10, 2327–2334.

Linker-Israeli, M., Deans, R.J., Wallace, D.J., Prehn, J., Ozeri-Chen, T. and Klinenberg, J.R. (1991). Elevated levels of endogenous IL-6 in systemic lupus erythematosus. J. Immunol. 147, 117–123.

Lord, K., Abdollahi, A., Thomas, S., et al. (1991). Leukemia inhibitory factor and interleukin-6 trigger the same immediate early response, including tyrosine phosphorylation, upon induction of myeloid leukemia differentiation. Mol. Cell. Biol. 11, 4371–4379.

Lotz, M. and Guerne, P.-A. (1991). Interleukin-6 induces the synthesis of tissue inhibitor of metalloproteinases-1/erythroid potentiating activity (TIMP-1/EPA). J. Biol. Chem. 266, 2017–2020.

Lu, C. and Kerbel, R.S. (1993). Interleukin-6 undergoes transition from paracrine growth inhibitor to autocrine stimulator during human melanoma progression. J. Cell. Biol. 120, 1281–1288.

Luger, T.A., Krutmann, J., Kirnbauer, R., et al. (1989a). IFN-beta2/IL-6 augments the activity of human natural killer cells. J. Immunol. 143, 1206–1209.

Luger, T.A., Schwarz, T., Krutmann, J., et al. (1989b). Interleukin-6 is produced by epidermal cells and plays an important role in the activation of human T-lymphocytes and natural killer cells. Ann. N.Y. Acad. Sci. 557, 405–414.

Lust, J.A., Donovan, K.A., Kline, M.P., Greipp, P.R., Kyle, R.A. and Maihle, N.J. (1992). Isolation of an mRNA encoding a soluble form of the human interleukin-6 receptor. Cytokine 4, 96–100.

Mackiewicz, A., Schooltink, H., Heinrich, P. and Rose-John, S. (1992). Complex of soluble human IL-6 receptor/IL-6 up-regulates expression of acute-phase proteins. J. Immunol. 149, 2021–2027.

Madhok, R., Crilly, A., Watson, J. and Capell, H. (1993). Serum interleukin 6 levels in rheumatoid arthritis: correlations with clinical and laboratory indices of disease activity. Ann. Rheum. Dis. 52, 232–234.

Maier, R., Ganu, V. and Lotz, M. (1993). Interleukin-11, an inducible cytokine in human articular chondrocytes and synoviocytes, stimulates the production of the tissue inhibitor of metalloproteinases. J. Biol. Chem. 268, 21527–21532.

Malejcyzk, J., Malejcyzk, M., Urbanski, A. and Luger, T. (1992). Production of natural killer cell activity augmenting factor (interleukin-6) by human epiphyseal chondrocytes. Arthritis Rheum. 35, 706–713.

Malter, J.S. (1989). Identification of an AUUUA-specific messenger RNA binding protein. Science 246, 664–666.

Masashi, M., Inoue, M., Wei, S., et al. (1996). STAT3 activation is a critical step in gp130-mediated terminal differentiation and growth arrest of a myeloid cell line. Proc. Natl Acad. Sci. USA 93, 3963–3966.

Matsuda, T., Hirano, T., Nagasawa, S. and Kishimoto, T. (1989). Identification of alpha-2-macroglobulin as a carrier protein for IL-6. J. Immunol. 142, 148–152.

May, L.T., Torcia, G., Cozzolino, F., et al. (1986). Interleukin-6 gene expression in human endothelial cells: RNA start sites, multiple IL-6 proteins and inhibition of proliferation. Biochem. Biophys. Res. Commun. 159, 991–998.

May, L.T., Ghrayeb, J., Santhanam, U., et al. (1988a). J. Biol. Chem. 263, 7760–7766.

May, L.T., Santhanam, U., Tatter, S.B., Bhardwaj, N., Ghrayeb, J. and Sehgal, P.B. (1988b). Phosphorylation of secreted forms of human β2-interferon/hepatocyte stimulating factor/interleukin-6. Biochem. Biophys. Res. Commun. 152, 1144–1150.

May, L., Ghrayeb, J., Santhanam, U., et al. (1988c). Synthesis and secretion of multiple forms of beta2-interferon/B-cell differentiation factor-2/hepatocyte stimulating factor by human fibroblasts and monocytes. J. Biol. Chem. 263, 7760–7766.

May, L.T., Santhanam, U., Tatter, S.B., Ghrayeb, J. and Sehgal, P.B. (1989). Multiple forms of human interleukin-6: phosphoglycoproteins secreted by many different tissues. Ann. N.Y. Acad. Sci. 557, 114–121.

May, L., Viguet, H., Kenney, J., Ida, N., Allison, A. and Sehgal, P. (1992). High levels of "complexed" interleukin-6 in human blood. J. Biol. Chem. 267, 19698–19704.

Miles, S.A., Rezai, A.R. and Salzar-Gonzales, J.F. et al. (1990). AIDS Kaposi sarcoma derived cells produce and respond to interleukin-6. Proc. Natl Acad. Sci. USA 87, 4068.

Miles, S.A., Martinez-Maza, O., Rezai, A., et al. (1992). Oncostatin M as a potent mitogen for AIDS-Kaposi's sarcoma-derived cells. Science 255, 1432–1434.

Miyaura, C., Onozaki, K. and Akiyama, Y. et al. (1988). Recombinant human IL-6 (B cell stimulatory factor 2) is a potent inducer of differentiation of mouse myeloid leukemia cells (M1). FEBS Lett. 234, 17–21.

Morrone, G., Ciliberto, G., Oliviero, S., et al. (1988). Recombinant interleukin-6 regulates the transcriptional activation of a set of human acute phase genes. J. Biol. Chem. 263, 12554–12558.

Mule, J., McKintosh, J., Jablons, D. and Rosenberg, S. (1990). Antitumor activity of recombinant interleukin 6 in mice. J. Exp. Med. 171, 629–636.

Mule, J.J., Marcus, S.G., Yang, J.C., Weber, J.S. and Rosenberg, S.A. (1992). Clinical application of IL6 in cancer therapy. Res. Immunol. 143, 777–779.

Mullberg, J., Schooltink, H., Stoyan, T., et al. (1993). The soluble interleukin-6 receptor is generated by shedding. Eur. J. Immunol. 23, 473–480.

Mullberg, J., Oberthur, W., Lottspeich, F., et al. (1994). The soluble human IL-6 receptor–mutational characterization of the proteolytic cleavage site. J. Immunol. 152, 4958–4968.

Muraguchi, A., Hirano, T., Tang, B., et al. (1988). The essential role of B cell stimulatory factor 2 (BSF-2/IL-6). for the terminal differentiation of B cells. J. Exp. Med. 167, 332–344.

Murakami, M., Narazaki, M., Hibi, M., et al. (1991). Critical cytoplasmic region of the interleukin 6 signal transducer gp130 is conserved in the cytokine receptor family. Proc. Natl Acad. Sci. USA 88, 11349–11353.

Murakami, M., Hibi, M., Nakagawa, N., et al. (1993). IL-6-induced homodimerization of gp130 and associated activation of a tyrosine kinase. Science 260, 1808–1810.

Naitoh, Y., Fukata, J., Tominaga, T., et al. (1988). Interleukin-6 stimulates the secretion of adrenocorticotropic hormone in conscious, freely-moving rats. Biochem. Biophys. Res. Commun. 155, 1459–1463.

Nakafuku, M., Satoh, T. and Kaziro, Y. (1992). Differentiation factors, including nerve growth factor, fibroblast growth factor, and interleukin-6, induce an accumulation of an active ras. GTP complex in rat pheochromocytoma PC12 cells. J. Biol. Chem. 267, 19448–19454.

Nakajima, K. and Wall, R. (1991). Interleukin-6 signals activating junB and TIS11 gene transcription in a B-cell hybridoma. Mol. Cell. Biol. 11, 1409–1418.

Nakajima, T., Kinoshita, S., Sasagawa, T., et al. (1993). Phosphorylation at threonine-235 by a ras-dependent mitogen-activated protein kinase cascade is essential for transcription factor NF-IL6. Proc. Natl Acad. Sci. USA 90, 2207–2211.

Narazaki, M., Yasukawa, K., Saito, T., et al. (1993). Soluble forms of the interleukin-6 signal-transducing receptor component gp130 in human serum possessing a potential to inhibit signals through membrane-anchored gp130. Blood 82, 1120–1126.

Navarro, S., Debili, N., Bernaudin, J.-F., Vainchenker, W. and Doly, J. (1989). Regulation of the expression of IL-6 in human monocytes. J. Immunol. 142, 4339–4345.

Nawata, Y., Eugi, E.M., Lee, S.W. and Allison, A.C. (1989). IL-6 is the principal factor produced by synovia of patients with rheumatoid arthritis that induces B-lymphocytes to secrete immunoglobulins. Ann. N.Y. Acad. Sci. 557, 230–238.

Nesbitt, J.E. and Fuller, G.M. (1992a). Dynamics of interleukin-6 internalization and degradation in rat hepatocytes. J. Biol. Chem. 267, 5739–5742.

Nesbitt, J.E. and Fuller, G.M. (1992b). Differential regulation of interleukin-6 receptor and gp130 gene expression in rat hepatocytes. Mol. Biol. Cell. 3, 103–112.

Ng, S.B., Tan, Y.H. and Guy, G.R. (1994). Differential induction of the interleukin-6 gene by tumor necrosis factor and interleukin-1. J. Biol. Chem. 269, 19021–19027.

Nijsten, M., deGroot, E., TenDuuis, H., Klensen, H., Hack, C. and Aarden, L. (1987). Serum levels of interleukin-6 and acute phase responses. Lancet 2, 921.

Northemann, W., Braciak, T.A., Hattori, M., Lee, F. and Fey, G.H. (1989). Structure of the rat interleukin 6 gene and its expression in macrophage-derived cells. J. Biol. Chem. 264, 16072–16082.

Novick, D., Engelmann, H., Wallach, D. and Rubinstein, M. (1989). Soluble cytokine receptors are present in normal human urine. J. Exp. Med. 170, 1409–1414.

Novick, D., Shulman, L.M., Chen, L. and Revel, M. (1992). Enhancement of interleukin 6 cytostatic effect on human breast carcinoma cells by soluble IL-6 receptor from urine and reversion by monoclonal antibody. Cytokine 4, 6–11.

Okada, M., Kitahara, M., Kishimoto, S., Matsuda, T., Hirano,

T. and Kishimoto, T. (1988). IL-6/BSF-2 functions as a killer helper factor in the in vitro induction of cytotoxic T cells. J. Immunol. 141, 1543 1549.

Okano, A., Suzuki, C., Takatsuki, F., et al. (1989). Effects of interleukin-6 on hematopoiesis in bone marrow-transplanted mice. Transplantation. 47, 738–740.

Patchen, M.L., MacVittie, T.J., Williams, J.L., Schwartz, G.N. and Souza, L.M. (1991). Administration of interleukin-6 stimulates multilineage hematopoiesis and accelerates recovery from radiation-induced hematopoietic depression. Blood 77, 472–480.

Perlmutter, D.H. (1989). IFNbeta2/IL-6 is one of several cytokines that modulate acute phase gene expression in human hepatocytes and human macrophages. Ann. N.Y. Acad. Sci. 557, 332–341.

Perlstein, R., Whitnall, M., Abrams, J., Mougey, E. and Neta, R. (1993). Synergistic roles of interleukin-6, interleukin-1 and tumor necrosis factor in the adrenocorticotropin response to bacterial lipopolysaccharide in vivo. Endocrinology 132, 946–952.

Pietzko, D., Zohlnhofer, D., Graeve, L., et al. (1993). The hepatic interleukin-6 receptor. J. Biol. Chem. 268, 4250–4258.

Podor, T.J., Jirik, F.R., Loskutoff, D.J., Carson, D.A. and Lotz, M. (1989). Human endothelial cells produce IL-6. Lack of responses to exogenous IL-6. Ann. N.Y. Acad. Sci. 557, 374–385.

Poli, V., Mancini, F.P. and Cortese, R. (1990). IL-6DBP, a nuclear protein involved in interleukin-6 signal transduction, defines a new family of leucine zipper proteins related to C/EBP. Cell 63, 643–653.

Porgador, A., Tzehoval, E. and Katz, A., et al. (1993). IL-6 gene transfection into 3LL tumor cells suppresses the malignant phenotype and confers immuno-protective competence against parental metastatic cells. Cancer Res. 53, 3679.

Poupart, P., De Wit, L. and Content, J. (1984). Induction and regulation of the 26-kDs protein in the absence of synthesis of β-interferon in RNA in human cells. Eur. J. Biochem. 143, 15–21.

Poupart, P., Vandenabeele, P., Cayphas, S., et al. (1987). B cell growth modulating and differentiating activity of recombinant human 26-kd protein (BSF-2, HuIFN-beta2, HPGF). EMBO J. 6, 1219–1224.

Prowse, K.R. and Baumann, H. (1988). Hepatocyte-stimulating factor, beta-2 interferon, and interleukin-1 enhance expression of the rat alpha-1-acid glycoprotein gene via a distal upstream regulatory region. Mol. Cell. Biol. 8, 42.

Ray, A. and Prefontaine, K.E. (1994). Physical association and functional antagonism between the p65 subunit of transcription factor NF-kB and the glucocorticoid receptor. Proc. Natl Acad. Sci. USA 91, 752–756.

Ray, A., Sassone-Corsi, P. and Sehgal, P.B. (1989). A multiple cytokine- and second messenger-responsive element in the enhancer of the human interleukin-6 gene: similarities with c-fos gene regulation. Mol. Cell. Biol. 9, 5537–5547.

Ray, A., LaForge, K.S. and Sehgal, P.B. (1990). On the mechanism for efficient repression of the interleukin-6 promoter by glucocorticoids: enhancer, TATA box, and RNA start site (Inr motif) occlusion. Mol. Cell. Biol. 10, 5736–5746.

Reis, L.F.L., Le, J., Hirano, T., Kishimoto, T. and Vilcek, J. (1988). Antiviral action of tumor necrosis factor in human fibroblasts is not mediated by B cell stimulatory factor 2/IFN-beta 2, and is inhibited by specific antibodies to IFN-beta. J. Immunol. 140, 1566–1570.

Rennick, D., Jackson, J., Yang, G., Wideman, J., Lee, F. and Hudak, S. (1989). Interleukin-6 interacts with interleukin-4 and other hematopoietic growth factors to selectively enhance the growth of megakaryocytes, erythroid, myeloid, and multipotential progenitor cells. Blood 73, 1828–1835.

Resnitzky, D., Tiefenbrun, N., Berissi, H. and Kimchi, A. (1992). Interferons and interleukin 6 suppress phosphorylation of the retinoblastoma protein in growth-sensitive hematopoietic cells. Proc. Natl Acad. Sci. USA 89, 402–406.

Revel, M. (1989). Host defense against infections and inflammations: role of the multifunctional IL-6/IFN-beta2 cytokine. Experientia 45, 549–557.

Richards, C.D. and Agro, A. (1994). Interaction between oncostatin-M, interleukin-1 and prostaglandin E2 in induction of IL-6 expression in human fibroblasts. Cytokine 6, 40–47.

Richards, C.D. and Saklatvala, J. (1991). Molecular cloning and sequence of porcine Interleukin-6 cDNA and expression of mRNA in synovial fibroblasts in vitro. Cytokine 3, 269–276.

Richards, C.D. and Shoyab, M. (1992). The role of oncostatin-M in the acute phase response. In "Acute Phase Proteins: Molecular Biology, Biochemistry and Clinical Applications" (eds. A. Mackiewicz, I. Kushner and H. Baumann), chapter 18, pp. 321–327. CRC Press, Boca Raton, FL.

Richards, C.D., Gauldie, J. and Baumann, H. (1991). Cytokine control of acute phase protein expression. Eur. Cyt. Net. 2, 89–98.

Richards, C.D., Brown, T.J., Shoyab, M., Baumann, H. and Gauldie, J. (1992). Recombinant oncostatin-M stimulates the production of acute phase proteins in hepatocytes in vitro. J. Immunol. 148, 1731–1736.

Richards, C.D., Shoyab, M., Brown, T.J. and Gauldie, J. (1993). Selective regulation of tissue inhibitor of metalloproteinases (TIMP-1) by oncostatin M in fibroblasts in culture. J. Immunol. 150, 5596–5603.

Richards, C.D., Langdon, C., Pennica, D. and Gauldie, J. (1996). Murine cardiotrophin-1 stimulates the acute-phase response in rat hepatocytes and H35 hepatoma cells. J. Inter. Cytokine Res. 16, 69–75.

Ripperger, J.A., Fritz, S., Richter, K., Hocke, G.M., Lottspeich, F. and Fey, G.H. (1995). Transciption factors STAT3 and STAT5b are present in rat liver nuclei late in an acute phase response and bind interleukin-6 response elements. J. Biol. Chem. 270, 29998–30006.

Ritchie, D.G. and Fuller, G.M. (1983). Hepatocyte-stimulating factor: a monocyte-derived acute-phase regulatory protein. Ann. N.Y. Acad. Sci. 408, 490.

Rose, T.M. and Bruce, A.G. (1991). Oncostatin M is a member of the cytokine family which includes LIF, GM-CSF and IL-6. Proc. Natl Acad. Sci. USA 88, 8641–8645.

Rose-John, S. and Heinrich, P.C. (1994). Soluble receptors for cytokines and growth factors: generation and biological function. Biochem. J. 300, 281–290.

Rose-John, S., Schooltink, H., Lenz, D., et al. (1990). Studies on the structure and regulation of the human hepatic interleukin-6 receptor. Eur. J. Biochem. 190, 79–83.

Sachs, L., Lotem, J. and Shabo, Y. (1989). The molecular regulators of macrophage and granulocyte development. Ann. N.Y. Acad. Sci. 557, 417–435.

Sadowski, H.B., Shuai, K., Darnell, Jr., J.E. and Gilman, M.Z. (1993). A common nuclear signal transduction pathway activated by growth factor and cytokine receptors. Science 261, 1739–1744.

Saito, M., Yoshida, K., Hibi, M., Taga, T. and Kishimoto, T. (1992). Molecular cloning of a murine IL-6 receptor-associated signal transducer, gp130, and its regulated expression in vivo. J. Immunol. 148, 4066–4071.

Santhanam, U., Ray, A. and Sehgal, P.B. (1991). Repression of the interleukin 6 gene promoter by p53 and the retinoblastoma susceptibility gene product. Proc. Natl Acad. Sci. USA 88, 7605–7609.

Sato, T., Ito, A. and Mori, Y. (1990). Interleukin 6 enhances the production of tissue inhibitor of metalloproteinases (TIMP) but not that of matrix metalloproteinases by human fibroblasts. Biochem. Biophys. Res. Commun. 170, 824–829.

Satoh, T., Nakamura, S., Taga, T., et al. (1988). Induction of neural differentiation in PC12 cells by B cell stimulatory factor 2/interleukin 6. Mol. Cell. Biol. 8, 3546–3549.

Satoh, T., Nakafuku, M. and Kaziro, Y. (1992). Function of ras as a molecular switch in signal transduction. J. Biol. Chem. 267, 24149–24152.

Savino, R., Lahm, A., Giorgio, M., Cabibbo, A., Tramontano, A. and Ciliberto, G. (1993). Saturation mutagenesis of the human interleukin 6 receptor-binding site: implications for its three-dimensional structure. Proc. Natl Acad. Sci. USA 90, 4067–4071.

Savino, R., Lahm, A., Salvati, A.L., et al. (1994). Generation of interleukin-6 receptor antagonists by molecular-modeling guided mutagenesis of residues important for gp130 activation. EMBO J. 13, 1357–1367.

Schaefer, T.S., Sanders, L.K. and Nathans, D. (1995). Cooperative transcriptional activity of Jun and STAT3β, a short form of STAT3. Proc. Natl Acad. Sci. USA 92, 9097–9101.

Schindler, R., Mancilla, J., Endres, S., Ghorbani, R., Clark, S.C. and Dinarello, C.A. (1990). Correlations and interactions in the production of interleukin-6 (IL-6)., IL-1, and tumor necrosis factor (TNF) in human blood mononuclear cells: IL-6 suppresses IL-1 and TNF. Blood 75, 40–47.

Schooltink, H., Schmitz-Van de Leur, H., Heinrich, P.C. and Rose-John, S. (1992). Up-regulation of the interleukin-6-signal transducing protein (gp130) by interleukin-6 and dexamethasone in HepG2 cells. FEBS Lett. 297, 263–265.

Schreiber, G. (1987). Synthesis, processing and secretion of plasma proteins by the liver (and other organs) and their regulation. Plasma Proteins 5, 293–363.

Schreiber, G., Tsykin, A., Aldred, A.R., et al. (1989). The acute phase response in the rodent. Ann. N.Y. Acad. Sci. 557, 61–86.

Scuderi, P. (1990). Differential effects of copper and zinc on human peripheral blood monocyte cytokine secretion. Cell. Immunol. 126, 391–405.

Seckinger, P., Kaufmann, M.-T. and Dayer, J.-M. (1990). An interleukin 1 inhibitor affects both cell-associated interleukin 1-induced T cell proliferation and PGE2/collagenase production by human dermal fibroblasts and synovial cells. Immunobiol. 180, 316–327.

Sehgal, P.B., Zilberstein, A., Ruggieri, R.-M., et al. (1986). Human chromosome 7 carries the beta2 interferon gene. Proc. Natl Acad. Sci. USA 83, 5219–5222.

Sehgal, P.B., May, L.T., Tamm, I. and Vilcek, J. (1987). Human beta2 interferon and B-cell differentiation factor BSF-2 are identical. Science 235, 731–732.

Sehgal, P.B., Helfgott, D.C., Santhanam, U., *et al.* (1988). Regulation of the acute phase and immune responses in viral disease: enhanced expression of the β2-interferon/hepatocyte-stimulating factor/interleukin-6 gene in virus-infected human fibroblasts. J. Exp. Med. 167, 1951–1956.

Shaw, G. and Kamen, R.A. (1986). Conserved AU sequence from the 3′ untranslated region of GM-CSF mRNA mediates selective mRNA degradation. Cell 46, 659–667.

Shuai, K., Horvath, C.M., Tsai Huang, L.H., Qureshi, S.A., Cowburn, D. and Darnell, J.E., Jr. (1994). Interferon activation of the transcription factor Stat91 involves dimerization through SH2-phosphotyrosyl peptide interactions. Cell 76, 821–828.

Snyers, L. and Content, J. (1992). Enhancement of IL-6 receptor beta chain (gp130) expression by IL-6, IL-1 and TNF in human epithelial cells. Biochem. Biophys. Res. Commun. 185, 902–908.

Spergel, J.M. and Chen-Kiang, S. (1991). Interleukin-6 enhances a cellular activity which functionally substitutes for E1A in transactivation. Proc. Natl Acad. Sci. USA 88, 6472–6476.

Spergel, J.M., Hsu, W., Akira, S., Thimmappaya, B., Kishimoto, T. and Chen-Kiang, S. (1992). NF-IL6, a member of the C/EBP family, regulates E1A-responsive promoters in the absence of E1A. J. Virol. 66, 1021–1030.

Stahl, N., Boulton, T.G., Farruggella, T., *et al.* (1994). Association and activation of Jak-Tyk kinases by CNTF-LIF-OSM-IL-6 beta receptor components. Science 263, 92–95.

Starnes, H.F., Jr., Pearce, M.K., Tewari, A., Yim, J.H., Zou, J.-C. and Abrams, J.S. (1990). Anti-IL-6 monoclonal antibodies protect against lethal *Escherichia coli* infection and lethal tumor necrosis factor-alpha challenge in mice. J. Immunol. 145, 4185–4191.

Stepien, H., Agro, A., Crossley, J., Padol, I., Richards, C. and Stanisz, A. (1993). Immunomodulatory properties of diazepam-binding inhibitor: effect on human interleukin-6 secretion, lymphocyte proliferation and natural killer cell activity in vitro. Neuropeptides 25, 207–211.

Suematsu, S., Matsuda, T., Aozasa, K., *et al.* (1989). IgG1 plasmacytosis in interleukin 6 transgenic mice. Proc. Natl Acad. Sci. USA 86, 7547–7551.

Suematsu, S., Matsusaka, T., Matsuda, T., *et al.* (1992). Generation of plasmacytomas with the chromosomal translocation t(12;15) in interleukin 6 transgenic mice. Proc. Natl Acad. Sci. USA 89, 232–235.

Taga, T. and Kishimoto, T. (1992). Cytokine receptors and signal transduction. FASEB J. 6, 3387–3396.

Taga, T., Kawanishi, Y., Hardy, R.R., Hirano, T. and Kishimoto, T. (1987). Receptors for B-cell stimulatory factor 2: quantitation, specificity, distribution, and regulation of their expression. J. Exp. Med. 166, 967–981.

Taga, T., Hibi, M., Hirata, Y., *et al.* (1989). Interleukin-6 triggers the association of its receptor with a possible signal transducer, gp130. Cell 58, 573–581.

Taga, T., Narazaki, M., Yasukawa, K., *et al.* (1992). Functional inhibition of hematopoietic and neurotrophic cytokines by blocking the interleukin 6 transducer gp130. Proc. Natl Acad. Sci. USA 89, 10998–11001.

Takai, Y., Wong, G., Clark, S., Burakoff, S. and Herman, S. (1988). B cell stimulatory-2 is involved in the differentiation of cytotoxic T-lymphocytes. J. Immunol. 140, 508–512.

Takatsuki, F., Okana, A., Suzuki, C., *et al.* (1988). Human recombinant IL-6/B cell stimulatory factor 2 augments murine antigen-specific antibody responses in vitro and in vivo. J. Immunol. 141, 3072–3077.

Tanabe, O., Akira, S., Kamiya, T., Wong, G., Hirano, T. and Kishimoto, T. (1989). Genomic structure of the murine IL-6 gene. J. Immunol. 141, 3875–3880.

Tilg, H., Trehu, E., Atkins, M.B., Dinarello, C.A. and Mier, J.W. (1994). Interleukin-6 (IL-6) as an anti-inflammatory cytokine: induction of circulating IL-1 receptor antagonist and soluble tumor necrosis factor receptor p55. Blood 83, 113–118.

Tilg, H., Dinarello, C.A. and Mier, J.W. (1997). IL-6 and APPs: anti-inflammatory and immunosuppressive mediators. Immunol. Today 18, 428–432.

Tosato, G. and Pike, S.E. (1989). A monocyte-derived B cell growth factor is IFN-beta2/BSF-2/IL-6a. Ann. N.Y. Acad. Sci. 557, 181–190.

Travis, J. and Salvesen, G.S. (1983). Human plasma proteinase inhibitors. Annu. Rev. Biochem. 52, 665–709.

Tsuchiya, Y., Hattori, M., Hayashida, K., Ishibashi, H., Okubo, H. and Sakaki, Y. (1987). Sequence analysis of the putative regulatory region of rat alpha2-macroglobulin gene. Gene 57, 73–80.

Ulich, T.R., Yin, S., Guo, K., Yi, E.S., Remick, D. and del Castillo, J. (1991). Intratracheal injection of endotoxin and cytokines. Am. J. Pathol. 138, 1097–1101.

Ulich, T.R., Guo, K., Yin, S., *et al.* (1992). Endotoxin-induced cytokine gene expression in vivo. Am. J. Pathol. 141, 61–68.

Van Dam, M., Mullberg, J., Schooltink, H., *et al.* (1993). Structure–function analysis of interleukin-6 utilizing human/murine chimeric molecules. Involvement of two separate domains in receptor binding. J. Biol. Chem. 268, 15285–15290.

Van Oers, M.H.J., Van Der Heyden, A.A.P.A.M. and Aarden, L.A. (1988). Interleukin 6 (IL-6). in serum and urine of renal transplant recipients. Clin. Exp. Immunol. 71, 314–319.

Van Snick, J. (1990). Interleukin-6: an overview. Annu. Rev. Immunol. 8, 253–278.

Van Snick, J., Cayphas, S., Szikora, J.-P., Renauld, J.-C., Roost, E. and Simpson, R. (1988). cDNA cloning of murine interleukin-HP1: homology with human interleukin 6. Eur. J. Immunol. 18, 193–197.

Waage, A., Slupphaug, G. and Shalaby, R. (1990). Glucocorticoids inhibit the production of IL6 from monocytes, endothelial cells and fibroblasts. Eur. J. Immunol. 20, 2439–2443.

Walther, Z., May, L. and Sehgal, P. (1988). Transcriptional regulation of the Interferon-B2/B cell differentiation factor BSF-2/hepatocyte-stimulating factor gene in human fibroblasts by other cytokines. J. Immunol. 140, 974–977.

Weber, J., Yang, J.C., Topalian, S.L., *et al.* (1993). Phase I trial of subcutaneous interleukin-6 in patients with advanced malignancies. J. Clin. Oncol. 11, 499–506.

Wegenka, U., Buschmann, J., Lutticken, C., Heinrich, P. and Horn, F. (1993). Acute-phase response factor, a nuclear factor binding to acute-phase response elements, is rapidly activated by interleukin-6 at the posttranslational level. Mol. Cell. Biol. 13, 276–288.

Woloski, B.M.R.N.J., Smith, E.M., Meyer, W.J., III, Fuller, G.M. and Blalock, J.E. (1985). Corticotropin-releasing activity of monokines. Science 230, 1035–1037.

Won, K.-A. and Baumann, H. (1990). The cytokine response element of the rat alpha1-acid glycoprotein gene is a complex of several interacting regulatory sequences. Mol. Cell. Biol. 10, 3965–3978.

Won, K.-A., Campos, S.P. and Baumann, H. (1993). Experimental systems for studying hepatic acute phase response. In "Acute Phase Proteins: Molecular Biology, Biochemistry and Clinical Applications" (eds. A. Mackiewicz, I. Kushner and H. Baumann), chapter 14, pp. 255–271. CRC Press, Boca Raton, FL.

Wong, G.G. and Clark, S.C. (1988). Multiple actions of interleukin 6 within a cytokine network. Immunol. Today 9, 137–139.

Xing, Z., Braciak, T., Jordana, M., Croitoru, K., Graham, F.L. and Gauldie, J. (1994). Adenovirus-mediated cytokine gene transfer at tissue sites. Overexpression of IL-6 induces lymphocytic hyperplasia in the lung. J. Immunol. 153, 4059–4069.

Yamasaki, K., Taga, T., Hirata, Y., et al. (1988). Cloning and expression of the human interleukin-6 (BSF-2/IFNbeta 2) receptor. Science 241, 825–828.

Yasukawa, K., Hirano, T., Watanabe, Y., et al. (1987). Structure and expression of human B cell stimulatory factor-2 (BSF-2/Il-6) gene. EMBO J. 6, 2939–2945.

Yin, T. and Yang, Y.-C. (1994). Mitogen-activated protein kinases and ribosomal S6 protein kinases are involved in signaling pathways shared by interleukin-11, interleukin-6, leukemia inhibitory factor, and oncostatin M in mouse 3T3-L1 cells. J. Biol. Chem. 269, 3731–3738.

Yonish-Rouach, E., Resnitzky, D., Lotem, J., Sachs, L., Kimchi, A. and Oren, M. (1991). Wild-type p53 induces apoptosis of myeloid leukaemic cells that is inhibited by interleukin-6. Nature 352, 345–347.

Yoshida, K., Taga, T., Saito, M., et al. (1996). Targeted disruption of gp130, a common signal transducer for the interleukin 6 family of cytokines, leads to myocardial and hematological disorders. Proc. Natl Acad. Sci. USA 93, 407–411.

Yoshizaka, K., Matsuda, T. and Nishimoto, N. et al. (1989). Pathogenic significance of interleukin-6 (IL-6/BSF-2) in Castleman's disease. Blood 74, 1360–1367.

Yuan, J., Wegenka, U.M., Lutticken, C., et al. (1994). The signalling pathways of interleukin-6 and gamma interferon converge by the activation of different transcription factors which bind to common responsive DNA elements. Mol. Cell. Biol. 14, 1657–1668.

Zhang, Y., Lin, J.-X. and Vilcek, J. (1988). Synthesis of interleukin 6 (interferon-beta2/B cell stimulatory factor 2) in human fibroblasts is triggered by an increase in intracellular cyclic AMP. J. Biol. Chem. 263, 6177–6182.

Zhang, Y., Lin, J.-X. and Vilcek, J. (1990). Interleukin-6 induction by tumor necrosis factor and interleukin-1 in human fibroblasts involves activation of a nuclear factor binding to a kB-like sequence. Mol. Cell. Biol. 10, 3818–3823.

Zhong, Z., Wen, Z. and Darnell, J.E., Jr., (1994a). Stat3 and Stat4: members of the family of signal transducers and activators of transcription. Proc. Natl Acad. Sci. USA 91, 4806–4810.

Zhong, Z., Wen, Z. and Darnell, J.E., Jr., (1994b). Stat3: a STAT family member activated by tyrosine phosphorylation in response to epidermal growth factor and interleukin-6. Science 264, 95–98.

Zilberstein, A., Ruggieri, R., Korn, J.H. and Revel, M. (1986). Structure and expression of cDNA and genes for human interferon-beta-2, a distinct species inducible by growth-stimulatory cytokines. EMBO J. 5, 2529–2537.

Zohlnhoefer, D., Graeve, L., Rose-John, S., Schooltink, H., Dittrich, E. and Heinrich, P.C. (1992). The hepatic interleukin-6 receptor. Down-regulation of the IL-6 binding subunit (gp80) by its ligand. FEBS Lett. 306, 219–222.

7. Interleukin-7

Anthony E. Namen *and* Anthony R. Mire-Sluis

1. Introduction

The process of hematopoiesis by which mature functional blood cells are continuously generated has been an object of intense study for decades. It has become clear that the regulation of hematopoietic development is a very complex phenomenon which involves multiple interactions between the developing hematopoietic cells and their immediate microenvironment. These environmental interactions are now known to include complex cell surface contributions from a stromal network and the elaboration of a growing number of soluble growth factors. The early investigations into hematopoiesis and its regulation yielded new information primarily concerning development of the various cells of the myeloid lineage and led to the isolation and characterization of a number of new soluble regulatory growth factors.

Information concerning the regulation of lymphoid development lagged behind the progress being made with myeloid lineage cells. This was certainly not due to

a lack of interest, as the cells of the lymphoid lineage mediate both the cell-mediated and the humoral immune responses, a necessary component of the maintenance of good health. The primary obstacle to the study of lymphoid development was the stringent *in vitro* growth requirements of lymphoid progenitors and the lack of a suitable source of cells for study.

A key element to beginning to unravel the particulars of early B cell development was the establishment of an *in vitro* culture system which allowed for the growth of large numbers of normal B cell precursors (Whitlock *et al.*, 1984). A stromal cell line derived from such a culture system was found to secrete a growth factor which alone could support the growth of B cell progenitors (Namen *et al.*, 1988a). Subsequently, this led to the isolation and cloning of this growth factor, which is now known as interleukin-7 (IL-7). The initial studies of the activities of IL-7 revealed a potent proliferative stimulus of pro-B and pre-B cells from normal bone marrow. Subsequent tissue distribution studies demonstrated that a high level of IL-7 mRNA was present in the thymus and it was soon discovered that IL-7 was also a potent growth promoter for T cell progenitors. IL-7 has since emerged as a major factor in the process of both B cell and T cell development. As studies with IL-7 have continued to progress, additional biological activities have begun to emerge. IL-7 stimulates the growth of immature and mature T cells and promotes the expansion and effector function of cytolytic T cells and their precursors. Additionally, IL-7 enhances lymphokine-activated killer (LAK) cell activity in peripheral blood and can stimulate the antitumor abilities of monocytes and macrophages.

Future studies utilizing IL-7 alone or in concert with other known growth factors will almost certainly reveal new unsuspected effects and provide potentially new avenues for therapeutic strategies. The lymphoid-restoring capacity of IL-7 as well as the cytotoxic and LAK cell-promoting properties would seem to indicate that IL-7 may play an important therapeutic role in adoptive immunotherapy and bone marrow transplantation following irradiation and/or chemotherapy.

2.   IL-7 Gene

2.1   IL-7 GENE SEQUENCE

A stromal cell line derived from a long-term bone marrow culture was originally identified as a source of soluble IL-7 (Namen *et al.*, 1988a). This cell line was utilized as a source of conditioned medium for the biochemical purification and sequencing of native murine IL-7. The same cell line was also utilized to provide a source of mRNA for constructing a cDNA library. Screening of this library by direct expression enabled the cloning of a single cDNA which encoded biologically active murine IL-7 (Namen *et al.*, 1988b). A comparison of the amino acid sequence predicted by the cloned cDNA and the N-terminal sequence derived from the purified native murine IL-7 showed them to be identical.

The cDNA encoding biologically active human IL-7 was obtained from the human hepatoma cell line, SK-Hep by hybridization with the homologous murine clone (Goodwin *et al.*, 1989), and is shown in Figure 7.1. Subsequent to the cloning of cDNAs encoding both human and murine IL-7, experiments to help define the structure and regulation of the IL-7 gene were carried out (Lupton *et al.*, 1990). A high degree of homology (81%) exists between the coding regions of human and murine IL-7. The 3′ noncoding region exhibits a 63% homology and the 5′ noncoding region a 73% homology, indicating a significant degree of conservation in both the coding and the flanking regions.

The human IL-7 gene is quite large, spanning at least 33 kbp. The 5′ noncoding regions of both the human and murine IL-7 gene are unusual in that the normal eukaryocytic promoter elements are noticeably absent. There is no evidence for the commonly observed TATA or CAAT sequences for at least 1200 bp upstream of the primary transcription initiation site. The 5′ regulatory region does, however, contain eight potential initiation of transcription sites which are situated upstream of the primary initiation site. Most of these upstream initiation sites result in frame shifts and/or premature stop signals and a biologically inactive protein fragment. It is possible that these alternate initiation sites are involved in a complex regulatory mechanism to tightly control IL-7 expression. This is indicated by the observation that removal of the 5′ noncoding regions from the IL-7 cDNA clone resulted in increased expression and production of mature functional IL-7 protein following transfection into COS cells.

The human gene does contain a potential SP1 binding site (GGGCGG) approximately 60 bp upstream of the primary initiation site. Approximately 720 bp upstream of the primary initiation site is a potential E12 binding site. In addition, five other sequences related to the E12 binding site are present in the 5′ noncoding region and are highly conserved in both the human and murine genes. These sites very possibly represent binding regions for members of the recently defined "helix–loop–helix" class of DNA binding regulatory proteins. The 700 bp region extending upstream from the primary initiation site contains 42 CpG dinucleotides. Such clusters of CpG dinucleotides have been shown to occur in many "housekeeping" and tissue-specific genes (Bird, 1987), and are involved in the regulation of transcription (Rachal *et al.*, 1989). In general, the regulation of IL-7 expression appears not to involve the commonly observed eukaryocytic control elements. Instead, it seems likely that a complex interplay of multiple initiation sites and less well defined regulatory sequences serves to tightly control the expression of IL-7. The tight control of IL-7 expression is evidenced by the observation that only very low levels of IL-7 are produced by all of the known native unmodified sources of IL-7. Moreover, those sources are generally refractory to the usual outside stimuli conventionally used to boost cytokine production. High levels of IL-7 production in cell lines has only been achieved following genetic modification of the IL-7 controlling regions. Clearly, further experimentation will be needed to more closely define the very interesting and unusual control elements regulating the expression of the IL-7 gene.

2.2   IL-7 GENE STRUCTURE

The human IL-7 gene consists of six exons and five introns distributed over at least 33 kbp (Lupton *et al.*, 1990), and is illustrated in Figure 7.2. The length of intron 2 is unknown, but consists of at least 15 kbp. The human IL-7 cDNA contains an open reading frame spanning 531 nucleotides, encoding a protein of 177 amino acids with a calculated molecular mass of 17 400

1101 GTCAGATAAATGATAGTCGTTATTATTATCGCTGTTGTTACTGGTTTACATTATCCACCTTCATCTAAGCACCCTTTCTGCAGAATAGCAGAAACCAAAC
TAATGTAGCAAATAAGCTACATAATTCAAGCCCAGGAAAAGTTAACATTTCAGTGGCATGCATTCAAGACGAATAGTTTGATTTATTAGCCAATTCAGA
TAAATGTGCACGTGGAAGTCATAGTTAAATATTATCGTCAGTTTCCACGTCCTGCGTTTAATTTGGGGTTTGATTTTCCAAATACAACACTTACCAGATT
AGGTGGACCCACAGGATTATTTTTCCTTGAGGTCTCACCTGAGCAGGTGCATGTACAGCAGACGGAGCAGAAAGAGACTGATTAGAGAGGTTGGAGTGGT
AGAGGGCGTGACCCTCTTAATCATTCTTCACTTCCTTTTTTAAAAGACGACTTGGCATCGTCCACCACATCCGCGGCAACGCCTCCTTGGTGTCGTCCGC
TTCCAATAACCCAGCTTGCGTCCTGCACACTTGTGGCTTCCGTGCACACATTAACAACTCATGGGTCTAGCTCCCAGTCGCCAAGCGTTGCCAAGGCGTT
GAGAGATCATCTGGGAAGTCTTTTACCCAGAATTGCTTTGATTCAGGCCAGCTGGTTTTTCTGCGGTGATTCGGAAATTCGCGAATTCCTCTGGTCCTCA
TCCAGGTGCGCGGGAAGCAGGTGCCCAGGAGAGAGGGGATAATGAAGATTCCATGCTGATGATCCCAAAGATTGAACCTGCAGACCAAGCGCAAAGTAGA
AACTGAAAGTACACTGCTGGCGGATCCTACGGAAGTTATGGAAAAGGCAAAGCGCAGAGCCACGCCGTAGTGTGTGCCGCCCCCCTTGGGATGGATGAAA
CTGCAGTCGCGGCGTGGGTAAGAGGAACCAGCTGCAGAGATCACCCTGCCCAACACAGACTCGGCAACTCCGCGGAAGACCAGGGTCCTGGGAGTGACTA
TGGGCGGTGAGAGCTTGCTCCTGCTCCAGTTGCGGTCATCATGACTACGCCCGCCTCCCGCAGACCATGTTCCATgtaagcgctcttctccctttattt
actgtcaaatttagTTTCTTTTAGGTATATCTTTGGACTTCCTCCCCTGATCCTTGTTCTGTTGCCAGTAGCATCATCTGATTGTGATATTGAAGGTAAA
GATGGCAAACAATATGAGAGTGTTCTAATGGTCAGCATCGATCAATTATTGgtatgtgattattttgtttt   2372

2373 ttttatgttatttattacagGACAGCATGAAAGAAATTGGTAGCAATTGCCTGAATAATGAATTTAACTTTTTTAAAAGACATATCTGTGATGCTAATAA
Ggtaatgataattatttggag   2493

2494 ctgacttttttcctataatagGAAGGTATGTTTTTATTCCGTGCTGCTCGCAAGTTGAGGCAATTTCTTAAAATGAATAGCACTGGTGATTTTGATCTCCA
CTTATTAAAAGTTTCAGAAGGCACAACAATACTGTTGAACTGCACTGGCCAGgtaag   2650

2651 aatgtgactttgtttttaagGTTAAAGGAAGAAAACCAGCTGCCCTGGGTGAAGCCCAACCAACAAAGAGTTTGgtgagaataattgtataatt   2744

2745 tttaaaactctattctctagGAAGAAAATAAATCTTTAAAGGAACAGAAAAAACTGAAATGACTTGTGTTTCCTAAAGAGACTATTACAAGAGATAAAAA
CTTGTTGGAATAAAATTTTGATGGGCACTAAAGAACACTGAAAAAATATGGAGTGGCAATAGAGAAACACGAACTTTAGCTGCATCCTCCAAGAATCTATC
TGCTTATGCAGTTTTTCAGAGTGGAATGCTTCCTAGAAGTTACTGAATGCACCATGGTCAAAACGGATTAGGGCATTTGAGAAATGCATATTGTATTACT
AGAAGATGAATACAAACAATGGAAACTGAATGCTCCAGTCAACAAACTATTTCTTATATATGTGAACATTTATCAATCAGTATAATTCTGTACTGATTTT
TGTAAGACAATCCATGTAAGGTATCAGTTGCAATAATACTTCTCAAAAATGTTTAAATATTTCAAGACATTAAATCTATGAAGTATATAATGGTTTCAAA
GATTCAAAATTGACATTGCTTTACTGTCAAAATAATTTTATGGCTCACTATGAATCTATTATACTGTATTAAGAGTGAAAATTGTCTTCTTCTGTGCTGG
AGATGTTTTAGAGTTAACAATGATATATGGATAATGCCGGTGAGAATAAGAGAGTCATAAACCTTAAGTAAGCAACAGCATAACAAGGTCCAAGATACCT
AAAAAGAGATTTCAAGAGATTTAATTAATCATGAATGTGTAACACAGTGCCTTCAATAAATGGTATAGCAAATG
AGAGGGCGTGACCCTCTTAATCATTCTTCACTTCCTTTTTTAAAAGACGACTTGGCATCGTCCACCACATCCGCGGCAACGCCTCCTTGGTGTCGTCCGC
TTCCAATAACCCAGCTTGCGTCCTGCACACTTGTGGCTTCCGTGCACACATTAACAACTCATGGGTCTAGCTCCCAGTCGCCAAGCGTTGCCAAGGCGTT
GAGAGATCATCTGGGAAGTCTTTTACCCAGAATTGCTTTGATTCAGGCCAGCTGGTTTTTCTGCGGTGATTCGGAAATTCGCGAATTCCTCTGGTCCTCA
TCCAGGTGCGCGGGAAGCAGGTGCCCAGGAGAGAGGGGATAATGAAGATTCCATGCTGATGATCCCAAAGATTGAACCTGCAGACCAAGCGCAAAGTAGA
AACTGAAAGTACACTGCTGGCGGATCCTACGGAAGTTATGGAAAAGGCAAAGCGCAGAGCCACGCCGTAGTGTGTGCCGCCCCCCTTGGGATGGATGAAA
CTGCAGTCGCGGCGTGGGTAAGAGGAACCAGCTGCAGAGATCACCCTGCCCAACACAGACTCGGCAACTCCGCGGAAGACCAGGGTCCTGGGAGTGACTA
TGGGCGGTGAGAGCTTGCTCCTGCTCCAGTTGCGGTCATCATGACTACGCCCGCCTCCCGCAGACCATGTTCCATgtaagcgctcttctccctttatt
actgtcaaattagTTTCTTTTAGGTATATCTTTGGACTTCCTCCCCTGATCCTTGTTCTGTTGCCAGTAGCATCATCTGATTGTGATATTGAAGGTAAAG
ATGGCAAACAATATGAGAGTGTTCTAATGGTCAGCATCGATCAATTATTGgtatgtgattattttgtttttttttatgttatttattacagGACAGCATGA
AAGAAATTGGTAGCAATTGCCTGAATAATGAATTTAACTTTTTTAAAAGACATATCTGTGATGCTAATAAGgtaatgataattatttggag
tcctataatagGAAGGTATGTTTTTATTCCGTGCTGCTCGCAAGTTGAGGCAATTTCTTAAAATGAATAGCACTGGTGATTTTGATCTCCACTTATTAAA
AGTTTCAGAAGGCACAACAATACTGTTGAACTGCACTGGCCAGgtaagaatgtgactttgtttttaagGTTAAAGGAAGAAAACCAGCTGCCCTGGGTGA
AGCCCAACCAACAAAGAGTTTGgtgagaataattgtataattttttaaaactctattctctagGAAGAAAATAAATCTTTAAAGGAACAGAAAAAACTGAA
TGACTTGTGTTTCCTAAAGAGACTATTACAAGAGATAAAAACTTGTTGGAATAAAATTTTGATGGGCACTAAAGAACACTGAAAAAATATGGAGTGGCAAT
ATAGAAACACGAAC
TTTAGCTGCATCCTCCAAGAATCTATCTGCTTATGCAGTTTTTCAGAGTGGAATGCTTCCTAGAAGTTACTGAATGCACCATGGTCAAAACGGATTAGGG
CATTTGAGAAATGCATATTGTATTACTAGAAGATGAATACAAACAATGGAAACTGAATGCTCCAGTCAACAAACTATTTCTTATATATGTGAACATTTAT
CAATCAGTATAATTCTGTACTGATTTTTGTAAGACAATCCATGTAAGGTATCAGTTGCAATAATACTTCTCAAAAATGTTTAAATATTTCAAGACATTAA
ATCTATGAAGTATATAATGGTTTCAAAGATTCAAAATTGACATTGCTTTACTGTCAÂAATAATTTTATGGCTCACTATGAATCTATTATACTGTATTAAG
AGTGAAAATTGTCTTCTTCTGTGCTGGAGATGTTTTAGAGTTAACAATGATATATGGATAATGCCGGTGAGAATAAGAGAGTCATAAACCTTAAGTAAGC
AACAGCATAACAAGGTCCAAGATACCTAAAAAGAGATTTCAAGAGATTTAATTAATCATGAATGTGTAACACAGTGCCTTCAATAAATGGTATAGCAAATG
TTTTGACATGAAAAAGGACAATTTCAAAAAAATAAAATAAAATAAAAATAAATTCACCTAGTCTAAGGATGCTAAACCTTAGTACTGAGTTACATTGTC
ATTTATATAGATTATAACTGTCTAAATAAGTTTGCAATTTGGGAGATATATTTTTAAGATAATAATATATGTTTACCTTTTAATTAATGAAATATCTGTA
TTTAATTTTGACACTATATCTGTATATAAAAATATTTTCATACAGCATTACAAATTGCTTACTTTGGAATACATTTCTCCTTTGATAAAATAAATGAGCTA
TGTATTAACACTGCCAGATTCAGTTAAATAAATCTCAACAGAATTTTTAAGGTGAGATTTTTAATACTTCACTGCTCTTTAATTTTCTACTTTCATTGAAT 3916

Figure 7.1   Nucleotide sequences from the human IL-7 gene. Intron sequences are indicated in *lower case letters*. A potential CAAT sequence, E12 binding site, SP1 binding site and polyadenylation signals (AATAAA) are *underlined*.

Da. This includes a 25-amino-acid signal sequence which is absent from the mature IL-7 protein. The 5′ untranslated region of the murine cDNA contains 548 nucleotides, while the analogous region of the human cDNA spans a region of 384 nucleotides. Conversely, the 3′ untranslated region of the human cDNA contains 658 nucleotides and is larger than the corresponding region of the murine cDNA (579 nucleotides). The murine cDNA contains an open reading frame encoding 154 amino acids. Twenty five of the amino acids comprise a signal sequence, which is identical in length to the human signal sequence. The human gene contains a 54 bp exon (exon 5) which is completely absent from the murine IL-7 gene. The artificial deletion of exon 5 from

the human cDNA has no effect on the biological activity of the protein produced from the altered cDNA. The biological significance of exon 5 is at present completely unknown. The chromosomal location of the human gene has been identified as chromosome 8, in region q12-13 (Sutherland *et al.*, 1989). The corresponding location of the murine gene has not been defined.

Northern blotting analysis was used to examine the tissue distribution of IL-7-specific transcripts. In murine tissue the highest level of expression observed was in the thymus, with somewhat lower levels of expression observed in the spleen and kidney. In the thymus, four different sizes of transcripts were detected corresponding to 2.9, 2.6, 1.7, and 1.5 kb. All of the four different sizes

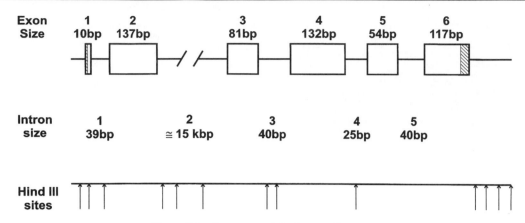

Figure 7.2   Structure of the human IL-7 gene.

of thymic transcripts have been shown to be derived by alternative splicing and polyadenylation of a single core transcript (Cosman *et al.*, 1989). Of particular interest was the presence of alternative splicing in the 5′ untranslated region of the transcript, which is quite rare. As mentioned earlier, this could relate to the regulatory control inherent to the 5′ noncoding region. Murine splenic tissue contains transcripts of 2.9, 2.6, and 1.7 kb, while kidney tissue exhibited two faint bands at 2.9 and 2.6 kb. Northern blot analysis of human spleen and thymus demonstrated the presence of two sizes of IL-7 transcripts. Human splenic tissue displayed a single 2.4 kb transcript while the human thymus exhibited a 2.4 kb and a 1.8 kb transcript. In contrast to murine tissue, IL-7 transcripts were expressed at a much higher level in human spleen than in human thymus. This species difference in IL-7 transcripts may relate to the age of the thymic tissue. In the murine tissue the thymic samples were derived from adult mice and the human thymic tissue was obtained from an infant. The significance of this disparity remains unknown.

3.   IL-7 Protein

The amino acid sequence of human IL-7 is shown in Figure 7.3. Native mature murine IL-7 consists of 129 amino acids giving rise to a predicted molecular mass of approximately 14 900 Da (Namen *et al.*, 1988b). Purified native murine IL-7 exhibits a molecular mass of 25 000 Da as determined by SDS-PAGE. The presence of two *N*-linked glycosylation sites at amino acids 69 and 90 probably account for at least part of the discrepancy between the native and the predicted molecular masses. Recombinant murine IL-7 produced by mammalian expression exhibits a molecular mass of 25 000 Da, very similar to the purified native molecule.

The coding region for the corresponding human cDNA contains 456 nucleotides, which would result in a

mature protein of 152 amino acids and a predicted molecular mass of 17 400 Da. Size analysis by SDS-PAGE of human recombinant IL-7 produced by mammalian expression gave rise to protein bands of 28, 24, and 20 kDa (Cosman *et al.*, 1989). The human IL-7 molecule contains three potential *N*-linked glycosylation sites. When human recombinant IL-7 was treated with *N*-glycanase and reanalyzed by SDS-PAGE, the molecular mass of the human IL-7 was reduced to 20 kDa. This confirms that at least a portion of the discrepancy between the predicted and observed molecular masses can be attributed to glycosylation. Owing to the lack of a suitable cellular source, native mature human IL-7 has never been purified to homogeneity, but presumably would exhibit the same molecular mass as the recombinant mammalian IL-7 species.

Human and murine IL-7 exhibit a high degree of homology at the amino acid level (approximately 60%) with the exception of an additional exon coding for 19 amino acids which are present in human IL-7. This additional exon has not been observed in any of the murine cDNAs isolated to date. Deletion of exon 5 (which encodes the additional 19 amino acids) had no effect on the biological activity of the human IL-7. Additionally, the extra 19 amino acids do not confer any species specificity as the murine and human IL-7s are fully cross-reactive on either murine or human cells. Not surprisingly, the isoelectric points are very similar, with a value for both IL-7s of approximately 9.0.

Both murine and human IL-7 contain six highly conserved cysteine residues, which are presumably involved in disulfide linkages. This is supported by the observations that both murine and human IL-7 bioactivity are highly resistant to both SDS (1%) or mild heat treatment as long as the disulfide linkages are maintained. Even brief treatment with reducing agents destroys all biological activity in a nonreversible manner. Likewise, both species of IL-7 are quite resistant to extremes of pH (from 2.0 to 10.0). Purified native or

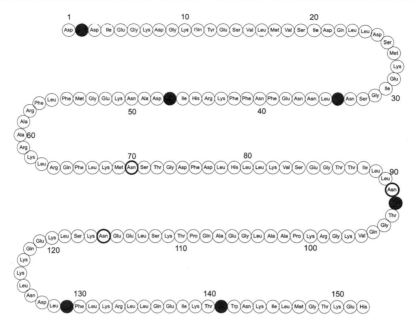

Figure 7.3   The amino acid sequence of mature human IL-7. The *N*-linked glycosylation sites are shown by dark circles and the six conserved cysteine residues are shaded.

recombinant IL-7 is quite stable at 4°C, and in our hands usually retains biological activity for at least several months. Storage at –70°C usually allows preservation of activity for in excess of one year.

While it is clear that at least some disulfide bridges are essential for bioactivity, the nature of the cysteine pairing is unknown. Both murine and human IL-7 contain six cysteine residues that are highly positionally conserved, underscoring the importance of the cysteine pairing. Similarly, no information is available concerning the secondary or tertiary structure of the molecule, but it is clear that IL-7 monomers contain full biological activity.

4. Cellular Sources

IL-7 was originally detected in supernatants produced from a bone marrow stromal cell line (Namen *et al.*, 1988a), and it is now well established that many stromal cell lines derived from bone marrow secrete IL-7 (Gimble *et al.*, 1989). The levels of IL-7 produced, however, are uniformly rather low in unmodified cell lines. IL-7 production by stromal cells appears to be tightly regulated, as no native high-producing cell lines have been described. Moreover, most IL-7-producing cell lines are refractory to inductive stimuli, and only modest increases in IL-7 production were achieved by stimulation with lipopolysaccharide. Transcripts encoding IL-7 have been detected in murine spleen, thymus, kidney, and bone marrow and in human spleen and thymus (Cosman *et al.*, 1989). Murine thymic stromal cell lines (Sakata *et al.*, 1990) have been established which constitutively produce IL-7, as well as thymic epithelial cell lines (Gutierrez and Palacios, 1991) and fetal liver cell lines (Gunji *et al.*, 1991). It has also been demonstrated that murine and human keratinocytes produce detectable levels of IL-7 (Heufler *et al.*, 1993). It appears in general that many stromal/epithelial cells or cell lines can produce IL-7 but that IL-7 production is notably absent in hematopoietic or lymphoid cells.

5. Biological Activities

Owing to the large amount of data produced by murine experiments, the section on murine IL-7 has been incorporated into this section as indicative of the probable biological role of human IL-7.

5.1 B CELLS

The first biological activity described for IL-7 was the stimulation of proliferation of murine pro-B and pre-B cells (Namen *et al.*, 1988a) in the absence of stromal cells. While the initial studies showed that B cell progenitors could respond to IL-7 alone, it is now clear that other factors, such as stem cell factor (SCF) and flt3 ligand can synergize with IL-7 to promote pre-B cell growth (McNiece *et al.*, 1991; Yasunaga *et al.*, 1995; Namikawa *et al.*, 1996; Ray *et al.*, 1996; Veiby *et al.*, 1996). Both IL-7 and stem cell factor are produced by bone marrow stromal cells and both molecules are probably involved in the regulation of B cell development. Almost certainly, as experimentation proceeds, new soluble factors and cell surface molecules will appear which impact the biological effects of IL-7. IL-7 does not appear to mediate the maturation of pre-B cells to mature functional B cells (Takeda, 1989), a process which seems to be under the control of undefined stromal cell factors and/or cell surface molecules. The growth response to the IL-7 during B cell ontogeny is restricted to cells expressing CD34 (Dittel and LeBien, 1995). Several known factors, which include transforming growth factor-β (Lee et al., 1989) and interleukin-1α (Suda et al., 1989), have been shown to downregulate the proliferative activity of IL-7 on pre-B cells. This most likely represents a control mechanism to regulate the expansion of the bone marrow pre-B cell compartment, which appears to be the primary effect of IL-7 on B lineage cells.

In contrast, mature surface IgM-positive B cells usually do not respond to IL-7 and also usually lack detectable receptors for IL-7 (Park et al., 1990). There is, however, one report that anti-Ig-activated B cells could proliferate in response to IL-7 (Joshi and Choi, 1991). The reasons for this discrepancy are not clear, but may be a function of the activation state of the cell populations tested. It is quite clear that at present the major action of IL-7 on cells of the B lineage is the expansion of the early B cell precursor population. This has been confirmed in experiments utilizing IL-7 gene-deleted mice whereby bone marrow B lymphopoiesis was blocked at the transition from pro-B to pre-B cells (von Freeden-Jeffry et al., 1995). In IL-7 transgenic mice there is an accumulation of immature B cells (Fisher et al., 1995; Mertsching et al., 1995). A role for IL-7 has been delineated in the differentiation of B cells during B lymphopoiesis (Corcoran et al., 1996). B cell proliferation is not sufficient to complete V–D–J rearrangement and distinct IL-7-induced signals are required for completion (Candeias et al., 1997a,b).

5.2   T CELLS

IL-7 has proved to have a much broader range of effects on T lineage cells. It has been shown that IL-7 can directly stimulate the proliferation of murine or fetal thymocytes independently of the cytokines IL-2, IL-4, or IL-6 (Conlon et al., 1989). Confirmation of the vital role of IL-7 in thymocyte/T cell development arose from studies depriving mice of IL-7 using neutralizing antibodies whereby thymic cellularity was reduced by 99% (Sudershan et al., 1995). Gene knockout mice exhibited a similar phenomenon (von Frieden-Jeffry et al., 1995). Data from transgenic mice overexpressing IL-7 illustrate that IL-7 enhances thymocyte proliferation but not thymocyte differentiation (Mertsching et al., 1995).

Transgenic mice have been produced utilizing the IL-7 cDNA under the control of the immunoglobulin heavy-chain enhancer and the kappa light-chain promoter (Samaridas et al., 1991). The mice exhibited increased levels of pre-B cells in the bone marrow and B cells in the spleen. The levels of T cells and thymocytes were elevated in the spleen and thymus, respectively, with no observable changes in the number of myeloid cells in the bone marrow or periphery. There were no abnormalities detected in the functional capabilities of the B cell or T cell compartments. These results are consistent with the studies involving the in vivo administration of IL-7 and anti-IL-7.

A second study also utilized the IL-7 cDNA, this time under the control of the immunoglobulin heavy-chain promoter and enhancer (Rich et al., 1993), with quite different results. A perturbation of the T cell compartment was observed, with a significant reduction in the levels of CD4$^+$ and CD8$^+$ cells in the thymus. Curiously, expression of the IL-7 cDNA in the skin resulted in a T cell infiltrate in the dermal region and the development of a progressive alopecia, hyperkeratosis, and finally exfoliation. Additionally, these transgenic animals exhibited lymphoproliferative disorders with the development of B cell and T cell lymphomas. The differences in the results of these two studies and those by Mertsching et al. (1995) are very striking, and the reasons for the discrepancies are unclear. Further study is obviously warranted to resolve these differences, but they do illustrate the potent in vivo effects of IL-7 depending upon the microenvironmental location and the level of expression of IL-7.

A number of independent studies have addressed the question of which of the various subpopulations of thymocytes proliferate in response to IL-7. Exogenous IL-7 has been shown to mediate the expansion of CD3$^-$CD4$^-$CD8$^-$ thymocyte precursors in fetal thymic organ cultures (Vissinga et al., 1992) and to maintain the viability of highly fractionated populations of CD3$^-$CD4$^-$CD8$^-$ thymocytes in vitro (Suda and Zlotnik, 1991). Studies by Conlon et al. (1989) and Suda and Zlotnik (1991) established that the CD3$^+$CD4$^-$CD8$^-$ population of adult or fetal thymocytes proliferates in response to IL-7. The results of these studies and several others indicate that IL-7 is a potent growth factor for primarily primitive T cell precursors. The largest expansion involves the CD3$^+$CD4$^-$CD8$^-$ precursor population, with more modest increases in cell number in the more primitive CD3$^-$CD4$^-$CD8$^-$ population. The association of IL-7 with early thymocyte growth is also suggested by the levels of IL-7 messenger RNA that occur during fetal thymic development (Wiles et al., 1992). Detectable IL-7 mRNA first appears at day 12 of development in the murine fetal thymus, peaks at days 14 to 16, and then declines as birthing approaches. The IL-7 protein is detectable at day 13 of development. Moreover, blocking antibodies to IL-7 have been shown to severely limit the expansion of thymocytes in fetal thymic organ cultures. While IL-7 is certainly a major factor in thymic development, the details regulating the effects of IL-7 in the thymus are still being explored. It has been shown that transforming growth factor-β can downregulate the proliferative effects of IL-7 on early thymocytes (Chantry *et al.*, 1989). Another study has shown that IL-7 is requisite for the IL-1-stimulated proliferation of thymocytes (Herbelin *et al.*, 1992), and is involved in the synergistic activities of granulocyte-macrophage colony-stimulating factor and tumor necrosis factor. Still another study demonstrated that exposure of fetal thymic lobes to IL-7 resulted in a marked expansion of the developing thymocytes with little or no development of cytolytic function (Widmer, 1990). When these cells were subsequently exposed to IL-2 they rapidly developed cytolytic function. This would seem to indicate that a major role for IL-7 in thymic development is the expansion of thymic precursors which could then differentiate in response to the complex interplay of additional factors in the microenvironment of the thymus.

In contrast to data from transgenic mice, several other studies have shown that IL-7 may also exhibit differentiative effects on early thymocyte precursors (Appasamy, 1992). Exposure of murine fetal liver to IL-7 stimulates expression of messenger RNA encoding the T cell receptor α, ß and γ chains. Others have shown that IL-7 can promote the differentiation of human thymic precursors into T cell receptor-positive γ-δ cells (Groh *et al.*, 1990). Similar experiments with murine fetal thymocytes (Watanabe *et al.*, 1991) also demonstrated a promotion of the development of γ-δ-bearing T cells. IL-7 seems to occupy a special role in the development of γ-δ T cells with the capability of both a proliferative and a differentiative role. The mechanisms of T cell development are extremely complex and still not well understood, but it is now becoming clear that IL-7 plays a major and still emerging role in the process.

In contrast to B-lineage cells, where IL-7 effects only the precursor pool, IL-7 has potent effects on mature peripheral T cells as well as developing primitive T cells (Webb *et al.*, 1997). IL-7 has been shown to be a potent growth factor on mature T cells, particularly in the presence of a costimulatory signal. Morrissey and colleagues (1989) demonstrated that IL-7 and concanavalin A stimulated the proliferation, expression of receptors for IL-2 on T cells, and the synthesis and secretion of IL-2 from T cells. Similar results have been shown with IL-7 and a number of different costimuli. These include IL-7 and anti-T-cell receptor monoclonal antibodies (Armitage *et al.*, 1990), the adhesion molecules CD2 and CD28 (Costello *et al.*, 1993), phorbol myristate acetate (PMA) (Chazen *et al.*, 1989), and phytohemagglutinin (PHA) (Welch *et al.*, 1989). Generally, on mature T cells IL-7 appears to require a second signal for maximal mitogenic activity. An interesting and perhaps physiologically relevant observation is that adhesion molecules can provide the second signal. Since the stromal cells known to produce IL-7 also express adhesion molecules and/or their ligands, it is interesting to speculate that this particular combination may have an important role in T cell function *in vivo*. The potential involvement of IL-2 in the IL-7-driven proliferation of mature T cells appears to depend upon the costimulatory signal. Costimulation with concanavalin A and IL-7 results in the production of IL-2 and the expression of receptors for IL-2. Costimulation with anti-T-cell receptor monoclonal antibodies or anti-CD28 antibodies in the presence of IL-7 promotes the expression of IL-2 receptors but elicits only a weak production of IL-2. Finally, the combination of IL-7 and PHA, PMA, or anti-CD2 stimulates IL-2 receptor expression in the absence of detectable IL-2 synthesis. The results are very complex and probably reflect the differences in experimental design and cell populations examined. While IL-2 may be involved in IL-7-driven proliferation of T cells in some instances, there is no evidence for the involvement of other T cell stimulators such as IL-4 or IL-6. There is general agreement between experimental studies that the stimulation of T cells can involve both the CD4[+] and the CD8[+] compartments. IL-12 is able to synergize with IL-7 during T cell activation (Mehrota *et al.*, 1995). During activation, IL-7 enhances CD23 expression (Carini and Fratazzi, 1996) and modulates expression of IL-4 and IFN-γ (Dokter *et al.*, 1994; Borger *et al.*, 1996).

5.3   CYTOTOXIC T CELLS AND LAK CELLS

IL-7 has been shown to enhance the generation of cytotoxic T cells (Alderson *et al.*, 1990) and lymphokine-activated killer cells (Lynch *et al.*, 1990; Dadmarz *et al.*, 1994). This enhancement appears to be independent of IL-2. Another study demonstrated that IL-7 induces cytotoxicity in mixed lymphocyte reactions and against virally infected targets (Hickman *et al.*, 1990). Neutralizing antiserum to IL-7 could partially reverse the cytotoxic responses in mixed lymphocyte cultures, while IL-2 neutralizing antibodies had no effect. Naume and Esperick (1991) have utilized purified CD56[+] cells and shown a marked proliferation in response to IL-7, without an IL-2 involvement. These cells also generated substantial levels of LAK activity following incubation in IL-7, although the levels were less than those generated in response to IL-2. This has been confirmed in a study using LAK cell activity against melanoma cells (Schadendorff *et al.*, 1994).

5.4   MYELOPOIESIS AND MONOCYTE ACTIVATION

Most of the biological effects of IL-7 have centered around its enhancement and/or differentiation of lymphoid cells. The effects of IL-7 have now been extended to include the myeloid lineage. Although IL-7 has no effect in myelopoiesis alone, in combination with either stem cell factor (Fahlman *et al.*, 1994) or the colony-stimulating factors (Jacobsen *et al.*, 1994a,b), IL-7 induces macrophage and granulocyte/macrophage colonies. Human monocytes purified from peripheral blood respond to IL-7 treatment by secreting IL-1α, IL-1β, IL-6, and tumor-necrosis factor α (Alderson *et al.*, 1991). Additionally, the IL-7-treated monocytes exhibited a significant tumoricidal activity against human melanoma cells. A second study demonstrated that IL-7 treatment stimulated expression of the gene for macrophage inflammatory protein-1β (Zeigler *et al.*, 1991). These studies suggest a role for IL-7 in the mediation of the inflammatory immune response through the activation of peripheral monocytes.

5.5   IL-7 *IN VIVO*

The availability of recombinant IL-7 and the gene encoding IL-7 have prompted a number of IL-7 experiments *in vivo* to better understand its physiological role. When normal mice were injected subcutaneously with IL-7, significant increases in the cellularity of bone marrow were due to marked increases in pre-B cells (Morrissey *et al.*, 1991a). There was a 2-fold increase in B cells in the spleen and lymph node over the same time period. In addition there were increased levels of both CD4[+] and CD8[+] T cells in the spleen and thymus. All cell numbers returned to normal upon the withdrawal of IL-7 treatment. Similarly, in mice rendered leukopenic by the administration of cyclophosphamide, subcutaneous IL-7 treatment dramatically accelerated the regeneration of pre-B cells in the bone marrow and B cells in both the spleen and lymph node (Morrissey *et al.*, 1991b). An accelerated recovery of CD4[+] and CD8[+] T cells was also observed in the spleen and lymph node. Another study by Grabstein and colleagues (1993) explored the effects of the administration *in vivo* of an anti-IL-7 neutralizing monoclonal antibody. A profound inhibition of B cell progenitors was observed in the bone marrow, manifesting itself at the pro-B cell stage. Anti-IL-7 treatment also substantially reduced thymic cellularity with a 3- to 4-fold reduction of cells in the thymus. The remaining thymocytes exhibited a normal subset distribution, indicating that anti-IL-7 treatment affected all of the major thymic subpopulations. These studies suggest that IL-7 plays a pivotal role in lymphoid development, affecting both the B cell and the T cell compartments. In both of these studies (administration of IL-7 or neutralization of IL-7), no significant effects were observed on myeloid development.

A separate study utilizing intraperitoneal administration of IL-7 yielded significantly different results (Damia *et al.*, 1992). In normal mice there was no change in overall bone marrow cellularity, but there was

a decrease in the numbers of myeloid progenitors observed. Increases in B cells, T cells, NK cells, and null cells were observed in the spleen, along with an increase in myeloid progenitors, suggesting the stimulation of a trafficking response. This study utilized a 20-fold higher dosage of IL-7 (10 μg, twice daily) and a different route of administration, which may contribute to the differing results.

In agreement with this, a study by Grzegorzewski and colleagues in 1994 showed that no changes in bone marrow cellularity occurred on administration of IL-7, again suggesting that changes seen in myelopoiesis result largely from emigration of myeloid progenitors from the bone marrow to peripheral organs.

It has also been confirmed that administration of IL-7 to normal mice induces a pronounced leukocytosis in spleen and lymph node with B cells, T cells, NK cells, and macrophages, but here CD4:CD8 ratios were dramatically reduced (Komschlies *et al.*, 1994).

6.   *IL-7 Receptors*

The initial characterization of the murine IL-7 receptor (IL-7R) was carried out utilizing a long-term IL-7-dependent cell line (IxN/2b) and purified iodinated recombinant murine IL-7. The binding profiles were nonlinear, exhibiting a high ($K_a \sim 5 \times 10^9$ M^{-1}) and a low ($K_a \sim 1 \times 10^6$ M^{-1}) affinity component, suggesting a two-chain structure. Approximately 20% of the 2500 IL-7 receptors on the IxN/2b cell line exhibited functional high-affinity binding which was rapidly saturable (Park *et al.*, 1990). Subsequently, the cDNAs encoding both the putative human and murine IL-7Rs were cloned and characterized (Goodwin *et al.*, 1990). Both cDNAs encode for a 439-amino-acid transmembrane protein sequence in addition to a signal sequence consisting of 20 amino acids. The receptors exhibit a quite high level of homology (64% identity) at the amino acid level and are associated with the γ chain of the IL-2 receptor (Noguchi *et al.*, 1993). Therefore, IL-7R consists of two chains, the IL-7R-specific chain (IL-7Rα) and the γ chain of the IL-2 receptor (Kawahara *et al.*, 1994; Kondo *et al.*, 1994; He *et al.*, 1995; Ziegler *et al.*, 1995). A schematic representation of the IL-7R is shown in Figure 7.4. When expressed in mammalian cell lines, cDNA for IL-7Rα gave rise to binding profiles indistinguishable from profiles on native cell lines. The cDNA contains an extracellular region which consists of 219 amino acids, followed by a hydrophobic 25-amino-acid transmembrane region, terminating in an intracellular cytoplasmic region of 195 amino acids. Interestingly, the six extracellular cysteines and the four intracellular cysteines are highly conserved between the murine and human receptor molecules, although the potential disulfide linkages and their relation to biological activity

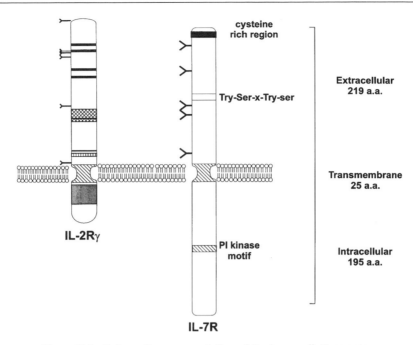

Figure 7.4   Schematic representation of the human IL-7 receptor.

are at present unknown. Both murine and human IL-7 bind to both the murine and human IL-7Rα with similar profiles and affinities.

Expression of the human IL-7Rα cDNA in a mammalian cell line gave rise to both high- and low-affinity binding species due to the presence or absence of the IL-2R γ chain (Kondo *et al.*, 1994). During expression of the human IL-7Rα cDNA it was observed that three species of mRNA encoding the IL-7Rα were produced (Goodwin *et al.*, 1990). Two of the observed mRNA molecules encode membrane-bound forms of the IL-7Rα which are capable of exhibiting both high- and low-affinity binding. One of the membrane-bound forms is full-length, containing full extracellular, transmembrane, and intracellular regions. The second membrane-bound species contains full-length extracellular and transmembrane regions, but contains a shorter, truncated cytoplasmic region. The binding characteristics of the two forms appear to be the same, but it is not known whether the two membrane-bound forms transmit the same biological signals following receptor occupancy.

The predicted molecular mass of the IL-7Rα chain is approximately 50 kDa. The observed molecular mass of the expressed IL-7Rα, or the native receptor in a number of different cell lines, was determined to comprise a major species at approximately 75 kDa and a minor species at approximately 160 kDa (Park *et al.*, 1990). Presumably, the higher than predicted molecular mass is due to glycosylation in the extracellular region. It has been suggested that the two observed receptor species may be due to dimerization and that the dimerization

process may generate alternative binding affinities. This remains to be determined.

The third mRNA species observed following expression of the human IL-7Rα cDNA encodes a soluble secreted form of the receptor which is truncated at the beginning of the transmembrane region. The secreted form of the receptor binds efficiently to IL-7 in solution and is reminiscent of the soluble receptors which have been identified for IL-2, IL-4, IL-6, IFN-γ, LIF, and tumor necrosis factor (Aggarwal and Pocsik, 1992). It has been proposed that the soluble forms of the cytokine receptors mediate a clearance mechanism to regulate the amount of active circulating cytokines.

Finally, in addition to the aforementioned forms of the IL-7R, a novel low-affinity receptor for IL-7 has been identified by Armitage and colleagues (1992); its relationship to the IL-2 receptor γ chain is elusive. This receptor binds IL-7 with an affinity 100- to 1000-fold less than that demonstrated previously. This low-affinity competable type of binding was observed on a number of primary cells and cell lines and included both B and T cells, pre-B-cells, and monocytic cells. These same binding studies demonstrated high numbers of low-affinity sites, up to approximately 100 000 sites per cell. The function of these peculiar low-affinity IL-7 sites is not known. It is interesting, though, that a monocytic cell line (THP-1) which expresses high levels of low-affinity sites and no high-affinity sites exhibited a dose-dependent decrease in proliferation when exposed to IL-7. Since IL-7 provides a positive growth signal to lymphoid cells, it is interesting to speculate that IL-7 may promote lymphoid growth at the expense of other hematopoietic cells (such as monocytic or myeloid cells), perhaps even at the developmental stage where cells become committed to the various hematopoietic lineages. This, however, is speculation and remains to be shown.

The distribution of IL-7 receptors closely parallels the observed range of IL-7-responsive cells, and receptors have been shown to occur on pre-B cells, thymocytes, mature T cells, and bone marrow macrophages, but have not been observed on mature IgM-displaying B cells. Transcripts encoding IL-7Rα have been observed in human and murine spleen (Goodwin *et al.*, 1990) and thymus, as well as in murine fetal liver. The transcript displayed a molecular mass of approximately 3.5 kDa.

It has been documented that the human and murine IL-7Rα genes both contain eight exons. Sequence comparisons with known receptor structures established that IL-7Rα merits inclusion in the hematopoietin receptor superfamily (Pleiman *et al.*, 1991). The IL-7Rα contains the canonical Trp-Ser-X-Trp-Ser region near the transmembrane region which is indicative of the hematopoietin superfamily. The typical cysteine-rich region near the amino terminus of the IL-7Rα contains a single cysteine rather than the usual two pairs. However, IL-7Rα now joins a growing list in this superfamily, which is now known to include IL-2R, IL-3R, IL-4R, IL-6R, EPO-R, G-CSF-R, and the GM-CSF-R. It was also observed by Goodwin and colleagues (1990) that differential splicing of a single mRNA species encoding IL-7Rα could lead to the production of both a transmembrane-bound receptor and a soluble secreted receptor. The alternative splicing event occurred at exon 6 (which encodes the transmembrane region).

The 5′ upstream region of the murine IL-7Rα gene contains several identifiable regulatory elements. The common eukaryotic promoter sequences TATA and CAATT have been identified (Pleiman *et al.*, 1991). Additionally, sequences suggestive of AP-1 and AP-2 sites as well as several glucocorticoid receptor binding sites are present. Particularly interesting is the presence of sequences quite similar to those contained in interferon response elements, which specifically bind interferon regulatory factors 1 and 2. Given that IL-7 is a known potent inducer of cytotoxic T cells, it is interesting to speculate on the potential and perhaps interactive roles of interferon and IL-7 in viral elimination.

7. IL-7 Signal Transduction

The hematopoietin receptor superfamily is now known to contain at least 10 members, but none of the members has been shown to exhibit intrinsic tyrosine kinase activity. Accordingly, it is likely that receptor expression alone may not be sufficient to trigger a biological response. In the case of the IL-7R, both resting and activated T cells express high-affinity receptor sites, but primarily activated T cells proliferate strongly in response to IL-7 (Grabstein *et al.*, 1990), while resting T cells respond by expressing activation antigens (Armitage *et al.*, 1990). It is likely that molecules associate with the IL-7R in activated T cells to mediate a full biological response.

In primary human B cell precursors and B cell lines it has been shown that IL-7 specifically stimulates tyrosine phosphorylation of multiple phosphoproteins, inositol phospholipid turnover, and DNA synthesis, resulting in proliferation (Uckun *et al.*, 1991; Seckinger and Fougereau, 1994). Similar results have been demonstrated with T cell lymphoblastic leukemias (Didbirdik *et al.*, 1991). IL-7 stimulation in human primary thymocytes and T cells results in tyrosine phosphorylation but no corresponding phosphatidylinositol turnover (Roifman *et al.*, 1993). This could be due to differences in the cell populations studied, or the conditions of stimulation, as T cell stimulation with IL-7 commonly utilizes a comitogen. c-*fos*, n-*myc*, and c-*myc* are transcriptionally upregulated in response to stimulation with IL-7 in normal primary B cell precursors (Morrow *et al.*, 1992; Imoto *et al.*, 1996). It has been demonstrated that there is a functional association of the IL-7Rα with the *src* family tyrosine kinases $p53/56^{lyn}$ and $p59^{fyn}$ in pre-B cells (Venkitaraman and Cowling, 1992; Seckinger and Fougereau, 1994) and T cells (Page *et al.*, 1995).

Phosphatidyl 3-kinase associates with the IL-7Rα chain through a canonical sequence Tyr(P)-X-X-Met surrounding Tyr-449 and is activated via tyrosine phosphorylation (Dadi *et al.*, 1994; Venkitaraman and Cowling, 1994). The IL-7Rα also mediates activation of the kinases JAK1 and JAK3 in murine T cells (Foxwell *et al.*, 1995) and STAT 5 and STAT 1 in B cell leukemia cells (van der Plas *et al.*, 1996). IL-7 induces differential association of PI3 kinase with IRS-1 and IRS-2 in thymocytes (Sharfe and Roifman, 1997). While it is clear that the IL-7R α chain is important for IL-7 signal transduction, the IL-2R γ chain is also critical for full IL-7 signal transduction (Kawahara *et al.*, 1994; Ziegler *et al.*, 1995).

8. Clinical Applications

8.1 GENE TRANSFER

There is a substantial amount of data accumulating to suggest that IL-7 could be useful in treatments involving tumor immunity and gene transfer. Studies by Lynch and colleagues (1991) showed that IL-7 was a potent inducer of antitumor cytotoxic lymphocytes which exhibited an enhanced immunotherapeutic efficacy in cellular adoptive immunotherapy. Antitumor lymphocytes could be expanded greatly *in vitro* in the presence of IL-7 with no loss of tumor specificity. The cells could be cultured for many months in the presence of IL-7 and retain the ability to mediate rejection of ultraviolet- or methylcholanthrene-induced fibrosarcomas. A similar study by Jicha and colleagues (1991) indicated that IL-7 was at least as effective as IL-2 at inducing antitumor immunity useful in adoptive immunotherapy. Several other studies have indicated that IL-7 is more effective than other cytokines at inducing a specific cytotoxic response against a weakly immunogenic viral peptide (Kos and Mullbacher, 1992) and a murine fibrosarcoma (Jicha *et al.*, 1992). These results are in agreement with a host of *in vitro* studies suggesting a potentially powerful role for IL-7 in the generation of cytotoxic T cells and lymphokine-activated killer cells. IL-7 alone, or in combination with other cytokines, appears to be a likely candidate for future studies in adoptive immunotherapy, particularly since IL-7 has now been shown also to stimulate monocytic antitumor responses (Alderson *et al.*, 1991).

A number of studies have now been published utilizing gene transfer of IL-7 into a variety of tumor cell lines. Hock and colleagues (1991) transfected the IL-7 gene into a plasmacytoma cell line which mediated complete rejection of the injected tumor cells through a CD4$^+$ T cell-mediated mechanism. Interestingly, an effective immune response was necessary for tumor rejection, as IL-7 transfection could not induce tumor elimination in nude or SCID mice. Aoki and colleagues (1992) demonstrated that when the IL-7 gene was introduced into a murine glioma cell line, the cell line had a much diminished tumorgenicity when injected into animals. Similarly, an IL-7-transfected murine fibrosarcoma was rejected following adoptive transfer (McBride *et al.*, 1992). Sharma and colleagues (1997) transferred non-small-cell lung cancer cells with similar results to cell lines. All of the IL-7-transfected tumor cell lines stimulated the infiltration of T cells and macrophages, both cell types which are known to exhibit cytotoxicity in response to IL-7. Of particular interest, these studies also demonstrated that when the animals were challenged secondarily with non-IL-7-transfected tumor cells, the cells were also rejected, indicating that the animals were endowed with a specific antitumor memory capability. These studies clearly show the effectiveness of IL-7 gene transfer in mediating a specific antitumor response with the resultant rejection of the tumor. Tumor cells transfected with IL-7 are potent vaccines (Cayeux *et al.*, 1995) and trials have been carried out with IL-7 gene transfer in melanoma and renal cell carcinoma (Schmidt *et al.*, 1994; Schadendorf *et al.*, 1995).

8.2   IMMUNORECONSTITUTION/ STIMULATION

Another potentially useful area for IL-7 therapy is in regeneration or stimulation of the lymploid compartment in immunocompromised situations. Intensive radiation and chemotherapy prior to transplantation is particularly effective at inducing lymphopenia, with regeneration of the lymphoid compartment lagging behind regeneration of the other hematopoietic lineages. Since IL-7 appears to have a major role in inducing lymphoid development, IL-7 could have a wide clinical application in stimulating lymphopoiesis. IL-7 has been shown to stimulate lymphopoiesis post bone marrow transplantation in murine models (Boerman *et al.*, 1995; Mathur *et al.*, 1995; Bolotin *et al.*, 1996). In addition, stimulation of the immune system by IL-7 may prove valuable in fighting infection (Kasper *et al.*, 1995; Tantawichien *et al.*, 1996).

8.3   IL-7 AND DISEASE

While IL-7 would appear to be a valuable therapeutic in tumor immunity and lymphopenic situations, a cautionary note must be exercised as to the effects of IL-7 on B cell and T cell leukemias. IL-7 has been shown to promote the proliferation of acute lymphoblastic leukemias (Elder *et al.*, 1990; Touw *et al.*, 1990; Smiers *et al.*, 1995), B cell lines (Benjamin *et al.*, 1994; Renard *et al.*, 1995), and T cell lymphomas (Foss *et al.*, 1994). It is not clear whether IL-7 has any kind of causative role in leukemogenesis or whether the IL-7 response is a remnant of normal development. Elevated levels of IL-7 have been found in Hodgkin disease, juvenile arthritis, psoriasis and multiple myeloma suggesting a role in these diseases (Trumper *et al.*, 1994; de Benedetti *et al.*, 1995; Foss *et al.*, 1995; Kroning *et al.*, 1997; Bonifati *et al.*, 1997). As has occurred with other known growth factors, as IL-7 proceeds through the stages of clinical applications, unforeseen effects, whether beneficial or harmful, are sure to appear. However, the potential therapeutic usefulness in tumor immunity and lymphopenic situations clearly warrants introduction into clinical trials.

9.   *References*

Aggarwal, B.B. and Pocsik, E. (1992). Cytokines: from clone to clinic. Arch. Biochem. Biophys. 292, 335–359.

Alderson, M.R., Sassenfeld, H.M. and Widmer, M.B. (1990). Interleukin 7 enhances cytolytic T lymphocyte generation and induces lymphokine-activated killer cells from human peripheral blood. J. Exp. Med. 172, 577–587.

Alderson, M.R., Tough, T.W., Zeigler, S.F., *et al.* (1991). Interleukin 7 induces cytokine secretion and tumoricidal activity by human peripheral blood monocytes. J. Exp. Med. 173, 923–930.

Aoki, T., Tashiro, K., Miyatake, S.-I., *et al.* (1992). Expression of murine interleukin 7 in a murine glioma cell line results in reduced tumorigenicity in vivo. Proc. Natl Acad. Sci. USA 89, 3850–3854.

Appasamy, P.M. (1992). IL-7 induced T cell receptor-gamma gene expression by pre-T cells in murine fetal liver cultures. J. Immunol. 149, 1649–1656.

Armitage, R.J., Namen, A.E., Sassenfeld, H.M., *et al.* (1990). Regulation of human T cell proliferation by IL-7. J. Immunol. 144, 938–941.

Armitage, R.J., Ziegler, S.F., Friend, D., *et al.* (1992). Identification of a novel low-affinity receptor for human interleukin-7. Blood 79, 1738–1744.

Benjamin, D., Sharma, V., Knobloch, T.J., Armitage, R.J., Dayton, M.A. and Goodwin, R.G. (1994). B cell IL-7. Human B cell lines constitutively secrete IL-7 and express IL-7 receptors. J. Immunol. 152(10), 4749–4757.

Bhatia, S.K., Tygrett, L.T., Grabstein, K.H. and Waldschmidt, T.J. (1995). The effect of in vivo IL-7 deprivation on T cell maturation. J. Exp. Med. 181(4), 1399–1409.

Boerman, O.C., Gregorio, T.A., Grzegorzewski, K.J., *et al.* (1995). Recombinant human IL-7 administration in mice affects colony-forming units-spleen and lymphoid precursor cell localization and accelerates engraftment of bone marrow transplants. J. Leukocyte Biol. 58(2), 151–158.

Bolotin, E., Smogorzewska, M., Smith, S., Widmer, M. and Weinberg, K. (1996). Enhancement of thymopoiesis after bone marrow transplant by in vivo interleukin-7. Blood 88(5), 1887–1894.

Bonifati, C., Trento, E., Cordiali-Fei, P., *et al.* (1997). Increased interleukin-7 concentrations in lesional skin and in the sera of patients with plaque-type psoriasis. Clin. Immunol. Immunopathol. 83(1), 41–44.

Borger, P., Kauffman, H.F., Postma, D.S. and Vellenga, E. (1996). IL-7 differentially modulates the expression of IFN-gamma and IL-4 in activated human T lymphocytes by transcriptional and post-transcriptional mechanisms. J. Immunol. 156(4), 1333–1338.

Candeias, S., Muegge, K. and Durum, S.K. (1997a). IL-7 receptor and VDJ recombination: trophic versus mechanistic actions. Immunity 6(5), 501–508.

Candeias, S., Peschon, J.J., Muegge, K. and Durum, S.K. (1997b). Defective T-cell receptor gamma gene rearrangement in interleukin-7 receptor knockout mice. Immunol. Lett. 57(1–3), 9–14.

Carini, C. and Fratazzi, C. (1996). CD23 expression in activated human T cells is enhanced by interleukin-7. Int. Arch. Allergy Immunol. 110(1), 23–30.

Cayeux, S., Beck, C., Aicher, A., Dorken, B. and Blankenstein, T. (1995). Tumor cells cotransfected with interleukin-7 and B7.1 genes induce CD25 and CD28 on tumor-infiltrating T lymphocytes and are strong vaccines. Eur. J. Immunol. 25(8), 2325–2331.

Chantry, D., Turner, M. and Feldman, M. (1989). Interleukin 7 (murine pre-B cell growth factor/lymphopoetin 1) stimulates thymocyte growth: Regulation by transforming growth factor beta. Eur. J. Immunol. 19, 783–786.

Chazen, G.D., Pereira, G.M., Legros, G., *et al.* (1989). Interleukin 7 is a T cell growth factor. Proc. Natl Acad. Sci. USA 86, 5923–5927.

Conlon, P.J., Morrissey, P.J., Norday, R.P., *et al.* (1989). Murine thymocytes proliferate in direct response to interleukin-7. Blood 74, 1368–1373.

Corcoran, A.E., Smart, F.M., Cowling, R.J., Crompton, T., Owen, M.J. and Venkitaraman, A.R. (1996). The interleukin-7 receptor alpha chain transmits distinct signals for proliferation and differentiation during B lymphopoiesis. EMBO J. 15(8), 1924–1932.

Cosman, D., Namen, A.E., Lupton, S., *et al.* (1989). Interleukin 7: a lymphoid growth factor active on T and B cell progenitors. Lymphokine Receptor Interactions 179, 229–236.

Costello, R., Mawas, C. and Olive, D. (1993). Differential immunosuppressive effects of metabolic inhibitors on T lymphocyte activation. Eur. Cytokine Netw. 4, 139–145.

Dadi, H., Ke, S. and Roifman, C.M. (1994). Activation of phosphatidylinositol-3 kinase by ligation of the interleukin-7 receptor is dependent on protein tyrosine kinase activity. Blood 84(5), 1579–1586.

Dadmarz, R., Bockstoce, D.C. and Golub, S.H. (1994). Interleukin-7 selectively enhances natural kill cytotoxicity mediated by the CD56 bright natural killer subpopulation. Lymphokine Cytokine Res. 13(6), 349–357.

Damia, G., Komschlies, K.L., Faltynek, C., *et al.* (1992). Administration of recombinant human interleukin-7 alters the frequency and number of myeloid progenitors in the bone marrow and spleen of mice. Blood 79, 1121–1129.

De Benedetti, F., Massa, M., Pignatti, P., Kelley, M., Faltynek, C.R. and Martini, A. (1995). Elevated circulating interleukin-7 levels in patients with systemic juvenile rheumatoid arthritis. J. Rheumatol. 22(8), 1581–1585.

Didbirdik, I., Langlie, M.C. and Ledbetter, J.A. (1991). Engagement of interleukin 7 receptor stimulates tyrosine phosphorylation, phosphoinositol turnover and clonal proliferation of human T lineage lymphoblastic leukemia cells. Blood 78, 564–573.

Dittel, B.N. and LeBien, T.W. (1995). The growth response to IL-7 during normal human B cell ontogeny is restricted to B-lineage cells expressing CD34. J. Immunol. 154(1), 58–67.

Dokter, W.H., Sierdsema, S.J., Esselink, M.T., Halie, M.R. and Vellenga, E. (1994). Interleukin-4 mRNA and protein in activated human T cells are enhanced by interleukin-7. Exp. Hematol. 22(1), 74–79.

Elder, M., Ottmann, O.G., Hannsen-Hagge, T.E., *et al.* (1990). Effects of recombinant human IL-7 on blast cell proliferation in acute lymphoblastic leukemia. Leukemia 4, 533–540.

Fahlman, C., Blomhoff, H.K., Veiby, O.P., McNiece, I.K. and Jacobsen, S.E. (1994). Stem cell factor and interleukin-7 synergize to enhance early myelopoiesis in vitro. Blood 84, 1450–1456.

Fisher, A.G., Burdet, C., Bunce, C., Merkenschlager, M. and Ceredig, R. (1995). Lymphoproliferative disorders in IL-7 transgenic mice: expansion of immature B cells which retain macrophage potential. Int. Immunol. 7(3), 415–423.

Foss, F.M., Koc, Y., Stetler-Stevenson, M.A., *et al.* (1994). Costimulation of cutaneous T-cell lymphoma cells by interleukin-7 and interleukin-2: potential autocrine or paracrine effectors in the Sezary syndrome. J. Clin. Oncol. 12(2), 326–335.

Foss, H.D., Hummel, M., Gottstein, S., *et al.* (1995). Frequent expression of IL-7 gene transcripts in tumor cells of classical Hodgkin's disease. Am. J. Pathol. 146(1), 33–39.

Foxwell, B.M., Beadling, C., Guschin, D., Kerr, I. and Cantrell, D. (1995). Interleukin-7 can induce the activation of Jak 1, Jak 3 and STAT 5 proteins in murine T cells. Eur. J. Immunol. 25(11), 3041–3046.

Gimble, J.M., Pietrangeli, C., Henley, A., *et al.* (1989). Characterization of murine bone marrow and spleen-derived stromal cells: analysis of leukocyte marker and growth factor mRNA transcript level. Blood 74, 303–311.

Goodwin, R.G., Lupton, S., Schmierer, A.E., *et al.* (1989). Human interleukin 7: molecular cloning and growth factor activity on human and murine B lineage cells. Proc. Natl Acad. Sci. USA 86, 302–306.

Goodwin, R.G., Friend, D.J., Ziegler, S.F., et al. (1990). Cloning of the human and murine IL-7 receptors: demonstration of a soluble form and homology to a new receptor superfamily. Cell 60, 941–951.

Grabstein, K.H., Namen, A.E., Shanebeck, K., et al. (1990). Regulation of human T cell proliferation by IL-7. J. Immunol. 144, 3015–3019.

Grabstein, K.H., Waldschmidt, T.J. and Finkelman, F.D. (1993). Inhibition of murine B and T lymphopoesis in vivo by an anti interleukin 7 monoclonal antibody. J. Exp. Med. 178.

Groh, V., Fabbi, M. and Strominger, J.L. (1990). Maturation of differentiation of human thymocyte precursors in vitro? Proc. Natl Acad. Sci. USA 87, 5973–5977.

Grzegorzewski, K., Komschlies, K.L., Mori, M., et al. (1994). Administration of recombinant human interleukin-7 to mice induces the exportation of myeloid progenitor cells from the bone marrow to peripheral sites. Blood 83(2), 377–385.

Gunji, Y., Sudo, T., Suda, J., et al. (1991). Support of early B-cell differentiation in mouse fetal liver by stromal cells and Interleukin 7. Blood 77, 2612–2617.

Gutierrez, J.C. and Palacios, R. (1991). Heterogeneity of thymic epithelial cells in promoting T lymphocyte differentiation in vivo. Proc. Natl Acad. Sci. USA 88, 642–646.

He, Y.W., Adkins, B., Furse, R.K. and Malek, T.R. (1995). Expression and function of the gamma c subunit of the IL-2, IL-4, and IL-7 receptors. Distinct interaction of gamma c in the IL-4 receptor. J. Immunol. 154(4), 1596–1605.

Herbelin, A., Machavoine, F., Schnieder, E., et al. (1992). Interleukin 7 is requisite for IL-1 induced thymocyte proliferation. Involvement of IL-7 in the synergistic effects of granulocyte macrophage colony-stimulating factor or tumor necrosis factor with IL-1. J. Immunol. 148, 99–105.

Heufler, C., Topar, G., Grasseger, A., et al. (1993). Interleukin 7 is produced by murine and human keratinocytes. J. Exp. Med. 178, 1109–1114.

Hickman, C.J., Crim, J.A., Mostowski, H.S., et al. (1990). Regulation of human cytotoxic T lymphocyte development by IL-7. J. Immunol. 145, 2415–2420.

Hock, H., Dorsch, M., Diamanstein, T., et al. (1991). Interleukin 7 induces CD4$^+$ T cell dependent tumor rejection. J. Exp. Med. 174, 1291–1298.

Hozumi, K., Kondo, M., Nozaki, H., et al. (1994). Implication of the common gamma chain of the IL-7 receptor in intrathymic development of pro-T cells. Int. Immunol. 6(9), 1451–1454.

Imoto, S., Hu, L., Tomita, Y., Phuchareon, J., Ruther, U. and Tokuhisa, T. (1996). A regulatory role of c-Fos in the development of precursor B lymphocytes mediated by interleukin-7. Cell. Immunol. 169(1), 67–74.

Jacobsen, F.W., Veiby, O.P. and Jacobsen, S.E. (1994a). IL-7 stimulates CSF-induced proliferation of murine bone marrow macrophages and Mac-1$^+$ myeloid progenitors in vitro. J. Immunol. 153(1), 270–276.

Jacobsen, F.W., Rusten, L.S. and Jacobsen, S.E. (1994b). Direct synergistic effects of interleukin-7 on in vitro myelopoiesis of human CD34$^+$ bone marrow progenitors. Blood 84(3), 775–779.

Jicha, D.L., Mule, J.J. and Rosenberg, S.A. (1991). Interleukin 7 generates antitimor cytotoxic T lymphocytes against murine sarcomas with efficacy in cellular adoptive immunotherapy. J. Exp. Med. 174, 1511–1515.

Jicha, D.L., Schwarz, S., Mule, J.J., et al. (1992). Interleukin 7 mediates the generation and expansion of murine allosensitized and antitumor CTL. Cell. Immunol. 141, 71–83.

Joshi, P.C. and Choi, Y.S. (1991). Human interleukin 7 is a B cell growth factor for activated B cells. Eur. J. Immunol. 21, 681–686.

Kasper, L.H., Matsuura, T. and Khan, I.A. (1995). IL-7 stimulates protective immunity in mice against the intracellular pathogen, Toxoplasma gondii. J. Immunol. 155(10), 4798–4804.

Kawahara, A., Minami, Y. and Taniguchi, T. (1994). Evidence for a critical role for the cytoplasmic region of the interleukin 2 (IL-2) receptor gamma chain in IL-2, IL-4, and IL-7 signalling. Mol. Cell. Biol. 14(8), 5433–5440.

Komschlies, K.L., Gregorio, T.A., Gruys, M.E., Back, T.C., Faltynek, C.R. and Wiltrout, R.H. (1994). Administration of recombinant human IL-7 to mice alters the composition of B-lineage cells and T cell subsets, enhances T cell function, and induces regression of established metastases. J. Immunol. 152(12), 5776–5784.

Kondo, M., Takeshita, T., Higuchi, M., et al. (1994). Functional participation of the IL-2 receptor gamma chain in IL-7 receptor complexes. Science 263(^{a}5152), 1453–1454.

Kos, F.J. and Mullbacher, A. (1992). Induction of primary antiviral cytotoxic T cells by an in vitro stimulation with short synthetic peptide and interleukin 7. Eur. J. Immunol. 22, 3183–3187.

Kroning, H., Tager, M., Thiel, U., et al. (1997). Overproduction of IL-7, IL-10 and TGF-beta 1 in multiple myeloma. Acta Haematol. 98(2), 116–118.

Lee, G., Namen, A.E., Gillis, S., et al. (1989). Normal B cell precursors responsive to recombinant murine IL-7 and inhibition of IL-7 activity by transforming growth factor beta. J. Immunol. 142, 3875–3883.

Lupton, S.D., Gimpel, S., Jerzy, R., et al. (1990). Characterization of the human and murine IL-7 genes. J. Immunol. 144, 3592–3601.

Lynch, D.H. and Miller, R.E. (1990). Induction of murine lymphokine activated killer cells by recombinant IL-7. J. Immunol. 145, 1983–1990.

Lynch, D.H., Namen, A.E. and Miller, R.E. (1991). In vivo evaluation of the effects of interleukin 2, 4 and 7 on enhancing immunotherapeutic efficacy of anti-tumor cytotoxic T lymphocytes. Eur. J. Immunol. 21, 2977–2985.

Mathur, A., Vallera, D.A., Taylor, P.A., et al. (1995). Effect of IL-7 or IL-4 on reconstitution of donor lymphoid cells in congenic murine bone marrow transplantation. Bone Marrow Transplant. 16(1), 119–124.

McBride, W.H., Thacker, J.D., Comora, S., et al. (1992). Genetic modification of a murine fibrosarcoma to produce IL-7 stimulates host cell infiltration and tumor immunity. Cancer Res. 52, 3931–3938.

McNiece, I.K., Langley, K.E. and Zsebo, K.M. (1991). The role of recombinant stem cell factor in early B cell development. Synergistic interaction with IL-7. J. Immunol. 146, 3785–3790.

Mehrotra, P.T., Grant, A.J. and Siegel, J.P. (1995). Synergistic effects of IL-7 and IL-12 on human T cell activation. J. Immunol. 154(10), 5093–5102.

Mertsching, E., Burdet, C. and Ceredig, R. (1995). IL-7 transgenic mice: analysis of the role of IL-7 in the differentiation of thymocytes in vivo and in vitro. Int. Immunol. 7(3), 401–414.

Morrissey, P.J., Goodwin, R.G., Nordan, R.P., et al. (1989). Recombinant interleukin 7, pre-B cell growth factor, has costimulatory activity on purified mature T cells. J. Exp. Med. 169, 707–716.

Morrissey, P.J., Conlon, P., Charrier, K., et al. (1991a). Administration of IL-7 to normal mice stimulates B-lymphopoesis and peripheral lymphadenopathy. J. Immunol. 147, 561–568.

Morrissey, P.J., Conlon, P., Braddy, S., et al. (1991b). Administration of IL-7 to mice with cyclophosphamide-induced lymphopenia accelerates lymphocyte repopulation. J. Immunol. 146, 1547–1552.

Morrow, M.A., Lee, G., Gillis, S., et al. (1992). Interleukin 7 induces n-myc and c-myc expression in normal precursor B lymphocytes. Genes Dev. 6, 61–70.

Namen, A.E., Schmierer, A.E., March, C.J., et al. (1988a). B cell precursor growth-promoting activity. Purification and characterization of a growth factor active on lymphoid progenitors. J. Exp. Med. 167, 988–1002.

Namen, A.E., Lupton, S., Hjerrild, K., et al. (1988b). Stimulation of B cell progenitors by cloned murine interleukin 7. Nature 333, 571–573.

Namikawa, R., Muench, M.O., de Vries, J.E. and Roncarolo, M.G. (1996). The FLK2/FLT3 ligand synergizes with interleukin-7 in promoting stromal-cell-independent expansion and differentiation of human fetal pro-B cells in vitro. Blood 87(5), 1881–1890.

Naume, B. and Esperick, T. (1991). Effects of IL-7 and IL-2 on highly enriched CD56+ natural killer cells. A comparative study. J. Immunol. 147, 2208–2214.

Noguchi, M., Nakamura, Y., Russell, S.M., et al. (1993). Interleukin 2 receptor gamma chain: a functional component of the interleukin-7 receptor. Science 262: 1877–1880.

Page, T.H., Lali, F.V. and Foxwell, B.M. (1995). Interleukin-7 activates p56lck and p59fyn, two tyrosine kinases associated with the p90 interleukin-7 receptor in primary human T cells. Eur. J. Immunol. 25(10), 2956–2960.

Park, L.S., Friend, D.J., Schmierer, A.E., et al. (1990). Murine interleukin 7 receptor. Characterization on an IL-7 dependent cell line. J. Exp. Med. 171, 1073–1089.

Pleiman, C.M., Gimpel, S.D., Park, L.S., et al. (1991). Organization of the murine and human interleukin-7 receptor genes: two mRNA's generated by differential splicing and presence of a type I interferon-inducible promoter. Mol. Cell. Biol. 11, 3052–3059.

Rachal, M.J., Yoo, H., Becker, F.F., et al. (1989). In vitro DNA cytosine methylation of cis-regulatory elements modulates C-Ha-ras promoter activity in vivo. Nucleic Acids Res. 13, 5135–5143.

Ray, R.J., Paige, C.J., Furlonger, C., Lyman, S.D. and Rottapel, R. (1996). Flt3 ligand supports the differentiation of early B cell progenitors in the presence of interleukin-11 and interleukin-7. Eur. J. Immunol. 26(7), 1504–1510.

Renard, N., Duvert, V., Matthews, D.J., et al. (1995). Proliferation of MIELIKI a novel t(7;9) early pre-B acute lymphoblastic leukemia cell line is inhibited concomitantly by IL-4 and IL-7. Leukemia 9(7), 1219–1226.

Rich, B.E., Campos-Torres, J., Tepper, R.I., et al. (1993). Cutaneous lymphoproliferation and lymphomas in Interleukin 7 transgenic mice. J. Exp. Med. 77, 305–316.

Roifman, C.M., Wang, G.W., Freedman, M., et al. (1992). IL-7 receptor mediates tyrosine phosphorylation but does not activate the phosphatidylinositol-phospholipase C gamma-1 pathway. J. Immunol. 148, 1136–1142.

Sakata, T., Iwagami, S., Tsuruta, Y., et al. (1990). Constitutive expression of interleukin 7 mRNA and production of IL-7 by a cloned murine thymic stromal cell line. J. Leukocyte Biol. 48, 205–212.

Samaridas, J., Casorati, G., Traunecker, A., et al. (1991). Development of lymphocytes in interleukin-7 transgenic mice. Eur. J. Immunol. 21, 453–460.

Schadendorf, D., Bohm, M., Moller, P., Grunewald, T. and Czarnetzki, B.M. (1994). Interleukin-7 induces differential lymphokine-activated killer cell activity against human melanoma cells, keratinocytes, and endothelial cells. J. Invest. Dermatol. 102(6), 838–842.

Schadendorf, D., Czarnetzki, B.M. and Wittig, B. (1995). Interleukin-7, interleukin-12, and GM-CSF gene transfer in patients with metastatic melanoma. J. Mol. Med. 73(9), 473–477.

Schmidt-Wolf, I.G., Huhn, D., Neubauer, A. and Wittig, B. (1994). Interleukin-7 gene transfer in patients with metastatic colon carcinoma, renal cell carcinoma, melanoma, or with lymphoma. Hum. Gene. Ther. 5(9), 1161–1168.

Seckinger, P. and Fougereau, M. (1994). Activation of src family kinases in human pre-B cells by IL-7. J. Immunol. 153(1), 97–109.

Sharfe, N. and Roifman, C.M. (1997). Differential association of phosphatidylinositol 3-kinase with insulin receptor substrate (IRS)-1 and IRS-2 in human thymocytes in response to IL-7. J. Immunol. 159(3), 1107–1114.

Sharma, S., Wang, J., Huang, M., et al. (1996). Interleukin-7 gene transfer in non-small-cell lung cancer decreases tumor proliferation, modifies cell surface molecule expression, and enhances antitumor reactivity. Cancer Gene. Ther. 3(5), 302–313.

Smiers, F.J., van Paassen, M., Pouwels, K., et al. (1995). Heterogeneity of proliferative responses of human B cell precursor acute lymphoblastic leukemia (BCP-ALL) cells to interleukin 7 (IL-7): no correlation with immunoglobulin gene status and expression of IL-7 receptor or IL-2/IL-4/IL-7 receptor common gamma chain genes. Leukemia 9(6), 1039–1045.

Suda, T. and Zlotnick, A. (1991). IL-7 maintains the T cell precursor potential of CD3−CD4−CD8− thymocytes. J. Immunol. 146, 3068–3073.

Suda, T., Okada, S., Suda, J., et al. (1989). A stimulatory effect of recombinant murine interleukin 7 (IL-7) on B-cell colony formation and an inhibitory effect of IL-1. Blood 74, 1936–1941.

Sutherland, G.R., Baker, E., Fernandez, K.E., et al. (1989). The gene for human interleukin 7 (IL-7) is at 8 q 12-13. Hum. of bone marrow pro-B and pro-T lymphocyte clones and fetal thymocyte clones. Proc. Natl Acad. Sci. USA 86, 1634–1638.

Tantawichien, T., Young, L.S. and Bermudez, L.E. (1996). Interleukin-7 induces anti-Mycobacterium avium activity in human monocyte-derived macrophages. J. Infect. Dis. 174(3), 574–582.

Touw, I., Pouwels, K. and Agthoven, T.N. (1990). Interleukin 7 is a growth factor of precursor B and T acute lymphoblastic leukemia. Blood 75, 2097–2101.

Trumper, L., Jung, W., Dahl, G., Diehl, V., Gause, A. and Pfreundschuh, M. (1994). Interleukin-7, interleukin-8, soluble TNF receptor, and p53 protein levels are elevated in the serum of patients with Hodgkin's disease. Ann. Oncol. (5, supplement 1), 93–96.

Uckun, F.M., Didbirdik, I., Smith, R., et al. (1991). Interleukin 7 receptor ligation stimulates tyrosine phosphorylation, inositol phospholipid turnover and clonal proliferation of human B cell precursors. Proc. Natl Acad. Sci. USA 88, 3589–3593.

van der Plas, D.C., Smiers, F., Pouwels, K., Hoefsloot, L.H., Lowenberg, B. and Touw, I.P. (1996). Interleukin-7 signaling in human B cell precursor acute lymphoblastic leukemia cells and murine BAF3 cells involves activation of STAT1 and STAT5 mediated via the interleukin-7 receptor alpha chain. Leukemia 10(8), 1317–1325.

Veiby, O.P., Lyman, S.D. and Jacobsen, S.E. (1996). Combined signaling through interleukin-7 receptors and flt3 but not c-kit potently and selectively promotes B-cell commitment and differentiation from uncommitted murine bone marrow progenitor cells. Blood 88(4), 1256–1265.

Venkitaraman, A.R. and Cowling, R.J. (1992). Interleukin 7 receptor functions by recruiting the tyrosine kinase p59^{fyn} through a segment of its cytoplasmic tail. Proc. Natl Acad. Sci. USA 89, 12083–12086.

Venkitaraman, A.R. and Cowling, R.J. (1994). Interleukin-7 induces the association of phosphatidylinositol 3-kinase with the alpha chain of the interleukin-7 receptor. Eur. J. Immunol. 24(9), 2168–2174.

Vissinga, C.S., Fatur-Saunders, D.J. and Takei, F. (1992). Dual role of IL-7 in the growth and differentiation of immature thymocytes. Exp. Hematol. 20, 998–1003.

von Freeden-Jeffry, U., Vieira, P., Lucian, L.A., McNeil, T., Burdach, S.E. and Murray, R. (1995). Lymphopenia in interleukin (IL)-7 gene-deleted mice identifies IL-7 as a nonredundant cytokine. J. Exp. Med. 181(4), 1519–1526.

Watanabe, Y., Sudo, T., Minato, N., et al. (1991). Interleukin 7 preferentially supports the growth of gamma delta T cell receptor-bearing T cells from fetal thymocytes in vitro. Int. Immunol. 3, 1067–1075.

Webb, L.M., Foxwell, B.M. and Feldmann, M. (1997). Interleukin-7 activates human naive CD4+ cells and primes for interleukin-4 production. Eur. J. Immunol. 27(3), 633–640.

Welch, P.A., Namen, A.E., Goodwin, R.G., et al. (1989). Human IL-7: a novel T cell growth factor. J. Immunol. 143, 3562–3566.

Whitlock, C.A., Robertson, D. and Witte, O.N. (1984). Murine B lymphopoesis in long term culture. J. Immunol. Methods 6, 353–374.

Widmer, M.B., Morrissey, P.J., Namen, A.E., et al. (1990). Interleukin 7 stimulates growth of fetal thymic precursors of cytolytic cells: induction of effector function by interleukin 2 and inhibition by interleukin 4. Int. Immunol. 2(11), 1055–1061.

Wiles, M.V., Ruiz, P. and Imhof, B.A. (1992). Interleukin 7 expression during mouse thymus development. Eur. J. Immunol. 22, 1037–1042.

Yasunaga, M., Wang, F., Kunisada, T., Nishikawa, S. and Nishikawa, S. (1995). Cell cycle control of c-kit$^+$IL-7R$^+$ B precursor cells by two distinct signals derived from IL-7 receptor and c-kit in a fully defined medium. J. Exp. Med. 182(2), 315–323.

Zeigler, S.F., Tough, T.W., Franklin, T.L., et al. (1991). Induction of macrophage inflammatory protein-1 beta gene expression in human monocytes by lipopolysaccharide and IL-7. J. Immunol. 147, 2234–2239.

Ziegler, S.E., Morella, K.K., Anderson, D., et al. (1995). Reconstitution of a functional interleukin (IL)-7 receptor demonstrates that the IL-2 receptor gamma chain is required for IL-7 signal transduction. Eur. J. Immunol. 25(2), 399–404.

8. Interleukin-8

Ivan J. D. Lindley

1. Introduction

Interleukin-8 (IL-8) was discovered by several groups at about the same time, resulting in a multitude of names for the factor, all based on the biological activities that allowed its initial identification. However, prior to these studies, the cDNA had been cloned from a differential screen of stimulated human peripheral blood leukocytes, and was named 3-10C (Schmid and Weissman 1987). IL-8 is a member of the *chemo*tactic cyto*kine*, or *chemokine* family (Lindley *et al.*, 1993), and is in fact the most studied member of that family to date. This C-X-C chemokine was identified in 1986–1987, as a soluble factor present in supernatants after endotoxin stimulation of monocytes. The bioactivities by which IL-8 was identified all involved its effect on activation of polymorphonuclear leukocytes (neutrophils), particularly chemotaxis or granule release.

IL-8 exerts its biological activity through two high-affinity glycosylated receptors, designated IL-8RA and IL-8RB (or I and II). These two receptors are now known as CXC chemokine receptors 1 and 2 (CXCR1 and CXCR2). They are members of the seven-transmembrane domain G-protein-coupled receptor family and share 77% homology at the amino acid level. Although IL-8 can bind to both receptors with high affinity, only the type I receptor is specific; the type II receptor is promiscuous, and can also bind some other C-X-C chemokines (GRO, NAP-2, ENA-78). Neither receptor can bind the unrelated chemotactic agents leukotriene B_4 (LTB_4), platelet-activating factor (PAF), the complement factor C5a, or the formyl peptide f-Met-Leu-Phe. Signal transduction from the receptor is achieved via GTP-binding proteins, and involves activation of protein kinase C (PKC) and mobilization of intracellular calcium. During cell activation by IL-8, the receptor–ligand complex is internalized, and the receptor is then recycled back to the surface.

1.1 SYNONYMS

Synonymous names include neutrophil-activating factor (NAF) (Walz *et al.*, 1987; Lindley *et al.*, 1988); monocyte-derived neutrophil chemotactic factor (MDNCF; Yoshimura *et al.*, 1987); monocyte-derived neutrophil-activating peptide and lymphocyte-derived neutrophil-activating peptide (MONAP and LYNAP; Schröder *et al.*, 1987; Gregory *et al.*, 1988); granulocyte chemotactic factor (GCP; Van Damme *et al.*, 1988), monocyte-derived chemotaxin (MOC; Kownatski *et al.*, 1988); and lung carcinoma derived chemotactin (LUCT) (Suzuki *et al.*, 1989). Subsequently, the name neutrophil-activating peptide 1 (NAP-1) was adopted by some groups, but the molecule was finally named Interleukin-8 (IL-8). In many publications the authors have attempted to avoid confusion by using their own nomenclature combined with the term IL-8 (e.g., NAP-1/IL-8).

1.2   THE CHEMOKINE FAMILY

The chemokines are a family of low-molecular-mass pro-inflammatory proteins with considerable homology at the amino acid level (Oppenheim *et al.*, 1991), and with potent chemotactic activity for leukocytes, both *in vitro* and *in vivo*. The family is separated into two major subfamilies, based on a structural motif within the molecule, a group of four conserved cysteine residues forming a pair of disulfide bridges which are essential for the activity of the molecule. The C-X-C, or α-chemokines, possess an amino acid between the first two cysteine residues, while the C-C, or β-chemokines, lack an amino acid at this position. IL-8 is a representative C-X-C chemokine, while, for instance, monocyte chemotactic peptide-1 (MCP-1) or RANTES (regulated on activation, normal T cell-expressed and secreted) are representative members of the C-C chemokine subfamily. There is also a third type of chemokine, the so-called "C-chemokines", which lack the first and third cysteines in the motif. To date, the only member of this family to be identified is lymphotactin (Kennedy *et al.*, 1995).

There are clear functional differences between the two main subfamilies. Broadly speaking, the C-X-C chemokines tend to exert their chemotactic and activating effects on neutrophils, and also on other cell types including mononuclear cells, but not on monocytes. C-C chemokines, however, frequently have activity on mononuclear cells in general and monocytes in particular (e.g., MCP-1) and some have marked effects also on granulocytes.

2.   *The Cytokine Gene*

2.1   GENOMIC ORGANIZATION OF THE IL-8 GENE

Human IL-8 is encoded by a single mRNA transcript of approximately 1.8 kb (Schmid and Weissman, 1987) (see Figure 8.1). The IL-8 gene comprises four exons and three introns, covering 5.25 kbp of DNA (Mukaida *et al.*, 1989), and is situated, in a cluster with the other C-X-C

chemokines, on chromosome 4q12-21. The overall structure of the gene is shown schematically in Figure 8.2.

2.2   TRANSCRIPTIONAL CONTROL

The structure of the 5′ flanking region of the IL-8 gene contains many known regulatory elements, including binding sites for NFκB, NF-IL-6, AP-1, AP-2, AP-3, interferon regulatory factor-1 and glucocorticoid response element (Mukaida *et al.*, 1989). Transcriptional activation occurs after treatment of producer cells with a variety of stimuli, of which the classical ones are IL-1α and IL-1β, TNF-α and bacterial endotoxin, but which also include phorbol-12-myristate 13-acetate (PMA), all-*trans* retinoic acid (ATRA), and reactive oxygen and nitrogen intermediates (Remick and Villarete, 1996), including nitric oxide (Andrew *et al.*, 1995).

It has been suggested that the AP-1 and octamer binding motifs are dispensable for IL-8 gene activation, while the NFκB and NF-IL-6 binding sites appear to be sufficient (Kunsch *et al.*, 1994). Recent studies now indicate that the region spanning the nucleotides −94 to −70 relative to the transcription start site of the IL-8 gene is essential for both induction and repression by certain stimuli (Stein and Baldwin 1993; Kunsch *et al.*, 1994; Mukaida *et al.*, 1994; Olivera *et al.*, 1994), mediated mainly by the transcription factor complexes NF-IL-6 and NFκB. NF-IL-6 was originally identified as an IL-1-induced *trans*-activator of the human IL-6 gene and belongs to the CCAAT/enhancer binding proteins (C/EBP) family of transcription factors (Isshiki *et al.*, 1990). Several members of the NFκB family, such as p50 (NFKB1), p65 (RelA), c-Rel, and p52 (NFKB2), have been shown to bind the NFκB motif of the IL-8 promoter. While p50, p65 and c-Rel are able to bind these sequences efficiently, NF-IL-6 binds to its own binding site very weakly. However, binding of NFκB to its site results in strong cooperative binding of NF-IL-6 to the adjacent site (Stein and Baldwin, 1993).

IL-8 is also regulated at the post-transcriptional level. In the 3′-flanking region, the IL-8 gene contains the repetitive ATTTA motif, which is responsible for destabilization of various cytokine mRNAs (Shaw and Kamen, 1986).

3.   *The Protein*

IL-8 is produced as a 99-amino-acid (99-aa) nonglycosylated peptide, the N-terminal end of which comprises a signal sequence, cleaved enzymatically upon release from the producer cell (Figure 8.3). Mature IL-8 appears to exist in multiple forms, arising from processing at the N-terminus which is thought to be

CTCCATAAGGCACAAACTTTCAGAGACAGCAGAGCACACAAGCTTCTAGGACAAGAGCCA

GGAAGAAACCACCGGAAGGAACCATCTCACTGTGTGTAAAC

ATGACTTCCAAGCTGGCCGTGGCTCTCTTGGCAGCCTTCCTGATTTCTGCAGCTCTGTGT

MetThrSerLysLeuAlaValAlaLeuLeuAlaAlaPheLeuIleSerAlaAlaLeuCys

GAAGGTGCAGTTTTGCCAAGGAGTGCTAAAGAACTTAGATGTCAGTGCATAAAGACATAC

GluGlyAlaValLeuProArgSerAlaLysGluLeuArgCysGlnCysIleLysThrTyr

TCCAAACCTTTCCACCCCAAATTTATCAAAGAACTGAGAGTGATTGAGAGTGGACCACAC

SerLysProPheHisProLysPheIleLysGluLeuArgValIleGluSerGlyProHis

TGCGCCAACACAGAAATTATTGTAAAGCTTTCTGATGGAAGAGAGCTCTGTCTGGACCCC

CysAlaAsnThrGluIleIleValLysLeuSerAspGlyArgGluLeuCysLeuAspPro

AAGGAAAACTGGGTGCAGAGGGTTGTGGAGAAGTTTTTGAAGAGGGCTGAGAATTCATAA

LysGluAsnTrpValGlnArgValValGluLysPheLeuLysArgAlaGluAsnSer.

AAAAATTCATTCTCTGTGG

TATCCAAGAATCAGTGAAGATGCCAGTGAAACTTCAAGCAAATCTACTTCAACACTTCAT

GTATTGTGTGGGTCTGTTGTAGGGTTGCCAGATGCAATACAAGATTCCTGGTTAAATTTG

AATTTCAGTAAACAATGAATAGTTTTTCATTGTACCATGAAATATCCAGAACATACTTAT

ATGTAAAGTATTATTTATTTGAATCTACAAAAAACAACAAATAATTTTTAAATATAAGGA

TTTTCCTAGATATTGCACGGGAGAATATACAAATAGCAAAATTGGGCCAAGGGCCAAGAG

AATATCCGAACTTTAATTTCAGGAATTGAATGGGTTTGCTAGAATGTGATATTTGAAGCA

TCACATAAAAATGATGGGACAATAAATTTTGCCATAAAGTCAAATTTAGCTGGAAATCCT

GGATTTTTTTCTGTTAAATCTGGCAACCCTAGTCTGCTAGCCAGGATCCACAAGTCCTTG

TTCCACTGTGCCTTGGTTTCTCCTTTATTTCTAAGTGGAAAAAGTATTAGCCACCATCTT

ACCTCACAGTGATGTTGTGAGGACATGTGGAAGCACTTTAAGTTTTTTCATCATAACATA

AATTATTTTCAAGTGTAACTTATTAACCTATTTATTATTTATGTATTTATTTAAGCATCA

AATATTTGTGCAAGAATTTGGAAAAATAGAAGATGAATCATTGATTGAATAGTTATAAAG

ATGTTATAGTAAATTTATTTTATTTTAGATATTAAATGATGTTTTATTAGATAAATTTCA

ATCAGGGTTTTTAGATTAAACAAACAAACAATTGGGTACCCAGTTAAATTTTCATTTCAG

ATATACAACAAATAATTTTTTAGTATAAGTACATTATTGTTTATCTGAAATTTTAATTGA

ACTAACAATCCTAGTTTGATACTCCCAGTCTTGTCATTGCCAGCTGTGTTGGTAGTGCTG

TGTTGAATTACGGAATAATGAGTTAGAACTATTAAAACAGCCAAAACTCCACAGTCAATA

TTAGTAATTTCTTGCTGGTTGAAACTTGTTTATTATGTACAAATAGATTCTTATAATATT

ATTTAAATGACTGCATTTTTAAATACAAGGCTTTATATTTTTAACTTTAAAAAAAACCGG

Figure 8.1   IL-8 cDNA and amino acid sequence. The amino acids encoding the signal sequence are underlined. (From Matsushima et al., 1988.) Genbank accession number of cDNA Y00787.

controlled by specific proteases. The 77-aa (or endothelial) form is the major product of endothelial cells (Gimbrone et al., 1989), but monocytes produce mostly the 72-aa form, together with smaller amounts of 77-aa, 70-aa and 69-aa forms (Lindley et al., 1988). The conversion of 77-aa to 72-aa form can also be achieved by thrombin and neutrophils (Hébert et al., 1990). The 72-aa form appears to be rather more potent than the 77-aa form in a number of assays, and the naturally occurring shorter forms are more potent still

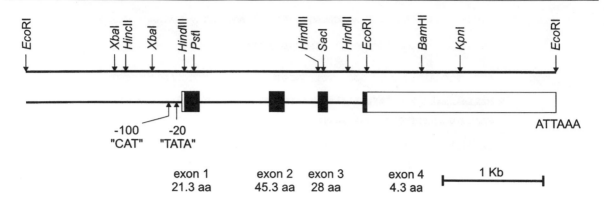

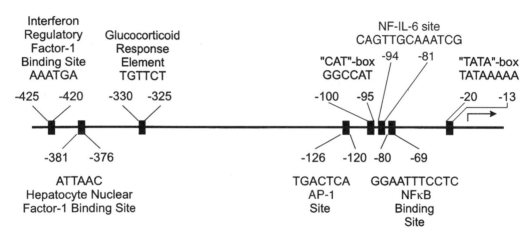

Figure 8.2   Gene structure of the IL-8 gene.

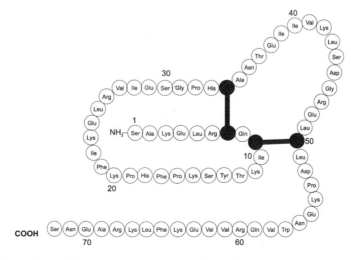

Figure 8.3   Amino acid sequence of IL-8. Cysteine residues are shown as filled circles.

(Clark-Lewis *et al.*, 1991). However, the activities of the variants seem to be qualitatively similar.

Several features of the molecule appear to be very important for activity. The two Cys–Cys bridges, which are a major feature in the chemokine family, are essential for activity. In the N-terminal region, the ELR (Glu-Leu-Arg) sequence is needed for functional activation of the receptor; C-X-C chemokines which do not possess this motif (e.g., platelet factor-4 (PF-4), γ-IP10) cannot activate neutrophils (Clark-Lewis *et al.*, 1991). Furthermore, the angiogenic effect of IL-8 and C-X-C chemokines reported by Koch and colleagues (1992) and Strieter and colleagues (1995) also requires this motif, while the non-ELR members of the C-X-C subfamily (e.g., PF-4) have been shown to antagonize the effect, resulting in anti-angiogenesis (Maione *et al.*, 1990). However, in another system, the ELR-containing C-X-C chemokines gro-α and gro-β have been demonstrated to have an inhibitory effect on angiogenesis (Cao *et al.*, 1995), suggesting that other, additional factors or motifs may be involved. The C-terminal α-helix, from amino acid 54 (Lys) to the end of the molecule, has been recognized as the binding site for the glycosaminoglycans which are present on the endothelium of the post-capillary venule, and to which IL-8 and other chemokines bind. This binding is thought to produce a solid-state gradient for recognition by rolling leukocytes, to promote haptotactic transmigration out of the blood vessel (Rot, 1992; Witt and Lander, 1994). Sequential truncation of this sequence has been shown to progressively reduce activity on neutrophils (Lindley *et al.*, 1990).

Amino acid substitutions have demonstrated the importance of other locations in the molecule in determining receptor binding specificity; mutation of Leu-25 to the conserved tyrosine present in C-C chemokines converts IL-8 to a monocyte chemoattractant which binds to the C-C chemokine receptor CKR-1 (Lusti-Narasimhan *et al.*, 1995). However, mutation of this leucine to cysteine has little effect on the activity of IL-8, and does not induce binding to CC CKR-1 (Lusti-Narasimhan *et al.*, 1996).

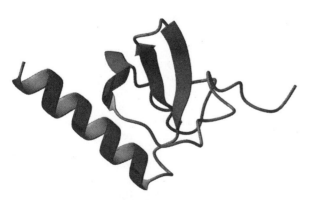

Figure 8.4   Three dimensional structure of IL-8.

IL-8 has been shown to be a noncovalent homodimer both in crystal form (Baldwin *et al.*, 1991), and in solution at high concentrations (Figure 8.4) (Clore *et al.*, 1990). However, it appears that IL-8 exerts its bioactivity as a monomer, since a single amino acid modification (Leu-25 to *N*-methylleucine) which prevented association of monomers did not affect receptor binding or activity in functional assays (Rajarathnam *et al.*, 1994).

4. Cellular Sources and Production

The variety of cell sources of IL-8 and the inducing stimuli have been dramatically extended since the time of the original reports; the current list of producer cells includes monocytes, macrophages, endothelial cells (Strieter *et al.*, 1989; Ueno *et al.*, 1996), lymphocytes (Gregory *et al.*, 1988; Schröder *et al.*, 1988), epithelial cells (see Table 8.1), smooth-muscle cells (Wang *et al.*, 1991), kidney mesangial cells (Brown *et al.*, 1991), chondrocytes (van Damme *et al.*, 1990; Lotz *et al.*, 1992), synovial cells (Golds *et al.*, 1989; Koch *et al.*, 1991), hepatocytes (Thornton *et al.*, 1990), fibroblasts (Schröder *et al.*, 1990; Tamura *et al.*, 1992, Tobler *et al.*, 1992), keratinocytes (Larsen *et al.*, 1989a; Gillitzer *et al.*, 1991; Chabot-Fletcher *et al.*, 1994; Li *et al.*, 1996), astrocytes (Aloisi *et al.*, 1992), neutrophils (Cassatella *et al.*, 1992, 1993), and many tumor cells including melanoma (Andrew *et al.*, 1995; Gutman *et al.*, 1996; Harant *et al.*, 1996), ovarian carcinoma (Harant *et al.*, 1995), and lung carcinoma cells (Hotta *et al.*, 1990). A more complete listing of sources of IL-8 is shown in Table 8.1.

For most of these "newer" sources, the inflammatory cytokines IL-1α or IL-1β, tumor necrosis factor-α (TNF-α), or lipopolysaccharide (LPS) provide powerful induction, although some cells have been shown to produce IL-8 also in response to other stimuli, for instance hypoxia–reoxygenation (Metinko *et al.*, 1992; Karakurum *et al.*, 1994; Kukielka *et al.*, 1995), adherence (Kasahara *et al.*, 1991; Strieter *et al.*, 1992), irradiation with UV-B (Singh *et al.*, 1995), viral infection (Van Damme *et al.*, 1989; Choi and Jacoby 1992; Arnold *et al.*, 1995), mycobacterial infection (Friedland *et al.*, 1992), double-stranded RNA (Oliveira *et al.*, 1992), or treatment with thrombin (Ueno *et al.*, 1996). Other cytokines and growth factors can induce IL-8 production, including IL-2 (Wei *et al.*, 1994), IL-7 (Standiford *et al.*, 1992), M-CSF (Hashimoto *et al.*, 1996), and GM-CSF (Takahashi *et al.*, 1993). Additionally, monocyte Fcγ receptor cross-linking has been shown to induce IL-8 (Marsh *et al.*, 1995). Stimulation with retinoic acid has been shown to upregulate IL-8 expression in some cells (Zhang *et al.*, 1992; Yang *et al.*, 1993; Harant *et al.*, 1995), and this agent can synergize strongly with TNF-α by a mechanism involving the transcription factor NFκB (Harant *et al.*, 1996).

Table 8.1   Cellular sources of IL-8

Monocyte/macrophage	Walz *et al.* (1987), Peveri *et al.* (1988), Schröder *et al.* (1987), Kristensen *et al.* (1991), Metinko *et al.* (1992), Anttila *et al.* (1992), De Waal Malefyt *et al.* (1991), Marsh *et al.* (1995)
Neutrophils	Cassatella *et al.* (1992, 1993)
T lymphocytes	Gregory *et al.* (1988), Schröder *et al.* (1988)
Epithelial cells	
cornea	Elner *et al.* (1991)
gastric	Crabtree *et al.* (1993, 1994)
kidney	Schmouder *et al.* (1992)
liver	Thornton *et al.* (1990)
lung	Standiford *et al.* (1990a)
retina	Elner *et al.* (1990)
thymus	Galy and Spits (1991)
thyroid	Weetman *et al.* (1992)
Endothelial cells	Sica *et al.* (1990), Kristensen *et al.* (1991), Strieter *et al.* (1988, 1989), Schröder and Christophers (1989), Gimbrone *et al.* (1989), Kaplanski *et al.* (1994), Ueno *et al.* (1996)
Keratinocytes	Gillitzer *et al.* (1991), Chabot-Fletcher *et al.* (1994), Li *et al.* (1996)
Fibroblasts	
gingiva	Tamura *et al.* (1992)
lung	Tobler *et al.* (1992)
dermal	Schröder *et al.* (1990)
Astrocytes	Aloisi *et al.* (1992)
Chondrocytes	Van Damme *et al.* (1990), Lotz *et al.* (1992)
Eosinophils	Braun *et al.* (1993), Yousefi *et al.* (1995)
Mesangial cells	Brown *et al.* (1991), Zoja *et al.* (1991), Abbott *et al.* (1991)
Mesothelial cells	Goodman *et al.* (1992)
Smooth-muscle cells	Wang *et al.* (1991)
Synovial fibroblasts	Golds *et al.* (1989)
Synovial macrophages	Koch *et al.* (1991)
Tumor cells	
astrocytoma	Van Meir *et al.* (1992)
gastric carcinoma	Yasumoto *et al.* (1992), Crabtree *et al.* (1994)
fibrosarcoma	Mukaida *et al.* (1992)
glioblastoma	Van Meir *et al.* (1992)
lung carcinoma	Hotta *et al.* (1990)
melanoma	Andrew *et al.* (1995), Colombo *et al.* (1992), Harant *et al.* (1996), Gutman *et al.* (1996)
osteosarcoma	Van Damme *et al.* (1989)
ovarian carcinoma	Harant *et al.* (1995)
renal carcinoma	Abruzzo *et al.* (1992)

The variety of factors which affect cytokine-induced IL-8 production is also quite extensive, but appears to be cell-specific. IL-8 can be downregulated by dexamethasone (Anttila *et al.*, 1992a; Mukaida *et al.*, 1994), the immunosuppressants cyclosporin A (Zipfel *et al.*, 1991; Wechsler *et al.*, 1994) and FK506 (Mukaida *et al.*, 1994), oxygen radical scavengers (DeForge *et al.*, 1992), interferons (Oliveira *et al.*, 1992, 1994; Cassatella *et al.*, 1993; Schnyder-Candrian *et al.*, 1995), IL-10 (de Waal Malefyt *et al.*, 1991; Kasama *et al.*, 1994; Wang *et al.*, 1994), and IL-4 (Standiford *et al.*, 1990b; Wertheim *et al.*, 1993). In keratinocytes, fibroblasts and PBMC, treatment with 1,25-dihydroxy vitamin D_3 reverses the IL-8-inducing effect of IL-1 (Larsen *et al.*, 1991), although not in endothelial cells. Type-I interferons can also inhibit IL-8 production in hematopoietic and bone marrow stromal cells (Aman *et al.*, 1993) and melanoma cells (Singh *et al.*, 1996).

5.   Biological Activities

IL-8 was originally characterized and isolated by its ability to induce neutrophil activation, observed as degranulation, shape change, and chemotaxis. Exocytosis occurs from specific granules and, after pretreatment with cytochalasin B, from azurophilic granules, measured by elastase or β-glucuronidase release (Peveri *et al.*, 1988). Activation is also witnessed in neutrophils by stimulation of the respiratory burst, reflecting activation of NADPH oxidase and resulting in reduction of molecular oxygen to superoxide (Thelen *et al.*, 1988). A further activity of IL-8 is the regulation of adhesion molecule expression on the neutrophil cell surface, inducing changes vital for cell migration *in vivo*.

Neutrophils transported by the bloodstream express constitutively on their surface the adhesion molecule L-selectin, and, as they come into contact with an area of activated endothelium, they slow down as they stick to the endothelial surface. The selectin binding forces are relatively weak and allow rolling of the neutrophils under the shear stress caused by the blood flow. However, it is hypothesized that when they come into contact with a solid-state gradient of IL-8 bound to the endothelium (Rot, 1992, 1993; Rot *et al.*, 1996), signaling through the IL-8 receptor causes the shedding of L-selectin, with a concomitant upregulation on the neutrophil surface of members of the integrin family of adhesion molecules, LFA-1 and Mac-1 (CD11a/CD18 and CD11b/CD18, respectively). These integrins bind very strongly to their intercellular adhesion molecule (ICAM) counterligands on the endothelium and stop the rolling action, allowing the cell to transmigrate by a haptotactic mechanism (Rot, 1993) through the endothelium toward the source of the chemoattractant.

However, although neutrophils are still thought by many to be the major target cell for this chemokine, the effects of IL-8 are not confined to neutrophils. It has also been reported to be chemotactic for GM-CSF or IL-3-primed eosinophils, and to have triggering effects on IL-3-primed basophils, inducing the release of histamine and the lipid mediator leukotriene C_4 (LTC$_4$) (Dahinden *et al.*, 1989). Also, there are reports of both direct and indirect chemotactic activity for T cells (Larsen *et al.*, 1989b; Zachariae *et al.*, 1992; Babi *et al.*, 1996; Taub *et al.*, 1996), although there is also evidence to the contrary (Roth *et al.*, 1995). Other reported activities of this pleiotropic cytokine include inhibition of IL-4-induced IgE production by, and growth of, B cells (Kimata *et al.*, 1992, 1995; Kimata and Lindley 1994), angiogenesis (Smith *et al.*, 1994; Strieter *et al.*, 1995), and mitogenesis and chemotaxis of keratinocytes (Tuschil *et al.*, 1992). These diverse activities have suggested a role for IL-8 in a variety of diseases, including rheumatoid arthritis, ulcerative colitis, cystic fibrosis, alcoholic hepatitis, psoriasis, ischemia–reperfusion injury, and allergy. Furthermore, IL-8 has been recently shown to stimulate the release of functional hematopoietic progenitors from the bone marrow to the peripheral blood (Laterveer *et al.*, 1995, 1996), and to suppress myeloid progenitor proliferation (Daly *et al.*, 1995), raising the possibility of a regulatory role for IL-8 in hematopoiesis.

Local administration of IL-8 *in vivo*, by either the intradermal or intraperitoneal routes, produces a marked neutrophil infiltrate at the site of administration, and can also induce plasma leakage (Colditz *et al.*, 1990). Interestingly, it has been shown that mast cells play a role in this process, as the neutrophil recruitment response is impaired in mast cell-deficient mice, but can be corrected by reconstitution with mast cells from wild-type mice (Rot, 1992). A single intravenous administration of IL-8 is followed by a rapid granulocytopenia, and then a longer-lasting granulocytosis (Van Damme *et al.*, 1988).

A novel physiological role for chemokines in general, and IL-8 in particular, is suggested by the presence of biologically relevant amounts of IL-8 in normal human sweat, produced by the epithelium of the eccrine sweat glands (Jones *et al.*, 1995). The presence of IL-8 has also been shown in human milk and colostrum (Rot *et al.*, 1993). Additionally, IL-8 appears to have further roles related to reproduction, witnessed by enhanced levels in the amniotic fluid (Romero *et al.*, 1991; Saito *et al.*, 1993) and reproductive tissues (Saito *et al.*, 1994; Osmers *et al.*, 1995) before parturition. Its production is at least partially under the control of progesterone (Ito *et al.*, 1994; Kelly *et al.*, 1992, 1994). The presence of IL-8 at this time may be explained by its ability to induce cervical ripening (Kanayama *et al.*, 1995).

6.   *The IL-8 Receptors*

IL-8 exerts its biological activity through two high-affinity receptors, designated IL-8RA and IL-8RB. In the new unified chemokine receptor nomenclature, these receptors are termed CXCR1 and CXCR2, respectively. They are members of the seven-transmembrane domain G-protein-coupled receptor family (rhodopsin receptor superfamily) (Holmes *et al.*, 1991; Gao *et al.*, 1993) and share 77% homology at the amino acid level (see Figure 8.5). Upon ligand binding, this type of receptor transduces signals by activating the GDP-bound G-protein (GTP/GDP-binding protein), which exchanges GTP for GDP. The GTP-bound protein then activates effector proteins and hydrolysis of the GTP occurs, yielding the GDP form of the G-protein, which is recycled by forming a complex with unoccupied receptors.

6.1   CHARACTERIZATION AND CLONING OF THE IL-8 RECEPTORS

The two high-affinity receptors were identified on the surface of neutrophils by binding studies, which indicated receptor numbers in the range of 20 000–75 000 sites per cell (Besemer *et al.*, 1989; Samanta *et al.*, 1989; Grob *et al.*, 1990). After ligand binding, the ligand-receptor complex is internalized, and the receptor is re-expressed on the cell surface 10 minutes later (Samanta *et al.*, 1990).

The cDNA encoding IL-8RA was cloned from a neutrophil expression library by binding of ^{125}I-labelled IL-8 (Holmes *et al.*, 1991) and the DNA sequence predicts a translation product of 350 amino acids with five potential glycosylation sites and seven hydrophobic transmembrane domains. The predicted molecular mass is ~40 kDa, but owing to glycosylation the actual molecular mass determined by cross-linking studies is 55–69 kDa (Samanta *et al.*, 1989). The cDNA for IL-8RB was isolated from the HL-60 cell line, and the

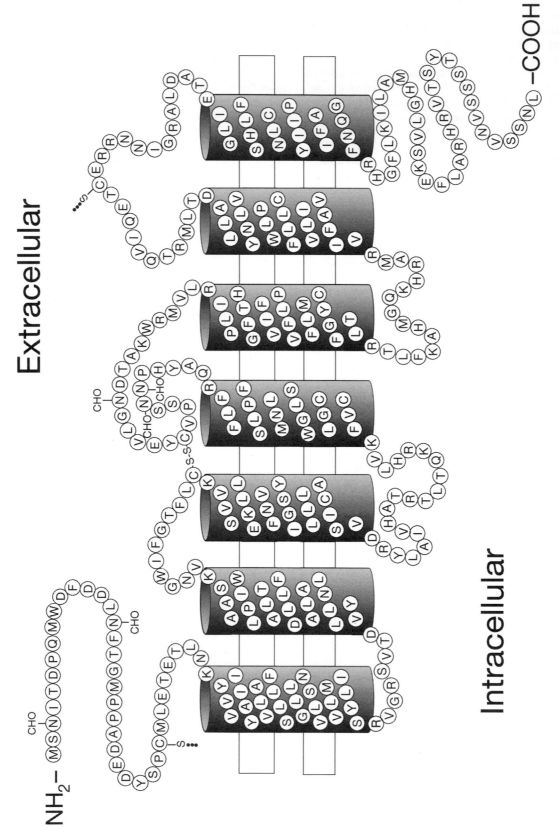

Figure 8.5   Proposed membrane topography of the IL-8RA (CXCR1). Membrane-spanning α-helices are defined on the basis of hydropathy analysis. Potential N-linked glycosylation sites are indicated by CHO, and disulfide bonds by S-S. From *Immunology Today*, 1994, 15: 170.

predicted protein product is slightly longer than IL-8RA, having 355 amino acids, but only one potential glycosylation site (Murphy and Tiffany, 1991).

Although IL-8 can bind to both receptors with high affinity (K_d = 2 nM) (Holmes *et al.*, 1991; Lee *et al.*, 1992; Cerretti *et al.*, 1993), only the IL-8RA receptor is IL-8-specific; the IL-8RB is promiscuous, and can also bind other C-X-C chemokines, gro-α/MGSA and NAP-2 (Lee *et al.*, 1992), and also gro-β and gro-γ. Binding studies using IL-8RA/IL-8RB chimeric molecules show that the specificity of binding is located at least partially in the N-terminus of the receptor, and in fact a peptide comprising the N-terminal extracellular region has been shown to bind IL-8 with low affinity (La Rosa *et al.*, 1992; Gayle *et al.*, 1993). Neither receptor can bind the unrelated chemotactic agents LTB$_4$, platelet-activating factor (PAF), the complement factor C5a, or the formyl peptide f-Met-Leu-Phe.

A further type of IL-8 receptor has been identified on red blood cells, and has been shown to be another promiscuous chemokine receptor, which, unlike other known chemokine receptors, binds both C-C and C-X-C chemokines (Horuk *et al.*, 1993a). It is expressed at about 5000–15 000 sites per cell, and binds chemokines with $K_d \sim 5$ nM (Neote *et al.*, 1993). It was originally thought to be a "sink" to remove circulating IL-8 and other chemokines from the blood, as it appeared not to be coupled to the signal transduction apparatus of the cell (Darbonne *et al.*, 1991). However, it has now been shown to be the Duffy blood group antigen, or DARC (Duffy antigen receptor for chemokines), which is also the molecule used by the malaria parasite to enter red blood cells (Horuk *et al.*, 1993b). DARC has recently been identified on several cell types, including the endothelial cells lining postcapillary venules in kidney and spleen (Hadley *et al.*, 1994; Peiper *et al.*, 1995) and in the brain (Horuk *et al.*, 1996). Such cellular distribution points to a more fundamental role than that of a sink for excess circulating chemokines.

Recent binding competition studies, in which binding of chemokines to mouse erythrocytes was also demonstrated for the first time, have shown that there are at least four functional subdivisions in the chemokine superfamily. C-C and ELR-containing C-X-C chemokines bind to human erythrocytes with high affinity, but non-ELR C-X-C chemokines exhibit different behavior; they are unable to displace ELR C-X-C chemokines but show a low affinity interaction with the C-C chemokine binding sites. The C chemokine lymphotactin was unable to displace either C-C or C-X-C chemokines (Szabo *et al.*, 1995).

Potential counterparts of chemokine receptors are also encoded by some viruses, and include the human cytomegalovirus proteins US27, US28, and UL33 (Chee *et al.*, 1990), and an open reading frame in Herpes saimiri virus ECRF3 (Nicholas *et al.*, 1992). Although ECRF3 has only 30% identity with the cloned chemokine receptors, it has 44% identity with IL-8RB in the N-terminal region. When expressed in *Xenopus* oocytes, ECRF-3 produced a Ca^{2+} efflux in response to the C-X-C chemokines IL-8, MGSA, and NAP-2, while the C-C chemokines produced little or no effect on the transfected oocytes (Ahuja and Murphy, 1993). The existence of virus-encoded chemokine receptors of this type suggests that chemokines play an important role in host defense against viral infection, and that the virus may express the receptor as a biological countermeasure.

6.2 GENOMIC ORGANIZATION AND REGULATION OF IL-8 RECEPTORS

The IL-8RA and IL-8RB genes have been localized to chromosomal region 2q34-35, where they span ~4 and 12 kbp of DNA, respectively. In each case, the open reading frame resides on a single exon. However, the 5′ untranslated regions differ greatly; for IL-8RA, it is composed of two exons, while for IL-8RB, alternative splicing of its 11 exons produces seven distinct mRNA species in neutrophils (Ahuja *et al.*, 1994). Neutrophils express two mRNA species for IL-8RA, of 2.0 and 2.4 kb, which arise by the use of two different polyadenylation signals. Primer extension studies have revealed two major transcription start sites for IL-8RA and eleven for IL-8RB.

Use of reporter gene constructs in HL-60, U-937, and Jurkat cell lines showed that strong constitutive promoter activity is contained within the regions extending 300 bp upstream from exon 1 of IL-8RA and 81 bp upstream from exon 3 of IL-8RB, although these two regions have only limited sequence homology and do not contain any classical promoter elements. However, a region 643 bp upstream from exon 1 of IL-8RB conferred only low levels of constitutive expression on the reporter gene. A conserved TATA sequence is located 47 bp upstream of the 5′ end of exon 1 of IL-8RB (Ahuja *et al.*, 1994).

7. *Signal Transduction*

Signal transduction from the receptor is achieved via GTP-binding proteins (G-proteins), and involves activation of protein kinase C (PKC) and mobilization of intracellular calcium. During cell activation by IL-8, the receptor–ligand complex is internalized, and the receptor is then recycled back to the surface.

In human neutrophils, receptors with bound IL-8 interact with G-proteins, as indicated by the ability of *Bordetella pertussis* toxin to inhibit neutrophil responses (Thelen *et al.*, 1988). Via the activation of a phosphotidyl inositol-specific phospholipase C, two second messengers are produced, diacylglycerol and inositol 1,4,5-trisphosphate (IP$_3$) (Pike *et al.*, 1992).

Diacylglycerol activates PKC, while IP_3 releases Ca^{2+} from intracellular stores, resulting in a transient rise in intracellular calcium ion concentration (Baggiolini et al., 1992; Smith et al., 1992), which is necessary for the induction of exocytosis and the respiratory burst.

8. Clinical Implications

IL-8 is frequently found at high levels in diseases associated with neutrophil influx. For instance, during the course of rheumatoid arthritis, IL-8 can be detected in the synovial fluid of the rheumatoid joint (Brennan et al., 1990), in some cases at levels which correlate well with the number of cells in the fluid (Peichl et al., 1991), suggesting a strong connection between the local disease process and the chemokine. Furthermore, synovial cells from rheumatoid arthritis patients have been shown to produce IL-8 (Seitz et al., 1991; Bas et al., 1992; Loetscher et al., 1994). In the periphery, an autoimmune response is mounted, resulting in circulating anti-IL-8 IgG at levels which reflect the disease severity (Peichl et al., 1992). In the case of ulcerative colitis, a similar picture is seen; elevated levels of IL-8 present in tissue from inflamed areas of the gut, and an anti-IL-8 IgG autoimmune response apparent in the periphery (Mahida et al., 1992; Raab et al., 1993). The role of these anti-IL-8 antibodies is unknown, but they are present at low but detectable levels even in healthy individuals (Sylvester et al., 1992), and may represent a defense mechanism against free IL-8 in the periphery which has been released from a major site of inflammation. High levels of circulating IL-8 in rabbits, induced by repeated intravenous injection, have been shown to produce lung damage, inducing lung pathology reminiscent of adult respiratory distress syndrome (Rot, 1991), and a natural defense against such a dangerous event may be vital.

IL-8 appears to play a role in airway inflammation, being associated particularly with cystic fibrosis (Dean et al., 1993; Dai et al., 1994), asthma (Shute et al., 1997), adult respiratory distress syndrome (ARDS) (Jorens et al., 1992), and infant respiratory distress syndrome (IRDS) (Little et al., 1995). Anti-IL-8 autoantibodies have also been found in ARDS (Kurdowska et al., 1996) and asthma (Shute et al., 1997).

Another area where IL-8 is thought to play a major role is hypoxia–reperfusion injury, participating in neutrophil-mediated myocardial injury (Kukielka et al., 1995). The importance of IL-8 in such conditions is underlined by studies where anti-IL-8 antibodies have been shown to greatly reduce tissue damage and infarct size in animal models of hypoxia–reperfusion (Sekido et al., 1993). In several other in vivo acute inflammation models, administration of anti-IL-8 antibodies also reduces neutrophil-mediated tissue damage and neutrophil infiltration (Broaddus et al., 1994; Harada et

al., 1994; Folkesson et al., 1995) and prevents proteinurea in experimental acute immune complex-induced glomerulonephritis (Wada et al., 1994).

IL-8 also appears to be involved in Helicobacter pylori-related gastroduodenal disease. IL-8 expression is increased in the gastric mucosal epithelium during H. pylori-associated inflammatory events (Crabtree et al., 1994; Crabtree and Lindley 1994), and a mucosal anti-IL-8 IgA response is apparent, with concentrations of antibody correlated to IL-8 levels (Crabtree et al., 1993). It is possible that this local IgA production is a host response designed to limit mucosal damage associated with a chronic bacterial infection.

Expression of IL-8 has also been detected in a variety of inflammatory skin diseases, including psoriasis (Gillitzer et al., 1991; Nickoloff et al., 1991; Schröder, 1992; Bruch-Gerharz et al., 1996; Kulke et al., 1996), palmoplantar pustulosis (Antilla et al., 1992b), and allergic contact dermatitis (Griffiths et al., 1991). Furthermore, anti-IL-8 antibodies have been shown to suppress the delayed-type hypersensitivity reaction in a rabbit model (Larsen et al., 1995). Taken together, these observations suggest an important mediator role for IL-8 in the pathomechanism of dermatological diseases.

9. References

Abbott, F., Ryan, J.J., Ceska, M., Matsushima, K., Sarraf, C.E. and Rees, A.J. (1991). Interleukin-1β stimulates human mesangial cells to synthesize and release interleukins-6 and -8. Kidney Int. 40, 597–605.

Abruzzo, L.V., Thornton, A.J., Liebert, M., et al. (1992). Cytokine-induced gene expression of interleukin-8 in human transitional cell carcinomas and renal cell carcinomas. Am. J. Pathol. 140, 365–373.

Ahuja, S.K. and Murphy, P.M. (1993). Molecular piracy of mammalian interleukin-8 receptor type B by herpesvirus saimiri. J. Biol. Chem. 268, 20691–20694.

Ahuja, S.K., Shetty, A., Tiffanny, H.L. and Murphy, P.M. (1994). Comparison of the genomic organization and promoter function for human interleukin-8 receptors A and B. J. Biol. Chem. 269, 26381–26389.

Aloisi, F., Caré, A., Borsellino, G., et al. (1992). Production of hemolymphopoietic cytokines (IL-6, IL-8, colony-stimulating factors) by normal human astrocytes in response to IL-1β and tumor necrosis factor-α. J. Immunol. 149, 2358–2366.

Aman, M.J., Rudolph, G., Goldschmitt, J., et al. (1993). Type-I interferons are potent inhibitors of interleukin-8 production in hematopoietic and bone marrow stromal cells. Blood 82, 2371–2378.

Andrew, P.J., Harant, H. and Lindley, I.J.D. (1995). Nitric oxide regulates IL-8 expression in melanoma cells at the transcriptional level. Biochem. Biophys. Res. Commun. 214, 949–956.

Anttila, H.S.I., Reitamo, S., Ceska, M. and Hurme, M. (1992a). Signal transduction pathways leading to the production of IL-8 by human monocytes are differentially regulated by dexamethasone. Clin. Exp. Immunol. 89, 509–512.

Anttila, H.S.I., Reitamo, S., Erkko, P., Ceska, M., Moser, B. and Baggiolini, M. (1992b). Interleukin-8 immunoreactivity in the skin of healthy subjects and patients with palmoplantar pustulosis and psoriasis. J. Invest. Dermatol. 98, 96–101.

Arnold, R., König, B., Galatti, H., Werchau, H. and König, W. (1995). Cytokine (IL-8, IL-6, TNF-α) and soluble TNF receptor-I release from human peripheral blood mononuclear cells after respiratory syncytial virus infection. Immunology 85, 364–372.

Babi, L.E.S., Moser, B., Soler, M.T.P., et al. (1996). The interleukin-8 receptor B and CXC chemokines can mediate transendothelial migration of human skin homing T cells. Eur. J. Immunol. 26, 2056–2061.

Baggiolini, M., Walz, A. and Kunkel, S.L. (1989). Neutrophil activating peptide 1/interleukin 8, a novel cytokine that activates neutrophils. J. Clin. Invest. 84, 1045–1049.

Baggiolini, M., Imboden, P. and Detmers, P. (1992). Neutrophil activation and the effects of interleukin-8/neutrophil activating peptide-1 (IL-8/NAP-1). In "Cytokines" (eds. M. Baggiolini and C. Sorg), Vol. 4, pp. 1–17. Karger, Basel.

Baldwin, E.T., Weber, I.T., St. Charles, R., et al. (1991). Crystal structure of interleukin-8: symbiosis of NMR and crystallography. Proc. Natl Acad. Sci. USA 88, 502–506.

Bas, S., Vischer, T.L., Ceska, M., et al. (1992). Production of neutrophil activating peptide-1 (NAP-1/IL-8) by blood and synovial mononuclear cells from patients with arthritis. Clin. Exp. Rheumatol. 10, 137–141.

Besemer, J., Huijber, A. and Kuhn, B. (1989). Specific binding, internalization and degradation of human neutrophil activating factor by human polymorphonuclear leukocytes. J. Biol. Chem. 264, 17409–17415.

Braun, R.K., Franchini, M., Erard, F., et al. (1993). Human peripheral blood eosinophils produce and release interleukin-8 on stimulation with calcium ionophore. Eur. J. Immunol. 23, 956–960.

Brennan, F.M., Zachariae, C.O.C., Chantry, D., et al. (1990). Detection of interleukin 8 biological activity in synovial fluids from patients with rheumatoid arthritis and production of interleukin 8 mRNA by isolated synovial cells. Eur. J. Immunol. 20, 2141–2144.

Broaddus, V.C., Boylan, A.M., Hoeffel, J.M., et al. (1994). Neutralization of IL-8 inhibits neutrophil influx in a rabbit model of endotoxin-induced pleurisy. J. Immunol. 152, 2960–2967.

Brown, Z., Strieter, R.M., Chensue, M., et al. (1991). Cytokine-activated human mesangial cells generate the neutrophil chemoattractant interleukin 8. Kidney Int. 40, 86–90.

Bruch-Gerharz, D., Fehsel, K., Suschek, C., Michel, G., Ruzicka, T. and Kolb-Bachofen, V. (1996). A proinflammatory activity of interleukin-8 in human skin: Expression of the inducible nitric oxide synthase in psoriatic lesions and cultured keratinocytes. J. Exp. Med. 184, 2007–2012.

Cao, Y., Chen, C., Weatherbee, J.A., Tsang, M. and Folkman, J. (1995). gro-β a C-X-C chemokine, is an angiogenesis inhibitor that suppresses the growth of Lewis Lung carcinoma in mice. J. Exp. Med. 182, 2069–2077.

Cassatella, M.A., Bazzoni, F., Ceska, M., Ferro, I., Baggiolini, M. and Berton, G. (1992). IL-8 production by human polymorphonuclear leukocytes: the chemoattractant formylmethionyl-leucyl-phenylalanine induces the gene expression and release of IL-8 through a pertussis toxin-sensitive pathway. J. Immunol. 148, 3216–3220.

Cassatella, M.A., Guasparri, I., Ceska, Bazzoni, F. and Rossi, F. (1993). Interferon-gamma inhibits interleukin-8 production by human polymorphonuclear leukocytes. Immunology 78, 177–184.

Cerretti, D.P., Kozlosky, C.J., Vanden Bos, T.Y., Nelson, N., Gearing, D.P. and Beckmann, M.P. (1993). Molecular characterization of receptors for human interleukin-8, GRO/melanoma growth stimulatory activity and neutrophil activating peptide-2. Mol. Immunol. 30, 359–367.

Chabot-Fletcher, M., Breton, J., Lee, J., Young, P. and Griswold, D.E. (1994). Interleukin-8 production is regulated by protein kinase C in human keratinocytes. J. Invest. Dermatol. 103, 509–515.

Chee, M.S., Satchwell, S.C., Preddie, E., Weston, K.M. and Barrell, B.G. (1990). Human cytomegalovirus encodes three G protein-coupled receptor homologues. Nature 344, 774–777.

Choi, A.M. and Jacoby, D.B. (1992). Influenza virus A infection induces IL-8 gene expression in human airway epithelial cells. FEBS Lett. 309, 327–329.

Clark-Lewis, I., Schumacher, C., Baggiolini, M. and Moser, B. (1991). Structure–activity relationships of interleukin-8 determined using chemically synthesized analogs. Critical role of NH$_2$-terminal residues and evidence for uncoupling of neutrophil chemotaxis, exocytosis, and receptor binding activities. J. Biol. Chem. 266, 23128–23134.

Clore, M.C., Appella, E., Yamada, M., Matsushima, K. and Gronenborn, A.M. (1990). Three-dimensional structure of interleukin 8 in solution. Biochemistry 29, 1689–1696.

Colditz, I.G., Zwahlen, R.D. and Baggiolini, M. (1990). Neutrophil accumulation and plasma leakage induced in vivo by neutrophil-activating peptide-1. J. Leukocyte Biol. 48, 129–137.

Colombo, M.P., Maccalli, C., Mattei, S., Melani, C., Radrizzani, M. and Parmiani, G. (1992). Expression of cytokine genes, including IL-6, in human malignant melanoma cell lines. Melanoma Res. 2, 181–189.

Crabtree, J.E. and Lindley, I.J.D. (1994). Mucosal Interleukin-8 and Helicobacter pylori-associated gastroduodenal disease. Eur. J. Gastroenterol. Hepatol. 6, 533–538.

Crabtree, J.E., Peichl, P., Wyatt, J.I., Stachl, U. and Lindley, I.J.D. (1993). Gastric interleukin-8 and IgA IL-8 autoantibodies in Helicobacter pylori infection. Scand. J. Immunol. 37, 65–70.

Crabtree, J.E., Wyatt, J.I., Peichl, P., et al. (1994). Interleukin-8 expression in Helicobacter pylori infected, normal and neoplastic gastroduodenal mucosa. J. Clin. Pathol. 47, 61–66.

Dahinden, C., Kurimoto, Y., de Weck, A., Lindley, I., Dewald, D. and Baggiolini, M. (1989). The neutrophil-activating peptide NAF/NAP-1 induces histamine and leukotriene release by interleukin-3 primed basophils. J. Exp. Med. 170, 1787–1792.

Dai, Y., Dean, T.P., Church, M.K., Warner, J.O. and Shute, J.K. (1994). Desensitisation of neutrophil responses by systemic interleukin 8 in cystic fibrosis. Thorax 49, 867–871.

Daly, T.J., LaRosa, G.J., Dolich, S., Maione, T.E., Cooper, S. and Broxmeyer, H.E. (1995). High activity suppression of myeloid progenitor proliferation by chimeric mutants of interleukin 8 and platelet factor 4. J. Biol. Chem. 270, 23282–23292.

Darbonne, W.C., Rice, G.C., Mohler, M.A., et al. (1991) Red blood cells are a sink for interleukin-8, a leukocyte chemotaxin. J. Clin Invest. 88, 1362–1369.

Dean, T.P., Dai, Y., Shute, J.K., Church, M.K. and Warner, J.O. (1993). Interleukin-8 concentrations are elevated in bronchoalveolar lavage, sputum, and sera of children with cystic fibrosis. Pediatr. Res. 34, 159–161.

DeForge, L., Fantone, J.C., Kenney, J.S. and Remick, D.G. (1992). Oxygen radical scavengers selectively inhibit interleukin-8 production in human whole blood. J. Clin. Invest. 90, 2123–2129.

DeForge, L.E., Preston, A.M., Takeuchi, E., Kenney, J., Boxer, L.A. and Remick, D.G. (1993). Regulation of IL-8 gene expression by oxidant stress. J. Biol. Chem. 268, 25568–25576.

de Waal Malefyt, R., Abrams, J., Bennett, B., Figdor, C.G. and de Vries, J.E. (1991). Interleukin 10 (IL-10) inhibits cytokine synthesis by human monocytes: an autoregulatory role of IL-10 produced by monocytes. J. Exp. Med. 174, 1209–1220.

Elner, V.M., Strieter, R.M., Pavilack, M.A., et al. (1991a). Human corneal interleukin-8: IL-1 and TNF-induced gene expression and secretion. Am. J. Pathol. 139, 977–988.

Elner, V.M., Strieter, R.M., Elner, S.G., Baggiolini, M., Lindley, I. and Kunkel, S.L. (1991b). Neutrophil chemotactic factor (IL-8) gene expression by cytokine-treated retinal pigment epithelial cells. Am. J. Pathol. 136, 745–750.

Folkesson, H.G., Matthay, M.A., Hebert, C.A. and Broaddus, V.C. (1995). Acid aspiration-induced lung injury in rabbits is mediated by interleukin-8-dependent mechanisms J. Clin. Invest. 96, 107–116.

Friedland, J.S., Remick, D.G., Shattrock, R. and Grifin, G.E. (1992). Secretion of interleukin-8 following phagocytosis of Mycobacterium tuberculosis by human monocyte cell lines. Eur. J. Immunol. 22, 1373–1378.

Galy, A.H.M. and Spits, H. (1991). IL-1, IL-4 and IFN-gamma differentially regulate cytokine production and cell surface molecular expression in cultured human thymic epithelial cells. J. Immunol. 147, 3823–3830.

Gao, J.L., Kuhns, D.B., Tiffany, H.L., et al. (1993). Structure and functional expression of the human macrophage inflammatory protein 1 alpha/RANTES receptor. J. Exp. Med. 177, 1421–1427.

Gayle, R.B., Sleath, P.R., Birks, C.W., et al. (1993). The importance of the amino terminus of the interleukin-8 receptor in ligand interactions. J. Biol. Chem. 268, 7283–7289.

Gillitzer, R., Berger, R., Mielke, V., Muller, C., Wolff, K. and Stingl, G. (1991). Upper keratinocytes of psoriatic skin lesions express high levels of NAP-1/IL-8 mRNA in situ. J. Invest. Dermatol. 97, 73–78.

Gimbrone, M.J., Obin, M.S., Brock, A.F., et al. (1989). Endothelial IL-8: a novel inhibitor of leukocyte-endothelial interactions. Science 246, 1601–1603.

Golds, E.E., Mason, P. and Nyirkos, P. (1989). Inflammatory cytokines induce synthesis and secretion of gro protein and a neutrophil chemotactic factor but not β_2-microglobulin in human synovial cells and fibroblasts. Biochem. J. 259, 585–588.

Goodman, R.B., Wood, R.G., Martin, T.R., Hanson-Painton, O. and Kinasewitz, G. (1992). Cytokine-stimulated human mesothelial cells produce chemotactic activity for neutrophils including NAP-1/IL-8. J. Immunol. 148, 457–465.

Gregory, H., Young, J., Schroeder, J.M., Mrowietz, U. and Christophers, E. (1988). Structure determination of a human lymphocyte derived neutrophil activating peptide (LYNAP). Biophys. Biochem. Res. Commun. 151, 883–890.

Griffiths, C.E.M., Barker, J.N.W.N., Kunkel, S. and Nickoloff, B.J. (1991). Modulation of leukocyte adhesion molecules, a T-cell chemotaxin (IL-8) and a regulatory cytokine (TNF-α) in allergic contact dermatitis (rhus dermatitis). Br. J. Dermatol. 124, 519–526.

Grob, P.M., David, E., Warren, T.C., DeLeon, R.P., Farina, P.R. and Hoaman, C.A. (1990). Characterization of a receptor for human monocyte-derived neutrophil chemotactic factor/interleukin-8. J. Biol. Chem. 14, 8311–8316.

Gutman, M., Singh, R.K., Xie, K., Bucana, C.D. and Fidler, I.J. (1996). Regulation of interleukin-8 expression in human melanoma cells by the organ environment. Cancer Res. 55, 2470–2475.

Hadley, T.J., Lu, Z.H., Wasniowska, K., et al. (1994). Postcapillary venule endothelial cells in kidney express a multispecific chemokine receptor that is structurally and functionally identical to the erythroid isoform, which is the Duffy blood group antigen. J. Clin. Invest. 94, 985–991.

Harada, A., Sekido, N., Akahoshi, T., Wada, T., Mukaida, N. and Matsushima, K. (1994). Essential involvement of interleukin-8 (IL-8) in acute inflammation. J. Leukocyte Biol. 56, 559–564.

Harant, H., Lindley, I., Uthman, A., et al. (1995). Regulation of interleukin-8 gene expression by all-trans retinoic acid. Biochem. Biophys. Res. Commun. 210, 898–906.

Harant, H., Andrew, P.J., de Martin, R., Foglar, E., Dittrich, C. and Lindley, I.J.D. (1996). Synergistic activation of interleukin-8 gene transcription by all-trans retinoic acid and tumor necrosis factor-α involves the transcription factor NF–κB. J. Biol. Chem. 271, 26954-26961.

Hashimoto, S., Yoda, M., Yamada, M., Yanai, N., Kawashima, T. and Motoyoshi, K. (1996). Macrophage colony stimulating factor induces interleukin-8 production in human monocytes. Exp. Hematol. 24, 123–128.

Hébert, C.A., Luscinskas, F.W., Kiely, J.-M., et al. (1990). Endothelial and leukocyte forms of IL-8: conversion by thrombin and interactions with neutrophils. J. Immunol. 145, 3033–3040.

Holmes, W., Lee, J., Kuang, W.J., Rice, G.C. and Wood, W.I. (1991). Structure and functional expression of a human IL-8 receptor. Science 253, 1278–1280.

Horuk, R., Colby, T.J., Darbonne, W.C., Schall, T.J., and Neote, K. (1993a). The human erythrocyte inflammatory peptide (chemokine) receptor. Biochemical characterization, solubilization, and development of a binding assay for the soluble receptor. Biochemistry 32, 5733–5738.

Horuk, R., Chitnis, C.E., Darbonne, W.C., et al. (1993b) A receptor for the malarial parasite Plasmodium vivax: the erythrocyte chemokine receptor. Science 261, 1182–1184.

Horuk, R., Martin, A., Hesselgesser, J., et al. (1996).The Duffy antigen receptor for chemokines: structural analysis and expression in the brain. J. Leukocyte Biol. 59, 29–38.

Hotta, K., Hayashi, K., Ishikawa, J., et al. (1990). Coding region structure of interleukin-8 gene of human lung giant cell carcinoma LU65C cells that produce LUCT/interleukin-8. Homogeneity in interleukin-8 genes. Immunol. Lett. 24, 165–170.

Isshiki, H., Akira, S., Tanabe, O., et al. (1990). Constitutive and interleukin-1 (IL-1) inducible factors interact with the IL-1 responsive element in the IL-6 gene. Mol. Cell. Biol. 10, 2757–2764.

Ito, A., Imada, K., Sato, T., Kubo, T., Matsushima, K. and Mori, Y. (1994). Suppression of interleukin 8 production by progesterone in rabbit uterine cervix. Biochem. J. 301, 183–186.

Jones, A.P., Webb, L.M.C., Anderson, A.O., Leonard, E.J. and Rot, A. (1995). Normal human sweat contains interleukin-8. J. Leukocyte Biol. 57, 434–437.

Jorens, P.G., Van Damme, J., De Backer, W., et al. (1992). Interleukin-8 in the bronchoalveolar lavage fluid from patients with the adult respiratory distress syndrome (ARDS) and patients at risk for ARDS. Cytokine 4, 592–597.

Kanayama, N., el-Maradny, E., Halim, A., et al. (1995). Urinary trypsin inhibitor suppresses premature cervical ripening. Eur. J. Obstet. Gynecol. Reprod. Biol. 60, 181–186.

Kaplanski, G., Farnarier, C., Kaplanski, S., et al. (1994). Interleukin-1 induces Interleukin-8 secretion from endothelial cells by a juxtacrine mechanism. Blood 84, 4242–4248.

Karakurum, M., Shreeniwas, R., Chen, J., et al. (1994). Hypoxic induction of interleukin-8 gene expression in human endothelial cells. J. Clin. Invest. 93, 1564–1570.

Kasahara, K., Strieter, R.M., Chensue, S.W., Standiford, T.J. and Kunkel, S.L. (1991). Mononuclear cell adherence induces neutrophil chemotactic factor/interleukin-8 gene expression. J. Leukocyte Biol. 50, 287–295.

Kasama, T., Strieter, R.M., Lukacs, N.W., Burdick, M.D. and Kunkel, S.L. (1994). Regulation of neutrophil-derived chemokine expression by IL-10. J. Immunol. 152, 3559–3569.

Kelly, R.W., Leask, R. and Calder, A.A. (1992). Chorio-decidual production of interleukin-8 and mechanism of parturition. Lancet 339, 776–777.

Kelly, R.W., Illingworth, P., Baldie, G., Leask, R., Brouwer, S. and Calder, A.A. (1994). Progesterone control of interleukin-8 production in endometrium and chorio-decidual cells underlines the role of the neutrophil in menstruation and parturition. Hum. Reprod. 9, 253–258.

Kennedy, J., Kelner, G.S., Kleyensteuber, S., et al. (1995). Molecular cloning and functional characterization of human lymphotactin. J. Immunol. 155, 203–209.

Kimata, H. and Lindley, I. (1994). Interleukin-8 differentially modulates interleukin-4 and interleukin-2-induced human B-cell growth. Eur. J. Immunol. 24, 3237–3240.

Kimata, H., Yoshida, A., Ishiola, C., Lindley, I. and Mikawa, H. (1992). IL-8 selectively inhibits IgE production induced by IL-4 in human B-cells. J. Exp. Med. 176, 1227–1231.

Kimata, H., Fujimoto, M., Lindley, I. and Furusho, K. (1995). IL-8 inhibits the interleukin-4 induced but not the spontaneous growth of human B-cells via mechanisms that may involve protein kinase C. Biochem. Biophys. Res. Commun. 207, 1044–1050.

Koch, A., Polverini, P.J., Kunkel, S.L., et al. (1991). Synovial tissue macrophages are a source of the chemotactic cytokine IL-8. J. Immunol. 147, 2187–2195.

Koch, A., Polverini, P.J., Kunkel, S.L., et al. (1992). IL-8 as a macrophage-derived mediator of angiogenesis. Science 258, 1798–1801.

Kownatzki, E., Uhrich, S. and Grüninger, G. (1988). Functional properties of a human neutrophil chemotactic factor derived from human monocytes. Immunobiology 177, 352–362.

Kristensen, M.S., Paludan, K., Larsen, C.G., et al. (1991). Quantitative determination of IL-1α-induced IL-8 mRNA levels in cultured human keratinocytes, dermal fibroblasts, endothelial cells and monocytes. J. Invest. Dermatol. 97, 506–510.

Kukielka, G.L., Smith, C.W., LaRosa, G.J., et al. (1995). Interleukin-8 gene induction in the myocardium after ischemia and reperfusion in vivo. J. Clin. Invest. 95, 89–103.

Kulke, R., Tödt-Pingel, I., Rademacher, D., Röwert, J., Schröder, J.-M. and Christophers, E. (1996). Colocalized expression of gro-α and IL-8 mRNA is restricted to the suprapapillary layers of psoriatic lesions. J. Invest. Dermatol. 106, 526–530.

Kunsch, C., Lang, R.K., Rosen, C.A. and Shannon, M.F. (1994). Synergistic transcriptional activation of the IL-8 gene by NFκB p65 (Rel A) and NF-IL-6. J. Immunol. 153, 153–164.

Kurdowska, A., Miller, E.J., Noble, J.M., et al. (1996). Anti-IL-8 autoantibodies in alveolar fluid from patients with the adult respiratory distress syndrome. J. Immunol. 157, 2699–2706.

LaRosa, G.J., Thomas, K.M., Kaufman, M.E., et al. (1992). Amino terminus of the interleukin-8 receptor is a major determinant of receptor subtype specificity. J. Biol. Chem. 267, 25402–25406.

Larsen, C., Anderson, A.O., Oppenheim, J.J., and Matsushima, K. (1989a). Production of IL-8 by human dermal fibroblasts and keratinocytes in response to IL-1 or TNF. Immunology 68, 31–36.

Larsen, C., Anderson, A.O., Appella, E., Oppenheim, J.J., and Matsushima, K. (1989b). The neutrophil-activating protein (NAP-1) is also chemotactic for T-lymphocytes. Science 243, 1464–1466.

Larsen, C.G., Kristensen, M., Paludan, K., et al. (1991). 1,25(OH)$_2$-D$_3$ is a potent regulator of interleukin-1 induced interleukin-8 expression and production. Biochem. Biophys. Res. Commun. 176, 1020–1026.

Larsen, C.G., Thomsen, M.K., Gesser, B., et al. (1995). The delayed-type hypersensitivity reaction is dependent on IL-8. Inhibition of a tuberculin skin reaction by an anti-IL-8 monoclonal antibody. J. Immunol. 155, 2151–2157.

Laterveer, L., Lindley, I.J.D., Hamilton, M.S., Willemze, R. and Fibbe, W.E. (1995). IL-8 induces rapid mobilization of hematopoietic stem cells with radioprotective capacity and long-term myelolymphoid repopulating ability. Blood 85, 2269–2275.

Laterveer, L., Lindley, I.J.D., Heemskerk, D.P., et al. (1996). Rapid mobilization of hematopoietic progenitor cells in rhesus monkeys by a single intravenous injection of interleukin-8. Blood 87, 781–788.

Lee, J., Horuk, R., Rice, G.C., Bennett, G.L., Camerato, T. and Wood, W.I. (1992). Characterization of two high affinity human IL-8 receptors. J. Biol. Chem. 267, 16283–16287.

Li, J., Ireland, G.W., Farthing, P.M. and Thornhill, M.H. (1996). Epidermal and oral keratinocytes are induced to produce RANTES and IL-8 by cytokine stimulation. J. Invest. Dermatol. 106, 661–666.

Lindley, I., Aschauer, H., Seifert, J.M., et al. (1988). Synthesis and expression in E. coli of the gene encoding monocyte-derived neutrophil-activating factor: Biological equivalence between natural and recombinant neutrophil-activating factor. Proc. Natl Acad. Sci. USA 85, 9199–9203.

Lindley, I., Aschauer, H., Lam, C., Besemer, J. and Rot, A. (1990). In vitro and in vivo activity of mutagenised versions of recombinant human NAP-1/IL-8 and identification of functionally important domains. In "Molecular and Cellular Biology of Cytokines" (eds. J.J. Oppenheim, M.C. Powanda, M.J. Kluger and C.A. Dinarello), pp. 345–350. Wiley-Liss, New York.

Lindley, I., Westwick, J. and Kunkel, S.L. (1993). Nomenclature announcement—the chemokines. Immunol. Today 14, 24.

Little, S., Dean, T., Bevin, S., *et al.* (1995). Role of elevated plasma soluble ICAM-1 and bronchial lavage fluid IL-8 levels as markers of chronic lung disease in premature infants. Thorax 50, 1073–1079.

Loetscher, P., Dewald, B., Baggiolini, M. and Seitz, M. (1994). Monocyte chemoattractant protein-1 and interleukin 8 production by rheumatoid synoviacytes. Effects of anti-rheumatic drugs. Cytokine 6, 162–170.

Lotz, M., Terkeltaub, R. and Villiger, P.M. (1992). Cartilage and joint inflammation: regulation of IL-8 expression by human articular chondrocytes. J. Immunol. 148, 466–473.

Lusti-Narasimhan, M., Power, C.A., Allet, B., *et al.* (1995). Mutation of Leu^{25} and Val^{27} introduces CC chemokine activity into interleukin-8. J. Biol. Chem. 270, 2716–2722.

Lusti-Narasimhan, M., Chollet, A., Power, C.A., Allet, B., Proudfoot, A.E.I. and Wells, T.N.C. (1996). A molecular switch of chemokine receptor selectivity. J. Biol. Chem. 271, 3148–3153.

Mahida, Y., Ceska, M., Effenberger, F., Kurlak, L., Lindley, I. and Hawkey, C.J. (1992). Enhanced synthesis of neutrophil activating peptide 1/IL-8 in active ulcerative colitis. Clin. Sci. 82, 273–275.

Maione, T.E., Gray, G.S., Petro, J., *et al.* (1990). Inhibition of angiogenesis by recombinant human platelet factor 4 and related peptides. Science 247, 77–79.

Marsh, C.B., Gadek, J.E., Kindt, G.C., Moore, S.A. and Wewers, M.D. (1995). Monocyte Fcγ receptor cross-linking induces IL-8 production. J. Immunol. 155, 3161–3167.

Matsushima, K., Morishita, K., Yoshimura, T., *et al.* (1988). Molecular cloning of a human monocyte-derived neutrophil chemotactic factor (MDNCF) and the induction of MDNCF mRNA by interleukin-1 and tumor necrosis factor. J. Exp. Med. 167, 1883–1893.

Metinko, A.P, Kunkel, S.L, Standiford, T.J, and Strieter, R.M, (1992). Anoxia-hyperoxia induces monocyte-derived interleukin-8. J. Clin. Invest. 90, 791–798.

Mukaida, N., Shiroo, M. and Matsushima, K. (1989). Genomic structure of the human monocyte-derived neutrophil chemotactic factor IL-8. J. Immunol. 143, 1366–1371.

Mukaida, N., Mahe, Y. and Matsushima, K. (1990). Cooperative interaction of NF-kB and cis-regulatory enhancer binding protein-like factor binding elements in activating the IL-8 gene by pro-inflammatory cytokines. J. Biol. Chem. 265, 21128–21133.

Mukaida, N., Gussella, G.L., Kasahara, T., *et al.* (1992). Molecular analysis of the inhibition of interleukin-8 production by dexamethasone in a human fibrosarcoma cell line. Immunology 75, 674–679.

Mukaida, N., Okamoto, S.I., Ishikawa, Y. and Matsushima, K. (1994). Molecular mechanism of interleukin-8 gene expression. J. Leukocyte Biol. 56, 554–558.

Murphy, P.M. and Tiffany, H.L. (1991). Cloning of complementary DNA encoding a functional human interleukin-8 receptor. Science 253, 1280–1283.

Neote, K., Darbonne, W.C., Ogez, J., Horuk, R., Colby, T.J. and Schall, T.J. (1993). Identification of a promiscuous inflammatory peptide receptor on the surface of red blood cells. J. Biol. Chem. 268, 12247–12249.

Nicholas, J., Cameron, K.R. and Honess, R.W. (1992). Herpesvirus saimiri encodes homologues of G protein-coupled receptors. Nature 355, 362–365.

Nickoloff, B.J., Karabin, G.D., Barker, J.N.W.N., *et al.* (1991). Cellular localization of interleukin-8 and its inducer, tumor necrosis factor-α in psoriasis. Am. J. Pathol. 138, 129–140.

Oliveira, I.C., Sciavolino, P.J., Lee, T.H. and Vilcek, J. (1992). Downregulation of interleukin-8 gene expression in human fibroblasts: unique mechanism of transcriptional inhibition by interferon. Proc. Natl Acad. Sci. USA 89, 9049–9053.

Oliveira, I.C., Mukaida, N., Matsushima, K. and Vilcek, J. (1994). Transcriptional inhibition of the interleukin-8 gene by interferon is mediated by the NFκB site. Mol. Cell. Biol. 14, 5300–5308.

Oppenheim, J.J., Zachariae, C.O.C., Mukaida, N. and Matsushima, K. (1991). Properties of the novel proinflammatory supergene intercrine cytokine family. Annu. Rev. Immunol. 9, 617–648.

Osmers, R.G., Blaser, J., Kuhn, W. and Tschesche, H. (1995). Interleukin-8 synthesis and the onset of labor. Obstet. Gynecol. 86, 223-229.

Peichl, P., Ceska, M., Effenberger, F., Haberhauer, G., Broell, H. and Lindley, I.J.D. (1991). Presence of NAP-1/IL-8 in synovial fluids indicates a possible pathogenic role in rheumatoid arthritis. Scand. J. Immunol. 34, 333–339.

Peichl, P., Ceska, M., Broell, H. Effenberger, F. and Lindley, I.J.D. (1992). NAP-1/IL-8 acts as an autoantigen in rheumatoid arthritis. Ann. Rheum. Dis. 51, 19–22.

Peiper, S.C., Wang, Z.X., Neote, K., *et al.* (1995). The Duffy antigen/receptor for chemokines (DARC) is expressed in endothelial cells of Duffy negative individuals who lack the erythrocyte receptor. J. Exp. Med. 181, 1311–1317.

Peveri, P., Walz, A., Dewald, B. and Baggiolini, M. (1988). A novel neutrophil-activating factor produced by human mononuclear phagocytes. J. Exp. Med. 167, 1547–1559.

Pike, M.C., Costello, K.M. and Lamb, K.M. (1992). IL-8 stimulates phosphotidylinositol-4-phosphate kinase in human polymorphonuclear leukocytes. J. Immunol. 148, 3158–3164.

Raab, Y., Gerdin, B., Ahlstedt, S and Hällgren, R. (1993). Neutrophil mucosal involvement is accompanied by enhanced local production of interleukin-8 in ulcerative colitis. Gut 34, 1203–1206.

Rajarathnam, K., Sykes, B.D., Kay, C.M., *et al.* (1994). Neutrophil activation by monomeric interleukin-8. Science 264, 90–92.

Remick, D.G. and Villarete, L. (1996). Regulation of cytokine gene expression by reactive oxygen and reactive nitrogen intermediates. J. Leukocyte Biol. 59, 471–475.

Romero, R., Ceska, M., Avila, C., Mazor, M., Benhke, E. and Lindley, I. (1991). Neutrophil attractant/activating peptide-1/interleukin-8 in term and preterm parturition. Am. J. Obstet. Gynecol. 165, 813–820.

Rot, A. (1991). Some aspects of NAP-1 pathophysiology: lung damage caused by a blood-borne cytokine. Adv. Exp. Med. Biol. 305, 127–135.

Rot, A. (1992). Endothelial cell binding of NAP-1/IL-8: role in neutrophil emigration. Immunol. Today 13, 291–294.

Rot, A. (1993). Neutrophil attractant/activation protein-1 (interleukin-8) induces in vitro neutrophil migration by haptotactic mechanism. Eur. J. Immunol. 23, 303–306.

Rot, A., Jones, A.P. and Webb, L.M.C. (1993). Some aspects of NAP-1 pathophysiology II: chemokine secretion by exocrine glands. Adv. Exp. Med. Biol. 351, 77–85.

Rot, A., Hub, E., Middleton, J., *et al.* (1996). Some aspects of IL-8 pathophysiology III: chemokine interaction with endothelial cells. J. Leukocyte Biol. 59, 39–44.

Roth, S.J., Carr, M.W. and Springer, T.A. (1995). CC chemokines, but not the CXC chemokines interleukin-8 and interferon-γ inducible protein-10, stimulate transendothelial chemotaxis of T lymphocytes. Eur. J. Immunol. 25, 3482–3488.

Saito, S., Kasahara, T., Kato, Y., Ishihara, Y. and Ichijo, M. (1993). Elevation of amniotic fluid interleukin 6 (IL-6), IL-8 and granulocyte colony stimulating factor (G-CSF) in term and preterm parturition. Cytokine 5, 81–88.

Saito, S., Kasahara, T., Sakakura, S., et al. (1994). Detection and localization of interleukin-8 mRNA and protein in human placenta and decidual tissues. J. Reprod. Immunol. 27, 161–172.

Samanta, A.K., Oppenheim, J.J. and Matsushima, K. (1989). Identification and characterization of specific receptors for monocyte-derived neutrophil chemotactic factor (MDNCF) on human neutrophils. J. Exp. Med. 169, 1185–1189.

Samanta, A.K., Oppenheim, J.J. and Matsushima, K. (1990). Interleukin-8 (monocyte-derived neutrophil chemotactic factor) dynamically regulates its own receptor expression on human neutrophils. J. Biol. Chem. 265, 183–189.

Schmid, J. and Weissman, C. (1987). Induction of mRNA for a serine protease and a J-thromboglobulin-like protein in mitogen-stimulated human leukocytes. J. Immunol. 139, 250–256.

Schmouder, R.L., Strieter, R.M., Wiggins, R.C., Chensue, S.W. and Kunkel, S.L. (1992). In vitro and in vivo interleukin-8 production in human renal cortical epithelia. Kidney Int. 41, 191–198.

Schnyder-Candrian, S., Strieter, R.M., Kunkel, S.L. and Walz, A. (1995). IFN-α and IFN-γ downregulate the production of IL-8 and ENA-78 in human monocytes. J. Leukocyte Biol. 57, 929–935.

Schröder, J.-M. (1992). Generation of NAP-1 and related peptides in psoriasis and other inflammatory skin diseases. In "Cytokines" (eds. M. Baggiolini and C. Sorg), Vol. 4, pp. 54–76. Karger, Basel.

Schröder, J.-M. and Christophers, E. (1989). Secretion of novel and homologous neutrophil-activating peptides by LPS-stimulated endothelial cells. J. Immunol. 142, 244–251.

Schröder, J.-M., Mrowietz, U., Morita, E. and Christophers, E. (1987). Purification and partial biochemical characterisation of a human monocyte-derived, neutrophil-activating peptide that lacks interleukin 1 activity. J. Immunol. 139, 3474–3483.

Schröder, J.-M., Mrowietz, U. and Christophers, E. (1988). Purification and partial biochemical characterisation of a human lymphocyte-derived peptide with potent neutrophil-stimulating activity. J. Immunol. 140, 3534–3540.

Schröder, J.-M., Sticherling, M., Henneicke, H.H., Preissner, W.C. and Christophers, E. (1990). IL-1α or tumor necrosis factor-α stimulate release of three NAP-1/IL-8 related neutrophil chemotactic proteins in human dermal fibroblasts. J. Immunol. 144, 2223–2232.

Seitz, M., Dewald, B., Gerber, N. and Baggiolini, M. (1991). Enhanced production of neutrophil-activating peptide-1/interleukin-8 in rheumatoid arthritis. J. Clin. Invest. 87, 463–467.

Sekido, N., Mukaida, N., Harada, A., Nakanishi, I., Watanabe, Y. and Matsushima, K. (1993). Prevention of lung reperfusion injury in rabbits by a monoclonal antibody against IL-8. Nature 365, 654–657.

Shaw, G. and Kamen, R. (1986). A conserved AU sequence from the 3′ untranslated region of GM-CSF mRNA mediates selective mRNA degradation. Cell 46, 659–667.

Shute, J.K., Vrugt, B., Lindley, I.J.D., et al. (1997). Free and complexed interleukin-8 in blood and bronchial mucosa in asthma. Am. J. Resp. Crit. Care Med. 155, 1877–1883.

Sica, A., Matsushima, K., Van Damme, J., et al. (1990). IL-1 transcriptionally activates the neutrophil chemotactic factor/IL-8 gene in endothelial cells. Immunology 69, 548–553.

Singh, R., Gutman, M., Reich, R. and Bar-Eli, M. (1995). UVB irradiation promotes tumorigenic and metastatic properties in primary cutaneous melanoma via induction of IL-8. Cancer Res. 54, 3242–3247.

Singh, R., Gutman, M., Llansa, M. and Fidler, I.J. (1996). Interferon-β prevents the upregulation of interleukin-8 expression in human melanoma cells. J. Interferon Cytokine Res. 16, 577–584.

Smith, D., Polverini, P.J., Kunkel, S.L., et al. (1994). Inhibition of interleukin 8 attenuates angiogenesis in bronchogenic carcinoma. J. Exp. Med. 179, 1409–1415.

Smith, R.J., Sam, L.M., Leach, K.L. and Justen, J.M. (1992). Postreceptor events associated with human neutrophil activation by interleukin-8. J. Leukocyte Biol. 52, 17–26.

Standiford, T.J., Kunkel, S.L., Basha, M.A., et al. (1990a). Interleukin-8 gene expression by a pulmonary epithelial cell line. A model for cytokine networks in the lung. J. Clin. Invest. 86, 1945–1953.

Standiford, T.J., Strieter, R.M., Chensue, S.W., Westwick, J., Kasahara, K. and Kunkel, S.L. (1990b). IL-4 inhibits the expression of IL-8 from stimulated human monocytes. J. Immunol. 145, 1435–1439.

Standiford, T.J., Strieter, R.M., Allen, R.M., Burdick, M.D. and Kunkel, S.L. (1992). IL-7 upregulates the expression of IL-8 from resting and stimulated human blood monocytes. J. Immunol. 149, 2035–2039.

Stein, B. and Baldwin, A.S., Jr (1993). Distinct mechanisms for regulation of the interleukin-8 gene involve synergism and cooperativity between C/EBP and NF-κB. Mol. Cell. Biol. 13, 7191–7198.

Strieter, R., Kunkel, S.L., Showell, H.J. and Marks, R.M. (1988). Monokine-induced gene expression of a human endothelial cell-derived neutrophil chemotactic factor. Biochem. Biophys. Res. Commun. 156, 1340–1345.

Strieter, R., Kunkel, S.L., Showell, H.J., et al. (1989). Endothelial cell gene expression of a neutrophil chemotactic factor by TNF-alpha, LPS and IL-1 beta. Science 243, 1467–1469.

Strieter, R., Kasahara, K., Allen, R.M., et al. (1992). Cytokine-induced neutrophil-derived interleukin-8. Am. J. Pathol. 141, 397–407.

Strieter, R., Polverini, P.J., Kunkel, S.L., et al. (1995). The functional role of the ELR motif in CXC chemokine-mediated angiogenesis. J. Biol. Chem. 270, 27348–27357.

Suzuki, K., Miyasaka, H., Ota, H., et al. (1989). Purification and partial primary sequence of a chemotactic protein for polymorphonuclear leukocytes derived from human lung giant cell carcinoma LU65C cells. J. Exp. Med. 169, 1085–1092.

Sylvester, I., Yoshimura, T., Sticherling, M., et al. (1992). Neutrophil attractant protein-1-immunoglobulin G immune complexes and free anti-NAP-1 antibody in normal human serum. J. Clin Invest. 90, 471–472.

Szabo, M.C., Soo, K.S., Zlotnik, A. and Schall, T.J. (1995). Chemokine class differences in binding to the Duffy antigen-erythrocyte chemokine receptor. J. Biol. Chem. 270, 25348–25351.

Takahashi, G.W., Andrews, D.F., Lilly, M.B., Singer, J.W. and Alderson, M.R. (1993). Effect of granulocyte-macrophage colony-stimulating factor and interleukin-3 on interleukin-8 production by human neutrophils and monocytes. Blood 81, 357–364.

Tamura, M., Tokuda, M., Nagaoka, S. and Takada, H. (1992). Lipopolysaccharides of *Bacteroides intermedius* (*Prevotella intermedia*) and *Bacteroides* (*Porphyromonas*) *gingivalis* induce interleukin-8 gene expression in human gingival fibroblast cultures. Infect. Immun. 60, 4932–4937.

Taub, D.D., Anver, M., Oppenheim, J.J., Longo, D.L. and Murphy, W. (1996). T-lymphocyte recruitment by interleukin-8. IL-8-induced degranulation of neutrophils releases potent chemoattractants for human T lymphocytes both in vitro and in vivo. J. Clin. Invest. 97, 1931–1941.

Thelen, M., Peveri, P., Kernen, P., von Tscharner, V., Walz, A. and Baggiolini, M. (1988). Mechanism of neutrophil activation by NAF, a novel monocyte-derived peptide agonist. FASEB J. 2, 2702–2706.

Thornton, A.J., Strieter, R.M., Lindley, I., Baggiolini, M. and Kunkel, S.L. (1990). Cytokine-induced gene expression of neutrophil chemotactic factor/IL-8 in human hepatocytes. J. Immunol. 144, 2609–2613.

Tobler, A., Meier, R., Seitz, M., *et al.* (1992). Glucocorticoids downregulate gene expression of GM-CSF, NAP-1/IL-8, and IL-6, but not of M-CSF in human fibroblasts. Blood 79, 45–51.

Tuschil, A., Lam, C., Haslberger, A. and Lindley, I. (1992). IL-8 stimulates calcium transients and promotes epidermal cell proliferation. J. Invest. Dermatol. 99, 294–298.

Ueno, A., Murakami, K., Yamanouchi, K., Watanabe, M. and Kondo, T. (1996). Thrombin stimulates production of interleukin-8 in human umbilical vein endothelial cells. Immunology 88, 76–81.

Van Damme, J., Van Beeumen, J., Opdenakker, G. and Billiau, A. (1988). A novel NH$_2$-terminal sequence-characterized human monokine possessing neutrophil chemotactic, skin-reactive, and granulocytosis-promoting activity. J. Exp. Med. 167, 1364–1376.

Van Damme, J., Decock, B., Conings, R., Lenaerts, J.P., Opdenakker, G. and Billiau, A. (1989). The chemotactic activity for granulocytes produced by virally infected fibroblasts is identical to monocyte-derived interleukin-8. Eur. J. Immunol. 19, 1189–1194.

Van Damme, J., Bunning, R.A.D., Conings, R., Graham, R., Russell, C. and Opdenakker, G. (1990). Characterization of granulocyte chemotactic activity from human cytokine-stimulated chondrocytes as interleukin-8. Cytokine 2, 106-111.

Van Meir, E., Ceska, M., Effenberger, F., *et al.* (1992). Interleukin-8 is produced in neoplastic and infectious diseases of the human central nervous system. Cancer Res. 52, 4297–4305.

Villarete, L. and Remick, D.G. (1995). NO regulation of IL-8 expression in human endothelial cells. Biochem. Biophys. Res. Commun. 211, 671–676.

Wada, T., Tomosugi, N., Naito, T., *et al.* (1994). Prevention of proteinurea by the administration of anti-interleukin-8 antibody in experimental acute immune complex-induced glomerulonephritis. J. Exp. Med. 180, 1135–1140.

Walz, A., Peveri, P., Aschauer, H. and Baggiolini, M. (1987). Purification and amino acid sequencing of NAF, a novel neutrophil-activating factor produced by monocytes. Biochem. Biophys. Res. Commun. 149, 755–761.

Wang, J.M., Sica, A., Peri, G., *et al.* (1991). Expression of monocyte chemotactic protein and interleukin-8 by cytokine-activated human vascular smooth muscle cells. Arterioscler. Thromb. 11, 1166–1174.

Wang, P., Wu, P., Anthes, J.C., Siegel, M.I., Egan, R.W. and Billah, M.M. (1994). Interleukin-10 inhibits interleukin-8 production in human neutrophils. Blood 83, 2678–2683.

Wechsler, A.S., Gordon, M.C., Dendorfer, U. and LeClair, K.P. (1994). Induction of IL-8 expression uses the CD28 costimulatory pathway. J. Immunol. 153, 2515–2523.

Weetman, A.P., Bennett, G.L. and Wong, W.L.T. (1992). Thyroid follicular cells produce interleukin-8. J. Clin. Endocrinol. Metab. 75, 328–330.

Wei, S., Liu, J.H., Blanchard, D.K. and Djeu, J.Y. (1994). Induction of IL-8 gene expression in human polymorphonuclear neutrophils by recombinant IL-2. J. Immunol. 152, 3630–3636.

Wertheim, W.A., Kunkel, S.L., Standiford, T.J., *et al.* (1993). Regulation of neutrophil-derived IL-8: the role of prostaglandin E2, dexamethasone and IL-4. J. Immunol. 151, 2166–2175.

Witt, D.P. and Lander, A.D. (1994). Differential binding of chemokines to glycosaminoglycan subpopulations. Current Biol. 4, 394–400.

Yang, K.D., Cheng, S.-N., Wu, N.-C. and Shaio, M.-F. (1993). Induction of interleukin-8 expression in neuroblastoma cells by retinoic acid: implication of leukocyte chemotaxis and activation. Pediatr. Res. 34, 720–724.

Yasumoto, K., Okamoto, S., Mukaida, N., Murakami, S., Mai, M. and Matsushima, K. (1992). Tumor necrosis factor-α and interferon-γ synergistically induce interleukin-8 production in a human gastric cancer cell line through acting concurrently on AP-1 and NF-kB-like binding sites of the interleukin-8 gene. J. Biol. Chem. 267, 22506–22511.

Yoshimura, T., Matsushima, K., Tanaka, S., *et al.* (1987). Purification of a human monocyte-derived neutrophil chemotactic factor that has peptide sequence similarity to other host defense cytokines. Proc. Natl Acad. Sci. USA 84, 9233–9237.

Yousefi, S., Hemmann, S., Weber, M., *et al.* (1995). IL-8 is expressed by human peripheral blood eosinophils. J. Immunol. 154, 5481–5490.

Zachariae, C.O.C., Jinquan, T., Nielsen, V., Kaltoft, K. and Thestrup-Pedersen, K. (1992). Phenotypic determination of T-lymphocytes responding to chemotactic stimulation from fMLP, IL-8, human IL-10, and epidermal lymphocyte chemotactic factor. Arch. Dermatol. Res. 284, 333–338.

Zipfel, P.F., Bialonski, A. and Skerka, C. (1991). Induction of members of the IL-8/NAP-1 family in human T-lymphocytes is suppressed by cyclosporin A. Biochem. Biophys. Res. Commun. 181, 179–183.

Zhang, Q.-Y., Hammerberg, C., Baldassare, J.J., *et al.* (1992). Retinoic acid and phorbol ester synergistically upregulate IL-8 expression and specifically modulate protein kinase C-ε in human skin fibroblasts. J. Immunol. 149, 1402–1408.

Zoja, C., Ming Wang, J., Bettoni, S., *et al.* (1991). Interleukin-1β and tumor necrosis factor-α induce gene expression and production of leukocyte chemotactic factors, colony-stimulating factors and interleukin-6 in human mesangial cells. Am. J. Pathol. 138, 991–1003.

9. Interleukin-9

Jean-Christophe Renauld

1. Introduction

Originally isolated in the mouse on the basis of its T cell growth factor activity, interleukin 9 (IL-9) seemed to be characterized by a narrow specificity for some T-helper clones. However, like many cytokines, IL-9 is presently known to interact with various biological targets and shows many similarities with other factors regarding the regulation of its expression as well as its cellular activation pathways.

The first isolation of IL-9 was made possible by its ability to sustain the long-term growth of murine T cell clones in the absence of feeder cells and antigen. Taking advantage of such factor-dependent cell lines, the protein was purified to homogeneity and provisionally designated P40, based on its apparent size in gel filtration (Uyttenhove et al., 1988). A P40 cDNA, isolated using oligonucleotides deduced from the protein sequence, allowed for production of the recombinant protein and further characterization of its biological activities (Van Snick et al., 1989).

Independently, Hültner and coworkers observed that activated splenocytes produced a factor able to enhance the proliferation of mast cell lines in response to IL-3 or IL-4 (Hültner et al., 1989; Moeller et al., 1989). This activity was partially purified and called mast cell growth enhancing activity (MEA). In addition, high levels of MEA were also found to be produced by T lymphocytes of the T_H2 subset derived by Schmitt and colleagues, who had observed that these cells produced a T cell growth factor distinct from IL-2 and IL-4, called TCGF-III (Moeller et al., 1990). The molecular cloning of a murine P40 cDNA and the availability of recombinant protein led to the conclusion that the same factor was responsible for all these biological activities (Hültner et al., 1990).

Human IL-9 cDNA was identified by cross-hybridization with a mouse probe (Renauld et al., 1990a) and by expression cloning of a factor stimulating the proliferation of a human megakaryoblastic leukemia (Yang et al., 1989). More recently, IL-9 activities were described on normal hematopoietic progenitors (Donahue et al., 1990), human T cells (Houssiau et al., 1993), B cells (Dugas et al., 1993; Petit-Frère et al.,

1993), fetal thymocytes (Suda *et al.*, 1990a), thymic lymphomas (Vink *et al.*, 1993), and murine neuronal cell lines (Mehler *et al.*, 1993).

2. *The IL-9 Gene: Structure and Expression*

The human IL-9 gene is a single-copy gene located on chromosome 5, in the 5q31–35 region (Modi *et al.*, 1991). Interestingly, this region also contains various growth factor genes and growth factor receptor genes such as IL-3, IL-4, IL-5, CSF-1, and CSF-1R and has been shown to be deleted in a series of hematological disorders. In addition, a linkage has recently been described between markers in this region and a gene controlling total serum IgE concentration (Marsh *et al.*, 1994). Radiation hybrid mapping analysis has recently located the IL-9 gene between the IL-3 and the EGR-1 (early growth response-1) genes (Warrington *et al.*, 1992).

The human IL-9 gene consists of five exons and four introns spreading over approximately 4 kb of DNA (Renauld *et al.*, 1990b). The first exon contains a 5′ untranslated region of 30 bp and the last exon contains a 3′ untranslated region of 164 bp including four ATTTA motifs that are thought to modulate the stability of the mRNAs. The structure and sequence of the IL-9 gene is shown in Figures 9.1 and 9.2.

The promoter of the IL-9 gene contains several consensus sequences such as a classical TATA box located 22–24 nucleotides upstream the transcription start, and potential recognition sites for several tetradecanoyl-phorbol-13-acetate (TPA)-inducible transcription factors such as AP-1 and AP-2, which could provide a structural basis for the induction of IL-9 expression by phorbol esters. A putative recognition site for IRF-1 (interferon regulatory factor-1) was also identified in the promoter but its physiological relevance remains more elusive (Renauld *et al.*, 1990b). Additional consensus motifs were described in the 5′ flanking region of the human IL-9 promoter (SP1, NFκB, octamer, AP-3, AP-5, glucocorticoid-responsive element, cAMP response element, and others) and it was suggested that the NFκB site and cAMP response element could be involved in the constitutive expression of IL-9 by human T-cell leukemia virus-1 (HTLV-1)-transformed T cells (Kelleher *et al.*, 1991).

3. *Characterization of the IL-9 Protein*

The cDNA for human IL-9 was independently cloned from activated peripheral T cells by hybridization with the mouse probe (Renauld *et al.*, 1990a) and from an HTLV-I infected T cell line by expression cloning of a factor active on a megakaryoblastic leukemia cell line

```
EXON 1    TCTGCAAGCGAGCTCCAGTCCGCTGTCAAG ATG CTT CTG GCC ATG GTC CTT ACC TCT GCC CTG CTC
                                        Met Leu Leu Ala Met Val Leu Thr Ser Ala Leu Leu
                                        -18
          CTG TGC TCC GTG GCA GGC CAG GGG TGT CCA ACC TTG GCG GGG ATC CTG GAC ATC AAC TTC
          Leu Cys Ser Val Ala Gly Gln Gly Cys Pro Thr Leu Ala Gly Ile Leu Asp Ile Asn Phe
                          1                                              10
          CTC ATC AAC AAG ATG CAG
          Leu Ile Asn Lys Met Gln
                          20

EXON 2    GAA GAT CCA GCT TCC AAG TGC CAC TGC AGT GCT AAT
          Glu Asp Pro Ala Ser Lys Cys His Cys Ser Ala Asn
                                          30

EXON 3    GTG ACC AGT TGT CTC TGT TTG GGC ATT CCC TCT
          Val Thr Ser Cys Leu Cys Leu Gly Ile Pro Ser
                                  40

EXON 4    GAC AAC TGC ACC AGA CCA TGC TTC AGT GAG AGA CTG TCT CAG ATG ACC AAT ACC ACC ATG
          Asp Asn Cys Thr Arg Pro Cys Phe Ser Glu Arg Leu Ser Gln Met Thr Asn Thr Thr Met
                              50                                          60
          CAA ACA AGA TAC CCA CTG ATT TTC AGT CGG GTG AAA AAA TCA GTT GAA GTA CTA AAG AAC
          Gln Thr Arg Tyr Pro Leu Ile Phe Ser Arg Val Lys Lys Ser Val Glu Val Leu Lys Asn
                              70                                          80
          AAC AAG TGT CCA
          Asn Lys Cys Pro

EXON 5    TAT TTT TCC TGT GAA CAG CCA TGC AAC CAA ACC ACG GCA GGC AAC GCG CTG ACA TTT CTG
          Tyr Phe Ser Cys Glu Gln Pro Cys Asn Gln Thr Thr Ala Gly Asn Ala Leu Thr Phe Leu
                                                  100
          AAG AGT CTT CTG GAA ATT TTC CAG AAA GAA AAG ATG AGA GGG ATG AGA GGC AAG ATA TGA
          Lys Ser Leu Leu Glu Ile Phe Gln Lys Glu Lys Met Arg Gly Met Arg Gly Lys Ile  *
              110                                    120                         126
          AGATGAAATATTATTTATCCTATTTATTAAATTTAAAAAGCTTTCTCTTTAAGTTGCTACAATTTAAAAATCAAGTAAG
          CTACTCTAAATCAGTATCAGTTGTGATTATTTGTTTAACATTGTATGTCTTTATTTTGAAATAAATACAT
```

Figure 9.1   DNA sequences of the human IL-9 cDNA. The five exons of the IL-9 gene are indicated. Amino acids corresponding to the signal peptide are underlined as well as nucleotide motifs involved in mRNA stability (ATTTA) and the polyadenylation signal (AATAAA).

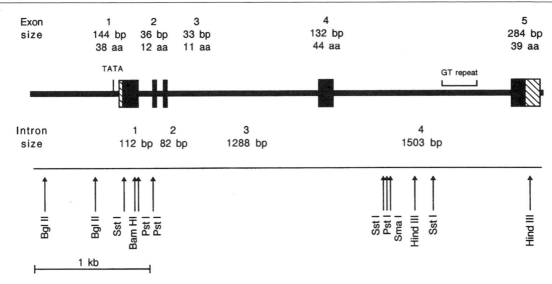

Figure 9.2   Structure of the human IL-9 gene. The five exons are indicated, with hatched sections indicating the untranslated regions.

(Yang *et al.*, 1989). Its sequence is shown in Figure 9.3. The deduced protein sequence contains 144 residues including a typical signal peptide of 18 amino acids and is characterized by four potential *N*-linked glycosylation sites, ten cysteines and a strong predominance of cationic residues, which explains the elevated pI (~10) of the protein. The deglycosylated protein has a molecular mass of approximately 14 kDa and the glycosylated material can have a molecular mass of 25–40 kDa.

The original bioassay for human IL-9 is based on the proliferation of the Mo7E cell line, isolated from a child with acute megakaryoblastic leukemia (Yang *et al.*, 1989). This cell line is also responsive to Steel factor, GM-CSF, and IL-3 which are more potent stimulators of these cells (Hendrie *et al.*, 1991) and render the measurement of IL-9 in mixtures of cytokines difficult without the use of specific neutralizing anti-IL4 antibodies. While human and mouse IL-9 are equally active in a Mo7E proliferation assay, human IL-9 is not active on murine cells. A new bioassay for human IL-9 was, however, described using murine TS1 cells transfected with the human IL-9 receptor cDNA (Renauld *et al.*, 1992). Among the human factors described so far, only LIF/HILDA and insulin were shown to promote the proliferation of this transfected murine cell line (Uyttenhove *et al.*, 1988; Van Damme *et al.*, 1992).

4.   Cellular Sources and Regulation of IL-9 Expression

The regulation of human IL-9 expression has been studied at the RNA level using freshly isolated peripheral blood mononuclear cells (PBMC). No IL-9 message could be detected in these cells in the absence of any stimulation or after activation of B cells or monocytes. By contrast, T cell-specific mitogens such as phytohemagglutinin (PHA) or anti-CD3 mAb induced a substantial IL-9 expression, which was further enhanced by addition of phorbol myristate acetate (PMA). Sorting experiments confirmed that IL-9 was preferentially produced by T cell-enriched lymphocyte populations, and, more specifically, by CD4+ T cells (Renauld *et al.*,

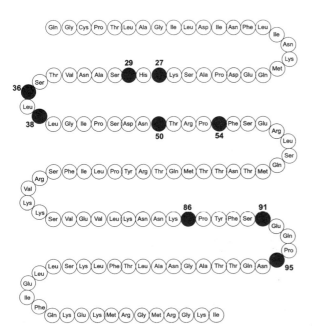

Figure 9.3   The amino acid sequence of human IL-9. Cysteine residues are shaded.

1990b). More recently, T cell activation by anti-CD3 antibodies and B7-1-mediated costimulation was also shown to result in IL-9 production. In this system, the "memory" T cell subset (CD4⁺CD45R0⁺ T cells) was found to be responsible of most of the IL-9 expression, although "naive" T cells (CD4⁺CD45R0⁺ T cells) were also able to express IL-9 under certain circumstances (Houssiau et al., 1995).

The IL-9 expression induced by T cell activation was completely abrogated by cycloheximide, an inhibitor of protein synthesis, indicating the involvement of secondary signals in this process. Further experiments identified IL-2 as a required mediator for IL-9 induction in T cells since anti-IL-2R mAb blocked IL-9 expression after stimulation with PMA and anti-CD3 mAb (Houssiau et al., 1992). However, the inhibitory activity of this mAb was not restricted to IL-9, as IL-4, IL-6, and IL-10 production was also dependent on IL-2 in this system, suggesting that the blockade of IL-9 expression could be secondary to the blockade of the production of these cytokines. Recent experiments have indeed demonstrated that IL-4 and IL-10 are both required for optimal IL-9 expression. In addition, kinetic experiments showed that IL-9 expression appears in the late stages of T cell activation while IL-2 is produced rapidly after stimulation and IL-4 and IL-10 at intermediate times, thereby suggesting the existence of a cytokine cascade during T cell activation (Houssiau et al., 1995).

5. Biological Activities of IL-9

So far, the biological activities of IL-9 have been studied mainly in murine systems, where mast cells were found to be one of the major targets of this cytokine. Surprisingly, in the human, no IL-9 activity has yet been described on mast cells and the main targets include hematopoietic progenitors, preactivated T cells, and B cells.

5.1 IL-9 AND HEMATOPOIETIC PROGENITORS

The first evidence for an involvement of IL-9 in the hematopoietic system was provided by the identification and cloning of the human protein as a growth factor for the megakaryoblastic leukemia Mo7E, a cell line displaying early markers of differentiation such as CD33 and CD34, and markers for bipotent erythromega-karyoblastic hematopoietic precursors (Yang et al., 1989; Hendrie et al., 1991). This finding prompted several groups to evaluate the activity of IL-9 on normal hematopoietic precursors. Although IL-9 did not seem to be active on megakaryoblastic precursors, it was found to support the clonogenic maturation of erythroid progenitors in the presence of erythropoietin (Donahue et al., 1990). This activity was confirmed by several

groups and reproducibly observed with highly purified progenitors after sorting for CD34⁺ cells and T cell depletion (Birner et al., 1992; Lu et al., 1992; Sonoda et al., 1992; Schaafsma et al., 1993).

By contrast, granulocyte or macrophage colony formation (CFU-GM, CFU-G, or CFU-M) was usually not influenced by IL-9. However, Schaafsma and colleagues observed that IL-9 also promoted some granulocytic as well as monocytic colony (CFU-GM) growth from CD34⁺CD2⁻ progenitors in two out of eight different bone marrow donors (Schaafsma et al., 1993). Experiments comparing the effects of IL-9 on fetal and adult progenitors have shown that IL-9 is more effective on fetal cells of the erythroid lineage and that addition of IL-9 to cultures of fetal progenitors induced maturation of CFU-Mix and CFU-GM (Holbrook et al., 1991). The observation that murine day-15 fetal thymocytes (Suda et al., 1990a), but not adult thymocytes (Suda et al., 1990b), respond synergistically to IL-9 and IL-2 further supports the hypothesis that the spectrum of activity of IL-9 is larger on fetal progenitors. The mechanism responsible for this wider spectrum of action and the physiological role of IL-9 in the regulation of hematopoiesis remain to be determined.

5.2 T CELLS

In the human system, unlike other T cell growth factors such as IL-2, IL-4 or IL-7, IL-9 did not induce any proliferation of freshly isolated T cells, either alone or in synergy with other cytokines or T cell costimuli. However, significant proliferation could be induced by IL-9 when PBMC were activated for only 10 days with PHA, IL-2, and irradiated allogeneic feeder cells, thereby indicating that responses to IL-9 might require previous activation (Houssiau et al., 1993; Lehrnbecher et al., 1994).

Experiments using human T cell clones derived either from established PHA-stimulated T cell lines or by direct cloning from freshly isolated purified T cells showed that most of these clones proliferated in response to IL-9, irrespective of their CD4 or CD8 phenotype. In addition, tumor-specific cytolytic T cell clones responded to IL-9 by proliferation or increased survival, indicating that the IL-9 responsiveness was not restricted to a particular T cell subset (Houssiau et al., 1993).

5.3 B CELLS

The ability of IL-9 to modulate immunoglobulin production by human B cells has been investigated by Dugas and colleagues (1993). Although not active by itself, IL-9 synergized with suboptimal doses of IL-4 for the IgG production by Staphylococcus aureus (SAC)-activated semipurified B cells by increasing the number of immunoglobulin-secreting cells. However, IL-9 does not potentiate the IL-4-induced activation of purified B cells stimulated by anti-CD40 antibodies (F. Brière, personal communication).

6. The IL-9 Receptor

The human IL-9 receptor cDNA, which was isolated by using the mouse cDNA as a probe, encodes a 522-amino-acid protein with 53% identity with the mouse molecule (Renauld *et al.*, 1992). The extracellular domain, composed of 233 amino acids, contains a WSEWS motif and 6 cysteine residues whose positions indicate that the IL-9 receptor is a member of the hematopoietin receptor superfamily (Bazan, 1990) (Figure 9.4). The cytoplasmic domain, less conserved than the extracellular region, is significantly larger in the human receptor (231 versus 177 residues in the mouse).

As already observed for many cytokine receptors, IL-9 receptor messenger RNAs encoding a putative soluble receptor have been identified as a result of alternative splicing. Another frequent alternative splicing of the human gene generates an intriguing heterogeneity in the 5′ untranslated region of the mRNA and introduces some short open reading frames that might represent an additional level in the regulation of IL9-R translation, as

suggested for many genes involved in cell growth (Kozak, 1991; Kermouni *et al.*, 1995). This phenomenon, as well as the use of alternative polyadenylation signals, is responsible for the multiple bands observed in northern blots, particularly in human cells, and raises the possibility of a post-transcriptional regulation of the IL-9 receptor expression (Renauld *et al.*, 1992).

The human genome contains at least four IL-9R pseudogenes with ~90% homology with the IL-9R gene, which is located in the subtelomeric region of chromosomes X and Y (Chang *et al.*, 1994; Kermouni *et al.*, 1995). *IL-9R* was thus the first gene to be identified in the long arm pseudoautosomal region and turned out to be a unique tool for study of this particular region of the genome. Using a polymorphism in the coding region of this gene, Vermeesch and colleagues recently showed that *IL-9R* is expressed both from X and Y and escapes X inactivation (Vermeesch *et al.*, 1997).

7. Signal Transduction

The amino acid sequence of the cytoplasmic domain of the IL-9 receptor did not provide much relevant information regarding signal transduction, except for a high percentage of serine and proline residues, as already

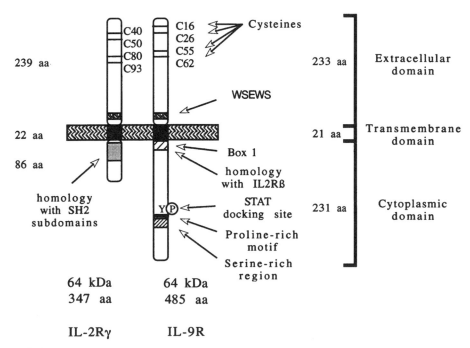

Figure 9.4   Schematic structure of the IL-9 receptor complex. Horizontal lines in the extracellular part of the receptors correspond to the four conserved cysteines of the hematopoietic receptor family, which are indicated with their respective positions. The WSXWS motif is shown for both receptors. The transmembrane domains are shown as black boxes. Hatched or dotted boxes represent the major structural features of the receptors as indicated in the figure.

observed for most cytokine receptors. Some sequence homology with other receptors was noted proximally to the transmembrane domain. As previously shown for many cytokine receptors (IL-4R, IL-7R, IL-3R, EPOR, IL-2Rβ, G-CSFR), a Pro-X-Pro sequence was found upstream from a cluster of hydrophobic residues that partially fits a consensus sequence designated Box 1 (Murakami *et al.*, 1991). Downstream from this motif, a striking homology was observed with the β chain of the IL-2 receptor and with the erythropoietin receptor. Interestingly, these two receptors interact with the cytokines that have been shown to synergize with IL-9 for the proliferation of fetal thymocytes and erythroid progenitors, respectively. A schematic view of the IL-9 receptor complex is shown in Figure 9.3.

Recently, additional evidence has been provided for a link between the IL-2 receptor system and the IL-9 receptor. Both receptors were indeed found to require a common chain, IL-2Rγ also designated γ$_c$, for signal transduction since an antibody directed against this protein inhibited the activity of IL-9 (Kitamura *et al.*, 1995). Interestingly, γ$_c$ does not seem to be involved in the affinity of the IL-9 receptor, although cross-linking experiments demonstrated a direct interaction between IL-9 and γ$_c$. In addition, IL-9 and IL-2 seem to trigger very similar signal transduction pathways. Upon IL-9 binding, both JAK1 and JAK3 become phosphorylated and catalytically active (Russel *et al.*, 1994). These kinases are likely to be responsible for IL-9R phosphorylation on one out of its five tyrosine residues. This single phosphorylated residue acts as a docking site for STAT1, STAT3 and STAT5, three transcription factors that, after phosphorylation by the JAK kinases associated to the receptor, form hetero- or homodimers and migrate to the nucleus (Demoulin *et al.*, 1996). Although several signal transduction studies on other cytokine receptors have shown that activation of STAT transcription factors is often dispensable for cell growth regulation, mutation of the single phosphorylated tyrosine of the IL-9R abolished both STAT activation and cell growth control by IL-9, including protection against apoptosis and positive as well as negative effects on proliferation (Demoulin *et al.*, 1996).

Besides the JAK/STAT pathway, IL-9 induces the activation of an adaptor protein called 4PS/IRS2, a feature shared with IL-4 signal transduction, where this pathway was shown to be critical for growth regulation (Keegan *et al.*, 1994; Yin *et al.*, 1995; Demoulin *et al.*, 1996). Phosphorylation of 4PS/IRS2 is not dependent on the phosphorylation of the IL-9 receptor, contrasting with the IL-4 system in which 4PS/IRS2 associates with the IL-4 receptor through a phosphotyrosine residue. Preliminary observations suggest that 4PS/IRS2 and JAK1 activation require the same region of the IL-9 receptor (Demoulin *et al.*, unpublished data) and these two molecules were shown to be associated upon IL-9 stimulation (Yin *et al.*, 1995). By contrast, IL-9 did not induce or enhance the phosphorylations of the serine-threonine kinases RAF1 or MAP-K (Miyazawa *et al.*, 1992).

8.   *Murine IL-9*

The purification of mouse IL-9 was made possible by the use of stable factor-dependent T cell lines derived from normal antigen-dependent clones. The purified protein was characterized by an elevated pI (~10) and a high level of glycosylation (Uyttenhove *et al.*, 1988). In parallel with the cDNA cloning (Van Snick *et al.*, 1989), a complete sequencing of the purified protein has been achieved (Simpson *et al.*, 1989), confirming the deduced amino acid sequence of the mature protein. Like its human counterpart, mouse IL-9 consists of 144 residues with a typical signal peptide of 18 amino acids. The overall identity with the human reached 69% at the nucleotide level and 55% at the protein level and four potential *N*-linked glycosylation sites were noted in both sequences. The ten cysteines are also perfectly matched in both mature proteins.

The mouse IL-9 gene does not seem to be linked to the same gene cluster as its human counterpart, since it was localized on mouse chromosome 13 (Mock *et al.*, 1990) while the IL-3, IL-4, IL-5, and GM-CSF genes are located on chromosome 11. The murine and human genes share a similar organization. The five exons are identical in size for both species and show homology levels ranging from 56% to 74% (Renauld *et al*, 1990b). Contrasting with the introns, which show no significant sequence homology, 3′ and 5′ flanking sequences exhibit a high level of identity between the human and murine genes, pointing to the importance of these regions in the transcriptional or post-transcriptional regulation of IL-9 expression.

Analysis of IL-9 expression by murine T-helper clones has shown a strict correlation with the T$_H$2 phenotype (Gessner *et al.*, 1993). *In vivo*, IL-9 production was observed during infections by helminth parasites such as *Trichinella spiralis* (Grencis *et al.*, 1991) or *Trichuris muris*, for which resistance was found to correlate with the production of IL-9 in mesenteric lymph nodes (Else *et al.*, 1992). During *Heligmosomoides polygyrus* infection, the patterns of secretion of IL-3 and IL-9 can clearly distinguish mice with fast- or slow-responder phenotypes. In slow-responder mice, the initial IL-3 and IL-9 production declines to background levels by week 6 post infection. By contrast, fast responders continue to secrete high amounts of these cytokines, thereby raising the hypothesis that IL-3 and IL-9 facilitate expulsion of adult worms (Wahid *et al.*, 1994). In this system, IL-9 expression appeared to be an early event in the immune response (Madden *et al.*, 1992; F. Finkelman, personal communication), in contrast to the kinetics of IL-9 expression *in vitro*. Further experiments are needed to investigate in more detail the mechanisms and cell types responsible for IL-9 expression in parasite infections.

Mast cells definitely represent one of the major biological targets of IL-9 in the murine system. Permanent bone marrow-derived mast cell lines (BMMC) such as L138.8A were found to proliferate in response to a factor that was initially designated mast cell growth enhancing activity (MEA) (Hültner *et al.*, 1989) and was eventually identified as IL-9 (Hültner *et al.*, 1990). The proliferative activity of IL-9, alone or in synergy with IL-3 or Steel factor, was confirmed on other permanent murine mast cell lines such as MC-6, H7, and MC-9 (Williams *et al.*, 1990; and unpublished data from our laboratory).

In addition to its proliferative activity, IL-9 also seems to be a potent regulator of mast cell effector molecules. Thus, mast cell-specific proteases of the mMCP family such as mMCP-1, mMCP-2 and mMCP-4 were found to be induced by IL-9 in mast cell lines cultured in the presence of Steel factor (Eklund *et al.*, 1993). Moreover, expression of the α chain of the high-affinity IgE receptor is also upregulated by IL-9, indicating that this factor could be a key mediator of mast cell differentiation (Louahed *et al.*, 1995). IL-9 was also shown to induce IL-6 secretion by mast cell lines (Hültner *et al.*, 1990; Hültner and Moeller, 1990).

For hematopoietic progenitors, IL-9 displayed an erythroid burst-promoting activity as described in the human. Surprisingly, this effect turned out to be dependent on the presence of T cells (Williams *et al.*, 1990).

The ability of IL-9 to modulate immunoglobulin production was confirmed with mouse B cells by Petit-Frère and colleagues (1993). IL-9 synergized with suboptimal doses of IL-4 for the IgE and IgG1 production by LPS-activated semi-purified B cells by increasing the number of immunoglobulin-secreting cells. The role of IL-9 in this model did not correspond to a simple upregulation of the IL-4 responsiveness as IL-9 did not affect the IL-4-induced CD23 expression by B cells. Although these experiments did not rule out the possibility that the effect of IL-9 observed on murine B cells was mediated by accessory cells, their *in vivo* relevance is underlined by the observation that IL-9 transgenic mice show a significant increase in total as well as antigen-specific serum immunoglobulin titers (J.-C. Renauld, unpublished observations).

The results obtained with murine T cells differ in several ways from those reported in the human model (Uyttenhove *et al.*, 1988; Schmitt *et al.*, 1989). Murine T cells require months of *in vitro* culture before responding to IL-9. Moreover, when murine T cells become responsive to IL-9, factor-dependent lines are readily derived that grow independently of their antigen and feeder cells, while no such long-term proliferation has yet been obtained in the human system. However,

these discrepancies could simply be related to the distinct experimental systems used and, in particular, to the differences in the stimulation procedures. Human IL-9-responsive T cells are strongly activated with PHA, IL-2, and feeder cells on a weekly basis, and this strong stimulation was found to be crucial for IL-9 responses. By contrast, a much less potent, antigen-specific, stimulation is used in the murine model. This difference could explain why freshly raised murine T cell clones do not respond to IL-9. On the other hand, the acquisition of IL-9-responsiveness by murine T cells is linked to a cryptic *in vitro* transformation process as indicated by the fact that IL-9 dependent T cells become tumorigenic after transfection with the IL-9 cDNA (Uyttenhove *et al.*, 1991). The fact that no permanent IL-9-dependent human T cell lines or clones have been derived so far suggests that such an *in vitro* transformation process does not take place under the conditions used in the human model. Taken together, these results indicate that different mechanisms may render T cells responsive to IL-9: potent activation of freshly isolated cells as observed in the human model, or progressive dysregulation resulting from long-term *in vitro* culture as observed in the mouse model.

Finally, a role for IL-9 in the differentiation of neuronal cells has been suggested by studies on immortalized murine embryonic hippocampal progenitor cell lines, which show little evidence of morphological maturation. In combination with bFGF and TGFα, IL-9 enhanced neurite outgrowth as well as other morphological modifications, and conferred some electrical excitability to these cells (Mehler *et al.*, 1993).

The murine IL-9 receptor is a 468-amino-acid polypeptide that binds IL-9 with high affinity (K_d ~100 pM) and was detected at the surface of T cells, mast cells, and macrophage cell lines (Druez *et al.*, 1990; Renauld *et al.*, 1992). The murine protein has a 53% identity with the human receptor. The extracellular region is particularly conserved with 67% identity, while the cytoplasmic domain is significantly larger in the human receptor (231 versus 177 residues). In contrast with the human IL-9R gene, the mouse gene is a single-copy gene located on chromosome 11 (Vermeesch *et al.*, 1997). It consists of nine exons and eight introns, sharing many characteristics with other genes encoding cytokine receptors (Renauld *et al.*, unpublished data).

9. *Putative Involvement of IL-9 in Pathology*

Analysis of IL-9 transgenic mice constitutively expressing high levels of IL-9 has shed some light on the *in vivo* activity of IL-9 (Renauld *et al.*, 1994). Although the normal T cell development did not appear to be affected, about 5% of these mice spontaneously developed T cell lymphomas. Moreover, the IL-9 transgenic mice were highly susceptible to chemical mutagenesis as all transgenic animals developed such tumors after low doses of irradiation or injections of a mutagen (*N*-methyl- *N*-nitrosourea) that were totally innocuous in control mice. The growth-promoting activity of IL-9 for T cell tumors was further investigated *in vitro* using other models of thymic lymphomas. In these experiments, it was found that IL-9 significantly stimulated the *in vitro* proliferation of primary lymphomas induced by either chemical mutagenesis in DBA/2 mice or by x-ray radiation in B6 mice (Vink *et al.*, 1993). Moreover, recent experiments have pointed to the potent antiapoptotic activity of IL-9 for dexamethasone-treated lymphoma cell lines (Renauld *et al.*, 1995).

In the human, an association between dysregulated IL-9 expression and lymphoid malignancies were initially suggested by the observation that lymph nodes from patients with Hodgkin and large-cell anaplastic lymphomas constitutively produce IL-9 (Merz *et al.*, 1991; Trümper *et al.*, 1993). Similar observations were reported *in vitro* for Hodgkin cell lines (Merz *et al.*, 1991; Gruss *et al.*, 1992). Moreover, the demonstration of an autocrine loop for the *in vitro* growth of one of these Hodgkin cell lines (Gruss *et al.*, 1992) suggests a potential involvement of IL-9 in this disease.

Another characteristic of IL-9 expression is its association with HTLV-1, a retrovirus involved in adult T-cell leukemia. It was indeed observed that HTLV-1-transformed T cells produce IL-9 constitutively (Yang *et al.*, 1989; Kelleher *et al.*, 1991). Although it is not yet clear which protein of HTLV-1 is responsible for this induction, Kelleher and colleagues suggested an implication of the Tax trans-activator through a NFκB consensus site in the IL-9 promoter (Kelleher *et al.*, 1991). An autocrine loop may also play a role in HTLV-1 leukemias, as illustrated by the *cis/trans*-activation of the IL-9 receptor gene by insertion of the HTLV-1 LTR in one leukemia cell line (Kubota *et al.*, 1996). Interestingly, in another system of T cell transformation by murine polytropic retroviruses, viral infection also resulted in IL-9 expression (Flubacher *et al.*, 1994).

10. References

Bazan, F. (1990). Structural design and molecular evolution of a cytokine receptor superfamily. Proc. Natl Acad. Sci. USA 87, 6934–6938.

Birner, A., Hültner, L., Mergenthaler, H.G., Van Snick, J. and Dörmer, P. (1992). Recombinant murine Interleukin 9 enhances the erythropoietin-dependent colony formation of human BFU-E. Exp. Hematol. 20, 541–545.

Chang, M.S., Engel, G., Benedict, C., Basu, R. and McNimch, J. (1994). Isolation and characterization of the human interleukin-9 receptor gene. Blood 83, 3199–3205.

Demoulin, J.B., Uyttenhove, C., Van Roost, E., *et al.* (1996). A single tyrosine of the Interleukin-9 (IL-9) receptor is required for STAT activation, antiapoptotic activity, and growth regulation by IL-9. Mol. Cell. Biol. 16. [In press]

Donahue, R.E., Yang, Y.C. and Clark, S.C. (1990). Human P40 T cell growth factor (interleukin 9) supports erythroid colony formation. Blood 75, 2271–2275.

Druez, C., Coulie, P., Uyttenhove, C. and Van Snick, J. (1990). Functional and biochemical characterization of mouse P40/IL-9 receptors. J. Immunol. 145, 2494–2499.

Dugas, B., Renauld, J.-C., Bonnefoy, J.Y., *et al.* (1993). Interleukin 9 potentiates the interleukin 4 induced immunoglobulin production by normal human B lymphocytes. J. Immunol. 23, 1687–1692.

Eklund, K.K., Ghildyal, N., Austen, F.K. and Stevens, R.L. (1993). Induction by IL-9 and suppression by IL-3 and IL-4 of the levels of chromosome 14-derived transcripts that encode late-expressed mouse mast cell proteases. J. Immunol. 151, 4266–4273.

Else, K., Hültner, L. and Grencis, R. (1992). Cellular immune responses to the murine nematode parasite *Trichuris muris*. II. Differential induction of Th cell subsets in resistant mice versus susceptible mice. Immunology 75, 232–237.

Flubacher, M.M., Bear, S.E. and Tsichlis, P.N. (1994). Replacement of interleukin-2 (IL-2)-generated mitogenic signals by a mink cell focus-forming (MCF) or xenotropic virus-induced IL-9-dependent autocrine loop: implications for MCF virus-induced leukemogenesis. J. Virol. 68, 7709–7716.

Gessner, A., Blum, H. and Röllinghoff, M. (1993). Differential regulation of IL-9 expression after infection with *Leishmania major* in susceptible and resistant mice. Immunobiology 189, 419–435.

Grencis, R.K., Hültner, L. and Else, K.J. (1991). Host protective immunity to *Trichinella spiralis* in mice: activation of Th cell subsets and lymphokine secretion in mice expressing different response phenotypes. Immunology 74, 329–332.

Gruss, H.J., Brach, M.A., Drexler, H.G., Bross, K.J. and Herrman, F. (1992). Interleukin 9 is expressed by primary and cultured Hodgkin and Reed–Sternberg cells. Cancer Res. 52, 1026–1031.

Hendrie, P., Miyazawa, K., Yang, Y.C., Langefeld, C. and Broxmeyer, H. (1991). Mast cell growth factor (c-kit ligand) enhances cytokine stimulation and proliferation of the human factor-dependent cell line Mo7E. Exp. Hematol. 19, 1031–1037.

Holbrook, S.T., Ohls, R.K., Schribler, K.R., Yang, Y.C. and Christensen, R.D. (1991). Effect of interleukin 9 on clonogenic maturation and cell cycle status of foetal and adult hematopoietic progenitors. Blood 77, 2129–2134.

Houssiau, F., Renauld, J.-C., Fibbe, W. and Van Snick, J. (1992). IL-2 dependence of IL-9 expression in human T lymphocytes. J. Immunol. 148, 3147–3151.

Houssiau, F., Renauld, J.-C., Stevens, M., Lehmann, F., Coulie, P.G. and Van Snick, J. (1993). Human T cell lines and clones respond to IL-9. J. Immunol. 150, 2634–2640.

Houssiau, F., Schandené, L., Stevens, M., *et al.* (1995). A cascade of cytokines is responsible for IL-9 expression in human T cells. Involvement of IL-2, IL-4 and IL-10. J. Immunol. 154, 2624–2630.

Hültner, L. and Moeller, J. (1990). Mast cell growth enhancing activity (MEA) stimulates IL-6 production in a mouse bone marrow-derived mast cell line and a malignant subline. Exp. Hematol. 18, 873–877.

Hültner, L., Moeller, J., Schmitt, E., *et al.* (1989). Thiol-sensitive mast cell lines derived from mouse bone marrow respond to a mast cell growth-enhancing activity different from both IL-3 and IL-4. J. Immunol. 142, 3440–3446.

Hültner, L., Druez, C., Moeller, J., *et al.* (1990). Mast cell growth enhancing activity (MEA) is structurally related and functionally identical to the novel mouse T cell growth factor P40/TCGFIII (interleukin 9). Eur. J. Immunol. 20, 1413–1416.

Keegan, A., Nelms, K., Wang, L.M., Pierce, J. and Paul, W. (1994). Interleukin-4 receptor: signaling mechanisms. Immunol. Today 15, 423–432.

Kelleher, K., Bean, K., Clark, S.C., *et al.* (1991). Human interleukin-9: genomic sequences, chromosomal location, and sequences essential for its expression in human T cell leukemia virus (HTLV)-I-transformed human T cells. Blood 77, 1436–1441.

Kermouni, A., Van Roost, E., Arden, K. C., *et al.* (1995). The IL-9 receptor gene: genomic structure, chromosomal localization in the pseudoautosomal region of the long arm of the sex chromosomes and identification of IL-9R pseudogenes at 9qter, 10pter, 16pter and 18pter. Genomics 29, 371–382.

Kitamura, Y., Takeshita, T., Kondo, M., *et al.* (1995). Sharing of the IL-2 receptor γ chain with the functional IL-9 receptor complex. Int. Immunol. 7, 115 120.

Kubota, S., Siomi, H., Hatanaka, M. and Pomerantz, R.J. (1996). Cis/trans-activation of the interleukin-9 receptor gene in an HTLV-I-transformed human lymphocytic cell. Oncogene 12, 1441–1447.

Kozak, M. (1991). An analysis of vertebrate mRNA sequences: intimations of translational control. J. Cell. Biol. 115, 887–903.

Lehrnbecher, T., Poot, M., Orscheschek, K., Sebald, W., Feller, A.C. and Merz, H. (1994). Interleukin 7 as interleukin 9 drives phytohemagglutinin-activated T cells through several cell cycles; no synergism between interleukin 7, interleukin 9 and interleukin 4. Cytokine 6, 279–284.

Lemoli, R.M., Fortuna, A., Tafuri, A., *et al.* (1996). Interleukin-9 stimulates the proliferation of human myeloid leukemic cells. Blood 87, 3852–3859.

Louahed, J., Kermouni, A., Van Snick, J. and Renauld, J.-C. (1995). IL-9 induces expression of granzymes and high affinity IgE receptor in murine T helper clones. J. Immunol. 154, 5061–5070.

Lu, L., Leemhuis, T., Srour, E. and Yang, Y.C. (1992). Human Interleukin 9 stimulates proliferation of CD34^{+++}DR$^+$CD33$^-$ erythroid progenitors in normal human bone marrow in the absence of serum. Exp. Hematol. 20, 418–424.

Madden, K., Urban, K., Ziltener, H., Schrader, J., Finkelman, F. and Katona, I. (1991). Antibodies to IL3 and IL4 suppress helminth-induced intestinal mastocytosis. J. Immunol. 147, 1387–1391.

Marsh, D.G., Neely, J.D., Breazeale, D.R., *et al.* (1994). Linkage analysis of IL-4 and other chromosome 5q31.1 markers and total serum immunoglobulin E concentrations. Science 264, 1152–1156.

Mehler, M.F., Rozental, R., Dougherty, M., Spray, D.C. and Kessler, J.A. (1993). Cytokine regulation of neuronal differentiation of hippocampal progenitor cells. Nature 362: 62–65.

Merz, H., Houssiau, F., Orscheschek, K., *et al.* (1991). Interleukin-9 expression in human malignant lymphomas: unique association with Hodgkin's disease and large cell anaplastic lymphoma. Blood 78, 1311–1317.

Miyazawa, K., Hendrie, P., Kim, Y.J., *et al.* (1992). Recombinant human interleukin-9 induces protein tyrosine phosphorylation and synergizes with Steel factor to stimulate proliferation of the human factor-dependent cell line, Mo7E. Blood 80, 1685–1692.

Mock, B.A., Krall, M., Kozak, C.A., *et al.* (1990). IL9 maps to mouse chromosome 13 and human chromosome 5. Immunogenetics 31, 265–270.

Modi, W.S., Pollock, D.D., Mock, B.A., Banner, C., Renauld, J.-C. and Van Snick, J. (1991). Regional localization of the human glutaminase (GLS) and interleukin-9 (IL9) genes by in situ hybridization. Cytogenet. Cell. Genet. 57, 114–116.

Moeller, J., Hültner, L., Schmitt, E. and Dörmer, P. (1989). Partial purification of a mast cell growth-enhancing activity and its separation from IL-3 and IL-4. J. Immunol. 142, 3447–3451.

Moeller, J., Hültner, L., Schmitt, E., Breuer, M. and Dörmer, P. (1990). Purification of MEA, a mast cell growth-enhancing activity, to apparent homogeneity and its partial amino acid sequencing. J. Immunol. 144, 4231–4234.

Murakami, M., Narazaki, M., Hibi, M., *et al.* (1991). Proc. Natl Acad. Sci. USA 88, 11349–11353.

Petit-Frère, C., Dugas, B., Braquet, P. and Mencia-Huerta, J.M. (1993). Interleukin-9 potentiates the interleukin-4-induced IgE and IgG1 release from murine B lymphocytes. Immunology 79, 146–151.

Renauld, J.-C., Goethals, A., Houssiau, F., Van Roost, E. and Van Snick, J. (1990a). Cloning and expression of a cDNA for the human homolog of mouse T cell and mast cell growth factor P40. Cytokine 2, 9–12.

Renauld, J.-C., Goethals, A., Houssiau, F., Merz, H., Van Roost, E. and Van Snick, J. (1990b). Human P40/IL-9: expression in activated CD4$^+$ T cells, genomic organization, and comparison with the mouse gene. J. Immunol. 144, 4235–4241.

Renauld, J.-C., Druez, C., Kermouni, A., *et al.* (1992). Expression cloning of the murine and human interleukin 9 receptor cDNAs. Proc. Natl Acad. Sci. USA 89, 5690–5694.

Renauld, J.-C., van der Lugt, N., Vink, A., *et al.* (1994). Thymic lymphomas in interleukin 9 transgenic mice. Oncogene 9, 1327–1332.

Renauld, J.-C., Vink, A., Louahed, J. and Van Snick, J. (1995). IL-9 is a major anti-apoptotic factor for mouse thymic lymphomas. Blood. [In press]

Russel, S.M., Johnston, J.A., Noguchi, M., *et al.* (1994). Interaction of IL-2Rβ and γc chains with Jak1 and Jak3: implications for XSCID and XCID. Science 266, 1042–1045.

Schmitt, E., van Brandwijk, R., Van Snick, J., Siebold, B. and Rüde, E. (1989). TCGFIII/P40 is produced by naive murine CD4$^+$ T cells but is not a general T cell growth factor. Eur. J. Immunol. 19, 2167–2170.

Schmitt, E., Beuscher, H.U., Huels, C., *et al.* (1991). IL-1 serves as a secondary signal for IL-9 expression. J. Immunol. 147, 3848–3854.

Schaafsma, M.R., Falkenburg, J.H., Duinkerken, N., *et al.* (1993). Interleukin-9 stimulates the proliferation of enriched human erythroid progenitor cells: additive effect with GM-CSF. Ann. Hematol. 66, 45–49.

Simpson, R.J., Moritz, R.L., Gorman, J.J. and Van Snick, J. (1989). Complete amino acid sequence of a new murine T-cell growth factor P40. Eur. J. Biochem. 183, 715–722.

Sonoda, Y., Maekawa, T., Kuzuyama, Y., Clark, S.C. and Abe, T.

(1992). Human interleukin-9 supports formation of a subpopulation of erythroid bursts that are responsive to interleukin-3. Am. J. Hematol. 41, 84-91.

Suda, T., Murray, R., Fischer, M., Tokota, T. and Zlotnik, A. (1990a). Tumor necrosis factor-α and P40 induce day 15 murine fetal thymocyte proliferation in combination with IL-2. J. Immunol. 144, 1783–1787.

Suda, T., Murray, R., Guidos, C. and Zlotnik, A. (1990b). Growth-promoting activity of IL-1α, IL-6 and tumor necrosis factor-α in combination with IL-2, IL-4, or IL-7 on murine thymocytes. J. Immunol. 144, 3039–3045.

Trümper, L.H., Brady, G., Bagg, A., et al. (1993). Single cell analysis of Hodgkin and Reed–Sternberg cells: molecular heterogeneity of gene expression and p53 mutations. Blood 81, 3097–3115.

Uyttenhove, C., Simpson, R.J. and Van Snick, J. (1988). Functional and structural characterization of P40, a mouse glycoprotein with T-cell growth factor activity. Proc. Natl Acad. Sci. USA 85, 6934–6938.

Uyttenhove, C., Druez, C., Renauld, J.-C., Herin, M., Noel, H. and Van Snick, J. (1991). Autonomous growth and tumorigenicity induced by P40/interleukin 9 cDNA transfection of a mouse P40-dependent T cell line. J. Exp. Med. 173, 519–522.

Van Damme, J., Uyttenhove, C., Houssiau, F., Put, W., Proost, P. and Van Snick, J. (1992). Human growth factor for murine interleukin-9 responsive T cell lines: co-induction with IL-6 in fibroblasts and identification as LIF/HILDA. Eur. J. Immunol. 22, 2801–2808.

Van Snick, J., Goethals, A., Renauld, J.-C., et al. (1989). Cloning and characterization of a cDNA for a new mouse T cell growth factor (P40). J. Exp. Med. 169, 363–368.

Vermeesch, J.R., Petit, P., Kermouni, A., Renauld, J.-C., Van Den Berghe, H. and Marynen, P. (1997). The IL-9 receptor gene, located in the Xq/Yq pseudoautosomal region, has an autosomal origin and escapes X inactivation. Hum. Mol. Genet. 6, 1–8.

Vink, A., Renauld, J.-C., Warnier, G. and Van Snick, J. (1993). Interleukin 9 stimulates in vitro growth of mouse thymic lymphomas. Eur. J. Immunol. 23, 1134–1138.

Wahid, F.N., Behnke, J.M., Grencis, R.K., Else, K.J. and Ben-Smith, A.W. (1994). Immunological relationships during primary infection with Heligmosomoides polygyrus: Th2 cytokines and primary response phenotype. Parasitology 108, 461–471.

Warrington, J., Bailey, S., Armstrong, E., et al. (1992). A radiation hybrid map of 18 growth factor, growth factor receptor, hormone receptor, or neurotransmitter receptor genes on the distal region of the long arm of chromosome 5. Genomics 13, 803–808.

Williams, D.E., Morrissey, P.J., Mochizuki, D.Y., et al. (1990). T cell growth factor P40 promotes the proliferation of myeloid cell lines and enhances erythroid burst formation by normal murine bone marrow cells in vitro. Blood 76, 906–911.

Yang, Y., Ricciardi, S., Ciarletta, A., Calvetti, J., Kelleher, K. and Clark, S.C. (1989). Expression cloning of a cDNA encoding a novel human hematopoietic growth factor: human homologue of murine T-cell growth factor P40. Blood 74, 1880–1884.

Yin, T., Keller, S.R., Quelle, F.W., et al. (1995). Interleukin-9 induces tyrosine phosphorylation of insulin receptor substrate-1 via Jak tyrosine kinases. J. Biol. Chem. 270, 20497–20502.

10. Interleukin-10

René de Waal Malefyt

1. Introduction

T cell clones have been divided into different subsets on the basis of the pattern of lymphokines which they are able to produce (Mossman *et al.*, 1986). T helper 1 clones (T_H1) are able to produce interleukin-2 (IL-2), interferon-γ (IFN-γ) and lymphotoxin, whereas T helper 2 (T_H2) clones secrete IL-4, IL-5, and IL-6. This subdivision proved functionally relevant, since T_H1 cells mediate cellular immune responses and T_H2 cells provide help for B cells in humoral immune responses (Mosmann, 1991; Mosmann and Coffman, 1989). It has been reported that T_H1 and T_H2 responses are mutually exclusive and that cytokines play an important role in directing these responses (Parish, 1972). For example, IFN-γ inhibits the proliferation of T_H2 cells and antagonizes many functions of the T_H2 cytokine IL-4 (Gajewski and Fitch, 1988). On the basis of these data, Mossman and colleagues proposed that T_H2 clones might produce a factor that could inhibit the proliferation

and/or cytokine production by T_H1 clones and set up an assay to test this hypothesis. They were successful in determining that supernatants of activated T_H2 clones could inhibit the production of IFN-γ by T_H1 clones and designated this activity cytokine synthesis inhibitory factor (CSIF) (Fiorentino *et al.*, 1989). The cDNA encoding this factor was identified from a library of a Con A-activated murine T_H2 clone by expression cloning, and this cDNA was subsequently used to isolate the human homologue from an activated T_H0 library by cross-hybridization (Moore *et al.*, 1990; Vieira *et al.*, 1991). Expression of these cDNAs allowed the biological characterization of the protein and revealed multiple activities on different cell types including mast cell growth factor activity (MCGFIII) and B cell-derived T cell growth factor activity (B-TCGF) (Suda *et al.*, 1991; Thompson-Snipes *et al.*, 1991). Subsequently this factor was renamed interleukin-10 (IL-10). At least two herpes viruses (Epstein–Barr virus (EBV) and equine herpes virus type 2 (EHV2)) have been shown to harbor analogs of the IL-10 gene (Moore *et al.*,

1990; Rode *et al.*, 1993; Vieira *et al.*, 1991). These analogs (viral IL-10 , vIL-10) are highly conserved and the expressed EBV-derived gene, called BCRFI, (*Bam*HI C fragment, rightward reading frame I), shares many, but not all, activities with human IL-10 (Hsu *et al.*, 1990). Several review articles have been written on the biology and activities of IL-10 (de Waal Malefyt *et al.*, 1992; Howard and O'Garra, 1992; Moore *et al.*, 1991, 1993; Mosmann and Moore, 1991) and a volume in the Molecular Biology Intelligence Unit series has been dedicated to it (de Vries and de Waal Malefyt, 1994).

2.   *The Cytokine Gene*

The sequence of the human IL-10 gene is given in Figure 10.1 and its structure is shown in Figure 10.2. The human IL-10 gene is located on chromosome 1q (Kim *et al.*, 1992). The gene spans ~4.7 kb of genomic DNA and consists of five exons separated by four introns. The first exon contains the 5′ untranslated sequence and the coding sequences for the 18-amino-acid signal peptide and 37 amino acids comprising the N-terminus of the mature protein. The fifth exon contains sequences encoding 30 amino acids of the C terminus and the 3′ untranslated sequence that has a ~300 bp repetitive element of the human *Alu* family (Vieira *et al.*, 1991). The intron/exon junctions do not disrupt amino acid codons, and introns are flanked by gt-ag sequences. A TATA box is located at 84 bp 5′ of the ATG codon. The start site of transcription has not yet been determined. A second TATA box is located 735 bp 5′ of the first. However, the significance of this second TATA box is unknown at present. CCAAT sequence motifs are found 148 bp, 363 bp, and 864 bp 5′ of the first TATA box. Putative AP-1 binding sites are located at positions − 29 bp and −695 bp 5′ of the TATA sequence. Several consensus binding sites for NF-IL-6 are located at −338 bp, −532 bp, −777 bp, −1504 bp, −1551 bp,

```
3813                                                       ccaatcatttttgcttacgat

       gcaaaaattgaaaactaagtttattagagaggttagagaaggaggagctctaagcagaaaaaatcctgtgccgggaaacc

       ttgattgtggcttttttaatgaatgaagaggcctccctgagcttacaatataaaaggggggacagagaggtgaaggtctaca

       catcaggggggttgctcttgcaaaaccaaaccacaagacagacttgcaaaagaaggcATG CAC AGC TCA GCA CTG

       CTC TGT TGC CTG GTC CTC CTG ACT GGG GTG AGG GCC AGC CCA GGC CAG GGC ACC CAG TCT       Exon 1

       GAG AAC AGC TGC ACC CAC TTC CCA GGC AAC CTG CCT AAC ATG CTT CGA GAT CTC CGA GAT

       GCC TTC AGC AGA GTG AAG ACT TTC TTT                                         4221


5187   CAA ATG AAG GAT CAG CTG GAC AAC TTG TTG TTA AAG GAG TCC TTG CTG GAG GAC TTT AAG   5247   Exon 2


5547   GGT TAC CTG GGT TGC CAA GCC TTG TCT GAG ATG ATC CAG TTT TAC CTG GAG GAG GTG ATG

       CCC CAA GCT GAG AAC CAA GAC CCA GAC ATC AAG GCG CAT GTG AAC TCC CTG GGG GAG AAC       Exon 3

       CTG AAG ACC CTC AGG CTG AGG CTA CGG CGC TGT                                  5700


6710   CAT CGA TTT CTT CCC TGT GAA AAC AAG AGC AAG GCC GTG GAG CAG GTG AAG AAT GCC TTT       Exon 4

       AAT AAG                                                                     6776


7851   CTC CAA GAG AAA GGC ATC TAC AAA GCC ATG AGT GAG TTT GAC ATC TTC ATC AAC TAC ATA

       GAA GCC TAC ATG ACA ATG AAG ATA CGA AAC TGA         gacatcagggtggcgactctatagact

       ctaggacataaattagaggtctccaaaatcggatctggggctctgggatagctgacccagcccccttgagaaaccttatt

       gtacctctcttatagaatatttattacctctgatacctcaacccccatttctatttatttactgagcttctctgtgaac

       gatttagaaagaagcccaatattataattttttttcaatatttattattttcacctgtttttaagctgtttccatagggt

       gacacactatggtatttgagtgtttttaagataaattataagttacataagggaggaaaaaaaatgttctttggggagcc

       aacagaagcttccattccaagcctgaccacgctttctagctgttgagctgttttccctgacctccctctaatttatctt       Exon 5

       gtctctgggcttggggcttcctaactgctacaaatactcttaggaagagaaaccagggagcccctttgatgattaattc

       tttataacaacctaaatttggttctaggccgggcggtggctcacgcctgtaatcccagcactttgggaggctgaggc

       gggtggatcacttgaggtcaggagttcctaaccagcctggtcaacatggtgaaacccgtctctactaaaaatacaaaa

       attagccgggcatggtggcgcgcacctgtaatcccagctacttgggaggctgaggcaagagaattgcttgaacccagga

       gatggaagttgcagtgagctgatatcatgccccctgtactccagcctgggtgacagagcaagactctgtctcaaaaaaat

       aaaaataaaaataaatttggttctaatagaactcagttttaactagaatttattcaattcctctgggaatgttacattg

       tttgtctgtcttcatagcagattttaattttgaataaataaatgtatcttattcacatc*                   8978
```

Figure 10.1   Sequence of human IL-10 gene. Exon nucleotides are in capitals. CAT, TATA, and AATAA sequences are underlined and the position of the codon representing the amino-terminus of the mature protein is boxed. The start of the poly(A) tail is indicated by an asterisk.

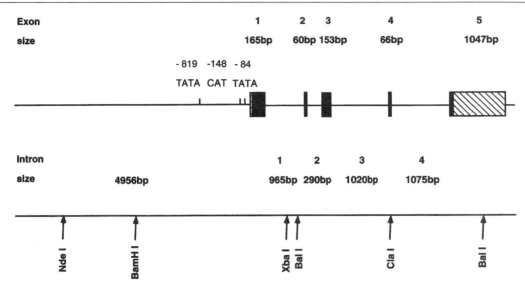

Figure 10.2   Genomic organization of human IL-10 gene. Intron/exon sizes and restriction sites are indicated.

−2085 bp, −2793 bp, −2894 bp, and −2937 bp 5′ of the TATA sequence. A glucocorticoid response element and a cyclic AMP response element, which are conserved in the promoter of the murine IL-10 gene, are located at −250 bp and −311 bp 5′ of the TATA sequence. The significance of these sites is currently under investigation. A 6-fold and an 11-fold repeat of the sequence CACA is located at positions −3942 bp and −1109 bp 5′ of the TATA sequence. Transcription of the IL-10 gene yields a mRNA of approximately 1.6 kb. The hIL-10 cDNA sequence and the hIL-10 gene sequence are deposited in the GenBank database under accession numbers M57627 and U16720. Shorter versions of the 5′ untranslated sequence were also deposited in GenBank by others under accession numbers X73536, X78437, Z30175, and U06844.

3.   The Protein

The human IL-10 gene encodes a protein of 178 amino acids which includes 18 amino acids of hydrophobic signal sequence (Figure 10.3). The mature hIL-10 protein consists therefore of 160 amino acids and Ser-1 was confirmed as N-terminal residue by protein sequencing (Windsor *et al.*, 1993). The predicted molecular mass of hIL-10 is 18 647 Da and it runs as a single species with an apparent molecular mass of 17 kDa in SDS-PAGE (Vieira *et al.*, 1991). hIL-10 contains four cysteine residues which form two intramolecular disulfide bonds with Cys-12 pairing to Cys-108 and Cys-62 to Cys-114 (Windsor *et al.*, 1993). hIL-10 contains one possible *N*-linked glycosylation site at Asn-115, but this site is unoccupied (Trotta and Windsor, 1994). Isoelectric focusing–PAGE of CHO-derived hIL-10

showed that the isoelectric point (pI) of IL-10 was 7.92, which is close to its predicted value of 7.8 (Windsor *et al.*, 1995). hIL-10 is stable at −20°C in phosphate buffered saline (PBS), pH 7.0, and can be reconstituted in PBS with full biological activity following lyophilization in ammonium bicarbonate pH 8.5. However, acidic treatment (below pH 5.5) will result in biologically inactive material (Windsor *et al.*, 1995). hIL-10 has a high degree of α-helicity. The protein contains 61% α-helix, 7% β-sheet, 4% β-turn, and 28% random structure. The α-helical regions were predicted to constitute at least four and possibly six independent segments (Windsor *et al.*, 1995). These predictions are consistent with the four-α-helix bundle motif which is found in a number of cytokines (Shanafelt *et al.*, 1991). In solution, IL-10 exists as a noncovalently linked homodimer. Gel filtration has indicated an apparent molecular mass of 37 kDa, which is consistent with this dimeric structure. The homodimer dissociates in 1.5 M guanidine-HCl or at pH values less than 4.5 (Trotta and Windsor, 1994). hIL-10 has been crystallized and the resulting IL-10 crystals were tetragonal of space group P43 21 2. The x-ray crystal structure has been determined at 2 Å and 1.8 Å resolution (Walter and Nagabhushan, 1995; Zidanov *et al.*, 1995). The molecule is a tight dimer made of two interpenetrating subunits forming a V-shaped structure. Each half of the structure consists of six α-helices, four originating from one subunit and two from the other. The overall structure resembles that of IFN-γ very closely. The human IL-10 protein is active on murine cells but the murine cytokine is not active on human cells. hIL-10 has 90% homology at the amino acid level with viral IL-10 and 73% with murine IL-10. The greatest divergences between these IL-10 species are found in the signal sequence and the first 21 amino acids of the N-terminus (Vieira *et al.*, 1991).

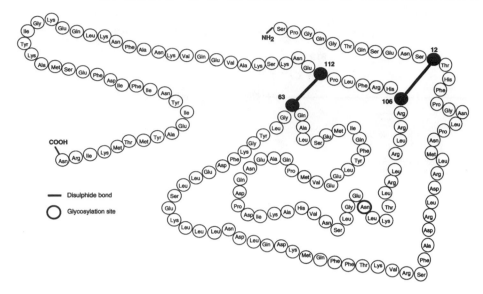

Figure 10.3   Secondary structure of human IL-10 protein.

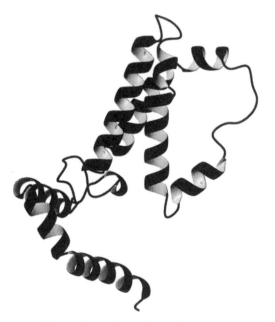

Figure 10.4   Three dimensional structure of human IL-10.

4. Cellular Sources and Production

hIL-10 is produced by a number of different cell types. Lipopolysaccharide (LPS)-activated monocytes (de Waal Malefyt *et al.*, 1991a), *Staphylococcus aureus* Cowan I (SAC)-activated B cells (Vieira *et al.*, 1991), B cells immediately following infection with EBV (Burdin *et al.*, 1993), EBV-transformed B cell lines (Vieira *et al.*, 1991; Benjamin *et al.*, 1994; Stewart *et al.*, 1994), Burkitt lymphoma (Vieira *et al.*, 1991; Benjamin *et al.*, 1992),

acquired immune deficiency syndrome (AIDS) lymphoma (Emilie *et al.*, 1993), and activated T cells and T cell clones (Yssel *et al.*, 1992) have been shown to produce IL-10. IL-10 protein has also been detected in the sera of patients with active non-Hodgkin lymphoma (Blay *et al.*, 1993), multiple myeloma (Merville *et al.*, 1992), melanoma (Becker *et al.*, 1994), or septicemia (Marchant *et al.*, 1994b), in cerebrospinal fluid of patients with bacterial meningitis (Frei *et al.*, 1993), and in ascites or biopsies of patients with ovarian and other intra-abdominal cancers (Gotlieb *et al.*, 1992); and IL-10 has been shown to be produced by human carcinoma cell lines (Gastl *et al.*, 1993) and melanoma cells (Chen *et al.*, 1994) *in vitro*. Elevated levels of IL-10 production were detected in diseases such as malaria (Peyron *et al.*, 1994), rheumatoid arthritis (Katsikis *et al.*, 1994), AIDS (Chehimi *et al.*, 1994; Clerici *et al.*, 1994), lepromatous leprosy (Yamamura *et al.*, 1991), visceral leishmaniasis (Kala azar) (Ghalib *et al.*, 1993; Karp *et al.*, 1993b), and lymphatic filariasis (King *et al.*, 1993). High levels of IL-10 were also produced by host-reactive T cell clones isolated from SCID patients that were successfully reconstituted with allogeneic fetal hematopoietic stem cells (Bacchetta *et al.*, 1994). In the mouse, it has been demonstrated that keratinocytes produce IL-10 following UV irradiation; in humans this is still controversial (Enk and Katz, 1992; Rivas and Ullrich, 1992; Kang *et al.*, 1994; Teunissen *et al.*, 1994).

Although IL-10 was initially characterized as a murine T_H2-derived cytokine, human T_H0, T_H1 and T_H2 T cell clones are all able to produce significant levels of IL-10 following activation (Yssel *et al.*, 1992; Del Prete *et al.*, 1993). In addition, T cells isolated from peripheral blood and belonging to CD4[+], CD8[+], CD45RO[+] (memory) or CD45RA[+] (naive) subsets, as well as CD4[+] cord blood T cells were able to produce IL-10 following activation (Yssel *et al.*, 1992). Production of IL-10 by memory T cells was ~5–20 times higher than that by naive T cells. The highest levels of IL-10 production by T cells or T cell clones were induced following activation of the cells by anti-CD3 mAbs and phorbol myristate acetate (PMA) or antigen. Relatively low levels of IL-10 were produced following activation of the T cell clones by Ca^{2+} ionophore and PMA, a stimulus that can induce high levels of IFN-γ, IL-2, and IL-4 secretion (Yssel *et al.*, 1992). IL-10 was produced relatively late after activation by both T cells and monocytes. mRNA levels were detectable 8 hours after activation and maximal levels could be observed at 24 hours (de Waal Malefijt *et al.*, 1991; Yssel, *et al.*, 1992). Several cytokines have been shown to affect the production of IL-10 by monocytes. Tumor necrosis factor (TNF-α) is able to induce the expression of IL-10 and enhance LPS-induced expression of IL-10 (Wanidworanun and Strober, 1993, van der Poll *et al.*, 1994), whereas IFN-γ, IL-4, IL-13, and IL-10 itself have been shown to inhibit IL-10 production by LPS-activated monocytes (Chomarat *et al.*, 1993; de Waal Malefijt, *et al.*, 1991, 1993a). Thus IL-10 has autoregulatory activities (de Waal Malefijt *et al.*, 1991).

5. Biological Activities

IL-10 has multiple biological activities and affects many different cell types. These include monocytes/macrophages, T cells, B cells, natural killer (NK) cells, neutrophils, endothelial cells and peripheral blood mononuclear cells (PBMC).

5.1 IN VITRO ACTIVITIES

5.1.1 Monocytes/Macrophages

Monocytes cultured *in vitro*, in the presence of IL-10 have a decreased ability to adhere to plastic surfaces, and detach and round up. In addition, these cells have few granules and a poorly developed rough endoplasmic reticulum, indicating that these cells are not extremely active in protein synthesis (de Waal Malefyt *et al.*, 1994). IL-10 inhibits the production of cytokines including IL-1α, IL-1β, IL6, IL-8, IL-10 itself, IL-12 p35, IL-12 p40, IFN-α, granulocyte-macrophage colony-stimulating factor (GM-CSF), granulocyte colony-stimulating factor (G-CSF), macrophage colony stimulating factor (M-CSF), macrophage inflammatory protein (MIP-1α), and TNF-α by LPS-activated human monocytes (de Waal Malefijt *et al.*, 1991; Chin and Kostura, 1993; D'Andrea *et al.*, 1993; Ralph *et al.*, 1992; Gruber *et al.*, 1994). However, IL-10 does not inhibit the production of all monokines, since it enhances the expression of IL-1

receptor antagonist (IL-1Ra) (de Waal Malefijt *et al.*, 1991; Cassatella *et al.*, 1994; Jenkins *et al.*, 1994).

The production of proinflammatory cytokines and chemokines by activated granulocytes and eosinophils was also inhibited by IL-10 at both transcriptional and post-transcriptional levels (Cassatella *et al.*, 1993, Kasama *et al.*, 1994, Takanaski *et al.*, 1994; Wang *et al.*, 1994). However, cytokine production and vascular cell adhesion molecule (VCAM) expression by endothelial cells was not affected by IL-10 (Pugin *et al.*, 1993; Sironi *et al.*, 1993). IL-10 downregulates the spontaneous expression of MHC class II antigens, ICAM-1 (CD54), B7 (CD80), and B70 (CD86), as well as the expression of these antigens on monocytes following their induction by IFN-γ or IL-4 (de Waal Malefijt *et al.*, 1991, 1994; Kubin *et al.*, 1994; Willems *et al.*, 1994).

In contrast, IL-10 upregulates the expression of FcγRI (CD64) on human monocytes to levels that are comparable to those induced by IFN-γ (te Velde *et al.*, 1992). The enhanced expression of CD64 by IL-10 correlated with enhanced antibody-dependent cell-mediated cytotoxicity (ADCC) activity by monocytes (te Velde *et al.*, 1992). On the other hand, it has also been reported that preactivation of monocytes and alveolar macrophages by LPS in the presence of IL-10 inhibited cytotoxicity against tumor cells (Nabioullin *et al.*, 1994). IL-10 inhibited the migration inhibitory factor (MIF)-, IFN-γ, or LPS-inducible production of nitric oxide (NO) in murine macrophages (Bogdan *et al.*, 1991; Cunha *et al.*, 1992; Gazzinelli *et al.*, 1992b; Wu *et al.*, 1993). This production of NO was dependent on endogenously produced TNF-α and the inhibitory effects of IL-10 were for the most part mediated by inhibition of this TNF-α (Flesch *et al.*, 1994; Oswald *et al.*, 1992b). In these assays, IL-10 synergized with IL-4 and transforming growth factor-β (TGF-β) for inhibition of NO production (Oswald *et al.*, 1992a). The inhibition of NO production, which plays an important role in monocyte cytotoxicity toward intracellular parasites, resulted in the enhanced survival of *Toxoplasma gondii*, *Leishmania donovani*, *Trypanosoma cruzi*, *Listeria monocytogenes*, *Mycobacterium bovis* and the fungus *Candida albicans* (Cenci *et al.*, 1993; Flesch *et al.*, 1994; Frei *et al.*, 1993; Gazzinelli *et al.*, 1992a,b; Wu *et al.*, 1993).

IL-10 strongly inhibits antigen presentation by monocytes/macrophages, resulting in abrogated proliferative responses or cytokine production by the responding T cells or T cell clones (de Waal Malefyt *et al.*, 1991; Fiorentino *et al.*, 1991a). The downregulation of MHC class II antigens, costimulatory molecules such as CD54, CD80, and CD86, as well as the inhibition of monokine production, including that of IL-12, by IL-10 contributes to the inhibition of the antigen presenting/accessory cell function (de Waal Malefyt, *et al.*, 1991, 1994; Kubin, *et al.*, 1994; Murphy *et al.*, 1994). IL-10 not only inhibits antigen-specific responses toward protein antigens, but also antigen-specific responses toward allo-antigens (Bejarano *et al.*, 1992; Caux *et al.*, 1994; Peguet *et al.*, 1994).

5.1.2   T Cells

Several direct effects of IL-10 on T cells have been described. IL-10 specifically inhibits the production of IL-2 and TNF-α by human T cells or T cell clones following antigen-presenting cell (APC)-independent activation. The production of IFN-γ, IL-4, and IL-5 was not affected in this system (de Waal Malefyt *et al.*, 1993). However, IL-5 production was inhibited by IL-10 following activation of T cells via CD28 and PMA (Schandene *et al.*, 1994). IL-10 has chemotactic activity on human CD8$^+$ T cells but it inhibits the IL-8 mediated chemotactic response of CD4$^+$ T cells (Jinquan *et al.*, 1993). In the mouse, IL-10 enhances the proliferation of murine thymocytes and it augmented the cytotoxic activity of murine CD8$^+$ T cells (MacNeil *et al.*, 1990; Chen and Zlotnik, 1991).

   IL-10 inhibited apoptosis of human T cells and T cell clones induced following growth factor deprivation or infectious mononucleosis (Taga *et al.*, 1993, 1994). The enhanced survival of cells in the presence of IL-10 has also been described for human monocytes, mouse macrophages, splenic B cells, and mast cells (Go *et al.*, 1990; Fiorentino *et al.*, 1991; Thompson-Snipes *et al.*, 1991; de Waal Malefyt *et al.*, 1994). The enhanced survival of splenic B cells was associated with an enhanced expression of the bcl-2 protein (Levy and Brouet, 1994). On the other hand, IL-10 induced apoptosis of human chronic lymphocytic leukemia (B-CLL) and these cells showed decreased levels of bcl-2 (Fluckiger *et al.*, 1994).

5.1.3   B Cells

B cell activating activities of IL-10 have been described, which is in line with the findings that the EBV-derived BCRF1 gene (vIL-10) plays an important, but possibly not essential, role in viral infection and B cell transformation (Burdin *et al.*, 1993; Miyazaki *et al.*, 1993; Swaminathan *et al.*, 1993). IL-10 is a costimulator for the proliferation of human B cell precursors and B cells following stimulation by anti-IgM mAbs, SAC particles, or cross-linking of their CD40 antigen (Rousset *et al.*, 1992; Saeland *et al.*, 1993). IL-10 synergizes with IL-2 for the proliferation of CD40-activated B cells, which might be due to an IL-10-induced enhancement of high-affinity IL-2 receptor expression on the B cells (Fluckiger *et al.*, 1993). IL-10 also induces B cell differentiation. B cells activated by SAC or anti-CD40 mAbs produce large amounts of IgM, IgG1, IgG2, IgG3, and IgA in the presence of IL-10 (Rousset *et al.*, 1992; Nonoyama *et al.*, 1993b). In addition, IL-10 may act as a switch factor for the production of IgA (in combination with TGF-β) or IgG1 and IgG3 by anti-CD40-activated naive sIgD$^+$ B cells (Defrance *et al.*,

1992; Briere *et al.*, 1994b). On the other hand, IL-10 has been reported to inhibit the production of IgE by IL-4-activated PBMC. However, this inhibitory effect of IL-10 is probably mediated through the effects of IL-10 on monocyte activation, since production of IgE by purified B cells and IL-4 was not inhibited by IL-10 (Punnonen *et al.*, 1993). In contrast, IL-10 inhibited IL-5-induced immunoglobulin production toward thymus-independent, but not to thymus-dependent antigens in mice (Pecanha *et al.*, 1992, 1993).

5.1.4   NK Cells

IL-10 inhibited the enhanced production of IFN-γ and TNF-α by IL-2-activated human NK cells when monocytes were present. This was an indirect effect of IL-10, mediated by inhibition of monokines (IL-12, IL-1). The cytotoxicity of human NK cells was not affected by IL-10 (Hsu *et al.*, 1992; D'Andrea *et al.*, 1993).

5.1.5   Precursor Cells

IL-10 by itself has no effect on hematopoiesis, but in combination with other cytokines it enhanced the growth of mast cells, megakaryocytes, megakaryocyte-mixed colonies and Thy1^{lo} Sca1$^+$ stem cells from mouse bone marrow (Thompson-Snipes *et al.*, 1991; Rennick *et al.*, 1994b) and that of B cell precursors from human fetal bone marrow (Saeland *et al.*, 1993). IL-10 also induced the reversible expression of mast cell protease-1 and -2 genes in mouse mast cell lines (Ghildyal *et al.*, 1992a,b, 1993). In a culture system of bone marrow stroma isolated from 5-fluorouracil-treated mice, IL-10 inhibited the *in vitro* formation of bone through the inhibition of endogenous TGF-β which acts to control the osteogenic differentiation. The inhibition of the outgrowth and differentiation of osteogenic precursors allowed hematopoiesis to occur as measured by the generation of granulocyte-macrophage colonies (CFU-GM) up to 8 weeks after onset of the cultures (Van Vlasselaer *et al.*, 1993, 1994).

5.2   *In Vivo* Activities

5.2.1   Inflammation

In vivo effects of IL-10 have been observed in animal models of inflammation, autoimmunity, tolerance, and parasite infections. IL-10 rescued mice from LPS- or SEB-induced toxic shock (Bean *et al.*, 1993; Gerard *et al.*, 1993; Howard *et al.*, 1993). Survival of mice following LPS-induced shock correlated with reduced levels of circulating TNF-α (Marchant *et al.*, 1994a). Endogenous IL-10 production had a protective effect in this model since mice treated with anti-IL-10 mAbs from birth were highly susceptible to LPS-induced shock and had elevated levels of TNF-α, IL-6, and IFN-γ in their circulation (Ishida *et al.*, 1993).

   IL-10 inhibited the initiation and effector phases of

cellular (T_H1) immune responses *in vivo*. IL-10 was identified as the product of keratinocytes which inhibited sensitization of a delayed-type hypersensitivity (DTH) response and it was shown to inhibit priming for contact hypersensitivity reactions by inducing hapten-specific tolerance (Enk and Katz, 1992; Rivas and Ullrich, 1992; Enk *et al.*, 1994). IL-10 acted in synergy with IL-4 to inhibit tuberculin type DTH reactions of BALB/c mice which had healed their *Leishmania major* infection (Powrie and Coffman, 1993, Powrie *et al.*, 1993a,b). Although IFN-γ and TNF-α production by draining lymph node cells was inhibited in mice that received the combination of IL-10 and IL-4, administration of anti-IFN-γ and anti-TNF-α antibodies could not reverse the protective effects of IL-10, indicating that the inhibition of this type of DTH may be through downregulation of other effector mechanisms.

Transfer of CD45RB^{high} T cells from BALB/c mice, which produce high levels of IFN-γ and TNF-α (T_H1), into C.B-17 *scid* mice results in spontaneous development of inflammatory bowel disease (IBD) (Powrie *et al.*, 1993a,b, 1994a,b). The induction of colitis in this model could be prevented by administration of CD45RB^{low} cells (which produce IL-4 and IL-10), treatment with anti-IFN-γ or anti-TNF-α antibodies, or administration of IL-10 (Powrie *et al.*, 1994). IBD did develop 4 weeks after the last IL-10 treatment, indicating that IL-10 did not appear to induce a long-lasting modulation of the immune response. Mice treated with IL-10 had lower levels of IFN-γ and TNF-α mRNA in the colon.

IL-10 inhibited IgG or IgA immune complex-induced lung injury in rats. This protective effect of IL-10 was evident by reductions in tissue damage, lower amounts of effector cells in bronchial alveolar lavage (BAL) fluids and strong reductions of TNF-α in BAL fluids (Mulligan *et al.*, 1993).

The inhibitory effects of IL-10 on inflammatory responses is clearly demonstrated in IL-10-deficient mice. These mice develop a severe chronic inflammatory bowel desease which probably results from the absence of the normal suppressive effect of IL-10 on immune-inflammatory responses to enteric antigens (Kuhn *et al.*, 1993). The inflammatory reaction in the gut is characterized by altered MHC class II expression by intestinal epithelial cells and increased numbers of mononuclear cells, including activated T cells, IgA- and IgG-producing plasma cells, neutrophils, and eosiniphils within focal gut lesions. High levels of TNF-α could be demonstrated at the site of the lesions, and in cultures of intestinal tissue IL-1 and IFN-γ were spontaneously produced (Rennick *et al.*, 1994a). Diseased mice were successfully treated by administration of IL-10, and the onset of disease could be delayed in young animals. However, treatment with either anti-TNF, anti-IL-6 or anti-IFN-γ mAbs was not effective (Rennick *et al.*, 1994a). IL-10-deficient mice also showed abnormal immune-inflammatory reactions upon antigenic challenge. Keyhole limpet hemocyanin (KLH) DTH reactions were exacerbated with a prolonged chronic phase which led to tissue damage and scar formation (Rennick *et al.*, 1994a). This correlated with the enhanced generation of T_H1 cells as determined by strongly enhanced production of IFN-γ by *in vitro* restimulated spleen cultures. Reactions to irritants and contact allergens were also increased in both magnitude and duration in these mice. Application of croton oil led to extensive hemorrhage and necrosis of the ears due to production of high levels of TNF-α and could be inhibited by coadministration of anti-TNF mAbs (Rennick *et al.*, 1994a). T_H2 responses developed normally in these mice. However, in mice challenged with *Nippostrongylus brasiliensis* a T_H1 response, as determined by high IFN-γ production, occurred in addition to a T_H2 response (Kuhn *et al.*, 1993), indicating the lack of normal cross-regulation between polarized T_H1 and T_H2 responses (Mosmann and Moore, 1991) and confirming the notion that IL-10 by itself does not support the generation of T_H2 responses (Hsieh *et al.*, 1992).

5.2.2 Autoimmunity

IL-10 has been tested in models of autoimmunity, including experimental autoimmune encephalomyelitis (EAE), diabetes, and systemic lupus erythematosus (SLE). In Lewis rats, systemic administration of IL-10 during the initiation phase of the disease markedly suppressed the induction of EAE, which coincided with an enhanced autoantibody production and sustained T cell proliferation to myelin basic protein as well as diminution of central nervous system (CNS) infiltration (Rott *et al.*, 1994). In a murine adoptive transfer model of EAE it was demonstrated that the high levels of IL-2, IL-4, IL-6, and IFN-γ mRNA which were present during the acute phase disappeared concomitantly with the induction of high levels of IL-10 mRNA, which remained elevated during the recovery/chronic phase of the disease (Kennedy *et al.*, 1992). These results extend *in vitro* observations that IL-10 inhibited the antigen-specific proliferation of encephalitogenic T_H1 clones (van der Veen and Stohlman, 1993). IL-10, administered daily from 8–9 weeks after birth, also inhibited the spontaneous onset of diabetes in non-obese diabetic (NOD) mice (Pennline *et al.*, 1994). However, mice did become diabetic 1 month after cessation of the treatment. Although some insulitis was detectable in the pancreas of these IL-10-treated mice, islet destruction and inhibition of insulin production were not observed (Pennline *et al.*, 1994). Insulitis without tissue destruction was also observed in IL-10 transgenic mice which expressed IL-10 under control of the insulin promoter (Wogensen *et al.*, 1993). However, when IL-10 transgenes were back-crossed with NOD mice, their offspring became diabetic at 8–10 weeks of age, indicating that local production of IL-10 accelerated insulitis and tissue destruction and had different effects from the systemic administration of IL-10 (Wogensen et al., 1994). The local production of IL-10 by the pancreas of IL-10 transgenes could also not prevent allograft rejection of fetal or adult pacreati transplanted in MHC-incompatible hosts or protect transgenic pancreatic islets expressing lymphocytic choriomeningitis virus (LCMV) viral antigens from destruction after sensitization by LCMV infection (Lee et al., 1994). The onset of spontaneous systemic lupus erythematosus, which develops in NZB/W F1 mice, was delayed upon treatment of these mice with neutralizing anti-IL-10 mAbs and accelerated by administration of IL-10. IL-10 suppressed the production of TNF-α, which has a protective effect in this case, since coadministration of anti-TNF-α and anti-IL-10 mAbs did not affect the kinetics of induction of disease (Ishida et al., 1994).

5.2.3   Transplantation

Murine models of organ rejection and graft-versus-host disease (GVHD) have shown that several conditioning regimens, including administration of anti-CD4, anti-CD8 or anti-CD40 mAbs, donor-derived spleen, or T_H2 cells, could establish tolerance and inhibit LPS-induced mortality during GVHD which coincided with increased levels IL-10 mRNA expression or IL-10 secretion (Takeuchi et al., 1992; Abramowicz et al., 1993; Hancock et al., 1993; Fowler et al., 1994; Gorczynski and Wojcik, 1994). However, high levels of IL-10 mRNA have also been observed in spleens of mice undergoing acute or chronic GVHD and, as mentioned earlier, IL-10 transgenic fetal or adult pacreati transplanted in MHC-incompatible hosts were rejected (Allen et al., 1993; De Wit et al., 1993; Garlisi et al., 1993; Lee et al., 1994). Administration of IL-10 in murine models of GVHD has not been studied extensively enough to allow conclusions on its efficacy (Roncarolo, 1994).

5.2.4   Parasitic Infections

IL-10 plays an important role in certain parasitic infections. The in vitro effects of IL-10, which include inhibition of production of NO, an important mediator in macrophage cytotoxicity against parasites, and inhibition of production of IFN-γ, which is a potent macrophage activator, are important in vivo in determining disease susceptibility (Silva et al., 1992). Mice susceptible to T. cruzi infections produced detectable levels of IL-10 upon infection, whereas resistant mice failed to do so (Reed and Sher, 1994). These responses could be modulated by anti-IL-10 or anti-IFN-γ mAbs. Neutralization of endogenous IFN-γ resulted in the appearance of IL-10 and the development of fatal infections in resistant mice, whereas neutralization of IL-10 production in susceptible mice prevented disease and death. The most likely producer of

IL-10 in this model was the macrophage, but what determines the balance between IL-10 and IFN-γ production in resistant versus susceptible mice is not clear. Schistosoma mansoni infection of mice induces the production of T_H2 cytokines and downregulation of T_H1 cytokines, and results at later stages in anergic responses towards parasite as well as other antigens (Pearce et al., 1991). High levels of IL-10 expression correlated with the observed inhibition of antigen-specific responses, and anti-IL-10 mAbs could restore protective T_H1 responses (Sher et al., 1991; Villanueva et al., 1994). This IL-10 was produced by T cells, macrophages, and (following induction by a schistosomal egg antigen) B cells (Vella et al., 1992; Murphy et al., 1993; Velupillai and Harn, 1994). Thus, IL-10, IFN-γ, and IL-12 play an important role in the balanced regulation of parasitic (protozoan, helminth, but also fungal) infections by their effects on the development and effector stages of protective T_H1 responses (Vella and Pearce, 1992, 1994; Flores et al., 1993; Romani et al., 1994; Wagner et al., 1994).

6.   The Receptor

Receptors for IL-10 have recently been identified and characterized (Ho et al., 1993; Tan et al., 1993; Liu et al., 1994). ^{125}I-labeled hIL-10 specifically bound to a variety of mouse and human hematopoietic cell lines with an observed affinity of 200–250 pM. Only several hundred receptors per cell were detected and cross-linking studies indicated a molecular mass of 90–120 kDa for hIL-10R (Tan et al., 1993). A cDNA clone encoding a human IL-10-binding protein was isolated from a library of a Burkitt lymphoma cell line (Liu et al., 1994). This cDNA encodes a protein of 578 amino acids, including a putative signal sequence of 21 amino acids, with a predicted molecular mass of 61 kDa. The difference between the predicted and observed molecular mass of the hIL-10R can be explained by the presence and use of six potential glycosylation sites. The extracellular portion of the IL-10R can be considered as two segments of ~110 amino acids that are similar to the size of the immunoglobulin-like ligand-binding domains of the growth hormone receptor. Expression of the extracellular domain of the receptor indicated that this soluble receptor is a glycosylated monomeric protein with a molecular mass of 35–45 kDa which is able to antagonize the effects of human IL-10. In solution a single complex was formed between two IL-10 dimers and four soluble receptor monomers (Tan et al. 1995). The overall structure of the hIL-10R places it in the family of class II cytokine receptors, which also contains the interferon-γ receptor (IFNR) (Bazan, 1990). Recently, CRFB4, another previously known member of the class II cytokine receptor family, was identified as a second chain of the IL-10 receptor

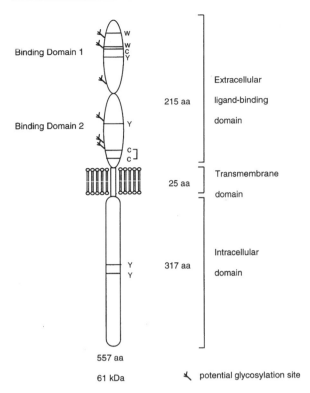

Binding Domain 1

Binding Domain 2

215 aa

25 aa

317 aa

557 aa

61 kDa

Extracellular

ligand-binding

domain

Transmembrane

domain

Intracellular

domain

⅃ potential glycosylation site

Figure 10.5   Structure of human IL-10 receptor. Indicated extracellular amino acids are characteristic for class II cytokine receptor sequences and two tyrosine residues are located in the intracellular domain.

complex. This chain serves as an accessory chain essential to the active IL-10 receptor complex and is involved in the initiation of IL-10 induced signal transduction (Kotenko *et al.*, 1997). Interestingly, IL-10 and IFN-γ antagonize each other's functions in several biological systems but both have been shown to induce expression of CD64 (FcγRI) on human monocytes (te Velde *et al.*, 1992). The molecular mechanism for induction of CD64 by IFN-γ and IL-10 involves tyrosine phosphorylation and formation of protein–DNA complexes between p91 and the IFN-γ response region in the promoter of CD64, indicating that IL-10 and IFN-γ signaling pathways may share either common receptor subunits or intracellular activation pathways (Larner *et al.*, 1993). The hIL-10R was found to be expressed by cells of hematopoietic origin, including monocytes, B cells, NK cells, and T cells. Activation of human T cell clones resulted in downregulation of hIL-10R mRNA expression (Liu *et al.*, 1994). It is not yet clear whether this also results in downregulation of hIL-10R protein expression at the cell membrane and whether this has any functional consequences. The gene encoding the hIL-10R is located on chromosome 11 and the sequence of the cDNA is deposited in the GenBank database under accession number U00672.

7.   Signal Transduction

Stimulation with IL-10 will lead to tyrosine phosphorylation of the kinases JAK1 and TYK2, but not of JAK2 or JAK3 in mouse Ba/F3 cells transfected with IL-10 receptor, human T cells, and monocytes (Finbloom and Winestock 1995; Ho *et al.*, 1995). In addition, IL-10 induced the tyrosine phosphorylation and assembly of STAT1α and STAT3 complexes in human T cells, monocytes, and mouse macrophages (Lehmann *et al.*, 1994; Finbloom and Winestock 1995). Tyrosine phosphorylation and DNA binding of STAT1 to the IFN-γ response region was also involved in the induction of CD64 by human monocytes and basophils (Larner *et al.*, 1993).

Activation of monocytes by IL-10 has also been shown to inhibit activation of NFκB, phosphatidylinositol 3-kinase (PI3-k), and p70 S6 kinase (Wang *et al.*, 1995; Crawley *et al.*, 1996). Inhibition of PI3-k and p70 S6 kinase seem to be involved in the inhibition of proliferation by IL-10 of D36 cells (Crawley *et al.*, 1996).

8.   Murine IL-10

The gene for murine IL-10 has been cloned and was assigned to chromosome 1 (Kim *et al.*, 1992). The mIL-10 gene is composed of five exons arrayed over 5.5 kb of genomic DNA. The sizes of the four introns are respectively 885, 365, 1008, and 1622 base pairs. The promoter contains several possible transcriptional control sequences including a NFAT/AP1 binding site (274), a NFκB-like recognition site (1074), an IL-6 responsive element (758), and a cAMP responsive element (1208) (Kim *et al.*, 1992). Transcription of the mIL-10 gene generates primarily a single mRNA of 1.4 kb. The gene codes for a 178-amino-acid protein that is modified to a 157-amino-acid mature protein (Moore *et al.*, 1990). mIL-10 has an *N*-glycosylation site near the N-terminus, which is resposible for the heterogeneity in molecular mass of 17, 19, and 21 kDa polypeptides. mIL-10 has five cysteine residues, which form two intramolecular disulfide bridges (C^9–C^{105}/C^{59}–C^{111}) and one unpaired residue (C^{146}). The predicted molecular mass is 18 453 Da and its isoelectric point is 7.8. Mouse IL-10 is produced by macrophages, T helper 2 cells, Ly1 (CD5) B cells, and keratinocytes (Fiorentino *et al.*, 1991a,b; Enk and Katz, 1992; O'Garra *et al.*, 1992).

mIL-10 binds with high affinity to specific cell surface receptors on hematopoietic cells (K_d = 70 pM). The mIL-10R cDNA encodes an open reading frame of 576 amino acids, which is detected on the cell surface as a 90–110 kDa protein (Ho *et al.*, 1993). The IL-10R gene is expressed as a 3.6 kb mRNA species. The IL-10R protein consists of a signal sequence of 16 amino acids, an extracellular domain of 222 amino acids, and a cytoplasmic domain of 314 amino acids. The intracellular domain contains four tyrosine residues. Mutational analysis of murine IL-10 receptor in Ba/F3 cells indicated that different regions of the cytoplasmic domain mediated proliferative and differentiation responses (Ho *et al.*, 1995).

9.   *Cytokine in Disease and Therapy*

IL-10 has potent anti-inflammatory activities *in vitro* and in animal models of acute and chronic inflammation as discussed. The role of IL-10 in inflammatory processes is being investigated in human diseases and evidence for an immunoregulatory role of IL-10 has been described in rheumatoid arthritis. Spontaneous IL-10 production was detected in synovial membrane biopsies and cultures. This endogenously produced IL-10 was active in suppressing TNF-α and IL-1β production; however, addition of exogenous IL-10 resulted in a further inhibition of these pro-inflammatory cytokines and an induction of IFN-γ, indicating that IL-10 treatment may be a possible therapeutic in this disease (Katsikis *et al.*, 1994).

The observation that IL-10 could inhibit allogeneic primary mixed leukocyte reactions provided a rationale for the study of IL-10 in the induction and maintenance of transplantation tolerance (Bejarano *et al.*, 1992). Host-reactive T cell clones, isolated from SCID patients, that were successfully reconstituted with allogeneic fetal hematopoietic stem cells, produced high levels of IL-10 following activation that suppressed the proliferation of these clones (Bacchetta *et al.*, 1994). In addition, freshly isolated PBMC of these patients secreted high levels of IL-10 *in vivo*, produced by host-derived monocytes and donor-derived T cells. IL-10 was active *in vivo* since HLA-DR expression could be upregulated on donor-derived monocytes after brief incubation (Bacchetta *et al.*, 1994). IL-10 mRNA expression has been detected in biopsies of patients who underwent heart or kidney transplantation (Merville *et al.*, 1993; Cunningham *et al.*, 1994). However, the significance of these findings in relation to acceptance or rejection of the graft still needs further investigation.

IL-10 has been shown to induce the production of IgA from anti-CD40-activated PBMC isolated from patients suffering from IgA deficiency. These patients were able to produce IL-10, although at reduced levels (Briere *et al.*, 1994a). IL-10 also induced immunoglobulin production in anti-CD40- or SAC-activated B cells from patients suffering from common variable immunodeficiency (CVI) (Nonoyama *et al.*, 1993a; Zielen *et al.*, 1993). However, a role for IL-10 in the etiology of this disease is not clear.

IL-10 also plays a role in viral immunity. EBV carries

its own version of IL-10 and induces hIL-10 production in the B cells it infects (Burdin *et al.*, 1993). This IL-10 may facilitate transformation and viral latency owing to its B cell stimulatory activities, although BCRF1-deleted viruses are still able to transform B cells (Miyazaki *et al.*, 1993; Swaminathan *et al.*, 1993; Stewart *et al.*, 1994). It may also affect cytotoxic T cell responses directed against the virus (Stewart and Rooney, 1992). IL-10 inhibited the replication of HIV in human monocyte-derived macrophages (Montaner *et al.*, 1994; Saville *et al.*, 1994; Schuitemaker *et al.*, 1994; Weissman *et al.*, 1994). Depending on the differentiation state of the monocytes, IL-10 inhibited the processing of viral proteins when administered before infection and it inhibited virus assembly when administered after infection. These effects may be related to the inhibitory effects of IL-10 on the production of TNF-α and IL-6 by infected monocytes which enhance HIV replication at a post-transcriptional level (Schuitemaker *et al.*, 1994; Weissman *et al.*, 1994). It has been reported that the viral protein gp120 is able to induce IL-10 expression by PBMC (Ameglio *et al.*, 1994). On the other hand, elevated levels of IL-10 production have been observed during end-stage disease and occurred concomitantly with diminished cellular responses against viral and mycobacterial antigens and lymphocytic chorio-meningitis virus (LCMV) suggesting that a switch from T_H1 to T_H2 responses may be a critical factor in the etiology of HIV infection (Clerici and Shearer, 1993; Chehimi *et al.*, 1994; Clerici *et al.*, 1994; Sieling and Modlin, 1994).

IL-10 plays a role in human infectious diseases such as mycobacterial infections (*M. lepra*, *M. tuberculosis*) and leishmaniasis. These diseases have in common that the clinical manifestations represent a spectrum of symptoms in which pathology is related to the level of cell-mediated immunity. The bacilli are cleared by a strong cellular response (tuberculoid form), whereas they survive when the patient develops a humoral response (lepromatous form). This phenomenon can be explained by the amount and type of cytokines that are produced. IL-10 mRNA was present in the lepromatous lesions and decreased in patients with reversal reactions (Yamamura *et al.*, 1991, 1992). IL-10 was also detected in the pleural fluid of patients suffering from tuberculosis pleuritis (Barnes *et al.*, 1993). The major cell producing IL-10 was the monocyte, which could be activated directly by mycobacterial cell wall products (Barnes *et al.*, 1992; Mutis *et al.*, 1993; Sieling *et al.*, 1993). Neutralization of endogenously produced IL-10 enhanced mycobacterium-specific proliferation of patient PBMC *in vitro* and increased IFN-γ, GM-CSF, IL-12, and TNF-α production (Sieling *et al.*, 1993; Sieling and Modlin, 1994). In addition, IL-10 inhibited the induction of CD1 by monocytes and the CD1-restricted proliferation of CD4⁻CD8⁻ mycobacterial specific T cell clones (Sieling and Modlin, 1994). A second infectious disease in man that is characterized by a dominant T_H2 response and a role of IL-10 involvement is leishmaniasis. It has been demonstrated that PBMC, lymph nodes, and spleen of patients suffering from visceral leishmaniasis and diffuse cutaneous leishmaniasis have elevated levels of IL-10 production, and activation of patients PBMC in the presence of anti-IL-10 mAbs resulted in antigen-specific proliferation and IFN-γ production (Dittmar *et al.*, 1993; Ghalib *et al.*, 1993; Karp *et al.*, 1993a; Carvalho *et al.*, 1994). As mentioned before, a third example of elevated IL-10 levels and a role for IL-10 in suppressing cellular immune responses to a parasitic infection leading to pathology and a disturbed T_H1/T_H2 balance is lymphatic filariasis (King *et al.*, 1993). Clearly, antagonists of IL-10 production or function could be of benefit to the treatment of these diseases.

A phase I clinical randomized, double-blind, placebo-controlled, parallel group study has been performed (Chernoff *et al.*, 1994). Healthy volunteers were administered 1, 10, or 25 µg/kg body weight IL-10 intravenously and were monitored for 96 hours. IL-10 was well tolerated and no adverse symptoms were reported or clinically relevant differences detected in blood chemistry, urinalysis, hemoglobin concentration, platelet count, coagulation parameters, complement components, immunoglobulin concentrations, neutrophil superoxide production, or electrocardiograms. Antibodies to IL-10 were not detected 21 days after administration. IL-10 was detectable in serum with peak concentrations of 14, 208, and 505 ng/ml, respectively, and the half-life of IL-10 was determined as ~3 h. IL-10 was active as determined by its effects on whole-blood cytokine production. IL-10 inhibited the production of IL-6, IL-1β, and TNF-α following *ex vivo* stimulation of whole blood with LPS. The conclusion of this study was that a single injection of rhIL-10 in healthy humans is safe (Chernoff *et al.*, 1994). Clinical trials which will address the immunosuppressive or anti-inflammatory activities of IL-10 are under way.

Much has been learned of the biology of IL-10 in just four years since its discovery. The most dramatic changes induced by IL-10 are its potent deactivating effects on monocytes/macrophages, including downregulation of MHC class II molecules, accessory molecules, and cytokine and chemokine production, which lead to inhibition of antigen presentation/accessory functions and result in abrogation of antigen-specific T cell responses and production of T cell cytokines. In addition, IL-10 has direct inhibitory effects on proliferation and production of IL-2 and TNF-α by T cells. Together with the observations that IL-10 is produced by T_H0, T_H1, and T_H2 subsets, and that its production by monocytes and T cells is relatively late after activation, these results indicate that IL-10 plays an important role as natural dampener of immune and inflammatory reactions. These characteristics of IL-10, in combination with the absence of toxicity observed in the

phase I clinical trial, indicate that IL-10 could be a potentially useful therapeutic in acute and chronic inflammatory reactions, including inflammatory bowel disease, rheumatoid arthritis, and sepsis, in autoimmune diseases, in prevention of graft-versus-host disease, and in the induction as well as maintenance of transplantation tolerance. The results of studies with animal models have already indicated that IL-10 may be effective in these applications. On the other hand, antagonists of IL-10 production or function may be useful in the reversal of strongly polarized T_H2 responses which lead to pathology as observed in infectious diseases and parasitic and viral infections. Finally, IL-10 may have direct beneficial effects in B cell disorders such as CVI and B-CLL. Studies on the production of IL-10 and its possible role in a number of disease states will continue to be investigated in the coming years as well as the more basic biology on the position of IL-10 in the cytokine network. In addition, elucidation of receptor structure and intracellular signal transduction pathways will provide important clues to the often synergistic or antagonistic effects of cytokines at the molecular level.

10. *References*

Abramowicz, D., Durez, P., Gerard, C., *et al.* (1993). Neonatal induction of transplantation tolerance in mice is associated with in vivo expression of IL-4 and -10 mRNAs. Transplant Proc. 25, 312–313.

Allen, R.D., Staley, T.A. and Sidman, C.L. (1993). Differential cytokine expression in acute and chronic murine graft versus host diseases. Eur. J. Immunol. 23, 333–337.

Ameglio, F., Capobianchi, M.R., Castilletti, C., *et al.* (1994). Recombinant gp120 induces IL-10 in resting peripheral blood mononuclear cells; correlation with the induction of other cytokines. Clin. Exp. Immunol. 95, 455–458.

Bacchetta, R., Bigler, M., Touraine, J.L., *et al.* (1994). High levels of interleukin 10 production in vivo are associated with tolerance in SCID patients transplanted with HLA mismatched hematopoietic stem cells. J. Exp. Med. 179, 493–502.

Barnes, P.F., Chatterjee, D., Abrams, J.S. and Modlin, R. (1992). Cytokine production induced by *Mycobacterium tuberculosis* lipoarabinomannan. Relationship to its chemical structure. J. Immunol. 149, 541–547.

Barnes, P.F., Lu, S., Abrams, J.S., Wang, E., Yamamura, M. and Modlin, R.L. (1993). Cytokine production at the site of disease of human tuberculosis. Infect. Immun. 61, 3482–3489.

Bazan, J.F. (1990). Structural design and molecular evolution of a cytokine receptor superfamily. Proc. Natl Acad. Sci. USA 87, 6934–6939.

Bean, A.G., Freiberg, R.A., Andrade, S., Menon, S. and Zlotnik, A. (1993). Interleukin 10 protects mice against staphylococcal enterotoxin B-induced lethal shock. Infect. Immun. 61, 4937–4939.

Becker, J.C., Czerny, C. and Brocker, E.-B. (1994). Maintenance of clonal anergy by endogenously produced IL-10. Int. Immunol. 6, 1605–1612.

Bejarano, M.T., de, W.M.R., Abrams, J.S., *et al.* (1992). Interleukin 10 inhibits allogeneic proliferative and cytotoxic T cell responses generated in primary mixed lymphocyte cultures. Int. Immunol. 4, 1389–1497.

Benjamin, D., Knobloch, T.J. and Dayton, M.A. (1992). Human B-cell interleukin-10: B-cell lines derived from patients with acquired immunodeficiency syndrome and Burkitt's lymphoma constitutively secrete large quantities of interleukin-10. Blood 80, 1289–1298.

Benjamin, D., Park, C.D. and Sharma, V. (1994). Human B cell interleukin 10. Leuk. Lymphoma. 12, 205–210.

Blay, J.Y., Burdin, N., Rousset, F., *et al.* (1993). Serum interleukin-10 in non-Hodgkin's lymphoma: a prognostic factor. Blood 82, 2169–2174.

Bogdan, C., Vodovotz, Y. and Nathan, C. (1991). Macrophage deactivation by interleukin 10. J. Exp. Med. 174, 1549–1555.

Briere, F., Bridon, J.-M., Chevet, D., *et al.* (1994a). Interleukin-10 induces B lymphocytes from IgA deficient patients to secrete IgA. J. Clin. Invest. 94, 97–104.

Briere, F., Servet, D.C., Bridon, J.M., Saint, R.J. and Banchereau, J. (1994b). Human interleukin 10 induces naive surface immunoglobulin D$^+$ (sIgD$^+$) B cells to secrete IgG1 and IgG3. J. Exp. Med. 179, 757–762.

Burdin, N., Peronne, C., Banchereau, J. and Rousset, F. (1993). Epstein–Barr virus transformation induces B lymphocytes to produce human interleukin 10. J. Exp. Med. 177, 295–304.

Carvalho, E.M., Bacellar, O., Brownell, C., Regis, T., Coffman, R.L. and Reed, S.G. (1994). Restoration of IFN-γ production and lymphocyte proliferation in visceral leishmaniasis. J. Immunol. 152, 5949–5956.

Cassatella, M.A., Meda, L., Bonora, S., Ceska, M. and Constantin, G. (1993). Interleukin 10 (IL-10) inhibits the release of proinflammatory cytokines from human polymorphonuclear leukocytes. Evidence for an autocrine role of tumor necrosis factor and IL-1 beta in mediating the production of IL-8 triggered by lipopolysaccharide. J. Exp. Med. 178, 2207–2211.

Cassatella, M.A., Meda, L., Gasperini, S., Calzetti, F. and Bonora, S. (1994). Interleukin 10 (IL-10) upregulates IL-1 receptor antagonist production from lipopolysaccharide-stimulated human polymorphonuclear leukocytes by delaying mRNA degradation. J. Exp. Med. 179, 1695–1699.

Caux, C., Massacrier, C. and Vandervliet, B. (1994). Interleukin-10 inhibits alloreaction induced by human dendritic cells. Int. Immunol. [In press]

Cenci, E., Romani, L., Mencacci, A., *et al.* (1993). Interleukin-4 and interleukin-10 inhibit nitric oxide-dependent macrophage killing of *Candida albicans*. Eur. J. Immunol. 23, 1034–1038.

Chehimi, J., Starr, S.E., Frank, I., *et al.* (1994). Impaired interleukin-12 production in human immunodeficiency virus-infected patients. J. Exp. Med. 179, 1361–1366.

Chen, Q., Daniel, V., Maher, D.W. and Hersey, P. (1994). Production of IL-10 by melanoma cells: examination of its role in immunosuppression mediated by melanoma. Int. J. Cancer. 56, 755–760.

Chen, W.-F. and Zlotnik, A. (1991). IL-10: a novel cytotoxic T cell differentiation factor. J. Immunol. 147, 528–534.

Chernoff, A.E., Granowitz, E.V., Shapiro, L., *et al.* (1994). A phase I study of interleukin-10 in healthy humans: safety and effects on cytokine production. In "Interleukin-10" (eds. J. E. de Vries and R. de Waal Malefyt), pp. 151–161. R.G. Landes, Austin TX.

Chin, J. and Kostura, M.J. (1993). Dissociation of IL-1 beta synthesis and secretion in human blood monocytes stimulated with bacterial cell wall products. J. Immunol. 151, 5574–5585.

Chomarat, P., Rissoan, M.C., Banchereau, J. and Miossec, P. (1993). Interferon gamma inhibits interleukin 10 production by monocytes. J. Exp. Med. 177, 523–527.

Clerici, M. and Shearer, G.M. (1993). A Th1 to Th2 switch is a critical step in the etiology of HIV infection. Immunol. Today 14, 107–111.

Clerici, M., Wynn, T.A., Berzofsky, J.A., *et al.* (1994). Role of interleukin-10 in T helper cell dysfunction in asymptomatic individuals infected with the human immunodeficiency virus. J. Clin. Invest. 93, 768–775.

Crawley, J.B., Williams, L.M., Mander, T., Brennan, F.M. and Foxwell, B.M.J. (1996). Interleukin-10 stimulation of phosphatidylinositol 3-kinase and p70 S6 kinase is required for the proliferative but not the antiinflammatory effects of the cytokine. J. Biol. Chem. 271, 16357–16362.

Cunha, F.Q., Moncada, S. and Liew, F.Y. (1992). Interleukin-10 (IL-10) inhibits the induction of nitric oxide synthase by interferon-gamma in murine macrophages. Biochem. Biophys. Res. Commun. 182, 1155–1159.

Cunningham, D.A., Dunn, M.J., Yacoub, M.H. and Rose, M.L. (1994). Local production of cytokines in the human cardiac allograft. Transplantation 57, 1333–1337.

D'Andrea, A., Aste, A.M., Valiante, N.M., Ma, X., Kubin, M. and Trinchieri, G. (1993). Interleukin 10 (IL-10) inhibits human lymphocyte interferon gamma-production by suppressing natural killer cell stimulatory factor/IL-12 synthesis in accessory cells. J. Exp. Med. 178, 1041–1048.

Defrance, T., Vanbervliet, B., Briere, F., Durand, I., Rousset, F. and Banchereau, J. (1992). Interleukin 10 and transforming growth factor beta cooperate to induce anti-CD40-activated naive human B cells to secrete immunoglobulin A. J. Exp. Med. 175, 671–682.

Del Prete, G., De Carli, M., Almerigogna, F., Giudizi, M.G., Biagiotti, R. and Romagnani, S. (1993). Human IL-10 is produced by both type 1 helper (Th1) and type 2 helper (Th2) T cell clones and inhibits their antigen-specific proliferation and cytokine production. J. Immunol. 150, 353–360.

de Vries, J.E. and de Waal Malefyt, R. (eds.) "Interleukin-10" R.G. Landes, Austin TX.

de Waal Malefijt, R., Abrams, J., Bennett, B., Figdor, C.G. and de Vries, J.E. (1991a). Interleukin 10 (IL-10) inhibits cytokine synthesis by human monocytes: an autoregulatory role of IL-10 produced by monocytes. J. Exp. Med. 174, 1209–1220.

de Waal Malefyt, R., Haanen, J., Spits, H., *et al.* (1991b). Interleukin 10 (IL-10) and viral IL-10 strongly reduce antigen-specific human T cell proliferation by diminishing the antigen-presenting capacity of monocytes via downregulation of class II major histocompatibility complex expression. J. Exp. Med. 174, 915–924.

de Waal Malefyt, R., Yssel, H., Roncarolo, M.G., Spits, H. and de Vries, J.E. (1992). Interleukin-10. Curr. Opin. Immunol. 4, 314-320.

de Waal Malefyt, R., Figdor, C.G., Huijbens, R., *et al.* (1993a). Effects of IL-13 on phenotype, cytokine production, and cytotoxic function of human monocytes. Comparison with IL-4 and modulation by IFN-gamma or IL-10. J. Immunol. 151, 6370–6381.

de Waal Malefyt, R., Yssel, H. and E., de Vries, J.E. (1993b). Direct effects of IL-10 on subsets of human CD4+ T cell clones and resting T cells. Specific inhibition of IL-2 production and proliferation. J. Immunol. 150, 4754–4765.

de Waal Malefyt, R., Figdor, C.G. and de Vries, J.E. (1994). Regulation of human monocyte functions by interleukin-10. In "Interleukin-10" (eds. J.E. de Vries and R. de Waal Malefyt), pp. 39–54. R.G. Landes, Austin TX.

De Wit, D., Van Mechelen, M., Zanin, C., et al. (1993). Preferential activation of Th2 cells in chronic graft-versus-host reaction. J. Immunol. 150, 361–366.

Dittmar, G., Tapia, F.G. and Sanchez, M.A. (1993). Determination of cytokine profile in American cutaneous leishmaniasis using PCR. Clin. Exp. Immunol. 91, 500–505.

Emilie, D., Galanaud, P., Raphael, M. and Joab, I. (1993). Interleukin-10 and acquired immunodeficiency syndrome lymphomas. Blood 81, 1106–1107.

Enk, A.H. and Katz, S.I. (1992). Identification and induction of keratinocyte-derived IL-10. J. Immunol. 149, 92–95.

Enk, A.H., Saloga, J., Becker, D., Bipmadzadeh, M. and Knop, J. (1994). Induction of hapten-specific tolerance by interleukin 10 in vivo. J. Exp. Med. 179, 1397–1402.

Finbloom, D.S. and Winestock, K.D. (1995). IL-10 induces the tyrosine phosphorylation of tyk2 and jak1 and the differential assembly of STAT1α and STAT3 complexes in human T cells and monocytes. J. Immunol. 155, 1079–1090.

Fiorentino, D.F., Bond, M.W. and Mosmann, T.R. (1989). Two types of mouse T helper cells IV. Th2 clones secrete a factor that inhibits cytokine production by Th1 clones. J. Exp. Med. 170, 2081–2095.

Fiorentino, D.F., Zlotnik, A., Viera, P., et al. (1991a). IL-10 acts on the antigen-presenting cell to inhibit cytokine production by Th1 cells. J. Immunol. 146, 3444–3451.

Fiorentino, D.F., Zlotnik, A., Mosmann, T.R., Howard, M. and O'Garra, A. (1991b). IL-10 inhibits cytokine production by activated macrophages. J. Immunol. 147, 3815–3820.

Flesch, I.E.A., Hess, J.H., Oswald, I.P. and Kaufmann, S.H.E. (1994). Growth inhibition of Mycobacterium bovis by IFN-γ stimulated macrophages: regulation by endogenous tumor necrosis factor-α and by IL-10. Int. Immunol. 6, 693–700.

Flores, V.P., Chikunguwo, S.M., Harris, T.S. and Stadecker, M.J. (1993). Role of IL-10 on antigen-presenting cell function for schistosomal egg-specific monoclonal T helper cell responses in vitro and in vivo. J. Immunol. 151, 3192–3198.

Fluckiger, A.C., Garrone, P., Durand, I., Galizzi, J.P. and Bancherau, J. (1993). Interleukin 10 (IL-10) upregulates functional high affinity IL-2 receptors on normal and leukemic B lymphocytes. J. Exp. Med. 178, 1473–1481.

Fluckiger, A.C., Durand, I. and Banchereau, J. (1994). Interleukin 10 induces apoptotic cell death of B-chronic lymphocytic leukemia cells. J. Exp. Med. 179, 91–99.

Fowler, D.H., Kurasawa, K., Husebekk, A., Cohen, P.A. and Gress, R.E. (1994). Cells of Th2 cytokine phenotype prevent LPS-induced lethality during murine graft-versus-host reaction: regulation of cytokines and CD8+ lymphoid engraftment. J. Immunol. 152, 1004–1013.

Frei, K., Nadal, D., Pfister, H.W. and Fontana, A. (1993). Listeria meningitis: identification of a cerebrospinal fluid inhibitor of macrophage listericidal function as interleukin 10. J. Exp. Med. 178, 1255–1261.

Gajewski, T.F. and Fitch, F.W. (1988). Anti-proliferative effect of IFN-gamma in immune regulation. I. IFN-gamma inhibits the proliferation of Th2 but not of Th1 murine helper T lymphocyte clones. J. Immunol. 140, 4245–4252.

Garlisi, C.G., Pennline, K.J. and Smith, S.R. (1993). Cytokine gene expression in mice undergoing chronic graft versus host disease. Mol. Immunol. 30, 669–677.

Gastl, G.A., Abrams, J.S., Nanus, D.M., et al. (1993). Interleukin-10 production by human carcinoma cell lines and its relationship to interleukin-6 expression. Int. J. Cancer. 55, 96–101.

Gazzinelli, R.T., Oswald, I.P., Hieny, S., James, S.L. and Sher, A. (1992a). The microbicidal activity of interferon-gamma-treated macrophages against Trypanosoma cruzi involves an L-arginine-dependent, nitrogen oxide-mediated mechanism inhibitable by interleukin-10 and transforming growth factor-beta. Eur. J. Immunol. 22, 2501–2506.

Gazzinelli, R.T., Oswald, I.P., James, S.L. and Sher, A. (1992b). IL-10 inhibits parasite killing and nitrogen oxide production by IFN-gamma-activated macrophages. J. Immunol. 148, 1792–1796.

Gerard, C., Bruyns, C., Marchant, A., et al. (1993). Interleukin 10 reduces the release of tumor necrosis factor and prevents lethality in experimental endotoxemia. J. Exp. Med. 177, 547–550.

Ghalib, H.W., Piuvezam, M.R., Skeiky, Y.A., et al. (1993). Interleukin 10 production correlates with pathology in human Leishmania donovani infections. J. Clin. Invest. 92, 324–329.

Ghildyal, N., McNeil, H.P., Gurish, M.F., Austen, K.F. and Stevens, R.L. (1992a). Transcriptional regulation of the mucosal mast cell-specific protease gene, MMCP-2, by interleukin 10 and interleukin 3. J. Biol. Chem. 267, 8473–8477.

Ghildyal, N., McNeil, H.P., Stechschulte, S., et al. (1992b). IL-10 induces transcription of the gene for mouse mast cell protease-1, a serine protease preferentially expressed in mucosal mast cells of Trichinella spiralis-infected mice. J. Immunol. 149, 2123–2129.

Ghildyal, N., Friend, D.S., Nicodemus, C.F., Austen, K.F. and Stevens, R.L. (1993). Reversible expression of mouse mast cell protease 2 mRNA and protein in cultured mast cells exposed to IL-10. J. Immunol. 151, 3206–3214.

Go, N.F., Castle, B.E., Barrett, R., et al. (1990). Interleukin-10 (IL-10), a novel B cell stimulating factor: unresponsiveness of X-chromosome-linked immunodeficiency B cells. J. Exp. Med. 172, 1625–1631.

Gorczynski, R.M. and Wojcik, D. (1994). A role for nonspecific (cyclosporin A) or specific (monoclonal antibodies to ICAM-1, LFA-1, and IL-10) immunomodulation in the prolongation of skin allografts after antigen-specific pretransplant immunization or transfusion. J. Immunol. 152, 2011–2019.

Gotlieb, W.H., Abrams, J.S., Watson, J.M., Velu, T.J., Berek, J.S. and Martinez-Maza, O. (1992). Presence of interleukin-10 (IL-10) in the ascites of patients with ovarian and other intra-abdominal cancers. Cytokine 4, 385–390.

Gruber, M.F., Williams, C.C. and Gerrard, T.L. (1994). Macrophage-colony-stimulating factor expression by anti-CD45 stimulated human monocytes is transcriptionally up-regulated by IL-1 beta and inhibited by IL-4 and IL-10. J. Immunol. 152, 1354–1361.

Hancock, W., Mottram, P.L., Purcell, L.J., Han, W.R., Pietersz, G.A. and McKenzie, I.F. (1993). Prolonged survival of mouse cardiac allografts after CD4 or CD8 monoclonal antibody therapy is associated with selective intragraft cytokine protein expression: interleukin (IL)-4 and IL-10 but not IL-2 or interferon-gamma. Transplant Proc. 25, 2937–2938.

Ho, A.S., Liu, Y., Khan, T.A., Hsu, D., Bazan, F.J. and Moore, K.W. (1993). A receptor for interleukin-10 is related to interferon receptors. Proc. Natl Acad. Sci. USA 90, 11267–11271.

Ho, A.S.-Y. and Moore, K.W. (1994). Interleukin-10 and its receptor. Ther. Immunol. 1, 173–185.

Ho, A.S., Wei, S.H., Mui, A.L, Mijayima, A. and Moore, K.W. (1995). Functional regions of the mouse interleukin-10 receptor cytoplasmic domain. Mol. Cell. Biol. 15, 5043–5053.

Howard, M. and O'Garra, A. (1992). Biological properties of interleukin 10. Immunol. Today 13, 198–200.

Howard, M., Muchamuel, T., Andrade, S. and Menon, S. (1993). Interleukin 10 protects mice from lethal endotoxemia. J. Exp. Med. 177, 1205–1208.

Hsieh, C.S., Heimberger, A.B., Gold, J.S., O'Garra, A. and Murphy, K.M. (1992). Differential regulation of T helper phenotype development by interleukins 4 and 10 in an alpha beta T-cell-receptor transgenic system. Proc. Natl Acad. Sci. USA 89, 6065–6069.

Hsu, D.H., Moore, K.W. and Spits, H. (1992). Differential effects of IL-4 and IL-10 on IL-2-induced IFN-gamma synthesis and lymphokine-activated killer activity. Int. Immunol. 4, 563–569.

Hsu, D.W., de Waal Malefyt, R., Fiorentino, D.F., et al. (1990). Expression of IL-10 activity by Epstein–Barr virus protein BCRFI. Science 250, 830–832.

Ishida, H., Hastings, R., Thompson, S.L. and Howard, M. (1993). Modified immunological status of anti-IL-10 treated mice. Cell. Immunol. 148, 371–384.

Ishida, H., Muchamuel, T., Sakaguchi, S., Andrade, S., Menon, S. and Howard, M. (1994). Continuous administration of anti-interleukin 10 antibodies delays onset of autoimmunity in NZB/W F1 mice. J. Exp. Med. 179, 305–310.

Jenkins, J.K., Malyak, M. and Arend, W.P. (1994). The effects of interleukin-10 on interleukin-1 receptor antagonist and interleukin-1-beta production in human monocytes and neutrophils. Lymphokine Cytokine Res. 13, 84–89.

Jinquan, T., Larsen, C.G., Gesser, B., Matsushima, K. and Thestrup, P.K. (1993). Human IL-10 is a chemoattractant for CD8+ T lymphocytes and an inhibitor of IL-8-induced CD4+ T lymphocyte migration. J. Immunol. 151, 4545–4551.

Kang, K., Hammerberg, C., Meunier, L. and Cooper, K.D. (1994). CD11b+ macrophages that infiltrate human epidermis after in vivo ultraviolet exposure potently produce IL-10 and represent the major secretory source of epidermal IL-10 protein. J. Immunol. 153, 5256–5265.

Karp, C.L., el, S.S., Wynn, T.A., et al. (1993a). In vivo cytokine profiles in patients with kala-azar. Marked elevation of both interleukin-10 and interferon-gamma. J. Clin. Invest. 91, 1644–1648.

Karp, C.L., el-Safi, S.H., Wynn, T.A., et al. (1993b). In vivo cytokine profiles in patients with kala azar. Marked elevation of both interleukin-10 and interferon-gamma. J. Clin. Invest. 91, 1644–1648.

Kasama, T., Strieter, R.M., Lukacs, N.W., Burdick, M.D. and Kunkel, S.L. (1994). Regulation of neutrophil-derived chemokine expression by IL-10. J. Immunol. 152, 3559–3569.

Katsikis, P.D., Chu, C.Q., Brennan, F.M., Maini, R.N. and Feldmann, M. (1994). Immunoregulatory role of interleukin 10 in rheumatoid arthritis. J. Exp. Med. 179, 1517–1527.

Kennedy, M.K., Torrance, D.S., Picha, K.S. and Mohler, K.M. (1992). Analysis of cytokine mRNA expression in the central nervous system of mice with experimental autoimmune encephalomyelitis reveals that IL-10 mRNA expression correlates with recovery. J. Immunol. 149, 2496–2505.

Kim, J.M., Brannan, C.I., Copeland, N.G., Jenkins, N.A., Khan, T.A. and Moore, K.W. (1992). Structure of the mouse IL-10 gene and chromosomal localization of the mouse and human genes. J. Immunol. 148, 3618–3623.

King, C.L., Mahanty, S., Kumaraswami, V., et al. (1993). Cytokine control of parasite-specific anergy in human lymphatic filariasis. Preferential induction of a regulatory T helper type 2 lymphocyte subset. J. Clin. Invest. 92, 1667–1673.

Kotenko, S.V., Krause, C.D., Izotova, L.S., Pollack, B.P., Wu, W. and Pestka, S. (1997). Identification and functional characterization of a second chain of the interleukin-10 receptor complex. J. 16, 5894–5903.

Kubin, M., Kamoun, M. and Trinchieri, G. (1994). Interleukin-12 synergizes with B7/CD28 interaction in inducing efficient proliferation and cytokine production of human T cells. J. Exp. Med. 180, 212–222.

Kuhn, R., Lohler, J., Rennick, D., Rajewsky, K. and Muller, W. (1993). Interleukin-10-deficient mice develop chronic enterocolitis. Cell 75, 263–274.

Larner, A.C., David, M., Feldman, G.M., et al. (1993). Tyrosine phosphorylation of DNA binding proteins by multiple cytokines. Science 261, 1730–1733.

Lee, M.S., Wogensen, L., Shizuru, J., Oldstone, M.B. and Sarvetnick, N. (1994). Pancreatic islet production of murine interleukin-10 does not inhibit immune-mediated tissue destruction. J. Clin. Invest. 93, 1332–1338.

Lehmann, J., Seegert, D., Strehlow, I., Schindler, C., Lohmann-matthes, M.-L. and Decker, T. (1994). IL-10 induced factors belonging to the p91 family of proteins bind to IFN-responsive promoter elements. J. Immunol. 153, 167–172.

Levy, Y. and Brouet, J.C. (1994). Interleukin-10 prevents spontaneous death of germinal center B cells by induction of the bcl-2 protein. J. Clin. Invest. 93, 424–428.

Liu, Y., Wei, S.H., Ho, A.S., de, W.M.R. and Moore, K.W. (1994). Expression cloning and characterization of a human IL-10 receptor. J. Immunol. 152, 1821–1829.

MacNeil, I.A., Suda, T., Moore, K.W., Mosmann, T.R. and Zlotnik, A. (1990). IL-10: a novel growth cofactor for mature and immature T cells. J. Immunol. 145, 4167–4173.

Marchant, A., Bruyns, C., Vandenabeele, P., et al. (1994a). Interleukin-10 controls interferon-γ and tumor necrosis factor production during experimental endotoxemia. Eur. J. Immunol. 24, 1167–1171.

Marchant, A., Deviere, J., Byl, B., De Groote, D., Vincent, J.L. and Goldman, M. (1994b). Interleukin-10 production during septicaemia. Lancet 343, 707–708.

Merville, P., Rousset, F., Banchereau, J., Klein, B. and Betaille, R. (1992). Serum interleukin-10 in early stage multiple myeloma. Lancet 340, 1544–1545.

Merville, P., Pouteil, N.C., Wijdenes, J., Potaux, L., Touraine, J.L. and Banchereau, J. (1993). Detection of single cells secreting IFN-gamma, IL-6, and IL-10 in irreversibly rejected human kidney allografts, and their modulation by IL-2 and IL-4. Transplantation 55, 639–646.

Miyajima, A.M., Hara, T. and Kitamura, T. (1992). Common subunits of cytokine receptors and the functional redundancy of cytokines. Trends Biochem. Sci. 17, 378–385.

Miyazaki, I., Cheung, R.K. and Dosch, H.M. (1993). Viral interleukin 10 is critical for the induction of B cell growth transformation by Epstein–Barr virus. J. Exp. Med. 178, 439–447.

Montaner, L.J., Griffin, P. and Gordon, S. (1994). Interleukin-10 (IL-10) inhibits initial reverse transcription of HIV-1 and mediates a virostatic latent state in primary blood-derived human macrophages in vitro. J. Gen. Virol. [In press]

Moore, K.W., Vieira, P., Fiorentino, D.F., Trounstine, M.L., Khan, T.A. and Mosmann, T.R. (1990). Homology of cytokine synthesis inhibitory factor (IL-10) to the Epstein Barr Virus gene BCRFI. Science 248, 1230–1234.

Moore, K.W., Rousset, F. and Banchereau, J. (1991). Evolving principles in immunopathology: interleukin 10 and its relationship to Epstein–Barr virus protein BCRF1. Springer Semin. Immunopathol. 13, 157–166.

Moore, K.W., O'Garra, A., de Waal Malefyt, R., Vieira, P. and Mosmann, T.R. (1993). Interleukin-10. Annu. Rev. Immunol. 11, 165–190.

Mosmann, T.R. (1991). Cytokine secretion patterns and cross-regulation of T cell subsets. Immunol. Res. 10, 183–188.

Mosmann, T.R. and Coffman, R.L. (1989). Th1 and Th2 cells: different patterns of lymphokine secretion lead to different functional properties. Annu. Rev. Immunol. 7, 145–173.

Mosmann, T.R. and Moore, K.W. (1991). The role of IL-10 in cross-regulation of Th1 and Th2 responses. Immunol. Today 12, A49–A53.

Mosmann, T.R., Cherwinski, H., Bond, M., Giedlin, M. and Coffman, R.L. (1986). Two types of mouse helper T cell clone. I. Definition according to profile of lymphokine activities and secreted proteins. J. Immunol. 136, 2348–2355.

Mulligan, M.S., Jones, M.L., Vaporciyan, A.A., Howard, M.C. and Ward, P.A. (1993). Protective effects of IL-4 and IL-10 against immune complex-induced lung injury. J. Immunol. 151, 5666–5674.

Murphy, E., Hieny, S., Sher, A. and O'Garra, A. (1993). Detection of in vivo expression of interleukin-10 using a semi-quantitative polymerase chain reaction method in Schistosoma mansoni infected mice. J. Immunol. Methods. 162, 211-223.

Murphy, E.E., Terres, G., Macatonia, S.E., et al. (1994). B7 and interleukin-12 cooperate for proliferation and IFN-γ production by mouse Th1 clones that are unresponsive to B7 costimulation. J. Exp. Med. 180, 223–231.

Mutis, T., Kraakman, E.M., Cornelisse, Y.E., et al. (1993). Analysis of cytokine production by Mycobacterium-reactive T cells. Failure to explain Mycobacterium leprae-specific nonresponsiveness of peripheral blood T cells from lepromatous leprosy patients. J. Immunol. 150, 4641–4651.

Nabioullin, R., Sone, S., Mizuno, K., et al. (1994). Interleukin-10 is a potent inhibitor of tumor cytotoxicity by human monocytes and alveolar macrophages. J. Leukocyte Biol. 55, 437–442.

Nonoyama, S., Farrington, M., Ishida, H., Howard, M. and Ochs, H.D. (1993a). Activated B cells from patients with common variable immunodeficiency proliferate and synthesize immunoglobulin. J. Clin. Invest. 92, 1282–1287.

Nonoyama, S., Hollenbaugh, D., Aruffo, A., Ledbetter, J.A. and Ochs, H.D. (1993b). B cell activation via CD40 is required for specific antibody production by antigen-stimulated human B cells. J. Exp. Med. 178, 1097–1102.

O'Garra, A., Chang, R., Go, N., Hastings, R., Haughton, G. and Howard, M. (1992). Ly-1 B (B1) cells are the main source of B-cell derived IL-10. Eur. J. Immunol. 22, 711–718.

Oswald, I.P., Gazzinelli, R.T., Sher, A. and James, S.L. (1992a). IL-10 synergizes with IL-4 and transforming growth factor-beta to inhibit macrophage cytotoxic activity. J. Immunol. 148, 3578–3582.

Oswald, I.P., Wynn, T.A., Sher, A. and James, S.L. (1992b). Interleukin 10 inhibits macrophage microbicidal activity by blocking the endogenous production of tumor necrosis factor alpha required as a costimulatory factor for interferon gamma-induced activation. Proc. Natl Acad. Sci. USA 89, 8676–8680.

Parish, C.R. (1972). The relationship between humoral and cell-mediated immunity. Transplant. Rev. 13, 35–66.

Pearce, E.J., Caspar, P., Grzych, J.M. and Sher, A. (1991). Downregulation of Th1 cytokine production accompanies induction of Th2 responses by a parasitic helminth, Scistosoma mansoni. J. Exp. Med. 173, 2713–2716.

Pecanha, L.M., Snapper, C.M., Lees, A. and Mond, J.J. (1992). Lymphokine control of type 2 antigen response. IL-10 inhibits IL-5- but not IL-2-induced Ig secretion by T cell-independent antigens. J. Immunol. 148, 3427–3432.

Pecanha, L.M., Snapper, C.M., Lees, A., Yamaguchi, H. and Mond, J.J. (1993). IL-10 inhibits T cell-independent but not T cell-dependent responses in vitro. J. Immunol. 150, 3215–3223.

Peguet, N.J., Moulon, C., Caux, C., Dalbiez, G.C., Banchereau, J. and Schmitt, D. (1994). Interleukin-10 inhibits the primary allogeneic T cell response to human epidermal Langerhans cells. Eur. J. Immunol. 24, 884–891.

Pennline, K.J., Roque-Gaffney, E. and Monahan, M. (1994). Recombinant human IL-10 prevents the onset of diabetes in the non-obese diabetic mouse. Clin. Immunol. Immunopathol. 71, 169–175.

Peyron, F., Burdin, N., Ringwald, P., Vuillez, J.P., Rousset, F. and Banchereau, J. (1994). High levels of circulating IL-10 in human malaria. Clin. Exp. Immunol. 95, 300–303.

Powrie, F. and Coffman, R.L. (1993). Inhibition of cell-mediated immunity by IL4 and IL10. Res. Immunol. 144, 639–643.

Powrie, F., Menon, S. and Coffman, R.L. (1993a). Interleukin-4 and interleukin-10 synergize to inhibit cell-mediated immunity in vivo. Eur. J. Immunol. 23, 3043–3049.

Powrie, F., Leach, M.W., Mauze, S. and Coffman, R.L. (1993b). Phenotypically distinct subsets of CD4$^+$ T cells induce or protect from chronic intestinal inflammation in C. B-17 scid mice. Int. Immunol. 5, 1461–1471.

Powrie, F., Correa, O.R., Mauze, S. and Coffman, R.L. (1994a). Regulatory interactions between CD45RBhigh and CD45RBlow CD4$^+$ T cells are important for the balance between protective and pathogenic cell-mediated immunity. J. Exp. Med. 179, 589–600.

Powrie, F., Leach, M.W., Mauze, S., Menon, S., Barcomb Caddle, L. and Coffman, R.L. (1994b). Inhibition of Th1 responses prevents inflammatory bowel disease in scid mice reconstituted with CD45RB^{hi} CD4$^+$ T cells. Immunity 1, 553–562.

Pugin, J., Ulevitch, R.J. and Tobias, P.S. (1993). A critical role for monocytes and CD14 in endotoxin-induced endothelial cell activation. J. Exp. Med. 178, 2193–2200.

Punnonen, J., de Waal Malefyt, R., van Vlasselaer, P., Gauchat, J.F. and de Vries, J. (1993). IL-10 and viral IL-10 prevent IL-4-induced IgE synthesis by inhibiting the accessory cell function of monocytes. J. Immunol. 151, 1280–1289.

Ralph, P., Nakoinz, I., Sampson, J.A., et al. (1992). IL-10, T lymphocyte inhibitor of human blood cell production of IL-1 and tumor necrosis factor. J. Immunol. 148, 808–814.

Reed, S.G. and Sher, A. (1994). Regulatory function of IL-10 in experimental parasitic infections. In "Interleukin-10" (eds. J.E. de Vries and R. de Waal Malefyt), pp. 71–79. R.G. Landes, Austin TX.

Rennick, D., Berg, D., Kuhn, R. and Muller, W. (1994a). Interleukin-10 deficient mice. In "Interleukin-10" (eds. J.E. de Vries and R. de Waal Malefyt), pp. 143–150. R.G. Landes, Austin TX.

Rennick, D., Hunte, B., Dang, W., Thompson, S.L. and Hudak, S. (1994b). Interleukin-10 promotes the growth of megakaryocyte, mast cell, and multilineage colonies: analysis with committed progenitors and Thy1^{lo}Sca1$^+$ stem cells. Exp. Hematol. 22, 136–141.

Rivas, J.M. and Ullrich, S.E. (1992). Systemic suppression of delayed-type hypersensitivity by supernatants from UV-irradiated keratinocytes. An essential role for keratinocyte-derived IL-10. J. Immunol. 149, 3865–3871.

Rode, H.J., Janssen, W., Rosen, W.A., et al. (1993). The genome of equine herpesvirus type 2 harbors an interleukin 10 (IL10)-like gene. Virus Genes 7, 111–116.

Romani, L., Puccetti, P., Mencacci, A., et al. (1994). Neutralization of IL-10 up-regulates nitric oxide production and protects susceptible mice from challenge with Candida albicans. J. Immunol. 152, 3514–3521.

Roncarolo, M.-G. (1994). Interleukin-10 and transplantation tolerance. In "Interleukin-10" (eds. J.E. de Vries and R. de Waal Malefyt), pp. 115–122. R.G. Landes, Austin TX.

Rott, O., Fleischer, B. and Cash, E. (1994). Interleukin-10 prevents experimental allergic encephalomyelitis in rats. Eur. J. Immunol. 24, 1434–1440.

Rousset, F., Garcia, E., Defrance, T., et al. (1992). Interleukin 10 is a potent growth and differentiation factor for activated human B lymphocytes. Proc. Natl Acad. Sci. USA 89, 1890–1893.

Saeland, S., Duvert, V., Moreau, I. and Banchereau, J. (1993). Human B cell precursors proliferate and express CD23 after CD40 ligation. J. Exp. Med. 178, 113–120.

Saville, M.W., Taga, K., Foli, A., Broder, S., Tosato, G. and Yarchoan, R. (1994). Interleukin-10 suppresses human immunodeficiency virus-1 replication in vitro in cells of the monocyte/macrophage lineage. Blood 83, 3591–3599.

Schandene, L., Alonso, V.C., Willems, F., et al. (1994). B7/CD28-dependent IL-5 production by human resting T cells is inhibited by IL-10. J. Immunol. 152, 4368–4374.

Schuitemaker, H., Kootstra, N.A. and Miedema, F. (1994). IL-10 and HIV replication. In "Interleukin-10" (eds. J.E. de Vries and R. de Waal Malefyt), pp. 103–114. R.G. Landes, Austin TX.

Shanafelt, A.B., Miyajima, A., Kitamura, T. and Kastelein, R.A. (1991). The amino-terminal helix of GM-CSF and IL-5 governs high affinity binding to their receptors. EMBO J. 10, 4105–4112.

Sher, A., Fiorentino, D., Caspar, P., Pearce, E. and Mosmann, T.

(1991). Production of IL-10 by CD4$^+$ T lymphocytes correlates with down-regulation of Th1 cytokine synthesis in helminth infection. J. Immunol. 147, 2713–2716.

Sieling, P.A., Abrams, J.S., Yamamura, M., et al. (1993). Immunosuppressive roles for IL-10 and IL-4 in human infection. In vitro modulation of T cell responses in leprosy. J. Immunol. 150, 5501–5510.

Sieling, P.A. and Modlin, R.L. (1994). IL-10 in mycobacterial infection. In "Interleukin-10" (eds. J.E. de Vries and R. de Waal Malefyt), pp. 181–192. R.G. Landes, Austin TX.

Silva, J.S., Morrissey, P.J., Grabstein, K.H., Mohler, K.M., Anderson, D. and Reed, S.G. (1992). Interleukin 10 and interferon gamma regulation of experimental Trypanosoma cruzi infection. J. Exp. Med. 175, 169–174.

Sironi, M., Munoz, C., Pollicino, T., et al. (1993). Divergent effects of interleukin-10 on cytokine production by mononuclear phagocytes and endothelial cells. Eur. J. Immunol. 23, 2692–2695.

Stewart, J.P. and Rooney, C.M. (1992). The interleukin-10 homolog encoded by Epstein–Barr virus enhances the reactivation of virus-specific cytotoxic T cell and HLA-unrestricted killer cell responses. Virology 191, 773–782.

Stewart, J.P., Behm, F.G., Arrand, J.R. and Rooney, C.M. (1994). Differential expression of viral and human interleukin-10 (IL-10) by primary B-cell tumors and B-cell lines. Virology 200, 724–732.

Suda, T., Mac, N.I., Fischer, M., Moore, K.W. and Zlotnik, A. (1991). Identification of a novel thymocyte growth factor derived from B cell lymphomas. Adv. Exp. Med. Biol. 292, 115–120.

Swaminathan, S., Hesselton, R. and Sullivan, J. (1993). Epstein–Barr virus recombinants with specifically mutated BCRF1 genes. J. Virol. 67, 7406–7413.

Taga, K., Cherney, B. and Tosato, G. (1993). IL-10 inhibits apoptotic cell death in human T cells starved of IL-2. Int. Immunol. 5, 1599–1608.

Taga, K., Chretien, J., Cherney, B., Diaz, L., Brown, M. and Tosato, G. (1994). Interleukin-10 inhibits apoptotic cell death in infectious mononucleosis T cells. J. Clin. Invest. 94, 251–260.

Takanaski, S., Nonaka, R., Xing, Z., O'Byrne, P., Dolovich, J. and Jordana, M. (1994). Interleukin-10 inhibits LPS induced survival and cytokine production by human peripheral blood eosinophils. J. Exp. Med. 180, 711–715.

Takeuchi, T., Lowry, R.P. and Konieczny, B. (1992). Heart allografts in murine systems. The differential activation of Th2-like effector cells in peripheral tolerance. Transplantation 53, 1281–1294.

Tan, J.C., Indelicato, S.R., Narula, S.K., Zavodny, P.J. and Chou, C.C. (1993). Characterization of interleukin-10 receptors on human and mouse cells. J. Biol. Chem. 268, 21053–21059.

Tan, J.C., Braun, S., Rong, H., et al. (1995). Characterization of recombinant extracellular domain of human interleukin-10 receptor. J. Biol. Chem. 270, 12906–12911.

Teunissen, M.B.M., Koomen, C.W. and de Waal Malefyt, R. (1994). Inability of human keratinocytes to synthesize interleukin-10. J. Invest. Dermatol. 102, 632.

te Velde, A.A., de Waal Malefijt, R., Huijbens, R.J., de Vries, J.E. and Figdor, C.G. (1992). IL-10 stimulates monocyte Fc gamma R surface expression and cytotoxic activity. Distinct regulation of antibody-dependent cellular cytotoxicity by IFN-gamma, IL-4, and IL-10. J. Immunol. 149, 4048–4052.

Thompson-Snipes, L., Dhar, V., Bond, M.W., Mosmann, T.R., Moore, K.W. and Rennick, D. (1991). Interleukin-10: a novel stimulatory factor for mast cells and their progenitors. J. Exp. Med. 173, 507–510.

Trotta, P.P. and Windsor, W.T. (1994). Physicochemical and structural properties of Interleukin-10. In "Interleukin-10" (eds. J.E. de Vries and R. de Waal Malefyt), pp. 11–18. R.G. Landes, Austin TX.

van der Poll, T., Jansen, J., Levi, M., ten Cate, H., ten Cate, J.W. and van Deventer, S.J.H. (1994). Regulation of interleukin-10 release by tumor necrosis factor in humans and chimpanzees. J. Exp. Med. 180, 1985–1988.

van der Veen, R.C. and Stohlman, S.A. (1993). Encephalitogenic Th1 cells are inhibited by Th2 cells with related peptide specificity: relative roles of interleukin (IL)-4 and IL-10. J. Neuroimmunol. 48, 213–220.

Van Vlasselaer, P., Borremans, B., Van Den Heuvel, R., Van Gorp, U. and de Waal Malefyt, R. (1993). Interleukin-10 inhibits the osteogenic activity of mouse bone marrow. Blood 82, 2361–2370.

Van Vlasselaer, P., Borremans, B., van Gorp, U., Dasch, J.R. and de Waal Malefyt, R. (1994). Interleukin 10 inhibits transforming growth factor-beta (TGF-beta) synthesis required for osteogenic commitment of mouse bone marrow cells. J. Cell. Biol. 124, 569–577.

Vella, A.T. and Pearce, E.J. (1992). CD4$^+$ Th2 response induced by *Schistosoma mansoni* eggs develops rapidly, through an early, transient, Th0-like stage. J. Immunol. 148, 2283–2290.

Vella, A.T. and Pearce, E.J. (1994). *Schistosoma mansoni* egg primed Th0 and Th2 cells: failure to downregulate IFN-g production following in vitro culture. Scand. J. Immunol. 39, 12–18.

Vella, A.T., Hulsebosch, M.D. and Pearce, E.J. (1992). *Schistosoma mansoni* eggs induce antigen-responsive CD44-hi T helper 2 cells and IL-4-secreting CD44-lo cells. Potential for T helper 2 subset differentiation is evident at the precursor level. J. Immunol. 149, 1714–1722.

Velupillai, P. and Harn, D.A. (1994). Oligosaccharide-specific induction of interleukin 10 production by B220$^+$ cells from schistosome-infected mice: a mechanism for regulation of CD4$^+$ T-cell subsets. Proc. Natl Acad. Sci. USA 91, 18–22.

Vieira, P., de Waal Malefyt, R., Dang, W., *et al.* (1991). Isolation and expression of human cytokine synthesis inhibitory factor (CSIF) cDNA clones: homology to Epstein Barr virus open reading frame BCRFI. Proc. Natl Acad. Sci. USA 88, 1172–1176.

Villanueva, P., Harris, T.S., Ricklan, D.E. and Stadecker, M.J. (1994). Macrophages from schistosomal egg granulomas induce unresponsiveness in specific cloned th-1 lymphocytes in-vitro and down-regulate schistosomal granulomatous-disease in-vivo. J. Immunol. 152, 1847–1855.

Wagner, D., Maroushek, N.M., Brown, J.F. and Czuprynski, C.J. (1994). Treatment with anti-interleukin-10 monoclonal antibody enhances early resistance to but impairs complete clearance of *Listeria monocytogenes* infection in mice. Infect. Immun. 62, 2345–2349.

Walter, M.R., and Nagabhushan, T.L. (1995). Crystal structure of interleukin-10 reveals an interferon gamma-like fold. Biochemistry 34, 12118–12125.

Wang, P., Wu, P., Anthes, J.C., Siegel, M.I., Egan, R.W. and Billah, M.M. (1994). Interleukin-10 inhibits interleukin-8 production in human neutrophils. Blood 83, 2678–2683.

Wang, P., Wu, P., Siegel, M.I., Egan, R.W. and Billah, M.M. (1995). Interleukin-10 inhibits nuclear factor kappa B (NF kappa B0) activation in human monocytes. IL-10 and IL-4 suppress cytokine synthesis by different mechanisms. J. Biol. Chem. 270, 9558–9563.

Wanidworanun, C. and Strober, W. (1993). Predominant role of tumor necrosis factor-alpha in human monocyte IL-10 synthesis. J. Immunol. 151, 6853–6861.

Weissman, D., Poli, G. and Fauci, A.S. (1994). Interleukin-10 blocks HIV replication in macrophages by inhibiting the autocrine loop of TNFα and IL-6 induction of virus. AIDS Res. Hum. Retroviruses. [In press]

Willems, F., Marchant, A., Delville, J.P., *et al.* (1994). Interleukin-10 inhibits B7 and intercellular adhesion molecule-1 expression on human monocytes. Eur. J. Immunol. 24, 1007–1009.

Windsor, W.T., Syto, R., Tsarbopoulos, A., *et al.* (1993). Disulfide bond assignments and secondary structure analysis of human and murine interleukin 10. Biochemistry 32, 8807–8815.

Windsor, W.T., Murgolo, N.J. and Syto, R. (1995). Structural and biological stability of the human Interleukin-10 homodimer. [Submitted]

Wogensen, L., Huang, X. and Sarvetnick, N. (1993). Leukocyte extravasation into the pancreatic tissue in transgenic mice expressing interleukin 10 in the islets of Langerhans. J. Exp. Med. 178, 175–185.

Wogensen, L., Lee, M.S. and Sarvetnick, N. (1994). Production of interleukin 10 by islet cells accelerates immune-mediated destruction of beta cells in nonobese diabetic mice. J. Exp. Med. 179, 1379–1384.

Wu, J., Cunha, F.Q., Liew, F.Y. and Weiser, W.Y. (1993). IL-10 inhibits the synthesis of migration inhibitory factor and migration inhibitory factor-mediated macrophage activation. J. Immunol. 151, 4325–4332.

Yamamura, M., Uyemura, K., Deans, R.J., *et al.* (1991). Defining protective responses to pathogens: cytokine profiles in leprosy lesions. Science 254, 277–279.

Yamamura, M., Wang, X., Ohmen, J.D., Bloom, B.R. and Modlin, R.L. (1992). Cytokine patterns of immunologically mediated tissue damage. J. Immunol. 149, 1470–1475.

Yssel, H., De Waal Malefyt, R., Roncarolo, M.G., *et al.* (1992). IL-10 is produced by subsets of human CD4$^+$ T cell clones and peripheral blood T cells. J. Immunol. 149, 2378–2384.

Zidanov, A., Schalk-Hihi, C., Gustchina, A., Tsang, M., Weatherbee, J. and Wlodawer, A. (1995). Crystal structure of interleukin-10 reveals the functional dimer with an unexpected topological similarity to interferon gamma. Structure 3, 591–601.

Zielen, S., Bauscher, P., Hofmann, D. and Meuer, S.C. (1993). Interleukin 10 and immune restoration in common variable immunodeficiency. Lancet 342, 750–751.

11. Interleukin-11

Paul F. Schendel *and* Katherine J. Turner

1. Introduction

Interleukin-11 (IL-11) was first identified as a soluble protein factor produced by a primate bone marrow stromal cell line in response to IL-1 (Paul *et al.*, 1990). The factor stimulated the growth of an IL-6-dependent mouse plasmacytoma cell line (T-1165). The gene encoding this new activity was identified by expression cloning, and its human homolog was isolated from an IL-1 and phorbol-ester-stimulated lung fibroblast line (Paul *et al.*, 1990). The protein encoded by this human cDNA was given the name recombinant human interleukin-11 (rhIL-11). Although named interleukin IL-11, it does not seem to be made by leukocytes. Rather

IL-11 is produced by a number of cell types in response to agents like IL-1, phorbol-esters, or transforming growth factor-β (TGF-β). Its effects are seen on a diverse set of cell types. This causes rhIL-11 to have pleiotropic effects both *in vitro* and *in vivo*. These effects have been the subject of intense investigation for the past six years. Much of this work will be summarized in this review. Other reviews or portions of this information have also been published recently (Kawashima and Takiguchi, 1992; Paul and Schendel, 1992; Quesniaux *et al.*, 1993; Du and Williams, 1994, 1995; Goldman, 1995; Turner and Clark, 1995).

2.   The IL-11 Gene

The human IL-11 cDNA contains an open reading frame of 597 nucleotides, as shown in Figure 11.1. The human genomic sequence of IL-11 has also been isolated and partially characterized. The gene spans approximately 7 kbp in length and is composed of five exons and four introns. An illustration of the IL-11 genomic structure is shown in Figure 11.2. By in-situ hybridization, the genomic sequence maps to the long arm of chromosome 19 at 19q13.3-13.4 (McKinley *et al.*, 1992). Alternative utilization of two polyadenylation sites in the 3′ noncoding region of the human IL-11 gene generates two distinct mRNA species of 1.5 and 2.5 kb that encode the same functional protein. Both

transcripts contain several copies of ATTTA in the 3′ noncoding region which have been implicated in regulating the stability of growth factor mRNAs (Shaw and Kamen, 1986). Although the 5′ flanking region of the human IL-11 gene shares no sequence similarity with other cytokine genes, it contains a number of potentially important transcriptional control sequences. These include sites for AP1, CTF/NF1, EF/C, SP1, and several copies of interferon-inducible elements. In addition, a sequence (ACATGGCAAAACCC) with significant similarity to an IL-1-responsive element found in the 5′ noncoding region of the IL-6 gene is located in the 3′ flanking region of the human IL-11 gene (McKinley *et al.*, 1992).

Recently the murine IL-11 gene has been isolated and is located on chromosome 7. The murine and human IL-11 homologs are 86% and 88% identical at the DNA and protein levels, respectively (Morris *et al.*, 1996).

3.   The IL-11 Protein

The IL-11 gene encodes a 199-amino-acid polypeptide. The first portion of the protein comprises a secretory signal sequence. This peptide is removed during transport from the cell to yield the active form of the factor. The precise site at which the protein is cleaved to release mature IL-11 *in vivo* is not known, but the recombinant factor expressed in either COS-1 cells or

691	<u>CCAGGCCCCC**TATATA**ACCCCCCAGGCGTCCACACTCCCT</u>	730
731	<u>CACTGCCGCGGGCCCTGCTGCTCAGGGCACATGCCTCCCCTCCCCAGCCGCGGGC</u>	
	<u>CCAGCTGACCCTCGGGGCTCCCCCGGCAGCGGACAGGGAAGGGTTAAAGGCCCCC</u>	Exon 1
	<u>GGCTCCCTGCCCCCTGCCCTGGGGAACCCCTGGCCCTGTGGGGAC</u>ATGAACT	892
2231	GTGTTTGCCGCCTGGTCCTGGTCGTGCTGAGCCTGTGGCCAGATACAG	
	CTGTCGCCCCTGGGCCACCACCTGGCCCCCCTCGAGTTTCCCCAGACCC	Exon 2
	TCGGGCCGAGCTGGACAGCACCGTGCTCCTGACCCGCTCTCTCCTGGCG	
	GACACGCGGCAGCTGGCTGCACAGCTG	2403
2603	AGGGACAAATTCCCAGCTGACGGGGACCACAACCTGGATTCCCTGCCC	Exon 3
	ACCCTGGCCATGAGTGCAGGGGCACTGGGAGCTCTACAG	2689
2801	CTCCCAGGTGTGCTGACAAGGCTGCGAGCGGACCTACTGTCCTACCTGC	
	GGCACGTGCAGTGGCTGCGCCGGGCAGGTGGCTCTTCCCTGAAGACCCT	Exon 4
	GGAGCCCGAGCTGGGCACCCTGCAGGCCCGACTGGACCGGCTGCTGCGC	
	CGGCTGCAGCTCCTG	2962
4994	ATGTCCCGCCTGGCCCTGCCCCAGCCACCCCCGGACCCGCCGGCGCCCC	
	CGCTGGCGCCCCCCTCCTCAGCCTGGGGGGGCATCAGGGCCGCCCACGC	
	CATCCTGGGGGGGCTGCACCTGACACTTGACTGGGCCGTGAGGGGACT	Exon 5
	GCTGCTGCTGAAGACTCGGCTGTGA<u>CCCGGGGGCCCAAAGCCACCACCGTCCT</u>	
	<u>TCAAAGCCAGATCTT**ATTTA**TTT**ATTTA**TTTCAGTACTGGGGGCGAAACAGC</u>	
	<u>CAGGTGATCCCCCCGCCATTATCTCCCCCTAGTTAGAGACAGTCCTTCCGTGAGG</u>	
	<u>CCTGGGGGGGCATCTGTGCCTT**ATTTA**TACTT**ATTTA**</u>	5391

Figure 11.1   Nucleotide sequence of the human IL-11 gene (exons only). The 5′ and 3′ untranslated regions are underlined containing the "TATA" and several "ATTTA" motifs shown in bold. For simplicity, exon 5 is truncated at nucleotide position 5391. (Data from McKinley *et al.*, 1992)

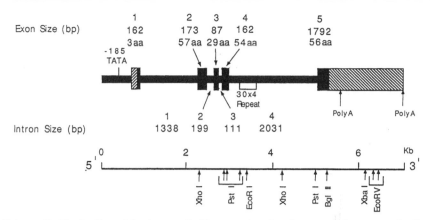

Figure 11.2   Schematic illustration of the human IL-11 gene showing the exon/intron organization and a partial restriction cleavage map. The coding and noncoding regions are indicated by solid and hatched boxes, respectively.

CHO cells is formed by proteolytic cleavage after amino acid 21. This yields a protein of 178 amino acids containing proline at its N-terminus (see Figure 11.3). Signal sequence processing next to a proline moiety is highly unusual and may account for the low level of expression seen in both COS and CHO cells.

The mature form of IL-11 has several unusual characteristics. Among these are the following:

- The protein is highly basic with a +7 charge at neutral pH. Within the basic amino acids is a preponderance of arginine residues the result of which is a protein with a pI>11.
- The amino acid composition of IL-11 contains 23 mol% leucine giving the protein considerable hydrophobic character. It also contains 12 mol% proline.

Most of these proline residues are concentrated at the N-terminus of the mature form of the protein and in a region located about two-thirds of the way through the protein. The internal proline-rich region probably forms an important structural loop between helical segments implicated in the binding of the factor to its receptor, while the N-terminal region can be deleted without major effect on activity. The amino acid composition is also unusual in that it contains no cysteine and thus the protein has no disulfide bonds to stabilize its structure.

- The protein sequence contains no N-linked glycosylation sites and the analysis of rhIL-11 expressed in COS-1 cells has not revealed any O-linked carbohydrate. Thus IL-11 may be naturally nonglycosylated.

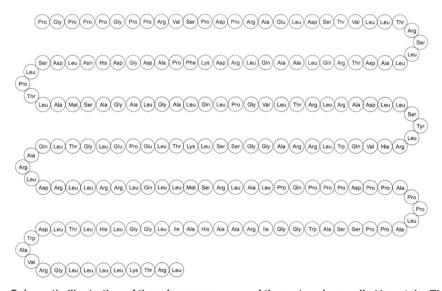

Figure 11.3   Schematic illustration of the primary sequence of the mature human IL-11 protein. The molecule contains no disulfide bonds or N-linked glycosylation sites.

The rhIL-11 protein elutes as a monomer upon native size exclusion chromatography with a mobility consistent with a globular protein of about 19 kDa. By contrast, its mobility during denaturing SDS-polyacrylamide gel electrophoresis is less than predicted. Rather than having the expected mobility of a 19 kDa protein, rhIL-11 has the mobility of a protein of 23 kDa. This reduced mobility is probably a reflection of its positive charge at neutral pH.

At present, the three-dimensional structure of rhIL-11 is uncertain. A four-helix bundle structure similar to that determined for several other cytokines (Boulay and Paul, 1993: Kaushansky and Karplus, 1993) has been proposed (Czupryn et al., 1995a) (see Figure 11.4). The results of various indirect tests of this model, such as alanine scanning of the proposed helix A and D regions and chemical modification of methionine and lysine residues, have been consistent with the model (Czupryn et al., 1995a,b). The model predicts two binding sites for interaction with the IL-11 receptor. Site I, which is comprised of the C-terminal helix (helix D) and the region around Met-59 may interact with the IL-11-specific subunit of the receptor (Hilton et al., 1994), while site II, which includes residues from helix A, may interact with the protein gp130 known to be involved in IL-11 signaling (Yin et al., 1993).

The integrity of the rIL-11 molecule in solution can be compromised by several forms of chemical instabilities (N. Warne, personal communication). Dilute solutions of

rhIL-11 are often found to contain less activity than expected because the protein has bound tenaciously to the walls of the tube in which it is stored. This is especially true at basic pH and when glass tubes are used. The presence of two Asp-Pro peptide sequences within the IL-11 protein renders the molecule sensitive to hydrolysis at these sites. This is the major form of instability under acidic conditions. Oxidation of Met-59 can also occur under oxidizing conditions. This alteration leads to a significant loss in activity. Finally, deamidation of Asn-50 has also been detected, but this change does not lead to a major change in the activity.

Despite the absence of disulfide bonds, the rhIL-11 structure is surprisingly heat stable. At neutral pH and in moderate salt solutions such as phosphate-buffered saline (PBS), the protein remains soluble up to a temperature of 80°C. On the other hand, exposure to hydrophobic surfaces, gas–liquid interfaces, or mechanical shear can cause structural changes that manifest themselves in protein aggregation and precipitation (N. Warne, personal communication).

4. Cellular Sources of IL-11

Expression of IL-11 has been noted from several cell types, including different bone marrow stromal fibroblasts (Paul et al., 1990), lung fibroblasts (Paul et al., 1990; Kawashima et al., 1991; Elias et al., 1994), glioblastoma cells (Murphy et al., 1995), melanoma cell lines (Paglia et al., 1995), a human trophoblast line TPA-30-1 (Paul et al., 1990), human osteosarcoma (Horowitz et al., 1993) and primary osteoblasts (Romas et al., 1996), and megakaryoblastic cell lines such as CMK and Meg-J (Kobayashi et al., 1993), as well as a human thyroid carcinoma cell line (Tohyama et al., 1992). The IL-11 transcript is not detectable in various T cell lines, or in lectin-stimulated peripheral blood cells (Paul et al., 1990). With the various fibroblast cell lines, treatment with IL-1 or phorbol esters substantially upregulates IL-11 expression. A recent analysis of expression of IL-11 by lung epithelial cells reveals that TGF-β synergizes with IL-1 in augmenting IL-11 protein and mRNA production. Interestingly, in these cultures, respiratory syncytial and parainfluenza viruses and rhinovirus are potent stimulators of IL-11 expression in a time- and dose-dependent manner (Elias et al., 1994, Einarrson et al., 1996) Similarly, in human articular chondrocytes and synoviocytes, addition of TGF-β, IL-1β, as well as phorbol ester induces IL-11 expression (Maier et al., 1993). IL-11 is also expressed by normal human trabecular bone cells enriched in osteoblasts after induction with TGF-β, or IL-1, but not lipopolysaccharide (LPS) or parathyroid hormone (Horowitz et al., 1993). The cytokine-regulated expression of IL-11 in mesenchymal cells from various tissues raises the possibility that IL-11 may play

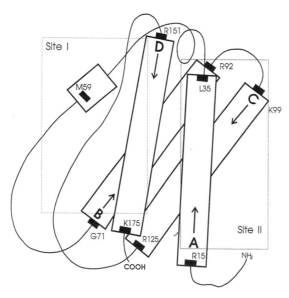

Figure 11.4   Schematic illustration of the proposed tertiary structure of human IL-11 (Czupryn et al., 1995a). The solid boxes represent proposed helical regions comprising the four-helix bundle structure. The dashed boxes represent the proposed surfaces that interact with the IL-11 receptor. Site I is proposed to interact with the IL-11-specific receptor subunit. Site II is proposed to interact with gp130.

important roles in controlling the growth, development, and function of cells outside of the hematopoietic system. This prediction is further supported by the growing number of *in vivo* studies showing effects of rhIL-11 with different types of target cells.

5. In Vitro Effects of IL-11

5.1 HEMATOPOIETIC EFFECTS OF IL-11

As the hematopoietic activity of IL-11 was one of the first activities noted for rhIL-11, it seems logical to begin a discussion of its *in vitro* activities with this topic.

5.1.1 Hematopoietic Stem and Multipotential Progenitor Cells

IL-11 is a pleiotropic cytokine which acts on multiple hematopoietic cell types in a synergistic manner with other cytokines. The effects of rhIL-11 on primitive hematopoietic progenitor cells were first recognized by Ogawa and colleagues in a blast colony assay system using bone marrow or spleen cells from mice pretreated with 5-fluorouracil (Musashi *et al.*, 1991b; Tsuji *et al.*, 1992). In this assay rhIL-11 has no activity alone, but synergizes with IL-3, IL-4, or steel factor (SF) to increase the number and kinetics of appearance of murine blast and CFU-GEMM colonies. This activity has also been noted for IL-6, G-CSF, and IL-12 and is attributed to the ability of these cytokines to trigger quiescent stem cells into cycle (Ikebuchi *et al.*, 1987, 1988; Hirayama *et al.*, 1994). rhIL-11 also synergizes with IL-3 or SF to promote the growth of very primitive lymphohemato-poietic progenitors in either methylcellulose or liquid culture (Hirayama *et al.*, 1992; Neben *et al.*, 1994, 1996). In the liquid culture system, the combination of rhIL-11 and SF supports the survival of long-term repopulating stem cells for a period of 6 days and increases the number of committed myeloid progenitors over 10 000-fold (Neben *et al.*, 1994, 1996). Similar results have been observed in murine long-term bone marrow cultures (LTBMC) supplemented with rhIL-11 (Du *et al.*, 1995; Neben *et al.*, 1994). In the human system, rhIL-11 acts synergistically with either IL-3 or SF to support blast cell colony formation from isolated CD34$^+$ cells (Leary *et al.*, 1992). rhIL-11 alone has no effect on colony formation from human CD34$^+$ HLA-DR$^-$ cells and stimulates a small number of CFU-GM derived from purified CD34$^+$ cells. However, two factor combinations of rhIL-11 with SF, IL-3, or GM-CSF in the presence of erythropoietin (EPO) results in a synergistic or additive increase in the number of CFU-C from both purified cell populations (Lemoli *et al.*, 1993). The effects of rhIL-11 have also been studied in stromal cell containing long-term human cell cultures (Keller *et*

al., 1993; Du *et al.*, 1995). In these cultures, rhIL-11 stimulates myeloid growth from enriched cultures of CD34$^+$, HLA-DR$^-$ bone marrow cells but has no effects on the more primitive CD34$^+$, HLA-DR$^-$ populations. Interestingly, a major effect of rhIL-11 in these cultures is to inhibit adipogenesis and to increase the adherent cell populations consisting of stromal cells and macrophages (Keller *et al.*, 1993).

5.1.2 Megakaryocytopoiesis

Like the effects of rhIL-11 on primitive hematopoietic progenitor populations, the action of rhIL-11 mega-karyocyte colony-forming cells is best observed in cultures containing combinations of cytokines. In a serum-containing fibrin clot system, IL-11 is inactive alone, but acts synergistically with IL-3 to promote the formation of murine megakaryocyte colony formation (Paul *et al.*, 1990). Similar synergistic effects of IL-3 and rhIL-11 on megakaryocyte growth are observed in semisolid and liquid cultures of murine and human bone marrow cells, in which rhIL-11 also increases the size of the megakaryocyte colonies as well as the size and ploidy of individual megakaryocytes (Burstein *et al.*, 1992; Teramura *et al.*, 1992; Yonemura *et al.*, 1992; Weich *et al.*, 1997). In the presence of IL-3, rhIL-11 stimulates significant BFU-MEG colony formation derived from immature CD34$^+$ DR$^-$ human bone marrow cells (Bruno *et al.*, 1991). On more mature IL-11-responsive target cells, the need for other growth factors diminishes. When highly purified CD41$^+$ bone marrow cells are cultured in rhIL-11 alone, the ploidy distribution increases to greater values (Teramura *et al.*, 1992). Similarly rhIL-11 alone increases the production of acetylcholinesterase in cultures of murine bone marrow cells, and can act as an autocrine growth factor for some human megakaryoblastic cell lines (Kobayashi *et al.*, 1993), suggesting that IL-11 has a role in maturation and differentiation. Combined, these results suggest that IL-11 has a role in multiple stages of the megakaryocyte differentiation pathway.

5.1.3 Erythropoiesis

rhIL-11 stimulates erythropoiesis *in vitro* at several stages along the differentiation pathway. In analyzing the effects of rhIL-11 on murine erythroid cell development, Quesniaux and colleagues found that IL-11 acts in concert with SF and EPO to promote macroscopic erythroblast bursts containing erythroid (E) progenitors capable of forming CFU-E-like colonies with high frequency (Quesniaux *et al.*, 1992). When rhIL-11 is combined with IL-3, a significant number of hemoglobinized bursts are observed in the absence of exogenously added EPO, suggesting a role for IL-11 in late stages of erythroid development as well. The effects of rhIL-11 on burst-forming units erythroid (BFU-E) formation have been confirmed using purified human bone marrow CD34$^+$ cells in the presence of EPO (Lemoli *et al.*, 1993).

5.1.4   Myelopoiesis and Macrophage Effects

rhIL-11 may play a role in the maturation and/or activation of monocytes and monocyte progenitors. Ogawa and colleagues reported that the replating of murine blast cell colonies grown in IL-3 alone into secondary cultures containing only rhIL-11 and EPO yields significant numbers of monocyte/macrophage colonies (Musashi *et al.*, 1991a). IL-11 stimulates significant myeloid colony formation from unfractionated human bone marrow at 100 ng/ml and from purified CD34$^+$ HLA-DR$^+$ bone marrow cells at 10 ng/ml. There is no effect of rhIL-11 alone on the more primitive CD34$^+$ HLA-DR$^-$ cell population (Keller *et al.*, 1993). In human long-term bone marrow cultures, addition of rhIL-11 stimulates the growth of myeloid progenitors and dramatically increases the numbers of macrophages in the adherent stromal cell layer (Keller *et al.*, 1993).

rhIL-11 has also been shown to affect gene expression in cells of the myeloid lineage. rhIL-11 treatment can increase the expression of IL-6 mRNA when added to cultures of human monocytes (Anderson *et al.*, 1992). It can also downregulate expression of tumor necrosis factor (TNF), IL-1, interferon-γ (γ-INF), and IL-12 in LPS-stimulated mouse peritoneal macrophages (Trepicchio *et al.*, 1996).These profound effects are directly attributable to rhIL-11 and are not mediated through the induction of other known anti-inflammatory cytokines such as TGF-β or IL-10 (Trepicchio *et al.*, 1996).

5.1.5   Lymphopoiesis

One of the earliest recognized effects of rhIL-11 in culture is its ability to increase the number of sheep red blood cell (SRBC)-specific plaque-forming cells in murine splenocyte cultures (Paul *et al.*, 1990; Yin *et al.*, 1992a). The activity of rhIL-11 in the splenocyte culture system is dramatically decreased by depletion of helper (CD4$^+$) T cells but not CD8$^+$ T cells, raising the possibility that the effects of IL-11 may be indirect.

In the human system, Anderson and colleagues have further defined the role of IL-11 on lymphopoiesis by testing rhIL-11 *in vitro* on fractionated cell populations. rhIL-11 produces a significant enhancement of Ig secretion without increasing DNA synthesis when B cells stimulated with pokeweed mitogen, irradiated T cells, and monocytes are cocultured with it (Anderson *et al.*, 1992). Interestingly, the purified helper T cell subset (CD4$^+$ 45RA$^-$) responds to rhIL-11 by increased proliferation and expression of IL-6, raising the possibility that some of the effects of IL-11 might be due to IL-6. However, neutralizing anti-IL-6 antibody only partially blocks IL-11-stimulated immunoglobulin secretion. Since IL-6 has been demonstrated to play an important role in B cell differentiation and immunoglobulin secretion, it will be important to discriminate its role from the role of IL-11.

5.1.6   Effects of IL-11 on Leukemia and Lymphoma Cells

In general, the effects of rhIL-11 on various cell lines and primary human leukemic samples parallel its effects with normal hematopoietic progenitors. rhIL-11 stimulates directly the proliferation of certain murine plasmacytoma cell lines such as T1165 and B9 in a dose-dependent manner, although the cells are much less responsive to rhIL-11 than they are to IL-6. However, rhIL-11 is not a growth factor for freshly isolated human myeloma cells and does not increase proliferation or enhance immunoglobulin production (Paul and Schendel, 1992; Zhang *et al.*, 1994). Interestingly, IL-10 has been shown to induce IL-11 receptor and IL-11 responsiveness in some human myeloma cell lines (Lu *et al.*, 1995). rhIL-11 alone has little if any effects when tested on a large panel of cell lines derived from patients with acute myeloblastic and lymphoblastic leukemias, and lymphoma. However, in combination with IL-3, GM-CSF, or SF, IL-11 can trigger acute myelogenous leukemia blast cells into S-phase and promote proliferation (Hu *et al.*, 1993; Lemoli *et al.*, 1995).

5.2   EFFECTS OF IL-11 ON NONHEMATOPOIETIC CELLS

5.2.1   Effect of IL-11 on Adipocytes

Besides affecting hematopoietic cells, rhIL-11 has significant activity in a variety of other culture systems. rhIL-11 directly inhibits the activity of a lipoprotein lipase in the murine 3T3-L1 preadipocyte cell line (Kawashima *et al.*, 1991) and blocks adipogenesis in H-1/A cells, a murine stromal cell line derived from a long-term bone marrow culture (Ohsumi *et al.*, 1991). In long-term human bone marrow cultures, rhIL-11 also blocks differentiation of preadipocytes and inhibits adipose accumulation in the adherent stromal layer (Keller *et al.*, 1993). The ability of rhIL-11 to influence stromal cell differentiation by preventing adipogenesis may well complement its stimulatory effects with hematopoietic stem cells in regulating hematopoiesis and the bone marrow microenvironment.

5.2.2   Effect of IL-11 on Hepatocytes and the Acute-phase Response

Like IL-6 and leukemia inhibitory factor (LIF), IL-11 can modulate gene expression in hepatocytes. rhIL-11 alone induces expression of acute-phase proteins such as fibrinogen, α_1-antitrypsin, and α_2-macroglobulin, in both primary rat hepatocytes and H-35, a rat hepatic cell line. The response to rhIL-11 is enhanced by dexamethasone in primary rat hepatocytes, but the magnitude of the effect is less than that seen with IL-6 (Baumann and Schendel, 1991). rhIL-11 also enhances the expression of the IL-1-regulated proteins including complement C3, haptoglobin, and hemopexin and has no effect on albumin synthesis. In contrast to IL-6, treatment of human HepG2 hepatoma cells with rhIL-11 elicits a very weak induction of haptoglobin mRNA and a large increase in the levels of microsomal heme-oxygenase, the rate-limiting enzyme in heme catabolism (Fukuda and Sassa, 1993). Induction of heme-oxygenase may also be important in protecting the host from oxidative stimuli by increasing the levels of bile pigments, which are antioxidants.

5.2.3   Effect of IL-11 on Neuronal Cells

Cytokines such as LIF and ciliary neurotrophic factor (CNTF) have also been shown to play an important role in the development of the nervous system. Thus, it should not be too surprising that IL-11 also has some activity on neuronal cells. rhIL-11 induces neuronal differentiation in murine embryonic hippocampal progenitor cell lines, which can be assessed by morphology as well as by an increase in mature neurofilament protein and development of tetradoxin-insensitive action potentials (Mehler *et al.*, 1993). However, at concentrations greater than 100 ng/ml, rhIL-11 has no effect on cholecystokinin, somatostatin, enkephalin, and vasoactive intestinal polypeptide mRNA expression in cultured sympathetic neurons and only weakly induces expression of substance P mRNA, suggesting that the receptor for IL-11 is expressed at very low levels on sympathetic neurons (Fann and Patterson, 1994).

5.2.4   Effect of IL-11 on Osteoclasts

The pleiotropic nature of IL-11 is further revealed by its action on osteoclastogenesis. rhIL-11 stimulates the development of osteoclasts in isolated murine calvaria cell cultures, as well as in cocultures of bone marrow cells and osteoblastic cells derived from calvaria (Passeri *et al.*, 1993). These results are consistent with the finding that the IL-11 receptor is present on murine calvaria cells and several osteoblast-like cell lines (Bellido *et al.*, 1996). Addition of rhIL-11 to bone marrow cells together with 10 nM 1,25-dihydroxyvitamin $D_3(1,25(OH)_2D_3)$, increases the number of osteoclasts and the number of nuclei per osteoclast, compared with $1,25(OH)_2D_3$ alone. In these cultures, the effects of $1,25(OH)_2D_3$ and parathyroid hormone on osteoclast induction are blocked by the addition of a neutralizing IL-11 antibody (Passeri *et al.*, 1993). In contrast to IL-6, the effects of IL-11 are independent of the estrogen status of the bone marrow donor. The finding that IL-11 is induced in osteoblastic cells by TGF-β and IL-1 suggests a potentially important role for IL-11 in bone remodeling.

5.2.5   Effect of IL-11 on Chondrocytes and Synoviocytes

IL-11 may also have potential effects on cartilage. In human articular chondrocytes and synoviocytes, rhIL-11 stimulates the production of tissue inhibitor of metalloproteinases (TIMP) and does not increase stromolysin activity or affect chondrocyte proliferation (Maier *et al.*, 1993). These results imply IL-11 may have a protective role in connective-tissue biology by inhibiting the breakdown of extracellular matrix which often accompanies inflammatory processes of the joint.

5.3   *IN VIVO* BIOLOGICAL ACTIVITIES

The biological effects of human IL-11 are not strongly species restricted. To date rhIL-11 has been administered to two strains of mice, two strains of rats, Golden Syrian hamsters, guinea-pigs, rabbits, beagle dogs, cynomolgus monkeys, and humans. In all cases the human protein is active and, while the dose–response relationship is different from species to species, the biological effects of the protein are remarkably consistent. When the murine homologue was cloned, it was more active per unit mass in mice than was the human protein, but it did not elicit any additional biological responses (C. Wood, personal communication).

5.3.1   Hematological Effects of rhIL-11 Treatment

5.3.1.1   *Effects in Normal Animals*

Despite the variety of *in vitro* effects reported for IL-11, administration of the rhIL-11 protein to animals produces only a limited set of detectable changes. In animals with unperturbed bone marrow, treatment with rhIL-11 causes a dose-related increase in circulating platelets (Bree *et al.*, 1991; Cairo *et al.*, 1993, 1994; Neben *et al.*, 1993; Yonemura *et al.*, 1993). While an increase in platelet number following rhIL-11 treatment is observed consistently, the magnitude of the change varies from species to species and animal to animal. The kinetics of the platelet response is also variable from species to species. In mice, platelet counts begin to rise after as little as 3 days of treatment and begin to fall soon after treatment is concluded (Neben *et al.*, 1993). In larger animals, there is a 5–7-day lag between initiation of rhIL-11 treatment and the observation of an increase in circulating platelet number. After rhIL-11 treatment is stopped, platelet numbers remain elevated for 5–7-days before beginning to drop back toward baseline (Bree *et al.*, 1991; Turner and Clark, 1995).

Along with the effects seen on circulating platelets, rhIL-11 treatment of mice causes an increase in the number of MEG-CFC in the bone marrow and spleen; an increase in the number, size, and ploidy of megakaryocytes in the spleen; and an increase in the ploidy, but no change in the number, of megakaryocytes in the bone marrow (Neben *et al.*, 1993). Similar effects are seen in rats (Cairo *et al.*, 1993, 1994; Yonemura *et al.*, 1993).

In contrast to its effects on cells of the megakaryocyte lineage, rhIL-11 treatment of normal rodents and primates has little effect on the number of circulating leukocytes (Bree *et al.*, 1991; Cairo *et al.*, 1994; Neben *et al.*, 1993; Yonemura *et al.*, 1993). Treatment with rhIL-11 can cause an increase in GM-CFC in the spleen and bone marrow (Cairo *et al.*, 1993, 1994) but no associated granulocytosis or monocytosis is seen. Similarly, no consistent effect of rhIL-11 on lymphocyte number or CDC4/8 ratio has been reported. While treatment with rhIL-11 can mobilize some bone marrow cells to the periphery, the use of G-CSF in combination with rhIL-11 dramatically increases the number of GM-CFC mobilized compared to the effect of either cytokine alone (Leonard *et al.*, 1993; Hastings *et al.*, 1994). The concomitant use of rhIL-11 with G-CSF may also enhance G-CSF-mediated leukocytosis (Cairo *et al.*, 1994).

Treatment with rhIL-11 may also cause a decrease in the animal's hematocrit. This effect is especially noticeable in humans and in large animals (Bree *et al.*, 1991; Gordon *et al.*, 1993; Ault *et al.*, 1994). The drop in hematocrit is detectable within a day of initiation of rhIL-11 treatment and continues for about a week. Thereafter the hematocrit stabilizes. This phenomenon has been studied carefully in humans and seems to be caused by plasma volume expansion (Ault *et al.*, 1994). Human volunteers treated with rhIL-11 retained more fluid and sodium than a cohort receiving a placebo control. The treated volunteers had about a 20% increase in their plasma volumes while their hematocrits fell by a similar amount. This effect is blunted by using a mild diuretic, and is readily reversible following cessation of rhIL-11 treatment. The mechanism by which rhIL-11 causes sodium and water retention is unknown at present.

5.3.1.2 *Effects in Myelosuppressed Animals*

The effects of rhIL-11 on hematological recovery in myelosuppressed mice have been reported from several laboratories (Du *et al.*, 1993a,b; Hangoc *et al.*, 1993; Leonard *et al.*, 1994; Maze *et al.*, 1994). The results of these studies are more variable than those seen in normal animals. This variability may result from the different myelosuppressive regimens employed having variable effects on the number of hematopoietic progenitors that survive, on the cytokine environment that is produced, or on both.

In lethally irradiated mice receiving bone-marrow and spleen cell support, treatment with rhIL-11 reduces the depth of the platelet nadir and stimulates the rate of neutrophil and platelet recovery significantly, but it has no significant effect on red cell recovery (Du *et al.*, 1993a,b). In mice treated with radiation and carboplatin, rhIL-11 again reduces the depth of the platelet nadir and stimulates platelet recovery. In this case red cell recovery is also stimulated, while no effect on neutrophil numbers is seen (Leonard *et al.*, 1994). For mice treated with cyclophosphamide, rhIL-11 treatment does not reduce either platelet or neutrophil nadirs, but accelerates the recovery of both cell types (Hangoc *et al.*, 1993). Finally, rhIL-11 can help sustain the levels of circulating platelets, red cells, and total white cells in BCNU-treated mice (Maze *et al.*, 1994). Despite the variability in response, platelet recovery has consistently been improved by rhIL-11 treatment in all of these models.

In addition to its hematological effects, rhIL-11 treatment of myelosuppressed mice produces significant increases in the number of hematopoietic progenitors found in the bone-marrow and spleen. In radiation/carboplatin-treated animals, where rhIL-11 promotes the recovery of circulating platelets, the number of CFU-MEG found in treated animals 2 weeks post chemotherapy is markedly higher than in controls (Leonard *et al.*, 1994). This result suggests that rhIL-11 not only stimulates megakaryocyte development and platelet production, but also supports the general growth and development of the whole megakaryocyte lineage. In addition to the effects on CFU-MEG, the numbers of CFU-GM, CFU-E, and CFU-GEMM are higher in rhIL-11-treated animals than in controls in all of the models reported to date. These results demonstrate both the specific effects of IL-11 on megakaryocyte development and the pleiotropic effects of IL-11 on general hematopoietic recovery in myelosuppressed animals. This conclusion is further supported by the observation that many patients naturally have elevated levels of IL-11 in their serum following myeloablative therapy or bone marrow transplantation (Suen *et al.*, 1994; Chang *et al.*, 1996).

5.3.2   Effects of IL-11 on Serum Proteins

In addition to its effects on blood cells and their precursors, rhIL-11 alters the level of several serum proteins in treated animals. Consistent with the observation that rhIL-11 induces acute-phase proteins in liver cells in culture (Baumann and Schendel, 1991) the concentration of acute-phase proteins increases in the blood of rhIL-11 treated people (Gordon *et al.*, 1993; Ault *et al.*, 1994; Kaye *et al.*, 1994). These proteins include ferritin, haptoglobin, C-reactive protein, and fibrinogen. Unexpectedly, the concentration of von Willebrand factor (vWF) also increases significantly (Kaye *et al.*, 1994). The source of the increased vWF is not understood at present.

5.3.3   Effects of IL-11 on Epithelial Cells

Much of the early work on rhIL-11 focused on its effects on hematopoiesis. Recently it has been recognized that this cytokine has effects on several other cell types. One such observation, first reported by D. Williams, is the effect of rhIL-11 on epithelial cells. Mice treated with both 5-fluorouracil (5-FU) and sublethal irradiation die of sepsis associated with infection by bacteria that enter the system through damaged intestinal crypts. Treatment with rhIL-11 allows the epithelial cells to maintain their integrity and thus allows most of the animals to survive the treatment (Du *et al.*, 1994). The mechanism by which IL-11 reduces epithelial damage is not clear, but it seems to be independent of its effects on circulating blood cells.

The initial observation by Williams and his coworkers have been extended to include models of chemotherapy and chemical damage. Hamsters treated with 5-FU develop severe mucositis after a week to 10 days. This epithelial damage can be strikingly reduced by the administration of rhIL-11 to the animals at the time of chemotherapy or shortly thereafter (Sonis *et al.*, 1995). In this model, rhIL-11 both prevents the mucosal damage and accelerates its healing. This is in contrast to the effects of other growth factors such as TGF-β, which accelerate healing but do not protect against the development of the mucositis lesions (Sonis *et al.*, 1994).

Chemical damage, such as acetic acid burns, has been used as a model of inflammatory damage in the large intestine of the rat (MacPherson and Pfeiffer, 1978). This model was used to determine whether rhIL-11 could affect the healing of inflammatory lesions. As in the case of radiation and chemotherapy, the colonic epithelial cells of animals receiving rhIL-11 are more resistant to the damage induced by acetic acid than are the controls. The architecture of the colonic walls is retained and the amount of inflammation is dramatically reduced in the rhIL-11-treated animal (Keith *et al.*, 1994). Similar results have been seen using TNB-induced damage (Qiu *et al.*, 1996).

While the mechanism by which IL-11 modulates epithelial damage is still unknown, the results are striking. As reviewed above, IL-11 is known to alter gene expression in monocytes and cells closely related to monocytes such as osteoclasts. Since monocytes play a central role in the inflammatory response, IL-11 may modulate inflammatory reactions by its effects on monocyte gene expression (Trepicchio *et al.*, 1996). It is also known that rhIL-11 can interact directly with epithelial cells in culture to alter their cell cycle characteristics and to modify gene expression (Peterson *et al.*, 1996). A combination of these two effects may lead to the epithelial protection seen in radiation, chemotherapy, and inflammation models.

6. *The IL-11 Receptor*

The initial observation that IL-11 stimulated the growth of T1165 cells in a manner similar to IL-6 prompted numerous studies to compare the biochemical and biological activities of the two cytokines and their receptors. In part, the overlapping activities are due to the finding that IL-6 and IL-11, like LIF, oncostatin M, and CNTF, utilize a common signal-transducing

receptor subunit gp130 (Yin *et al.*, 1993; Fourcin *et al.*, 1994; Zhang *et al.*, 1994). The best-studied member of this complex receptor family is the high-affinity IL-6 receptor, which is composed of a heterodimeric complex consisting of a unique IL-6 binding α chain, gp80, in association with a homodimer of gp130. The structure of the IL-11 receptor is less well understood. Murine and human IL-11 binding receptor α chains have been cloned (Hilton *et al.*, 1994; Cherel *et al.*, 1995). Figure 11.5a shows an illustration of the known subunits of the human IL-11 receptor complex. The relationship of the IL-11 receptor complex with other members of the gp130 family is depicted in Figure 11.5b. The human IL-11 receptor, cloned from a placental cDNA library, has 82% sequence homology with the murine receptor. The murine IL-11 receptor, cloned from an adult mouse liver cDNA library, has 24% sequence identity to the α chain of the IL-6 receptor and contains the characteristic residues found in most hematopoietin receptors. IL-11 binding studies in the factor-dependent Ba/F3 cell line, transfected with the IL-11 receptor α chain in the presence or absence of gp130, confirm the previously published preliminary binding studies (Yin *et al.*, 1992b). When expressed alone in the Ba/F3 cell line, IL-11 receptor binds rhIL-11 with a low affinity of approximately 10 nM. Coexpression of gp130 and the IL-11 receptor α chain generates a high-affinity complex of 300–800 pM, similar to that found on 3T3-L1 cells (Hilton *et al.*, 1994, Yin *et al.*, 1992b).

7. *IL-11 Signal Transduction*

The utilization of gp130 as a common signal transducer by the IL-11, IL-6, LIF, CNTF, and oncostatin M receptors in part explains the biological overlap among these cytokines. To identify additional downstream intracellular components of the IL-11 signaling transduction pathway, tyrosine phosphorylation studies have been performed by several investigators (Yin and Yang, 1993; Berger *et al.*, 1994; Yin *et al.*, 1994). Following rhIL-11 treatment of the IL-11-dependent B9-TY1 murine plasmacytoma cell line, early response genes such as *tis*11, *tis*21, and *jun*B are rapidly activated (Yin and Yang, 1993). More recently, rhIL-11 has been shown to activate mitogen-activated protein kinase (MAPK) in mouse 3T3L1 cells and induce the formation of a Grb2, Fyn, and JAK2 complex, thus potentially serving as intermediate signaling molecules to link IL-11 receptor activation with the Ras/MAPK system (Yin *et al.*, 1994; Wang *et al.*, 1995). The protein tyrosine phosphatase Syp has also been reported to associate with gp130 and JAK2 after stimulation of 3T3-L1 cells with rhIL-11 (Fuhrer *et al.*, 1995).

The effects of IL-6 and IL-11 on tyrosine phosphorylation patterns of the JAK-TYK tyrosine

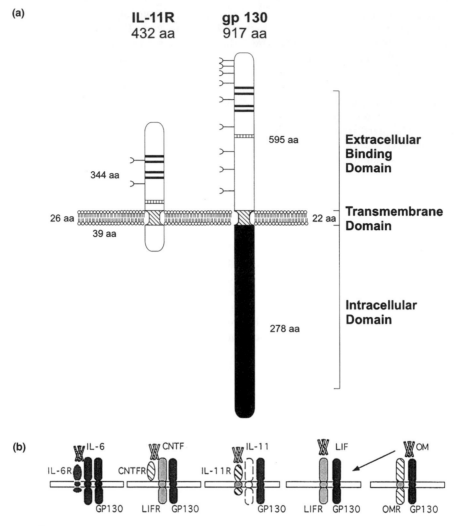

Figure 11.5 (a) Schematic illustration of the human IL-11 receptor as currently defined. The heavy black horizontal lines represent the conserved cysteines in the extracellular region. Potential *N*-linked glycosylation sites and the "WSXWS" motif are indicated by the half-circles and the hatched regions, respectively. (b) Schematic illustration of the gp130 family of hematopoietin receptors, including IL-6, CNTF, IL-11, LIF, and oncostatin M (OM). The receptor chain indicated for IL-11 by the dashed line is speculation based on the predicted structures of other members of this receptor family.

kinases have recently been examined in human and murine plasma-cell tumor lines (Berger *et al.*, 1994). rhIL-11 induces similar patterns of JAK-TYK tyrosine phosphorylation as does IL-6 in the murine B9 hybridoma and T10 plasmacytoma cell lines. Both IL-6 and IL-11 are mitogens that can support the growth of these cell lines. Interestingly, the patterns of tyrosine phosphorylation of the JAK-TYK kinases and the transcription factor STAT91 were different in response to IL-6 in two human myeloma cell lines. Treatment of these two cell lines with IL-11 did not result in any tyrosine-phosphorylated proteins, consistent with its lack of biological activity in human myeloma (see Section 5) (Berger *et al.*, 1994).

8.   *Clinical Implications*

The hematopoietic effects of rhIL-11 have prompted clinical studies in the use of IL-11 to treat patients suffering from severe thrombocytopenia. Not only does rhIL-11 stimulate platelet production, it also stimulates the production of megakaryocytes and their progenitors. This could be especially useful in very high-dose chemotherapy where most megakaryocytes are destroyed. An additional benefit of rhIL-11 treatment for patients undergoing chemotherapy may be a reduction in the severity of mucositis they experience. rhIL-11 is currently in clinical trials treating thrombocytopenia caused by chemotherapy with and without stem cells support. Results from phase I clinical trials (Gordon *et al.*, 1996; Champlin *et al.*, 1994) have demonstrated that the treatment is well tolerated and causes an increase in platelet count in the chemotherapy study and an earlier return to platelet transfusion independence in the transplantation study. Masked, placebo-controlled trials have been completed and have demonstrated that rhIL-11 treatment can reduce the proportion of patients who require platelet transfusion following chemotherapy (Tepler *et al.*, 1996; Isaacs *et al.*, 1997). Other trials are currently underway to measure the effects of IL-11 on mucositis.

The anti-inflammatory effects of rhIL-11 in several models of inflammatory bowel disease also suggest the possible therapeutic effects of treatment with this cytokine for these conditions. Clearly inflammatory bowel disease is a complex condition but the results obtained in animal models were striking enough to warrant human clinical trials. These trials are currently underway.

9. Summary

As reviewed in this article, interleukin-11 has been found to be a multifunctional cytokine. It is part of a family of cytokines which includes IL-6. All members of this family utilize the gp130 signal-transducing chain as part of their receptor, and they all share many functional similarities. While in some ways typical of this family, IL-11 also has some interesting and unique features. Unlike IL-6, IL-11 is not made by T cells; it is only rarely mitogenic; and its effects are often demonstrated as synergistic enhancements of the activity of other growth factors. The actual molecular mechanism of this synergy is not known, nor is the reason why some cells are caused to divide by IL-11 while most are not. Explaining these different signal transduction pathways is one of the future challenges for IL-11 research.

To date, the normal physiological role of IL-11 is unclear, but its expression pattern and distribution suggest that it may be involved in regulating both the bone marrow microenvironment and some kinds of inflammatory responses. Its effects on bone cells, adipocytes, and hematopoietic stem cells make it an ideal candidate for a factor that controls the architecture and cell composition of the bone marrow cavity. On the other hand, its expression in response to IL-1 and TGF-β in synoviocytes and lung fibroblasts; its effects on TNF, IL-1, and γ-INF expression in macrophages; its stimulation of cells to produce TIMP which may in turn inhibit release of TNFα from inflammatory macrophages (Gearing *et al.*, 1994); and its effects on the severity and progression of inflammation in the intestine following chemical damage, all suggest a potential role in reducing the adverse effects of inflammatory reactions.

IL-11 is being developed as a biopharmaceutical. Its hematopoietic effects suggest that it may be useful in treating patients suffering from severe thrombocytopenia. This condition is sometimes the result of prolonged or intensive chemotherapy. It may also result from genetic defects or bone marrow disease. Clinical trials are currently underway to assess the utility of IL-11 in the treatment of these conditions.

In addition, the potential for using IL-11 in the treatment of inflammatory conditions such as inflammatory bowel disease seems great. Its ability to reduce inflammation without completely inhibiting it would seem to make IL-11 an ideal candidate for the treatment of the detrimental effects of a necessary bodily defense mechanism. Clinical trials to assess this potential are underway.

10. References

Anderson, K.C., Morimoto, C., Paul, S.R., *et al.* (1992). Interleukin-11 promotes accessory cell-dependent B-cell differentiation in humans. Blood 80, 2797–2804.

Ault, K.A., Mitchell, J., Knowles, C., *et al.* (1994). Recombinant human interleukin eleven (Neumega™ rhIL-11 Growth Factor) increases plasma volume and decreases urine sodium excretion in normal human subjects. Blood 84 (supplement I), 276a. [Abstract]

Baumann, H. and Schendel, P. (1991). Interleukin-11 regulates the hepatic expression of the same plasma protein genes as interleukin-6. J. Biol. Chem. 266, 20424–20427.

Bellido, T., Stahl, N., Farruggella, T.J., Borba, V., Yancopoulos, G.D. and Manolagas, S.C. (1996). Detection of receptors for interleukin-6, interleukin-11, leukemia inhibitory factor, oncostatin M, and ciliary neurotrophic factor in bone marrow stromal/osteoblastic cells. J. Clin. Invest. 97, 431-437.

Berger, L.C., Hawley, T.S., Lust, J.A., Goldman, S.J. and Hawley, R.G. (1994). Tyrosine phosphorylation of JAK-TYK kinases in malignant plasma cell lines growth-stimulated by interleukins 6 and 11. Biochem. Biophys. Res. Commun. 202, 596–605.

Boulay, J.-L. and Paul, W.E. (1993). Hematopoietin sub-family classification based on size, gene organization and sequence homology. Current Biol. 3, 573–581.

Bree, A., Schlerman, F., Timony, G., McCarthy, K. and Stoudimire, J. (1991). Pharmacokinetics and thrombopoietic effects of recombinant human interleukin-11 (rhIL-11) in nonhuman primates and rodents. Blood 78 (supplement I), 132a. [Abstract]

Bruno, E., Briddel, R.A., Cooper, R.J. and Hoffman, R. (1991). Effects of recombinant interleukin 11 on human megakaryocyte progenitor cells. Exp. Hematol. 19, 378–381.

Burger, R. and Gramatzki, M. (1993). Responsiveness of the interleukin (IL)-6-dependent cell line B9 to IL-11. J. Immunol. Methods. 158, 147–148.

Burstein, S.A., Mei, R.-L., Henthorn, J., Friese, P. and Turner, K. (1992). Leukemia inhibitory factor and interleukin-11 promote maturation of murine and human megakaryocytes *in vitro*. J. Cell. Physiol. 153, 305–312.

Cairo, M.S., Plunkett, J.M., Nguyen, A.,Schendel, P. and van de Ven, C. (1993). Effect of Interleukin-11 with and without granulocyte colony-stimulating factor on *in vivo* neonatal rat hematopoiesis: induction of thrombocytosis by interleukin-11 and enhancement of neutrophilia by interleukin-11 and granulocyte colony-stimulating factor. Pediatr. Res. 34, 56–61.

Cairo, M.S., Plunkett, J.M., Schendel, P. and van de Ven, C. (1994). The combined effects of interleukin-11, stem cell factor, and granulocyte colony-stimulating factor on newborn rat hematopoiesis: significant enhancement of the absolute neutrophil count. Exp. Hematol. 22, 1118–1123.

Champlin, R.E., Mehra, R., Kaye, J.A., *et al.* (1994). Recombinant human Interleukin-eleven (rhIL-11) following autologous BMT for breast cancer. Blood 84 (supplement I), 395a. [Abstract]

Chang, B.M., Suen, Y., Meng, G., *et al.* (1996). Differential mechanisms in the regulation of endogenous levels of thrombopoietin and interleukin-11 during thrombocytopenia: insight into the regulation of platelet production. Blood 88, 3354–3362.

Cherel, M., Sorel, M., Lebeau, B., *et al.* (1995). Molecular cloning of two isoforms of a receptor for the human hematopoietic cytokine interleukin-11. Blood 86, 2534–2540.

Czupryn, M.J., McCoy, J.M. and Scoble, H.A. (1995a). Structure–function relationships in human interleukin-11: identification of regions involved in activity by chemical modification and site-directed mutagenesis. J. Biol. Chem. 270, 985–987.

Czupryn, M., Bennett, F., Dube, J., *et al.* (1995b). Alanine-scanning mutagenesis of human interleukin-11: identification of regions important for biological activity. Ann. N.Y. Acad. Sci. 762, 152–164.

Du, X.X. and Williams, D.A. (1994). Interleukin-11: a multi-functional growth factor derived from the hematopoietic microenvironment. Blood 83, 2023–2030.

Du, X.X., Keller, D., Maze, R. and Williams, D.A. (1993a). Comparative effects of in vivo treatment using interleukin-11 and stem cell factor on reconstitution in mice after bone marrow transplantation. Blood 82, 1016–1022.

Du, X.X., Neben, T., Goldman, S. and Williams, D.A. (1993b). Effects of recombinant human interleukin-11 on hematopoietic reconstitution in transplant mice: acceleration of recovery of peripheral blood neutrophils and platelets. Blood 81, 27–34.

Du, X.X., Doerschuk, C.M., Orazi, A. and Williams, D.A. (1994). A bone marrow stromal-derived growth factor, interleukin-11, stimulates recovery of small intestinal mucosal cells after cytoablative therapy. Blood 83, 33–37.

Du, X.X., Scott, Z.X., Yang, R., *et al.* (1995). Interleukin-11 stimulates multilineage progenitors, but not stem cells, in murine and human long-term marrow cultures. Blood 86, 128–134.

Du, X. and Williams, D.A. (1997). Review: Interleukin-11: Review of molecular, cell biology and clinical use. Blood 87, 3897–3908.

Einarsson, O., Geba, G.P., Zhu, Z., Landry, M. and Elias, J.A. (1996). Interleukin-11: stimulation in vivo and in vitro by respiratory viruses and induction of airways hyper-responsiveness. J. Clin. Invest. 97, 915–924.

Elias, J.A., Zheng, T., Einarsson, O., *et al.* (1994). Epithelial interleukin-11: regulation by cytokines, respiratory syncytial virus, and retinoic acid. J. Biol. Chem. 269, 22261–22268.

Fann, M.-J. and Patterson, P.H. (1994). Neuropoietic cytokines and activin A differentially regulate the phenotype of cultured sympathetic neurons. Proc. Natl Acad. Sci. USA 91, 43–47.

Fourcin, M., Chevalier, S., Lebrun, J.-J., *et al.* (1994). Involvement of gp130/interleukin-6 receptor transducing component in interleukin-11 receptor. Eur. J. Immunol. 24, 277–280.

Fuhrer, D.K., Feng, G.S. and Yang, Y.C. (1995). Syp associates with gp130 and Janus kinase 2 in response to interleukin-11 and 3T3-L1 mouse preadipocytes. J. Biol. Chem. 270, 24826–24830.

Fukuda, Y. and Sassa, S. (1993). Effect of interleukin-11 on the levels of mRNAs encoding heme oxygenase and haptoglobin in human HEPG2 hepatoma cells. Biochem. Biophys. Res. Commun. 193, 297–302.

Gearing, A.J.H., Beckett, P., Christodoulou, M., *et al.* (1994). Processing of tumour necrosis factor-α precursor by metalloproteinases. Nature 370, 555–557.

Goldman, S.J. (1995). Preclinical biology of interleukin-11: a multifunctional hematopoietic cytokine with potent thrombopoietic activity. Stem Cells 13, 462–471.

Gordon, M.S., Sledge, G.W., Battiato, L., *et al.* (1993). The *in vivo* effects of subcutaneously (SC) administered recombinant human interleukin-11 (Neumega™ rhIL-11 Growth Factor; rhIL-11) in women with breast cancer (BC). Blood 82 (supplement I), 498a. [Abstract]

Gordon, M.S., Hoffman, R., Battiato, L., *et al.* (1994). Recombinant human Interleukin-eleven (Neumega™ rhIL-11 Growth Factor; rhIL-11) prevents severe thrombocytopenia in breast cancer patients receiving multiple cycles of cyclophosphamide (C) and doxorubicin (A) chemotherapy. Proc. Am. Soc. Clin. Oncol. 13, 133. [Abstract 326]

Gordon, M.S., McCaskill-Stevens, W.J., Battiato, L.A., *et al.* (1996). A phase 1 trial of recombinant human interleukin-11 (Neumega rhIL-11 growth factor) in women with breast cancer receiving chemotherapy. Blood 87, 3615–3624.

Hangoc, G., Yin, T., Cooper, S., Schendel, P., Yang, Y.-C. and Broxmeyer, H.E. (1993). In vivo effects of recombinant interleukin-11 on myelopoiesis in mice. Blood 81, 965–972.

Hastings, R., Kaviani, M.D., Schlerman, F., Hitz, S., Bree, A. and Goldman, S.J. (1994). Mobilization of hematopoietic progenitors by rhIL-11 and rhG-CSF in non-human primates. Blood 84 (supplement I), 23a. [Abstract]

Hilton, D.J., Hilton, A.A., Raicevic, A., *et al.* (1994). Cloning of a murine IL-11 receptor α-chain; requirement for gp130 for high affinity binding and signal transduction. EMBO 13, 4765–4775.

Hirayama, F., Shih, J.-P., Awgulewitsch, A., Warr, G.W., Clark, S.C. and Ogawa, M. (1992). Clonal proliferation of murine lymphohemopoietic progenitors in culture. Proc. Natl Acad. Sci. USA 89, 5907–5911.

Hirayama, F., Katayama, N., Neben, S., *et al.* (1994). Synergistic relationship between interleukin-12 and steel factor in support of proliferation of murine lymphohematopoietic progenitors in culture. Blood 83, 92–98.

Horowitz, M., Fields, A., Zamparo, J., Heyden, D., Tang, W. and Elias, J. (1993). IL-11 secretion and its regulation in bone cells. J. Bone Miner. Res. 1, 107A:S143

Hu, J.P., Cesano, A., Santoli, D., Clark, S.C. and Hoang, T. (1993). Effects of interleukin-11 on the proliferation and cell cycle status of myeloid leukemic cells. Blood 81, 1586–1592.

Ikebuchi, K., Wong, G.G., Clark, S.C., Ihle, J.N., Hirai, Y. and Ogawa, M. (1987). Interleukin-6 enhancement of interleukin-3-dependent proliferation of multipotential hemopoietic progenitors. Proc. Natl Acad. Sci. USA 84, 9035–9039.

Ikebuchi, K., Clark, S.C., Ihle, J.N., Souza, L.M. and Ogawa, M. (1988). Granulocyte colony stimulating factor enhances interleukin-3-dependent proliferation of multipotential hemopoietic progenitors. Proc. Natl Acad. Sci. USA 85, 3445–3449.

Isaacs, C., Robert, N.J., Bailey, F.A., et al. (1997). Randomized placebo-controlled study of recombinant human interleukin-11 to prevent chemotherapy-induced thrombocytopenia in patients with breast cancer receiving dose-intensive cyclophosphamide and doxorubicin. J. Clin. Oncol. 15, 3368–3377.

Kaushansky, K. and Karplus, P.A. (1993). Hematopoietic growth factors: understanding functional diversity in structural terms. Blood 82, 3229–3240.

Kawashima, I. and Takiguchi, Y. (1992). Interleukin-11: a novel stroma-derived cytokine. Prog. Growth Factor Res. 4, 191–206.

Kawashima, I., Ohsumi, J., Mita-Honjo, K., et al. (1991). Molecular cloning of cDNA encoding adipogenesis inhibitory factor and identify with interleukin-11. Fed. Eur. Biochem. Soc. 283, 199–202.

Kaye, J.A., Loewy, J., Blume, J., et al. (1994). Recombinant human interleukin eleven (Neumega™ rhIL-11 Growth Factor) increases plasma von Willebrand factor and fibrinogen concentrations in normal human subjects. Blood 84 (supplement I), 276a. [Abstract]

Keith, J.C., Albert, L., Sonis, S.T., Pfeiffer, C.J. and Schaub, R.G. (1994). IL-11, a pleiotropic cytokine: exciting new effects of IL-11 on gastrointestinal mucosal biology. Stem Cells 12 (supplement 1), 79–90.

Keller, D.C., Du, X.X., Srour, E.F., Hoffman, R. and Williams, D.A. (1993). Interleukin-11 inhibits adipogenesis and stimulates myelopoiesis in human long-term marrow cultures. Blood 82, 1428–1435.

Kobayashi, S., Teramura, M., Sugawara, I., Oshimi, K. and Mizoguchi, H. (1993). Interleukin-11 acts as an autocrine growth factor for human megakaryoblastic cell lines. Blood 81, 889–893.

Leary, A.G., Zeng, H.Q., Clark, S.C. and Ogawa, M. (1992). Growth factor requirements for survival in G_0 and entry into the cell cycle of primitive human hemopoietic progenitors. Proc. Natl Acad. Sci. USA 89, 4013–4017.

Lemoli, R.M., Fogli, M., Fortuna, A., et al. (1993). Interleukin-11 stimulates the proliferation of human hematopoietic CD34+ and CD34+ CD33−DR− cells and synergizes with stem cell factor, interleukin-3, and granulocyte-macrophage colony-stimulating factor. Exp. Hematol. 21, 1668–1672.

Lemoli, R.M., Fogli, M., Fortuna, A., et al. (1995). Interleukin-11 (IL-11) acts as a synergistic factor for the proliferation of human myeloid leukaemic cells. Br. J. Haematol. 91, 319–326.

Leonard, J.P., Quinto, C.M., Kozitza, M.K., Neben, T.Y. and Goldman, S.J. (1994). Recombinant human interleukin-11 stimulates multilineage hematopoietic recovery in mice after a myelosuppressive regimen of sublethal irradiation and carboplatin. Blood 83, 1499–1506.

Lu, Z.Y., Gu, Z.J., Zhang, X.G., et al. (1995). Interleukin-10 induces interleukin-11 responsiveness in human myeloma cell lines. FEBS Lett. 377, 515–518.

MacPherson, B. and Pfeiffer, C.J. (1978). Experimental production of diffuse colitis in rats. Digestion 17, 135–150.

Maier, R., Ganu, V. and Lotz, M. (1993). Interleukin-11, an inducible cytokine in human articular chondrocytes and synoviocytes, stimulates the production of the tissue inhibitor of metalloproteinases. J. Biol. Chem. 268, 21527–21532.

Mason, L., Timony, G., Perkin, C., et al. (1993). Human recombinant interleukin-11 promotes muturation of bone marrow megakaryocytes in nonhuman primates: an electron microscopic study. Blood 82 (supplement 1) 69a. [Abstract]

Maze, R., Moritz, T. and Williams, D.A. (1994). Increased survival and multilineage hematopoietic production from delayed and severe myelosuppressive effects on a nitrosourea with recombinant interleukin-11. Cancer Res. 54, 4947–4951.

McKinley, D., Wu, Q., Yang-Feng, T. and Yang, Y.-C. (1992). Genomic sequence and chromosomal location of human interleukin-11 gene (IL11). Genomics 13, 814–819.

Mehler, M.F., Rozental, R., Dougherty, M., Spray, D.C. and Kessler, J.A. (1993). Cytokine regulation of neuronal differentiation of hippocampal progenitor cells. Nature 362, 62–65.

Morris, J.C., Neben, S., Bennett, F., et al. (1996). Molecular cloning and characterization of murine interleukin-11. Exp. Hematol. 24, 1369–1376.

Murphy, G.M., Bitting, L., Majewska, A., Schmidt, K., Song, Y. and Wood, C.R. (1995). Expression of interleukin-11 and its encoding mRNA by glioblastoma cells. Neurosci. Lett. 196, 153–156.

Musashi, M., Yang, Y.-C., Paul, S.R., Clark, S.C., Sudo, T. and Ogawa, M. (1991a). Direct and synergistic effects of interleukin 11 on murine hemopoiesis in culture. Proc. Natl Acad. Sci. USA 88, 765–769.

Musashi, M., Clark, S.C., Sudo, T., Urdal, D.L. and Ogawa, M. (1991b). Synergistic interactions between interleukin-11 and interleukin-4 in support of proliferation of primitive hematopoietic progenitors of mice. Blood 78, 1448–1451.

Neben, T.Y., Loebelenz, J., Hayes, L., et al. (1993). Recombinant human interleukin-11 stimulates megakaryocytopoiesis and increases peripheral platelets in normal and splenectomized mice. Blood 81, 901–908.

Neben, S., Donaldson, D., Sieff, C., et al. (1994). Synergistic effects of interleukin-11 with other growth factors on the expansion of murine hematopoietic progenitors and maintenance of stem cells in liquid culture. Exp. Hematol. 22, 353–359.

Neben, S., Donaldson, D., Fitz, L., et al. (1996). Interleukin-4 (IL-4) in combination with IL-11 or IL-6 reverses the inhibitory effect of IL-3 on early B lymphocyte development. Exp. Hematol. 24, 783–789.

Ohsumi, J., Mijadai, I., Ishikawa-Ohsumi, H., Sakakibara, S., Mito-Honjo, K. and Takiguchi, T. (1991). Adipogenesis inhibitory factor. A novel inhibitory regulator of adipose conversion in bone marrow. FEBS Lett. 288, 13–16.

Paglia, D., Oran, A., Lu, C., Kerbel, R.S., Sauder, D.N. and McKenzie, R.C. (1995). Expression of leukemia inhibitory factor and interleukin-11 by human melanoma cell lines: LIF, IL-6, and IL-11 are not coregulated. J. Interferon Cytokine Res. 15, 455–460.

Passeri, G., Girasole, G., Knutson, S., Yang, Y.-C., Manolagas, S.C. and Jilka, R.L. (1993). Interleukin-11 (IL-11): a new cytokine with osteoclastogenic and bone resorptive properties and a critical role in PTH- and $1,25(OH)_2D_3$-induced osteoclast development. J. Bone Miner. Res. 1, 110A:S110.

Paul, S.R. and Schendel, P. (1992). The cloning and biological characterization of recombinant human interleukin 11. Int. J. Cell Cloning 10, 135–143.

Paul, S.R., Bennett, F., Calvetti, J.A., *et al.* (1990). Molecular cloning of a cDNA encoding interleukin 11, a stromal cell-derived lymphopoietic and hematopoietic cytokine. Proc. Natl Acad. Sci. USA 87, 7512–7516.

Peterson, R.L., Bozza, M.M. and Dorner, A.J. (1996). Interleukin-11 induces intestinal epithelial cell growth arrest through effects of retinoblastoma protein phosphorylation. Am. J. Pathol. 149, 895–902.

Qiu, B.S., Pfeiffer, C.J. and Keith, J.C. (1996). Protection by recombinant human interleukin-11 against experimental TNB-induced colitis in rats. Dig. Dis. Sci. 41(8), 1625–1630.

Quesniaux, V.F.J., Clark, S.C., Turner, K. and Fagg, B. (1992). Interleukin-11 stimulates multiple phases of erythopoiesis *in vitro*. Blood 80, 1218–1223.

Quesniaux, V.F.J., Mayer, P., Liehl, E., Turner, K., Goldman, S.J. and Fagg, B. (1993). Review of a novel hematopoietic cytokine, interleukin-11. Int. Rev. Exp. Pathol. 34A, 205–214.

Quinto, C.M., Leonard, J.P., Kozitza, M.K. and Goldman, S.J. (1993). Synergistic interactions between rhIL-11 and G-CSF in the mobilization of hematopoietic progenitors. Blood 82 (supplement 1), 369a. [Abstract]

Romas, E., Udagawa, N., Zhou, H., *et al.* (1996). The role of gp130-mediated signals in osteoclast development: regulation of interleukin-11 production by osteoblasts and distribution of its receptor in bone marrow cultures. J. Exp. Med. 183, 2581–2591.

Shaw, G. and Kamen, R.A. (1986). Conserved AU sequence from the 3′ untranslated region of GM-CSF mRNA mediates selective mRNA degradation. Cell 46, 659–667.

Sonis, S.T., Lindquist, L., Van Vugt, A., *et al.* (1994). Prevention of chemotherapy-induced ulcerative mucositis by transforming growth factor-β3. Cancer Res. 54, 1135–1138.

Sonis, S., Muska, A., O'Brien, J., Van Vugt, A., Langer-Safer, P. and Keith, J. (1995). Alteration in the frequency, severity and duration of chemotherapy-induced mucositis in hamsters by interleukin-11. Oral Oncol. Eur. J. Cancer 31B, 261–266.

Suen, Y., Chang, M., Killen, R., Cameron, K., Rosenthal, J. and Cairo, M. (1994). Physiological role of endogenous IL-11 during thrombocytopenia post myeloablative therapy (MAT) and BMT. Blood 84 (supplement 1), 91a. [Abstract]

Tepler, I., Elias, L., Smith, J.W., *et al.* (1996). A randomized placebo-controlled trial of recombinant human interleukin-11 in cancer patients with severe thrombocytopenia due to chemotherapy. Blood 87, 3607–3614.

Teramura, M., Kobayashi, S., Hoshino, S., Oshimi, K. and Mizoguchi, H. (1992). Interleukin-11 enhances human megakaryocytopoiesis *in vitro*. Blood 79, 327–331.

Tohyama, K., Yoshida, Y., Ohashi, K., *et al.* (1992). Production of multiple growth factors by a newly established human thyroid carcinoma cell line. Jpn J. Cancer Res. 83, 153–158.

Trepicchio, W.L., Bozza, M., Pedneault, G. and Dorner, A.J. (1996). Recombinant human interleukin-11 attenuates the inflammatory response through downregulation of proinflammatory cytokine release and nitric oxide production. J. Immunol. 157, 3627–3634.

Tsuji, K., Lyman, S.D., Sudo, T., Clark, S.C. and Ogawa, M. (1992). Enhancement of murine hematopoiesis by synergistic interactions between steel factor (ligand for c-*kit*), interleukin-11, and other early acting factors in culture. Blood 79, 2855–2860.

Turner, K.J. and Clark, S.C. (1995). Interleukin-11: biological and clinical perspectives. In "Hematopoietic Growth Factors in Clinical Applications" (eds. R. Mertelsmann and F. Herrmann), pp. 315–336. Marcel Dekker, New York.

Wang, X.Y., Fuhrer, D.K., Marshall, M.S. and Yang, Y.C. (1995). Interleukin-11 induces complex formation of Grb2, Fyn, and JAK2 in 3T3L1 cells. J. Biol. Chem. 270, 27999–28002.

Weich, N.S., Wang, A., Fitzgerald, M., *et al.* (1998). Recombinant human interleukin-11 directly promotes megakaryocytopoiesis *in vitro*. Blood [In press]

Yin, T. and Yang, Y.-C. (1993). Protein tyrosine phosphorylation and activation of primary response genes by interleukin 11 in B9-TY1 cells. Cell Growth Differ. 4, 603–609.

Yin, T., Schendel, P. and Yang, Y.-C. (1992a). Enhancement of *in vitro* and *in vivo* antigen-specific antibody responses by interleukin 11. J. Exp. Med. 175, 211–216.

Yin, T., Miyazawa, K. and Yang, Y.-C. (1992b). Characterization of interleukin-11 receptor and protein tyrosine phosphorylation induced by interleukin-11 in mouse 3T3-L1 cells. J. Biol. Chem. 267, 8347–8351.

Yin, T., Taga, T., Tsang, M.L., Yasukawa, K., Kishimoto, T. and Yang, T.C. (1993). Interleukin-11 mediated TF-1 cell proliferation is associated with interleukin-6 signal transducer, gp130. J. Immunol. 151, 2555–2561.

Yin, T., Yasukawa, K., Taga, T., Kishimoto, T. and Yang, Y.-C. (1994). Identification of a 130-kilodalton tyrosine-phosphorylated protein induced by interleukin-11 as JAK2 tyrosine kinase, which associates with gp130 signal transducer. Exp. Hematol. 22, 467–472.

Yonemura, Y., Kawakita, M., Masuda, T., Fujimoto, K., Kato, K. and Takatsuki, K. (1992). Synergistic effects of interleukin 3 and interleukin 11 on murine megakaryopoiesis in serum-free culture. Exp. Hematol. 20, 1011–1016.

Yonemura, Y., Kawakita, M., Masuda, T., Fujimoto, K. and Takatsuki, K. (1993). Effect of recombinant human interleukin-11 on rat megakaryopoiesis and thrombopoiesis *in vivo*: comparative study with interleukin-6. Br. J. Haematol. 84, 16–23.

Zhang, X.-G., Gu, J.-J., Lu, Z.-Y., *et al.* (1994). Ciliary neurotropic factor, interleukin 11, leukemia inhibitory factor, and oncostatin M are growth factors for human myeloma cell lines using the interleukin 6 signal transducer GP130. J. Exp. Med. 177, 1337–1342.

12. Interleukin-12

Richard Chizzonite, Ueli Gubler, Jeanne Magram *and* Alvin S. Stern

1. Introduction

Interleukin-12 (IL-12) is a potent modulator of natural killer (NK) and T cell functions. Through the activities on these two cell types, the cytokine provides an important link between natural resistance (NK cells) and adaptive immune responses (T cells) (Locksley, 1993; Trinchieri, 1993). IL-12 is unusual among the known cytokines because it is a heterodimer composed of two subunits (p35 and p40) that represent unrelated gene products. The two subunits have to be coexpressed in order to yield secreted, bioactive IL-12 (Gubler *et al.*, 1991; Wolf *et al.*, 1991). IL-12 was discovered independently by two research teams. Investigators at the Wistar Institute and Genetics Institute first designated the cytokine natural killer cell stimulatory factor (NKSF), while investigators at Hoffmann-La Roche proposed the term cytotoxic lymphocyte maturation factor (CLMF). The early

Cytokines
ISBN 0–12–498340–5

observation that B cell lines could directly contribute to an IL-2-independent proliferation of NK cells ultimately led to the identification of NKSF through its ability to induce the synthesis of IFN-γ in resting peripheral blood mononuclear cells (PBMC) (Kobayashi *et al.*, 1989). CLMF was identified by its ability to synergize with IL-2 in augmenting cytotoxic lymphocyte responses (Stern *et al.*, 1990). Through the purification of the natural proteins (Kobayashi *et al.*, 1989; Stern *et al.*, 1990) and the subsequent cloning of their mRNAs (Gubler *et al.*, 1991; Wolf *et al.*, 1991), it became clear that NKSF and CLMF were identical cytokines; hence, the common name IL-12 was proposed (Gubler *et al.*, 1991).

2. The IL-12 Genes

2.1 Sequences of the Human IL-12 p35 and p40 Subunit cDNAs

The sequences for the p40 and p35 cDNAs are shown in Figures 12.1 and 12.2. The cDNAs for both IL-12 subunits were originally cloned from EBV-transformed induced human B cell lines NC-37 (Gubler *et al.*, 1991) and RPMI 8866 (Wolf *et al.*, 1991). In both cases, the natural IL-12 protein was purified (Kobayashi *et al.*, 1989; Stern *et al.*, 1990) and partial amino acid sequence information from each respective subunit was used to design oligonucleotides for the screening of the cDNA libraries (Gubler *et al.*, 1991; Wolf *et al.*, 1991). Sequencing of the isolated cDNAs led to the conclusion that the p35 and p40 mRNAs represented unrelated gene products. The p35 mRNA is approximately 1.4 kb, whereas the p40 mRNA is approximately 2.4 kb. Although the human p35 mRNA sequence has two in-frame AUG codons (see Figure 12.2), there is good evidence that the second AUG codon is the one used for translational initiation: (i) an mRNA containing only the second AUG gives rise to functional p35 protein; (ii) the amino acid sequence following the second AUG has all the characteristics of a standard signal peptide, whereas the peptide sequence after the first (upstream) AUG codon does not (Gubler *et al.*, 1991; Wolf *et al.*, 1991); and, finally, (iii) the corresponding 5′ region in the murine p35 mRNA does not contain this upstream in-frame AUG and cannot be translated into the corresponding peptide sequence (Schoenhaut *et al.*, 1992). Both IL-12 subunit transcripts contain several copies of the RNA destabilizing sequence motif ATTTA in their 3′ noncoding regions (Caput *et al.*, 1986; Shaw and Kamen, 1986).

In the B cell lines used for the cDNA cloning, both subunit mRNAs are induced with similar kinetics from low or undetectable levels to readily detectable levels within a few hours (Gubler *et al.*, 1991; Wolf *et al.*, 1991). Both lipopolysaccharide (LPS) and *Staphylococcus*

aureus Cowan I strain (SAC) are potent inducers of the p40 transcript in normal PBMC; however, under these conditions, the p35 mRNA is expressed in an almost constitutive fashion (D'Andrea *et al.*, 1992, 1993).

2.2 Sequences of the Murine IL-12 p35 and p40 Subunit cDNAs

Since human recombinant IL-12 is inactive on murine lymphocytes (Schoenhaut *et al.*, 1992), it became important to obtain recombinant murine IL-12 for biological studies in mice. Using cross-hybridization, cDNAs encoding the murine IL-12 subunit homologs were isolated from cDNA libraries that had been generated from mRNA derived from pokeweed mitogen-stimulated splenocytes. The human subunit cDNAs were first used under conditions of reduced hybridization stringency to isolate partial murine genomic clones encoding bona fide exons; subsequently, these murine genomic probes were used to screen the murine cDNA libraries (Schoenhaut *et al.*, 1992). Human and murine p40 mRNAs are approximately 70% identical in sequence, whereas the p35 mRNA sequences are only about 60% identical. By northern blot analysis, the murine p35 mRNA is approximately 1.5 kb and the murine p40 mRNA approximately 2.6 kb. Similar to the human mRNAs, the 3′ noncoding regions of both murine IL-12 subunit mRNAs also contain several copies of the ATTTA destabilizing motif.

Some data are available on the expression of the murine subunit mRNAs in different tissues or cell types and under different induction conditions. Similar to the human IL-12 p35 subunit mRNA, very little regulation of the expression of murine p35 mRNA in macrophages has been observed (Romani *et al.*, 1994). In contrast, murine p40 mRNA is induced in macrophages by LPS or heat-killed *Listeria* (Reiner *et al.*, 1994). Treatment of the macrophage cell line J774 with IFN-γ was shown to induce expression of the p40 mRNA (Yoshida *et al.*, 1994). Studies on murine tissues demonstrated the presence of the p40 mRNA in spleen, cultured splenocytes, and thymus. Treatment of splenocytes with pokeweed mitogen results in a decrease in p40 mRNA levels. The p35 mRNA is present in spleen, and treatment with pokeweed mitogen has no effect on its steady-state levels. p35 mRNA is also expressed in nonhemopoietic tissues, i.e., lung and brain (Schoenhaut *et al.*, 1992), although the relevance of these findings is not known. IL-12 p35 and p40 mRNAs were also detected in fetal and adult thymus, and in two thymic epithelial cell lines (Godfrey *et al.*, 1994).

2.3 Genomic Structure

The IL-12 subunits p40 and p35 are encoded by separate and unrelated genes. Analysis of the human genes has

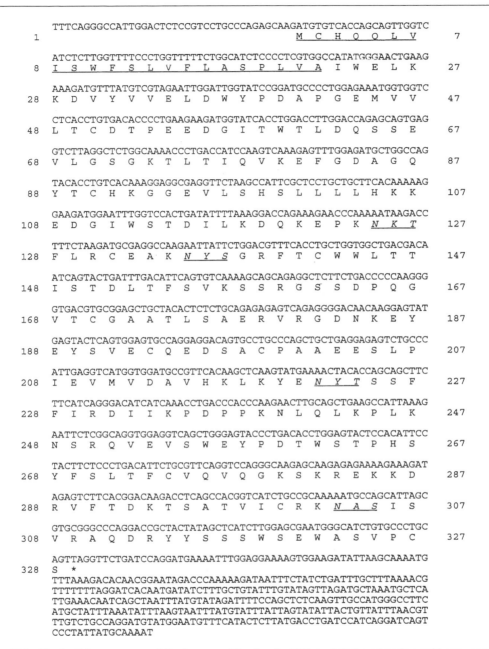

```
     TTTCAGGGCCATTGGACTCTCCGTCCTGCCCAGAGCAAGATGTGTCACCAGCAGTTGGTC
  1                                             M  C  H  Q  Q  L  V     7

     ATCTCTTGGTTTTCCCTGGTTTTTCTGGCATCTCCCCTCGTGGCCATATGGGAACTGAAG
  8  I  S  W  F  S  L  V  F  L  A  S  P  L  V  A  I  W  E  L  K    27

     AAAGATGTTTATGTCGTAGAATTGGATTGGTATCCGGATGCCCCTGGAGAAATGGTGGTC
 28  K  D  V  Y  V  V  E  L  D  W  Y  P  D  A  P  G  E  M  V  V    47

     CTCACCTGTGACACCCCTGAAGAAGATGGTATCACCTGGACCTTGGACCAGAGCAGTGAG
 48  L  T  C  D  T  P  E  E  D  G  I  T  W  T  L  D  Q  S  S  E    67

     GTCTTAGGCTCTGGCAAAACCCTGACCATCCAAGTCAAAGAGTTTGGAGATGCTGGCCAG
 68  V  L  G  S  G  K  T  L  T  I  Q  V  K  E  F  G  D  A  G  Q    87

     TACACCTGTCACAAAGGAGGCGAGGTTCTAAGCCATTCGCTCCTGCTGCTTCACAAAAAG
 88  Y  T  C  H  K  G  G  E  V  L  S  H  S  L  L  L  L  H  K  K   107

     GAAGATGGAATTTGGTCCACTGATATTTTAAAGGACCAGAAAGAACCCAAAAATAAGACC
108  E  D  G  I  W  S  T  D  I  L  K  D  Q  K  E  P  K  N  K  T   127

     TTTCTAAGATGCGAGGCCAAGAATTATTCTGGACGTTTCACCTGCTGGTGGCTGACGACA
128  F  L  R  C  E  A  K  N  Y  S  G  R  F  T  C  W  W  L  T  T   147

     ATCAGTACTGATTTGACATTCAGTGTCAAAAGCAGCAGAGGCTCTTCTGACCCCCAAGGG
148  I  S  T  D  L  T  F  S  V  K  S  S  R  G  S  S  D  P  Q  G   167

     GTGACGTGCGGAGCTGCTACACTCTCTGCAGAGAGAGTCAGAGGGGACAACAAGGAGTAT
168  V  T  C  G  A  A  T  L  S  A  E  R  V  R  G  D  N  K  E  Y   187

     GAGTACTCAGTGGAGTGCCAGGAGGACAGTGCCTGCCCAGCTGCTGAGGAGAGTCTGCCC
188  E  Y  S  V  E  C  Q  E  D  S  A  C  P  A  A  E  E  S  L  P   207

     ATTGAGGTCATGGTGGATGCCGTTCACAAGCTCAAGTATGAAAACTACACCAGCAGCTTC
208  I  E  V  M  V  D  A  V  H  K  L  K  Y  E  N  Y  T  S  S  F   227

     TTCATCAGGGACATCATCAAACCTGACCCACCCAAGAACTTGCAGCTGAAGCCATTAAAG
228  F  I  R  D  I  I  K  P  D  P  P  K  N  L  Q  L  K  P  L  K   247

     AATTCTCGGCAGGTGGAGGTCAGCTGGGAGTACCCTGACACCTGGAGTACTCCACATTCC
248  N  S  R  Q  V  E  V  S  W  E  Y  P  D  T  W  S  T  P  H  S   267

     TACTTCTCCCTGACATTCTGCGTTCAGGTCCAGGGCAAGAGCAAGAGAGAAAAGAAAGAT
268  Y  F  S  L  T  F  C  V  Q  V  Q  G  K  S  K  R  E  K  K  D   287

     AGAGTCTTCACGGACAAGACCTCAGCCACGGTCATCTGCCGCAAAAATGCCAGCATTAGC
288  R  V  F  T  D  K  T  S  A  T  V  I  C  R  K  N  A  S  I  S   307

     GTGCGGGCCCAGGACCGCTACTATAGCTCATCTTGGAGCGAATGGGCATCTGTGCCCTGC
308  V  R  A  Q  D  R  Y  Y  S  S  S  W  S  E  W  A  S  V  P  C   327

     AGTTAGGTTCTGATCCAGGATGAAAATTTGGAGGAAAAGTGGAAGATATTAAGCAAAATG
328  S  *
     TTTAAAGACACAACGGAATAGACCCAAAAAGATAATTTCTATCTGATTTGCTTTAAAACG
     TTTTTTTAGGATCACAATGATATCTTTGCTGTATTTGTATAGTTAGATGCTAAATGCTCA
     TTGAAACAATCAGCTAATTTATGTATAGATTTTCCAGCTCTCAAGTTGCCATGGGCCTTC
     ATGCTATTTAAATATTTAAGTAATTTATGTATTTATTAGTATATTACTGTTATTTAACGT
     TTGTCTGCCAGGATGTATGGAATGTTTCATACTCTTATGACCTGATCCATCAGGATCAGT
     CCCTATTATGCAAAAT
```

Figure 12.1   Nucleotide sequence of the human p40 subunit cDNA and deduced amino acid sequence for the preform of the p40 protein. The signal peptide is underlined and the predicted N-linked glycosylation sites are underlined and italicized.

demonstrated that they are unlinked and reside on chromosomes 5q31-33 and 3p12-3q13.2, respectively (Sieburth *et al.*, 1992). Further analysis of the human gene structure has not yet been published.

Although the chromosomal location of the mouse genes has yet to be determined, the genomic structure of these genes has been analyzed (J. Magram, unpublished data). The sequence of the p40 gene which aligns with the cDNA spans approximately 13 kbp and is divided into eight exons (see Figure 12.3). The first two introns are significantly larger than the others and constitute a large portion of the gene. The first exon consists of noncoding DNA since the initiating ATG is encoded in the second exon. Sequence analysis upstream from the first exon reveals the presence of a putative TATA site and a CAT box approximately 47 and 109 nucleotides, respectively, 5′ to the first nucleotide encoded by the known cDNA

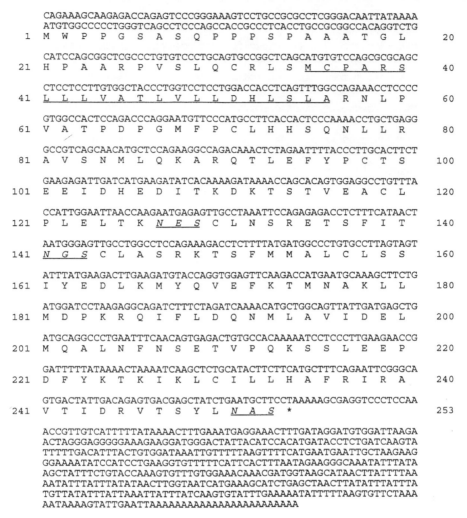

```
      CAGAAAGCAAGAGACCAGAGTCCCGGGAAAGTCCTGCCGCGCCTCGGGACAATTATAAAA
      ATGTGGCCCCCTGGGTCAGCCTCCCAGCCACCGCCCTCACCTGCCGCGGCCACAGGTCTG
  1   M  W  P  P  G  S  A  S  Q  P  P  P  S  P  A  A  A  T  G  L      20

      CATCCAGCGGCTCGCCCTGTGTCCCTGCAGTGCCGGCTCAGCATGTGTCCAGCGCGCAGC
 21   H  P  A  A  R  P  V  S  L  Q  C  R  L  S  M  C  P  A  R  S      40

      CTCCTCCTTGTGGCTACCCTGGTCCTCCTGGACCACCTCAGTTTGGCCAGAAACCTCCCC
 41   L  L  L  V  A  T  L  V  L  L  D  H  L  S  L  A  R  N  L  P      60

      GTGGCCACTCCAGACCCAGGAATGTTCCCATGCCTTCACCACTCCCAAAACCTGCTGAGG
 61   V  A  T  P  D  P  G  M  F  P  C  L  H  H  S  Q  N  L  L  R      80

      GCCGTCAGCAACATGCTCCAGAAGGCCAGACAAACTCTAGAATTTTACCCTTGCACTTCT
 81   A  V  S  N  M  L  Q  K  A  R  Q  T  L  E  F  Y  P  C  T  S     100

      GAAGAGATTGATCATGAAGATATCACAAAAGATAAAACCAGCACAGTGGAGGCCTGTTTA
101   E  E  I  D  H  E  D  I  T  K  D  K  T  S  T  V  E  A  C  L     120

      CCATTGGAATTAACCAAGAATGAGAGTTGCCTAAATTCCAGAGAGACCTCTTTCATAACT
121   P  L  E  L  T  K  N  E  S  C  L  N  S  R  E  T  S  F  I  T     140

      AATGGGAGTTGCCTGGCCTCCAGAAAGACCTCTTTTATGATGGCCCTGTGCCTTAGTAGT
141   N  G  S  C  L  A  S  R  K  T  S  F  M  M  A  L  C  L  S  S     160

      ATTTATGAAGACTTGAAGATGTACCAGGTGGAGTTCAAGACCATGAATGCAAAGCTTCTG
161   I  Y  E  D  L  K  M  Y  Q  V  E  F  K  T  M  N  A  K  L  L     180

      ATGGATCCTAAGAGGCAGATCTTTCTAGATCAAAACATGCTGGCAGTTATTGATGAGCTG
181   M  D  P  K  R  Q  I  F  L  D  Q  N  M  L  A  V  I  D  E  L     200

      ATGCAGGCCCTGAATTTCAACAGTGAGACTGTGCCACAAAAATCCTCCCTTGAAGAACCG
201   M  Q  A  L  N  F  N  S  E  T  V  P  Q  K  S  S  L  E  E  P     220

      GATTTTTATAAAACTAAAATCAAGCTCTGCATACTTCTTCATGCTTTCAGAATTCGGGCA
221   D  F  Y  K  T  K  I  K  L  C  I  L  L  H  A  F  R  I  R  A     240

      GTGACTATTGACAGAGTGACGAGCTATCTGAATGCTTCCTAAAAAGCGAGGTCCCTCCAA
241   V  T  I  D  R  V  T  S  Y  L  N  A  S  *                      253

      ACCGTTGTCATTTTTATAAAACTTTGAAATGAGGAAACTTTGATAGGATGTGGATTAAGA
      ACTAGGGAGGGGGAAAGAAGGATGGGACTATTACATCCACATGATACCTCTGATCAAGTA
      TTTTTGACATTTACTGTGGATAAATTGTTTTTAAGTTTTCATGAATGAATTGCTAAGAAG
      GGAAAATATCCATCCTGAAGGTGTTTTTCATTCACTTTAATAGAAGGGCAAATATTTATA
      AGCTATTTCTGTACCAAAGTGTTTGTGGAAACAAACGATGGTAAGCATAACTTATTTTAA
      AATATTTATTTATATAACTTGGTAATCATGAAAGCATCTGAGCTAACTTATATTTATTTA
      TGTTATATTTATTAAATTATTTATCAAGTGTATTTGAAAAATATTTTTAAGTGTTCTAAA
      AATAAAAGTATTGAATTAAAAAAAAAAAAAAAAAAAAAAAAA
```

Figure 12.2   Nucleotide sequence of the human p35 subunit cDNA and deduced amino acid sequence for the preform of the p35 protein. The signal peptide following the second AUG codon is underlined and the predicted N-linked glycosylation sites are underlined and italicized.

(Schoenhaut *et al.*, 1992). Since the true initiation of transcription has not yet been identified, these sites are potentially aligned to function correctly as bona fide promoter elements.

The p35 subunit coding sequence spans 7.4 kb of genomic DNA and like the p40 gene contains two introns which are larger than the rest. Interestingly, it also contains two extremely small introns of 72 and 100 bp. The p35 gene consists of seven exons as shown in Figure 12.4. In contrast to the p40 gene, sequence analysis upstream from exon 1 does not reveal any putative promoter elements. This most likely indicates one of two possibilities: (i) since the initiation of transcription is unknown, the sequence obtained may not represent the promoter region (i.e. there may be additional noncoding exons yet to be identified) or (ii) the p35 promoter may not be a classical promoter.

3.   The Protein

Among the family of interleukin proteins that have been characterized to date, IL-12 is structurally unique. It is a heterodimeric protein comprising two disulfide-linked subunits (Kobayashi *et al.*, 1989; Stern *et al.*, 1990; Podlaski *et al.*, 1992) with neither IL-12 subunit alone displaying significant activity over the range of concentrations at which the IL-12 heterodimer is active (Gubler *et al.*, 1991; Wolf *et al.*, 1991). However, it has been proposed that the larger subunit (p40) may interact with the receptor or stabilize the smaller subunit (p35) in a conformation that allows binding to the receptor (Podlaski *et al.*, 1992). The p40 subunit (molecular mass 40 kDa) of human IL-12 is composed of 306 amino acids (Figure 12.5a) and contains 10 cysteine residues, while the p35 subunit (molecular mass 35 kDa) consists of 197

p40 Genomic Structure

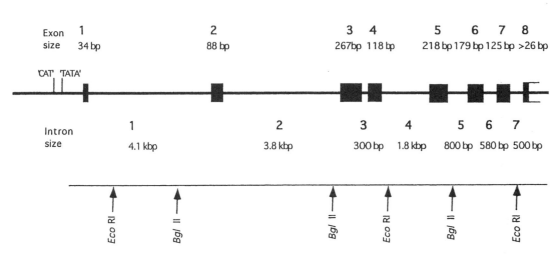

Figure 12.3   Genomic structure of the p40 subunit of IL-12. Exon/intron boundaries and locations were determined by PCR and sequencing. This diagram illustrates the alignment with the known cDNA and therefore contains the entire coding sequence but is not complete for noncoding sequences. Putative CAT and TATA boxes were determined by sequencing. Some known restriction sites are shown (↑).

p35 Genomic Structure

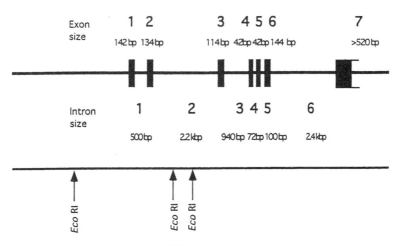

Figure 12.4   Genomic structure of the p35 subunit of IL-12. Exon/intron boundaries and location were determined by PCR and sequencing. This diagram illustrates the alignment with the known cDNA and therefore contains the entire coding sequence but is not complete for noncoding sequences. EcoRI sites are shown (↑).

amino acids (Figure 12.5b) and has 7 cysteine residues (Gubler *et al.*, 1991; Wolf *et al.*, 1991). p35 contains six intramolecular S—S bonds: Cys^{15}–Cys^{88}, Cys^{42}–Cys^{174}, and Cys^{63}–Cys^{101}. The intermolecular disulfide bond is between p35 Cys-74 and p40 Cys-177. p40 contains 8 intramolecular S—S bonds: Cys^{-28}–Cys^{68}, Cys^{109}–Cys^{120}, Cys^{148}–Cys^{171}, and Cys^{278}–Cys^{305}. (Tangarone *et al.*, 1995). It has also been reported that the free sulfhydryl group found in the recombinant p40 protein is

cysteinylated or paired with the sulfur in thioglycolic acid (Tangarone *et al.*, 1995). IL-12 had been initially shown to be a glycoprotein by its adsorption to immobilized *Lens culinaris* agglutinin (Kobayashi *et al.*, 1989). Furthermore, the deduced amino acid sequence of the subunits revealed four potential *N*-linked glycosylation sites on the 40 kDa structure (Figure 12.5a) and three putative *N*-linked sites on the 35 kDa subunit (Gubler *et al.*, 1991; Wolf *et al.*, 1991). It has been demonstrated

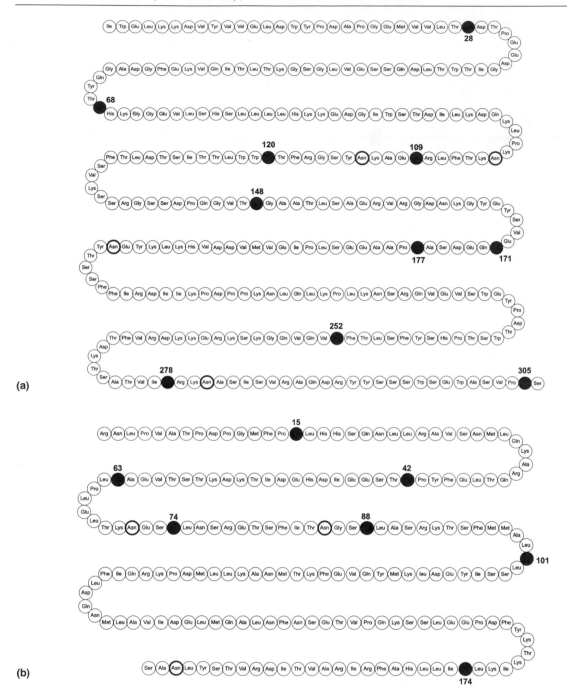

Figure 12.5   Primary protein structure of (a) the 40 kDa and (b) 35 kDa subunits of IL-12. The complete amino acid sequence of the subunits were predicted from the cloned cDNA of the subunits (Gubler *et al.*, 1991; Wolf *et al.*, 1991). Potential *N*-linked oligosaccharide sites are shown as bold circles. Cysteine residues are shown shaded.

(Podlaski *et al.*, 1992) that at least one of the sites on the 40 kDa subunit is glycosylated (Asn-200), whereas a second is not (Asn-103). Chemical and enzymatic analyses of the human IL-12 heterodimer, isolated from the human lymphoblastoid cell line NC-37 demonstrated that the p40 and p35 subunits contain 10% and 20% carbohydrate, respectively (Podlaski *et al.*, 1992).

The 35 kDa subunit of IL-12 shares homology with IL-6, G-CSF, and chicken myelomonocytic growth factor (Merberg *et al.*, 1992), and has, like most other cytokines, an α-helix-rich structure. The 40 kDa subunit is not homologous to other cytokines, but belongs to the hemopoietin receptor family and most resembles the IL-6 receptor and the ciliary neurotrophic factor receptor

(Gearing *et al.*, 1991; Schoenhaut *et al.*, 1992). However, there is no evidence for the existence of membrane-associated forms of IL-12 or either of its subunits.

The predicted amino acid sequence of the murine IL-12 p40 subunit shares 70% identity to human IL-12 p40, whereas the human and mouse p35 subunits share 60% amino acid sequence identity (Schoenhaut *et al.*, 1992). Murine IL-12 is fully active on human lymphocytes, but, as discussed above, human IL-12 is inactive on murine lymphocytes. Amino acid differences in the p35 subunits of human and murine IL-12 are critical for the observed species specificity of IL-12 (Schoenhaut *et al.*, 1992).

Recombinant IL-12 protein has been purified from the medium of cultures of CHO cells transfected with IL-12 cDNAs. The protein backbone has a molecular mass of 57.2 kDa, although the glycosylated protein is approximately 70 kDa as determined by analytical ultracentrifugation (E.H. Braswell and A.S., unpublished data). Based on isoelectric focusing gels, there are three major bands in the pI range of 4.5–5.3. Although long-term storage of the bulk protein in a liquid state has not been thoroughly investigated, it has been found that the protein is stable in pH 7 buffer free of calcium, magnesium, and potassium salts, at concentrations of approximately 1 mg/ml at –70°C.

4. Sources of IL-12

The primary cellular source of IL-12 production is the monocyte/macrophage. In addition, B cells produce IL-12 but at significantly lower levels (D'Andrea *et al.*, 1992). These observations derive from analyses of a variety of both human and mouse cells and cell lines.

When cultures of human PBMC are incubated with SAC, LPS, or killed *Mycobacterium tuberculosis*, a dramatic increase in IL-12 production is observed (D'Andrea *et al.*, 1992). SAC is the most potent inducer. Although PBMC appear to secrete low levels of IL-12 in the absence of inducers, the presence of minimal amounts of endotoxin in the medium could not be excluded in these studies (D'Andrea *et al.*, 1992). The cell types in these cultures which appear to produce the IL-12 are comprised of both adherent monocytes/macrophages and nonadherent cells. The nonadherent cells which are most likely producing the IL-12 are B lymphocytes, since phytohemagglutinin (PHA)-activated PBMC (PHA blasts; mostly activated T cells) and NK cells do not produce detectable IL-12. PBMC fail to enhance IL-12 secretion in response to phorbol esters (D'Andrea *et al.*, 1992). In addition, levels of IL-12 secreted are not elevated in response to a variety of cytokines including IL-1α, IL-1β, IL-2, IL-4, IL-6, IFN-γ, IFN-β, TNF-α, TNF-β or GM-CSF (D'Andrea *et al.*, 1992).

Analysis of human IL-12 produced by a number of *in vitro* EBV-transformed cell lines demonstrated that most of these lines secrete both the free p40 chain and the heterodimer (D'Andrea *et al.*, 1992). As opposed to PBMC, production of both of these molecules was significantly increased upon stimulation by phorbol ester (D'Andrea *et al.*, 1992). In contrast, no detectable IL-12 was produced by Burkitt lymphoma-derived cell lines. Also, no IL-12 could be detected in several T cell lines, myeloid leukemic cell lines, or other solid tumor cell lines, in the absence or presence of stimulation by phorbol ester (D'Andrea *et al.*, 1992).

In the murine system, evidence also points to the monocyte/macrophage as the primary source for the production of IL-12 (D'Andrea *et al.*, 1992): (i) production of IL-12 was observed in *Listeria monocytogenes*-stimulated spleen cells derived from severe combined immunodeficiency (SCID) mutant mice which lack B and T cells (Tripp *et al.*, 1993) and (ii) production of IL-12 was also detected from *Listeria monocytogenes*-stimulated peritoneal macrophages *in vitro* and was negatively regulated by IL-10 (Hsieh *et al.*, 1993).

5. Biological Activities

5.1 EFFECTS ON CYTOLYTIC CELLS

One of the initially described activities of IL-12 was its ability to activate spontaneously cytotoxic human NK/lymphokine-activated killer (NK/LAK) cells to become cytolytic (Kobayashi *et al.*, 1989). The original observation has now been confirmed and extended by numerous laboratories using recombinant IL-12 on both human (Lieberman *et al.*, 1991; Wolf *et al.*, 1991; Chehimi *et al.*, 1992, 1993; Gately *et al.*, 1992; Naume *et al.*, 1992, 1993b; Robertson *et al.*, 1992; Zeh *et al.*, 1993) and murine NK cells (Schoenhaut *et al.*, 1992; Gately *et al.*, 1994b). Activation of CD56⁺ NK cells by overnight incubation *in vitro* with IL-12 resulted in enhanced killing of NK-sensitive and NK-resistant tumor target cells (Chehimi *et al.*, 1992; Robertson *et al.*, 1992), antibody-coated tumor target cells (Lieberman *et al.*, 1991; Robertson *et al.*, 1992), and virus-infected fibroblasts or T cells (Chehimi *et al.*, 1992, 1993). The effects of IL-12 are independent of IL-2, IFN-α, IFN-β, IFN-γ, and TNF-α (Robertson *et al.*, 1992; Chehimi *et al.*, 1993). As with human NK activity, mouse IL-12 induced NK activity of murine spleen cells *in vitro* (Schoenhaut *et al.*, 1992). Furthermore, enhanced NK activity has been demonstrated in the spleens, livers, lungs, and peritoneal cavities of mice treated *in vivo* with recombinant murine IL-12 (Gately *et al.*, 1994b).

LAK cell activity has been observed following incubation of human peripheral blood lymphocytes (PBL) with IL-12 for longer times (Gately *et al.*, 1992; Naume *et al.*, 1992, 1993b; Zeh *et al.*, 1993; Rossi *et al.*, 1994). The primary cell type stimulated in these cultures is a CD56$^+$ NK cell (Gately et al., 1992; Naume et al., 1992, 1993b) and this induction can be partially inhibited by antibodies to TNF-α, but antibodies to IL-2 or IFN-γ had little or no effect on IL-12-induced activation (Gately et al., 1992; Naume et al., 1992). The maximum NK/LAK activity induced by optimal concentrations of IL-12 is lower than that which can be achieved with IL-2 (Kobayashi et al., 1989; Chehimi et al., 1992; Gately et al., 1992; Naume et al., 1992; Robertson et al., 1992; Cesano et al., 1993). Combinations of IL-12 and IL-2 can result in synergistic (Stern et al., 1990; Gately et al., 1992; Rossi et al., 1994), additive (Gately et al., 1992; Cesano et al., 1993), less than additive (Chehimi et al., 1992), or inhibitory (Gately et al., 1992; Zeh et al., 1993) effects on enhancement of NK/LAK activity depending on the cytokine concentrations and length of culture.

IL-12 has also been shown to upregulate the cell surface expression of adhesion/activation molecules and cytokine receptors on cytolytic cells, including CD2, CD11a, CD54, CD56, CD69, CD71, HLA-DR, the 75 kDa TNF receptor, and receptors for IL-2 (α and β subunits), IL-4, and IL-12 (Robertson et al., 1992; Gerosa et al., 1993; Naume et al., 1992, 1993b; Rabinowich et al., 1993; Jewett and Bonavida, 1994).

In addition to its effects on NK/LAK cells, IL-12 can also facilitate the induction of specific human CTL responses to weakly immunogenic allogeneic melanoma tumor cells (Wong et al., 1988; Gately et al., 1992). The induction of CTL required both IL-12 and irradiated melanoma cells, and the resulting CD3$^+$ T cells specifically lysed allogeneic cells (Gately et al., 1992). Both IL-12 and IL-2 were active in this assay, but the concentration of IL-12 necessary to induce a half-maximal response was approximately 10-fold lower than the concentration of IL-2 (Gately et al., 1992). IL-12 can also enhance the lytic activity of purified human CD8$^+$ T cells stimulated by immobilized anti-CD3 (Mehrotra et al., 1993). Consistent with these in vitro results, it has now been shown that IL-12 administered in vivo to mice injected with allogeneic cells can enhance specific CD8$^+$ CTL responses (Gately et al., 1994b). The amount of IL-12 necessary to induce this specific CTL response in vivo was 10- to 100-fold higher than that required to augment NK activity (Gately et al., 1994b).

5.2 EFFECTS ON CELL PROLIFERATION

IL-12 has been shown to stimulate the proliferation of T cells in various short-term in vitro assays. In contrast to other cytokines (e.g., IL-2 and IL-7), IL-12 stimulates minimal proliferation of resting PBMC but can induce proliferation of lymphocytes that are activated by a number of stimuli (Kobayashi et al., 1989; Stern et al., 1990; Gately et al., 1991; Gubler et al., 1991; Wolf et al., 1991; Bertagnolli et al., 1992;

Perussia et al., 1992; Andrews et al., 1993; Zeh et al., 1993). The growth-promoting effects of IL-12 on activated T cells have been demonstrated using T cells isolated from PBMC, T cell clones, CTL lines, and tumor infiltrating lymphocytes (TILs) (Gately et al., 1991; Bertagnolli et al., 1992; Perussia et al., 1992; Andrews et al., 1993; Zeh et al., 1993). IL-12 stimulates the proliferation of CD4$^+$ and CD8$^+$ TCR-$\alpha\beta^+$ T cells and TCR-$\gamma\delta^+$ lymphoblasts (Gately et al., 1991; Perussia et al., 1992). This effect of IL-12 is independent of IL-2 (Gately et al., 1991; Bertagnolli et al., 1992; Perussia et al., 1992; Andrews et al., 1993). The maximum proliferation induced by IL-12 on PHA-activated lymphocytes is approximately one-half that induced by IL-2 but similar to that obtained with either IL-4 or IL-7 (Gately et al., 1991). However, the concentration of IL-12 needed to obtain half-maximal proliferation was substantially less than the concentrations needed for IL-2, IL-4, or IL-7 (Gately et al., 1991). Combinations of IL-12 with suboptimal amounts of IL-2 resulted in additive proliferation of activated lymphocytes (Gately et al., 1991; Bertagnolli et al., 1992; Perussia et al., 1992; Andrews et al., 1993). However, at higher concentrations of IL-2, IL-12 has no effect or inhibits IL-2 induced proliferation depending on the cell type (Perussia et al., 1992). Recently, it has been reported that the highest level of T cell proliferation, comparable to that observed with IL-2, is obtained by combining IL-12 with activation of the B7-CD28 pathway (Kubin et al., 1994a; Murphy et al., 1994). The bioactivity of IL-12 is frequently measured by its ability to stimulate proliferation of activated T cells in vitro (Gately and Chizzonite, 1992).

Several groups have also demonstrated that IL-12 can induce the proliferation of NK cells (Gately et al., 1991; Naume et al., 1992, 1993b; Perussia et al., 1992; Robertson et al., 1992). IL-12 can directly stimulate the proliferation of CD56$^+$ NK cells, but the level of proliferation induced was only 10% of the level induced by IL-2 and approximately 50% of the IL-7-induced proliferative responses (Naume et al., 1992). IL-12 had either an additive (Naume et al., 1992; Perussia et al., 1992) or inhibitory (Robertson et al., 1992; Perussia et al., 1992) effect on IL-2-induced NK cell proliferation depending on the culture conditions used. The inhibitory effect of IL-12 on IL-2-induced NK proliferation was blocked by neutralizing anti-TNF antibodies (Perussia et al., 1992). In contrast to the inhibitory effects of IL-4 on IL-2- and IL-7-induced NK cell proliferation, combinations of IL-4 and IL-12 resulted in synergistic proliferation of NK cells (Nagler et al., 1988; Stotter et al., 1991; Robertson et al., 1992; Naume et al., 1993b); these results suggest that IL-12-induced signaling on NK cells may differ from that of IL-2 or IL-7. In general, IL-12 by itself is a weaker inducer of NK cell proliferation than of T cell proliferation.

5.3   EFFECTS ON CYTOKINE INDUCTION

Incubation of human resting or activated PBL with IL-12 results in a dose-dependent induction of IFN-γ (Chan *et al.*, 1991, 1992; Kobayashi *et al.*, 1989; Wolf *et al.*, 1991; Perussia *et al.*, 1992; Naume *et al.*, 1993a; Wu *et al.*, 1993) from both T and NK cells (Chan *et al.*, 1991; Perussia *et al.*, 1992). Resting PBL and neonatal T cells require the presence of accessory cells to produce IFN-γ in response to IL-12 (Chan *et al.*, 1991; Wu *et al.*, 1993), but there is no accessory cell requirement for activated T or NK cells to respond to IL-12 (Chan *et al.*, 1991). The effect of IL-12 is largely independent of the induction of IL-2 since anti-IL-2 antibodies only slightly reduce the IL-12-stimulated production of IFN-γ (Chan *et al.*, 1991). Synergistic induction of IFN-γ has also been observed with combinations of IL-12 plus other stimuli including IL-2 (Kobayashi *et al.*, 1989; Chan *et al.*, 1991, 1992; Wolf *et al.*, 1991; Wu *et al.*, 1993), IL-1 and TNF-α (Wu *et al.*, 1993), PHA, phorbol ester, or anti-CD3 antibodies (Chan *et al.*, 1991). In contrast IL-4 (Kiniwa *et al.*, 1992), transforming growth factor beta (TGF-β) (Chan *et al.*, 1991) and IL-10 (D'Andrea *et al.*, 1993) inhibit IL-12-induced production of IFN-γ from human lymphoid cells. IL-6 does not synergize with IL-12 to induce IFN-γ (Wu *et al.*, 1993).

Murine IL-12 can also induce/enhance IFN-γ secretion from murine spleen cells (Schoenhaut *et al.*, 1992; Gazzinelli *et al.*, 1993; Tripp *et al.*, 1993) and naive T cells that were antigen-primed *in vitro* (Hsieh *et al.*, 1993). IL-12-stimulated IFN-γ secretion can be further enhanced by exposure of the cells concurrently to IL-2 (Hsieh *et al.*, 1993) or TNF-α (Gazzinelli *et al.*, 1993; Tripp *et al.*, 1993). In contrast, IL-10 inhibits IL-12 plus TNF-α-induced production of IFN-γ from murine NK cells (Tripp *et al.*, 1993). Mice injected with murine IL-12 produce high serum levels of IFN-γ, and liver lymphoid cells from these animals spontaneously secrete IFN-γ *ex vivo* (Gately *et al.*, 1994b; Morris *et al.*, 1994).

IL-12 has been shown to induce the secretion of low amounts of TNF-α from alloactivated (Perussia *et al.*, 1992) or resting NK cells (Naume *et al.*, 1992), but substantially more TNF-α is secreted by IL-2-induced than by IL-12-induced NK cells (Naume *et al.*, 1992). Incubation of human NK cells with IL-12 plus IL-2 results in an additive induction of TNF-α (Perussia *et al.*, 1992). Human NK cells also produce low levels of GM-CSF and IL-8 in response to IL-12 (Naume *et al.*, 1993a). Following injection of mice with IL-12, the levels of mRNA for splenic TNF-α and IL-10 are increased, as are the number of spleen cells secreting IL-10 (Morris *et al.*, 1994).

5.4   EFFECT ON REGULATION OF T$_H$1/T$_H$2 CELLS

Helper T cells can be divided into a T$_H$1 subset that secretes IFN-γ and IL-2, resulting in cell-mediated immune responses; a T$_H$2 subset that produces IL-4, IL-5 and IL-10, facilitating humoral immune responses; and a T$_H$0 subset that secretes both IFN-γ and IL-4 (Scott, 1993; Trinchieri, 1993). There is now strong evidence accumulating that a unique property of IL-12 is the regulation of the induction of the T$_H$1 subsets of murine T helper cells (Hsieh *et al.*, 1993; Seder *et al.*, 1993; Schmitt *et al.*, 1994a,b) and human T helper cells (Romagnani, 1992; Manetti *et al.*, 1993; Trinchieri, 1993). In an *in vitro* model utilizing ovalbumin-specific αβ T cell receptor CD4⁺ T cells, recombinant murine IL-12, but not a number of other cytokines, preferentially induced the development of T$_H$1 cells from naive murine T cells (Hsieh *et al.*, 1993). IL-12 has also been shown to directly effect naive CD4⁺ T cells (Seder *et al.*, 1993; Schmitt *et al.*, 1994a) and murine T$_H$1 but not T$_H$2 cell clones (Germann *et al.*, 1993; Yanagida *et al.*, 1994). In cultures of T$_H$2 clones incubated with IL-4 and IL-12, the effects of IL-4 were dominant and T$_H$2 cells were induced (Hsieh *et al.*, 1993; Seder *et al.*, 1993; Schmitt *et al.*, 1994a). IFN-γ is necessary for most of the IL-12-mediated effects on T$_H$1 cells but it cannot replace IL-12 (Hsieh *et al.*, 1993; Seder *et al.*, 1993; Schmitt *et al.*, 1994b). TGF-β inhibited IL-12-induced development of T$_H$1 cells (Schmitt *et al.*, 1994b). In addition to promoting the commitment of naive T cells to the T$_H$1 pathway, IL-12 has been shown to serve as an important costimulus with the B7-CD28 pathway for the activation of fully differentiated murine T$_H$1 cells (Murphy *et al.*, 1994).

The effect of IL-12 on T$_H$1/T$_H$2 cell development has also been observed on human cells (Romagnani, 1992; Manetti *et al.*, 1993; Trinchieri, 1993). CD4⁺ T cell lines, activated *in vitro* with *Dermatophagoides pteronysinus* group 1 antigen, generally exhibit a T$_H$2-like phenotype producing IL-4 but little or no IFN-γ (Parronchi *et al.*, 1991). However, cell lines generated in the presence of human IL-12 exhibit a T$_H$0 profile (secreting both IFN-γ and IL-4) or T$_H$1-like cytokine profile (secreting INF-γ but little IL-4) (Manetti *et al.*, 1993). Conversely, purified protein derivative-specific T cell lines which normally exhibit a T$_H$1-like phenotype develop into T$_H$0-like cell lines if generated in the presence of anti-IL-12 antibody. In contrast to murine T$_H$2 clones (Germann *et al.*, 1993; Yanagida *et al.*, 1994), activation of cloned human T$_H$2 cells in the presence of IL-12 resulted in conversion to the T$_H$0 or T$_H$1 phenotype (Manetti *et al.*, 1994).

5.5   EFFECT ON IgE PRODUCTION

IL-12 markedly inhibits IL-4-stimulated IgE production but not pokeweed mitogen-stimulated synthesis of IgG, IgM, or IgA from human PBMC *in vitro* (Kiniwa *et al.*, 1992). The effect of IL-12 was mediated in part by an IFN-γ-dependent and in part by an IFN-γ-independent mechanism (Kiniwa et al., 1992). Similar effects have now also been demonstrated in vivo (Finkelman et al., 1994; Morris et al., 1994; see Section 5.7). IL-12 does not inhibit IgE responses by B cells that have already switched to membrane IgE expression (Morris et al., 1994). In addition, administration of IL-12 can inhibit IgE responses in helminth-infected mice during a primary but not during a secondary response (Finkelman et al., 1994). These results suggest that IL-12 can induce a T_H1 response during a primary response but is much less effective in reversing established T_H2 responses.

Simultaneous administration of IL-12 and the hapten–protein conjugate TNP-keyhole limpet hemocyanin (KLH) suppressed the production of anti-TNP antibodies of the IgG1 isotype, which is associated with T_H2-type responses (McKnight et al., 1994). In the same animals, IL-12 caused a modest increase in the levels of anti-TNP antibodies of the IgG2a isotype, which is associated with T_H1-type responses (McKnight et al., 1994). Thus, under certain circumstances, administration of IL-12 may lead to an increase of humoral, as well as cell-mediated, immunity.

5.6   EFFECT ON HEMATOPOIESIS AND LYMPHOID PROGENITOR CELLS

Although IL-12 alone cannot stimulate murine hematopoietic stem and progenitor cells in vitro, synergistic colony formation was found when IL-12 was combined with other hematopoietic cytokines (Jacobsen et al., 1993; Ploemacher et al., 1993a,b; Hirayama et al., 1994). Using single-cell cloning assays, IL-12 was shown to have a direct effect on stem cells (Jacobsen et al., 1993; Hirayama et al., 1994). In the presence of various cytokines (e.g., stem cell factor (SCF), IL-3, and IL-11), IL-12 enhanced formation of myeloid, lympho-hematopoietic progenitors (pre-B), megakaryocytic, erythroid, and mast cell colonies but not eosinophil colonies (Jacobsen et al., 1993; Ploemacher et al., 1993a,b; Hirayama et al., 1994). The effects of IL-12 were not reversed by anti-IFN-γ antibodies. Recent data have also demonstrated that human IL-12 can synergize with SCF and IL-3 to induce formation of mixed, erythroid, and myeloid colonies from human progenitor cells in peripheral blood and bone marrow (Bellone and Trinchieri, 1994). In the presence of NK cells, IL-12 inhibited formation of hematopoietic colonies through the induction of TNF-α and IFN-γ, cytokines which suppress hematopoietic colony formation (Bellone and Trinchieri, 1994). It is possible that the inhibitory effects of IL-12 on murine bone marrow hematopoiesis in vivo may be mediated through an NK cell mechanism.

IL-12 may also influence T cell development in the thymus. Addition of IL-12 to fetal thymic organ cultures led to a significant reduction in total cell number but there was an increase in the number of αβ $TCR^+CD4^-CD8^+$ thymocytes (Godfrey et al., 1994). In combination with IL-2 and IL-4, IL-12 stimulated the proliferation of isolated $CD3^+CD4^-CD8^+$ and $CD3^-CD4^-CD8^-$ thymocytes. In addition, IL-12 plus SCF induced proliferation of $CD3^-CD4^-CD8^-$ triple negative $CD44^+CD25^+$ pro-T cells.

5.7   IMMUNOMODULATORY EFFECTS IN NORMAL MICE

The majority of biological activities of IL-12 originally elucidated in vitro can also be demonstrated in vivo. Administration of mouse IL-12 to normal mice over 2 days resulted in enhanced NK activity in the spleen, liver, lungs, and peritoneal cavity in a dose-dependent manner. Administration of IL-12 also enhanced specific CTL responses in mice which were immunized with allogeneic splenocytes in the footpads and then given daily injections of IL-12, beginning on the day of immunization. The lytic activity was mediated by $CD8^+$ T cells (Gately et al., 1994b). Dose–response studies indicated that higher doses of IL-12 were required to enhance specific CTL responses than to increase NK lytic activity (Gately et al., 1994b).

IL-12-treated mice also developed splenomegaly, which was shown histologically to be due largely to increased extramedullary hematopoiesis (Gately et al., 1994b). In contrast to this marked extramedullary hematopoiesis, IL-12 suppressed hematopoiesis in the bone marrow. When mice were allowed to rest for 1–2 weeks without further treatment, the IL-12-induced hematological changes reverted to normal or near normal.

In contrast to the enhancement of cell-mediated immunity observed in IL-12-treated mice, IL-12 tends to suppress humoral immunity, particularly IgE responses. When mice which had been injected with goat anti-mouse IgD antibody (GaMδ) were treated with IL-12, the IL-12 suppressed GaMδ-induced serum IgG1 and IgE responses by >98% and IgG2a and IgG3 responses to a lesser extent (Morris et al., 1994). These effects correlated with the ability of IL-12 to promote T_H1-type cytokine responses (enhanced IFN-γ and reduced IL-4 and IL-3) and to inhibit T_H2-type responses. Unexpectedly, IL-12 also enhanced the expression of IL-10 mRNA in the spleens of both normal and GaMδ-treated mice. Since IL-10 can inhibit IL-12 synthesis (D'Andrea et al., 1993), this may represent a negative feedback loop through which levels of endogenous IL-12 are normally regulated. Treatment of the GaMδ-injected mice with neutralizing anti-IFN-γ antibody largely reversed the IL-12-mediated suppression of IgG1 synthesis but not of IgE synthesis and resulted in enhancement of IgG2a and IgG3 responses by IL-12.

5.8 ANTAGONISTIC EFFECTS OF IL-12 p40 SUBUNIT HOMODIMER

Disulfide-linked heterodimeric IL-12 is necessary for the biological activity since neither the p40 nor the p35 subunit alone can substitute for IL-12. However, culture supernatants containing only murine p40 inhibit the biological activity of murine IL-12 in several *in vitro* assays (Mattner *et al.*, 1993). Although both p40 monomer and a disulfide-linked p40 homodimer are present in such culture supernatants, the p40 homodimer is substantially more potent than the monomer in inhibiting IL-12 activity (Gately and Brunda, 1994; Gillessen *et al.*, 1994). Purified mouse homodimeric p40 produced a dose-dependent inhibition of mouse IL-12-induced (i) proliferation of mouse Con A blasts, (ii) IFN-γ secretion by mouse splenocytes, and (iii) activation of mouse NK cells. However, mouse p40 homodimer did not inhibit human IL-12-induced proliferation of human PHA-lymphoblasts, even though mouse IL-12 is equally active on human cells. Likewise, a disulfide-linked homodimer of human p40 is an antagonist of IL-12 receptor binding and biological activity on human cells (Ling *et al.*, 1994). Lymphoid cells that have been induced to secrete IL-12 heterodimer also produce a substantial excess of p40 (D'Andrea *et al.*, 1992; Podlaski *et al.*, 1992). At present it is not known whether the excess p40 includes p40 homodimer which could act as a physiological regulator of IL-12 activity.

6. *IL-12 Receptor*

6.1 EXPRESSION OF IL-12 RECEPTOR

Using flow cytometry, IL-12 receptors (IL-12R) have been detected on resting human NK cells (Desai *et al.*, 1992) but not on resting human T cells (Desai *et al.*, 1992), B cells (Desai *et al.*, 1992), or monocytes (M. Gately, unpublished results). IL-12R are upregulated on activated CD4$^+$ and CD8$^+$ T cells (Desai *et al.*, 1992) and on activated NK cells (Naume *et al.*, 1993a; Desai *et al.*, 1992), but not on activated B cells (Desai *et al.*, 1992). Several cytokines, including IL-2 and IL-12 itself, have been shown to upregulate expression of the IL-12R (Desai *et al.*, 1992; Naume *et al.*, 1993a). Using radiolabeled IL-12, initial equilibrium binding studies suggested that activated human T cells displayed a single affinity class of IL-12R (Chizzonite *et al.*, 1992). However, more detailed analyses using a more biologically active and lower radiospecific activity preparation of IL-12 indicated that PHA-activated human T-lymphoblasts, a human T cell clone, and a human NK-like cell line possess three affinity classes of binding sites with apparent dissociation constants of 5–20 pM, 50–200 pM, and 2–6 nM, respectively (Chizzonite *et al.*, 1994; Chua *et al.*, 1994).

Equilibrium binding studies with ^{125}I-murine IL-12 on concanavalin A-activated murine splenocytes also identified three affinity classes of IL-12 binding sites with apparent dissociation constants of 40 pM, 200 pM, and 7 nM (Chizzonite *et al.*, 1993). High- or low-affinity ^{125}I-IL-12 binding sites were not detected on unactivated splenocytes on mouse B cell, fibroblast, or macrophage cell lines (Chizzonite *et al.*, 1993).

6.2 STRUCTURE OF IL-12 RECEPTOR ON HUMAN AND MURINE LYMPHOBLASTS

Affinity cross-linking of surface-bound ^{125}I-IL-12 to activated human lymphoblasts identified a major complex of ~210–280 kDa, suggesting that IL-12 was bound to one or more protein(s) of aggregate size of 140–210 kDa (Chizzonite *et al.*, 1992). Anti-IL-12 antibodies also immunoprecipitated a complex of 210–280 kDa that was produced by cross-linking unlabeled IL-12 to ^{125}I-labeled lymphoblast surface proteins (Chizzonite *et al.*, 1992). Cleavage of this complex with reducing agent and analysis by SDS-PAGE identified a 110 kDa IL-12-binding protein. An 85 kDa protein also coprecipitated with the IL-12-110 kDa protein complex, but it is not known whether the association was specific or nonspecific. Affinity cross-linking studies also demonstrated that ^{125}I-IL-12 was bound in a complex of 210–280 kDa on the surface of murine activated splenocytes, similar to the complex identified from human lymphoblasts (Chizzonite *et al.*, 1993). These data suggested that the IL-12R is composed of at least two subunits: a ~110 kDa IL-12 binding protein and an additional protein.

6.3 STRUCTURE OF AN IL-12 RECEPTOR COMPONENT

Using a nonneutralizing monoclonal antibody raised against the human IL-12 receptor complex on PHA-blasts, a cDNA encoding an IL-12 receptor subunit was isolated by expression cloning (Chua *et al.*, 1994). This receptor subunit is a 662-amino acid type I transmembrane protein, with an extracellular domain of 516 amino acids and a cytoplasmic tail of 91 amino acids (Figure 12.6). The protein sequence as deduced from the cDNA predicts six *N*-linked glycosylation sites within the extracellular domain. When expressed in COS cells, the receptor size is approximately 100 kDa. It is a member of the cytokine receptor superfamily of proteins and is most closely related to gp130 and the receptors for LIF and G-CSF. When expressed in COS cells, this IL-12 receptor subunit binds the IL-12 ligand only with low affinity (2–5 nM). The low-affinity binding on COS cells is mediated by receptor dimers or oligomers; however, their

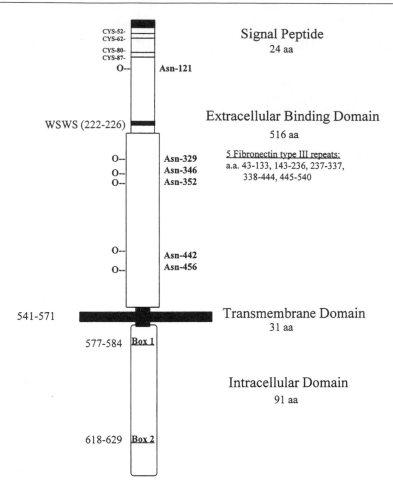

Figure 12.6   Schematic outline of the structure of the human low-affinity IL-12 receptor β₁ subunit (IL-12Rβ₁). Signal peptide is shown shaded. The two pairs of disulfide bonds and the WSXWS motif conserved among the cytokine receptor superfamily members are shown. The fibronectin type III repeats in the extracellular domain are listed. O— (Asn-#) indicates the predicted *N*-linked glycosylation sites. The locations of the cytoplasmic box 1 and 2 motifs are shown.

formation is not IL-12 dependent. Receptor monomers on COS cells do not bind IL-12 with a measurable affinity. The rather short cytoplasmic tail of the receptor shows the box 1 and box 2 motifs (Figure 12.6) that are also found in other signaling members of the receptor superfamily (Murakami *et al.*, 1991). Thus, this cloned IL-12 receptor component appears to be a β-type receptor subunit (Stahl and Yancopoulos, 1993). There is experimental support for the notion that it is indeed involved in IL-12 signaling: polyclonal antibodies raised against the receptor protein specifically inhibit the IL-12-induced proliferation of PHA-blasts (Chua *et al.*, 1994). More recent work has identified a second IL-12 receptor chain. This IL-12 receptor subunit is also a member of the cytokine receptor superfamily of proteins and within that family is most closely related to gp130. The two known IL-12 receptor subunits are now classified as IL-12Rβ₁ (previously IL-12Rβ) and IL-12Rβ₂ (newly identified

subunit). Coexpression of IL-12Rβ₁ and β₂ leads to the formation of a functional, high-affinity IL-12 receptor complex (Gubler and Presky, 1996).

6.4   Interaction between IL-12 Subunits and IL-12 Receptor

Anti-p40 subunit specific antibodies and p40 homodimer block ¹²⁵I-IL-12 binding to both high- and low-affinity IL-12 receptors on human and mouse lymphoblasts (Gately *et al.*, 1994, Chizzonite *et al.*, 1991, 1994) and to low-affinity recombinant receptor expressed in COS cells (Chizzonite *et al.*, 1994; Gillessen *et al.*, 1994). However, 20C2, an anti-human IL-12 monoclonal antibody (mAb) that binds to an epitope composed of both the p35 and p40 IL-12 subunits, blocks ¹²⁵I-IL-12 binding to the high-affinity receptor but not the low-affinity receptor on human lymphoblasts (Chizzonite *et al.*, 1994; Gillessen *et al.*, 1994). In addition, mAb 20C2 does not block IL-12 binding to recombinant IL-12 receptor expressed on COS cells. A combination of the anti-p40 reagents (either anti-p40 antibody or p40 homodimer) and mAb 20C2 is synergistic at inhibiting human IL-12-induced proliferation of human lymphoblasts. These data suggest that (i) the p40 subunit is responsible for IL-12 binding to its receptor and (ii) the p35 subunit is required for signaling.

6.5   IL-12 RECEPTOR SIGNALING

Thus far, only a few reports investigating IL-12 signaling events have appeared in the literature. In mitogen-activated human T cells, but not in resting T cells, IL-12 induces rapid tyrosine phosphorylation of a 44 kDa protein. This protein is a member of the MAP kinase family. The IL-12 induced increase in phosphorylation of this MAP kinase was shown to result in enhanced kinase activity (Pignata *et al.*, 1994). In purified NK cells, IL-12 induced rapid phosphorylation and enhanced autophosphorylation activity of p56^{lck} (Pignata *et al.*, 1993). In PHA-activated lymphoblasts, IL-12 was reported to induce the phosphorylation of JAK2 and TYK2 (Bacon *et al.*, 1995a) and of STAT3 and STAT4 (Bacon *et al.*, 1995b; Jacobson *et al.*, 1995).

7.   *Clinical Implications*

7.1   ACTIVITY IN INFECTIOUS DISEASE MODELS

IL-12 has been shown to have therapeutic effects in several murine models of infectious disease caused by intracellular pathogens. The initial observation that IL-12 might be therapeutically useful was demonstrated in a murine leishmaniasis model (Heinzel *et al.*, 1993; Sypek *et al.*, 1993). Previously it had been shown that different strains of mice varied in their susceptibility to infection by *Leishmania major* (Heinzel *et al.*, 1989). Some strains, such as C57BL/6, generate a predominantly T_H1-type cytokine response to infection and resolve the infection, while others, such as BALB/c, form a predominantly T_H2-type cytokine response and succumb to infection. Administration of IL-12 to *Leishmania*-infected BALB/c mice during the first week after infection resulted in a substantial reduction of the parasite burden and cure (Heinzel *et al.*, 1993; Sypek *et al.*, 1993). Furthermore, mice that had been cured of *Leishmania major* infection by IL-12 were immune to reinfection (Heinzel *et al.*, 1993). IL-12-induced resistance to infection was correlated with decreased production of IL-4 by draining lymph node cells cultured with *Leishmania* antigen and preserved or increased

production of IFN-γ, i.e., a shift from a T_H2-type cytokine response to a T_H1-type response. The curative effect of IL-12 in this model could be abolished by simultaneous administration of neutralizing anti-IFN-γ antibody (Heinzel *et al.*, 1993), although comparable curative effect could not be achieved by administration of recombinant IFN-γ (Sadick *et al.*, 1990).

IL-12 was also found to inhibit both primary and secondary pulmonary granuloma formation induced by *Schistosoma mansoni* eggs. The mechanism of this inhibition was shown to involve an IL-12-induced shift from a predominantly T_H2-type cytokine response to a T_H1-type response in the primary granulomas or to a T_H0-type response in the secondary granulomas (Wynn *et al.*, 1994). In addition to promoting resistance to infectious agents by enhancing T_H1-type responses, IL-12 has been shown to increase resistance to intracellular pathogens by a T cell-independent mechanism involving IL-12-induced IFN-γ production by NK cells followed by IFN-γ-mediated activation of macrophages to display enhanced microbicidal activity (Gazzinelli *et al.*, 1993; Tripp *et al.*, 1993). Thus in T and B cell-deficient SCID mice infected with *Toxoplasma gondii*, treatment with IL-12 significantly prolonged survival by a mechanism dependent on NK cells and IFN-γ (Gazzinelli *et al.*, 1993). IL-12 treatment has also been shown to protect normal mice against acute infection with a lethal dose of *Toxoplasma gondii* (Khan *et al.*, 1994) and to reduce the bioburden in the brains and livers of mice infected with *Cryptococcus neoformans* (Clemons *et al.*, 1994). In the *Cryptococcus* infection model, IL-12 was also shown to synergize with the antifungal agent fluconazole in reducing infection (Clemons *et al.*, 1994).

Studies using neutralizing anti-IL-12 antibodies have indicated that IL-12 produced endogenously in response to infection plays an important role in the host defense against infection by intracellular pathogens. Administration of anti-IL-12 to *Leishmania*-infected but normally resistant C57BL/6 mice exacerbated infection (Sypek *et al.*, 1993), and giving anti-IL-12 to mice injected with *Schistosoma mansoni* eggs led to a dramatic increase in the size of egg-induced pulmonary granulomas (Wynn *et al.*, 1994). In both these models, the effects of anti-IL-12 correlated with a shift from a predominantly T_H1-type cytokine response to a T_H2-type response. In either normal or SCID mice infected with a sublethal dose of *Listeria monocytogenes*, injection of anti-IL-12 resulted in increased *Listeria* burden in the spleen, decreased macrophage I-A expression, and death of the mice; administration of IFN-γ together with the anti-IL-12 reversed all of these effects (Tripp *et al.*, 1994). Administering anti-IL-12 to normal mice infected with *Toxoplasma gondii* at the LD$_{50}$ was also shown to decrease survival if the antibody was given at the time of infection (Khan *et al.*, 1994). Thus, endogenous IL-12 appears to play an essential role in the host defense against a variety of intracellular pathogens in murine models. Consistent with this observation, IL-12 levels were significantly higher in pleural fluids from human patients with tuberculous pleuritis than in serum from the same patients or in malignant pleural effusions (Zhang *et al.*, 1994). In addition, anti-IL-12 antibody partially suppressed the proliferation of pleural fluid mononuclear cells induced *in vitro* by *Microbacteria tuberculosis* (Zhang *et al.*, 1994). These results suggest that IL-12 may play a role in the immune response to *M. tuberculosis* infection in man.

The therapeutic effects of IL-12 in *Toxoplasma*-infected SCID mice suggested that IL-12 might have utility in treating opportunistic infections in patients lacking normal T cell function, including HIV-infected patients. The possible use of IL-12 in HIV-infected patients is also supported by the observations that (i) PBMC from HIV-infected patients are deficient in their ability to produce IL-12 *in vitro* but produce normal or elevated levels of IL-1, TNF, and IL-6 (Chehimi *et al.*, 1994); (ii) PBMC from HIV-infected patients can respond to IL-12 in assays of IL-12-induced IFN-γ secretion and enhancement of NK lytic activity (Chehimi *et al.*, 1992); and (iii) IL-12 restored the ability of PBMC from HIV-infected patients to secrete IL-2 and proliferate in response to a variety of stimuli, including a pool of Env peptides (Clerici *et al.*, 1993). The observation that PBMC from HIV-infected patients are deficient in their ability to produce IL-12 but can respond to it suggests that endogenous IL-12 appears to play an essential role in the host defense against infection by a number of intracellular pathogens. The observation that administration of IL-12 can have therapeutic effects against such pathogens in T cell-deficient mice provides a compelling rationale for testing the possible therapeutic effects of IL-12 in HIV-associated opportunistic infections. The possibility that IL-12, through its effects in promoting T_H1-mediated immunity, might slow the course of HIV infection is more speculative.

In some infectious disease models, IL-12 has shown detrimental rather than beneficial effects. In mice infected with lymphocytic choriomeningitis virus (LCMV), administration of low doses of IL-12 (1–10 ng/day) caused a small but significant decrease in the splenic and renal LCMV titers, whereas treatment with a high dose of IL-12 (1 μg/day) resulted in approximately a 2 log increase in viral titers as compared to untreated controls (Orange *et al.*, 1994). The adverse effects of high-dose IL-12 in this model were associated with the disappearance of CD8$^+$ T cells from the spleen, blood, and lymph nodes of IL-12-treated mice and with enhanced serum IFN-γ and TNF levels. These adverse effects were not seen in mice infected with murine cytomegalovirus and treated with IL-12 (Orange *et al.*, 1994). In at least some nematode infections, T_H2-type cytokine responses may be protective while T_H1-type responses are detrimental. Consistent with this, administration of IL-12 to mice infected with

Nippostrongylus brasiliensis enhanced adult worm survival of the mice and egg production if given during a primary infection, but survival was only enhanced if IL-12 was given for the first time during a secondary infection (Finkelman *et al.*, 1994). This correlated with the ability of IL-12 to shift the cytokine profile from a T_H2-like to a T_H1-like response if given during the primary infection and to a T_H0-like response if given only during the secondary infection. The ability of IL-12 to enhance worm survival and fecundity was IFN-γ dependent (Finkelman *et al.*, 1994).

An additional use of IL-12 in infectious diseases, and potentially also in oncology, is its use as a vaccine adjuvant. This was first suggested by studies in the murine leishmaniasis model in which susceptible BALB/c mice that had been immunized with soluble *Leishmania* antigen (SLA) plus IL-12 were protected against a subsequent infection with *Leishmania major*, whereas mice immunized with SLA alone were not (Afonso *et al.*, 1994). Protection was correlated with the development of a T_H1-type cytokine response which was dependent on both NK cells and IFN-γ (Afonso *et al.*, 1994). However, immunization with SLA plus IFN-γ resulted in only partial protection (Scott, 1991). The potential utility of IL-12 as a vaccine adjuvant is also suggested by studies in the murine schistosomiasis model in which mice were immunized with schistosome eggs with and without IL-12 (Wynn *et al.*, 1994). Mice that received IL-12 were almost completely protected from pulmonary granulomatous responses following intravenous challenge with schistosome eggs, and this was correlated with enhanced production of IFN-γ and decreased production of IL-4, IL-5, and IL-13, as assessed by measurement of cytokine mRNA levels in lung tissue.

7.2 Antimetastatic and Antitumor Effects in Mice

As noted above, administration of IL-12 to normal mice can promote T_H1-type cytokine responses, increase the lytic activity of NK/LAK cells, augment specific CTL responses, induce the production of IFN-γ, and enhance macrophage tumoricidal activity. Given these actions, it is not surprising that IL-12 has been found to exert potent antimetastatic and antitumor effects in a number of murine tumor models.

7.2.1   Activity in Animal Models

The antimetastatic effects of IL-12 were first demonstrated in an experimental pulmonary metastasis model using B16F10 melanoma cells (Brunda *et al.*, 1993a). Treatment with IL-12 beginning on one day after intravenous injection of melanoma cells resulted in dose-dependent inhibition of the number of pulmonary metastases. In this model, a substantial antimetastatic effect could be seen even when the initiation of IL-12 treatment was delayed until day 7 after tumor cell injection. Similar results were observed in an MC-38 experimental pulmonary metastasis model in which treatment with IL-12 was begun on day 10 after injection of tumor cells (Nastala *et al.*, 1994). Treatment with IL-12 also reduced the number of experimental hepatic metastases of M5076 reticulum cell sarcoma, resulting in increased survival (Brunda *et al.*, 1993a). In addition to its efficacy in experimental metastasis models, IL-12 has been shown to inhibit spontaneous hepatic metastases of M5076 reticulum cell sarcoma (Gately *et al.*, 1994a) and spontaneous pulmonary metastases of Lewis lung carcinoma (Stern *et al.*, 1994). In the hepatic metastasis model, IL-12 was effective when treatment was initiated at day 28 following subcutaneous injection of tumor cells, indicating that it could inhibit established micrometastases (Brunda *et al.*, 1993a).

IL-12 has also been shown to exert potent antitumor effects against a number of established, subcutaneous tumors of a variety of types including B16F10 melanoma, Renca renal adenocarcinoma, M5076 reticulum cell sarcoma, MCA-105 and MCA-207 sarcomas, Lewis lung carcinoma, and MC-38 colon adenocarcinoma (Brunda *et al.*, 1993a; O'Toole *et al.*, 1993; Brunda, 1994; Brunda and Gately, 1994; Gately *et al.*, 1994a; Nastala *et al.*, 1994; Stern *et al.*, 1994). The antitumor effects of IL-12 were initially demonstrated in the B16F10 melanoma model in which administration of IL-12 daily, five days per week, beginning 7 or 14 days after tumor cell inoculation, resulted in dose-dependent inhibition of tumor growth and increased length of survival (Brunda *et al.*, 1993a). However, complete regressions were not observed in this model, and tumor growth resumed rapidly when IL-12 therapy was discontinued. In other tumor models, including Renca renal carcinoma, MCA-105 and MCA-207 sarcomas, and MC-38 colon adenocarcinoma, complete regressions have been observed following peritumoral or systemic administration of IL-12 (Brunda *et al.*, 1993a,b; Gately and Brunda, 1994; Nastala *et al.*, 1994).

7.2.2   Mechanism of Action

The mechanism by which IL-12 exerts its antitumor effects is not fully understood and may vary with the tumor type and location. IL-12 has not been found to have any direct antiproliferative activity on tumor lines *in vitro* and, thus, it is likely that the antitumor effects of IL-12 are mediated through its actions on cells of the immune system. IL-12 displayed full antitumor activity in NK-deficient beige mice inoculated with B16F10 melanoma cells and in Renca or MCA-207 tumor-bearing mice that had been depleted of NK cells by treatment with anti-asialo-GM1 antibody (Brunda *et al.*, 1993a; Natala *et al.*, 1994), suggesting that NK cells do not constitute an essential component of the mechanism by which IL-12 exerts its antitumor effects in these three

tumor models. However, in both the B16F10 and Renca models the antitumor efficacy of IL-12 was substantially reduced in T cell-deficient nude mice, indicating that T cells play a critical role in mediating the antitumor activity of IL-12 against these tumors (Brunda *et al.*, 1993a). IFN-γ plays an important role in mediating the antitumor effects of IL-12 (Natala *et al.*, 1994). Treatment of mice with neutralizing anti-IFN-γ antibody significantly reduced the antitumor activity of IL-12 in the MC-38 colon carcinoma (Natala *et al.*, 1994). However, it appears that the antitumor activity of IL-12 is not due solely to the direct effects of IL-12-induced IFN-γ on tumor cells, since administration of IL-12 to normal or tumor-bearing nude mice resulted in serum IFN-γ levels approximately 8-fold higher than in euthymic controls (M.J. Brunda *et al.*, personal communication). Thus induction of IFN-γ is necessary but not sufficient for the antitumor action of IL-12. Furthermore, administration of recombinant mouse IFN-γ at the maximum nonlethal dose to Renca tumor-bearing mice did not result in antitumor effects comparable to those achieved with IL-12 therapy (M.J. Brunda *et al.*, personal communication). It may be that sustained local production of IFN-γ induced by IL-12 is much more effective in achieving antitumor effects than periodic administration of high doses of IFN-γ systemically, or IL-12 may exert antitumor effects through the combination of inducing IFN-γ and exerting other actions that are not mimicked by administration of IFN-γ alone. Possible mechanisms by which IFN-γ may contribute to the antitumor effects of IL-12 include activation of macrophages and enhancement of T cell-mediated immunity, either via direct effects on T cells or via upregulation of antigen expression by tumor cells. Both macrophages (Brunda and Gately, 1994; Tahara *et al.*, 1994) and CD8+ T cells (Nastala *et al.*, 1994) have been reported to infiltrate tumors in mice treated with IL-12.

7.3   PHARMACOKINETIC AND TOXICOLOGICAL PROPERTIES

The pharmacokinetic properties of human IL-12 have been examined in both rats and rhesus monkeys (Gately *et al.*, 1994a). Following intravenous bolus injection of IL-12 (50 μg/kg), to Sprague–Dawley rats, the majority (95%) of the IL-12 was eliminated from the serum with a half-life of 2.7 h. When human IL-12 (42.5 μg/kg) was administered to rhesus monkeys by 40-min intravenous infusion, the mean elimination serum half-life was 14 h. In rhesus monkeys given IL-12 by the subcutaneous route, the bioavailability was 17–29%. Thus, compared to cytokines such as IFN-α (Wills and Spiegel, 1985) and IL-2 (Gustavson *et al.*, 1989), IL-12 has a relatively long serum half-life. On the other hand, the bioavailability of IL-12 is low in comparison to that of IFN-α (Wills *et al.*,

1984). These properties may reflect its high molecular mass in comparison to other cytokines and suggest that less frequent dosing compared to cytokines with lower molecular masses (e.g., INF-α and IL-2) may be possible.

Studies to define the toxicities which may be associated with IL-12 therapy have been performed in normal and tumor-bearing mice and primates. The primary toxicities observed in normal mice treated with murine IL-12 (doses of 0.1–10 µg/day for up to 2 weeks) were hematological toxicities, hepatotoxicity, and skeletal muscle degeneration (Gately et al., 1994a; Lipman et al., 1994). Daily administration of 1 µg of IL-12 for 7 days led to severe anemia with red blood cell counts dropping by ~50%. Both lymphopenia and neutropenia were also observed and were probably related to margination of leukocytes onto vascular endothelium and migration into tissues such as the liver. A mild thrombocytopenia was seen but was not associated with any bleeding tendencies. Histological studies showed that the splenomegaly resulting from IL-12 administration was largely due to extensive extramedullary hematopoiesis involving the erythroid, myeloid, and megakaryocytic lineages. On the other hand, the bone marrow was hypoplastic, with a loss of mature neutrophils and of red blood cell precursors. Histological examination of livers of mice receiving high doses of IL-12 revealed marked Kupffer cell hyperplasia and mononuclear cell infiltrates in the hepatic parenchyma. Flow cytometry studies indicated that these infiltrates are composed predominantly of macrophages, NK cells, and $CD8^+$ T cells (Gately et al., 1994b). Lymphoid infiltrates in the liver were initially associated with occasional necrosis of isolated hepatocytes, but with continued administration of high doses of IL-12 this progressed to areas of coagulative necrosis with marked elevation of serum transaminases. Muscle toxicity from an unknown mechanism was seen in mice receiving IL-12 at doses of 1 to 10 µg/day and was characterized by visibly white muscle at necropsy, muscle necrosis and calcification, and elevations of serum muscle enzymes beginning after about 5 days of treatment. Normal mice given high doses of IL-12 displayed ascites and pleural effusions, but in contrast to prior studies with IL-2 (Gately et al., 1988), no pulmonary edema was observed and there was minimal or no lymphoid infiltration into the lung.

An initial evaluation of the toxicological profile of IL-12 in primates was performed in squirrel monkeys (Saimiri sciureus) (Gately et al., 1994a; Sarmiento et al., 1994). Monkeys received daily subcutaneous injections of IL-12 at doses ranging from 0.1 to 50 µg/kg per day for 14 days. The major dose-related findings were mild temperature elevation, mild to moderate anemia, leukocytosis, hypoproteinemia, hypoalbuminemia, hypophosphatemia, hypocalcemia, generalized lymph node enlargement and splenomegaly, and, at the highest dose

only, vascular leak. No hepatotoxicity or muscle degeneration was observed. Histopathology demonstrated dose-dependent thymic cortical atrophy, splenic lymphoid hyperplasia with histiocytosis and extramedullary hematopoiesis of the red pulp, hepatic Kupffer cell hypertrophy and hyperplasia, trilineage bone marrow hyperplasia, and reactive hyperplasia of the lymph nodes. Monkeys receiving IL-12 at doses of 0.1 or 1 µg/kg did not display any signs of toxicity, although IL-12 caused increased LAK activity and lectin-dependent cytotoxicity at these doses (Gately et al., 1994a; Sarmiento et al., 1994).

Besides the toxicities described above, there are additional concerns which apply to the use of IL-12 in cases of sepsis or autoimmune disease. Recent data suggest that IL-12 may play a role in the pathogenesis of septic shock. In a recent study in the generalized Shwartzman reaction in mice, the results suggest that endotoxin induces production of IL-12, which in turn stimulates the secretion of IFN-γ, resulting in priming of macrophages and other cell types for a lethal response to the endotoxin challenge (Ozmen et al., 1994). In another recent study, injection of anti-IL-12 antibodies resulted in a significant reduction in the levels of serum IFN-γ observed following intraperitoneal injection of endotoxin (Heinzel et al., 1994). Thus, IL-12 produced in response to endotoxin appears to regulate the production of IFN-γ, a cytokine believed to play an important role in endotoxin-induced pathology (Doherty et al., 1992). These results suggest that IL-12 should not be used in the case of sepsis.

Overall, toxicological results to date suggest no major toxicities that would preclude the use of IL-12 in man for the treatment of malignancies. In the squirrel monkey studies, evidence of immunomodulatory activity was observed at a dose of IL-12 (0.1 µg/kg per day) that was 500-fold below the dose of IL-12 associated with severe toxicity. Clinical studies will determine whether these results will translate into an acceptable therapeutic index in man.

8.   References

Afonso, L.C.C., Scharton, T.M., Vieira, L.Q., Wysocka, M., Trinchieri, G. and Scott, P. (1994). The adjuvant effect of interleukin-12 in a vaccine against Leishmania major. Science 263, 235–237.

Andrews, J.V.R., Schoof, D.D., Bertagnolli, M.M., Peoples, G.E., Goedegebuure, P.S. and Eberlein, T.J. (1993). Immunomodulatory effects of interleukin-12 on human tumor-infiltrating lymphocytes. J. Immunother. 14, 1–10.

Bacon, C.M., McVicar, D.W., Ortaldo, J.R., Rees, R.C., O'Shea, J.J. and Johnston, J.A. (1995a). IL-12 induces tyrosine phosphorylation of JAK2 and TYK2: differential use of Janus family kinases by IL-2 and IL-12. J. Exp. Med. 181, 399–404.

Bacon, C.M., Petricoin, E.F., Ortaldo, J.R., *et al.* (1995b). IL-12 induces tyrosine phosphorylation and activation of STAT4 in human lymphocytes. Proc. Natl Acad. Sci. USA 92, 7307–7311.

Bellone, G. and Trinchieri, G. (1994). Dual stimulatory and inhibitory effect of NK cell stimulatory factor/IL-12 on human hematopoiesis. J. Immunol. 153, 930–937.

Bertagnolli, M.M., Lin, B.-Y., Young, D. and Hermann, S.H. (1992). IL-12 augments antigen-dependent proliferation of activated T lymphocytes. J. Immunol. 149, 3778–3783.

Bonnema, J.D., Rivlin, K.A., Ting, A.T., Schoon, R.A., Abraham, R.T. and Leibson, P.J. (1994). Cytokine-enhanced NK cell-mediated cytotoxicity. Positive modulatory effects of IL-2 and IL-12 on stimulus-dependent granule exocytosis. J. Immunol. 152, 2098–2104.

Brunda, M.J. (1994). Interleukin-12. J. Leukocyte Biol. 55, 280–288.

Brunda, M.J. and Gately, M.K. (1994). Antitumor activity of interleukin-12. Clin. Immunol. Immunopathol. 71, 253–255.

Brunda, M.J. and Rosenbaum, D. (1984). Modulation of murine natural killer cell activity *in vitro* and *in vivo* by recombinant human interferons. Cancer Res. 44, 597–601.

Brunda, M.J., Luistro, L., Warrier, R., Hubbard, B., Wolf, S.F. and Gately, M.K. (1993b). Antitumor activity of interleukin-12 (IL-12). Proc. Am. Assoc. Cancer Res. 34, 464. [Abstract]

Brunda, M.J., Luistro, L., Warrier, R.R., *et al.* (1993a). Antitumor and antimetastatic activity of interleukin-12 against murine tumors. J. Exp. Med. 178, 1223–1230.

Brunda, M.J., Luistro, L., Hendrzak, J.A., Fountoulakis, M., Garotta, G. and Gately, M.K. (1994). Interleukin-12: biology and preclinical studies of a new anti-tumor cytokine. In "The Biology of Renal Cell Carcinoma" (ed. R.M. Bukowski). [In press]

Brunda, M.J., Luistro, L., Hendrzak, J.A., Fountoulakis, M., Garotta, G. and Gately, M.K. (1995). Interferon gamma is necessary but not sufficient to mediate the antitumor effect of interleukin-12. J. Immunother. 17, 71–77.

Caput, D., Beutler, B., Hartog, K., Thayer, R., Brown-Shimer, S. and Cerami, A. (1986). Identification of a common nucleotide sequence in the 3′ untranslated region of mRNA molecules specifying inflammatory mediators. Proc. Natl Acad. Sci. USA 83, 1670–1674.

Cesano, A., Visonneau, S., Clark, S.C. and Santoli, D. (1993). Cellular and molecular mechanisms of activation of MHC nonrestricted cytotoxic cells by IL-12. J. Immunol. 151, 2943–2957.

Chan, S.H., Perussia, B., Gupta, J.W., *et al.* (1991). Induction of interferon γ production by natural killer cell stimulatory factor: characterization of the responder cells and synergy with other inducers. J. Exp. Med. 173, 869–879.

Chan, S.H., Kobayashi, M., Santoli, D., Perussia, B. and Trinchieri, G. (1992). Mechanisms of IFN-γ induction by natural killer cell stimulatory factor (NKSF/IL-12). Role of transcription and mRNA stability in the synergistic interaction between NKSF and IL-2. J. Immunol. 148, 92–98.

Chehimi, J., Starr, S.E., Frank, I., *et al.* (1992). Natural killer (NK) cell stimulatory factor increases the cytotoxic activity of NK cells from both healthy donors and human immunodeficiency virus-infected patients. J. Exp. Med. 175, 789–796.

Chehimi, J., Valiante, N.M., D'Andrea, A., *et al.* (1993). Enhancing effect of natural killer cell stimulatory factor (NKSF/interleukin-12) on cell-mediated cytotoxicity against tumor-derived and virus-infected cells. Eur. J. Immunol. 23, 1826–1830.

Chehimi, J., Starr, S.E., Frank, I., *et al.* (1994). Impaired interleukin 12 production in human immunodeficiency virus-infected patients. J. Exp. Med. 179, 1361–1366.

Chizzonite, R., Truitt, T., Desai, B.B., *et al.* (1992). IL-12 receptor. I. Characterization of the receptor on phytohemagglutinin-activated human lymphoblasts. J. Immunol. 148, 3117–3124.

Chizzonite, R., Truitt, T., Griffin, M., *et al.* (1993). Initial characterization of the IL-12 receptor (IL-12R) on concanavalin-A activated mouse splenocytes. J. Cell. Biochem. 17B, 73. [Abstract]

Chizzonite, R., Truitt, T., Nunes, P., *et al.* (1994). High and low affinity receptors for Interleukin-12 (IL-12) on human T cells: evidence for a two subunit receptor by IL-12 and anti-receptor antibody binding. Cytokine 6, A82a. [Abstract]

Chua, A.O., Chizzonite, R., Desai, B.B., *et al.* (1994). Expression cloning of a human IL-12 receptor component: a new member of the cytokine receptor superfamily with strong homology to gp130. J. Immunol. 153, 128–136.

Clemons, K.V., Brummer, E. and Stevens, D.A. (1994). Cytokine treatment of central nervous system infection: efficacy of interleukin-12 alone and synergy with conventional antifungal therapy in experimental cryptococcosis. Antimicrob. Agents Chemother. 38, 460–464.

Clerici, M., Lucey, D.R., Berzofsky, J.A., *et al.* (1993). Restoration of HIV-specific cell-mediated immune responses by interleukin-12 *in vitro*. Science 262, 1721–1724.

D'Andrea, A., Rengaraju, M., Valiante, N.M., *et al.* (1992). Production of natural killer cell stimulatory factor (interleukin-12) by peripheral blood mononuclear cells. J. Exp. Med. 176, 1387–1398.

D'Andrea, A., Aste-Amezaga, M., Valiante, N.M., Ma, X., Kubin, M. and Trinchieri, G. (1993). Interleukin 10 (IL-10) inhibits human lymphocyte interferon γ production by suppressing natural killer cell stimulatory factor/IL-12 synthesis in accessory cells. J. Exp. Med. 178, 1041–1048.

Desai, B.B., Quinn, P.M., Wolitzky, A.G., Mongini, P.K.A., Chizzonite, R. and Gately, M.K. (1992). IL-12 receptor. II. Distribution and regulation of receptor expression. J. Immunol. 148, 3125–3132.

de Waal Malefyt, R., Figdor, C.G., Huijbens, R., *et al.* (1993). Effects of IL-13 on phenotype, cytokine production, and cytotoxic function of human monocytes. Comparison with IL-4 and modulation by IFN-γ or IL-10. J. Immunol. 151, 6370–6381.

Doherty, G.M., Lange, J.R., Langstein, H.N., Alexander, H.R., Buresh, C.M. and Norton, J.A. (1992). Evidence for IFN-γ as a mediator of the lethality of endotoxin and tumor necrosis factor-α. J. Immunol. 149, 1666–1670.

Doherty, T.M., Kastelein, R., Menon, S., Andrade, S. and Coffman, R.L. (1993). Modulation of murine macrophage function by IL-13. J. Immunol. 151, 7151–7160.

Finkelman, F.D., Madden, K.B., Cheever, A.W., *et al.* (1994). Effects of interleukin 12 on immune responses and host protection in mice infected with intestinal nematode parasites. J. Exp. Med. 179, 1563–1572.

Gately, M.K. (1993). Interleukin-12: a recently discovered cytokine with potential for enhancing cell-mediated immune responses to tumors. Cancer Invest. 11, 500–506.

Gately, M.K. and Brunda, M.J. (1994). Interleukin-12: a pivotal regulator of cell-mediated immunity. In "Cytokines: Interleukins and Their Receptors" (eds. R. Kurzrock and M. Talpaz). Kluwer Academic, Norwell, MA. [In press]

Gately, M.K. and Chizzonite, R. (1992). Measurement of human and mouse interleukin 12. In "Current Protocols in Immunology", vol. 1 (eds. J.E. Coligan, A.M. Kruisbeek, D.H. Margulies, E.M. Shevach and W. Strober), pp. 6.16.1.–6.16.8.

Gately, M.K., Anderson, T.D. and Hayes, T.J. (1988). Role of asialo-GM1-positive lymphoid cells in mediating the toxic effects of recombinant IL-2 in mice. J. Immunol. 141, 189–200.

Gately, M.K., Desai, B.B., Wolitzky, A.G., et al. (1991). Regulation of human lymphocyte proliferation by a heterodimeric cytokine, IL-12 (cytotoxic lymphocyte maturation factor). J. Immunol. 147, 874–882.

Gately, M.K., Wolitzky, A.G., Quinn, P.M. and Chizzonite, R. (1992). Regulation of human cytolytic lymphocyte responses by interleukin-12. Cell. Immunol. 143, 127–142.

Gately, M.K., Gubler, U., Brunda, M.J., et al. (1994a). Interleukin-12: a cytokine with therapeutic potential in oncology and infectious diseases. Ther. Immunol. 1, 187–196.

Gately, M.K., Warrier, R.R., Honasoge, S., et al. (1994b). Administration of recombinant IL-12 to normal mice enhances cytolytic lymphocyte activity and induces production of IFN-γ in vivo. Int. Immunol. 6, 157–167.

Gazzinelli, R.T., Heiny, S., Wynn, T.A., Wolf, S. and Sher, A. (1993). Interleukin 12 is required for the T-lymphocyte-independent induction of interferon γ by an intracellular parasite and induces resistance in T-cell-deficient hosts. Proc. Natl Acad. Sci. USA 90, 6115–6119.

Gearing, D.P. and Cosman, D. (1991). Homology of the p40 subunit of natural killer cell stimulatory factor (NKSF) with the extracellular domain of the interleukin-6 receptor. Cell 66, 9–10.

Germann, T., Gately, M.K., Schoenhaut, D.S., et al. (1993). Interleukin-12/T cell stimulating factor, a cytokine with multiple effects on T helper type 1 (T_H1) but not on T_H2 cells. Eur. J. Immunol. 23, 1762–1770.

Germann, T., Bongartz, M., Dlugonska, H., et al. (1995). Interleukin-12 profoundly up-regulates the synthesis of antigen-specific complement-fixing IgG2a, IgG2b and IgG3 antibody subclasses in vivo. Eur. J. Immunol. 25, 823–829.

Gerosa, F., Tommasi, M., Benati, C., et al. (1993). Differential effects of tyrosine kinase inhibition in CD69 antigen expression and lytic activity induced by rIL-2, IL-12, and rIFN-α in human NK cells. Cell. Immunol. 150, 382–390.

Gillessen, S., Carvajal, D., Ling, P., et al. (1995). Mouse interleukin-12 p40 Homodimer: A potent IL-12 antagonist. Eur. J. Immunol. 25, 200–206.

Godfrey, D.I., Kennedy, J., Gately, M.K., Hakimi, J., Hubbard, B.R. and Zlotnik, A. (1994). IL-12 influences intrathymic T cell development. J. Immunol. 152, 2729–2735.

Gubler, U. and Presky, D. (1996). Molecular biology of IL-12 receptors. Ann. N.Y. Acad. Sci. 795, 36–40.

Gubler, U., Chua, A.O., Schoenhaut, D.S., et al. (1991). Coexpression of two distinct genes is required to generate secreted, bioactive cytotoxic lymphocyte maturation factor. Proc. Natl Acad. Sci. USA 88, 4143–4147.

Gustavson, L.E., Nadeau, R.W. and Oldfild, N.F. (1989). Pharmacokinetics of Teceleukin (recombinant human interleukin-2) after intravenous or subcutaneous administration to patients with cancer. J. Biol. Resp. Modifiers 8, 440–449.

Heinzel, F.P., Sadick, M.D., Holaday, B.J., Coffman, R.L. and Locksley, R.M. (1989). Reciprocal expression of interferon γ or interleukin 4 during the resolution or progression of murine leishmaniasis. Evidence for expansion of distinct helper T cell subsets. J. Exp. Med. 169, 59–72.

Heinzel, F.P., Schoenhaut, D.S., Rerko, R.M., Rosser, L.E. and Gately, M.K. (1993). Recombinant interleukin 12 cures mice infected with Leishmania major. J. Exp. Med. 177, 1505–1509.

Heinzel, F.P., Rerko, R.M., Ling, P., Hakimi, J. and Schoenhaut, D.S. (1994). Interleukin 12 is produced in vivo during endotoxemia and stimulates synthesis of interferon-γ. Infect. Immun. 62, 4244–4249.

Hirayama, F., Katayama, N., Neben, S., et al. (1994). Synergistic interaction between interleukin-12 and steel factor in support of proliferation of murine lympho-hematopoietic progenitors in culture. Blood 83, 92–98.

Hsieh, C.-S., Macatonia, S.E., Tripp, C.S., Wolf, S.F., O'Garra, A. and Murphy, K.M. (1993). Development of T_H1 CD4$^+$ T cells through IL-12 produced by Listeria-induced macrophages. Science 260, 547–549.

Jackson, J.D., Yan, Y., Brunda, M.J., Kelsey, L.S. and Talmadge, J.E. (1995). IL-12 enhances peripheral hematopoiesis in vivo. Blood 85, 2371–2376.

Jacobsen, S.E.W., Veiby, O.P. and Smeland, E.B. (1993). Cytotoxic lymphocyte maturation factor (interleukin 12) is a synergistic growth factor for hematopoietic stem cells. J. Exp. Med. 178, 413–418.

Jacobson, N.G., Szabo, S.J., Weber-Nordt, R.M., et al. (1995). IL-12 signaling in T helper type 1 (Th1) cells involves tyrosine phosphorylation of signal transducer and activator of transcription (STAT)3 and STAT4. J. Exp. Med. 181, 1755–1762.

Jewett, A. and Bonavida, B. (1994). Activation of the human immature natural killer cell subset by IL-12 and its regulation by endogenous TNF-α and IFN-γ secretion. Cell. Immunol. 154, 273–286.

Khan, I.A., Matsuura, T. and Kasper, L.H. (1994). Interleukin-12 enhances murine survival against acute toxoplasmosis. Infect. Immun. 62, 1639–1642.

Kiniwa, M., Gately, M., Gubler, U., Chizzonite, R., Fargeas, C. and Delespesse, G. (1992). Recombinant interleukin-12 suppresses the synthesis of IgE by interleukin-4 stimulated human lymphocytes. J. Clin. Invest. 90, 262–266.

Kips, J.C., Brusselle, G.G., Peleman, R.A., et al. (1996). Interleukin-12 inhibits allergen induced airway hyper-responsiveness in mice. Am. J. Respir. Crit. Care Med. 153, 535–539.

Kobayashi, M., Fitz, L., Ryan, M., et al. (1989). Identification and purification of natural killer cell stimulatory factor (NKSF), a cytokine with multiple biological effects on human lymphocytes. J. Exp. Med. 170, 827–845.

Kubin, M., Kamoun, M. and Trinchieri, G. (1994a). Interleukin 12 synergizes with B7/CD28 interaction in inducing efficient proliferation and cytokine production of human T cells. J. Exp. Med. 180, 211–222.

Kubin, M., Chow, J.M. and Trinchieri, G. (1994b). Differential regulation of interleukin-12 (IL-12), tumor necrosis factor α, and IL-1β production in human myeloid leukemia cell lines and peripheral blood mononuclear cells. Blood 83, 1847–1855.

Lieberman, M.D., Sigal, R.K., Williams, N.N. II and Daly, J.M. (1991). Natural killer cell stimulatory factor (NKSF) augments natural killer cell and antibody-dependent tumoricidal response against colon carcinoma cell lines. J. Surg. Res. 50, 410–415.

Ling, P., Gately, M.K., Gubler, U., *et al.* (1995). A homodimer of the IL-12 p40 subunit binds to the IL-12 receptor but does not mediate biologic activity. J. Immunol. 154, 116–127.

Lipman, J.M., Hall, L.B., Gately, M.K. and Anderson, T.D. (1994). Toxicologic profile of recombinant murine and human IL-12 in CD-1 mice. Toxicol. Pathol. [In press]

Locksley, R.M. (1993). Interleukin 12 in host defense against microbial pathogens. Proc. Natl Acad. Sci. USA 90, 5879–5880.

Macatonia, S.E., Hsieh, C.-S., Murphy, K.M. and O'Garra, A. (1993). Dendritic cells and macrophages are required for Th1 development of CD4+ T cells from αβ TCR transgenic mice: IL-12 substitution for macrophages to stimulate IFN-γ production is IFN-γ-dependent. Int. Immunol. 5, 1119–1128.

Malefyt, R.dW., Figdor, C.G., Huijbens, R., *et al.* (1993). Effects of IL-13 on phenotype, cytokine production, and cytotoxic function of human monocytes. Comparison with IL-4 and modulation by IFN-γ or IL-10. J. Immunol. 151, 6370–6381.

Manetti, R., Parronchi, P., Giudizi, M.G., *et al.* (1993). Natural killer cell stimulatory factor (interleukin 12 [IL-12]) induces T helper type 1 (Th1)-specific immune responses and inhibits the development of IL-4-producing Th cells. J. Exp. Med. 177, 1199–1204.

Manetti, R., Gerosa, F., Giudizi, M.G., *et al.* (1994). Interleukin 12 induces stable priming for interferon γ (IFN-γ) production during differentiation of human T helper (Th) cells and transient IFN-γ production in established T$_H$2 cell clones. J. Exp. Med. 179, 1273–1283.

Mattner, F., Fischer, S., Guckes, S., *et al.* (1993). The interleukin-12 subunit p40 specifically inhibits effects of the interleukin-12 heterodimer. Eur. J. Immunol. 23, 2202–2208.

Mayor, S.E., O'Donnell, M.A. and Clinton, S.K. (1994). Interleukin-12 (IL-12) immunotherapy of experimental bladder cancer. Proc. Am. Assoc. Cancer Res. 35, 474. [Abstract]

McKnight, A.J., Zimmer, G.J., Fogelman, I., Wolf, S.F. and Abbas, A.K. (1994). Effects of IL-12 on helper T cell-dependent immune responses in vivo. J. Immunol. 152, 2172–2179.

Mehrotra, P.T., Wu, D., Crim, J.A., Mostowski, H.S. and Siegel, J.P. (1993). Effects of IL-12 on the generation of cytotoxic activity in human CD8+ T lymphocytes. J. Immunol. 151, 2444–2452.

Merberg, D.M., Wolf, S.F. and Clark, S.C. (1992). Sequence similarity between NKSF and the IL-6/G-CSF family. Immunol. Today 13, 77–78.

Morris, S.C., Madden, K.B., Adamovicz, J.J., *et al.* (1994). Effects of IL-12 on in vivo cytokine gene expression and Ig isotype selection. J. Immunol. 152, 1047–1056.

Murakami, M., Narazaki, M., Hibi, M., *et al.* (1991). Critical cytoplasmic region of the interleukin-6 signal transducer gp130 is conserved in the cytokine receptor family. Proc. Natl Acad. Sci. USA 88, 11349–11353.

Murphy, E.E., Terres, G., Macatonia, S.E., *et al.* (1994). B7 and interleukin 12 cooperate for proliferation and interferon γ production by mouse T helper clones that are unresponisve to B7 costimulation. J. Exp. Med. 180, 223–231.

Nagler, A., Lanier, L.L. and Phillips, J.H. (1988). The effects of IL-4 on human natural killer cells. A potent regulator of IL-2 activation and proliferation. J. Immunol. 141, 2349–2351.

Nastala, C.L., Edington, H.D., McKinney, T.G., *et al.* (1994). Recombinant interleukin-12 (IL-12) administration induces tumor regression in association with interferon-gamma production. J. Immunol. 153, 1697–1706.

Naume, B., Gately, M. and Espevik, T. (1992). A comparative study of IL-12 (cytotoxic lymphocyte maturation factor)-, IL-2-, and IL-7-induced effects on immunomagnetically purified CD56+ NK cells. J. Immunol. 148, 2429–2436.

Naume, B., Johnsen, A.-C., Espevik, T. and Sundan, A. (1993a). Gene expression and secretion of cytokines and cytokine receptors from highly purified CD56+ natural killer cells stimulated with interleukin-2, interleukin-7 and interleukin-12. Eur. J. Immunol. 23, 1831–1838.

Naume, B., Gately, M.K., Desai, B.B., Sundan, A. and Espevik, T. (1993b). Synergistic effects of interleukin 4 and interleukin 12 on NK cell proliferation. Cytokine 5, 38–46.

O'Toole, M., Wolf, S.F., O'Brien, C., Hubbard, B. and Herrmann, S. (1993). Effect of in vivo IL-12 administration on murine tumor cell growth. J. Immunol. 150, 294A.

Orange, J.S., Wolf, S.F. and Biron, C.A. (1994). Effects of IL-12 on the response and susceptibility to experimental viral infections. J. Immunol. 152, 1253–1264.

Ozmen, L., Pericin, M., Hakimi, J., *et al.* (1994). IL12, IFN-γ and TNF-α are the key cytokines of the generalized Shwartzman reaction. J. Exp. Med. [In press]

Parronchi, P., Macchia, D., Piccinni, M.P., *et al.* (1991). Allergen- and bacterial antigen-specific T-cell clones established from atopic donors show a different profile of cytokine production. Proc. Natl Acad. Sci. USA 88, 4538–4542.

Perussia, B., Chan, S.H., D'Andrea, A., *et al.* (1992). Natural killer (NK) cell stimulatory factor or IL-12 has differential effects on the proliferation of TCR-αβ+, TCR-γδ+ T lymphocytes, and NK cells. J. Immunol. 149, 3495–3502.

Pignata, C., Prasad, K.V.S., Robertson, M.J., Levine, H., Rudd, C.E. and Ritz, J. (1993). FcgammaRIIIA-mediated signaling involves *src*-family *lck* in human natural killer cells. J. Immunol. 151, 6794–6800.

Pignata, C., Sanghera, J.S., Cossette, L., Pelech, S.L. and Ritz, J. (1994). Interleukin-12 induces tyrosine phosphorylation and activation of 44-kD mitogen-activated protein kinase in human T cells. Blood 83, 184–190.

Ploemacher, R.E., van Soest, P.L., Boudewijn, A. and Neben, S. (1993a). Interleukin-12 enhances interleukin-3 dependent multilineage hematopoietic colony formation stimulated by interleukin-11 or steel factor. Leukemia 7, 1374–1380.

Ploemacher, R.E., van Soest, P.L., Voorwinden, H. and Boudewijn, A. (1993b). Interleukin-12 synergizes with interleukin-3 and steel factor to enhance recovery of murine hemopoietic stem cells in liquid culture. Leukemia 7, 1381–1388.

Podlaski, F.J., Nanduri, V.B., Hulmes, J.D., *et al.* (1992). Molecular characterization of interleukin 12. Arch. Biochem. Biophys. 294, 230–237.

Rabinowich, H., Herberman, R.B. and Whiteside, T.L. (1993). Differential effects of IL12 and IL2 on expression and function of cellular adhesion molecules on purified human natural killer cells. Cell. Immunol. 152, 481–498.

Reiner, S.L., Zheng, S., Wang, Z.-E., Stowring, L. and Locksley, R.M. (1994). *Leishmania* promastigotes evade interleukin-12 induction by macrophages and stimulate a broad range of cytokines from CD4⁺ T cells during initiation of infection. J. Exp. Med. 179, 447–456.

Robertson, M.J., Soiffer, R.J., Wolf, S.F., *et al.* (1992). Responses of human natural killer (NK) cells to NK cell stimulatory factor (NKSF): cytolytic activity and proliferation of NK cells are differentially regulated by NKSF. J. Exp. Med. 175, 779–788.

Romagnani, S. (1992). Induction of T$_H$1 and T$_H$2 responses: a key role for the "natural" immune response? Immunol. Today 13, 379–381.

Romani, L., Menacci, A., Tonnetti, L., *et al.* (1994). Interleukin-12 but not interferon gamma production correlates with induction of T helper type-1 phenotype in murine candidiasis. Eur. J. Immunol. 24, 909–915.

Rossi, A.R., Pericle, F., Rashleigh, S., Janiec, J. and Djeu, J.Y. (1994). Lysis of neuroblastoma cell lines by human natural killer cells activated by interleukin-2 and interleukin-12. Blood 83, 1323–1328.

Sadick, M.D., Heinzel, F.P., Holaday, B.J., Pu, R.T., Dawkins, R.S. and Locksley, R.M. (1990). Cure of murine leishmaniasis with anti-interleukin 4 monoclonal antibody. Evidence for a T cell-dependent, interferon-γ-independent mechanism. J. Exp. Med. 171, 115–124.

Salcedo, T.W., Azzoni, L., Wolf, S.F. and Perussia, B. (1993). Modulation of perforin and granzyme messenger RNA expression in human natural killer cells. J. Immunol. 151, 2511–2520.

Sarmiento, U.M., Riley, J.H., Knaack, P.A., *et al.* (1994). Biologic effects of recombinant human interleukin-12 in squirrel monkeys (*Sciureus saimiri*). Lab. Invest. 71, 862–873.

Schmitt, E., Hoehn, P., Germann, T. and Rüde, E. (1994a). Differential effects of interleukin-12 on the development of naive mouse CD4⁺ T cells. Eur. J. Immunol. 24, 343–347.

Schmitt, E., Hoehn, P., Huels, C., *et al.* (1994b). T helper type 1 development of naive CD4⁺ T cells requires the coordinate action of interleukin-12 and interferon-γ and is inhibited by transforming growth factor-β. Eur. J. Immunol. 24, 793–798.

Schoenhaut, D.S., Chua, A.O., Wolitzky, A.G., *et al.* (1992). Cloning and expression of murine IL-12. J. Immunol. 148, 3433–3440.

Scott, P. (1991). IFN-γ modulates the early development of T$_H$1 and T$_H$2 responses in a murine model of cutaneous leishmaniasis. J. Immunol. 147, 3149–3155.

Scott, P. (1993). IL-12: initiation cytokine for cell-mediated immunity. Science 260, 496–497.

Seder, R.A., Gazzinelli, R., Sher, A. and Paul, W.E. (1993). Interleukin 12 acts directly on CD4⁺ T cells to enhance priming for interferon γ production and diminishes interleukin 4 inhibition of such priming. Proc. Natl Acad. Sci. USA 90, 10188–10192.

Shaw, G. and Kamen, R. (1986). A conserved AU sequence from the 3′ untranslated region of GM-CSF mRNA mediates selective mRNA degradation. Cell 46, 659–667.

Sieburth, D., Jabs, E.W., Warrington, J.A., *et al.* (1992). Assignment of NKSF/IL 12, a unique cytokine composed of two unrelated subunits, to chromosomes 3 and 5. Genomics 14, 59–62.

Stahl, N. and Yancopoulos, G. (1993). The alphas, betas and kinases of cytokine receptor complexes. Cell 74, 587–590.

Stern, A.S., Podlaski, F.J., Hulmes, J.D., *et al.* (1990). Purification to homogeneity and partial characterization of cytotoxic lymphocyte maturation factor from human B-lymphoblastoid cells. Proc. Natl Acad. Sci. USA 87, 6808–6812.

Stern, L.L., Tarby, C.M, Tamborini, B. and Truitt, G.A. (1994). Preclinical development of IL-12 as an anticancer drug: comparison to IL-2. Proc. Am. Assoc. Cancer Res. 35, 520. [Abstract]

Stotter, H., Custer, M.C., Bolton, E.S., Guedez, L. and Lotze, M.T. (1991). IL-7 induces human lymphokine-activated killer cell activity and is regulated by IL-4. J. Immunol. 146, 150–155.

Sypek, J.P., Chung, C.L., Mayor, S.E.H., *et al.* (1993). Resolution of cutaneous leishmaniasis: interleukin 12 initiates a protective T helper type 1 immune response. J. Exp. Med. 177, 1797–1802.

Tahara, H., Zeh, H.J., Storkus, W.J., III, *et al.* (1994). Fibroblasts genetically engineered to secrete IL-12 can suppress tumor growth and induce antitumor immunity to a murine melanoma *in vivo*. Cancer Res. 54, 182–189.

Talmadge, J.E., Herberman, R.B., Chirigos, M.A., *et al.* (1985). Hyporesponsiveness to augmentation of murine natural killer cell activity in different anatomical compartments by multiple injections of various immunomodulators including recombinant interferons and interleukin 2. J. Immunol. 135, 2483–2491.

Tangarone, B., Vath, J., Nickbarg, E., Yu, W., Harris, A. and Scoble, H. (1995). The disulfide bond structure of recombinant human interleukin-12 (rhIL-12). Protein Sci. 4 [supplement 2], 150.

Trinchieri, G. (1993). Interleukin-12 and its role in the generation of T$_H$1 cells. Immunol. Today 14, 335–337.

Trinchieri, G., Kubin, M., Bellone, G. and Cassatella, M.A. (1993). Cytokine cross-talk between phagocytic cells and lymphocytes: relevance for differentiation/activation of phagocytic cells and regulation of adaptive immunity. J. Cell. Biochem. 53, 301–308.

Tripp, C.S., Wolf, S.F. and Unanue, E.R. (1993). Interleukin 12 and tumor necrosis factor α are costimulators of interferon production by natural killer cells in severe combined immunodeficiency mice with listeriosis, and interleukin 10 is a physiologic antagonist. Proc. Natl Acad. Sci. USA 90, 3725–3729.

Tripp, C.S., Gately, M.K., Hakimi, J., Ling, P. and Unanue, E.R. (1994). Neutralization of IL-12 decreases resistance to *Listeria* in SCID and C.B-17 mice. Reversal by IFN-γ. J. Immunol. 152, 1883–1887.

Wills, R.J. and Spiegel, H.E. (1985). Continuous intravenous infusion pharmacokinetics of interferon to patients with leukemia. J. Clin. Pharmacol. 25, 616–619.

Wills, R.J., Dennis, S., Spiegel, H.E., Gibson, D.M. and Nadler, P.I. (1984). Interferon kinetics and adverse reactions after intravenous, intramuscular, and subcutaneous injection. Clin. Pharmacol. Ther. 35, 722–727.

Wolf, S.F., Temple, P.A., Kobayashi, M., *et al.* (1991). Cloning of cDNA for natural killer cell stimulatory factor, a heterodimeric cytokine with multiple biologic effects on T and natural killer cells. J. Immunol. 146, 3074–3081.

Wong, H.L., Wilson, D.E., Jenson, J.C., Familletti, P.C., Stremlo, D.L. and Gately, M.K. (1988). Characterization of a factor(s) which synergizes with recombinant interleukin 2 in promoting allogeneic human cytolytic T-lymphocyte responses in vitro. Cell. Immunol. 111, 39–54.

Wu, C.-Y., Demeure, C., Kiniwa, M., Gately, M. and Delespesse, G. (1993). IL-12 induces the production of IFN-γ by neonatal human CD4 T cells. J. Immunol. 151, 1938–1949.

Wynn, T.A., Eltoum, I., Oswald, I.P., Cheever, A.W. and Sher, A. (1994). Endogenous interleukin 12 (IL-12) regulates granuloma formation induced by eggs of *Schistosoma mansoni* and exogenous IL-12 both inhibits and prophylactically immunizes against egg pathology. J. Exp. Med. 179, 1551–1561.

Yanagida, T., Kato, T., Igarashi, O., Inoue, T. and Nariuchi, H. (1994). Second signal activity of IL-12 on the proliferation and IL-2R expression of T helper cell-1 clone. J. Immunol. 152, 4919–4928.

Yoshida, A., Koide, Y., Uchijima, M. and Yoshida, T.O. (1994). IFN-gamma induces IL-12 mRNA expression by a murine macrophage cell line, J774. Biochem. Biophys. Res. Com. 198, 3, 857–861.

Zeh, H.J., III, Hind, S., Storkus, W.J. and Lotze, M.T. (1993). Interleukin-12 promotes the proliferation and cytolytic maturation of immune effectors: implications for the immunotherapy of cancer. J. Immunother. 14, 155–161.

Zhang, M., Gately, M.K., Wang, E., *et al.* (1994). Interleukin 12 at the site of disease in tuberculosis. J. Clin. Invest. 93, 1733–1739.

13. Interleukin-13

David J. Matthews *and* Robin E. Callard

1. Introduction

Interleukin-13 (IL-13) is an anti-inflammatory cytokine produced by activated T_H2-like T cells. Human IL-13 was identified by a subtraction cloning approach from anti-CD28-activated T cells (Minty *et al.*, 1993) and by cloning the human homolog of the mouse *p600* gene (Zurawski *et al.*, 1993). The mouse IL-13 gene (or *P600*) has been identified for some time but it has failed to show any biological activity on mouse T and B cells and has not been fully investigated (Cherwinski *et al.*, 1987; Morgan *et al.*, 1992).

In human B cells, IL-13 upregulates the expression of many surface molecules, enhances B cell proliferation induced by anti IgM antibodies or CD40L, and is a switch factor for IgE and IgG4 (Samaridis *et al.*, 1991; Punnonen *et al.*, 1993; DeFrance *et al.*, 1994). IL-13 has profound effects on monocytes that include changes in morphology, surface antigen expression, antibody-dependent cellular cytotoxicity, and cytokine synthesis (McKenzie *et al.*, 1993a; Minty *et al.*, 1993; de Waal Malefyt *et al.*, 1993a,b; Zurawski and de Vries, 1994a,b).

In general IL-13 behaves very like IL-4, although in many instances it is less potent. The similarity between these two cytokines appears to stem from them sharing the IL-4Rα chain as a receptor component (Callard *et al.*, 1996; Hilton *et al.*, 1996). However, an important distinction between the two cytokines is that IL-13 does not act on or bind to T cells (de Waal Malefyt *et al.*, 1995).

In mice, IL-13 modulates monocyte and macrophage functions again in a similar manner to that of IL-4 (Doherty *et al.*, 1993); however, it has no activity on murine B or T cells (Zurawski and de Vries, 1994).

Cytokines
ISBN 0–12–498340–5

2.   The Cytokine Gene

The human *IL-13* gene spans a 4.6 kb region located on chromosome 5q23-31 that encodes a 1.3 kb mRNA transcript (McKenzie *et al.*, 1993a,b; Minty *et al.*, 1993) (Figure 13.1).

2.1   GENOMIC ORGANIZATION OF THE IL-13 GENE

The *IL-13* gene is located 12 kb upstream of the *IL-4* gene and consists of four exons and three introns (McKenzie *et al.*, 1993a,b; Minty *et al.*, 1993; Smirnov *et al.*, 1995) (Figure 13.2). Multiple forms of IL-13 have been identified with no detectable difference in biological activity (McKenzie *et al.*, 1993a). One subset encode proteins with a Gly-to-Asp substitution at amino acid position 61. Many cDNAs encode an extra Gln residue at position 98, apparently owing to the presence of an alternate 3′ splice/acceptor site found at the 5′ end of exon 4. Exon 1 encodes 44 amino acid residues and contains the 5′ untranslated region. Exons 2 and 3 encode 18 and 35 amino acids, respectively. Exon 4 encodes the C-terminal 35 amino acid residues and contains the 3′ untranslated region. The length of the introns 1, 2, and 3 are 1055 bp, 251 bp and 345 bp, respectively. Using the nomenclature of McKenzie *et al.*, (1993b), the TATA box is located at genomic position 697–701, and the polyadenylation sequence at position 4055–4060. The cap site has not yet been identified, but a number of putative cap sites are located within the region 735–765.

2.2   TRANSCRIPTIONAL CONTROL

Factors that bind to the *hIL-13* promoter have not been identified at this time, but a number of regulatory sequence motifs have been identified in the 5′ upstream region and in intron 1 (McKenzie *et al.*, 1993b; Smirnov *et al.*, 1995). The TATA box is located at position 697 and 701 and putative binding sites for AP-1, AP-2 and AP-3, NF-IL-6, and IFN-responsive elements and PU.1 have also been identified at positions 625–631, 743–750, 635–642, 259–267, and 436–443, 523–530, 589–596, respectively.

3.   The Protein

The precursor hIL-13 molecule is 132 amino acids long and the mature protein begins at the first N-terminal Gly and is 112 amino acids long (Minty *et al.*, 1993) (Figure 13.3). The predicted molecular mass is 12 300 Da but the gylcosylated form has a molecular mass of 17 000 Da. The unglycosylated recombinant form of IL-13 migrates at a molecular mass of 9000 Da on SDS-PAGE. There are four potential *N*-linked glycosylation sites at Asn residues 18, 29, 37, and 52, and two disulfide bonds between Cys-28 and Cys-56 and between Cys-44 and Cys-70 (Minty *et al.*, 1993; Callard and Gearing, 1994).

675   ACTTGGGCCTATAAAAGCTGCCACAAGAGCCCAAGCCACAAGCCACCCAGCCTATG   726

727   CATCCGCTCCTCAATCCTCTCCTGTTGGCACTGGGCCTCATGGCGCTTTTGTTGACC
ACGGTCATTGCTCTCACTTGCCTTGGCGGCTTTGCCTCCCCAGGCCCTGTGCCT   Exon 1
CCCTCTACAGCCCTCAGGGAGCTCATTGAGGAGCTGGTCAACATCACCCAGAAC   903
CAGAAG

1960   GCTCCGCTCTGCAATGGCAGCATGGTATGGAGCATCAACCTGACAGCTGGCATG   2013   Exon 2

2266   TACTGTGCAGCCCTGGAATCCCTGATCAACGTGTCAGGCTGCAGTGCCATCGAG
AAGACCCAGAG GATGCTGAGCGGATTCTGCCCGCACAAGGTCTCAGCTGGG   2370   Exon 3

2717   CAGTTTTCCAGCTTGCATGTCCGAGACACCAAAATCGAGGTGGCCCAGTTTGTA
AAGGACCTGCTCTTACATTTAAAGAAACTTTTTCGCGAGGGACGGTTCAACTGAA
ACTTCGAAAGCATCATTATTTGCAGAGACAGGACCTGACTATTGAAGTTGCAGATTCA
TTTTTCTTTCTGATGTCAAAAATGTCTTGGGTAGGCGGGAAGGAGGGTTAGGGAGGG
GTAAAATTCCTTAGCTTAGACCTCAGCCTGTGCTGCCCGTCTTCAGCCTAGCCGACCT
CAGCCTTCCCCTTGCCCAGGGCTCAGCCTGGTGGGCCTCCTCTGTCCAGGGCCCTGAG
CTCGGTGGACCCAGGGATGACATGTCCCTACACCCCTCCCCTGCCCTAGAGCACACTG
TAGCATTACAGTGGGTGCCCCCCTTGCCAGACATGTGGTGGGACAGGGACCCACTTC   Exon 4
ACACACAGGCAACTGAGGCAGACAGCAGCTCAGGCACACTTCTTCTTGGTCTTATTTA
TTATTGTGTGTTATTTAAATGAGTGTGTTTGTCACCGTTGGGGATTGGGGAAGACTGT
GGCTGCTGGCACTTGGAGCCAAGGGTTCAGAGACTCAGGGCCCCAGCACTAAAGCAG
TGGACCCCAGGAGTCCCTGGTAATAAGTACTGTGTACAGAATTCTGCTACCTCACTGG
GGTCCTGGGGCCTCGGAGCCTCATCCGAGGCAGGGTCAGGAGAGGGGCAGAACAGC
CGCTCCTGTCTGCCAGCCAGCAGCCAGCTCTCAGCCAACGAGTAATTTATTGTTTTTC
CTCGTATTTAAATATTAAATATGTTAGCAAAGAGTTAATATATAGAAGGGTACCTTGA
ACACTGGGGGAGGGGACATTGAACAAGTTGTTTCATTGACTATCAAACTGAAGCCAG
AAATAAAAGTTGGTGACAGATAGGCCTGAT   3651

Figure 13.1   The sequence of the human *IL-13* gene. The amino acid coding regions of the four exons are indicated in bold. Untranslated regions of exon 1 and 4 are underlined and the first mature amino acid is underlined in bold. The TATA box and the polyadenylation signal are asterisked.

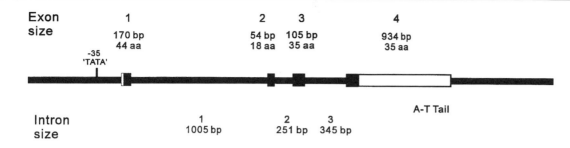

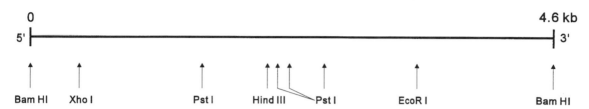

Figure 13.2   Gene structure of the *IL-13* gene. The coding regions of each exon are indicated as filled black boxes and untranslated regions as empty boxes.

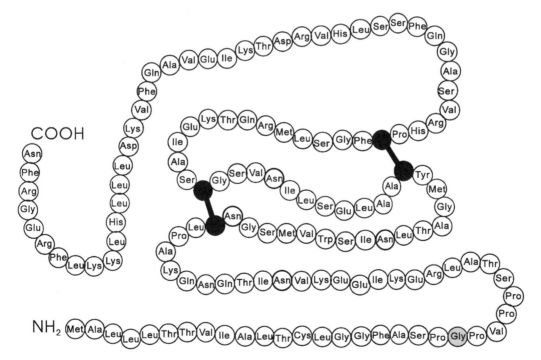

Figure 13.3   The primary and secondary structure of the *IL-13* protein. Cysteine residues are indicated as solid black circles; the shaded circle denotes the first amino acid of the mature protein; and possible glycosylation sites indicated by thick circles.

3.1   STRUCTURE

IL-13 is a member of the hematopoietin cytokine family that includes IL-2, IL-4, GH, GM-CSF, and M-CSF. The members of this family share similar structural homologies. All have a four-helix bundle structure consisting of four antiparallel α-helices (Diederichs *et al.*, 1991; Bazan, 1992; McKay, 1992; Powers *et al.*, 1992). Circular diochroism (CD) analysis of mIL-13 shows it to have the characteristics of a highly α-helical protein

(Zurawski et al., 1993). The 3D structure of IL-13 has not been determined. Computer modeling based on conserved structural regions from members of the hematopoietic cytokine family have produced two structural models for IL-13 (Bamborough et al., 1994). The models suggest that IL-13 may be more like GM-CSF than other cytokines of this family in terms of the length of the helical regions but similar to IL-4 in terms of lengths of loops.

4.   Cellular Sources and Production

Human IL-13 is produced by both CD4[+]- and CD8[+]-activated T cells and can be induced by either antigen-specific or polyclonal stimuli (Minty et al., 1993; Zurawski and de Vries, 1994; de Waal Malefyt et al., 1995). CD4[+] T cell clones with T_H0, T_H1, and T_H2 characteristics can produce IL-13 (Zurawski and de Vries, 1994; de Waal Malefyt et al., 1995). Naive human CD4[+]45RO[-] T cells activated by the cross-linking of TCR/CD3 produce IL-13 and IFN-γ but not IL-4 (Brinkmann and Kristofic, 1995). After activation of CD4[+] T cells, IL-13 mRNA production can be detected within 2 h and maximum production of the protein is reached by 6 h and continues for at least 24 h, but may continue to be produced for up to 7 days is some conditions (Zurawski and de Vries, 1994; de Waal Malefyt et al., 1995). A study of cytokine production in human tonsils found IL-13 localized to the extrafollicular, T cell-rich area (Andersson et al., 1994).

Human B cells can synthesize IL-13. Expression of the IL-13 gene was detected by RT-PCR analysis in both malignant and EBV-transformed B lymphocytes (Fior et al., 1994) and small amounts of IL-13 protein can be detected by ELISA in EBV culture supernatants (de Waal Malefyt et al., 1995).

Mast cells produce IL-13 when activated through their IgE receptors. Murine bone marrow-derived mast cells and human mast cell lines produce IL-13 in response to IgE or specific antigen or in response to phorbol myristate acetate (PMA) and calcium ionophore (Burd et al., 1995). Bone marrow-derived mouse mucosal-like mast cells constitutively express IL-13, but bone marrow-derived connective-tissue mast cells require activation by IL-3 or via FcϵRI (Marietta et al., 1996).

5.   Biological Activity

5.1   T Cells

IL-13 has no biological activity on T cells and T cells do not appear to express the IL-13R (Zurawski and de Vries, 1994; de Waal Malefyt et al., 1995).

5.2   B Cells

IL-13 is a human B cell growth and differentiation factor with very similar properties to that of IL-4. B cell responses to IL-13 differ from those to IL-4 in their magnitude, and B cell activation, proliferation, and immunoglobulin production responses to IL-13 are less than for IL-4 (Cocks et al., 1993; Punnonen et al., 1993; Defrance et al., 1994; Punnonen and de Vries, 1994; Matthews et al., 1995).

5.2.1   B Cell Activation

IL-13 will induce B cells, from peripheral blood, spleen, and tonsil, to increase the expression of a variety of surface antigens that includes CD23, CD71, CD72, IgM, and MHC class II (Punnonen et al., 1993; Defrance et al., 1994; Zurawski and de Vries, 1994; Matthews et al., 1995).

5.2.2   B Cell Proliferation

Human B cell proliferation can be induced by costimulation with IL-13 and anti-IgM, anti-CD40, or the CD40L (Cocks et al., 1993; McKenzie et al., 1993a; Banchereau et al., 1994; Defrance et al., 1994). IL-13 can support long-term B cell growth when CD40 is cross-linked with monoclonal anti-CD40 antibody presented by a murine L cells transfected with FcγRII/CDw32 (Banchereau et al., 1994). Comparisons of B cell proliferation responses to IL-13 found that signals via CD40 are more effective at supporting proliferation than through sIgM (McKenzie et al., 1993a; Fluckiger et al., 1994; Callard et al., 1996).

5.2.3   IgE Production

In humans, IL-13 induces IgE production (McKenzie et al., 1993a) and is a switch factor for IgE and IgG4 (Punnonen et al., 1993; Punnonen and de Vries, 1994). B cells in the presence of the activated CD4[+] T cell clone B21 and IL-13 can increase the production of IgM and IgG but not IgA (McKenzie et al., 1993a). Purified splenic or PBMC IgD[+] B cells stimulated by CD40L and IL-13 produce IgM, IgG4, and IgE (Cocks et al., 1993; Punnonen et al., 1993). Immature human fetal bone marrow B cells (sμ[+], CD10[+], CD19[+]) have also been shown to produce IgM, IgG4, and IgE in response to IL-13 and activated CD4[+] T cells or the CD40L (Punnonen and de Vries, 1994).

5.3   Monocytes and Macrophages

IL-13 is an anti-inflammatory cytokine. Responses of human monocytes to IL-13 are similar to those elicited by IL-4, although they are generally weaker. Monocytes change their morphology in response to IL-13 and develop long processes, aggregate homotypically, adhere strongly to the substrates, and migrate in a chemotactic manner (McKenzie et al., 1993a; Magazin et al., 1994).

IL-13 also modulates the expression of surface antigens on human monocytes. The expression of CD11b, CD11c, CD13, CD18, CD23, CD29, CD49e (VLA-5), mannose receptor, and MHC class II is increased, but the expression of CD14, CD16, CD32, and CD64 is decreased (de Waal Malefyt *et al.*, 1993; Doyle *et al.*, 1994; Cosentino *et al.*, 1995).

In monocytes IL-13 can also inhibit the production of pro-inflammatory cytokines and antibody-dependent cellular cytotoxicity (ADCC) induced by IL-10 and IFN-γ (de Waal Malefyt *et al.*, 1993; Minty *et al.*, 1993; Cosentino *et al.*, 1995). Other anti-inflammatory activities include the induction of 15-lipoxygenase (Nassar *et al.*, 1994), the induction of IL-1Ra (Muzio *et al.*, 1994; Yanagawa *et al.*, 1995), the upregulation of thrombomodulin (Herbert *et al.*, 1993), the inhibition of nitric oxide production (Doyle *et al.*, 1994), and inhibition of tissue factor (Herbert *et al.*, 1993).

Separate mechanisms appear to mediate the inhibition of monocyte responses to LPS by IL-13. Pretreatment of peripheral blood monocytes with GM-CSF or M-CSF abrogates IL-13 inhibition of lipopolysaccharide (LPS)-induced TNF-α but not of IL-1β production (Hart *et al.*, 1995). Moreover, pretreatment of peripheral blood mononuclear cells with IL-4 or IL-13 can augment the LPS induction of IL-12 and TNF-α but not IL-1β or IL-10 production (D'Andrea *et al.*, 1995).

The expression of Lsk, a Csk-like tyrosine kinase, is induced by IL-13 and may mediate anti-inflammatory activity by inhibiting the activation of src-like kinases (Musso *et al.*, 1994).

5.4 OTHER CELL TYPES

Mast cells respond to IL-13 by inducing c-*fos* expression and increasing ICAM-1 expression, but proliferation of mast cell lines and expression of CD117 is inhibited (Nilsson and Nilsson, 1995).

In the presence of IL-13, endothelial cells (HUVEC) specifically express VCAM-1. As a consequence, IL-13 promotes the adhesion of eosinophils, but not neutrophils to the endothelial monolayer (Sironi *et al.*, 1994; Bochner *et al.*, 1995). Consistently with its anti-inflammatory properties, IL-13 inhibits TNF-α and IFN-γ induced production of the chemokine RANTES in human umbilical vein endothelial cells (HUVECs) (Marfaing Koka *et al.*, 1995) and the procoagulant activity by HUVECs induced by LPS (Herbert *et al.*, 1993).

IL-13 inhibits the production of IFN-γ by highly purified NK cells but, in contrast, will induce large granular lymphocytes to produce IFN-γ (Minty *et al.*, 1993; Zurawski and de Vries, 1994). Polymorphonuclear cells upregulate the expression of type I and type II IL-1 decoy receptors in response to IL-13 (Colotta *et al.*, 1994). The expression and activity of aminopeptidase

N/CD3 and dipeptidase IV/CD26 are increased by IL-13 in renal cell carcinomas and tubular epithelial cells (Riemann *et al.*, 1995). Osteoblast chemotaxis is stimulated by IL-13 (Lind *et al.*, 1995) and IL-13 inhibits bone reabsorption by suppressing cyclooxygenase-2-dependent prostaglandin synthesis of osteoblasts (Onoe *et al.*, 1996). In human mesangial cells, IL-13 acts by inhibiting inducible nitric-oxide synthase expression (Saura *et al.*, 1996).

5.5 HEMATOPOIESIS

Studies of cytokine expression in human bone marrow by RT-PCR failed to detect the expression of IL-13 (Cluitmans *et al.*, 1995), and northern blot analysis of mouse bone marrow failed to detect the expression of mIL-13R (Hilton *et al.*, 1996). IL-13 induces immature human B cell progenitors to express IgE (Punnonen and de Vries, 1994) and inhibits leukemic human B cell precursor proliferation (Fluckiger *et al.*, 1994). IL-13 suppresses the propagation of human macrophage progenitors (Sakamoto *et al.*, 1995). Murine hematopoietic progenitor cells (Lin⁻, Sca-1⁺) proliferate in response to IL-13 (Jacobsen *et al.*, 1994); however, IL-13 knockout mice have normal numbers and distribution of hematopoietic lineages (Kirshna *et al.*, 1995).

6. *IL-13 Receptor*

The human IL-13 receptor has not been fully characterized but appears to be composed of at least two components, the IL4Rα chain (CDw124) and an IL-13 binding chain (Callard *et al.*, 1996; Caput *et al.*, 1996; Hilton *et al.*, 1996) (Figure 13.4). Evidence suggesting that IL-4 and IL-13 share a receptor component arose when the IL-4 antagonist IL4.Y124D was also found to inhibit cellular responses to IL-13 (Aversa *et al.*, 1993; Zurawski *et al.*, 1993; Tony *et al.*, 1994). IL4.Y124D can bind almost normally to the IL-4R but causes only weak activation and therefore acts as a competitive inhibitor to IL-4 (Kruse *et al.*, 1992, 1993; Zurawski *et al.*, 1993; Tony *et al.*, 1994). Proliferation of the human cell line TF-1 and B cell responses to IL-13 are inhibited by IL4.Y124D (Aversa *et al.*, 1993; Zurawski *et al.*, 1993; Tony *et al.*, 1994). These reports argue that IL-4 and IL-13 share a receptor subunit. However, IL-13 does not bind to T cells or to the IL-4Rα subunit (Zurawski *et al.*, 1993; Zurawski and de Vries, 1994). Therefore, IL-4 and IL-13 have distinct receptors which have shared component(s).

Although IL-13 does not bind to IL-4Rα (Zurawski *et al.*, 1993), this subunit appears to be a component of the IL-13R. IL-13 can cross-compete for IL-4 binding on human, monkey, and mouse tissue types (Obiri *et al.*,

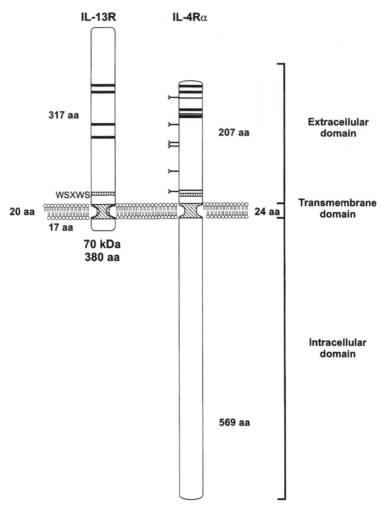

Figure 13.4   The putative structure of the human IL-13 receptor. Thick lines indicate the four conserved cysteines of the hemopoietic cytokine receptor family. Vertical bars donate the WSXWS motif. The hatched transmembrane domain is indicated.

1995; Vita *et al.*, 1995; Zurawski *et al.*, 1995; Hilton *et al.*, 1996) and a subset of blocking IL4Rα antibodies inhibit IL-13 activity (Lefort *et al.*, 1995; Lin *et al.*, 1995; Obiri *et al.*, 1995; Zurawski *et al.*, 1995). The IL4Rα appears to act as an affinity enhancer for the IL-13R (Hilton *et al.*, 1996) and form a major signaling subunit of the receptor (Welham *et al.*, 1995; Keegan *et al.*, 1996). The properties of the IL4Rα chain are addressed more fully in the chapter on IL-4.

The common receptor γ chain (γ_c) has been discounted as a likely component of the IL-13R (He *et al.*, 1995; Matthews *et al.*, 1995). B cells from patients with X-linked severe combined immunodeficiency (X-SCID), a human genetic disease cause by mutations in γ_c, respond normally to IL-4 and IL-13. Blocking antibodies directed against murine γ_c inhibited mouse mast cell line MC/9 responses to IL-4 but not IL-13, and transfection studies with γ_c in the B9 plasmacytoma cell line (γ_c^-) failed to influence responses to IL-13 (He *et*

al., 1995). From these experiments it can be concluded that γ_c is not essential for IL-13 responses and is not likely to be a component of the IL-13 receptor.

An IL-13R binding chain has been cloned in humans (Caput *et al.*, 1996) and distinct IL-13 binding chain has also been cloned in mice (Hilton *et al.*, 1996). The human IL-13R is mRNA transcript is approximately 1.4 kb long and has a short 3′ untranslated region of 103 bases. The protein is 380 amino acids long with a putative signal peptide of 26 amino acids, a single membrane spanning domain, and a short cytoplasmic tail. The extracellular domain has four potential sites for *N*-linked glycosylation, a WSXWS motif near the membrane domain and a cytokine binding domain with four conserved cysteines characteristic of the hematopoietic cytokine receptor superfamily. The protein migrates on SDS-PAGE gels at ~70 kDa and consistently with previous reports indicating that the IL-13R migrates on SDS-PAGE at between 60 and 70 kDa

(Vita *et al.*, 1995; Zurawski *et al.*, 1995). The IL-13 receptor is expressed in large amounts on human renal cell carcinoma cell lines (RCC) but at low levels on human B cells and monocytes, and does not cross-react with anti γ_c or anti-IL4Rα antibodies (Caput *et al.*, 1996; Obiri *et al.*, 1995). The IL-13 receptor described in mice (Hilton *et al.*, 1996) is not the human homolog of that described by Caput and colleagues (1996). At this time the relationship of the IL-13 receptor described in mice to that in humans is not clear.

The IL-13R complex is also an IL-4 receptor. In cells that respond only to IL-4 but not to IL-13, the γ_c has been found to be essential for IL-4 receptor activation. Cells that do not express γ_c but do respond to IL-13, such as X-SCID B cells, RCC lines, and fibroblasts, also respond to IL-4 (Matthews *et al.*, 1995; Obiri *et al.*, 1995). Evidence in support of two IL-4 receptors on human B cells has also been reported on the basis of binding studies (Foxwell *et al.*, 1989) and the differential regulation of sIgM and CD23 by IL-4 (Rigley *et al.*, 1991). The recent cloning of the mouse IL-13R has provided further evidence that the IL-13R complex is also an IL-4 receptor (Hilton *et al.*, 1996).

7. Signal Transduction

Signal transduction via the IL-13 receptor is very similar to that for the IL-4 receptor, presumably owing to the contribution of the IL4Rα subunit. IL-13 induces the phosphorylation of JAK1 (Welham *et al.*, 1995; Izuhara *et al.*, 1996; Keegan *et al.*, 1996), a p170 now identified as IRS-2 or 4PS (Sun *et al.*, 1995; Wang *et al.*, 1995; Welham *et al.*, 1995; Keegan *et al.*, 1996), the p85 subunit of PI3-kinase, and the IL4Rα (Smertz-Bertling and Duschl, 1995; Welham *et al.*, 1995). IL-13 stimulates similar transcription factor complexes to those that IL-4 does and activates NF-IL4 (Kohler *et al.*, 1994; Lin *et al.*, 1995). Human monocytes release intracellular calcium, increase cAMP levels, and activate protein kinase C (PKC) in response to IL-13 (Sozzani *et al.*, 1995). IL-13 does not activate JAK3 (Lin *et al.*, 1995; Welham *et al.*, 1995; Keegan *et al.*, 1996). JAK3 has been shown to be a signaling component of γ_c (Russell *et al.*, 1994; Malabarba *et al.*, 1995). The absence of JAK3 activation by IL-13 is therefore consistent with the IL-13R not utilizing the γ_c chain.

8. Murine IL-13

The murine *IL-13* gene has been mapped to the central region of chromosome 11 that is syntenic with human chromosomes 5q and 17p (McKenzie *et al.*, 1993b). Genetic linkage mapping determined that the *mIL-13* locus is linked to the *AdraI*, *IL-3*, *Myhs*, and *Trp53* loci.

The genomic organization of *mIL-13* is similar to that of humans and consists of four exons and three introns spanning 4.3 kb of DNA. Exon 1 contains the 5′ untranslated region and the first 47 amino acids. Exons 2 and 3 encode 18 and 35 amino acids, respectively. Exon 4 encodes the C-terminal 31 amino acid residues and contains the 3′ untranslated region. The introns 1, 2, and 3 are 1258 bp, 576 bp, and 311 bp long, respectively.

IL-13 has IL-4-like effects on mouse macrophages and inhibits the production of pro-inflammatory cytokines and macrophage cytotoxicity (Doherty *et al.*, 1993; Zurawski and de Vries, 1994). Macrophages derived *in vitro* from bone marrow cultured with M-CSF (MM0) or GM-CSF (GMM0) respond to IL-13. MM0 and GMM0 macrophage survival is promoted by IL-13; however, IL-13 enhances the expression of MHC class I and MHC class II antigens on MM0 but not on GMM0. The expression of FcγR is decreased in both types of macrophages but FcεR is unaffected by IL-13. The generation of NO induced by LPS or IFN-γ is inhibited by IL-13 but the generation of NO by MM0 macrophages is unaffected by IL-13.

The differences between IL-13 activity in mouse and human are significant. Murine IL-13 has no activity on mouse B cells, suggesting that the regulation of IgE production is different between mouse and man. IL-13 knockout mice have normal numbers and distribution of hematopoietic lineages (Kirshna *et al.*, 1995), indicating that its presence is not crucial for hematopoiesis.

The mIL-13R has been cloned (Hilton *et al.*, 1996) and there are two hybridizing transcripts of 2.2 kb and 5.2 kb in length. Northern blot analysis found that the IL-13R mRNA is expressed in a wide variety of tissues including spleen, liver, thymus, heart, and lung, but not in bone marrow or muscle. On its own, mIL-13R binds IL-13 with a low affinity ($K_D \approx 2$–10 pM) and is converted to a high-affinity receptor ($K_D \approx 75$ pM) by IL-4Rα. The IL-13R is predicted to encode 424 amino acid residues. The extracellular region of the protein contains an immunoglobulin-like domain (amino acids 27–117) and a typical hematopoietin receptor domain (118–340 amino acids) which includes the characteristic WSXWS motif. The cytoplasmic tail is 60 amino acid residues long. Cross-competition binding studies on CTLL-2 cells transfected with IL-13R found that IL-4 and IL-13 compete for receptor occupancy, thereby providing compelling evidence that the IL-13R is also an IL-4 receptor (Hilton *et al.*, 1996). The IL-13R is not expressed in mouse B cells.

9. IL-13 in Disease and Therapy

IL-13 is a newly discovered cytokine and its role in disease has not been explored as thoroughly as it has for IL-4. The IL-13 receptor complex appears to be a second

IL-4 receptor (Callard et al.,1996; Hilton et al., 1996) and has a subset of IL-4 activities. This may have important implications for IgE production in allergy and other diseases associated with IL-4 (see chapter on IL-4). The IL-4R and IL-13Rs may have separate functional properties that are reflected in different disease states and therefore require specific theraputic intervention.

9.1 ASSOCIATED PATHOLOGIES

IL-13 has been found expressed in bronchoalveolar lavage cells of atopic patients but not in normal controls, and allergen-challenged asthmatic and rhinitic patients induced a significant increase of IL-13 production (Huang et al., 1995). This suggests that IL-13 is involved in the regulation of allergen-induced late-phase inflammatory responses.

The spontaneous production of IgE and IgG4 in patients suffering from nephrotic syndrome has been attributed to IL-13 rather than IL-4 (Kimata et al., 1995).

9.2 IL-13 AS AN ANTITUMOR AGENT

IL-13 has an antiproliferative effect on a number of leukemic cell lines in vitro (Fluckiger et al., 1994) and IL-13 can mediate antitumor activity in vivo (Lebel-Binay et al., 1995). In a mouse tumor model the in vivo production of IL-13 near to the tumor elicited an antitumor immunological response by indirectly stimulated macrophage and NK cell activity.

9.3 IL-13 AS A THERAPEUTIC AGENT

HIV replication in cultured human monocytes and macrophages is inhibited by the addition of IL-13 (Montaner et al., 1993; Denis and Ghadirian, 1994; Mikovits et al., 1994); however, replication was found to be increased in cultures chronically treated with the cytokine (Mikovits et al., 1994).

In a rat model for autoimmune encephalomyelitis, treatment with IL-13 suppressed the disease. The anti-inflammatory action of IL-13 suppressed monocyte and macrophage activation with no undesirable effects on T and B cell function (Cash et al., 1994).

9.4 THERAPEUTIC POTENTIAL OF ANTI-IL-4R ANTIBODIES AND IL4.Y124D

The IL-4 antagonist IL4.Y124D is also an efficient antagonist of IL-13 (Aversa et al., 1993; Zurawski et al., 1993; Tony et al., 1994), and a subset of blocking IL-4Rα antibodies also inhibit responses to IL-13 (Lefort et al., 1995; Lin et al., 1995; Obiri et al., 1995; Zurawski et al., 1995). These antagonists may prove to have therapeutic value in controlling IgE production by B cells. However, the peripheral blood of allergic individuals contains long-lived allergen-specific B cells which have already switched to IgE and are not sensitive to IL-4 or IL-13 (Dolecek et al., 1995). Thus, the therapeutic use of such antagonists requires investigation.

10. References

Andersson, J., Abrams, J., Bjork, L., et al. (1994). Concomitant in vivo production of 19 different cytokines in human tonsils. Immunology 83, 16–24.

Aversa, G., Punnonen, J., Cocks, B.G., et al. (1993). An interleukin 4 (IL-4) mutant protein inhibits both IL-4 or IL-13-induced human immunoglobulin G4 (IgG4) and IgE synthesis and B cell proliferation: support for a common component shared by IL-4 and IL-13 receptors. J. Exp. Med. 178, 2213–2218.

Bamborough, P., Duncan, D. and Richards, W.G. (1994). Predictive modelling of the 3-D structure of interleukin-13. Protein Eng. 7, 1077–1082.

Banchereau, J., Bazan, F., Blanchard, D., et al. (1994). The CD40 antigen and its ligand. Annu. Rev. Immunol. 12, 881–922.

Bazan, J.F. (1992). Unraveling the structure of IL-2 [letter]. Science 257, 410–413.

Bochner, B.S., Klunk, D.A., Sterbinsky, S.A., Coffman, R.L. and Schleimer, R.P. (1995). IL-13 selectively induces vascular cell adhesion molecule-1 expression in human endothelial cells. J. Immunol. 154, 799–803.

Brinkmann, V. and Kristofic, C. (1995). TCR-stimulated naive human CD4+ 45RO− T cells develop into effector cells that secrete IL-13, IL-5, and IFN-gamma, but no IL-4, and help efficient IgE production by B cells. J. Immunol. 154, 3078–3087.

Burd, P.R., Thompson, W.C., Max, E.E. and Mills, F.C. (1995). Activated mast cells produce interleukin 13. J. Exp. Med. 181, 1373–1380.

Callard, R.E., Matthews, D.J., and Hibbert, L. (1996). IL-4 and IL-13 receptors: are they one and the same? Immunol. Today 17, 108–110.

Caput, D., Laurent, P., Kaghad, M., et al. (1996). Cloning and characterisation of a specific interleukin (IL)-13 binding protein structurally related to the IL-5 receptor α chain. J. Biol. Chem. 271, 16981–16986.

Cash, E., Minty, A., Ferrara, P., Caput, D., Fradelizi, D. and Rott, O. (1994). Macrophage-inactivating IL-13 suppresses experimental autoimmune encephalomyelitis in rats. J. Immunol. 153, 4258–4267.

Cherwinski, H.M., Schumacher, J.H., Brown, K.D. and Mosmann, T.R. (1987). Two types of mouse helper T cell clone. III. Further differences in lymphokine synthesis between Th1 and Th2 clones revealed by RNA hybridization, functionally monospecific bioassays, and monoclonal antibodies. J.Exp.Med. 166, 1229–1244.

Cluitmans, F.H., Esendam, B.H., Landegent, J.E., Willemze, R. and Falkenburg, J.H. (1995). Constitutive in vivo cytokine and hematopoietic growth factor gene expression in the bone marrow and peripheral blood of healthy individuals. Blood 85, 2038–2044.

Cocks, B.G., de Waal Malefyt, R., Galizzi, J.P., de Vries, J.E. and Aversa, G. (1993). IL-13 induces proliferation and differentiation of human B cells activated by the CD40 ligand. Int. Immunol. 5, 657–663.

Colotta, F., Re, F., Muzio, M., et al. (1994). Interleukin-13 induces expression and release of interleukin-1 decoy receptor in human polymorphonuclear cells. J. Biol. Chem. 269, 12403–12406.

Cosentino, G., Soprana, E., Thienes, C.P., Siccardi, A.G., Viale, G. and Vercelli, D. (1995). IL-13 down-regulates CD14 expression and TNF-α secretion in normal human monocytes. J. Immunol. 155, 3145–3151.

D'Andrea, A., Ma, X., Aste Amezaga, M., Paganin, C. and Trinchieri, G. (1995). Stimulatory and inhibitory effects of interleukin (IL)-4 and IL-13 on the production of cytokines by human peripheral blood mononuclear cells: priming for IL-12 and tumor necrosis factor alpha production. J. Exp. Med. 181, 537–546.

Defrance, T., Carayon, P., Billian, G., et al. (1994). Interleukin 13 is a B cell stimulating factor. J. Exp. Med. 179, 135–143.

de Vries, J.E. (1994). Novel fundamental approaches to intervening in IgE-mediated allergic diseases. J. Invest. Dermatol. 102, 141–144.

de Vries, J.E. and Zurawski, G. (1995). Immunoregulatory properties of IL-13: its potential role in atopic disease. Int. Arch. Allergy Immunol. 106, 175–179.

de Waal Malefyt, R., Figdor, C.G. and de Vries, J.E. (1993a). Effects of interleukin 4 on monocyte functions: comparison to interleukin 13. Res. Immunol. 144, 629–633.

de Waal Malefyt, R., Figdor, C.G., Huijbens, R., et al. (1993b). Effects of IL-13 on phenotype, cytokine production, and cytotoxic function of human monocytes. Comparison with IL-4 and modulation by IFN-gamma or IL-10. J. Immunol. 151, 6370–6381.

de Waal Malefyt, R., Abrams, J.S., Zurawski, S.M., et al. (1995). Differential regulation of IL-13 and IL-4 production by human CD8+ and CD4+T$_h$0, T$_h$1 and T$_h$2 T cell clones and EBV-transformed B cells. Int. Immunol. 7, 1405–1416.

Diederichs, K., Boone, T. and Karplus, P.A. (1991). Novel fold and putative receptor binding site of granulocyte-macrophage colony-stimulating factor. Science 254, 1779–1782.

Doherty, T.M., Kastelein, R., Menon, S., Andrade, S. and Coffman, R.L. (1993). Modulation of murine macrophage function by IL-13. J. Immunol. 151, 7151–7160.

Dolecek, C., Steinberger, P., Susani, M., Kraft, D., Valenta, R. and Boltz-Nitulescu, G. (1995). Effects of IL-4 and IL-13 on total and allergen specific IgE production by cultured PBMC from allergic patients determined with recombinant pollen allergens. Clin. Exp. Allergy. 25, 879–889.

Doyle, A.G., Herbein, G., Montaner, L.J., et al. (1994). Interleukin-13 alters the activation state of murine macrophages in vitro: comparison with interleukin-4 and interferon-gamma. Eur. J. Immunol. 24, 1441–1445.

Fior, R., Vita, N., Raphael, M., et al. (1994). Interleukin-13 gene expression by malignant and EBV-transformed human B lymphocytes. Eur. Cytokine Netw. 5, 593–600.

Fluckiger, A.C., Briere, F., Zurawski, G., Bridon, J.M. and Banchereau, J. (1994). IL-13 has only a subset of IL-4-like activities on B chronic lymphocytic leukaemia cells. Immunology 83, 397–403.

Hart, P.H., Ahern, M.J., Smith, M.D. and Finlay Jones, J.J. (1995). Regulatory effects of IL-13 on synovial fluid macrophages and blood monocytes from patients with inflammatory arthritis. Clin. Exp. Immunol. 99, 331–337.

He, Y.W., Adkins, B., Furse, R.K. and Malek, T.R. (1995). Expression and function of the gamma c subunit of the IL-2, IL-4, and IL-7 receptors. Distinct interaction of gamma c in the IL-4 receptor. J. Immunol. 154, 1596–1605.

Herbert, J.M., Savi, P., Laplace, M.C., et al. (1993). IL-4 and IL-13 exhibit comparable abilities to reduce pyrogen-induced expression of procoagulant activity in endothelial cells and monocytes. FEBS Lett. 328, 268–270.

Hilton, D.J., Zhang, J., Metcalf, D., Alexander, W.S., Nicola, N.A. and Willson, T.A. (1996). Cloning and characterisation of a binding subunit of the interleukin 13 receptor that is also a component of the interleukin 4 receptor. Proc. Natl Acad. Sci. USA 93, 497–501.

Huang, S., Xiao, H., Kleine-Tebbe, J., et al. (1995). IL-13 expression at the sites of allergen challenge in patients with asthma. J. Immunol. 155, 2688–2694.

Izuhara, K., Heike, T., Otsuka, T., et al. (1996). Signal transduction pathway of interleukin-4 and interleukin-13 in human B cells derived from X-linked severe combined immunodeficiency patients. J. Biol. Chem. 271, 619–622.

Jacobsen, S.E., Okkenhaug, C., Veiby, O.P., Caput, D., Ferrara, P. and Minty, A. (1994). Interleukin 13: novel role in direct regulation of proliferation and differentiation of primitive hematopoietic progenitor cells. J. Exp. Med. 180, 75–82.

Keegan, A.D., Johnston, J.A., Tortolani, P.J., et al. (1996). Similarities and differences in the signal transduction by interleukin 4 and interleukin 13: analysis of Janus kinase activation. Proc. Natl Acad. Sci. USA 92, 7681–7685.

Kimata, H., Fujimoto, M. and Furusho, K. (1995). Involvement of IL-13 but not IL-4 in spontaneous IgE and IgG4 production in nephrotic syndrome. Eur. J. Immunol. 25, 1497–1501.

Kirshna, M., Murray, R., Zurawski, G. and McKenzie, A.N. (1995). Studies of infectious disease models in interleukin-13 deficient mice. Cytokine 7, 637.

Kodelja, V. and Goerdt, S. (1994). Dissection of macrophage differentiation pathways in cutaneous macrophage disorders and in vitro. Exp. Dermatol. 3, 257–268.

Kohler, I., Alliger, P., Minty, A., et al. (1994). Human interleukin-13 activates the interleukin-4-dependent transcription factor NF-IL4 sharing a DNA binding motif with an interferon-gamma-induced nuclear binding factor. FEBS Lett. 345, 187–192.

Lebel-Binay, S., Laguerre, B., Quintin-Colonna, F., et al. (1995). Experimental gene therapy of cancer using tumour cells engineered to secrete interleukin-13. Eur. J. Immunol. 25, 2340–2348.

Lefort, S., Vita, N., Reeb, R., Caput, D. and Ferrara, P. (1995). IL-13 and IL-4 share signal transduction elements as well as receptor components in TF-1 cells. FEBS Lett. 366, 122–126.

Lin, J.X., Migone, T.S., Tsang, M., et al. (1995). The role of shared receptor motifs and common Stat proteins in the generation of cytokine pleiotropy and redundancy by IL-2, IL-4, IL-7, IL-13, and IL-15. Immunity 2, 331–339.

Lind, M., Deleuran, B., Yssel, H., Fink Eriksen, E. and Thestrup Pedersen, K. (1995). IL-4 and IL-13, but not IL-10, are chemotactic factors for human osteoblasts. Cytokine 7, 78–82.

Magazin, M., Guillemot, J.C., Vita, N. and Ferrara, P. (1994). Interleukin-13 is a monocyte chemoattractant. Eur. Cytokine Netw. 5, 397–400.

Malabarba, M.G., Kirken, R.A., Rui, H., *et al.* (1995). Activation of JAK3, but not JAK1, is critical to interleukin-4 (IL4) stimulated proliferation and requires a membrane-proximal region of IL4 receptor alpha. J. Biol. Chem. 270, 9630–9637.

Marfaing Koka, A., Devergne, O., Gorgone, G., *et al.* (1995). Regulation of the production of the RANTES chemokine by endothelial cells. Synergistic induction by IFN-gamma plus TNF-alpha and inhibition by IL-4 and IL-13. J. Immunol. 154, 1870–1878.

Marietta, E.V., Chen, Y. and Weis, J.H. (1996). Modulation of expression of the anti-inflammatory cytokines interleukin-13 and interleukin-10 by interleukin-3. Eur. J. Immunol. 26, 49–56.

Matthews, D.J., Clark, P.A., Herbert, J., *et al.* (1995). Function of the interleukin-2 (IL-2) receptor gamma-chain in biologic responses of X-linked severe combined immunodeficient B cells to IL-2, IL-4, IL-13, and IL-15. Blood 85, 38–42.

McKenzie, A.N., Culpepper, J.A., de Waal Malefyt, R., *et al.* (1993a). Interleukin 13, a T-cell-derived cytokine that regulates human monocyte and B-cell function. Proc. Natl Acad. Sci. USA 90, 3735–3739.

McKenzie, A.N., Li, X., Largaespada, D.A., *et al.* (1993b). Structural comparison and chromosomal localization of the human and mouse IL-13 genes. J. Immunol. 150, 5436–5444.

Minty, A., Chalon, P., Derocq, J.M., *et al.* (1993). Interleukin-13 is a new human lymphokine regulating inflammatory and immune responses. Nature 362, 248–250.

Montaner, L.J., Doyle, A.G., Collin, M., *et al.* (1993). Interleukin 13 inhibits human immunodeficiency virus type 1 production in primary blood-derived human macrophages in vitro. J. Exp. Med. 178, 743–747.

Morgan, J., Dolganov, G., Robbins, S., Hinton, L. and Lovett, M. (1992). The selective isolation of novel cDNAs encoded by the regions surrounding the human interleukin 4 and 5 genes. Nucleic Acids Res. 20, 5173–5179.

Musso, T., Varesio, L., Zhang, X., *et al.* (1994). IL-4 and IL-13 induce Lsk, a Csk-like tyrosine kinase, in human monocytes. J. Exp. Med. 180, 2383–2388.

Muzio, M., Re, F., Sironi, M., *et al.* (1994). Interleukin-13 induces the production of interleukin-1 receptor antagonist (IL-1ra) and the expression of the mRNA for the intracellular (keratinocyte) form of IL-1ra in human myelomonocytic cells. Blood 83, 1738–1743.

Nassar, G.M., Morrow, J.D., Roberts, L.J., Lakkis, F.G. and Badr, K.F. (1994). Induction of 15-lipoxygenase by interleukin-13 in human blood monocytes. J. Biol. Chem. 269, 27631–27634.

Nilsson, G. and Nilsson, K. (1995). Effects of interleukin (IL)-13 on immediate-early response gene expression, phenotype and differentiation of human mast cells. Comparison with IL-4. Eur. J. Immunol. 25, 870–873.

Obiri, N.I., Debinski, W., Leonard, W.J. and Puri, R.K. (1995). Receptor for interleukin 13. Interaction with interleukin 4 by a mechanism that does not involve the common gamma chain shared by receptors for interleukins 2, 4, 7, 9, and 15. J. Biol. Chem. 270, 8797–8804.

Onoe, Y., Miyaura, C., Kaminakayashiki, T., *et al.* (1996). IL-13 and IL-4 inhibit bone resorption by suppressing cyclooxygenase-2-dependent prostaglandin synthesis in osteoblasts. J. Immunol. 156, 758–764.

Powers, R., Garrett, D.S., March, C.J., Frieden, E.A., Gronenborn, A.M. and Clore, G.M. (1992). Three-dimensional solution structure of human interleukin-4 by multidimensional heteronuclear magnetic resonance spectroscopy. Science 256, 1673–1677.

Punnonen, J. and de Vries, J.E. (1994). IL-13 induces proliferation, Ig isotype switching, and Ig synthesis by immature human fetal B cells. J. Immunol. 152, 1094–1102.

Punnonen, J., Aversa, G., Cocks, B.G., *et al.* (1993). Interleukin 13 induces interleukin 4-independent IgG4 and IgE synthesis and CD23 expression by human B cells. Proc. Natl Acad. Sci. USA 90, 3730–3734.

Renard, N., Duvert, V., Banchereau, J. and Saeland, S. (1994). Interleukin-13 inhibits the proliferation of normal and leukemic human B-cell precursors. Blood 84, 2253–2260.

Riemann, D., Kehlen, A. and Langner, J. (1995). Stimulation of the expression and the enzyme activity of aminopeptidase N/CD13 and dipeptidylpeptidase IV/CD26 on human renal cell carcinoma cells and renal tubular epithelial cells by T cell-derived cytokines, such as IL-4 and IL-13. Clin. Exp. Immunol. 100, 277–283.

Russell, S.M., Johnston, J.A., Noguchi, M., *et al.* (1994). Interaction of IL-2R beta and gamma c chains with Jak1 and Jak3: implications for XSCID and XCID. Science 266, 1042–1045.

Sakamoto, O., Hashiyama, M., Minty, A., Ando, M. and Suda, T. (1995). Interleukin-13 selectively suppresses the growth of human macrophage progenitors at the late stage. Blood 85, 3487–3493.

Saura, M., Martinez Dalmau, R., Minty, A., Perez Sala, D. and Lamas, S. (1996). Interleukin-13 inhibits inducible nitric oxide synthase expression in human mesangial cells. Biochem. J. 313, 641–646.

Sironi, M., Sciacca, F.L., Matteucci, C., *et al.* (1994). Regulation of endothelial and mesothelial cell function by interleukin-13: selective induction of vascular cell adhesion molecule-1 and amplification of interleukin-6 production. Blood 84, 1913–1921.

Smerz-Bertling, C. and Duschl, A. (1995). Both interleukin 4 and interleukin 13 induce tyrosine phosphorylation of the 140-kDa subunit of the interleukin 4 receptor. J. Biol. Chem. 270, 966–970.

Smirnov, D.V., Smirnova, M.G., Korobko, V.G. and Frolova, E.I. (1995). Tandem arrangement of human genes for interleukin-4 and interleukin-13: resemblance in their organization. Gene 155, 277–281.

Sun, X.J., Wang, L., Zhang, Y., *et al.* (1995). Role of IRS-2 in insulin and cytokine signalling. Nature 377, 173.

Sozzani, P., Cambon, C., Vita, N., *et al.* (1995). Interleukin-13 inhibits protein kinase C-triggered respiratory burst in human monocytes. Role of calcium and cyclic AMP. J. Biol. Chem. 270, 5084–5088.

Tony, H.P., Shen, B.J., Reusch, P. and Sebald, W. (1994). Design of human interleukin-4 antagonists inhibiting interleukin-4-dependent and interleukin-13-dependent responses in T-cells and B-cells with high efficiency. Eur. J. Biochem. 225, 659–665.

Vita, N., Lefort, S., Laurent, P., Caput, D. and Ferrara, P. (1995). Characterization and comparison of the interleukin 13 receptor with the interleukin 4 receptor on several cell types. J. Biol. Chem. 270, 3512–3517.

Wang, L.-M., Michieli, P., Lie, W., *et al.* (1995). The insulin receptor substrate-1-related 4PS substrate but not the interleukin-2 gamma chain is involved in interleukin-13-mediated signal transduction. Blood 86, 4218–4227.

Welham, M.J., Learmonth, L., Bone, H. and Schrader, J.W. (1995). Interleukin-13 signal transduction in lympho-hemopoietic cells. Similarities and differences in signal transduction with interleukin-4 and insulin. J. Biol. Chem. 270, 12286–12296.

Yanagawa, H., Sone, S., Haku, T., *et al.* (1995). Contrasting effect of interleukin-13 on interleukin-1 receptor antagonist and proinflammatory cytokine production by human alveolar macrophages. Am. J. Respir. Cell Mol. Biol. 12, 71–76.

Yano, S., Sone, S., Nishioka, Y., Mukaida, N., Matsushima, K. and Ogura, T. (1995). Differential effects of anti-inflammatory cytokines (IL-4, IL-10 and IL-13) on tumoricidal and chemotactic properties of human monocytes induced by monocyte chemotactic and activating factor. J. Leukocyte Biol. 57, 303–309.

Zurawski, G. and de Vries, J.E. (1994a). Interleukin 13 elicits a subset of the activities of its close relative interleukin 4. Stem. Cells Dayt. 12, 169–174.

Zurawski, G. and de Vries, J.E. (1994b). Interleukin 13, an interleukin 4-like cytokine that acts on monocytes and B cells, but not on T cells. Immunol. Today 15, 19–26.

Zurawski, S.M., Vega, F., Jr, Huyghe, B. and Zurawski, G. (1993). Receptors for interleukin-13 and interleukin-4 are complex and share a novel component that functions in signal transduction. EMBO J. 12, 2663–2670.

Zurawski, S.M., Chromarat, P., Djossou, O., *et al.* (1995). The primary binding subunit of the human Interleukin-4 receptor is also a component of the interleukin-13 receptor. J. Biol. Chem. 270, 13869–13878.

14. Interleukin-14

Claudia Ballaun

Preface

This chapter summarizes the literature regarding a human B-cell growth factor, originally reported by Ambrus et al. (1993), who designated this factor as IL-14. However, a recent publication from the group (Ambrus et al., 1996) contains a retraction of the originally published sequence, thus calling into question the exact nature of this protein.

1.   Introduction

Interleukin-14 (IL-14) is a member of a group of molecules called human B cell growth factors (BCGFs) (Kehrl et al., 1984; Kishimoto and Hirano, 1988). It was identified during the study of human B cell development, which is a multistep maturation process, involving activation, proliferation, and differentiation into plasma cells accompanied by coordinated acquisition and loss of B-lineage differentiation/activation antigens and surface factors. IL-14 was first described as a high-molecular-mass B cell growth factor (HMW-BCGF) owing to its ability to induce proliferation of activated B cells and its inability to stimulate resting B cells. (Ambrus and Fauci, 1985; Ambrus et al., 1985; Delfraissay et al., 1986). HMW-BCGF has also been found to inhibit antibody synthesis or secretion by B cells (Ambrus et al., 1990). In 1993, Ambrus and colleagues identified a cDNA for the human high-molecular-mass B cell growth factor, which they termed interleukin 14 (Ambrus et al., 1993). This cytokine has an apparent molecular mass of 50–60 kDa and is produced by malignant B cells, as well as by normal and malignant T cells (Ambrus and Fauci, 1985; Kishimoto et al., 1985; Sahasrabuddhe et al., 1984; Delfraissy et al., 1986; Ambrus et al., 1987). Only cells of the B cell lineage bear receptors for IL-14 (Ambrus et al., 1988; Uckun et al., 1987, 1989).

2.   The Cytokine Gene

After isolation of the human cDNA clone for IL-14 from phytohemagglutinin (PHA)-stimulated Namalva cells, the complete sequence was reported (Ambrus et al., 1993; GenBank accession number L15344). The 1.8 kb cDNA of IL-14 has a short (72 bp) 5′ untranslated region followed by a start signal that does not contain an ideal Kozak sequence. There is no further information available concerning chromosomal localization or transcriptional control of the IL-14 gene.

3. *The Protein*

The coding sequence comprises 1491 bases, encoding a protein of 497 amino acids of a predicted molecular mass of 53.1 kDa. There are three potential *N*-glycosylation sites, which is consistent with the observation that natural IL-14 is glycosylated (Ambrus and Fauci, 1985). At the amino-terminus is a 15-amino-acid signal peptide, as expected for a secreted protein (Figure 14.1) IL-14 has 8% overall sequence homology with the complement protein Bb, consistent with antigenic similarities between these two proteins (Peters *et al.*, 1988; Ambrus *et al.*, 1991). There is as yet no more information about protein structure and biochemical features of the IL-14 protein.

3.1 Cellular Sources and Biological Activities

IL-14 is produced by normal T cells, T cell clones, and T-lineage as well as B-lineage lymphoma cell lines (Ambrus and Fauci, 1985; Ambrus *et al.*, 1985).

This cytokine delivers a proliferative signal to B cells which have been activated *in vivo* by *Staphylococcus aureus* Cowan I (SAC), or anti-μ *in vitro* (Ambrus and Fauci, 1985; Ambrus *et al.*, 1985; Vazquez *et al.*, 1987). Owing to these biological features, IL-14 is similar to the Bb activation fragment from factor B of the alternative pathway of complement, although much higher concentrations of Bb than of IL-14 are required for optimal B cell proliferation. Furthermore, IL-14, but not Bb, increases cytoplasmic cAMP and cytoplasmic Ca^{2+}, leading to enhanced receptor expression.

IL-14 is present in the effusion fluids of non-Hodgkin lymphoma (NHL-B) patients with high tumor cell burdens. Freshly isolated NHL-B cells from patients

```
   1  caacaccttc agaaataatc ctttgggtga tctcttgtca atcatttgtg
  51  caggctagag aggcacctgt gaATGATAAG GCTACTGAGA AGCATCATTG
 101  GCCTGGTCCT GGCACTACCA AAGGGCAGGG GAAGCGATGC CCAAGGGGCT
 151  CCTGACCAGC ACATCATCCC ACGCAAAAAC ATTCTCCAGG TCCCTTGTTC
 201  CAGGCAGGAA ATCCCCAGCT CTGAGCGCCC TGCCAGGGCT CTGCCTAGGG
 251  ACACCTTTTC CAGGTCTAGA GAATCAAAGG AGCCTCCAGA GCAGCTAGGA
 301  GGGCCTGAGC TGACCAAGCA AGCCCTGCTC ACAAGACAAA TGCAGTCAAG
 351  ACCTGGGTGT ATTACTTGTC TTGAGCTCTG AAGGGCAGGG AGGGGTCTGA
 401  GCCTCAAATC AGACAGAGAA ATGCTCAAGT CACTTCTGCC AACTCACTGT
 451  GATGGCAGCT ACAGATGACA GCCCCTCTCA AGACTCTTCA GCTCACAGAC
 501  AAGCCACTGA CTTCATCTGT ACACACCCCC ATCCCCAATG CAAGCCCACT
 551  GTACACTTAC AGGTATAAAT GCATTTGCAA GGCCTTGCAA AATGCCCTAT
 601  GTACGTAAAA CTGACCCACA AAATCCAAAA TTGCAAGTGC CAGATGCCAG
 651  CCAGGTCAGA ACATCCTGGC TTCAGCAATG GGCTGCTCAG CATGGGAGCC
 701  TTTTATGGGC CAGGCCTGGC TGGGCTGCCG CTCCCTTCCC AGCATGACCC
 751  AACACCAGGC TCTCTAGGCC CTGGCGGAGG TGGGCTCTTG AGGCCCAGTC
 801  TGGCCTGATG CTTCTGTGCT CGGTGCTCCT GGGTAGCAAG GCGCTTCTGT
 851  GACCCTGGGG GAGCTGGGTG CTTGAGCCCC AGGCCCCTCT GGCCTCCTCT
 901  CAGGGCCACT GTCAGTGAGG GAGCCCTGGC CACCAGCACT CAGGTCCTGT
 951  ACCCTCTTGT TCAGGTCATT GCGCTCTGTC TGCAGTGCCC GGCACAGCTT
1001  CTCCAGCCGT TGGATTTTTA CCTGCAGGCC CTCCAGTTCT TTATCCCGGA
1051  CTGTTTTCTC CTCAGCCATC TCAAGCAGGG CCTTGTTGCT GCTCTCCCAC
1101  CGGGACCGGT ACATGGTGGT TTCTTTCTCC AGCTTCTTGA TCTTCTTAGT
1151  CATCTTTTCC ATCTCCTGCT TGAATGTGGT GAATACCTCG CTGCTTTTGG
1201  AAAGTGTGTT CTGGAACTCC TCAAACTTCT CTGTGTATAG GGCAAGCTGT
1251  TGCTTCAGGT GGGTCTCTTG CTGCTTCATC AGCTCACACA TCCTCTGGGA
1301  CTCTACTGCC TCTTTCAGGA GAAAATCCTT CTCCCGCTGG TGCCGCTCCT
1351  CTGCCTCCTT TAGCATCTCC TGGGCCTGCT GGAGCTTGGC ATCCACCAGC
1401  TGCTGTTGTA GGTCCTTGTG TTTGAAGACT TTGTCGATAT GCTCCTCGCG
1451  CAGCTCATAC TGCTCAATCA GCTTCTTGAG CCTCTCAGCC AGCTCCATGT
1501  TCTCTTGGCG CAGCTTGGAG TTGCGCTCAT TGTGCTGTTC CATCTGCAGC
1551  tgaatgtcat tcagtgtcac ctggaagtgc gaggtcacct ccttgcgctt
1601  ctcctcctcc tcccgggccc gctgcacacc ttcttccttg agggagcggt
1651  tgtgcccgtg cagctcacgg cataggctct caagcttgct gcgggccagg
1701  acggcttgct gtgctcaccg cgcaggtggt ccttctcttg caccagctgg
1751  ctctgctttt tctgtaggag cttcatctgc ttctgtgaat tcc
```

Figure 14.1   Nucleotide sequence of the IL-14 cDNA. The 3′ and 5′ untranslated regions are indicated in lower case letters; capital letters display translated sequences. Bold letters denote the start and stop codons. Sequence encoding the signal peptide is underlined, and potential *N*-glycosylation sites are double underlined. Translation of this cDNA sequence using commercial software (Genetic Computer Group, Maddison, Wisconsin, USA) does not give rise to the amino acid sequence of IL-14 published by Ambrus *et al.* (1993); who in fact retracted this sequence in a later publication (Ambrus *et al.*, 1996).

express IL-14 mRNA, secrete IL-14, and proliferate in response to IL-14, while exogenous IL-14 is also able to stimulate NHL-B cell proliferation *in vitro*. It may therefore act as an autocrine as well as a paracrine growth factor on NHL-B cell lines and presumably causes NHL-B cells to escape from the normal growth regulation (Ford *et al.*, 1995). Thus, one might use antisense oligonucleotides based on IL-14 for a therapeutic approach for aggressive (intermediate and high-grade) NHL-B. IL-14 stimulates chronic lymphocytic leukemia cell (CLL) colony formation in the absence of costimulants, leading to the hypothesis that this cytokine is an important hematopoietic growth factor which regulates the proliferative activity of clonogenic "mature" leukemic B cells in CLL (Uckun *et al.*, 1989).

4. *The Receptor*

Functional IL-14 receptors were detected on activated B cells (Butler *et al.*, 1984) and leukemic B cells from 60% of CLL cases (Uckun *et al.*, 1989) by use of ^{125}I-labeled purified IL-14 and a monoclonal antibody (BA-5) reactive with the ligand-binding site of the IL-14 receptor. Cross-linking studies of radioactively labeled IL-14 to its receptor on activated B cells or leukemic CLL B cells revealed a ~90 kDa receptor molecule.

5. *References*

Ambrus, J.L., Jr and Fauci, A.S. (1985). Human B lymphoma cell line producing B cell growth factor. J. Clin. Invest. 75, 732–739.

Ambrus, J.L., Jr, Jurgensen, C., Brown, E. and Fauci, A. (1985). Purification to homogeneity of a high-molecular-weight human B-cell growth factor: demonstration of specific binding to activated B cells and development of a monoclonal antibody to this factor. J. Exp. Med. 162, 1319–1335.

Ambrus, J.L., Jr, Jurgensen, C., Bowen, D.L., *et al.* (1987). The activation, proliferation, and differentiation of human B lymphocytes. In "Mechanisms of Lymphocyte Activation and Immune Regulation", Vol. 213 (eds. S. Gupta, W.E. Paul and A.S. Fauci), pp. 163–175. Plenum Press, New York.

Ambrus, J.L., Jr, Jurgensen, C., Brown, E.J,., MacFarland, P. and Fauci, A.S. (1988). Identification of a receptor for high molecular weight human B-cell growth factor. J. Immunol. 141, 660–665.

Ambrus, J.L., Jr, Chesky, L., Stephany, D., McFarland, P., Mostowski, H. and Fauci, A. (1990). Functional studies examining the subpopulation of human B lymphocytes responding to high molecular weight B cell growth factor. J. Immunol. 145, 3949–3955.

Ambrus, J.L., Jr, Chesky, L., Chused, T., *et al.* (1991). Intracellular signalling events associated with the induction of proliferation of normal human B lymphocytes by two different antigenically related human B cell growth factors and the complement factor Bb. J. Biol. Chem. 266, 3702–3708.

Ambrus, J.L., Jr, Pippin, J., Joseph, A., *et al.* (1993). Identification of a cDNA for a human high-molecular-weight B-cell growth factor. Proc. Natl Acad. Sci. USA 90, 6330–6334.

Ambrus, J.L., Pippin, J., Joseph, A., *et al.* (1996). Correction. Proc. Natl Acad. Sci. USA 93, 8154.

Butler, J.L., Ambrus, J.L., Jr and Fauci, A.S. (1984). Characterization of monoclonal B cell growth factor (BCGF) produced by a human T-T hybridoma. J. Immunol. 133, 251–255.

Delfraissy, J., Wallen, C., Vazquez, A., Dugas, B., Dormant, J. and Galanaud, P. (1986). B cell hyperactivity in systemic lupus erythematosus: selectively enhanced responsiveness to a high molecular weight B cell growth factor. Eur. J. Immunol. 16, 1251–1256.

Ford, R., Tamayo, A., Martin, B., *et al.* (1995). Identification of B-cell growth factors (interleukin-14; high molecular weight B cell growth factors) in effusion fluids from patients with aggressive B-cell lymphomas. Blood 86, 283–293.

Kehrl, J.H., Muraguchi, A., Butler, J., Falkoff, R. and Fauci, A. (1984). Human B cell activation, proliferation, and differentiation. Immunol. Rev. 78, 75–96.

Kishimoto, T. and Hirano, T. (1988). Molecular regulation of B lymphocyte response. Annu. Rev. Immunol. 6, 485–512.

Kishimoto, T., Yoshizaki, K., Okada, M., *et al.* (1985). Growth and differentiation factors and activation of human B cells. In "Lymphokines", Vol. 10 (eds. M.H. Schreier and K.A. Smith), p. 15. Academic Press, New York.

Peters, M., Ambrus, J., Jr, Fauci, A. and Brown, E. (1988). The Bb fragment of complement factor B acts as a B cell growth factor. J. Exp. Med. 168, 1225–1235.

Sahasrabuddhe, C.G., Morgan, J., Sharma, S., *et al.* (1984). Evidence for an intracellular precursor for human B cell growth factor. Proc. Natl Acad. Sci. USA 81, 7902–7906.

Uckun, F.M., Fauci, A.S., Heerema, N.A., *et al.* (1987). B-cell growth factor receptor expression and B-cell growth factor response of leukemic B-cell precursors and B-lineage lymphoid progenitor cells. Blood 70, 1020–1034.

Uckun, F.M., Fauci, A.S., Chandan-Langlie, M., Myers, D.E. and Ambrus, J.L. (1989). Detection and characterization of human high molecular weight B cell growth factor receptors on leukemic B cells in chronic lymphocytic leukemia. J. Clin. Invest. 84, 1595–1608.

Vazquez, A., Auffredou, M.-T., Gerard, J.-P., Delfraissy, J.-F. and Galanaud, P. (1987). Sequential effect of a high molecular weight B cell growth factor and of interleukin 2 on activated human B cells. J. Immunol. 139, 2344–2348.

15. Interleukin-15

Claudia Ballaun

1. Introduction

Interleukin-15 (IL-15) was originally discovered as a T cell stimulatory activity present in the culture supernatant of a simian kidney epithelial cell line (CV-1/EBNA). Biologically active IL-15 was detected in a highly enriched fraction of the concentrated culture supernatant, being capable of supporting proliferation of an IL-2-dependent murine CTLL T cell line (Grabstein *et al.*, 1994). This novel mitogen shares many activities with IL-2, including growth stimulation of T cells, B cells and natural killer (NK) cells, generation of cytolytic effector cells, induction of IgM, IgA, and IgG$_1$ synthesis, and generation of human primary antigen-specific *in vitro* response (Carson *et al.*, 1994; Giri *et al.*, 1994; Grabstein *et al.*, 1994; Armitage *et al*, 1995). IL-15 binds to the β chain of the IL-2 receptor, but requires the IL-2 receptor γ chain for correct internalization. Despite these common features of IL-15 and IL-2, they share no significant sequence homology. There are also differences in receptor usage, since IL-15 uses an additional binding component termed the IL-15 receptor α chain (Grabstein *et al.*, 1994; Anderson *et al.*, 1995a; Giri *et al.*, 1995a,b). Considering its properties, IL-15 is able to substitute for its "sister" cytokine IL-2 as a T cell activator under conditions where only small amounts of IL-2 are present.

2. The Cytokine Gene

The human IL-15 cDNA was obtained by hybridization of the simian IL-15 cDNA to the human stromal cell line IMTLH cDNA library (Grabstein *et al.*, 1994; GenBank-EMBL accession number U14407). It contains a 5′ 316 bp noncoding region preceding an open reading frame of 486 bp and a 400 bp 3′ noncoding region (Figure 15.1). The human gene is mapped to the human chromosome 4q31. Several other human growth factors and cytokine genes map to the same arm of human chromosome 4, including epidermal growth factor, fibroblast growth factor, the α-chemokines, and other cytokines, including IL-2. The organization of the IL-2 and IL-15 gene is very similar and characteristic of the helical cytokine family (Anderson *et al.*, 1995; Bazan *et al.*, 1992). The murine gene for IL-15 consists of eight exons spanning at least 34 kb and is localized to the central region of mouse chromosome 8 (Anderson *et al.*, 1995). The human contiguous IL-15 sequence of 15 kb

```
   1  tgtccggcgc cccccgggag ggaactgggt ggccgcaccc tcccggctgc
  51  ggtggctgtc gcccccacc  ctgcagccag gactcgatgg agaatccatt
 101  ccaatatatg gccatgtggc tctttggagc aatgttccat catgttccat
 151  gctgctgctg acgtcacatg gagcacagaa atcaatgtta gcagatagcc
 201  agcccataca agatcgtatt gtattgtagg aggcatcgtg gatggatggc
 251  tgctggaaac cccttgccat agccagctct tcttcaatac ttaaggattt
 301  accgtggctt tgagtaATGA GAATTTCGAA ACCACATTTG AGAAGTATTT
 351  CCATCCAGTG CTACTTGTGT TTACTTCTAA ACAGTCATTT TCTAACTGAA
 401  GCTGGCATTC ATGTCTTCAT TTTGGGCTGT TTCAGTGCAG GGCTTCCTAA
 451  AACAGAAGCC AACTGGGTGA ATGTAATAAG TGATTTGAAA AAAATTGAAG
 501  ATCTTATTCA ATCTATGCAT ATTGATGCTA CTTTATATAC GGAAAGTGAT
 551  GTTCACCCCA GTTGCAAAGT AACAGCAATG AAGTGCTTTC TCTTGGAGTT
 601  ACAAGTTATT TCACTTGAGT CCGGAGATGC AAGTATTCAT GATACAGTAG
 651  AAAATCTGAT CATCCTAGCA AACAACAGTT TGTCTTCTAA TGGGAATGTA
 701  ACAGAATCTG GATGCAAAGA ATGTGAGGAA CTGGAGGAAA AAAATATTAA
 751  AGAATTTTTG CAGAGTTTTG TACATATTGT CCAAATGTTC ATCAACACTT
 801  CTtgattgca attgattctt tttaaagtgt ttctgttatt aacaaacatc
 851  actctgctgc ttagacataa caaaacactc ggcatttaaa atgtgctgtc
 901  aaaacaagtt tttctgtcaa gaagatgatc agaccttgga tcagatgaac
 951  tcttagaaat gaaggcagaa aaatgtcatt gagtaatata gtgactatga
1001  acttctctca gacttacttt actcatttt  ttaatttatt attgaaattg
1051  tacatatttg tggaataatg taaaatgttg aataaaaata tgtacaagtg
1101  ttgtttttta agttgcactg atattttacc tcttattgca aaatagcatt
1151  tgtttaaggg tgatagtcaa attatgtatt ggtggggctg ggtaccaatg
1201  ct
```

Figure 15.1   Nucleotide sequence of the human IL-15 cDNA. Nucleotides are numbered according to the sequence published by Grabstein *et al.* (1994). The coding sequence is given in capital letters, untranslated sequences are indicated in lower case letters. The "tata"-box and the start- and stop-codons are indicated in bold letters. The human genomic IL-15 nucleotide sequence has been deposited under Acc. No. X91233.

contains six protein-coding exons and the flanking intron sequences (EMBL Acc. No. X91233). The overall size of the gene is estimated to be at least 32 kb (Krause *et al.*, 1996).

3.   The Protein

The full-length human cDNA clone encodes a 162-amino acid (aa) precursor polypeptide containing an unusually long 48-aa leader sequence producing a 114-aa mature protein after cleavage. The IL-15 protein has an apparent molecular mass of 14–18 kDa. Computer modeling studies for IL-15 predicts strong helical structures for regions containing aa1–15, 18–57, 65–78, 97–114, and support a four-helix bundle-like structure for this protein. IL-15 therefore belongs to the family of four-helix bundle-type cytokines, which bind to receptors of the hematopoietin superfamily. A three-dimensional-model (distance geometry-based homology modeling package: FOLDER) suggests two disulfide cross-links: Cys[35]–Cys[85] and Cys[42]–Cys[88], where the latter is analogous to the only disulfide bridge in IL-2. There are two potential N-glycosylation sites at the carboxy-terminus of the IL-15 protein at Asn-79 and Asn-112 (Figure 15.2).

4.   Cellular Sources and Production

In distinct contrast to IL-2, IL-15 appears to be expressed in a variety of tissues and cell types. Expression of IL-15 mRNA was detected in several human tissues, being most abundant in placenta and skeletal muscle, adherent peripheral blood mononuclear cells, and epithelial and fibroblast cell lines. Activated peripheral blood T lymphocytes, a rich source of IL-2 and IFN-γ, express no detectable IL-15 mRNA, nor do B lymphoblastoid cell lines; instead IL-15 appears to be synthesized by monocytes/macrophages (Grabstein *et al.*, 1994). Human bone marrow (BM)-derived stromal cells express IL-15 transcripts, and supernatants from long-term BM stromal cell cultures contain IL-15 protein (Mrozek *et al.*, 1996). The differences in expression patterns, especially the expression of IL-15 in nonlymphoid tissues, suggests that IL-15 may have other functions *in vivo*, in addition to promoting the growth and functional activation of lymphocytes.

5.   Biological Activities

Purified recombinant IL-15 stimulates the proliferation of CTLL cells and phytohemagglutinin (PHA)-stimulated peripheral blood T lymphocytes, to the same

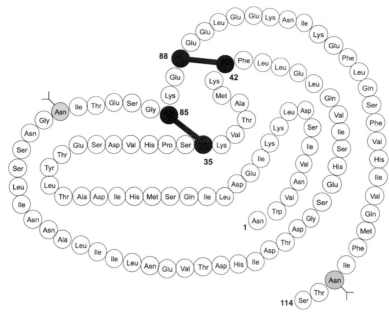

Figure 15.2   Deduced amino acid sequence of the human IL-15 protein. Two potential *N*-glycosylation sites are indicated. Cysteine residues, important for disulfide cross-links are blocked circled.

extent as IL-2 and with similar potency (Grabstein *et al.*, 1994). It is also reported that IL-15 stimulates tumor-derived activated T cells in culture (TDAC, also referred to as tumor infiltrating lymphocytes) in a dose-dependent manner, alone or in the presence of IL-2. IL-15 doubled the level of secreted IFN-α and GM-CSF, implying that this cytokine is capable of stimulating the growth of TDAC that have antitumor activity (Lewko *et al.*, 1995).

Like IL-2, IL-15 induces proliferation of activated, but not resting, B cells and induces secretion of IgM, IgG, and IgA in combination with recombinant CD40-ligand (Armitage *et al.*, 1995). It was also shown that IL-15 stimulates locomotion of human T lymphocytes, causing them to accelerate and to move directionally, implying that this cytokine is a more powerful chemoattractant than IL-8 or MIP-1α (Wilkinson and Liew, 1995).

IL-15 is able to induce CD34⁺ hematopoietic progenitor cells to differentiate into CD3⁻ CD56⁺ NK cells (Mrozek *et al.*, 1996). CD56*dim* NK cells display a weak proliferative but strong cytotoxic response to IL-15 indistinguishable from IL-2. Its ability to induce NK cytotoxic activity is not mediated by IL-2, because neutralizing antibodies to IL-2 did not inhibit the activation of NK cells by IL-15. Moreover, results of preliminary murine models suggest that IL-15 is also a potent enhancer of NK activity *in vivo* (Carson *et al.*, 1994). In fact, resting human NK cells express IL-15 Rα mRNA and further, picomolar amounts of IL-15 can sustain NK cell survival for up to 8 days in the absence of serum and other certain cytokines (Carson *et al.*, 1997).

The combination of IL-2 with IL-15 never produced an additive or synergistic response, demonstrating that IL-2 and IL-15 are mutually redundant. In granulated metrial gland (GMG) cells which differentiate *in situ* in the mouse pregnant uterus into NK-like cells, IL-15 and its cognate receptor are expressed and moreover IL-15 is able to induce the expression of the cytolytic mediators perforin and granzymes (Ye *et al.*, 1996).

Corresponding to its high expression in skeletal muscle tissue, IL-15 was found to stimulate muscle-specific myosin heavy-chain accumulation by differentiated myocytes and muscle fibers in culture (Quinn *et al.*, 1995).

In human skin, constitutive expression of IL-15 is confined to the dermis. Following UVB treatment, IL-15 mRNA was induced in epidermal sheets and enhanced in dermal sheets. The fact that skin cells can secrete UVB-inducible IL-15 supports a pro-inflammatory T cell activating role for UVB radiation (Mohamadzadeh *et al*, 1995).

6.   *The Receptor*

Cell surface receptors for IL-15 are expressed on a variety of T cells, such as murine antigen-dependent T cell clones, and human and murine T cell lines, which also express IL-2 receptors (IL-2R). Low numbers of high-affinity IL-15 receptors (IL-15R) were detected on NK cells and activated peripheral blood mononuclear cells (PBMC), while activation of fresh monocytes and

T cell lines leads to expression of a higher number of receptors for IL-15. In addition to cells of lymphoid origin, receptors for IL-15 are also detected on fresh human venous endothelial cells, which do not bind IL-2. The human IL-2R β chain alone is not able to bind IL-15 under physiological conditions and the γ chain is required in addition to the β chain for IL-15 binding (Giri *et al.*, 1994). The fact that IL-15 was able to bind to B-LCLs (Epstein–Barr virus-transformed B lymphoblastoid cell lines) established from X-SCID (X-linked severe combined immunodeficiency) patients lacking cell surface expression of the γ chain suggested the existence of another IL-15-specific binding molecule (Kumaki *et al.*, 1995). Indirect evidence for an additional binding site for IL-15 has been also derived from the observation that IL-2 only partially inhibits IL-15 binding to the human NK cell line YT (Giri *et al.*, 1994). Another indication for the existence of additional IL-15 binding molecules was seen in murine cells, which express fully functional IL-2Rs but were not able to bind or respond to human recombinant or simian-derived IL-15 (Giri *et al.*, 1994). A distinct murine IL-15R α chain (Giri *et al.*, 1995a,b) and the human analogue (Anderson *et al.*, 1995) have been characterized. The IL-15R α chain is structurally related to the IL-2R α chain, defining a new cytokine receptor family, but in contrast to the IL-2R α chain it contains just one Sushi domain. This Sushi domain seems to be an essential element for specific IL-15 binding to its receptor. The cytoplasmic domain of the IL-15R α chain is dispensable for IL-15 signaling, as is true for the IL-2R α chain (Anderson *et al.*, 1995) (Figure 15.3). A biologically functional heterotrimeric IL-15R complex, consisting of IL-2Rβ, IL-2Rγ, and IL-15Rα, capable of mediating IL-15 responses, was generated through reconstruction experiments in a murine myeloid cell. IL-15Rα alone displays a 1000-fold higher affinity for IL-15 binding than does IL-2Rα for IL-2 (Anderson *et al.*, 1995). Such a high-affinity binding protein might either serve to remove IL-15 from effective action or, conversely, serve as a source of presentation of IL-15 to neighboring populations of cells capable of responding to it. The distribution of IL-15 and IL-15R α chain mRNA suggests that IL-15 may have biological activities distinct from IL-2, as well as additional activities outside the immune system.

7.   Signal Transduction

For signal transduction in B cells, IL-15 requires the β and γ chains of the IL-2 receptor complex, but does not use the IL-2R α chain. Capacity for signal transduction was shown by induction of mRNA expression for c-*myc*, c-*fos*, and c-*jun* protooncogenes (Giri *et al.*, 1994).

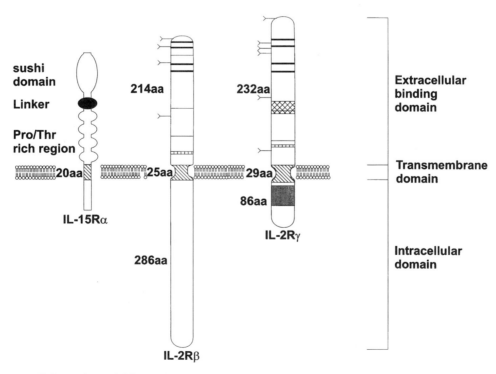

Figure 15.3   Schematic model for the IL-15 receptor. The three components of the IL-15 receptor, IL-2Rβ,γ, and IL-15Rα are outlined. The various regions of the IL-15Rα, the Pro/Thr-rich region (proline/threonine-rich region), and the sushi domain (short consensus repeats) are indicated.

Studies of the role of shared receptor motifs and common STAT (signal transducer and activation of transcription) proteins revealed that IL-15 activates STAT3 and STAT5 proteins, highlighting the redundancy of cytokine action mediated by IL-2, IL-7, and IL-15 (Lin *et al.*, 1995). IL-15, as well as IL-2, IL-7, and IL-4, rapidly stimulates the tyrosine phosphorylation of the signaling molecules insulin receptor substrate IRS-1 and IRS-2 in human peripheral blood T cells, NK cells, and lymphoid cell lines, indicating that IRS-1 and IRS-2 may have important roles in T lymphocyte activation (Johnston *et al.*, 1995). The requirement for IL-2R γ chain participation in IL-15 binding and signaling is significant in view of the finding that a mutation in the IL-2R γ chain results in X-linked severe combined immunodeficiency in humans (Noguchi *et al.*, 1993). Considering that the β chain and the γ chain of the IL-2R are physically associated with the tyrosine kinases JAK1 and JAK3 (Russell *et al.*, 1994), and kinases of the *src*-family with the β chain of the IL-2R (Taniguchi and Minami, 1993), the utilization of IL-2Rβ,γ by IL-15 implies some common signaling pathways of IL-15 and IL-2. Another distinct signal transduction pathway has been reported in mast cells. IL-15 induces mast cell proliferation in the absence of IL-2Rα,β and does not induce the IL-2Rγ, as demonstrated by transfectants of these cells with a truncated mutant of IL-2Rγ. Moreover, the mast cell IL-15 receptors recruit JAK2 and STAT5 instead of JAK1/3 and STAT3/5, that are activated in T cells (Tagaya *et al.*, 1996).

8.   Cytokine in Disease and Therapy

IL-15 triggers the proliferation of freshly isolated leukemic cells obtained from patients with chronic lymphoproliferative disorder (CLD) but not normal resting B lymphocytes (Trentin *et al.*, 1996). This observation proposes that leukemic B cell proliferation is probably better related to the presence of a distinct receptor structure (IL-2R subunits) that is fully functional to transduce stimulatory signals, rather than to differences in terms of B cell subsets.

By its ability to recruit and activate T lymphocytes into the membrane by chemoattraction during a chronic inflammatory disease (rheumatoid arthritis), it has been demonstrated that IL-15 plays a biological role in a pathological setting (McInnes *et al.*, 1996). Current therapies directed against T cells in rheumatoid arthritis demonstrate encouraging improvements in disease activity. IL-15 might be a useful target for future therapeutic intervention in rheumatoid arthritis.

The new cytokine IL-15 may also play an important role in antitumor immunity with a possible use for cancer immunotherapy. *In vitro* studies have shown that IL-15 is able to induce CTL and lymphokine-activated killer (LAK) cell activity in peripheral blood mononuclear cells against known tumor targets from normal donors. Gamero and colleagues (1995) demonstrated that IL-15 is able to induce LAK activity in lymphocytes from metastatic melanoma patients to kill autologous melanoma tumor cells from fresh tumor biopsies.

9.   References

Anderson, D.M., Kumaki, S., Ahdieh, M., *et al.* (1995a). Functional characterization of the human interleukin-15 receptor α chain and close linkage of IL15RA and IL2RA genes. J. Biol. Chem. 270, 29862–29869.

Anderson, D.M., Johnson, L., Glaccum, M.B., *et al.* (1995b). Chromosomal assignment and genomic structure of IL-15. Genomics 25, 701–706.

Armitage, R.J., MacDuff, B.M., Eisenman, J., Paxton R. and Grabstein, K.H. (1995). IL-15 has stimulatory activity for the induction of B cell proliferation and differentiation. J. Immunol. 154, 483–490.

Bazan, J.F. (1992). Unraveling the structure of IL-2. Science 257, 410–413.

Carson, W.E., Giri, J.G., Lindemann, M.J., *et al.* (1994). Interleukin 15 (IL-15) is a novel cytokine that activates human natural killer cells via components of the IL-2 receptor. J. Exp. Med. 180, 1395–1403.

Carson, W.E., Fehninger, T.A., Haldar, S., Eckhert, K., *et al.* (1997). A potential role for interleukin-15 in the regulation of human natural killer cell survival. J. Clin. Invest. 99(5), 937–943.

Gamero, A., Ussery, D., Reintgen, D., Puleo, C. and Djeu, J. (1995). Interleukin 15 induction of lymphokine-activated killer cell function against autologous tumor cells in melanoma patient lymphocytes by a CD18-dependent, perforin-related mechanism. Cancer Res. 55, 4988–4994.

Giri, J.G., Ahdieh, M., Eisenman, J., *et al.* (1994). Utilization of the β and γ chains of the IL-2 receptor by the novel cytokine IL-15. EMBO J. 13, 2822–2830.

Giri, J.G., Kumaki, S., Ahdieh, M., *et al.* (1995a). Identification and cloning of a novel IL-15 binding protein that is structurally related to the alpha chain of the IL-2 receptor. EMBO J. 14, 3654–3663.

Giri, J.G., Anderson, D., Kumaki, S., Park, L.S., Grabstein, K. and Cosman, D. (1995b). IL-15, a novel T cell growth factor that shares activities and receptor components with IL-2. J. Leukocyte Biol. 57, 763–766.

Grabstein, K., Eisenman, J., Shanebeck, K., *et al.* (1994). Cloning of a T cell growth factor that interacts with the β chain of the interleukin-2 receptor. Science 264, 965–968.

Johnston, J., Wang, L.-M., Hanson, E., *et al.* (1995). Interleukins 2, 4, 7, and 15 stimulate tyrosine phosphorylation of insulin receptor substrates 1 and 2 in T cells. J. Biol. Chem. 270, 28527–28530.

Krause, H., Jandrig, B., Wernicke, C., Bulfone-Paus, S., *et al.* (1996). Genomic structure and chromosomal localisation of the human interleukin 15 gene (IL-15). Cytokine 8(9), 667–674.

Kumaki, S., Ochs, H.D., Timour, M., *et al.* (1995). Characterization of B-cell lines established from two X-linked severe combined immunodeficiency patients: Interleukin-15 binds to the B cells but is not internalized efficiently. Blood 86, 1428–1436.

Lewko, W.M., Smith, T.L., Bowman, D.J., Good, R.W, and Oldham, R.K. (1995). Interleukin 15 and the growth of tumor derived activated T-cells. Cancer Biother. 10, 13–20.

Lin, J.-X., Migone, T.-S., Tsang, M., et al. (1995). The role of shared receptor motifs and common STAT proteins in the generation of cytokine pleiotropy and redundancy by IL-2, IL-4, IL-7, IL-13, and IL-15. Immunity 4, 331–339.

McInnes, I., Al-Mughales, J., Field, M., et al. (1996). The role of interleukin-15 in T-cell migration and activation in rheumatoid arthritis. Nature Med. 2, 175–182.

Mrozek, E., Anderson, P. and Caligiuri, M. (1996). Role of interleukin-15 in the development of human CD56$^+$ natural killer cells from CD34$^+$ hematopoietic progenitor cells. Blood 87, 2632–2640.

Mohamadzadeh, M., Takashima, A., Dougherty, I., Knop, J., Bergstresser, P. and Cruz, P., Jr (1995). Ultraviolet B radiation up-regulates the expression of IL-15 in human skin. J. Immunol. 155, 4492–4496.

Noguchi, M., Yi, H., Rosenblatt, H.M., et al. (1993). Interleukin 2 receptor γ chain mutation results in X-linked severe immunodeficiency in humans. Cell 73, 147–157.

Quinn, L.S., Haugk, K.L. and Grabstein, K.H. (1995). Interleukin-15: a novel cytokine for skeletal muscle. Endocrinology 136, 3669–3672.

Russell, A., Johnston, J., Noguchi, M., et al. (1994). Interaction of IL-2Rβ and γ chains with Jak1 and Jak3: implications for XSCID and SCID. Science 266, 1042–1045.

Tagaya, Y., Burton, J.D., Miyamoto, Y. and Waldmann, T.A. (1996). Identification of a novel receptor/signal transduction pathway for IL-15/T in mast cells. EMBO J. 15(8), 4928–4939.

Taniguchi, T. and Minami, Y. (1993). The IL-2/IL-2 receptor system: a current overview. Cell 73, 5–8.

Trentin, L., Cerutti, A., Zambello, R., et al. (1996). Interleukin-15 promotes the growth of leukemic cells of patients with B-cell chronic lymphoproliferative disorders. Blood 87, 3327–3335.

Wilkinson, P.C. and Liew, F.Y. (1995). Chemoattraction of human blood T lymphocytes by interleukin-15. J. Exp. Med. 181, 1255–1259.

Ye, W., Zheng, L.H., Young, J.D. and Lin, C.C. (1996). The involvement of interleukin (IL)-15 in regulating the differentiation of granulated metrial gland cells in mouse pregnant uterus. J. Exp. Med. 184(6), 2405–2410.

16. Interleukin-16

Anthony R. Mire-Sluis

Interleukin-16 (IL-16) was originally described as a lymphocyte chemoattractant factor (LCF) (Center and Cruikshank, 1982; Cruikshank and Center, 1982). IL-16 is produced from CD8$^+$ T cells induced by lectins, histamine or serotonin (Berman et al., 1985; van Epps et al., 1983; Laberge et al., 1996). IL-16 is also produced from CD4$^+$ T cells on stimulation with lectins, antigen, or anti CD3 antibodies (Center et al., 1996) as well as from eosinophils (Lim et al., 1996) epithelial cells (Bellini et al., 1993; Center et al., 1996) and mast cells (Rumsaeng et al., 1997).

The cDNA for IL-16 was originally found to result in mRNA that contains 3′ AUUUA sequences, two polyadenylation sites but no leader sequence (Cruikshank et al., 1994). IL-16 mRNA appears to be present constitutively in both subsets of T cells and has a half-life of 2 h (Labege et al., 1995, 1996), although only CD8 cells appear to contain preformed biologically active tetrameric IL-16 (Labege et al., 1996). The preformed IL-16 in CD8 cells is released on activation without the requirement of de novo protein synthesis. The chromosomal location of the IL-16 gene is on 15q26.1 (Center et al., 1996).

IL-16 protein has been suggested to consist of a 56 kDa homotetramer of 14–17 kDa 130-amino-acid monomers and has a pI of 9.1 (Cruikshank and Center, 1982; Cruikshank et al., 1994). There appears to be some controversy whether the monomers result from a much larger precursor that could arise from a different cDNA than that described and an 80 kDa protein has been identified in nonactivated CD4 cells that cross-reacts with antisera to the 14 kDa monomer (Bazan and Schall, 1996; Bannert et al., 1996; Center et al., 1996; Cruikshank et al., 1996). Recloning of the IL-16 cDNA revealed a 2.6 kb mRNA producing a deduced 67–80 kDa protein proform of IL-16 that is subsequently cleaved to lower molecular weight forms as yet not clearly defined (Baier et al., 1997) (Figures 16.1 and 16.2).

IL-16 chemoattracts CD4 T cells, monocytes and eosinophils (Cruikshank and Center, 1982; Cruikshank et al., 1987; Rand et al., 1991). IL-16 is also an activating factor for CD4 T cells, inducing the expression of the IL-2 receptor and HLA-DR surface antigen (Cruikshank et al., 1987). IL-16 also stimulates HLA-DR expression on monocytes (Cruikshank et al., 1987) and eosinophil adhesion molecule expression (Wan et al., 1995).

The receptor for IL-16 is the 55 kDa CD4 molecule (Berman et al., 1958; Cruikshank et al., 1991). CD4 contains an extracellular domain containing four immunoglobulin-like domains and a short 37-amino-acid intracellular domain (Figure 16.3) (Maddon et al., 1985).

CD4 is present on a subpopulation of T cells as well as eosinophils and monocytes (Reinherz et al., 1979; Steward et al., 1986; Lucey et al., 1989). It has been shown that the binding of IL-16 to CD4 mediates its biological activity (Rand et al., 1991; Ryan et al., 1995).

Ligation of CD4 with IL-16 on T cells results in inositol lipid turnover, increases in intracellular calcium, and autophosphorylation of p56^{lck} (Cruikshank et al., 1991; Ryan et al., 1995).

IL-16 has been shown to be produced in inflammatory conditions such as asthma and sarcoidosis (Bellini et al., 1993; Cruikshank et al., 1995; Center et al., 1996; Laberge et al., 1997). IL-16 may have a useful role in AIDS as it has been shown to prevent HIV binding to CD4 and infecting cells (Baier et al., 1995), is increased in nonprogressive HIV infection (Scala et al., 1997) and is able to repress HIV-1 promoter activity (Maciaszek et al., 1997) and mRNA expression (Zhou et al., 1997).

```
                                                                                                                         120
CTGCTGCTACCACAGGAAGACACAGCAGGGAGAAGCCCTAGTGCCTCTGCCGGCTGCCCAGGACCTGGTATCGGCCCACAGACCAAGTCCTCCACAGAGGGCGAGCCAGGGTGGAGAAGA
                                                                                                                         240
GCCAGCCCAGTGACCCAAACATCCCCGATAAAACACCCACTGCTTAAGAGGCAGGCTCGGATGGACTATAGCTTTGATACCACAGCCGAAGACCCTTGGGTTAGGATTTCTGACTGCATC
                                                                                                                         360
AAAAAACTTATTTAGCCCCATCATGAGTGAGAACCATGGCCACATGCCTCTACAGCCCAATGCCAGCCTGAATGAAGAAGAAGGGACACAGGGCCACCCAGATGGGACCCCACCAAAGCTG
                                                                                                                         480
GACACCGCCAATGGCACTCCCAAAGTTTACAAGTCAGCAGACAGCAGCACTGTGAAGAAAGGTCCTCCTGTGGCTCCCAAGCCAGCCTGGTTTCGCCAAAGCTTGAAAGGTTTGAGGAAT
                                                                                                                         600
CGTGCTTCAGACCCAAGAGGGCTCCCTGATCCTGCCTTGTCCACCCAGCCAGCACCTGCTTCCAGGGAGCACCTAGGATCACACATCCGGGCCTCCTCCTCCTCCTCCTCCATCAGGCAG
                                                                                                                         720
AGAATCAGCTCCTTTGAAACCTTTGGCTCCTCTCAACTGCCTGACAAAGGAGCCCAGAGACTGAGCCTCCAGCCCTCCTCCGGGGAGGCAGCAAAACCTCTTGGGAAGCATGAGGAGGA
                                                                                                                         840
CGGTTTTCTGGACTCTTGGGGCGAGGGGCTGCACCCACTCTTGTGCCCCAGCAGCCTGAGCAAGTACTGTCCTCGGGGTCCCCTGCAGCCTCCGAGGCCAGAGACCCAGGCGTGTCTGAG
                                                                                                                         960
TCCCCTCCCCCAAGGCGGCAGCCCAATCAGAAAACTCTCCCCCCTGGCCCGGACCCGCTCCTAAGGCTGCTGTCAACACAGGCTGAGGAATCTCAAGGCCCAGTGCTCAAGATGCCTAGC
                                                                                                                        1080
CAGCGAGCACGGAGCTTCCCCCTGACCAGGTCCCAGTCCTGTGAGACGAAGCTACTTGACGAAAAGACCAGCAAACTCTATTCTATCAGCAGCCAAGTGTCATCGGCTGTCATGAAATCC
                                                                                                                        1200
TTGCTGTGCCTTCCATCTTCTATCTCCTGTGCCCAGACTCCCTGCATCCCCAAGGAAGGGGCATCTCCAACATCATCATCCAACGAAGACTCAGCTGCAAATGGTTCTGCTGAAACATCT
                                                                                                                        1320
GCCTTGGACACAGGGTTCTCGCTCAACCTTTCAGAGCTGAGAGAATATACAGAGGGTCTCACGGAAGCCAAGGAAGACGATGATGGGGACCACAGTTCCCTTCAGTCTGGTCAGTCCGTT
                                                                                                                        1440
ATCTCCCTGCTGAGCTCAGAAGAATTAAAAAAACTCATCGAGGAGGTGAAGGTTCTGGATGAAGCAACATTAAAGCAATTAGACGGCATCCATGTCACCATCTTACACAAGGAGGAAGGT
                                                                                                                        1560
GCTGGTCTTGGGTTCAGCTTGGCAGGAGGAGCAGATCTAGAAAACAAGGTGATTACGGTTCACAGAGTGTTTCCAAATGGGCTGGCCTCCCAGGAAGGGACTATTCAGAAGGGCAATGAG
                                                                                                                        1680
GTTCTTTCCATCAACGGCAAGTCTCTCAAGGGGACCACGCACCATGATGCCTTGGCAATCCTCCGCCAAGCTCGAGAGCCCAGGCAAGCTGTGATTGTCACAAGGAAGCTGACTCCAGAG
                                                                                                                        1800
GCCATGCCTGACCTCAACTCCTCCACTGACTCTGCAGCCTCAGCCTCTGCAGCCAGTGATGTTTCTGTAGAATCTACAGCAGAGGCCACAGTCTGCACGGTGACACTGGAGAAGATGTCG
                                                                                                                        1920
GCAGGGCTGGGCTTCAGCCTGGAAGGAGGGAAGGGCTCCCTACACGGAGACAAGCCTCTCACCATTAACAGGATTTTCAAAGGAGCAGCCTCAGAACAAAGTGAGACAGTCCAGCCTGGA
                                                                                                                        2040
GATGAAATCTTGCAGCTGGGTGGCACTGCCATGCAGGGCCTCACACGGTTTGAAGCCTGGAACATCATCAAGGCACTGCCTGATGGACCTGTCACGATTGTCATCAGGAGAAAAAGCCTC
                                                        2076
CAGTCCAAGGAAACCACAGCTGCTGGAGACTCCTAG
```

Figure 16.1   Nucleotide sequence of the IL-16 precursor cDNA. The two possible initiation codons are underlined. The ATG that was originally believed to mark the beginning of the 390 bp IL-16 coding region is marked by a black bar. The location of the splice junction where in some cDNAs short intervening sequences have been found is identified by a triangle. With kind permission from M. Baier.

```
                                                                                       MetAspTyrSerPheAspThrThrAlaGluAspProTrpValArgIleSerAspCysIle
LysAsnLeuPheSerProIleMetSerGluAsnHisGlyHisMetProLeuGlnProAsnAlaSerLeuAsnGluGluGluGlyThrGlnGlyHisProAspGlyThrProProLysLeu
AspThrAlaAsnGlyThrProLysValTyrLysSerAlaAspSerSerThrValLysLysGlyProProValAlaProLysProAlaTrpPheArgGlnSerLeuLysGlyLeuArgAsn
ArgAlaSerAspProArgGlyLeuProAspProAlaLeuSerThrGlnProAlaProAlaSerArgGluHisLeuGlySerHisIleArgAlaSerSerSerSerSerIleArgGln
ArgIleSerSerPheGluThrPheGlySerSerGlnLeuProAspLysGlyAlaGlnArgLeuSerLeuGlnProSerSerGlyGluAlaAlaLysProLeuGlyLysHisGluGluGly
ArgPheSerGlyLeuLeuGlyArgGlyAlaAlaProThrLeuValProGlnGlnProGluGlnValLeuSerSerGlySerProAlaAlaSerGluAlaArgAspProGlyValSerGlu
SerProProProArgArgGlnProAsnGlnLysThrLeuProProGlyProAspProLeuLeuArgLeuLeuSerThrGlnAlaGluGluSerGlnGlyProValLeuLysMetProSer
GlnArgAlaArgSerPheProLeuThrArgSerGlnSerCysGluThrLysLeuLeuAspGlyLysThrSerLysLeuTyrSerIleSerSerGlnValSerSerAlaValMetLysSer
LeuLeuCysLeuProSerSerIleSerCysAlaGlnThrProCysIleProLysGluHlyAlaSerProThrSerSerSerAsnGluAspSerAlaAlaAsnGlySerAlaGluThrSer
AlaLeuAspThrGlyPheSerLeuAsnLeuSerGluLeuArgGluTyrThrGluGlyLeuThrGluAlaLysGluAspAspAspGlyAspHisSerSerLeuGlnSerGlyGlnSerVal
IleSerLeuLeuSerSerGluGluLeuLysLysLeuIleGluGluValLysValLeuAspGluAlaThrLeuLysGlnLeuAspGlyIleHisValThrIleLeuHisLysGluGluGly
AlaGlyLeuGlyPheSerLeuAlaGlyGlyAlaAspLeuGluAsnLysValIleThrValHisArgValPheProAsnGlyLeuAlaSerGlnGluGlyThrIleGlnLysGlyAsnGlu
ValLeuSerIleAsnglyLysSerLeuLysGlyThrThrHisHisAspAlaLeuAlaIleLeuArtGlnAsnAlaArgGluProArgGlnAlaValIleValThrArgLysLeuThrProGlu
AlaMetProAspLeuAsnSerSerThrAspSerAlaAlaSerAlaSerAlaAlaSerAspValSerValGluSerThrAlaGluAlaThrValCysThrValThrLeuGluLysMetSer
AlaGlyLeuGlyPheSerLeuGluGlyGlyLysGlySerLeuHisGlyAspLysProLeuThrIleAsnArgIlePheLysGlyAlaAlaSerGluGlnSerGluThrValGlnProGly
AspGluIleLeuGlnLeuGlyGlyThrAlaMetGlnGlyLeuThrArgPheGluAlaTrpAsnIleIleLysAlaLeuProAspGlyProValThrIleValIleArgArgLysSerLeu
GlnSerLyGluThrThrAlaAlaGlyAspSer
```

Figure 16.2   Deduced amino acid sequence of the IL-16 precursor cDNA. Two possible amino acids starts and a putative cleavage site at Asp 510 are underlined. With kind permission from M. Baier.

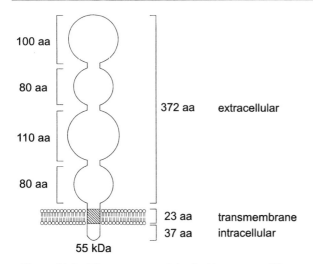

100 aa

80 aa

372 aa   extracellular

110 aa

80 aa

23 aa   transmembrane
37 aa   intracellular

55 kDa

Figure 16.3   The structure of the IL-16 receptor, CD4.

References

Baier, M., Werner, A., Bannert, N., Metzner, K. and Kurth, R. (1995). HIV suppression by interleukin-16. Nature 371(6557), 561. [Letter]

Baier, M., Bannert, N., Werner, A., Lang, K. and Kurth, R. (1997). Molecular cloning, sequence, expression, and processing of the interleukin 16 precursor. Proc. Natl Acad. Sci. USA 94(10), 5273–5277.

Bannert, N., Baier, M., Werner, A. and Kurth, R. (1996). Interleukin-16 or not? Nature 381(6577), 30. [Letter]

Bazan, J.F. and Schall, T.J. (1996). Interleukin-16 or not? Nature 381(6577), 29–30. [Letter]

Bellini, A., Yoshimura, H., Vittori, E., Marini, M. and Mattoli, S. (1993). Bronchial epithelial cells of patients with asthma release chemoattractant factors for T lymphocytes. J. Allergy Clin. Immunol. 92(3), 412–424.

Berman, J.S., Cruikshank, W.W., Center, D.M., Theodore, A.C. and Beer, D.J. (1985). Chemoattractant lymphokines specific for the helper/inducer T-lymphocyte subset. Cell. Immunol. 95, 105–112.

Center, D.M. and Cruikshank, W. (1982). Modulation of lymphocyte migration by human lymphokines. I. Identification and characterization of chemoattractant activity for lymphocytes from mitogen-stimulated mononuclear cells. J. Immunol. 128(6), 2563–2568.

Center, D.M., Kornfeld, H. and Cruikshank, W. (1996). Interleukin 16 and its function as a CD4 ligand. Immunol. Today 17, 476–478.

Cruikshank, W. and Center, D.M. (1982). Modulation of lymphocyte migration by human lymphokines. II. Purification of a lymphotactic factor (LCF). J. Immunol. 128(6), 2569–2574.

Cruikshank, W.W., Berman, J.S., Theodore, A.C., Bernardo, J. and Center, D.M. (1987). Lymphokine activation of T4⁺ T lymphocytes and monocytes. J. Immunol. 138, 3817–3825.

Cruikshank, W., Greenstein, J.L., Theodore, A.C. and Center, D.M. (1991). Lymphocyte chemoattractant factor induces CD4-dependent intracytoplasmic signaling in lymphocytes. J. Immunol. 146(9), 2928–2934.

Cruikshank, W., Center, D.M., Nisar, N., et al. (1994). Molecular and functional analysis of a lymphocyte chemoattractant factor: association of biologic function with CD4 expression. Proc. Natl Acad. Sci. USA 91, 5109–5113.

Cruikshank, W.W., Long, A., Tarpy, R.E., et al. (1995). Early identification of interleukin-16 (lymphocyte chemoattractant factor) and macrophage inflammatory protein 1 alpha (MIP1 alpha) in bronchoalveolar lavage fluid of antigen-challenged asthmatics. Am. J. Respir. Cell Mol. Biol. 13(6), 738–747.

Cruikshank, W., Kornfeld, H., Berman, J., Chupp, G., Keane, J. and Center, D. (1996). Biological activity of interleukin-16. Nature 382(6591), 501–502. [Letter]

Laberge, S., Cruikshank, W.W., Kornfeld, H. and Center, D.M. (1995). Histamine-induced secretion of lymphocyte chemoattractant factor from CD8⁺ T cells is independent of transcription and translation. Evidence for constitutive protein synthesis and storage. J. Immunol. 155, 2902–2910.

Laberge, S., Cruikshank, W.W., Beer, D.J. and Center, D.M. (1996). Secretion of IL-16 (lymphocyte chemoattractant factor) from serotonin-stimulated CD8⁺ T cells in vitro. J. Immunol. 156(1), 310–315.

Laberge, S., Ernst, P., Ghaffar, O., et al. (1997). Increased expression of interleukin-16 in bronchial mucosa of subjects with atopic asthma. Am. J. Respir. Cell Mol. Biol. 17(2), 193–202.

Lim, K.G., Wan, H.C., Bozza, P.T., et al. (1996). Human eosinophils elaborate the lymphocyte chemoattractants IL-16 lymphocyte chemoattractant factor and RANTES. J. Immunol. 156(7), 2566–2570.

Lucey, D.R., Dorsky, D.I., Nicholson-Weller, A. and Weller, P.F. (1989). Human eosinophils express CD4 protein and bind human immunodeficiency virus 1 gp120. J. Exp. Med. 169, 327–331.

Maciaszek, J.W., Parada, N.A., Cruikshank, W.W., et al. (1997). IL-16 represses HIV-1 promoter activity. J. Immunol. 158(1), 5–8.

Maddon, P.J., Littman, D.R., Godfrey, M., Maddon, D.E., Chess, L. and Axel, R. (1985). The isolation and nucleotide sequence of a cDNA encoding the T cell surface protein T4: a new member of the immunoglobulin gene family. Cell 42, 93–104.

Rand, T.H., Cruikshank, W.W., Center, D.M. and Weller, P.F. (1991). CD4-mediated stimulation of human eosinophils: lymphocyte chemoattractant factor and other CD4-binding ligands elicit eosinophil migration. J. Exp. Med. 133, 1521–1528.

Reinherz, E.L., Kung, P.C., Goldstein, G. and Schlossman, S.F. (1979). Separation of functional subsets of human T cells by a monoclonal antibody. Proc. Natl Acad. Sci. USA 76, 4061–4065.

Rumsaeng, V., Cruikshank, W.W., Foster, B., et al. (1997). Human mast cells produce the CD4⁺ T lymphocyte chemo-attractant factor, IL-16. J. Immunol. 159(6), 2904–2910.

Ryan, T.C., Cruikshank, W.W., Kornfeld, H. and Collins, T.L. (1995). The CD4-associated tyrosine kinase p56lck is required for lymphocyte chemoattractant factor-induced T lymphocyte migration. J. Biol. Chem. 270(29), 17081–17086.

Scala, E., D'Offizi, G., Rosso, R., et al. (1997). C-C chemokines, IL-16, and soluble antiviral factor activity are increased in cloned T cells from subjects with long-term nonprogressive HIV infection. J. Immunol. 158(9), 4485–4492.

Steward, S.J., Fujimoto, J. and Levy, R. (1986). Human T lymphocytes and monocytes for the same Leu-3 (T4) antigen. J. Immunol. 136(3), 773–778.

van Epps, D.E., Potter, J.W. and Durant, D.A. (1983). Production of a human T lymphocyte chemotactic factor by T cell subpopulations. J. Immunol. 130(6), 2722–2731.

Wan, H.C., Lazarovits, A.I., Cruikshank, W.W., Kornfeld, H., Center, D.M. and Weller, P.F. (1995). Expression of alpha 4 beta 7 integrin on eosinophils and modulation of alpha 4-integrin-mediated eosinophil adhesion via CD4. Int. Arch. Allergy Immunol. 107(1–3), 343–344.

Zhou, P., Goldstein, S., Devadas, K., Tewari, D. and Notkins, A.L. (1997). Human CD4$^+$ cells transfected with IL-16 cDNA are resistant to HIV-1 infection: inhibition of mRNA expression [see comments]. Nat. Med. 3(6), 659–664.

17. Granulocyte Colony-stimulating Factor

MaryAnn Foote, Helmut Hasibeder, Robin Campbell, T. Michael Dexter *and* George Morstyn

1. Introduction

Hematopoiesis is controlled by specific growth factors, glycoproteins that act on cells at various stages in the hematopoietic cascade to produce mature hematopoietic cells. One factor that has been isolated, purified, cloned, and produced in commercial quantities is granulocyte colony-stimulating factor (G-CSF), a protein that acts on the neutrophil lineage to stimulate the proliferation, differentiation, and activation of committed progenitor cells and functionally active mature neutrophils. A distinctive property of G-CSF that distinguishes it from other colony-stimulating factors and facilitated its purification, molecular cloning, and large-scale production in prokaryotic cells is its ability to induce terminal differentiation of a murine leukemic cell line (WEHI-3B). After observing that serum from endotoxin-treated mice was capable of causing the differentiation of a WEHI-3B myelomonocytic leukemic cell line, Metcalf (1980) named the molecules GM-DF or granulocyte-macrophage differentiating factor. Further analysis

showed that this serum contained G-CSF as well as granulocyte-macrophage colony-stimulating factor (GM-CSF). Nicola and colleagues (1983) were able to further purify G-CSF from medium conditioned by lung tissue of endotoxin-treated mice. This G-CSF was able to stimulate WEHI-3B^{D+} cells as well as normal cells, supporting the formation of numerous small, neutrophil-containing colonies at a concentration similar to that needed for WEHI-3B differentiation (Nicola, 1989). Subsequently, murine G-CSF was partially purified as a protein and was shown to have differentiation-inducing activity for WEHI-3B^{D+} as well as granulocyte colony-stimulating activity in bone-marrow cells (Nicola *et al.*, 1983). Other researchers, notably Asano and colleagues (1980) and Welte and colleagues (1985), found several human carcinoma cells that constitutively produce colony-stimulating factors, and one of these factors was purified to apparent homogeneity from the conditioned medium of bladder carcinoma 5637 cells (Welte *et al.*, 1985) or squamous carcinoma CHU-2 (Nomura *et al.*, 1986). As the purified colony-stimulating factor could selectively stimulate specific neutrophilic granulocyte-colony formation from bone marrow cells, it was concluded that this factor was the human counterpart to mouse G-CSF. The protein initially identified as G-CSF was also called CSF-β and pluripoietin (pCSF).

The study of G-CSF progressed to the purification and molecular cloning of both murine and human forms and then to the first clinical trials of G-CSF in cancer patients (Bronchud *et al.*, 1987; Gabrilove *et al.*, 1988a,b; Morstyn *et al.*, 1988, 1989). Because of its unique biological activities, recombinant human (rh)G-CSF is used for the reversal or amelioration of neutropenias of various causes, for increasing cancer chemotherapy doses, and for hematopoietic stem cell transplantation.

2. *The Cytokine Gene*

The mouse G-CSF gene is on murine chromosome 11, the same chromosome that encodes for GM-CSF and IL-3 (Nagata, 1989). The human gene is on chromosome 17 (Platzer, 1989) which is homologous to mouse chromosome 11. In both human and mouse, one chromosomal gene exists per haploid genome. It is worth noting that in humans the colony-stimulating factors do not represent a gene "family", and while genes for IL-3, GM-CSF, monocyte colony-stimulating factor (M-CSF),

and M-CSF receptor are all located at a rather small region of human chromosome 5q, the gene encoding for G-CSF is on chromosome 17.

2.1 GENE SEQUENCE

The chromosomal genes for G-CSF were isolated from human and mouse gene libraries and characterized by nucleotide sequence analysis. The encoding regions of human and mouse G-CSF genes are approximately 2500 nucleotides (Nagata *et al.*, 1986; Tsuchiya *et al.*, 1987).

The genes encoding the amino acid sequence of G-CSF were isolated independently from different tissue sources. The first, Souza and colleagues (1986), isolated a G-CSF cDNA from a bladder carcinoma cell line that coded for a predicted amino acid sequence of 174 amino acids. Nagata and colleagues (1986) reported a G-CSF cDNA from a squamous cell line that encoded for a polypeptide of 177 amino acids. The nucleotide sequence of G-CSF cDNA is given in Figure 17.1 (Nagata *et al.*, 1986; Souza *et al.*, 1986).

Murine and human G-CSF genes do not show any close homology to any other known gene sequences. There is, however, some rather weak sequence homology to the human IL-6 gene (also known as B cell stimulatory factor-2, hepatocyte-stimulating factor, or IFN-β2) and a rather strong similarity in exon size and distribution (but not in intron size) (Yasukawa *et al.*, 1987).

```
GGGGACAGGCTTGAGAATCCCAAAGGAGAGGGGCAAAGGACACTGCCCCCGCAAGTCTGCCAGAGCAGAGAGGGAGACCCCGACTCAGCTGCCACTTCCC      100

CACAGGCTCGTGCCGCTTCCAGGCGTCTATCAGCGGCTCAGCCTTTGTTCAGCTGTTCTGTTCAAACACTCTGGGGCCATTCAGGCCTGGGTGGGGCAGC      200

GGGAGGAAGGGAGTTTGAGGGGGGCAAGGCGACGTCAAAGGAGGATCAGAGATTCCACAATTTCACAAAACTTTCGCAAACAGCTTTTTGTTCCAACCCC      300

CCTGCATTGTCTTGGACACCAAATTTGCATAAATCCTGGGAAGTTATTACTAAGCCTTAGTCGTGGCCCCAGGTAATTTCCTCCCAGGCCTCCATGGGGT      400

                                                          -30
                                   *                       Met Ala Gly Pro Ala Thr Gln
TATGTATAAAGGGCCCCCTAGAGCTGGGCCCCAAAAACAGCCCGGAGCCTGCAGCCCAGCCCCACCCAGACCC ATG GCT GGA CCT GCC ACC CAG      493

    -20
Ser Pro Met Lys Leu Met A
AGC CCC ATG AAG CTG ATG G GTGAGTGTCTTGGCCCAGGATGGGAGAGCCGCCTGCCCTGGCATGGGAGGGAGGCTGGTGTGACAGAGGGGCTG      586

GGGATCCCCGTTCTGGGAATGGGGATTAAAGGCACCCAGTGTCCCCGAGAGGGCCTCAGGTGGTAGGGAACAGCATGTCTCCTGAGCCCGCTCTGTCCCC      686

                   -10                                          -1 +1
         la Leu Gln Leu Leu Leu Trp His Ser Ala Leu Trp Thr Val Gln Glu Ala Thr Pro Leu Gly Pro Ala Ser Ser
AG CC CTG CAG CTG CTG CTG TGG CAC AGT GCA CTC TGG ACA GTG CAG GAA GCC ACC CCC CTG GGC CCT GCC AGC TCC      762

    10                            20                            30
Leu Pro Gln Ser Phe Leu Leu Lys Cys Leu Glu Gln Val Arg Lys Ile Gln Gly Asp Gly Ala Ala Leu Gln Glu
CTG CCC CAG AGC TTC CTG CTC AAG TGC TTA GAG CAA GTG AGG AAG ATC CAG GGC GAT GGC GCA GCG CTC CAG GAG      837

Lys Leu Val Ser Glu
AAG CTG GTG AGT GAG GTGGGTGAGAGGGCTGTGGAGGGAAGCCCGGTGGGGAGAGCTAAGGGGGATGGAACTGCAGGGCCAACATCCTCTGGAAG      932

GGACATGGGAGAATATTAGGAGCAGTGGAGCTGGGGAAGGCTGGGAAGGGACTTGGGGAGGAGGACCTTGGTGGGGACAGTGCTCGGGAGGGCTGGCTGG      1032

GATGGGAGTGGAGGCATCACATTCAGGAGAAAGGGCAAGGGCCCCTGTGAGATCAGAGAGTGGGGGTGCAGGGCAGAGAGGAACTGAACAGCCTGGCAGG      1132

ACATGGAGGGAGGGGAAAGACCAGAGAGTCGGGGAGGGACCCCGGGAAGGAGCGGCGACCCGGCCACGGCGAGTCTCACTCAGCATCCTTCCATCCCCAG      1230
```

```
                40                              50                              60
Cys Ala Thr Tyr Lys Leu Cys His Pro Glu Glu Leu Val Leu Leu Gly His Ser Leu Gly Ile Pro Trp Ala Pro
TGT GCC ACC TAC AAG CTG TGC CAC CCC GAG GAG CTG GTG CTG CTC GGA CAC TCT CTG GGC ATC CCC TGG GCT CCC    1305

                70
Leu Ser Ser Cys Pro Ser Gln Ala Leu Gln Leu
CTG AGC AGC TGC CCC AGC CAG GCC CTG CAG CTG GTGAGTGTCAGGAAAGGATAAGGCTAATGAGGAGGGGGAAGGAGAGGAGGAACACC    1394

                                                                                          Ala Gly Cys
CATGGGCTCCCCCATGTCTCCAGGTTCCAAGCTGGGGGCCTGACGTATCTCAGGCAGCACCCCCTAACTCTTCCGCTCTGTCTCACAG GCA GGC TGC    1491

                80                              90
Leu Ser Gln Leu His Ser Gly Leu Phe Leu Tyr Gln Gly Leu Leu Gln Ala Leu Glu Gly Ile Ser Pro Glu Leu
TTG AGC CAA CTC CAT AGC GGC CTT TTC CTC TAC CAG GGG CTC CTG CAG GCC CTG GAA GGG ATC TCC CCC GAG TTG    1566

100                             110                             120
Gly Pro Thr Leu Asp Thr Leu Gln Leu Asp Val Ala Asp Phe Ala Thr Thr Ile Trp Gln Gln
GGT CCC ACC TTG GAC ACA CTG CAG CTG GAC GTC GCC GAC TTT GCC ACC ACC ATC TGG CAG CAG GTGAGCCTTGTTGGGC    1645

AGGGTGGCCAAGGTCGTGCTGGCATTCTGGGCACCACAGCCGGGCCTGTGTATGGGCCCTGTCCATGCTGTCAGCCCCCAGCATTTCCTCATTTGTAATA    1745

                                                                  130
                                Met Glu Glu Leu Gly Met Ala Pro Ala Leu Gln Pro Thr
ACGCCCACTCAGAAGGGCCCAACCACTGATCACAGCTTTCCCCCACAG ATG GAA GAA CTG GGA ATG GCC CCT GCC CTG CAG CCC ACC    1832

                140                             150
Gln Gly Ala Met Pro Ala Phe Ala Ser Ala Phe Gln Arg Arg Ala Gly Gly Val Leu Val Ala Ser His Leu Gln
CAG GGT GCC ATG CCG GCC TTC GCC TCT GCT TTC CAG CGC CGG GCA GGA GGG GTC CTG GTT GCC TCC CAT CTG CAG    1907

160                             170
Ser Phe Leu Glu Val Ser Tyr Arg Val Leu Arg His Leu Ala Gln Pro
AGC TTC CTG GAG GTG TCG TAC CGC GTT CTA CGC CAC CTT GCC CAG CCC TGA GCCAAGCCCTCCCCATCCCATGTATTTATCTC    1990

TATTTAATATTTATGTCTATTTAAGCCTCATATTTAAAGACAGGGAAGAGCAGAACGGAGCCCCAGGCCTCTGTGTCCTTCCCTGCATTTCTGAGTTTCA    2090

TTCTCCTGCCTGTAGCAGTGAGAAAAAGCTCCTGTCCTCCCATCCCCTGGACTGGGAGGTAGATAGGTAAATACCAAGTATTTATTACTATGACTGCTCC    2190

CCAGCCCTGGCTCTGCAATGGGCACTGGGATGAGCCGCTGTGAGCCCCTGGTCCTGAGGGTCCCCACCTGGGACCCTTGGAGAGTATCAGGTCTCCCACGT    2290

GGGAGACAAGAAATCCCTGTTTAATATTTAAACAGCAGTGTTCCCCATCTGGGTCCTTGCACCCCTCACTCTGGCCTCAGCCGACTGCACAGCGGCCCCT    2390

GCATCCCCTTGGCTGTGAGGCCCCTGGACAAGCAGAGGTGGCCAGAGCTGGGAGGCATGGCCCTGGGGTCCCACGAATTTGCTGGGGAATCTCGTTTTTC    2490


TTCTTAAGACTTTTGGGACATGGTTTGACTCCCGAACATCACCGACGTGTCTCCTGTTTTTCTGGGTGGCCTCGGGACACCTGCCCTGCCCCCACGAGGG    2590

TCAGGACTGTGACTCTTTTTAGGGCCAGGCAGGTGCCTGGACATTTGCCTTGCTGGATGGGGACTGGGGATGTGGGAGGGAGCAGACAGGAGGAATCATG    2690

TCAGGCCTGTGTGTGAAAGGAAGCTCCACTGTCACCCTCCACCTCTTCACCCCCCACTCACCAGTGTCCCCTCCACTGTCACATTGTAACTGAACTTCAG    2790

TATAATAAAGTGTTTGCCTCCAGTCACGTCCTTCCTCCTTCTTGAGTCCAGCTGGTGCCTGGCCAGGGGCTGGGGAGGTGGCTGAAGGGTGGGAGAGGCC    2890

AGAGGGAGGTCGGGGAGGAGGTCTGGGGAGGAGGTCCAGGGAGGAGGAGGAAAGTTCTCAAGTTCGTCTGACATTCATTCCGTTAGCACATATTTATCTG    2990

AGCACCTACTCTGTGCAGACGCTGGGCTAAGTGCTGGGGACACAGCAGGGAACAAGGCAGACATGGAATCTGCACTCGAG    3070
```

Figure 17.1   The nucleotide sequence of the G-CSF gene. The dashed box represents the consensus decanucleotide also found in other CSF promoters. The TATA sequence is overlined; the polyadenylation signal is underlined. The putative site for transcription initiation is indicated by an asterisk. The boxed nucleotides depict the start of the mature portion of the protein (Nagata *et al*., 1986b).

2.2 INTRONS AND EXONS

The chromosomal genes for murine and human G-CSF consist of five mRNA-encoding exons separated by four introns (Nagata, 1989). At the end of exon 2 of the human G-CSF gene, two 5′ splice-donor sequences are arranged in tandem, and two different human G-CSF mRNAs can be produced using alternative sequences. The five exons are illustrated in Figure 17.2a. Although coding regions of the exons are highly conserved between murine and human, as is the 5′ flanking region, there is little sequence homology in the intron. Unlike the human gene, the murine gene does not contain a second splice donor site in exon 2, so that only one spliced mRNA is produced. The single-copy murine gene for G-CSF has been mapped to the distal half of chromosome 11 (homologous to human chromosome 17) close to the *erb*B and *erb*B-2 loci (Buckberg *et al*., 1988).

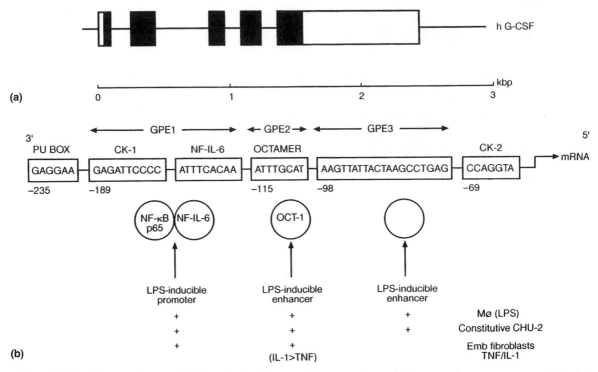

Figure 17.2 **(a) Genomic organization of G-CSF genes. Exons are shown in boxes with coding segments in black and noncoding segments in white. (Adapted from Metcalf and Nicola, 1995.) (b) Promoter and enhancer elements in the human G-CSF gene 5′ of exon 1. Nucleotides are labeled relative to the transcriptional initiation site. Recognized sequencer elements that correspond to nuclear factor binding sites are boxed and labeled. Plus signs indicate the elements that are active in LPS-activated macrophages and embryonic fibroblasts stimulated with TNF or IL-1, and that are constitutively active in CHU-2 cells. (Adapted from Metcalf and Nicola, 1995.)**

2.3 CHROMOSOME LOCATION

The gene for human G-CSF is located at 17q21-22 (Simmers *et al.*, 1987) and is approximately 2.5 kbp (Nagata, 1989). This region corresponds to a translocation between chromosomes 15 and 17 (t915;17) (q23 q21)) that is commonly associated with acute promyelocytic leukemia (Larson *et al.*, 1984). It remains to be determined how closely the break point is associated with the central region of the G-CSF gene. The protooncogenes for c-*erb*B-2 and c-*erb*A, and the nerve growth factor receptor gene are located at the same segment of chromosome 17 (Nicola, 1989).

2.4 TRANSCRIPTIONAL CONTROL OF THE G-CSF GENE

The human and mouse G-CSF genes are highly homologous for approximately 300 bp upstream of the transcription initiation site. This region is sufficient for essential and inducible (by lipopolysaccharide (LPS), tumor necrosis factor (TNF), and IL-1) expression of G-CSF by macrophages (Figure 17.2b). Within this region are several potential regulatory elements including the PU

BO, CK-1, NF-IL-6, OCTAMER (thought to be involved in the transcriptional control of immunoglobulin genes), and the CK-2. Using reporter constructs in embryonic fibroblasts, Shannon and colleagues (1992) found that the CK-1, NF-IL-6, and OCTAMER sequences only were required for induction of G-CSF by TNF and IL-1, with the OCTAMER sequence contributing to the differential response to IL-1 compared with TNF. Nishizawa and Nagata (1990) found that in macrophages, the CK-1 element was necessary for LPS inducibility. In addition, the OCTAMER sequence did not confer inducibility on G-CSF genes in macrophages (Nishizawa and Nagata, 1990), or on any other cell types other than the embryonic fibroblasts, which suggests that its action is cell specific (Shannon *et al.*, 1992).

3. The Protein

3.1 AMINO ACID SEQUENCE

The amino acid identities between pairs of aligned sequences of human, murine, bovine, and canine G-CSFs are between 70% and 80% (Kuga *et al.*, 1989). Murine recombinant G-CSF and nonglycosylated G-CSF differ in that there is a fifth cysteine residue found in the human but not the murine molecule (Kuga *et al.*, 1989). The murine G-CSF has a serine at this position, and when substituted in the human G-CSF causes no decrease in biological activity. The cysteine at position 17 in the murine G-CSF does not form a disulfide linkage and is not accessible to modification by iodoacetic acid (Lu *et al.*, 1989). Biological activity with a substitution at position 17 with alanine is at least as good as that of the unaltered recombinant protein. The amino acid sequence for human G-CSF is given in Figure 17.3. As mentioned earlier, the differences between the squamous cell and bladder carcinoma cell human G-CSF is due to an alternative mRNA splicing site with three additional amino acids between residues 32 and 33 of the 174-amino-acid polypeptide; these amino acids diminish the biological activity of the longer squamous cell line-derived G-CSF (Osslund and Boone, 1994). Both murine and human G-CSF are synthesized with a 30-amino-acid leader sequence typical of secreted proteins (Nagata *et al.*, 1986; Souza *et al.*, 1986). The fact that nonglycosylated G-CSF has an extra methionine at the N-terminus compared with native G-CSF does not seem to have any effect on the biological or physical properties of the molecule (Souza *et al.*, 1986).

The native protein has been calculated to have an apparent molecular mass of 18.6 kDa (Platzer, 1989), 18.8 kDa (Souza *et al.*, 1986), or 19.6 kDa (Welte *et al.*, 1985; Nomura *et al.*, 1986; Oheda *et al.*, 1988; Lu *et al.*, 1989) as measured by sodium dodecyl polyacrylamide gel electrophoresis (SDS-PAGE).

Under appropriate conditions, the nonglycosylated form of G-CSF is fully biologically active and is stable for >1 year at 4°C (Welte *et al.*, 1985; Nomura *et al.*, 1986; Oheda *et al.*, 1988). The protein molecule, however, is highly hydrophobic, and the hydrophilic *O*-linked sugar chain may increase the solubility and *in vitro* stability of the glycosylated form at neutral pH (Oheda *et al.*, 1990). See Section 3.5 for more details on the stability of the commercial forms.

3.2 CARBOHYDRATES

CHO cell-derived rhG-CSF (glycosylated G-CSF) and native G-CSF are *O*-glycosylated (Nagata *et al.*, 1986; Souza *et al.*, 1986), whereas *E. coli*-derived rhmet-G-CSF is not glycosylated. The sugar chains are the reason for the differences in molecular mass. It is known that the presence and type of glycosylation of the recombinant protein depends on the cellular source (e.g., yeast, mammalian, or bacterial) (Morstyn and Burgess, 1988). Glycosylation by CHO is heterogeneous; that is, not all the sugar chains are identical. Not all structural components of native human G-CSF have been identified, but it is known that its sugar composition is not identical to that of glycosylated CHO-derived G-CSF (Oheda *et al.*, 1988, 1990).

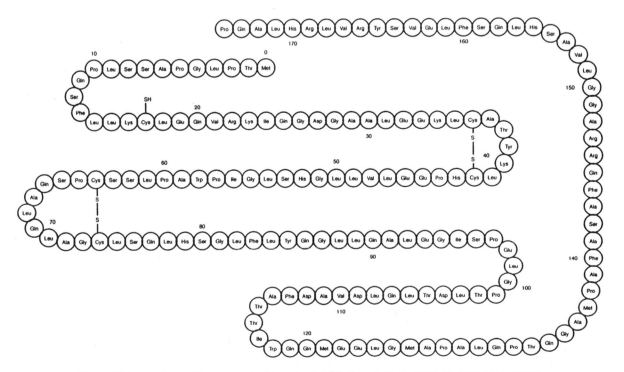

Figure 17.3   Amino acid sequence of human G-CSF (Souza *et al.*, 1986; Nagata *et al.*, 1986).

The lack of binding of G-CSF to concanavalin A (Nicola and Metcalf, 1981) suggests that it does not have mannose-containing carbohydrates. Further work showed a reduction in the charge heterogeneity in isoelectric focusing after treatment with neuraminidase, suggesting sialic acid-containing *O*-glycosylation (Nicola and Metcalf, 1984).

Crystallographic studies have shown that the sugar chain is attached to the C-D loop of G-CSF, at a distance from the active biological sites (Kuga *et al.*, 1989; Osslund and Boone, 1994). The actual glycosylation site is the threonine 134 position. Although the glycosylation does not seem to have a role in the biological function of the molecule, it may partially protect the molecule from proteolytic degradation. The observation that a limited proteolytic degradation of nonglycosylated G-CSF results in the cleavage of the molecule near threonine 134 points to a role of glycosylation for proteolytic protection, and indicates that the residues along this portion of the protein structure may serve as handles for proteolytic degradation.

3.3 SECONDARY STRUCTURE

Secondary structure prediction analyses show that the cytokine family members have substantial helical infrastructures. The protein's secondary structure has abundant α-helices with a small amount of β-sheet pleating. Native G-CSF appears to exist in two forms: type a with 177 amino acids and type b with 174 amino acids. Type b appears to be more active than type a (Asano, 1991).

3.4 TERTIARY AND QUATERNARY STRUCTURES

The atomic structure of nonglycosylated G-CSF has been determined by x-ray crystallography at high resolution (Hill *et al.*, 1993). The crystal structure confirmed the predicted topology and showed that nonglycosylated G-CSF has four antiparallel bundles with crossing angles of 18°. Besides the core bundle, there is a short 3-to-10 helix immediately after the first disulfide bond, and two long loops and one short loop connecting the helical core bundle. The first helix of nonglycosylated G-CSF begins at residue 10, implying that the first 10 residues, including the initiating methionine, are not part of the core structure.

The quaternary structure is given in Figure 17.4. The protein has two disulfide bridges (36, 42; and 64, 74). The kinetics of the formation of the two disulfide bonds of nonglycosylated G-CSF differ dramatically. The first disulfide bond of nonglycosylated G-CSF (between 36 and 42) is between the first helix of the helical bundle and the short fifth helix, E. In contrast, the second disulfide bond is positioned at the carboxy-terminal end

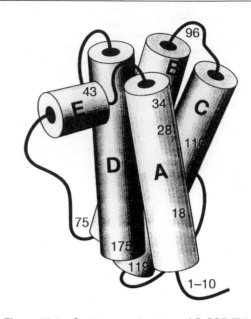

Figure 17.4   Quaternary structure of G-CSF. This is a schematic representation of non-glycosylated G-CSF with its four antiparallel α-helices arranged to form a helical bundle. The bundle has constituents labeled A, B, C, and D. There is a fifth helix, E, which is a short 3-to-10 type helix. The helices are connected with two long loops and one short loop. The relative positions of the amino acids are numbered and indicated. (Adapted from Osslund and Boone, 1994.)

of the flexible E to B loop (between 64 and 74). In the process of folding and oxidation of nonglycosylated G-CSF, the protein rapidly forms an intermediate that is a partially oxidized form of nonglycosylated G-CSF containing a single disulfide bond. The rate-limiting step in the formation of the active form of nonglycosylated G-CSF is the formation of the second disulfide bond (Lu *et al.*, 1992); and the difference in the rates may be due to the greater mobility of the second disulfide loop.

There is a structural relationship between G-CSF and related cytokines (IL-2, IL-3, IL-6, erythropoietin (EPO), and GM-CSF). Using x-ray crystallographic techniques, it has been found, however, that nonglycosylated G-CSF is most closely related to human growth hormone (HGH) (Osslund and Boone, 1994). Nonglycosylated G-CSF, HGH, and GM-CSF each have four helices, and, while the structure of GM-CSF has a similar topology, it is different in the folding and positioning of the helices within its core structure. In G-CSF, the amino-terminus and two extended loops are not part of the core structure.

3.5 STABILITY AND STORAGE

Under appropriate conditions, r-metHuG-CSF is fully biologically active and is stable for >1 year at 4°C (Welte *et al.*, 1985; Nomura *et al.*, 1986; Oheda *et al.*, 1988). The protein molecule, however, is highly hydrophobic, and the hydrophilic *O*-linked sugar chain may increase the solubility and *in vitro* stability of the glycosylated form at neutral pH (Oheda *et al.*, 1990).

The nonglycosylated G-CSF package insert (marketed under the name Neupogen® in the United States) gives the storage temperature as +2 to +8°C; in the Japanese nonglycosylated G-CSF package insert (marketed under the name Gran®), the package insert gives the storage temperature as <10°C. It is thought that the commercial product is stable at room temperature for 1 week (as given in the European package insert of Neupogen®), although the package insert for the United States gives a stability at room temperature of 24 h.

The glycosylated G-CSF package insert (marketed as Neutrogin® in Japan and Granocyte® in Europe) also places the storage temperature at +2 to +8°C for Europe and <10°C for Japan. After reconstitution of the lyophilized powder, the commercial product is stable for 24 h at room temperature.

4. *Cellular Sources and Production*

Various cell types in the human body are known to produce G-CSF, including stromal cells, macrophages, endothelial cells, fibroblasts, and monocytes (Zsebo *et al.*, 1986; Devlin *et al.*, 1987; Rennick *et al.*, 1987; Morstyn and Burgess, 1988; Vellenga *et al.*, 1988; Ernst *et al.*, 1989; Platzer, 1989). Since these cell types are widely distributed in the body, it is possible that G-CSF participates in the production and functional enhancement of neutrophils that occur in response to local infection or other causes. It is thought that there are two types of G-CSF response to infection, either direct activation of G-CSF production or induction of monocytes and macrophages to release G-CSF (Wong *et al.*, 1985). In response to minor infection, microorganisms release lipopolysaccharides (LPS, endotoxin), which results in the direct production of G-CSF, or indirectly by first inducing the production of IL-1 which in turn initiates G-CSF production. This G-CSF results in the localization and activation of circulating neutrophils and macrophages responsible for eliminating the microorganism. However, if the initial infection becomes established, because of the failure of the local response to suppress the infection, the infection provokes a massive local production of G-CSF that enters the circulation. In addition, there is also a gradual entry of the microbial products into the circulation, which results in the amplification of the host response by indirectly stimulating the production of G-CSF via the production of IL-1 and by directly stimulating the production of G-CSF. This general increase in systemic G-CSF stimulates the production of mature cells in the marrow and spleen, and ultimately results in the local accumulation of huge numbers of neutrophils and macrophages that eliminate

the infectious microorganisms (Metcalf and Nicola, 1995).

Macrophages recovered from the lungs of patients with pneumonia have demonstrated the ability of these cells to produce G-CSF spontaneously and in higher concentrations than those produced by normal alveolar macrophages. (Priming with endotoxin was necessary to obtain detectable G-CSF levels in normal alveolar macrophages.) Similar findings have been observed in animal models of pneumonia in rats. G-CSF levels were detectable in the normal lung but not in the serum. After infection, lung G-CSF levels increased rapidly, preceding an increase in serum levels of G-CSF (Metcalf and Nicola, 1995).

The blood levels of endogenous G-CSF increase in a variety of pathological conditions, such as infection and exposure to endotoxins (Metcalf, 1988; Watari *et al.*, 1989; Kawakami *et al.*, 1990; Cebon *et al.*, 1991, 1994; Morstyn *et al.*, 1991). In patients with cyclic neutropenia there are marked fluctuations (Yujiri *et al.*, 1992), and in patients with autoimmune neutropenia peripheral absolute neutrophil count (ANC) changes in parallel with serum level of endogenous G-CSF, with a delay of 4–5 h (Omori *et al.*, 1992).

5. *Biological Activities*

Granulocyte colony-stimulating factor reduces neutrophil maturation time from 5 days to 1 day, leading to a rapid release of mature neutrophils from the bone marrow into the peripheral circulation (Lord *et al.*, 1989). Neutrophils treated with G-CSF have normal survival (Bronchud *et al.*, 1988; Lord *et al.*, 1989) and enhanced chemotaxis through increased binding of f-Met-Leu-Phe (fMLP) (Colgan *et al.*, 1992).

Granulocyte colony-stimulating factor works on progenitor cells as well as on mature cells. Its primary action is on granulocyte progenitors (CFU-G), but at high concentrations G-CSF acts also on progenitors for granulocyte-macrophages (CFU-GM) and macrophages (CFU-M) (Platzer, 1989; Ogawa, 1993). Granulocyte colony-stimulating factor stimulates the differentiation and activation of mature neutrophils. Its action on the neutrophil lineage causes rapid, dose-dependent increases in cell numbers, while small or no effects are seen on monocytes or eosinophils (Lindemann *et al.*, 1989; Morstyn *et al.*, 1988; Lieschke *et al.*, 1989).

The existing data suggest that, within the human body, G-CSF is the primary factor mediating neutrophilic response to infection and neutropenia (Cebon *et al.*, 1991, 1994). In cases of gram-negative and fungal infections, G-CSF levels are elevated in the blood (Kawakami *et al.*, 1990). The highest G-CSF levels are found in neutropenic patients and are correlated with fever (Cebon *et al.*, 1991, 1994).

Mice lacking endogenous G-CSF have chronic neutropenia and impaired neutrophil function (Lieschke *et al.*, 1994). This suggests that G-CSF is indispensable for maintaining the normal quantitative balance of neutrophil production during "steady state" granulopoiesis *in vivo*, and indicates that G-CSF has a role in "emergency" granulopoiesis during infection.

In addition, G-CSF can act synergistically with other CSFs and growth factors, to promote an enhanced proliferative response. However, these synergistic responses vary in magnitude depending on the concentrations of CSFs used (Metcalf and Nicola, 1995).

6.   Receptors

Receptors have been identified for G-CSF (Nicola, 1989; Demetri and Griffin, 1991). Figure 17.5 is an illustration of the receptor.

The G-CSF receptor is expressed on cells of the neutrophil lineage from myeloblasts to the mature neutrophil as well as on subsets of cells of the monocyte lineage. Receptors for G-CSF also have been found on nonhematopoietic cell lines (Mazanet and Griffin, 1992).

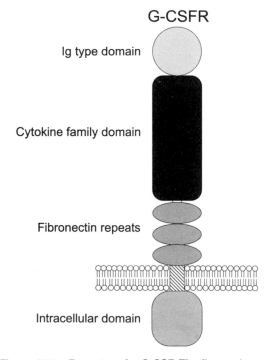

G-CSFR

Ig type domain

Cytokine family domain

Fibronectin repeats

Intracellular domain

Figure 17.5   Receptors for G-CSF. The figure shows a schematic representation of the structure of a nonglycosylated G-CSF receptor. A ribbon diagram of nonglycosylated G-CSF has been inserted between two molecules of nonglycosylated G-CSF receptors, suggesting that signal transduction is related to the binding of the ligand to two molecules of the receptor. (Adapted from Osslund and Boone, 1994.)

The receptors are single, high-affinity receptors, specific to G-CSF (Nicola, 1989). Receptor density varies from 50 to 500 per cell, and the receptor affinity (K_d) equals approximately 100 pM. Studies in mice have shown that the number of G-CSF receptors increases as cells mature (Nicola, 1989; Demetri and Griffin, 1991).

With regard to the biochemical structure, the receptor is a 130–150 kDa glycoprotein that can be downregulated by LPS, f-MLP, or GM-CSF (Nicola, 1989).

The human G-CSF receptor is encoded by a single gene on chromosome 1p35-p34.3 (Inazawa *et al.*, 1991) and contains 17 exons that span approximately 16.5 kbp of DNA (Seto *et al.*, 1992). The G-CSF receptor gene does not contain a classical TATA box, but the major transcription initiation site is 20 bp upstream of the cap site with a minor initiation site at 10 bp upstream of the cap site (Seto *et al.*, 1992).

The murine and human G-CSF receptors have a single transmembrane domain of 813 and 812 amino acids, respectively. Within the extracellular domain of the G-CSF receptors, there are at least three regions that have statistically significant identity to other proteins (Osslund and Boone, 1994). These are the immunoglobulin-like domain, three fibronectin type-3 repeat units, and the WSXWS motif common to the cytokine receptor family.

7.   Signal Transduction

The signal transduction activities of G-CSF are not well known. It appears that G-CSF stimulation of cells does not change the resting transmembrane electric potential, intracellular concentration of unbound calcium ions, or cytosolic pH (Tkatch and Tweardy, 1993). The primary G-CSF interaction with its receptor results in the rapid internalization of receptor and ligand followed by the slow degradation of internalized G-CSF (Nicola *et al.*, 1986; Walker *et al.*, 1985).

When quiescent IL-3-deprived NFS-60 cells were exposed to 500 IU/ml nonglycosylated G-CSF, they showed a rapid increase in p42 mitogen-activated protein (MAP) kinase activity (Bashey *et al.*, 1994). The activity was concentration dependent and appeared within 2–5 min after nonglycosylated G-CSF was added; the effect remained elevated at 60 min, but diminished significantly at 150 min. The increase in enzymatic activity paralleled the appearance of the slower-migrating forms of both p42 and p44 MAP kinase on immunoblots of cell lysates. The rapid stimulation of MAP kinases suggests that the activation is a result of nonglycosylated G-CSF binding to these cells, rather than a result of autocrine growth factor secretion in response to the presence of nonglycosylated G-CSF. Exposure of growth factor-deprived NFS-60 cells to nonglycosylated G-CSF led to a rapid increase of p21^{ras} in GTP-bound form with maximum activation within 5 min of exposure. At 60 min, approximately 28% of the $p21^{ras}$ was GTP bound. In a further experiment, Bashey and colleagues assessed whether MAP kinase activation is a universal component of G-CSF-receptor (G-CSF-R) signal transduction. Activation of these enzymes was assessed in cell types that showed nonproliferative responses to G-CSF. Non-glycosylated G-CSF-induced granulocytic differentiation in the 32DC13(G) and HL-60 cell lines occurred in the absence of detectable activation of p42 MAP kinase. Much work remains to be done in the area of signal transduction and, as understanding of G-CSF signaling increases, therapeutic opportunities may evolve.

8. Clinical Applications

Reviews of clinical experience with rhG-CSF can be found in several papers, notably by Hollingshead and Goa (1991), Lieschke and Burgess (1992), Steward (1993), Frampton et al. (1994), and Welte et al. (1996).

Phase III trials have demonstrated the beneficial effect of rhG-CSF on neutropenia following standard-dose chemotherapy. Two randomized, placebo-controlled, double-blind studies involving more than 300 patients with small-cell lung cancer receiving cyclophos-phamide–Adriamycin®–etoposide (CAE) chemotherapy showed that nonglycosylated rhG-CSF significantly decreased the incidence, severity, and duration of severe neutropenia (Crawford et al., 1991; Trillet-Lenoir et al., 1993). In other randomized, placebo-controlled, double-blind trials, nonglycosylated rhG-CSF allowed increases in dose intensity of doxorubicin (Bronchud et al., 1989) and cyclophosphamide–Adriamycin®–5-fluorouracil (CAF) chemotherapy (Demetri et al., 1991).

In more recent studies, new chemotherapeutic concepts or new agents such as paclitaxel (Taxol®) have been tested with rhG-CSF as a standard adjunct. Compared with prior experience without rhG-CSF, in a study of paclitaxel with nonglycosylated rhG-CSF the incidence, depth, and duration of neutropenia was reduced (Reichmann et al., 1993). In a study to determine the maximum tolerated doses and principal toxicities of a combination of paclitaxel and cisplatin, doses of these drugs could be increased with the support of nonglycosylated rhG-CSF, but severe peripheral neuropathy and/or severe myalgias became the predominant toxicities (Rowinsky et al., 1993). In a study undertaken to define an escalated dose schedule of MVAC (methotrexate, vinblastine, doxorubicin, and cisplatin) with the support of nonglycosylated rhG-CSF, the delivered relative dose intensity was 33% higher than the previously reported one without hematopoietic support; leukopenia and thrombocytopenia became dose limiting (Seidman et al., 1993).

A randomized phase III trial in patients with non-Hodgkin lymphoma (NHL) showed that nonglyco-sylated rhG-CSF significantly improved delivery of full-dose chemotherapy compared with control patients (Pettengell et al., 1992).

rhG-CSF alone or in combination with chemotherapy is an effective agent for recruiting peripheral blood progenitor cells (PBPCs) with long-term reconstituting ability (Dürhsen et al., 1988; Sheridan et al., 1990, 1992; Chao et al., 1993; Hohaus et al., 1993; Kawano et al., 1993; Schwinger et al., 1993; Sato et al., 1994). In a historically controlled study, nonglycosylated rhG-CSF-generated PBPC when used in conjunction with autologous bone marrow transplantation and nonglycosylated rhG-CSF accelerated recovery of neutrophil and platelet count (Sheridan et al., 1992). Use of nonglycosylated rhG-CSF for mobilization of PBPCs resulted in a significantly accelerated time to recovery of granulocytes when compared with nonmobilized PBPC recipients who did not receive mobilized PBPCs in a study of 85 patients with relapsed Hodgkin disease (Chao et al., 1993). The use of mobilized PBPC resulted in a significantly accelerated time to platelet engraftment when compared with nonmobilized PBPC recipients. There was a statistically significant reduction of costs in patients who received nonglycosylated rhG-CSF-mobilized PBPC.

Results from a recent phase III study reported by Schmitz and colleagues demonstrated that nonglycosylated rhG-CSF-mobilized PBPCs were significantly more effective than bone marrow in accelerating platelet recovery and neutrophil recovery in lymphoma patients receiving high-dose chemotherapy (Schmitz et al., 1996).

Placebo-controlled studies have shown that rhG-CSF accelerates neutrophil recovery following autologous bone marrow transplantation (BMT). An ANC recovery to $>0.5 \times 10^9/l$ was achieved in 14 days or less in studies with rhG-CSF (Peters et al., 1989; Sheridan et al., 1989; Taylor et al., 1989). In a 2-year, multicenter, randomized, vehicle-controlled, single-blind, dose-ranging trial, glycosylated rhG-CSF was administered subcutaneously in 121 patients with nonmyeloid malignancies. A dose–response effect was apparent, and neutrophil recovery was significantly accelerated while infectious complications were reduced (Harousseau, 1992; Linch, 1992; Linch et al., 1993). In a prospective, randomized trial of high-dose chemotherapy and autologous transplantation (Stahel et al., 1994), nonglycosylated rhG-CSF at 10 or 20 µg/kg per day reduced the median time to neutrophil recovery (10 days versus 18 days) compared with a control group not receiving the cytokine. The median number of days with fever (1 day versus 4 days) and neutropenic fever also were significantly reduced. Days on intravenous antibiotics and the duration of hospitalization were also shorter in the cytokine-treated group, although the two end points did not reach statistical significance.

rhG-CSF has been shown to increase neutrophil counts in some patients with moderate aplastic anemia but, in general, patients with very severe hypoplasia do not respond to growth factors, and their use is still quite experimental (Kojima *et al.*, 1991; Bessho *et al.*, 1992). In a Japanese study of 20 children with severe or intermediate aplastic anemia, a dose of $400 \mu g/m^2$ per day increased the neutrophil count in 12 patients. Increasing doses to as much as $1200 \mu g/m^2$ were administered to 5 patients who did not respond to the initial dose, and 3 of the 5 showed an increase in ANC (Kojima *et al.*, 1991). In another small Japanese study, 5 patients were given nonglycosylated rhG-CSF 75–500 µg/day by subcutaneous injection or by intravenous infusion (Sonoda *et al.*, 1992). All five responded with increased neutrophil counts and two of the patients also showed improvement in their anemia.

Studies have been done in patients with myelodysplastic syndromes (MDS), where treatment with nonglycosylated rhG-CSF was associated with a sustained improvement in neutrophil function, in the absence of increased adherence or impaired chemotaxis (Negrin *et al.*, 1990). A phase II study showed the efficacy of glycosylated rhG-CSF, with most patients responding to intravenous doses of 2 or 5 µg/kg per day (Yoshida *et al.*, 1991). In a phase III randomized study involving 102 patients with RAEB or RAEB-t subtypes of MDS, nonglycosylated rhG-CSF was shown to be efficacious in increasing ANC (Greenberg *et al.*, 1993), although imbalances in patient characteristics made it difficult to define the overall benefit, if any, of the use of G-CSF in this setting.

A phase III trial of nonglycosylated rhG-CSF in severe chronic neutropenia (SCN) patients has shown long-term efficacy (>200 patient-years experience) and tolerance, and the hematological and clinical benefits were sustained during maintenance treatment (Dale *et al.*, 1990, 1993). In one study, when children with SCN did not respond to treatment with rhGM-CSF, they were switched to treatment with nonglycosylated rhG-CSF with a resulting increase in neutrophil counts (Welte *et al.*, 1990). (For a recent review, see Welte and Dale, 1996).

The potential of CSFs to ameliorate the myelo-suppression of antiviral and anti-infective therapies has been investigated in HIV setting. Nonglycosylated rhG-CSF has been shown to allow delivery of full doses to treat cytomegalovirus (CMV) infection in advanced HIV infection (Jacobsen *et al.*, 1992). Nonglycosylated rhG-CSF given for 2 weeks at doses of 0.3–3.6 µg/kg per day increased neutrophil numbers 9-fold and maintained this increase during concomitant therapy with EPO and zidovudine (Miles *et al.*, 1991). Use of nonglycosylated rhG-CSF permitted some patients to receive full doses of antiviral therapy and these patients had preserved or improved neutrophil function. In a phase I/II trial, glycosylated rhG-CSF 0.4–10 µg/kg per day subcuta-

neously at low doses was effective in ameliorating zidovudine-induced neutropenia (Van der Wouw *et al.*, 1991). While most patients required 2.0 µg/kg per day, 2 of 12 patients required 0.4 µg/kg per day. Administration of nonglycosylated rhG-CSF has been shown to permit the greater use of myelosuppressive medications without the potentially life-threatening complications of neutropenia (Hermans *et al.*, 1995.) In a small study, 10 consecutive patients at risk of developing severe neutropenia because of the use of antiviral drugs were given nonglycosylated rhG-CSF at 5 µg/kg per week, in two divided doses, for 6 months (Balbiano *et al.*, 1994). Even at low intermittent doses, nonglycosylated rhG-CSF was able to prevent severe neutropenia and neutropenic fever. A prospective, controlled study of non-glycosylated rhG-CSF has shown that rhG-CSF prevented severe neutropenia and reduced the incidence of severe bacterial infection, and hospitalization for bacterial infections (Kuritzkes *et al.*, 1997).

In a phase I open-label trial of community acquired pneumonia (CAP), 31 patients were treated with non-glycosylated rhG-CSF, at doses as great as 600 µg/day for up to 10 days in conjunction with antibiotics (Kuhlberg and van der Meer, 1996). Doses >150 µg/day increased neutrophil counts up to 3 times the normal value and were not associated with any clinically evident toxicity. Overall, pulmonary symptoms and signs improved or remained normal for the majority of the patients regardless of the dose. The study was too small to determine whether there was any clinical benefit and clinical outcomes (i.e., duration of fever, antibiotic use, hospitalization) had no apparent relation to dose. In a subsequent phase III trial, treatment with rhG-CSF plus antibiotics was compared with antibiotic treatment alone in 756 patients with moderate CAP (Nelson *et al.*, 1996). Preliminary analysis of the study showed that the primary end point, time to resolution of morbidity (i.e., absence of clinical signs and symptoms), was not met. The analysis, however, did demonstrate a benefit in patients receiving nonglycosylated rhG-CSF in resolving radiographic signs of pneumonia, as well as clinical benefit, i.e., reduction of incidence of end-organ failure and incidence of acute respiratory distress syndrome.

Clinical response to rhG-CSF in patients with pneumonia and severe sepsis was reported by Wunderink and colleagues (1995). Mortality at day 15 and day 29 and resolution of end-organ failure on day 29 were assessed in a double-blind, placebo-controlled, random-ized study in which placebo or nonglycosylated rhG-CSF was administered at 300 µg/day intravenously for 5 days. Mortality at day 15 and day 29 was less in the cytokine-treated group compared with the placebo-treated group. These studies suggest there may be clinical benefit for rhG-CSF in severely ill patients with pneumonia.

The effects of nonglycosylated rhG-CSF in neonatal sepsis were studied in a randomized, placebo-controlled, phase I–II trial (Gillan *et al.*, 1994). Forty-two neonates, gestational age 26–42 weeks, with presumed bacterial sepsis in the first 3 days of life were administered nonglycosylated rhG-CSF at 1, 5, or 10 µg/kg per day. There was a significant increase in their peripheral blood neutrophil count 24 h after administration of the 5 and 10 µg/kg per day doses, a dose-dependent increase in neutrophil storage pool, and increase in functional activity of neutrophils. In the pneumonia and neonatal settings, further trials are required to define any benefit.

9. References

Asano, S. (1991). Human granulocyte colony-stimulating factor: its basic aspects and clinical applications. Am. J. Pediatr. Hematol./Oncol. 13, 400–413.

Asano, S., Sato, N., Mori, M., *et al.* (1980). Detection and assessment of human granulocyte-macrophage colony-stimulating factor (GM-CSF) producing tumours by heterotransplantation into nude mice. Br. J. Cancer 41, 689–694.

Balbiano, R., Degioanni, M., Valle, M., *et al.* (1994). Prevention of severe neutropenia in AIDS patients with intermittent, low-dose G-CSF (filgrastim). Int. Conf. AIDS. 10, 223. [abstract No. PB0323]

Bashey, A., Healy, L. and Marshall, C.J. (1994). Proliferative but not nonproliferative responses to granulocyte colony-stimulating factor are associated with rapid activation of the p21ras/MAP kinase signalling pathway. Blood 83, 949–957.

Bessho, M., Toyoda, A., Itoh, Y., *et al.* (1992). Trilineage recovery by combination therapy with recombinant human granulocyte colony-stimulating factor (rhG-CSF) and erythropoietin (rhEPO) in severe aplastic anemia. Br. J. Haematol. 80, 409–411.

Bronchud, M.H., Scarffe, J.H., Thatcher, N., *et al.* (1987). Phase I/II study of recombinant human granulocyte colony-stimulating factor in patients receiving intensive chemotherapy for small cell lung cancer. Br. J. Cancer 56, 809–813.

Bronchud, M.H., Potter, M.R., Morgenstern, G., *et al.* (1988). In vitro and in vivo analysis of the effects of recombinant human granulocyte colony-stimulating factor in patients. Br. J. Cancer 58, 64–69.

Bronchud, M.H., Howell, A., Crowther, D., *et al.* (1989). Phase I/II study of recombinant human granulocyte colony-stimulating factor to increase the intensity of treatment with doxorubicin in patients with advanced breast and ovarian cancer. Br. J. Cancer 60, 121–128.

Buckberg, A.M., Bedigan, H.G., Taylor, B.A., *et al.* (1988). Localization of Evi-2 to chromosome 11: linkage to other photo-oncogene and growth factor locii using interspecific backcross mice. Oncogene Res. 2, 149–165.

Cebon, J.S., Layton, J., Pavlovic, R., *et al.* (1991). Endogenous cytokine production in response to sepsis in neutropenic and non-neutropenic patients. Blood 78, 8a.

Cebon, J., Layton, J.E., Maher, D. and Morstyn, G. (1994). Endogenous haemopoetic growth factors in neutropenia and infection. Br. J. Haematol. 86, 265–274.

Chao, N.J., Schriber, J.R., Grimes, K., *et al.* (1993). Granulocyte colony-stimulating factor "mobilized" peripheral blood progenitor cells accelerate granulocyte and platelet recovery after high-dose chemotherapy. Blood 81, 2031–2035.

Colgan, S.P., Gasper, P.W., Thrall, M.A., Boone, T.C., Blancquaert, A.M. and Bruyninckx, W.J. (1992). Neutrophil function in normal and Chediak–Higashi syndrome cats following administration of recombinant canine granulocyte colony-stimulating factor. Exp. Hematol. 20, 1229–1234.

Crawford, J., Ozer, H., Stoller, R., *et al.* (1991). Reduction by granulocyte colony-stimulating factor of fever and neutropenia induced by chemotherapy in patients with small cell lung cancer. N. Engl. J. Med. 325, 164–170.

Dale, D.C., Hammond, W.P., Gabrilove, J., *et al.* (1990). Long term treatment of severe chronic neutropenia with recombinant human granulocyte colony-stimulating factor (r-metHuG-CSF). Blood 76, 545a.

Dale, D.C., Bonilla, M.A., Davis, M.S., *et al.* (1993). A randomized controlled phase III trial of recombinant human granulocyte colony-stimulating factor (Filgrastim) for treatment of severe chronic neutropenia. Blood 81, 2496–2502.

Demetri, G.D. and Griffin, J.D. (1991). Granulocyte colony-stimulating factor and its receptor. Blood 78, 2791–2802.

Demetri, G.D., Younger, J., McGuire, B.W., *et al.* (1991). Recombinant methionyl granulocyte-CSF (r-metHuG-CSF) allows an increase in the dose intensity of cyclophosphamide/doxorubicin/5-fluorouracil (CAF) in patients with advanced breast cancer. Proc ASCO 10, 70a.

Devlin, J.J., Devlin, P.E., Myamabo, K., *et al.* (1987). Expression of granulocyte colony-stimulating factor by human cell lines. J. Leukocyte Biol. 41, 302–306.

Dürhsen, U., Villeval, J.L., Boyd, J., Kannourakis, G., Morstyn, G. and Metcalf, D. (1988). Effects of recombinant human granulocyte colony-stimulating factor on hematopoietic progenitor cells in cancer patients. Blood 72, 2074–2081.

Ernst, T.J., Ritchie, A.R., Demetri, G.D., *et al.* (1989). Regulation of granulocyte colony-stimulating factor mRNA levels in human blood monocytes is mediated primarily at the posttranscriptional level. J. Biol. Chem. 264, 5700–5703.

Frampton, J.E., Lee, C.R. and Faulds, D. (1994). Filgrastim. A review of its pharmacological properties and therapeutic efficacy in neutropenia. Drug Evaluation 48, 731–760.

Gabrilove, J.L., Jakubowski, A., Sher, H., *et al.* (1988a). Effect of granulocyte colony-stimulating factor on neutropenia and associated morbidity due to chemotherapy for transitional cell carcinoma of the urothelium. N. Engl. J. Med. 111, 887–892.

Gabrilove, J.L., Jakubowski, A., Fain, K., *et al.* (1988b). Phase I study of granulocyte colony-stimulating factor in patients with transitional cell carcinoma of the urothelium. J. Clin. Invest. 82, 1454–1461.

Gillan, E.R., Christensen, R.D., Suen, Y., Ellis, R., van de Ven, C. and Cairo, M.S. (1994). A randomized, placebo-controlled trial of recombinant human granulocyte colony-stimulating factor administration in newborn infants with presumed sepsis: significant induction of peripheral and bone marrow neutrophilia. Blood 84, 1427–1433.

Greenberg, P., Taylor, K., Larson, R., *et al.* (1993). Phase III randomized multicenter trial of G-CSF vs observation for myelodysplastic syndromes (MDS). Blood 82, 196a.

Harousseau, J.L. (1992). Lenograstim after bone marrow transplantation: results of a European multicenter randomised study in 315 patients. Satellite Symposium to the 24th Congress of the ISH.

Hermans, P., Rozenbaum, W., Jou, A., *et al.* (1995). Filgrastim (r-metG-CSF) to treat neutropenia and support myelosuppressive medication dosing in HIV infection. In "5th European Conference on Clinical Aspects and Treatment of HIV Infection", 1995 September 27–29, Copenhagen. [Abstract 540]

Hill, C.P., Osslund, T.D. and Eisenberg, D.S. (1993). The structure of granulocyte-colony stimulating factor (r-hu-G-CSF) and its relationship to other growth factors. Proc. Natl Acad. Sci. USA 90, 5167–5171.

Hohaus, S., Goldschmidt, H., Ehrhardt, R. and Haas, R. (1993). Successful autografting following myeloablative conditioning therapy with blood stem cells mobilized by chemotherapy plus rhG-CSF. Exp. Hematol. 21, 508–514.

Hollingshead, L.M. and Goa, K.L. (1991). Recombinant granulocyte colony-stimulating factor (rG-CSF). A review of its pharmacological properties and prospective role in neutropenic conditions. Drug Evaluation 42, 300–330.

Inazawa, J., Fukunaga, R., Seto, Y., *et al.* (1991). Assignment of the human granulocyte colony-stimulating factor receptor gene (CSF3R) to chromosome 1 at region p35–p34.3. Genomics 10, 1075–1078.

Jacobsen, M.A., Stanley, H.D. and Heard, S.E. (1992). Ganciclovir with recombinant methionyl human granulocyte colony-stimulating factor for treatment of cytomegalovirus disease in AIDS patients. AIDS 6, 515–517.

Kawakami, M., Tsutsumi, H., Kumakawa, T., *et al.* (1990). Levels of serum granulocyte colony-stimulating factor in patients with infections. Blood 76, 1962–1964.

Kawano, Y., Takaue, Y., Watanabe, T., *et al.* (1993). Effects of progenitor cell dose and preleukapheresis use of human recombinant granulocyte colony-stimulating factor on the recovery of hematopoiesis after blood stem cell autografting in children. Exp. Hematol. 21, 103–108.

Kojima, S., Fukuda, M., Miyajima, Y., Matsuyama, T. and Horibe, K. (1991). Treatment of aplastic anemia in children with recombinant human granulocyte colony-stimulating factor. Blood 77, 937–941.

Kuga, T., Komatsu, Y., Yamaski, M., *et al.* (1989). Mutagenesis of human granulocyte colony stimulating factor. Biochem. Biophys. Res. Commun. 159, 103–111.

Kuhlberg, B.J. and van der Meer, J.W.M. (1996). "Infectious Diseases: The Role of Haematopoietic Growth Factors". Gardiner-Caldwell Communications Ltd. Cheshire, UK.

Kuritzkes, D., Parenti, D., Ward, D., Rachilis, A. and Jacobson, M. (1997). Filgrastim (r-metHuG-CSF) for prevention of severe neutropenia and associated clinical sequelae in HIV-infected patients: results of a 24-week, prospective, randomized, controlled trial. 4th Conference on Retroviruses and Opportunistic Infections. Washington, D.C.; abstract 365.

Larson, R.A., Kondo, K., Vardiman, J.W., Butler, A.E., Golombo, H.M. and Rowley, J.D. (1984). Evidence for a 15,17 translocation in every patient with acute promyelocytic leukemia. Am. J. Med. 76, 827–841.

Lieschke, G.J., Burgess, A.W. (1992). Granulocyte colony-stimulating factor and granulocyte-macrophage colony-stimulating factor (2). N. Engl. J. Med. 327, 28–35.

Lieschke, G.J., Cebon, J. and Morstyn, G. (1989). Characterization of the clinical effects after the first dose of bacterially synthesized recombinant human granulocyte-macrophage colony-stimulating factor. Blood 74, 2634–2643.

Lieschke, G.J., Grail, D., Hodgson, G., *et al.* (1994). Mice lacking granulocyte colony-stimulatiang factor have chronic neutropenia, granulocyte and macrophage progenitor cell deficiency, and impaired neutrophil mobilization. Blood 84, 1737–1746.

Linch, D.C. (1992). Lenograstim. A new glycosylated rHuG-CSF: pharmacology and clinical profile in bone marrow transplants. Br. J. Haematol. 82, 274–275.

Linch, D.C., Scarrffe, H., Proctor, S., *et al.* (1993). Randomised vehicle-controlled dose-finding study of glycosylated recombinant human granulocyte colony-stimulating factor after bone marrow transplantation. Bone Marrow Transplant. 11, 307–311.

Lindemann, A., Herrmann, F., Oster, W., *et al.* (1989). Hematologic effects of recombinant human granulocyte colony-stimulating factor in patients with malignancy. Blood 74, 2644–2651.

Lord, B.I., Bronchud, M.H., Owens, S., *et al.* (1989). The kinetics of human granulopoiesis following treatment with granulocyte colony-stimulating factor in vivo. Proc. Natl Acad. Sci. USA 86, 9499–9503.

Lu, H.S., Boone, T.C., Souza, L.M. and Lai, P.-H. (1989). Disulfide and secondary structures of recombinant human granulocyte colony stimulating factor. Arch. Biochem. Biophys. 268, 81–92.

Lu, H.S., Clogston, C.L., Narhi, L.O., Merewether, L.A., Pearl, W.R. and Boone, T.C. (1992). Folding and oxidation of recombinant human granulocyte colony-stimulating factor produced in *Escherischia coli*. Characterization of the disulfide-reduced intermediates and cysteine-serine analogs. J. Biol. Chem. 267, 8770–8777.

Mazanet, R. and Griffin, J.D. (1992). Hematopoietic growth factors. In "High-Dose Cancer Chemotherapy". (eds. J.O. Armitage and K.H. Antman), pp. 289–313. Williams & Wilkins, Baltimore, MD.

Metcalf, D. (1980). Clonal extinction of myelomonocytic leukaemia cells by serum from mice injected with endotoxin. Int. J. Cancer 25, 225–233.

Metcalf, D. (1988). "The Molecular Control of Blood Cells". Harvard University Press. Cambridge, MA.

Metcalf, D. and Nicola, N. (1995). "The Hemopoietic Colony-Stimulating Factors: From Biology to Clinical Applications". Cambridge University Press, Cambridge, UK.

Miles, S.A., Mitsuyasu, R.T., Moreno, J., *et al.* (1991). Combined therapy with recombinant granulocyte colony-stimulating factor and erythropoietin decreases hematologic toxicity from zidovudine. Blood 77, 2109–2117.

Morstyn, G. and Burgess, A.W. (1988). Hemopoietic growth factors: a review. Cancer Res. 45, 5624–5637.

Morstyn, G., Campbell, L., Souza, L.M., *et al.* (1988). Effect of granulocyte colony-stimulating factor on neutropenia induced by cytotoxic chemotherapy. Lancet 1, 667–672.

Morstyn, G., Campbell, L., Leischke, G., *et al.* (1989). Treatment of chemotherapy-induced neutropenia by subcutaneously administered granulocyte colony-stimulating factor with optimization of dose and duration of therapy. J. Clin. Oncol. 7, 1554–1562.

Morstyn, G., Cebon, J., Layton, J., *et al.* (1991). Cytokines in infections and as anticancer agents. Ann. Haematol. 29A, 96a.

Nagata, S. (1989). Gene structure and function of granulocyte colony-stimulating factor. BioEssays 10, 113–117.

Nagata, S., Tsuchiya, M., Asano, S., *et al.* (1986). Molecular cloning and expression of cDNA for human granulocyte colony-stimulating factor. Nature 319, 415–418.

Negrin, R.S., Haeuber, D.H., Nagler, A., *et al.* (1990). Maintenance treatment of patients with myelodysplastic syndromes using recombinant human granulocyte colony-stimulating factor. Blood 7, 36–43.

Nelson, S., Farkas, S., Fotheringham, N., Ho, H., Marrie, T. and Movahhed, H. (1996). Filgrastim in the treatment of hospitalized patients with community acquired pneumonia (CAP). Am. J. Respir. Crit. Care Med. 153, A535.

Nicola, N.A. (1989). Hemopoietic cell growth factors and their receptors. Annu. Rev. Biochem. 58, 45–77.

Nicola, N.A. and Metcalf, D. (1981). Biochemical properties of differentiation factors for murine myelomonocytic leukemic cells in organ conditioned media. Separation from colony stimulating factors. J. Cell. Physiol. 109, 253–264.

Nicola, N.A. and Metcalf, D. (1984). Differentiation induction in leukemic cells by normal growth regulators: molecular and binding properties of purified granulocyte colony-stimulating factor. In "Genes and Cancer" (eds. J.M. Bishop, J.D. Rowley and M. Greaves), pp. 591–610. Alan R. Liss, New York.

Nicola, N.A., Metcalf, D., Matsumoto, M. and Johnson, G.R. (1983). Purification of a factor inducing differentiation in murine myelomonocytic leukaemia cells: identification as granulocyte colony-stimulating factor (G-CSF). J. Biol. Chem. 258, 9017–9023.

Nicola, N.A., Vadas, M.A. and Lopez, A.F. (1986). Down-modulation of receptors for granulocyte colony-stimulating factor on human neutrophils by granulocyte-activating agents. J. Cell. Physiol. 128, 501–509.

Nishizawa, M. and Nagata, S. (1990). Regulatory elements responsible for inducible expression of granulocyte colony-stimulating factor gene in macrophages. Mol. Cell. Biol. 10, 2002–2011.

Nomura, H., Imazeki, I., Oheda, M., *et al.* (1986). Purification and characterization of human granulocyte colony stimulating factor (G-CSF). EMBO J. 5, 871–876.

Ogawa, M. (1993). Differentiation and proliferation of hematopoietic stem cells. Blood 81, 2844–2853.

Oheda, M., Hase, S., Ono, M. and Ikenaka, T. (1988). Structure of the sugar chains of recombinant human granulocyte colony-stimulating factor produced by Chinese hamster ovary cells. J. Biochem. 103, 544–546.

Oheda, M., Hasegawa, M., Hattori, K., *et al.* (1990). O-linked sugar chain of human granulocyte colony-stimulating factor protects it against polymerization and denaturation allowing it to retain its biological activity. Biol. Chem. 265, 11432–11435.

Omori, F., Okamura, S., Shimoda, K., *et al.* (1992). Levels of human serum granulocyte colony-stimulating factor under pathological conditions. Biotherapy 4, 147–153.

Osslund, T. and Boone, T.C. (1994). Biochemistry and structure of Filgrastim (r-metHuG-CSF). In "Filgrastim (r-metHuG-CSF) in Clinical Practice", (eds. G. Morstyn and T.M. Dexter), pp. 23–31. Marcel Dekker, New York.

Peters, W.P., Kurtzberg, J., Atwater, S., *et al.* (1989). Comparative effects of rHuG-CSF and rHuGM-CSF on hematopoietic reconstitution and granulocyte function following high dose chemotherapy and autologous bone marrow transplantation (ABMT). Proc ASCO 18, 18A.

Pettengell, R., Gurney, H., Radford, J., *et al.* (1992). Granulocyte colony-stimulating factor to prevent dose-limiting neutropenia in non-Hodgkin's lymphoma: a randomized controlled trial. Blood 80, 1430–1436.

Platzer, E. (1989). Human hematopoietic growth factors. Eur. J. Haematol. 42, 1–15.

Reichman, B.S., Seidman, A.D., Crown, J.P, *et al.* (1993). Paclitaxel and recombinant human granulocyte colony-stimulating factor as initial chemotherapy for metastatic breast cancer. J. Clin. Oncol. 11, 1943–1951.

Rennick, D., Yang, G., Gemmell, L., *et al.* (1987). Control of hemopoeisis by a bone marrow stroma cell clone: liopopolysaccharide- and interleukin-1 inducible production of colony-stimulating factors. Blood 69, 682–691.

Rowinsky, E.K., Chaudhry, V., Forastiere, A.A., *et al.* (1993). Phase I and pharmacologic study of paclitaxel and cisplatin with granulocyte colony stimulating factor: neuromuscular toxicity is dose-limiting. J. Clin. Oncol. 11, 2010–2020.

Sato, N., Sawada, K., Takahashi, T.A., *et al.* (1994). A time course study for optimal harvest of peripheral blood progenitor cells by granulocyte colony-stimulating factor in healthy volunteers. Exp. Hematol. 22, 973–978.

Schmitz, N., Linch, D.C., Dreger, P., *et al.* (1996). Randomised trial of filgrastim-mobilised peripheral blood progenitor cell transplantation versus autologous bone marrow transplantation in lymphoma patients. Lancet 347, 353–367.

Schwinger, W., Mache, C., Urban, C., Beaurfort, E. and Töglhofer (1993). Single dose of filgrastim (rhG-CSF) increases the number of hemopoietic progenitors in the peripheral blood of adult volunteers. Bone Marrow Transplant. 11, 489–492.

Seidman, A.D., Scher, H.I., Gabrilove, J.L., *et al.* (1993). Dose-intensification of MVAC with recombinant granulocyte colony-stimulating factor in the treatment of advanced urothelial cancer. J. Clin. Oncol. 11, 408–414.

Seto, Y., Fukunaga, R. and Nagata, S. (1992). Chromosomal gene organization of the human granulocyte colony-stimulating factor receptor. J. Immunol. 148, 259–266.

Shannon, M.F., Coles, L.S., Fielke, R.K., *et al.* (1992). Three essential promoter elements mediate tumour necrosis factor and interleukin-1 activation of granulocyte-colony stimulating factor gene. Growth Factors 7, 181–193.

Sheridan, W.P., Morstyn, G., Wolf, M., *et al.* (1989). Granulocyte colony-stimulating factor and neutrophil recovery after high-dose chemotherapy and autologous bone marrow transplantation. Lancet 2, 891–895.

Sheridan, W.P., Juttner, C., Szer, J., *et al.* (1990). Granulocyte colony-stimulating factor (G-CSF) in peripheral blood stem cell (PBSC) and bone marrow transplantation. Blood 76, S1.

Sheridan, W.P., Begley, C.G., Juttner, C.A., *et al.* (1992). Effect of peripheral-blood progenitor cells mobilised by non-glycosylated G-CSF (G-CSF) on platelet recovery after high-dose chemotherapy. Lancet 339, 640–644.

Simmers, R.N., Webber, L.M., Shannon, M.F., *et al.* (1987). Localization of the G-CSF gene on chromosome 17 proximal to the breakpoint in the T(15;17) in acute promyelocytic leukemia. Blood 70, 330–332.

Souza, L.M., Boone, T.C., Gabrilove, J., *et al.* (1986). Recombinant human granulocyte colony-stimulating factor: effects on normal and leukemic myeloid cells. Science 232, 61–65.

Stahel, R.A., Jost, L.M., Cerny, T., *et al.* (1994). Randomized study of recombinant human granulocyte colony-stimulating factor after high-dose chemotherapy and autologous bone marrow transplantation for high-risk lymphoid malignancies. J. Clin. Oncol. 12, 1931–1938.

Steward, W.P. (1993). Granulocyte and granulocyte-macrophage colony-stimulating factor. Lancet 342, 153–157.

Taylor, K.M., Jagganath, S., Spitzer, G., et al. (1989). Recombinant human granulocyte colony-stimulating factor hastens granulocyte recovery after high-dose chemotherapy and autologous bone marrow transplantation in Hodgkin's disease. J. Clin. Oncol. 7, 1791–1799.

Tkatch, L.S. and Tweardy, D.J. (1993). Human granulocyte colony-stimulating factor (G-CSF), the premier granulopoietin: biology, clinical utility, and receptor structure and function. Lymphokine Cytokine Res. 12, 477–488.

Trillet-Lenoir, V., Green, J., Manegold, C., et al. (1993). Recombinant granulocyte colony stimulating factor reduces the infectious complications of cytotoxic chemotherapy. Eur. J. Cancer 29A, 319–324.

Tsuchiya, M., Kaziro, Y. and Nagata, S. (1987). The chromosomal gene structure for murine granulocyte colony-stimulating factor. Eur. J. Biochem. 165, 7–12.

Van der Wouw, P.A., van Leeuwen, R., van Oers, R.H., et al. (1991). Effects of recombinant human granulocyte colony-stimulating factor on leucopenia in zidovudine-treated patients with AIDS and AIDS related complex, a phase I/II study. Br. J. Haematol. 78, 319–324.

Vellenga, E., Rambaldi, A., Ernst, T.J., Ostapovicz, D. and Griffin, J.D. (1988). Independent regulation of M-CSF and G-CSF gene expression in human monocytes. Blood 71, 1529–1532.

Walker, F., Nicola, N.A., Metcalf, D. and Burgess, A.W. (1985). Hierarchial down-modulation of hemopoietic growth factor receptors. Cell 43, 269–276.

Watari, K., Asano, S., Shirafuji, N., et al. (1989). Serum granulocyte colony-stimulating factor levels in healthy volunteers and patients with various disorders as estimated by enzyme immunoassay. Blood 73, 117–122.

Welte, K. and Dale, D. (1996). Pathophysiology and treatment of severe chronic neutropenia. Ann Hematol. 72, 158–165.

Welte, K., Platzer, E., Lu, L., et al. (1985). Purification and biochemical characterization of human pluripotent hematopoietic colony-stimulating factor. Proc. Natl Acad. Sci. USA 82, 1526–1530.

Welte, K., Ziedler, C., Reiter, A., et al. (1990). Differential effects of granulocyte colony-stimulating factor and granulocyte-macrophage colony-stimulating factor in children with severe congenital neutropenia. Blood 75, 1056–1063.

Wong, G.G., Witek, J.S., Temple, P.A., et al. (1985). Human GM-CSF: molecular cloning of the complementary DNA and purification of the natural and recombinant proteins. Science 228, 810–815.

Wunderink, R.G., Leeper, K.V., Schein, R.M.H., et al., (1995). Clinical response to Filgrastim (r-metHuG-CSF) in pneumonia with severe sepsis. Am. J. Resp. Crit. Care Med. 153, A123.

Yasukawa, K., Hirno, T., Watanabe, Y., et al. (1987). Structure and expression of human B cell stimulating factor-2 (BSF-2/IL-6) gene. EMBO J. 6, 2939–2945.

Yoshida, Y., Hirashima, K., Asano, S., et al. (1991). A phase II trial of recombinant human granulocyte colony-stimulating factor in the myelodysplastic syndromes. Br. J. Haematol. 78, 378–384.

Yujiri, T., Shinohara, K., Kurimoto, F., et al. (1992). Fluctuations in serum cytokine levels in patients with cyclic neutropenia. Am. J. Hematol. 39, 144–145.

Zsebo, K.M., Cohen, A.M., Murdock, D.C., et al. (1986). Recombinant human granulocyte colony-stimulating factor: molecular and biological characterization. Immunobiology 172, 175–184.

18. Macrophage Colony-stimulating Factor

Robert G. Schaub, Joseph P. Sypek, James C. Keith, Jr, David H. Munn, Matthew L. Sherman, Andrew J. Dorner *and* Marc B. Garnick

1. Introduction

Macrophage colony-stimulating factor (M-CSF, also known as CSF-1) is a hematopoietic growth factor which supports the survival, proliferation and differentiation of cells of the monocyte lineage (Tushinski *et al.*, 1982; Clark and Kamen, 1987; Becker *et al.*, 1988). M-CSF is produced in various cells, including monocytes, smooth-muscle cells, endothelial cells, and fibroblasts as a disulfide-linked homodimer (Rambaldi *et al.*, 1987; Takahashi *et al.*, 1988; Rettenmier and Sherr, 1989; Praloran *et al.*, 1990; Clinton *et al.*, 1992). The

molecular mass of M-CSF varies owing to the synthesis of two distinct M-CSF precursor proteins and the post-transcriptional addition of glycosyl residues and chondroitin sulfate to the molecule (Ralph and Sampson-Johannes, 1990; Price *et al.*, 1992; Kimura *et al.*, 1994). M-CSF was originally identified by its ability to selectively support macrophage colony formation in semisolid medium by progenitor cells found in normal bone marrow (Metcalf, 1970, 1985; Stanley *et al.*, 1983). It is one of the few cytokine/growth factors found in measurable concentrations in various tissues including plasma, urine, cerebrospinal fluid, and placenta

(Ralph and Sampson-Johannes, 1990). cDNA clones for M-CSF were successfully isolated and the biological properties of the recombinant protein (rhM-CSF) confirmed the biological activity of the natural factors (Wong *et al.*, 1987; Kawasaki *et al.*, 1985). In addition, the availability of the recombinant protein has permitted evaluation of M-CSF as a therapeutic agent in the treatment of cancer, infectious diseases, and atherosclerosis.

2. The Cytokine Gene

Human M-CSF is encoded by a single 20 kb gene containing 10 exons. The gene for M-CSF was thought to reside on the long arm of chromosome 5 at band q33.1 (Pettenati *et al.*, 1987). Recently, however, the human M-CSF gene was reassigned to the short arm of chromosome 1 at band p13-p21 (Morris *et al.*, l991; Saltman *et al.*, 1992; Landegent *et al.*, 1992). The molecular cloning of the cDNA for M-CSF was identified through a structural approach (Clark and Kamen, 1987). The natural protein was purified to homogeneity, the amino-terminal sequence was determined, and degenerate oligonucleotides were synthesized on the basis of the predicted nucleotide sequence. These oligonucleotides were used as hybridization probes to screen a human cDNA library. The original cDNA clone represented a 1.6 kb mRNA, although subsequently a 4 kb message was found to be the predominant transcript (Wong *et al.*, 1987). This 4 kb mRNA species contained an insertion of 894 nucleotides in the protein coding region and a substantially longer 3′ noncoding sequence relative to the 1.6 kb mRNA. These differences result from alternative splicing of the M-CSF transcript. The 894-nucleotide insert preserves the M-CSF reading frame in the 4 kb RNA compared to the 1.6 kb. The 4 kb mRNA encodes a 554-amino-acid precursor protein and the 1.6 kb mRNA encodes a 256-amino-acid precursor protein which lacks amino acid residues 150–447 encoded by the 894 nucleotide insert (Figures 18.1 and 18.2). A third M-CSF cDNA which encodes a 438-amino-acid protein has also been described (Pandit *et al.*, 1992).

3. The Protein

The two precursor proteins share amino acid residues −32 to +149. Residues 149 to 331 and 149 to 447 are

-570 ctgcagaggaagaaggggggctgccggcaaacctgctgactcaggctccacgagggagcaagtaacactggactcctttcg
gcactccgagaatggggtgggggcgtcttcaaaggatttccctcccttcccagtgcttgtccctgctctcggtccgttttt
ctgctaagatttggggattttcagggcctggagggaaagtcccttgggacgatcatagagcgctagcactgaatcagcct
ggagagcgcggaaggaaagggtcggtccgcagagggcgcggggaaggcaggtggggacgcggtggagcccgcgctcgtt
tgctgaaggcttggaagtgcagcgcagaagacagagggtgactaggaagacgcgcgaacggggctggccggccggcgggt
gggggaggggaggcggggaaggcggctgagtgggcctctggagtgtgtgtgtctgtgtcagtgtgtgtgtgtgtgtgtgt
tatgtgtgtgtctggcgcctggccaggtgatttcc cataaa ccacatgccccccagtcctctc ttaaaa ggctgtgccg
agggctggccagtgaggctcggcccgggaaagtgaaagtttgcctgggtcctctcggcgccagagccgctctccgcatc
ccaggacagcggtgcggccctcggccgggggcgcccactccccaccagccagcgagcgagcgagcgaggcggccga
cgcgcccggccgggacccagctgcccgtATGACCGCGCCGGGCGCCGCCGGGCGCTGCCCTCCCACGACATGGCTGGGCT
CCCTGCTGTTGTTGGTCTGTCTCCTGGCGAGCAGGAGTATCACCGAGGAGGTGTCGGAGTACTGTAGCCACATGATTGGG
AGTGGACACCTGCAGTCTCTGCAGCGGCTGATTGACAGTCAGATGGAGACCTGCTGCCAAATTACATTTGAGTTTGTAGA
CCAGGAACAGTTGAAAGATCCAGTGTGCTACCTTAAGAAGGCATTTCTCCTGGTACAAGACATAATGGAGGACACCATGC
GCTTCAGAGATAACACCCCCAATGCCATCGCCATTGTGCAGCTGCAGGAACTCTCTTTGAGGCTGAAGAGCTGCTTCACC
AAGGATTATGAAGAGCATGACAAGGCCTGCGTCCGAACTTTCTATGAGACACCTCTCCAGTTGCTGGAGAAGGTCAAGAA
TGTCTTTAATGAAACAAAGAATCTCCTTGACAAGGACTGGAATATTTTCAGCAAGAACTGCAACAACAGCTTTGCTGAAT
GCTCCAGCCAAGATGTGGTGACCAAGCCTGATTGCAACTGCCTGTACCCCAAAGCCATCCCTAGCAGTGACCCGGCCTCT
GTCTCCCCTCATCAGCCCCTCGCCCCCTCCATGGCCCCTGTGGCTGGCTTGACCTGGGAGGACTCTGAGGGAACTGAGGG
CAGCTCCCTCTTGCCTGGTGAGCAGCCCCTGCACACAGTGGATCCAGGCAGTGCCAAGCAGCGGCCACCCAGGAGCACCT
GCCAGAGCTTTGAGCCGCCAGAGACCCCAGTTGTCAAGGACACGACCATCGGTGGCTCACCACAGCCTCGCCCCTCTGTC
GGGGCCTTCAACCCCGGGATGGAGGATATTCTTGACTCTGCAATGGGCACTAATTGGGTCCCAGAAGAAGCCTCTGGAGA
GGCCAGTGAGATTCCCGTACCCCAAGGGACAGAGCTTTCCCCCTCCAGGCCAGGAGGGGGCAGCATGCAGACAGAGCCCG
CCAGACCCAGCAACTTCCTCTCAGCATCTTCTCCACTCCCTGCATCAGCAAAGGGCCAACAGCCGGCAGATGTAACTGGT
ACAGCCTTGCCCAGGGTGGGCCCCGTGAGGCCCACTGGCCAGGACTGGAATCACACCCCCCAGAAGACAGACCATCCATC
TGCCCTGCTCAGAGACCCCCCGGAGCCAGGCTCTCCCAGGATCTCATCACTGCGCCCCCAGGGCCTCAGCAACCCCTCCA
CCCTCTCTGCTCAGCCACAGCTTTCCAGAAGCCACTCCTCGGGCAGCGTGCTGCCCCTTGGGGAGCTGGAGGGCAGGAGG
AGCACCCAGGGATCGGAGGAGCCCCGCAGAGCCAGAAGGAGGACCAGCAAGTGAAGGGGCAGCCAGGCCCCTGCCCCGTTT
TAACTCCGTTCCTTTGACTGACACAGGCCATGAGAGGCAGTCCGAGGGATCCTCCAGCCCGCAGCTCCAGGA GTCTGTCT
TCCACCTGCTGGTCCCCAGTGTCATCCTGGTCTTGCTGGCCGTCGGAGGCCTCTTGTTCTAC AGGTGGAGGCGGCGGAGC
CATCAAGAGCCTCAGAGAGCGGATTCTCCCTTGGAGCAACCAGAGGGCAGCCCCCTGACTCAGGATGACAGACAGGTGGA
ACTGCCAGTGtagagggaattctaag   **1856**

Figure 18.1   The nucleotide sequence of the M-CSF cDNA. Negative numbers refer to the 32 amino acid signal sequence (underlined). Shorter translation products occur owing to alternative splicing of exon 6. Primary translation products are membrane spanning proteins that are modified posttranslationally to yield secreted forms by proteolytic cleavage on the amino terminal side of the transmembrane domain. For complete cDNAs see Aggerwal and Gutterman, 1991.

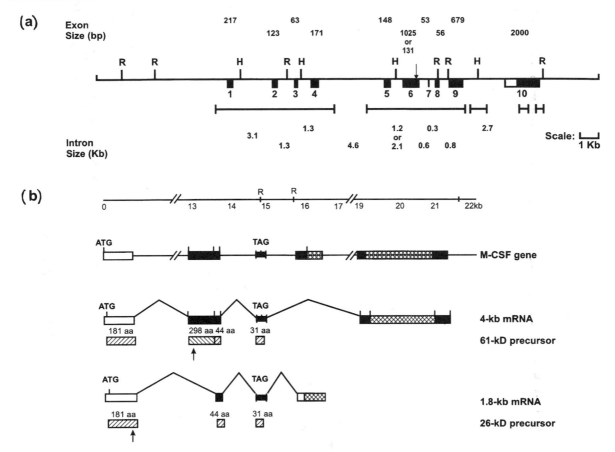

Figure 18.2   (a) Organization of the human M-CSF gene. Three overlapping genomic clones containing the entire gene are indicated at the top of the figure. The 10 exons are shown as solid boxes; the open box indicates the presumed location of the first half of exon 10. The location of the exon-6 5′ splice site used by pCSF-17-type transcripts is shown by an arrow. The map indicates EcoRI (R) and HindIII (H) restriction sites and sequenced regions of the gene are delineated below. (b) The two M-CSF mRNAs arise from alternate gene splicing pathways as illustrated. The M-CSF gene consists of many exons distributed over 22 kb of DNA. The first 750 bases of either mRNA result from identical splicing together of many small exon sequences (not shown) that are located in the first half of the gene (12.5 kb). The differences in the structure of the two mRNAs result from alternative splicing of the four remaining exons (boxes). The positions of the initiation (ATG) and termination (TAG) codons are as indicated. The primary translation products of the two mRNAs have identical amino- and carboxyl-terminal domains (indicated by hatched boxes) but differ by the 298-residue domain (indicated by the cross-hatched boxes) as shown. The arrows indicate the locations where carboxyl terminal processing ultimately occurs to yield the related but distinct M-CSF polypeptides encoded by either mRNA species. (From Clark and Kamen, 1987.)

missing in the two smaller precursor proteins (Figure 18.3). The transmembrane and carboxyl terminal residues are shared by all three variants (Pandit et al., 1992) (Figure 18.4). The biologically active form of M-CSF is a homodimer derived from the amino-terminal region. Each monomer contains three intramolecular disulfide bonds at Cys^7–Cys^{90}, Cys^{48}–Cys^{139} and Cys^{102}–Cys^{146} and one interchain bond at Cys^{31}–Cys^{31} (Pandit et al., 1992). The precursor proteins undergo post-translational processing including N-linked and O-linked glycosylation (438 and 554 forms only) (Ralph and Sampson-Johannes, 1990; Price et al., 1992; Kimura et al., 1994). The N-linked sites are fully occupied and are found at amino acids

122 and 140 (H. Scoble, personal communication). Many O-linked sites occur on M-CSF. Those which have been identified occur on amino acids 153, 167, 172, 174, and 182. An additional O-linked site has been localized to amino acids 194, 200, or 201 (H. Scoble, personal communication). Glycosylation, however, is not required for biological activity (Pandit et al., 1992). The structure of recombinant M-CSF dimer has been determined by x-ray crystallography (Figure 18.5). The molecule contains two bundles of four α-helices with an interchain disulfide bond. The monomer has a structural similarity to both G-CSF and human growth hormone, suggesting similar receptor binding determinants (Pandit et al., 1992).

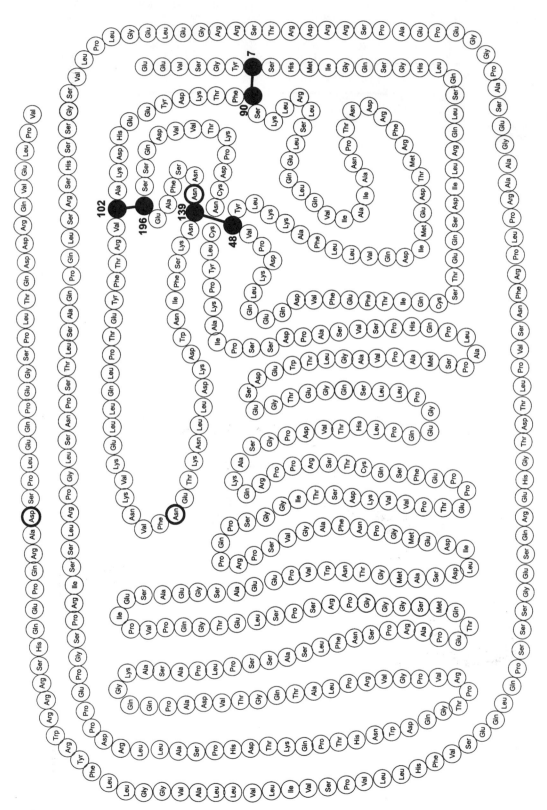

Figure 18.3   Predicted amino acid sequence of M-CSF. The positions of the potential sites for asparagine-linked glycosylation are indicated by heavy circles and cysteine residues are filled.

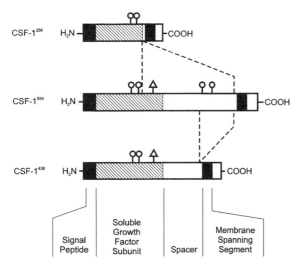

Figure 18.4   The biologically active form of M-CSF is a homodimer derived from the amino-terminal region. The precursor proteins undergo post-translational processing including *N*-linked and *O*-linked glycosylation (438 and 554 forms only). The *N*-linked sites are fully occupied and are found at amino acids 122 and 140. (From Rettenmier and Sherr, 1989.)

4.   Cellular Sources and Production

M-CSF gene expression is constitutive or inducible in a variety of cell types including fibroblasts, smooth-muscle cells, vascular endothelial cells, bone marrow stromal cells, T cells, uterine cells, hepatocytes and cells of the monocyte/macrophage lineage (Rambaldi *et al.*, 1987; Takahashi *et al.*, 1988; Praloran *et al.*, 1990; Clinton *et al.*, 1992; Tsukui *et al.*, 1992). M-CSF is also expressed in malignant cell lines (Horiguchi *et al.*, 1988). Gene expression can be induced by endotoxin, IL-1α, tumor necrosis factor-α (TNF-α), interferon-γ (IFN-γ), granulocyte-macrophage colony-stimulating factor (GM-CSF), phorbol esters, and progesterone/estrogen (Horiguchi *et al.*, 1987; Pollard *et al.*, 1987; Rambaldi *et al.*, 1987; Sherman *et al.*, 1990; Clinton *et al.*, 1992). M-CSF is found in detectable concentrations in blood, urine, cerebrospinal fluid, and fetal amniotic fluid (Ralph and Sampson-Johannes, 1990). Plasma or serum concentrations range from 1.8 to 7.1 ng/ml. Elevation of M-CSF concentration has been reported in trauma, infections, certain malignancies, and pregnancy (Ralph and Sampson-Johannes, 1990).

5.   Cellular and Animal Biological Activity

The role of M-CSF in the proliferation, differentiation and activation of cells of the monocyte-macrophage lineage has been described in detail (for review see, Stanley, 1984; Rettenmier *et al.*, 1988; Ralph and Sampson-Johannes, 1990; Sherr, 1990; Crosier and Clark, 1992). The cloning and expression of recombinant human M-CSF (rhM-CSF) has provided

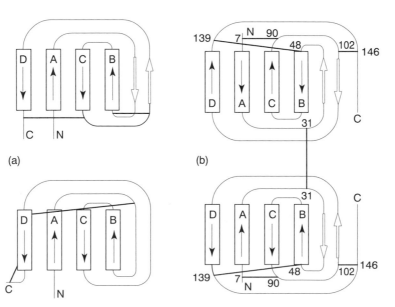

Figure 18.5   The structure of recombinant M-CSF dimer has been determined by x-ray crystallography. The molecule contains two bundles of four α-helices with an interchain disulfide bond. Helices are shown as β-strands as arrows in two views (a). A topological structure showing all disulfide bonds is shown in (b). The monomer has structural similarity to both G-CSF and human growth hormone, suggesting similar receptor binding determinants. (From Pandit *et al.*, 1992.)

sufficient material to permit detailed *in vitro* and *in vivo* evaluation of the role of M-CSF in various physiological processes. This section will focus on selected aspects of M-CSF bioactivity.

5.1 Hematopoietic Effects of M-CSF

M-CSF has been administered to rabbits, cynomolgus monkeys, and nonhuman primates by constant intravenous infusion (CIVI) or subcutaneous injection. In rabbits, infusion of rhM-CSF at 100 or 300 µg/kg per day resulted in a transient decrease in platelets for the first week after the initiation of treatment. In addition, rhM-CSF-treated rabbits developed a sustained decrease in white blood cell count that was primarily due to a decrease in circulating neutrophils. Changes in monocyte counts in rabbits have been more variable, ranging from no effect to significant monocytosis (Stoudemire and Garnick, 1990; Schaub *et al.*, 1994). In cynomolgus monkeys, doses of 10 or 25 µg/kg administered by CIVI for 14 days had no consistent effect on hematology. Doses of 50 and 100 µg/kg per day increased monocyte count by 400% and promonocytes by almost 900% after 4–6 days of treatment (Garnick and Stoudemire, 1990). A dose-dependent reduction in platelet count was associated with the increased monocyte count in monkeys. At a dose of 100 µg/kg per day, platelet counts decreased by 60% (Garnick and Stoudemire, 1990). A qualitatively similar monocyte and platelet response was observed in cynomolgus monkeys treated with rhM-CSF at daily subcutaneous doses of 200 µg/kg per day (Garnick and Stoudemire, 1990). In humans, an increase in monocyte count and a decrease in platelet count were seen at doses greater than 30 µg/kg per day (VandePol and Garnick, 1991).

5.2 Antitumor Activity of M-CSF

5.2.1 *In Vitro*

rhM-CSF treatment of monocytes increases tumoricidal activity through poorly understood mechanisms which may involve the release of cytotoxic cytokines such as TNF, oxygen free radicals or nitric oxide from activated macrophages (Garnick and Stoudemire, 1990). rhM-CSF treated monocytes and macrophages are more tumoricidal when a monoclonal antibody directed against the tumor target cell (antibody-dependent cellular cytotoxicity, ADCC) is present (Munn and Cheung, 1989). rhM-CSF-treated human peripheral blood monocytes were more effective in mediating ADCC against colon cancer targets at lower effector-to-target cell ratios and with lower concentrations of anti-colon-cancer antibodies than untreated monocytes, while rhGM-CSF and IL-3 did not enhance ADCC in this system (Mufson *et al.*, 1989). Macrophage cell surface

antigens such as CD16, Leu-11, Ia, LFA3, CR3, and CD14 increase after rhM-CSF treatment (Falk *et al.*, 1988). CD16, a low-affinity receptor for IgG, may be important in rhM-CSF-mediated ADCC. rhM-CSF has significant effects on proliferation of leukemic blast cells from patients with acute myelogenous leukemia (AML) (Pebusque *et al.*, 1988; Suzuki *et al.*, 1988). In addition, rhM-CSF was found to induce terminal differentiation of some, but not all, peripheral blood AML cells (Miyauchi *et al.*, 1988a,b).

5.2.2 *In Vivo*

The antitumor effects of rhM-CSF have been evaluated in a variety of murine tumor models. Administration of rhM-CSF (4 days of 20 000 U/day × 2 cycles) starting 21 days after inoculation of mice with 50 000 B16 melanoma cells resulted in a reduction of the median number of metastases from 65 in the control mice to 1 in the rhM-CSF treated mice (Garnick and Stoudemire, 1990). Administration of rhM-CSF (50 or 200 µg daily in 2 divided doses) had no antitumor effect on 3-day pulmonary or hepatic metastases from methylcholanthrene (MCA)-induced sarcomas. rhM-CSF alone did not reduce B16 pulmonary or hepatic metastases. The combination of rhM-CSF and anti-B16 melanoma antibody, while not reducing pulmonary metastases, did reduce hepatic metastases at a dose of 200 µg daily (Bock *et al.*, 1991). rhM-CSF alone at doses of 1, 30, and 100 µg/kg per day did not reduce the growth of Lewis lung carcinoma in C57BL mice (Lu *et al.*, 1991; Teicher *et al.*, 1995). However, the combination of rhM-CSF with radiation resulted in incremental increases in tumor growth delay at all three doses of rhM-CSF. Treatment of Lewis lung carcinoma-bearing mice with the immuno-modulatory cytokine IL-12 (4.5 µg/kg) and M-CSF (30 or 100 µg/kg) resulted in increased growth delay of the primary tumor compared with treatment with IL-12 alone and a parallel decrease in lung metastases. When the combination of rhM-CSF and rmIL-12 was added to fractionated radiation, there was a markedly increased tumor growth delay of 30.6 days and a reduction of tumor metastases with the most clinically relevant radiation doses (2 or 3 Gy × 5 days) (Teicher *et al.*, 1995). In a more recent study, B16F10 murine melanoma cells were transduced with a retroviral vector containing genes encoding neomycin resistance and M-CSF (Walsh *et al.*, 1995). Mice given a mixture of B16F10 and M-CSF$^+$ cells had an 80% survival rate at 8 weeks and survived at least twice as long as mice given B16F10 tumor and M-CSF$^-$ cells (0% survival at 8 weeks).

5.3 Anti-infectious Disease Activity of M-CSF

Initial investigations evaluating M-CSF as an antifungal or antimicrobial agent used *in vitro* exposure of both murine and human monocyte-macrophages with M-CSF. Such experiments generally resulted in observations noting enhanced microbistatic/microbicidal effects (Karbassi et al., 1987; Lee and Warren, 1987; Cheers et al., 1989; Wang et al., 1989; Ho et al., 1990; Cenci et al., 1991; Rose et al., 1991; Newman and Gootee, 1992; Brummer et al., 1994; Brummer and Stevens, 1994; Khemani et al., 1995). M-CSF synergizes with fluconazole for fungal killing and with IFN-γ against Leishmania and Mycobacterium avium (Ho et al., 1991; Rose et al., 1991). The M-CSF-mediated antimicrobial effects in macrophages do not appear to be due to the products of an oxidative burst, nitric oxide production (Brummer et al., 1994; Brummer and Stevens, 1994) or the presence of TNF-α (Rose et al., 1991). Additional studies indicated that augmented expression of mannose receptors on the M-CSF-treated macrophages was, in part, responsible for the observed enhanced antifungal activity and that this activity was independent of IFN-α/β (Karbassi et al., 1987).

Enhanced microbistatic/microbicidal effects against C. neoformans (Nassar et al., 1994) and H. capsulatum (Khemani et al., 1995) have been observed ex vivo in murine bronchoalveolar macrophages derived from mice treated with M-CSF at doses of 2.5 mg/kg and against Candida albicans in human peripheral blood-derived monocyte/macrophages derived from nonleukopenic bone marrow transplant patients receiving a 2 h intravenous infusion of M-CSF (Khjwaja et al., 1991). Peripheral blood monocytes from guinea-pigs treated with M-CSF also have enhanced cytolytic activity against herpes simplex virus-infected target cells ex vivo (Ho et al., 1991).

Prophylactic or therapeutic administration of M-CSF has been shown to enhance survival to infection or reduce microbial burden in a number of infectious disease animal models (Cenci et al., 1991; Ho et al., 1991; Kayashima et al., 1991; Doyle et al., 1992; Gregory et al., 1992; Gregory and Wing, 1993; Vitt et al., 1994). In contrast, in C. albicans-infected mice, M-CSF exacerbates disease and caused a significantly earlier death in one study (Hume and Denkins, 1992). It has also been observed that the therapeutic administration of M-CSF to Listeria-infected mice had no effect on the replication of the organism (Gregory et al., 1992). In several studies, M-CSF has been observed to enhance the replication of HIV in vitro in monocytes/macrophages (Gendleman et al., 1988; Koyanagi et al., 1988; Kalter et al., 1991; Kitano et al., 1991). Overall these studies suggest that the major effect of M-CSF is on the macrophage and not directly on the retrovirus. This effect appears to be related to M-CSF stimulating an increase in the expression of the CD4 receptor on the macrophage, the expression of which is known to correlate with an enhanced susceptibility of these cells to become infected with HIV (Bergamini et al., 1994). Further studies have shown that while M-CSF may augment the replication of HIV in monocyte/macrophages, it has little affect on the in vitro efficacy of zidovudine (AZT) or antiviral activity by other dideoxynucleosides (Perno et al., 1992, 1994). However, the antiviral activity of compounds that inhibit viral binding, such as dextran sulfate and soluble CD4, is dramatically reduced by M-CSF (Bergamini et al., 1994).

5.4 M-CSF AND PREGNANCY

Most of the data concerning the role of CSFs in pregnancy have come from experiments in rodents. Location and distribution of CSFs and their receptors during rodent pregnancy are detailed elsewhere (Arceci et al., 1989; Regenstreif and Rossant, 1989; Kanazaki et al., 1991). Initially, the CSF family was observed to induce the growth of a mixed population of cells in the placenta in general (Athanassakis et al., 1987) and, later, in ectoplacental core trophoblasts in particular (Armstrong and Chaouat, 1989). Pollard and colleagues (1987) have shown that M-CSF, secreted by uterine gland cells, was increased 1000-fold within the first days of pregnancy, while Uzumake and colleagues (1989) have found the c-fms receptor on invasive trophoblast cells. In human placenta, the expression and localization of mRNA for M-CSF have been demonstrated in mesenchymal cells of the chorionic villous stroma (Kanazaki et al., 1992). Daiter and colleagues (1992) have also shown a trimester-dependent distribution of mRNA for M-CSF. M-CSF appeared in cytotrophoblasts lining the villous core and in the cytotrophoblastic core in the first trimester, in villous mesenchymal cells in the second trimester, and in cells lining the villous vessels in the third trimester. In addition, circulating levels of M-CSF during pregnancy are higher than those of nonpregnant women (Yong et al., 1992).

The elevated levels of M-CSF seen in pregnancy may be due to the stimulation of uterine glandular epithelial cells by ovarian hormones (Hanamura et al., 1988). M-CSF synthesized by normal uterine glandular epithelial cells in response to ovarian hormones during pregnancy appears to stimulate receptor-bearing trophoblasts of the adjacent placenta (Pollard et al., 1987; Bartocci et al., 1986).

5.5 M-CSF IN LIPID METABOLISM AND ATHEROSCLEROSIS

M-CSF gene expression and immunoreactive protein has been demonstrated within the endothelium, smooth-muscle cells, and macrophages associated with atherosclerotic lesions (Clinton et al., 1992; Rosenfeld et al., 1992). The presence of M-CSF within the arterial lesion has led to speculation that the protein may play a role in promoting atherosclerosis by enhancing monocyte recruitment and proliferation, foam-cell development, and the release of monocyte-derived growth factors for smooth-muscle cells and pathogenic cytokines (Shyy *et al.*, 1993). The importance of M-CSF as a maintenance factor for monocyte/macrophages has also led to the speculation that M-CSF may be important in maintaining long-term survival of macrophages within lesions and that necrosis of foam cells in advanced lesions may occur because of a deficiency in local M-CSF production (Clinton *et al.*, 1992; Rosenfeld *et al.*, 1992; Schaub *et al.*, 1994). Many of the *in vivo* studies performed with M-CSF in hypercholesterolemic rabbits would suggest that M-CSF can promote changes in lipid metabolism which can significantly reduce the development of atherosclerosis in animals models.

5.5.1   M-CSF Decreases Plasma Cholesterol

M-CSF decreased plasma total cholesterol by 43% when administered by intravenous bolus to normo-cholesterolemic nonhuman primates and by 25% in normocholesterolemic New Zealand White rabbits when administered by constant intravenous infusion (Stoudemire and Garnick, 1990). Low-density lipoprotein (LDL) clearance was increased from 2.07 pools to 3.75 pools per day when M-CSF was administered at 100 µg/kg per day (Stoudemire and Garnick, 1990). M-CSF administration to hyper-cholesterolemic rabbits also reduced plasma cholesterol by 30–73% when M-CSF was given at doses from 100 µg/kg to 1 mg/kg per day by continuous intravenous infusion or by intravenous bolus (Schaub *et al.*, 1994, 1995). Most of the decrease in plasma cholesterol was due to a fall in LDL-cholesterol. As with normo-cholesterolemic rabbits, ^{125}I-labeled LDL was cleared more rapidly from plasma in M-CSF-treated animals (1.23 pools per day compared to 0.43 pools in non-treated hypercholesterolemic rabbits) (Stoudemire and Garnick, 1990). In addition, M-CSF treatment of hypercholesterolemic rabbits increased bile salt concentration and total cholic acid excretion by 50%, suggesting that M-CSF increases biliary excretion of cholesterol (Schaub *et al.*, 1995).

5.5.2   M-CSF Promotes Reverse Cholesterol Transport from Peripheral Tissue

M-CSF stimulated the uptake and degradation of oxidized and acetylated LDL in cultured macrophages (Yamada *et al.*, 1992). In rabbits with tissue cholesterol pre-labeled with [³H]cholesterol, M-CSF treatment decreased plasma cholesterol by 21% while increasing the specific activity of plasma HDL, suggesting movement of cholesterol from tissue to plasma (Yamada *et al.*, 1992). Daily treatment with M-CSF also increases HDL particle size, which is consistent with increased cholesterol accumulation onto the HDL particle and reverse cholesterol transport from tissue to blood (Schaub *et al.*, 1995).

5.5.3   M-CSF Decreases Atherosclerotic Lesion Formation

M-CSF treatment of hypercholesterolemic rabbits was found to decrease aortic cholesterol and cholesteryl ester content, luminal surface lesion frequency in the thoracic and abdominal aorta, and macrophage foam cell development, and to enhance regression of atherosclerotic lesions (Inoue *et al.*, 1992; Schaub *et al.*, 1994).

6.   M-CSF Receptor

The receptor for M-CSF is a member of the tyrosine kinase class of growth factor receptors and is encoded by the c-*fms* protooncogene (Sherr *et al.*, 1985) (Figure 18.6). The gene for c-*fms* is located on chromosome 5 at q33.2-q33.3 (Roussel *et al.*, 1983). The receptor is an integral transmembrane protein which contains an extracellular binding domain with five immunoglobulin-like repeats with 8–12 potential *N*-linked carbohydrate binding sites, a transmembrane segment, and a cytoplasmic tyrosine kinase domain. Signal transduction results from M-CSF binding to a single high-affinity site consisting of the first three immunoglobulin domains in each of two c-*fms* molecules to induce dimerization. Dimerization induces rapid transphosphorylation of tyrosine residues on nearby receptor molecules as well as phosphorylation of intracellular proteins believed to be involved in signal transduction following receptor–ligand interaction (Wang *et al.*, 1993). Activation is terminated by lysosomal degradation of receptor and ligand. c-*fms* is found on mononuclear phagocytes, placental trophoblasts, and smooth-muscle cells in atherosclerotic lesions (Rettenmier *et al.*, 1988; Inaba *et al.*, 1992).

7.   Signal Transduction

Binding of M-CSF to c-*fms* results in receptor dimerization and in upregulation of protein tyrosine kinase activity. Three sites of autophosphorylation have been identified. These are tyrosines 699, 708 and 809 (Sherr, 1990). These phosphorylated sites serve as important docking sites for a variety of signal transduction molecules containing the Src homology domains (SHR) (Courtneidge *et al.*, 1993; Heldin, 1995). The c-*myc* gene appears to be essential in the cellular proliferation of M-CSF-stimulated cells. Activation of p21*ras* and the *ets* gene family are necessary for c-*myc* induction (Roussel, 1994). Other signaling or regulatory factors associated with M-CSF binding to c-*fms* are the SH2 domain-containing tyrosine phosphatase SHPTP1, Grb2, phosphatidylinositol 3-kinase, and fyn-associated protein tyrosine kinase activity (Shurtleff *et al.*,

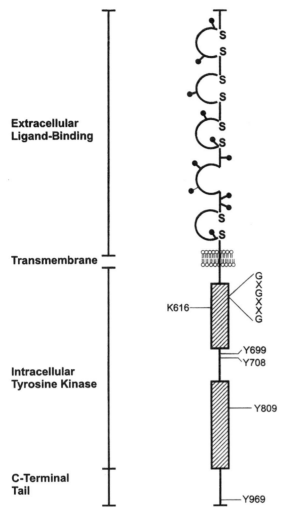

Figure 18.6   The receptor for M-CSF. The receptor is an integral transmembrane protein which contains an extracellular binding domain with five immunoglobulin-like repeats with 8–12 potential *N*-linked carbohydrate binding sites, a transmembrane segment and a cytoplasmic tyrosine kinase domain. (From Sherr, 1990.)

estimated to be 15 kDa (Das and Stanley, 1982). Two full-length cDNA clones of the murine protein have been isolated from L929 cells (Delamarter *et al.*, 1987; Ladner *et al.*, 1988). The murine gene maps to chromosome 3 and contains ten exons which are transcribed in two RNAs as a result of alternative use of exons 9 and 10 (Ladner *et al.*, 1988; Daiter and Pollard, 1992). These two clones both produce a 520-amino-acid protein which is similar to the large human M-CSF precursor and exhibits over 80% amino acid identity in the growth factor region with human M-CSF (Rettenmier *et al.*, 1988). Murine M-CSF has *in vitro* and *in vivo* activities similar to those reported for human M-CSF (Ralph and Sampson-Johannes, 1990).

9.   M-CSF in Therapy

9.1   CANCER THERAPY

9.1.1   Urinary-derived M-CSF

Urinary-derived M-CSF has been studied in clinical trials in Japan (Motoyoshi and Takaku, 1990). These studies have evaluated the ability of M-CSF to ameliorate the myelosuppression associated with chronic neutropenia of childhood; to enhance bone marrow engraftment following allogeneic, syngeneic, and autologous bone marrow transplantation (BMT) in patients with a variety of hematological and solid tumors; and to evaluate myelorestoration following intensive cancer chemotherapy and in individuals with myelodysplasia. In these studies, M-CSF was found to accelerate production of neutrophils and platelets after chemotherapy and BMT and to improve survival rate after BMT (Motoyoshi and Takaku, 1990).

9.1.2   Recombinant M-CSF

Several phase I studies of glycosylated M-CSF produced by mammalian expression have been published. Patients with metastatic melanoma have been treated with doses of M-CSF ranging from 10 to 120 μg/kg per day over two 7-day continuous intravenous infusion regimens (VandePol and Garnick, 1991; Jakubowski *et al.*, 1996). Monocytosis was seen at doses greater than 30 μg/kg per day. Doses of 80 μg/kg per day increased absolute peripheral monocyte counts almost 6-fold and increased percentage peripheral monocytes 4-fold. M-CSF was well tolerated with no evidence of systemic toxicity. There was a mild dose-related decrease in platelets which was reversible when treatment was discontinued. Antitumor activity was noted in one patient who demonstrated a complete response 3 months after therapy was discontinued. Patients with metastatic solid tumors refractory to conventional therapy were treated with M-CSF at doses of 50, 100, and 150 μg/kg per day by continuous intravenous infusion for 7–14 days with 1–2-week

1990; Li and Chen, 1995; Chen *et al.*, 1996). The activation of fyn-kinase may be associated with the development of adherence capacity in macrophages (Li and Chen, 1995), while SHPTP1 and Grb2 may have a role as negative regulators of M-CSF signaling (Chen *et al.*, 1996).

8.   Murine M-CSF

Murine M-CSF was originally purified from L-cell-conditioned media (Stanley and Heard, 1977) and described as a glycosylated disulfide-linked dimer of 70 kDa. The unglycosylated monomeric protein was intervals between treatment. Doses of 150 μg/kg produced dose-limiting grade 4 thrombocytopenia. No objective clinical responses were observed (Cole *et al.*, 1994). Three phase I studies of unglycosylated *E. coli*-derived rhM-CSF administered intravenously have been reported. A phase I study or rhM-CSF administered by rapid intravenous infusion was performed in 14 patients with advanced cancer. No clinical or laboratory toxicity was seen; a peripheral blood monocytosis was observed at the higher dose levels. Antitumor activity was seen in 2 patients with metastatic leiomyosarcoma (Redman *et al.*, 1992). A second phase I trial using this route of administration was conducted in 20 patients with refractory malignancies. Mild decreases in platelet counts were observed (Zamkoff *et al.*, 1992). A third phase I trial in 23 patients demonstrated decreases in platelet counts and increases in absolute monocyte counts; several patients also developed a variety of ophthalmological side-effects (Sanda *et al.*, 1992). One patient with metastatic renal cell carcinoma had a complete response (Sanda *et al.*, 1992). Cohorts of 4–7 patients with malignancy received M-CSF by subcutaneous administration at doses of 0.1–25.6 mg/m^2 per day on days 1–5 and 8–12 of 2-week cycles which were repeated at 28-day intervals. Forty-four patients received 88 cycles of M-CSF. Toxicity included thrombocytopenia and iritis at a dose of 25.6 mg/m^2 per day. No clinical response was reported (Bukowski *et al.*, 1994). A phase I study of M-CSF in combination with mouse R24 monoclonal antibody against CD3 ganglioside was conducted in 19 patients with metastatic melanoma. Monocytosis, as well as transient thrombocytopenia were observed in the majority of patients. Antitumor activity was observed in three patients who then received a second course of therapy (Minasian *et al.*, 1995). A phase I study of M-CSF in combination with the murine D612 monoclonal antibody was conducted in 14 patients with gastrointestinal cancers. Increases in CD16$^+$ monocytes were demonstrated after M-CSF treatment (Saleh *et al.*, 1995). The combination of M-CSF and IFN-γ has also been evaluated (Weiner *et al.*, 1994). M-CSF was administered at doses of 10–140 μg/kg per day by continuous intravenous infusion for 14 days. IFN-γ was administered subcutaneously at doses of 0.05–0.1 mg/m^2 per day on days 8–14 of the M-CSF regimen. The maximum tolerated dose (MTD) of M-CSF was 120 μg/kg per day. The coadministration of IFN-γ did not alter the MTD compared to the single agent use of M-CSF. There was a dose-dependent increase in peripheral monocytes and a decrease in platelet counts at M-CSF doses exceeding 40 μg/kg per day. A partial clinical response was reported for a patient with metastatic renal cell carcinoma and minor responses were reported in patients with a diffuse/follicular lymphoma, metastatic renal cell carcinoma and metastatic thymoma. A phase II trial of M-CSF administered by rapid intravenous infusion was undertaken in patients with metastatic soft-tissue sarcoma. One partial response was observed in a patient with metastatic leiomyosarcoma of the small bowel (Momin *et al.*, 1994).

9.2   Infectious Diseases

Phase I and phase II studies have been completed evaluating M-CSF in patients with invasive *Candida* and *Aspergillus* fungal infections (Nemunaitis *et al.*, 1991) using daily dosages of M-CSF ranging from 100 to 2000 μg/m^2 per day in combination with conventional antifungal therapy (amphotericin, fluconazole). Six patients had resolution of their infections, 12 were not evaluable for their response, and 6 did not respond. Ten patients survived 100 days after the initiation of M-CSF therapy and 14 died. Long-term follow-up of 46 consecutive BMT patients who were given M-CSF, administered 0–28 days after determination of progressive fungal disease, showed that survival of patients who received M-CSF was greater than that of historical patients (27% vs 5%). In a phase II study (Schiller *et al.*, 1994), the ability of either M-CSF (2000 μg/m^2 intravenously over 2 h for 4–30 days) or placebo to augment standard antifungal therapy was evaluated in 25 patients. Long-term survival showed a 19% mortality reduction in patients treated with M-CSF (58% mortality with placebo vs 47% with M-CSF).

9.3   Cholesterol Lowering And Atherosclerosis

Homozygous familial hypercholesterolemic patients have received M-CSF by continuous intravenous infusion for 14 days at doses of 10, 20, 40, or 60 μg/kg per day. Dose-dependent decreases in both total cholesterol and LDL-cholesterol were observed (Sherman *et al.*, 1995). At a dose of 60 μg/kg per day, mean decreases of 29% and 33% in total cholesterol and LDL-cholesterol, respectively, were seen. One patient treated at 60 μg/kg per day had a maximum decrease in LDL-cholesterol of 48%.

9.4   Pharmacokinetics Of M-CSF

The pharmacokinetics of M-CSF have been evaluated in rodents, nonhuman primates and humans (Bartocci *et al.*, 1987; Garnick and Stoudemire, 1990; Stoudemire and Garnick; 1990, Redman *et al.*, 1992; Zamkoff *et al.*, 1992; Bauer *et al.*, 1994). Macrophages appear to selectively clear M-CSF by receptor-mediated endocytosis in the mouse. Greater than 94% of injected M-CSF has been found to be cleared by the liver and splenic macrophages (Bartocci *et al.*, 1987). The role of the liver and spleen in M-CSF clearance has been further elucidated through biodistribution studies which have found that spleen and liver show the highest accumulation of ^{125}I-M-CSF, while less than 2% of the injected dose was recovered in the kidney and other organs (Garnick and Stoudemire, 1990). A slightly different clearance mechanism has been reported for *E. coli*-derived recombinant M-CSF (a nonglycosylated 49 kDa form). This lower-molecular-mass variant of M-CSF has a first-order elimination by the kidney (Bauer *et al.*, 1994). These authors suggest that, since intact M-CSF has been isolated from urine, glomerular filtration and renal excretion could be a primary mechanism of M-CSF elimination. However, they do not discuss a major molecular mass and charge difference which exists between endogenous M-CSF and CHO-derived recombinant M-CSF compared to the *E. coli*-derived material. These molecular mass and charge differences would make glomerular filtration of this molecule unlikely. Our recent unpublished observation, using immunohistochemistry, that M-CSF protein is localized constitutively in the uroepithelium suggests that epithelial synthesis and release is a more likely source of urinary M-CSF. Plasma clearance curves of M-CSF are best fitted to an exponential function of the form $C(T) = Ae^{-at}$. At a dose of 100 µg/kg, the pharmacokinetic parameters derived from this function indicate a clearance rate of 22 ml/h and 3.2 ml/h and a plasma half-life of approximately 6.4 h and 2.4 h in the primate and rat, respectively (Stoudemire and Garnick, 1990). The clearance rate of M-CSF was dose dependent. As the dose increased from 5 to 1000 µg/kg there was a decrease in the clearance rate from 0.34 to 0.013 ml/min and an increase in plasma half-life from 23 to 578 min in the rat (Stoudemire and Garnick, 1990). Carrageenan injection has been shown to reduce this saturable clearance, suggesting that macrophages are responsible for this mechanism (Bauer *et al.*, 1994). To evaluate the effect of monocytosis and M-CSF-mediated c-*fms* induction on the pharmacokinetic profile of M-CSF, the plasma clearance of M-CSF was evaluated in primates and rats following a 7-day treatment regimen which elevated monocytes 10-fold over baseline. Serum half-life was reduced to 2 h from 6.2 h in primates and to 3 min from 2.4 h in rats (Garnick and Stoudemire, 1990). In the rat studies, this elevated clearance rate was returned to normal by treatment with anti-c-*fms* receptor antibody (Timony, unpublished observation). Similar observations have been made in human studies which have reported lower blood levels of M-CSF after multiple doses of M-CSF (Redman *et al.*, 1992; Zamkoff *et al.*, 1992; Bukowski *et al.*, 1994).

10. Summary

Macrophage colony-stimulating factor is a pleiotropic cytokine with a variety of biological activities including activation and enhancement of survival of cells of the monocyte/macrophage lineage, increase of macrophage response to infectious microorganisms, anti-tumor activity, and enhancement of cholesterol metabolism. The availability of recombinant M-CSF has permitted extensive investigation of M-CSF in a variety of myelodysplastic conditions, cancer, infectious disease, and hypercholesterolemia. With the exception of lowering plasma cholesterol, M-CSF as a single agent has demonstrated mixed response in clinical studies. However, M-CSF has demonstrated greater efficacy when combined with other microbicidal or anticancer therapies. These observations support a potential role of M-CSF as a pharmaceutical agent in atherosclerosis, oncology, and infectious diseases.

11. References

Aggerwal, B.B. and Gutterman, J.V. (eds.) (1991). Human cytokines. Blackwell, Oxford, UK.

Arceci, R.J., Shanahan, F., Stanley, E.R. and Pollard, J.W. (1989). Temporal expression and location of colony-stimulating factor 1 (CSF-1) and its receptor in the female reproductive tract are consistent with CSF-1-regulated placental development. Proc. Natl Acad. Sci. USA 86, 8818–8822.

Armstrong, D.T.A. and Chaouat, L.G. (1989). Effects of lymphokines and immune complexes on murine placental cell growth in vitro. Biol. Reprod. 40, 466–474.

Athanassakis, I., Bleackley, R.C., Paetkau, V., Guibert, L., Barr, P.J. and Wegmann, T.G. (1987). The immunostimulatory effects of T cells and T-cell lymphokines on murine fetally derived placental cells. J. Immunol. 138, 37–44.

Bartocci, A., Pollard, J.W. and Stanley, E.R. (1986). Regulation of colony-stimulating factor-1 during pregnancy. J. Exp. Med. 164, 956–961.

Bartocci, A., Mastrogiannis, D.S., Migliorati, G., Stockert, R.J., Wolkoff, W.A. and Stanley, E.R. (1987). Macrophages specifically regulate the concentration of their own growth factor in the circulation. Proc. Natl Acad. Sci. USA 84, 6179–6183.

Bauer, R.J., Gibbons, J.A., Bell, D.P., Luo, Z.P. and Young, J.D. (1994). Nonlinear pharmacokinetics of recombinant macrophage colony-stimulating factor (M-CSF) in rats. J. Pharmacol. Exp. Ther. 268, 152–158.

Becker, S.J., Devlin, R.B. and Haskill, J.S. (1988). Differential production of tumor necrosis factor (TNF), macrophage colony stimulating factor (CSF1) and interleukin-1 (IL1) by human alveolar macrophages. J. Leukocyte Biol. 45, 353–361.

Bergamini, A., Perno, C.F., Dini, L., *et al.* (1994). Macrophage colony-stimulating factor enhances the susceptibility of macrophages to infection by human immunodeficiency virus and reduces the activity of compounds that inhibit virus binding. Blood 84, 3405–3412.

Bock, S.N., Cameron, R.B., Kragel, P., Mule, J.J. and Rosenberg, S.A. (1991). Biological and antitumor effects of recombinant human macrophage colony-stimulating factor in mice. Cancer Res. 51, 2649–2654.

Brugger, W., Kreutz, M. and Andreesen, R. (1991). Macrophage colony-stimulating factor is required for human monocyte survival and acts as a cofactor for their terminal differentiation to macrophages in vitro. J. Leukocyte Biol. 49, 483–488.

Brummer, E., Nassar, F. and Stevens, D.A. (1994). Effects of macrophage colony-stimulating factor on anticryptococcal activity of bronchoalveolar macrophages: synergy with fluconazole killing. Antimicrob. Agents Chemother. 38, 2158–2161.

Brummer, E. and Stevens, D.A. (1994). Macrophage colony-stimulating factor induction of enhanced macrophage anticryptococcal activity: synergy with fluconazole for killing. J. Infect. Dis. 170, 173–179.

Bukowski, R.M., Budd, G.T., Gibbons, J.A., *et al.* (1994). Phase I trial of subcutaneous recombinant macrophage colony-stimulating factor: clinical and immunomodulatory effects. J. Clin. Oncol. 12, 97–106.

Cebon, J., Layton, J.E., Maher, D. and Morstyn, G. (1994). Endogenous hemopoietic growth factors in neutropenia and infection. Br. J. Hematol. 86, 265–274.

Cenci, E., Bartocci, A., Pucetti, P., Mocci, S., Stanley, E.R. and Bistoni, F. (1991). Macrophage stimulating-factor in murine candidiasis:serum and tissue levels during infection and protective effect of exogenous administration. Infect. Immun. 59, 868–872.

Cheers, C. and Stanley, E.R. (1988). Macrophage production during listeriosis:CSF-1 and CSF-1 binding to cells in genetically resistant and susceptible mice. Infect. Immun. 56, 2972–2978.

Cheers, C. and Young, A.M. (1987). Serum colony stimulating activity and colony forming cells in murine brucellosis: relationship to immunopathology. Microb. Path. 3, 185–194.

Cheers, C., Hill, M., Haigh, A.M. and Stanley, E.R. (1989). Stimulation of macrophage phagocytic but not bacterial activity by colony-stimulating factor 1. Infect. Immun. 57, 1512–1516.

Chen, H.E., Chang, S., Trub, T. and Neel, B.G. (1996). Regulation of colony-stimulating factor 1 receptor signaling by the SH2 domain-containing tyrosine phosphatases SHPTP1. Mol. Cell. Biol. 16, 3685–3697.

Clark, S. and Kamen, R. (1987). The human hematopoietic colony stimulating factors. Science 236, 1229–1237.

Clinton, S.K., Underwood, R., Hayes, L., Sherman, M.L., Kufe, D.W. and Libby, P. (1992). Macrophage colony-stimulating factor gene expression in vascular cells and in experimental and human atherosclerosis. Am. J. Pathol. 140, 301–316.

Cole, D.J., Sanda, M.G., Yang, J.C., *et al.* (1994). Phase I trial of recombinant human macrophage colony-stimulating factor administered by continuous intravenous infusion in patients with metastatic cancer. J. Natl Cancer Inst. 86, 39–45.

Courtneidge, S.A., Dhand, R., Pilat, D., Twamley, G.M., Waterfield, M.D. and Roussel, M.F. (1993). Activation of Src family kinases by colony stimulating factor-1 and their association with its receptor. EMBO J. 12, 943–950.

Crosier, P.S. and Clark, S.C. (1992). Basic biology of the hematopoietic growth factors. Semin. Oncol. 19, 349–361.

Daiter, E. and Pollard, J.W. (1992). Colony stimulating factor-1 (CSF-1) in pregnancy. Reprod. Med. Rev. 1, 83–97.

Daiter, E., Pampfer, S., Ueng, Y., Bared, D., Stanley, E.R. and Pollard, J.W. (1992). Expression of colony-stimulating factor-1 in the human uterus and placenta. J. Clin. Endocrinol. Metab. 74, 850–858.

Das, S.K. and Stanley, E.R. (1982). Structure–function studies of a colony stimulating factor (CSF-1). J. Biol. Chem. 257, 13679–13684.

Delamarter, J.F., Hession, C., Semon, D., Gough, N.M., Rothenbuhler, R. and Mermod, J.J. (1987). Nucleotide sequence of a cDNA encoding murine CSF-1 (macrophage-CSF). Nucleic Acids Res. 15, 2389–2390.

Doyle, A.G., Halliday, W.J., Barnett, C.J., Dunn, T.L. and Hume, D.A. (1992). Effect of recombinant macrophage colony-stimulating factor 1 on immunopathology of experimental brucellosis in mice. Infect. Immun. 60, 1465–1472.

Evan, R., Kamdar, S.J., Fuller, J.A. and Krupke, D.M. (1995). The potential role of the macrophage colony-stimulating factor, CSF-1, in inflammatory responses: characterization of macrophage cytokine gene expression. J. Leukocyte Biol. 58, 99–107.

Falk, L.A., Wahl, L.M. and Vogel, S.N. (1988). Analysis of Ia antigen expression in macrophage-colony stimulating factor. J. Immunol. 140, 2652–2660.

Filonzi, E.L., Zoellner, H., Stanton, H. and Hamilton, J.A. (1993). Cytokine regulation of granulocyte-macrophage colony stimulating factor and macrophage colony-stimulating factor production in human arterial smooth muscle cells. Atherosclerosis 99, 241–252.

Garnick, M.B. and Stoudemire, J.B. (1990). Preclinical and clinical evaluation of recombinant human macrophage colony stimulating factor (rhM-CSF). Int. J. Cell Cloning 8, 356–373.

Gendleman, H.E., Orenstein, J.M., Martin, M.A., *et al.* (1988). Efficient isolation and propagation of human immunodeficiency virus on recombinant colony-stimulating factor 1-treated monocytes. J. Exp. Med. 167, 1428–1441.

Gregory, S.H. and Wing, E.J. (1993). Macrophage colony stimulating factor and the enhanced migration of monocytes are essential in primary but not secondary host defenses to *Listeria* organisms. J. Infect. Dis. 168, 934–942.

Gregory, S.H., Wing, E.J., Tweardy, D.J., Shadduck, R.K. and Lin, H.-S. (1992). Primary listerial infections are exacerbated in mice administered neutralizing antibody to macrophage colony-stimulating factor. J. Immunol. 149, 188–193.

Hamilton, J.A., Filonzi, E.L. and Ianches, G. (1993). Regulation of macrophage colony-stimulating factor (M-CSF) production in cultured human synovial fibroblasts. Growth Factors 9, 157–165.

Hanamura, T., Motoyoshi, K., Yoshida, K., *et al.* (1988). Quantification and identification of human monocyte colony-stimulating factor in human serum by enzyme linked immunosorbent assay. Blood 72, 886–892.

Heldin, C.-H. (1995). Dimerization of cell surface receptors in signal transduction. Cell 80, 213–223.

Ho, J.L., Reed, S.G., Wick, E.A. and Giodana, M. (1990). Granulocyte-macrophage and macrophage colony-stimulating factors activate intramacrophage killing of *Leishmania mexicana amazonensis*. J. Infect. Dis. 162, 224–230.

Ho, R.J.Y., Chong, K.T. and Merigan, T.C. (1991). Antiviral activity and dose optimum of recombinant macrophage colony-stimulating factor on herpes simplex genitalis in guinea pigs. J. Immunol. 146, 3578–3582.

Horiguchi, J., Warren, M.K. and Kufe, D. (1987). Expression of the macrophage-specific colony-stimulating factor in human monocytes treated with granulocyte-macrophage colony-stimulating factor. Blood 69, 1259–1261.

Horiguchi, J., Sherman, M.L., Sampson-Johannes, A., Weber, B.L. and Kufe, D.W. (1988). CSF-1 and C-FMS gene expression in human carcinoma cell lines. Biochem. Biophys. Res. Commun. 157, 395–401.

Hume, D.A. and Denkins, Y. (1992). The deleterious effect of macrophage colony-stimulating factor (CSF-1) on the pathology of experimental candidiasis in mice. Lymphokine Cytokine Res. 11, 92–98.

Inaba, T., Yamada, N., Gotoda, T., et al. (1992). Expression of M-CSF receptor encoded by c-fms on smooth muscle cells derived from arteriosclerotic lesion. J. Biol. Chem. 267, 5693–5699.

Inoue, I., Inaba, T., Motoyoshi, K., et al. (1992). Macrophage colony stimulating factor prevents the progression of atherosclerosis in Watanabe heritable hyperlipidemic rabbits. Atherosclerosis 93, 245–254.

Jakubowski, A.A., Bajorin, D.F., Templeton, M.A., et al. (1996). A Phase 1 study of continuous infusion of recombinant macrophage colony-stimulating factor in patients with metastatic melanoma. Clin. Cancer Res. 2, 295–302.

Kalter, D.C., Nakamura, M., Turpin, J.A., et al. (1991). Enhanced HIV replication in macrophage colony-stimulating factor-treated monocytes. J. Immunol. 146, 298–306.

Kanazaki, H., Crainie, M., Lin, H., et al. (1991). The in situ expression of granulocyte-macrophage colony-stimulating factor (GM-CSF) mRNA at the maternal-fetal interface. Growth Factors 5, 69–74.

Kanazaki, H., Yui, J., Iwai, M. and Imai, K. (1992). The expression and localization of mRNA for colony-stimulating factor (CSF-1) in human term placenta. Hum. Reprod. 7, 563–567.

Karbassi, A., Becker, J.M., Foster, J.S. and Moore, R.N. (1987). Enhanced killing of Candida albicans by murine macrophages treated with macrophage colony-stimulating factor: evidence for augmented expression of mannose receptors. J. Immunol. 139, 417–421.

Kawasaki, E.S., Ladner, M.B., Wang, A.M., et al. (1985). Molecular cloning of a complementary DNA encoding human macrophage-specific colony-stimulating factor (CSF-1). Science 230, 291–296.

Kayashima, S., Tsuru, S., Shinomiya, N., et al. (1991). Effects of macrophage colony-stimulating factor on reduction of viable bacteria and survival of mice during Listeria monocytogenes infection: characteristics of monocyte subpopulations. Infect. Immun. 59, 4677–4680.

Khemani, S., Brummer, E. and Stevens, D.A. (1995). In vivo and in vitro effects of macrophage colony-stimulating factor (M-CSF) on bronchoalveolar macrophages for antihistoplasmal activity. Int. J. Immunopharmacol. 17, 49–53.

Khjwaja, A., Johnson, B., Addison, I.E., et al. (1991). In vivo effects of macrophage colony-stimulating factor on human monocyte function. Br. J. Haematol. 77, 25–31.

Kimura, F., Suzu, S., Nakamura, Y., et al. (1994). Structural analysis of proteoglycan macrophage colony-stimulating factor. J. Biol. Chem. 269, 19751–19756.

Kitano, K., Abboud, C.N., Ryan, D.H., Quan, S.G., Baldwin, G.C. and Golde, D.W. (1991). Macrophage-active colony-stimulating factors enhance human immunodeficiency virus type 1 infection in bone marrow stem cells. Blood 77, 1699–1705.

Koyanagi, Y., O'Brien, W.A., Zhao, J.Q., Golde, D.W., Gasson, J.C. and Chen, I.S.Y. (1988). Cytokines alter production of HIV-1 from primary mononuclear phagocytes. Science 241, 1673–1675.

Ladner, M.B., Martin, G.A., Noble, J.A., et al. (1988). cDNA cloning and expression of murine macrophage colony-stimulating factor from L929 cells. Proc. Natl Acad. Sci. USA 85, 6706–6710.

Landegent, J.E., Kluck, P.M.C., Bolk, M.W.J. and Willemze, R. (1992). The human macrophage colony-stimulating factor gene is localized at chromosome 1 band p21 and not at 5q33.1. Hematology 64, 110–111.

Lee, M.T. and Warren, M.K. (1987). CSF-1 induced resistance to viral infection in murine macrophages. J. Immunol. 138, 3019–3022.

Li, Y. and Chen, B. (1995). Differential regulation of fyn-associated protein tyrosine kinase activity by macrophage colony-stimulating factor (M-CSF) and granulocyte-macrophage colony-stimulating factor (GM-CSF). J. Leukocyte Biol. 57, 484–490.

Lieschke, G.J., Stanley, E., Grail, D., et al. (1994). Mice lacking both macrophage- and granulocyte-macrophage colony-stimulating factor have macrophages and coexistent osteopetrosis and severe lung disease. Blood 84, 27–35.

Lu, L., Shen, R.-N., Lin, Z.-H., Aukerman, S.L., Ralph, P. and Broxmeyer, H.E. (1991). Anti-tumor effects of recombinant human macrophage colony-stimulating factor, alone or in combination with local irradiation, in mice inoculated with Lewis lung carcinoma cells. Int. J. Cancer. 47, 143–147.

Magee, D.M., Williams, D.M., Wing, E.J., Bleicker, C.A. and Schachter, J. (1991). Production of colony-stimulating factors during pneumonia caused by Chlamydia trachomatis. Infect. Immun. 59, 2370–2375.

Metcalf, D. (1970). Studies on colony formation in vitro by mouse bone marrow cells. II. Action of colony stimulating factor. J. Cell. Physiol. 76, 89–99.

Metcalf, D. (1985). The granulocyte-macrophage colony stimulating factors. Science 229, 16–22.

Minasian, L.M., Yao, T.J., Steffens, T.A., et al. (1995). A phase 1 study of anti-GD3 ganglioside monoclonal antibody R24 and recombinant human macrophage-colony stimulating factor in patients with metastatic melanoma. Cancer 75, 2251–2257.

Miyauchi, J., Wang, C., Kelleher, C.A., et al. (1988a). The effects of recombinant CSF-1 on the blast cells of acute myeloblastic leukemia in suspension culture. J. Cell. Physiol. 135, 55–62.

Miyauchi, J., Kelleher, C.A. Wong, G.G., et al. (1988b). The effects of combinations of the recombinant growth factors GM-CSF, G-CSF, IL-3 and CSF-1. Leukemia 2, 382–387.

Momin, F.A., Zalupski, M., Heilbrun, L.K., et al. (1994). Phase II trial of recombinant human macrophage colony-stimulating factor in metastatic soft tissue sarcoma. J. Immunother. 16, 224–228.

Morris, S.W., Valentine, M.B., Shapiro, D.N., et al. (1991). Reassignment of the human CSF1 gene to chromosome 1p13–p21. Blood 78, 2013–2020.

Motoyoshi, K. and Takaku, F. (1990). Human monocyte colony-stimulating factor (hM-CSF), Phase I/II clinical studies. Hum. Growth Factors Clin. Appl. 161–171.

Mufson, R.A., Aghajanian, T., Wong, G., Woodhouse, C. and Morgan, A.C. (1989). Macrophage colony stimulating factor enhances monocyte and macrophage antibody-dependent cell-mediated cytotoxicity. Cell. Immunol. 119, 182–192.

Munn, D.H. and Cheung, N.-K.V. (1989). Antibody-dependent antitumor cytotoxicity by human monocytes cultured with recombinant macrophage colony stimulating factor. J. Exp. Med. 170, 136–148.

Nassar, F., Brummer, E. and Stevens, D.A. (1994). Effect of in vivo macrophage colony-stimulating factor on fungistasis of bronchoalveolar and peritoneal macrophages against *Cryptococcus neoformans*. Antimicrob. Agents Chemother. 38, 2162–2164.

Nemunaitis, J., Meyers, J.D., Buckner, C.D., *et al.* (1991). Phase I trial of recombinant human macrophage colony-stimulating factor in patients with invasive fungal infections. Blood 78, 907–913.

Nemunaitis, J., Shannon-Dorcy, K., Appelbaum, F.R., *et al.* (1993). Long-term follow-up of patients with invasive fungal disease who received adjunctive therapy with recombinant human macrophage colony-stimulating factor. Blood 82, 1422–1427.

Newman, S.L. and Gootee, L. (1992). Colony-stimulating factors activate human macrophages to inhibit intracellular growth of *Histoplasma capsulatum*. Infect. Immun. 60, 4593–4597.

Orme, I.M., Roberts, A.D., Griffin, J.P. and Abrams, J.S. (1993). Cytokine secretion by CD4 T lymphocytes acquired in response to *Mycobacterium tuberculosis* infection. J. Immunol. 151, 518–525.

Pandit, J., Bohm, A., Jancarik, J., Halenbeck, R., Koths, K. and Kim, S.H. (1992). Three dimensional structure of dimeric human recombinant macrophage colony-stimulating factor. Science 258, 1358–1362.

Pebusque, M.J., Lopez, M., Torres, H., Carotti, A., Guilbert, L. and Mannoni, P. (1988). Growth response of human myeloid leukemia cells to colony stimulating factors. Exp. Hematol. 16, 360–366.

Perno, C.-F., Cooney, D.A., Gao, W.-Y., *et al.* (1992). Effects of bone marrow stimulatory cytokines on human immunodeficiency virus replication and the antiviral activity of dideoxynucleosides in cultures of monocyte/macrophages. Blood 80, 995–1003.

Perno, C.-F., Aquaro, S., Rosenwirth, B., *et al.* (1994). In vitro activity of inhibitors of late stages of the replication of HIV in chronically infected macrophages. J. Leukocyte Biol. 56, 381–386.

Petros, W.P., Rabinowitz, J., Stuart, A.R., Gupton, C., Alderman, E.M. and Peters, W.P. (1994). Elevated endogenous serum macrophage colony-stimulating factor in the early stage of fungemia following bone marrow transplantation. Exp. Hematol. 22, 582–586.

Pettenati, M.J., LeBeau, M.M. and Lemons, R.S. (1987). Assignment of CSF-1 to 5q33.1: evidence for clustering of genes regulating hematopoiesis and for their involvement in the deletion of the long arm of chromosome 5 in myeloid disorders. Proc. Natl Acad. Sci. USA 84, 2970–2974.

Pollard, J.W., Bartocci, A., Arceci, R., Orlofsky, A., Ladner, M.B. and Stanley, R. (1987). Apparent role of the macrophage growth factor, CSF-1, in placental development. Nature 330, 484–486.

Praloran, V., Gascan, H., Papin, S., Chevalier, S., Trossaert, M., and Boursier, M.C. (1990). Inducible production of macrophage colony-stimulating factor (CSF-1) by malignant and normal human T cells. Leukemia 4, 411–414.

Price, L.K.H., Choi, H.U., Rosenberg, L. and Stanley, E.R. (1992). The predominant form of secreted colony stimulating factor-1 is a proteoglycan. J. Biol. Chem. 267, 2190–2199.

Ralph, P. and Sampson-Johannes, A. (1990). Macrophage growth and stimulating factor, M-CSF. Hematopoietic Growth Factors Transfus. Med. 43–63.

Rambaldi, A., Young, D.C. and Griffin, J.D. (1987). Expression of the M-CSF (CSF-1) gene by human monocytes. Blood 69, 1409–1413.

Redman, B.G., Flaherty, L., Chou, T.H., *et al.* (1992). Phase I trial of recombinant macrophage colony-stimulating factor by rapid intravenous infusion in patients with cancer. J. Immunother. 12, 50–54.

Regenstreif, L.J. and Rossant, J. (1989). Expression of the c-fms proto-oncogene and the cytokine, CSF-1, during mouse embryogenesis. Dev. Biol. 133, 284–294.

Rettenmier, C.W. and Sherr, C.J. (1989). The Mononuclear Phagocyte Colony Stimulating Factor (CSF-1, M-CSF) in Hematopoietic Growth Factors (ed. D.W. Golde), Hematology/Oncology Clinics of North America 3, 479. W.B Saunders, Philadelphia, PA.

Rettenmier, C.W., Roussel, M.F. and Ashmun, R.A. (1987). Synthesis of membrane bound CSF-1 in NIH 3T3 cells transformed by cotransfection of the human CSF-1 and c-fms (CSF-1 receptor) genes. Mol. Cell. Biol. 7, 2378–2387.

Rettenmier, C.W., Roussel, M.F. and Sherr, C.J. (1988). The colony-stimulating factor 1 (CSF1) receptor (c-fms proto-oncogene products) and its ligand. J. Cell. Sci. Suppl. 9, 27–34.

Rose, R.M., Fuglestad, J.M. and Remington, L. (1991). Growth inhibition of *Mycobacterium avium* complex in human alveolar macrophages by the combination of recombinant macrophage colony-stimulating factor and interferon-gamma. Am. J. Respir. Cell Mol. Biol. 4, 248–254.

Rosenfeld, M.E., Yla-Herttuala, S., Lipton, B.A., Ord, V.A., Witztum, J.L. and Steinberg, D. (1992). Macrophage colony-stimulating factor mRNA and protein in atherosclerotic lesions of rabbits and humans. Am. J. Pathol. 140, 291–300.

Roussel, M.F. (1994). Signal transduction by the macrophage-colony stimulating factor receptor (CSF-1R). J. Cell. Sci. Suppl. 18, 105–108.

Roussel, M.F., Sherr, C.J., Barker, P.E. and Ruddle, F.H. (1983). Molecular cloning of the c-fms locus and its assignment to human chromosome 5. J. Virol. 48, 770–773.

Saito, S., Motoyoshi, K., Ichijo, M., Saito, M. and Takaku, F. (1992). High serum human macrophage colony-stimulating factor level during pregnancy. Int. J. Hematol. 55, 219–225.

Saleh, M.N., Goldman, S.J., LoBuglio, A.F., *et al.* (1995). CD16[+] monocytes in patients with cancer: spontaneous elevation and pharmacologic induction by recombinant human macrophage colony-stimulating factor. Blood 85, 2910–2917.

Saltman, D.L., Dolganov, G.M., Hinton, L.M. and Lovett, M. (1992). Reassignment of the human macrophage colony stimulating factor gene to chromosome 1p13–21. Biochem. Biophys. Res. Commun. 182, 1139–1143.

Sanda, M.G., Yang, J.C., Topalian, S.L., *et al.* (1992). Intravenous administration of recombinant human macrophage colony-stimulating factor to patients with metastatic cancer: a phase 1 study. J. Clin. Oncol. 10, 1643–1649.

Schaub, R.G., Bree, M.P., Hayes, L.L., *et al.* (1994). Recombinant human macrophage colony-stimulating factor reduces plasma cholesterol and carrageenan granuloma foam cell formation in Watanabe heritable hyperlipidemic rabbits. Arterioscler. Thromb. 14, 70–76.

Schaub, R.G., Donnelly, L., Parker, T.S., Clinton, S.K. and Garnick, M.B. (1995). The protective role of the macrophage in atherogenesis: insight from using M-CSF. In "Atherosclerosis X" (eds. F.P. Woodford, J. Davignon and A. Sniderman), pp. 537–544. Elsevier, Amsterdam.

Schiller, G., O'Neill, C., Lee, M., et al. (1994). A phase II study of placebo versus recombinant human macrophage colony-stimulating factor to augment antifungal therapy in patients with invasive Candida or Aspergillus fungal infection. Blood 84 (supplement 1), 504A.

Scholl, S.M., Pallud, C., Beuvon, F., et al. (1994). Anti-colony stimulating factor-1 antibody staining in primary breast adenocarcinomas correlates with marked inflammatory cell infiltrates and prognosis. J. Natl Cancer Inst. 86, 120–126.

Sherman, M.L., Weber, B.L., Datta, R. and Kufe, D.W. (1990). Transcriptional and posttranscriptional regulation of macrophage-specific colony stimulating factor gene expression by tumor necrosis factor. J. Clin. Invest. 85, 442–447.

Sherman, M.L., Stein, E.A., Isaacsohn, J., Lees, R.S. and Garnick, M.B. (1995). Role of recombinant human macrophage colony stimulating factor (rhM-CSF) in cholesterol metabolism and atherosclerosis. In "Atherosclerosis X" (eds. F.P. Woodford, J. Davignon and A. Sniderman), pp. 598–602. Elsevier, Amsterdam.

Sherr, C. (1990). Colony-stimulating factor-1 receptor. Blood 75, 1–12.

Sherr, C.J., Rettenmier, C.W., Sacca, R., Roussel, M.F., Look, A.T. and Stanley, E.R. (1985). The c-fms proto-oncogene product is related to the receptor for the mononuclear phagocyte growth factor CSF-1. Cell 41, 665–676.

Shimada, M., Inaba, T., Shimano, H., et al. (1992). Platelet-derived growth factor BB-dimer suppresses the expression of macrophage colony-stimulating factor in human vascular smooth muscle cells. J. Biol. Chem. 267, 15455–15458.

Shiratsuchi, H., Johnson, J.L. and Ellner, J.J. (1991). Bidirectional effects of cytokines on the growth of Mycobacterium avium within human macrophages. J. Immunol. 146, 3165–3170.

Shurtleff, S.A., Downing, J.R., Rock, C.O., Hawkins, S.A., Roussel, M.F. and Sherr, C.J. (1990). Structural features of the colony-stimulating factor 1 receptor that affect its association with phosphatidylinositol 3-kinase. EMBO J. 9, 2415–2421.

Shyy, Y.-J., Wickham, L.L., Hagan, J.P., et al. (1993). Human monocyte colony-stimulating factor stimulates the gene expression of monocyte chemotactic protein-1 and increases the adhesion of monocytes to endothelial monolayer. J. Clin. Invest. 92, 1745–1751.

Stanley, E.R. (1984). The macrophage colony-stimulating factor, CSF-1. Methods Enzymol. 116, 5026–5032.

Stanley, E.R. and Heard, P.M. (1977). Factors regulating macrophage production and growth. Purification of some properties of the colony stimulating factor from medium conditioned by mouse L cells. J. Biol. Chem. 252, 4305–4312.

Stanley, E.R., Guilbert, L.J. and Tushinski, R.J. (1983). CSF-1, a mononuclear phagocyte lineage-specific hemopoietic growth factor. J. Cell. Biochem. 21, 151–159.

Stoudemire, J.B. and Garnick, M.B. (1990). M-CSF. In vivo evaluations of recombinant human macrophage colony-stimulating factor. In "Cytokines in Hemopoiesis, Oncology, and AIDS" (eds. Freund/Kink/Welte), pp. 468–469. Springer-Verlag, Berlin, Heidelberg.

Suzuki, T., Nagata, K., Murohashi, I. and Nara, N. (1988). Effect of recombinant human M-CSF on the proliferation of leukemic blast progenitors in AML patients. Leukemia 2, 358–362.

Takahashi, M., Yeong-Man, H. and Setsuko, Y. (1988). Macrophage colony-stimulating factor is produced by human T lymphoblastoid cell line, CEM-ON: identification by amino-terminal amino acid sequence analysis. Biochem. Biophys. Res. Commun. 152, 1401–1409.

Teicher, B.A., Gulshan, A., Menon, K. and Schaub, R.G. (1995). In vivo studies with interleukin-12 alone and in combination with macrophage-colony stimulating factor and/or fractionated radiation treatment. Int. J. Cancer 63, 1–5.

Tsukui, T., Kikuchi, K., Mabuchi, A., et al. (1992). Production of macrophage colony-stimulating factor by adult murine parenchymal liver cells (hepatocytes). J. Leukocyte Biol. 52, 383–389.

Tushinski, R.J., Oliver, I.T. and Guilbert, L.J. (1982). Survival of mononuclear phagocytes depends on a lineage-specific growth factor that the differentiation cells selectively destroy. Cell 28, 71–81.

Uzumake, H., Okabe, T., Sasaki, N., et al. (1989). Identification and characterization of receptors for granulocyte colony-stimulating factors on human placenta and trophoblast cells. Proc. Natl Acad. Sci. USA 86, 9323–9326.

VandePol, C.J. and Garnick, M.B. (1991). Clinical applications of recombinant macrophage-colony stimulating factor (rhM-CSF). Biotech. Ther. 2, 231–239.

Vitt, C.R., Fidler, J.M., Ando, D., Zimmerman, R.J. and Aukerman, S.L. (1994). Antifungal activity of recombinant human macrophage colony-stimulating factor in models of acute and chronic candidiasis in the rat. J. Infect. Dis. 169, 369–374.

Walsh, P., Dorner, A., Duke, R.C., Su, L.-J. and Glode, M.L. (1995). Macrophage colony-stimulating factor complementary DNA: a candidate for gene therapy in metastatic melanoma. J. Natl Cancer Inst. 87, 809–816.

Wang, M., Freidman, H. and Djeu, J.Y. (1989). Enhancement of human monocyte function against Candida albicans by the colony-stimulating factors (CSF): IL-3, granulocyte-macrophage-CSF and macrophage-CSF. Immunology 143, 671–677.

Wang, Z., Myles, G.M., Brandt, C.S., Lioubin, M.N. and Rohrschnieider, L. (1993). Identification of the ligand-binding regions in the macrophage colony-stimulating factor receptor extracellular domain. Mol. Cell. Biol. 13, 5348–5359.

Weiner, L.M., Holmes, M., Catolano, R.B., Dovnarsky, M., Padavic, K. and Alpaugh, R.K. (1994). Phase I trial of recombinant macrophage colony-stimulating factor and recombinant gamma-interferon: toxicity, monocytosis, and clinical effects. Cancer Res. 54, 4084–4090.

Wong, G.G., Temple, P.A. and Leary, A.C. (1987). Human CSF-1; molecular cloning and expression of 4 kb cDNA encoding the human urinary protein. Science 235, 1504–1508.

Yamada, N., Ishibashi, S., Shimano, H., et al. (1992). Role of monocyte colony-stimulating factor in foam cell generation. Proc. Soc. Exp. Biol. Med. 200, 240–244.

Yong, K., Salooja, N., Donahue, R.E., Hegde, U. and Linch, D.C. (1992). Human macrophage colony-stimulating factor levels are elevated in pregnancy and in immune thrombocytopenia. Blood 180, 2897–2902.

Zamkoff, K.W., Hudson, J., Groves, E.S., Childs, A., Konrad, M. and Rudolph, A.R. (1992). A phase I trial of recombinant human macrophage colony-stimulating factor by rapid intravenous infusion in patients with refractory malignancy. J. Immunother. 11, 103–110.

19. Granulocyte-Macrophage Colony-stimulating Factor

S. Devereux *and* D.C. Linch

1. Introduction

Granulocyte-macrophage colony-stimulating factor (GM-CSF) is one of the four classical colony-stimulating factors discovered in the 1960s following the introduction of semi-solid bone marrow culture techniques (Bradley and Metcalf, 1966; Pluznik and Sachs, 1965). The cDNA for human GM-CSF was isolated by expression cloning from the Mo leukemic cell line (Wong *et al.*, 1985) and a T lymphocyte line (Lee *et al.*, 1985) and the genomic sequence was ascertained shortly afterwards (Kaushansky *et al.*, 1986; Mitayake *et al.*, 1985). The availability of recombinant material over the last decade has allowed extensive characterization of the biochemical, biological, and clinical properties of GM-CSF which are summarized in this chapter. Although originally identified because of its effects on hematopoietic progenitors, GM-CSF has proved to be pleiotropic in activity with many of its most important effects being exerted on fully differentiated cells.

2. The Cytokine Gene

2.1 GENOMIC ORGANIZATION

The human GM-CSF gene is located on chromosome 5 (q25-q31) within 10 kb of the interleukin-3 (IL-3) gene (Yang *et al.*, 1988) and in the same region as the genes for IL-5, IL-4, and the M-CSF receptor (Van Leeuwe *et al.*,

1989). This finding is of interest because this region is deleted in some cases of myelodysplasia and acute myeloid leukemia (Huebner *et al.*, 1985). The human GM-CSF gene is 2.5 kb in size, contains three introns (Kaushansky *et al.*, 1986), and shows considerable homology to its murine counterpart, particularly in the 5′ untranslated region (Figure 19.1) (Mitayake *et al.*, 1985).

2.2 REGULATION OF GENE EXPRESSION

GM-CSF is produced in response to immunological and inflammatory signals in a range of cell types. Expression is controlled at the transcriptional level through constitutive and inducible cis-acting elements and post-transcriptionally by changes in mRNA half-life mediated through AUUUA sequences in the 3′ untranslated region of the gene.

2.2.1 GM-CSF Promoter–Enhancer

Following T cell activation there is a rapid induction of a number of cytokine genes including GM-CSF. Several

conserved regulatory elements have been identified in the 5′ untranslated region of the GM-CSF gene (Figures 19.1 and 19.2). These include the CLE0 (conserved lymphokine element 0)/PB-1 (purine box 1) sequence which contains direct repeats of the CATT(A/T) motif and is responsible for cyclosporin A-sensitive GM-CSF transcription in T cells stimulated with phorbol ester and calcium ionophore (Nimer *et al.*, 1990). A protein complex similar to NF-AT (nuclear factor-activated T cells) present in activated but not resting T cells binds to this element and contains the fos/jun heterodimer AP-1 and an Ets family member Elf-1 (Fraser *et al.*, 1994; Wang *et al.*, 1994).

Further conserved elements have been defined in the GM-CSF gene including CLE2/GC and CLE1 motifs (also known as cytokine consensus elements, CK-1 and CK-2). The CLE2/CK-2 element, which is also found in the IL-3 gene, binds a factor induced mainly by phorbol ester designated NF-GM2 that is immunologically identical to NFκB and the GC element which binds constitutive factors such as Sp1 (Tsuboi *et al.*,

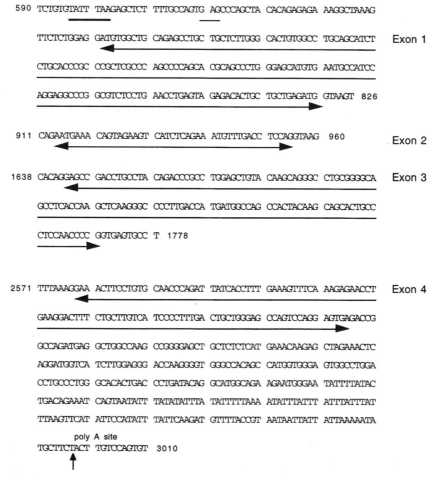

Figure 19.1   Sequence of the GM-CSF gene. Nucleotide (nt) sequence of portions of the GM-CSF gene showing TATA box at nt 597–603 (bold underlined), transcription initiation site nt 620–622 (underlined), and exons (double-headed arrows). The polyadenylation site is indicated by a vertical arrow at nt 2998.

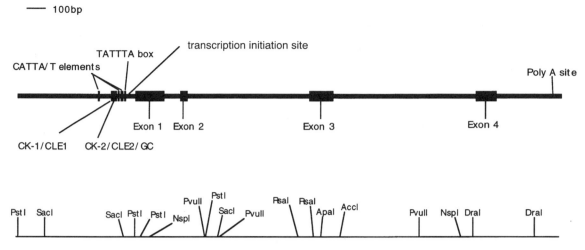

Figure 19.2   GM-CSF genomic organization. Genomic organization of GM-CSF receptor gene showing the location of regulatory sequences and selected restriction enzyme sites.

1991). Studies in which ERK-1 was overexpressed in Jurkat cells show that MAP kinases play a role in the formation of both NF-AT and NFκB transcription factors (Park and Levitt, 1993). Further upstream the CLE1/CK-1 element, which is also found in the IL-3, G-CSF, and IL-2 genes, binds a ubiquitous nuclear protein NF-GMa which is induced in tumor necrosis factor (TNF)-stimulated fibroblasts (Shannon et al., 1988, 1990). In addition, a strong cyclosporin-sensitive enhancer element has been identified 3 kb upstream of the GM-CSF gene. These sites bind the inducible transcription factor AP-1 and may contribute to the coordinate expression of IL-3 and GM-CSF in activated T cells (Cockerill et al., 1993).

2.2.2   Post-transcriptional Control

In common with other growth factor and protooncogene RNAs, AU-rich sequences have been identified in the 3′ untranslated region which mediate rapid RNA degradation (Shaw and Kamen, 1986; Wilson and Treisman, 1988). Following stimulation of fibroblasts or T cells, a protein factor is induced which binds to this element and increases mRNA half-life (Malter et al., 1990). The mechanism of action of the destabilizing AU element is unknown but it may promote deadenylation of mRNA and removal of the 5′ cap, allowing degradation by cellular exonuclease in a 5′ → 3′ fashion (Mulhrad et al., 1994).

3.   The GM-CSF Protein

Human GM-CSF protein contains 144-amino-acid (aa) residues including a 17-aa signal peptide that is removed during secretion (Gough et al., 1985; Lee et al., 1985;

Wong et al., 1985) (Figure 19.3). Despite more than 50% homology between the proteins there is no functional cross-reactivity between human and murine GM-CSF. Native GM-CSF is heavily glycosylated with two potential N-linked and several O-linked sites, resulting in a molecular mass of 14.5–34 kDa (Cebon et al., 1990). Compared to bacterially derived, nonglycosylated recombinant GM-CSF, glycosylated forms have a lower specific activity and bind the receptor with a lower affinity (Cebon et al., 1990). When human GM-CSF separated on the basis of the number of N-linked glycosylation sites occupied was administered to rats, it was shown that the 2N type had a 5-fold longer half-life than the 0N type (Okamoto et al., 1991). Parallel in vitro studies demonstrated that the 2N GM-CSF had a 200-fold lower biological activity in vitro. Changes in glycosylation status may also result in altered immunogenicity of recombinant GM-CSF from different sources (Gribben et al., 1990). There are two intrachain disulfide bonds in human GM-CSF, C54-C96, and C88-C121, but only the first is necessary for biological activity (Shanafelt and Kastlein, 1989). The isoelectric point of human GM-CSF is 4.2 (Nicola et al., 1979).

The crystal structure of human GM-CSF has been determined (Diederichs et al., 1991; Walter et al., 1992), revealing a tertiary structure similar to many other cytokines with four α-helices arranged in an "up–up, down–down" topology (Figure 19.4).

A range of methods have been employed to elucidate structure–function relationships. These have included the synthesis of truncated GM-CSF molecules (Clark-Lewis et al., 1988), construction of interspecies hybrids (Kaushansky et al., 1989; Shanafelt et al., 1991), site-directed mutagenesis (Hercus et al., 1994; Lopez et al., 1992), and mapping the epitopes to which monoclonal antibodies bind (C.B. Brown et al., 1990, 1994;

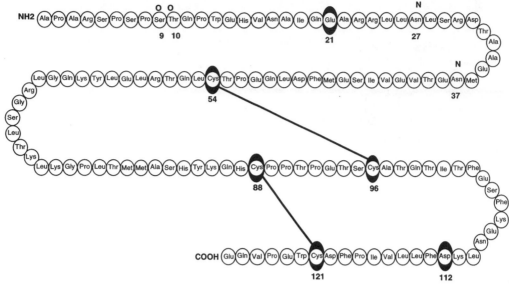

Figure 19.3   GM-CSF protein sequence. Protein sequence of GM-CSF showing intrachain disulfide bonds and location of Glu-21 and Asp-112, which are critical for binding to receptor β and α subunits. The *N*-linked glycosylation sites (Asn-27 and Asn-37) and known *O*-linked sites (Ser-9 and Thr-10) are marked N and O.

Figure 19.4   3D structure of human GM-CSF. Ribbon diagram of GM-CSF structure generated from the coordinates using the RasMac molecular graphics program. The up–up, down–down topology of α-helices is evident.

Kanakura *et al.*, 1991). These studies have shown that the first and third helices are involved in the interaction with the affinity converting/signal transducing β subunit of the receptor and that Glu-21 is critical in this regard. Residues in the fourth helix, particularly Asp-112 are involved in interactions with the ligand binding α subunit of the GM-CSF receptor (Hercus *et al.*, 1994).

4.   Cellular Sources and Production

A wide range of cell types produce GM-CSF either constitutively or in response to a variety of immunological and inflammatory stimuli. GM-CSF production occurs in T lymphocytes stimulated via the antigen receptor or phytohemagglutinin (Cline and Golde, 1974; Wong *et al.*, 1985), or IL-1 (Herrmann *et al.*, 1988). Endotoxin stimulates its production in B lymphocytes (Pluznik *et al.*, 1989). Other hematopoietic cells are capable of producing GM-CSF: cross-linking of the Fcε receptor or calcium ionophore causes mast cells to release IL-3 and GM-CSF (Plaut *et al.*, 1989) and monocytes express GM-CSF mRNA in response to phagocytosis and inflammatory stimuli such as TNF (Thorens *et al.*, 1987). A number of other cells of mesenchymal origin including endothelial cells and fibroblasts produce GM-CSF in response to TNF, whilst mesothelial cells produce GM-CSF after stimulation with TNF and epidermal growth factor (EGF), and osteoblasts (Horowitz *et al.*, 1989) can be induced to secrete GM-CSF in response to parathormone and endotoxin. Both mRNA and protein for GM-CSF and its receptor have been demonstrated in fallopian tube epithelial cells, leading to speculation that it may play a role in the normal fertilization/implantation process (Zhao and Chegini, 1994).

5.   Biological Activities

The actions of GM-CSF have been studied extensively *in vitro* as well as *in vivo*, in animals and man. As with other cytokines, its effects are to stimulate the survival, function, proliferation, and differentiation of target cells. These effects may be direct or may serve to enhance or "prime" responses to other agonists. Although earlier studies focused on its actions on immature cells, there is now a large amount of information on the effects of GM-CSF on differentiated cells.

5.1 IN VITRO ACTIONS

5.1.1 Effects on Survival

In the absence of stromal cells or cytokines, primary hematopoietic progenitor cells and factor-dependent cell lines die rapidly by apoptosis (Williams et al., 1990). Signaling through the IL-3/GM-CSF receptor can suppress apoptotic cell death in murine BaF3 cells transfected with the relevant human receptor subunits, an effect that seems to depend on the Ras/MAP kinase pathway (Kinoshita et al., 1995). GM-CSF also stimulates cell membrane functions necessary for maintenance of viability including glucose transport (Hamilton et al., 1988) and the Na^+/K^+ antiporter (Vairo and Hamilton, 1988).

5.1.2 Effects on Function

GM-CSF has been shown to enhance a wide range of effector functions in differentiated cells of neutrophil monocyte and eosinophil lineages.

Following exposure to GM-CSF, neutrophils undergo a rapid shape change associated with membrane ruffling (Lopez et al., 1986). There is upregulation of a number of cell surface molecules including the fmet-leu-phe (FMLP) (Weisbart et al., 1986) and inositol triphosphate receptors (Bradford et al., 1992) and integrins both in vitro (Arnaout et al., 1986) and in vivo (Devereux et al., 1989). Release of secondary granule products also occurs in vitro (Kaufman et al., 1989; Lopez et al., 1986) and in vivo (Devereux et al., 1990), and there is mRNA synthesis and secretion of various cytokines, including IL-1, G-CSF, and M-CSF (Lindemann et al., 1988, 1989). A number of other neutrophil functions are primed or enhanced by GM-CSF including superoxide generation (Weisbart et al., 1985) and release of platelet-activating factor (PAF) (Wirthmueller et al., 1989) in response to chemotactic factors and 5-lipoxygenase activation and release of arachidonate and leukotriene B_4 in response to PAF (McColl et al., 1991). Other effects on neutrophil function include enhanced uptake and killing of microorganisms (Fleischmann et al., 1986; Villalta and Kierszenbaum, 1986), increased phagocytosis of IgA-opsonized particles (Weisbart et al., 1988), and enhanced antibody-dependent cytotoxicity (Vadas et al., 1983). Neutrophil migration in vitro is reduced (Gasson et al., 1984) as is skin window migration in patients receiving GM-CSF treatment (Addison et al., 1989).

GM-CSF exerts similar effects on eosinophil function. Survival, cytotoxicity to schistosome ova, and calcium ionophore-induced generation of leukotriene C_4 have been shown to be enhanced by GM-CSF (Lopez et al., 1986; Silberstein et al., 1986).

GM-CSF exerts analogous effects on monocyte/ macrophage function. There is increased adhesion (Gamble et al., 1989), expression of a range of cytokines (Oster et al., 1989; Wing et al., 1989), as well as priming of superoxide responses (Coleman et al., 1988). Killing of tumor cells (Grabstein et al., 1986) and intracellular parasites is enhanced (Weiser et al., 1987) and there is augmentation of monocyte antigen-presenting function (Fischer et al., 1988). In addition, the dendritic/ Langerhans cell lineage, now recognized to be related to but distinct from monocytes, is profoundly influenced by GM-CSF in respect of both maturation and functional attributes (Heufler et al., 1988; Witmer et al., 1987). This property of GM-CSF is currently being explored with a view to generating antigen-presenting cells for immunotherapy (Romani et al., 1996)

5.1.3 Effects on Proliferation and Differentiation

GM-CSF was originally defined as a factor that supports the growth of granulocyte, macrophage, and eosinophil progenitors in semisolid bone marrow culture (Metcalf et al., 1986). Like IL-3 it can supply burst-promoting activity in erythroid cultures (Sieff et al., 1985) and there is some effect on megakaryocyte differentiation (Mazur et al., 1987). In combination with TNF, the proliferation and differentiation of dendritic Langerhans cells from a $CD34^+$ progenitor is promoted (Caux et al., 1992). Peripheral blood monocytes can be induced to differentiate into dendritic cells when exposed to GM-CSF, IL-4, and TNF (Zhou and Tedder, 1996).

In addition to its effects on hematopoietic cells, GM-CSF has been reported to cause the proliferation of other cell types including oligodendrocytes (Baldwin et al., 1993) and other cells originating from the neural crest (Dedhar et al., 1988). An effect on cultured endothelial cells has been reported (Bussolino et al., 1989) but not reproduced (Yong et al., 1991).

5.1.4 In Vivo Effects

The first clinical study using GM-CSF in patients with HIV infection showed the expected rise in neutrophil, monocyte, and eosinophil levels over a period of several days (Groopman et al., 1987). Paralleling the rise in the numbers of mature cells in the peripheral blood, increased numbers of progenitor cells are also observed (Socinski et al., 1988), a finding that has been exploited clinically. In vivo consequences of functional activation of mature cells are also observed. Within the first hour of administration of GM-CSF, circulating levels of neutrophils, monocytes, and eosinophils fall transiently, a phenomenon that is associated with upregulation of surface adhesion molecules, neutrophil degranulation, and sequestration in the lungs (Devereux *et al.*, 1987, 1989, 1990). Neutrophil migration into skin windows is depressed during GM-CSF administration (Addison *et al.*, 1989), whilst cytotoxicity of monocytes to tumor targets is increased (Wing *et al.*, 1989). Over a period of time serum cholesterol levels fall in individuals receiving GM-CSF, probably as a consequence of monocyte/macrophage activation (Nimer *et al.*, 1988).

5.2　PHYSIOLOGICAL ROLE OF GM-CSF

When initially described it was imagined that GM-CSF would function as a regulator of basal hematopoiesis. However, the fact that circulating levels of GM-CSF do not rise during neutropenia (Cebon *et al.*, 1994) and that transgenic mice with a targeted disruption of the GM-CSF gene are hematologically normal (Dranoff *et al.*, 1994) suggests that this is not the case. The *in vitro* effects of GM-CSF on mature cells point rather to a function as an inflammatory mediator and in particular as a modulator of antigen-presenting cell function (Morrissey *et al.*, 1987). The major abnormality in GM-CSF knockout mice is pulmonary alveolar proteinosis, a syndrome in which there is defective clearance of pulmonary proteins by alveolar macrophages. It has been suggested that a key role of GM-CSF is to regulate the uptake of particulate or opsonized material and this may also account in some measure for its effects on antigen presentation and phagocytosis (Dranoff and Mulligan, 1994).

6.　The GM-CSF Receptor

In common with other cytokines, GM-CSF acts on target cells through specific surface membrane receptors that are present at low density. The cDNAs for two human GM-CSF receptor components have been cloned (Gearing *et al.*, 1989; Hayashida *et al.*, 1990), an α subunit that binds GM-CSF with low affinity and a β chain that does not bind ligand but cooperates with the α chain to increase receptor affinity by reducing the rate of ligand dissociation from the complex. Ligand binding α chains have also been identified for interleukins-3 and -5. Both appear to share and compete for the same signal transducing GM-CSFRβ_c subunit (Figure 19.5).

6.1　MOLECULAR GENETICS OF THE GM-CSF RECEPTOR

The human GM-CSFRα gene has been mapped to the pseudo-autosomal region of the sex chromosomes (Gough *et al.*, 1990), extends over 45 kb, and contains 12 introns. Although abnormalities of the X chromosome are found in human leukemias, no abnormalities in the GM-CSFRα gene have been reported (M.A. Brown *et al.*, 1994; Wagner *et al.*, 1994). The gene for human GM-CSFRβ_c maps to chromosome 22 (Shen *et al.*, 1992). No abnormalities in the intracytoplasmic domain of the human β_c have been reported in human leukemia, although like the α subunit it appears to be quite polymorphic (Freeburn *et al.*, 1996).

The GM-CSF, IL-3, and IL-5 receptor α and β subunits belong to the hemopoietin receptor family. All are single-chain glycoproteins with a single transmembrane (TM) domain, lack intrinsic enzymatic activity and show conserved structural features including two paired extracellular cysteine residues, the amino acid motif Trp-Ser-X-Trp-Ser (WSXWS) close to the TM region (Bazan, 1990), and a proline-rich sequence in the cytoplasmic domain (Oneal and Yulee, 1993).

6.2　GM-CSF RECEPTOR ISOFORMS

Several isoforms of the GM-CSFRα subunit arising by alternative splicing have been identified. These include a soluble receptor lacking a transmembrane domain (Ashworth and Kraft, 1990; Raines *et al.*, 1991), a receptor with a serine-rich intracytoplasmic tail (Crosier *et al.*, 1991), and an isoform containing a 34-amino-acid insertion between the TM region and the WSXWS motif (Devereux *et al.*, 1993). In addition, alternatively spliced isoforms in the 5′ untranslated portion of the receptor transcript have been identified which appear to affect translational efficiency (Chopra *et al.*, 1996). The soluble isoform of the GM-CSF receptor antagonizes the effects of GM-CSF in semisolid cultures (C.B. Brown *et al.*, 1995) although the functional significance of the other isoforms is at present unclear. A GM-CSFRβ_c isoform with a truncated intracytoplasmic tail has been described, the message for which is highly expressed in some acute myelogenous leukemia (AML) blasts (Gale *et al.*, 1993).

6.3　GM-CSF RECEPTOR SUBUNIT STRUCTURE AND INTERACTIONS

The structure of the GM-CSF receptor subunits and their interactions with ligand have been inferred by comparison with the crystal structure of the growth hormone receptor (Cunningham *et al.*, 1991; De Vos *et al.*, 1992). This model predicts separate interactions between GM-CSFR α subunit and ligand, GM-CSR β_c subunit and ligand, and GM-CSFR α and β_c subunits. The fact that IL-3 and IL-5 can compete with GM-CSF for β_c subunits suggests that the α/β_c complex is not preformed but assembles following binding of GM-CSF to the α subunit. GM-CSF binds to isolated receptor α subunits with a low affinity and does not result in any cellular response other than increased glucose uptake

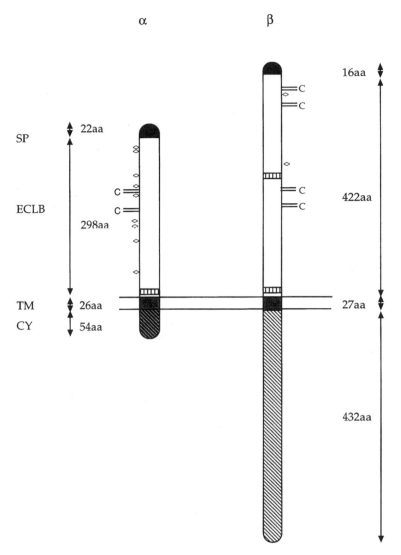

Figure 19.5   The GM-CSF receptor and signal transduction. Schematic representation of GM-CSF receptor α and β subunits. The signal peptide (SP), extracellular ligand binding (ECLB), transmembrane (TM), and cytoplasmic (CY) regions are indicated and number of amino acids (aa) is shown. Conserved cysteine residues are marked C and potential glycosylation sites indicated with a diamond. The WSXWS motif is shaded with vertical bars.

(Ding *et al.*, 1994). Mutation of either residue Glu-21 of GM-CSF or His-367 of the β_c subunit abolishes high-affinity binding, implying that a direct association occurs at this site between the two molecules. This interaction is not, however, critical for signal transduction, which still occurs if GM-CSF concentrations are appropriately increased. Thus it would appear that it is the interaction between the α and β_c subunits, which presumably occurs because of a ligand-induced change in α chain conformation, that is the initiating event in signal transduction.

Activation through ligand-induced homo- or hetero-oligomerization of receptor subunits has also been demonstrated for several of the other cytokines (Davis *et al.*, 1993; Stahl and Yancopoullos, 1993; Watowich *et al.*, 1992) and appears to be a common theme in transmembrane signaling. The IL-6 receptor complex, for example, may form a hexamer consisting of two molecules each of the IL-6 ligand, the IL6R α chain, and the signal-transducing β subunit Gp130 (Paeonessa *et al.*, 1995). The precise composition of the GM-CSFR complex is as yet unknown but there is some evidence that two alternatives, of differing affinity, may assemble depending on the availability of each subunit. In conditions of α subunit excess over β_c, dual high- and low-affinity GM-CSF binding is observed, whereas, when α subunits are limiting, a single intermediate-affinity receptor results (Budel *et al.*, 1993; Wheadon *et al.*, 1995). A recent study in which cross-linking studies were performed using an epitope-tagged β subunit suggested that preformed β homodimers are activated by ligand-induced association with the α subunit (Muto *et al.*, 1996; Freeburn *et al.*, 1996).

7.   Signal Transduction

Despite the fact that neither receptor subunit possesses intrinsic kinase activity, GM-CSF binding is associated with the rapid tyrosine phosphorylation of a number of substrates (Kanakura *et al.*, 1990). Activation of the Janus kinase (JAK2), probably as a result of dimerization-induced phosphorylation, appears to be a critical event in GM-CSF signaling (Quelle *et al.*, 1994). Substrates of activated JAK2 include MGF-STAT5 (mammary gland factor, signal transducer and activator of transcription 5) which was originally identified as responsible for the lactogenic response to prolactin. MGF-STAT5 activated by GM-CSF stimulation is able to bind IFN-γ activation sites (Gouilleux *et al.*, 1995). Other proteins tyrosine-phosphorylated following GM-CSF stimulation include the β_c subunit itself, shc which activates p21 ras and the Raf/MAP kinase pathway (Dorsch *et al.*, 1994) as well as the src-related kinases lyn and yes which are found in association with activated phosphatidylinositol-3-kinase (Corey *et al.*, 1993).

Relationships between structure and receptor function have been studied by expressing subunits in hematopoietic and nonhematopoietic cell lines from other species. Studies in *Xenopus* oocytes reveal that signaling through the GM-CSFR α subunit stimulates glucose uptake (Ding *et al.*, 1994), although this is not associated with protein phosphorylation. Other cellular responses to GM-CSF, however, appear to require both α and β_c subunits since transfer of both into murine cell lines is required to confer responsiveness to human GM-CSF. Experiments in the murine pro-B cell line BaF3, where the human α subunit can interact with the murine β_c subunit, showed that the membrane proximal residues of the α subunit (346–382) are sufficient for cell proliferation (Polotskaya *et al.*, 1994). Removal of residues 382–400 had no effect on growth or translocation of protein kinase C to the membrane but did prevent tyrosine phosphorylation of the β_c subunit when this was cotransfected. It seems that the α subunit confers signaling specificity since, despite using the same signal-transducing β_c subunit, IL-3- and GM-CSF-induced mitogenic responses are associated with protein tyrosine phosphorylation, whilst IL-5 stimulated proliferation is associated with dephosphorylation (Mire-Sluis *et al.*, 1995).

A similar deletion analysis of the β_c subunit has

enabled mapping of domains required for cellular responses to GM-CSF. BaF3 cells transfected with GM-CSFR subunits growing in serum required only the membrane-proximal portion of the β_c subunit for proliferation, whilst the distal portion of receptor was needed for tyrosine phosphorylation. Subsequent studies have shown that it is the membrane-proximal region that binds JAK2, induces expression of c-myc, the serine threonine kinase PIM-1, and an SH2 domain containing the negative regulator CIS (cytokine-inducible SH2) that binds to phosphorylated sites on the distal portion of the receptor (Yoshimura *et al.*, 1995). The distal domain is required for the majority of tyrosine phosphorylation events and activation of the Ras/MAP pathway (Quelle *et al.*, 1994; Sato *et al.*, 1993). Studies in BaF3 cells transfected with the same deleted β_c subunits indicate that signaling through the proximal portion induces DNA synthesis whereas the distal portion suppresses apoptosis through the Ras/MAP kinase pathway (Kinoshita *et al.*, 1995). This is consistent with the finding that signaling through the proximal portion of the receptor is insufficient to support long-term growth in serum-free medium (Sakamaki and Yonehara, 1994). Serum is presumably required to complement the lack of a distal Ras/MAP activating domain and prevent apoptosis by signaling through another pathway.

8.   Murine GM-CSF

Murine GM-CSF was first purified to homogeneity from lung conditioned medium in 1977 (Burgess *et al.*, 1977) and partial sequencing of this material subsequently allowed the construction of oligonucleotide probes and isolation of cDNAs from murine lung (Gough *et al.*, 1984) and T lymphocyte libraries (Gough *et al.*, 1985). The murine GM-CSF gene localizes to chromosome 11, closely linked to that for other cytokines including IL-3 (Barlow *et al.*, 1987). The regulatory elements of the murine GM-CSF gene are closely homologous to those found in the human gene and GM-CSF secretion is regulated at both the transcriptional and post-transcriptional levels (Bickel *et al.*, 1988). Murine GM-CSF contains 141 amino acids including a 17-aa signal peptide that is removed during secretion. There is no functional cross-reactivity between human and murine GM-CSF.

Cellular sources and biological activities of murine GM-CSF are broadly similar to those of the human molecule. Differences exist in receptor biology, however, as there is both a common β_c subunit shared between murine GM-CSF, IL-3, and IL-5 (AIC2B) and an IL-3-specific β subunit (AIC2A) (Kitamura *et al.*, 1991). The gene for the murine GM-CSF receptor α is found on autosome 19 (Disteche *et al.*, 1992) and the β subunit genes are on chromosome 15 (Gorman *et al.*, 1992).

Transgenic mice with a disrupted β_c gene show the same pulmonary abnormality as GM-CSF knockout mice but in addition are unable to mount an eosinophilia following parasitic infection, presumably because IL-5 signaling is also disrupted (Nishinakamura *et al.*, 1995). Mice with both the IL-3-specific β subunit and the β_c subunit disrupted show no additional abnormalities (Nishinakamura *et al.*, 1996a). The pulmonary abnormalities in β subunit knockout mice may be corrected by bone marrow transplantation (Nishinakamura *et al.*, 1996b).

The *in vivo* effects of GM-CSF in mice are as predicted from *in vitro* data, namely a rise in neutrophil, monocyte, and eosinophil numbers over a period of days (Metcalf *et al.*, 1986). Several studies have examined the effects of chronically elevated GM-CSF levels. Mice expressing a transgene resulting in sustained systemic elevation of GM-CSF levels develop macrophage infiltration in the retina, lens, and pleural and peritoneal cavities and die prematurely with muscle wasting, possibly due to abnormal macrophage activation (Lang *et al.*, 1987). When abnormal GM-CSF expression is restricted to hematopoietic tissue in mice transplanted with retrovirally transduced marrow cells, a lethal but polyclonal myeloproliferative syndrome develops (Johnson *et al.*, 1989).

9. GM-CSF in Disease and Therapy

9.1 GM-CSF IN DISEASE STATES

A number of groups have examined the expression of GM-CSF or its receptor at RNA or protein level in a variety of disease states with a view to understanding pathogenesis or to use as a disease marker. The difficulty with these studies in the main is that basal levels of circulating GM-CSF are very low and cannot be detected by most assay systems. As noted earlier, GM-CSF levels do not rise in neutropenia, but a local increase in GM-CSF levels can be found in inflammatory states. For example, in early histiocytosis X lesions, bronchial epithelia express high levels of GM-CSF protein which may be involved in the recruitment of CD1a-positive immunostimulatory cells (Tazi *et al.*, 1996). The presence of GM-CSF in rheumatoid synovial fluid suggests that it may play a part in the inflammatory process in this disease (Xu *et al.*, 1989). Elevated serum GM-CSF levels have also been found in eosinophilic states such as Hodgkin disease (Endo *et al.*, 1995), and episodic eosinophil-myalgia syndrome (Bochner *et al.*, 1991).

There has been much interest in the possible role of GM-CSF and other growth factors as an autocrine stimuli in myeloid malignancies. Certainly AML blast cells do express GM-CSF receptors and can synthesize and respond to GM-CSF (Young and Griffin, 1986). It is generally felt, however, that AML blasts do not express GM-CSF constitutively but may be induced to do so *in vitro* by other factors such as IL-1 or TNF (Lowenberg and Touw, 1993). Autonomous growth of AML blasts in culture, possibly mediated through autocrine mechanisms (Bradbury *et al.*, 1992), is sometimes seen and is associated with a poor prognosis (Lowenberg *et al.*, 1993). However, the fact that GM-CSF and other growth factors may be administered to patients with AML without obviously causing expansion of the leukemic clone does suggest that growth factors do not play a major role in the pathogenesis of leukemia.

9.2 CLINICAL USE OF GM-CSF

Clinical trials with recombinant GM-CSF commenced rapidly following the availability of purified recombinant material. Both *E. coli*-derived nonglycosylated and glycosylated formulations from yeast and CHO cells have been produced, the availability varying from country to country. In the United Kingdom only the *E. coli* nonglycosylated product is licensed at present.

9.2.1 Dosage and Side-effects

The maximum recommended daily dose of GM-CSF is $10\,\mu g/kg$ per day by subcutaneous or intravenous injection. At higher doses than this a syndrome of erythroderma, weight gain edema with pleural and pericardial effusions, and ascites has been reported which is presumably related to endothelial damage, phagocyte activation, and secondary cytokine production (Lieschke *et al.*, 1989). Bone pain, worse with rapid administration to patients with high white cell counts, and transient dyspnea and hypoxia are sometimes seen and are also likely to be due to activation of mature phagocytes. In general the toxicity profile of GM-CSF is worse than that for G-CSF, although at the recommended dosage side-effects are not so severe as to prevent its use. In the circulation, GM-CSF has a biphasic half-life with an initial $t_{1/2}$ of 10 min followed by a secondary $t_{1/2}$ of 85 min (Herrmann *et al.*, 1989).

9.2.2 Clinical Studies

The first clinical studies with GM-CSF were performed on patients with HIV infection, with the aim of improving cytopenia and reducing the number of infections (Groopman *et al.*, 1987). An increase in bone marrow cellularity and the number of circulating neutrophils, eosinophils, and monocytes was documented but there was no effect on infective episodes. Subsequent enthusiasm for the use of GM-CSF in HIV infection was tempered by the finding that HIV replication is enhanced in monocytes exposed to GM-CSF (Koyangani *et al.*, 1988). GM-CSF does enhance the anti-retroviral properties of AZT, however (Perno *et al.*, 1989), and its use to support AZT therapy is now a licensed indication.

Subsequent clinical trials focused on the use of GM-CSF to accelerate recovery after conventional chemotherapy and after ablative chemo/radiotherapy and bone marrow transplantation. Whilst these studies have consistently shown improvements in neutrophil (but not platelet) recovery, definite clinical benefits have been harder to document and have been significant in only a few randomized studies. In one such randomized study in which GM-CSF or placebo was administered after modified COP-BLAM III therapy for non-Hodgkin lymphoma, the use of GM-CSF was associated with reductions in febrile days, antibiotic use, and time in hospital (Gerhartz et al., 1993). In addition, an improved complete remission rate was reported in high-risk patients, but this analysis was not performed on an intention-to-treat basis.

Despite the theoretical risk of stimulating the leukemic clone, several groups have explored the use of GM-CSF as an adjuvant to chemotherapy in myeloid malignancies in an attempt to increase the number of cycling cells and thus augment the effectiveness of therapy. Results have been contradictory, one study reporting a lower survival and remission rate (Estey et al., 1992) whilst others were more encouraging (Bernell et al., 1994; Bettelheim et al., 1991).

GM-CSF has also been used in a variety of other clinical situations including the congenital and cyclical neutropenias, myelodysplastic syndromes, and aplastic anemia. Again, no real clinical benefit could be demonstrated in these disorders although, in contrast, G-CSF had a markedly beneficial effect in patients with severe congenital (Bonilla et al., 1989; Welte et al., 1990) and cyclical neutropenia (Hammond et al., 1989).

As discussed earlier, GM-CSF does increase the number of progenitor cells in the peripheral blood, an effect that is amplified after myelosuppressive chemotherapy. When these are used instead of bone marrow as support after myeloablative chemotherapy, a striking reduction in the length of neutropenia and thrombocytopenia is observed (Gianni et al., 1989). Such significant reductions in the period of cytopenia are associated with real improvements in the morbidity and mortality of these procedures, which are consequently becoming much more common.

The effects of GM-CSF on mature phagocyte function have encouraged studies in a number of infectious diseases. No definite benefit has been demonstrated using GM-CSF, with the possible exception of invasive fungal infection in neutropenic patients, where it may be a useful adjunct to other therapy (Bodey et al., 1993).

Several groups are examining the use of GM-CSF as an immune adjuvant, exploiting its properties as an enhancer of antigen-presenting cell function. In one such study, irradiated murine melanoma cells engineered by transduction with a retroviral vector to secrete GM-CSF and were used as "tumor vaccine" (Dranoff et al., 1993). Injection into mice with small pre-existing wild-type tumor burdens resulted in tumor eradication and persisting immunity to rechallenge. This approach is now being tested in human studies. GM-CSF has also been shown to enhance the humoral immune response when fused to specific B cell lymphoma idiotypes (Tao and Levy, 1993) and may prove of value as an adjuvant in more conventional immunization strategies.

10. References

Addison, I., Johnson, B., Devereux, S., et al. (1989). Granulocyte-macophage colony stimulating factor may inhibit neutrophil migration in vivo. J. Exp. Immunol. 76, 149.

Arnaout, M., Wang, E., Clark, S. and Sieff, C. (1986). Human recombinant granulocyte-macrophage colony-stimulating factor increases cell to cell adhesion and surface expression of adhesion promoting glycoproteins on mature granulocytes. J. Clin. Invest. 78, 597.

Ashworth, A. and Kraft, A. (1990). Cloning of a potentially soluble receptor for human GM-CSF. Nucleic Acids Res. 18, 7178.

Baldwin, G.C., Benveniste, E.N., Chung, G.Y., et al. (1993). Identification and characterization of a high-affinity granulocyte-macrophage colony-stimulating factor receptor on primary rat oligodendrocytes. Blood 82, 3279.

Barlow, D.P., Bucan, M., Lehrach, H., et al. (1987). Close genetic and physical linkage between the murine hematopoietic growth-factor genes GM-CSF and multi-CSF (IL3). EMBO J. 6, 617.

Bazan, J. (1990). Structural design and molecular evolution of a cytokine receptor superfamily. Proc. Natl Acad. Sci. USA 87, 6934.

Bernell, P., Kimby, E. and Hast, R. (1994). Recombinant human granulocyte-macrophage colony-stimulating factor in combination with standard induction chemotherapy. Leukemia 8, 1631.

Bettelheim, P., Valent, P., Andreeff, M., et al. (1991). Recombinant human granulocyte-macrophage colony-stimulating factor in combination with standard induction chemotherapy in denovo acute myeloid-leukemia. Blood 77, 700.

Bickel, M., Mergenhagen, S.E. and Pluznik, D.H. (1988). Post-transcriptional control of murine granulocyte macrophage-colony stimulating factor (GM-CSF). Exp. Hematol. 16, 414.

Bochner, B.S., Friedman, B., Krishnaswami, G., et al. (1991). Episodic eosinophilia-myalgia like syndrome in a patient without L-tryptophan use—association with eosinophil activation and increased serum levels of granulocyte-macrophage colony stimulating factor. J. Allergy Clin. Immunol. 88, 629.

Bodey, G.P., Anaissie, E., Gutterman, J. and Vadhanraj, S. (1993). Role of granulocyte-macrophage colony-stimulating factor as adjuvant therapy for fungal infection in patients with cancer. Clin. Infect. Dis. 17, 705.

Bonilla, M.A., Gillio, A.P., Ruggeiro, M., et al. (1989). Effects of recombinant human granulocyte colony-stimulating factor on neutropenia in patients with congenital agranulocytosis. N. Engl. J. Med. 320, 1574.

Bradbury, D., Rogers, S., Reilly, I.A.G., *et al.* (1992). Role of autocrine and paracrine production of granulocyte-macrophage colony-stimulating factor and interleukin-1-beta in the autonomous growth of acute myeloblastic-leukemia cells—studies using purified CD34-positive cells. Leukemia 6, 562.

Bradford, P.G., Jin, Y., Hui, P. and Wang, X. (1992). IL-3 and GM-CSF induce the expression of the inositol trisphosphate receptor in K562 myeloblast cells. Biochem. Biophys. Res. Commun. 187, 438.

Bradley, T. and Metcalf, D. (1966). The growth of mouse bone marrow cells in vitro. J. Cell Comp. Physiol. 66, 287.

Brown, C.B., Hart, C.E., Curtis, D.M., *et al.* (1990). Two neutralizing monoclonal antibodies against human granulocyte-macrophage colony-stimulating factor recognize the receptor binding domain of the molecule. J. Immunol. 144, 2184.

Brown, C.B., Pihl, C.E. and Kaushansky, K. (1994). Mapping of human granulocyte-macrophage-colony-stimulating-factor domains interacting with the human granulocyte-macrophage-colony-stimulating-factor-receptor alpha-sub-unit. Eur. J. Biochem. 225, 873.

Brown, C.B., Beaudry, P., Laing, T.D., *et al.* (1995). In vitro characterization of the human recombinant soluble granulocyte-macrophage colony-stimulating factor receptor. Blood 85, 1488.

Brown, M.A., Harrisonsmith, M., Deluca, E., *et al.* (1994). No evidence for GM-CSF receptor α chain gene mutation in AML-M2 leukemias which have lost a sex-chromosome. Leukemia 8, 1774.

Budel, L., Hoogerbrugge, H., Pouwels, K., *et al.* (1993). Granulocyte-macrophage colony-stimulating factor receptors alter their binding characteristics during myeloid maturation through up-regulation of the affinity converting β subunit (KH97). J. Biol. Chem. 14, 10154.

Burgess, A., Camakaris, J. and Metcalf, D. (1977). Purification of and properties of colony stimulating factors from mouse lung conditioned medium. J. Biol. Chem. 252, 1998.

Bussolino, F., Wang, J., Defilippi, P., *et al.* (1989). Granulocyte and granulocyte-macrophage colony-stimulating factors induce human endothelial cells to migrate and proliferate. Nature 337, 471.

Caux, C., Dezutter, D.C., Schmitt, D. and Banchereau, J. (1992). GM-CSF and TNF-alpha cooperate in the generation of dendritic Langerhans cells. Nature 360, 258.

Cebon, J., Nicola, N., Ward, M., *et al.* (1990). Granulocyte-macrophage colony stimulating factor from human lymphocytes. The effect of glycosylation on receptor binding and biological activity. J. Biol. Chem. 265, 4483.

Cebon, J., Layton, J., Maher, D. and Morstyn, G. (1994). Endogenous haemopoietic growth factors in neutropenia and infection. Br. J. Haematol. 86, 265.

Chopra, R., Kendall, G., Gale, R., *et al.* (1996). Expression of two alternatively spliced forms of the 5′ untranslated region of the GM-CSF receptor α chain mRNA. Exp. Hematol. 24, 755.

Clark-Lewis, I., Lopaez, A., To, L., *et al.* (1988). Structure function studies of human granulocyte-macrophage colony-stimulating factor. J. Immunol. 141, 881.

Cline, M. and Golde, D. (1974). Production of colony stimulating activity by human lymphocytes. Nature 248, 703.

Cockerill, P.N., Shannon, M.F., Bert, A.G., *et al.* (1993). The granulocyte-macrophage colony-stimulating factor/

interleukin 3 locus is regulated by an inducible cyclosporin A-sensitive enhancer. Proc. Natl Acad. Sci. USA 90, 2466.

Coleman, D.L., Chodakewitz, J.A., Bartiss, A.H. and Mellors, J.W. (1988). Granulocyte-macrophage colony-stimulating factor enhances selective effector functions of tissue-derived macrophages. Blood 72, 573.

Corey, S., Eguinoa, A., PuyanaTheall, K., *et al.* (1993). Granulocyte macrophage-colony stimulating factor stimulates both association and activation of phosphoinositide 30H-Kinase and src-related tyrosine kinase(s) in human myeloid derived cells. EMBO J. 12, 2681.

Crosier, K., Wong, G., Mathey-Prevot, B., *et al.* (1991). A functional isoform of the human granulocyte-macrophage colony stimulating factor receptor has an unusual cytoplasmic domain. Proc. Natl Acad. Sci. USA 88, 7744.

Cunningham, B., Ultsch, M., de Vos, A., *et al.* (1991). Dimerisation of the extracellular domain of the human growth hormone receptor by a single hormone molecule. Science 254, 821.

Davis, S., Aldrich, T., Stahl, N., *et al.* (1993). LIFrβ and gp130 as heterodimerising transducers of the tripartite CNTF receptor. Science 260, 1805.

Dedhar, S., Gaboury, L., Galloway, P. and Eaves, C. (1988). Human granulocyte-macrophage colony-stimulating factor is a growth factor active on a variety of cell types of nonhemopoietic origin. Proc. Natl Acad. Sci. USA 85, 9253.

Devereux, S., Linch, D., Campos-Costa, D., *et al.* (1987). Transient leucopenia induced by granulocyte macrophage colony stimulating factor. Lancet ii, 1523.

Devereux, S., Bull, H., Campos-Costa, D., *et al.* (1989). Granulocyte-macrophage colony-stimulating factor induced changes in cellular adhesion molecule expression and adhesion to endothelium: in vitro and in vivo studies in man. Br. J. Haematol. 71, 323.

Devereux, S., Porter, J., Hoyes, K., *et al.* (1990). Secretion of neutrophil secondary granules occurs during granulocyte-macrophage colony-stimulating factor induced margination. Br. J. Haematol. 74, 17.

Devereux, S., Wagner, H., Khwaja, A., *et al.* (1993). A novel isoform of the granulocyte-macrophage colony-stimulating factor receptor. Br. J. Haematol. 84(supplement 1), 79.

De Vos, A., Ultsch, M. and Kossiakoff, A. (1992). Human growth hormone and extracellular domain of its receptor: Crystal structure of the complex. Science 255, 306.

Diederichs, K., Boone, T. and Karplus, P.A. (1991). Novel fold and putative receptor-binding site of granulocyte-macrophage colony-stimulating factor. Science 254, 1779.

Ding, D.-H., Rivas, C., Heaney, M., *et al.* (1994). The α subunit of the human granulocyte-macrophage colony-stimulating factor receptor signals for glucose transport via a phosphorylation independent pathway. Proc. Natl Acad. Sci. USA 91, 2537.

Disteche, C.M., Brannan, C.I., Larsen, A., *et al.* (1992). The human pseudoautosomal GM-CSF receptor alpha-subunit gene is autosomal in mouse. Nature Genetics 1, 333.

Dorsch, M., Hock, H. and Diamantstein, T. (1994). Tyrosine phosphorylation of SHC is induced by IL-3, IL-5 and GM-CSF. Biochem. Biophys. Res. Commun. 200, 562.

Dranoff, G. and Mulligan, R. (1994). Activities of granulocyte-macrophage colony-stimulating factor revealed by gene transfer and gene knockout studies. Stem Cells 12(supplement), 173.

Dranoff, G., Jaffee, E., Lazenby, A., *et al.* (1993). Vaccination with irradiated tumor cells engineered to secrete murine granulocyte-macrophage colony-stimulating factor stimulates potent, specific, and long-lasting anti-tumor immunity. Proc. Natl Acad. Sci. USA 90, 3539.

Dranoff, G., Crawford, A.D., Sadelain, M., *et al.* (1994). Involvement of granulocyte-macrophage colony-stimulating factor in pulmonary homeostasis. Science 264, 713.

Endo, M., Usuki, K., Kitazume, K., *et al.* (1995). Hypereosinophilic syndrome in Hodgkins disease with increased granulocyte-macrophage colony-stimulating factor. Ann. Hematol. 71, 313.

Estey, E., Thall, P.F., Kantarjian, H., *et al.* (1992). Treatment of newly diagnosed acute myelogenous leukemia with granulocyte-macrophage colony-stimulating factor (GM-CSF) before and during continuous-infusion high-dose Ara-C plus daunorubicin—comparison to patients treated without GM-CSF. Blood 79, 2246.

Fischer, H.G., Frosch, S., Reske, K. and Reske, K.A. (1988). Granulocyte-macrophage colony-stimulating factor activates macrophages derived from bone marrow cultures to synthesis of MHC class II molecules and to augmented antigen presentation function. J. Immunol. 141, 3882.

Fleischmann, J., Golde, D.W., Weisbart, R.H. and Gasson, J.C. (1986). Granulocyte-macrophage colony-stimulating factor enhances phagocytosis of bacteria by human neutrophils. Blood 68, 708.

Fraser, J.K., Tran, S., Nimer, S.D. and Gasson, J.C. (1994). Characterization of nuclear factors that bind to a critical positive regulatory element of the human granulocyte-macrophage colony-stimulating factor promoter. Blood 84, 2523.

Freeburn, R., Gale, R., Wagner, H. and Linch, D. (1996). The beta subunit common to the GM-CSF, IL-3 and IL-5 receptors is highly polymorphic but pathogenic point mutations in patients with acute myeloid leukemia (AML) are rare. Leukemia 10, 123.

Freeburn, R.W., Gale, R.E. and Linch, D.C. (1996). The β_c chain of the GM-CSF receptor can dimerise in the absence of specific α chains and ligand. Blood 88 (supplement 1), 2635.

Gale, R., Chopra, R., Freeburn, R. and Linch, D. (1993). A truncated form of the beta common chain for the human GM-CSF, IL-3 and IL-5 receptors. Blood 82 (supplement 1), 722a.

Gamble, J.R., Elliott, M.J., Jaipargas, E., *et al.* (1989). Regulation of human monocyte adherence by granulocyte-macrophage colony-stimulating factor. Proc. Natl Acad. Sci. USA 86, 7169.

Gasson, J., Weisbart, R., Kaufman, S., *et al.* (1984). Purified human granulocyte-macrophage colony-stimulating factor: Direct action on neutrophils. Science 226, 1339.

Gearing, D.P., King, J.A., Gough, N.M. and Nicola, N.A. (1989). Expression cloning of a receptor for human granulocyte-macrophage colony-stimulating factor. EMBO J. 8, 3667.

Gerhartz, H.H., Engelhard, M., Meusers, P., *et al.* (1993). Randomized, double-blind, placebo-controlled, phase III study of recombinant human granulocyte-macrophage colony-stimulating factor as adjunct to induction treatment of high-grade malignant non-Hodgkin's lymphomas. Blood 82, 2329.

Gianni, A., Siena, S., Bregni, M., *et al.* (1989). Granulocyte-macrophage colony-stimulating factor to harvest circulating haemopoietic stem cells for autotransplantation. Lancet ii, 580.

Gorman, D.M., Itoh, N., Jenkins, N.A., *et al.* (1992). Chromosomal localization and organization of the murine genes encoding the beta-subunits (AIC2A and AIC2B) of the interleukin-3, granulocyte/macrophage colony-stimulating factor, and interleukin-5 receptors. J. Biol. Chem. 267, 15842.

Gough, N., Gough, J., Metcalf, D., *et al.* (1984). Molecular cloning of cDNA encoding a murine haemopoietic growth regulator, granulocyte-macrophage colony stimulating factor. Nature 309, 763.

Gough, N., Metcalf, D., Gough, J., *et al.* (1985). Structure and expression of the mRNA for murine granulocyte-macrophage colony stimulating factor. EMBO J. 4, 645.

Gough, N., Gearing, D., Nicola, N., *et al.* (1990). Localisation of the human GM-CSF receptor gene to the X-Y pseudoautosomal region. Nature 345, 734.

Gouilleux, F., Pallard, C., DusanterFourt, I., *et al.* (1995). Prolactin, growth hormone, erythropoietin and granulocyte-macrophage colony-stimulating factor induce MGF-Stat5 DNA binding activity. EMBO J. 14, 2005.

Grabstein, K., Urdal, D., Tushinski, R., *et al.* (1986). Induction of tumoricidal activity by granulocyte-macrophage colony-stimulating factor. Science 232, 506.

Gribben, J.G., Devereux, S., Thomas, N.S., *et al.* (1990). Development of antibodies to unprotected glycosylation sites on recombinant human GM-CSF. Lancet 335, 434.

Groopman, J.E., Mitsuyasu, R.T., DeLeo, M.J., *et al.* (1987). Effect of recombinant human granulocyte-macrophage colony-stimulating factor on myelopoiesis in the acquired immunodeficiency syndrome. N. Engl. J. Med. 317, 593.

Hamilton, J., Vairo, G. and Lingelbach, S. (1988). Activation and proliferation signals in murine macrophages: stimulation of glucose uptake by hemopoietic growth factors and other agents. J. Cell. Physiol. 134, 405.

Hammond, W.T., Price, T.H., Souza, L.M. and Dale, D.C. (1989). Treatment of cyclic neutropenia with granulocyte colony-stimulating factor. N. Engl. J. Med. 320, 1306.

Hayashida, K., Kitamura, T., Gorman, D.M., *et al.* (1990). Molecular cloning of a second subunit of the receptor for human granulocyte-macrophage colony-stimulating factor (GM-CSF): reconstitution of a high-affinity GM-CSF receptor. Proc. Natl Acad. Sci. USA 87, 9655.

Hercus, T.R., Cambareri, B., Dottore, M., *et al.* (1994). Identification of residues in the first and fourth helices of human granulocyte-macrophage colony-stimulating factor involved in biologic activity and in binding to the alpha- and beta-chains of its receptor. Blood 83, 3500.

Herrmann, F., Oster, W., Meuer, S.C., *et al.* (1988). Interleukin-1 stimulates lymphocytes-T to produce granulocyte-monocyte colony-stimulating factor. J. Clin. Invest. 81, 1415.

Herrmann, F., Schulz, G., Lindemann, A., *et al.* (1989). Hematopoietic responses in patients with advanced malignancy treated with recombinant granulocyte-macrophage colony-stimulating factor. J. Clin. Oncol. 7, 159.

Heufler, C., Koch, F. and Schuler, G. (1988). Granulocyte-macrophage colony-stimulating factor and interleukin-1 mediate the maturation of murine epidermal Langerhans cells into potent immunostimulatory dendritic cells. J. Exp. Med. 167, 700.

Horowitz, M.C., Coleman, D.L., Flood, P.M., *et al.* (1989). Parathyroid hormone and lipopolysaccharide induce murine osteoblast-like cells to secrete a cytokine indistinguishable from granulocyte-macrophage colony-stimulating factor. J. Clin. Invest. 83, 149.

Huebner, K., Isobe, M., Croce, C., et al. (1985). The human gene encoding GM-CSF is at 5q21-q32, the chromosomal region deleted in the 5q- anomaly. Science 230, 1282.

Johnson, G., Gonda, T., Metcalf, D., et al. (1989). A lethal myeloproliferative disorder in mice transplanted with bone marrow cells infected with a retrovirus expressing granulocyte-macrophage colony-stimulating factor. J. Cell. Biol. 77, 35.

Kanakura, Y., Druiker, B., Cannistra, S., et al. (1990). Signal transduction of the human granulocyte-macrophage colony-stimulating factor and interleukin-3 receptors involves tyrosine phosphorylation of a common set of cytoplasmic proteins. Blood 76, 706.

Kanakura, Y., Cannistra, S.A., Brown, C.B., et al. (1991). Identification of functionally distinct domains of human granulocyte-macrophage colony-stimulating factor using monoclonal antibodies. Blood 77, 1033.

Kaufman, F., DiPersio, J. and Gasson, J. (1989). Effect of human GM-CSF on neutrophil degranulation in vitro. Exp. Hematol. 17, 800.

Kaushansky, K., O Hara, P., Berkner, K., et al. (1986). Genomic cloning, characterisation and multilineage growth promoting activity of human granulocyte-macrophage colony-stimulating factor. Proc. Natl Acad. Sci. USA 83.

Kaushansky, K., Shoemaker, S., Alfaro, S. and Brown, C. (1989). Hematopoietic activity of granulocyte-macrophage colony-stimulating factor is dependent upon two distinct regions of the molecule: functional analysis based upon the activities of interspecies hybrid growth factors. Proc. Natl Acad. Sci. USA 86, 1213.

Kinoshita, T., Yokata, T., Arai, K. and Miyajima, A. (1995). Suppression of apoptotic death in hematopoietic cells by signalling through the IL-3/GM-CSF receptors. EMBO J. 14, 266.

Kitamura, T., Hayashida, K., Sakamaki, K., et al. (1991). Reconstitution of functional receptors for human granulocyte/macrophage colony-stimulating factor (GM-CSF): evidence that the protein encoded by the AIC2B cDNA is a subunit of the murine GM-CSF receptor. Proc. Natl Acad. Sci. USA 88, 5082.

Koyangani, Y., O'Brien, W., Zao, J., et al. (1988). Cytokines alter production of HIV-1 from primary mononuclear phagocytes. Science 241, 1673.

Lang, R.A., Metcalf, D., Cuthbertson, R.A., et al. (1987). Transgenic mice expressing a hematopoietic growth-factor gene (GM-CSF) develop accumulations of macrophages, blindness, and a fatal syndrome of tissue-damage. Cell 51, 675.

Lee, F., Yokata, T., Otsuka, T., et al. (1985). Isolation of cDNA for a human granulocyte-macrophage colony-stimulating factor by functional expression in mammalian cells. EMBO J. 82, 4360.

Lieschke, G.J., Maher, D., Cebon, J., et al. (1989). Effects of bacterially synthesized recombinant human granulocyte-macrophage colony-stimulating factor in patients with advanced malignancy. Ann. Intern. Med. 110, 357.

Lindemann, A., Riedel, D., Oster, W., et al. (1988). Granulocyte-macrophage colony-stimulating factor induces interleukin-1 production by human polymorphonuclear neutrophils. J. Immunol. 140, 837.

Lindemann, A., Riedel, D., Oster, W., et al. (1989). Granulocyte-macrophage colony-stimulating factor induces cytokine secretion by human polymorphonuclear leukocytes. J. Clin. Invest. 83, 1308.

Lopez, A.F., Williamson, D.J., Gamble, J.R., et al. (1986). Recombinant human granulocyte-macrophage colony-stimulating factor stimulates in vitro mature human neutrophil and eosinophil function, surface-receptor expression, and survival. J. Clin. Invest. 78, 1220.

Lopez, A.F., Shannon, M.F., Hercus, T., et al. (1992). Residue 21 of human granulocyte-macrophage colony-stimulating factor is critical for biological activity and for high but not low affinity binding. EMBO J. 11, 909.

Lowenberg, B. and Touw, I. (1993). Hematopoietic growth factors and their receptors in acute leukemia. Blood 81, 281.

Lowenberg, B., Vanputten, W.L.J., Touw, I.P., et al. (1993). Autonomous proliferation of leukemic-cells in vitro as a determinant of prognosis in adult acute myeloid-leukemia. N. Engl. J. Med. 328, 614.

Malter, J., McCroroy, W., Wilson, M. and Gillis, P. (1990). Adenosine-ridine binding factor requires metals for binding to granulocyte-macrophage colony-stimulating factor. Enzyme 44, 203.

Mazur, E.M., Cohen, J.L., Wong, G.G. and Clark, S.C. (1987). Modest stimulatory effect of recombinant human GM-CSF on colony growth from peripheral-blood human megakaryocyte progenitor cells. Exp. Hematol. 15, 1128.

McColl, S.R., Krump, E., Naccache, P.H., et al. (1991). Granulocyte-macrophage colony-stimulating factor increases the synthesis of leukotriene B4 by human neutrophils in response to platelet-activating factor. Enhancement of both arachidonic acid availability and 5-lipoxygenase activation. J. Immunol. 146, 1204.

Metcalf, D., Burgess, A., Johnson, G., et al. (1986). In vitro actions on hemopoietic cells of recombinant murine GM-CSF purified after production in Escherichia coli: comparison with purified native GM-CSF. J. Cell. Physiol. 128, 421.

Mire-Sluis, A., Page, L.A., Wadhwa, M. and Thorpe, R. (1995). Evidence for a signaling role for the alpha-chains of granulocyte-macrophage colony-stimulating factor (GM-CSF), interleukin-3 (IL-3), and IL-5 receptors—divergent signaling pathways between GM-CSF/IL-3 and IL-5. Blood 86, 2679.

Mitayake, S., Otsuka, T., Yokata, T., et al. (1985). Structure of the chromosomal gene for granulocyte-macrophage colony-stimulating factor: comparison of the mouse and human genes. EMBO J. 4, 2561.

Morrissey, P., Bressler, L., Park, L., et al. (1987). Granulocyte-macrophage colony-stimulating factor augments the primary antibody response by enhancing the function of antigen presenting cells. J. Immunol. 139.

Mulhrad, D., Decker, C. and Parker, R. (1994). Deadenylation of the unstable mRNA encoded by yeast MFA2 gene leads to decapping followed by 5' to 3' digestion of the transcript. Genes Dev. 8, 855.

Muto, A., Watanabe, S., Miyajima, A., et al. (1996). The β subunit of the human granulocyte-macrophage colony-stimulating factor receptor forms a homodimer and is activated by association with the α subunit. J. Exp. Med. 183, 1911.

Nicola, N.A., Metcalf, D., Johnson, G.R. and Burgess, A.W. (1979). Separation of functionally distinct human granulocyte-macrophage colony-stimulating factors. Blood 54, 614.

Nimer, S., Champlin, R. and Golde, D. (1988). Serum cholesterol lowering activity of granulocyte-macrophage colony stimulating factor. J. Am. Med. Assoc. 260, 3297.

Nimer, S., Fraser, J., Richards, J., *et al.* (1990). The repeated sequence CATT(A/T) is required for granulocyte-macrophage colony-stimulating factor promoter activity. Mol. Cell. Biol. 10, 6084.

Nishinakamura, R., Nakayama, N., Hirabayashi, Y., *et al.* (1995). Mice deficient for the IL-3/GM-CSF/IL-5 beta c receptor exhibit lung pathology and impaired immune response, while beta (IL3) receptor-deficient mice are normal. Immunity 2, 211.

Nishinakamura, R., Miyajima, A., Mee, P.J., *et al.* (1996a). Hematopoiesis in mice lacking the entire granulocyte-macrophage colony-stimulating factor/interleukin-3/interleukin-5 functions. Blood 88, 2458.

Nishinakamura, R., Wiler, R., Dirksen, U., *et al.* (1996b). The pulmonary alveolar proteinosis in granulocyte-macrophage colony-stimulating factor/interleukins-3/5 beta-C receptor-deficient mice is reversed by bone-narrow transplantation. J. Exp. Med. 183, 2657.

Okamoto, M., Nakai, M., Nakayama, C., *et al.* (1991). Purification and characterization of three forms of differently glycosylated recombinant human granulocyte-macrophage colony-stimulating factor. Arch. Biochem. Biophys. 286, 562.

Oneal, K.D. and Yulee, L.Y. (1993). The proline-rich motif (Prm)—a novel feature of the cytokine hematopoietin receptor superfamily. Lymphokine Cytokine Res. 12, 309.

Oster, W., Lindemann, A., Mertelsmann, R. and Herrmann, F. (1989). Granulocyte-macrophage colony stimulating factor and multilineage CSF recruit human monocytes to express granulocyte-CSF. Blood 73, 64–67.

Paeonessa, G., Grazizni, R., De Serio, A., *et al.* (1995). Two distinct and independent sites on IL-6 trigger gp130 dimer formation and signalling. EMBO J. 14, 1942.

Park, J.H. and Levitt, L. (1993). Overexpression of mitogen-activated protein kinase (ERK1) enhances T-cell cytokine gene expression: role of AP1, NF-AT, and NF-KB. Blood 82, 2470.

Perno, A., Kongsuwan, R., Cooney, D., *et al.* (1989). Replication of human immunodeficiency virus in monocytes: granulocyte-macrophage colony-stimulating factor (GM-CSF) potentiates viral production yet enhances the antiviral effect mediated by 3'-azido-2'3'-dideoxythymidine (AZT) and other dideoxynucleotide congeners of thymidine. J. Exp. Med. 169, 933.

Plaut, M., Pierce, J.H., Watson, C.J., *et al.* (1989). Mast cell lines produce lymphokines in response to cross-linkage of Fc epsilon RI or to calcium ionophores. Nature 339, 64.

Pluznik, D. and Sachs, L. (1965). The cloning of normal mast cells in cell culture. J. Cell Comp. Physiol. 66, 319.

Pluznik, D.H., Bickel, M. and Mergenhagen, S.E. (1989). B lymphocyte derived hematopoietic growth factors. Immunol. Invest. 18, 103.

Polotskaya, A., Zhao, Y., Lilly, M.B. and Kraft, A.S. (1994). Mapping the intracytoplasmic regions of the alpha granulocyte-macrophage colony-stimulating factor receptor necessary for cell growth regulation. J. Biol. Chem. 269, 14607.

Quelle, F.W., Sato, N., Witthuhn, B.A., *et al.* (1994). JAK2 associates with the beta (c) chain of the receptor for granulocyte-macrophage colony-stimulating factor, and its activation requires the membrane-proximal region. Mol. Cell. Biol. 14, 4335.

Raines, M.A., Liu, L.D., Quan, S.G., *et al.* (1991). Identification and molecular-cloning of a soluble human granulocyte macrophage colony-stimulating factor receptor. Proc. Natl Acad. Sci. USA 88, 8203.

Romani, N., Reider, D., Heuer, M., *et al.* (1996). Generation of mature dendritic cells from human blood—an improved method with special regard to clinical applicability. J. Immunol. Methods 196, 137.

Sakamaki, K. and Yonehara, S. (1994). Serum alleviates the requirement of the granulocyte-macrophage colony-stimulating factor (GM-CSF)-induced Ras activation for proliferation of BaF3 cells. FEBS Lett. 353, 133.

Sato, N., Sakamaki, K., Terada, N., *et al.* (1993). Signal transduction by the high-affinity GM-CSF receptor: two distinct cytoplasmic regions of the common beta subunit responsible for different signaling. EMBO J. 12, 4181.

Shanafelt, A. and Kastlein, R. (1989). Identification of critical regions in mouse granulocyte-macrophage colony-stimulating factor by scanning deletion analysis. Proc. Natl Acad. Sci. USA 86, 4872.

Shanafelt, A.B., Johnson, K.E. and Kastlein, R.A. (1991). Identification of critical amino acid residues in human and mouse granulocyte-macrophage colony-stimulating factor and their involvement in species specificity. J. Biol. Chem. 266, 13804.

Shannon, M., Gamble, J. and Vadas, M. (1988). Nuclear proteins interacting with the promoter region of the human granulocyte-macrophage colony-stimulating factor gene. Proc. Natl Acad. Sci. USA 85, 674.

Shannon, M., Pell, L., Lenardo, M., *et al.* (1990). A novel tumor necrosis factor responsive transcription factor which recognises a regulatory element in hemopoietic growth factor genes. Mol. Cell. Biol. 10, 2950–2956.

Shaw, G. and Kamen, R. (1986). A conserved AU sequence from the 3' untranslated region of GM-CSF messenger-RNA mediates selective messenger-RNA degradation. Cell 46, 659.

Shen, Y., Baker, E., Callen, D.F., *et al.* (1992). Localization of the human GM-CSF receptor beta-chain gene (CSF2RB) to chromosome 22q12.2-]q13.1. Cytogenet. Cell Genet. 61, 175.

Sieff, C., Emerson, S., Donahue, R., *et al.* (1985). Human recombinant granulocyte-macrophage colony-stimulating factor: a human multilineage hemopoietin. Science 230, 1171.

Silberstein, D.S., Owen, W.F., Gasson, J.C., *et al.* (1986). Enhancement of human eosinophil cytotoxicity and leukotriene synthesis by biosynthetic (recombinant) granulocyte-macrophage colony-stimulating factor. J. Immunol. 137, 3290.

Socinski, M.A., Cannistra, S.A., Elias, A., *et al.* (1988). Granulocyte-macrophage colony stimulating factor expands the circulating haemopoietic progenitor cell compartment in man. Lancet 1, 1194.

Stahl, N. and Yancopoullos, G. (1993). The alphas, betas and kinases of cytokine receptor complexes. Cell 74, 587.

Tao, M. and Levy, R. (1993). Idiotype/granulocyte-macrophage colony-stimulating factor fusion protein as a vaccine for B-cell lymphoma. Nature 362, 755.

Tazi, A., Bonay, M., Bergeron, A., *et al.* (1996). Role of granulocyte-macrophage colony-stimulating factor (GM-CSF) in the pathogenesis of adult pulmonary histiocytosis-X. Thorax 51, 611.

Thorens, B., Mermod, J.J. and Vassalli, P. (1987). Phagocytosis and inflammatory stimuli induce GM-CSF messenger-RNA in macrophages through posttranscriptional regulation. Cell 48, 671.

Tsuboi, A., Sugimoto, K., Yodoi, J., et al. (1991). A nuclear factor NF-GM2 that interacts with a regulatory region of the GM-CSF gene essential for its induction in responses to T-cell activation: purification from human T-cell leukemia line Jurkat cells and similarity to NF-kappa B. Int. Immunol. 3, 807.

Vadas, M., Nicola, N. and Metcalf, D. (1983). Activation of antibody dependent cell mediated cytotoxicity of human neutrophils and eosinophils by separate colony stimulating factors. J. Immunol. 130, 795.

Vairo, G. and Hamilton, J. (1988). Activation and proliferation signals in murine macrophages: stimulation of Na⁺, K⁺ ATPase activity by hemopoietic growth factors and other agents. J. Cell. Physiol. 134, 13.

Van Leeuwe, B., Martinson, M., Webb, G. and Young, I. (1989). Molecular organisation of the cytokine gene cluster, involving the human IL-3, IL-4, IL-5 and GM-CSF genes, on human chromosome 5. Blood 73, 1142.

Villalta, F. and Kierszenbaum, F. (1986). Effects of human colony-stimulating factor on the uptake and destruction of a pathogenic parasite (Trypanosoma cruzi) by human neutrophils. J. Immunol. 137, 1703.

Wagner, H., Gale, R., Freeburn, R., et al. (1994). Analysis of mutations in the GM-CSFr α coding sequence in patients with acute myeloid leukaemia and haematologically normal individuals by RT-PCR-SSCP. Leukaemia 8, 1527.

Walter, M.R., Cook, W.J., Ealick, S.E., et al. (1992). 3-dimensional structure of recombinant human granulocyte-macrophage colony-stimulating factor. J. Mol. Biol. 224, 1075.

Wang, C.Y., Bassuk, A.G., Boise, L.H., et al. (1994). Activation of the granulocyte-macrophage colony-stimulating factor promoter in T cells requires cooperative binding of Elf-1 and AP-1 transcription factors. Mol. Cell. Biol. 14, 1153.

Watowich, S., Yoshimura, A., Longmore, G., et al. (1992). Homodimerisation and constitutive activation of the erythropoietin receptor. Proc. Natl Acad. Sci. USA 89, 2140.

Weisbart, R., Golde, D., Clark, S., et al. (1985). Human granulocyte macrophage colony-stimulating factor is a neutrophil activator. Nature 314, 361.

Weisbart, R., Kacena, A., Schuh, A. and Golde, D. (1988). GM-CSF induces human neutrophil IgA mediated phagocytosis by an IgA receptor activation mechanism. Nature 332, 647.

Weisbart, R.H., Golde, D.W. and Gasson, J.C. (1986). Biosynthetic human GM-CSF modulates the number and affinity of neutrophil f-Met-Leu-Phe receptors. J. Immunol. 137, 3584.

Weiser, W.Y., Van, N.A., Clark, S.C., et al. (1987). Recombinant human granulocyte/macrophage colony-stimulating factor activates intracellular killing of Leishmania donovani by human monocyte-derived macrophages. J. Exp. Med. 166, 1436.

Welte, K., Zeidler, C., Reiter, A., et al. (1990). Differential effects of granulocyte-macrophage colony-stimulating factor and granulocyte colony-stimulating factor in children with severe congenital neutropenia. Blood 75, 1056.

Wheadon, H., Devereux, S., Khwaja, A. and Linch, D. (1995). The molecular basis of the intermediate and high affinity GM-CSF receptors. Br. J. Haematol. 89 (supplement 1), 32a.

Williams, G., Smith, C., Spooncer, E., et al. (1990). Haemopoietic colony-stimulating factors promote cell survival by suppressing apoptosis. Nature 343, 76.

Wilson, T. and Treisman, R. (1988). Removal of poly(A) and consequent degradation of c-fos mRNA facilitated by 3′ AU rich sequences. Nature 336, 396.

Wing, E.J., Magee, D.M., Whiteside, T.L., et al. (1989). Recombinant human granulocyte/macrophage colony-stimulating factor enhances monocyte cytotoxicity and secretion of tumor necrosis factor alpha and interferon in cancer patients. Blood 73, 643.

Wirthmueller, U., De, W.A. and Dahinden, C.A. (1989). Platelet-activating factor production in human neutrophils by sequential stimulation with granulocyte-macrophage colony-stimulating factor and the chemotactic factors C5A or formyl-methionyl-leucyl-phenylalanine. J. Immunol. 142, 3213.

Witmer, P.M., Olivier, W., Valinsky, J., et al. (1987). Granulocyte/macrophage colony-stimulating factor is essential for the viability and function of cultured murine epidermal Langerhans cells. J. Exp. Med. 166, 1484.

Wong, G., Witek, J., Temple, P., et al. (1985). Human GM-CSF: molecular cloning of the complementary DNA and purification of the natural and recombinant proteins. Science 228, 810.

Xu, W.D., Firestein, G.S., Taetle, R., et al. (1989). Cytokines in chronic inflammatory arthritis.2. Granulocyte-macrophage colony-stimulating factor in rheumatoid synovial effusions. J. Clin. Invest. 83, 876.

Yang, Y.C., Kovacic, S., Kriz, R., et al. (1988). The human genes for GM-CSF and IL-3 are closely linked in tandem on chromosome-5. Blood 71, 958.

Yong, K., Cohen, H., Khwaja, A., et al. (1991). Lack of effect of granulocyte-macrophage and granulocyte colony-stimulating factors on cultured human endothelial cells. Blood 77, 1675.

Yoshimura, A., Ohkubo, T., Kiguch, T., et al. (1995). A novel cytokine-inducible gene CIS encodes an SH2 containing protein that binds to tyrosine phosphorylated interleukin-3 and erythropoietin receptors. EMBO J. 14, 2816.

Young, D.C. and Griffin, J.D. (1986). Autocrine secretion of gm-csf in acute myeloblastic-leukemia. Blood 68, 1178.

Zhao, Y. and, Chegini, N. (1994). Human fallopian tube expresses granulocyte-macrophage colony stimulating factor (GM-CSF) and GM-CSF alpha and beta receptors and contain immunoreactive GM-CSF protein. J. Clin. Endocrinol. Metab. 79, 662.

Zhou, L.J. and Tedder, T.F. (1996). Cd14(+) blood monocytes can differentiate into functionally mature CD83(+) dendritic cells. Proc. Natl Acad. Sci. USA 93, 2588.

20. Leukemia Inhibitory Factor

Douglas J. Hilton *and* Nicholas M. Gough

1. Introduction

Leukemia inhibitory factor (LIF), a 180-amino-acid-residue glycoprotein, was purified and cloned on the basis of its ability to induce the differentiation of the monocytic leukemia cell line M1 (Gearing *et al.*, 1987; Gough *et al.*, 1988a; Hilton *et al.*, 1988a,b). LIF was, however, discovered independently by a number of groups on the basis of different biological functions. As a result, LIF is known by a variety of alternative names including differentiation-inducing factor (D-factor or DIF; Tomida *et al.*, 1984; Yamamoto *et al.*, 1980), macrophage/granulocyte inducer type 2 (MGI-2; Lipton and Sachs, 1981; Hilton *et al.*, 1988a), differentiation-inhibiting activity (DIA; Smith *et al.*, 1988), differentiation-retarding factor (DRF; Koopman and Cotton, 1984), human interleukin for DA-1a cells (HILDA; Godard *et al.*, 1988; Moreau *et al.*, 1986, 1988), growth-stimulatory activity for TS-1 cells (GATS; Van Damme *et al.*, 1992), hepatocyte-stimulating factor type II and type III (HSF-II and HSF-III; Baumann *et al.*, 1987a; Baumann and Wong, 1989), cholinergic neuronal differentiation factor (CDF; Yamamori *et al.*, 1989), and melanocyte-derived lipoprotein lipase inhibitor (MLPLI; Mori *et al.*, 1989), and possibly osteoclast-activating factor (OAF; Abe *et al.*, 1986).

LIF exhibits a remarkable range of biological effects *in vitro* and when overexpressed *in vivo*. Many of these actions are, in fact, shared by other cytokines, notably IL-6, IL-11, oncostatin M (OSM) and ciliary neurotrophic factor (CNTF) (Bruce *et al.*, 1992; Hilton, 1992; Yang, 1993). The redundancy of LIF action is highlighted in mice that are incapable of producing LIF. While the females are incapable of producing pups because of a defect in blastocyst implantation, LIF-deficient mice nonetheless develop apparently normally (Stewart *et al.*, 1992).

The overlapping biological effects of LIF, CNTF, OSM, IL-6, and IL-11 are explained in part by the existence of shared receptor components. The LIF receptor is composed of two proteins, the LIF receptor α chain and gp130 (Gearing *et al.*, 1991, 1992a). The LIF receptor α chain and gp130 are also components of the OSM and CNTF receptors, while gp130 but not the LIF receptor α chain forms a part of the IL-6 and IL-11 receptors (Davis *et al.*, 1993; Gearing *et al.*, 1991, 1992a,b; Murakami *et al.*, 1993; Stahl *et al.*, 1993; Yin *et al.*, 1994).

2.   *The Cytokine Gene*

LIF is encoded by a single-copy gene found on chromosome 22q12 in the human (Sutherland *et al.*, 1989). The human LIF gene is tightly linked to the gene for the related cytokine oncostatin-M. The genes for LIF and OSM are separated by approximately 10 kbp and lie in the same transcriptional orientation (Giovannini *et al.*, 1993a,b). The intron/exon structures of the LIF and OSM genes are also similar, suggesting that they arose by gene duplication. Rearrangements of chromosome 22 are found in certain tumors, for example, meningioma (Budarf *et al.*, 1989; Pergolizzi and Erster, 1994). Given the role of LIF in regulating proliferation and differentiation of various cell types, some interest has focused on whether the LIF gene is altered in such tumors.

The murine and human LIF genes contain three exons and two introns (Stahl *et al.*, 1990). Exon 1 encodes the 5′ untranslated region and the first 6 amino acids of the leader sequence, exon 2 encodes the remaining 16 residues of the leader sequence and 53 residues of the mature LIF protein, while exon 3 encodes the C-terminal 137 residues of the mature protein and an extensive, approximately 3.2 kb 3′ untranslated region (Gough *et al.*, 1992; Stahl *et al.*, 1990; Figures 20.1 and 20.2).

Cross-species nucleotide sequence comparisons of the 5′ flanking region of the LIF gene revealed high interspecies similarity that extended up to 270 bp 5′ of the major transcriptional start site. All four TATA boxes are conserved and there are large blocks of conserved nucleotides surrounding the major and minor transcriptional start sites. Indeed, the region defined in the murine gene as the minimal essential promoter is almost entirely conserved. In contrast, the negative regulatory element found in the murine gene falls immediately 5′ of the conserved region and shows a low level of sequence similarity with the corresponding region of the human gene (Stahl *et al.*, 1990; Wilson *et al.*, 1992).

3.   *The Protein*

LIF is a basic, 180-amino-acid-residue glycoprotein (Figure 20.3). Human LIF contains three disulfide bonds (Cys^{12}–Cys^{134}, Cys^{18}–Cys^{131}, and Cys^{60}–Cys^{163}). A high degree of sequence similarity is observed between LIF from various species (Gearing *et al.*, 1987; Gough *et al.*, 1988a; Wilson *et al.*, 1992; Yamamori *et al.*, 1989). LIF is also related to oncostatin-M (Rose and Bruce, 1991), CNTF (Lin *et al.*, 1989; Stockli *et al.*, 1989), and cardiotropin (Pennica *et al.*, 1995), and more distantly to G-CSF and IL-6 (Bazan, 1991). Seven potential *N*-glycosylation sites are found in the human LIF amino acid sequence and LIF is in fact heavily glycosylated. From the sequence of the cDNA, the predicted molecular mass of LIF is approximately 20 kDa. LIF purified from a variety of mammalian sources migrates with an apparent molecular mass of 32–67 kDa; however, deglycosylation with *N*-glycosidase results in a reduction in the apparent molecular mass to 20–25 kDa (Gascan *et al.*, 1989; Hilton *et al.*, 1988b). Evidence also exists that human LIF is glycosylated on serine and/or threonine residues; however, this is not extensive (Gascan *et al.*, 1989). The glycosylation state of LIF does not appear to effect its biological activity. *Escherichia coli*-derived LIF is not glycosylated, yet its biological potency is indistinguishable from mammalian LIF and hyperglycosylated yeast-derived LIF (Gough *et al.*, 1988b; Hilton *et al.*, 1988b). Likewise, *E. coli*-derived LIF and mammalian LIF are cleared from the circulation of mice with a similar half-life and exhibit a similar tissue fate (Hilton *et al.*, 1991b).

LIF is a stable molecule. The following treatments do not irreversibly diminish the biological activity of LIF: pH 2.0–pH 11.0; 8 M urea; 6 M guanidine-HCl; a variety of ionic, nonionic and zwitterionic detergents at 1% w/v or v/v, including SDS, Triton-X100, Tween 20, β-octylglucoside, CHAPS, 3.10 and 3.12; iododination on tyrosine residues (Godard *et al.*, 1992; Hilton *et al.*, 1988b; O.J. Hilton, M.J. Layton and N.A. Nicola, unpublished observations). The biological activity of LIF is destroyed by reduction or by treatment with proteases. In addition, murine LIF and, to a lesser extent, human LIF are sensitive to derivatization of amines using reagents such as maleic anhydride. LIF is stable for extended periods at −20°C and when sterile at 4°C. Storage of low concentrations of LIF requires a carrier to maintain biological activity, presumably by reducing absorption to the surface of the storage vessel. Suitable carriers include 0.02% (v/v) Tween-20 or protein such as bovine serum albumin or calf serum (D.J. Hilton, M.J. Layton and N.A. Nicola, unpublished observations; Hilton *et al.*, 1988b).

The tertiary structure of murine LIF has been solved using x-ray crystallography (Robinson *et al.*, 1994). LIF, like many other cytokines, is folded as a four-α–helical

```
 435                            ttataa ttttatcaat caaattctta gaagagggaa aaagtctgtt
 481 ctccccaccc tccccctca ctcgtccccc cccttcactc tcactttctt ccattcataa
 541 tttcctatga tgcacctcaa acaacttcct ggactgggga tcccggctaa atataqctgt
 601 ttctgtctta caacacaggc tccagtatat aaatcaggca aattccccat ttgagcATGA Exon 1
 661 ACCTCTGAAA ACTGCCGGCA TCTGAGGTTT CCTCCAAGGC CCTCTGAAGT GCAGCCCATA
 721 ATGAAGGTCT TGGCGGCAG

2471               GAGTTGTGCC CCTGCTGTTG GTTCTGCACT GGAAACATGG GGCGGGGAGC Exon 2
2521 CCCCTCCCCA TCACCCCTGT CAACGCCACC TGTGCCATAC GCCACCCAT

3343                                        TACACAGC CCAGGGGGAG Exon 3
3361 CCGTTCCCCA ACAACCTGGA CAAGCTATGT GGCCCCAACG TGACGGACTT CCCGCCCTTC
3421 CACGCCAACG GCACGGAGAA GGCCAAGCTG GTGGAGCTGT ACCGCATAGT CGTGTACCTT
3481 GGCACCTCCC TGGGCAACAT CACCCGGGAC CAGAAGATCC TCAACCCCAG TGCCCTCAGC
3541 CTCCACAGCA AGCTCAACGC CACCGCCGAC ATCCTGCGAG GCCTCCTTAG CAACGTGCTG
3601 TGCCGCCTGT GCAGCAAGTA CCACGTGGGC CATGTGGACG TGACCTACGG CCCTGACACC
3661 TCGGGTAAGG ATGTCTTCCA GAAGAAGAAG CTGGGCTGTC AACTCCTGGG GAAGTATAAG
3721 CAGATCATCG CCGTGTTGGC CCAGGCCTTC TAGCAGGAGG TCTTGAAGTG TGCTGTGAAC
3781 CGAGGGATCT CAGGAGTTGG GTCCAGATGT GGGGGCCTGT CCAAGGGTGG CTGGGCCCAG
3841 GGCATCGCTA AACCCAAATG GGGGCTGCTG GCTGACCCCG AGGGTGCCTG GCCAGTCCAC
3901 TCCACTCTGG GCTGGGCTGT GATGAAGCTG AGCAGAGTGG AAACTTCCAT AGGGAGGGAG
3961 CTAGAAGAAG GTGCCCCTTC CTCTGGGAGA TTGTGGACTG GGGAGCGTGG GCTGGACTTC
4021 TGCCTCTACT TGTCCCTTTG GCCCCTTGCT CACTTTGTGC AGTGAACAAA CTACACAAGT
4081 CATCTACAAG AGCCCTGACC ACAGGGTGAG ACAGCAGGGC CCAGGGGAGT GGACCAGCCC
4141 CCAGCAAATT ATCACCATCT GTGCCTTTGC TGCCCCTTAG GTTGGGACTT AGGTGGGCCA
4201 GAGGGGCTAG GATCCCAAAG GACTCCTTGT CCCCTAGAAG TTTGATGAGT GGAAGATAGA
4261 GAGGGGCCTC TGGGATGGAA GGCTGTCTTC TTTTGAGGAT GATCAGAGAA CTTGGGCATA
4321 GGAACAATCT GGCAGAAGTT TCCAGAAGGA GGTCACTTGG CATTCAGGCT CTTGGGGAGG
4381 CAGAGAAGCC ACCTTCAGGC CTGGGAAGGA AGACACTGGG AGGAGGAGAG GCCTGGAAAG
4441 CTTTGGTAGG TTCTTCGTTC TCTTCCCCGT GATCTTCCCT GCAGCCTGGG ATGGCCAGGG
4501 TCTGATGGCT GGACCTGCAG CAGGGGTTTG TGGAGGTGGG TAGGGCAGGG GCAGGTTGCT
4561 AAGTCAGGTG CAGAGGTTCT GAGGGACCCA GGCTCTTCCT CTGGGTAAAG GTCTGTAAGA
4621 AGGGGCTGGG GTAGCTCAGA GTAGCAGCTC ACATCTGAGG CCCTGGGAGG TCTTGTGAGG
4681 TCACACAGAG GTACTTGAGG GGGACTGGAG GCCGTCTCTG GTCCCCAGGG CAAGGGAACA
4741 GCAGAACTTA GGGTCAGGGT CTCAGGGAAC CCTGAGCTCC AAGCGTGCTG TGCGTCTGAC
4801 CTGGCATGAT TTCTATTTAT TATGATATCC TATTTATATT AACTTATTGG TGCTTTCAGT
4861 GGCCAAGTTA ATTCCCCTTT CCCTGGTCCC TACTCAACAA AATATGATGA TGGCTCCCGA
4921 CACAAGCGCC AGGGCCAGGG CTTAGCAGGG CCTGGTCTGG AAGTCGACAA TGTTACAAGT
4981 GGAATAAGCC TTACGGGTGA AGCTCAGAGA AGGGTCGGAT CTGAGAGAAT GGGGAGGCCT
5041 GAGTGGGAGT GGGGGGCCTT GCTCCACCCC CATCCCCTAC TGTGACTTGC TTTAGCGTGT
5101 CAGGGTCCAG GCTGCAGGGG CTGGGCCAAT TTGTGGGAGAG GCCGGGTGCC TTTCTGTCTT
5161 GCTTCCAGGG GGCTGGTTCA CACTGTTCTT GGGCGCCCCA GCATTGTGTT GTGAGGCGCA
5221 CTGTTCCTGG CAGATATTGT GCCCCCTGGA GCAGTGGGCA AGACAGTCCT TGTGGCCCAC
5281 CCTGTCCTTG TTTCTGTGTC CCCATGCTGC CTCTGAAATA GCGCCCTGGA ACAACCCTGC
5341 CCCTGCACCC AGCATGCTCC GACACAGCAG GGAAGCTCCT CCTGTGGCCC GGACACCCAT
5401 AGACGGTGCG GGGGGCCTGG CTGGGCCAGA CCCCAGGAAG GTGGGGTAGA CTGGGGGGAT
5461 CAGCTGCCCA TTGCTCCCAA GAGGAGGAGA GGGGAGGCTGC AGACGCCTGG GACTCAGACC
5521 AGGAAGCTGT GGGCCCTCCT GCTCCACCCC CATCCCACTC CCACCCATGT CTGGGCTCCC
5581 AGGCAGGGAA CCCGATCTCT TCCTTTGTGC TGGGGCCAGG CGAGTGGAGA AACGCCCTCC
5641 AGTCTGAGAG CAGGGGAGGG AAGGAGGCAG CAGAGTTGGG GCAGCTGCTC AGAGCAGTGT
5701 TCTGGCTTCT TCTCAAACCC TGAGCGGGCT GCCGGCCTCC AAGTTCCTCC GACAAGATGA
5761 TGGTACTAAT TATGGTACTT TTCACTCACT TTGCACCTTT CCCTGTCGCT CTCTAAGCAC
5821 TTTACCTGGA TGGCGCGTGG GCAGTGTGCA GGCAGGTCCT GAGGCCTGGG GTTGGGGTGG
5881 AGGGTGCGGC CCGGAGTTGT CCATCTGTCC ATCCCAACAG CAAGACGAGG ATGTGGCTGT
5941 TGAGATGTGG GCCACACTCA CCCTTGTCCA GGATGCAGGG ACTGCCTTCT CCTTCCTGCT
6001 TCATCCGGCT TAGCTTGGGG CTGGCTGCAT TCCCCCAGGA TGGCTTCGAG AAAGACAAAC
6061 TTGTCTGGAA ACCAGAGTTG CTGATTCCAC CCGGGGGGCC CGGCTGACTC GCCCATCACC
6121 TCATCTCCCT GTGGACTTGG GAGCTCTGTG CCAGGCCCAC CTTGCGGCCC TGGCTCTGAG
6181 TCGCTCTCCC ACCCAGCCTG GACTTGGCCC CATGGGACCC ATCCTCAGTG CTCCCTCCAG
6241 ATCCCGTCCG GCAGCTTGGC GTCCACCCTG CACAGCATCA CTGAATCACA GAGCCTTTGC
6301 GTGAAACAGC TCTGCCAGGC CGGGGAGTGG GTTTCTCTTC CCTTTTTATC TGCTGGTGTG
6361 GACCACACCT GGGCCTGGCC GGAGGAAGAG AGAGTTTACC AAGAGAGATG TCTCCGGGCC
6421 CTTATTTATT ATTTAAACAT TTTTTTAAAA AGCACTGCTA GTTTACTTGT CTCTCCTCCC
6481 CATCGTCCCC ATCGTCCTCC TTGTCCCTGA CTTGGGGCAC TTCCACCCTG ACCCAGCCAG
6541 TCCAGCTCTG CCTTGCCGGC TCTCCAGAGT AGACATAGTG TGTGGGGTGG GAGCTCTGGC
6601 ACCCGGGGAG GTAGCATTTC CCTGCAGATG GTACAGATGT TCCTGCCTTA GAGTCATCTC
6661 TAGTTCCCCA CCTCAATCCC GGCATCCAGC CTTCAGTCCC GCCCACGTGC TAGCTCCGTG
6721 GGCCCACCGT GCGGCCTTAG AGGTTTCCCT CCTTCCTTTC CACTGAAAAG CACATGGCCT
6781 TGGGTGACAA ATTCCTCTTT GATGAATGTA CCCTGTGGGG ATGTTTCATA CTGACAGATT
6841 ATTTTTATTT ATTCAATGTC ATATTTAAAA TATTTATTTT TTATACCAAA TGAATACTTT
6901 TTTTTTTAAG AAAAAAAAGA GAAATGAATA AAGAATCTAC TCTTGG
```

Figure 20.1   Sequence of the human LIF gene. The sequence of the human LIF gene was obtained from the GenBank database (accession number HUMALIFA) and is numbered according to Stahl *et al.* (1990). The promoter region is shown in lower case with important elements underlined. The exons are shown in upper case, with the nucleotides encoding the LIF protein shown in bold. The first amino acid of mature LIF is underlined

bundle, with the helices arranged in an up–up, down–down configuration (Figure 20.4; Robinson *et al.*, 1994). LIF may be classed as a long-chain cytokine owing to its long helices, long overhand A–B and C–D loops, and an additional fifth helix A loop (Purvis and Mabbutt, 1997). In this regard LIF most resembles G–CSF.

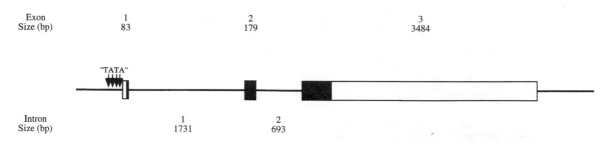

Figure 20.2   Schematic structure of the human LIF gene. The intron/exon structure of the human LIF gene. The exons are shown as boxes, with the coding regions shown in black. The numbers and size of the exons are shown above the figure, while the number and size of the introns are shown below.

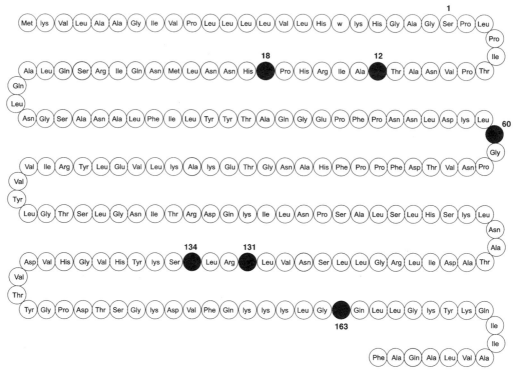

Figure 20.3   Amino acid sequence of LIF. Cysteine residues are shaded.

4. Cellular Sources and Production

In vitro, LIF is produced and secreted by a wide variety of primary tissues and cell lines (see Aloisi *et al.*, 1994; Anegon *et al.*, 1990; Gascan and Lemetayer, 1991; Greenfield *et al.*, 1993; Mezzasoma *et al.*, 1993). In general, most cells seem to constitutively secrete a small amount of LIF *in vitro*. Transcription of the LIF gene, and as a result synthesis and secretion of the protein, is elevated in most cells and tissues upon stimulation with bacterial products such as lipopolysaccharide (LPS) and in response to inflammatory cytokines such as IL-6, IL-1, and G-CSF (Brown *et al.*, 1994; Derigs and Boswell, 1993; Wetzler *et al.*, 1991).

Production of LIF *in vivo* appears to be associated with infection and inflammation (Kreisberg *et al.*, 1993;

Lecron *et al.*, 1993; Waring *et al.*, 1992, 1993b; Wesselingh *et al.*, 1994). In a survey of sera from a range of healthy volunteers and patients with a variety of diseases, a correlation was observed between the amount of circulating LIF and the severity of the infection or inflammation. In sera from healthy volunteers LIF was, in general, undetectable. Sera from patients with mild infections or inflammation often contained low levels of LIF, while the sera from patients with septicemia, especially those with septic shock, contained high concentrations of LIF (2–200 ng/ml, Waring *et al.*, 1992). LIF has also been found localized to sites of inflammation, for example, synovial fluid from arthritic joints (Campbell *et al.*, 1993; Carroll and Bell, 1993; Hamilton *et al.*, 1993; Ishimi *et al.*, 1992; Lotz *et al.*, 1992; Waring *et al.*, 1993b).

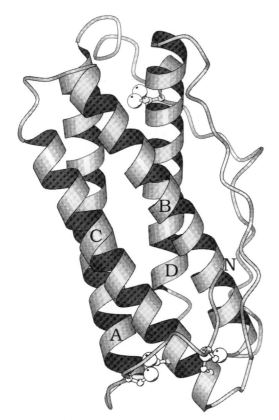

Figure 20.4   Three-dimensional structure of murine LIF.

Pathological consequences of LIF production by certain tumors have also been described. The human melanoma cell line SEKI secretes high levels of LIF and when injected into mice causes a severe, rapid, and fatal cachexia (Mori et al., 1989, 1991). Similar effects are observed in mice that have been experimentally manipulated to increase LIF levels (Metcalf and Gearing, 1989a,b; Metcalf et al., 1990).

Physiologically, a most important and interesting site of LIF production is the uterus (Bhatt et al., 1991; Kojima et al., 1994; Stewart et al., 1992; Yang et al., 1994). Production of LIF by the uterus is under strict control and peaks at the time of blastocyst formation, just prior to implantation. Most work in this area has been performed in mice, in a series of elegant studies by Stewart and colleagues (Bhatt et al., 1991; Stewart et al., 1992; see also Croy et al., 1991). The site of LIF production was found to be the endometrial glands (Bhatt et al., 1991). LIF secretion in the mouse peaks at approximately day 4 of gestation; however, production of LIF was found to be independent of the presence of an embryo since pseudopregnant mice produced LIF at a similar time after mating (Bhatt et al., 1991). Moreover, LIF production was delayed in mice in which development of the embryo had been delayed either naturally by the presence of suckling pups or artificially

by bilateral ovariectomy and administration of progesterone. Production of LIF did, however, occur in these mice upon commencement of normal embryonic development triggered by withdrawal of suckling pups or β-estradiol injection (Bhatt et al., 1991). These studies clearly demonstrate that LIF production is under exquisite maternal control; however, the mechanisms regulating production remain unclear. Recent evidence illustrates that LIF may also play a role in blastocyst implantation in other mammals, including humans (Kojima et al., 1994; Yang et al., 1994; reviewed in Robertson et al., 1994).

5.   Biological Activity

5.1   HEMOPOIETIC SYSTEM

LIF was originally defined on the basis of its ability to induce the macrophage differentiation and suppress the clonogenicity of the murine monocytic leukemia cell line M1 (Gearing et al., 1987; Gough et al., 1988b; Hilton et al., 1988a,b; Ichikawa, 1969, 1970; Lotem and Sachs, 1992; Maeda et al., 1977; Metcalf et al., 1988; Tomida et al., 1984; Yamamoto et al., 1980). After 4 days of LIF treatment, M1 cells differentiate into mature macrophages. These cells cease proliferation, exhibit a low nuclear to cytoplasmic ratio and numerous vacuoles, and are capable of migrating through semisolid agar (Metcalf, 1989; Metcalf et al., 1988). LIF treatment also results in an increase in the synthesis of macrophage markers such as Mac-1 (CD11b), FCγRII (CD32), c-fms (the M-CSF receptor), and lysozyme (see for example, Krystosek and Sachs, 1976). Like LIF, OSM and IL-6 also induce macrophage differentiation of M1 cells (Metcalf, 1989; Rose and Bruce, 1991; Shabo et al., 1988). LIF also suppresses the clonogenicity of the human leukemic cell lines HL-60 and U937 (Maekawa et al., 1990). The effect of LIF on these human lines is not as marked as its actions on M1 cells, with little evidence of morphological differentiation observed.

In addition to its effects on differentiation, LIF also stimulates the survival and proliferation of certain factor-dependent hemopoietic cell lines. Notably, LIF has been purified and cloned on the basis of its ability to stimulate the proliferation of murine DA1a cells and human TF-1 cells (Gascan et al., 1989; Godard et al., 1988; Moreau et al., 1986, 1987, 1988; Van Damme et al., 1992). Expression of the LIF receptor α chain and gp130 in BaF3 cells also allows these cells to grow in LIF (Gearing et al., 1994).

Alone, LIF has little or no effect on the proliferation and differentiation of primary hemopoietic progenitors from the bone marrow, spleen, or fetal liver (Metcalf et al., 1988). In combination with IL-3, but not GM-CSF, LIF enhances the formation of megakaryocyte colonies (Burstein et al., 1992; Debili et al., 1993; Metcalf et al.,

1991; Waring et al., 1993a; Warren et al., 1993). LIF also acts with IL-3 to synchronize the proliferation of primitive multipotential hemopoietic progenitors, termed blast-CFC (Leary et al., 1990). Similarly, LIF has been claimed to augment the survival and stimulate the proliferation of hemopoietic stem cells (Dick et al., 1991; Fletcher et al., 1990, 1991a,b; Verfaillie and McGlave, 1991). Although this effect of LIF is not observed universally (Schaafsma et al., 1992; Szilvassy and Cory, 1994), it has led to the inclusion of LIF, with other cytokines, in cultures of stem cells used for retroviral infection prior to reconstitution of the hemopoietic system of compromised recipients (Dick et al., 1991; Fletcher et al., 1990, 1991a,b; Moore et al., 1992).

5.2   EMBRYONIC STEM CELLS

The blastocyst of mammalian embryos consists of an outer trophoectoderm layer which contributes to the ectoplacental cone, giant cells, and extraembryonic ectoderm (Gardner et al., 1973), and an inner cell mass (ICM) which gives rise to tissues of the fetus proper, the fetal membranes (amnion, allantois, and yolk sac), and the extraembryonic endoderm (Gardner and Johnson, 1975; Gardner and Rossant, 1979; Rossant, 1975a,b, 1976). The ICM may be explanted and cultured in vitro. In the absence of any factors, the cultured ICM rapidly differentiates (Hogan and Tilly, 1978). In the presence of purified LIF or conditioned medium containing LIF, the ICM proliferates indefinitely and retains its undifferentiated phenotype (Evans and Kaufman, 1981; Magnuson et al., 1982; Martin, 1981; Smith et al., 1988; Smith and Hooper, 1987; Williams et al., 1988). Remarkably, the concentration of LIF required to maintain these cells in an undifferentiated state is identical to that necessary to induce macrophage differentiation of M1 leukemic cells.

The "cell lines" that result from the outgrowth of the ICM in LIF have been termed embryonic stem (ES) cells. An important feature of ES cells cultured in LIF is that they may be injected into the blastocoel of recipient blastocysts, where they become integrated with the host ICM. If chimeric blastocysts are then placed into the uterus of a foster mother, embryos develop normally and the ES cells contribute to the formation of all tissues, including the germ-line (Bradley et al., 1984; Nagy et al., 1990; Nichols et al., 1990; Pease et al., 1990; Pease and Williams, 1990; Smith et al., 1988; Williams et al., 1988). Indeed, if germ-line chimeric animals are bred, a proportion of their offspring will contain a genetic contribution from the ES cell line in every cell (Bradley et al., 1984; Nagy et al., 1990). Since ES cells cultured in LIF may also be manipulated genetically in vitro, either by introduction of a transgene or by mutation of specific genes by homologous recombination, the effect of such genetic changes may also be studied within the context of the entire animal (reviewed by Capecchi, 1993, 1994; Robertson, 1991).

Embryonic carcinoma (EC) cells are derived from germ cell tumors often in 129 mice. These cells exhibit some similarities to ES cells, since they are capable of differentiating in vitro into a wide range of cell types and have a limited ability to contribute to normal embryogenesis when combined with the ICM of a normal recipient blastocyst. Interestingly, normal primordial germ cells may also form ES cell lines. The differentiation of both EC cells and germ cell-derived ES cells is inhibited by LIF (Brown et al., 1992; Dolci et al., 1993; Matsui et al., 1992; Mummery et al., 1990; Nichols et al., 1990; Pesce et al., 1993; Pruitt and Natoli, 1992; Takagi, 1993).

5.3   OTHER CELLS

5.3.1   Neuronal Cells

LIF exerts two broad actions on neuronal cells—first to stimulate a switch in neurotransmitter phenotype (Bamber et al., 1994; Banner and Patterson, 1994; Fan and Katz, 1993; Fann and Patterson, 1993; Kalberg et al., 1993; Ludlam and Kessler, 1993; Ludlam et al., 1994; Michikawa et al., 1992; Shadiack et al., 1993; Sun et al., 1994), and second to enhance the survival of nerves, especially after explant or injury (Banner and Patterson, 1994; Cheema et al., 1994a,b; Kotzbauer et al., 1994; Martinou et al., 1992; Murphy et al., 1993; Richards et al., 1992; Thaler et al., 1994; Ure and Campenot, 1994; Zurn and Werren, 1994). Evidence is also emerging that LIF may control the generation of mature astrocytes (Nishiyama et al., 1993; Yoshida et al., 1993) and oligodendrocytes (Barres et al., 1993; Mayer et al., 1994).

Patterson and colleagues demonstrated that LIF was capable of inducing a switch in the neurotransmitter phenotype of sympathetic nerves from adrenergic to cholinergic (Fukada, 1985; Yamamori et al., 1989). This switch was accompanied by a reduction in the expression of tyrosine hydroxylase and hence catecholamine expression, while increasing the expression of choline acetyltransferase (Fukada, 1985; Nawa et al., 1991; Ure et al., 1992; Yamamori et al., 1989). LIF also enhances expression of peptide neurotransmitters including somatostatin, substance-P, vasoactive intestinal polypeptide (VIP), cholecystokinin, and enkephalin (Fann and Patterson, 1993; Nawa et al., 1991).

To survive in vitro, cultures of neuronal precursors and mature nerves require the addition of cytokines termed neurotropic factors. LIF enhances the survival of neuronal tissue from a variety of sources or following injury (Banner and Patterson, 1994; Cheema et al., 1994a,b; Kotzbauer et al., 1994; Martinou et al., 1992; Murphy et al., 1993; Richards et al., 1992; Thaler et al., 1994; Ure and Campenot, 1994; Zurn and Werren,

1994). Many of the actions of LIF on nerve cells are shared by the cytokine CNTF (Kotzbauer *et al.*, 1994; Thaler *et al.*, 1994; Zurn and Werren, 1994), reviewed in Patterson (1992, 1994), Patterson and Nawa (1993), Yamamori (1992).

5.3.2 Muscle Cells

In addition to its capacity to support the survival of neuronal cells, LIF enhances the survival and proliferation of muscle cells (Austin *et al.*, 1992; Austin and Burgess, 1991). LIF also acts as a chemotactic factor for muscle precursor cells (Robertson *et al.*, 1993). Since LIF is often produced at the site of injury, the capacity to attract progenitor cells, to enhance their survival, and to stimulate their proliferation suggests that LIF might play a role in muscle repair following injury.

5.3.3 Hepatocytes

In response to injury and infection, hepatocytes increase the expression of a set of proteins known as the acute–phase proteins. These proteins include C-reactive protein, serum amyloid A, the protease inhibitors α_1-antichymotrypsin and α_1-antitrypsin, fibrinogen, α_1-acidglycoprotein, haptoglobin, and complement factor C3 (reviewed by Baumann, 1989). Acute-phase protein synthesis is at least in part controlled by cytokines produced at the site of injury, which enter the circulation and act on the liver. The identity of these important regulators has been elucidated primarily using hepatocyte cell lines, such as HepG2, that retain the ability to express acute-phase proteins (Baumann *et al.*, 1980, 1983, 1984, 1986, 1987a,b; Baumann and Eldredge, 1982; Baumann and Muller-Eberhard, 1987). In the presence of dexamethasone, cytokines such as LIF, IL-6, IL-11, and OSM have been found to be powerful enhancers of acute–phase protein synthesis (Baumann *et al.*, 1988, 1989, 1993; Baumann and Schendel, 1991; Gauldie *et al.*, 1987, 1989; Kordula *et al.*, 1991; Marinkovic *et al.*, 1989; Piquet-Pellorce *et al.*, 1994; Richards *et al.*, 1993). The effects of each cytokine on acute-phase protein synthesis are similar quantitatively and qualitatively (Baumann *et al.*, 1980, 1983, 1984, 1986, 1987a,b, 1988, 1989; Baumann and Muller-Eberhard, 1987; Gauldie *et al.*, 1987, 1989; Marinkovic *et al.*, 1989).

5.3.4 Adipocytes

3T3-L1 is a preadipocytic cell line. Under appropriate culture conditions 3T3-L1 cells form mature adipocytes that accumulate fat and express enzymes such as lipoprotein lipase (Green and Kehinde, 1975; Green and Meuth, 1974). LIF inhibits the expression of lipoprotein lipase in 3T3-L1 adipocytes (Mori *et al.*, 1989). The effect of LIF on adipocyte differentiation does not, however, appear to be as dramatic as that of TNF (Marshall *et al.*, 1994). As mentioned in Sections 4 and 5.4, LIF reduces the amount of subcutaneous fat *in vivo*,

and when produced by tumor cells may be responsible for some cases of cachexia (Metcalf and Gearing, 1989a,b; Metcalf *et al.*, 1990; Mori *et al.*, 1991).

5.3.5 Osteoblasts, Osteoclasts and Chondrocytes

Osteoclast activation is measured by examining bone resorption in cultures that contain osteoblasts, macrophages, and fibroblasts, in addition to osteoclasts. Evidence exists that the action of LIF in activating osteoclasts is indirect and occurs via osteoblasts. First, no increase in bone resorption is stimulated by LIF in pure cultures of osteoclasts—the action requires the presence of contaminating osteoblasts. Second osteoblasts but not osteoclasts express receptors for LIF at their cell surface (Allan *et al.*, 1990). LIF appears to be produced within the bone microenvironment, by osteoblasts and macrophages as well as by synoviocytes and chondrocytes (Campbell *et al.*, 1993; Greenfield *et al.*, 1993; Hamilton *et al.*, 1993; Lotz *et al.*, 1992; Marusic *et al.*, 1993). LIF also effects a range of other functions of primary osteoblasts and osteoblastic cell lines. Notably, LIF acts as a mitogen and modulates the expression of a range of genes including tissue plasminogen inhibitor and alkaline phosphatase (Abe *et al.*, 1986; Allan *et al.*, 1990; Evans *et al.*, 1994; Hakeda *et al.*, 1991; Ishimi *et al.*, 1992; Lotz *et al.*, 1992; Lowe *et al.*, 1991; Noda *et al.*, 1990; Reid *et al.*, 1990; Richards *et al.*, 1993; Rodan *et al.*, 1990; Van Beek *et al.*, 1993). Cartilage, as well as bone formation, is also affected by LIF; LIF increases proteoglycan resorption in porcine cartilage (Carroll and Bell, 1993).

5.4 ELEVATION OF LIF

LIF levels have been experimentally elevated in animals by three means: (i) injection or infusion of purified LIF (Barnard *et al.*, 1994; Cheema *et al.*, 1994a,b; Cornish *et al.*, 1993; Metcalf *et al.*, 1990; Moran *et al.*, 1994; Waring *et al.*, 1993a); (ii) engraftment with nonleukemic hemopoietic cell lines engineered to secrete LIF (Metcalf and Gearing, 1989a,b; Waring *et al.*, 1993a); and (iii) the generation of transgenic mice in which LIF is expressed from a strong promoter (Bamber *et al.*, 1994; Conquet *et al.*, 1992; Shen *et al.*, 1994). In each type of study the consequences of increased LIF levels appear to recapitulate, at least in part, the effects of LIF observed *in vitro*.

Injection of LIF or LIF-producing hemopoietic cells into mice results in a complex pathology. Among the most dramatic effects observed is a rapid weight loss. Mice lose 30% of their body weight after 3 days of LIF injection. Weight loss can be attributed to an almost complete disappearance of subcutaneous body fat, correlating with the action of LIF on adipocytes *in vitro* (Metcalf and Gearing, 1989a,b; Metcalf *et al.*, 1990); see also Section 5.3.4 (Marshall *et al.*, 1994; Mori *et al.*, 1989). In mice engrafted with LIF-producing hemopoietic cells which home to the bone marrow, there is a remarkable proliferation of osteoblasts and increased deposition of bone, to such an extent that hemopoiesis is almost nonexistent in the long bones (Metcalf and Gearing, 1989a,b). There is a compensatory increase in extramedullary hemopoiesis, especially in the spleen. Again the effect of LIF on osteoblasts *in vivo* is consistent with actions defined by *in vitro* studies (Section 5.3.5). Although it has not been directly examined, evidence also exists that mice with elevated LIF levels undergo an acute-phase response; notably there is a reduction in the ratio of serum albumin to serum calcium and an increase in the erythrocyte sedimentation rate—both classic signs of an acute-phase reaction (Metcalf *et al.*, 1990).

Mice with elevated LIF levels also contain increased numbers of platelets, megakaryocytes, and megakaryocyte progenitor cells (Burstein *et al.*, 1992; Metcalf and Gearing, 1989b; Metcalf *et al.*, 1990; Waring *et al.*, 1993a). These effects are of the same order as observed for IL-6 (Debili *et al.*, 1993; Imai *et al.*, 1991; Koike *et al.*, 1990; Warren *et al.*, 1989; Williams *et al.*, 1992) and IL-11 (Burstein *et al.*, 1992; Yonemura *et al.*, 1992). Similar effects have been observed in rabbits (Moran *et al.*, 1994) and primates (Farese *et al.*, 1994) injected with LIF. The action of LIF on platelets and megakaryocytes occurs at low doses, thus when 20–200 ng of LIF is injected into mice once daily, weight loss and other adverse effects are not observed, while significant elevation of platelets and megakaryocyte is retained (Metcalf *et al.*, 1990).

Transgenic mice generated to express LIF and nerve growth factor (NGF) from an insulin promoter exhibit a number of interesting neuronal effects in the pancreas (Bamber *et al.*, 1994). First, the number of neurons is increased owing to the action of NGF, while there is a marked increase in the number of cholinergic neurons at the expense of adrenergic neurons owing to the action of LIF (Bamber *et al.*, 1994). This result demonstrates that LIF effects neurotransmitter phenotype *in vivo*. LIF also enhances neuronal survival following axotomy. Application of LIF at the site of nerve damage results in retrograde transport of LIF and increases the survival of damaged nerves (Cheema *et al.*, 1994a,b; Rao *et al.*, 1993).

5.5   Analysis of LIF-deficient Mice

Mice incapable of producing LIF have been generated by disrupting the LIF gene in ES cells using homologous recombination (Escary *et al.*, 1993; Rao *et al.*, 1993; Stewart *et al.*, 1992). The resultant LIF-deficient mice appear to develop and behave normally. Interestingly, while males are fertile, female LIF-deficient mice never produce pups. The defect in these mice does not appear to be in the germ cells or in fertilization; rather it is at the

level of blastocyst implantation (Stewart *et al.*, 1992). Blastocysts do not implant in the uterus in the absence of LIF. An effect on blastocyst implantation is consistent with the tight regulation of LIF production in the uterus during pregnancy (see Section 4) (Bhatt *et al.*, 1991; Stewart *et al.*, 1992). A similar effect is likely in humans, since LIF is also produced by the endometrium (Kojima *et al.*, 1994). Whether certain forms of human infertility are due to mutations at the LIF locus is, however, unclear.

While the inability of female mice to produce offspring is the most dramatic phenotype associated with an inability to produce LIF, other, more subtle, effects are also observed. For example, mice are claimed to contain reduced numbers of multipotential hemopoietic stem cells in the bone marrow and spleen and additionally to show defects in thymocyte stimulation (Escary *et al.*, 1993). In both cases there appears to be a gene dosage effect, since mice with a single normal LIF allele show a phenotype intermediate to wild-type mice and LIF-deficient mice (Escary *et al.*, 1993). Additional studies have suggested that while the neurotransmitter switch that occurs during embryonic development proceeds normally in LIF-deficient mice, changes occurring in response to nerve injury, such as enhanced VIP (vasoactive intestinal peptide) expression are defective (Rao *et al.*, 1993).

5.6   Natural Inhibitors of LIF Activity

One inhibitor of LIF action has been described, a soluble form of the LIF receptor α chain (Layton *et al.*, 1992) (see Section 6). In the mouse a secreted form of the low-affinity LIF receptor α chain is generated by alternative splicing (Gearing *et al.*, 1991; Tomida *et al.*, 1994). This soluble receptor binds murine LIF with a relatively low affinity (K_d = 1–2 nM), yet the concentrations present in normal mouse serum are sufficient to inhibit the biological activity of moderate concentrations of LIF *in vitro* (Layton *et al.*, 1992, 1994a,b). In mice, the serum level of soluble LIF receptor is approximately 1 μg/ml; during pregnancy, however, the levels of soluble LIF receptor rise dramatically to 30–100 μg/ml, suggesting that during this time placing a limit on the systemic action of LIF is especially important (Layton *et al.*, 1992; Tomida *et al.*, 1993; Yamaguchi-Yamamoto *et al.*, 1993). Serum LIF binding proteins have also been found in other species but have not yet been detected in humans (M.J. Layton and N.A. Nicola, personal communication).

6.   *The Receptor*

Studies using ^{125}I-LIF revealed that all LIF-responsive cells express high-affinity receptors for LIF on their cell surface (Allan *et al.*, 1990; Godard *et al.*, 1992; Hendry *et al.*, 1992; Hilton and Nicola, 1992; Hilton *et al.*, 1991a, 1992; Qiu *et al.*, 1994; Rodan *et al.*, 1990; Williams *et al.*, 1988; Yamamoto-Yamaguchi *et al.*, 1986). Receptor numbers are in general low, 50–3000 per cell, and exhibit an apparent dissociation constant (K_d) between 20 and 100 pM. Certain cell types express low-affinity receptor (K_d = 1–2 nM) in addition to high-affinity receptors. The interaction between LIF and its low- and high-affinity receptors is governed by a similar association rate constant (k_a = 5 × 10^8–10^9 min^{-1} M^{-1}); however, the dissociation rate constant is different (k_d=0.5–1.0 min^{-1} M^{-1} vs 0.001–0.0002 min^{-1} M^{-1}) (Godard *et al.*, 1992; Hilton and Nicola, 1992; Hilton *et al.*, 1988c, 1991a).

The molecular basis for high- and low-affinity LIF receptors was elucidated with the cloning of the two components of the LIF receptor (Figure 20.5). A low-affinity LIF receptor α chain was cloned by expression in COS cells, and was found to be a member of the hemopoietin receptor family, which includes the receptors for growth hormone, prolactin, erythropoietin, thrombopoietin, G-CSF, GM-CSF, IL-2, IL-3, IL-4, IL-5, IL-6, IL-7, IL-9, IL-11, IL-12, IL-13, IL-15, OSM, and CNTF (Bazan, 1990a,b; Gearing *et al.*, 1991). The defining features of the hemopoietin receptor family are found in the extracellular region within the 200-amino-acid residue hemopoietin receptor domain. These features include two pairs of conserved cysteines and the 5-amino-acid motif Trp-Ser-Xaa-Trp-Ser (WSXWS). Members of the hemopoietin receptors often have other modular domains within their extracellular region,

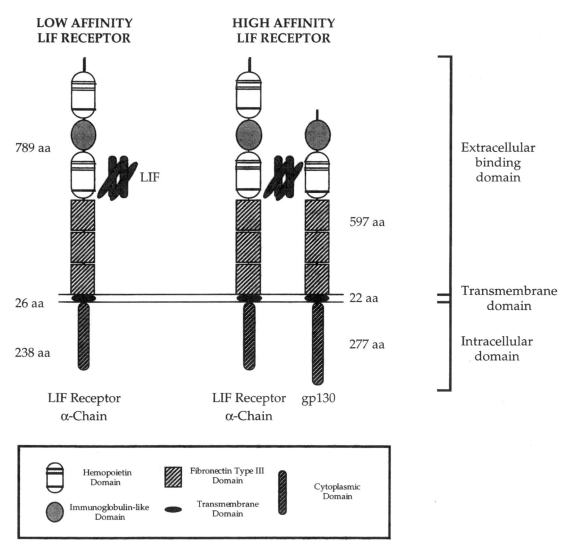

LOW AFFINITY LIF RECEPTOR

HIGH AFFINITY LIF RECEPTOR

789 aa

LIF

26 aa

238 aa

597 aa

22 aa

277 aa

Extracellular binding domain

Transmembrane domain

Intracellular domain

LIF Receptor α-Chain

LIF Receptor α-Chain   gp130

Hemopoietin Domain

Immunoglobulin-like Domain

Fibronectin Type III Domain

Transmembrane Domain

Cytoplasmic Domain

Figure 20.5   Schematic structure of the LIF receptor. The low-affinity LIF receptor comprises the LIF receptor α chain alone. The generation of a high-affinity receptor capable of signal transduction requires the presence of gp130. The positions of the conserved cysteine residues and the WSXWS motif within the hemopoietin receptor domain are shown by thin and thick lines, respectively.

notably immunoglobulin-like domains and fibronectin-type III like domains (Figure 20.5). The low-affinity LIF receptor α chain contains two hemopoietin domains, an immunoglobulin-like domain and three fibronectin type III domains (Gearing et al., 1991). A secreted form of the LIF receptor, produced by alternative splicing of the LIF receptor mRNA, has been identified in mouse but not man (Gearing et al., 1991; Layton et al., 1992; Tomida et al., 1994; Yamaguchi-Yamamoto et al., 1993). Soluble LIF receptor may be found in relatively high levels in mouse sera and increases 20- to 30-fold during pregnancy (Layton et al., 1992; Yamaguchi-Yamamoto et al., 1993). As described in Section 5.6, the soluble receptor may function as an antagonist of LIF action and therefore limit the effect of LIF at local sites of production.

Expression of the LIF receptor α chain in isolation fails to generate a high-affinity receptor capable of signal transduction (Gearing et al., 1991). A fully functional high-affinity receptor was found to require coexpression of the LIF receptor α chain and a second member of the hemopoietin receptor family gp130 (Gearing et al., 1992a,b). gp130 itself cannot bind to LIF, but rather it interacts with the low-affinity complex of LIF and the LIF receptor α chain as well as a 150 kDa protein to form a high-affinity receptor capable of signal transduction (Heymann et al., 1996).

The LIF receptor α chain and gp130 are components of other cytokine receptors (reviewed by Hirano et al., 1994; Kishimoto et al., 1992, 1994). Indeed, gp130 was initially isolated as the affinity-converting and signal-transducing subunit of the IL-6 receptor (Hibi et al., 1990; Taga et al., 1989). Subsequently, it has been shown to play a similar role in the IL-11 receptor (Hilton et al., 1994; Nishimoto et al., 1994; Yin et al., 1994; Zhang et al., 1994). The generation of high-affinity IL-6, and presumably IL-11, receptors requires binding to an α chain with a short cytoplasmic tail and the recruitment of two molecules of gp130, which become disulfide-linked (Murakami et al., 1993). Like the IL-6 and IL-11 receptors, the receptor for CNTF contains a low-affinity α chain, in this case with a GPI-anchor tethering it to the plasma membrane and gp130, but also requires the LIF receptor α chain (Davis et al., 1993; Ip et al., 1992; Stahl et al., 1993; Zhang et al., 1994). One form of the OSM receptor appears to be composed of the same subunits as the LIF receptor, that is, the LIF receptor α chain and gp130. Unlike LIF, however, OSM binds first to gp130 with low affinity and then interacts with the LIF receptor α chain to generate a high-affinity receptor (Gearing and Bruce, 1992; Gearing et al., 1992a). On the basis of receptor structure, LIF and OSM would be expected to exhibit identical biological activities. OSM, however, acts upon cells on which LIF is inactive. One such OSM-specific biological effect is as a mitogen for Kaposi sarcoma cells (Miles et al., 1992; Nair et al., 1992). This suggests that there may also be an OSM-specific receptor. Other groups of cytokines share receptor subunits. GM-CSF, IL-3, and IL-5 share a common β chain, while IL-2, IL-4, IL-7, IL-9, IL-13, and IL-15 each use the γ chain of the IL-2 receptor. This phenomenon is likely, in part, to explain why cytokines often regulate an overlapping group of biological processes (reviewed by Kishimoto et al., 1994; Nicola and Metcalf, 1991).

The cross-species binding and activity of human and murine LIF are particularly interesting (Ichibara et al., 1997). Murine LIF does not bind to the human LIF receptor α chain and interacts only very weakly to the combination of human LIF receptor α chain and human gp130 (Layton et al., 1994a,b; R. Clark, M.J. Layton, D.J. Hilton and N.A. Nicola, unpublished observation). In contrast, human LIF interacts with the murine LIF receptor α chain, whether membrane-bound or soluble, with a higher affinity than murine LIF (K_d= 10–20 pM versus 1–2 nM) and, indeed, with a higher affinity than it interacts with the human LIF receptor α chain (Layton et al., 1994a,b). These properties have been exploited to map the residues in human LIF that allow it to bind to its own receptor α chain and to bind to the murine LIF receptor α chain with an unusually high affinity. The residues that are responsible for both properties are one and the same and appear to lie on one surface of the LIF molecule (Owczarek et al., 1993).

7.   Signal Transduction

As described above, ligand-induced dimerization is an essential event in the formation of high-affinity receptors capable of signal transduction. Although the cytoplasmic domains of the LIF receptor and other members of the hemopoietin receptor family contain no intrinsic kinase domain, tyrosine phosphorylation occurs rapidly after receptor occupation (Amaral et al., 1993; Boulton et al., 1994; Ip et al., 1992; Lord et al., 1991; Thoma et al., 1994; Starr et al., 1997). Members of the interferon and hemopoietin receptor families have been found to interact with two classes of cytoplasmic receptor tyrosine kinases, the src family of kinases and the JAK kinases (reviewed by Darnell et al., 1994; Ihle, 1994; Ihle et al., 1994a,b). In the case of LIF, three members of the JAK kinase family, JAK1, JAK2, and TYK2, have been found to interact with the LIF receptor and to be phosphorylated and presumably activated upon stimulation with LIF (Boulton et al., 1994; Stahl et al., 1994). In addition, in ES cells the src kinase member hck appears to interact with the receptor and to be rapidly phosphorylated and activated; moreover, production by ES cells of an activated mutant of hck renders these cells hypersensitive to LIF (Ernst et al., 1994). In the case of the JAK kinases, one group of likely downstream substrates are the STAT (signal transducers and activators of transcription) family of transcription factors. These molecules contain SH2 domains responsible for interaction with tyrosine phosphate residues and DNA-binding domains (reviewed by Darnell *et al.*, 1994). Three STATs have been implicated in LIF signal transduction (Boulton *et al.*, 1994; Lamb *et al.*, 1994; Raz *et al.*, 1994), STAT1, STAT3 and STAT5B, also known as the acute-phase response factor (APRF) (Kuropatwinski *et al.*, 1997). These molecules are also activated to varying extents in response to IL-6, CNTF, IFN-α, IFN-γ, platelet-derived growth factor (PDGF), and epidermal growth factor (EGF) (Boulton *et al.*, 1994; Fu and Zhang, 1993; Fujitani *et al.*, 1994; Lamb *et al.*, 1994; Matsuda and Hirano, 1994; Raz *et al.*, 1994; Ruff-Jamison *et al.*, 1994; Shual *et al.*, 1993; Wegenka *et al.*, 1994; Zhong *et al.*, 1994). Activation of STATs involves tyrosine phosphorylation, possibly by activated JAKs, and their migration from the cytoplasm to the nucleus where they increase the rate of transcription of genes with the appropriate target sequences in their promoters (reviewed by Darnell *et al.*, 1994). Ras appears to be a point of convergence for LIF signaling (Schwiemann *et al.*, 1997).

It is likely that signal transduction through the LIF receptor is more complex than a simple linear cascade of receptor dimerization leading to activation of JAKs, which in turn phosphorylate and activate STATs, which then alter transcription. Initial studies have revealed that a different region of the LIF receptor is required for stimulating BaF3 cell proliferation as compared with enhancing hepatic acute-phase synthesis (Baumann *et al.*, 1994). This suggests that the multitude of biological effects mediated by LIF rely on distinct, though possibly overlapping, signal transduction pathways.

8. Murine Cytokine

For the most part, the biology of LIF has been studied using mouse models and has not been found to differ from the activity of human LIF. Therefore the biology of murine LIF is described more thoroughly in other sections and the most important aspects of the murine LIF gene are described here.

LIF is encoded by a single-copy gene found on chromosome 11A1 in the mouse (Kola *et al.*, 1990). The murine and human LIF genes contain three exons and two introns (Stahl *et al.*, 1990). In the mouse, but not the human or any other species, an alternative first exon has been described that is claimed to yield a protein in which the first four residues of the leader sequence are altered. It should be noted, however, that in order to generate the new signal sequence identified by Rathjen and colleagues, it is necessary to invoke an unprecedented splicing event (Gough *et al.*, 1992; Rathjen *et al.*, 1990). This putative minor change in a region of the protein which is presumably cleaved within the lumen of the endoplasmic reticulum during

translation is also claimed to direct the mature LIF protein to a matrix-bound rather than a soluble extracellular location (Rathjen *et al.*, 1990). The mechanism by which this occurs has not been clarified.

Nucleotide sequence analysis of the mouse LIF gene revealed four TATA-like elements within a region of 340 bp 5′ of the translational start codon of exon 1 (Stahl *et al.*, 1990). A major transcriptional start site adjacent to the most proximal of these TATA boxes has been located (Stahl *et al.*, 1990), while a minor transcriptional start site was identified adjacent to the second most distal element (Stahl *et al.*, 1990).

The minimal essential region of the 5′ flanking region able to function as a promoter of transcription of a reporter gene (the bacterial chloramphenicol acetyltransferase (CAT gene) has been found to consist of the major start site of transcription (nucleotide +1), the proximal TATA box (nucleotide −31) and an additional 50–72 nucleotides (Stahl and Gough, 1993).

A negative regulatory element was identified between positions −249 and −360. The negative influence of this region on promoter function could be overridden by the SV40 viral enhancer or, more interestingly by the sequence −660 to −860 of the murine LIF gene, suggesting the existence of an enhancer-like element in this region (Stahl and Gough, 1993).

9. Cytokine in Disease and Therapy

LIF is yet to be used clinically. Perhaps the most promising potential clinical use of LIF is in the elevation of platelet levels (reviewed by Metcalf, 1991). Although LIF appears capable of shortening the period of thrombocytopenia following platelet destruction in animal models (Farese *et al.*, 1994; Akiyama *et al.*, 1997a,b), the clinical use of LIF in this setting may be restricted for two reasons. First, the recently cloned thrombopoietin increases platelet numbers more dramatically than LIF or the related cytokines IL-6 and IL-11 (Burstein *et al.*, 1992; de Sauvage *et al.*, 1994; Debili *et al.*, 1993; Farese *et al.*, 1994; Imai *et al.*, 1991; Kaushansky *et al.*, 1994; Koike *et al.*, 1990; Lok *et al.*, 1994; Metcalf and Gearing, 1989b; Metcalf *et al.*, 1990; Moran *et al.*, 1994; Waring *et al.*, 1993a; Warren *et al.*, 1989; Wendling *et al.*, 1994; Williams *et al.*, 1992; Yonemura *et al.*, 1992); and second, although platelet elevation occurs in mice at LIF doses that are well tolerated (Metcalf *et al.*, 1990), LIF, like IL-6 and IL-11, is extremely pleiotropic (see Section 5; reviewed by Gough and Williams, 1989; Hilton, 1992; Hilton and Gough, 1991; Kurzrock *et al.*, 1991; Metcalf, 1991, 1992; Ryffel, 1993; Van Vlasselaer, 1992). Clinical use of LIF in a local rather than systemic manner requires development of more advanced delivery systems. If these delivery systems are developed, application of LIF to the repair of bone, nerve, and muscle tissue following injury might be envisaged.

Clinical applications might also be found for LIF *in vitro*. Recent work suggests that LIF enhances the survival and retroviral infection of hemopoietic stem cells *in vitro*, thus LIF may find a role in protocols for gene therapy or hemopoietic reconstitution following marrow-ablative therapy (Dick *et al.*, 1991; Fletcher *et al.*, 1990, 1991a,b).

The remarkable affinity of the soluble murine LIF receptor for human LIF also suggests that the receptor might be used as an antagonist in situations where LIF production is pathological (Layton *et al.*, 1994a,b). One situation where localized LIF production has adverse affects is seen in arthritis; administration of the soluble LIF receptor directly into the joint may therefore be of therapeutic value.

10. Acknowledgments

D.J.H. was supported by a Queen Elizabeth II Postdoctoral Fellowship from the Australian Research Council. The authors' work is supported by the Anti-Cancer Council of Victoria, Melbourne, Australia, AMRAD Operations Pty Ltd, Melbourne, Australia, The National Health and Medical Research Council, Canberra , Australia, The J.D. and L. Harris Trust, The National Institutes of Health, Bethesda, Maryland, USA (grant CA-22556), and the Australian Federal Government Cooperative Research Centres Program.

11. References

Abe, E., Tanaka, H., Ishimi, Y., *et al.* (1986). Differentiation-inducing factor purified from conditioned medium of mitogen-treated spleen cell cultures stimulates bone resorption. Proc. Natl Acad. Sci. USA 83, 5958–5962.

Akiyama, Y., Kajimura, N., Matsuzaki, J., *et al.* (1997a). In vivo effect of recombinant human leukemia inhibitory factor in primates. Jpn. J. Cancer Res. 88(6), 578–583.

Akiyama, Y., Kikuchi, Y., Matsuzaki, J., *et al.* (1997b). Protective effect of recombinant human leukemia inhibitory factor against thrombocytopenia in carboplatin-treated mice. Jpn. J. Cancer Res. 88(6), 584–589.

Allan, E.H., Hilton, D.J., Brown, M.A., *et al.* (1990). Osteoblasts display receptors for and responses to leukemia-inhibitory factor. J. Cell. Physiol. 145, 110–119.

Aloisi, F., Rosa, S., Testa, U., *et al.* (1994). Regulation of leukemia inhibitory factor synthesis in cultured human astrocytes. J. Immunol. 152, 5022–5031.

Amaral, M.C., Miles, S., Kumar, G. and Nel, A.E. (1993). Oncostatin-M stimulates tyrosine protein phosphorylation in parallel with the activation of p42MAPK/ERK-2 in Kaposi's cells. Evidence that this pathway is important in Kaposi cell growth. J. Clin. Invest. 92, 848–857.

Anegon, I., Moreau, J.F., Godard, A., *et al.* (1990). Production of human interleukin for DA cells (HILDA)/leukemia inhibitory factor (LIF) by activated monocytes. Cell. Immunol. 130, 50–65.

Austin, L. and Burgess, A.W. (1991). Stimulation of myoblast proliferation in culture by leukaemia inhibitory factor and other cytokines. J. Neuro. Sci. 101, 193–197.

Austin, L., Bower, J., Kurek, J. and Vakakis, N. (1992). Effects of leukaemia inhibitory factor and other cytokines on murine and human myoblast proliferation. J. Neuro. Sci. 112, 185–191.

Bamber, B.A., Masters, B.A., Hoyle, G.W., Brinster, R.L. and Palmiter, R.D. (1994). Leukemia inhibitory factor induces neurotransmitter switching in transgenic mice. Proc. Natl Acad. Sci. USA 91, 7839–7843.

Banner, L.R. and Patterson, P.H. (1994). Major changes in the expression of the mRNAs for cholinergic differentiation factor/leukemia inhibitory factor and its receptor after injury to adult peripheral nerves and ganglia. Proc. Natl Acad. Sci. USA 91, 7109–7113.

Barnard, W., Bower, J., Brown, M.A., Murphy, M. and Austin, L. (1994). Leukemia inhibitory factor (LIF) infusion stimulates skeletal muscle regeneration after injury: injured muscle expresses lif mRNA. J. Neuro. Sci. 123, 108–113.

Barres, B.A., Schmid, R., Sendtner, M. and Raff, M.C. (1993). Multiple extracellular signals are required for long-term oligodendrocyte survival. Development 118, 283–295.

Baumann, H. (1989). Hepatic acute phase reaction in vivo and in vitro. In Vitro Cell. Dev. Biol. 25, 115–126.

Baumann, H. and Eldredge, D. (1982). Dexamethasone increases the synthesis and secretion of a partially active fibronectin in rat hepatoma cells. J. Cell Biol. 95, 29–40.

Baumann, H. and Muller-Eberhard, U. (1987). Synthesis of hemopexin and cysteine protease inhibitor is coordinately regulated by HSF-II and interferon-beta 2 in rat hepatoma cells. Biochem. Biophys. Res. Commun. 146, 1218–1228.

Baumann, H. and Schendel, P. (1991). Interleukin-11 regulates the hepatic expression of the same plasma protein genes as interleukin-6. J. Biol. Chem. 266, 20424–20427.

Baumann, H. and Wong, G.G. (1989). Hepatocyte-stimulating factor III shares structural and functional identity with leukemia-inhibitory factor. J. Immunol. 143, 1163–1167.

Baumann, H., Gelehrter, T.D. and Doyle, D. (1980). Dexamethasone regulates the program of secretory glycoprotein synthesis in hepatoma tissue culture cells. J. Cell Biol. 85, 1–8.

Baumann, H., Firestone, G.L., Burgess, T.L., Gross, K.W., Yamamoto, K.R. and Held, W.A. (1983). Dexamethasone regulation of alpha 1-acid glycoprotein and other acute phase reactants in rat liver and hepatoma cells. J. Biol. Chem. 258, 563–570.

Baumann, H., Jahreis, G.P., Sauder, D.N. and Koj, A. (1984). Human keratinocytes and monocytes release factors which regulate the synthesis of major acute phase plasma proteins in hepatic cells from man, rat, and mouse. J. Biol. Chem. 259, 7331–7342.

Baumann, H., Hill, R.E., Sauder, D.N. and Jahreis, G.P. (1986). Regulation of major acute-phase plasma proteins by hepatocyte-stimulating factors of human squamous carcinoma cells. J. Cell Biol. 102, 370–383.

Baumann, H., Onorato, V., Gauldie, J. and Jahreis, G.P. (1987a). Distinct sets of acute phase plasma proteins are stimulated by separate human hepatocyte-stimulating factors and monokines in rat hepatoma cells. J. Biol. Chem. 262, 9756–9768.

Baumann, H., Richards, C. and Gauldie, J. (1987b). Interaction among hepatocyte-stimulating factors, interleukin 1, and glucocorticoids for regulation of acute phase plasma proteins in human hepatoma (HepG2) cells. J. Immunol. 139, 4122–4128.

Baumann, H., Prowse, K.R., Won, K.A., Marinkovic, S. and Jahreis, G.P. (1988). Regulation of acute phase protein genes by hepatocyte-stimulating factors, monokines and glucocorticoids. Tokai J. Exp. Clin. Med. 13, 277–292.

Baumann, H., Won, K.A. and Jahreis, G.P. (1989). Human hepatocyte-stimulating factor-III and interleukin-6 are structurally and immunologically distinct but regulate the production of the same acute phase plasma proteins. J. Biol. Chem. 264, 8046–8051.

Baumann, H., Ziegler, S.F., Mosley, B., Morella, K.K., Pajovic, S. and Gearing, D.P. (1993). Reconstitution of the response to leukemia inhibitory factor, oncostatin M, and ciliary neurotrophic factor in hepatoma cells. J. Biol. Chem. 268, 8414–8417.

Baumann, H., Symes, A.J., Comeau, M.R., et al. (1994). Multiple regions within the cytoplasmic domains of the leukemia inhibitory factor receptor and gp130 cooperate in signal transduction in hepatic and neuronal cells. Mol. Cell. Biol. 14, 138–146.

Bazan, J.F. (1990a). Haemopoietic receptors and helical cytokines. Immunol. Today 11, 350–354.

Bazan, J.F. (1990b). Structural design and molecular evolution of a cytokine receptor superfamily. Proc. Natl Acad. Sci. USA 87, 6934–6938.

Bazan, J.F. (1991). Neuropoietic cytokines in the hematopoietic fold. Neuron 7, 197–208.

Bhatt, H., Brunet, L.J. and Stewart, C.L. (1991). Uterine expression of leukemia inhibitory factor coincides with the onset of blastocyst implantation. Proc. Natl Acad. Sci. USA 88, 11408–11412.

Boulton, T.G., Stahl, N. and Yancopoulos, G.D. (1994). Ciliary neurotrophic factor/leukemia inhibitory factor/interleukin 6/oncostatin M family of cytokines induces tyrosine phosphorylation of a common set of proteins overlapping those induced by other cytokines and growth factors. J. Biol. Chem. 269, 11648–11655.

Bradley, A., Evans, M., Kaufman, M.H. and Robertson, E. (1984). Formation of germ-line chimaeras from embryo-derived teratocarcinoma cell lines. Nature 309, 255–256.

Brown, G.S., Brown, M.A., Hilton, D., Gough, N.M. and Sleigh, M.J. (1992). Inhibition of differentiation in a murine F9 embryonal carcinoma cell subline by leukemia inhibitory factor (LIF). Growth Factors 7, 41–52.

Brown, M.A., Metcalf, D. and Gough, N.M. (1994). Leukaemia inhibitory factor and interleukin 6 are expressed at very low levels in the normal adult mouse and are induced by inflammation. Cytokine 6, 300–309.

Bruce, A.G., Linsley, P.S. and Rose, T.M. (1992). Oncostatin M. Prog. Growth Factor Res. 4, 157–170.

Budarf, M., Emanuel, B.S., Mohandas, T., Goeddel, D.V. and Lowe, D.G. (1989). Human differentiation-stimulating factor (leukemia inhibitory factor, human interleukin DA) gene maps distal to the Ewing sarcoma breakpoint on 22q. Cytol. Cell Genet. 52, 19–22.

Burstein, S.A., Mei, R.L., Henthorn, J., Friese, P. and Turner, K. (1992). Leukemia inhibitory factor and interleukin-11 promote maturation of murine and human megakaryocytes in vitro. J. Cell. Physiol. 153, 305–312.

Campbell, I.K., Waring, P., Novak, U. and Hamilton, J.A. (1993). Production of leukemia inhibitory factor by human articular chondrocytes and cartilage in response to interleukin-1 and tumor necrosis factor alpha. Arthritis Rheum. 36, 790–794.

Capecchi, M.R. (1993). Mouse genetics. YACs to the rescue. Nature 362, 205–206.

Capecchi, M.R. (1994). Targeted gene replacement. Sci. Am. 270, 52–59.

Carroll, G.J. and Bell, M.C. (1993). Leukaemia inhibitory factor stimulates proteoglycan resorption in porcine articular cartilage. Rheum. Int. 13, 5–8.

Cheema, S.S., Richards, L., Murphy, M. and Bartlett, P.F. (1994a). Leukemia inhibitory factor prevents the death of axotomised sensory neurons in the dorsal root ganglia of the neonatal rat. J. Neurosci. Res. 37, 213–218.

Cheema, S.S., Richards, L.J., Murphy, M. and Bartlett, P.F. (1994b). Leukemia inhibitory factor rescues motoneurones from axotomy-induced cell death. Neuroreport 5, 989–992.

Conquet, F., Peyrieras, N., Tiret, L. and Brulet, P. (1992). Inhibited gastrulation in mouse embryos overexpressing the leukemia inhibitory factor. Proc. Natl Acad. Sci. USA 89, 8195–8199.

Cornish, J., Callon, K., King, A., Edgar, S. and Reid, I.R. (1993). The effect of leukemia inhibitory factor on bone in vivo. Endocrinology 132, 1359–1366.

Croy, B.A., Guilbert, L.J., Browne, M.A., et al. (1991). Characterization of cytokine production by the metrial gland and granulated metrial gland cells. J. Reprod. Immunol. 19, 149–166.

Darnell, J., Jr, Kerr, I.M. and Stark, G.R. (1994). Jak-STAT pathways and transcriptional activation in response to IFNs and other extracellular signaling proteins. Science 264, 1415–1421.

Davis, S., Aldrich, T.H., Stahl, N., et al. (1993). LIFR beta and gp130 as heterodimerizing signal transducers of the tripartite CNTF receptor. Science 260, 1805–1808.

de Sauvage, F.J., Hass, P.E., Spencer, S.D., et al. (1994). Stimulation of megakaryocytopoiesis and thrombopoiesis by the c-Mpl ligand. Nature 369, 533–538.

Debili, N., Masse, J.M., Katz, A., Guichard, J., Breton-Gorius, J. and Vainchenker, W. (1993). Effects of the recombinant hematopoietic growth factors interleukin-3, interleukin-6, stem cell factor, and leukemia inhibitory factor on the mega-karyocytic differentiation of CD34+ cells. Blood 82, 84–95.

Derigs, H.G. and Boswell, H.S. (1993). LIF mRNA expression is transcriptionally regulated in murine bone marrow stromal cells. Leukemia 7, 630–634.

Dick, J.E., Kamel-Reid, S., Murdoch, B. and Doedens, M. (1991). Gene transfer into normal human hematopoietic cells using in vitro and in vivo assays. Blood 78, 624–634.

Dolci, S., Pesce, M. and De Felici, M. (1993). Combined action of stem cell factor, leukemia inhibitory factor, and cAMP on in vitro proliferation of mouse primordial germ cells. Mol. Reprod. Dev. 35, 134–139.

Ernst, M., Gearing, D.P. and Dunn, A.R. (1994). Functional and biochemical association of Hck with the LIF/IL-6 receptor signal transducing subunit gp130 in embryonic stem cells. EMBO J. 13, 1574–1584.

Escary, J.L., Perreau, J., Dumenil, D., Ezine, S. and Brulet, P. (1993). Leukaemia inhibitory factor is necessary for maintenance of haematopoietic stem cells and thymocyte stimulation. Nature 363, 361–364.

Evans, D.B., Gerber, B. and Feyen, J.H. (1994). Recombinant human leukemia inhibitory factor is mitogenic for human bone-derived osteoblast-like cells. Biochem. Biophys. Res. Commun. 199, 220–226.

Evans, M.J. and Kaufman, M.H. (1981). Establishment in culture of pluripotential cells from mouse embryos. Nature 292, 154–156.

Fan, G. and Katz, D.M. (1993). Non-neuronal cells inhibit catecholaminergic differentiation of primary sensory neurons: role of leukemia inhibitory factor. Development 118, 83–93.

Fann, M.J. and Patterson, P.H. (1993). A novel approach to screen for cytokine effects on neuronal gene expression. J. Neurochem. 61, 1349–1355.

Farese, A.M., Myers, L.A. and MacVittie, T.J. (1994). Therapeutic efficacy of recombinant human leukemia inhibitory factor in a primate model of radiation-induced marrow aplasia. Blood 84, 3675–3678.

Fletcher, F.A., Williams, D.E., Maliszewski, C., Anderson, D., Rives, M. and Belmont, J.W. (1990). Murine leukemia inhibitory factor enhances retroviral-vector infection efficiency of hematopoietic progenitors. Blood 76, 1098–1103.

Fletcher, F.A., Moore, K.A., Ashkenazi, M., et al. (1991a). Leukemia inhibitory factor improves survival of retroviral vector-infected hematopoietic stem cells in vitro, allowing efficient long-term expression of vector-encoded human adenosine deaminase in vivo. J. Exp. Med 174, 837–845.

Fletcher, F.A., Moore, K.A., Williams, D.E., Anderson, D., Maliszewski, C. and Belmont, J.W. (1991b). Effects of leukemia inhibitory factor (LIF) on gene transfer efficiency into murine hematolymphoid progenitors. Adv. Exp. Med. Biol. 292, 131–138.

Fu, X.Y. and Zhang, J.J. (1993). Transcription factor p91 interacts with the epidermal growth factor receptor and mediates activation of the c-fos gene promoter. Cell 74, 1135–1145.

Fujitani, Y., Nakajima, K., Kojima, H., Nakae, K., Takeda, T. and Hirano, T. (1994). Transcriptional activation of the IL-6 response element in the junB promoter is mediated by multiple Stat family proteins. Biochem. Biophys. Res. Commun. 202, 1181–1187.

Fukada, K. (1985). Purification and partial characterization of a cholinergic neuronal differentiation factor. Proc. Natl Acad. Sci. USA 82, 8795–8799.

Gardner, R.L. and Johnson, M.H. (1975). Investigation of cellular interaction and deployment in the early mammalian embryo using interspecific chimaeras between the rat and mouse. Ciba Found. Symp. 0, 183–200.

Gardner, R.L. and Rossant, J. (1979). Investigation of the fate of 4–5 day post-coitum mouse inner cell mass cells by blastocyst injection. J. Embryol. Exp. Morphol. 52, 141–152.

Gardner, R.L., Papaioannou, V.E. and Barton, S.C. (1973). Origin of the ectoplacental cone and secondary giant cells in mouse blastocysts reconstituted from isolated trophoblast and inner cell mass. J. Embryol. Exp. Morphol. 30, 561–572.

Gascan, H. and Lemetayer, J. (1991). Induction of leukemia inhibitory factor secretion by interleukin-1 in a human T lymphoma cell line. Lymphocyte Cytokine Res. 10, 115–118.

Gascan, H., Godard, A., Ferenz, C., et al. (1989). Characterization and NH2-terminal amino acid sequence of natural human interleukin for DA cells: leukemia inhibitory factor. Differentiation inhibitory activity secreted by a T lymphoma cell line. J. Biol. Chem. 264, 21509–21515.

Gauldie, J., Richards, C., Harnish, D., Lansdorp, P. and Baumann, H. (1987). Interferon beta 2/B-cell stimulatory factor type 2 shares identity with monocyte-derived hepatocyte-stimulating factor and regulates the major acute phase protein response in liver cells. Proc. Natl Acad. Sci. USA 84, 7251–7255.

Gauldie, J., Richards, C., Northemann, W., Fey, G. and Baumann, H. (1989). IFN beta 2/BSF2/IL-6 is the monocyte-derived HSF that regulates receptor-specific acute phase gene regulation in hepatocytes. Ann. N.Y. Acad. Sci. 557, 46–58.

Gearing, D.P. and Bruce, A.G. (1992). Oncostatin M binds the high-affinity leukemia inhibitory factor receptor. New Biol. 4, 61–65.

Gearing, D.P., Gough, N.M., King, J.A., et al. (1987). Molecular cloning and expression of cDNA encoding a murine myeloid leukaemia inhibitory factor (LIF). EMBO J. 6, 3995–4002.

Gearing, D.P., Thut, C.J., VandenBos, T., et al. (1991). Leukemia inhibitory factor receptor is structurally related to the IL-6 signal transducer, gp130. EMBO J. 10, 2839–2848.

Gearing, D.P., Comeau, M.R., Friend, D.J., et al. (1992a). The IL-6 signal transducer, gp130: an oncostatin M receptor and affinity converter for the LIF receptor. Science 255, 1434–1437.

Gearing, D.P., VandenBos, T., Beckmann, M.P., et al. (1992b). Reconstitution of high affinity leukaemia inhibitory factor (LIF) receptors in haemopoietic cells transfected with the cloned human LIF receptor. Ciba Found. Symp. 167, 245–255.

Gearing, D.P., Ziegler, S.F., Comeau, M.R., et al. (1994). Proliferative responses and binding properties of hematopoietic cells transfected with low-affinity receptors for leukemia inhibitory factor, oncostatin M, and ciliary neurotrophic factor. Proc. Natl Acad. Sci. USA 91, 1119–1123.

Giovannini, M., Djabali, M., McElligott, D., Selleri, L. and Evans, G.A. (1993a). Tandem linkage of genes coding for leukemia inhibitory factor (LIF) and oncostatin M (OSM) on human chromosome 22. Cytol. Cell Genet. 64, 240–244.

Giovannini, M., Selleri, L., Hermanson, G.G. and Evans, G.A. (1993b). Localization of the human oncostatin M gene (OSM) to chromosome 22q12, distal to the Ewing's sarcoma breakpoint. Cytol. Cell Genet. 62, 32–34.

Godard, A., Gascan, H., Naulet, J., et al. (1988). Biochemical characterization and purification of HILDA, a human lymphokine active on eosinophils and bone marrow cells. Blood 71, 1618–1623.

Godard, A., Heymann, D., Raher, S., et al. (1992). High and low affinity receptors for human interleukin for DA cells/leukemia inhibitory factor on human cells. Molecular characterization and cellular distribution. J. Biol. Chem. 267, 3214–3222.

Gough, N.M. and Williams, R.L. (1989). The pleiotropic actions of leukemia inhibitory factor. Cancer Cells 1, 77–80.

Gough, N.M., Gearing, D.P., King, J.A., et al. (1988a). Molecular cloning and expression of the human homologue of the murine gene encoding myeloid leukemia-inhibitory factor. Proc. Natl Acad. Sci. USA 85, 2623–2627.

Gough, N.M., Hilton, D.J., Gearing, D.P., et al. (1988b). Biochemical characterization of murine leukaemia inhibitory factor produced by Krebs ascites and by yeast cells. Blood Cells 14, 431–442.

Gough, N.M., Wilson, T.A., Stahl, J. and Brown, M.A. (1992). Molecular biology of the leukaemia inhibitory factor gene. Ciba Found. Symp. 167, 24–38.

Green, H. and Kehinde, O. (1975). An established preadipose cell line and its differentiation in culture. II. Factors affecting the adipose conversion. Cell 5, 19–27.

Green, H. and Meuth, M. (1974). An established pre-adipose cell line and its differentiation in culture. Cell 3, 127–133.

Greenfield, E.M., Gornik, S.A., Horowitz, M.C., Donahue, H.J. and Shaw, S.M. (1993). Regulation of cytokine expression in osteoblasts by parathyroid hormone: rapid stimulation of interleukin-6 and leukemia inhibitory factor mRNA. J. Bone Miner. Res. 8, 1163–1171.

Hakeda, Y., Sudo, T., Ishizuka, S., et al. (1991). Murine recombinant leukemia inhibitory factor modulates inhibitory effect of 1,25 dihydroxyvitamin D3 on alkaline phosphatase activity in MC3T3-E1 cells. Biochem. Biophys. Res. Commun. 175, 577–582.

Hamilton, J.A., Waring, P.M. and Filonzi, E.L. (1993). Induction of leukemia inhibitory factor in human synovial fibroblasts by IL-1 and tumor necrosis factor-alpha. J. Immunol. 150, 1496–1502.

Hendry, I.A., Murphy, M., Hilton, D.J., Nicola, N.A. and Bartlett, P.F. (1992). Binding and retrograde transport of leukemia inhibitory factor by the sensory nervous system. J. Neurosci. 12, 3427–3434.

Heymann, D., Godard, A., Raher, S., et al. (1996) Leukemia inhibitory factor (LIF) and oncostatin M (OSM) high affinity binding require additional receptor subunits besides GP130 and GP190. Cytokine 8(3), 197–205.

Hibi, M., Murakami, M., Saito, M., Hirano, T., Taga, T. and Kishimoto, T. (1990). Molecular cloning and expression of an IL-6 signal transducer, gp130. Cell 63, 1149–1157.

Hilton, D.J. (1992). LIF: lots of interesting functions. Trends Biochem. Sci. 17, 72–76.

Hilton, D.J. and Gough, N.M. (1991). Leukemia inhibitory factor: a biological perspective. J. Cell. Biochem. 46, 21–26.

Hilton, D.J. and Nicola, N.A. (1992). Kinetic analyses of the binding of leukemia inhibitory factor to receptor on cells and membranes and in detergent solution. J. Biol. Chem. 267, 10238–10247.

Hilton, D.J., Nicola, N.A., Gough, N.M. and Metcalf, D. (1988a). Resolution and purification of three distinct factors produced by Krebs ascites cells which have differentiation-inducing activity on murine myeloid leukemic cell lines. J. Biol. Chem. 263, 9238–9243.

Hilton, D.J., Nicola, N.A. and Metcalf, D. (1988b). Purification of a murine leukemia inhibitory factor from Krebs ascites cells. Anal. Biochem. 173, 359–367.

Hilton, D.J., Nicola, N.A. and Metcalf, D. (1988c). Specific binding of murine leukemia inhibitory factor to normal and leukemic monocytic cells. Proc. Natl Acad. Sci. USA 85, 5971–5975.

Hilton, D.J., Nicola, N.A. and Metcalf, D. (1991a). Distribution and comparison of receptors for leukemia inhibitory factor on murine hemopoietic and hepatic cells. J. Cell. Physiol. 146, 207–215.

Hilton, D.J., Nicola, N.A., Waring, P.M. and Metcalf, D. (1991b). Clearance and fate of leukemia-inhibitory factor (LIF) after injection into mice. J. Cell. Physiol. 148, 430–439.

Hilton, D.J., Nicola, N.A. and Metcalf, D. (1992). Distribution and binding properties of receptors for leukaemia inhibitory factor. Ciba Found. Symp. 167, 227–239.

Hilton, D.J., Hilton, A.A., Raicevic, A., et al. (1994). Cloning of a murine IL-11 receptor alpha-chain; requirement for gp130 for high affinity binding and signal transduction. EMBO J. 13, 4765–4775.

Hirano, T., Matsuda, T. and Nakajima, K. (1994). Signal transduction through gp130 that is shared among the receptors for the interleukin 6 related cytokine subfamily. Stem Cells 12, 262–277.

Hogan, B. and Tilly, R. (1978). In vitro development of inner cell masses isolated immunosurgically from mouse blastocysts. I. Inner cell masses from 3.5-day p.c. blastocysts incubated for 24 h before immunosurgery. J. Embryol Exp. Morphol. 45, 93–105.

Ichihara, M., Hara, T., Kim, H., Murate, T. and Miyajima, A. (1997). Oncostatin M and leukemia inhibitory factor do not use the same functional receptor in mice. Blood 90(1), 165–173.

Ichikawa, Y. (1969). Differentiation of a cell line of myeloid leukemia. J. Cell. Physiol. 74, 223–234.

Ichikawa, Y. (1970). Further studies on the differentiation of a cell line of myeloid leukemia. J. Cell. Physiol. 76, 175–184.

Ihle, J.N. (1994). The Janus kinase family and signaling through members of the cytokine receptor superfamily. Proc. Soc. Exp. Biol. Med. 206, 268–272.

Ihle, J.N., Witthuhn, B., Tang, B., Yi, T. and Quelle, F.W. (1994a). Cytokine receptors and signal transduction. Baillières Clinical Haematology 7, 17–48.

Ihle, J.N., Witthuhn, B.A., Quelle, F.W., et al. (1994b). Signaling by the cytokine receptor superfamily: JAKs and STATs. Trends Biochem. Sci. 19, 222–227.

Imai, T., Koike, K., Kubo, T., et al. (1991). Interleukin-6 supports human megakaryocytic proliferation and differentiation in vitro. Blood 78, 1969–1974.

Ip, N.Y., Nye, S.H., Boulton, T.G., et al. (1992). CNTF and LIF act on neuronal cells via shared signaling pathways that involve the IL-6 signal transducing receptor component gp130. Cell 69, 1121–1132.

Ishimi, Y., Abe, E., Jin, C.H., et al. (1992). Leukemia inhibitory factor/differentiation-stimulating factor (LIF/D-factor): regulation of its production and possible roles in bone metabolism. J. Cell. Physiol. 152, 71–78.

Kalberg, C., Yung, S.Y. and Kessler, J.A. (1993). The cholinergic stimulating effects of ciliary neurotrophic factor and leukemia inhibitory factor are mediated by protein kinase C. J. Neurochem. 60, 145–152.

Kaushansky, K., Lok, S., Holly, R.D., et al. (1994). Promotion of megakaryocyte progenitor expansion and differentiation by the c-Mpl ligand thrombopoietin. Nature 369, 568–571.

Kishimoto, T., Hibi, M., Murakami, M., Narazaki, M., Saito, M. and Taga, T. (1992). The molecular biology of interleukin 6 and its receptor. Ciba Found. Symp. 167, 5–16.

Kishimoto, T., Taga, T. and Akira, S. (1994). Cytokine signal transduction. Cell 76, 253–262.

Koike, K., Nakahata, T., Kubo, T., et al. (1990). Interleukin-6 enhances murine megakaryocytopoiesis in serum-free culture. Blood 75, 2286–2291.

Kojima, K., Kanzaki, H., Iwai, M., et al. (1994). Expression of leukemia inhibitory factor in human endometrium and placenta. Biol. Reprod. 50, 882–887.

Kola, I., Davey, A. and Gough, N.M. (1990). Localization of

the murine leukemia inhibitory factor gene near the centromere on chromosome 11. Growth Factors 2, 235–240.

Koopman, P. and Cotton, R.G. (1984). A factor produced by feeder cells which inhibits embryonal carcinoma cell differentiation. Characterization and partial purification. Exp. Cell. Res. 154, 233–242.

Kordula, T., Rokita, H., Koj, A., Fiers, W., Gauldie, J. and Baumann, H. (1991). Effects of interleukin-6 and leukemia inhibitory factor on the acute phase response and DNA synthesis in cultured rat hepatocytes. Lymphocyte Cytokine Res. 10, 23–26.

Kotzbauer, P.T., Lampe, P.A., Estus, S., Milbrandt, J. and Johnson, E., Jr (1994). Postnatal development of survival responsiveness in rat sympathetic neurons to leukemia inhibitory factor and ciliary neurotrophic factor. Neuron 12, 763–773.

Kreisberg, R., Detrick, M.S. and Moore, R.N. (1993). Opposing effects of tumor necrosis factor alpha and leukemia inhibitory factor in lipopolysaccharide-stimulated myelopoiesis. Infect. Immun. 61, 418–422.

Krystosek, A. and Sachs, L. (1976). Control of lysozyme induction in the differentiation of myeloid leukemic cells. Cell 9, 675–684.

Kuropatwinski, K.K., De-Imus, C., Gearing, D., Baumann, H. and Mosley, B. (1997). Influence of subunit combinations on signaling by receptors for oncostatin M, leukemia inhibitory factor, and interleukin-6. J. Biol. Chem. 272(24), 15135–15144.

Kurzrock, R., Estrov, Z., Wetzler, M., Gutterman, J.U. and Talpaz, M. (1991). LIF: not just a leukemia inhibitory factor. Endocrine Rev. 12, 208–217.

Lamb, P., Kessler, L.V., Suto, C., et al. (1994). Rapid activation of proteins that interact with the interferon gamma activation site in response to multiple cytokines. Blood 83, 2063–2071.

Layton, M.J., Cross, B.A., Metcalf, D., Ward, L.D., Simpson, R.J. and Nicola, N.A. (1992). A major binding protein for leukemia inhibitory factor in normal mouse serum: identification as a soluble form of the cellular receptor. Proc. Natl Acad. Sci. USA 89, 8616–8620.

Layton, M.J., Lock, P., Metcalf, D. and Nicola, N.A. (1994a). Cross-species receptor binding characteristics of human and mouse leukemia inhibitory factor suggest a complex binding interaction. J. Biol. Chem. 269, 17048–17055.

Layton, M.J., Owczarek, C.M., Metcalf, D., et al. (1994b). Complex binding of leukemia inhibitory factor to its membrane-expressed and soluble receptors. Proc. Soc. Exp. Biol. Med. 206, 295–298.

Leary, A.G., Wong, G.G., Clark, S.C., Smith, A.G. and Ogawa, M. (1990). Leukemia inhibitory factor differentiation-inhibiting activity/human interleukin for DA cells augments proliferation of human hematopoietic stem cells. Blood 75, 1960–1964.

Lecron, J.C., Roblot, P., Chevalier, S., et al. (1993). High circulating leukaemia inhibitory factor (LIF) in patients with giant cell arteritis: independent regulation of LIF and IL-6 under corticosteroid therapy. Clin. Exp. Immunol. 92, 23–26.

Lin, L.F., Mismer, D., Lile, J.D., et al. (1989). Purification, cloning, and expression of ciliary neurotrophic factor (CNTF). Science 246, 1023–1025.

Lipton, J.H. and Sachs, L. (1981). Characterization of macrophage- and granulocyte-inducing proteins for normal and leukemic myeloid cells produced by the Krebs ascites tumor. Biochim. Biophys. Acta 4, 552–569.

Lok, S., Kaushansky, K., Holly, R.D., et al. (1994). Cloning and expression of murine thrombopoietin cDNA and stimulation of platelet production in vivo. Nature 369, 565–568.

Lord, K.A., Abdollahi, A., Thomas, S.M., et al. (1991). Leukemia inhibitory factor and interleukin-6 trigger the same immediate early response, including tyrosine phosphorylation, upon induction of myeloid leukemia differentiation. Mol. Cell. Biol. 11, 4371–4379.

Lotem, J. and Sachs, L. (1992). Regulation of leukaemic cells by interleukin 6 and leukaemia inhibitory factor. Ciba Found. Symp. 167, 80–88.

Lotz, M., Moats, T. and Villiger, P.M. (1992). Leukemia inhibitory factor is expressed in cartilage and synovium and can contribute to the pathogenesis of arthritis. J. Clin. Invest. 90, 888–896.

Lowe, C., Cornish, J., Callon, K., Martin, T.J. and Reid, I.R. (1991). Regulation of osteoblast proliferation by leukemia inhibitory factor. J. Bone Miner. Res. 6, 1277–1283.

Ludlam, W.H. and Kessler, J.A. (1993). Leukemia inhibitory factor and ciliary neurotrophic factor regulate expression of muscarinic receptors in cultured sympathetic neurons. Dev. Biol. 155, 497–506.

Ludlam, W.H., Zang, Z., McCarson, K.E., Krause, J.E., Spray, D.C. and Kessler, J.A. (1994). mRNAs encoding muscarinic and substance P receptors in cultured sympathetic neurons are differentially regulated by LIF or CNTF. Dev. Biol. 164, 528–539.

Maeda, M., Horiuchi, M., Numa, S. and Ichikawa, Y. (1977). Characterization of a differentiation-stimulating factor for mouse myeloid leukemia cells. Gann 68, 435–447.

Maekawa, T., Metcalf, D. and Gearing, D.P. (1990). Enhanced suppression of human myeloid leukemic cell lines by combinations of IL-6, LIF, GM-CSF and G-CSF. Int. J. Cancer 45, 353–358.

Magnuson, T., Epstein, C.J., Silver, L.M. and Martin, G.R. (1982). Pluripotent embryonic stem cell lines can be derived from tw5/tw5 blastocysts. Nature 298, 750–753.

Marinkovic, S., Jahreis, G.P., Wong, G.G. and Baumann, H. (1989). IL-6 modulates the synthesis of a specific set of acute phase plasma proteins in vivo. J. Immunol. 142, 808–812.

Marshall, M.K., Doerrler, W., Feingold, K.R. and Grunfeld, C. (1994). Leukemia inhibitory factor induces changes in lipid metabolism in cultured adipocytes. Endocrinology 135, 141–147.

Martin, G.R. (1981). Isolation of a pluripotent cell line from early mouse embryos cultured in medium conditioned by teratocarcinoma stem cells. Proc. Natl Acad. Sci. USA 78, 7634–7638.

Martinou, J.C., Martinou, I. and Kato, A.C. (1992). Cholinergic differentiation factor (CDF/LIF) promotes survival of isolated rat embryonic motoneurons in vitro. Neuron 8, 737–744.

Marusic, A., Kalinowski, J.F., Jastrzebski, S. and Lorenzo, J.A. (1993). Production of leukemia inhibitory factor mRNA and protein by malignant and immortalized bone cells. J. Bone Miner. Res. 8, 617–624.

Matsuda, T. and Hirano, T. (1994). Association of p72 tyrosine kinase with Stat factors and its activation by interleukin-3, interleukin-6, and granulocyte colony-stimulating factor. Blood 83, 3457–3461.

Matsui, Y., Zsebo, K. and Hogan, B.L. (1992). Derivation of

pluripotential embryonic stem cells from murine primordial germ cells in culture. Cell 70, 841–847.

Mayer, M., Bhakoo, K. and Noble, M. (1994). Ciliary neurotrophic factor and leukemia inhibitory factor promote the generation, maturation and survival of oligodendrocytes in vitro. Development 120, 143–153.

Metcalf, D. (1989). Actions and interactions of G-CSF, LIF, and IL-6 on normal and leukemic murine cells. Leukemia 3, 349–355.

Metcalf, D. (1991). The leukemia inhibitory factor (LIF). Int. J. Cell Cloning 9, 95–108.

Metcalf, D. (1992). Leukemia inhibitory factor—a puzzling polyfunctional regulator. Growth Factors 7, 169–173.

Metcalf, D. and Gearing, D.P. (1989a). Fatal syndrome in mice engrafted with cells producing high levels of the leukemia inhibitory factor. Proc. Natl Acad. Sci. USA 86, 5948–5952.

Metcalf, D. and Gearing, D.P. (1989b). A myelosclerotic syndrome in mice engrafted with cells producing high levels of leukemia inhibitory factor (LIF). Leukemia 3, 847–852.

Metcalf, D., Hilton, D.J. and Nicola, N.A. (1988). Clonal analysis of the actions of the murine leukemia inhibitory factor on leukemic and normal murine hemopoietic cells. Leukemia 2, 216–221.

Metcalf, D., Nicola, N.A. and Gearing, D.P. (1990). Effects of injected leukemia inhibitory factor on hematopoietic and other tissues in mice. Blood 76, 50–56.

Metcalf, D., Hilton, D. and Nicola, N.A. (1991). Leukemia inhibitory factor can potentiate murine megakaryocyte production in vitro. Blood 77, 2150–2153.

Mezzasoma, L., Biondi, R., Benedetti, C., et al. (1993). In vitro production of leukemia inhibitory factor (LIF) by Hep G2 hepatoblastoma cells. J. Biol. Regul. Homeostatic Agents 7, 126–132.

Michikawa, M., Kikuchi, S. and Kim, S.U. (1992). Leukemia inhibitory factor (LIF) mediated increase of choline acetyltransferase activity in mouse spinal cord neurons in culture. Neurosci. Lett. 140, 75–77.

Miles, S.A., Martinez-Maza, O., Rezai, A., et al. (1992). Oncostatin M as a potent mitogen for AIDS-Kaposi's sarcoma-derived cells. Science 255, 1432–1434.

Moore, K.A., Deisseroth, A.B., Reading, C.L., Williams, D.E. and Belmont, J.W. (1992). Stromal support enhances cell-free retroviral vector transduction of human bone marrow long-term culture-initiating cells. Blood 79, 1393–1399.

Moran, C.S., Campbell, J.H., Simmons, D.L. and Campbell, G.R. (1994). Human leukemia inhibitory factor inhibits development of experimental atherosclerosis. Art. Thromb. 14, 1356–1363.

Moreau, J.F., Bonneville, M., Peyrat, M.A., Jacques, Y. and Soulillou, J.P. (1986). Capacity of alloreactive human T clones to produce factor(s) inducing proliferation of the IL3-dependent DA-1 murine cell line. I. Evidence that this production is under IL2 control. Ann. Inst. Pasteur Immunol. 1, 25–37.

Moreau, J.F., Bonneville, M., Godard, A., et al. (1987). Characterization of a factor produced by human T cell clones exhibiting eosinophil-activating and burst-promoting activities. J. Immunol. 138, 3844–3849.

Moreau, J.F., Donaldson, D.D., Bennett, F., Witek-Giannotti, J., Clark, S.C. and Wong, G.G. (1988). Leukaemia inhibitory factor is identical to the myeloid growth factor human interleukin for DA cells. Nature 336, 690–692.

Mori, M., Yamaguchi, K. and Abe, K. (1989). Purification of a lipoprotein lipase-inhibiting protein produced by a melanoma cell line associated with cancer cachexia. Biochem. Biophys. Res. Commun. 160, 1085–1092.

Mori, M., Yamaguchi, K., Honda, S., et al. (1991). Cancer cachexia syndrome developed in nude mice bearing melanoma cells producing leukemia-inhibitory factor. Cancer Res. 51, 6656–6659.

Mummery, C.L., Feyen, A., Freund, E. and Shen, S. (1990). Characteristics of embryonic stem cell differentiation: a comparison with two embryonal carcinoma cell lines. Cell Differ. Dev. 30, 195–206.

Murakami, M., Hibi, M., Nakagawa, N., et al. (1993). IL-6-induced homodimerization of gp130 and associated activation of a tyrosine kinase. Science 260, 1808–1810.

Murphy, M., Reid, K., Brown, M.A. and Bartlett, P.F. (1993). Involvement of leukemia inhibitory factor and nerve growth factor in the development of dorsal root ganglion neurons. Development 117, 1173–1182.

Nagy, A., Gocza, E., Diaz, E.M., et al. (1990). Embryonic stem cells alone are able to support fetal development in the mouse. Development 110, 815–821.

Nair, B.C., DeVico, A.L., Nakamura, S., et al. (1992). Identification of a major growth factor for AIDS-Kaposi's sarcoma cells as oncostatin M. Science 255, 1430–1432.

Nawa, H., Nakanishi, S. and Patterson, P.H. (1991). Recombinant cholinergic differentiation factor (leukemia inhibitory factor) regulates sympathetic neuron phenotype by alterations in the size and amounts of neuropeptide mRNAs. J. Neurochem. 56, 2147–2150.

Nichols, J., Evans, E.P. and Smith, A.G. (1990). Establishment of germ-line-competent embryonic stem (ES) cells using differentiation inhibiting activity. Development 110, 1341–1348.

Nicola, N.A. and Metcalf, D. (1991). Subunit promiscuity among hemopoietic growth factor receptors. Cell 67, 1–4.

Nishimoto, N., Ogata, A., Shima, Y., et al. (1994). Oncostatin M, leukemia inhibitory factor, and interleukin 6 induce the proliferation of human plasmacytoma cells via the common signal transducer, gp130. J. Exp. Med. 179, 1343–1347.

Nishiyama, K., Collodi, P. and Barnes, D. (1993). Regulation of glial fibrillary acidic protein in serum-free mouse embryo (SFME) cells by leukemia inhibitory factor and related peptides. Neurosci. Lett. 163, 114–116.

Noda, M., Vogel, R.L., Hasson, D.M. and Rodan, G.A. (1990). Leukemia inhibitory factor suppresses proliferation, alkaline phosphatase activity, and type I collagen messenger ribonucleic acid level and enhances osteopontin mRNA level in murine osteoblast-like (MC3T3E1) cells. Endocrinology 127, 185–190.

Owczarek, C.M., Layton, M.J., Metcalf, D., et al. (1993). Inter-species chimeras of leukaemia inhibitory factor define a major human receptor-binding determinant. EMBO J. 12, 3487-3495.

Patterson, P.H. (1992). The emerging neuropoietic cytokine family: first CDF/LIF, CNTF and IL-6; next ONC, MGF, GCSF? Curr. Opin. Neurobiol. 2, 94–97.

Patterson, P.H. (1994). Leukemia inhibitory factor, a cytokine at the interface between neurobiology and immunology. Proc. Natl Acad. Sci. USA 91, 7833–7835.

Patterson, P.H. and Nawa, H. (1993). Neuronal differentiation factors/cytokines and synaptic plasticity. Cell 72, 123–137.

Pease, S. and Williams, R.L. (1990). Formation of germ-line chimeras from embryonic stem cells maintained with recombinant leukemia inhibitory factor. Exp. Cell Res. 190, 209–211.

Pease, S., Braghetta, P., Gearing, D., Grail, D. and Williams, R.L. (1990). Isolation of embryonic stem (ES) cells in media supplemented with recombinant leukemia inhibitory factor (LIF). Dev. Biol. 141, 344–352.

Pennica, D., King, K.L., Shaw, K.J., et al. (1995). Expression cloning of cardiotrophin 1, a cytokine that induces myocyte hypertrophy. Proc. Natl Acad. Sci. USA 92, 1142–1146.

Pergolizzi, R.G. and Erster, S.H. (1994). Analysis of chromosome 22 loci in meningioma. Alterations in the leukemia inhibitory factor (LIF) locus. Mol. Chem. Neuropathol. 21, 189–217.

Pesce, M., Farrace, M.G., Piacentini, M., Dolci, S. and De Felici, M. (1993). Stem cell factor and leukemia inhibitory factor promote primordial germ cell survival by suppressing programmed cell death (apoptosis). Development 118, 1089–1094.

Piquet-Pellorce, C., Grey, L., Mereau, A. and Heath, J.K. (1994). Are LIF and related cytokines functionally equivalent? Exp. Cell Res. 213, 340–347.

Pruitt, S.C. and Natoli, T.A. (1992). Inhibition of differentiation by leukemia inhibitory factor distinguishes two induction pathways in P19 embryonal carcinoma cells. Differentiation 50, 57–65.

Purvis, D.H. and Mabbutt, B.C. (1997). Solution dynamics and secondary structure of murine leukemia inhibitory factor: a four-helix cytokine with a rigid CD loop. Biochemistry 36(33), 10146–10154.

Qiu, L., Bernd, P. and Fukada, K. (1994). Cholinergic neuronal differentiation factor (CDF)/leukemia inhibitory factor (LIF) binds to specific regions of the developing nervous system in vivo. Dev. Biol. 163, 516–520.

Rao, M.S., Sun, Y., Escary, J.L., et al. (1993). Leukemia inhibitory factor mediates an injury response but not a target-directed developmental transmitter switch in sympathetic neurons. Neuron 11, 1175–1185.

Rathjen, P.D., Toth, S., Willis, A., Heath, J.K. and Smith, A.G. (1990). Differentiation inhibiting activity is produced in matrix-associated and diffusible forms that are generated by alternate promoter usage. Cell 62, 1105–1114.

Raz, R., Durbin, J.E. and Levy, D.E. (1994). Acute phase response factor and additional members of the interferon-stimulated gene factor 3 family integrate diverse signals from cytokines, interferons, and growth factors. J. Biol. Chem. 269, 24391–24395.

Reid, L.R., Lowe, C., Cornish, J., et al. (1990). Leukemia inhibitory factor: a novel bone-active cytokine. Endocrinology 126, 1416–1420.

Richards, L.J., Kilpatrick, T.J., Bartlett, P.F. and Murphy, M. (1992). Leukemia inhibitory factor promotes the neuronal development of spinal cord precursors from the neural tube. J. Neurosci. Res. 33, 476–484.

Richards, C.D., Shoyab, M., Brown, T.J. and Gauldie, J. (1993). Selective regulation of metalloproteinase inhibitor (TIMP-1) by oncostatin M in fibroblasts in culture. J. Immunol. 150, 5596–5603.

Robertson, E.J. (1991). Using embryonic stem cells to introduce mutations into the mouse germ line. Biol. Reprod. 44, 238–245.

Robertson, S.A., Seamark, R.F., Guilbert, L.J. and Wegmann, T.G. (1994). The role of cytokines in gestation. Crit. Rev. Immunol. 14, 239–292.

Robertson, T.A., Maley, M.A., Grounds, M.D. and Papadimitriou, J.M. (1993). The role of macrophages in skeletal muscle regeneration with particular reference to chemotaxis. Exp. Cell Res. 207, 321–331.

Robinson, R.C., Grey, L.M., Staunton, D., et al. (1994). The crystal structure and biological function of leukemia inhibitory factor: implications for receptor binding. Cell 77, 1101–1116.

Rodan, S.B., Wesolowski, G., Hilton, D.J., Nicola, N.A. and Rodan, G.A. (1990). Leukemia inhibitory factor binds with high affinity to preosteoblastic RCT-1 cells and potentiates the retinoic acid induction of alkaline phosphatase. Endocrinology 127, 1602–1608.

Rose, T.M. and Bruce, A.G. (1991). Oncostatin M is a member of a cytokine family that includes leukemia-inhibitory factor, granulocyte colony-stimulating factor, and interleukin 6. Proc. Natl Acad. Sci. USA 88, 8641–8645.

Rossant, J. (1975a). Investigation of the determinative state of the mouse inner cell mass. I. Aggregation of isolated inner cell masses with morulae. J. Embryol. Exp. Morphol. 33, 979–990.

Rossant, J. (1975b). Investigation of the determinative state of the mouse inner cell mass. II. The fate of isolated inner cell masses transferred to the oviduct. J. Embryol. Exp. Morphol. 33, 991–1001.

Rossant, J. (1976). Investigation of inner cell mass determination by aggregation of isolated rat inner cell masses with mouse morulae. J. Embryol. Exp. Morphol. 36, 163–174.

Ruff-Jamison, S., Zhong, Z., Wen, Z., Chen, K., Darnell, J., Jr and Cohen, S. (1994). Epidermal growth factor and lipopolysaccharide activate Stat3 transcription factor in mouse liver. J. Biol. Chem. 269, 21933–21935.

Ryffel, B. (1993). Pathology induced by leukemia inhibitory factor. Int. Rev. Exp. Pathol. 34, 69–72.

Schaafsma, M.R., Falkenburg, J.H., Duinkerken, N., et al. (1992). Human interleukin for DA cells (HILDA) does not affect the proliferation and differentiation of hematopoietic progenitor cells in human long-term bone marrow cultures. Exp. Hematol. 20, 6–10.

Schiemann, W.P., Bartoe, J.L. and Nathanson, N.M. (1997). Box 3-independent signaling mechanisms are involved in leukemia inhibitory factor receptor alpha- and gp130-mediated stimulation of mitogen-activated protein kinase. Evidence for participation of multiple signaling pathways which converge at Ras. J. Biol. Chem. 272(26), 16631–16636.

Shabo, Y., Lotem, J., Rubinstein, M., et al. (1988). The myeloid blood cell differentiation-inducing protein MGI-2A is interleukin-6. Blood 72, 2070–2073.

Shadiack, A.M., Hart, R.P., Carlson, C.D. and Jonakait, G.M. (1993). Interleukin-1 induces substance P in sympathetic ganglia through the induction of leukemia inhibitory factor (LIF). J. Neurosci. 13, 2601–2609.

Shen, M.M., Skoda, R.C., Cardiff, R.D., Campos-Torres, J., Leder, P. and Ornitz, D.M. (1994). Expression of LIF in transgenic mice results in altered thymic epithelium and apparent interconversion of thymic and lymph node morphologies. EMBO J. 13, 1375–1385.

Shual, K., Ziemiecki, A., Wilks, A.F., et al. (1993). Polypeptide signalling to the nucleus through tyrosine phosphorylation of Jak and Stat proteins. Nature 366, 580–583.

Smith, A.G. and Hooper, M.L. (1987). Buffalo rat liver cells produce a diffusible activity which inhibits the differentiation of murine embryonal carcinoma and embryonic stem cells. Dev. Biol. 121, 1–9.

Smith, A.G., Heath, J.K., Donaldson, D.D., *et al.* (1988). Inhibition of pluripotential embryonic stem cell differentiation by purified polypeptides. Nature 336, 688–690.

Stahl, J. and Gough, N.M. (1993). Delineation of positive and negative control elements within the promoter region of the murine leukemia inhibitory factor (LIF) gene. Cytokine 5, 386–393.

Stahl, J., Gearing, D.P., Willson, T.A., Brown, M.A., King, J.A. and Gough, N.M. (1990). Structural organization of the genes for murine and human leukemia inhibitory factor. Evolutionary conservation of coding and non-coding regions. J. Biol. Chem. 265, 8833–8841.

Stahl, N., Davis, S., Wong, V., *et al.* (1993). Cross-linking identifies leukemia inhibitory factor-binding protein as a ciliary neurotrophic factor receptor component. J. Biol. Chem. 268, 7628–7631.

Stahl, N., Boulton, T.G., Farruggella, T., *et al.* (1994). Association and activation of Jak-Tyk kinases by CNTF-LIF-OSM-IL-6 beta receptor components. Science 263, 92–95.

Starr, R., Novak, U., Willson, T.A., *et al.* (1997). Distinct roles for leukemia inhibitory factor receptor alpha-chain and gp130 in cell type-specific signal transduction. J. Biol. Chem. 272(32), 19982–19986.

Stewart, C.L., Kaspar, P., Brunet, L.J., *et al.* (1992). Blastocyst implantation depends on maternal expression of leukaemia inhibitory factor. Nature 359, 76–79.

Stockli, K.A., Lottspeich, F., Sendtner, M., *et al.* (1989). Molecular cloning, expression and regional distribution of rat ciliary neurotrophic factor. Nature 342, 920–923.

Sun, Y., Rao, M.S., Zigmond, R.E. and Landis, S.C. (1994). Regulation of vasoactive intestinal peptide expression in sympathetic neurons in culture and after axotomy: the role of cholinergic differentiation factor/leukemia inhibitory factor. J. Neurobiol. 25, 415–430.

Sutherland, G.R., Baker, E., Hyland, V.J., Callen, D.F., Stahl, J. and Gough, N.M. (1989). The gene for human leukemia inhibitory factor (LIF) maps to 22q12. Leukemia 3, 9–13.

Szilvassy, S.J. and Cory, S. (1994). Efficient retroviral gene transfer to purified long-term repopulating hematopoietic stem cells. Blood 84, 74–83.

Taga, T., Hibi, M., Hirata, Y., *et al.* (1989). Interleukin-6 triggers the association of its receptor with a possible signal transducer, gp130. Cell 58, 573–581.

Takagi, N. (1993). Variable X chromosome inactivation patterns in near-tetraploid murine EC × somatic cell hybrid cells differentiated in vitro. Genetica 88, 107–117.

Thaler, C.D., Suhr, L., Ip, N. and Katz, D.M. (1994). Leukemia inhibitory factor and neurotrophins support overlapping populations of rat nodose sensory neurons in culture. Dev. Biol. 161, 338–344.

Thoma, B., Bird, T.A., Friend, D.J., Gearing, D.P. and Dower, S.K. (1994). Oncostatin M and leukemia inhibitory factor trigger overlapping and different signals through partially shared receptor complexes. J. Biol. Chem. 269, 6215–6222.

Tomida, M., Yamamoto-Yamaguchi, Y. and Hozumi, M. (1984). Purification of a factor inducing differentiation of mouse myeloid leukemic M1 cells from conditioned medium of mouse fibroblast L929 cells. J. Biol. Chem. 259, 10978–10982.

Tomida, M., Yamamoto-Yamaguchi, Y. and Hozumi, M. (1993). Pregnancy associated increase in mRNA for soluble D-factor/LIF receptor in mouse liver. FEBS Lett. 334, 193–197.

Tomida, M., Yamamoto-Yamaguchi, Y. and Hozumi, M. (1994). Three different cDNAs encoding mouse D-factor/LIF receptor. J. Biochem. 115, 557–562.

Ure, D.R. and Campenot, R.B. (1994). Leukemia inhibitory factor and nerve growth factor are retrogradely transported and processed by cultured rat sympathetic neurons. Dev. Biol. 162, 339–347.

Ure, D.R., Campenot, R.B. and Acheson, A. (1992). Cholinergic differentiation of rat sympathetic neurons in culture: effects of factors applied to distal neurites. Dev. Biol. 154, 388–395.

Van Beek, E., Van der Wee-Pals, L., van de Ruit, M., Nijweide, P., Papapoulos, S. and Lowik, C. (1993). Leukemia inhibitory factor inhibits osteoclastic resorption, growth, mineralization, and alkaline phosphatase activity in fetal mouse metacarpal bones in culture. J. Bone Miner. Res. 8, 191–198.

Van Damme, J., Uyttenhove, C., Houssiau, F., Put, W., Proost, P. and Van Snick, J. (1992). Human growth factor for murine interleukin (IL)-9 responsive T cell lines: co-induction with IL-6 in fibroblasts and identification as LIF/HILDA. Eur. J. Immunol. 22, 2801–2808.

Van Vlasselaer, P. (1992). Leukemia inhibitory factor (LIF): a growth factor with pleiotropic effects on bone biology. Prog. Growth Factor Res. 4, 337–353.

Verfaillie, C. and McGlave, P. (1991). Leukemia inhibitory factor/human interleukin for DA cells: a growth factor that stimulates the in vitro development of multipotential human hematopoietic progenitors. Blood 77, 263–270.

Waring, P., Wycherley, K., Cary, D., Nicola, N. and Metcalf, D. (1992). Leukemia inhibitory factor levels are elevated in septic shock and various inflammatory body fluids. J. Clin. Invest. 90, 2031–2037.

Waring, P., Wall, D., Dauer, R., Parkin, D. and Metcalf, D. (1993a). The effects of leukaemia inhibitory factor on platelet function. Br. J. Haematol. 83, 80–87.

Waring, P.M., Carroll, G.J., Kandiah, D.A., Buirski, G. and Metcalf, D. (1993b). Increased levels of leukemia inhibitory factor in synovial fluid from patients with rheumatoid arthritis and other inflammatory arthritides. Arthritis. Rheum. 36, 911–915.

Warren, M.K., Conroy, L.B. and Rose, J.S. (1989). The role of interleukin 6 and interleukin 1 in megakaryocyte development. Exp. Hematal. 17, 1095–1099.

Warren, M.K., Guertin, M., Rudzinski, I. and Seidman, M.M. (1993). A new culture and quantitation system for megakaryocyte growth using cord blood CD34+ cells and the GPIIb/IIIa marker. Exp. Hematol. 21, 1473–1479.

Wegenka, U.M., Lutticken, C., Buschmann, J., *et al.* (1994). The interleukin-6-activated acute-phase response factor is antigenically and functionally related to members of the signal transducer and activator of transcription (STAT) family. Mol. Cell. Biol. 14, 3186–3196.

Wendling, F., Maraskovsky, E., Debili, N., *et al.* (1994). cMpl ligand is a humoral regulator of megakaryocytopoiesis. Nature 369, 571–574.

Wesselingh, S.L., Levine, B., Fox, R.J., Choi, S. and Griffin, D.E. (1994). Intracerebral cytokine mRNA expression during fatal and nonfatal alphavirus encephalitis suggests a predominant type 2 T cell response. J. Immunol. 152, 1289–1297.

Wetzler, M., Talpaz, M., Lowe, D.G., Baiocchi, G., Gutterman, J.U. and Kurzrock, R. (1991). Constitutive expression of leukaemia inhibitory factor RNA by human bone marrow stromal cells and modulation by IL-1, TNF-alpha, and TGF-beta. Exp. Hematol. 19, 347–351.

Williams, R.L., Hilton, D.J., Pease, S., et al. (1988). Myeloid leukaemia inhibitory factor maintains the developmental potential of embryonic stem cells. Nature 336, 684–687.

Williams, N., Bertoncello, I., Jackson, H., Arnold, J. and Kavnoudias, H. (1992). The role of interleukin 6 in megakaryocyte formation, megakaryocyte development and platelet production. Ciba Found. Symp. 167, 160–170.

Wilson, T.A., Metcalf, D. and Gough, N.M. (1992). Cross-species comparison of the sequence of the leukaemia inhibitory factor gene and its protein. Eur. J. Biochem. 204, 21–30.

Yamaguchi-Yamamoto, Y., Tomida, M. and Hozumi, M. (1993). Pregnancy associated increase in differentiation-stimulating factor (D-factor)/leukemia inhibitory factor (LIF)-binding substance(s) in mouse serum. Leuk. Res. 17, 515–522.

Yamamori, T. (1992). Molecular mechanisms for generation of neural diversity and specificity: roles of polypeptide factors in development of postmitotic neurons. Neurosci. Res. 12, 545–582.

Yamamori, T., Fukada, K., Aebersold, R., Korsching, S., Fann, M.J. and Patterson, P.H. (1989). The cholinergic neuronal differentiation factor from heart cells is identical to leukemia inhibitory factor. Science 246, 1412–1416.

Yamamoto, Y., Tomida, M. and Hozumi, M. (1980). Production by mouse spleen cells of factors stimulating differentiation of mouse myeloid leukemic cells that differ from the colony-stimulating factor. Cancer Res. 40, 4804–4809.

Yamamoto-Yamaguchi, Y., Tomida, M. and Hozumi, M. (1986). Specific binding of a factor inducing differentiation to mouse myeloid leukemic M1 cells. Exp. Cell Res. 164, 97–102.

Yang, Y.C. (1993). Interleukin 11: an overview. Stem Cells 11, 474–486.

Yang, Z.M., Le, S.P., Chen, D.B. and Harper, M.J. (1994). Temporal and spatial expression of leukemia inhibitory factor in rabbit uterus during early pregnancy. Mol. Reprod. Dev. 38, 148–152.

Yin, T., Yasukawa, K., Taga, T., Kishimoto, T. and Yang, Y.C. (1994). Identification of a 130-kilodalton tyrosine-phosphorylated protein induced by interleukin-11 as JAK2 tyrosine kinase, which associates with gp130 signal transducer. Exp. Hematol. 22, 467–472.

Yonemura, Y., Kawakita, M., Masuda, T., Fujimoto, K., Kato, K. and Takatsuki, K. (1992). Synergistic effects of interleukin 3 and interleukin 11 on murine megakaryopoiesis in serum-free culture. Exp. Hematol. 20, 1011–1016.

Yoshida, T., Satoh, M., Nakagaito, Y., Kuno, H. and Takeuchi, M. (1993). Cytokines affecting survival and differentiation of an astrocyte progenitor cell line. Dev. Brain Res. 76, 147–150.

Zhang, X.G., Gu, J.J., Lu, Z.Y., et al. (1994). Ciliary neurotropic factor, interleukin 11, leukemia inhibitory factor, and oncostatin M are growth factors for human myeloma cell lines using the interleukin 6 signal transducer gp130. J. Exp. Med. 179, 1337–1342.

Zhong, Z., Wen, Z. and Darnell, J., Jr (1994). Stat3: a STAT family member activated by tyrosine phosphorylation in response to epidermal growth factor and interleukin-6. Science 264, 95–98.

Zurn, A.D. and Werren, F. (1994). Development of CNS cholinergic neurons in vitro: selective effects of CNTF and LIF on neurons from mesencephalic cranial motor nuclei. Dev. Biol. 163, 309–315.

21. The Steel Factor/kit Ligand/ Stem Cell Factor

Stewart D. Lyman, Anthony R. Mire-Sluis *and* Brian Gliniak

1. Introduction

The discovery of the kit ligand/stem cell factor/Steel factor/mast cell growth factor is intimately linked with the study of two very well known mutations in mice known as *W* (dominant white spotting) and *Sl* (Steel). The *W* mutation was first described in the early part of the twentieth century and was found to be a pleiotrophic mutation (reviewed in Russell, 1979; Silvers, 1979a). Mice afflicted with mutations at the *W* locus were originally identified because of a change in their pigmentation, which is generally seen as a nonpigmented white spot on their bellies. Inheritance of the spotting defect was dominant in crosses with normally pigmented animals. In addition to the pigmentation defect, detailed examination of these mice showed that they also suffered from defects in germ cell development (manifested as reproductive difficulties) and in hematopoiesis (characterized by a macrocytic anemia). Over the years a number of alleles of the W locus were described, all of which manifested themselves with this same pleiotrophic phenotype. Genetic analysis showed that all of these alleles mapped to the same location on mouse chromosome 5.

A second mutation in mice that had a phenotype virtually identical to that observed in W mice was discovered in the 1950s. Despite the similarities in phenotype, this new mutation, termed Steel (Sl), was found to map to mouse chromosome 10 (reviewed in Russell, 1979; Silvers, 1979b). The fact that two different mutations on different chromosomes led to the same complex phenotype that affected pigmentation, germ cells, and hematopoiesis led researchers to conclude that there would likely be some relationship between the proteins encoded at these loci. Elizabeth Russell, who had done much of the pioneering research on both of these mutations, suggested that W and Sl loci might encode a receptor and its cognate ligand (Russell, 1979). The identity of this putative receptor–ligand pair remained a mystery for another decade.

A breakthrough in this area came when it was shown that the W locus encoded a tyrosine kinase receptor protein known as c-kit (Chabot et al., 1988; Geissler et al., 1988). A segment of the c-kit gene had originally been identified as the transforming protein (v-kit) contained within the Hardy–Zuckerman 4 strain of feline sarcoma virus (Besmer et al., 1986). With the recognition that the W locus encoded a receptor, it was immediately suspected that the Sl locus would encode the ligand for that same receptor. A soluble form of the c-kit receptor was shown to bind to fibroblasts from normal mice, but not to fibroblasts isolated from mice carrying a mutation at the Sl locus (Flanagan and Leder, 1990). Cloning of the gene encoded at the Sl locus was accomplished simultaneously by three different groups (see below). Once the ligand had been cloned, its biological activities were determined; the various names given to the protein reflect the variety of its biological activities. One group referred to the protein as mast cell growth factor, since it stimulated the proliferation of murine mast cell lines (D.E. Williams et al., 1990). A second group referred to the protein as stem cell factor, since the protein stimulated colony formation from primitive hematopoietic cells (Zsebo et al., 1990b). A third group named the protein kit ligand, which reflected the fact that the protein cloned is in fact a ligand for the c-kit tyrosine kinase receptor (Huang et al., 1990). The protein has also been referred to as multipotential growth factor since it functions on so many different cell types (Hooper, 1990). In this review we will use the name Steel factor for this protein, since it has been shown to be encoded at the $Steel$ locus on mouse chromosome 10 (Copeland et al., 1990; Huang et al., 1990; Zsebo et al., 1990a).

2.   The Gene

2.1   Cloning of the Steel Factor Gene

As mentioned above, purification of the Steel factor protein and the cloning of a cDNA encoding this factor were reported simultaneously by three different groups, each of which had discovered a different source of the factor (Huang et al., 1990; Martin et al., 1990; D.E. Williams et al., 1990) (Figure 21.1). All three groups used a similar approach in that they first purified the protein from medium conditioned by a cell line, obtained N-terminal amino acid sequence, and then made degenerate oligonucleotide primers based on the protein sequence in order to isolate cDNA clones by the polymerase chain reaction (PCR). D.E. Williams and coworkers (1990) originally detected an activity in medium conditioned by a bone marrow stromal cell line, isolated from normal mice, that stimulated the proliferation of a mast cell line (Boswell et al., 1990). A stromal cell line from $Steel$ mice was unable to provide this stimulation, suggesting that the activity resulted from a protein that was a product of the $Steel$ locus. The protein was subsequently shown to be both a ligand for the c-kit tyrosine kinase receptor and to be encoded at the $Steel$ locus (Anderson et al., 1990; Copeland et al., 1990; D.E. Williams et al., 1990).

A second group, working independently, isolated the Steel factor protein from medium conditioned by a Buffalo rat liver cell line. Their biological assay was detection of an activity that stimulated the proliferation of hematopoietic precursor cells present in the marrow of mice treated with 5-fluorouracil. Analysis of the protein showed it to be both a ligand for the c-kit tyrosine kinase receptor and to be encoded at the $Steel$ locus (Martin et al., 1990; Zsebo et al., 1990a,b).

A third group, also working independently, purified their Steel factor protein from medium conditioned by BALB/c 3T3 cells and assayed its activity on bone marrow-derived mast cells. This group also showed that the protein was a ligand for the c-kit tyrosine kinase receptor and was encoded at the $Steel$ locus (Huang et al., 1990).

2.2   Chromosomal Location of the Steel Factor Gene

The human gene encoding the Steel factor has been localized to human chromosome 12q22-24 (Anderson et al., 1991; Geissler et al., 1991) in a region that is syntenic with the corresponding region of mouse chromosome 10.

```
   1   ccgcctcgcgccgagactagaagcgctgcggaagcagggacagtggagagggcgctgcgctcgggctacccaatgcgtggactatctgcc   90

  91   gccgctgttcgtgcaatatgctqgaqctccagaacagctaaacggagtcgccacaccactgtttgtgctggatcgcagcgctgcctttcc   180

 181   ttATGAAGAAGACACAAACTTGGATTCTCACTTGCATTTATCTTCAGCTGCTCCTATTTAATCCTCTCGTCAAAACTGAAGGGATCTGCA   270

 271   GGAATCGTGTGACTAATAATGTAAAAGACGTCACTAAATTGGTGGCAAATCTTCCAAAAGACTACATGATAACCCTCAAATATGTCCCCG   360

 361   GGATGGATGTTTTGCCAAGTCATTGTTGGATAAGCGAGATGGTAGTACAATTGTCAGACAGCTTGACTGATCTTCTGGACAAGTTTTCAA   450

 451   ATATTTCTGAAGGCTTGAGTAATTATTCCATCATAGACAAACTTGTGAATATAGTGGATGACCTTGTGGAGTGCGTGAAAGAAAACTCAT   540

 541   CTAAGGATCTAAAAAAATCATTCAAGAGCCCAGAACCCAGGCTCTTTACTCCTGAAGAATTCTTTAGAATTTTTAATAGATCCATTGATG   630

 631   CCTTCAAGGACTTTGTAGTGGCATCTGAAACTAGTGATTGTGTGGTTTCTTCAACATTAAGTCCTGAGAAAGATTCCAGAGTCAGTGTCA   720

 721   CAAAACCATTTATGTTACCCCCTGTTGCAGCCAGCTCCCTTAGGAATGACAGCAGTAGCAGTAATAGGAAGGCCAAAAATCCCCCTGGAG   810

 811   ACTCCAGCCTACACTGGGCAGCCATGGCATTGCCAGCATTGTTTTCTCTTATAATTGGCTTTGCTTTTGGAGCCTTATACTGGAAGAAGA   900

 901   GACAGCCAAGTCTTACAAGGGCAGTTGAAAATATACAAATTAATGAAGAGGATAATGAGATAAGTATGTTGCAAGAGAAAGAGAGAGAGT   990

 991   TTCAAGAAGTGTAAattgtggcttgtatcaacactgttactttcgtacattggctggtaacagttcatgttttgcttcataaatgaagcag   1080

1081   ctttaaacaaattcatattctgtctggagtgacagaccacatctttatctgttcttgctacccatgactttatatggatgattcagaaat   1170

1171   tggaacagaatgttttactgtgaaactggcactgaattaatcatctataaagaagaacttgcatggagcaggactctattttaaggactg   1260

1261   cgggacttgggtctcatttagaacttgcagctgatgttggaagagaaagcacgtgtctcagactgcatgtaccatttgcatggctccaga   1350

1351   aatgtctaaatgctgaaaaaacacctagctttattcttcagatacaaactgcag                                       1404
```

Figure 21.1   Gene sequence of the human SCF gene. Adapted from Martin et al., 1990.

2.3 ALTERNATIVE SPLICING OF STEEL FACTOR mRNAs

The cloned Steel factor gene was shown to encode a transmembrane protein that underwent proteolytic cleavage to generate a soluble form of the protein (Figure 21.2). Alternative splicing of Steel factor mRNAs has been reported in both the mouse and the human proteins and has been suggested to be a method of regulating the generation of soluble versus membrane-bound forms of the protein (Anderson et al., 1990, 1991; Flanagan et al., 1991; Huang et al., 1992; Lyman and Williams, 1991). The Steel factor protein can undergo cleavage at either of two alanine residues (residues 164 or 165 that are found within the extracellular domain of the protein) to generate a soluble form of the protein that is biologically active (Anderson et al., 1990; Lu et al., 1991; Martin et al., 1990). This site is considered to be the primary cleavage site within the protein (Majumdar et al., 1994).

Alternative splicing of mRNAs encoding the Steel factor has been demonstrated in both mouse and human genes (Anderson et al., 1990, 1991; Flanagan et al., 1991; Huang et al., 1992; Lyman and Williams, 1991). This splicing regulates the presence or absence of the sixth exon that encodes 28 amino acids and, as discussed above, contains the primary proteolytic cleavage site (Huang et al., 1992; Majumdar et al., 1994). The ratio of protein isoforms containing this sixth exon varies from tissue to tissue (Huang et al., 1992). The presence of the sixth exon is not required for proteolytic cleavage of Steel

factor. Data have been presented indicating that there is a secondary cleavage site present in a stretch of amino acids encoded by exon 7 at a site just upstream of the transmembrane domain. This cleavage site can also be used to generate a soluble form of the ligand (Majumdar et al., 1994). As discussed above, the cell-bound form of the Steel factor appears to be required for normal development in mice since a mutation (Sl^d) that eliminates the membrane-bound form of the factor, but still makes a biologically active soluble form, results in developmental abnormalities (Brannan et al., 1991; Flanagan et al., 1991). It is still unclear, however, how the alternative splicing of Steel factor mRNAs is regulated in a tissue-specific fashion, what regulates the proteolytic cleavage of the protein, and how these effects are manifested physiologically (McNeice and Briddel, 1995).

2.4 STRUCTURE OF THE GENOMIC LOCUS

The genomic locus encoding the Steel factor has been cloned from human sources (Martin et al., 1990) (Figure 21.2). The human gene contains at least eight exons that contain the entire coding region of the protein. Those intron:exon boundaries that have been identified are seen to occur at identical positions within the rat, human, and murine genes.

Steel factor is structurally related to M-CSF and flt3

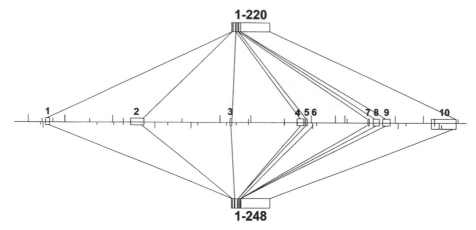

Figure 21.2   Alternative splicing of the human SCF gene. With kind permission of Vann Parker, Amgen.

ligand (Bazan, 1991a; Lyman *et al.*, 1993, 1994b; Hannum *et al.*, 1994) and this is reflected in similar gene structure (Ladner *et al.*, 1988; Martin *et al.*, 1990) and alternative splicing in the sixth exon (Cerretti *et al*, 1988; Lyman *et al.*, 1995).

3.   *The Protein*

3.1   STRUCTURE OF THE PROTEIN

The primary translation product of the Steel factor gene is a type 1 transmembrane protein; that is, the N-terminus of the protein is located outside of the cell (Figure 21.3). The human protein contains 273 amino acids; the first 25 of these amino acids constitute a signal peptide that is removed in the mature protein. In conventional numbering, the first amino acid of the mature protein is amino acid number 1, and therefore the signal peptide is numbered −25 to −1. As a transmembrane protein, the Steel factor is biologically active on the cell surface (Anderson *et al.*, 1990). As mentioned above, the Steel factor protein shares structural similarities to CSF-1 (Bazan, 1991b) and the flt3 ligand (Hannum *et al.*, 1994; Lyman *et al.*, 1993, 1994b). There are four cysteine residues that are conserved between all three of these proteins; in the case of the Steel factor, it has been shown that these cysteine residues form two intramolecular disulfide bonds that establish the three-dimensional structure of the protein (Lu *et al.*, 1991). The Steel factor protein is likely to form a four-helix bundle, since the related CSF-1 protein has been shown by x-ray crystallography to form a four-helix bundle with a structure similar to granulocyte-macrophage colony stimulating factor (GM-CSF) (Pandit *et al.*, 1992) (Figures 21.4 and 21.5).

Epitope mapping suggests that three regions in the first, third, and fourth helices of the protein are responsible for receptor binding (Matous *et al*, 1996;

Mendiaz *et al*, 1996). The Steel factor protein exists as a noncovalent dimer in continual monomer exchange, with a pI of 5.0 (Arakawa *et al.*, 1991; Lu *et al*, 1995, 1996).

The molecular mass of the protein depends on a number of factors, such as the addition of *N*-linked and *O*-linked carbohydrates. Analysis of the human Steel factor produced by Chinese hamster ovary (CHO) cells showed it was glycosylated at asparagines 65 and 120, whereas asparagines 72 and 93 appear to be nonglycosylated (Langley *et al.*, 1992). *O*-Linked sites of glycosylation were at serine 142 and at threonine residues 143 and 155. The CHO-derived material is about 30% sugars by mass (Arakawa *et al.*, 1991).

3.2   LOCATION OF INTRAMOLECULAR DISULFIDE BONDS

Two intramolecular disulfide bonds help to establish the secondary structure of the Steel protein: cysteine 4 is linked to cysteine 89, and cysteine 43 to cysteine 138 (Lu *et al.*, 1991). Circular dichroism spectra of the Steel factor protein show that it has considerable secondary structure, including both α-helical and β-sheet (Arakawa *et al.*, 1991).

3.3   CLEAVAGE OF THE STEEL FACTOR PROTEIN

The mature Steel factor protein (from which the signal sequence has been removed) encodes a 248-amino-acid transmembrane protein (murine and human) that undergoes proteolytic cleavage to generate a soluble form of the protein that is 164–165 amino acids long (Martin *et al.*, 1990) and is biologically active (Anderson *et al.*, 1990; Huang *et al.*, 1990; Zsebo *et al.*, 1990b). The primary site for proteolytic cleavage is encoded within exon 6 (Martin *et al.*, 1990);

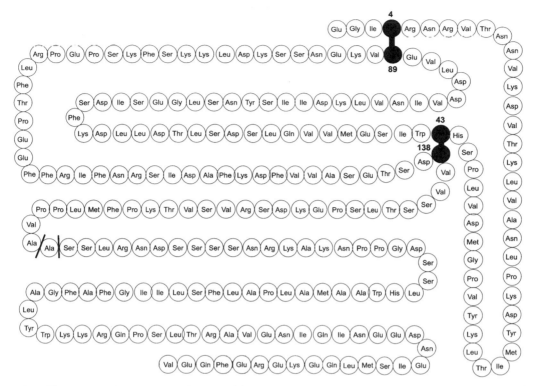

Figure 21.3   The amino acid sequence of human SCF protein. The lines bisecting the alanine residue indicate the variable end residues of the soluble steel factor. Cysteine residues are shaded. Adapted from Martin *et al.*, 1990.

however, mutagenesis experiments have shown that there is a secondary proteolytic cleavage site just upstream of the transmembrane region (Majumdar *et al.*, 1994). This secondary site is utilized only if the primary site is missing, which can occur by splicing out the sixth exon (Anderson *et al.*, 1990; Flanagan *et al.*, 1991; Toksoz *et al.*, 1992). The protease that is responsible for cleavage of the Steel protein has not been identified, and it is unknown whether it is the same protease that generates soluble forms of the structurally related proteins CSF-1 and flt3 ligand. These growth factors, like Steel factor, are expressed as transmembrane proteins and have been shown to undergo proteolytic cleavage to generate soluble forms (Cerretti *et al.*, 1988; Hannum *et al.*, 1994; Lyman *et al.*, 1993).

4.   Cellular Sources and Production

4.1   EXPRESSION OF THE STEEL FACTOR PROTEIN

Much of the information on where the Steel factor protein is expressed comes from in-situ hybridization studies performed in fetal and adult mice. It is clear from

these studies that the Steel protein (or at least mRNA encoding this protein) is widely expressed during embryogenesis. This widespread expression pattern suggests that the Steel factor effects the growth, survival, and/or differentiation of cells besides those three lineages (hematopoietic cells, germ cells, and melanocytes) shown to be affected in both *W* and *Sl* mutant mice. Compensatory mechanisms in these other tissues may prevent the deleterious effects like those observed in the tissues mentioned above. Cells expressing the Steel factor are frequently juxtaposed to cells expressing c-*kit*, that is, there are complementary patterns of expression of the ligand and its receptor (Keshet *et al.*, 1991; Matsui *et al.*, 1990; Motro *et al.*, 1991).

One primary role the Steel factor plays in embryogenesis is to establish a migratory gradient over which a number of c-*kit*-expressing cells move (e.g., primordial germ cells migrating into the developing gonad (Keshet *et al.*, 1991; Matsui *et al.*, 1990; Motro *et al.*, 1991) and melanocyte precursors migrating into the skin (Funasaka *et al.*, 1992)). Expression of the Steel factor in the gonad has been extensively examined in both normal and *Steel* mice in an effort to explain the patterns of sterility observed in some, but not all, *Sl* mutations (reviewed in Copeland *et al.*, 1990). The Steel protein is expressed in the follicular (granulosa) cells that

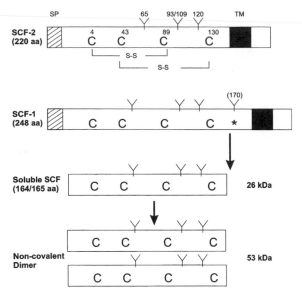

Figure 21.4   The structure of the SCF protein products, their sizes, the locations and numbers of cysteine residues that are involved in intramolecular disulfide bonds on sites of *N*-glycosylation (Y) or potential *N*-glycosylation (170) (Anderson *et al.*, 1990; Arakawa *et al.*, 1991; Flanagan *et al.*, 1991; Huang *et al.*, 1990; Langley *et al.*, 1992; Lu *et al.*, 1991; Martin *et al.*, 1990). All characteristics depicted are conserved between rat and human SCF, except for one site of *N*-glycosylation, which occurs at amino acid 93 in the human and at amino acid 109 in the rat. Note: all amino acids numbered on the basis of their position in the mature peptide. Adapted from Galli, Zsebo and Geissler (1994). *Advances in Immunology* 55, 1–96.

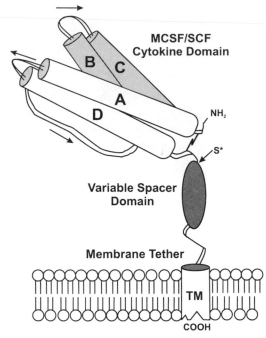

Figure 21.5   Proposed common tertiary structural frameworks of SCF. The four predicted helices of the SCF monomer are labled A to D. Allowed disulfide bridges (zig-zag lines; the second is occluded) could link cysteine residues 4–89 and 43–138 in SCF. From Bazan, 1991a.

surround the ovary as well as the Sertoli cells of the testis (Keshet *et al.*, 1991; Matsui *et al.*, 1990; Motro *et al.*, 1991). In the brain, the Steel factor is expressed in a number of regions including the olfactory bulb, the cerebral cortex, the hippocampus, the dorsal thalamus, the dentate gyrus, and the cerebellum (Keshet *et al.*, 1991; Matsui *et al.*, 1990; Motro *et al.*, 1991). Steel factor has also been implicated as playing a role in the generation of autonomic gut motility (Huizinga *et al.*, 1995; Maeda *et al.*, 1992).

In addition to the sources identified by hybridization studies, expression of the ligand has also been demonstrated either by northern blot analysis or by assaying for activity in the supernatants of various cell lines. Thus, Steel factor has been shown to be expressed by stromal cells (Aye *et al.*, 1992; McNiece *et al.*, 1991), by fibroblasts (Anderson *et al.*, 1990; Flanagan and Leder, 1990; Nocka *et al.*, 1990), and by endothelial cells, where expression was upregulated by treating the cells with thrombin (Aye *et al.*, 1992). Northern blot analysis also showed that the protein was expressed in fetal lung, heart, placenta, kidney, liver, brain, and visceral yolk sac (Matsui *et al.*, 1990).

4.2   Differential Expression of Steel Factor Isoforms

As outlined above, there are two primary isoforms of the Steel factor protein; one of these includes amino acids encoded by a sixth exon (248 amino acids), the other does not (220 amino acids). Although numerous studies have been directed at examining expression of the Steel factor (Keshet *et al.*, 1991; Matsui *et al.*, 1990; Motro *et al.*, 1991), few studies have been done to examine which isoforms are present in each tissue. In one such study (Huang *et al.*, 1992), an RNAase protection assay was used to show tissue-specific expression of the different isoforms. The primary isoform in fibroblasts was the 248-amino-acid form containing the sixth exon. The ratio of the form containing the exon to the form missing the exon was 26:1 in brain, 3:1 in bone marrow, 1.5:1 in spleen, and 1:2.6 in testis. The physiological significance of these altered isoform ratios is unknown, but presumably reflects a relative need by each tissue to produce a form capable of interacting with the appropriate c-*kit*-expressing cells.

4.3   Soluble Steel Factor Protein

Functional, recombinant Steel factor protein (consisting of the extracellular domain of the protein) has been produced in yeast (Anderson *et al.*, 1990) as well as *E. coli* and CHO cells (Langley *et al.*, 1992). Soluble naturally occurring Steel factor is thought to be generated by proteolytic cleavage of the protein expressed on the cell surface as described above. The Steel factor protein has been found to be normally expressed in human serum; circulating levels are in the range of 1–8 ng/ml (mean about 3 ng/ml) as determined by an ELISA assay (Langley *et al.*, 1993).

5. Biological Activities

Owing to the vast amount of research based on murine models, biological data derived from such sources have not been exclusively directed to the murine section of the chapter (Section 8).

5.1 GERM CELLS

Mice that are homozygous for mutations at the Dominant White Spotting (*W*) and Steel (*Sl*) loci in mice are characterized by reproductive deficiencies. Detailed histological analysis of *W* and *Sl* embryos showed that mutations at these loci lead to the developmental failure of germ cells during early embryogenesis in both testis and ovary (reviewed in Russell, 1979). The nature of these fertility problems depends upon the mutation associated with each specific allele. For example, in the homozygous condition the mutation associated with the Sl^d allele leads to sterility in both male and female mice, whereas the mutation in the Sl^{17H} and Sl^{pan} alleles lead to sterility only in male or only in female mice, respectively (reviewed in Copeland *et al.*, 1990). With the identification of the c-*kit* tyrosine kinase receptor and the Steel factor as the proteins encoded by the *W* and *Sl* loci, respectively (see above), a number of investigations were launched to study the expression of these proteins during germ cell development (Keshet *et al.*, 1991; Motro *et al.*, 1991; Orr-Urtreger *et al.*, 1990).

Primordial germ cells originate in the allantois and migrate to the developing genital ridges between days 9.5 and 11.5 (Bennett, 1956; Donovan *et al.*, 1986; Mintz and Russell, 1957). In-situ hybridization analysis showed that the primordial germ cells, which express c-*kit*, migrate to the genital ridge along a gradient of cells that express the Steel factor protein (Keshet *et al.*, 1991; Motro *et al.*, 1991). Once the germ cells have migrated into the genital ridge, the gradient of Steel factor expression along the pathway disappears. These data suggest that it is the expression of Steel factor protein along this migratory pathway that guides the germ cells to the genital ridge. The c-*kit* receptor is expressed in primordial germ cells of fetal gonads (Orr-Urtreger *et al.*, 1990).

A number of studies have been done that show that the Steel protein is required for the growth of murine primordial germ cells in culture (Dolci *et al.*, 1991; Godin *et al.*, 1991; Matsui *et al.*, 1991). Antibodies that block Steel factor (expressed on feeder cells) from binding to c-*kit* on the surface of the primordial germ cells were shown to inhibit the growth of the germ cells (Matsui *et al.*, 1991). These studies also revealed that a soluble form of Steel factor protein cannot substitute for membrane-bound Steel factor to promote the long-term survival of primordial germ cells in culture. The reason for this is unknown, but presumably reflects the need for a second accessory molecule(s) expressed only on the surface of the feeder cells. Steel factor has been shown to promote primordial germ cell survival by suppressing apoptosis (Pesce *et al.*, 1993).

Spermatogenesis is controlled by the release of hormones of the anterior pituitary gland. These hormones do not act directly on germ cells themselves, but stimulate the somatic cells in the gonad that then act to stimulate the germ cells. Steel factor has been shown by several groups to be produced by murine Sertoli cells (Rossi *et al.*, 1991; Tajima *et al.*, 1991), and analysis of Sertoli cells derived from Sl^d/Sl^d mice showed that a cell-bound form of Steel factor was required to support the growth of c-*kit*-expressing mast cells (Tajima *et al.*, 1991). Follicle stimulating hormone can induce levels of Steel factor mRNA in mouse Sertoli cells (Rossi *et al.*, 1993). Sertoli cells, however, do not express the c-*kit* receptor (Yoshinaga *et al.*, 1991).

In contrast, c-*kit* is expressed in murine germ cells, particularly in spermatogonia, the earliest stage of spermatogenesis in the mouse (Manova *et al.*, 1990; Sorrentino *et al.*, 1991). No c-*kit* was expressed in later spermatocytes (Yoshinaga *et al.*, 1991). Therefore, c-*kit* expression (and by implication Steel factor) is only required for certain intermediate steps of spermatogenesis, specifically those involved in division of type A (but not type B) spermatogonia (Yoshinaga *et al.*, 1991). Steel factor was able to stimulate thymidine incorporation by isolated germ cells that had been enriched for spermatogonia (Rossi *et al.*, 1993). An anti-c-*kit* monoclonal antibody (ACK2) that blocks Steel factor binding to the receptor was tested for its capacity to interfere with the development of male or female germ cells (Yoshinaga *et al.*, 1991). Intravenous injection of ACK2 into male mice led to the depletion of differentiating, but not nondifferentiating, type A spermatogonia. Interestingly, ACK2 had no apparent effect on oocyte maturation even though those cells express high levels of c-*kit*. ACK2 also did not effect Leydig cell function (measured as testosterone production) even though these cells also express high levels of c-*kit*.

A number of testicular germ cell tumors have been analyzed for c-*kit* expression; 80% of seminomas but only 7% of non-seminomas expressed this receptor (Strohmeyer *et al.*, 1991). Unfortunately, the tumor cells in this study were not tested for their capacity to proliferate in Steel factor. In another study, Steel factor and c-*kit* expression were shown to be dysregulated in a collection of primary male germ cell tumors (Murty *et al.*, 1992).

Analysis of c-*kit* and Steel factor expression in adult ovaries revealed a complementary pattern of expression (Keshet *et al.*, 1991; Motro *et al.*, 1991). The c-*kit* receptor is expressed in the oocytes, but not in the granulosa cells that support the oocytes. c-*kit* receptor expression has not been detected in late fetal ovaries; expression was maximal in ovaries of juvenile mice and declined in adult ovaries (Manova *et al.*, 1990). This time course of c-*kit* expression is consistent with the expression of c-*kit* in primordial and growing oocytes. The c-*kit* receptor has been shown to be expressed on primordial, growing, and full-grown oocytes both by northern blot analysis and by indirect immuno-fluorescence (Manova *et al.*, 1990). After fertilization of the egg, c-*kit* levels decline rapidly and are undetectable in blastocysts (Manova *et al.*, 1990).

Steel factor is produced by the granulosa cells that surround the developing oocytes, but not by the oocytes themselves (Keshet *et al.*, 1991; Motro *et al.*, 1991). Steel factor stimulated the proliferation of oocytes in an *in vitro* culture system, and this proliferation was inhibited by ACK2, which blocks binding of Steel factor to the receptor (Packer *et al.*, 1994). The ACK2 antibody has also been shown to inhibit the growth of primordial germ cells on a feeder layer of cells that produce Steel factor (Matsui *et al.*, 1991).

As outlined above, Steel factor is required for the survival *in vitro* of primordial germ cells. Analysis of the Steel–Dickie (Sl^d) mutation showed that animals carrying this allele are capable of making a soluble form of the protein that is biologically active, but not a membrane-bound form (Brannan *et al.*, 1991; Flanagan *et al.*, 1991). However, the soluble form of the Steel factor cannot support the long-term survival of primordial germ cells in culture (Dolci *et al.*, 1991), which is consistent with the observation that both Sl^d/Sl^d and Sl/Sl^d mice are sterile. The exact difference in the biological activity of the membrane-bound and soluble isoforms of the Steel factor is unknown but is being investigated. Membrane-bound Steel factor appears to be required for binding of germ cells (which express c-*kit*) to Sertoli cells. Germ cells do not bind to Sertoli cells isolated from Sl^d mice, but this binding activity can be rescued by ectopic expression of the transmembrane form of the Steel factor in these Sertoli cells (Marziali *et al.*, 1993). The relative expression of isoforms of the Steel factor (which differ in their capacity to undergo proteolytic cleavage) has been reported to be developmentally regulated (Marziali *et al.*, 1993).

5.2 Melanocytes

As outlined above, the most obvious phenotype of the pleiotrophic defects associated with the *Steel* or *W* loci are the effects on coat color. The pigmentation deficiencies are believed to be due to a failure of neural crest-derived melanocytes either to migrate to their proper location in the skin or to survive there following migration. Mice that are homozygous for mutant alleles at either the *Steel* or *W* loci (or at least at those alleles that are not homozygous lethal) are generally black-eyed animals with white fur. The black eyes distinguish these animals from true albinos, which have nonpigmented or pink eyes. The melanocyte lineage is derived from neural crest cells that migrate from the apex of the neural tube to the dorsal region of the somites (Mayer, 1970). Pigmented cells in the retina that give the eye its color, however, are derived from the neural tube and are not affected in *Sl* or *W* mice. In addition to the obvious effects on skin pigmentation, failure of melanocyte migration and/or establishment can manifest itself as other problems in affected mice. These include hearing deficiencies as a result of problems in the stria vascularis in the inner ear (Deol, 1970; Steel and Barkway, 1989) or night blindness resulting from problems associated with melanocytes that populate the choroid structure in the eye (Dräger, 1986; Dräger and Balkema, 1987).

In-situ hybridization studies have shown that melanocyte precursors (expressing c-*kit*) migrate along a gradient of Steel factor expression in a manner consistent with a chemotactic role for Steel factor (Keshet *et al.*, 1991). Murphy and coworkers (1992) showed that melanocyte precursor cells survive transiently, but do not proliferate, in response to Steel factor alone. The factor is required for maintenance, but not differentiation, of neural crest-derived melanocyte precursors. The transient nature of the requirement for Steel factor for melanocyte precursors in the neural crest has been confirmed (Morrison-Graham and Weston, 1993). In addition, Steel factor alone was not capable of stimulating melanin production *in vitro*. Nishikawa and coworkers (1991) reported that injection of ACK2 antibodies into pregnant mice on or about day 14.5 of gestation resulted in coat color abnormalities after birth. Therefore, interruption of interactions between Steel factor and c-*kit* during embryonic development in normal mice produced a phenotype similar to that seen in *W* or *Sl* mice and demonstrated the critical nature of these receptor-ligand interactions. An unexpected finding in these studies was the ability to modulate coat color in neonatal and adult, normally pigmented wild-type mice by injection of the anti-c-*kit* antibody (Nishikawa *et al.*, 1991). These experiments indicate that c-*kit*/Steel factor interactions are required not just during embryogenesis but throughout life to maintain the viability of melanocytes.

Melanocytes can readily be isolated from human foreskin tissue and propagated *in vitro* (Halaban *et al.*, 1987). Isolated melanocytes were shown to proliferate in response to Steel factor in combination with activators of protein kinase C (PKC), such as dbCAMP or TPA (Funasaka *et al.*, 1992). Neither Steel factor nor the activators of PKC alone were capable of sustaining melanocyte proliferation, although Steel factor had transient survival promoting effects as a single agent. This requirement for multiple stimulators was similar to that observed with other proteins that stimulate melanocyte proliferation (e.g, fibroblast growth factors and hepatocyte growth factor), which only act in synergy with stimulators of protein kinases A or C.

Rottapel and coworkers (1991) showed that the addition of Steel factor to melanocytes resulted in the rapid autophosphorylation of the c-*kit* receptor, which was followed by a cascade of tyrosine phosphorylations on a number of intracellular substrates. In contrast to the proliferative effects mentioned above, this phosphorylation was not dependent upon dbCAMP or TPA. Proteins phosphorylated in melanocytes appear to be different from those proteins phosphorylated in mast cells, suggesting the utilization of lineage-specific signal transduction pathways.

The c-*kit* receptor, although normally expressed in neonatal and adult melanocytes, is expressed in only a minority of cell lines established from human and murine melanomas (Lassam and Bickford, 1992; Zakut *et al.*, 1993). The lack of c-*kit* expression in the majority of the melanoma cell lines is likely due to a lack of transcription because the gene encoding this protein appears to be neither deleted nor rearranged (Zakut *et al.*, 1993). The level of expression of the c-*kit* receptor in those melanoma cell lines that do express the receptor has been shown to be high (Zakut *et al.*, 1993). Despite this, no constitutive c-*kit* autophosphorylation has been observed in a limited survey of fresh melanoma tissue, suggesting that autocrine or paracrine Steel factor/c-*kit* stimulation is not involved in the pathophysiology of this malignancy (Funasaka *et al.*, 1992). In addition, Steel factor did not stimulate the proliferation of freshly isolated melanoma cells in a manner similar to that observed in normal melanocytes (Funasaka *et al.*, 1992). In fact, Steel factor has been shown actually to inhibit the growth of those melanoma cell lines that do express c-*kit*, probably by slowing down the cell cycle (Zakut *et al.*, 1993).

There is a rare autosomal dominant genetic trait in humans known as piebald that manifests itself as a spotting defect similar to that seen in W and Sl mutant mice. Affected individuals have a white forelock as well as large nonpigmented patches on the chest and/or other locations. Not surprisingly, the piebald trait in humans has been shown to result from both missense and frameshift mutations in the c-*kit* tyrosine kinase receptor (Ezoe *et al.*, 1995; Fleischman, 1992a,b, 1993; Fleischman *et al.*, 1991; Giebel and Spritz, 1991; Spritz *et al.*, 1992, 1993), which in mice is encoded at the W locus (Chabot *et al.*, 1988; Geissler *et al.*, 1988). Affected individuals are believed to be heterozygous for defects in the c-*kit* protein; the dominant nature of the trait is believed to reflect dominant negative effects of the

mutant c-*kit* allele. Such dominant-negative effects are believed to result from the fact that receptor dimerization (induced by Steel factor dimers) is required for proper biological function. One possible case of homozygous piebald trait has been reported, but this was before the molecular nature of the defect was known (Hulten *et al.*, 1987).

Since pigmentation defects in mice carrying mutations at the W and Sl loci are often indistinguishable, it would be reasonable to suspect that at least some cases of piebald trait in humans would arise from mutations in the Steel factor gene, i.e., from a defect in the ligand instead of the receptor. However, no defects in the Steel gene have ever been reported in piebald humans and, in cases where the molecular cause of this trait has been identified, it has always been shown to result from mutations in the c-*kit* receptor. Piebald trait thus represents the human homologue of the W defect in mice.

5.3 HEMATOPOIETIC PROGENITOR CELLS

Hematopoiesis involves the continual proliferation, differentiation, and self-renewal of a population of stem cells into at least eight distinct lineages. Factors that stimulate tyrosine kinase receptors have been shown to play a major role in this process. For example, CSF-1, the ligand for c-*fms*, regulates the proliferation, differentiation and survival of cells in the monocyte/macrophage lineage (Stanley *et al.*, 1983). Also, the recently discovered flt3 ligand has been shown to stimulate the proliferation of both human and murine hematopoietic cells that were highly enriched for primitive progenitor and stem cells (Hannum *et al.*, 1994; Lyman *et al.*, 1993, 1994a).

A role for Steel factor in the regulation of hematopoiesis was first suggested by the analysis of naturally occurring W/W and Sl/Sl^d mutations in mice (Russell, 1979). As previously described, abnormalities in the hematopoietic system were one of the prominent phenotypes associated with these mutations. Specifically, the most significant defects were seen as a severe macrocytic anemia and a dramatic reduction of cells within the mast cell lineage. In addition, a reduction in hematopoietic stem cells was seen in the yolk sac and fetal liver of Sl/Sl^d mice (Chui and Russell, 1974; Ikuta and Weissman, 1992). However, mutations in the Sl or W locus did not disrupt every hematopoietic lineage, and hematopoiesis could occur in the absence of a functional Steel factor or c-*kit*. These results suggested a critical, although not essential, role for Steel factor in full lineage development and expansion of early hematopoietic progenitor populations.

The cloning of Steel factor made it possible to test the capacity of this protein to stimulate purified hematopoietic populations. A significant amount of *in vitro* data has been generated over the years which has shown that Steel factor can directly stimulate purified stem cell populations. Steel factor appears to have little activity alone on these cells but can synergize with numerous cytokines in a variety of *in vitro* assays to stimulate cell proliferation. For example, de Vries and coworkers (1991) demonstrated that Steel factor could synergize with IL-1 or IL-3 to stimulate the proliferation of purified stem cells in liquid culture. The development of multilineage colonies could also be induced with Steel factor plus erythropoietin, IL-3, IL-6, IL-11, or G-CSF (Ogawa *et al.*, 1991; Tsuji *et al.*, 1991, 1992). Moreover, Steel factor has been shown to enhance the development of high-proliferative-potential colony-forming cells (HPP-CFC) *in vitro* (Kriegler *et al.*, 1994; Lowry *et al.*, 1991; Muench *et al.*, 1992; Williams *et al.*, 1992). HPP-CFC are believed to represent some of the most primitive cells from the bone marrow that can be grown *in vitro* (Bartelmez *et al.*, 1989; Hodgson *et al.*, 1982). In these experiments, Steel factor was able to synergize with various combinations of CSF-1, IL-3, IL-1, and IL-6 to stimulate maximum HPP-CFC development.

Steel factor has also been shown to be a potent *in vitro* stimulator of primitive human hematopoietic cells. Several groups have shown that cells expressing the CD34 antigen are enriched for the most primitive progenitors. These cells can form HPP-CFC colonies, contain the long-term bone marrow culture-initiating cell, and reconstitute hematopoiesis *in vivo* (Berenson *et al.*, 1988, 1991; Brandt *et al.*, 1988; Gunji *et al.*, 1992; McNiece *et al.*, 1989; Sutherland *et al.*, 1989). Analysis of CD34$^+$ cells demonstrated that c-*kit* expression could be detected on a subset of this population (Briddell *et al.*, 1992; Gunji *et al.*, 1993). Consistent with these results, stimulation of CD34$^+$ bone marrow cells with Steel factor plus IL-3, GM-CSF, or G-CSF resulted in increased numbers of granulocyte-macrophage colonies, whereas the combination of Steel factor, IL-3, and erythropoietin produced macroscopic erythroid burst-forming units (BFU-E) (Bernstein *et al.*, 1991). Further analysis of a CD34$^+$ population depleted of the major histocompatibility class II locus HLA-DR and CD15 showed that Steel factor could synergize with GM-CSF and IL-3, or a recombinant GM-CSF/IL-3 fusion protein (PIXY321), to promote the formation of HPP-CFC colonies (Brandt *et al.*, 1992). Steel factor has also been shown to play a role in the survival of primitive progenitor cells (Brandt *et al.*, 1994). Culturing CD34$^+$, HLA-DR$^-$, c-*kit*$^+$ cells in the presence of Steel factor maintained the frequency of colony formation and suppressed apoptosis for 48 hours. Thus, depending on which cytokines Steel factor is paired with, Steel factor has the ability to promote proliferation, differentiation, or survival of the most primitive human hematopoietic cells.

The *in vitro* studies have clearly shown that Steel

factor requires additional cytokines to exert its maximum effect on cell proliferation. Administration of Steel factor as a single agent *in vivo*, however, has been shown to have dramatic effects on multiple lymphohematopoietic lineages. In mice, a transient expansion (3–5-fold) of hematopoietic stem cells was seen, in addition to mobilization out of the bone marrow compartment, as judged by a loss in radioprotective capacity (Bodine *et al.*, 1993; Fleming *et al.*, 1993). Stem cells isolated from the peripheral blood or spleens of these mice treated with Steel factor were found to confer an increased radioprotective capacity upon transfer to recipient animals and could completely repopulate the erythroid, myeloid, and lymphoid lineages of irradiated or *W/W^v* hosts (Bodine *et al.*, 1993). In baboons, a dose-dependent expansion of hematopoietic colony-forming cells of multiple lineages in both blood and bone marrow was observed in response to Steel factor (Andrews *et al.*, 1991, 1992a). Moreover, an increase in circulating hematopoietic progenitors that were capable of engrafting and rescuing lethally irradiated baboons was observed following Steel factor treatment (Andrews *et al.*, 1992b). In addition, a dose-dependent leukocytosis, reticulocytosis and increased marrow cellularity were seen. These effects were reversible, and normal cell levels were obtained 7–14 days after ceasing administration of Steel factor (Andrews *et al.*, 1992a). The dramatic effects of Steel factor reported with *in vivo* animal models have resulted in the recent introduction of Steel factor into the clinic (Costa *et al.*, 1993; Demetri *et al.*, 1993).

Both the *in vitro* and *in vivo* data have demonstrated that Steel factor can result in the expansion of multiple hematopoietic subpopulations. However, no convincing data have to date been presented that demonstrate a stem cell self-renewal activity associated with Steel factor. *In vitro* stimulation of isolated stem cells with Steel factor in combination with IL-3 or IL-1 resulted in the survival of day-14 CFU-S forming cells for up to 14 days in liquid culture (de Vries *et al.*, 1991). Likewise, the survival of Lin$^-$, c-*kit*$^+$ cells or Lin$^-$ progenitors isolated from 5-fluorouracil-treated mice could be maintained for 7 days in the presence of Steel factor or Steel factor plus IL-3 (Katayama *et al.*, 1993; Miura *et al.*, 1993). However, the ability to maintain long-term repopulating stem cells was not tested with secondary transplants in these studies. Bodine and coworkers (1992) reported that both short-term and long-term competitive repopulating cells isolated after 5-fluorouracil treatment could be maintained for 6 days *in vitro* with Steel factor. However, the cultured cells were not purified stem cells, and the presence of accessory cells that may have provided additional cytokines could not be ruled out. Li and Johnson (1994) demonstrated that the long-term repopulating ability of a stem cell-enriched population characterized by a Sca-1$^+$, Lin$^-$, Rh123^{lo} phenotype could be maintained for 10 days *in vitro* in the presence of Steel factor, but the frequency for long-term engraftment declined during *in vitro* culture in comparison to freshly isolated stem cells. The most consistent interpretation of the data, taken together, is that Steel factor can enhance *in vitro* survival, but not self-renewal, of primitive hematopoietic stem cells. It is also interesting to note that there is evidence that Steel factor acts as a major regulator of hematopoietic progenitor cell trafficking via its potent chemotactic/kinetic influence on these cells (Okumura *et al*, 1996).

As seen with the *in vivo* data, Steel factor can also stimulate the development of granulocyte-macrophage progenitors *in vitro*. Culturing of murine bone marrow cells in semisolid medium with Steel factor alone resulted in the development of small colonies containing blast cells and granulocytes (Metcalf and Nicola, 1991). However, Steel factor plus IL-3, GM-CSF, CSF-1, or G-CSF resulted in dramatically larger colonies containing a variety of differentiated myeloid cell types (Heyworth *et al.*, 1992; Metcalf, 1991). Furthermore, the presence of Steel factor was able to lower the concentration of IL-3 or G-CSF needed for optimal colony formation (Heyworth *et al.*, 1992).

5.4 ERYTHROPOIESIS

The role of Steel factor in the development and expansion of differentiated lineages has also been studied thoroughly. The hematological defects observed in *Sl* mice are seen most dramatically in several differentiation pathways. Specifically, the erythroid and mast cell lineages are most severely compromised, whereas less severe defects are observed in the megakaryocytic and granulocytic lineages (Ebbe *et al.*, 1973; Kitamura and Go, 1978; Ruscetti *et al.*, 1976; Russell, 1979). More recent work with a c-*kit* antagonist, the ACK2 monoclonal antibody, has confirmed the importance of Steel factor in both myeloid and erythroid development (Ogawa *et al.*, 1991). Injection of ACK2 into adult mice resulted in a dramatic reduction of hematopoietic progenitor cells from the bone marrow and the eventual disappearance of mature myeloid and erythroid cells. Interestingly, the IL-7-responsive B cells were not effected by ACK2 treatment. These results suggest that Steel factor is required for full development of the myeloid and erythroid lineages, but the absence of Steel factor can be compensated for in the lymphoid lineage.

A direct effect of Steel factor on erythropoiesis has been reported by numerous investigators (Anderson *et al.*, 1990; Martin *et al.*, 1990; D.E. Williams *et al.*, 1990). Steel factor alone has little effect on BFU-E colony formation but could synergize with erythropoietin, erythropoietin plus IL-3, and erythropoietin plus GM-CSF to enhance the number and size of BFU-E colonies (Broxmeyer *et al.*, 1991; Papayannopoulou *et al.*, 1993; Xiao *et al.*, 1992). Moreover, Steel factor alone or in combination with IL-3 or GM-CSF could induce erythroid colony formation in the absence of erythropoietin. Steel factor has also been shown to play a role in erythroid maturation. de Wolf and coworkers (1994) demonstrated that BFU-E colony forming cells were derived from CD34$^+$/CD36$^-$ human marrow cells, whereas the more mature cells capable of forming CFU-E colonies were derived from the CD34$^-$/CD36$^+$ cell population. Culturing of the CD34$^+$/CD36$^-$ cells for 7 days in erythropoietin plus Steel factor resulted in the transition of BFU-E colony-forming cells to CFU-E colony-forming cells. This transition was also accompanied by the appearance of cells with the CD34$^-$/CD36$^+$ phenotype. Consistent with the actions of Steel factor on erythroid cells *in vitro*, *in vivo* administration of Steel factor reversed the macrocytic anemia of *Sl/Sl*^{d} mice (Zsebo *et al.*, 1990a) and caused an increase in circulating erythrocytes in baboons (Andrews *et al.*, 1991). However, it appears that such responses are due to Steel factor acting by retarding differentiation of normal erythroid progenitors while enhancing their proliferation (Muta *et al.*, 1995).

Of potential clinical relevance is the observation that Steel factor could stimulate the *in vitro* erythroid development of marrow cells isolated from Diamond–Blackfan anemia (Abkowitz *et al.*, 1991; Bagnara *et al.*, 1991; Olivieri *et al.*, 1991). The primary defect in this congenital disease is a macrocytic anemia. Steel factor could synergize with IL-3 and erythropoietin to increase the size and number of BFU-E in cells from Diamond–Blackfan patients. However, differences in responsiveness to Steel factor were seen in BFU-E production, with some patients not responding at all. Thus, it appears that Steel factor or c-*kit* defects are not the primary cause of this disorder, but that Steel factor can enhance erythroid development in a subpopulation of patients. There is increasing evidence that the erythropoietic activity of Steel factor is mediated by a unique interaction between the Steel factor and erythropoietin receptors (Olweus *et al.*, 1995; H. Wu *et al.*, 1995).

5.5 MAST CELLS

The development of mature myeloid cells can also be enhanced by Steel factor. Most dramatic is the effect of Steel factor on the survival, proliferation, and differentiation of mast cells. Culturing of murine bone marrow cells in the presence of Steel factor will result in pure mast cell populations (Tsai *et al.*, 1991). Likewise, several groups have demonstrated that Steel factor induces human mast cell differentiation from unfractionated cord blood mononuclear cells (Mitsui *et al.*, 1993) or adult bone marrow (Galli *et al*, 1995; Valent *et al.*, 1992). Moreover, culturing of human CD34$^+$ cord blood cells in the presence of Steel factor and IL-3 under serum-free conditions results in the proliferation and survival of immature mast cells (Durand et al., 1994). These cells die if either Steel factor or IL-3 is removed. Data generated with mouse (Coleman et al., 1993; Wershil et al., 1992) and human (Bischoff and Dahinden, 1992; Columbo et al., 1992) mast cells indicate that soluble forms of Steel factor can also activate these cells and stimulate their degranulation. Activation of mast cells through the FcεRI can be significantly enhanced by low concentrations of Steel factor (Columbo et al., 1992). Steel factor activity on mast cells also includes induction of histamine release (Lukacs et al., 1996; Nakajima et al., 1992; Taylor et al., 1995), adherence (Dastych and Metcalf, 1994; Kinashi and Springer, 1994), and chemotaxis (Meininger et al., 1992).

5.6 MEGAKARYOCYTES

A role for Steel factor in megakaryocytopoiesis was first suggested in Sl/Sl^d mice in response to 5-FU (Hunt et al., 1992). After 5-FU treatment, platelet counts drop significantly and return to normal levels very slowly in comparison to wild-type controls. Hunt and coworkers (1992) demonstrated that the administration of Steel factor to these 5-FU-treated Sl/Sl^d mice could dramatically stimulate the recovery of platelet levels. In these experiments, the Steel factor-treated Sl/Sl^d mice recovered from the thrombocytopenia as quickly as their wild-type littermates. In addition, W/W^v mice did not demonstrate this enhanced recovery. Consistent with this observation, injection of Steel factor into untreated Sl/Sl^d mice resulted in an increase of steady-state platelet levels (Zsebo et al., 1990a).

The role of Steel factor on murine and human megakaryocyte proliferation and maturation has also been studied extensively in vitro as well as in vivo (Grossi et al., 1995). Alone, Steel factor has been shown to have little stimulatory effect on immature megakaryocyte progenitors. However, Steel factor can synergize with IL-3 or GM-CSF to enhance megakaryocyte colony formation of human CD34+ bone marrow cells (Briddell et al., 1991). Using partially purified mouse bone marrow after 5-FU treatment, Tanaka and coworkers (1992) demonstrated that Steel factor plus IL-3 significantly increased the number of both megakaryocyte and granulocyte-macrophage-megakaryocyte colonies. In contrast, IL-3 plus IL-6 enhanced only megakaryocyte colony formation. Steel factor also appears to play a role in megakaryocyte maturation. Debili and coworkers (1993) demonstrated that the megakaryocyte maturation of human CD34+ cells, as judged by cell polyploidy, could be enhanced when Steel factor was combined with IL-6 or IL-3. More recently, Steel factor has been shown to synergize with thrombopoietin to stimulate megakaryocyte development (Zeigler et al., 1994). Thrombopoietin alone could induce highly purified

murine stem cells to differentiate along the megakaryocytic lineage and this differentiation could be enhanced with Steel factor. Interestingly, the combination of thrombopoietin and Steel factor also led to the production of cells from other myeloid lineages, but no megakaryocytes were observed if thrombopoietin was omitted. Steel factor is able to induce Fc receptor expression on megakaryoblasts, suggesting that this factor can influence both the development and function of megakaryocytes (Kiss et al., 1996).

5.7 LYMPHOPOIETIC CELLS

Lymphopoiesis in adult W or Sl mutant mice appears to be normal, even though intrathymically expressed Steel factor appears to be important in the expansion of very immature thymocytes in vivo (Rodewald et al., 1995). The steady-state levels of circulating lymphocytes in the peripheral blood are normal as well as the number of B cells in the bone marrow and spleen. Only in certain W mutations can a perturbation in B cell development be observed (Landreth et al., 1994). In fetal W/W^x and W^x/W^x mice, a reduction of absolute numbers of pre-B cells was observed. Moreover, W/W^x- and W^x/W^x-derived cells from adult bone marrow or fetal liver were both significantly less able to form B lymphocytes in lethally irradiated recipients in comparison to wild-type controls. These results suggest that a lack of Steel factor or functional receptor can be compensated for, resulting in normal levels of B cells in the adult animal. However, under active development or B cell regeneration, a reduced capacity for B cell expansion may exist.

Murine B cell development in vivo is believed to involve the progressive maturation of specific subpopulations of cells. These populations have been characterized in vitro by the analysis of cell surface expression of specific antigens and the progressive rearrangement and expression of immunoglobulin (Ig) genes. The earliest cells, termed pro-B cells, are defined by the expression of CD45 (B220) antigen and the retention of Ig genes in germline configuration. Pre-B cells are more mature cells that retain B220 expression and progressively undergo rearrangement of the Ig light and heavy chain genes. However, a functional Ig protein product is not yet expressed on the cell surface. The most mature B cells are marked by the surface expression of specific Ig gene products (Hardy et al., 1991). This process is believed to require stage-specific interactions of the cells with stromal elements and certain cytokines.

Steel factor has been shown to be able to significantly enhance B cell growth in vitro. Steel factor plus IL-7 was able to increase the number and size of B cell colonies induced from murine bone marrow cells in comparison to IL-7 alone (McNiece et al., 1991). The morphology of these colonies revealed that they were primarily pre-B cells as judged by the B220+, surface Ig-negative phenotype. Likewise, Rolink and coworkers (1991) demonstrated that the growth of murine pre-B cell clones on stromal cells plus IL-7 could be inhibited by the ACK2 monoclonal antibody. However, stimulation of mature B cells by mitogens was unaffected by ACK2. These results suggest that Steel factor plays a role in early B cell expansion, but not in late-stage antigen-dependent B cell stimulation.

The role Steel factor plays in the maturation of the earliest pro-B cells remains unclear. *In vitro* studies have shown that Steel factor can synergize with IL-7 to induce the proliferation of pre-B cells in a stroma-independent liquid culture system (Billips *et al.*, 1992). However, pro-B cells require stroma contact for proliferation and maturation to the pre-B cells stage (Faust *et al.*, 1993). Even though these pro-B cells express the receptors for both Steel factor and IL-7, they fail to grow in the presence of either factor alone or in combination. Furthermore, pro-B cells have been shown to proliferate and mature on stroma isolated from *Sl* mice. Thus, Steel factor can synergize with IL-7 to enhance the expansion of the pre-B cell compartment but appears to be dispensable for the growth and maturation of earlier pro-B cells.

Recent studies have suggested a possible role for Steel factor in thymic T cell development, although no obvious T cell defects have been observed in *Sl* or *W* mutant mice. Both Steel factor and c-*kit* transcripts can be detected in cells isolated from murine thymus. Stromal cells isolated from fetal and adult thymi express Steel factor mRNA, whereas thymic precursor cells express c-*kit* mRNA (Palacios and Nishikawa, 1992). These c-*kit*[+] precursor cells, isolated from day-14 fetal livers, could fully reconstitute the T and B cell compartments of immunodeficient SCID mice. Likewise, de Vries and coworkers (1992) demonstrated that one c-*kit*[+] hematopoietic stem cell purified from adult mouse bone marrow could reconstitute the thymus of a sublethally irradiated recipient upon intrathymic transfer. The expression of both Steel factor and c-*kit* transcripts has also been reported in human thymic stromal cells and thymocytes (Wolf and Cohen, 1992).

An analysis has been done to determine which thymocytes respond to Steel factor *in vitro*. Thymocyte precursor cells are believed to have a CD3[-]/CD4[lo]/CD8[-] and Thy-1[lo] phenotype (Wu *et al.*, 1991). These cells commit to the T cell lineage and assume an immature triple-negative phenotype of CD3[-]/CD4[-]/CD8[-]. This triple-negative population has been shown to express high levels of c-*kit*, and Steel factor can synergize with IL-2, IL-3, or IL-7 to induce proliferation of these cells in suspension culture (Godfrey *et al.*, 1992; Morrissey *et al.*, 1994). Similar results have been reported with human triple-negative thymocytes (deCastro *et al.*, 1994). Using fetal thymus lobe submersion cultures, it was shown that both Steel factor and IL-7 could stimulate thymocytes to proliferate with comparable magnitudes within the thymic environment. The combination of Steel factor and IL-7 resulted in a dramatic synergy to expand thymocyte numbers. Phenotypic analysis of the stimulated populations suggested that Steel factor, in comparison to IL-7, stimulated the expansion of a less mature thymocyte population. Furthermore, thymocytes grown in Steel factor have low IL-2 receptor expression levels, whereas reculturing of these cells in the presence of IL-7 dramatically elevated the level of IL-2 receptor expression. These results are consistent with the hypothesis that Steel factor stimulates the growth of less mature thymocytes in comparison to IL-7.

It is clear from a large body of experimental evidence that Steel factor can stimulate cells within every hematopoietic lineage described, in addition to dendritic cell lineages (Santiago-Schwartz *et al.*, 1995). This stimulation usually requires other cytokines for maximal effect and to guide the eventual developmental outcome. However, the *Sl* and *W* mutations severely affect only the erythroid and mast cell lineages, suggesting that redundant factors with similar biological activities must exist to compensate for the lack of Steel factor or c-*kit* proteins. It is tempting to speculate that the newly discovered flt3 ligand may provide some of those compensatory activities. The flt3 ligand and receptor share many similar biological and structural similarities to Steel factor and c-*kit*. As more work is done to compare and contrast the actions of Steel factor with those of other hematopoietic factors, a better understanding of cytokine regulation of hematopoiesis should emerge.

6.   *The c-kit Receptor*

A number of genes were identified during the 1980s that were capable of causing malignant transformation, that is, uncontrolled growth when they were transfected and expressed in certain cells. Collectively, these genes came to be known as oncogenes, and their encoded proteins were referred to as transforming proteins. One of these oncogenes was v-*kit*, which was found within the Hardy–Zuckerman 4 strain of feline sarcoma virus (Besmer *et al.*, 1986). Analysis of the v-*kit* gene showed that it encoded a truncated version of a normal cellular gene denoted as c-*kit*, which was later cloned from mouse (Qiu *et al.*, 1988) and human (Yarden *et al.*, 1987) sources. The c-*kit* gene encoded a tyrosine kinase receptor for which no ligand had yet been discovered (Majumder *et al.*, 1988). The c-*kit* receptor has also been designated as CD117. The c-*kit* protein has a similar structure to several other tyrosine kinase receptors, including c-*fms*, flt3 (Matthews *et al.*, 1991; Rosnet *et al.*, 1991, 1993; Small *et al.*, 1994), and both of the receptors for PDGFA (designated as A and B) (Figure 21.6). All of these receptors have five Ig-like domains in

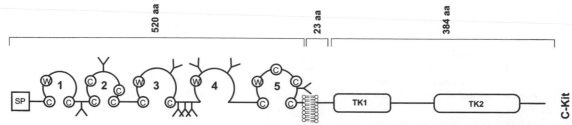

Figure 21.6   Structural representation of the c-*kit* protein. The protein is represented schematically and the Ig-like domains are shown as loops. The protein is drawn to scale and the designations of functional domains are indicated. Numbers 1 to 5 indicate the respective Ig-like domains. The locations of all the cysteine residues of the extracellular domains are indicated by circled C. Tryptophan residues that belong to the consensus structure of the Ig homology unit are shown by circled W. SP = signal peptide; TK = tyrosine kinase. (From Blochman *et al.* (1993). J. Biol. Clem. 268, 4399–4406.)

their extracellular regions, and a catalytic domain in their cytoplasmic portions that phosphorylates tyrosine residues in specific target proteins following activation of the receptor by ligand.

The c-*kit* gene was mapped to human chromosome 4q11-q12 (d'Auriol *et al.*, 1988). The genomic loci encoding the human (André *et al.*, 1992; Giebel *et al.*, 1992; Vandenbark *et al.*, 1992) and murine c-*kit* genes have each been cloned. The genomic loci have been evolutionarily conserved between species and share a very similar overall structure; each comprises 21 exons and spans a distance of over 70 kilobases.

Analysis of independently derived cDNA clones revealed that there were two different isoforms of the c-*kit*-encoded protein (Reith *et al.*, 1991). These c-*kit* receptor isoforms differ by four amino acids (glycine-asparagine-asparagine-lysine) that are either present or absent at a point just upstream of the transmembrane domain. The isoforms have been identified in both mouse and human c-*kit* proteins (Reith *et al.*, 1991). The different isoforms arise as a result of alternative splicing of mRNAs and the use of a cryptic splice donor site located at the 3′ end of exon 9 of the c-*kit* gene (Hayashi *et al.*, 1991). It is not clear whether biological differences occur as a result of ligand signaling via one c-*kit* isoform versus another, although data have been presented suggesting that ligand-independent constitutive phosphorylation occurs only in the isoform that is missing these four amino acids (Reith *et al.*, 1991).

Expression of the two different c-*kit* isoforms has been looked at in various leukemic lines. In one study, Crosier and coworkers (1993) examined isoform expression both in leukemic cell lines and in primary acute myeloid leukemias using an RNAase protection assay. Both isoforms appeared to be expressed in all of the cells examined, with the ratio of GNNK⁻ to GNNK⁺ isoforms being 10:1 to 15:1. A second study confirmed the expression of both isoforms in a series of acute myeloid leukemias, although a greater range of GNNK⁻ to GNNK⁺ isoform ratios was found, 1.3:1 to 12:1 (Piao *et al.*, 1994).

In addition to the isoforms discussed above, other variants have been seen in the c-*kit* receptor. Alternative splicing of mRNAs has been shown to insert an extra serine residue in the cytoplasmic domain at position 715, and a survey of cell lines and acute myeloid leukemias has shown that both of these isoforms are normally expressed (Crosier *et al.*, 1993). The significance of this serine isoform is unknown. A novel transcript that has the potential to encode a truncated receptor has been identified in mouse spermatids (Rossi *et al.*, 1992). This partial receptor contains only a small region of the cytoplasmic domain, and its physiological significance, if any, is unknown. A soluble version of the c-*kit* receptor has been reported in human serum at high concentration (324 ± 105 ng/ml) (Wypych *et al.*, 1995). The mechanism by which this soluble receptor is generated, as well as its physiological importance, if any, are unknown.

It has been shown that dimerization of the receptor is necessary for signal transduction (Philo *et al.*, 1996) and that the fourth immunoglobulin domain of the receptor couples ligand binding to signal transduction (Blechman *et al.*, 1995).

The exact defect in the c-*kit* receptor has now been identified at the molecular level for a number of alleles of the *W* locus (Bernstein *et al.*, 1991; Herbst *et al.*, 1992; Nocka *et al.*, 1990; Reith *et al.*, 1990; Tan *et al.*, 1990). Most of the alleles arise as a result of point mutations that are found in the cytoplasmic domain of the receptor; these changes result in a decrease in the tyrosine phosphorylating activity of the protein. However, in several cases the mutations appear to be of a regulatory instead of a structural nature, and result in the reduced expression of the c-*kit* receptor.

7.   Signal Transduction through the c-kit Receptor

The capacity of Steel factor to activate the c-*kit* receptor has been examined by several groups. Steel factor stimulated the autophosphorylation of c-*kit* on tyrosine residues on mast cells, and c-*kit* was the major phosphorylated protein detected in these cells (Rottapel *et al.*, 1991). In contrast to the results described above for the wild type c-*kit* protein, Steel factor did not stimulate the phosphorylation of c-*kit* in mast cells from mice homozygous for the W^{42} allele of c-*kit*; this mutation has been shown to eliminate the *in vitro* kinase activity of c-*kit* (Reith *et al.*, 1990).

Induction of receptor dimerization and tyrosine autophosphorylation of the c-*kit* receptor by Steel factor was examined to determine how this affects the binding of signal transduction proteins to the receptor (Blume-Jensen *et al.*, 1995; Rottapel *et al.*, 1991). The phosphorylated c-*kit* protein was found to be associated with the signaling proteins phosphatidylinositol 3'-kinase (PI3K), phospholipase C-γ1 (PLCγ1) (Shearman *et al.*, 1993), MAP kinases (Lev *et al*, 1991; Welham and Schrader, 1992), and JAK2 (Linnekin *et al.*, 1996; Weiler *et al.*, 1996). However, there was no detectable binding or phosphorylation of Ras GTPase-activating protein (GAP) to c-*kit* either before or after Steel factor treatment (Duronio *et al.*, 1992). Phosphorylation of c-*kit* apparently acts as a trigger for the formation of a complex with PI3K and PLCγ1. However, this results in activation of protein kinase C (PKC) and phosphorylation of the receptor on Ser-741 and Ser-746 (Blume-Jensen *et al.*, 1995). PKC therefore acts in a negative feedback loop by inhibiting Steel factor-induced receptor kinase activity and modulates responses to Steel factor. The mutant c-*kit* protein expressed in mast cells isolated from homozygous W^{42}/W^{42} mice (see above) also did not associate with PI3K, even after Steel factor treatment. Stimulation of the c-*kit* receptor results in the phosphorylation of a range of substrates, of which a 200 kDa substrate associates rapidly with c-*kit*, although its exact role in signal transduction remains unclear (Linnekin *et al.*, 1995).

8. Steel Factor from Other Species

The Steel (*Sl*) locus was originally described as a novel spotting mutation in mice (Sarvella and Russell, 1956). As described in the Introduction, the chromosomal location of the *Sl* locus on mouse chromosome 10 was an integral part of the biological story that led to the cloning of the gene encoded at that locus. Initial localization of the gene encoding the Steel factor protein showed that it was indeed encoded on mouse chromosome 10 and was deleted in some, but not all, *Sl* alleles (Copeland *et al.*, 1990; Huang *et al.*, 1990; Zsebo *et al.*, 1990a). Formal proof that the *Sl* locus actually encoded the Steel factor protein was obtained when the Steel–Dickie allele of *Sl* was molecularly cloned (Brannan *et al.*, 1991; Flanagan and Leder, 1990). The Steel factor gene has now been cloned from mouse (Anderson *et al.*, 1990; Huang *et al.*,

1990; Zsebo *et al.*, 1990a), rat (Martin *et al.*, 1990), chicken (Zhou *et al.*, 1993), and dog (Shull *et al.*, 1992).

The mouse gene contains at least eight exons that contain the entire coding region of the protein. Those intron:exon boundaries that have been identified are seen to occur at identical positions within the rat, human, and murine genes. In the case of the mouse protein, a ninth exon has been shown to be present and to encode the end of the cytoplasmic tail (Brannan *et al.*, 1992).

Mutations at the murine Steel locus (reviewed in Copeland *et al.*, 1990) have occurred spontaneously or have been induced by chemical mutagenesis, x-ray irradiation, or transgene insertion (Keller *et al.*, 1990). Deletion of most or all of the Steel factor gene has been found to be associated with the alleles Steel-J, Steel Grizzle Belly, Steel 8^H, Steel 10^H, and Steel 18^H (Copeland *et al.*, 1990). A molecular analysis of some alleles of the *Sl* locus has been done in an effort to determine how each particular allele reflects changes in the Steel factor protein or in its expression. In contrast to the results presented above, the *Sl* gene is not deleted in the alleles Steel Dickie (Sl^d), Steel Panda, and Steel^{17H} (Copeland *et al.*, 1990).

The formal possibility remained following the initial cloning of the *Sl* gene that what we have referred to in this review as the Steel factor protein was actually encoded at a locus adjacent to the actual Steel locus. Formal proof that the Steel locus really did encode the Steel protein was obtained by analysis of the most-studied of the *Sl* alleles, Sl^d. The defect in the case of Sl^d turned out to be a 4 kb intragenic deletion that removes the transmembrane region as well as the entire cytoplasmic tail of the protein (Brannan *et al.*, 1991; Flanagan *et al.*, 1991). The resulting protein is not membrane-bound like the wild-type protein, but it is biologically active. This finding suggested that the membrane-bound form of the ligand was absolutely required for normal biological function *in vivo*. The molecular defect responsible for the Sl^{17H} mutation has also been identified (Brannan *et al.*, 1992). The unadulterated murine protein contains 273 amino acids; the first 25 of these amino acids constitute a signal peptide that is removed in the mature protein.

Rat Steel factor protein contains four potential sites for the attachment of *N*-linked sugars, at asparagine residues 65, 72, 109, and 120 (Lu *et al.*, 1991). Steel factor purified from medium conditioned by buffalo rat liver cells was found to be *N*-glycosylated in a somewhat heterogeneous fashion. Asparagine 120 appears to be used in all of the molecules, asparagine 65 and asparagine 109 in some of the molecules, and asparagine 72 in none of the molecules. Sequencing of the purified rat Steel protein revealed that the serine residue at position 142 and threonine residues at positions 143 and 155 were blocked (Zsebo *et al.*, 1990a). Because the protein appears to contain *O*-linked sugars, it is likely that they are attached through one or more of these sites.

It has been reported by two groups working independently (Chabot et al., 1988; Geissler et al., 1988) that the gene encoding the murine c-*kit* tyrosine kinase receptor mapped to mouse chromosome 5 and was encoded at the *W* locus.

9. Clinical Implications

9.1 STEEL FACTOR AND CANCER

The expression of Steel factor and/or its receptor has been shown to result in the autocrine growth of a wide number of different tumor types, including neuroblastoma (Beck et al., 1995a,b), glioma (Stanulla et al., 1995), testicular tumors (Strohmeyer et al., 1995), hemangioma (Meininger et al., 1995), small cell lung tumors (Krystal et al., 1996; Papadimitriou et al., 1995), and melanocytic tumors (Papadimitriou et al., 1995; Takahashi et al., 1995). In fact, not only has Steel factor been shown to increase survival of malignant hematopoietic cells, but it also protects leukemic cells from chemotherapy and tumor cells from radiotherapy (Hassan and Zander, 1996; Hassan and Freund, 1995; Shui et al., 1995).

9.2 STEEL FACTOR AND MAST CELLS/ALLERGY

Steel factor has been shown to cause a respiratory distress syndrome in mice following intravenous administration (Lynch et al., 1992). This syndrome results from degranulation of mast cells in the lungs and is characterized by breathing difficulties. The effects of the Steel factor on mast cell degranulation have also been implicated in some of the deleterious effects seen with this protein in phase I clinical trials (Costa et al., 1993; Demetri et al., 1993). Injection of recombinant Steel factor results in wheal and flare mast cell reactions at the site of injection, increased dermal mast cell density at distances from the site, and increased levels of urinary histamine metabolites (Costa et al., 1996). Steel factor has also been shown to be involved with the recruitment and activation of mast cells/eosinophils in allergic airway inflammation and inflammatory bowel disease (Bischoff et al., 1996; Louis et al., 1995; Lukacs et al., 1996a,b).

9.3 CLINICAL APPLICATIONS OF STEEL FACTOR

In preclinical trials Steel factor can protect against lethal radiation (Fleming et al., 1993), elicit mutilineage hematopoietic responses and increases in marrow cellularity (Bodine et al., 1993, 1996; Hassan et al., 1996; Holyoake et al., 1996; Molineux et al., 1991), and increase the number of circulating peripheral blood stem cells (Molineux et al., 1991; Andrews et al., 1992a, 1995). Steel factor appears to function in synergy with G-CSF in mobilizing peripheral blood progenitor cells and may prove to have clinical potential in this manner (Andrews et al., 1995; Briddell et al., 1993; Glaspy et al., 1996).

The discovery of the Steel factor provided the last piece of the puzzle to a mystery that had existed since the early part of the century. It is now clear that the Steel locus encodes a protein with extraordinary pleiotrophic effects on the growth of not just hematopoietic cells, but also germ cells and melanocytes. How the factor functions in each of these settings is still a subject of great research interest. It is not yet apparent whether the Steel protein will be a clinically useful molecule, and this reflects the balance of the positive effects of the protein on hematopoietic progenitor cells and the apparent negative effects (in a clinical setting) on mast cells.

10. Acknowledgments

We thank our colleagues Doug Williams, Dave Cosman, Phil Morrissey, and John Sims for their critical reading of this review, and Anne Bannister for expert editorial assistance.

11. References

Abkowitz, J.L., Sabo, K.M., Nakamoto, B., et al. (1991). Diamond–Blackfan anemia: in vitro response of erythroid progenitors to the ligand for c-*kit*. Blood 78, 2198–2202.

Anderson, D.M., Lyman, S.D., Baird, A., et al. (1990). Molecular cloning of mast cell growth factor, a hematopoietin that is active in both membrane bound and soluble forms. Cell 63, 235–243.

Anderson, D.M., Williams, D.E., Tushinski, R., et al. (1991). Alternate splicing of mRNAs encoding human mast cell growth factor and localization of the gene to chromosome 12q22-q24. Cell Growth Differ. 2, 373–378.

André, C., Martin, E., Cornu, F., Hu, W.-X., Wang, X.-P. and Galibert, F. (1992). Genomic organization of the human c-*kit* gene: evolution of the receptor tyrosine kinase subclass III. Oncogene 7, 685–691.

Andrews, R.G., Knitter, G.H., Bartelmez, S.H., et al. (1991). Recombinant human stem cell factor, a c-*kit* ligand, stimulates hematopoiesis in primates. Blood 78, 1975–1980.

Andrews, R.G., Bartelmez, S.H., Knitter, G.H., et al. (1992a). A c-*kit* ligand, recombinant human stem cell factor, mediates reversible expansion of multiple CD34+ colony-forming cell types in blood and marrow of baboons. Blood 80, 920–927.

Andrews, R.G., Bensinger, W.I., Knitter, G.H., et al. (1992b). The ligand for c-kit, stem cell factor, stimulates the circulation of cells that engraft lethally irradiated baboons. Blood 80, 2715–2720.

Andrews, R.G., Briddell, R.A., Knitter, G.H., Rowley, S.D., Appelbaum, F.R. and McNiece, I.K. (1995). Rapid engraftment by peripheral blood progenitor cells mobilized by recombinant human stem cell factor and recombinant human granulocyte colony-stimulating factor in nonhuman primates. Blood 85(1), 15–20.

Arakawa, T., Yphantis, D.A., Lary, J.W., et al. (1991). Glycosylated and unglycosylated recombinant-derived human stem cell factors are dimeric and have extensive regular secondary structure. J. Biol. Chem. 266, 18942–18948.

Aye, M.T., Hashemi, S., Leclair, B., et al. (1992). Expression of stem cell factor and c-kit mRNA in cultured endothelial cells, monocytes and cloned human bone marrow stromal cells (CFU-RF). Exp. Hematol. 20, 523–527.

Bagnara, G.P., Zauli, G., Vitale, L., et al. (1991). In vitro growth and regulation of bone marrow enriched CD34+ hematopoietic progenitors in Diamond–Blackfan anemia. Blood 78, 2203–2210.

Bartelmez, S.H., Bradley, T.R., Bertoncello, I., et al. (1989). Interleukin 1 plus interleukin 3 plus colony-stimulating factor 1 are essential for clonal proliferation of primitive myeloid bone marrow cells. Exp. Hematol. 17, 240–245.

Bazan, J.F. (1991a). Genetic and structural homology of stem cell factor and macrophage colony-stimulating factor. Cell 65, 9–10.

Bazan, J.F. (1991b). Neuropoietic cytokines in the hematopoietic fold. Neuron 7, 197–208.

Beck, D., Gross, N. and Beretta-Brognara, C. (1995a). Effects of stem cell factor and other bone marrow-derived growth factors on the expression of adhesion molecules and proliferation of human neuroblastoma cells. Eur. J. Cancer 31A(4), 467–470.

Beck, D., Gross, N., Brognara, C.B. and Perruisseau, G. (1995b). Expression of stem cell factor and its receptor by human neuroblastoma cells and tumors. Blood 86(8), 3132–3138.

Bennett, D. (1956). Developmental analysis of a mutant with pleiotrophic effects in the mouse J. Morphol. 98, 199–234.

Berenson, R.J., Andrews, R.G., Bensinger, W.I., et al. (1988). Antigen CD34+ marrow cells engraft lethally irradiated baboons. J. Clin. Invest. 81, 951–955.

Berenson, R.J., Bensinger, W.I., Hill, R.S., et al. (1991). Engraftment after infusion of CD34+ marrow cells in patients with breast cancer or neuroblastoma. Blood 77, 1717–1722.

Bernstein, I.D., Andrews, R.G. and Zsebo, K.M. (1991). Recombinant human stem cell factor enhances the formation of colonies by CD34+ and CD34+lin− cells, and the generation of colony-forming cell progeny from CD34+lin− cells cultured with interleukin-3, granulocyte colony-stimulating factor, or granulocyte-macrophage colony-stimulating factor. Exp. Hematol. 77, 2316–2321.

Besmer, P., Murphy, J.E., George, P.C., et al. (1986). A new acute transforming feline retrovirus and relationship of its oncogene v-kit with the protein kinase gene family. Nature 320, 415–421.

Billips, L.G., Petitte, D., Dorshkind, K., Narayanan, R., Chiu, C.-P. and Landreth, K.S. (1992). Differential roles of stromal cells, interleukin-7, and kit-ligand in the regulation of B lymphopoiesis. Blood 79, 1185–1192.

Bischoff, S.C. and Dahinden, C.A. (1992). c-kit ligand: a unique potentiator of mediator release by human lung mast cells. J. Exp. Med. 175, 237–244.

Bischoff, S.C., Schwengberg, S., Wordelmann, K., Weimann, A., Raab, R. and Manns, M.P. (1996). Effect of c-kit ligand, stem cell factor, on mediator release by human intestinal mast cells isolated from patients with inflammatory bowel disease and controls. Gut 38(1), 104–114.

Blechman, J.M., Lev, S., Barg, J., et al. (1995). The fourth immunoglobulin domain of the stem cell factor receptor couples ligand binding to signal transduction. Cell 80(1), 103–113.

Blume-Jensen, P., Wernstedt, C., Heldin, C.H. and Ronnstrand, L. (1995). Identification of the major phosphorylation sites for protein kinase C in kit/stem cell factor receptor in vitro and in intact cells. J. Biol. Chem. 270(23), 14192–14200.

Bodine, D.M., Orlic, D., Birkett, N.C., Seidel, N.E. and Zsebo, K.M. (1992). Stem cell factor increases colony-forming unit-spleen number in vitro in synergy with interleukin-6, and in vivo in Sl/Sl[d] mice as a single factor. Blood 79, 913–919.

Bodine, D.M., Seidel, N.E., Zsebo, K.M. and Orlic, D. (1993). In vivo administration of stem cell factor to mice increases the absolute number of pluripotent hematopoietic stem cells. Blood 82, 445–455.

Bodine, D.M., Seidel, N.E. and Orlic, D. (1996). Bone marrow collected 14 days after in vivo administration of granulocyte colony-stimulating factor and stem cell factor to mice has 10-fold more repopulating ability than untreated bone marrow. Blood 88(1), 89–97.

Boswell, H.S., Mochizuki, D.Y., Burgess, G.S., et al. (1990). A novel mast cell growth factor (MCGF-3) produced by marrow-adherent cells that synergizes with interleukin 3 and interleukin 4. Exp. Hematol. 18, 794–800.

Brandt, J., Baird, N., Lu, L., Srour, E. and Hoffman, R. (1988). Characterization of a human hematopoietic progenitor cell capable of forming blast cell containing colonies in vitro. J. Clin. Invest. 82, 1017–1027.

Brandt, J., Briddell, R.A., Srour, E.F., Leemhuis, T.B. and Hoffman, R. (1992). Role of c-kit ligand in the expansion of human hematopoietic progenitor cells. Blood 79, 634–641.

Brandt, J.E., Bhalla, K. and Hoffman, R. (1994). Effects of interleukin-3 and c-kit ligand on the survival of various classes of human hematopoietic progenitor cells. Blood 83, 1507–1514.

Brannan, C.I., Lyman, S.D., Williams, D.E., et al. (1991). Steel–Dickie mutation encodes c-Kit ligand lacking transmembrane and cytoplasmic domains. Proc. Natl Acad. Sci. USA 88, 4671–4674.

Brannan, C.I., Bedell, M.A., Resnick, J.L., et al. (1992). Developmental abnormalities in Steel[17H] mice result from a splicing defect in the steel factor cytoplasmic tail. Genes Dev. 6, 1832–1842.

Briddell, R.A., Bruno, E., Cooper, R.J., Brandt, J.E. and Hoffman, R. (1991). The effect of c-kit ligand on in vitro human megakaryocytopoiesis. J. Clin. Invest. 78, 2854–2859.

Briddell, R.A., Broudy, V.C., Bruno, E., Brandt, J.E., Srour, E.F. and Hoffman, R. (1992). Further phenotypic characterization and isolation of human hematopoietic progenitor cells using a monoclonal antibody to the c-kit receptor. Blood 79, 3159–3167.

Briddell, R.A., Hartley, C.A., Smith, K.A. and McNiece, I.K. (1993). Recombinant rat stem cell factor synergizes with recombinant human granulocyte-colony stimulating factor in vivo in mice to mobilize peripheral blood progenitor cells that have enhanced repopulating ability. Blood 82, 1720–1728.

Broxmeyer, H.E., Cooper, S., Lu, L., et al. (1991). Effect of murine mast cell growth factor (c-kit proto-oncogene ligand) on colony formation by human marrow hematopoietic progenitor cells. Blood 77, 2142–2149.

Cerretti, D.P., Wignall, J., Anderson, D., *et al.* (1988). Human macrophage-colony stimulating factor: alternate RNA and protein processing from a single gene. Mol. Immunol. 25, 761–770.

Chabot, B., Stephenson, D.A., Chapman, V.M., Besmer, P. and Bernstein, A. (1988). The proto-oncogene c-*kit* encoding a transmembrane tyrosine kinase receptor maps to the mouse *W* locus. Nature 335, 88–89.

Chui, D.H. and Russell, E.S. (1974). Fetal erythropoiesis in steel mutant mice. I. A morphological study of erythroid cell development in fetal liver. Dev. Biol. 40, 256–269.

Coleman, J.W., Holliday, M.R., Kimber, I., Zsebo, K.M. and Galli, S.J. (1993). Regulation of mouse peritoneal mast cell secretory function by stem cell factor, IL-3 or IL-4. J. Immunol. 150, 556–562.

Columbo, M., Horowitz, E.M., Botana, L.M., *et al.* (1992). The human recombinant c-*kit* receptor ligand, rhSCF, induces mediator release from human cutaneous mast cells and enhances IgE-dependent mediator release from both skin mast cells and peripheral blood basophils. J. Immunol. 149, 599–608.

Copeland, N.G., Gilbert, D.J., Cho, B.C., *et al.* (1990). Mast cell growth factor maps near the Steel locus on mouse chromosome 10 and is deleted in a number of Steel alleles. Cell 63, 175–183.

Costa, J.J., Demetri, G.D., Hayes, D.F., Merica, E.A., Menchaca, D.M. and Galli, S.J. (1993). Increased skin mast cells and urine methyl histamine in patients receiving recombinant methionyl human stem cell factor. Proc. Am. Assoc. Cancer Res. 34, 211.

Costa, J.J., Demetri, G.D., Harris, T.J., *et al.* (1996). Recombinant human stem cell factor (kit ligand) promotes human mast cell and melanocyte hyperplasia and functional activation in vivo. J. Exp. Med. 183(6), 2681–2686.

Crosier, P.S., Ricciardi, S.T., Hall, L.R., Vitas, M.R., Clark, S.C. and Crosier, K.E. (1993). Expression of isoforms of the human receptor tyrosine kinase c-*kit* in leukemic cell lines and acute myeloid leukemia. Blood 82, 1151–1158.

Dastych, J. and Metcalf, D.D. (1994). Stem cell factor induces mast cell adhesion to fibronectin. J. Immunol. 152, 213–219.

d'Auriol, L., Mattei, M.G., Andre, C. and Galibert, F. (1988). Localization of the human c-kit protooncogene on the q11-q12 region of chromosome 4. Hum. Genet. 78, 374–376.

Debili, N., Masse, J.M., Katz, A., Guichard, J., Breton-Gorius, J. and Vainchenker, W. (1993). Effects of the recombinant hematopoietic growth factors interleukin-3, interleukin-6, stem cell factor, and leukemia inhibitory factor on the megakaryocytic differentiation of CD34+ cells. Blood 82, 84–95.

deCastro, C.M., Denning, S.M., Langdon, S., *et al.* (1994). The c-*kit* proto-oncogene receptor is expressed on a subset of human CD3−CD4−CD8− (triple-negative) thymocytes. Exp. Hematol. 22, 1025–1033.

Demetri, G., Costa, J., Hayes, D., *et al.* (1993). A phase I trial of recombinant methionyl human stem cell factor (SCF) in patients with advanced breast carcinoma pre- and post-chemotherapy with cyclophosphamide and doxorubicin. Proc. Am. Assoc. Clin. Oncol. 12, A367.

Deol, M.S. (1970). The relationship between abnormalities of pigmentation and of the inner ear. Proc. Roy. Soc. Lond. 175, 201–217.

de Vries, P., Brasel, K.A., Eisenman, J.R., Alpert, A.R. and Williams, D.E. (1991). The effect of recombinant mast cell growth factor on purified murine hematopoietic stem cells. J. Exp. Med. 173, 1205–1211.

de Vries, P., Brasel, K.A., McKenna, H.J., Williams, D.E. and Watson, J.D. (1992). Thymus reconstitution by c-*kit*-expressing hematopoietic stem cells purified from adult mouse bone marrow. J. Exp. Med. 176, 1503–1509.

de Wolf, J.Th.M., Muller, E.W., Hendriks, D.H., Halie, R.M. and Vellenga, E. (1994). Mast cell growth factor modulates CD36 antigen expression on erythroid progenitors from human bone marrow and peripheral blood associated with ongoing differentiation. Blood 84, 59–64.

Dolci, S., Williams, D.E., Ernst, M.K., *et al.* (1991). Requirement for mast cell growth factor for primordial germ cell survival in culture. Nature 352, 809–811.

Donovan, P.J., Stott, D., Cairns, L.A., Heasman, J. and Wylie, C.C. (1986). Migratory and postmigratory mouse primordial germ cells behave differently in culture. Cell 44, 831–838.

Dräger, U.C. (1986). Albinism and visual pathways. N. Engl. J. Med. 314, 1636–1638.

Dräger, U.C. and Balkema, G.W. (1987). Does melanin do more than protect from light? Neurosci. Res. Suppl. 6, S75–S86.

Durand, B., Migliaccio, G., Yee, N.S., *et al.* (1994). Long-term generation of human mast cells in serum-free cultures of CD34+ cord blood cells stimulated with stem cell factor and interleukin-3. Blood 84, 3667–3674.

Duronio, V., Welham, M.J., Abraham, S., Dryden, P. and Schrader, J.W. (1992). p21ras activation via hemopoietin receptors and c-kit requires tyrosine kinase activity but not tyrosine phosphorylation of p21 ras GTPase activating protein. Proc. Natl Acad. Sci. USA 89, 1587–1591.

Ebbe, S., Phalen, E. and Stohlman, F.J. (1973). Abnormal megakaryocytopoiesis in *Sl*/*Sl*^d mice. Blood 42, 865.

Ezoe, K., Holmes, S.A., Ho, L., *et al.* (1995). Novel mutations and deletion of the *KIT* (steel factor receptor) gene in human piebaldism. Am. J. Hum. Genet. 56, 58–66.

Faust, E.A., Saffran, D.C., Toksoz, D., Williams, D.A. and Witte, O.N. (1993). Distinctive growth requirements and gene expression patterns distinguish progenitor B cells from pre-B cells. J. Exp. Med. 177, 915–923.

Flanagan, J.G. and Leder, P. (1990). The *kit* ligand: a cell surface molecule altered in Steel mutant fibroblasts. Cell 63, 185–194.

Flanagan, J.G., Chan, D.C. and Leder, P. (1991). Transmembrane form of the *kit* ligand growth factor is determined by alternate splicing and is missing in the *Sl*^d mutant. Cell 64, 1025–1035.

Fleischman, R.A. (1992a). Effect of the c-kit codon 584 Phe→Leu substitution demonstrated in human piebaldism. Am. J. Hum. Genet. 51, 677–678.

Fleischman, R.A. (1992b). Human piebald trait resulting from a dominant negative mutant allele of the c-*kit* membrane receptor gene. J. Clin. Invest. 89, 1713–1717.

Fleischman, R.A. (1993). From white spots to stem cells: the role of the Kit receptor in mammalian development. Trends Genet. 9, 285–290.

Fleischman, R.A., Saltman, D.L., Stastny, V. and Zneimer, S. (1991). Deletion of the c-*kit* protooncogene in the human developmental defect piebald trait. Proc. Natl Acad. Sci. USA 88, 10885–10889.

Fleming, W.H., Alpern, E.J., Uchida, N., Ikuta, K. and

Weissman, I.L. (1993). Steel factor influences the distribution and activity of murine hematopoietic stem cells *in vivo*. Proc. Natl Acad. Sci. USA 90, 3760–3769.

Funasaka, Y., Boulton, T., Cobb, M., *et al.* (1992). c-Kit-kinase induces a cascade of protein tyrosine phosphorylation in normal human melanocytes in response to mast cell growth factor and stimulates mitogen-activated protein kinase but is down-regulated by melanomas. Mol. Biol. Cell. 3, 197–209.

Galli, S.J., Tsai, M., Wershil, B.K., Tam, S.Y. and Costa, J.J. (1995). Regulation of mouse and human mast cell development, survival and function by stem cell factor, the ligand for the c-kit receptor. Int. Arch. Allergy Immunol. 107(1–3), 51–53.

Geissler, E.N., Ryan, M.A. and Houseman, D.E. (1988). The dominant-white spotting (*W*) locus of the mouse encodes the c-*kit* proto-oncogene. Cell 55, 185–192.

Geissler, E.N., Liao, M., Brook, J.D., *et al.* (1991). Stem cell factor (*SCF*), a novel hematopoietic growth factor and ligand for c-kit tyrosine kinase receptor, maps on human chromosome 12 between 12q14.3 and 12qter. Somat. Cell Mol. Genet. 17, 207–214.

Giebel, L.B. and Spritz, R.A. (1991). Mutation of the *KIT* (mast/stem cell growth factor receptor) protooncogene in human piebaldism. Proc. Natl Acad. Sci. USA 88, 8696–8699.

Giebel, L.B., Strunk, K.M., Holmes, S.A. and Spritz, R.A. (1992). Organization and nucleotide sequence of the human *KIT* (mast/stem cell growth factor receptor) proto-oncogene. Oncogene 7, 2207–2217.

Glaspy, J., Davis, M.W., Parker, W.R., Foote, M. and McNiece, I. (1996). Biology and clinical potential of stem-cell factor. Cancer Chemother. Pharmacol. 38 (supplement), S53–S57.

Godfrey, D.I., Zlotnik, A. and Suda, T. (1992). Phenotypic and functional characterization of c-kit expression during intrathymic T cell development. J. Immunol. 149, 2281–2285.

Godin, I., Deed, R., Cooke, J., Zsebo, K., Dexter, M. and Wylie, C.C. (1991). Effects of the *steel* gene product on mouse primordial germ cells in culture. Nature 352, 807–809.

Gokkel, E., Grossman, Z., Ramot, B., Yarden, Y., Rechavi, G. and Givol, D. (1992). Structural organization of the murine c-*kit* proto-oncogene. Oncogene 7, 1423–1429.

Grossi, A., Vannucchi, A.M., Bacci, P., *et al.* (1995). In vivo administration of stem cell factor enhances both proliferation and maturation of murine megakaryocytes. Haematologica 80(1), 18–24.

Gunji, Y., Nakamura, M., Hagiwara, T., *et al.* (1992). Expression and function of adhesion molecules on human hematopoietic stem cells: CD34[+] LFA-1[-] cells are more primitive than CD34[+] LFA-1[+] cells. Blood 80, 429–436.

Gunji, Y., Nakamura, M., Osawa, H., *et al.* (1993). Human primitive hematopoietic progenitor cells are more enriched in KIT[low] cells than in KIT[high] cells. Blood 82, 3283–3289.

Halaban, R., Ghosh, S. and Baird, A. (1987). bFGF is the putative natural growth factor for human melanocytes. In Vitro Cell. Dev. Biol. 23, 47–52.

Hannum, C., Culpepper, J., Campbell, D., *et al.* (1994). Ligand for FLT3/FLK2 receptor tyrosine kinase regulates growth of haematopoietic stem cells and is encoded by variant RNAs. Nature 368, 643–648.

Hardy, R.R., Carmack, C.E., Shinton, S.A., Kemp, J.D. and Hayakawa, K. (1991). Resolution and characterization of pro-B and pre-pro-B cell stages in normal mouse bone marrow. J. Exp. Med. 173, 1213–1225.

Hassan, H.T. and Freund, M. (1995). Human stem cell factor protects CD34 positive human myeloid leukaemia cells from chemotherapy-induced apoptosis. Eur. J. Cancer 31A(11), 1883–1884. [Letter]

Hassan, H.T. and Zander, A. (1996). Stem cell factor as a survival and growth factor in human normal and malignant hematopoiesis. Acta Haematol. 95(3–4), 257–262.

Hassan, H.T., Biermann, B. and Zander, A.R. (1996). Maintenance and expansion of erythropoiesis in human long-term bone marrow cultures in presence of erythropoietin plus stem cell factor and interleukin-3 or interleukin-11. Eur. Cytokine Netw. 7(2), 129–136.

Hayashi, S.-I., Kunisada, T., Ogawa, M., Yamaguchi, K. and Nishikawa, S.-I. (1991). Exon skipping by mutation of an authentic splice site of c-*kit* gene in *W/W* mouse. Nucleic Acids Res. 19, 1267–1271.

Herbst, R., Shearman, M.S., Obermeier, A., Schlessinger, J. and Ullrich, A. (1992). Differential effects of *W* mutations on p145[c-kit] tyrosine kinase activity and substrate interaction. J. Biol. Chem. 267, 13210–13216.

Heyworth, C.M., Whetton, A.D., Nicholls, S., Zsebo, K. and Dexter, T.M. (1992). Stem cell factor directly stimulates the development of enriched granulocyte-macrophage colony-forming cells and promotes the effects of other colony-stimulating factors. Blood 80, 2230–2236.

Hodgson, G.S., Bradley, T.R. and Radley, J.M. (1982). The organization of hemopoietic tissue as inferred from the effects of 5-fluorouracil. Exp. Hematol. 10, 26–35.

Holyoake, T.L., Freshney, M.G., McNair, L., *et al.* (1996). Ex vivo expansion with stem cell factor and interleukin-11 augments both short-term recovery posttransplant and the ability to serially transplant marrow. Blood 87(11), 4589–4595.

Hooper, C. (1990). Multipotent growth factor: a new muse for hematopoietics. J. NIH Res. 2, 54–58.

Huang, E., Nocka, K., Beier, D.R., *et al.* (1990). The hematopoietic growth factor KL is encoded by the *Sl* locus and is the ligand of the c-*kit* receptor, the gene product of the *W* locus. Cell 63, 225–233.

Huang, E.J., Nocka, K.H., Buck, J. and Besmer, P. (1992). Differential expression and processing of two cell associated forms of the kit-ligand: KL-1 and KL-2. Mol. Biol. Cell 3, 349–362.

Huizinga, J.D., Thuneberg, L., Klüppel, M., Malysz, J., Mikkelsen, H.B. and Bernstein, A. (1995). *W/kit* gene required for interstitial cells of Cajal and for intestinal pacemaker activity. Nature 373, 347–349.

Hulten, M.A., Honeyman, M.M., Mayne, A.J. and Tarlow, M.J. (1987). Homozygosity in piebald trait. J. Med. Genet. 24, 568–571.

Hunt, P., Zsebo, K.M., Hokom, M.M., *et al.* (1992). Evidence that stem cell factor is involved in the rebound thrombocytosis that follows 5-fluorouracil treatment. Blood 80, 904–911.

Ikuta, K. and Weissman, I.L. (1992). Evidence that hematopoietic stem cells express mouse c-*kit* but do not depend on steel factor for their generation. Proc. Natl Acad. Sci. USA 89, 1502–1506.

Katayama, N., Clark, S.C. and Ogawa, M. (1993). Growth factor requirement for survival in cell-cycle dormancy of primitive murine lymphohematopoietic progenitors. Blood 81, 610–616.

Keller, J.R., Ortiz, M. and Ruscetti, F.W. (1995). Steel factor (c-kit ligand) promotes the survival of hematopoietic stem/progenitor cells in the absence of cell division. Blood 86(5), 1757–1764.

Keller, S.A., Liptay, S., Hajra, A. and Meisler, M.H. (1990). Transgene-induced mutation of the murine steel locus. Proc. Natl Acad. Sci. USA 87, 10019–10022.

Keshet, E., Lyman, S.D., Williams, D.E., et al. (1991). Embryonic RNA expression patterns of the c-kit receptor and its cognate ligand suggest multiple functional roles in mouse development. EMBO J. 10, 2425–2435.

Kinashi, T. and Springer, T.A. (1994). Steel factor and c-kit regulate cell–matrix adhesion. Blood 83, 1033–1041.

Kiss, C., Surrey, S., Schreiber, A.D., Schwartz, E. and McKenzie, S.E. (1996). Human c-kit ligand (stem cell factor) induces platelet Fc receptor expression in megakaryoblastic cells. Exp. Hematol. 24(10), 1232–1237.

Kitamura, Y. and Go, S. (1978). Decreased production of mast cells in Sl/Sl^t anemic mice. Blood 53, 492–497.

Kriegler, A.B., Verschoor, S.M., Bernardo, D. and Bertoncello, I. (1994). The relationship between different high proliferative potential colony-forming cells in mouse bone marrow. Exp. Hematol. 22, 432–440.

Krystal, G.W., Hines, S.J. and Organ, C.P. (1996). Autocrine growth of small cell lung cancer mediated by coexpression of c-kit and stem cell factor. Cancer Res. 56(2), 370–376.

Ladner, M.B., Martin, G.A., Noble, J.A., et al. (1988). cDNA cloning and expression of murine macrophage colony-stimulating factor from L929 cells. Proc. Natl Acad. Sci. USA 85, 6706–6710.

Landreth, K.S., Kincade, P.W., Lee, G. and Harrison, D.E. (1994). B lymphocyte precursors in embryonic and adult W anemic mice. J. Immunol. 132, 2724–2729.

Langley, K.E., Wypych, J., Mendiaz, E.A., et al. (1992). Purification and characterization of soluble forms of human and rat stem cell factor recombinantly expressed by Escherichia coli and by Chinese hamster ovary cells. Arch. Biochem. Biophys. 295, 21–28.

Langley, K.E., Bennett, L.G., Wypych, J., et al. (1993). Soluble stem cell factor in human serum. Blood 81, 656–660.

Lassam, N. and Bickford, S. (1992). Loss of c-kit expression in cultured melanoma cells. Oncogene 7, 51–56.

Lev, S., Givol, D. and Yarden, Y. (1991). A specific combination of substrates is involved in signal transduction by the kit-encoded receptor. EMBO J. 10, 647–654.

Li, C.L. and Johnson, G.R. (1994). Stem cell factor enhances the survival but not the self-renewal of murine hematopoietic long-term repopulating cells. Blood 84, 408–414.

Linnekin, D., Keller, J.R., Ferris, D.K., Mou, S.M., Broudy, V. and Longo, D.L. (1995). Stem cell factor induces phosphorylation of a 200 kDa protein which associates with c-kit. Growth Factors 12(1), 57–67.

Linnekin, D., Weiler, S.R., Mou, S., et al. (1996). JAK2 is constitutively associated with c-Kit and is phosphorylated in response to stem cell factor. Acta Haematol. 95(3–4), 224–228.

Louis, R., Tilkin, P., Poncelet, M., et al. (1995). Regulation of histamine release from human bronchoalveolar lavage mast cells by stem cell factor in several respiratory diseases. Allergy 50(4), 340–348.

Lowry, P.A., Zsebo, K.M., Deacon, D.H., Eichman, C.E. and Quesenberry, P.J. (1991). Effects of rrSCF on multiple cytokine responsive HPP-CFC generated from SCA+Lin−

murine hematopoietic progenitors. Exp. Hematol. 19, 994–996.

Lu, H.S., Clogston, C.L., Wypych, J., et al. (1991). Amino acid sequence and post-translational modification of stem cell factor isolated from buffalo rat liver cell-conditioned medium. J. Biol. Chem. 266, 8102–8107.

Lu, H.S., Chang, W.-C., Mendiaz, E.A., Mann, M.B., Langley, K.E. and Hsu, Y.-R. (1995). Spontaneous dissociation–association of monomers of the human-stem-cell-factor dimer. Biochem. J. 305, 563–568.

Lu, H.S., Jones, M.D., Shieh, J.H., et al. (1996). Isolation and characterization of a disulfide-linked human stem cell factor dimer. Biochemical, biophysical, and biological comparison to the noncovalently held dimer. J. Biol. Chem. 271(19), 11309–11316.

Lukacs, N.W., Kunkel, S.L., Strieter, R.M., et al. (1996a). The role of stem cell factor (c-kit ligand) and inflammatory cytokines in pulmonary mast cell activation. Blood 87(6), 2262–2268.

Lukacs, N.W., Strieter, R.M., Lincoln, P.M., et al. (1996b). Stem cell factor (c-kit ligand) influences eosinophil recruitment and histamine levels in allergic airway inflammation. J. Immunol. 156(10), 3945–3951.

Lyman, S.D. and Williams, D.E. (1991). In "Blood Cell Growth Factors: Their Present and Future Use in Hematology and Oncology". Proceedings of the Beijing Symposium, August, 1991 (ed. M.J. Murphy, Jr), pp. 183–193. AlphaMed Press, Dayton, OH.

Lyman, S.D., James, L., VandenBos, T., et al. (1993). Molecular cloning of a ligand for the flt3/flk-2 tyrosine kinase receptor: a proliferative factor for primitive hematopoietic cells. Cell 75, 1157–1167.

Lyman, S.D., Brasel, K., Rousseau, A.-M. and Williams, D.E. (1994a). The flt3 ligand: a hematopoietic stem cell factor whose activities are distinct from the steel factor. Stem Cells 12, 99–110.

Lyman, S.D., James, L., Johnson, L., et al. (1994b). Cloning of the human homologue of the murine flt3 ligand: a growth factor for early hematopoietic progenitor cells. Blood 83, 2795–2801.

Lyman, S.D., James, L., Escobar, S., et al. (1995). Identification of soluble and membrane-bound isoforms of the murine flt3 ligand generated by alternative splicing of mRNAs. Oncogene 10, 149–157.

Lynch, D.H., Jacobs, C., DuPont, D., et al. (1992). Pharmacokinetic parameters of recombinant mast cell growth factor (rMGF). Lymphokine Cytokine Res. 11, 233–243.

Maeda, H., Yamagata, A., Nishikawa, S., et al. (1992). Requirement of c-kit for development of intestinal pacemaker system. Development 116, 369–375.

Majumdar, M.K., Feng, L., Medlock, E., Toksoz, D. and Williams, D.A. (1994). Identification and mutation of primary and secondary proteolytic cleavage sites in murine stem cell factor cDNA yields biologically active, cell-associated protein. J. Biol. Chem. 269, 1237–1242.

Majumder, S., Brown, K., Qiu, F.-H. and Besmer, P. (1988). c-kit protein, a transmembrane kinase: identification in tissues and characterization. Mol. Cell. Biol. 8, 4896–4903.

Manova, K., Nocka, K., Besmer, P. and Bachvarova, R.F. (1990). Gonadal expression of c-kit encoded at the W locus of the mouse. Development 110, 1057–1069.

Martin, F.H., Suggs, S.V., Langley, K.E., et al. (1990). Primary structure and functional expression of rat and human stem cell factor cDNAs. Cell 63, 203–211.

Marziali, G., Lazzaro, D. and Sorrentino, V. (1993). Binding of germ cells to mutant Sl^d Sertoli cells is defective and is rescued by expression of the transmembrane form of the c-*kit* ligand. Dev. Biol. 157, 182–190.

Matous, J.V., Langley, K. and Kaushansky, K. (1996). Structure–function relationships of stem cell factor: an analysis based on a series of human–murine stem cell factor chimera and the mapping of a neutralizing monoclonal antibody. Blood 88, 437–444.

Matsui, Y., Zsebo, K.M. and Hogan, B.L.M. (1990). Embryonic expression of a haematopoietic growth factor encoded by the *Sl* locus and the ligand for c-kit. Nature 347, 667–669.

Matsui, Y., Toksoz, D., Nishikawa, S., *et al.* (1991). Effect of Steel factor and leukaemia inhibitory factor on murine primordial germ cells in culture. Nature 353, 750–752.

Matthews, W., Jordan, C.T., Wiegand, G.W., Pardoll, D. and Lemischka, I.R. (1991). A receptor tyrosine kinase specific to hematopoietic stem and progenitor cell-enriched populations. Cell 65, 1143–1152.

Mayer, T.C. (1970). A comparison of pigment cell development in albino, steel, and dominant spotting mutant mouse embryos. Dev. Biol. 23, 297–309.

McNiece, I.K. and Briddell, R.A. (1995). Stem cell factor. J. Leukocyte Biol. 58(1), 14–22.

McNiece, I.K., Stewart, F.M., Deacon, D.M., *et al.* (1989). Detection of a human CFC with a high proliferative potential. Blood 74, 609–612.

McNiece, I.K., Langley, K.E. and Zsebo, K.M. (1991). The role of recombinant stem cell factor in early B cell development. Synergistic interaction with IL-7. J. Immunol. 146, 3785–3790.

Mendiaz, E.A., Chang, D.G., Boone, T.C., *et al.* (1996). Epitope mapping and immunoneutralization of recombinant human stem-cell factor. Eur. J. Biochem. 239(3), 842–849.

Meininger, C.J., Yano, H., Rottapel, R., Bernstein, R., Zsebo, K.M. and Zetter, B.R. (1992). The c-kit receptor ligand functions as a mast cell chemoattractant. Blood 79, 958–964.

Meininger, C.J., Brightman, S.E., Kelly, K.A. and Zetter, B.R. (1995). Increased stem cell factor release by hemangioma-derived endothelial cells. Lab. Invest. 72(2), 166–173.

Metcalf, D. (1991). Lineage commitment of hemopoietic progenitor cells in developing blast cell colonies: influence of colony-stimulating factors. Proc. Natl Acad. Sci. USA 88, 11310–11314.

Metcalf, D. and Nicola, N.A. (1991). Direct proliferative actions of stem cell factor on murine bone marrow cells *in vitro*: effects of combination with colony-stimulating factors. Proc. Natl Acad. Sci. USA 88, 6239–6243.

Mintz, B. and Russell, E.S. (1957). Gene induced embryological modifications of primordial germ cells in the mouse. J. Exp. Zool. 134, 207–237.

Mitsui, H., Furitsu, T., Dvorak, A.M., *et al.* (1993). Development of human mast cells from umbilical cord blood cells by recombinant human murine c-kit ligand. Proc. Natl Acad. Sci. USA 90, 735–739.

Miura, N., Okada, S., Zsebo, K.M., Miura, Y. and Suda, T. (1993). Rat stem cell factor and IL-6 preferentially support the proliferation of c-kit-positive murine hemopoietic cells rather than their differentiation. Exp. Hematol. 21, 143–149.

Molineux, G., Migdalska, A., Szmitkowski, M., Zsebo, K. and

Dexter, T.M. (1991). The effects of hematopoiesis of recombinant stem cell factor (ligand for c-kit) administered in vivo to mice either alone or in combination with granulocyte colony-stimulating factor. Blood 78, 961.

Morrison-Graham, K. and Weston, J.A. (1993). Transient steel factor dependence by neural crest-derived melanocyte precursors. Dev. Biol. 159, 346–352.

Morrissey, P.J., McKenna, H., Widmer, M.B., *et al.* (1994). Steel factor (c-kit ligand) stimulates the *in vitro* growth of immature CD3$^-$/CD4$^-$/CD8$^-$ thymocytes: synergy with IL-7. Cell. Immunol. 157, 118–131.

Motro, B., van der Kooy, D., Rossant, J., Reith, A. and Bernstein, A. (1991). Contiguous patterns of c-*kit* and *steel* expression: analysis of mutations at the *W* and *Sl* loci. Development 113, 1207–1221.

Muench, M.O., Schneider, J.G. and Moore, M.A.S. (1992). Interactions among colony-stimulating factors, IL-1β, IL-6, and *kit*-ligand in the regulation of primitive murine hematopoietic cells. Exp. Hematol. 20, 339–349.

Murphy, M., Reid, K., Williams, D.E., Lyman, S.D. and Bartlett, P.F. (1992). Steel factor is required for maintenance, but not differentiation, of melanocyte precursors in the neural crest. Dev. Biol. 153, 396–401.

Murty, V.V.S., Houldsworth, J., Baldwin, S., *et al.* (1992). Allelic deletions in the long arm of chromosome 12 identify sites of candidate tumor suppressor genes in male germ cell tumors. Proc. Natl Acad. Sci. USA 89, 11006–11010.

Muta, K., Krantz, S.B., Bondurant, M.C. and Dai, C.H. (1995). Stem cell factor retards differentiation of normal human erythroid progenitor cells while stimulating proliferation. Blood 86(2), 572–580.

Nakajima, K., Hirai, K., Yamaguchi, M., *et al.* (1992). Stem cell factor has histamine releasing activity in rat connective tissue-type mast cells. Biochem. Biophys. Res. Commun. 183, 1076–1082.

Nishikawa, S., Kusakabe, M., Yoshinaga, K., *et al.* (1991). *In utero* manipulation of coat color formation by a monoclonal anti-c-*kit* antibody: two distinct waves of c-*kit*-dependency during melanocyte development. EMBO J. 10, 2111–2118.

Nocka, K., Tan, J.C., Chiu, E., *et al.* (1990). Molecular bases of dominant negative and loss of function mutations at the murine c-*kit*/white spotting locus: W^{37}, W^v, W^{41} and W. EMBO J. 9, 1805–1813.

Ogawa, M., Matsuzaki, Y., Nishikawa, S., *et al.* (1991). Expression and function of c-*kit* in hemopoietic progenitor cells. J. Exp. Med. 174, 63–71.

Okumura, N., Tsuji, K., Ebihara, Y., *et al.* (1996). Chemotactic and chemokinetic activities of stem cell factor on murine hematopoietic progenitor cells. Blood 87(10), 4100–4108.

Olivieri, N.F., Grunberger, T., Ben-David, Y., *et al.* (1991). Diamond–Blackfan anemia: heterogenous response of hematopoietic progenitor cells in vitro to the protein product of the *Steel* locus. Blood 78, 2211–2215.

Olweus, J., Terstappen, L.W., Thompson, P.A. and Lund-Johansen, F. (1996). Expression and function of receptors for stem cell factor and erythropoietin during lineage commitment of human hematopoietic progenitor cells. Blood 88(5), 1594–1607.

Orr-Urtreger, A., Aviv, A., Zimmer, Y., Givol, D., Yarden, Y. and Lonai, P. (1990). Developmental expression of c-*kit*, a proto-oncogene encoded by the *W* locus. Development 109, 911–923.

Packer, A.I., Hsu, Y.C., Besmer, P. and Bachvarova, R.F.

(1994). The ligand of the c-*kit* receptor promotes oocyte growth. Dev. Biol. 161, 194–205.

Palacios, R. and Nishikawa, S.-I. (1992). Developmentally regulated cell surface expression and function of c-*kit* receptor during lymphocyte ontogeny in the embryo and adult mice. Development 115, 1133–1147.

Pandit, J., Bohm, A., Jancarik, J., Halenbeck, R., Koths, K. and Kim, S.-H. (1992). Three-dimensional structure of dimeric human recombinant macrophage colony-stimulating factor. Science 258, 1358–1362.

Papadimitriou, C.A., Topp, M.S., Serve, H., *et al.* (1995). Recombinant human stem cell factor does exert minor stimulation of growth in small cell lung cancer and melanoma cell lines. Eur. J. Cancer 31A(13–14), 2371–2378.

Papayannopoulou, T., Brice, M. and Blau, C.A. (1993). *Kit* ligand in synergy with interleukin-3 amplifies the erythropoietin-independent, globin-synthesizing progeny of normal human burst-forming units-erythroid in suspension cultures: physiologic implications. Blood 81, 299–310.

Pesce, M., Farrace, M.G., Piacentini, M., Dolci, S. and De Felici, M. (1993). Stem cell factor and leukemia inhibitory factor promote primordial germ cell survival by suppressing programmed cell death (apoptosis). Development 118, 1089–1094.

Philo, J.S., Wen, J., Wypych, J., Schwartz, M.G., Mendiaz, E.A. and Langley, K.E. (1996). Human stem cell factor dimer forms a complex with two molecules of the extracellular domain of its receptor, Kit. J. Biol. Chem. 271(12), 6895–6902.

Piao, X., Curtis, J.E., Minkin, S., Minden, M.D. and Bernstein, A. (1994). Expression of the *Kit* and *KitA* receptor isoforms in human acute myelogenous leukemia. Blood 83, 476–481.

Qiu, F., Ray, P., Brown, K., *et al.* (1988). Primary structure of c-*kit*: relationship with the CSF-1/PDGF receptor kinase family—oncogenic activation of v-*kit* involves deletion of extracellular domain and C terminus. EMBO J. 7, 1003–1011.

Reith, A.D., Rottapel, R., Giddens, E., Brady, C., Forrester, L. and Bernstein, A. (1990). *W* mutant mice with mild or severe developmental defects contain distinct point mutations in the kinase domain of the c-*kit* receptor. Genes Dev. 4, 390–400.

Reith, A.D., Ellis, C., Lyman, S.D., *et al.* (1991). Signal transduction by normal isoforms and *W* mutant variants of the Kit receptor tyrosine kinase. EMBO J. 10, 2451–2459.

Rodewald, H.R., Kretzschmar, K., Swat, W. and Takeda, S. (1995). Intrathymically expressed c-kit ligand (stem cell factor) is a major factor driving expansion of very immature thymocytes in vivo. Immunity 3(3), 313–319.

Rolink, A., Streb, M., Nishikawa, S.-I. and Melchers, F. (1991). The c-kit-encoded tyrosine kinase regulates the proliferation of early pre-B cells. Eur. J. Immunol. 21, 2609–2612.

Rosnet, O., Marchetto, S., deLapeyriere, O. and Birnbaum, D. (1991). Murine *Flt3*, a gene encoding a novel tyrosine kinase receptor of the PDGFR/CSF1R family. Oncogene 6, 1641–1650.

Rosnet, O., Schiff, C., Pébusque, M.-J., *et al.* (1993). Human *FLT3/FLK2* gene: cDNA cloning and expression in hematopoietic cells. Blood 82, 1110–1119.

Rossi, P., Albanesi, C., Grimaldi, P. and Geremia, R. (1991). Expression of the mRNA for the ligand of c-kit in mouse sertoli cells. Biochem. Biophys. Res. Commun. 176, 910–914.

Rossi, P., Marziali, G., Albanesi, C., Charlesworth, A., Geremia,

R. and Sorrentino, V. (1992). A novel c-kit transcript, potentially encoding a truncated receptor, originates within a kit gene intron in mouse spermatids. Dev. Biol. 152, 203–207.

Rossi, P., Dolci, S., Albanesi, C., Grimaldi, P., Ricca, R. and Geremia, R. (1993). Follicle-stimulating hormone induction of Steel factor (SLF) mRNA in mouse Sertoli cells and stimulation of DNA synthesis in spermatogonia by soluble SLF. Dev. Biol. 155, 68–74.

Rottapel, R., Reedijk, M., Williams, D.E., *et al.* (1991). The *Steel/W* transduction pathway: Kit autophosphorylation and its association with a unique subset of cytoplasmic signaling proteins is induced by the Steel factor. Mol. Cell. Biol. 11, 3043–3051.

Ruscetti, F.N., Boggs, D.R., Torok, B.J. and Boggs, S.S. (1976). Reduced blood and marrow neutrophils and granulocytic colony-forming cells in Sl/Sl^d mice. Proc. Soc. Exp. Biol. Med. 152, 398.

Russell, E.S. (1979). Hereditary anemias of the mouse: a review for geneticists. Adv. Gen. 20, 357–459.

Santiago-Schwarz, F., Laky, K. and Carsons, S.E. (1995). Stem cell factor enhances dendritic cell development. Adv. Exp. Med. Biol. 378, 7–11.

Sarvella, P.A. and Russell, L.B. (1956). Steel, a new dominant gene in the house mouse. J. Hered. 47, 123–128.

Shadle, P.J., Allen, J.I., Geier, M.D. and Koths, K. (1989). Detection of endogenous macrophage colony-stimulating factor (M-CSF) in human blood. Exp. Hematol. 17, 154–159.

Shearman, M.S., Herbst, R., Schlessinger, J. and Ullrich, A. (1993). Phosphatidylinositol 3′-kinase associates with p145 c-kit as part of a cell type characteristic multimeric signaling complex. EMBO J. 12, 3817–3826.

Shui, C., Khan, W.B., Leigh, B.R., Turner, A.M., Wilder, R.B. and Knox, S.J. (1995). Effects of stem cell factor on the growth and radiation survival of tumor cells. Cancer Res. 55(15), 3431–3437.

Shull, R.M., Suggs, S.V., Langley, K.E., Okino, K.H., Jacobsen, F.W. and Martin, F.H. (1992). Canine stem cell factor (c-kit ligand) supports the survival of hematopoietic progenitors in long-term canine marrow culture. Exp. Hematol. 20, 1118–1124.

Silvers, W.K. (1979a). In "The Coat Colors of Mice: A Model for Mammalian Gene Action and Interaction", pp. 206–242. Springer Verlag, New York.

Silvers, W.K. (1979b). In "The Coat Colors of Mice: A Model for Mammalian Gene Action and Interaction", pp. 243–267. Springer Verlag, New York.

Small, D., Levenstein, M., Kim, E., *et al.* (1994). STK-1, the human homolog of Flk-2/Flt-3, is selectively expressed in CD34+ human bone marrow cells and is involved in the proliferation of early progenitor/stem cells. Proc. Natl Acad. Sci. USA 91, 459–463.

Sorrentino, V., Giorgi, M., Geremia, R., Besmer, P. and Rossi, P. (1991). Expression of the c-*kit* proto-oncogene in the murine male germ cells. Oncogene 6, 149–151.

Spritz, R.A., Giebel, L.B. and Holmes, S.A. (1992). Am. J. Hum. Genet. 51, 678.

Spritz, R.A., Holmes, S.A., Berg, S.Z., Nordlund, J.J. and Fukai, K. (1993). A recurrent deletion in the KIT (mast/stem cell growth factor receptor) proto-oncogene is a frequent cause of human piebaldism. Hum. Mol. Genet. 2, 1499–1500.

Stanley, E.R., Guilbert, L.J., Tushinski, R.J. and Bartelmez, S.H. (1983). CSF-1: a mononuclear phagocyte lineage specific hemopoietic growth factor. J. Cell. Biochem. 21, 151–159.

Stanulla, M., Welte, K., Hadam, M.R. and Pietsch, T. (1995). Coexpression of stem cell factor and its receptor c-Kit in human malignant glioma cell lines. Acta Neuropathol. Berl. 89(2), 158–165.

Steel, K.P. and Barkway, C. (1989). Another role for melanocytes: their importance for normal stria vascularis development in the mammalian inner ear. Development 107, 453–463.

Strohmeyer, T., Peter, S., Hartmann, M., et al. (1991). Expression of the hst-1 and c-kit protooncogenes in human testicular germ cell tumors. Cancer Res. 51, 1811–1816.

Strohmeyer, T., Reese, D., Press, M., Ackermann, R., Hartmann, M. and Slamon, D. (1995). Expression of the c-kit proto-oncogene and its ligand stem cell factor (SCF) in normal and malignant human testicular tissue. J. Urol. 153(2), 511–515.

Sutherland, H.J., Eaves, C.J., Eaves, A.C., Dragowska, W. and Lansdorp, P.M. (1989). Characterization and partial purification of human marrow cells capable of initiating long-term hematopoiesis in vitro. Blood 74, 1563–1570.

Tajima, Y., Onoue, H., Kitamura, Y. and Nishimune, Y. (1991). Biologically active kit ligand growth factor is produced by mouse Sertoli cells and is defective in Sl^d mutant mice. Development 113, 1031–1035.

Takahashi, H., Saitoh, K., Kishi, H. and Parsons, P.G. (1995). Immunohistochemical localisation of stem cell factor (SCF) with comparison of its receptor c-Kit proto-oncogene product (c-KIT) in melanocytic tumours. Virchows Arch. 427(3), 283–288.

Tan, J.C., Nocka, K., Ray, P., Traktman, P. and Besmer, P. (1990). The dominant W^{42} spotting phenotype results from a missense mutation in the c-kit receptor kinase. Science 247, 209–212.

Tanaka, R., Koike, K., Imai, T., et al. (1992). Stem cell factor enhances proliferation, but not maturation, of murine megakaryocytic progenitors in serum-free culture. Blood 80, 1743–1749.

Taylor, A.M., Galli, S.J. and Coleman, J.W. (1995). Stem-cell factor, the kit ligand, induces direct degranulation of rat peritoneal mast cells in vitro and in vivo: dependence of the in vitro effect on period of culture and comparisons of stem-cell factor with other mast cell-activating agents. Immunology 86(3), 427–433.

Toksoz, D., Zsebo, K.M., Smith, K.A., et al. (1992). Support of human hematopoiesis in long-term bone marrow cultures by murine stromal cells selectively expressing the membrane-bound and secreted forms of the human homolog of the steel gene product, stem cell factor. Proc. Natl Acad. Sci. USA 89, 7350–7354.

Tsai, M., Takeishi, T., Thompson, H., et al. (1991). Induction of mast cell proliferation, maturation and heparin synthesis by rat c-kit ligand, stem cell factor. Proc. Natl Acad. Sci. USA 88, 6382–6386.

Tsuji, K., Zsebo, K.M. and Ogawa, M. (1991). Enhancement of murine blast cell colony formation in culture by recombinant rat stem cell factor, ligand for c-kit. Blood 78, 1223–1229.

Tsuji, K., Lyman, S.D., Sudo, T., Clark, S.C. and Ogawa, M. (1992). Enhancement of murine hematopoiesis by synergistic interactions between Steel factor (ligand for c-kit),

interleukin-11, and other early acting factors in culture. Blood 79, 2855–2860.

Valent, P., Spanblöchl, E., Sperr, W.R., et al. (1992). Induction of differentiation of human mast cells from bone marrow and peripheral blood mononuclear cells by recombinant human stem cell factor/kit-ligand in long-term culture. Blood 80, 2237–2245.

Vandenbark, G.R., deCastro, C.M., Taylor, H., Dew-Knight, S. and Kaufman, R.E. (1992). Cloning and structural analysis of the human c-kit gene. Oncogene 7, 1259–1266.

Weiler, S.R., Mou, S., DeBerry, C.S., et al. (1996). JAK2 is associated with the c-kit proto-oncogene product and is phosphorylated in response to stem cell factor. Blood 87(9), 3688–3693.

Welham, M.J. and Schrader, J.W. (1992). Steel factor-induced tyrosine phosphorylation in murine mast cells. J. Immunol. 149, 2772–2783.

Wershil, B.K., Tsai, M., Geissler, E.N., Zsebo, K.M. and Galli, S.J. (1992). The rat c-kit ligand, stem cell factor, induces c-kit receptor-dependent mouse mast cell activation in vivo. Evidence that signaling through the c-kit receptor can induce expression of cellular function. J. Exp. Med. 175, 245–255.

Williams, D.E., Eisenman, J., Baird, A., et al. (1990). Identification of a ligand for the c-kit proto-oncogene. Cell 63, 167–174.

Williams, N., Bertoncello, I., Kavnoudias, H., Zsebo, K. and McNiece, I. (1992). Recombinant rat stem cell factor stimulates the amplification and differentiation of fractionated mouse stem cell populations. Blood 79, 58–64.

Wolf, S.S. and Cohen, A. (1992). Expression of cytokines and their receptors by human thymocytes and thymic stromal cells. Immunology 77, 362–368.

Wu, H., Klingmuller, U., Besmer, P. and Lodish, H.F. (1995). Interaction of the erythropoietin and stem-cell-factor receptors. Nature 377(6546), 242–246.

Wu, L., Antica, M., Johnson, G.R., Scollay, R. and Shortman, K. (1991). Developmental potential of the earliest precursor cells from the adult mouse thymus. J. Exp. Med. 174, 1617–1627.

Wypych, J., Bennett, L.G., Schwartz, M.G., et al. (1995). Soluble Kit receptor in human serum. Blood 85, 66–73.

Xiao, M., Leemhuis, T., Broxmeyer, H.E. and Lu, L. (1992). Influence of combinations of cytokines on proliferation of isolated single cell-sorted human bone marrow hematopoietic progenitor cells in the absence and presence of serum. Exp. Hematol. 20, 276–279.

Yarden, Y., Kuang, W.-J., Yang-Feng, T., et al. (1987). Human proto-oncogene c-kit: a new cell surface receptor tyrosine kinase for an unidentified ligand. EMBO J. 6, 3341–3351.

Yoshinaga, K., Nishikawa, S., Ogawa, M., et al. (1991). Role of c-kit in mouse spermatogenesis: identification of spermatogonia as a specific site of c-kit expression and function. Development 113, 689–699.

Zakut, R., Perlis, R., Eliyahu, S., et al. (1993). KIT ligand (mast cell growth factor) inhibits the growth of KIT-expressing melanoma cells. Oncogene 8, 2221–2229.

Zeigler, F.C., de Sauvage, F., Widmer, H.R., et al. (1994). In vitro megakaryocytopoietic and thrombopoietic activity of c-mpl ligand (TPO) on purified murine hematopoietic stem cells. Blood 84, 4045–4052.

Zhou, J.-H., Ohtaki, M. and Sakurai, M. (1993). Sequence of a cDNA encoding chicken stem cell factor. Gene 127, 269–270.

Zsebo, K.M., Williams, D.A., Geissler, E.N., *et al.* (1990a). Stem cell factor is encoded at the *Sl* locus of the mouse and is the ligand for the c-*kit* tyrosine kinase receptor. Cell 63, 213–224.

Zsebo, K.M., Wypych, J., McNiece, I.K., *et al.* (1990b). Identification, purification, and biological characterization of hemopoietic stem cell factor from buffalo rat liver-conditioned medium. Cell 63, 195–201.

22. flt3 Ligand

Claudia Ballaun

1. Introduction

A number of ligands for tyrosine kinase receptors play a role in regulating the proliferation and differentiation of cells in the hematopoietic system. These include flt3 ligand, CSF-1 (MCSF, c-*fms*), and Steel factor (mast cell growth factor, stem cell factor, c-*kit* ligand), which interact with their respective tyrosine kinase receptors flt3(/flk2) (FMS-like tyrosine kinase 3/fetal liver kinase 2), c-*fms*, and c-*kit*. The human and murine ligands for flt3 were cloned independently by two groups (Lyman *et al.*, 1993, 1994; Hannum *et al.*, 1994). The cDNA for the flt3 ligand encodes a type I transmembrane protein that undergoes proteolytic cleavage to generate a soluble factor. Both the membrane-bound form and the soluble protein are biologically active. The flt3 ligand also has structural similarity to the ligands for c-*kit* and c-*fms*. The human and murine flt3 ligands are 72% identical at the amino acid level, and both stimulate the proliferation of human and murine cells expressing flt3 receptors. Cells expressing flt3 ligand are of myeloid, B cell, and T cell lineages at various stages of differentiation. This mitogen stimulates proliferation of hematopoietic precursor cells (CD34$^+$), and seems to function, together with its receptor, as a growth factor receptor–ligand system on hematopoietic stem and/or progenitor cells.

2. The Cytokine Gene

Lyman and colleagues (1993) constructed a fusion protein consisting of the extracellular domain of the murine flt3 receptor followed by the Fc portion of human IgG. By use of this soluble receptor molecule in binding studies, they identified a murine cell line, P7B-0.3A.4, which expresses the cell surface-bound form of the flt3 ligand. cDNA clones were then isolated from an expression library of these cells and further analyzed. The human analog was identified by screening a human T cell library with the murine flt3 ligand cDNA (Lyman *et al.*, 1994). Simultaneously Hannum and colleagues (1994) purified the murine flt3 ligand, produced by mouse thymic stromal cells, to homogeneity and partially sequenced this soluble form of the flt3 ligand protein. Complementary DNAs were isolated and hybridized to a human cDNA library from a thymic stromal cell line. The various mouse and human cDNAs found by Hannum and colleagues define a family of flt3 ligand molecules sharing a common *Amino*-terminus of 135 amino acids (including the signal peptide). The main difference between these mouse and human flt3 ligand cDNAs is caused by divergent carboxy-termini. This is analogous to the variable chain-length region of Steel factor and CSF-1, which may link the receptor binding domains of these cytokines to the membrane. Analysis of several clones discovered by Lyman and colleagues (1995a; Genbank acc.no.:U29874) show an open reading frame of 705 bp flanked by 79 bp of 5′ noncoding sequence and 321 bp of 3′noncoding sequence (Figure 22.1). The open reading frame encodes a type I transmembrane protein of 235 amino acids. The search for the murine flt3 ligand revealed two individual clones differing from nucleotide 521 onward, generating different splice forms of the flt3 ligand. One clone appeared to represent the authentic flt3 ligand sequence, as it shared homology throughout its length with human flt3 ligand sequence, and was therefore chosen for further studies. This murine flt3 ligand cDNA consists of 829 bp, where the open reading frame of 696 bp is flanked by 31 bp of 5′ noncoding region and 102 bp of 3′ noncoding region (Lyman et al., 1993). Genetic mapping studies localized the gene encoding the flt3 ligand to the maximal portion of mouse chromosome 7 and to human chromosome 19q13.3 (Lyman et al., 1995a).

2.1 flt3 LIGAND SPLICE VARIANTS

Alternative splicing is one method to regulate the generation of soluble versus membrane-bound protein forms. PCR techniques have revealed additional flt3 ligand isoforms. One splice variant contains an extra exon introducing a stop codon in the extracellular domain, which should give rise to a soluble flt3 ligand protein. This introduction of a sixth exon is similar to the splicing mechanism used by the Steel factor, with

the difference that exon 6 of the Steel factor cDNA generates a primary proteolytic cleavage site, leading to the generation of the soluble protein. In the murine system there exists an extra flt3 ligand isoform. It encodes a protein with a longer transmembrane domain containing one additional intron (intron 5H), generated by a failure of splicing, resulting in a nonsoluble form of the protein which is still biologically active (Lyman et al., 1995a). In the case of flt3 ligand the mechanisms of regulation of alternative mRNA splicing, resulting in the production of membrane bound or soluble forms, are as yet unknown.

3. The Protein

Analysis of the amino acid sequence (Figure 22.2) indicates that the human flt3 ligand protein has an N-terminal signal peptide of 26 amino acids (aa), followed by a 156-aa extracellular domain, a 23-aa transmembrane domain and a 30-aa cytoplasmic domain. There are two potential N-glycosylation sites in the extracellular domain (aa 100, aa 123). The murine analog consists of 231 aa, where the first 27 aa represent the N-terminal signal peptide, followed by a 161-aa extracellular domain, a 22-aa transmembrane domain and a 21-aa cytoplasmic domain. The predicted molecular mass of the murine protein is 23.16 kDa with an estimated pI of 8.17. Natural flt3 ligand protein purified from a stromal cell line, is shown to be a 65 kDa nondisulfide-linked homodimeric glycoprotein comprised of 30 kDa

```
   1   ggggcatgag ggtccgagac ttgttcttct gtcccttcca agacccggcg
       acaggaggca tgaggggccc ccggccgaaA TGACAGTGCT GGCGCCAGCC        Exon 1
       TGGAGCCCAA CA                                           112

1131   ACCTATCTCC TCCTGCTGCT GCTGCTGAGC TCGGGACTCA GTGGGACCCA
       GGACTGCTCC TTCCAACACA GCCCCATCTC CTCCGACTTC GCTGTCAAAA        Exon 2
       TCCGTGAGCT G                                           1241

1586   TCTGACTACC TGCTTCAAGA TTACCCAGTC ACCGTGGCCT CCAACCTGCA        Exon 3
       GGAC                                                   1639

1864   TAGGAGGAGC TCTGCGGGGG CCTCTGGCGG CTGGTCCTGG CACAGCGCTG
       GATGGAGCGG CTCAAGACTG TCGCTGGGTC CAAGATGCAA GGCTTGCTGG        Exon 4
       AGCGCGTGAA CACGGAGATA CACTTTGTCA CCAAATGTGC CTTTCAG   2007

4305   CCCCCCCCCA GCTGTCTTCG CTTCGTCCAG ACCAACATCT CCCGCCTCCT
       GCAGGAGACC TCCGAGCAGC TGGTGGCGCT GAAGCCCTGG ATCACTCGCC        Exon 5
       AGAACTTCTC CCGGTGCCTG GAGCTGCAGT GTCAGCCCG              4443

5085   TAGAGACGGT GTTTCACCGT GTCAGCCAGG ATGGTCTCGA TCTCCTGACC        Exon 6
       TCGTGATCTG CCCGCCTCGG CCTCCCAAAG TGCTAGGATT ACAGAT     5180

5693   ACTCCTCAAC CCTGCCACCC CCATGGAGTC CCCGGCCCCT GGAGGCCACA
       GCCCCGACAG CCCCGCAGCC CCCTCTGCTC CTCCTACTGC TGCTGCCCGT        Exon 7
       GGGCCTCCTG CTGCTGGCCG CTGCCTGGTG CCTGCACTGG CAGAGGACGC
       GGCGGAGGAC ACCCCGCCCT GGGGAGCAG                        5871

5947   GTGCCCCCCG TCCCCAGTCC CCAGGACCTG CTGCTTGTGG AGCACTGACc        Exon 8
       tggccaaggc ctcatcctgg                                  6015
```

Figure 22.1   Nucleotide sequence of the human flt3 ligand gene. Nucleotide numbering corresponds to the sequence published by Lyman et al. (1995a). The start and stop codons are indicated by bold letters, untranslated regions by lower-case letters.

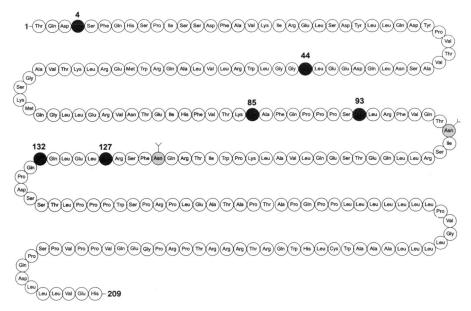

Figure 22.2   Amino acid sequence of the human flt3 ligand. Potential *N*–glycosylation sites are indicated and cysteine residues are full circled.

subunits, each containing 12 kDa of N- and O-linked sugars. (McClanahan *et al.*, 1996). The human flt3 ligand is 72% identical to the murine analog at the amino acid level, and conserves many of the features of the murine protein, including the number of glycosylation sites (human aa 100, aa 123; murine aa 127, aa 152), key cysteine residues (human aa 4, aa 44, aa 85, aa 132; murine aa 31, aa 71, aa 112, aa 156), and splice junctions. The main differences between the human and mouse molecules occurs at the C-terminus. The flt3 ligand is similar to Steel factor and CSF-1 regarding size of the extracellular and cytoplasmic domain, and the four conserved cysteine residues. The overall structure of these proteins appears to be similar and suggests that flt3 ligand is also a four-helix-bundle protein (Lyman *et al.*, 1993).

A carboxy-terminal valine residue appears to be a critical determinant for extracellular cleavage of membrane-bound cytokines into a soluble biologically active growth factor, as has been reported for TGF-α, Steel factor, and CSF-1 (Bosenberg *et al.*, 1992; Anderson *et al.*, 1990; Stanley *et al.*, 1983). The murine and human flt3 ligands have different amino acids at their C-termini but neither of these residues is a valine. Although the flt3 ligand appears to have a similar overall structure to Steel factor and CSF-1 (Lyman *et al.*, 1993), and also generates a soluble form, it seems that flt3 ligand uses either a histidine residue (human flt3 ligand) or a proline residue (mouse flt3 ligand) for proteolytic cleavage, or presumably generates its soluble form by alternative splicing (Lyman *et al.*, 1995a).

4.   *Cellular Sources*

A comprehensive study of mRNA expression by 110 leukemia-lymphoma cell lines revealed widespread distribution of flt3 ligand, whereas flt3 receptor mRNA expression was basically limited to immature B cell lines and monocytic cell lines (Meierhoff *et al.*, 1995). This expression pattern of the flt3 ligand versus its receptor suggests that restrictive expression of the receptor may be the dominant factor in regulating ligand–receptor interactions. Although most cell lines express some amount of flt3 ligand mRNA, only very little flt3 ligand protein is actually made, with T cells and stromal cells being the major producers (McClanahan *et al.*, 1996).

Northern blot analysis of flt3 ligand RNA showed that virtually all hematopoietic cell lines assayed were positive. However, not all of the tested cell lines had detectable ligand expressed on the cell surface, nor was any ligand detectable in medium conditioned by these cells (myeloid cell line HL-60, monocytic cell line THP1, and pre B-ALL cell lines EU-1, LAZ-221, and 697) (Brasel *et al.*, 1995). These findings could be explained by incomplete translation of the mRNA, lack of cell surface expression of flt3 ligand protein, or possibly sequestration of newly produced ligand by receptors in the cytosol.

5.   *Biological Activities*

The hematopoietic activities of flt3 ligand have been assessed in murine and human systems. The murine and human flt3 ligands can each stimulate the proliferation of either murine or human cells at similar concentrations, showing no apparent species-specific restriction in activity. Flt3 ligand shows relatively few effects by itself on the proliferation and differentiation of hematopoietic cells, but exhibits a potent costimulatory activity in enhancing the proliferation of progenitor cells of multiple lineages. In the murine system, flt3 ligand induced [^{3}H]thymidine incorporation by AA4.1$^+$Sca1$^+$Lin fetal liver cells, a population which contains totipotent stem cells and myeloid-erythroid progenitors (Lyman et al., 1993). Furthermore, the response of AA4.1$^+$ cells was increased synergistically by combining flt3 ligand with Steel factor. Similar results were obtained with c-*kit*$^+$ stem cells. But it has also been demonstrated that flt3 ligand alone is incapable of supporting colony formation by Thy^{lo}Sca1$^+$ stem cells (Hannum et al., 1994). The failure of flt3 ligand to induce clonal growth of these primitive cells is not surprising, owing to their requirement for signaling by multiple factors (Heimfeld et al., 1991). It is also reported that flt3 ligand by itself can stimulate the proliferation of human CD34$^+$ progenitor cells (Lyman et al., 1993, 1994; Hannum et al., 1994), which is in contrast to the murine system where it has been shown that flt3 ligand does not support colony formation in accessory cell-depleted cultures (Hudak et al., 1995).

The combination of flt3 ligand with IL-6, IL-11, or G-CSF supports the proliferation of primitive hematopoietic progenitor cells, including lymphohematopoietic progenitors that are capable of differentiation along both myeloid and B-lymphoid lineages (Hirayama et al., 1995; Yonemura et al., 1997; Molineux et al., 1997). In this regard, flt3 ligand is similar to Steel factor, although the number and size of the colonies supported by flt3 ligand-containing factor combinations are smaller than those supported by combinations containing Steel factor. If bone marrow microvascular endothelial cells were used as support stroma cultures, flt3 ligand in combination with IL-3, IL-6, and G-CSF generated a greater number of progenitors than the Steel factor-containing cultures (Shapiro et al., 1996). For the expansion of human bone marrow-derived mononuclear cells flt3 ligand is a more potent growth factor than Steel factor in the presence of IL-3, GM-CSF and erythropoietin (Koller el al., 1996). Another difference between these two related hematopoietic growth factors is their function on pre-B cells and on mast cells. The flt3 ligand by itself maintains the proliferation of B-cell progenitors and their maturation to B220$^+$ cells, while Steel factor alone does not. The opposite takes place concerning stimulation of mast cells; where Steel factor is able to stimulate but flt3 ligand is not. The flt3 ligand has been shown to synergize with Steel factor in supporting the proliferation of B cell progenitors (Hirayama et al., 1995) In a more detailed study it has been demonstrated that flt3 ligand failed to affect CD43lowB220$^+$ pre-B cells or CD43$^+$B220$^+$ pro-B cells whether used alone or in combination with stem cell factor (SCF) or IL-7. In striking contrast, flt3 ligand was a potent cofactor for the CD43$^+$B220low progenitor cells, interacting with either IL-7 and/or SCF to stimulate their growth. When this subset was further divided on the basis of expression of heat-stable antigen (CD24), the flt3 ligand responsive cells were contained only within the CD24$^-$ subset. Since the CD24$^-$ subset was the most immature of the B cell populations studied, these data suggest that flt3 ligand costimulates the expansion of very primitive B cell progenitors (Hunte et al., 1996). The comparison of both growth factors on hematopoietic progenitors were also shown in clonal colony assays. Both factors synergize with IL3-granulocyte-macrophage colony-stimulating factor fusion protein (Pixy321) to induce granulocytic-monocytic (GM) and high proliferative potential (HPP) colonies, and synergized with Pixy321+erythropoietin to induce multipotent granulocytic–erythroid–monocytic–megakaryocytic colonies. *Ex vivo* expansion studies with isolated CD34$^+$ bone marrow stem cells from normal donors showed that flt3 ligand alone supported maintenance of both GM and HPP progenitors for 3 to 4 weeks *in vitro* (McKenna et al., 1995).

The manner in which flt3 ligand acts has been elucidated by Ohishi and colleagues (1996). This cytokine enhances the rate of growth of IL3-dependent colonies by shortening the time for each progenitor in the colonies to divide. Cell cycle analysis demonstrated that shortening of the cell cycle time induced by flt3 ligand is mainly due to alteration in the G$_1$ phase that hematopoietic progenitors go through.

6.   *The Receptor*

The human flt3/flk2 cDNA encodes a receptor-type III tyrosine kinase of 993 amino acids (Rosnet et al., 1993b) with strong similarity to the corresponding mouse flt3/flk2 protein and to the subclass III receptor tyrosine kinases (RTK) PDGF-R (platelet derived growth factor receptor), c-*kit*, and c-*fms* (Birg et al., 1994). The flt3 gene is closely linked to the *flt1* gene, encoding the receptor for vascular endothelial growth factor; both genes are located in the mouse on chromosome 5, and in humans on chromosomal band 13q12 (Rosnet et al., 1993a).

There are conflicting reports concerning flt3 receptor expression. Brasel and colleagues (1995) reported that flt3 receptor expression is restricted to human B cell and myelomonocytic cell lines, although not all cell lines of these two lineages are flt3 receptor positive. These findings are in contrast to those of Da Silva and colleagues (1994), who detected receptor expression only by B lymphoid cell lines. Another difference concerns the human pre-B cell line 1E8. The full-length flt3 receptor cDNA was originally cloned from these cells

(Rosnet *et al.*, 1993b). However, Brasel and colleagues could detect neither cell surface expression nor specific flt3 receptor RNA expression by this cell line, and Da Silva and colleagues found only a smaller form (2 kb) of the mRNA encoding the receptor. Extended hybridization studies on flt3 receptor expression, using a larger probe (2.8 kb), covering extracellular regions, transmembrane domain, and intracellular tyrosine kinase domains, confirmed the findings of Brasel and colleagues that, in addition to the pre-B cell lines, myelomonocytic cell lines were flt3 receptor positive (Meierhoff *et al.*, 1995). These discrepancies are possibly due to the use of different cell lines, methods, or probes for hybridization. One explanation might be differential splicing within the extracellular domain of the flt3 receptor. Differential splicing of extracellular domains may confer promiscuous ligand interaction, resulting in different ligand specificities, as has been reported for other RTK genes.

7. Cytokine in Disease and Therapy

Blood serum levels of flt3 ligand are very low in normal individuals: only 12% of normal individuals had flt3 ligand serum levels above 100 pg/ml. In contrast, 86% of samples from patients with Fanconi anemia and 100% of samples from patients with acquired aplastic anemia had plasma or serum levels above 100 pg/ml. Concentrations of flt3 ligand are therefore specifically elevated to a level that may be physiologically relevant in hematopoietic disorders (Lyman *et al.*, 1995a).

Studies published by Scopes and colleagues (1995), who investigated the effect of the human flt3 ligand on the committed progenitor colony formation of normal bone marrow and bone marrow from aplastic anemia (AA) and three Diamond–Blackfan anemia (DBA) patients, revealed that flt3 ligand has no effect on AA and DBA bone marrows, and on the production of erythroid progenitor colonies.

In contrast to restricted flt3 receptor expression in CD34+ stem cells, it has been demonstrated that flt3 RNA is overexpressed in 100% of B-lineage acute leukemias (ALL), in 11 of 12 cases of acute myeloid leukemias (AML), and 3 of 11 T cell acute leukemias (T-ALL). Stimulation of these patient samples with flt3 ligand resulted in autophosphorylation of the flt3 receptor, suggesting the receptor is functional in these cells. These data suggest that overexpression of flt3 receptor may be involved in the maintenance/proliferation of malignant clones in cases of acute leukemia (Carow *et al.*, 1996).

Nevertheless, the combination of flt3 ligand, Steel factor, GM-CSF, and TNF-α enhanced the *ex vivo* generation of functional dendritic cells from mobilized CD34+ blood progenitors for anticancer therapy up to 5-fold. In practice, the stimulation of CD34+ cells in a blood cell autograft provided by the four above-mentioned growth factors should permit *ex vivo* generation of approximately 4×10^{10} dendritic cells in an adult patient. These new findings provide advantageous tools for the large-scale generation of dendritic cells that are potentially usable for clinical protocols of immunotherapy or vaccination in patients undergoing cancer treatment (Siena *et al.*, 1995).

8. References

Anderson, D.M., Lyman, S.D., Baird, A., *et al.* (1990). Molecular cloning of mast cell growth factor (MGF), a hematopoietin that is active in both membrane bound and soluble forms. Cell 63, 235–243.

Birg, F., Rosnet, O., Carbuccia, N. and Birnbaum, D. (1994). The expression of FMS, KIT and FLT3 in hematopoietic malignancies. Leuk. Lymphoma 13, 223–227.

Bosenberg, M.W., Pandiella, A. and Massague, J. (1992). The cytoplasmic carboxy-terminal amino acid specifies cleavage of membrane TGF-α into soluble growth factor. Cell 71, 1157–1165.

Brasel, K., Escobar, S., Anderberg, R., De Vries, P., Gruss, H.-J. and Lyman, S.D. (1995). Expression of the flt3 receptor and its ligand on hematopoietic cells. Leukemia 9, 1212–1218.

Carow, C., Levenstein, M., Kaufmann, S., *et al.* (1996). Expression of the hematopoietic growth factor receptor flt3 (STK-1/FLK-2) in human leukemias. Blood 87, 1089–1096.

Da Silva, N., Hu, L.B., Ma, W., Rosnet, O., Birnbaum, D. and Drexler, H.G. (1994). Expression of the flt3 gene in human leukemia-lymphoma cell lines. Leukemia 8, 885–888.

Hannum, C., Culpepper, J., Campbell, D., *et al.* (1994). Ligand for flt3/flk2 receptor tyrosine kinase regulates growth of haematopoietic stem cells and is encoded by variant RNAs. Nature 368, 643–648.

Heimfeld, S., Hudak, S., Weissman, I. and Rennick, D. (1991). The in vitro response of phenotypically defined mouse stem cells and myeloerythroid progenitors to single or multiple growth factors. Proc. Natl Acad. Sci. USA 88, 9902–9906.

Hirayama, F., Lyman, S.D., Clark, S.C., and Ogawa, M. (1995). The flt3 ligand supports proliferation of lymphohematopoietic progenitors and early B-lymphoid progenitors. Blood 85, 1762–1768.

Hudak, S., Hunte, B., Culpepper, J., *et al.* (1995). FLT3/FLK2 ligand promotes the growth of murine stem cells and the expansion of colony-forming cells and spleen colony-forming units. Blood 10, 2747–2755.

Hunte, B., Hudak, S., Campbell, D., Xu, Y. and Rennick, D. (1996). Flk2/flt3 ligand is a potent cofactor for the growth of primitive B cell progenitors. J. Immunol. 156, 489–496.

Koller, M.R., Oxender, M., Brott, D.A. and Palsson, B.O. (1996). Flt-3 ligand is more potent than c-kit ligand for the synergistic stimulation of ex vivo hematopoietic cell expansion. J. Hematother. 5, 449–459.

Lyman, S.D., James, L., Vanden Bos, T., *et al.* (1993). Molecular cloning of a ligand for the flt3/flk2 tyrosine kinase receptor: a proliferative factor for primitive hematopoietic cells. Cell 75, 1157–1167.

Lyman, S.D., James, L., Johnson, L., *et al.* (1994). Cloning of the human homologue of the murine flt3 ligand: a growth factor for early hematopoietic progenitor cells. Blood 10, 2795–2801.

Lyman, S.D., Stocking, K., Davison, B., Fletcher, F., Johnson, L. and Escobar, S. (1995a). Structural analysis of human and murine flt3 ligand genomic loci. Oncogene 11, 1165–1172.

Lyman, S.D., James, L., Escobar, S., *et al.* (1995b). Identification of soluble and membrane-bound isoforms of the murine flt3 ligand generated by alternative splicing of mRNAs. Oncogene 10, 149–157.

Lyman, S.D., Seaberg, M., Hanna, R., *et al.* (1995c). Plasma/serum levels of flt3 ligand are low in normal individuals and highly elevated in patients with Fanconi anemia and acquired aplastic anemia. Blood 86, 4091–4096.

McClanahan, T., Culpepper, J., Campbell, D., Wagner, J., *et al.* (1996). Biochemical and genetic characterization of multiple splice variants of the flt-3 ligand. Blood 88, 3371–3382.

McKenna, H., DeVries, P., Brasel, K., Lyman, S. and Williams, D. (1995). Effect of flt3 ligand on the ex vivo expansion of human CD34$^+$ hematopoietic progenitor cells. Blood 86, 3413–3420.

Meierhoff, G., Dehmal, U., Gruss, H.-J., *et al.* (1995). Expression of flt3 receptor and flt3 ligand in human leukemia-lymphoma cell lines. Leukemia 9, 1368–1372.

Molineux, G., McCrea, C., Yan, X.Q., Kerzic, P. and McNiece, I. (1997). Flt-3 ligand synergizes with granulocyte colony-stimulating factor to increase neutrophil numbers and to mobilize peripheral blood stem cells with long-term repopulating potential. Blood 89, 3998–4004.

Ohishi, K., Katayama, N., Itoh, R., *et al.* (1996). Accelerated cell-cycling of hematopoietic progenitors by the flt3 ligand that is modulated by transforming growth factor-β. Blood 87, 1718–1727.

Rosnet, O., Stephenson, D., Mattei, M.G., *et al.* (1993a). Close physical linkage of the flt1 and flt3 genes on chromosome 13 in man and chromosome 5 in mouse. Oncogene 8, 173–179.

Rosnet, O., Schiff, C., Pébusque, M.-J., *et al.* (1993b). Human flt3/flk2 gene: cDNA cloning and expression in hematopoietic cells. Blood 82, 1110–1119.

Scopes, J., Daly, S., Ball, S., McGuckin, C., Gordon–Smith, E. and Gibson, F. (1995). The effect of human flt3 ligand on committed progenitor cell production from normal, aplastic anaemia and Diamond–Blackfan anaemia bone marrow. Br. J. Haematol. 91, 544–550

Shapiro, F., Pytowski, B., Rafii, S., *et al.* (1996). The effects of Flk-2/flt3 ligand as compared with c-kit ligand on short-term and long-term proliferation of CD34$^+$ hematopoietic progenitors elicited from human fetal liver, umbilical cord blood, bone marrow, and mobilised peripheral blood. J. Hematother. 5, 655–662.

Siena, S., Di Nicola, M., Bregni, M., *et al.* (1995). Massive ex vivo generation of functional dendritic cells from mobilized CD34$^+$ blood progenitors for anticancer therapy. Exp. Hematol. 23, 1463–1471.

Stanley, E.R., Guilbert, L.J., Tushinski, R.J. and Bartelmez, S.H. (1983). CSF-1: A mononuclear phagocyte lineage specific hemopoietic growth factor. J. Cell Biochem. 21, 151–159.

Yonemura, Y., Ku, H., Lyman, S.D. and Ogawa, M. (1997). In vitro expansion of hematopoietic progenitors and maintenance of stem cells: comparison between flt-3/flk-2 ligand and kit ligand. Blood 89, 1915–1921.

23. Thrombopoietin

Claudia Ballaun

1. Introduction

The successful cloning of thrombopoietin (TPO) began with the identification of the murine retrovirus myeloproliferative leukemia virus. In 1986, Wendling and colleagues described a transforming viral complex that induced a myeloproliferative syndrome in mice (Wendling *et al.*, 1986). The responsible virus was ultimately cloned and the transforming gene, v-*mpl*, was identified in 1990 (Souyri *et al.*, 1990) and the human cellular homolog, c-*mpl*, in 1992 (Vigon *et al.*, 1992). c-*mpl* encodes a cell surface protein belonging to the highly conserved cytokine receptor family, and is specifically involved in the regulation of megakaryocytopoiesis. The involvement of c-*mpl* in thrombopoiesis was recently confirmed when "knockout" mice lacking a functional c-*mpl* gene were generated. These mice were found to have dramatically reduced numbers of circulating platelets and bone marrow megakaryocytes (Gurney *et al.*, 1994). Thereupon scientists searched for the c-*mpl* ligand, using various strategies. Four articles published in *Nature* (de Sauvage *et al.*, 1994; Kaushansky *et al.*, 1994; Lok *et al.*, 1994; Wendling *et al.*, 1994) and one article in *Cell* (Bartley *et al.*, 1994) announced the long-awaited confirmation that thrombopoietin, the proposed lineage-specific cytokine responsible for regulation of platelets, actually exists. Thrombopoietin stimulates the proliferation and differentiation of megakaryocytic progenitor cells, is essential for the full maturation of megakaryocytes, and acts to enhance platelet production and to speed recovery after cytoreductive therapies. Platelets are the component of blood that initiates clotting reactions and are generated from pluripotent hematopoietic stem cells in a multistep process called thrombopoiesis. Platelets are released from progenitor cells called megakaryocytes which, during maturation, form characteristically large cells with highly polyploid nuclei. Erythrocytes pass through a parallel multistep process (erythropoiesis), and they arc also generated from a pluripotent progenitor cell, which in turn is derived from a bone marrow stem cell. Thus, it is not surprising that thrombopoietin has a high degree of homology with erythropoietin (EPO), the main differentiation factor for erythrocytes. Synonymous terms are megakaryocyte growth and development factor (MGDF) and mpl ligand (ML).

Cytokines
ISBN 0–12–498340–5

2. *The Cytokine Gene*

Three different strategies were used to obtain cDNA clones for TPO. De Sauvage and colleagues (1994) used affinity chromatography based on immobilized recombinant c-*mpl* receptor to purify TPO from the plasma of irradiated pigs, obtained the N-terminal amino acid sequence, and cloned the corresponding cDNA. An almost identical strategy was used by Bartley and colleagues (1994), except that canine TPO was first purified and cloned. Another approach was taken by Lok and colleagues (1994) and Kaushansky and colleagues (1994). They mutagenized growth-dependent murine hematopoietic cell lines expressing the murine c-*mpl* gene, and then selected for autonomously growing clones. One of the clones selected in this way turned out to produce thrombopoietin as an autocrine factor that sustained their growth *in vitro*. Finally, Kato and colleagues (1994) and Kuter and colleagues (1994) used conventional protein fractionation methods to purify a plasma protein from thrombocytopenic rats or sheep, respectively, capable of promoting megakaryocyte

differentiation. The TPO gene spans approximately 6.2 kb and is organized in six coding exons (Figure 23.1), with intron/exon boundaries that precisely correspond to the intron/exon boundaries of human and murine erythropoietin (Foster *et al.*, 1994 (Genbank accession number L36051 (gene) and L36052 (cDNA)); Bartley *et al.*, 1994 (Genbank accession number U11025); Ogami *et al.*, 1995 (rat cDNA, EMBL accession number D3220)). The human TPO cDNA clone consists of 1774 nucleotides followed by a poly(A)$^+$ tail. It contains an open reading frame of 1059 nucleotides, which predicts a primary translation product of 353 amino acids (De Sauvage *et al.*, 1994).

The human gene is localized to chromosome 3q26-28 (Eaton *et al.*, 1994; Foster *et al.*, 1994; Sohma *et al.*, 1994; Gurney *et al.*, 1995a; Suzukawa *et al.*, 1995), a region of the long arm of chromosome 3 previously associated with elevated platelet counts and increased bone marrow megakaryocytes in patients with acute nonlymphocytic leukemia (Rowley and Potter, 1976; Pintado *et al.*, 1985; Jenkins *et al.*, 1989). The gene of the murine TPO is located on mouse chromosome 16 (Chang *et al.*, 1995).

```
  92     tcctaccaat ctgctcccca gagggctgcc tgctgtgcac ttgggtcctg gagcccttct
         ccacccg                                                                   158    Exon 1

1827     gatagattct tcacccttgg tccgcctttg ccccacccta ctctgcccag aagtgcaaga
         gcctaagccg cctccatggc cccaggaagg attcaggggga gaggccccaa acagggagcc
         acgccagcca gacacccgg ccagaATGGA GCTGACTG                                  1984    Exon 2

2216     AATTGCTCCT CGTGGTCATG CTTCTCCTAA CTGCAAGGCT AACGCTGTCC AGCCCGGCTC
         CTCCTGCTTG TGACCTCCGA GTCCTCAGTA AACTGCTTCG TGACTCCCAT GTCCTTCACA         Exon 3
         GCAGACTG                                                                  2343

2630     AGCCAGTGCC CAGAGGTTCA CCCTTTGCCT ACACCTGTCC TGCTGCCTGC TGTGGACTTT         Exon 4
         AGCTTGGGAG AATGGAAAAC CCAGATG                                             2716

4649     GAGGAGACCA AGGCACAGGA CATTCTGGGA GCAGTGACCC TTCTGCTGGA GGGAGTGATG
         GCAGCACGGG GACAACTGGG ACCCACTTGC CTCTCATCCC TCCTGGGGCA GCTTTCTGGA         Exon 5
         CAGGTCCGTC TCCTCCTTGG GGCCCTGCAG AGCCTCCTTG GAACCCAG                      4816

5053     CTTCCTCCAC AGGGCAGGAC CACAGCTCAC AAGGATCCCA ATGCCATCTT CCTGAGCTTC
         CAACACCTGC TCCGAGGAAA GGTGCGTTTC CTGATGCTTG TAGGAGGGTC CACCCTCTGC
         GTCAGGCGGG CCCCACCCAC CACAGCTGTC CCCAGCAGAA CCTCTCTAGT CCTCACACTG
         AACGAGCTCC CAAACAGGAC TTCTGGATTG TTGGAGACAA ACTTCACTGC CTCAGCCAGA
         ACTACTGGCT CTGGGCTTCT GAAGTGGCAG CAGGGATTCA GAGCCAAGAT TCCTGGTCTG         Exon 6
         CTGAACCAAA CCTCCAGGTC CCTGGACCAA ATCCCCGGAT ACCTGAACAG GATACACGAA
         CTCTTGAATG GAACTCGTGG ACTCTTTCCT GGACCCTCAC GCAGGACCCT AGGAGCCCCG
         GACATTTCCT CAGGAACATC AGACACAGGC TCCCTGCCAC CCAACCTCCA GCCTGGATAT
         TCTCCTTCCC CAACCCATCC TCCTACTGGA CAGTATACGC TCTTCCCTCT TCCACCCACC
         TTGCCCACCC CTGTGGTCCA GCTCCACCCC CTGCTTCCTG ACCCTTCTGC TCCAACGCCC
         ACCCCTACCA GCCCTCTTCT AAACACATCT TACACCCACT CCCAGAATCT GTCTCAGGAA
         GGGTAAggtt ctcagacact gccgacatca gcattgtctc gtgtacagct cccttccctg         stop 5719
         cagggcgccc ctgggagaca actggacaag atttcctact ttctcctgaa acccaaagcc
         ctggtaaaag ggatacacag gactgaaaag ggaatcattt ttcactgtac attataaacc
         ttcagaagct attttttttaa gctatcagca atactcatca gagcagctag ctctttggtc
         tattttctgc agaaatttgc aactcactga ttctcaacat gctctttttc tgtgataact
         ctgcaaagac ctgggctggc ctggcagttg aacagaggga gagactaacc ttgagtcaga
         aaacagagga agggtaattt cctttgcttc aaattcaagg ccttccaacg ccccatctcc
         ctttactatc attctcagtg ggactctgat cccatattct taacagatct ttactcttga
         gaaatgaata agctttctct cagaatgctg tccctataca ctagacaaaa ctgagcctgt
         ataaggaata aatgggagcg ccgaaaagct ccctaaaaag caagggaaag atgtt            6306
```

Figure 23.1   Nucleotide sequence of the human TPO gene. Coding exon regions are indicated by upper-case letters, flanking sequences by lower-case letters. Nucleotide numbering is according to the sequence published by Foster *et al.* (1994). The start codon and the stop codon are highlighted by bold letters. The consensus polyadenylation signal is underlined.

2.1 ALTERNATIVE SPLICING OF TPO

Two alternative splice forms of TPO have been identified. They differ by the presence (TPO) or absence (TPO-2) of a four-amino-acid insertion at positions 112 to 115. The four amino acids are encoded at the 5′ junction of exon 6 in the human gene. The TPO-2 mRNA encodes a protein which is poorly secreted. This four-amino-acid deletion within TPO-2 would be predicted to alter the loop between α-helices C and D in the proposed structure of TPO. A similar deletion has been introduced into EPO, resulting in an inhibition of secretion of this protein. These results suggest that the level of TPO expression could potentially be modulated by altering the proportion of TPO mRNA that is spliced to encode each of the forms (Gurney *et al.*, 1995a).

2.2 REGULATION OF TPO GENE TRANSCRIPTION

The mechanism by which the production of TPO is upregulated when there is a need for more platelets remains to be elucidated. Two models of TPO gene regulation have been proposed. The first one suggests that TPO serum levels are maintained solely by platelet uptake and metabolism of TPO. The second hypothesis proposes that platelet levels are recognized, resulting in increased or decreased levels of TPO gene expression. Studies by McCarty and colleagues (1995) support the second model by demonstrating that TPO-specific mRNA levels in the spleen and marrow derived from thrombocytopenic mice are inversely related to platelet counts. Stoffel and colleagues (1996) support the first model, demonstrating that TPO activity during thrombocytopenia is not caused by regulation at the level of TPO mRNA, but by a regulation at the post-transcriptional level and/or through absorption and metabolism by platelets. This model is also established by Fielder and colleagues (1996). They used c-mpl⁻/⁻ mice to analyze the mechanisms leading to the increased concentration of circulating TPO in response to low platelet counts. In contrast to platelets from the c-mpl⁻/⁻ mice, platelets from normal mice are capable of binding, internalizing, and degrading TPO. The absence of receptors in the c-mpl⁻/⁻ mice leads to a slower rate of clearance from the plasma and a longer initial half-life of TPO.

Another interesting model of transcriptional TPO gene regulation is presented by Shivdasani and colleagues (1995). They generated mice lacking the hematopoietic subunit (p45) of the heterodimeric erythroid transcription factor NF-E2. This results in a complete, and apparently late block in megakaryocyte maturation, such that circulating platelets are undetectable *in vivo*. Though platelets are absent in these NF-E2 deficient mice, serum levels of TPO are not elevated. Thus, as an essential factor for megakaryocyte maturation and

platelet production, NF-E2 must regulate critical target genes independently of the action of thrombopoietin. One explanation for the inconsistency observed in this study of the above-mentioned "platelet count/TPO production" relationship can be found in the pleiotropic effects of TPO. This cytokine appears to exert substantial physiological effects on many aspects of megakaryocyte maturation, including cell size, DNA ploidy, and platelet production. These multiple effects are likely to be mediated by genes that are activated following a thrombopoietin-mediated signaling cascade. It seems possible that a subset of genes responsive to this pathway are also regulated through NF-E2, particularly late in megakaryocyte development.

3. *The Protein*

The human TPO cDNA encodes a polypeptide identical to that predicted from the human genomic sequence and with a similar overall size, two-domain structure, and significant sequence conservation with the murine TPO (Figure 23.2). The predicted human TPO protein is a polypeptide of 353 amino acids (aa) (rat TPO 305 aa; murine TPO 335 aa), including a 21-aa highly hydrophobic signal peptide, a 152-aa domain with homology to EPO (23% identity), and a dibasic Arg-Arg, representing a potential protease cleavage site (aa 153–154). This protease recognition site is followed by a 177-aa carbohydrate-rich domain (Figure 23.2). This second domain is highly enriched in serine, threonine, and proline, contains several potential N-glycosylation sites (depending on the species: human 6; murine 7; porcine 6), and bears no recognizable homology with other known protein sequences (Foster *et al.*, 1994). TPO, like EPO, contains four cysteine residues (aa 7, 29, 85, 151), three of which are conserved. Site-directed mutagenesis experiments have shown that the first and the last cysteines of EPO form a disulfide bond essential for EPO activity (Wang *et al.*, 1983). By analogy, the first and last cysteine residues of TPO also form a critical disulfide bond essential for exhibiting TPO activity. The mature polypeptide has a predicted molecular mass of 35.5 kDa; however, it is anticipated that mature TPO will have a much greater molecular mass, since both the recombinant murine and human proteins expressed in yeast have been observed to be heavily glycosylated and to be composed of as much as 50% carbohydrate. Bartley and colleagues (1994) purified 25 kDa and 31 kDa forms of TPO, which may be products of proteolytic processing at the C-terminus. This interpretation is consistent with the fact that expression of a truncated human cDNA, encoding amino acids 1–195, yields a biologically active protein.

Comparison of the predicted amino acid sequences of the mature TPO from human, mouse and pig showed a

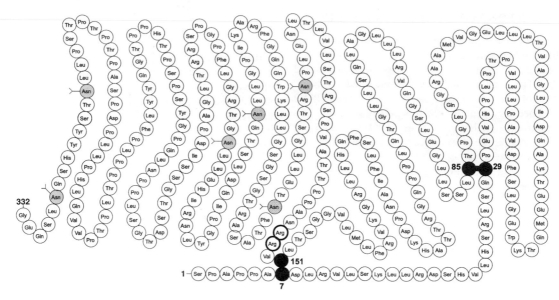

Figure 23.2 Deduced amino acid sequence of the human TPO. Potential *N*-glycosylation sites are indicated; cysteine residues creating disulfide bridges are shown as filled circles. The proteolytic cleavage site assembled by the dibasic Arg-Arg pair is indicated in bold circles.

72% identity between mouse and human, 68% identity of mouse versus pig, and 73% identity between human and pig. The homology is substantially greater in the amino-terminal half of TPO, being 81% to 85% homologous between any two species (Gurney *et al.*, 1995a).

4. Cellular Sources

Initial northern blotting studies using murine thrombopoietin cDNA as a probe revealed that the mRNA is expressed in a number of tissues (Lok *et al.*, 1994; de Sauvage *et al.*, 1994; Bartley *et al.*, 1994). Although liver is the predominant site of expression, lower amounts are easily detectable in kidney, smooth muscle, and spleen. Reverse transcriptase PCR demonstrated that both endothelial cells and fibroblasts contain readily detectable thrombopoietin transcripts.

5. Biological Activities

5.1 IN VITRO ACTIVITIES

TPO functions primarily as a differentiation factor, with limited ability to promote megakaryocyte progenitor growth by itself (Kaushansky *et al.*, 1995). The generation of megakaryocyte colonies (CFU-Mk) in semisolid matrices, reflecting megakaryocyte progenitor proliferation and maturation, was most potently stimulated when thrombopoietin was used in combination with early-acting growth factors such as interleukin-3 or the c-*kit* ligand (Banu *et al.*, 1995; Broudy *et al.*, 1995). Although IL-3 is able to induce early stages of megakaryocytic development in the absence of TPO, the latter is required for full megakaryocyte maturation. In liquid culture, TPO augments terminal differentiation of megakaryocytes, as demonstrated by a significant increase in their DNA content (Broudy *et al.*, 1995), as well as expression of the platelet-specific differentiation antigens gpIb (Kaushansky *et al.*, 1994; Morgan *et al.*, 1994) and gpIIb/IIIa (de Sauvage *et al.*, 1994; Bartley *et al.*, 1994; Papayannopoulou *et al.*, 1994). Electron microscopy studies showed that TPO, but not IL-3, IL-6, or IL-11, is able to complete the program of megakaryocyte differentiation required to produce mature platelets (Kaushansky *et al.*, 1995). TPO also acts in synergy with other pluripotent cytokines on hematopoietic stem cells to augment development of erythroid and myeloid progenitors (Kaushansky *et al.*, 1997).

5.2 IN VIVO ACTIVITIES

There are conflicting reports about the *in vivo* activities of TPO, possibly owing to species-specific differences between murine and human TPO protein used in the studies. Miyazaki and colleagues (1994) found a modest increase in platelet counts 8 days after the administration of human TPO to mice. Similar results were obtained by de Sauvage and colleagues (1994), who injected recombinant murine TPO into mice and detected a modest increase (20%) in circulating platelets and a 40% increase in incorporation of ^{35}S into platelets. In contrast, when TPO was administered to normal Balb/c mice for 6 days, a profound increase in marrow and splenic colony-forming units megakaryocyte (CFU-Mk) and megakaryocytes, and a 4-fold increase in platelet counts was observed. Splenic colony-forming units erythroid (CFU-E), marrow and splenic burst forming units erythroid (BFU-E), and marrow and splenic colony forming units granulocyte-macrophage (CFU-GM) were reproducibly increased up to 3-fold (Kaushansky *et al.*, 1994). This effect on CFU-GM was also noted by Levin and colleagues (1980) after the induction of acute thrombocytopenia using antiplatelet antiserum. Injection of recombinant TPO into nonhuman primates increases platelet counts by stimulating endoreduplication and megakaryocyte formation from marrow progenitor cells (Harker *et al.*, 1996). Additional support for the concept that thrombopoietin plays an important role in megakaryopoiesis *in vivo* arises from studies of mice genetically engineered to lack the c-*mpl* receptor. These mice display platelet counts that are 15% of the level found in normal mice and reduced numbers of megakaryocytes in the marrow. Myelopoiesis, erythropoiesis, and lymphopoiesis appear to be normal in the c-*mpl*-deficient mice (Methia *et al.*, 1993).

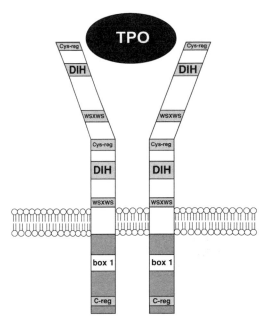

Figure 23.3   Model for the TPO receptor. The homologous regions of hematopoietic cytokine receptors, the four conserved cysteine residues (Cys-reg), the dimer interface homology regions (DIH), and the common WSXWS motif and the carboxy-terminal region (C-reg) are indicated.

6. *The Receptor*

The murine myeloproliferative leukemia virus (MPLV) is a replication-defective retrovirus that induces an acute myeloproliferative disorder after inoculation of mice in the presence of helper murine leukemia virus. The envelope gene of MPLV contains the oncogene v-*mpl*, which has characteristics of a transmembrane protein. Both the murine and the human cellular homologs have been cloned (Vigon *et al.*, 1992, 1993). Since MPLV was found to transform multiple hematopoietic lineages *in vitro*, and the structure of c-*mpl* shares features with members of the hematopoietin cytokine receptor family, it was postulated that c-*mpl* encodes the receptor for a hematopoietic growth factor. The extracellular part of c-*mpl* consists of two hematopoietin domains, which in turn contain a conserved array of cysteine residues, a dimer interface homology region (Alexander *et al.*, 1995), and a WSXWS box near the transmembrane region (Figure 23.3). A box 1/2 motif but no tyrosine kinase consensus sequence has been identified in the intracellular region of the receptor protein. C-*mpl* displays an overall sequence similarity that is most closely related to IL-3Rβ (Vigon *et al.*, 1992).

Transcripts specific for c-*mpl* are found in primitive hematopoietic cells, including CD34[+] cells, in megakaryocytes, and in platelets but not in lymphocytes, neutrophils, or monocytes (Methia *et al.*, 1993).

7. *Signal Transduction*

The thrombopoietin receptor (c-*mpl*) fits the general model of signal transduction developed for other members of the cytokine receptor superfamily. TPO binding leads to activation of tyrosine kinases, including JAK2 and TYK2, in platelets (Miyakawa *et al.*, 1995; Ezumi *et al.*, 1995) and two factor-dependent hematopoietic cell lines (Sattler *et al.*, 1995), whereas only JAK2 is stimulated in human megakaryoblastic leukemia-derived cell lines (M07e, mumpl-UT7), which express c-*mpl* and proliferate in response to TPO (Pallard *et al.*, 1995; Sasaki *et al.*, 1995; Tortolani *et al.*, 1995). Drachman and colleagues (1995) demonstrated that a direct interaction of JAK2 and c-*mpl* in a hematopoietic murine cell line, BaF3/mpl, was detectable only after 20 min of ligand stimulation and increased at 60 min. This rather delayed event is in marked contrast to the early phosphorylation of JAK2, and they suggest that JAK2 becomes phosphorylated prior to binding the cytoplasmic domain of c-*mpl* and that another kinase may first be activated by c-*mpl*, which in turn might be responsible for JAK2 phosphorylation.

TPO stimulation also results in increased phosphotyrosine content of known signaling molecules, such as Shc and Shc-associated protein (Sasaki *et al.*, 1995), but in contrast phospholipase C-γ and phosphatidylinositol 3-kinase displayed little and no tyrosine phosphorylation, respectively (Drachman *et al.*, 1995). TPO also induces phosphorylation of adaptor molecules, such as Vav, serine/threonine kinases Raf-1 and mitogen-activated protein (MAP) kinases Erk-1 and Erk-2 (Yamada *et al.*, 1995) and further activates Ras, MAP kinase kinase, and Pim-1 (Nagata and Todoro 1995).

There are differing reports concerning activation of signal transducer and activator of transcription proteins (STAT proteins). The phosphorylation of the STAT proteins has been shown to be essential for homo- or heterodimerization, translocation into the nucleus, and formation of DNA binding complexes. TPO stimulates tyrosine phosphorylation of STAT3 in platelets (Ezumi *et al.*, 1995), and STAT1 and STAT3 in human megakaryoblastic CMK cells (Gurney *et al.*, 1995b). In contrast to these observations are results published by Pallard and colleagues (1995). They demonstrated that TPO activates a STAT5-like transcriptional factor in murine UT7 cells expressing c-*mpl* (mumpl UT7), in a very rapid and transient manner, but not STAT1, STAT2, STAT3, or STAT4 proteins. The activation of STAT5 by TPO in these cells is shared by GM-CSF and EPO, but not IFN-γ or stem cell factor (SCF). The transcriptional factor STAT5 can be constitutively activated in cells expressing the oncogenic form of a cytokine receptor such as v-*mpl* (Pallard *et al.*, 1995). Miyakawa and colleagues (1996) demonstrated that TPO induces tyrosine phosphorylation of STAT3 and STAT5 in human platelets. Furthermore it has been shown that, in the cell line Ba/F3-mMPL, the closely related proteins STAT5A and STAT5B are both activated by TPO stimulation and are capable of heterodimerisation (Drachman *et al.*, 1997).

Taken together, these observations demonstrate that TPO induces the activation of at least two distinct signaling pathways, a specific JAK2-TYK2/STAT signaling cascade and a common Shc/Vav/Ras/Raf-1/MAP kinase kinase/MAP kinase signaling cascade.

By analogy to the G-CSF receptor, EPO receptor, prolactin receptor, and growth hormone receptor, it has been shown that c-*mpl* functions through receptor homodimerization and is mediated by conserved hematopoietin receptor dimer interface domains (Alexander *et al.*, 1995). The function of most hematopoietin receptors has been defined by the ability of the receptors to control proliferation of hematopoietic cells. To achieve a proliferative cell response, two structural motifs (box1 and box2) in the cytoplasmic domain are required (Hirano *et al.*, 1990). Morella and colleagues (1995) and Gurney and colleagues (1995b) demonstrated that the membrane-proximal box1 sequence motif is critical for c-*mpl* signaling. A second region of the c-*mpl* intracellular domain located at the C-terminus is required for tyrosine phosphorylation of Shc and induction of c-*fos* mRNA.

8.   Cytokine in Disease and Therapy

Platelets are necessary for blood clotting, and when their numbers are very low a patient is at serious risk of death from catastrophic hemorrhage.

The possibility of producing recombinant thrombopoietin for use as a therapeutic treatment for thrombocytopenia, for example, following bone-marrow failure resulting from chemotherapy and radiation, is an attractive proposition. There are certainly convincing arguments for the development of a recombinant protein approach to stimulating platelet production. First, the existing treatments for thrombocytopenia rely on transfusions of platelet concentrates prepared from recently donated blood. Second, platelets have a relatively short shelf-life of 5 days. Third, platelets carry surface antigens and, if repeated transfusions are required, there is a potential risk of the formation of platelet allo-antibodies. The use of recombinant EPO for increasing red-blood-cell counts in anemia, and of recombinant G-CSF for increasing neutrophil counts in chemotherapy-induced neutropenia, has been particularly successful. First promising studies by Broudy and colleagues (1995) demonstrated that thrombopoietin actually has the capacity to reduce the time for platelet recovery (3 weeks to 12 days) of mice treated by chemotherapy and radiation. Cwirla *et al.* (1997) found a peptide agonist (14aa) with high affinity to the TPO receptor, that increases platelet count by stimulating megakaryocytopoiesis. When administered to normal mice this peptide may therefore serve as useful lead compound for the development of a therapeutically effective agent.

9.   References

Alexander, W.S., Metcalf, D. and Dunn, A.R. (1995). Point mutations within a dimer interface homology domain of c-mpl induce constitutive receptor activity and tumorigenicity. EMBO J. 14, 5569–5578.

Banu, N., Wang, J., Deng, B., Groopman, J. and Avraham, H. (1995). Modulation of megakaryocytopoiesis by thrombopoietin: The c-*mpl* ligand. Blood 86, 1331–1338.

Bartley, T., Bogenberger, J., Hunt, P., *et al.* (1994). Identification and cloning of a megakaryocyte growth and development factor that is a ligand for the cytokine receptor Mpl. Cell 77, 1117–1124.

Broudy, V.C., Lin, N. and Kaushansky K. (1995). Thrombopoietin (c-*mpl* ligand) acts synergistically with erythropoietin, stem cell factor, and interleukin-III to enhance murine megakaryocyte colony growth and increases megakaryocyte ploidy in vitro. Blood 85, 1719–1726.

Chang, M.S., Hsu, R.Y., McNinch, J., Copeland, N. and Jenkins, N.A. (1995). The gene for murine megakaryocyte growth and development factor (thrombopoietin, Thpo) is located on mouse chromosome 16. Genomics 26, 636–637.

Cwirla, S., Balasubramanian, P., Duffin, D.J., *et al.* (1997). Peptide agonist of the thrombopoietin receptor as potent as the natural cytokine. Science 276, 1696–1699.

De Sauvage, F.J., Hass, P.E., Spencer, S.D., *et al.* (1994). Stimulation of megakaryocytopoiesis and thrombopoiesis by the c-*mpl* ligand. Nature 369, 533–538.

Drachman, J.G., Griffin, J.D. and Kaushansky, K. (1995). The c-*mpl* ligand (thrombopoietin) stimulates tyrosine phosphorilation of JAK2, Shc, and c-*mpl*. J. Biol. Chem. 270, 4979–4982.

Drachman, J., Sabath, D.F., Fox, N.E. and Kaushansky, K. (1997). Thrombopoietin signal transduction in purified murine megakaryocytes. Blood 89, 483–492.

Eaton, D., Gurney, A., Malloy, B., *et al.* (1994). Biological activity of human thrombopoietin (TPO), the c-*mpl* ligand, TPO variants and the chromosomal localization of TPO. Blood 84, 241a.

Ezumi, Y., Takayama, H. and Okuma, M. (1995). Thrombopoietin, c-MPL ligand induces tyrosine phosphorylation of Tyk2, JAK2, and STAT3, and enhances agonists-induced aggregation in platelets in vitro. FEBS Lett. 374, 48–52.

Fielder, P., Gurney, A., Stefanich, E., *et al.* (1996). Regulation of thrombopoietin levels by c-mpl-mediated binding to platelets. Blood 87, 2154–2161.

Foster, D., Sprecher, C., Grant, F., *et al.* (1994). Human thrombopoietin: gene structure, cDNA sequence, expression, and chromosomal localization. Proc. Natl Acad. Sci. USA 91, 13023–13027.

Gurney, A., Carver-Moore, K., de Sauvage, F. and Moore, M. (1994). Thrombocytopenia in c-*mpl*-deficient mice. Science 265, 1445–1447.

Gurney, A., Kuang, W.-J., Xie, M.-H., Malloy, B., Eaton, D. and de Sauvage, F. (1995a). Genomic structure, chromosomal localization, and conserved alternative splice forms of thrombopoietin. Blood 85, 981–988.

Gurney, A., Wong, S., Henzel, W. and de Sauvage, F. (1995b). Distinct regions of c-*mpl* cytoplasmic domain are coupled to JAK-STAT signal transduction pathway and Shc phosphorylation. Proc. Natl Acad. Sci. USA 92, 5292–5296.

Harker, L.A., Hunt, P., Marzec, U., *et al.* (1996). Regulation of platelet production and function by megacaryocyte growth and development factor in nonhuman primates. Blood 87, 1833–1844.

Hirano, T., Matsuda, T. and Nakajiama, K. (1990). Signal transduction through gp130 that is shared among the receptors for the interleukin-6 related cytokine subfamily. Cell 12, 262–271.

Jenkins, R., Tefferi, A., Solberg, L. and Dewald, G. (1989). Acute leukemia with abnormal thrombopoiesis and inversions on chromosome 3. Cancer Genet. Cytogenet. 39, 167–179.

Kato, T., Iwamatsu, A, Shimada, Y., *et al.* (1994). Purification and characterization of thrombopoietin derived from thrombocytopenic rat plasma. Blood 84, 329a.

Kaushansky, K., Lok, S., Holly, R.D., *et al.* (1994). Promotion of megacaryocyte progenitor expansion and differentiation by the c-*mpl* ligand thrombopoietin. Nature 369, 568–571.

Kaushansky, K., Broudy, V.C., Lin, N., *et al.* (1995). Thrombopoietin, the Mpl-ligand is essential for full megakaryocyte development. Proc. Natl Acad. Sci. USA 92, 3234–3238.

Kaushansky, K. (1997). Thrombopoietin; Understanding and manipulating platelet production. Annu. Rev. Med. 48, 1–11.

Kuter, D., Beeler, D. and Rosenberg, R. (1994). The purification of megapoietin: a regulator of megakaryocyte growth and platelet production. Proc. Natl Acad. Sci. USA 91, 11104–11108.

Levin, J., Levin, F.C. and Metcalf, D. (1980). The effects of acute thrombocytopenia on megakaryocyte-CFC and granulocytemacrophage-CFC in mice: studies of bone marrow and spleen. Blood 56, 274–283.

Lok, S., Kaushansky, K., Holly, R.D., *et al.* (1994). Cloning and expression of murine thrombopoietin cDNA and stimulation of platelet productin *in vivo*. Nature 369, 565–568.

McCarty, J., Sprugel, K., Fox, N., Sabath, D. and Kaushansky, K. (1995). Murine thrombopoietin mRNA levels are modulated by platelet count. Blood 86, 3668–3675.

Methia, N., Louache, F., Vainchenker, W. and Wendling, F. (1993). Oligodeoxynucleotides antisense to the proto-oncogenec-*mpl* specifically inhibit in vitro megakaryo-cytopoiesis. Blood 82, 1395–1401.

Miyakawa, Y., Oda, A., Druker, B.J., *et al.* (1995). Recombinant thrombopoietin induces rapid protein tyrosine phosphorylation of Janus kinase 2 and Shc in human blood platelets. Blood 86, 23–27.

Miyakawa, Y., Oda, A., Druker, B.J., *et al.* (1996). Thrombo-poietin induces tyrosine phosphorylation of Stat3 and Stat5 in human blood platelets. Blood 87, 439–446.

Miyazaki, H., Kato, T., Ogami, K., *et al.* (1994). Isolation and cloning of a novel human thrombopoietic factor. Exp. Hematol. 22, 838a.

Morella, K., Bruno, E., Kumaki, S., *et al.* (1995). Signal transduction by the receptors for thrombopoietin (c-*mpl*) and interleukin-3 in hematopoietic and nonhematopoietic cells. Blood 86, 557–571.

Morgan, D., Soslau, G. and Brodsky, I. (1994). Differential effects of thrombopoietin (mpl) on cell lines MB-02 and HU-3 derived from patients with megakaryoblastic leukemia. Blood 84, 330a.

Nagata, Y. and Todokoro, K. (1995). Thrombopoietin induces at least two distinct signaling pathways. FEBS Lett. 377, 497–501.

Ogami, K., Shimada, Y., Sohma, Y., *et al.* (1995). The sequence of rat cDNA encoding thrombopoietin. Gene 158, 309–310.

Pallard, C., Gouilleux, F., Benit, L., *et al.* (1995). Thrombopoietin activates a STAT5-like factor in hematopoietic cells. EMBO J. 14, 2847–2856.

Papayannopoulou, T., Brice, M. and Kaushansky, K. (1994). The influence of Mpl-ligand in the development of megakaryocytes from CD34+ cells isolated from bone marrow, peripheral blood. Blood 84, 32a.

Pintado, T., Ferro, M.T., San Ramon, C., Mayayo, M. and Larana, J.G. (1985). Clinical correlations of the 3q21;q26 cytogenic anomaly. Cancer 55, 535–541.

Rowley, J.D. and Potter, D. (1976). Chromosomal banding patterns in acute nonlymphocytic leukemia. Blood 47, 705–721.

Sasaki, K., Odai, H., Hanatono, Y., *et al.* (1995). TPO/c-mpl ligand induces tyrosine phosphorylation of multiple cellular proteins including proto-oncogene products, vav and c-cbl, and ras signaling molecules. Biochem. Biophys. Res. Commun. 216, 338–347.

Sattler, M., Durstin, M.A., Frank, D.A., *et al.* (1995). The thrombopoietin receptor c-*mpl* activates JAK2 and TYK2 tyrosine kinases. Exp. Hematol. 23, 1040–1048.

Shivdasani, R.A:, Rosenblatt, M.F., Zucker-Franklin, D., *et al.*

(1995). Transcription factor NF-E2 is required for platelet formation independent of the action of thrombopoietin/MGDF in megakaryocyte development. Cell 81, 695–704.

Sohma, Y., Akahori, H., Seki, N., et al. (1994). Molecular cloning and chromosomal localization of the human thrombopoietin gene. FEBS Lett. 353, 57–61.

Souyri, M., Vigon, I., Penciolelli, J.-F., Heard, J.M., Tambourin, P. and Wendling, F. (1990). A putative truncated cytokine receptor gene transduced by the myeloproliferative leukemia virus immortalizes hematopoietic progenitors. Cell 63, 1137–1147.

Stoffel, R., Wiestner, A. and Skoda, R. (1996). Thrombopoietin in thrombocytopenic mice: Evidence against regulation at the mRNA level and for a direct regulatory role of platelets. Blood 87, 567–573.

Suzukawa, K., Satoh., Taniwaki, M., Yokota, J. and Morishita, K. (1995). The human thrombopoietin gene is located on chromosome 3q26.33-q27, but is not transcriptionally activated in leukemia cells with 3q21 and 3q26 abnormalities (3q21q26 syndrome). Leukemia 9, 1328–1331.

Tortolani, P.J., Johnston, J.A., Bacon, C.H., et al. (1995). TPO induces tyrosine phosphorylation and activation of the Janus kinase 2. Blood 85, 3444–3451.

Vigon, I., Mornon, J.P., Cocault, L., et al. (1992). Molecular cloning and characterization of MPL, the human homologue of the v-mpl oncogene: identification of a member of the hematopoietic growth factor receptor super family. Proc. Natl Acad. Sci. USA 89, 5640–5644.

Vigon, I., Florindo, C., Fichelson, S., et al. (1993). Characterization of the murine Mpl proto-oncogene, a member of the hematopoietic cytokine receptor family: molecular cloning, chromosomal location and evidence for a function in cell growth. Oncogene 8, 2607–2615.

Wang, F., Kung, C.K. and Goldwasser, E. (1983) Some chemical properties of human erythropoietin. Endocrinology 116, 2286–2292.

Wendling, F., Varlet, P., Charon, M. and Tambourin, P. (1986). A retrovirus complex inducing an acute myeloproliferative leukemia disorder in mice. Virology 149, 242–246.

Wendling, F., Maraskovsky, E., Debili, N., et al. (1994). c-mpl ligand is a humoral regulator of megakaryocytopoiesis. Nature 369, 571–574.

Yamada, M., Komatsu, N., Okada, K., Kato, T., Miyazaki, H. and Miura, Y. (1995). Thrombopoietin induces tyrosine phosphorylation and activation of mitogen-activated protein kinases in a human thrombopoietin dependent cell line. Biochem. Biophys. Res. Commun. 217, 230–237.

24. Tumor Necrosis Factor and Lymphotoxin

Rudi Beyaert *and* Walter Fiers

1. Introduction

Today's knowledge of tumor necrosis factor (TNF) is the culmination of over two centuries of clinical observations and experimentation. It was initially observed that cancer patients who experienced a severe infection sometimes had a regression of their tumor (Nauts, 1989). Many examples were described where an induced infection was used successfully as a cancer treatment. The best-documented early clinical studies were carried out around the turn of the century by W.B. Coley, a surgeon at the New York Cancer Hospital. He found that treatment with a mixture of bacterial extracts, later known as "Coley's toxins", was as effective as inoculations of live bacteria

(Coley, 1894). Although many successful cases were reported, the approaches of Coley did not catch on in the cancer field in view of the considerable progress made by surgery and by radiotherapy at that time, and because of the toxicity of the regimen. Fortunately, experimental and preclinical research continued. Fifty years later, Shear and his colleagues demonstrated that *Serratia marcescens* extracts induced hemorrhagic necrosis when administered to mice bearing transplanted sarcomas (Shear *et al.*, 1943). Later, the active component was identified as lipopolysaccharide (LPS) derived from the cell wall of Gram-negative bacteria. It was also established that mice primed with *Bacillus Calmette–Guérin* (BCG) for 2–3 weeks become hypersensitive to LPS. This led to

Cytokines
ISBN 0–12–498340–5

the key finding by Old and his colleagues at the Memorial Sloan-Kettering Cancer Center in New York that mice primed with BCG and then treated with LPS release within hours in their serum a factor which causes hemorrhagic tumor necrosis in subcutaneous MethA sarcoma-transplanted tumors (Carswell *et al.*, 1975). This factor, which turned out to be a protein, was given the name "tumor necrosis factor". Both the initial exposure to BCG and the subsequent LPS challenge were required for optimal production of the TNF activity, and both were known to stimulate macrophage activity. *In vitro* TNF production by isolated macrophages was soon established. TNF was directly cytotoxic to some tumor cell lines *in vitro*, the prototype of which is the mouse fibrosarcoma line L929; this provided a quantitative assay system *in vitro*. Protein purification and partial sequencing led to the cloning of both human and murine TNF (Pennica *et al.*, 1984; Fransen *et al.*, 1985; Marmenout *et al.*, 1985; Shirai *et al.*, 1985; Wang *et al.*, 1985). At the present time, TNF has been purified from a wide range of other animal species also (reviewed in Fiers, 1992).

Independent research led to the discovery of the same factor. Cerami and his coworkers at the Rockefeller Institute had been studying the severe wasting or cachexia observed in chronic infection or in cancer (reviewed in Beutler and Cerami, 1988). Their studies led to the identification of a serum factor, named "cachectin", which suppressed lipoprotein lipase activity in an adipocyte cell line. When this factor was purified and partially sequenced, it turned out to be identical to the previously cloned murine TNF.

A third route of research led to the discovery of lymphotoxin-α (LT-α). Unlike TNF, LT-α is exclusively made by stimulated T-lymphocytes. In 1984, Aggarwal and colleagues isolated human LT-α and cloned the gene (Gray *et al.*, 1984). The mature polypeptide turned out to be 30% homologous with human TNF, appears to act at the same receptor, and has been called TNF-β, while the "classical" TNF would become TNF-α. Remarkably, a membrane-associated form of LT has been discovered, consisting of a heterotrimeric complex containing two LT-α monomers together with a 33 kDa transmembrane protein (Androlewicz *et al.*, 1992). The gene encoding this LT-α-associated protein was cloned (Browning *et al.*, 1993). Since this protein (p33) forms a complex with LT-α, is structurally related to LT-α, and lies next to the TNF/LT-α locus in the genome, it was given the name LT-β. As the function and biology of LT-β are currently poorly characterized (reviewed in Ware *et al.*, 1995), the latter will not be discussed in detail, and we will continue to use the term "LT" for LT-α, except when we specifically refer to LT-β.

The availability of pure recombinant TNF in large quantities allowed a detailed study of its biological properties. TNF has turned out to be a cytokine with very diverse biological activities that might explain its role in various physiological and pathological phenomena, such as infection, inflammation and immunomodulation, cancer, cachexia, and lethal septic shock. The reader is also referred to some reviews which focus in more detail on the different aspects of the biology of TNF (Beutler and Cerami, 1989; Vassalli, 1992; Sidhu and Bollon, 1993; Beyaert and Fiers, 1994; Heller and Krönke, 1994; Vandenabeele *et al.*, 1995a; Fiers *et al.*, 1996; Argiles *et al.*, 1997; Wallach *et al.*, 1997). The literature on LT is less voluminous. In general, TNF and LT display similar spectra of activities in *in vitro* systems, although LT is often less potent (Browning and Ribolini, 1989) or apparently has partial agonistic activity (Andrews *et al.*, 1990).

2.   *The TNF and LT Genes*

2.1   GENE SEQUENCE AND STRUCTURE

Both TNF and LT genes were found to be encoded by a single gene of about 3 kilobases (kb) in size (Figure 24.1(a,b)). The gene for LT-β is about 2 kb (Figure 24.1(c)). Localization of the TNF and LT genes on the human genome led to the surprising finding that they were closely linked and mapped in the middle of the major histocompatibility complex (MHC), on the short arm of the human chromosome 6. More exactly, the TNF-LT gene cluster is about 210 kb away from the HLA-B locus, toward the centromere and the distal end of the class III loci (Spies *et al.*, 1991). The polyadenylation site of the LT gene is separated from the transcription/initiation site of the TNF gene by only approximately 1 kb. The LT-β gene is localized within 2 kb of the TNF gene. Both human TNF and LT genes are made up of four exons (Figure 24.2(a,b)). Also the LT-β gene is contained within four exons in an arrangement very similar to that of TNF and LT, except for being oriented in the opposite direction (Figure 24.2(c)). In each case, the last exon encodes the majority (over 80%) of the secreted protein and these sequences are approximately 56% homologous. The position of the intron/exon junction linking the last large exon, which encodes essentially all of the extracellular domain and most likely the receptor-binding region, is completely conserved in all three genes. The 5′ regions of the TNF and LT genes are also somewhat homologous and there are short sequences which are identical in the putative promoter regions. Based on these observations it is likely that these genes arose from a common ancestral sequence through a tandem duplication event.

2.2   GENE REGULATION

The regulatory elements mainly responsible for transcriptional activation of TNF and LT have been studied by Jongeneel and colleagues (Jongeneel, 1992). It turns out that elements responsive to the NFκB

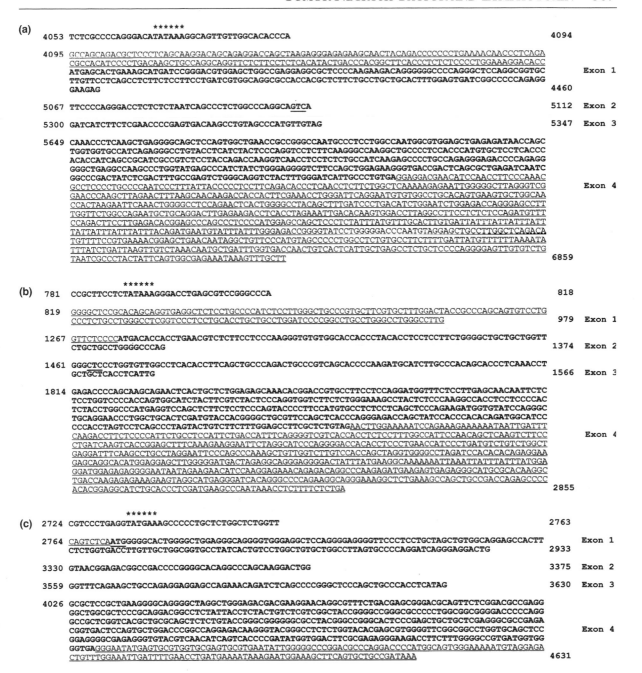

Figure 24.1 Genomic sequence of human TNF (a) and LT-α (b), according to Nedospasov *et al.* (1986), and of LT-β (c), according to Browning *et al.* (1993). The four exons of each gene are indicated. Untranslated regions are underlined. The position of the amino terminal residue of the mature protein is bold underlined. Asterisks indicate the TATA box.

transcription factor are especially important to confer LPS inducibility. However, more factors must be involved to explain the selective activation and expression of the TNF gene. For example, truncated versions of the TNF promoter, lacking NFκB-binding motifs, are active in non-macrophage cell lines (Kruys *et al.*, 1992). The mRNAs of TNF and LT, like several other cytokine mRNAs, have AU-rich sequences in the untranslated 3′ sequence of the mRNA, which decrease the stability of the mRNA (Caput *et al.*, 1986). This sequence may represent recognition sites for specific mRNA-processing proteins. A ribonuclease has been isolated from mouse macrophages that specifically destabilizes mRNA containing this UUAUUUAU sequence in the 3′ region (Beutler and

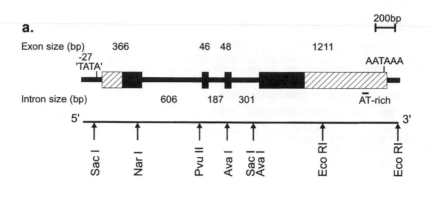

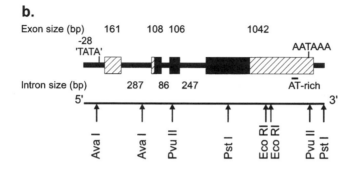

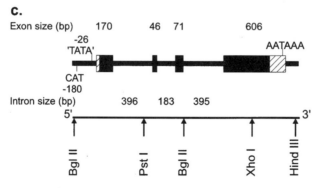

Figure 24.2   Schematic representation of the gene structure of human TNF (a), LT-α (b), and LT-β (c), as determined by the DNA sequence shown in Figure 24.1. The sizes of the exons and introns are denoted in bp above and below each figure. Some special restriction sites are indicated in the lower part of each figure. Hatched sections indicate the untranslated regions.

Cerami, 1988). Furthermore, this octamer also produces a translational block (Han *et al.*, 1990). Remarkably, this AU-rich motif is lacking in LT-β (Browning *et al.*, 1993). A factor has been identified that is able to overcome the inhibitory influence of the TNF 3′ untranslated region (Kruys *et al.*, 1992). Its action depends on the presence of sequences found in the TNF 5′ untranslated region.

3.   *Protein Structure*

Mature secreted human TNF consists of 157 amino acids (aa), whereas human LT contains 171 residues. Human

LT-β contains 240 amino acids. The mature LT protein is preceded by a "classical" 34-aa signal sequence, while the mature TNF is preceded by a 76-aa presequence, the first part of which corresponds to an intracellular domain when TNF is present as a 26 kDa membrane-bound form (see below). The homology between LT and TNF at the amino acid level amounts to 28% (Figure 24.3). LT-β is a transmembrane protein with a short 15-aa N-terminal cytoplasmic domain, and a stretch of 30 hydrophobic amino acids which acts as a membrane-anchoring domain (Browning *et al.*, 1993). LT-β is 21% identical to TNF and 24% to LT (Browning *et al.*, 1993). In contrast to LT, TNF contains a disulfide bridge (Cys^{145}–Cys^{177}),

```
            -76                                                    -27
hTNF        MSTESMIRDV ELAEEALPKK TGGPQGSRRC LFLSLFSFLI VAGATTLFCL
hLT-α       .......... ...MTPPER LFLPRVCGTT LHLLLLGLLL ..........
hLT-β       ..MGALGLEG RGGRLQGRGS LLLAVAGATS LVTLLLAVPI TVLAV.....

            -26                         ↓1                          15
hTNF        LHFGVIGPQR EEFPRDLSLI SPLAQAVRSS SRT....PSD K.....PVAH
hLT-α       ....VLLPGA QGLP.GVGLT PSAAQTARQH PKMHLAHSTL K.....PAAH
hLT-β       ..LALVPQDQ GGLVTETADP GAQAQQGLGF QKLPEEEPET DLSPGLPAAH

            16                                                      64
hTNF        VVANPQAEGQ LQWLNRRANA LLANGVELRD .NQLVVPSEG LYLIYSQVLF
hLT-α       LIGDPSKQNS LLWRANTDRA FLQDGFSLSN .NSLLVPTSG IYFVYSQVVF
hLT-β       LIGAPLKGQG LGWETTKEQA FLTSGTQFSD AEGLALPQDG LYYLYCLVGY

            65                                                      104
hTNF        KGQGCP.... ..STHVLLTH TISRIAVSYQ TKVNLL.... SAIKSPCQRE
hLT-α       SGKAYSPKAT ..SSPLYLAH EVQLFSSQYP FHVPLL.... SSQKMVY...
hLT-β       RGRAPPGGGD PQGRSVTLRS SLYRAGGAYG PGTPELLLEG AETVTPVLDP

            105                                                     154
hTNF        TPEGAEAKPW YEPIYLGGVF QLEKGDRLSA EINRPDYLDF AESGQVYFGI
hLT-α       ...PGLQEPW LHSMYHGAAF QLTQGDQLST HTDGIPHLVL SPST.VFFGA
hLT-β       ARRQGYGPLW YTSVGFGGLV QLRRGERVYV NISHPDMVDF AR.GKTFFGA

            155
hTNF        IAL.
hLT-α       FAL.
hLT-β       VMVG
```

Figure 24.3   Comparison of the amino acid sequence of human TNF (top), LT-α (middle), and LT-β (bottom). The numbering refers to human TNF. The sequences are optimally aligned; identical residues are shaded. LT-α and LT-β each contain a single *N*-glycosylation site (single underlined). Human TNF contains two cysteines involved in a disulfide bond (double underlined). Basic residues (K or R), where trypsin can cleave, are indicated in bold. The start of mature TNF and LT-α is indicated by an arrow.

which does not seem to be required for activity (Figure 24.3). Under denaturing conditions, the molecular masses of human TNF, LT, and LT-β are approximately 17 kDa, 25 kDa, and 33 kDa, respectively. Both TNF and LT are homotrimers (Wingfield *et al.*, 1987; Lewit-Bentley *et al.*, 1988). The three subunits are noncovalently linked by secondary forces. The interaction is so strong that even at very high dilution there is no evidence whatsoever for dissociation, meaning that the trimer is also the biologically active form (Wingfield *et al.*, 1987). It is very likely that LT and LT-β associate to form a heteromeric complex with a trimeric structure similar to those of TNF and LT. The stoichiometry of the complex was believed to be $\alpha_1\beta_2$ (Androlewicz *et al.*, 1992), although a small portion of the complex may also exist as an $\alpha_2\beta_1$ form. Natural LT is a glycoprotein, while TNF is not. The isoelectric point of TNF is 5.6. In contrast to TNF (pH stability from 5–10), LT is rather acid-labile. TNF activity is destroyed by heating for 1 h at 70°C (Ruff and Gifford, 1981).

Well-diffracting crystals of TNF have been obtained (Lewit-Bentley *et al.*, 1988), and the three-dimensional structure was solved at both 2.9 Å and 2.6 Å resolutions

(Eck and Sprang, 1989; Jones *et al.*, 1989). The shape of the molecule resembles a triangular pyramid (Figure 24.4) in which each subunit consists of two β-pleated sheets, five antiparallel β strands in each. The three subunits are arranged edge to face. It is quite remarkable that this arrangement is also found in the capsid proteins of certain viruses, including picorna viruses and satellite tobacco necrosis virus. The outside β-sheet is rich in hydrophilic residues, while the inner sheet is largely hydrophobic and contains the C-terminal segment, which is located close to the central axis of the trimer. The first 8–10 residues on the amino acid terminus appear to be relatively unconstrained and are not crucial for activity. By mutational analysis, the active site has been located in the lower half of the trimeric cone, in the cleft between two subunits (Yamagishi *et al.*, 1990; Van Ostade *et al.*, 1991). Each active site formed between two subunits corresponds to a receptor-binding domain, implying three such binding sites in the TNF molecule. Also, the crystal structure of the extracellular domain of the p55 TNF receptor complexed to LT confirms that these regions are involved in receptor binding (Banner *et al.*, 1993). Interestingly, a number of TNF mutants were

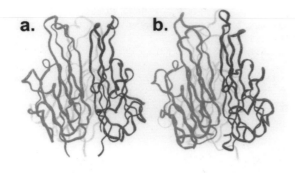

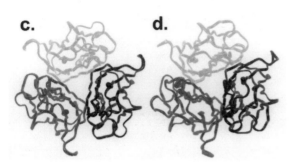

Figure 24.4   Quaternary structure of human LT-α ((a) and (c)) and human TNF ((b) and (d)) trimers. (a) and (b) are frontal views, (c) and (d) are top views.

generated with exclusive specificity for the p55 or p75 TNF receptors (Loetscher *et al.*, 1993; Van Ostade *et al.*, 1993). These mutants provide a physiological tool to independently activate one or the other receptor, even when they are present on the same cell, and enable distinction between the signals generated by each receptor (see Section 6).

The three-dimensional structure of human LT has been determined at 2.9 Å resolution (Eck *et al.*, 1992), and shows close similarity to the structure of TNF (Figure 24.4). Insertions and deletions in the two sequences correspond to loops near the top of the trimer.

4.   *Cellular Sources and Production*

Originally it was believed that TNF was strictly produced by monocytes and macrophages. Nowadays, it seems that, at least *in vitro*, many cell types can produce TNF after an appropriate stimulus (reviewed in Sidhu and Bollon, 1993). It should be mentioned that the intracellular TNF mRNA level may not be related to the amount of protein secreted: treatment with LPS increases gene expression 3-fold, intracellular TNF mRNA 100-fold, and TNF protein production 1000-fold (Beutler and Cerami, 1988). Besides LPS (endotoxin), which represents the main stimulus, viral, fungal, and parasital antigens, enterotoxin, mycobacterial cord factor, C5a

anaphylatoxin, immune complexes, interleukin (IL)-1, IL-2 and, in an autocrine manner, TNF itself, may trigger the synthesis of TNF.

TNF is expressed as a 26 kDa integral transmembrane precursor protein from which a 17 kDa mature TNF protein is released into the medium by proteolytic cleavage (Figure 24.5(a)) (Pennica *et al.*, 1984; Kriegler *et al.*, 1988). It was shown that synthetic metallo-proteinase inhibitors can specifically inhibit TNF processing and secretion at a post-translational step both *in vitro* and *in vivo* (Gearing *et al.*, 1994; McGeehan *et*

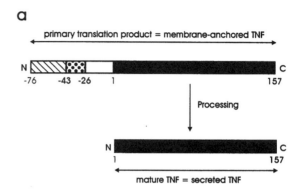

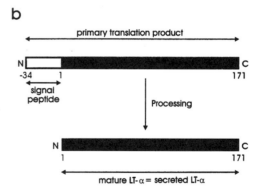

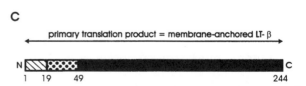

Figure 24.5   Schematic representation of the processing of human TNF (a), LT-α (b), and LT-β (c). The primary translation products of TNF and LT-β are membrane-anchored proteins. Their intracellular part is indicated by a hatched box, while their transmembrane region is stippled. TNF, unlike LT-β, can be processed to mature TNF, which is released into the medium. The primary translation product of LT-α is preceded by a "classical" signal sequence, which is cleaved off in the endoplasmic reticulum, after which LT-α is secreted.

al., 1994; Mohler *et al.*, 1994). Such drugs could be very useful in the management of septic shock and other TNF-associated pathologies. Cleavage does not always occur efficiently in all cell types, and often polypeptides of slightly different lengths can be observed (Cseh and Beutler, 1989). Recently, a TNF-converting enzyme has been identified (Black *et al.*, 1997; Moss *et al.*, 1997). The fact that the presequence is strongly maintained among different species indicates that it fulfills a very important biological function. Not only secreted TNF, but also the membrane-bound form is biologically active (Perez *et al.*, 1990; Decoster *et al.*, 1995). Membrane insertion may be an effective way to keep the action of TNF restricted to specific locations. Recently, the transmembrane form of TNF has been shown to be superior to soluble TNF in activating the TNF p75 receptor in various cell systems (Grell *et al.*, 1995).

Unlike TNF, LT is exclusively produced by T lymphocyte subsets, both of the $CD4^+$ and $CD8^+$ type, following antigenic stimulation in the context of class II and class I restriction, respectively (Paul and Ruddle, 1988). Some normal as well as Epstein–Barr virus-immortalized B lymphocytes can produce LT. LT synthesis can also be induced by IL-2 or IL-2 plus interferon-γ (IFN-γ), or some viruses, such as vesicular stomatitis virus (VSV) or herpes simplex-2 (HSV-2) (Paul and Ruddle, 1988).

As already discussed above, LT starts with a "classical" signal sequence which is cleaved off after secretion (Figure 24.5(b)). But LT can be retained at the surface by association with LT-β, which is an integral membrane glycoprotein (Figure 24.5(c)) (Browning *et al.*, 1991; Androlewicz *et al.*, 1992; Ware *et al.*, 1995). LT-α and LT-β expressions seem to parallel each other, and the expression of LT-β mRNA is increased by phorbol esters and IL-2.

There are also antagonists of TNF induction and synthesis. Glucocorticoids and prostaglandin E_2 (PGE_2) inhibit TNF synthesis, both at the transcriptional and at a post-transcriptional level (Beutler *et al.*, 1992; Seckinger and Dayer, 1992). The suppressive activity of PGE_2 is believed to be mediated by cAMP as a secondary messenger. It is of interest that after administration of TNF to experimental animals, both ACTH/glucocorticoids as well as PGE_2 are induced, and these mediators presumably function as negative feedback regulators. Other important physiological, antagonistic regulators of TNF synthesis are transforming growth factor-β (TGF-β) (Flynn and Palladino, 1992), IL-4 (Hart *et al.*, 1989), and especially IL-10 (Fiorentino *et al.*, 1991). The latter even prevents lethality in experimental endotoxemia (Gérard *et al.*, 1993). Also pharmacological agents can be used as antagonists of TNF synthesis. Reducing agents which are known to inhibit NFκB activation, e.g., *N*-acetylcysteine and glutathione, have been shown to inhibit TNF synthesis in some cell types (Roederer *et al.*, 1992). Pentoxifylline

and thalidomide selectively inhibit TNF production by monocytes/macrophages, without affecting IL-1 and IL-6 production (Waage *et al.*, 1990; Sampaio *et al.*, 1991). Cyclosporin A inhibits TNF production in macrophages at the translational level and in T cells and B cells at the transcriptional level (Goldfeld *et al.*, 1992). Recently, a new series of pyridinyl-imidazole compounds which act as specific inhibitors of the stress-activated p38 mitogen-activated protein (MAP) kinase, has been shown to inhibit the production of TNF and IL-1 from stimulated human monocytes (Lee *et al.*, 1994). Finally, lipoxygenase inhibitors have also been shown to suppress LPS-induced TNF biosynthesis in murine macrophages (Schade *et al.*, 1989).

5. *Biological Activities*

5.1 TNF ACTIVITIES *IN VITRO*

5.1.1 Cytotoxicity

Although cytotoxicity may not be the major activity of TNF *in vivo*, it was among the first activities attributed to the protein and, given its uniqueness, may be considered the hallmark of this cytokine. Although the early studies suggested that TNF was selectively cytostatic or cytotoxic on transformed cell lines and had no effect on normal cells in culture (Sugarman *et al.*, 1985; Fransen *et al.*, 1986a,b), subsequent studies showed that there may be some exceptions, such as endothelial cells, smooth-muscle cells, adipocytes, fibroblasts, and keratinocytes, which, under certain conditions, are inhibited by TNF (Sato *et al.*, 1986; Palombella and Vilček, 1989; Robaye *et al.*, 1991). The cytotoxic action of TNF can vary depending on the growth conditions, viz., the degree of confluence and differentiation (Kirstein *et al.*, 1986). Also, not all tumor cell lines are sensitive to the antiproliferative effect of TNF. This is not due to a difference in TNF receptor number, nor is there a correlation between susceptibility to TNF and embryological or histological origin of the tumor cell types. Many factors and conditions have been identified which increase the sensitivity of the cell to TNF, and even render TNF-resistant cells sensitive to TNF: interferon (Williamson *et al.*, 1983), high temperature (Ruff and Gifford, 1981), transcription or translation inhibitors (Ruff and Gifford, 1981), mitosis inhibitors (Darzynkiewicz *et al.*, 1984), topoisomerase II inhibitors (Alexander *et al.*, 1987), some protein kinase inhibitors (Beyaert *et al.*, 1993b), and LiCl (Beyaert *et al.*, 1989).

Depending on the type of target cell and on the presence of metabolic inhibitors, TNF can induce necrotic or apoptotic cell death (Schmid *et al.*, 1986; Dealtry *et al.*, 1987; Laster *et al.*, 1988; Grooten *et al.*, 1993). Necrosis is characterized by cell swelling, destruction of cell organelles, and cell lysis. In apoptosis, the cell shrinks, apoptotic bodies are formed, and in most cases specific internucleosomal DNA fragmentation is observed.

In murine cells, TNF and LT are nearly equipotent for inducing a cytotoxic response, whereas with many human tumor cells LT is often far less potent than TNF (by more than 100-fold). The reason for this difference is unclear, and cannot be explained by a difference in affinity of LT binding (Browning and Ribolini, 1989).

5.1.2   Differentiation

The process of differentiation can be affected by many cytokines, including TNF and LT. For example, TNF has been shown to reverse adipocyte differentiation, decreasing the expression of genes associated with lipogenesis to yield a more undifferentiated pre-adipocyte (Torti et al., 1985). In HL-60 cells, TNF induces monocytic differentiation (Peetre et al., 1986). The latter phenomenon is characterized by increased nitro-blue tetrazolium staining, nonspecific esterase activity, expression of monocyte-specific cell-surface antigens, and alterations in protooncogene expression (Krönke et al., 1987). LT also induces HL-60 differentiation along the monocytic lineage (Hemmi et al., 1987). IFN-α and IFN-γ, retinoic acid, and 1α,25-dihydroxyvitamin D$_3$ have been reported to be synergistic with TNF in the induction of differentiation (Tobler and Koeffler, 1987; Trinchieri et al., 1987a,b). This induction may also be involved in the direct antitumor activity of TNF.

5.1.3   Growth Stimulation

Besides being cytotoxic to many tumor cells, TNF has also been shown to be mitogenic for a number of normal cells, such as fibroblasts, smooth-muscle cells, T cells and B cells (Vilček et al., 1986; Kahaleh et al., 1988). Some tumor cells such as osteosarcoma and ovarian tumors are growth-stimulated by TNF (Kirstein and Baglioni, 1988; Wu et al., 1992). TNF can be cytotoxic and mitogenic at the same time, in that low doses of TNF induce DNA synthesis and enhance growth, but high doses are cytotoxic (Palombella and Vilček, 1989).

5.1.4   Antiviral Activity

TNF or LT treatment of cells mediates an antiviral effect against several viruses, such as VSV and HSV-2 (Wong and Goeddel, 1986). TNF has also been shown to lyse cells infected with different viruses (Koff and Fann, 1986; Rood et al., 1990). In contrast, TNF has been shown to enhance the infection rate and replication of human immunodeficiency virus (HIV) (Ito et al., 1989; Matsuyama et al., 1989). This enhancement is mediated by transcriptional upregulation of the HIV long terminal repeat as a result of TNF-induced activation of the transcription factor NFκB (Osborn et al., 1989).

5.1.5   Immune-modulatory and Pro-inflammatory Activity

Activated monocytes/macrophages are the major in vivo sources of endogenous TNF synthesis. TNF was found to activate monocytes and macrophages as well as to mediate their cytotoxic activity against TNF-sensitive tumor cells (Philip and Epstein, 1986; Decker et al., 1987). However, the main macrophage-activating factor is IFN-γ. The fact that TNF-resistant cells can also be lysed by activated macrophages indicates that TNF is not the only cytotoxic mediator released by macrophages. Cytotoxicity is also mediated by reactive oxygen species, such as peroxides and NO, which can be induced by TNF (Hoffman and Weinberg, 1987; Higuchi et al., 1990). TNF also causes the secretion of various cytokines, such as IL-1, IL-6, and IL-8. The main negative regulator of macrophage activation is TGF-β, which deactivates macrophages and reduces, for example, their capacity to release peroxides (Flynn and Palladino, 1992). Macrophages play a key role in the immune system by presenting antigens in an MHC class II context. The latter determinant is upregulated by IFN-γ and TNF, and this effect is again counteracted by TGF-β.

TNF very rapidly induces neutrophil adherence to endothelial cells (Gamble et al., 1985), activates phagocytosis (Klebanoff et al., 1986), and enhances specific antibody-dependent cellular cytotoxicity (Shalaby et al., 1985). The oxidative burst is increased 10-fold, but only under conditions of adherence (Schleiffenbaum and Fehr, 1990). TNF also increases adhesion of neutrophils to extracellular matrix proteins, such as fibrinogen, fibronectin, and even serum-coated plastic. The mechanism of the TNF-induced increase in adhesiveness seems to be based on a conformational rearrangement of the CD11a/CD18 and CD11b/CD18 integrins on the cell surface (Gamble et al., 1992).

Resting T cells and B cells appear to express no or only very low numbers of TNF receptors, and TNF upregulates its receptors after activation (Kehrl et al., 1987; Scheurich et al., 1987). TNF is produced by activated lymphocytes and can act as an autocrine growth factor (Ehrke et al., 1988; Yokota et al., 1988). This stimulation involves induction of high-affinity IL-2 receptors and synthesis of IL-2. TNF also provides a synergistic effect with IL-2 in the generation of lymphokine-activated killer (LAK) cells (Chouaib et al., 1988).

Natural killer (NK) cells and natural cytotoxic cells mediate their cytotoxicity at least partially by producing TNF (Wright and Bonavida, 1987). Activated NK cells propagated in vitro in the presence of IL-2 are dependent on endogenously produced TNF for their proliferation (Naume et al., 1991). TNF also enhances the cytotoxic activity of the NK cells, perhaps by stimulating endogenous TNF synthesis (Østensen et al., 1987).

TNF drastically alters the properties of endothelial

cells. The immune properties of the endothelial cells are increased as a result of TNF-induced MHC class I expression. TNF treatment leads to the surface expression of membrane-bound IL-1, as well as adhesion molecules, such as ELAM-1, ICAM-1, ICAM-2, and VCAM-1 (Collins et al., 1984; Kurt-Jones et al., 1987; Pober and Cotran, 1990). IL-8, a strong stimulator of polymorphonuclear neutrophils (PMNs), is secreted by TNF-treated endothelial cells (Strieter et al., 1989). The expression of these molecules results in the binding of granulocytes, lymphocytes, and monocytes to vascular endothelium, and play an important role in the development of an inflammatory state and migration of leukocytes into tumors. TNF treatment of an endothelial cell layer also leads to morphological changes, and these effects may contribute to increased microvascular permeability and to extravasation of neutrophils, lymphocytes, and monocytes (Brett et al., 1989). TNF also changes the properties of endothelium from anticoagulant to procoagulant, by increased production of plasminogen activator inhibitor and tissue factor procoagulant activity, and suppression of both plasminogen activators and thrombomodulin, a cell-surface cofactor required for the activation of protein C (van Hinsbergh et al., 1988; Lentz et al., 1991; Vassalli, 1992). Depending on the conditions, TNF can be toxic or angiogenic to endothelial cells (Leibovich et al., 1987; Robaye et al., 1991). Finally, TNF induces the synthesis and secretion of cytokines (GM-CSF, G-CSF, M-CSF, IL-6, IL-8, IL-1, TNF), as well as several low-molecular-mass mediators, such as PGE_2 and NO (Kilbourn and Belloni, 1990). The latter is implicated in TNF-induced hypotension.

Although TNF and LT have most of their properties in common, some data indicate that LT induces fewer pro-inflammatory activities on cultured endothelial cells than TNF in vitro. Indeed, LT leads to less neutrophil adherence and cytokine production (Broudy et al., 1987; Locksley et al., 1987).

5.1.6   Gene Induction

TNF induces many gene products involved in inflammation, tissue repair, hematopoiesis, immune response and antitumor effects. An exhaustive list of such TNF-responsive genes has been compiled (Fiers, 1993), including genes coding for transcription factors, growth factors, cytokines, and cell surface antigens. Some of the TNF-responsive genes code for so-called "TNF resistance proteins", which can inhibit TNF cytotoxicity. Indeed, the cytotoxic activity of TNF is enhanced up to 100-fold in the presence of transcription or translation inhibitors, such as actinomycin D or cycloheximide, respectively (Ruff and Gifford, 1981). Many cell lines are quite resistant to TNF, but become sensitive when they are also treated with actinomycin D (Fransen et al., 1986a). Examples of "TNF resistance proteins" are manganese superoxide dismutase (Wong and Goeddel,

1988), the zinc-finger protein A20 (Opipari et al., 1992), and the heat-shock protein hsp70 (Jäättelä et al., 1992).

5.1.7   Other in vitro Activities

Receptors for TNF are present on nearly all cell types, and it is therefore not unexpected that TNF exerts an effect on almost every cell type studied. Only a few more examples of particular interest are given below.

TNF induces the synthesis of collagenase, hyaluronic acid, plasminogen activator, and PGE_2 in synovial cells (Brenner et al., 1989). It also causes resorption of human articular cartilage and stimulates synthesis of plasminogen activator in human articular chondrocytes (Campbell et al., 1990).

In muscle cells, TNF stimulates glycolysis and glycogenolysis. There is an increase in glucose transport, a rise in fructose 2,6-bisphosphate, and an increased release of lactate (Sherry and Cerami, 1988; Zentella et al., 1993). TNF also causes a small drop in transmembrane potential (Tracey et al., 1986). Vascular, smooth-muscle cells release IL-1 and IL-6 after stimulation with TNF (Warner and Libby, 1989). Furthermore, induction of NO synthetase, NO production, and release of prostaglandins may contribute to TNF-induced blood vessel relaxation and hypotension (Geng et al., 1992).

TNF has been shown to stimulate the production of various hormones, including ACTH, growth hormone, and thyroid-stimulating hormone, in pituitary cells (Milenkovic et al., 1989).

Finally, TNF is cytotoxic to oligodendrocytes in culture, and mediates demyelination of cultured mouse spinal cord (Powell and Steinman, 1992). The latter findings have led to speculation about a role for TNF in multiple sclerosis.

5.2   TNF Activities IN VIVO

5.2.1   Antitumor Activity

A murine MethA sarcoma in a syngenic mouse was the first system in which antitumor activity of TNF was shown (Carswell et al., 1975). A single injection of 2 µg human TNF was sufficient to cause hemorrhagic necrosis within 24 h. The cure rate increased with increased frequency of administration and TNF dose, and often complete cure could be obtained. It should be mentioned that MethA sarcoma cells are completely resistant to TNF in vitro. Hence, the antitumor effect on MethA sarcoma cells in vivo is completely host-mediated (Palladino et al., 1987). Several lines of evidence indicate that the antitumor activity is mediated by vascular changes in the tumor. Only well-established subcutaneous tumors were subject to necrosis after TNF treatment, while small early tumors and intraperitoneal tumors were resistant (Manda et al., 1987). Moreover, there is local fibrin deposition and thrombus formation in the microcapillaries of the tumor tissue which lead to ischemic necrosis starting in the central part of the tumor (Nawroth *et al.*, 1988; Shimomura *et al.*, 1988). Finally, the antitumor effect of TNF could be prevented by the anticoagulant dicoumarol (Shimomura *et al.*, 1988). It should be stressed that the MethA sarcoma is quite unique in its *in vivo* high sensitivity to TNF. In contradistinction to most human tumors, the MethA sarcoma is fairly immunogenic, and this might explain the high susceptibility. The role of an immune response was further supported by Asher and coworkers (Asher *et al.*, 1987), who compared weakly immunogenic and nonimmunogenic syngenic tumors. In these experiments, only immunogenic tumors showed a response. Furthermore, hemorrhagic necrosis of a MethA tumor is much reduced in nude mice (Haranaka *et al.*, 1984). Hence, T cells presumably play an important role in the induction of hemorrhagic necrosis. It should be noted that infiltration of inflammatory cells in the tumor may also play an important role. The specific susceptibility of the microvascular system of a tumor to TNF remains intriguing. Whether this specificity is obtained through T cell-derived factors or tumor-derived factors, or is an intrinsic property of nascent tumor vasculature, remains unclear (Nawroth *et al.*, 1988; McIntosh *et al.*, 1990).

T cell-deficient nude mice bearing tumors derived from human malignant cell lines or from transplantable human xenografts offer another model system which has often been used to study the antitumor activity of TNF (Balkwill *et al.*, 1986, 1987; Nosoh *et al.*, 1987; Manetta *et al.*, 1989). Although there is no contribution from activated cytotoxic T cells in these mice, other potential contributory systems remain functional, such as activation of tumoricidal macrophages, natural killer cells, and neutrophils. For example, a marked influx of PMNs in the peritoneal cavity of nude mice carrying a human ovarian xenograft in the peritoneum has been demonstrated after intraperitoneal injection of TNF (Malik and Balkwill, 1992). Nevertheless, these studies also demonstrated that local perilesional or intratumoral treatment of subcutaneous tumors was much more efficient than intraperitoneal treatment, suggesting a direct antitumor cell activity in this model. This is further suggested by the observation that tumor cells for which a synergistic effect was seen *in vitro* between TNF and IFN-γ, TNF and LiCl, or TNF and staurosporine gave rise after intradermal injection to tumors in nude mice, which were now sensitive *in vivo* to the same synergistic combinations (Balkwill *et al.*, 1986; Beyaert *et al.*, 1989, 1993b). An even stronger argument comes from results obtained with a human TNF mutant which was specific for human p55 receptors and did not recognize murine TNF receptors, neither p55 nor p75. When these muteins were injected into nude mice carrying a human tumor xenograft, they induced an antitumor response as

effective as the one obtained with wild-type human TNF, which can also bind to murine cells. Again this effect became more pronounced in the presence of human IFN-γ (Van Ostade *et al.*, 1993).

In vitro studies have shown that the antiproliferative effect of TNF can be enhanced by many agents, such as cytokines (IFN, IL-1, IL-2), some cytotoxic drugs, and higher temperature (see also Section 5.1.1). A similar synergistic effect has been demonstrated *in vivo*. TNF and IFN-γ have shown either additive or synergistic antitumor activity against mouse and rat tumors as well as human ovarian tumor xenografts in nude mice (Brouckaert *et al.*, 1986; Gresser *et al.*, 1986; Balkwill *et al.*, 1987; Marquet *et al.*, 1987). Both direct and indirect effects were shown to be involved in the increased antitumor response. However, the toxicity of a combination treatment was fairly high. Also IFN-α enhances the antitumor activity of TNF against human breast carcinoma and renal cell carcinoma xenografts in mice (Balkwill *et al.*, 1986; Baisch *et al.*, 1990). With several tumor models there was a synergy with TNF and IL-2 therapy (Nishimura *et al.*, 1987; McIntosh *et al.*, 1988). Host immunity and tumor immunogenicity played an important role in tumor regression. TNF has been shown to enhance the cytotoxicity of IL-2-induced LAK (lymphocyte-activated killer) cells against tumors such as adenocarcinoma, sarcoma, and leukemia (Owen-Schaub *et al.*, 1988; Teichmann *et al.*, 1992). However, although induction of LAK activity can occur in athymic nude mice, there was no potentiation of the antitumor activity of TNF by IL-2 administration (Malik *et al.*, 1989). Cytotoxic drugs, such as cyclophosphamide, doxorubicin, adriamycin, etoposide, and staurosporine, enhanced the antitumor action of TNF *in vivo* (Alexander *et al.*, 1987; Regenass *et al.*, 1987; Krosnick *et al.*, 1989; Beyaert *et al.*, 1993b). Some of these combinations can also increase the systemic toxicity and suppress immune responses. For some tumor cells there is a dramatic synergy between TNF and LiCl, both *in vitro* and *in vivo*, without inducing toxicity. Animals carrying tumors susceptible to the two agents showed complete and lasting remissions when treated perilesionally with TNF and LiCl (Beyaert *et al.*, 1989). Finally, the effective synergy between TNF and hyperthermia should be recalled (Haranaka *et al.*, 1987; Watanabe *et al.*, 1988). The synergistic action is readily observed in tissue culture and presumably contributes to a direct antitumor activity *in vivo*. Furthermore, Fujimoto and collaborators have shown that a reduced blood flow induced by hyperthermia might be involved in tumor regression (Fujimoto *et al.*, 1992).

A number of studies have followed the ability of TNF to limit experimental metastasis. In some studies involving treatment with the combination of TNF and IFN-γ (Schultz and Altom, 1990) or TNF and IL-6 (Mulé *et al.*, 1990), important inhibition of experimental metastasis was observed. Treatment with TNF as such was mostly not as successful. In fact, TNF even promoted peritoneal metastasis of ovarian carcinoma in nude mice (Malik *et al.*, 1989). Two factors contributed to this phenomenon: increased adherence of tumor cells to the peritoneal surface, and stromal proliferation. Interestingly, cells secreting TNF show enhanced metastasis in nude mice (Malik *et al.*, 1990). Similarly, endogenous TNF or administered TNF enhanced the metastatic potential of circulating tumor cells in an experimental fibrosarcoma metastasis model in mice (Orosz *et al.*, 1993).

5.2.2 Synthesis of Other Mediators

TNF rapidly induces the synthesis of mediators such as prostaglandins (Kettelhut *et al.*, 1987), which contribute to the pathology of TNF but are also required for induction of protecting factors such as glucocorticoids (Takahashi *et al.*, 1993). IL-6 is also strongly and rapidly induced by TNF in a large number of cell types and released into the circulation. IL-6 is a very pleiotropic cytokine, acting on the immune system and on the neuroendocrine system and, perhaps most importantly of all, is one of the main inducers of the acute-phase response (Van Snick, 1990; Baumann and Gauldie, 1994). A sustained increase in IL-6 after injection of TNF correlated with TNF-induced lethality (Libert *et al.*, 1990; Brouckaert *et al.*, 1992). TNF-induced glucocorticoids can counteract the toxicity of TNF. Indeed, TNF becomes much more lethal when the feedback control, exerted by endogenous glucocorticoids, is inhibited by injecting the antagonistic drug RU38486. Again this correlates with high IL-6 levels persisting until death (Brouckaert *et al.*, 1992). Finally, sensitization of mice to TNF toxicity by tumors again correlates with high inducibility of IL-6 (Cauwels *et al.*, 1995). However, although the above observations may suggest a causal relationship between IL-6 levels and lethality, antibodies against IL-6 or IL-6 receptor provided only limited protection against TNF-induced lethality (Libert *et al.*, 1992). Moreover, mice deleted of functional IL-6 genes remained fully sensitive to TNF (Libert *et al.*, 1994).

Another important cytokine induced by TNF is IL-1. Although TNF and IL-1 have an almost overlapping spectrum of activities (Last-Barney *et al.*, 1988), the biological effects of the two cytokines are not identical. In contrast to TNF, a single injection of IL-1 is not lethal, even at quite high doses. However, coadministration of TNF and IL-1 leads to a highly synergistic toxicity (Okusawa *et al.*, 1988; Everaerdt *et al.*, 1989), suggesting that IL-1 sensitizes the animal to the toxic effect of TNF. Indeed, the lethality of TNF for healthy mice can be partially counteracted by repeated injections of IL-1 receptor antagonist (Everaerdt *et al.*, 1994). When mice are treated with IL-1 or low doses of TNF itself 12 h before a high dose of TNF, they become much more resistant to a normally lethal challenge with TNF (Wallach *et al.*, 1988; Libert *et al.*, 1991a;

Takahashi *et al.*, 1991). The protective effect of IL-1 or TNF pretreatment may operate via the liver (Libert *et al.*, 1991b), although the nature of the protective factor is not known at present.

6. Receptors

Like other cytokines, TNF acts via specific cell-surface receptors. These are present on nearly all cell types studied, except for erythrocytes and unstimulated lymphocytes. Two receptors have been cloned which differ in size and in binding affinity. Based on the molecular mass of the proteins, they are referred to as p55 and p75 (Dembic *et al.*, 1990; Gray *et al.*, 1990; Himmler *et al.*, 1990; Loetscher *et al.*, 1990; Nophar *et al.*, 1990; Schall *et al.*, 1990; Smith *et al.*, 1990a). The binding constant is about 2×10^{-10} M and the number of receptors varies from about 200 to 10 000 per cell. Both receptors bind TNF as well as LT. Of considerable interest is the fact that human TNF does not bind to murine p75 (Lewis *et al.*, 1991). A third receptor recognized by TNF, but not by LT, has been reported in liver extracts (Schwalb *et al.*, 1993). Further characterization and cloning are necessary before conclusions can be drawn regarding its significance.

The two receptor types consist of an extracellular domain which binds TNF and is homologous for 28%, a transmembrane region, and an intracellular part which is totally different and does not contain any recognizable structure associated with a particular function. Each extracellular domain contains four conserved, cysteine-rich repeats, about 38 to 42 amino acids in length (Figure 24.6). Most or all of the cysteines are fixed in cysteine bridges. Based on similarities in their extracellular domains, these receptors belong to a receptor superfamily including the low-affinity nerve growth factor receptor, the Fas antigen, the human B-lymphocyte activation molecule CD40, and the OX40 antigen found on activated T cells (Smith *et al.*, 1994; Baker and Reddy, 1996). The remarkable absence of homology between the intracellular regions of both TNF receptors suggests that they are involved in different functions or signal-transducing pathways. Most of the TNF responses known (NFκB activation, cytotoxicity, IL-6 induction, fibroblast proliferation) occur by activation of p55 (Engelmann *et al.*, 1990; Thoma *et al.*, 1990). However, TNF activities on T cells seem to be p75-mediated (Tartaglia *et al.*, 1991; Vandenabeele *et al.*, 1992). A contribution of p75 was also demonstrated in cytotoxicity (Shalaby *et al.*, 1990; Heller *et al.*, 1992; Grell *et al.*, 1993; Bigda *et al.*, 1994), in endothelial and neutrophil functions (Barbara *et al.*, 1994), as well as in inhibition of early hematopoiesis (Jacobsen *et al.*, 1994). Some of the latter data might be explained by the "ligand passing model" in which p75, having a 5-fold higher

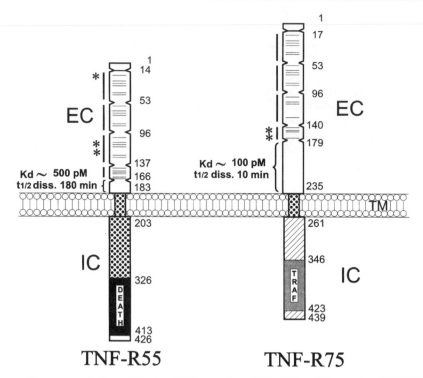

Figure 24.6   Schematic representation of the two TNF receptors (TNF-R55 = p55 receptor; TNF-R75 = p75 receptor). The extracellular (EC) domain, the transmembrane domain (TM), and the intracellular domain (IC) are indicated. The EC domain can be separated into four cysteine-rich repeats (horizontal lines show the disulfide bridges). The death domain has been shown, by mutational analysis, to be responsible for cytotoxic action, while the total IC domain of p55 is required for NO synthetase induction (Tartaglia *et al.*, 1993a). The TRAF-binding domain of p75 has been shown to be responsible for NFκB activation (Rothe *et al.*, 1994 and 1995b). Potential *N*-glycosylation sites are indicated by an asterisk.

affinity and fast dissociation rate, presents TNF to neighboring p55 molecules (Tartaglia *et al.*, 1993b). It has been reported that p75 can contribute, in HeLa cells, to the cytotoxic effect of TNF both by its own signaling and by regulating the access of TNF to p55 (Bigda *et al.*, 1994). In the case of a T cell-derived cell line, true cooperation between the two signaling pathways for the induction of apoptosis, and not "ligand passing", has been shown to occur (Vandenabeele *et al.*, 1995b).

Soluble TNF-binding proteins, which later were identified as extracellular domains of the TNF receptors, have been found in urine and serum (Nophar *et al.*, 1990). They arise by proteolytic cleavage from surface-bound receptors (Porteu *et al.*, 1991). At the amino terminus, 11 and 4 residues are missing from soluble p55 and p75, respectively. The cleavage at the C-terminus of the former is after residue 171 (Nophar *et al.*, 1990), while the C-terminal cleavage site for soluble p75 may be heterogenous. The protease responsible for TNF receptor shedding remains to be identified. Receptor shedding from cell lines can be induced by protein kinase C (PKC) activators, by *N*-formylmethionyl-leucyl-phenylalanine, by complement fragment 5A, by the calcium ionophore A23187, or by TNF itself (Porteu

and Nathan, 1990; Porteu and Hieblot, 1994). Remarkably, the cytoplasmic domain of p75 has been shown not to be absolutely essential for shedding (Brakebusch *et al.*, 1992; Crowe *et al.*, 1993). Soluble TNF receptors can affect TNF activity not only by interfering with its binding to cells but also by stabilizing the structure and preserving its activity, thus prolonging some of its effects (Aderka *et al.*, 1992; cf. Section 7).

The TNF receptors have half-lives between 30 min and 2 h, and it has been suggested that this rapid turnover is related to the high content of so-called "PEST" sequences (proline, glutamic acid, serine, threonine) in the intracellular domains. Ubiquitination may be involved in this degradation process, at least as far as p75 is concerned (Loetscher *et al.*, 1990). TNF receptors can be up- or downmodulated. Both type I IFNs and IFN-γ have been shown to increase or decrease the expression of cell surface receptors, depending on the cell type studied (Tsujimoto *et al.*, 1986; Ruggiero *et al.*, 1987; Aggarwal and Pandita, 1994). cAMP and analogs strongly enhance p75 expression in HL-60 cells and in some myeloid cells, indicating a regulatory role for protein kinase A (Hohmann *et al.*, 1991). On the other hand, activation of PKC leads to disappearance of the

TNF receptors from the cell surface (Unglaub et al., 1987). In the latter case the mechanism may be due to receptor shedding. Also, treatment with TNF leads to downmodulation of its receptors. TNF receptors are rapidly internalized and there is no recycling.

A receptor for LT-β has been described which is unable to bind TNF or LT-α (Crowe et al., 1994). The LT-β receptor also contains a cysteine-rich motif characteristic of the TNF receptor superfamily. The cytoplasmic region of the LT-β receptor has little sequence similarity with other members of the receptor family, which suggests that the mechanism used to signal cellular responses may be unrelated. Although the LT-β receptor lacks a typical death domain (see below), signaling through the LT-β receptor can induce the death of some adenocarcinoma tumor lines (Browning et al., 1996).

7. Signal Transduction

The primary trigger for signaling is clustering of the TNF receptors, which is brought about by binding to the trimeric TNF. The main argument in support of a clustering mechanism comes from the observation that overexpression of a fusion protein of the intracellular part of the p55 TNF receptor with chloramphenicol acetyltransferase, the latter enforcing trimerization, is sufficient to induce cell death (Vandevoorde et al., 1997). In addition, monoclonal antibodies to the TNF receptor can mimic the action of TNF (Engelmann et al., 1990; Espevik et al., 1990). The fact that a pentameric immunoglobulin (IgM) M monoclonal antibody is considerably more active in mimicking TNF action as compared to the bivalent IgG is further support for a clustering mechanism. Moreover, stoichiometric binding studies (Pennica et al., 1992) and three-dimensional resolution of ligand–receptor complexes which employed genetically engineered TNF receptor ectodomains (Banner et al., 1993) have confirmed that trimeric TNF is indeed able to bind up to three TNF receptors. Furthermore, it has been shown that interaction with TNF leads to the formation of disulfide-linked receptor aggregates (Grazioli et al., 1994). Following clustering, the ligand–receptor complex is internalized by clathrin-coated pits, and via endosomes and multivesicle bodies ends up in the lysosomes, where it is degraded (Mosselmans et al., 1988). Whether internalization and degradation are necessary to induce a biological response is still under debate (Decker et al., 1987; Smith et al., 1990b). The intracellular part of p55 can be divided in two domains: a membrane-proximal and a membrane-distal (i.e., carboxy-terminal) domain. The latter is largely homologous to the Fas intracellular domain, and has been referred to as the "death domain" (Tartaglia et al., 1993a,c). Both domains are important for NO synthase induction, but the death domain is sufficient for

cytotoxicity. Furthermore, TNF-induced activation of two distinct types of sphingomyelinase (SMase), a membrane-associated neutral SMase and an endosomal acidic SMase, is mediated by the membrane-proximal domain and the membrane distal (death) domain of p55, respectively (Wiegmann et al., 1994). In a yeast two-hybrid screening using the intracellular domain of p55 as a bait, Hsu and colleagues (1995) identified TRADD, a 34 kDa cytoplasmic protein containing a C-terminal death domain. TRADD and p55 interact through their death domains and TRADD is recruited to p55 in a TNF-dependent process (Hsu et al., 1995, 1996b). As observed for p55, overexpression of TRADD causes apoptosis and activation of NFκB, suggesting that TRADD is critically involved in p55 signal transduction. The recent discovery that TRADD interacts with TRAF2, FADD, and RIP (see below), suggests that TRADD may function as an adaptor to recruit other signaling proteins to p55 after TNF stimulation (Hsu et al., 1996a,b). TRAF2 was originally identified as a protein that associates with a C-terminal region in the cytoplasmic domain of p75, which is indispensable for p75-mediated signal transduction (Rothe et al., 1994). Binding of TRAF2 to p75 and p55/TRADD is required for NFκB activation (Rothe et al., 1995b; Hsu et al., 1996b). TRAF2 recruits the NFκB-inducing kinase NIK to the membrane, which in turn activates a recently indentifed IκB kinase complex through phosphorylation of two related IκB kinases (Maniatis, 1997). Two other proteins, known as cellular inhibitors of apoptosis (c-IAP1 and c-IAP2), are also recruited to the TNF receptors by binding to TRAF2 (Rothe et al., 1995a), but their physiological function is still not clear. FADD and RIP are two death domain proteins which were originally isolated based on their interaction with Fas (Boldin et al., 1995; Chinnaiyan et al., 1995; Stanger et al., 1995), another member of the TNF receptor family. Both proteins have been shown to mediate p55-induced apoptosis, while only RIP can also mediate p55-induced NFκB activation (Chinnaiyan et al., 1996; Hsu et al., 1996a,b). Evidence has also been provided that both receptors can be phosphorylated by an associated kinase (Darnay et al., 1994a,b; Beyaert et al., 1995), but its biological function remains unknown.

Recent studies implicate a group of cytoplasmic thiol proteases which are structurally related to the Coenorhabditis elegans protease CED3 and the mammalian IL-1β-converting enzyme (ICE) in the onset of cell death induced by TNF and various other apoptotic stimuli (Nicholson and Thornberry, 1997; Van de Craen et al., 1997). Specific peptides and virus-encoded proteins blocking protease functions were found to protect cells from TNF-mediated cytotoxicity. In addition, rapid cleavage of specific intracellular proteins, mediated by proteases of the CED3/ICE family, was observed shortly after TNF stimulation. The mechanism of activation of the proteases by TNF receptors has not yet been established, but the cloning of a novel CED3/ICE protease, MACHα or FLICE (Boldin *et al.*, 1996; Muzio *et al.*, 1996), that binds to the TNF receptor-associated protein FADD indicates that protease activation constitutes the most upstream enzymatic step in the cascade of signaling for the cytocidal effects of TNF.

Mitochondrial radical production (Schulze-Osthoff *et al.*, 1992, 1993; Goossens *et al.*, 1995) and a variety of enzymes, including phospholipases (Suffys *et al.*, 1991; Schütze *et al.*, 1992b; De Valck *et al.*, 1993; Beyaert *et al.*, 1993a), SMases (Wiegmann *et al.*, 1994), and protein kinases (Kaur *et al.*, 1989; Van Lint *et al.*, 1992; Beyaert *et al.*, 1993b, 1996; Beyaert *et al.*, 1997) may also participate in the cellular activities of TNF. Some of these enzymes may become activated by the proteolytic cleavage by CED3/ICE members. However, it cannot be excluded that distinct signaling routes, independently of MACHα/FLICE, are involved in TNF-induced cell death.

TNF induces the synthesis of a wide variety of proteins (see Section 5.1.6). Changes in gene expression result from the activation of transcription factors by TNF. The most important transcription factor involved in TNF-induced gene activation is undoubtedly the nuclear factor κB (NFκB). It has been shown to be essential for TNF-mediated induction of IL-6 (Zhang *et al.*, 1990) and the protective protein A20 (Laherty *et al.*, 1993). Activation of cytosolic NFκB involves phosphorylation and proteolytic degradation of the inhibitory subunit IκB, after which an active NFκB dimer can translocate to the nucleus (Henkel *et al.*, 1993; Naumann and Scheidereit, 1994; Maniatis, 1997), and subsequent nuclear transactivation of NFκB, which seems to be mediated by the p38 stress-activated MAP kinase (Beyaert *et al.*, 1996). In addition to NFκB, TNF has been reported also to activate or induce the transcription factors activation protein 1 (AP-1), nuclear factor IL-6 (NF-IL-6), cAMP-responsive element-binding protein (CREB), and possibly others (Zhang *et al.*, 1988; Brenner *et al.*, 1989). As already mentioned, gene expression is not required for TNF-induced cytotoxicity.

Signaling events leading to cell death may follow the following model. TNF binding induces the clustering of its cell surface receptors and the association of several molecules, which then initiate competing processes: transcription of protective genes versus a program of self-destruction involving activation of proteases, protein kinases, and phospholipases. This results in the formation of arachidonic acid, inositol phosphates, diacylglycerol, phosphatidic acid, and derivatives. These mediators in some way, either directly or through the release of other second messengers, provoke activation of other pathways, including mitochondrial radical production, finally resulting in cell death (reviewed in Beyaert and Fiers, 1994). Other models which might be more relevant for other TNF-induced activities have been described in other reviews (Larrick and Wright, 1990; Camussi *et al.*, 1991; Vilček and Lee, 1991; Schütze *et al.*, 1992a; Heller and Krönke, 1994; Fiers *et al.*, 1996; Wallach *et al.*, 1997).

8. Murine TNF and Murine LT

The genes for murine TNF (Fransen *et al.*, 1985) and murine LT (Li *et al.*, 1987) have been cloned and mapped to chromosome 17, where they are closely linked at about the same position of the MHC locus. Both genes consist of four exons and three introns. The murine TNF gene encodes a 235-amino-acid (aa) protein of which a presequence of 79 amino acids is cleaved off to obtain the mature murine TNF protein of 156 amino acids. There is a remarkably great homology in both the propeptide (86%) and mature protein (79%) regions between the amino acid sequences of murine and human TNF. This high sequence homology may be related to the nearly complete lack of species specificity of the biological action of TNF. It should be mentioned, however, that human TNF only binds to p55 in the murine system, while murine TNF binds to the two TNF receptors both in the murine and human system (Lewis *et al.*, 1991). Murine TNF is glycosylated at position 86 of the precursor sequence. Two cysteines, at positions 148 and 179 in the precursor, are involved in an intrasubunit disulfide bridge. Murine LT has a 33-aa signal peptide and a mature protein of 169 amino acids. There is one potential glycosylation site at residue 60. Murine LT is highly homologous to human LT (74%) and is 35% homologous to murine TNF (Paul and Ruddle, 1988).

On the basis of homology with the human TNF receptor clones, murine TNF receptor cDNAs have also been obtained (Lewis *et al.*, 1991). Murine p55 and p75 are 64% and 62% identical to their human counterparts, respectively. Again the two extracellular domains are homologous, having 20% identity, and there is no relationship between the intracellular domains. The p55 cDNA codes for a 454-aa precursor protein of the 425-aa mature receptor with a 23-aa transmembrane domain, a 183-aa extracellular domain and a 219-aa intracellular domain. The p75 cDNA codes for a 474-aa precursor protein of the 452-aa mature receptor with a 29-aa transmembrane domain, a 235-aa extracellular domain and a 188-aa intracellular domain. The binding constant is approximately 2×10^{-10} M, which is similar to that of the human TNF receptors. Unlike the TNF/LT genes, TNF receptors are not linked in the genome; p55 maps on murine chromosome 6, and p75 on murine chromosome 4.

9. TNF in Disease and Therapy

The main physiological role of TNF is undoubtedly activation of the first-line reaction of the organism to microbial, parasitic, viral, or mechanical stress. It has an important role in antibacterial resistance (Havell, 1989; Roll *et al.*, 1990) and may be important in the host resistance against leishmaniasis (Liew *et al.*, 1990) and *Plasmodium falciparum* (Butcher and Clark, 1990). Mice in which the p55 or p75 TNF receptor is deleted have been described (Rothe *et al.*, 1993). Deletion of the p55 receptor gene causes pronounced immunodeficiency in which animals show enhanced susceptibility to *Listeria monocytogenes*. However, "knockout" of p55 does not confer resistance to the lethal effect of LPS; only in galactosamine-treated animals does p55 gene deletion abrogate the sensitivity to LPS. Deletion of the p75 gene causes a minimal phenotype, in which scab formation fails to occur in response to repeated intradermal injection of TNF and there is modest resistance to the lethal effect of TNF. Rather unexpectedly, a developmental defect has been described in LT-α knockout mice (De Togni *et al.*, 1994). Deletion of the LT-α gene results in a distinctive phenotype, characterized by the absence of lymph nodes and Peyer's patches, and by disordered segregation of B cells and T cells in the spleen. Because the two TNF receptors (p55 and p75) cannot be implicated as essential participants in the ontogeny of lymph nodes and spleen (as discussed above), suspicion immediately centers on the heteromeric ligand LT-α/LT-β and its receptor. Indeed, LT-β-deficient mice reveal severe defects in organogenesis of the lymphoid system similar to LT-α-deficient mice, except that mesenteric and cervical lymph nodes are present (Alimzhanov *et al.*, 1997; Koni *et al.*, 1997).

The activity of TNF is tightly regulated at the levels of secretion and receptor expression. Additional regulatory mechanisms are provided by the concomitant action of different cytokines and the presence in biological fluids of specific inhibitory proteins, especially soluble cytokine receptors (see also Section 6). Abnormalities in the production of these substances might contribute to the pathophysiology of immune and neoplastic diseases. Besides their role in regulating cytokine activity *in vivo*, soluble cytokine receptors hold significant potential for therapeutic use as very specific anticytokine agents and as indicators in diagnosis and assessment of immune parameters in a variety of autoimmune and malignant diseases (Fernandez-Botran, 1991).

In cases of overreaction of the host or deficiency of a natural autoregulatory network, several deleterious effects of TNF can be observed. These are discussed in more detail below.

9.1   SEPSIS

TNF plays a central role in the pathophysiology of sepsis. Studies in rabbits, dogs, baboons, and human volunteers have demonstrated circulating levels of TNF (from 0.1 to 5 nM) after an endotoxin challenge (Beutler *et al.*, 1985;

Michie *et al.*, 1988). After administration of LPS, TNF appears in the plasma after 30 min and reaches maximal levels after 60–90 min. It is cleared rapidly and has a plasma half-life of 10–20 min in mammals. Passive immunization with antisera to TNF can partially protect mice and baboons from the lethal effects of endotoxin (Beutler *et al.*, 1985). Since high doses of dexamethasone can suppress not only the synthesis of TNF but also its activity, it is not surprising that corticosteroids blunt the shock response to LPS. However, pretreatment with corticosteroids is required for an effect. Similar results have been observed in the clinical treatment of septic shock (Bone *et al.*, 1987). It would be a gross simplification, however, to infer that all changes associated with sepsis are related to TNF. Other cytokines, like IL-1 and IL-6 (and LPS itself), might act synergistically with TNF during the induction of shock (Okusawa *et al.*, 1988; Rothstein and Schreiber, 1988; Waage *et al.*, 1989).

Administration of TNF results in metabolic changes similar to those seen in sepsis. TNF administration to rats and dogs results in hypotension, metabolic acidosis, and hemoconcentration, followed by hypoglycemia, hyperkalemia and respiratory arrest (Tracey *et al.*, 1986). In the circulation, marked increases of epinephrine, norepinephrine, cortisol, and glucagon are observed. The pattern of tissue injury resulting from TNF infusions is similar to the effect of LPS administration: pulmonary leukostasis and hemorrhage; hemorrhagic necrosis in the gut, adrenal gland, and pancreas; intravascular thrombosis, and acute tubular necrosis in the kidney. In humans, TNF infusions have been associated with increases in ACTH, cortisol and catecholamines, and increased plasma triglyceride levels (Michie *et al.*, 1988; Sherman *et al.*, 1988). These alterations are similar to the changes observed after endotoxin administration to normal human volunteers (Michie *et al.*, 1988).

9.2   CACHEXIA

Some diseases associated with wasting and cachexia have been reported to be characterized by chronically high circulating TNF levels. In this regard, a parasitic infection has been most frequently associated with high plasma TNF levels (Scuderi *et al.*, 1986). As already discussed in a previous section, TNF induces a catabolic state in adipocytes, increasing lipase activity while inhibiting the activity of lipogenic enzymes (Pekala *et al.*, 1983). When TNF is injected daily in mice, there is an anorexic effect: the animals stop eating and drinking, and there is an arrest of bowel movement (Takahashi *et al.*, 1991); but after a day or two the animals recover both their weight and physical fitness. In cancer cachexia, however, the state of anorexia persists. In addition, cancer patients who show severe wasting have no TNF in their serum. Various physiological and biochemical studies, which as a whole strongly argue that TNF is not the main mediator of cachexia, have been discussed in more detail by García-Martinez *et al.* (1997).

9.3 INFECTION

Malaria is undoubtedly the most important parasitic disease of man. About 1% of the patients develop cerebral malaria, which is often fatal, especially in children. It was found that plasmodial infection results in an increase in circulating TNF levels, and Grau and colleagues (1989) reported that treatment of infected mice with anti-TNF antibodies could protect against cerebral implications.

Not only parasitic and bacterial infections, but also some viral infections, can become more pathogenic or fatal due to TNF in circulation. For example, CD4$^+$ T cells latently infected by HIV can be stimulated to active viral replication by TNF. In children with HIV, elevated serum levels of TNF correlate with progressive encephalopathy (Mintz *et al.*, 1989). However, it is still not clear whether increased TNF levels in the sera of HIV patients are directly involved in the pathology of HIV or are only observed in relation to opportunistic infections.

9.4 (AUTO)IMMUNE RESPONSES, INFLAMMATION, AND OTHER PATHOPHYSIOLOGICAL PHENOMENA

Graft-versus-host disease in an animal model can be prevented or diminished by anti-TNF therapy or by treatments preventing the synthesis of endogenous TNF (Piguet *et al.*, 1987). In a retrospective study, it was shown that bone marrow transplantation patients with elevated serum TNF often developed major complications (Höller *et al.*, 1990). Renal allograft rejection has been correlated with raised TNF serum levels (Maury and Teppo, 1987). It may also be noted that immunosuppressive treatment with the monoclonal antibody OKT3, directed against activated T cells, has a toxic side-effect due to induction of TNF. This toxicity can be avoided by administration of drugs preventing endogenous TNF synthesis, such as steroids or pentoxifylline (Alegre *et al.*, 1991).

In a rat model for ischemia/reperfusion there was clear evidence for synthesis of TNF, which reached a peak level about 3 h post-reperfusion (Kunkel *et al.*, 1991). The pathology of the liver, and also of the lung, was considerably improved by treatment with anti-TNF antibodies. Furthermore, adult respiratory distress syndrome (ARDS) may be linked to an ischemic event and has been shown to correlate with increased TNF levels in the bronchoalveolar fluid (Grau, 1990).

A TNF involvement is also suspected in a number of other autoimmune and inflammatory conditions. In the case of rheumatoid arthritis, TNF is often present at the site of inflammation (Saxne *et al.*, 1988). Moreover,

transgenic mice carrying a 3'-modified human TNF gene show disregulated patterns of human TNF gene expression and develop chronic inflammatory polyarthritis that can be completely suppressed by treatment with antibodies against human TNF (Keffer *et al.*, 1991). It is quite possible that TNF plays a role in demyelination and oligodendrocyte toxicity in multiple sclerosis (Selmaj and Raine, 1988; Selmaj *et al.*, 1991). Patients with active psoriasis often have increased TNF and IL-6 levels in the psoriatic plaques, and even in their plasma (Nickoloff *et al.*, 1991). Furthermore, a role for endogenous TNF has been implicated in the development or exacerbation of psoriasis in manic-depressive patients who have been treated with lithium salts (Beyaert *et al.*, 1992).

9.5 HUMAN CANCER

Many clinical trials have been initiated with TNF (reviewed in Jones and Selby, 1989; Alexander and Rosenberg, 1991; Taguchi and Sohmura, 1991). Short bolus infusions lead to transient peak levels of about 10 ng/ml, but the plasma levels drop to undetectable levels in less than 3 h, the half-life being 10–40 min. The maximally tolerated dose in humans has been reported to be 200–800 µg/m^2. Fatigue, fever, chills, anorexia, headache, diarrhea, nausea, and vomiting were noted, but were not dose-related. The commoner, more serious, and more specific dose-limiting toxicities were hypotension, hepatotoxicity, and thrombocytopenia. Neither in phase I studies nor in limited phase II studies was there more than a very occasional response to treatment with TNF as a single agent. Some promising clinical results were obtained in cases where direct intratumor therapy was tried. For a number of different carcinomas, a response rate (complete and partial remission combined) of 42% was obtained (Taguchi and Sohmura, 1991). In a study involving patients with refractory malignant ascites, TNF was administered by intraperitoneal infusions; 22 out of 29 patients responded with complete or considerable resolution of their ascites (Räth *et al.*, 1991). In another phase I study of 15 patients with liver metastasis arising from colorectal, pancreatic, gastric, or hepatic cancers, including one case of undetermined origin, intratumor injection of TNF (100–300 µg) stopped tumor growth in all patients, with only mild toxicity (van der Schelling *et al.*, 1992).

The combination of different agents has been a cornerstone in the success of chemotherapy in the treatment of human malignancy. A similar philosophy is being applied to TNF and other biological agents. In preclinical studies, TNF has been combined with other cytokines and chemotherapeutic agents. Abbruzzese and colleagues (1990) reported a phase I study of intravenous TNF administration in combination with intramuscular IFN-γ in patients with advanced gastrointestinal malignancies. However, no objective responses were found in this study, although two patients had a minor response. Also, the maximally tolerated dose of TNF was reduced 2-fold to 5-fold by addition of IFN-γ. In another phase I trial, sequential intravenous administration of IL-2 and TNF to 31 patients with different metastatic malignancies showed only two cases of partial remission (Negrier *et al.*, 1992). Cytotoxic drugs have shown synergistic activity with TNF in preclinical studies. As a result, clinical trials of TNF and etoposide, a topoisomerase II inhibitor, in patients with advanced malignancy are in progress.

A different approach was followed in a very successful therapy of patients with metastatic melanoma or recurrent soft-tissue sarcoma based on the use of a triple combination of TNF, IFN-γ, and the alkylating agent melphalan. This procedure involved isolation perfusion of a limb, and allowed the administration of high doses of TNF without systemic side-effects after administration in the general circulation (Eggermont *et al.*, 1997).

10. Acknowledgments

Research in the authors' laboratory was supported by the Belgian IUAP and ASLK, as well as by an EC BIOMED2 program grant BMH4-CT96-0300. R.B. is a post-doctoral researcher with the FWO.

11. References

Abbruzzese, J.L., Levin, B., Ajani, J.A., *et al.* (1990). A phase II trial of recombinant human interferon-gamma and recombinant tumor necrosis factor in patients with advanced gastrointestinal malignancies: Results of a trial terminated by excessive toxicity. J. Biol. Response Mod. 9, 522–527.

Aderka, D., Engelmann, H., Maor, Y., Brakebusch, C. and Wallach, D. (1992). Stabilization of the bioactivity of tumor necrosis factor by its soluble receptors. J. Exp. Med. 175, 323–329.

Aggarwal, B.B. and Pandita, R. (1994). Both type I and type II interferons down-regulate human tumor necrosis factor receptors in human hepatocellular carcinoma cell line Hep G2. Role of protein kinase C. FEBS Lett. 337, 99–102.

Alegre, M.-L., Vandenabeele, P., Depierreux, M., *et al.* (1991). Cytokine release syndrome induced by the 145-2C11 anti-CD3 monoclonal antibody in mice: Prevention by high doses of methylprednisolone. J. Immunol. 146, 1184–1191.

Alexander, R.B., Isaacs, J.T. and Coffey, D.S. (1987). Tumor necrosis factor enhances the *in vitro* and *in vivo* efficacy of chemotherapeutic drugs targeted at DNA topoisomerase II in the treatment of murine bladder cancer. J. Urol. 138, 427–429.

Alexander, R.B. and Rosenberg, S.A. (1991). Tumor necrosis factor: clinical applications. In "Biologic Therapy of Cancer" (eds. V.T. DeVita, Jr, S. Hellman and S.A. Rosenberg), pp. 378–392. J.B. Lippincott, Philadelphia.

Alimzhanov, M.B., Kuprash, D.V., Kosco-Vilbois, M.H., *et al.* (1997). Abnormal development of secondary lymphoid tissues in lymphotoxin β-deficient mice. Proc. Natl Acad. Sci. USA 94, 9302–9307.

Andrews, J.S., Berger, A.E. and Ware, C.F. (1990). Characterization of the receptor for tumor necrosis factor (TNF) and lymphotoxin (LT) on human T lymphocytes. TNF and LT differ in their receptor binding properties and the induction of MHC class I proteins on a human CD4⁺ T cell hybridoma. J. Immunol. 144, 2582–2591.

Androlewicz, M.J., Browning, J.L. and Ware, C.F. (1992). Lymphotoxin is expressed as a heteromeric complex with a distinct 33-kDa glycoprotein on the surface of an activated human T cell hybridoma. J. Biol. Chem. 267, 2542–2547.

Argiles, J.M., Lopez-Soriano, J., Busquets, S. and Lopez-Soriano, F.J. (1997). Journey from cachexia to obesity by TNF. FASEB J. 11, 743–751.

Asher, A., Mulé, J.J., Reichert, C.M., Shiloni, E. and Rosenberg, S.A. (1987). Studies on the anti-tumor efficacy of systemically administered recombinant tumor necrosis factor against several murine tumors *in vivo*. J. Immunol. 138, 963–974.

Baisch, H., Otto, U. and Kloppel, G. (1990). Antiproliferative and cytotoxic effects of single and combined treatment with tumor necrosis factor α and/or α interferon on a human renal cell carcinoma xenotransplanted into nu/nu mice: cell kinetic studies. Cancer Res. 50, 6389–6395.

Baker, S.J. and Reddy, E.P. (1996). Transducers of life and death: TNF receptor superfamily and associated proteins. Oncogene 12, 1–9.

Balkwill, F.R., Lee, A., Aldam, G., *et al.* (1986). Human tumor xenografts treated with recombinant human tumor necrosis factor alone or in combination with interferons. Cancer Res. 46, 3990–3993.

Balkwill, F.R., Ward, B.G., Moodie, E. and Fiers, W. (1987). Therapeutic potential of tumor necrosis factor-α and γ-interferon in experimental human ovarian cancer. Cancer Res. 47, 4755–4758.

Banner, D.W., D'Arcy, A., Janes, W., *et al.* (1993). Crystal structure of the soluble human 55 kd TNF receptor-human TNFβ complex: implications for TNF receptor activation. Cell 73, 431–445.

Barbara, J.A.J., Smith, W.B., Gamble, J.R., *et al.* (1994). Dissociation of TNF-α cytotoxic and proinflammatory activities by p55 receptor- and p75 receptor-selective TNF-α mutants. EMBO J. 13, 843–850.

Baumann, H. and Gauldie, J. (1994). The acute phase response. Immunol. Today 15, 74–80.

Beutler, B. and Cerami, A. (1988). Tumor necrosis, cachexia, shock, and inflammation: a common mediator. Annu. Rev. Biochem. 57, 505–518.

Beutler, B. and Cerami, A. (1989). The biology of cachectin/TNF—a primary mediator of the host response. Annu. Rev. Immunol. 7, 625–655.

Beutler, B., Milsark, I.W. and Cerami, A.C. (1985). Passive immunization against cachectin/tumor necrosis factor protects mice from lethal effect of endotoxin. Science 229, 869–871.

Beutler, B., Han, J., Kruys, V. and Giroir, B.P. (1992). Coordinate regulation of TNF biosynthesis at the levels of transcription and translation. Patterns of TNF expression *in vivo*. In "Tumor Necrosis Factors. The Molecules and their Emerging Role in Medicine" (ed. B. Beutler), pp. 561–574. Raven Press, New York.

Beyaert, R. and Fiers, W. (1994). Molecular mechanisms of tumor necrosis factor-induced cytotoxicity: what we do understand and what we do not. FEBS Lett. 340, 9–16.

Beyaert, R., Vanhaesebroeck, B., Suffys, P., Van Roy, F. and Fiers, W. (1989). Lithium chloride potentiates tumor necrosis factor-mediated cytotoxicity *in vitro* and *in vivo*. Proc. Natl Acad. Sci. USA 86, 9494–9498.

Beyaert, R., Schulze-Osthoff, K., Van Roy, F. and Fiers, W. (1992). Synergistic induction of interleukin-6 by tumor necrosis factor and lithium chloride in mice: possible role in the triggering and exacerbation of psoriasis by lithium treatment. Eur. J. Immunol. 22, 2181–2184.

Beyaert, R., Heyninck, K., De Valck, D., Boeykens, F., Van Roy, F. and Fiers, W. (1993a). Enhancement of tumor necrosis factor cytotoxicity by lithium chloride is associated with increased inositol phosphate accumulation. J. Immunol. 151, 291–300.

Beyaert, R., Vanhaesebroeck, B., Heyninck, K., *et al.* (1993b). Sensitization of tumor cells to tumor necrosis factor action by the protein kinase inhibitor staurosporine. Cancer Res. 53, 2623–2630.

Beyaert, R., Vanhaesebroeck, B., Declercq, W., *et al.* (1995). Casein kinase-1 phosphorylates the p75 tumor necrosis factor receptor and negatively regulates tumor necrosis factor signaling for apoptosis. J. Biol. Chem. 270, 23293–23299.

Beyaert, R., Cuenda, A., Vanden Berghe, W., *et al.* (1996). The p38/RK mitogen-activated protein kinase pathway regulates interleukin-6 synthesis in response to tumour necrosis factor. EMBO J. 15, 1914–1923.

Beyaert, R., Kidd, V.J., Cornelis, S., *et al.* (1997). Cleavage of PITSLRE kinases by ICE/CASP-1 and CPP32/CASP-3 during apoptosis induced by tumor necrosis factor. J. Biol. Chem. 272, 11694–11697.

Bigda, J., Beletsky, I., Brakebusch, C., *et al.* (1994). Dual role of the p75 tumor necrosis factor (TNF) receptor in TNF cytotoxicity. J. Exp. Med. 180, 445–460.

Black, R.A., Rauch, C.T., Kozlosky, C.J., *et al.* (1997). A metalloproteinase disintegrin that releases tumour-necrosis factor-α from cells. Nature 385, 729–733.

Boldin, M.P., Mett, I.L., Varfolomeev, E.E., *et al.* (1995). Self-association of the "death domains" of the p55 tumor necrosis factor (TNF) receptor and Fas/APO1 prompts signaling for TNF and Fas/APO1 effects. J. Biol. Chem. 270, 387–391.

Boldin, M.P., Goncharov, T.M., Goltsev, Y.V. and Wallach, D. (1996). Involvement of MACH, a novel MORT1/FADD-interacting protease, in Fas/APO-1- and TNF receptor-induced cell death. Cell 85, 803–815.

Bone, R., Fisher, J., Clemmer, T., Slotman, G., Metz, C. and Balk, R. (1987). A controlled clinical trial of high-dose methylprednisolone in the treatment of severe sepsis and septic shock. N. Engl. J. Med. 317, 653–658.

Brakebusch, C., Nophar, Y., Kemper, O., Engelmann, H. and Wallach, D. (1992). Cytoplasmic truncation of the p55 tumour necrosis factor (TNF) receptor abolishes signalling, but not induced shedding of the receptor. EMBO J. 11, 943–950.

Brenner, D.A., O'Hara, M., Angel, P., Chojkier, M. and Karin, M. (1989). Prolonged activation of jun and collagenase genes by tumour necrosis factor-α. Nature 337, 661–663.

Brett, J., Gerlach, H., Nawroth, P., Steinberg, S., Godman, G. and Stern, D. (1989). Tumor necrosis factor/cachectin increases permeability of endothelial cell monolayers by a mechanism involving regulatory G proteins. J. Exp. Med. 169, 1977–1991.

Brouckaert, P.G.G., Leroux-Roels, G.G., Guisez, Y., Tavernier, J. and Fiers, W. (1986). *In vivo* anti-tumour activity of recombinant human and murine TNF, alone and in combination with murine IFN-gamma, on a syngeneic murine melanoma. Int. J. Cancer 38, 763–769.

Brouckaert, P., Everaerdt, B. and Fiers, W. (1992). The glucocorticoid antagonist RU38486 mimics interleukin-1 in its sensitization to the lethal and interleukin-6-inducing properties of tumor necrosis factor. Eur. J. Immunol. 22, 981–986.

Broudy, V.C., Harlan, J.M. and Adamson, J.W. (1987). Disparate effects of tumor necrosis factor-α/cachectin and tumor necrosis factor-β/lymphotoxin on hematopoietic growth factor production and neutrophil adhesion molecule expression by cultured human endothelial cells. J. Immunol. 138, 4298–4302.

Browning, J. and Ribolini, A. (1989). Studies on the differing effects of tumor necrosis factor and lymphotoxin on the growth of several human tumor lines. J. Immunol. 143, 1859–1867.

Browning, J.L., Androlewicz, M.J. and Ware, C.F. (1991). Lymphotoxin and an associated 33-kDa glycoprotein are expressed on the surface of an activated human T cell hybridoma. J. Immunol. 147, 1230–1237.

Browning, J.L., Ngam-ek, A., Lawton, P., *et al.* (1993). Lymphotoxin β, a novel member of the TNF family that forms a heteromeric complex with lymphotoxin on the cell surface. Cell 72, 847–856.

Browning, J.L., Miatkowski, K., Sizing, I., *et al.* (1996). Signaling through the lymphotoxin-β receptor induces the death of some adenocarcinoma tumor lines. J. Exp. Med. 183, 867–878.

Butcher, G.A. and Clark, I.A. (1990). The inhibition of *Plasmodium falciparum* growth *in vitro* by sera from mice infected with malaria or treated with TNF. Parasitology 101, 321–326.

Campbell, I.K., Piccoli, D.S., Roberts, M.J., Muirden, K.D. and Hamilton, J.A. (1990). Effects of tumor necrosis factor α and β on resorption of human articular cartilage and production of plasminogen activator by human articular chondrocytes. Arthritis Rheum. 33, 542–552.

Camussi, G., Albano, E., Tetta, C. and Bussolino, F. (1991). The molecular action of tumor necrosis factor-α. Eur. J. Biochem. 202, 3–14.

Caput, D., Beutler, B., Hartog, K., Thayer, R., Brown-Shimer, S. and Cerami, A. (1986). Identification of a common nucleotide sequence in the 3′-untranslated region of mRNA molecules specifying inflammatory mediators. Proc. Natl Acad. Sci. USA 83, 1670–1674.

Carswell, E.A., Old, L.J., Kassel, R.L., Green, S., Fiore, N. and Williamson, B. (1975). An endotoxin-induced serum factor that causes necrosis of tumors. Proc. Natl Acad. Sci. USA 72, 3666–3670.

Cauwels, A., Brouckaert, P., Grooten, J., Huang, S., Aguet, M. and Fiers, W. (1995). Involvement of IFN-γ in bacillus Calmette–Guérin-induced but not in tumor-induced sensitization to TNF-induced lethality. J. Immunol. 154, 2753–2763.

Chinnaiyan, A.M., O'Rourke, K., Tewari, M. and Dixit, V.M. (1995). FADD, a novel death domain-containing protein, interacts with the death domain of Fas and initiates apoptosis. Cell 81, 505–512.

Chinnaiyan, A.M., Tepper, C.G., Seldin, M.F., et al. (1996). FADD/MORT1 is a common mediator of CD95 (Fas/APO-1) and tumor necrosis factor receptor-induced apoptosis. J. Biol. Chem. 271, 4961–4965.

Chouaib, S., Bertoglio, J., Blay, J.Y., Marchiol-Fournigault, C. and Fradelizi, D. (1988). Generation of lymphokine-activated killer cells: synergy between tumor necrosis factor and interleukin 2. Proc. Natl Acad. Sci. USA 85, 6875–6879.

Coley, W.B. (1894). Treatment of inoperable malignant tumors with the toxi of erysipelas and the Bacillus prodigiosus. Trans. Am. Surg. Assoc. 12, 183–212.

Collins, T., Korman, A.J., Wake, C.T., et al. (1984). Immune interferon activates multiple class-II major histocompatibility complex genes and the associated invariant chain gene in human endothelial cells and dermal fibroblasts. Proc. Natl Acad. Sci. USA 81, 4917–4921.

Crowe, P.D., VanArsdale, T.L., Goodwin, R.G. and Ware, C.F. (1993). Specific induction of 80-kDa tumor necrosis factor receptor shedding in T lymphocytes involves the cytoplasmic domain and phosphorylation. J. Immunol. 151, 6882–6890.

Crowe, P.D., VanArsdale, T.L., Walter, B.N., et al. (1994). A lymphotoxin-β-specific receptor. Science 264, 707–710.

Cseh, K. and Beutler, B. (1989). Alternative cleavage of the cachectin/tumor necrosis factor propeptide results in a larger, inactive form of secreted protein. J. Biol. Chem. 264, 16256–16260.

Darnay, B.G., Reddy, S.A.G. and Aggarwal, B.B. (1994a). Identification of a protein kinase associated with the cytoplasmic domain of the p60 tumor necrosis factor receptor. J. Biol. Chem. 269, 20299–20304.

Darnay, B.G., Reddy, S.A.G. and Aggarwal, B.B. (1994b). Physical and functional association of a serine-threonine protein kinase to the cytoplasmic domain of the p80 form of the human tumor necrosis factor receptor in human histiocytic lymphoma U-937 cells. J. Biol. Chem. 269, 19687–19690.

Darzynkiewicz, Z., Williamson, B., Carswell, E.A. and Old, L.J. (1984). Cell cycle-specific effects of tumor necrosis factor. Cancer Res. 44, 83–90.

Dealtry, G.B., Naylor, M.S., Fiers, W. and Balkwill, F.R. (1987). DNA fragmentation and cytotoxicity caused by tumor necrosis factor is enhanced by interferon-γ. Eur. J. Immunol. 17, 689–693.

Decker, T., Lohmann-Matthes, M.L. and Gifford, G.E. (1987). Cell-associated tumor necrosis factor (TNF) as a killing mechanism of activated cytotoxic macrophages. J. Immunol. 138, 957–962.

Decoster, E., Vanhaesebroeck, B., Vandenabeele, P., Grooten, J. and Fiers, W. (1995). Generation and biological characterization of membrane-bound, uncleavable murine tumor necrosis factor. J. Biol. Chem. 270, 18473–18478.

Dembic, Z., Loetscher, H., Gubler, U., et al. (1990). Two human TNF receptors have similar extracellular, but distinct intracellular, domain sequences. Cytokine 2, 231–237.

De Togni, P., Goellner, J., Ruddle, N.H., et al. (1994). Abnormal development of peripheral lymphoid organs in mice deficient in lymphotoxin. Science 264, 703–707.

De Valck, D., Beyaert, R., Van Roy, F. and Fiers, W. (1993). Tumor necrosis factor cytotoxicity is associated with phospholipase D activation. Eur. J. Biochem. 212, 491–497.

Eck, M.J. and Sprang, S.R. (1989). The structure of tumor necrosis factor-α at 2.6 Å resolution. Implications for receptor binding. J. Biol. Chem. 264, 17595–17605.

Eck, M.J., Ultsch, M., Rinderknecht, E., de Vos, A.M. and Sprang, S.R. (1992). The structure of human lymphotoxin (tumor necrosis factor-β) at 1.9-Å resolution. J. Biol. Chem. 267, 2119–2122.

Eggermont, A.M., Schraffordt Koops, H., Klausner, J.M., et al. (1997). Isolation limb perfusion with tumor necrosis factor α and chemotherapy for advanced extremity soft tissue sarcomas. Sem. Oncol. 24, 547–555.

Ehrke, M.J., Ho, R.L.X. and Hori, K. (1988). Species-specific TNF induction of thymocyte proliferation. Cancer Immunol. Immunother. 27, 103–108.

Engelmann, H., Holtmann, H., Brakebusch, C., et al. (1990). Antibodies to a soluble form of a tumor necrosis factor (TNF) receptor have TNF-like activity. J. Biol. Chem. 265, 14497–14504.

Espevik, T., Brockhaus, M., Loetscher, H., Nonstad, U. and Shalaby, R. (1990). Characterization of binding and biological effects of monoclonal antibodies against a human tumor necrosis factor receptor. J. Exp. Med. 171, 415–426.

Everaerdt, B., Brouckaert, P., Shaw, A. and Fiers, W. (1989). Four different interleukin-1 species sensitize to the lethal action of tumour necrosis factor. Biochem. Biophys. Res. Commun. 163, 378–385.

Everaerdt, B., Brouckaert, P. and Fiers, W. (1994). Recombinant IL-1 receptor antagonist protects against TNF-induced lethality in mice. J. Immunol. 152, 5041–5049.

Fernandez-Botran, R. (1991). Soluble cytokine receptors: their role in immunoregulation. FASEB J. 5, 2567–2574.

Fiers, W. (1992). Precursor structures and structure-function analysis of TNF and lymphotoxin. In "Tumor Necrosis Factors. Structure, Function, and Mechanism of Action" (eds. B.B. Aggarwal and J. Vilček), pp. 79–92. Marcel Dekker, New York.

Fiers, W. (1993). Tumour necrosis factor. In "The Natural Immune System: Humoral Factors" (ed. E. Sim), pp. 65–119. IRL Press, Oxford.

Fiers, W., Beyaert, R., Boone, E., et al. (1996). TNF-induced intracellular signaling leading to gene induction or to cytotoxicity by necrosis or by apoptosis. J. Inflammation 47, 67–75.

Fiorentino, D.F., Zlotnik, A., Mosmann, T.R., Howard, M. and O'Garra, A. (1991). IL-10 inhibits cytokine production by activated macrophages. J. Immunol. 147, 3815–3822.

Flynn, R.M. and Palladino, M.A. (1992). TNF and TGF-β: the opposite sides of the avenue? In "Tumor Necrosis Factors. The Molecules and their Emerging Role in Medicine" (ed. B. Beutler), pp. 131–144. Raven Press, New York.

Fransen, L., Müller, R., Marmenout, A., et al. (1985). Molecular cloning of mouse tumour necrosis factor cDNA and its eukaryotic expression. Nucleic Acids Res. 13, 4417–4429.

Fransen, L., Ruysschaert, M.R., Van der Heyden, J. and Fiers, W. (1986a). Recombinant tumor necrosis factor: species specificity for a variety of human and murine transformed cell lines. Cell. Immunol. 100, 260–267.

Fransen, L., Van der Heyden, J., Ruysschaert, R. and Fiers, W. (1986b). Recombinant tumor necrosis factor: its effect and its synergism with interferon-γ on a variety of normal and transformed human cell lines. Eur. J. Cancer Clin. Oncol. 22, 419–426.

Fujimoto, S., Kobayashi, K., Takahashi, M., et al. (1992). Effects on tumour microcirculation in mice of misonidazole and tumour necrosis factor plus hyperthermia. Br. J. Cancer 65, 33–36.

Gamble, J.R., Harlan, J.M., Klebanoff, S.J. and Vadas, M.A. (1985). Stimulation of the adherence of neutrophils to umbilical vein endothelium by human recombinant tumor necrosis factor. Proc. Natl Acad. Sci. USA 82, 8667–8671.

Gamble, J.R., Smith, W.B. and Vadas, M.A. (1992). TNF modulation of endothelial and neutrophil adhesion. In "Tumor Necrosis Factors. The Molecules and Their Emerging Role in Medicine" (ed. B. Beutler), pp. 65–86. Raven Press, New York.

García-Martinez, C., Costelli, P., Lopez-Soriano, F.J. and Argiles, J.M. (1997). Is TNF really involved in cachexia? Cancer Invest. 15, 47–54.

Gearing, A.J.H., Beckett, P., Christodoulou, M., et al. (1994). Processing of tumour necrosis factor-α precursor by metalloproteinases. Nature 370, 555–557.

Geng, Y., Hansson, G.K. and Holme, E. (1992). Interferon-γ and tumor necrosis factor synergize to induce nitric oxide production and inhibit mitochondrial respiration in vascular smooth muscle cells. Circ. Res. 71, 1268–1276.

Gérard, C., Bruyns, C., Marchant, A., et al. (1993). Interleukin 10 reduces the release of tumor necrosis factor and prevents lethality in experimental endotoxemia. J. Exp. Med. 177, 547–550.

Goldfeld, A.E., Flemington, E.K., Boussiotis, V.A., et al. (1992). Transcription of the tumor necrosis factor α gene is rapidly induced by anti-immunoglobulin and blocked by cyclosporin A and FK506 in human B cells. Proc. Natl Acad. Sci. USA 89, 12198–12201.

Goossens, V., Grooten, J., De Vos, K. and Fiers, W. (1995). Direct evidence for tumor necrosis factor-induced mitochondrial reactive oxygen intermediates and their involvement in cytotoxicity. Proc. Natl Acad. Sci. USA 92, 8115–8119.

Grau, G.E. (1990). Implications of cytokines in immunopathology: experimental and clinical data. Eur. Cytokine Netw. 1, 203–210.

Grau, G.E., Piguet, P.F., Vassalli, P. and Lambert, P.H. (1989). Tumor-necrosis factor and other cytokines in cerebral malaria. Experimental and clinical data. Immunol. Rev. 112, 49–70.

Gray, P.W., Aggarwal, B.B., Benton, C.V., et al. (1984). Cloning and expression of cDNA for human lymphotoxin, a lymphokine with tumour necrosis activity. Nature 312, 721–724.

Gray, P.W., Barrett, K., Chantry, D., Turner, M. and Feldmann, M. (1990). Cloning of human tumor necrosis factor (TNF) receptor cDNA and expression of recombinant soluble TNF-binding protein. Proc. Natl Acad. Sci. USA 87, 7380–7384.

Grazioli, L., Casero, D., Restivo, A., Cozzi, E. and Marcucci, F. (1994). Tumor necrosis factor-driven formation of disulfide-linked receptor aggregates. J. Biol. Chem. 269, 22304–22309.

Grell, M., Scheurich, P., Meager, A. and Pfizenmaier, K. (1993). TR60 and TR80 tumor necrosis factor (TNF)-receptors can independently mediate cytolysis. Lymphokine Cytokine Res. 12, 143–148.

Grell, M., Douni, E., Wajant, H., et al. (1995). The transmembrane form of tumor necrosis factor is the prime activating ligand of the 80 kDa tumor necrosis factor receptor. Cell 83, 793–802.

Gresser, I., Belardelli, F., Tavernier, J., et al. (1986). Anti-tumor effects of interferon in mice injected with interferon-sensitive and interferon-resistant Friend leukemia cells. V.

Comparisons with the action of tumor necrosis factor. Int. J. Cancer 38, 771–778.

Grooten, J., Goossens, V., Vanhaesebroeck, B. and Fiers, W. (1993). Cell membrane permeabilization and cellular collapse, followed by loss of dehydrogenase activity: early events in tumour necrosis factor-induced cytotoxicity. Cytokine 5, 546–555.

Han, J., Brown, T. and Beutler, B. (1990). Endotoxin-responsive sequences control cachectin/tumor necrosis factor biosynthesis at the translational level. J. Exp. Med. 171, 465–475.

Haranaka, K., Satomi, N. and Sakurai, A. (1984). Antitumor activity of murine tumor necrosis factor (TNF) against transplanted murine tumors and heterotransplanted human tumors in nude mice. Int. J. Cancer 34, 263–267.

Haranaka, K., Sakurai, A. and Satomi, N. (1987). Antitumor activity of recombinant human tumor necrosis factor in combination with hyperthermia, chemotherapy, or immunotherapy. J. Biol. Resp. Modif. 6, 379–391.

Hart, P.H., Vitti, G.F., Burgess, D.R., Whitty, G.A., Piccoli, D.S. and Hamilton, J.A. (1989). Potential antiinflammatory effects of interleukin 4: suppression of human monocyte tumor necrosis factor α, interleukin 1, and prostaglandin E$_2$. Proc. Natl Acad. Sci. USA 86, 3803–3807.

Havell, E.A. (1989). Evidence that tumor necrosis factor has an important role in antibacterial resistance. J. Immunol. 143, 2894–2899.

Heller, R.A. and Krönke, M. (1994). Tumor necrosis factor receptor-mediated signaling pathways. J. Cell. Biol. 126, 5–9.

Heller, R.A., Song, K., Fan, N. and Chang, D.J. (1992). The p70 tumor necrosis factor receptor mediates cytotoxicity. Cell 70, 47–56.

Hemmi, H., Nakamura, T., Tamura, K., et al. (1987). Lymphotoxin: induction of terminal differentiation of the human myeloid leukemia cell lines HL-60 and THP-1. J. Immunol. 138, 664–666.

Henkel, T., Machleidt, T., Alkalay, I., Krönke, M., Ben-Neriah, Y. and Baeuerle, P.A. (1993). Rapid proteolysis of IκB-α is necessary for activation of transcription factor NF-κB. Nature 365, 182–185.

Higuchi, M., Higashi, N., Taki, H. and Osawa, T. (1990). Cytolytic mechanisms of activated macrophages. Tumor necrosis factor and L-arginine-dependent mechanisms act synergistically as the major cytolytic mechanisms of activated macrophages. J. Immunol. 144, 1425–1431.

Himmler, A., Maurer-Fogy, I., Krönke, M., et al. (1990). Molecular cloning and expression of human and rat tumor necrosis factor receptor chain (p60) and its soluble derivative, tumor necrosis factor-binding protein. DNA Cell. Biol. 9, 705–715.

Hoffman, M. and Weinberg, J.B. (1987). Tumor necrosis factor-α induces increased hydrogen peroxide production and Fc receptor expression, but not increased Ia antigen expression by peritoneal macrophages. J. Leukocyte Biol. 42, 704–707.

Hohmann, H.-P., Kolbeck, R., Remy, R. and van Loon, A.P.G.M. (1991). Cyclic AMP-independent activation of transcription factor NF-κB in HL60 cells by tumor necrosis factors α and β. Mol. Cell. Biol. 11, 2315–2318.

Höller, E., Kolb, H.J., Möller, A., et al. (1990). Increased serum levels of tumor necrosis factor α precede major complications of bone marrow transplantation. Blood 75, 1011–1016.

Hsu, H., Xiong, J. and Goeddel, D.V. (1995). The TNF receptor 1-associated protein TRADD signals cell death and NF-κB activation. Cell 81, 495–504.

Hsu, H., Huang, J., Shu, H.-B., Baichwal, V. and Goeddel, D.V. (1996a). TNF-dependent recruitment of the protein kinase RIP to the TNF receptor-1 signaling complex. Immunity 4, 387–396.

Hsu, H., Shu, H.-B., Pan, M.-G. and Goeddel, D.V. (1996b). TRADD-TRAF2 and TRADD-FADD interactions define two distinct TNF receptor 1 signal transduction pathways. Cell 84, 299–308.

Ito, M., Baba, M., Sato, A., et al. (1989). Tumor necrosis factor enhances replication of human immunodeficiency virus (HIV) in vitro. Biochem. Biophys. Res. Commun. 158, 307–312.

Jäättelä, M., Wissing, D., Bauer, P.A. and Li, G.C. (1992). Major heat shock protein hsp70 protects tumor cells from tumor necrosis factor cytotoxicity. EMBO J. 11, 3507–3512.

Jacobsen, F.W., Rothe, M., Rusten, L., et al. (1994). Role of the 75-kDa tumor necrosis factor receptor: inhibition of early hematopoiesis. Proc. Natl Acad. Sci. USA 91, 10695–10699.

Jones, A.L. and Selby, P. (1989). Tumour necrosis factor: clinical relevance. Cancer Surveys 8, 817–836.

Jones, E.Y., Stuart, D.I. and Walker, N.P.C. (1989). Structure of tumour necrosis factor. Nature 338, 225–228.

Jongeneel, C.V. (1992). The TNF and lymphotoxin promoters. In "Tumor Necrosis Factors. The Molecules and Their Emerging Role in Medicine" (ed. B. Beutler), pp. 539–559. Raven Press, New York.

Kahaleh, M.B., Smith, E.A., Soma, Y. and LeRoy, E.C. (1988). Effect of lymphotoxin and tumor necrosis factor on endothelial and connective tissue cell growth and function. Clin. Immunol. Immunopathol. 49, 261–272.

Kaur, P., Welch, W.J. and Saklatvala, J. (1989). Interleukin 1 and tumour necrosis factor increase phosphorylation of the small heat shock protein. Effects in fibroblasts, Hep G2 and U937 cells. FEBS Lett. 258, 269–273.

Keffer, J., Probert, L., Cazlaris, H., et al. (1991). Transgenic mice expressing human tumour necrosis factor: a predictive genetic model of arthritis. EMBO J. 10, 4025–4031.

Kehrl, J.H., Miller, A. and Fauci, A.S. (1987). Effect of tumor necrosis factor α on mitogen-activated human B cells. J. Exp. Med. 166, 786–791.

Kettelhut, I.C., Fiers, W. and Goldberg, A.L. (1987). The toxic effects of tumor necrosis factor in vivo and their prevention by cyclooxygenase inhibitors. Proc. Natl Acad. Sci. USA 84, 4273–4277.

Kilbourn, R.G. and Belloni, P. (1990). Endothelial cell production of nitrogen oxides in response to interferon γ in combination with tumor necrosis factor, interleukin-1, or endotoxin. J. Natl Cancer Inst. 82, 772–776.

Kirstein, M. and Baglioni, C. (1988). Tumor necrosis factor stimulates proliferation of human osteosarcoma cells and accumulation of c-myc messenger RNA. J. Cell. Physiol. 134, 479–484.

Kirstein, M., Fiers, W. and Baglioni, C. (1986). Growth inhibition and cytotoxicity of tumor necrosis factor in L929 cells is enhanced by high cell density and inhibition of mRNA synthesis. J. Immunol. 137, 2277–2280.

Klebanoff, S.J., Vadas, M.A., Harlan, J.M., et al. (1986). Stimulation of neutrophils by tumor necrosis factor. J. Immunol. 136, 4220–4225.

Koff, W.C. and Fann, A.V. (1986). Human tumor necrosis factor-alpha kills herpesvirus-infected but not normal cells. Lymphokine Res. 5, 215–221.

Koni, P.A., Sacca, R., Lawton, P., et al. (1997). Distinct roles in lymphoid organogenesis for lymphotoxins α and β revealed in lymphotoxin β-deficient mice. Immunity 6, 491–500.

Kriegler, M., Perez, C., DeFay, K., Albert, I. and Lu, S.D. (1988). A novel form of TNF/cachectin is a cell surface cytotoxic transmembrane protein: ramifications for the complex physiology of TNF. Cell 53, 45–53.

Krönke, M., Schlüter, C. and Pfizenmaier, K. (1987). Tumor necrosis factor inhibits MYC expression in HL-60 cells at the level of mRNA transcription. Proc. Natl Acad. Sci. USA 84, 469–473.

Krosnick, J.A., Mulé, J.J., McIntosh, J.K. and Rosenberg, S.A. (1989). Augmentation of antitumor efficacy by the combination of recombinant tumor necrosis factor and chemotherapeutic agents in vivo. Cancer Res. 49, 3729–3733.

Kruys, V., Kemmer, K., Shakhov, A., Jongeneel, V. and Beutler, B. (1992). Constitutive activity of the tumor necrosis factor promoter is canceled by the 3′ untranslated region in nonmacrophage cell lines; a trans-dominant factor overcomes this suppressive effect. Proc. Natl Acad. Sci. USA 89, 673–677.

Kunkel, S.L., Strieter, R.M., Chensue, S.W., Campbell, D.A. and Remick, D.G. (1991). The role of TNF in diverse pathologic processes. Biotherapy 3, 135–141.

Kurt-Jones, E.A., Fiers, W. and Pober, J.S. (1987). Membrane interleukin 1 induction on human endothelial cells and dermal fibroblasts. J. Immunol. 139, 2317–2324.

Laherty, C.D., Perkins, N.D. and Dixit, V.M. (1993). Human T cell leukemia virus type I Tax and phorbol 12-myristate 13-acetate induce expression of the A20 zinc finger protein by distinct mechanisms involving nuclear factor κB. J. Biol. Chem. 268, 5032–5039.

Larrick, J.W. and Wright, S.C. (1990). Cytotoxic mechanism of tumor necrosis factor-α. FASEB J. 4, 3215–3223.

Last-Barney, K., Homon, C.A., Faanes, R.B. and Merluzzi, V.J. (1988). Synergistic and overlapping activities of tumor necrosis factor-α and IL-1. J. Immunol. 141, 527–530.

Laster, S.M., Wood, J.G. and Gooding, L.R. (1988). Tumor necrosis factor can induce both apoptic and necrotic forms of cell lysis. J. Immunol. 141, 2629–2634.

Lee, J.C., Laydon, J.T., McDonnell, P.C., et al. (1994). A protein kinase involved in the regulation of inflammatory cytokine biosynthesis. Nature 372, 739–746.

Leibovich, S.J., Polverini, P.J., Shepard, H.M., Wiseman, D.M., Shively, V. and Nuseir, N. (1987). Macrophage-induced angiogenesis is mediated by tumour necrosis factor-α. Nature 329, 630–632.

Lentz, S.R., Tsiang, M. and Sadler, J.E. (1991). Regulation of thrombomodulin by tumor necrosis factor-α: comparison of transcriptional and posttranscriptional mechanisms. Blood 77, 542–550.

Lewis, M., Tartaglia, L.A., Lee, A., et al. (1991). Cloning and expression of cDNAs for two distinct murine tumor necrosis factor receptors demonstrate one receptor is species specific. Proc. Natl Acad. Sci. USA 88, 2830–2834.

Lewit-Bentley, A., Fourme, R., Kahn, R., et al. (1988). Structure of tumour necrosis factor by X-ray solution scattering and preliminary studies by single crystal X-ray diffraction. J. Mol. Biol. 199, 389–392.

Li, C.B., Gray, P.W., Lin, P.F., McGrath, K.M., Ruddle, F.H. and Ruddle, N.H. (1987). Cloning and expression of murine lymphotoxin cDNA. J. Immunol. 138, 4496–4501.

Libert, C., Brouckaert, P., Shaw, A. and Fiers, W. (1990). Induction of interleukin 6 by human and murine recombinant interleukin 1 in mice. Eur. J. Immunol. 20, 691–694.

Libert, C., Van Bladel, S., Brouckaert, P. and Fiers, W. (1991a). The influence of modulating substances on tumor necrosis factor and interleukin-6 levels after injection of murine tumor necrosis factor or lipopolysaccharide in mice. J. Immunother. 10, 227–235.

Libert, C., Van Bladel, S., Brouckaert, P., Shaw, A. and Fiers, W. (1991b). Involvement of the liver, but not of IL-6, in IL-1-induced desensitization to the lethal effects of tumor necrosis factor. J. Immunol. 146, 2625–2632.

Libert, C., Vink, A., Coulie, P., et al. (1992). Limited involvement of interleukin-6 in the pathogenesis of lethal septic shock as revealed by the effect of monoclonal antibodies against interleukin-6 or its receptor in various murine models. Eur. J. Immunol. 22, 2625–2630.

Libert, C., Takahashi, N., Cauwels, A., Brouckaert, P., Bluethmann, H. and Fiers, W. (1994). Response of interleukin-6-deficient mice to tumor necrosis factor-induced metabolic changes and lethality. Eur. J. Immunol. 24, 2237–2242.

Liew, F.Y., Li, Y. and Millott, S. (1990). Tumor necrosis factor-α synergizes with IFN-γ in mediating killing of Leishmania major through the induction of nitric oxide. J. Immunol. 145, 4306–4310.

Locksley, R.M., Heinzel, F.P., Shepard, H.M., et al. (1987). Tumor necrosis factors α and β differ in their capacities to generate interleukin 1 release from human endothelial cells. J. Immunol. 139, 1891–1895.

Loetscher, H., Pan, Y.E., Lahm, H.W., et al. (1990). Molecular cloning and expression of the human 55 kd tumor necrosis factor receptor. Cell 61, 351–359.

Loetscher, H., Stueber, D., Banner, D., Mackay, F. and Lesslauer, W. (1993). Human tumor necrosis factor α (TNFα) mutants with exclusive specificity for the 55-kDa or 75-kDa TNF receptors. J. Biol. Chem. 268, 26350–26357.

Malik, S.T.A., Griffin, D.B., Fiers, W. and Balkwill, F.R. (1989). Paradoxical effects of tumour necrosis factor in experimental ovarian cancer. Int. J. Cancer 44, 918–925.

Malik, S.T.A., Naylor, M.S., East, N., Oliff, A. and Balkwill, F.R. (1990). Cells secreting tumour necrosis factor show enhanced metastasis in nude mice. Eur. J. Cancer 26, 1031–1034.

Malik, S.T.A. and Balkwill, F.R. (1992). Antiproliferative and antitumor activity of TNF in vitro and in vivo. In "Tumor Necrosis Factors: Structure, Function, and Mechanism of Action" (eds. B.B. Aggarwal and J. Vilček), pp. 239–268. Marcel Dekker, New York.

Manda, T., Shimomura, K., Mukumoto, S., et al. (1987). Recombinant human tumor necrosis factor-α: evidence of an indirect mode of antitumor activity. Cancer Res. 47, 3707–3711.

Manetta, A., Podczaski, E., Zaino, R.J. and Satyaswaroop, P.G. (1989). Therapeutic effect of recombinant human tumor necrosis factor in ovarian carcinoma xenograft in nude mice. Gynecol. Oncol. 34, 360–364.

Maniatis, T. (1997). Catalysis by a multiprotein IκB kinase complex. Science 278, 818–819.

Marmenout, A., Fransen, L., Tavernier, J., et al. (1985). Molecular cloning and expression of human tumor necrosis factor and comparison with mouse tumor necrosis factor. Eur. J. Biochem. 152, 515–522.

Marquet, R.L., IJzermans, J.N.M., de Bruin, R.W.F., Fiers, W. and Jeekel, J. (1987). Anti-tumor activity of recombinant mouse tumor necrosis factor (TNF) on colon cancer in rats is promoted by recombinant rat interferon gamma; toxicity is reduced by indomethacin. Int. J. Cancer 40, 550–553.

Matsuyama, T., Hamamoto, Y., Soma, G., Mizuno, D., Yamamoto, N. and Kobayashi, N. (1989). Cytocidal effect of tumor necrosis factor on cells chronically infected with human immunodeficiency virus (HIV): enhancement of HIV replication. J. Virol. 63, 2504–2509.

Maury, C.P. and Teppo, A.M. (1987). Raised serum levels of cachectin/tumor necrosis factor α in renal allograft rejection. J. Exp. Med. 166, 1132–1137.

McGeehan, G.M., Becherer, J.D., Bast, R.C., Jr, et al. (1994). Regulation of tumour necrosis factor-α processing by a metalloproteinase inhibitor. Nature 370, 558–561.

McIntosh, J.K., Mulé, J.J., Merino, M.J. and Rosenberg, S.A. (1988). Synergistic antitumor effects of immunotherapy with recombinant interleukin-2 and recombinant tumor necrosis factor-α. Cancer Res. 48, 4011–4017.

McIntosh, J.K., Mulé, J.J., Travis, W.D. and Rosenberg, S.A. (1990). Studies of effects of recombinant human tumor necrosis factor on autochthonous tumor and transplanted normal tissue in mice. Cancer Res. 50, 2463–2469.

Michie, H.R., Manogue, K.R., Spriggs, D.R., et al. (1988). Detection of circulating tumor necrosis factor after endotoxin administration. N. Engl. J. Med. 318, 1481–1486.

Milenkovic, L., Rettori, V., Snyder, G.D., Beutler, B. and McCann, S.M. (1989). Cachectin alters anterior pituitary hormone release by a direct action in vitro. Proc. Natl Acad. Sci. USA 86, 2418–2422.

Mintz, M., Rapaport, R., Oleske, J.M., et al. (1989). Elevated serum levels of tumor necrosis factor are associated with progressive encephalopathy in children with acquired immunodeficiency syndrome. Am. J. Dis. Child. 143, 771–774.

Mohler, K.M., Sleath, P.R., Fitzner, J.N., et al. (1994). Protection against a lethal dose of endotoxin by an inhibitor of tumour necrosis factor processing. Nature 370, 218–220.

Moss, M.L., Jin, S.-L.C., Milla, M.E., et al. (1997). Cloning of a disintegrin metalloproteinase that processes precursor tumour-necrosis factor-α. Nature 385, 733–736.

Mosselmans, R., Hepburn, A., Dumont, J.E., Fiers, W. and Galand, P. (1988). Endocytic pathway of recombinant murine tumor necrosis factor in L-929 cells. J. Immunol. 141, 3096–3100.

Mulé, J.J., McIntosh, J.K., Jablons, D.M. and Rosenberg, S.A. (1990). Antitumor activity of recombinant interleukin 6 in mice. J. Exp. Med. 171, 629–636.

Muzio, M., Chinnaiyan, A.M., Kischkel, F.C., et al. (1996). FLICE, a novel FADD homologous ICE/CED-3-like protease, is recruited to the CD95 (Fas/Apo-1) death-inducing signaling complex. Cell 85, 817–827.

Naumann, M. and Scheidereit, C. (1994). Activation of NF-κB in vivo is regulated by multiple phosphorylations. EMBO J. 13, 4597–4607.

Naume, B., Shalaby, R., Lesslauer, W. and Espevik, T. (1991). Involvement of the 55- and 75-kDa tumor necrosis factor receptors in the generation of lymphokine-activated killer cell activity and proliferation of natural killer cells. J. Immunol. 146, 3045–3048.

Nauts, H.C. (1989). Bacteria and cancer—antagonisms and benefits. Cancer Surveys 8, 713–723.

Nawroth, P., Handley, D., Matsueda, G., *et al.* (1988). Tumor necrosis factor/cachectin-induced intravascular fibrin formation in meth A fibrosarcomas. J. Exp. Med. 168, 637–647.

Nedospasov, S.A., Shakhov, A.N., Turetskaya, R.L., *et al.* (1986). Tandem arrangement of genes coding for tumor necrosis factor (TNF-α) and lymphotoxin (TNF-β) in the human genome. Cold Spring Harbor Symp. Quant. Biol. 51, 611–624.

Negrier, M.S., Pourreau, C.N., Palmer, P.A., *et al.* (1992). Phase I trial of recombinant interleukin-2 followed by recombinant tumor necrosis factor in patients with metastatic cancer. J. Immunother. 11, 93–102.

Nicholson, D.W. and Thornberry, N.A. (1997). Caspases: Killer proteases. Trends Biochem. Sci. 22, 299–306.

Nickoloff, B.J., Karabin, G.D., Barker, J.N., *et al.* (1991). Cellular localization of interleukin-8 and its inducer, tumor necrosis factor-α, in psoriasis. Am. J. Pathol. 138, 129–140.

Nishimura, T., Ohta, S., Sato, N., Togashi, Y., Goto, M. and Hashimoto, Y. (1987). Combination tumor-immunotherapy with recombinant tumor necrosis factor and recombinant interleukin 2 in mice. Int. J. Cancer 40, 255–261.

Nophar, Y., Kemper, O., Brakebusch, C., *et al.* (1990). Soluble forms of tumor necrosis factor receptors (TNF-Rs). The cDNA for the type I TNF-R, cloned using amino acid sequence data of its soluble form, encodes both the cell surface and a soluble form of the receptor. EMBO J. 9, 3269–3278.

Nosoh, Y., Toge, T., Nishiyama, M., *et al.* (1987). Antitumor effects of recombinant human tumor necrosis factor against human tumor xenografts transplanted into nude mice. Jpn J. Surg. 17, 51–54.

Okusawa, S., Gelfand, J.A., Ikejima, T., Connolly, R.J. and Dinarello, C.A. (1988). Interleukin 1 induces a shock-like state in rabbits. Synergism with tumor necrosis factor and the effect of cyclooxygenase inhibition. J. Clin. Invest. 81, 1162–1172.

Opipari, A.W., Jr, Hu, H.M., Yabkowitz, R. and Dixit, V.M. (1992). The A20 zinc finger protein protects cells from tumor necrosis factor cytotoxicity. J. Biol. Chem. 267, 12424–12427.

Orosz, P., Echtenacher, B., Falk, W., Ruschoff, J., Weber, D. and Männel, D.N. (1993). Enhancement of experimental metastasis by tumor necrosis factor. J. Exp. Med. 177, 1391–1398.

Osborn, L., Kunkel, S. and Nabel, G.J. (1989). Tumor necrosis factor α and interleukin 1 stimulate the human immunodeficiency virus enhancer by activation of the nuclear factor κB. Proc. Natl Acad. Sci. USA 86, 2336–2340.

Østensen, M.E., Thiele, D.L. and Lipsky, P.E. (1987). Tumor necrosis factor-α enhances cytolytic activity of human natural killer cells. J. Immunol. 138, 4185–4191.

Owen-Schaub, L.B., Gutterman, J.U. and Grimm, E.A. (1988). Synergy of tumor necrosis factor and interleukin 2 in the activation of human cytotoxic lymphocytes: effect of tumor necrosis factor α and interleukin 2 in the generation of human lymphokine-activated killer cell cytotoxicity. Cancer Res. 48, 788–792.

Palladino, M.A., Jr, Shalaby, M.R., Kramer, S.M., *et al.* (1987). Characterization of the antitumor activities of human tumor

necrosis factor-α and the comparison with other cytokines: induction of tumor-specific immunity. J. Immunol. 138, 4023–4032.

Palombella, V.J. and Vilček, J. (1989). Mitogenic and cytotoxic actions of tumor necrosis factor in BALB/c 3T3 cells. Role of phospholipase activation. J. Biol. Chem. 264, 18128–18136.

Paul, N.L. and Ruddle, N.H. (1988). Lymphotoxin. Annu. Rev. Immunol. 6, 407–438.

Peetre, C., Gullberg, U., Nilsson, E. and Olsson, I. (1986). Effects of recombinant tumor necrosis factor on proliferation and differentiation of leukemic and normal hemopoietic cells *in vitro*. Relationship to cell surface receptor. J. Clin. Invest. 78, 1694–1700.

Pekala, P., Kawakami, M., Vine, W., Lane, M.D. and Cerami, A. (1983). Studies of insulin resistance in adipocytes induced by macrophage mediator. J. Exp. Med. 157, 1360–1365.

Pennica, D., Lam, V.T., Mize, N.K., *et al.* (1992). Biochemical properties of the 75-kDa tumor necrosis factor receptor. Characterization of ligand binding, internalization, and receptor phosphorylation. J. Biol. Chem. 267, 21172–21178.

Pennica, D., Nedwin, G.E., Hayflick, J.S., *et al.* (1984). Human tumour necrosis factor: precursor structure, expression and homology to lymphotoxin. Nature 312, 724–729.

Perez, C., Albert, I., DeFay, K., Zachariades, N., Gooding, L. and Kriegler, M. (1990). A nonsecretable cell surface mutant of tumor necrosis factor (TNF) kills by cell-to-cell contact. Cell 63, 251–258.

Philip, R. and Epstein, L.B. (1986). Tumour necrosis factor as immunomodulator and mediator of monocyte cytotoxicity induced by itself, γ-interferon and interleukin-1. Nature 323, 86–89.

Piguet, P.F., Grau, G.E., Allet, B. and Vassalli, P. (1987). Tumor necrosis factor/cachectin is an effector of skin and gut lesions of the acute phase of graft-vs.-host disease. J. Exp. Med. 166, 1280–1289.

Pober, J.S. and Cotran, R.S. (1990). The role of endothelial cells in inflammation. Transplantation 50, 537–544.

Porteu, F. and Hieblot, C. (1994). Tumor necrosis factor induces a selective shedding of its p75 receptor from human neutrophils. J. Biol. Chem. 269, 2834–2840.

Porteu, F. and Nathan, C. (1990). Shedding of tumor necrosis factor receptors by activated human neutrophils. J. Exp. Med. 172, 599–607.

Porteu, F., Brockhaus, M., Wallach, D., Engelmann, H. and Nathan, C.F. (1991). Human neutrophil elastase releases a ligand-binding fragment from the 75-kDa tumor necrosis factor (TNF) receptor. Comparison with the proteolytic activity responsible for shedding of TNF receptors from stimulated neutrophils. J. Biol. Chem. 266, 18846–18853.

Powell, M.B. and Steinman, L. (1992). The role of lymphotoxin and TNF in demyelinating diseases of the CNS. In "Tumor Necrosis Factors. The Molecules and Their Emerging Role in Medicine" (ed. B. Beutler), pp. 355–369. Raven Press, New York.

Räth, U., Kaufmann, M., Schmid, H., *et al.* (1991). Effect of intraperitoneal recombinant human tumour necrosis factor alpha on malignant ascites. Eur. J. Cancer 27, 121–125.

Regenass, U., Müller, M., Curschellas, E. and Matter, A. (1987). Anti-tumor effects of tumor necrosis factor in combination with chemotherapeutic drugs. Int. J. Cancer 39, 266–273.

Robaye, B., Mosselmans, R., Fiers, W., Dumont, J.E. and Galand, P. (1991). Tumor necrosis factor induces apoptosis (programmed cell death) in normal endothelial cells *in vitro*. Am. J. Pathol. 138, 447–453.

Roederer, M., Staal, F.J.T., Raju, P.A., Herzenberg, L.A. and Herzenberg, L.A. (1992). The interrelationship of tumor necrosis factor, glutathione, and AIDS. In "Tumor Necrosis Factor: Structure–function Relationship and Clinical Application" (ed T. Osawa and B. Bonavida), pp. 215–229. Karger, Basel.

Roll, J.T., Young, K.M., Kurtz, R.S. and Czuprynski, C.J. (1990). Human rTNFα augments anti-bacterial resistance in mice: potentiation of its effects by recombinant human rIL-1α. Immunology 69, 316–322.

Rood, P.A., Lorence, R.M. and Kelley, K.W. (1990). Serum protease inhibitor abrogation of Newcastle disease virus enhancement of cytolysis by recombinant tumor necrosis factors alpha and beta. J. Natl. Cancer Inst. 82, 213–217.

Rothe, J., Lesslauer, W., Lötscher, H., *et al.* (1993). Mice lacking the tumour necrosis factor receptor 1 are resistant to TNF-mediated toxicity but highly susceptible to infection by *Listeria monocytogenes*. Nature 364, 798–802.

Rothe, M., Wong, S.C., Henzel, W.J. and Goeddel, D.V. (1994). A novel family of putative signal transducers associated with the cytoplasmic domain of the 75 kDa tumor necrosis factor receptor. Cell 78, 681–692.

Rothe, M., Pan, M.-G., Henzel, W.J., Ayres, T.M. and Goeddel, D.V. (1995a). The TNFR2-TRAF signaling complex contains two novel proteins related to baculoviral inhibitor of apoptosis proteins. Cell 83, 1243–1252.

Rothe, M., Sarma, V., Dixit, V.M. and Goeddel, D.V. (1995b). TRAF2-mediated activation of NF-κB by TNF receptor 2 and CD40. Science 269, 1424–1427.

Rothstein, J.L. and Schreiber, H. (1988). Synergy between tumor necrosis factor and bacterial products causes hemorrhagic necrosis and lethal shock in normal mice. Proc. Natl Acad. Sci. USA 85, 607–611.

Ruff, M.R. and Gifford, G.E. (1981). Tumor necrosis factor. In "Lymphokines" (ed. E. Pick), vol. 2, pp. 235–272. Academic Press, New York.

Ruggiero, V., Latham, K. and Baglioni, C. (1987). Cytostatic and cytotoxic activity of tumor necrosis factor on human cancer cells. J. Immunol. 138, 2711–2717.

Sampaio, E.P., Sarno, E.N., Galilly, R., Cohn, Z.A. and Kaplan, G. (1991). Thalidomide selectively inhibits tumor necrosis factor α production by stimulated human monocytes. J. Exp. Med. 173, 699–703.

Sato, N., Goto, T., Haranaka, K., *et al.* (1986). Actions of tumor necrosis factor on cultured vascular endothelial cells: morphologic modulation, growth inhibition, and cytotoxicity. J. Natl Cancer Inst. 76, 1113–1121.

Saxne, T., Palladino, M.A., Heinegard, D., Tatal, N. and Wollheim, F.A. (1988). Detecting of tumor necrosis factor α but not tumor necrosis factor β in rheumatoid arthritis synovial fluid and serum. Arthritis Rheum. 31, 1041–1045.

Schade, U.F., Ernst, M., Reinke, M. and Wolter, D.T. (1989). Lipoxygenase inhibitors suppress formation of tumor necrosis factor *in vitro* and *in vivo*. Biochem. Biophys. Res. Commun. 159, 748–754.

Schall, T.J., Lewis, M., Koller, K.J., *et al.* (1990). Molecular cloning and expression of a receptor for human tumor necrosis factor. Cell 61, 361–370.

Scheurich, P., Thoma, B., Ücer, U. and Pfizenmaier, K. (1987). Immunoregulatory activity of recombinant human tumor necrosis factor (TNF)-α: induction of TNF receptors on human T cells and TNF-α-mediated enhancement of T cell responses. J. Immunol. 138, 1786–1790.

Schleiffenbaum, B. and Fehr, J. (1990). The tumor necrosis factor receptor and human neutrophil function. Deactivation and cross-deactivation of tumor necrosis factor-induced neutrophil responses by receptor down-regulation. J. Clin. Invest. 86, 184–195.

Schmid, D.S., Tite, J.P. and Ruddle, N.H. (1986). DNA fragmentation: manifestation of target cell destruction mediated by cytotoxic T-cell lines, lymphotoxin-secreting helper T-cell clones, and cell-free lymphotoxin-containing supernatant. Proc. Natl Acad. Sci. USA 83, 1881–1885.

Schultz, R.M. and Altom, M.G. (1990). Protective activity of recombinant murine tumor necrosis factor-α and interferon-γ against experimental murine lung carcinoma metastases. J. Interferon Res. 10, 229–236.

Schulze-Osthoff, K., Bakker, A.C., Vanhaesebroeck, B., Beyaert, R., Jacob, W.A. and Fiers, W. (1992). Cytotoxic activity of tumor necrosis factor is mediated by early damage of mitochondrial functions. Evidence for the involvement of mitochondrial radical generation. J. Biol. Chem. 267, 5317–5323.

Schulze-Osthoff, K., Beyaert, R., Vandevoorde, V., Haegeman, G. and Fiers, W. (1993). Depletion of the mitochondrial electron transport abrogates the cytotoxic and gene-inductive effects of TNF. EMBO J. 12, 3095–3104.

Schütze, S., Machleidt, T. and Krönke, M. (1992a). Mechanisms of tumor necrosis factor action. Semin. Oncol. 19 (supplement 4), 16–24.

Schütze, S., Potthoff, K., Machleidt, T., Berkovic, D., Wiegmann, K. and Krönke, M. (1992b). TNF activates NF-κB by phosphatidylcholine-specific phospholipase C-induced "acidic" sphingomyelin breakdown. Cell 71, 765–776.

Schwalb, D.M., Han, H.-M., Marino, M., *et al.* (1993). Identification of a new receptor subtype for tumor necrosis factor-α. J. Biol. Chem. 268, 9949–9954.

Scuderi, P., Sterling, K.E., Lam, K.S., *et al.* (1986). Raised serum levels of tumour necrosis factor in parasitic infections. Lancet ii, 1364–1365.

Seckinger, P. and Dayer, J.-M. (1992). Natural inhibitors of TNF. In "Tumor Necrosis Factors: Structure, Function, and Mechanism of Action" (eds. B.B. Aggarwal and J. Vilček), pp. 217–236. Marcel Dekker, New York.

Selmaj, K. and Raine, C.S. (1988). Tumor necrosis factor mediates myelin and oligodendrocyte damage *in vitro*. Ann. Neurol. 23, 339–346.

Selmaj, K., Raine, C.S., Cannella, B. and Brosnan, C.F. (1991). Identification of lymphotoxin and tumor necrosis factor in multiple sclerosis lesions. J. Clin. Invest. 87, 949–954.

Shalaby, M.R., Aggarwal, B.B., Rinderknecht, E., Svedersky, L.P., Finkle, B.S. and Palladino, M.A., Jr, (1985). Activation of human polymorphonuclear neutrophil functions by interferon-γ and tumor necrosis factor. J. Immunol. 135, 2069–2073.

Shalaby, M.R., Sundan, A., Loetscher, H., Brockhaus, M., Lesslauer, W. and Espevik, T. (1990). Binding and regulation of cellular functions by monoclonal antibodies against human tumor necrosis factor receptors. J. Exp. Med. 172, 1517–1520.

Shear, M.J., Turner, F.C. and Perrault, A. (1943). Chemical treatment of tumors. V. Isolation of the hemorrhage producing fraction from *Serratia marcescens* culture filtrate. J. Natl Cancer Inst. 4, 81–97.

Sherman, M.L., Spriggs, D.R., Arthur, K.A., Imamura, K., Frei, E., III, and Kufe, D.W. (1988). Recombinant human tumor necrosis factor administered as a five-day continuous infusion in cancer patients: phase I toxicity and effects on lipid metabolism. J. Clin. Oncol. 6, 344–350.

Sherry, B. and Cerami, A. (1988). Cachectin/tumor necrosis factor exerts endocrine, paracrine and autocrine control of inflammatory responses. J. Cell Biol. 107, 1269–1277.

Shimomura, K., Manda, T., Mukumoto, S., Kobayashi, K., Nakano, K. and Mori, J. (1988). Recombinant human tumor necrosis factor-alpha: thrombus formation is a cause of anti-tumor activity. Int. J. Cancer 41, 243–247.

Shirai, T., Yamaguchi, H., Ito, H., Todd, C.W. and Wallace, R.B. (1985). Cloning and expression in *Escherichia coli* of the gene for human tumour necrosis factor. Nature 313, 803–806.

Sidhu, R.S. and Bollon, A.P. (1993). Tumor necrosis factor activities and cancer therapy. A perspective. Pharmacol. Ther. 57, 79–128.

Smith, C.A., Davis, T., Anderson, D., *et al.* (1990a). A receptor for tumor necrosis factor defines an unusual family of cellular and viral proteins. Science 248, 1019–1023.

Smith, C.A., Farrah, T. and Goodwin, R.G. (1994). The TNF receptor superfamily of cellular and viral proteins: Activation, costimulation, and death. Cell 76, 959–962.

Smith, M.R., Munger, W.E., Kung, H.-F., Takacs, L. and Durum, S.K. (1990b). Direct evidence for an intracellular role for tumor necrosis factor-α. Microinjection of tumor necrosis factor kills target cells. J. Immunol. 144, 162–169.

Spies, T., Blanck, G., Bresnahan, M., Sands, J. and Strominger, J.L. (1991). A new cluster of genes within the human major histocompatibility complex. Science 243, 214–217.

Stanger, B.Z., Leder, P., Lee, T.H., Kim, E. and Seed, B. (1995). RIP: a novel protein containing a death domain that interacts with Fas/APO-1 (CD95) in yeast and causes cell death. Cell 81, 513–523.

Strieter, R.M., Kunkel, S.L., Showell, H.J., *et al.* (1989). Endothelial cell gene expression of a neutrophil chemotactic factor by TNF-α, LPS, and IL-1β. Science 243, 1467–1469.

Suffys, P., Beyaert, R., De Valck, D., Vanhaesebroeck, B., Van Roy, F. and Fiers, W. (1991). Tumour-necrosis-factor-mediated cytotoxicity is correlated with phospholipase-A$_2$ activity, but not with arachidonic acid release *per se*. Eur. J. Biochem. 195, 465–475.

Sugarman, B.J., Aggarwal, B.B., Hass, P.E., Figari, I.S., Palladino, M.A., Jr and Shepard, H.M. (1985). Recombinant human tumor necrosis factor-α: effects on proliferation of normal and transformed cells *in vitro*. Science 230, 943–945.

Taguchi, T. and Sohmura, Y. (1991). Clinical studies with TNF. Biotherapy 3, 177–186.

Takahashi, N., Brouckaert, P. and Fiers, W. (1991). Induction of tolerance allows separation of lethal and antitumor activities of tumor necrosis factor in mice. Cancer Res. 51, 2366–2372.

Takahashi, N., Brouckaert, P. and Fiers, W. (1993). Cyclooxygenase inhibitors prevent the induction of tolerance to the toxic effects of tumor necrosis factor. J. Immunother. 14, 16–21.

Tartaglia, L.A., Weber, R.F., Figari, I.S., Reynolds, C., Palladino, M.A., Jr and Goeddel, D.V. (1991). The two different receptors for tumor necrosis factor mediate distinct cellular responses. Proc. Natl Acad. Sci. USA 88, 9292–9296.

Tartaglia, L.A., Ayres, T.M., Wong, G.H.W. and Goeddel, D.V. (1993a). A novel domain within the 55 kd TNF receptor signals cell death. Cell 74, 845–853.

Tartaglia, L.A., Pennica, D. and Goeddel, D.V. (1993b). Ligand passing: the 75-kDa tumor necrosis factor (TNF) receptor recruits TNF for signaling by the 55-kDa TNF receptor. J. Biol. Chem. 268, 18542–18548.

Tartaglia, L.A., Rothe, M., Hu, Y.-F. and Goeddel, D.V. (1993c). Tumor necrosis factor's cytotoxic activity is signaled by the p55 TNF receptor. Cell 73, 213–216.

Teichmann, J.V., Ludwig, W.D. and Thiel, E. (1992). Cytotoxicity of interleukin 2-induced lymphokine-activated killer (LAK) cells against human leukemia and augmentation of killing by interferons and tumor necrosis factor. Leukocyte Res. 16, 287–298.

Thoma, B., Grell, M., Pfizenmaier, K. and Scheurich, P. (1990). Identification of a 60-kD tumor necrosis factor (TNF) receptor as the major signal transducing component in TNF responses. J. Exp. Med. 172, 1019–1023.

Tobler, A. and Koeffler, H.P. (1987). Differentiation of human acute myelogenous leukemia cells: therapeutic possibilities. Acta Haematol. 78, 127–135.

Torti, F.M., Dieckmann, B., Beutler, B., Cerami, A. and Ringold, G.M. (1985). A macrophage factor inhibits adipocyte gene expression: an *in vitro* model of cachexia. Science 229, 867–869.

Tracey, K.J., Beutler, B., Lowry, S.F., *et al.* (1986). Shock and tissue injury induced by recombinant human cachectin. Science 234, 470–474.

Trinchieri, G., Rosen, M. and Perussia, B. (1987a). Induction of differentiation of human myeloid cell lines by tumor necrosis factor in cooperation with 1-α,25-dihydroxyvitamin D$_3$. Cancer Res. 47, 2236–2242.

Trinchieri, G., Rosen, M. and Perussia, B. (1987b). Retinoic acid cooperates with tumor necrosis factor and immune interferon in inducing differentiation and growth inhibition of the human promyelocytic leukemic cell line HL-60. Blood 69, 1218–1224.

Tsujimoto, M., Feinman, R. and Vilček, J. (1986). Differential effects of type 1 IFN and IFN-γ on the binding of tumor necrosis factor to receptors in two human cell lines. J. Immunol. 137, 2272–2276.

Unglaub, R., Maxeiner, B., Thoma, B., Pfizenmaier, K. and Scheurich, P. (1987). Downregulation of tumor necrosis factor (TNF) sensitivity via modulation of TNF binding capacity by protein kinase C activators. J. Exp. Med. 166, 1788–1797.

Vandenabeele, P., Declercq, W., Vercammen, D., *et al.* (1992). Functional characterization of the human tumor necrosis factor receptor p75 in a transfected rat/mouse T cell hybridoma. J. Exp. Med. 176, 1015–1024.

Vandenabeele, P., Declercq, W., Beyaert, R. and Fiers, W. (1995a). Two tumour necrosis factor receptors: structure and function. Trends Cell Biol. 5, 392–399.

Vandenabeele, P., Declercq, W., Vanhaesebroeck, B., Grooten, J. and Fiers, W. (1995b). Both TNF receptors are required for TNF-mediated induction of apoptosis in PC60 cells. J. Immunol. 154, 2904–2913.

Van de Craen, M., Vandenabeele, P., Declercq, W., *et al.* (1997). Characterization of seven murine caspase family members. FEBS Lett. 403, 61–69.

Vandevoorde, V., Haegeman, G. and Fiers, W. (1997). Induced expression of trimerized intracellular domains of the human tumor necrosis factor (TNF) p55 receptor elicits TNF effects. J. Cell Biol. 137, 1627–1638.

van der Schelling, G.P., IJzermans, J.N., et al. (1992). A phase I study of local treatment of liver metastases with recombinant tumour necrosis factor. Eur. J. Cancer 28A, 1073–1078.

van Hinsbergh, V.W.M., Kooistra, T., van den Berg, E.A., Princen, H.M.G., Fiers, W. and Emeis, J.J. (1988). Tumor necrosis factor increases the production of plasminogen activator inhibitor in human endothelial cells in vitro and in rats in vivo. Blood 72, 1467–1473.

Van Lint, J., Agostinis, P., Vandevoorde, V., et al. (1992). Tumor necrosis factor stimulates multiple serine/threonine protein kinases in Swiss 3T3 and L929 cells. Implication of casein kinase-2 and extracellular signal-regulated kinases in the tumor necrosis factor signal transduction pathway. J. Biol. Chem. 267, 25916–25921.

Van Ostade, X., Tavernier, J., Prangé, T. and Fiers, W. (1991). Localization of the active site of human tumour necrosis factor (hTNF) by mutational analysis. EMBO J. 10, 827–836.

Van Ostade, X., Vandenabeele, P., Everaerdt, B., et al. (1993). Human TNF mutants with selective activity on the p55 receptor. Nature 361, 266–269.

Van Snick, J. (1990). Interleukin-6: an overview. Annu. Rev. Immunol. 8, 253–278.

Vassalli, P. (1992). The pathophysiology of tumor necrosis factors. Annu. Rev. Immunol. 10, 411–452.

Vilček, J. and Lee, T.H. (1991). Tumor necrosis factor. New insights into the molecular mechanisms of its multiple actions. J. Biol. Chem. 266, 7313–7316.

Vilček, J., Palombella, V.J., Henriksen-DeStefano, D., et al. (1986). Fibroblast growth enhancing activity of tumor necrosis factor and its relationship to other polypeptide growth factors. J. Exp. Med. 163, 632–643.

Waage, A., Brandtzaeg, P., Halstensen, A., Kierulf, P. and Espevik, T. (1989). The complex pattern of cytokines in serum from patients with meningococcal septic shock. Association between interleukin 6, interleukin 1, and fatal outcome. J. Exp. Med. 169, 333–338.

Waage, A., Slupphaug, G. and Shalaby, R. (1990). Glucocorticoids inhibit the production of IL6 from monocytes, endothelial cells and fibroblasts. Eur. J. Immunol. 20, 2439–2443.

Wallach, D., Holtmann, H., Engelmann, H. and Nophar, Y. (1988). Sensitization and desensitization to lethal effects of tumor necrosis factor and IL-1. J. Immunol. 140, 2994–2999.

Wallach, D., Boldin, M., Varfolomeev, E., et al. (1997). Cell death induction by receptors of the TNF family: Towards a molecular understanding. FEBS Lett. 410, 96–106.

Wang, A.M., Creasey, A.A., Ladner, M.B., et al. (1985). Molecular cloning of the complementary DNA for human tumor necrosis factor. Science 228, 149–154.

Ware, C.F., VanArsdale, T.L., Crowe, P.D. and Browning, J.L. (1995). The ligands and receptors of the lymphotoxin system. Curr. Top. Microbiol. Immunol. 198, 175–218.

Warner, S.J.C. and Libby, P. (1989). Human vascular smooth muscle cells. Target for and source of tumor necrosis factor. J. Immunol. 142, 100–109.

Watanabe, N., Niitsu, Y., Umeno, H., et al. (1988). Synergistic cytotoxic and antitumor effects of recombinant human tumor necrosis factor and hyperthermia. Cancer Res. 48, 650–653.

Wiegmann, K., Schütze, S., Machleidt, T., Witte, D. and Krönke, M. (1994). Functional dichotomy of neutral and acidic sphingomyelinases in tumor necrosis factor signaling. Cell 78, 1005–1015.

Williamson, B.D., Carswell, E.A., Rubin, B.Y., Prendergast, J.S. and Old, L.J. (1983). Human tumor necrosis factor produced by human B-cell lines: synergistic cytotoxic interaction with human interferon. Proc. Natl Acad. Sci. USA 80, 5397–5401.

Wingfield, P., Pain, R.H. and Craig, S. (1987). Tumour necrosis factor is a compact trimer. FEBS Lett. 211, 179–184.

Wong, G.H.W. and Goeddel, D.V. (1986). Tumour necrosis factors α and β inhibit virus replication and synergize with interferons. Nature 323, 819–822.

Wong, G.H.W. and Goeddel, D.V. (1988). Induction of manganous superoxide dismutase by tumor necrosis factor: Possible protective mechanism. Science 241, 941–944.

Wright, S.C. and Bonavida, B. (1987). Studies on the mechanism of natural killer cell-mediated cytotoxicity. VII. Functional comparison of human natural killer cytotoxic factors with recombinant lymphotoxin and tumor necrosis factor. J. Immunol. 138, 1791–1798.

Wu, S., Rodabaugh, K., Martinez Maza, O., et al. (1992). Stimulation of ovarian tumor cell proliferation with monocyte products including interleukin-1, interleukin-6, and tumor necrosis factor-α. Am. J. Obstet. Gynecol. 166, 997–1007.

Yamagishi, J., Kawashima, H., Matsuo, N., et al. (1990). Mutational analysis of structure–activity relationships in human tumor necrosis factor-alpha. Protein Eng. 3, 713–719.

Yokota, S., Geppert, T.D. and Lipsky, P.E. (1988). Enhancement of antigen- and mitogen-induced human T lymphocyte proliferation by tumor necrosis factor-α. J. Immunol. 140, 531–536.

Zentella, A., Manogue, K. and Cerami, A. (1993). Cachectin/TNF-mediated lactate production in cultured myocytes is linked to activation of a futile substrate cycle. Cytokine 5, 436–447.

Zhang, Y., Lin, J.-X., Yip, Y.K. and Vilček, J. (1988). Enhancement of cAMP levels and of protein kinase activity by tumor necrosis factor and interleukin 1 in human fibroblasts: role in the induction of interleukin 6. Proc. Natl Acad. Sci. USA 85, 6802–6805.

Zhang, Y., Lin, J.-X. and Vilček, J. (1990). Interleukin-6 induction by tumor necrosis factor and interleukin-1 in human fibroblasts involves activation of a nuclear factor binding to a κB-like sequence. Mol. Cell. Biol. 10, 3818–3823.

25. Interferons Alpha, Beta, and Omega

Anthony Meager

1. Introduction

1.1 OUTLINE OF DISCOVERY AND CHARACTERIZATION OF INTERFERONS (IFN)

The phenomenon of viral interference was first described nearly 60 years ago when Hoskins (1935) described the protective action of a neurotropic yellow fever virus against a viserotropic strain of the same virus in monkeys. Although viral interference was further investigated in the 1940s and 1950s, the underlying mechanism was not discovered until 1957 when Isaacs and Lindemann, working at The National Institute for Medical Research (London, UK), isolated a biologically active substance from virally-infected chicken cell cultures that, on transfer to fresh chicken cell cultures,

Cytokines
ISBN 0–12–498340–5

produced a protective antiviral effect (Isaacs and Lindemann, 1957). The word Interferon (IFN) was coined for this substance. Its discovery aroused considerable scientific and medical interest since by 1957 antibiotics were widely available to control bacterial infections, but, in stark contrast, viral diseases such as influenza, measles, polio, and smallpox were virtually untreatable. Interest was further heightened by many subsequent studies that demonstrated that IFN could be produced by human cells and was active against a broad spectrum of viruses (see Schlesinger, 1959, for an early review).

At that time, IFN was being hailed by the media as a wonder drug, but it soon became clear that IFN was being produced naturally in too small quantities for that extravagant claim to be immediately confirmed. In fact, the low production of IFN was to bedevil attempts both to characterize it molecularly and evaluate it clinically for many years following its discovery.

Although the protein nature of IFN was recognized at an early stage in its development (see Fantes, 1966, for an early review), it was only following the introduction of large-scale production methods in the 1970s (Cantell and Hirvonen, 1977) and the simultaneous development of efficient purification procedures (Knight, 1976; Rubinstein et al., 1978) that sufficient amounts of partially pure IFN protein became available for characterization and clinical use. Gradually, it became apparent that IFN was not a single protein and that there were likely to be different types of IFN molecules. However, despite progress in the area of purification and in initial characterization by sequencing N-terminal polypeptides, IFN proteins all but defied full characterization until the advent of recombinant DNA (rDNA) technology in the late 1970s. This technology, spurred on by the pharmaceutical industry's desire to produce pharmacologically active proteins cheaply, revealed that one type of human IFN, now designated IFN-α, was a mixture of several closely related proteins, termed subtypes, expressed from distinct chromosomal genes (Nagata et al., 1980). Second and third types of IFN, designated IFN-β and IFN-γ respectively, have subsequently been "cloned" (Taniguchi et al., 1980; Gray et al., 1982) but, unlike IFN-α, are single protein species. IFN-β is molecularly related to IFN-α subtypes but is antigenically distinct from them, whereas IFN-γ is both molecularly and antigenically distinct from either IFN-α subtypes or IFN-β. (For this reason, IFN-γ is considered separately elsewhere in this volume.) Finally, a fourth type of IFN, antigenically distinct from IFN-α and IFN-β, but molecularly related to both, has more recently been cloned and characterized. Rather untypically, this new IFN type has been designated IFN omega (IFN-ω) (Adolf, 1987).

The genes for IFN-α subtypes, IFN-β and IFN-ω are tandemly arranged on the short arm of chromosome 9.

They are only transiently expressed following induction by a variety of exogenous stimuli, including viruses. IFN-α, IFN-β and IFN-ω proteins are synthesized from their respective mRNAs for relatively short periods following gene activation and are secreted to act, via specific cell surface receptors, on other cells. Early studies on the characterization of IFN receptors indicated that IFN-α and IFN-β were likely to share a common receptor, but it has only been comparatively recently that such receptors have been cloned (Uzé et al., 1990; Novick et al., 1994).

IFN actions are initiated by activated receptors and cytoplasmic signal transduction pathways, which are now well characterized for the IFN-α/β receptor, and manifested following expression of a number of IFN-specific inducible genes. Induction of the antiviral state, which is dependent on such protein synthesis, may now be viewed as just one of the many activities attributed to IFN in general; these activities include inhibition of cell proliferation and immunomodulation (see Pestka et al., 1987, for a review).

1.2 Nomenclature

In the 1960s, two types of IFN were defined on the basis of the capacity of their antiviral activity to withstand acidification to pH 2. These were termed type I IFN for acid-stable IFN and type II IFN for acid-labile IFN. Type I IFN included IFN produced by virally infected leukocytes, alternatively known as leukocyte IFN, and IFN produced by virally infected human diploid fibroblasts, alternatively known as fibroblast IFN. Type II IFN, which was only produced by antigenically or mitogenically stimulated human peripheral blood mononuclear cells (PBMC), has often been referred to as immune IFN (Stewart, 1979).

Antigenic differences were described for leukocyte and fibroblast IFN and these were put on a molecular basis when knowledge of their respective N-terminal amino acid sequences became available (Allen and Fantes, 1980; Knight et al., 1980; Levy et al., 1980; Zoon et al., 1980). At that time, an international nomenclature committee (Stewart et al., 1980) reviewed the growing evidence for the existence of distinct molecular forms of IFN and introduced the Greek alphabetical system to apply to the then known antigenically distinct types of IFN. Leukocyte IFN was designated IFN-α, fibroblast IFN was designated IFN-β, and immune IFN became IFN-γ. However, complications immediately arose when it was revealed, following the cloning of several different leukocyte IFN complementary DNAs (cDNAs) (Nagata et al., 1980; Brack et al., 1981; Goeddel et al., 1981; Streuli et al., 1980), that leukocyte IFN was heterogeneous and contained many different, molecularly and antigenically related species, now commonly referred to as subtypes. The research group at

Hoffmann-La Roche labeled the subtypes produced in *E. coli* αA, αB, αC, αD, etc. (Evinger *et al.*, 1981; Rehberg *et al.*, 1982), distinguishing them from natural components of leukocyte IFN (Rubinstein *et al.*, 1981), while the Biogen group labeled these recombinant subtypes α1, α2, α3, α4, etc. (Streuli *et al.*, 1980), regrettably without an appropriate alphabetical–numerical correspondence: for example, αA ≡ α2, αD ≡ α1. However, the numerical system now prevails. When an IFN preparation is a mixture of IFN-α subtypes, e.g., leukocyte IFN, lymphoblastoid IFN, this is often designated IFN-αn.

The later cloning of cDNAs encoding IFN-α-like proteins (Capon *et al.*, 1985; Hauptmann and Swetly, 1985) initially led to the naming of this new IFN as IFN-α subclass II, with all of the earlier-characterized IFN-α subtypes being reclassified as IFN-α subclass I. This large, unwieldy nomenclature system has been superseded by the renaming of IFN-α II as IFN-ω; this has been generally accepted with the finding that IFN-ω is antigenically distinct from IFN-α and IFN-β proteins and thus qualifies as a separate type of IFN (Adolf, 1987).

Fortunately, the nomenclature for IFN-β has remained straightforward since there is only one protein species, at least in humans (Derynck *et al.*, 1980, 1981; Taniguchi *et al.*, 1980).

From the initial cloning of IFN-α cDNAs, there has been a plethora of reports on the cloning of new, and sometimes distinct, genomic and cDNA clones, and fairly disparate nomenclatures have arisen. Diaz and Allen (1993) therefore undertook the considerable task of compiling the IFN genes and genomic and cDNA clones from the literature and introduced an arabic-alphabetical system for naming IFN genes to enable their distinction from IFN proteins. Thus, IFN-α genes became IFNA genes with the addition of a numeral to denote subtype, i.e., IFNA1, IFNA2, etc. (Table 25.1). The IFN-β gene became IFNB and, since there is only one gene in humans, it is referred to as IFNB1. For IFN-ω genes, W has been used; hence IFNW1 (Table 25.1). Besides genes that are capable of being expressed and translated into IFN proteins, there are a number of pseudogenes which are unable to give rise to IFN proteins, and which in this new nomenclature system are designated by a P, e.g., IFNAP22, IFNWP2, or simply IFNP1 where pseudogenes are clearly IFN-like but cannot be definitely included in any one of the IFNA, IFNB or IFNW gene families (Table 25.1).

The IFN genomic and cDNA clones have been designated in a variety of ways, as illustrated in Table 25.1. Pseudogenic or non-translatable clones are normally prefixed with a Greek ψ. For the purposes of this chapter, and to reduce the complexity of naming IFN clones and proteins, the nomenclature system adopted by Weissmann and colleagues will be adhered to: IFN-α_1, α_2, α_4, α_5, etc.

2. IFN Genes

2.1 NUMBERS, STRUCTURE, AND LOCALIZATION

In humans, there are 14 nonallelic IFNA genes, one of which, IFNAP22, is a pseudogene (Table 25.1). In addition, there are a further four nonallelic pseudogenes that possibly also belong to the IFNA gene family. Probable allelic variants of certain IFNA genes, e.g., IFNA2, are also known to exist (Streuli *et al.*, 1980; Goeddel *et al.*, 1981; Dworkin-Rastl *et al.*, 1982; Emanuel and Pestka, 1993). This extensive family of IFNA genes are tandemly arranged on the short-arm of chromosome 9 (9p23) and span a region of approximately 400 kb (Owerbach *et al.*, 1981; Shows *et al.*, 1982; Slate *et al.*, 1982; Ullrich *et al.*, 1982). The IFNB gene and the IFNW gene/pseudogene family are also located in the same region of chromosome 9 (Meager *et al.*, 1979a,b; Owerbach *et al.*, 1981; Henry *et al.*, 1984; Capon *et al.*, 1985).

There is a high degree of homology among the IFNA genes, but these show much less homology to either IFNB or IFNW genes. Nevertheless, all of these IFN genes share the common feature of being intron-less (Taniguchi *et al.*, 1980; Goeddel *et al.*, 1981; Houghton *et al.*, 1981; Capon *et al.*, 1985), suggesting a very ancient origin of their common ancestral gene. It has been proposed that the primordial IFN gene arose some 500 million years ago, with the first split occurring around 400 million years ago to yield the first IFNA and IFNB genes (Wilson *et al.*, 1983). Since then the IFNA gene has evolved and duplicated many times to give rise to the multiple IFNA genes found in present-day animals and man (Mijàta and Hayashida, 1982; Gillespie and Carter, 1983). Around 100 million years ago, an IFNA gene appears to have diverged sufficiently from the main group to give rise to the IFNW gene family, which is present in most mammals except mice and dogs (Himmler *et al.*, 1987; Roberts *et al.*, 1992).

It is not clear what characteristic of IFNA and IFNW genes enabled the numerous reduplication events to occur in comparison to the nonexistent or more limited (some mammals have more than one IFNB gene, e.g., bovines (Wilson *et al.*, 1983)) reduplication of the IFNB gene (Ohlsson *et al.*, 1985). It is apparent that gene conversion, as a result of mismatch repair and unequal crossover, contributed significantly to the creation of distinct, but highly homologous, nonallelic IFNA (and IFNW) genes (De Maeyer and De Maeyer-Guignard, 1988). In most cases, both coding and noncoding regions have diverged, but in the case of IFNA13 the coding region has remained identical to that of IFNA1, although its 5′ and 3′ flanking regions have diverged (Todokoro *et al.*, 1984).

The structures of IFNA, IFNB, and IFNW genes are similar. Each gene has a 5′ regulatory promoter region upstream from the transcriptional start (cap) site, a

Table 25.1   Nomenclature of the human interferon genes and proteins

IFN genes; new symbols	Corresponding genomic clones (examples)	Corresponding cDNA clones (examples)	Corresponding proteins
IFN-α			
IFNA$_1$	IFN-α_1	IFN-α_1, LeIF-D	IFN-α_1 (D)
IFNA$_2$	λa_2, IFN-A	IFN-α_2, LeIF-A	IFN-α_2 (A)
IFNA$_4$	IFN-α_{4a}, IFN-α_{7b}	–	IFN-α_4
IFNA$_5$	IFN-α_5, IFN-α_{61}	LeIF-G	IFN-α_5 (G)
IFNA$_6$	IFN-α_6, LeIF-K	–	IFN-α_6 (K)
IFNA$_7$	IFN-α_7, LeIF-J	IFN-α_{J1}	IFN-α_7 (J1)
IFNA$_8$	IFN-α_8, IFNα_{B2}	LeIF-B, IFN-B	IFN-α_8 (B)
IFNA$_{10}$	ψIFN-α_{10}, ψLeIF-L	LeIF-C	IFN-α_{10} (L)
IFNA$_{13}$	IFN-α_{13} (similar to IFN-α_1)	–	IFN-α_{13}
IFNA$_{14}$	IFN-α_{14}, IFN-α_N	LeIF-H	IFN-α_{14} (H1)
IFNA$_{16}$	IFN-α_{16}, IFN-α_{WA}	IFN-αN (Gren)	IFN-α_{16} (WA)
IFNA$_{17}$	IFN-α_{17}, IFN-α_T	IFN-α_1, IFN-α_{88}	IFN-α_{17}
IFNA$_{21}$	IFN-α_{21}	LeIF-F	IFN-α_{21} (F)
IFNAP$_{22}$	ψIFN-α_{22}	ψLeIF-E	IFN-α_{22} (E)
IFN-β			
IFNB$_1$	IFN-β	IFN-β	IFN-β
IFN-ω			
IFNW$_1$	IFN-α_{II1}	IFN-αII$_1$, IFN-ω	IFN-ω
IFNWP$_2$	ψIFN-α_{II2}	–	–
IFNWP$_4$	ψIFN-α_{II4}	–	–
IFNWP$_5$	ψIFN-$\alpha_{\omega 5}$	–	–
IFNWP$_9$	ψIFN-α_{II9}	–	–
IFNWP$_{15}$	ψIFN-α_{II3} or 15	–	–
IFNWP$_{18}$	ψIFN-α_{IIM} or 18	–	–
IFNWP$_{19}$	ψIFN-α_{19}	–	–
IFN pseudogenes not included in α or ω gene families			
IFNP$_{11}$	ψIFN-α_{11}	–	–
IFNP$_{12}$	ψIFN-α_{12}	–	–
IFNP$_{20}$	ψIFN-α_{20}	–	–
IFNP$_{23}$	ψIFN-α? (closely linked to IFNA1)	–	–

Adapted and modified from Diaz and Allen (1993) and Allen and Diaz (1996), which see for specific references for genomic clones and cDNAs. Reproduced with permission of Mary Ann Liebert, Inc., New York, USA.

coding region containing a nucleotide sequence encoding a signal polypeptide of 21–23 mainly hydrophobic amino acids, which is typical for secreted proteins, and consecutively the sequence encoding the mature IFN protein, followed by the 3′ flanking noncoding region, which can vary in length up to 450 base pairs (bp) (Figure 25.1) (Derynck *et al.*, 1980, 1981; Nagata *et al.*, 1980; Streuli *et al.*, 1980; Taniguchi *et al.*, 1980; Degrave *et al.*, 1981; Goeddel *et al.*, 1981; Gross *et al.*, 1981; Lawn *et al.*, 1981a,b; Gren *et al.*, 1984; Capon *et al.*, 1985; Henco *et al.*, 1985). The 5′ flanking region contains a TATA or Hogness box, which delineates the boundary of the upstream promoter, approximately 30 bp from the cap site. Farther upstream are found a number of hexameric repeat sequences GAAANN, where N can be any base, which in their dimeric or multimeric forms act as binding sites for nuclear transcription factors and repressor molecules (Fujita *et al.*, 1985; Ryals *et al.*, 1985). (This area is covered in more detail in Section 2.2,

Inducers and Transcriptional Control). The 3′ flanking regions vary in length and contain several polyadenylation sites and thus can give rise to mRNAs of different lengths (Mantei and Weissmann, 1982; Henco *et al.*, 1985). They contain above-average numbers of the sequence motifs ATTA or TTATTTAT. Such sequences are common, however, in many other cytokine genes and other genes, such as protooncogenes, that are inducibly and transiently expressed. It has been proposed that these sequences contribute to the relative instability and short half-lives of IFN and cytokine mRNAs (Caput *et al.*, 1986; Shaw and Kamen, 1986).

2.2   INDUCERS AND TRANSCRIPTIONAL CONTROL

All IFN genes are normally silent and thus require some sort of stimulus to induce expression. A wide range of

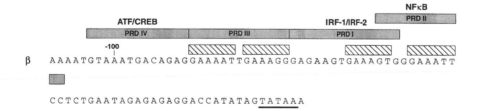

```
-220
AAAACAAAACATTTGAGAAACACGGCTCTAAACTCATGTAAAGAGTGCATGAAGGAAAGCAAAA
                                                        ******
ACAGAAATGGAAAGTGGCCCAGAAGCATTAAGAAAGTGGAAATCAGTATGTTCCCTATTTAAGG
CATTTGCAGGAAGCAAGGCCTTCAGAGAACCTAGAGCCCAAGGTTCAGAGTCACCCATCTCAGC
                          +1
AAGCCCAGAAGTATCTGCAATATCTACGATGGCCTCGCCCTTTGCTTTACTGATGGTCCTGGTG
GTGCTCAGCTGCAAGTCAAGCTGCTCTCTGGGCTGTGATCTCCCTGAGACCCACAGCCTGGATA
ACAGGAGGACCTTGATGCTCCTGGCACAAATGAGCAGAATCTCTCCTTCCTGTCTGATGGACAG
ACATGACTTTGGATTTCCCCAGGAGGAGTTTGATGGCAACCAGTTCCAGAAGGCTCCAGCCATC
TCTGTCCTCCATGAGCTGATCCAGATCTTCAACCTCTTTACCACAAAAGATTCATCTGCTGCTT
GGGATGAGGACCTCCTAGACAAATTCTGCACCGAACTCTACCAGCAGCTGAATGACTTGGAAGC
CTGTGTGATGCAGGAGGAGAGGGTGGGAGAAACTCCCCTGATGAATGCGGACTCCATCTTGGCT
GTGAAGAAATACTTCCGAAGAATCACTCTCTATCTGACAGAGAAGAAATACAGCCCTTGTGCCT
GGGAGGTTGTCAGAGCAGAAATCATGAGATCCCTCTCTTTATCAACAAACTTGCAAGAAAGATT
     560
AAGGAGGAAGGGAATAACATCTGGTCCAACATGAAAACAATTCTTATTGACTCATACACCAGGTC
ACGCTTTCATGAATTCTGTCATTTCAAAGACTCTCACCCCTGCTATAACTATGACCATGCTGAT
                   700
AAACTGATTTATCTATTTAAATATTTATTTAACTATTCATAAGATTTAAATTATTTTTGTTCAT
                                            800
ATAACGTCATGTGCACCTTTACACTGTGGTTAGTGTAATAAAACATGTTCCTTATATTTATATT
```

(a)
```
TACTCAAAAAAAA
```

(b)

Figure 25.1   (a) Nucleotide sequence of the chromosomal IFN-α_1 coding segment and its flanking regions. The coding sequence including the signal sequence is highlighted. The TATA box in the 5' flanking region is marked by asterisks. (b) A comparison of the promoter regions for IFN-α_1, IFN-β, and IFN-ω genes. The sequences upstream of the TATA box (underlined) of the IFN-α_1, IFN-β, and IFN-ω promoters are aligned. The positions of GAAANN hexamers are indicated by hatched boxes. The regions of the IFN-β promoter thought to be required for binding of transcription factors are indicated by named shaded boxes.

inducers, including viruses, bacteria, mycoplasma, endotoxins, double-stranded polynucleotides or RNA (dsRNA), and some cytokines, have been shown to efficiently activate transcription of IFN genes (Stewart, 1979; De Maeyer and De Maeyer-Guignard, 1988). Such inducers have in general the potential to induce expression of all IFNA, IFNB, and IFNW genes; however, there appears to be cell- and inducer-specific selectivity that governs the type and numbers of IFN genes expressed. For instance, virally induced human diploid fibroblasts produce mainly IFN-β and only a minor amount of IFN-α (Havell *et al.*, 1978), whereas virally induced PBMC produce mainly IFN-α plus IFN-ω and only a minor amount of IFN-β (Cantell and Hirvonen, 1977; Adolf *et al.*, 1990). This has to some extent been confirmed at the mRNA level (Shuttleworth *et al.*, 1983; Hiscott *et al.*, 1984). Differences have also been reported in the proportions of individual IFN-α subtypes produced by different cell types (Goren *et al.*, 1986; Finter, 1991; Greenway *et al.*, 1992), suggesting that IFNA genes may be differentially expressed. However, the way in which such differential expression is regulated is presently not understood.

Transcriptional control of IFNA genes resides in their 5′ flanking region, upstream from the cap site. Nucleotide deletions outside of position −117 from the cap site have little impact on transcriptional control, but deletions farther in eliminated induction, indicating that this region, −117 to −1, contained promoter regulatory elements (Ragg and Weissmann, 1983; Weidle and Weissmann, 1983). These have been further delineated as a purine-rich nucleotide tract between −109 and −64, containing hexameric repeats of GAAANN (GAAA G/C T/C), which appears to be necessary for inducible transcription and which has been termed the "virus-regulating element" (VRE) (Ryals *et al.*, 1985).

Similar studies involving 5′ deletions have been carried out with the IFNB gene, and it has been found that 5′ sequences within −110 to −1 contain regulatory elements that are required for induction by viruses and dsRNA. The minimum VRE has been localized to −74 to −37 with respect to the cap site (Goodbourn *et al.*, 1986; Goodbourn and Maniatis, 1988) and contains two positive virus-inducible elements, termed positive regulatory domains (PRDI, −77 to −64; PRDII, −66 to −55), and a negative regulatory domain (NRD, −57 to −37) (Figure 25.1) (Fujita *et al.*, 1985; Fan and Maniatis, 1989; Whittemore and Maniatis, 1990; Nourbakhsh *et al.*, 1993). In addition, the hexameric repeat sequences GAAANN (also present in IFNA genes) spanning from −110 to −65 contain variants of the PRDI sequence and two further regulatory elements, PRDIII (−90 to −78) and PRDIV (−104 to −91) have been identified that are required for a functional VRE in IFNB gene expression in mouse L cells (Dinter and Hauser, 1987; Du and Maniatis, 1992). PRDI and PRDIII act as binding sites for a nuclear transcription factor, designated "interferon regulatory factor-1" (IRF-1) (Miyamoto *et al.*, 1988), whose expression is transiently increased by virus infection and which appears to mediate the activation of transcription of the IFNB gene (Fujita *et al.*, 1988; Harada *et al.*, 1989; Xanthoudakis *et al.*, 1989). A second virus-inducible factor, designated "interferon regulatory

factor-2" (IRF-2), also binds to PRDI but suppresses, rather than activates, transcription (Harada *et al.*, 1989, 1990). PRDII is a binding site for the nuclear transcription factor NFκB (Clark and Hay, 1989; Fujita *et al.*, 1989a; Hiscott *et al.*, 1989; Lenardo *et al.*, 1989; Visvanathan and Goodbourn, 1989), which interacts with the major groove of the DNA (Thanos and Maniatis, 1992). Additionally, another protein, high-mobility group Y/1, also binds to PRDII, interacting with the minor groove of the DNA (Thanos and Maniatis, 1992). Both factors appear to be necessary for virus induction of the IFNB gene promoter. PRDIV contains a binding site for a protein of the cAMP response element binding protein (ATF/CREB) family of transcription factors (Du and Maniatis, 1992).

Viral induction of the IFNB gene is thought to occur following activation of pre-existing NFκB and by *de novo* synthesis of IRF-1; these nuclear transcription factors bind to the tandemly arranged PRDI and PRDII and act cooperatively to initiate/activate transcription (Leblanc *et al.*, 1990; Lenardo *et al.*, 1989; Visvanathan and Goodbourn, 1989; Fujita *et al.*, 1989b; Watanabe *et al.*, 1991). Reporter constructs containing PRDI supported by a simian virus 40 (SV40) enhancer, or (GAAAGT)$_4$, which contains the functional equivalent of dimeric PRDI (Näf *et al.*, 1991) are activated not only by virus but also by overexpression of IRF-1 (Näf *et al.*, 1991; MacDonald *et al.*, 1990). However, in most cell lines, overexpression of IRF-1 has led to poor induction of IFNB (and IFNA) genes (Harada *et al.*, 1990; Fujita *et al.*, 1989b) or none at all (MacDonald *et al.*, 1990; Reis *et al.*, 1992). This has been attributed to the repressive effect of IRF-2, a homologue of IRF-1, which also binds to PRDI (Harada *et al.*, 1989). In the undifferentiated murine embryonal carcinoma (stem) cell line P19, which is refractory to viral induction of IFNB (and IFNA) genes and which expresses neither IRF-1 nor IRF-2, overexpression of an introduced IRF-1 construct leads to activation of endogenous IFN genes and to activation of reporter plasmids with the IFNB promoter (Harada *et al.*, 1990). In addition, cell lines permanently transformed with an antisense IRF-1 expression plasmid exhibited strongly reduced IFNB gene inducibility that nevertheless could be restored by transient transformation with an IRF-1-overproducing expression plasmid (Reis *et al.*, 1992). However, the role of IRF-1 in virus-induced activation of the IFNB promoter remains controversial (Whiteside *et al.*, 1992; Pine *et al.*, 1990) and Ruffner and colleagues (1993) have shown that in murine embryonal stem cells devoid of both IRF-1 gene alleles (IRF-1 $^0/_0$) viral induction of IFNB was only slightly higher in control IRF-1 $^0/_+$ differentiated stem cells than that in IRF-1 $^0/_0$ differentiated cells. This suggests that while IRF-1 at high levels may elicit or enhance induction of IFNB under certain circumstances, it is not essential for viral induction. In cultured mouse fibroblasts devoid of IRF-1, IFNB induction by the synthetic dsRNA molecule poly(I):poly(C) was absent, whereas induction by Newcastle disease virus (NDV) was normal (Matsuyama et al., 1993). However, IFN induction in vivo by either virus or dsRNA has been found to be unimpaired in IRF-1 $^0/_0$ mice, indicating that IRF-1 is not essential (Reis et al., 1994). It has also become clear recently that the PRDI site can bind factors other than IRF-1 and IRF-2, and these may be more important for regulating IFNB gene activation (Whiteside et al., 1992; Keller and Maniatis, 1991).

In contrast, targeted disruption of the IRF-2 gene to yield mouse fibroblasts deficient in the repressor IRF-2 has been found to lead to upregulated induction of IFNB following NDV infection (Matsuyama et al., 1993). This suggests that IRF 2 negatively regulates or represses IFNB gene induction.

The induction of the IFNA1 gene appears to be regulated differently from that of the IFNB gene. IRF-1 is bound poorly by the equivalent PRDI site in the IFNA1 promoter and this promoter also lacks an NFκB site (Figure 25.1) (Miyamoto et al., 1988; MacDonald et al., 1990). The IFNA1 gene VRE does contain a hexameric repeat nucleotide sequence $(GAAATG)_4$, designated a "TG-sequence" (MacDonald et al., 1990), which appears to mediate virus inducibility when supported by an SV40 enhancer, but which does not respond to IRF-1 (Näf et al., 1991). It has, however, been reported that overexpression of IRF-1 can induce IFNA genes, at least under special circumstances (Harada et al., 1990; Au et al., 1992).

The IFNW gene, like IFNA and IFNB genes, is virus inducible and has structural features in its 5′ promoter region similar to those in IFNA/B promoters (Figure 25.1). In particular, hexameric repeat units are present, but are organized differently from those present in IFNA/B genes (Hansen et al., 1991; Roberts et al., 1992). However, the regulation of transcription of the IFNW gene has not been studied in detail.

3. IFN Proteins

3.1 STRUCTURAL FEATURES

3.1.1 IFN-α Subtypes

The 14 IFN-α subtypes are secreted proteins and as such are transcribed from mRNAs as precursor proteins, pre-IFN-α, containing N-terminal signal polypeptides of 23 mainly hydrophobic amino acids (Figure 25.2). The signal polypeptide is cleaved off before "mature" IFN-α molecules are secreted from the cell. From their cDNA sequences, mature IFN-α subtypes have been predicted to contain 166 amino acids (except IFN-α₂, 165 amino acids) (Mantei et al., 1980; Nagata et al., 1980; Streuli et al., 1980; Goeddel et al., 1981; Lawn et al., 1981a,b; Gren et al., 1984). The calculated molecular mass of

recombinant IFN-α subtypes is approximately 18.5 kDa, although apparent molecular masses of leukocyte-derived IFN-α subtypes in sodium dodecyl sulphate (SDS)–polyacrylamide gels vary between 17 and 26 kDa, possibly owing to variable processing of C-terminal amino acids (Levy et al., 1981) and post-translational modifications. The amino acid sequences of IFN-α subtypes are highly related, with complete identity at 85 of the 166 amino acid positions (Langer and Pestka, 1985; De Maeyer and De Maeyer-Guignard, 1988). This is illustrated in Figure 25.2, where the amino acid sequences of the subtypes are compared to an idealized consensus sequence. Many of the positions where amino acids differ from subtype to subtype are conservative substitutions. Interestingly, IFN-α subtypes contain four cysteine residues whose positions (1, 29, 99, and 139) are highly conserved (Figure 25.2). These four cysteines form disulfide bridges (1–99, 29–139) which induce folding of the IFN-α molecule (Figure 25.3) and whose integrity is essential for biological activity (Morehead et al., 1984). IFN-α subtypes are predicted to contain a high proportion (~60%) of α-helical regions and are folded to form globular proteins (Zoon and Wetzel, 1984). It has not yet proved possible to apply x-ray crystallographic techniques to IFN-α subtypes, or to human IFN-β, but the three-dimensional crystal structure of recombinant mouse IFN-β, which is approximately 60% related in amino acid sequences to its human counterpart, has been solved (Senda et al., 1992). This has revealed that mouse IFN-β has a structure which consists of five α-helices folded into a compact α-helical bundle (Figure 25.4). From comparative sequence analysis it is predicted that in all mammalian IFN-α and IFN-β proteins these five α-helical domains are conserved (Korn et al., 1994; Horisberger and Di Marco, 1995) and thus show similarities with many other cytokines, which also have α-helical bundle structures (Bazan, 1990).

With one exception, IFN-α subtypes do not contain recognition sites (Asn-X-Ser/Thr) for N-linked glycosylation (Henco et al., 1985); only IFN-α₁₄ contains two of these sites (Figure 25.2). Nevertheless, O-linked glycosylation may be possible in other IFN-α subtypes. For example, it has been found that natural IFN-α₂, purified from human leukocyte IFN, contains the disaccharide galactosyl-N-acetylgalactosamine in O-linkage to Thr-106 (Adolf et al., 1991a). However, since IFN-α₂ is the only IFN-α subtype with a theonine at position 106, it may represent the only O-glycosylated IFN-α protein (Figure 25.2). (N.B. Recombinant IFN-α subtypes produced by E. coli are nonglycosylated since bacteria lack the biosynthetic machinery to add sugar residues to polypeptides; all recombinant IFN-α₂ products used clinically are nonglycosylated.)

3.1.2 IFN-β

Pre-IFN-β contains 187 amino acids, of which 21 comprise the N-terminal signal polypeptide and 166

```
                  S1      S10        S20  S23 1       10        20        30        40        50
IFN-α Consensus   MALSFSLLMA VLVLSYKSIC SLG CDLPQTHSLG NRRALILLAQ MGRISPFSCL KDRHDFGFPQ EEFDGNQFQK
IFN -α₁           ..SP.A...V LV...C..S. ... ....E....D ...T.M.... .S...S.... M......... ..........
IFN -α₂           ...T.A..V. L....C..S. .V. .......... S..T.M.... .R...L.... .........- ..........
IFN -α₄           .......... .......... ... .......... .......... .R........ ......E... ....DK....
IFN -α₅           ...P.V.... LV..NC.... ... ........S. ...T.MIM.. .......... .......... ..........
IFN -α₆           ...P.A.... LV...C..S. ..D .......... H..TMM.... .R...L.... ......R... ..........
IFN -α₇           ..R......V .......... ... .........R .......... .......... ...E.R..E .....H....
IFN -α₈           ...T.Y..V. LV......FS ... .......... .......... .R........ ......E... ....DK....
IFN -α₁₀          .......... .........* ... .......T.R .......G. .......... ......RI.. ..........
IFN -α₁₃          ..SP.A...V LV...C..S. ... ....E....D ...T.M.... .S...S.... M......... ..........
IFN -α₁₄          ...P...M.. LV...C..S. ... .N.S.....N ...T.M.M.. .R........ ......E... ..........
IFN -α₁₆          .......... .......... ... .......... .......... .....H.... ...Y...... .V........
IFN -α₁₇          .......... .......... ... .......... .......... .......... ...P...L.. ..........
IFN -α₂₁          .......... .......... ... .......... .......... .......... .......... ..........
IFN -α₂₂          ...P...M.. LV...C..S. ... .N.S.....N ...T.MI... .......... .......... ..........
IFN -ω            .V.LLP..V. LPLCHCGPCG ..S .....N.G.L S.NT.V..H. .R.....L.. ...R..R... .MVK.S.L..
IFN -β            .TNKCL.QI. L.LCFSTTAL .MS YN.LGFLQRS SNFQCQK.LW QLNGRLEY.. ...MN.DI.E .IKQLQ....


                  60         70         80         90         100        110
IFN -α Consensus  AQAISVLHEM IQQTFNLFST KDSSAAWDES LLEKFYTELY QQLNDLEACV IQEVGVEETP
IFN -α₁           .P.......L ...I....T. .........D ..D..C.... .......... M..ER.G...
IFN -α₂           .ET.P..... ...I...... .........T ..D..Y.... .......... ...G..T...
IFN -α₄           .......... .......... ......L..T .......... .......... ..........
IFN -α₅           .......... .......... ......L...T ..D..Y.... .........M M........D.
IFN -α₆           .E.......V .......... ....V....R ..D.LY.... .......... M...W.GG..
IFN -α₇           T......... .......... E......EQ. .......... .......... ..........
IFN -α₈           .......... .......... ......L..T ..DE.YI..D .......VLC D.....I.S.
IFN -α₁₀          .......... .......... E......EQ. ........I. .......... I......II.
IFN -α₁₃          .P.......L ...I....T. .........D ..D..C...C .......... M..ER.G...
IFN -α₁₄          .......... M......... .N.......T .....YI..F ..M....... ..........
IFN -α₁₆          .....AF... .......... .........T ..D..YI..F .......... T.......IA
IFN -α₁₇          T......... .......... E......EQ. .......... .....N.... .....M....
IFN -α₂₁          .......... .......... ....T.EQ. ........N .....M.... ..........
IFN -α₂₂          .......... .......T..T ..D..Y.... .......... .........M M......D..
IFN -ω            .HVM...... L..I.S..H. ER.....NMT ..DQLH...H ...QH..T.L L.V..EG.SA
IFN -β            ED.ALTIY.. L.NI.AI.RQ DS..TG.N.T IV.NLLANV. H.I.H.KTVL EEKLEK.DFT


                  120        130        140        150        160        166
IFN- α Consensus  LMNEDSILAV RKYFQRITLY LTEKKYSPCA WEVVRAEIMR SFSFSTNLQK RLRRKD------
IFN -α₁           ...A...... K...R..... .......... .......... .L.L....E .....E
IFN -α₂           ..K....... .......... .K........ .......... ...L....E S...S.E
IFN -α₄           .......... .......... .I.R...... .......... .L........ ......
IFN -α₅           ...V...T.. .......... .......... .......... ...L.A...E .....E
IFN -α₆           .......... .......... .......... .......... ...S.R...E .....E
IFN -α₇           .....F.... .......... .M........ .......... .......K. G.....
IFN -α₈           ..Y....... .......... .......S.. .......... ...L.I.... ..KS.E
IFN -α₁₀          .......... .......... .I.R...... .......... .L........ ......
IFN -α₁₃          ...A.....L K...R..... .......... .......... .L.L....E .....E
IFN -α₁₄          .......... .......K... .M........ .......... .......... ......
IFN -α₁₆          .......... .......... .MG....... .......... .......... G.....
IFN -α₁₇          .......... .......... .......... .......... .L........ I....
IFN -α₂₁          ...V...... K......... .......... .......... ...L.KIF.E .....E
IFN -α₂₂          ...V...T.. .......... .......... .......... ...L.A...E .....E
IFN -ω            GAISSPA.TL .R...G.AV. .K....D.. ......M..K .LFL...M.E ...S.DADLGSS
IFN -β            RGKLM.S.HL KR.YG..LH. .KA.E..H.. .TI..V..L. N.Y.INR.TG Y..N
```

Figure 25.2 Amino acid sequence comparison of human IFN-α subtypes, IFN-β, and IFN-ω. All sequences are shown in comparison to an IFN-α consensus subtype sequence obtained by computer analysis of IFN-α subtype sequences. Signal sequences are marked S. Sites of *N*-linked glycosylation are underlined.

comprise the mature IFN-β protein (Derynck *et al.,* 1980, 1981; Taniguchi *et al.,* 1980; Houghton *et al.,* 1981). Although IFN-β is the same length as the majority of IFN-α subtypes, it shows only approximately 30% amino acid sequence relatedness with them (Figure 25.2) and is antigenically distinct. The IFN-β protein lacks the N-terminal Cys-1 residue present in IFN-α subtypes, but contains three other cysteines at positions 17, 31, and 141, the latter two corresponding to the disulfide bond pairing 29–139 in IFN-α subtypes. Replacement of Cys-17 by serine does not result in any loss of biological activity, whereas serine substitution of Cys-141 does (Mark *et al.,* 1981; Shepard *et al.,* 1981). As mentioned previously, on the basis of the three-dimensional structure of recombinant mouse IFN-β (Figure 25.4), human IFN-β is predicted to contain five α-helices and to fold up into an α-helical bundle structure (Senda *et al.,* 1992).

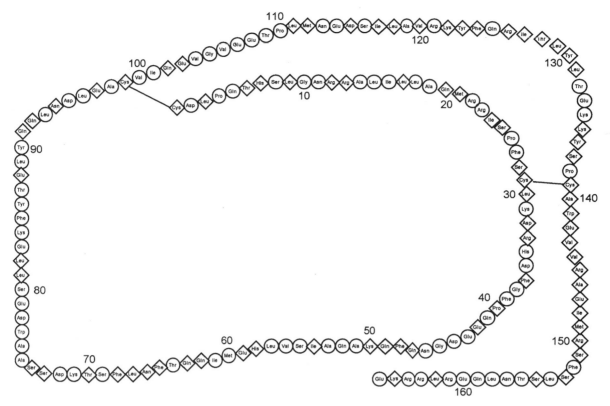

Figure 25.3   Consensus amino acid sequence of human IFN-α. Residues in squares are common to all known IFN-α subtypes.

Human IFN-β has one potential *N*-glycosylation site at Asn-80 (Taniguchi *et al.*, 1980) and *N*-linked oligosaccharides, primarily of the biantennary complex-type, are known to be attached to this site in natural IFN-β (Hosoi *et al.*, 1988). However, these may vary considerably depending on the producer cell type (Utsumi *et al.*, 1989).

3.1.3   IFN-ω

From cDNA sequence data, it was predicted that pre-IFN-ω contains 195 amino acids, the N-terminal 23 comprising the signal sequence and the remaining 172 the mature IFN-ω protein (Capon *et al.*, 1985; Hauptmann and Swetly, 1985). The amino acid sequence of IFN-ω is therefore six residues longer at the C-terminus than IFN-α or IFN-β proteins. However, it has been found that natural IFN-ω is heterogeneous at the N-terminus owing to variable cleavage of pre-IFN-ω; about 60% of mature IFN-ω molecules carry two additional N-terminal amino acids (Adolf, 1990; Shirono *et al.*, 1990). It is approximately 60% related to IFN-α subtype sequences, but only 30% related to that of IFN-β (Figure 25.2), and is antigenically distinct from both IFN-α and IFN-β (Adolf, 1990). Nevertheless, the four cysteines occur in the same notional positions, 1, 29,

99, and 139, as they do in IFN-α subtypes and it is likely that IFN-ω will have a similar α-helical bundle structure to those predicted for both IFN-α and IFN-β proteins (Senda *et al.*, 1992). IFN-ω has one potential site at Asn-78 for *N*-linked glycosylation and natural IFN-ω has been demonstrated to be a glycoprotein with biantennary complex oligosaccharides (containing neuraminic acid) attached at this site (Adolf, 1990; Adolf *et al.*, 1991b).

4.   Cellular Sources and Production

Type I IFNs (IFN-α, -β, and -ω) are produced by a variety of normal cell types responding to extracellular or intracellular stimuli (Stewart, 1979). IFN-α, as a mixture of subtypes, and IFN-ω may be produced together following viral infection of null lymphocytes or monocytes/macrophages (Cantell and Hirvonen, 1977; Adolf, 1990). The proportions of IFN-α subtypes may vary according to the type of virus used as inducer (Hiscott, 1984; Finter, 1991). However, production of IFN-β is usually restricted to double-stranded polynucleotide, e.g., poly-inosinic, poly-cytidylic acid, or

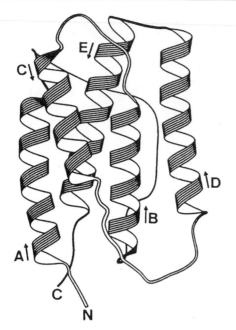

Figure 25.4   Ribbon drawing of the Cα structure of mIFN-β (Senda et al., 1992). The view is perpendicular to the helical axes. Helices A, B, C, D, and E, as well as NH₂- and COOH-termini, are labeled. The arrows indicate the direction of the helices. (Reprinted with permission of Horisberger and Di Marco (1995) Pharmac. Ther. 66, 535. Copyright Elsevier Science Ltd., Amsterdam, The Netherlands.)

virally-induced normal fibroblasts and other tissue cell types, e.g., epithelial cells (Meager et al., 1979; Stewart, 1979). In all the above cases, the amount of IFN secreted is dependent on the dose of the inducer.

Besides normal cells, a range of transformed and tumor-derived cell lines are known IFN producers, e.g., MG63 human osteosarcoma cell line (Meager et al., 1982), and the Namalwa B-lymphoblastoid cell line (Phillips et al., 1986). Generally speaking, adherent fibroblastic cell lines produce mainly IFN-β and only a minor quantity of IFN-α (Havell et al., 1978), whereas nonadherent myeloid or lymphoid cell lines produce mainly IFN-α and only a small amount of IFN-β (Cantell and Hirvonen, 1977; Shuttleworth et al., 1983; Zoon et al., 1992).

Actual production of IFNs lasts only a matter of a few hours following induction. This is due to IFN mRNA instability and the rapid shut-off of IFN gene transcription (Caput et al., 1986; Shaw and Kamen, 1986). Under conditions where IFN mRNA stability is increased, e.g., by blocking protein and RNA synthesis following induction, IFN production has been shown to be "superinduced" once the block on protein synthesis is removed (Meager et al., 1979; Stewart, 1979).

5.   Biological Activities Associated with IFN-α/β/ω

It is well established that the biological activities of IFNs are mostly dependent upon protein synthesis with selective subsets of proteins mediating individual activities. Antiviral, antiproliferative and immunomodulatory activities have been ascribed to IFN-α/β/ω (reviewed in Pestka et al., 1987; De Maeyer and De Maeyer-Guignard, 1988). The proteins and mechanisms involved in these activities are described below.

5.1   ANTIVIRAL ACTIVITY

5.1.1   Molecular Mechanisms

Despite there being vast numbers of viruses with different replication strategies, it appears that many viruses can be countered by relatively few IFN-inducible "antiviral" proteins (Samuel, 1987). One of the best-characterized of these is a family of enzymes collectively known as "2-5A synthetase" which, in the presence of dsRNA (often an intermediate of viral RNA synthesis) catalyses the formation of an unusual oligonucleotide, ppp (A2′p) nA (2-5A), which in turn activates an IFN-induced latent endonuclease, RNase L (Williams and Kerr, 1978; Wreschner et al., 1981; Ghosh et al., 1991; Hovanessian, 1991; Lengyel, 1982; Zhou et al., 1993). When activated, the RNase L degrades viral (and cellular mRNA) and therefore inhibits viral protein synthesis. Small RNA viruses (picornaviridae), e.g., Mengo virus and murine encephalomyocarditis virus (EMCV), whose replication is cytoplasmic are most inhibited by the induction of 2-5A synthetase (Lengyel, 1982; Rice et al., 1985; Chebath et al., 1987; Kumar et al., 1988). A further important IFN-induced "antiviral" protein is a dsRNA-dependent protein kinase, now designated PKR, which in the active form phosphorylates the peptide initiation factor, eIF2, involved in polyribosomal translation of mRNA (Miyamoto and Samuel, 1980; Gupta et al., 1982; Samuel, 1987). Phosphorylated eIF2 is inactive and thus viral protein synthesis is inhibited. This inhibition has been associated with the loss of replicating capacity of reoviruses and rhabdoviruses such as vesicular stomatitis virus (VSV).

The 2-5A synthetase and PKR antiviral mechanisms are rather general and potentially could affect a wide range of viruses. However, IFN can induce certain proteins that inhibit specifically one class of virus. For example, the IFN-inducible Mx proteins block the replication of influenza virus, probably by inhibiting the nuclear phase of viral transcription (mouse cells) or later cytoplasmic phases (human cells), without affecting the replication of many other viruses (Staeheli, 1990; Melén et al., 1992; Ronni et al., 1993).

In addition, since IFNs can impair various steps of viral replication, including penetration, uncoating and assembly of progeny virions as well as transcription and translation, there are likely to be several other antiviral proteins and mechanisms (De Maeyer and De Maeyer-Guignard, 1988). For instance, some viruses, e.g., herpes virus and certain retroviruses, appear to be inhibited at the relatively late stage of virus particle maturation and budding (Aboud and Hassan, 1983).

5.1.2 Defense Mechanisms of Viruses

In the course of evolution, many viruses have developed countermechanisms by which they can disrupt the antiviral mechanisms induced by IFN. Such countermechanisms often point to the significance of particular "antiviral" proteins. One of the main "targets" for several different viruses is the IFN-inducible PKR. The action of this kinase is overcome in adenovirus or Epstein–Barr virus (a member of the herpes virus family) by the production of small viral RNA molecules, VAI- and EBER-RNAs, respectively, which bind to PKR and block its activation by dsRNA (Clarke et al.; 1991; Ghadge et al., 1991; Mathews and Shenk, 1991). Reoviruses and vaccinia virus (a member of the pox virus family) produce viral proteins (sigma3 and SKI, respectively), that bind to dsRNA and thus reduce activation of PKR (Sen and Lengyel, 1992). Interestingly, if IFN-treated VSV-infected cells are co-infected by vaccinia virus, VSV replication is rescued, presumably partly by the inhibitory effect of SKI on PKR (Whitaker-Dowling and Youngner, 1983). Vaccinia virus also produces a nonfunctional protein analog of eIF2 which competes with the real eIF2 for phosphorylation by PKR and thus dilutes out the antiviral effect of activated PKR (Beattie et al., 1991). Other viruses, such as influenza, may activate latent cellular inhibitors of PKR activity, e.g., a 58 kDa protein (p58) (Lee et al., 1992).

The 2-5A synthetase–RNase L system can also be subverted. For example, EMCV, a picornavirus, can inactivate RNase L in several cell lines, but this inactivation is usually blocked by IFN treatment (Lengyel, 1982). Herpes viruses, in contrast, appear to inhibit RNase L activation by producing competing analogs of 2-5A (Cayley et al., 1984).

Some viruses even have the ability to block the transcription of IFN-inducible genes. The "early" E1a regulatory proteins of adenoviruses prevent the activation of ISGF3 by IFN, probably by inhibiting the transcription of the ISGF3γ subunit (Ackrill et al., 1991; Gutch and Reich, 1991; Kalvakolanu et al., 1991; Nevins, 1991). In the case of hepatitis B virus-infected cells, the so-called virus-specified "terminal protein" inhibits IFN-inducible gene expression (Foster et al., 1991).

5.2 ANTIPROLIFERATIVE ACTIVITY

The antiviral mechanisms induced by IFN are mediated by enzymes, e.g., 2-5A synthetase and PKR, whose activities have broad implications for cell growth and proliferation. Viral replication may be regarded as a form of pathological growth of a foreign, "cell-like", entity at the expense of a living cell. In the presence of IFN, enzymes are activated which curtail protein synthesis in general, but because viral protein synthesis is normally rapid, the inhibitory effect on viral replication appears more dramatic than on the slower and more complex cellular growth. Possibly, the "IFN system" was evolved more as a part of a complex, interactive network of intercellular mediators of cell growth and proliferation than as one for antiviral mechanisms. Recent investigations tend to support the role of IFNs in regulating cell growth. For example, if a mutant form of PKR that is unable to phosphorylate eIF2 is introduced into cells, they undergo neoplastic transformation (Koromilas et al., 1992; Lengyel, 1993; Meurs et al., 1993). This suggests that PKR normally acts as a "tumor-suppressor'" gene product. Therefore, one of the mechanisms by which IFN inhibits cell proliferation could be through its capacity to induce enhanced expression/activity of PKR. IFN-α has also been reported to inhibit cyclin-dependent CDK-2 kinase, which is responsible for phosphorylation of retinoblastoma (RB) protein, and this could contribute to antiproliferative activity (Kumar and Atlas, 1992; Resnitzky et al., 1992; Zhang and Kumar, 1994).

The 2-5A synthetase–RNase L system may also have antiproliferative and tumor suppressor activities. For instance, the levels of these two enzymes are high in growth-arrested cells: introduction of 2-5A-like oligoadenylates into proliferating cells also causes growth impairment (Sen and Lengyel, 1992; Lengyel, 1993; Zhou et al., 1993).

The IFN-stimulated increases in synthesis of PKR and 2-5A synthetase are dependent on IFN-inducible transcription factors, such as IRF-1 (ISGF2) (Miyamoto et al., 1988; Pine et al., 1990; Williams, 1991; Reis et al., 1992). The latter has a short half-life and thus transcription of IFN-inducible genes is rapidly repressed by the longer-lasting, inhibitory IRF-2 (Harada et al., 1989). If IRF-2 is overexpressed, cells become transformed as even low level constitutive production of PKR and 2-5A synthetase, which can regulate normal cell growth, is abrogated. This transformation was reversed by overexpressing IRF-1 (Harada et al., 1993), indicating that IRF-1 can be viewed as a pivotal player in the growth, regulatory, and tumor suppressor machinery. A variety of other IFN-induced mechanisms, including suppression of oncogenes (Contente et al., 1990), depletion of essential metabolites (Sekar et al., 1983), and increased cell rigidity (E. Wang et al., 1981), could also contribute to its antiproliferative activity.

The antiproliferative effects of IFNs in different tumor cell lines cultured in vitro is highly variable. Besides tumor cell lines, IFN-α/β have antiproliferative activity in hematopoietic precursor cells, e.g., of the myeloid lineage (Rigby *et al.*, 1985; De Maeyer and De Maeyer-Guignard, 1988, for review). They are also potent inhibitors of angiogenesis, the process whereby blood capillaries are formed to envasculate tissues (Sidky and Borden, 1987).

5.3 IMMUNOREGULATORY ACTIVITY

Besides activating intracellular processes, IFNs can also activate intercellular activities, especially within the immune system, which are an essential part of host defense against infectious and invasive diseases. Thus, IFNs can stimulate indirect antiviral and antitumor mechanisms, which in the main rest upon cellular differentiation and the induction of cytotoxic activity. For example, in the presence of antigen-specific antibodies, macrophages can effect cell-mediated cytotoxicity. Such antibody-dependent cell-mediated cytotoxicity (ADCC) is enhanced by IFN, possibly through an augmentation of immunoglobulin G (IgG)-Fc receptor (FcR) expression (Hokland and Berg, 1981; Vogel *et al.*, 1983; De Maeyer and De Maeyer-Guignard, 1988). In addition, another category of leukocytes comprising large granular lymphocytes and known as natural killer (NK) cells are activated, by unknown mechanisms, to kill virally-infected or tumor cell targets independently of major histocompatibility complex (MHC) antigen expression (Trinchieri and Perussia, 1984; Rager-Zisman and Bloom, 1985; De Maeyer and De Maeyer-Guignard, 1988).

IFNs can stimulate increased expression of class I MHC antigens, i.e., HLA-A, -B, -C, which are crucial for recognition of foreign antigen by cytotoxic T lymphocytes (CTL, CD8[+]); recognition of virally infected cells by CTL depends on class I MHC antigen presentation of viral antigens at the cell membrane (Heron *et al.*, 1978; Fellous *et al.*, 1979). IFN-α/β have sometimes been observed to increase class II MHC antigen expression, which is necessary to trigger both humoral and cell-mediated immunity, but probably play a lesser role than IFN-γ, which is the major class II MHC antigen inducer (Baldini *et al.*, 1986; Rhodes *et al.*, 1986; De Maeyer and De Maeyer-Guignard, 1988).

5.4 BIOLOGICAL ACTIVITIES *IN VIVO*

All of the biological activities so far described (see above) for IFNs have followed from *in vitro* experimentation. Here it is possible to pick and choose conditions that favor particular outcomes, e.g., the antiviral response, by adjusting doses of IFN, times of incubation, levels of virus challenge, and so on. Such experiments illustrate the range of biological activities of IFNs but cannot define their physiological roles. That IFNs have the potential for inducing antiviral and antitumor activity suggests their main role *in vivo* is to act as regulators of host defense mechanisms, and to prevent pathophysiological events occurring. Investigations in experimental animals have supported this likelihood. For example, injection of mice with anti-IFN-α/β antibody has been shown to increase their susceptibility to a range of virus infections (Virelizier and Gresser, 1978; Gresser, 1984). The earliest evidence for an antitumor effect of IFN-α/β came from inoculation of murine L1210 cells into mice. L1210 cells are sensitive to the antiproliferative action of IFN-α/β *in vitro* and *in vivo*, IFNα/β prevented tumor growth by these cells. However, when a clone of L1210 was isolated that was resistant to the antiproliferative action of IFN-α/β, there occurred a similar retardation of tumor growth upon IFN-α/β treatment to that observed with "sensitive" L1210 cells, suggesting that IFN was acting indirectly *in vivo* by a host-mediated mechanism (Gresser *et al.*, 1970, 1972). A similar conclusion was reached more recently using IFN-resistant B-cell lymphoma cells (Reid *et al.*, 1989). In the intervening years, many studies have been conducted confirming that IFN can act directly (e.g., human IFN-α₂ against a range of human tumor xenografts in nude mice where human IFN-α₂ has no activity on the murine immune system) and indirectly (reviewed by Balkwill, 1989). Although antitumor activity has been clearly demonstrated by the application of exogenous IFNs, it is not certain that endogenously produced IFNs are involved in countering tumor growth. However, some experimental evidence that endogenous IFN could play a role in host resistance to cancer or its spread has been obtained by treating mice with anti-IFN antibodies. Under these conditions, the intraperitoneal transplantability of six different experimental murine tumors was observed (Gresser, 1984).

IFN-α/β can inhibit the growth of hematopoietic progenitor cells *in vitro* (Rigby *et al.*, 1985) and this is also likely to occur *in vivo*. Such an occurrence is undesirable in most instances, but suppression of proliferation of bone marrow hematopoietic cell precursors has been turned to advantage in protecting tumor-bearing mice against the cytotoxicity of chemotherapeutic agents such as 5-fluorouracil (5-FU) (Stolfi *et al.*, 1983).

6. *Receptors*

6.1 CHARACTERIZATION

6.1.1 General Features

IFNs exercise their actions in cells via IFN-specific cell surface receptors. These receptors bind IFNs with high affinity (Aguet, 1980) and transduce the signal occasioned by ligand (IFN) binding across the cell membrane into the cytoplasm. IFN-α, IFN-β and IFN-ω share the same binding sites (Aguet *et al.*, 1984; Flores *et al.*, 1991), but IFN-γ binds to different sites (Branca and Baglioni, 1981; Aguet *et al.*, 1988). The binding of IFN-α and IFN-β to lymphoid cells and fibroblasts has been studied extensively and has been reviewed (Rubinstein and Orchansky, 1986; Branca, 1988; Langer and Pestka, 1988; Grossberg *et al.*, 1989), and results have generally demonstrated the presence of up to a few thousand complex, high-affinity receptors per cell. Chemical cross-linking studies with ^{125}I-labeled IFN-α or IFN-β to receptor-bearing cells have led to the identification of various IFN–receptor complexes, their molecular masses ranging from 80 to 300 kDa (Joshi *et al.*, 1982; Eid and Mogensen, 1983; Faltynek *et al.*, 1983; Raziuddin and Gupta, 1985; Thompson *et al.*, 1985; Hannigan *et al.*, 1986; Vanden Broecke and Pfeffer, 1988; Colamonici *et al.*, 1992). Such results have suggested that there are either multiple binding sites for IFN-α, IFN-β, and IFN-ω or that there are complex multichain receptors (Colamonici *et al.*, 1992; Hu *et al.*, 1993).

Although on human cells all the IFN-α, IFN-β, and IFN-ω proteins compete for common binding sites, individual IFN-α subtypes show different levels of activities on cells (Streuli *et al.*, 1981; Weck *et al.*, 1981; Rehberg *et al.*, 1982) which appear to correlate with their binding behavior to the cell surface (Uzé *et al.*, 1985). In particular, IFN-α_1 shows a much lower binding for human membrane receptors than either IFN-α_2 or IFN-α_8 (Uzé *et al.*, 1988). Interestingly, this differential binding of human IFN-α subtypes is not manifested in bovine cells, and all of the subtypes exhibit high specific activities (Yonehara *et al.*, 1983; Shafferman *et al.*, 1987). IFN-β and IFN-ω are also active in bovine cells (Capon *et al.*, 1985; Adolf *et al.*, 1990), but this cross-reactivity does not extend to mouse cells, a feature that has provided experimental systems in which to characterize IFN-receptors. Thus, somatic cell genetic studies with human × rodent hybrid cells containing various combinations of human chromosomes have provided evidence that the presence of human chromosome 21 confers sensitivity of such hybrid "rodent" cells to human IFN-α, IFN-β and IFN-ω (Tan *et al.*, 1973; Slate *et al.*, 1978; Epstein *et al.*, 1982; Raziuddin *et al.*, 1984). Further, it was demonstrated that antibodies raised against human chromosome 21-encoded cell surface proteins were able to block the binding and action of human IFN-α to human cells, indicating that this chromosome contained a gene(s) specifying the human IFN cell surface receptor (Shulman *et al.*, 1984).

6.1.2 Molecular Cloning of IFN Receptor Components

The elucidation of the full complement of components of the IFN-α/β/ω receptor has long been sought. One methodology used to isolate receptor cDNAs involves transfecting mouse cells with total human DNA and then selecting for cells sensitive to human IFN-α. After several attempts, this approach led successfully to the isolation of a 2.7 kb cDNA from a library constructed from human lymphoblastoid (Daudi) cells, which encoded an IFN-α binding protein (Uzé *et al.*, 1990) containing 557 amino acids (molecular mass 63 485 Da) including a signal sequence of 27 mainly hydrophobic amino acids. This protein has a structure typical of a transmembrane glycoprotein: a large N-terminal extracellular domain, which potentially could be highly glycosylated owing to a preponderance of N-linked glycosylation sites, a short hydrophobic transmembrane domain, and an intracellular or cytoplasmic tail (Figure 25.5). Its amino acid sequence shows little homology with any currently available sequences of proteins, including the sequence of the human IFN-γ receptor (Aguet *et al.*, 1988); however, the extracellular domain has been predicted to show structural similarities with the latter receptor and to a lesser extent with the so-called hematopoietin receptor supergroup (Bazan, 1990a,b). The gene coding for this putative IFN-α receptor has been mapped to chromosome 21.q22 (Lutfalla *et al.*, 1992), in confirmation of the earlier rodent × human hybrid cell data (Tan *et al.*, 1973; Slate *et al.*, 1978; Epstein *et al.*, 1982; Raziuddin *et al.*, 1984).

Although the cloned "IFN-α receptor" could be shown to confer sensitivity to human IFN-α_8 in transfected mouse cells (Uzé *et al.*, 1990), such cells were relatively insensitive to human IFN-α_2 and human IFN-β. These findings, together with those from anti-IFN-α receptor antibody blocking studies (Colamonici *et al.*, 1990; Revel *et al.*, 1991; Uzé *et al.*, 1991) and affinity cross-linking studies with IFN-α_2 (Colamonici *et al.*, 1992), have suggested that a second IFN-α receptor exists or another component is required besides the cloned IFN-α_8 binding protein, to complete the receptor complex. This hypothesis is further supported by a study (Soh *et al.*, 1994) in which introduction of a yeast artificial chromosome (YAC) containing a segment of human chromosome 21 into chinese hamster ovary (CHO) cells conferred a greatly increased response to both IFN-α_2 and IFN-α_8, as well as an increased response to IFN-β and IFN-ω, whereas the expression of the IFN-α_8 binding protein alone did not confer sensitivity (Revel *et al.*, 1991). However, these increased responses can be "knocked out" by disruption of the IFN-α_8 binding protein gene in the YAC, and then reconstituted by expression of the cDNA encoding the IFN-α_8 binding protein (Cleary *et al.*, 1994), suggesting that cell surface expression of this protein is required for a fully functional receptor (see also Hertzog *et al.*, 1994; Constantinescu *et al.*, 1994).

A second human IFN-α/β receptor, which is probably the additional component of the receptor complex referred to above, has been cloned (Novick *et al.*, 1994). The 1.5 kb cDNA encodes a 331-amino-acid protein,

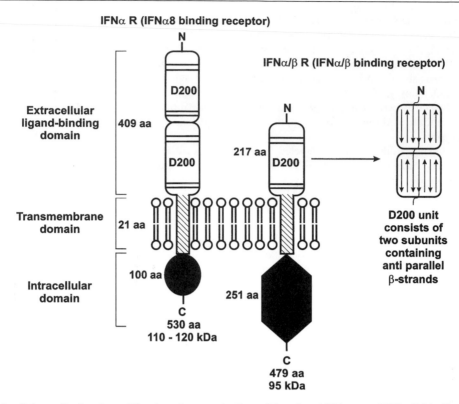

Figure 25.5   Schematic drawing of the domain organization of the clonal IFN-α_8- and IFN-α/β binding proteins.

including a signal sequence, which has the predicted structure of a transmembrane glycoprotein. The N-terminal ectodomain (217 amino acids) corresponds in sequence to a soluble 40 kDa IFN-α/β binding protein, p40, isolated from urine. This domain is linked to a transmembrane segment (21 amino acids) and a relatively small cytoplasmic domain of 67 amino acids (Figure 25.5). Overall, the primary sequence shows little homology with that of the previously cloned IFN-α_8 binding protein (Uzé *et al.*, 1990), but when the extracellular domains are compared, 23.4% relatedness is found (Novick *et al.*, 1994), suggesting that both of these IFN binding proteins belong to the same so-called class II cytokine receptor family (Uzé *et al.*, 1995). Two classes of cytokine receptor (class I and class II) have been proposed by Bazan (1990a,b), these being distinguished by the positions of cysteine pairs in the extracellular domain. The latter is comprised of fibronectin type III-like units containing around 200 amino acids and designated D200 (Uzé *et al.*, 1995). A schematic drawing of both IFN-α/β receptor chains is shown in Figure 25.5. Subsequently, it has been found that alternative splicing of the IFN-α/β receptor gene can produce a transcript encoding a long form of the receptor protein containing a larger cytoplasmic domain of 251 amino acids (Domanski *et al.*, 1995; Lutfalla *et al.*, 1995).

Mouse cells transfected with the cDNA encoding the

IFN-α/β binding protein bind IFN-α_2 but are insensitive to its effects, suggesting that an accessory protein, possibly the cloned IFN-α_8 binding protein, is required for signaling (Novick *et al.*, 1994; Constantinescu *et al.*, 1994). The findings that anti-p40 antiserum and a particular monoclonal antibody to the IFN-α_8 binding protein (Benoit *et al.*, 1993) both block the biological activity of IFN-$\alpha\beta\omega$ indicate that the IFN-α/β and IFN-α_8 binding proteins are in close proximity, and thus probably interact to form a high-affinity IFN-$\alpha/\beta/\omega$ receptor complex. The most likely scenario on present evidence is for a two-chain IFN-$\alpha/\beta/\omega$ receptor, comprising the cloned IFN-α_8 binding protein (Uzé *et al.*, 1990) and the long form of the IFN-α/β binding protein (Domanski *et al.*, 1995; Lutfalla *et al.*, 1995), each of which binds to some extent particular IFN types or IFN-α subtypes but which together more strongly bind all IFN-α, -β, and -ω species and function to transmit signals across the cell membrane. However, it is not completely ruled out that other cell surface components, e.g., membrane glycosphingolipids, are required for fully functional receptors (Colamonici *et al.*, 1992; Platanias *et al.*, 1994; Ghislain *et al.*, 1995; Uzé *et al.*, 1995) or, possibly, that there are alternative IFN receptors, e.g., the Epstein–Barr virus/complement C3d receptor as an IFN-α receptor on B-lymphocytes (Delcayre *et al.*, 1991). Vaccinia virus and other orthopoxviruses contain a gene B18R encoding a soluble type I IFN receptor which, unlike the class II cytokine type receptors, belongs to the immunoglobulin superfamily (Symons *et al.*, 1995; Colamonici *et al.*, 1995).

7. Signal Transduction

7.1 SIGNAL TRANSDUCTION

7.1.1 Molecular Mechanisms

The intracellular domains of the two cloned IFN-binding proteins are unrelated to the tyrosine kinase class of receptors, e.g., epidermal growth factor receptor (EGF R) and platelet-derived growth factor-receptor (PDGF-R), and are not predicted to have kinase activity of any sort (Uzé *et al.*, 1990; Novick *et al.*, 1994). However, it appears that the cytoplasmic domain of the IFN-α_8 and α/β binding proteins associate with nonreceptor tyrosine kinases TYK2 and Janus kinase 1 (JAK1), respectively, known to be involved in the signal transduction pathway of IFN-α/β and other cytokines (Novick *et al.*, 1994; Ghislain *et al.*, 1995; Ihle, 1995; Ihle and Kerr, 1995; Velasquez *et al.*, 1995). The current understanding of this pathway is as follows. After binding of IFN-$\alpha/\beta/\omega$ to their cognate receptors, the intracellular domains are phosphorylated by TYK2 and JAK1. These phosphorylated domains act as docking sites for the cytoplasmic STAT (signal transducers and activators of transcription) proteins p84/p91 (STAT1a/b) and p113 (STAT 2) (Ihle, 1996). The latter undergo tyrosine phosphorylation mediated by receptor-associated TYK2/JAK1, dimerize, translocate to the nucleus, and combine with a DNA binding protein, p48, to form the IFN-stimulated gene factor-3 (ISGF3) transcription factor complex (Schindler *et al.*, 1992; Velasquez *et al.*, 1992; Müller *et al.*, 1993; Platanias *et al.*, 1994; Shuai *et al.*, 1994; Gupta *et al.*, 1996; Yan *et al.*, 1996). Both TYK2 and JAK1 need to be reciprocally activated for signal transduction to occur, since cell mutants lacking either TYK2 or JAK1 are unresponsive to IFN-α (Ihle, 1995; Ihle and Kerr, 1995). ISGF3 binds to *cis*-acting IFN-stimulated response elements (ISRE), present in the promoter regions of IFN-inducible genes, to initiate their transcription (Williams, 1991). Targeted disruption of the STAT 1 gene in mice has shown that STAT 1 has an obligatory role in IFN-α and IFN-γ signaling (Durbin *et al.*, 1996; Meraz *et al.*, 1996).

The JAK1/TYK2 – ISGF3 pathway may not be the only "receptor-to-cell nucleus" signaling mechanism activated in IFN-stimulated cells. There has been some evidence to implicate protein kinase C (PKC) pathways as well (Reich and Pfeffer, 1990; Pfeffer *et al.*, 1991; C. Wang *et al.*, 1993). However, this remains controversial owing to the lack of specificity of kinase inhibitors used (Kessler and Levy, 1991; James *et al.*, 1992).

7.2 IFN-INDUCIBLE GENES

7.2.1 IFN-response Gene Sequences

IFN-inducible genes have a common regulatory nucleotide sequence (G/A)GGAAAN(N)GAAACT in their 5' flanking region and this type of sequence, which resembles the VRE sequences (Ryals *et al.*, 1985; Reid *et al.*, 1989) present in IFN genes, is designated interferon-stimulated response element (ISRE) (Williams, 1991). The resemblance between ISRE and VRE sequences probably accounts for the finding that many IFN-inducible genes are transcriptionally activated by virus infection or dsRNA, which also activate the transcription of IFN genes (Hug *et al.*, 1988; Wathelet *et al.*, 1988). As mentioned previously (see Section 5.1.3), IFN-receptor occupation activates cytoplasmic ISGF-3 and this complex is translocated to the nucleus and binds to ISRE of IFN-inducible genes as a transcriptional activator. In addition, a second factor, ISGF2, forms complexes with ISRE in IFN-stimulated cells. ISGF2 is a single, inducible phosphoprotein that has been shown to be identical to IRF-1 (Miyamoto *et al.*, 1988; Pine *et al.*, 1990; Williams, 1991; Reis *et al.*, 1992). The role of a third transcription factor, ISGF1, which is constitutively produced and requires only the central 9 bp core of ISRE for binding, remains to be fully defined (Kessler *et al.*, 1988). A number of other negative regulatory factors, including IRF2 (Harada *et al.*, 1989) and the ISGF2 (IRF1)/ISGF3γ-related "human interferon consensus sequence binding protein" (ICSBP) (Weisz *et al.*, 1992; Bovolenta *et al.*, 1994), which also bind to ISRE, are also probably involved in the regulation of transcription of IFN-inducible genes.

7.2.2 Proteins Induced by IFN

It is clear that the regulation of expression of IFN-inducible genes is complex and that the mechanisms that control their selective expression are not fully understood (see Taylor and Grossberg, 1990, for review). IFN-inducible proteins, whose number probably exceeds 20, include both those proteins induced early after IFN stimulation and those proteins that may be produced at later times, often in response to the actions of "early" IFN-inducible proteins (Sen and Lengyel, 1992). The full set of IFN-inducible proteins is probably not known, but several have been identified and characterized. Table 25.2 shows an incomplete list of IFN-inducible proteins together with their likely functions. Some of these proteins are not exclusively induced by IFN-$\alpha/\beta/\omega$; IFN-γ and certain other cytokines, e.g., tumor necrosis factor-α (TNF-α) often induce spectra of proteins that overlap with the set induced by IFN-$\alpha/\beta/\omega$ (Revel and Chebath, 1986; Rubin *et al.*, 1988; Wathelet *et al.*, 1992). It should be noted that IFN-inducible proteins tend also to be cell type-specific and thus not all proteins listed in Table 25.2 will be expressed in all cell types. In some cases, IFN-inducible proteins are completely absent from a cell before IFN stimulation, but in other cases

Table 25.2   IFN-inducible proteins

Protein	Function	Reference
2-5A synthetase	dsRNA-dependent synthesis of ppp(A2p)n-A [2-5A]; activator of RNase L	Revel and Chebath, 1986; Sen and Lengyel, 1992; Samuel, 1987; Staeheli, 1990.
dsRNA-activatable protein kinase (PKR)	Phosphorylation of peptide initiation factor eIF-2α	Revel and Chebath, 1986; Sen and Lengyel, 1992.
Class I MHC antigens (HLA-A, B, C) and β-microglobulin	Antigen presentation to cytotoxic T lymphocytes (CTL)	De Maeyer and De Maeyer Guignard, 1988; Heron et al., 1978.
Guanylate-binding proteins (GBP; γ67)	GTP, GDP binding	Schwemmle and Staeheli, 1994.
MxA	Specific inhibition of influenza virus replication	Ronni et al., 1993.
Metallothionein	Metal detoxification	Revel and Chebath, 1986.
Protein kinase C-ϵ (PKC- ϵ)	Serine/threonine protein phosphorylation	C. Wang et al., 1993.
Retinoblastoma (RB) gene product	Tumor suppressor protein	Kumar and Atlas, 1992.
15 kDa Ubiquitin cross-reactive protein	Targeting of structurally abnormal proteins for degradation	Loeb and Haas, 1992.
Vimentin	Intermediate filament network	Alldridge et al., 1989.
Tubulin	Cellular structural filaments	Fellous et al., 1982.
IRF1/ISGF2	Nuclear transcription factor	Sen and Lengyel, 1992.
IRF2	Nuclear repressor factor	Sen and Lengyel, 1992.
Interferon-inducible protein 35 (IFP35)	Leucine-zipper type transcription factor	Bange et al., 1994.
Interferon-inducible protein 56 (IFP56)	Unknown	Chebath et al., 1983.
Gene 200 cluster products	204 protein is nucleolar phosphoprotein	Choubey and Lengyel, 1992.
1-8U, 1-8D and 9-27 gene products	9-27 product is an RNA binding protein	Lawn et al., 1981a,b; Constantoulakis et al., 1993.
6-16 gene product	13 kDa hydrophobic protein of unknown function	Porter and Itzhaki, 1993.
Immunoglobulin Fc-receptor (FcR)	Binding of immunoglobulins	Hokland and Berg, 1981.
Intracellular 50 kDa Fcγ-binding protein	Unknown; binds IgG but not IgM, IgA or IgE	Thomas and Linch, 1991.
Thymosin B4	Induction of terminal transferase in B lymphocytes	Revel and Chebath, 1986.

they are being constitutively produced, their synthesis augmented by IFN.

8.   *Mouse IFN-α and IFN-β*

The genes for mouse IFN-α subtypes and mouse IFN-β (no functional mouse IFN-ω gene has been found) are located on mouse chromosome 4 (Dandoy *et al.,* 1984,

1985; De Maeyer and De Maeyer-Guignard, 1988, for review). These genes are, like their human counterparts, intronless and of comparable structure. Twelve mouse IFN-α genes or pseudogenes have been identified, of which the cDNAs for 10 different genes have been cloned and expressed (Langer and Pestka, 1985; De Maeyer and De Maeyer-Guignard, 1988). Mouse IFN-α subtype proteins contain 166 or 167 amino acids, or exceptionally 162 (mouse IFN-α_8), and the four cysteines at positions

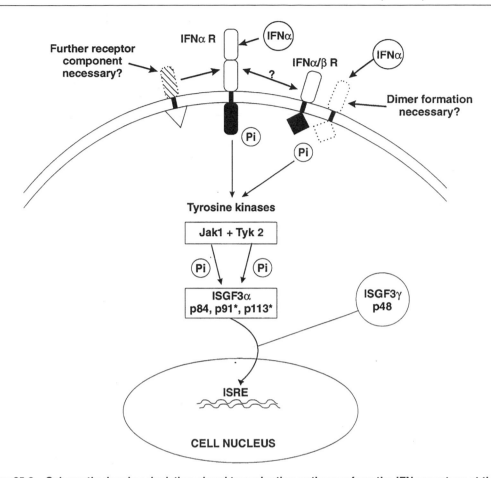

Figure 25.6   Schematic drawing depicting signal transduction pathways from the IFN receptors at the cell membrane to the cell nucleus. ISGF3 = interferon-stimulated gene factor-3. The ISGF3α complex contains three structurally related proteins p84, p91, and p113, and the ISGF3γ subunit is comprised of a single protein, p48. ISRE = interferon-stimulated response element, present in IFN-inducible genes. JAK1, JAK2 and TYK2, nonreceptor protein tyrosine kinases involved in the phosphorylation of ISGF3α proteins, p91 and p113. PKC = protein kinase C involved in the phosphorylation of nuclear transcription factor NFκB.

1, 29, 99, and 139, which are responsible for disulfide bond (Cys^1-Cys^{99}, Cys^{29}-Cys^{139}) formation in human IFN-α subtypes, are perfectly conserved. Most of the mouse IFN-α subtypes contain an N-linked glycosylation site at position 78 and thus are glycoproteins. In amino acid sequences, mouse IFN-α subtypes are about 40% homologous with their respective human counterparts (Langer and Pestka, 1985).

There is only a single-copy mouse IFN-β gene and this encodes the 161-amino-acid mature mouse IFN-β protein (Higashi *et al.*, 1983; De Maeyer and De Maeyer-Guignard, 1988). Mouse IFN-β contains only one cysteine and thus cannot form intramolecular disulfide bonds. It has three potential N-linked glycosylation sites and is heavily glycosylated when secreted from mouse fibroblasts; the molecular mass of the native glycoprotein is approximately 34 kDa compared to the predicted 17 kDa for the nonglycosylated counterpart (De Maeyer and De Maeyer-Guignard,

1988). The amino acid sequence of mouse IFN-β is about 48% related to that of human IFN-β. The three-dimensional structure of mouse IFN-β has been solved (Senda *et al.*, 1992) (see Section 3.1.1) and the protein has been shown to comprise five α-helices folded into a compact α-helical bundle (Figure 25.4).

Induction of transcription of mouse IFN-α subtype genes and the mouse IFN-β gene is probably regulated by transcription factor-binding nucleotide sequences present in the 5' noncoding promoter region, in a similar way to that of human IFN-α and IFN-β genes (see Section 2.2). For example, repeated GAAA-rich sequences are present in the 5' flanking regions of most mouse IFN-α subtype genes and these are likely to be important for virus-inducible transcription (Shaw *et al.*, 1983; Zwarthoff *et al.*, 1985). Inducers of mouse IFN-α and IFN-β synthesis, which include a number of viruses and double-stranded polynucleotides, are similar to those which induce human IFN-α, IFN-β, and IFN-ω production (Stewart, 1979; De Maeyer and De Maeyer-Guignard, 1988). Similarly, the type of IFN produced follows the pattern found among different human cell types: fibroblastic and epithelial cell lines produce mainly IFN-β, whereas leukocytes produce mainly IFN-α subtypes (De Maeyer and De Maeyer-Guignard, 1988).

The biological properties of mouse IFN-α and IFN-β are similar to those of human IFN-α, IFN-β and IFN-ω (see Section 5). Since mouse and human IFN-α subtypes are only 40% homologous, there is considerable species preference in biological activity, i.e., mouse IFN-α is weakly active in human cells and vice versa. Mouse IFN-β is also not active in human cells (Stewart, 1979).

Rather less is known regarding receptors for mouse IFN-α and IFN-β, than for the human counterparts, but it is probable that they comprise two or more chains, as is the case for the human IFN-α/β/ω receptors (Uzé et al., 1995). The mouse equivalent receptor chain to the IFN-$α_8$ binding protein (Uzé et al., 1990) has been cloned (Uzé et al., 1992). The gene for this mouse IFN-α/β receptor has been located to mouse chromosome 16 (Cheng et al., 1993). The mouse IFN-α/β receptor is 564 amino acids long and is divided into a large N-terminal extracellular domain (403 amino acids), a short hydrophobic transmembrane segment (20 amino acids), and a cytoplasmic domain (141 amino acids) (Uzé et al., 1992). The extracellular domain contains eight potential N-linked glycosylation sites and is predicted to exhibit the two-D200 domain structure of the human IFN-$α_8$ binding protein extracellular domain (Figure 25.5) (Uzé et al., 1995). Further mouse IFN-α/β receptors or components thereof await identification and characterization. Signal transduction via mouse IFN-α/β receptors is expected to involve the JAK1/TYK2–ISGF3 (STAT 1/2) pathway as outlined for human IFN-α/β/ω receptors (see Section 7.1). In STAT 1 gene-deleted mice there are no overt developmental abnormalities, but they display a complete lack of responsiveness to mouse IFN-α and IFN-γ (Durbin et al., 1996; Meraz et al., 1996). As a consequence, STAT 1 $^{-/-}$ mice are highly susceptible to infection by viruses and microbial pathogens. STAT 1 is therefore an obligatory mediator in the signal transduction pathway triggered by IFNs. Targeted disruption of the cloned mouse IFN-α/β receptor gave rise to a knockout with a similar phenotype (Müller et al., 1994). Such mice, lacking the IFN-α/β receptor, developed normally but were unable to respond to mouse IFN-α/β and thus unable to cope with viral infections.

9.   Clinical Uses of IFNs

9.1   GENERAL CONSIDERATIONS

The potent antiviral activity of IFN-α/β/ω together with their potential antitumor actions provided the impetus for large-scale manufacture of IFNs for the purpose of clinical evaluation in a variety of viral and malignant diseases. In the early 1970s, IFN production depended on pooled, human buffy coats (leukocytes) and thus only limited quantities could be made (Cantell and Hirvonen, 1977). Later in that decade, human lymphoblastoid cells (e.g., Namalwa), which could be grown to large culture volumes, became available for IFN production. By the 1980s, following the cloning of IFN-α and IFN-β, these IFN species were mass-produced by recombinant rDNA technology, leading to abundant availability of certain IFN-α subtypes, e.g., IFN-$α_2$ and "stabilized" IFN-β ser 17. There followed production of IFN-γ and IFN-ω by this means. Clinical usage of IFN-α preparations far exceeds that of IFN-β and IFN-ω because of early production difficulties with the latter types, though these are now solved.

At the beginning of the 1980s there was tremendous enthusiasm, both from manufacturers of IFNs and from clinicians, to evaluate the therapeutic potential of IFNs. However, early clinical trials had been poorly devised, were not "blinded", and often yielded only anecdotal evidence of success. It was only after many controlled, randomized studies had been conducted that it became apparent that IFNs in general, administered as a single agent, were not beneficial for the treatment of the majority of malignant diseases, including the major cancers (lung, breast, colon) of the developed world. The initial optimism all but vanished and was replaced in the mid-to-late 1980s by a more sober and realistic appreciation of the potential therapeutic value of IFNs. A number of general conclusions have been drawn, as follows. (i) IFN-α and IFN-β, and to a lesser extent IFN-γ, have antitumor activity in a small number of cancers, particularly in those that are relatively slow-growing and well-differentiated. (ii) There is no indication that heterogeneous IFN-α preparations containing mixtures of IFN-α subtypes (e.g., leukocyte IFN, lymphoblastoid IFN) have different clinical effects from those of homogenous, recombinant IFN-α subtype or IFN-β preparations. (iii) Continuous or intermittent high dosing appears to be required for antitumor efficacy. (iv) IFNs probably work best in patients with a minimal tumor burden (Balkwill, 1989).

A major concern that has emerged from clinical studies is that IFNs all generate a considerable number of undesirable, clinically observable, side-effects, including fever, chills, malaise, myalgia, headache, fatigue, and weight loss, and in certain cases these have been severe enough for treatment to be halted (Bottomly and Toy, 1985; Rohatiner et al., 1985; Goldstein and Laszlo, 1986). In addition, a variable proportion (1–40%) of patients treated with IFN-α or IFN-β, especially recombinant IFN-$α_2$ and recombinant IFN-β ser 17, develop neutralizing antibodies to the IFN species used (Rinehart et al., 1986; Antonelli et al., 1991), that in some instances have been associated with clinical "resistance" to IFN (Steis *et al.*, 1988; Öberg *et al.*, 1989; Freund *et al.*, 1989; Fossa *et al.*, 1992). A further important, but generally unrecognized, side-effect of IFN-α treatment is the possible induction of certain types of autoimmune disease (Feldmann *et al.*, 1989; Gutterman, 1994), probably mediated via IFN-induced upregulation of MHC antigen expression and generalized immunosuppression.

9.2   IFN TREATMENT OF MALIGNANT DISEASES

The most responsive cancer to IFN-α therapy is a very rare form of B-cell leukemia, known as "hairy cell" leukemia (HCL), in which a response rate up to 80% has been reported (Gutterman, 1994; Baron *et al.*, 1991; Vedantham *et al.*, 1992). In HCL patients, the "hairy cells" invade the spleen and bone marrow and the disease takes an indolent course. It has been shown convincingly that IFN-α therapy continued over several months leads to a clearance of "hairy cells" and in some patients a long-term remission is achieved. IFN-β ser 17 or IFN-γ were less effective against HCL (Saven and Piro, 1992). The IFN-α-induced mechanisms whereby clearance of "hairy cells" is achieved are not fully understood, but it is believed that a direct action of IFN-α leading to differentiation of "hairy cells" to a nonproliferating phenotype is involved (Vedantham *et al.*, 1992; Gutterman, 1994). Not all patients benefit greatly from IFN-α treatment and some develop neutralizing antibodies, particularly when IFN-α$_2$ is used (Steis *et al.*, 1988). When such neutralizing antibodies cause resistance to further IFN-α$_2$ treatment, clinical responses can be "rescued" by switching to a heterogeneous IFN-α preparation, e.g., leukocyte IFN-α (von Wussow *et al.*, 1991). However, on the whole, IFN-α therapy of HCL appears at least as effective and durable as chemotherapy with the drug pentostatin (2-deoxycoformycin) (Saven and Piro, 1992).

IFN-α therapy has also been shown to slow down the progression of chronic myelogenous leukemia (CML) (Baron *et al.*, 1991; Gutterman, 1994). In this malignant disease, leukemic cells grow slowly in the initial chronic, but benign, phase and persist for 2–4 years, but there follows a dramatic "blast crisis" producing rapidly proliferating myeloid leukemia cells and a fatal outcome. CML patients treated with IFN-α in the chronic phase often achieve durable remissions, associated with the elimination of leukemic cells bearing the so-called "Philadephia chromosome", sometimes lasting up to 8 years.

Other malignancies in which IFN-α therapy seems to work, although with generally a lower percentage of patients responding than in HCL and CML, include low-grade non-Hodgkin lymphoma, cutaneous T cell lymphoma, carcinoid tumors, renal cell carcinoma, squamous epithelial tumors of the head and neck, multiple myeloma, and malignant melanoma. In most of these cancers, complete responses are low compared to partial responses, but IFN-α may help with maintenance therapy of diseases in some cases, e.g., multiple myeloma (Mandelli *et al.*, 1990; Johnson and Selby, 1994).

The neovascularization of primary tumors is a crucial step in their development and thus the anti-angiogenic activity of IFN-α/β (Sidky and Borden, 1987) may have therapeutic value in certain early malignancies, e.g., primary melanoma (Gutterman, 1994). Kaposi sarcoma, often found in AIDS patients, has been regarded as an angiogenic tumor or angioproliferative disease, which may explain why IFN-α treatment can lead to regression of lesions in up to 40% of patients with this condition (De Wit *et al.*, 1988; Groopman and Scadden, 1989).

In preclinical systems, the combination of IFN therapy and conventional chemotherapy has appeared to offer greater chances of producing effective treatment of many cancers, but in clinical trials this strategy has produced mostly disappointing results (see Wadler and Schwartz, 1990, for review). This may be due to (i) the inability of preclinical models accurately to predict the clinical situation; (ii) the lack of understanding of the biochemical interactions and biological consequences of combining IFNs and chemotoxic agents; (iii) a failure to incorporate information on dose, scheduling, and sequence of administration of IFNs and chemotoxic agents into clinical trials.

9.3   IFN TREATMENT OF VIRAL DISEASES

Despite having proven antiviral activity *in vitro*, IFNs have not proved the hoped-for panacea for most common viral infections in man. IFN-α/β prevent the replication of common cold viruses (rhinoviruses and coronaviruses) in the test tube and when administered to volunteers in the form of a nasal spray, but cannot "cure" colds once they are established (Scott *et al.*, 1982; R.M. Douglas *et al.*, 1986; Turner *et al.*, 1986). IFN-α is only partially effective in preventing influenza virus infections (Treanor *et al.*, 1987).

Topical applications of IFN-α/β in the form of creams or ointments to herpes virus lesions, e.g., in herpes zoster (chickenpox), and genital warts (*Condyloma acuminatum*) caused by papilloma viruses have been investigated, but have given limited beneficial effects. However, when administered parenterally, i.e., by intramuscular or intravenous injection, greater beneficial effects of IFN-α/β on virally caused lesions and warts have been found, although not to an extent that IFN therapy has become the treatment of choice (Schneider *et al.*, 1987; J.M. Douglas *et al.*, 1990; Baron *et al.*, 1991; Gutterman, 1994). Another wart-like disease, juvenile laryngeal papilloma (JLP), which can severely obstruct the airways of young children, caused by the same papilloma virus types (6 and 11) as cause genital warts, has also been found to respond beneficially to IFN-α therapy. Disappointingly, IFN-α therapy appears neither curative nor of substantial value as an adjunctive agent in the long-term management of JLP (Healy *et al.*, 1988).

Probably the most successful application of IFN-α therapy to viral disease is in the treatment of chronic active hepatitis, caused by either hepatitis B or C viruses (Baron *et al.*, 1991; Gutterman, 1994). Up to about 40% of chronic active hepatitis B patients respond to IFN-α therapy; viral infectivity markers disappear and seroconversion and cure follow. It is interesting in the case of hepatitis B virus that viral activity is responsible for inhibiting the endogenous IFN system (Foster *et al.*, 1991), and thus the administration of exogenous IFN-α constitutes a replacement therapy. In hepatitis C virus infection, some serotypes of the virus are apparently more sensitive to IFN-α therapy than others and prolonged treatment may be necessary (>6 months) to prevent relapses occurring (Gutterman, 1994).

Both IFN-α and IFN-β have been shown to inhibit human immunodeficiency virus-1 (HIV-1) replication *in vitro* (Hartshorn *et al.*, 1987). However, *in vivo*, there is little evidence showing that IFN-α therapy has any long-term beneficial effect in asymptomatic HIV-1-positive individuals or AIDS patients (Friedland *et al.*, 1988; Lane *et al.*, 1990), except for limited regressions in Kaposi sarcoma lesions (De Wit *et al.*, 1988; Groopman and Scadden, 1989). Combination therapies for HIV-1-infected individuals involving IFN-α and antiviral drugs such as zidovudine (AZT) have also proved to be ineffective (Berglund, 1991).

9.4   IFN TREATMENT OF OTHER HUMAN DISEASES

As mentioned earlier, IFN-α/β inhibits hematopoiesis and therefore induces leukopenia in patients. This effect has generally been thought to be undesirable and it can lead to immunosuppression; however, it has proved useful for the treatment of diseases in which there is uncontrolled leukocytosis, e.g., thrombocytosis (markedly elevated platelet numbers), associated with various myeloproliferative diseases (Gisslinger *et al.*, 1989). Resistance to IFN-$α_2$ therapy has occurred in such patients when neutralizing antibodies to IFN-$α_2$ have developed, but successful retreatment with a heterogeneous IFN-α preparation, lymphoblastoid IFN (IFN-αN1) has been reported (Brand *et al.*, 1993).

The findings that production of IFN-α and IFN-γ was deficient in multiple sclerosis (MS) patients (Neighbor and Bloom, 1979) stimulated clinical trials to evaluate IFNs in MS. Rather unexpectedly, it has repeatedly been found that IFN-β, either natural fibroblast-derived or the later recombinant IFN-β ser 17 (IFN-β-1b), injected intrathecally, subcutaneously, or intramuscularly in patients with relapsing/remitting disease leads to a reduced rate of exacerbations of the disease and thus is possibly of clinical benefit in some patients (Jacobs *et al.*, 1981, 1987, 1993; The IFNB Multiple Sclerosis Study Group, 1993; Paty *et al.*, 1993). The IFN-β-induced mechanisms that contribute to this beneficial outcome are not known, but probably immunomodulatory actions are involved, e.g., suppression of growth and activity of autoreactive T lymphocytes in the central nervous system (Goodkin, 1994). It is unclear whether IFN-α would have a similar effect. However, the results with IFN-β treatment have been encouraging so far, although more follow-up of patients will be necessary to monitor any effects on the clinical progression of MS (Ebers, 1994).

10. *References*

Aboud, M. and Hassan, Y. (1983). Accumulation and breakdown of RNA-deficient intracellular virus particles in interferon-treated NIH 3T3 cells chronically producing Moloney murine leukaemia virus. J. Virol. 45, 489–495.

Ackrill, A.M., Foster, G.R., Laxton, C.D., Flavell, D.M., Stark, G.R. and Kerr, I.M. (1991). Inhibition of the cellular response to interferons by products of the adenovirus type 5 E1A oncogene. Nucleic Acids Res. 19, 4387–4393.

Adolf, G.R. (1987). Antigenic structure of human interferon ω1 (interferon alpha II): comparison with other human interferons. J. Gen. Virol. 68, 1669–1676.

Adolf, G.R. (1990). Monoclonal antibodies and enzyme immunoassays specific for human interferon (IFN) ω1: evidence that IFN-ω1 is a component of human leukocyte IFN. Virology 175, 410–417.

Adolf, G.R., Maurer-Fogy, I., Kalsner, I. and Cantell, K. (1990). Purification and characterisation of natural interferon ω1: two alternative cleavage sites for the signal peptidase. J. Biol. Chem. 265, 9290–9295.

Adolf, G.R., Kalsner, I., Ahorn, H., Maurer-Fogy, I. and Cantell, K. (1991a). Natural human interferon-α2 is O-glycosylated. Biochem. J. 276, 511–518.

Adolf, G.R., Frühbeis, B., Hauptmann, R., *et al.* (1991b). Human interferon ω1: isolation of the gene, expression in Chinese hamster ovary cells and characterization of the recombinant protein. Biochim. Biophys. Acta 1089, 167–174.

Aguet, M. (1980). High affinity binding of ^{125}I-labelled mouse interferon to a specific cell surface receptor. Nature 284, 459–461.

Aguet, M., Grobke, M. and Dreiding, P. (1984). Various human interferon α subclasses cross-react with common receptors: their binding activities correlate with their specific biological activities. Virology 132, 211–216.

Aguet, M., Dembic, Z. and Merlin, G. (1988). Molecular cloning and expression of the human interferon-γ receptor. Cell 55, 273–280.

Alldridge, L.C., O'Farrell, M.K. and Dealtry, G.B. (1989). Interferon β increases expression of vimentin at the mRNA and protein levels in differentiated embryonal carcinoma (PSMB) cells. Exp. Cell Res. 185, 387–393.

Allen, G. and Diaz, M.O. (1996). Nomenclature of the human

interferon proteins. J. Interferon and Cytokine Res. 16, 181–184.

Allen, G. and Fantes, K.H. (1980). A family of structural genes for human lymphoblastoid (leukocyte-type) interferon. Nature 287, 408–411.

Antonelli, G., Currenti, M., Turriziani, O. and Dianzani, F. (1991). Neutralizing antibodies to interferon-α: relative frequency in patients treated with different interferon preparations. J. Infect. Dis. 163, 882–885.

Au, W.-C., Raj, N.K.B., Pine, R. and Pitha, P.M. (1992). Distinct activation of murine interferon-α promoter region by IRF-1/ISGF-2 and virus infection. Nucleic Acids Res. 20, 2877–2884.

Baldini, L., Cortelezzi, A., Polli, N., et al. (1986). Human recombinant interferon α-2C enhances the expression of class II HLA antigens on hairy cells. Blood 67, 458–464.

Balkwill, F.R. (1989). "Cytokines in Cancer Therapy", pp. 8–53. Oxford University Press, Oxford.

Bange, F.-C., Vogel, U., Flohr, T., Kiekenbeck, M., Denecke, B. and Böttger, E.C. (1994). IFP35 is an interferon-induced leucine zipper protein that undergoes interferon-regulated cellular redistribution. J. Biol. Chem. 269, 1091–1098.

Baron, S., Tyring, S.K., Fleischmann, W.R.J., et al. (1991). The interferons: mechanisms of action and clinical applications. J. Am. Med. Assoc. 266, 1375–1383.

Bazan, J.F. (1990a). Shared architecture of the hormone binding domains in type I and II interferon receptors. Cell 61, 753–754.

Bazan, J.F. (1990b). Haemopoietic receptors and helical cytokines. Immunol. Today 11, 350–354.

Beattie, E., Tartaglia, J. and Paoletti, E. (1991). Vaccinia virus-encoded eIF-2 alpha homolog abrogates the antiviral effect of interferon. Virology 183, 419–422.

Benoit, P., Maguire, D., Plavec, I., Kocher, H., Tovey, M. and Meyer, F. (1993). A monoclonal antibody to recombinant human IFN-α receptor inhibits biologic activity of several species of human IFN-α, IFN-β and IFN-ω. J. Immunol. 150, 707–716.

Berglund, O., Engman, K., Ehrnst, A., et al. (1991). Combined treatment of symptomatic human immunodeficiency virus type 1 infection with native interferon α and zidovudine. J. Infect. Dis. 163, 710–715.

Bottomly, J.M. and Toy, J.L. (1985). Clinical side effects and toxicities of interferon, In "Interferon, vol. 4. In vivo and Clinical Studies" (eds. N.B. Finter and R.K. Oldham), pp.155–180. Elsevier, Amsterdam.

Bovolenta, C., Driggers, P.H., Marks, M.S., et al. (1994). Molecular interactions between interferon consensus sequence binding protein and members of the interferon regulatory factor family. Proc. Natl Acad. Sci. USA 91, 5046–5050.

Brack, C., Nagata, S., Mantei, N. and Weissmann, C. (1981). Molecular analysis of the human interferon-α gene family. Gene 15, 379–394.

Branca, A.A. (1988). The interferon receptors. In Vitro Cell Dev. Biol. 24, 155–165.

Branca, A.A. and Baglioni, C. (1981). Evidence that types I and II interferons have different receptors. Nature 294, 768–770.

Brand, C.M., Leadbeater, L., Lechner, K. and Gisslinger, H. (1993). Successful retreatment of an anti-interferon resistant polycythaemic vera patient with lymphoblastoid interferon-αN1 and in vitro studies on the specificity of the antibodies. Br. J. Haematol. 86, 216–218.

Cantell, K. and Hirvonen, S. (1977). Preparation of human leukocyte interferon for clinical use. Tex. Rep. Biol. Med. 35, 138–144.

Capon, D.J., Shepard, H.M. and Goeddel, D.V. (1985). Two distinct families of human and bovine interferon-α genes are coordinately expressed and encode functional polypeptides. Mol. Cell. Biol. 5, 768–779.

Caput, D., Beutler, B., Hartog, K., Thayer, R., Brown-Shimer, S. and Cerami, A. (1986). Identification of a common nucleotide sequence in the 3'-untranslated region of mRNA molecules specifying inflammatory mediators. Proc. Natl Acad. Sci. USA 83, 1670–1674.

Cayley, P.J., Davies, J.A., McCullagh, K.G. and Kerr, I.M. (1984). Activation of the ppp(A2'p)nA system in interferon-treated, herpes simplex virus-infected cells and evidence for novel inhibitors of the ppp(A2'p)nA-dependent RNase. Eur. J. Biochem. 143, 165–174.

Chebath, J., Merlin, G., Metz, R., Benech, P. and Revel, M. (1983). Interferon-induced 56,000 M_r protein and its mRNA in human cells: molecular cloning and partial sequence of the cDNA. Nucleic Acids Res. 11, 1213–1226.

Chebath, J., Benech, P., Revel, M. and Vigneron, M. (1987). Constitutive expression of (2'-5') oligo A synthetase confers resistance to picornavirus infection. Nature 330, 587–588.

Cheng, S., Lutfalla, G., Uzé, G., Chumakov, I. and Gardiner, K. (1993). GART, SON, IFNAR, and CRF-24 genes cluster on human chromosome 21 and mouse chromosome 16. Mammal Genome 3, 338–342.

Choubey, D. and Lengyel, P. (1992). Interferon action: nucleolar and nucleoplasmic localization of the interferon-inducible 72-kD protein that is encoded by the If: 204 gene from the gene 200 cluster. J. Cell Biol. 116, 1333–1341.

Clark, L. and Hay, R.T. (1989). Sequence requirement for specific interaction of an enhancer binding protein (EBP1) with DNA. Nucleic Acids Res. 17, 499–516.

Clarke, P.A., Schwemmle, M., Schickinger, J., Hilse, K. and Clemens, M. (1991). Binding of Epstein–Barr virus small RNA EBER-1 to the double-stranded RNA-activated protein kinase DAI. Nucleic Acids Res. 19, 243–248.

Cleary, C.M., Donnelly, R.J., Soh, J., Mariano, T.M. and Pestka, S. (1994). Knockout and reconstitution of a functional human type I interferon receptor complex. J. Biol. Chem. 269, 18747–18749.

Colamonici, O.R., D'Allesandro, F., Diaz, M.O., Gregory, L., Neckers, M. and Nordan, R. (1990). Characterization of three monoclonal antibodies that recognize the IFNα2 receptor. Proc. Natl Acad. Sci. USA 87, 7230–7234.

Colamonici, O.R., Pfeffer, L.M., D'Alessandro, F., et al. (1992). Multichain structure of the IFN-α receptor on hematopoietic cells. J. Immunol. 148, 2126–2132.

Colamonici, O., Domanski, P., Sweitzer, S., Larner, A. and Buller, R. (1995). Vaccinia virus B18R gene encodes a type I interferon-binding protein that blocks interferon α transmembrane signaling. J. Biol. Chem. 270, 15974–15978.

Constantinescu, S.N., Croze, E., Wang, C., et al. (1994). Role of interferon α/β receptor chain 1 in the structure and transmembrane signalling of the interferon α/β receptor complex. Proc. Natl Acad. Sci. USA 91, 9602–9606.

Constantoulakis, P., Campbell, M., Felber, B.K., Nasioulas, G., Afonina, E. and Parlakis, G.N. (1993). Inhibition of Rev-mediated HIV-1 expression by an RNA binding protein encoded by the interferon-inducible 9–27 gene. Science 259, 1314–1316.

Contente, S., Kenyon, K., Rimoldi, D. and Friedman, R.M. (1990). Expression of gene rrg is associated with reversion of NIH 3T3 transformed by LTR-c-H-ras. Science 249, 796–798.

Dandoy, F., Kelley, K., De Maeyer-Guignard, J., De Maeyer, E. and Pitha, P. (1984). Linkage analysis of the murine interferon-α locus on chromosome 4. J. Exp. Med. 160, 294–302.

Dandoy, F., De Maeyer, E., Bonhomme, F., Guenet, J. and De Maeyer-Guignard, J. (1985). Segregation of restriction fragment length polymorphism in an interspecies cross of laboratory and wild mice indicates tight linkage of the murine IFN-β gene to the murine IFN-α genes. J. Virol. 56, 216–220.

Degrave, W., Derynck, R., Tavernier, J., Haegeman, G. and Fiers, W. (1981). Nucleotide sequence of the chromosomal gene for human fibroblast (β1) interferon and of the flanking regions. Gene 14, 137–143.

Delcayre, A.X., Salas, F., Mathur, S., Kovats, K., Lotz, M. and Lernhardt, W. (1991). Epstein Barr virus/complement C3d receptor is an interferon α receptor. EMBO J. 10, 919–926.

De Maeyer, E. and De Maeyer-Guignard, J. (1988). "Interferons and Other Regulatory Cytokines." Wiley Interscience, New York.

Derynck, R., Content, J., De Clercq, E., et al. (1980). Isolation and structure of a human fibroblast interferon gene. Nature 285, 542–547.

Derynck, R., Devos, R., Remant, E., et al. (1981). Isolation and characterisation of a human fibroblast interferon gene and its expression in Escherichia coli. Rev. Infect. Dis. 3, 1186–1195.

De Wit, R., Schattenkerk, J.K., Boucher, C.A., Bakker, P.J., Veenhof, K.H. and Danner, S.A. (1988). Clinical and virological effects of high-dose recombinant interferon alpha in disseminated AIDS-related Kaposi's sarcoma. Lancet ii, 1214–1217.

Diaz, M.O. and Allen, G. (1993). Nomenclature of the human interferon genes. J. Interferon Res. 13, 443–444.

Dinter, H. and Hauser, H. (1987). Cooperative interaction of multiple DNA elements in the interferon-β promoter. Eur. J. Biochem. 166, 103–109.

Domanski, P., Witte, M., Kellum, M., et al. (1995). Cloning and expression of a long form of the β subunit of the interferon αβ receptor that is required for signalling. J. Biol. Chem. 270, 21606–21611.

Douglas, J.M., Eron, L.J., Judson, F.N., et al. (1990). A randomized trial of combination therapy with intralesional interferon alpha 2b and podophyllin versus podophyllin alone for the therapy of anogenital warts. J. Infect. Dis. 162, 52–59.

Douglas, R.M., Moore, B.W., Miles, H.B., et al. (1986). Prophylactic efficacy of intranasal alpha-2 interferon against rhinovirus infections in the family setting. N. Engl. J. Med. 314, 65–70.

Du, W. and Maniatis, T. (1992). An ATF/CREB binding site protein is required for virus induction of the human interferon β gene. Proc. Natl Acad. Sci. USA 89, 2150–2154.

Durbin, J., Hackenmiller, R., Simon, M. and Levy, D. (1996). Targeted disruption of the mouse Stat 1 gene results in compromised innate immunity to viral disease. Cell 84, 443–450.

Dworkin-Rastl, E., Dworkin, M.B. and Swetly, P. (1982). Molecular cloning of human alpha and beta genes from Namalwa cells. J. Interferon Res. 2, 575–585.

Ebers, G.C. (1994). Treatment of multiple sclerosis. Lancet 343, 275–278.

Eid, R. and Mogensen, K.E. (1983). Isolated interferon alpha-receptor complexes stabilised in vitro. FEBS Lett. 156, 157–160.

Emanuel, S.L. and Pestka, S. (1993). Human interferon-αA, -α2 and -α2 (Arg) genes in genomic DNA. J. Biol. Chem. 268, 12565–12569.

Epstein, C.J., McManus, N.H., Epstein, L.B., Branca, A.A., D'Alessandro, S.B. and Baglioni, C. (1982). Direct evidence that the gene product of the human chromosome 21 locus, IFRC, is the interferon-α receptor. Biochem. Biophys. Res. Commun. 107, 1060–1066.

Evinger, M., Maeda, S. and Pestka, S. (1981). Recombinant human leukocyte interferon produced in bacteria has antiproliferative activity. J. Biol. Chem. 256, 2113–2114.

Faltynek, C.R., Branca, A.A., McCandless, S. and Baglioni, C. (1983). Characterization of an interferon receptor on human lymphoblastoid cells. Proc. Natl Acad. Sci. USA 80, 3269–3273.

Fan, C.-M. and Maniatis, T. (1989). Two different virus-inducible elements are required for human β-interferon gene regulation. EMBO J. 8, 101–110.

Fantes, K.H. (1966). Purification, concentration and physico-chemical properties of interferons, In "Interferons" (ed. N.B. Finter), pp.118–180. North-Holland, Amsterdam.

Feldmann, M., Londei, M. and Buchan, G. (1989). Interferons and autoimmunity. In "Interferon 9" (ed. I. Gresser), pp. 73–90. Academic Press, New York.

Fellous, M., Kamoun, M., Gresser, I. and Bono, R. (1979). Enhanced expression of HLA antigens and α2-microglobulin on interferon-treated human lymphoid cells. Eur. J. Immunol. 9, 446–449.

Fellous, A., Ginzburg, I. and Littauer, U.Z. (1982). Modulation of tubulin mRNA levels by interferon in human lymphoblastoid cells. EMBO J. 1, 835–839.

Finter, N.B. (1991). Why are there so many subtypes of alpha-interferons. J. Interferon Res. Special Issue, Jan. 1991, 185–194.

Flores, I., Mirano, M. and Pestka, S. (1991). Human interferon omega (ω) binds to the α/β receptor. J. Biol. Chem. 266, 19875–19877.

Fossa, S.D., Lehne, G., Gunderson, R., Hjelmaas, U. and Holdener, E.E. (1992). Recombinant interferon α-2A combined with prednisone in metastatic renal-cell carcinoma: treatment results, serum interferon levels and the development of antibodies. Int. J. Cancer 50, 868–870.

Foster, G.M., Ackrill, A.M., Goldin, R.D., Kerr, I.M., Thomas, H.C. and Stark, G.R. (1991). Expression of the terminal protein region of hepatitis B virus inhibits cellular responses to interferons α and γ and double-stranded RNA. Proc. Natl Acad. Sci. USA 88, 2888–2892.

Freund, M., von Wussow, P., Diedrich, H., et al. (1989). Recombinant human interferon (IFN) alpha-2b in chronic myelogenous leukaemia: dose dependency of response and frequency of neutralizing anti-interferon antibodies. Br. J. Haematol. 72, 350–356.

Friedland, G.H., Klein, R.S., Saltzman, B.R. and The Interferon Alpha Study Group (1988). A randomized placebo-controlled trial of recombinant human interferon alpha 2a in patients with AIDS. J. Acquir. Immune Defic. Syndr. 1, 111–118.

Fujita, T., Ohno, S., Yasumitsu, H. and Taniguchi, T. (1985). Delimitation and properties of DNA sequences required for the regulated expression of human interferon-β gene. Cell 41, 489–496.

Fujita, T., Sakakibara, J., Sudo, Y., Miyamoto, M., Kimura, Y. and Taniguchi, T. (1988). Evidence for a nuclear factor(s), IRF-1, mediating induction and silencing properties to human IFN-β gene regulatory elements. EMBO J. 7, 3397–3405.

Fujita, T., Miyamoto, M., Kimura, Y., Hammer, J. and Taniguchi, T. (1989a). Involvement of a cis-element that binds an H2TF-1/NF-κB like factor(s) in virus-induced interferon-β expression. Nucleic Acids Res. 17, 3335–3346.

Fujita, T., Reis, L., Watanabe, N., Kimura, Y., Taniguchi, T. and Vilcek, J. (1989b). Induction of the transcription factor IRF-1 and interferon-β mRNAs by cytokines and activators of second messenger pathways. Proc. Natl Acad. Sci. USA 86, 9936–9940.

Fujita, T., Kimura, Y., Miyamoto, M., Barsoumian, E.L. and Taniguchi, T. (1989c). Induction of endogenous IFN-α and IFN-β genes by a regulatory transcription factor, IRF-1. Nature 337, 270–272.

Ghadge, G.D., Swaminathan, S., Katze, M.G. and Thimmapaya, B. (1991). Binding of the adenovirus VAI RNA to the interferon-induced 68-kDa protein kinase correlates with function. Proc. Natl Acad. Sci. USA 88, 7140–7144.

Ghislain, J., Sussman, G., Goelz, S., Ling, L.E. and Fish, E. (1995). Configuration of the interferon-α/β receptor complex determines the context of the biological response. J. Biol. Chem. 270, 21785–21792.

Ghosh, S.K., Kusari, J., Bandyopadhyay, S.K., Samanta, H., Kumar, R. and Sen, G.C. (1991). Cloning, sequencing and expression of two murine 2′-5′-oligoadenylate synthetases. J. Biol. Chem. 266, 15293–15299.

Gillespie, D. and Carter, W. (1983). Concerted evolution of human interferon α genes. J. Interferon Res. 3, 83–88.

Gisslinger, H., Ludwig, H., Linkesch, W., Chott, A., Fritz, E. and Radaszkiewicz, T.H. (1989). Long-term interferon therapy for thrombocytosis in myeloproliferative diseases. Lancet i, 634–637.

Goeddel, D.V., Yelverton, E., Ullrich, A., et al. (1980). Human leukocyte interferon produced by E. coli is biologically active. Nature 287, 411–416.

Goeddel, D.V., Leung, D.W., Dull, T.J., et al. (1981). The structure of eight distinct cloned human leukocyte cDNAs. Nature 290, 20–26.

Goldstein, D. and Laszlo, J. (1986). Interferon therapy in cancer: from imagination to interferon. Cancer Res. 46, 4315–4329.

Goodbourn, S. and Maniatis, T. (1988). Overlapping positive and negative regulatory domains of the human β-interferon gene. Proc. Natl Acad. Sci. USA 85, 1447–1451.

Goodbourn, S., Burstein, H. and Maniatis, T. (1986). The human β-interferon gene enhancer is under negative control. Cell 45, 601–610.

Goodkin, D.E. (1994). Interferon beta-1b. Lancet 344, 1057–1060.

Goren, T., Fischer, D.G. and Rubinstein, N. (1986). Human monocytes and lymphocytes produce different mixtures of α-interferon subtypes. J. Interferon Res. 6, 323–327.

Gray, P.W., Leung, D.W., Pennica, D., et al. (1982). Expression of human immune interferon cDNA in E. coli and monkey cells. Nature 295, 503–508.

Greenway, A.L., Overall, M.L., Sattayasai, N., et al. (1992). Selective production of interferon-alpha subtypes by cultured peripheral blood mononuclear cells and lymphoblastoid cell lines. Immunology 75, 182–188.

Gren, E., Berzin, V., Jansone, I., Tsimanis, A., Vishnevsky, Y. and Npsalons, U. (1984). Novel human leukocyte interferon subtype and structural comparison of α interferon genes. J. Interferon Res. 4, 609–617.

Gresser, I. (1984). Role of interferon in resistance to viral infection in vivo. In "Interferon 2: Interferons and the Immune System" (eds. J. Vilcek and E. De Maeyer), pp. 221–247. Elsevier, Amsterdam.

Gresser, I., Brouty-Boyé, D., Thomas, M.-T. and Macieira-Coelho, A. (1970). Interferon and cell division. I. Inhibition of the multiplication of mouse leukaemia L1210 cells in vitro by interferon preparations. Proc. Natl Acad. Sci. USA 66, 1052–1058.

Gresser, I., Maury, C. and Brouty-Boyé, D. (1972). Mechanism of antitumour effect of interferon in mice. Nature 239, 167–168.

Gresser, I., Belardelli, F., Maury, C., Maunoury, M.T. and Tovey, M.G. (1983). Injection of mice with antibody to interferon enhances the growth of transplantable murine tumours. J. Exp. Med. 158, 2095–2107.

Groopman, J.E. and Scadden, D.T. (1989). Interferon therapy for Kaposi's sarcoma associated with the acquired immunodeficiency syndrome (AIDS). Ann. Intern. Med. 10, 335–337.

Gross, G., Mayr, U., Bruns, W., Grosveld, F., Dahl, H.H.M. and Collins, J. (1981). The structure of a thirty-six kilobase region of the human chromosome including the fibroblast interferon gene IFN-β. Nucleic Acids Res. 9, 2495–2507.

Grossberg, S.E., Taylor, J.L. and Kushnaryov, V.M. (1989). Interferon receptors and their role in interferon action. Experientia 45, 508–513.

Gupta, S.L., Holmes, S.L. and Mehra, L.L. (1982). Interferon action against reovirus: activation of interferon-induced protein kinase in mouse L929 cells upon reovirus infection. Virology 120, 495–499.

Gupta, S., Yan, H., Wong, L., Ralph, S., Krolewski, J. and Schineller, C. (1996). The SH2 domains of Stat 1 and Stat 2 mediate multiple interactions in the transduction of IFNα signals. EMBO J. 15, 1075–1084.

Gutch, M.J. and Reich, N.C. (1991). Regression of the interferon signal transduction pathway by the adenovirus E1A oncogene. Proc. Natl Acad. Sci. USA 88, 7913–7917.

Gutterman, J.U. (1994). Cytokine therapeutics: lessons from interferon α. Proc. Natl Acad. Sci. USA 91, 1198–1205.

Hannigan, G.E., Lau, A.S. and Williams, B.R. (1986). Differential human interferon alpha receptor expression on proliferating and non-proliferating cells. Eur. J. Biochem. 157, 187–193.

Hansen, T.R., Leaman, D.W., Cross, J.C., Mathialagan, N., Bixby, J.A. and Roberts, R.M. (1991). The genes for the trophoblast interferons and the related interferon αII possess distinct 5′-promoter and 3′-flanking sequences. J. Biol. Chem. 266, 3060–3067.

Harada, H., Fujita, T., Miyamoto, M., et al. (1989). Structurally similar, but functionally distinct factors, IRF-1 and IRF-2, bind to the same regulatory elements of IFN and IFN-inducible genes. Cell 58, 729–739.

Harada, H., Willison, K., Sakakibara, J., Miyamoto, M., Fujita, T. and Taniguchi, T. (1990). Absence of type I IFN system in EC cells: transcriptional activator (IRF-1) and repressor (IRF-2) are developmentally regulated. Cell 63, 303–312.

Harada, H., Kitagawa, M., Tanaka, N., et al. (1993).

Anti-oncogenic and oncogenic potentials of interferon regulatory factors 1 and 2. Science 259, 971–974.

Hartshorn, K.L., Neumeyer, D., Vogt, M.W., Schooley, R.T. and Hirsh, M.S. (1987). Activity of interferons alpha, beta, and gamma against human immunodeficiency virus replication *in vitro*. AIDS Res. Human Retrovir. 3, 125–133.

Hauptmann, R. and Swetly, P. (1985). A novel class of human type I interferons. Nucleic Acids Res. 13, 4739–4749.

Havell, E.A., Hayes, T.G. and Vilcek, J. (1978). Synthesis of two distinct interferons by human fibroblasts. Virology 89, 330–334.

Healy, G.B., Gelber, R.D., Trowbridge, A.L., Grundfast, K.M., Ruben, R.J. and Price, K.N. (1988). Treatment of recurrent respiratory papillomatosis with human leukocyte interferon. N. Engl. J. Med. 319, 401–407.

Henco, K., Brosius, J., Fujisawa, A., *et al.* (1985). Structural relationship of human interferon-α genes and pseudogenes. J. Mol. Biol. 185, 227–260.

Henry, L., Sizun, J., Turleau, C., Boue, J., Azoulay, M. and Junien, C. (1984). The gene for human fibroblast interferon (IFB) maps to 9p21. Hum. Genet. 68, 67–70.

Heron, I., Hokland, M. and Berg, K. (1978). Enhanced expression of β2-microglobulin and HLA antigens on human lymphoid cells by interferon. Proc. Natl Acad. Sci. USA 75, 6215–6219.

Hertzog, P.J., Hwang, S.Y., Holland, K.A., Tymms, M.J., Iannello, R. and Kola, I. (1994). A gene on human chromosome 21 located in the region 21q22.2 to 21q22.3 encodes a factor necessary for signal transduction and antiviral response to type I interferons. J. Biol. Chem. 269, 14088–14093.

Higashi, Y., Sokawa, Y., Watanabe, Y., *et al.* (1983). Structure and expression of a cloned cDNA for mouse interferon-β. J. Biol. Chem. 258, 9522–9529.

Himmler, A., Hauptmann, R., Adolf, G.R. and Swetly, P. (1987). Structure and expression in *Escherichia coli* of canine alpha-interferon genes. J. Interferon Res. 7, 173–183.

Hiscott, J., Cantell, K. and Weissmann, C. (1984). Differential expression of human interferon genes. Nucleic Acids Res. 12, 3727–3746.

Hiscott, J., Alper, D., Cohen, L., *et al.* (1989). Induction of human interferon gene expression is associated with a nuclear-factor that interacts with the NFκB site of the human immunodeficiency virus enhancer. J. Virol. 65, 2557–2566.

Hokland, P. and Berg, K. (1981). Interferon enhances the antibody-dependent cellular cytotoxicity (ADCC) of human polymorphonuclear leukocytes. J. Immunol. 127, 1585–1588.

Horisberger, M. and Di Marco, S. (1995). Interferon-alpha hybrids. Pharmac. Ther. 66, 507–534.

Hoskins, M. (1935). A protective action of neurotropic against viserotropic yellow fever virus in *Macacus rhesus*. Am. J. Trop. Med. Hyg. 15, 675–680.

Hosoi, K., Utsumi, J., Kitagawa, T., Shimizu, H. and Kobayashi, S. (1988). Structural characterization of fibroblast human interferon-beta. J. Interferon Res. 8, 375–384.

Houghton, M., Jackson, I.J., Porter, A.G., *et al.* (1981). The absence of introns within a human fibroblast interferon gene. Nucleic Acids Res. 9, 247–266.

Hovanessian, A.G. (1991). Interferon-induced and double-stranded RNA-activated enzymes: a specific protein kinase and 2′,5′-oligoadenylate synthetases. J. Interferon Res. 11, 199–205.

Hu, R., Gan, Y., Liu, J., Miller, D. and Zoon, K.C. (1993). Evidence for multiple binding sites for several components of human lymphoblastoid interferon-α. J. Biol. Chem. 268, 12591–12595.

Hug, H., Costas, M., Staeheli, P., Aebi, M. and Weissmann, C. (1988). Organisation of the murine Mx gene and characterisation of its interferon-inducible promoter. Experientia 44, A23.

IFNB Multiple Sclerosis Study Group (1993). Interferon beta-1b is effective in relapsing-remitting multiple sclerosis. I. Clinical results of a multicenter, randomized, double-blind, placebo-controlled trial. Neurology 43, 655–661.

Ihle, J. (1995). Cytokine receptor signalling. Nature 377, 591–594.

Ihle, J. (1996). STATs: signal transducers and activators of transcription. Cell 84, 331–334.

Ihle, J. and Kerr, I. (1995). JAKs and STATs in signaling by the cytokine receptor superfamily. Trends Genet. 11, 69–74.

Isaacs, A. and Lindemann, J. (1957). Virus interference. I. The interferon. Proc. Roy. Soc. B. 147, 258–267.

Jacobs, L., O'Malley, J., Freeman, A. and Ekes, R. (1981). Intrathecal interferon reduces exacerbations of multiple sclerosis. Science 214, 1026–1028.

Jacobs, L., Salazar, A.M., Herndon, R., *et al.* (1987). Intrathecally administered natural human fibroblast interferon reduces exacerbations of multiple sclerosis: results of a multicenter, double-blind study. Arch. Neurol. 44, 589–595.

Jacobs, L., Cookfair, D., Rudick, R., *et al.* (1993). A phase III trial of intramuscular recombinant beta interferon as treatment for multiple sclerosis: current status. Ann. Neurol. 34, 310.

James, R.I., Menaya, J., Hudson, K., *et al.* (1992). Role of protein kinase C in induction of gene expression and inhibition of cell proliferation by interferon α. Eur. J. Biochem. 209, 813–822.

Johnson, P.W.M. and Selby, P.J. (1994). The treatment of multiple myeloma—an important MRC trial. Br. J. Cancer 70, 781–785.

Joshi, A.R., Sarkar, I.H. and Gupta, S.L. (1982). Interferon receptors. Crosslinking of human leukocyte alpha-2 to its receptor on human cells. J. Biol. Chem. 257, 13884–13887.

Kalvakolanu, D.V.R., Bandyopadhyay, S.K., Harter, M.L. and Sen, G.C. (1991). Inhibition of interferon-inducible gene expression by adenovirus E1A proteins: block in transcriptional complex formation. Proc. Natl Acad. Sci. USA 88, 7459–7463.

Keller, A.D. and Maniatis, T. (1991). Identification and characterisation of a novel repressor of interferon-β expression. Genes Dev. 5, 868–879.

Kessler, D.S. and Levy, D.E. (1991). Protein kinase activity required for an early step in interferon-α signalling. J. Biol. Chem. 266, 23471–23476.

Kessler, D.S., Levy, D.E., Pine, R. and Darnell, J.E.J. (1988). Two interferon-induced nuclear factors bind a single promoter element in interferon-stimulated genes. Proc. Natl Acad. Sci. USA 85, 8521–8525.

Knight, E. (1976). Interferon: purification and initial characterisation from diploid cells. Proc. Natl Acad. Sci. USA 73, 520–523.

Knight, E.J., Hunkapillar, M.W., Korant, B.D., Hardy, R.W.F. and Hood, L.E. (1980). Human fibroblast interferon: amino acid analysis and amino terminal amino acid sequence. Science 207, 525–526.

Korn, A., Rose, D. and Fish, E. (1994). Three-dimensional model of a human interferon alpha consensus sequence. J. Interferon Res. 14, 1–9.

Koromilas, A.E., Roy, S., Barber, G.N., Katze, M.G. and Sononberg, N. (1992). Malignant transformation by a mutant of the IFN-inducible dsRNA-dependent protein kinase. Science 257, 1685–1687.

Kumar, R., Choubey, D., Lengyel, P. and Sen, G.C. (1988). Studies on the role of the 2′-5′-oligoadenylate synthetase-RNase L pathway in β-interferon-mediated inhibition of encephalomyocarditis virus replication. J. Virol. 62, 3175–3181.

Kumar, R. and Atlas, F. (1992). Interferon α induces the expression of retinoblastoma gene product in human Burkitt lymphoma Daudi cells: role in growth regulation. Proc. Natl Acad. Sci. USA 89, 6599–6603.

Lane, H.C., Davey, V., Kovacs, J.A., et al. (1990). Interferon alpha in patients with asymptomatic human immunodeficiency virus (HIV) infection: a randomized, placebo-controlled trial. Ann. Intern. Med. 112, 805–811.

Langer, J.A. and Pestka, S. (1985). Structure of interferons. Pharmac. Ther. 27, 371–401.

Langer, J.A. and Pestka, S. (1988). Interferon receptors. Immunology Today 9, 393–400.

Lawn, R.M., Adelman, J., Dull, T.J., Gross, M., Goeddel, D. and Ullrich, A. (1981a). DNA sequence of two closely linked human leukocyte interferon genes. Science 212, 1159–1162.

Lawn, R.M., Gross, M., Houck, C.M., Franke, A.E., Gray, P.J. and Goeddel, D.V. (1981b). DNA sequence of a major human leukocyte interferon gene. Proc. Natl Acad. Sci. USA 78, 5435–5439.

Leblanc, J.-F., Cohen, L., Rodrigues, M. and Hiscott, J. (1990). Synergism between distinct enhanson domains in viral induction of human beta interferon gene. Mol. Cell. Biol. 10, 3987–3993.

Lee, T.G., Tomita, J., Hovanessian, A.G. and Katze, M.G. (1992). Characterization and regulation of the 58,000-dalton cellular inhibition of the interferon-induced, dsRNA-activated protein kinase. J. Biol. Chem. 267, 14238–14243.

Lenardo, M.J., Fan, C.-M., Maniatis, T. and Baltimore, D. (1989). The involvement of NF-κB in β-interferon gene regulation reveals its role as a widely inducible mediator of signal transduction. Cell 57, 287–294.

Lengyel, P. (1982). Biochemistry of interferons and their actions. Annu. Rev. Biochem. 51, 251–282.

Lengyel, P. (1993). Tumor-suppressor genes: news about the interferon connection. Proc. Natl Acad. Sci. USA 90, 5893–5895.

Levy, W.P., Shively, J., Rubinstein, M., Del Valle, U. and Pestka, S. (1980). Amino terminal amino acid sequence of human leukocyte interferon. Proc. Natl Acad. Sci. USA 77, 5102–5104.

Levy, W.P., Rubinstein, M., Shively, J., et al. (1981). Amino acid sequence of a human leukocyte interferon. Proc. Natl Acad. Sci. USA 78, 6186–6190.

Lewin, A.R., Reid, L.E., McMahon, M., Stark, G.R. and Kerr, I.M. (1991). Molecular analysis of a human interferon-inducible gene family. Eur. J. Biochem. 199, 417–423.

Loeb, K.R. and Haas, A.L. (1992). The interferon-inducible 15-kDa ubiquitin homolog conjugates to intracellular proteins. J. Biol. Chem. 267, 7806–7813.

Lutfalla, G., Gardiner, K., Proudhon, D., Vielh, E. and Uzé, G. (1992). The structure of the human interferon α/β receptor gene. J. Biol. Chem. 267, 2802–2809.

Lutfalla, G., Holland, S., Cinato, E., et al. (1995). Mutant U5A cells are complemented by an interferon-αβ receptor subunit generated by alternative processing of a new member of a cytokine receptor gene cluster. EMBO J. 14, 5100–5108.

MacDonald, N.J., Kuhl, D., Maguire, D., et al. (1990). Different pathways mediate virus inducibility of the human IFN-α1 and IFN-β genes. Cell 60, 767–779.

Mandelli, F., Avvisati, G., Amadori, S., et al. (1990). Maintenance treatment with recombinant interferon alpha 2b in patients with multiple myeloma responding to conventional induction chemotherapy. N. Engl. J. Med. 322, 1430–1434.

Mantei, N. and Weissmann, C. (1982). Controlled transcription of a human α-interferon gene introduced into mouse L-cells. Nature 297, 128–132.

Mantei, N., Schwartzstein, M., Streuli, M., Panem, S., Nagata, S. and Weissmann, C. (1980). The nucleotide sequence of a cloned human leukocyte interferon cDNA. Gene 10, 1–10.

Mark, D.F., Lu, S.D., Creasey, A.A., Yamamoto, R. and Lin, L.S. (1981). Site-specific mutagenesis of the human fibroblast interferon gene. Proc. Natl Acad. Sci. USA 81, 5662–5666.

Mathews, M.B. and Shenk, T. (1991). Adenovirus virus-associated RNA and translational control. J. Virol. 65, 5657–5662.

Matsuyama, T., Kimura, T., Kitagawa, M., et al. (1993). Targeted disruption of IRF-1 or IRF-2 results in abnormal type I IFN gene induction and aberrant lymphocyte development. Cell 75, 83–97.

Meager, A., Graves, H., Burke, D.C. and Swallow, D.M. (1979a). Involvement of a gene on chromosome 9 in interferon production. Nature 280, 493–495.

Meager, A., Graves, H., Shuttleworth, J. and Zucker, N. (1979b). Interferon production: variation in yields from human cell lines. Infect. Immun. 25, 658–663.

Meager, A., Graves, H., Walker, J.R., Burke, D.C., Swallow, D.M. and Westerveld, A. (1979c). Somatic cell genetics of human interferon production in human-rodent cell lines. J. Gen. Virol. 45, 309–321.

Meager, A., Shuttleworth, J., Just, M., Boseley, P. and Morser, J. (1982). The effect of hypertonic salt on interferon and interferon mRNA synthesis in human MG63 cells. J. Gen. Virol. 59, 177–181.

Melén, K., Ronni, T., Broni, B., Krug, R.M., von Bonsdorff, C.-H. and Julkunen, I. (1992). Interferon-induced Mx proteins form oligomers and contain a putative leucine zipper. J. Biol. Chem. 267, 25898–25992.

Meraz, M., White, J., Sheehan, K., et al. (1996). Targeted disruption of the Stat 1 gene in mice reveals unexpected physiologic specificity in the JAK-STAT signaling pathway. Cell 84, 431–442.

Meurs, E.F., Galabru, J., Barber, G.N., Katze, M.G. and Hovanessian, A.G. (1993). Tumor suppressor function of the interferon-induced double-stranded RNA-activated protein kinase. Proc. Natl Acad. Sci. USA 90, 232–236.

Mijàta, T. and Hayashida, H. (1982). Recent divergence from a common ancestor of human IFN-α genes. Nature 295, 165–168.

Miyamoto, N.G. and Samuel, C.E. (1980). Mechanism of interferon action. Interferon mediated inhibition of reovirus mRNA translation in the absence of detectable mRNA degradation but in the presence of protein phosphorylation. Virology 107, 461–475.

Miyamoto, M., Fumita, T., Kimura, Y., *et al.* (1988). Regulated expression of a gene encoding a nuclear factor, IRF-1, that specifically binds to IFN-β gene regulatory elements. Cell 54, 903–913.

Morehead, H., Johnston, P.D. and Wetzel, R. (1984). Roles of the 29–138 disulfide bond of subtype A of human α interferon in its antiviral activity and conformational stability. Biochemistry 23, 2500–2507.

Müller, M., Briscoe, J., Laxton, C., *et al.* (1993). The protein tyrosine kinase JAK1 complements defects in the IFN-α/β and -γ signal transduction. Nature 366, 129–135.

Müller, U., Steinhoff, U., Reis, L., *et al.* (1994). Functional role of type I and type II interferons in antiviral defense. Science 264, 1918–1924.

Näf, D., Hardin, S.E. and Weissmann, C. (1991). Multimerization of AAGTGA and GAAAGT generates sequences that mediate virus inducibility by mimicking an interferon promoter element. Proc. Natl Acad. Sci. USA 88, 1369–1373.

Nagata, S., Mantei, N. and Weissmann, C. (1980). The structure of one of the eight or more distinct chromosomal genes for human interferon-α. Nature 287, 401–408.

Neighbor, P.A. and Bloom, B.R. (1979). Absence of virus-induced lymphocyte suppression and interferon production in multiple sclerosis. Proc. Natl Acad. Sci. USA 76, 476–480.

Nevins, J.R. (1991). Transcriptional activation by viral regulatory proteins. Trends Biochem. Sci. 16, 435–439.

Nourbakhsh, M., Hoffmann, K. and Hauser, H. (1993). Interferon-β promoters contain a DNA element that acts as a position-independent silencer on the NF-κB site. EMBO J. 12, 451–459.

Novick, D., Cohen, B. and Rubinstein, M. (1994). The human interferon α/β receptor: characterisation and molecular cloning. Cell 77, 391–400.

Öberg, K., Alm, G., Magnusson, A., *et al.* (1989). Treatment of malignant carcinoid tumors with recombinant interferon alfa-2b: development of neutralizing interferon antibodies and possible loss of antitumor activity. J. Natl Cancer Inst. 81, 531–535.

Ohlsson, M., Feder, J., Cavalli-Sforza, L.L. and von Gabain, A. (1985). Close linkage of α and β interferons and infrequent duplication of β interferon in humans. Proc. Natl Acad. Sci. USA 82, 4473–4476.

Owerbach, D., Rutter, W.J., Shows, T.B., Gray, P., Goeddel, D.V. and Lawn, R. (1981). Leukocyte and fibroblast interferon genes are located on human chromosome 9. Proc. Natl Acad. Sci. USA 78, 3123–3127.

Paty, D.W., Li, D.K.B., UBC MS MRI Study Group and IFNB Multiple Sclerosis Study Group (1993). Interferon beta-1b is effective in relapsing-remitting multiple sclerosis. II. MRI analysis results of a multicenter, randomized, double-blind, placebo-controlled trial. Neurology 43, 662–667.

Pestka, S., Langer, J.A., Zoon, K.C. and Samuel, C.E. (1987). Interferons and their actions. Annu. Rev. Biochem. 56, 727–777.

Pfeffer, L.M., Eisenkraft, B.L., Reich, N.C., *et al.* (1991). Transmembrane signalling by interferon α involves diacylglycerol production and activation of ε isoform of protein kinase C in Daudi cells. Proc. Natl Acad. Sci. USA 88, 7988–7992.

Phillips, A.W., Finter, N.B., Burman, C.J. and Ball, G.D. (1986). Large scale production of human interferon from lymphoblastoid cells. Methods Enzymol. 119, 35–38.

Pine, R., Levy, D.E., Reich, N. and Darnell, J.E.J. (1990). Purification and cloning of interferon-stimulated gene factor 2 (ISGF2): ISGF2 (IRF-1) can bind to the promoters of both beta interferon and interferon-stimulated genes, but is not a primary transcriptional activator of either. Mol. Cell. Biol. 10, 2448–2457.

Platanias, L.C., Uddin, S. and Colamonici, O.R. (1994). Tyrosine phosphorylation of the α and β subunits of the α and β subunits of the type I interferon receptor. J. Biol. Chem. 269, 17761–17764.

Porter, A.C.G. and Itzhaki, J.E. (1993). Gene targeting in human somatic cells: complete inactivation of an interferon-inducible gene. Euro. J. Biochem. 218, 273–281.

Rager-Zisman, B. and Bloom, B.R. (1985). Interferons and natural killer cells. Br. Med. Bull. 41, 22–27.

Ragg, H. and Weissmann, C. (1983). Not more than 117 base pairs of 5′-flanking sequence are required for inducible expression of a human IFN-α gene. Nature 303, 439–442.

Raziuddin, A. and Gupta, S.L. (1985). Receptors for human interferon-alpha: two forms of interferon-receptor complexes identified by chemical cross-linking. In "The 2-5A System: Molecular and Clinical Aspects of the Interferon-Regulated Pathway" (eds. B.R.G. Williams and R.H. Silverman), pp. 219–240. Alan Liss, New York.

Raziuddin, A., Sarkar, F.H., Dutkowski, R., Shulman, L., Ruddle, F.H. and Gupta, S.L. (1984). Receptors for human α and β interferon but not for γ interferon are specified by human chromosome 21. Proc. Natl Acad. Sci. USA 81, 5504–5508.

Rehberg, E., Kelder, B., Hoal, E.G., and Pestka, S. (1982). Specific molecular activities of recombinant and hybrid leukocyte interferons. J. Biol. Chem. 257, 11497–11502.

Reich, N.C. and Pfeffer, L.M. (1990). Evidence for involvement of protein kinase C in the cellular response to interferon α. Proc. Natl Acad. Sci. USA 87, 8761–8765.

Reid, L.E., Brasnett, A.H., Gilbert, C.S., *et al.* (1989). A single DNA response element can confer inducibility by both α- and γ-interferons. Proc. Natl Acad. Sci. USA 86, 840–844.

Reid, T.R., Race, E.R., Wolff, B.H., Friedman, R.M., Merigan, T.C. and Basham, T.Y. (1989). Enhanced *in vivo* therapeutic response to interferon in mice with an *in vitro* interferon-resistant B-cell lymphoma. Cancer Res. 49, 4163–4169.

Reis, F.L., Ruffner, H., Stark, G., Aguet, M. and Weissmann, C. (1994). Mice devoid of interferon regulatory factor (IRF-1) show normal expression of type I interferon genes. EMBO J. 13, 4798–4806.

Reis, L.F.L., Harada, H., Wolchok, J.D., Taniguchi, T. and Vilcek, J. (1992). Critical role of a common transcription factor, IRF-1, in the regulation of IFN-β and IFN-inducible genes. EMBO J. 11, 185–193.

Resnitzky, D., Tiefenbrun, N., Berissi, H. and Kimchi, A. (1992). Interferons and interleukin-6 suppress phosphorylation of the retinoblastoma protein in growth-sensitive hematopoietic cells. Proc. Natl Acad. Sci. USA 89, 402–406.

Revel, M. and Chebath, J. (1986). Interferon-activated genes. Trends Biochem. Sci. 11, 166–170.

Revel, M., Cohen, B., Abramovich, C., Novick, D., Rubinstein, M. and Shulman, L. (1991). Components of the human type

I IFN receptor system. J. Interferon Res. 11 (supplement), 561.

Rhodes, J., Ivanyi, J. and Cozens, P. (1986). Antigen presentation by human monocytes: effects of modifying major histocompatibility complex class II antigen expression and interleukin 1 production by using recombinant interferons and corticosteroids. Eur. J. Immunol. 16, 370–375.

Rice, A.P., Duncan, R., Hershey, J.W.B. and Kerr, I.M. (1985). Double-stranded RNA-dependent protein kinase and 2-5A system are both activated in interferon-treated, encephalomyocarditis virus-infected HeLa cells. J. Virol. 54, 894–898.

Rigby, W.F.C., Ball, E.D., Guyre, P.M. and Fanger, M.W. (1985). The effect of recombinant-DNA-derived interferons on the growth of myeloid progenitor cells. Blood 65, 858–861.

Rinehart, J., Malspeis, L., Young, D. and Neidhart, J. (1986). Phase I/II trial of human recombinant β-interferon serine in patients with renal cell carcinoma. Cancer Res. 46, 5364–5367.

Roberts, R.M., Cross, J.C. and Leaman, D.W. (1992). Interferons as hormones of pregnancy. Endocrine Rev. 13, 432–452.

Rohatiner, A.Z., Prior, P., Burton, A., Balkwill, F. and Lister, T.A. (1985). Central nervous system toxicity of interferons. Prog. Exp. Tumor Res. 29, 197–202.

Ronni, T., Melén, K., Malygin, A. and Julkunen, I. (1993). Control of IFN-inducible MxA gene expression in human cells. J. Immunol. 150, 1715–1726.

Rubin, B.Y., Anderson, S.L., Lunn, R.M., et al. (1988). Tumour necrosis factor and IFN induce a common set of proteins. J. Immunol. 141, 1180–1184.

Rubinstein, M. and Orchansky, P. (1986). The interferon receptors. CRC Crit. Rev. Biochem. 21, 249–270.

Rubinstein, M., Rubinstein, S., Familetti, P.C., et al. (1978). Human leukocyte interferon purified to homogeneity. Science 202, 1289–1290.

Rubinstein, M., Levy, W.P., Moschera, J.A., et al. (1981). Human leukocyte interferon: isolation and characterisation of several molecular forms. Arch. Biochem. Biophys. 210, 307–318.

Ruffner, H., Reis, L.F., Näf, D. and Weissmann, C. (1993). Induction of type I interferon genes and interferon-inducible genes in embryonal stem cells devoid of interferon regulatory factor I. Proc. Natl Acad. Sci. USA 90, 11503–11507.

Ryals, J., Dierks, P., Ragg, H. and Weissmann, C. (1985). A 46-nucleotide promoter segment from an IFN-α gene renders an unrelated promoter inducible by virus. Cell 41, 497–507.

Samuel, C.E. (1987). Interferon induction of the antiviral state proteins induced by interferons and their possible roles in the antiviral mechanisms of action. In "Interferon Actions" (ed. L.M. Pfeffer), pp. 110–130. CRC Press, Boca Raton, FL.

Saven, A. and Piro, L.D. (1992). Treatment of hairy cell leukaemia. Blood 79, 1111–1120.

Schindler, C., Shuai, K., Prezioso, V.R. and Darnell, J.E.J. (1992). Interferon-dependent tyrosine phosphorylation of a latent cytoplasmic transcription factor. Science 257, 809–813.

Schlesinger, R.W. (1959). Interference between animal viruses. In "The Viruses", vol. III (eds. F.M. Barnet and W.M. Stanley) pp. 157–194. Academic Press, New York.

Schneider, A., Papendick, U., Gissmann, L. and De Villiers, E.M. (1987). Interferon treatment of human genital papilloma virus infection: importance of viral type. Int. J. Cancer 40, 610–614.

Schwemmle, M. and Staeheli, P. (1994). The interferon-induced 67-kDa guanylate-binding protein (hGBP1) is a GTPase that converts GTP to GMP. J. Biol. Chem. 269, 11299–11305.

Scott, G.M., Philpotts, R.J., Wallace, J., Gauci, C.L., Greiner, J. and Tyrell, D.A.J. (1982). Prevention of rhinovirus colds by human interferon alpha-2 from E. coli. Lancet ii, 186–188.

Sekar, V., Atmar, V.J., Joshi, A.R., Krim, M. and Kuehn, G. (1983). Inhibition of ornithine decarboxylase in human fibroblast cells by type I and type II interferons. Biochem. Biophys. Res. Commun. 114, 950–954.

Sen, G.C. and Lengyel, P. (1992). The interferon system: a bird's eye view of the biochemistry. J. Biol. Chem. 267, 5017–5020.

Senda, T., Shimazu, T., Matsuda, S., et al. (1992). Three-dimensional crystal structure of recombinant murine interferon-β. EMBO J. 11, 3193–3201.

Shafferman, A., Velan, B., Cohen, S., Leitner, M. and Grosfeld, H. (1987). Specific residues within an amino-terminal domain of 35 residues of interferon-α are responsible for recognition of the human interferon-α cell receptor and for triggering biological effects. J. Biol. Chem. 262, 6227–6237.

Shaw, G., Boll, W., Taira, H., Mantei, N., Lengyel, P. and Weissmann, C. (1983). Structure and expression of cloned murine IFN-α genes. Nucleic Acids Res. 11, 555–573.

Shaw, G.D. and Kamen, R. (1986). A conserved AU sequence from the 3′ untranslated region of GM-CSF mRNA mediators selective mRNA-degradation. Cell 46, 659–667.

Shepard, H.M., Leung, D., Stebbing, N. and Goeddel, D.V. (1981). A single amino acid change in IFN-β1 abolishes its antiviral activity. Nature 294, 563–565.

Shirono, H., Kono, K., Koga, J., Hayashi, S., Matsuo, A. and Hiratani, H. (1990). Existence and unique N-terminal sequence of alpha II (omega) interferon in natural leukocyte interferon preparation. Biochem. Biophys. Res. Commun. 168, 16–21.

Shows, T.B., Sakaguchi, A.Y., Naylor, S.L., Goeddel, D.V. and Lawn, R.M. (1982). Clustering of leukocyte and fibroblast genes on human chromosome 9. Science 218, 373–374.

Shuai, K., Horvath, C.M., Tsai Huang, L.H., Qureshi, S.A., Cowburn, D. and Darnell, J.E.J. (1994). Interferon activation of the transcription factor Stat91 involves dimerization through SH2-phosphotyrosyl peptide interactions. Cell 76, 821–828.

Shulman, L.M., Kamarck, M.E., Slate, D.L., et al. (1984). Antibodies to chromosome 21 coded cell surface components block binding of human α interferon but not γ interferon to human cells. Virology 137, 422–427.

Shuttleworth, J., Morser, J. and Burke, D.C. (1983). Expression of interferon-α and interferon-β genes in human lymphoblastoid (Namalwa) cells. Eur. J. Biochem. 133, 399–404.

Sidky, Y.A. and Borden, E.C. (1987). Inhibition of angiogenesis by interferons: effects on tumour- and lymphocyte-induced vascular responses. Cancer Res. 47, 5155–5161.

Slate, D.L., Shulman, L., Lawrence, J.B., Revel, M. and Ruddle, F.H. (1978). Presence of human chromosome 21 alone is sufficient for hybrid cell sensitivity to human interferon. J. Virol. 25, 319–325.

Slate, D.L., D'Eustachio, P., Pravtcheva, D., *et al.* (1982). Chromosomal location of a human α interferon gene family. J. Exp. Med. 155, 1019–1024.

Soh, J., Mariano, T.M., Lim, J.-Y., *et al.* (1994). Expression of a functional human type I interferon receptor in hamster cells: application of functional yeast artificial chromosome (YAC) screening. J. Biol. Chem. 269, 18102–18110.

Staeheli, P. (1990). Interferon-induced proteins and the antiviral state. Adv. Virus Res. 38, 147–200.

Steis, R.G., Smith, J.W., Urba, W.J., *et al.* (1988). Resistance to recombinant interferon alfa-2a in hairy-cell leukaemia associated with neutralizing anti-interferon antibodies. N. Engl. J. Med. 318, 1409–1413.

Stewart, W.E., II, (1979). "The Interferon System", pp. 134–183. Springer-Verlag, New York.

Stewart, W.E.I., II, Blalock, J.E., Burke, D.C., *et al.* (1980). Interferon nomenclature. Nature 286, 110.

Stolfi, R.L., Martin, D.S., Sawyer, R.C. and Spiegelman, S. (1983). Modulation of 5-fluorouracil-induced toxicity in mice with interferon or with the interferon inducer, polyinosinic-polycytidylic acid. Cancer Res. 43, 561–566.

Streuli, M., Nagata, S. and Weissmann, C. (1980). At least three human type α interferons: structure of α_2. Science 209, 1343–1347.

Streuli, M., Hall, A., Boll, W., Stewart, W.E., II, Nagata, S. and Weissmann, C. (1981). Target specificity of two species of human interferon-α produced in *Escherichia coli* and of hybrid molecules derived from them. Proc. Natl Acad. Sci. USA 78, 2848–2852.

Symons, J., Alcami, A. and Smith, G. (1995). Vaccinia virus encodes a soluble type I interferon receptor of novel structure and broad species specificity. Cell 81, 551–560.

Tan, Y.H., Tischfield, J. and Ruddle, F.H. (1973). The linkage of genes for the human interferon-induced antiviral protein and indophenol oxidase-B traits to chromosome G-21. J. Exp. Med. 137, 317–330.

Taniguchi, T., Mantei, N., Schwarzstein, M., Nagata, S., Muramatsu, M. and Weissmann, C. (1980). Human leukocyte and fibroblast interferons are structurally related. Nature 285, 547–549.

Taylor, J.L. and Grossberg, S.E. (1990). Recent progress in interferon research: molecular mechanisms of regulation, action, and virus circumvention. Virus Res. 15, 1–26.

Thanos, D. and Maniatis, T. (1992). The high mobility group protein HMGI (Y) is required for NF-κB-dependent virus induction of the human IFN-β gene. Cell 71, 777–789.

Thomas, N.S.B. and Linch, D.C. (1991). An intracellular 50kDa Fcγ-binding protein is induced in human cells by α-IFN. J. Immunol. 146, 1649–1654.

Thompson, M.R., Zhang, Z., Fournier, A. and Tan, Y.H. (1985). Characterization of human beta-interferon-binding sites on human cells. J. Biol. Chem. 360, 563–567.

Todokoro, K., Kioussis, D. and Weissmann, C. (1984). Two non-allelic human interferon α genes with identical coding regions. EMBO J. 3, 1809–1812.

Treanor, J.J., Betts, R.F., Erb, S.M., Roth, F.K. and Dolin, R. (1987). Intranasally administered interferon as prophylaxis against experimentally induced influenza A virus infection in humans. J. Infect. Dis. 156, 379–383.

Trinchieri, G. and Perussia, B. (1984). Human natural killer cells: biologic and pathologic aspects. Lab. Invest. 50, 489–503.

Turner, R.B., Fetton, A., Kosak, K., Kelsey, D.K. and Meschievitz, C.K. (1986). Prevention of experimental coronavirus colds with intranasal alpha-2b interferon. J. Infect. Dis. 154, 443–447.

Ullrich, A., Gray, A., Goeddel, D.V. and Dull, T.J. (1982). Nucleotide sequence of a portion of human chromosome 9 containing a leukocyte interferon gene cluster. J. Mol. Biol. 156, 467–486.

Utsumi, J., Mizuno, Y., Hosoi, K., *et al.* (1989). Characterization of four different mammalian-cell-derived recombinant human interferon-β1s. Eur. J. Biochem. 181, 545–553.

Uzé, G., Mogensen, K.E. and Aguet, M. (1985). Receptor dynamics of closely related ligands: "fast" and "slow" interferons. EMBO J. 4, 65–70.

Uzé, G., Bandu, M.-T., Eid, P., Grütter, M. and Mogensen, K.E. (1988). Electrostatic interactions in the cellular dynamics of the interferon-receptor complex. Eur. J. Biochem. 171, 683–691.

Uzé, G., Lutfalla, G. and Gresser, I. (1990). Genetic transfer of a functional human interferon α receptor into mouse cells: cloning and expression of its cDNA. Cell 60, 225–234.

Uzé, G., Lutfalla, G., Eid, P., *et al.* (1991). Murine tumor cells expressing the gene for the human interferon αβ receptor elicit antibodies in syngeneic mice to the active form of the receptor. Eur. J. Immunol. 21, 447–451.

Uzé, G., Lutfalla, G., Bandu, M.-T., Proudhon, D. and Mogensen, K. (1992). Behaviour of a cloned murine interferon alpha beta receptor expressed in homospecific or heterospecific background. Proc. Natl Acad. Sci. USA 889, 4774–4778.

Uzé, G., Lutfalla, G. and Mogensen, K.E. (1995). α and β interferons and their receptor and their friends and relations. J. Interferon Cytokine Res. 15, 3–26.

Vanden Broecke, C. and Pfeffer, L.M. (1988). Characterization of interferon-alpha binding sites on human cell lines. J. Interferon Res. 8, 803–811.

Vedantham, S., Gamliel, H. and Golomb, H.M. (1992). Mechanism of interferon action in hairy cell leukaemia: a model of effective cancer biotherapy. Cancer Res. 52, 1056–1066.

Velasquez, L., Fellous, M., Stark, G.R. and Pellegrini, S. (1992). A protein tyrosine kinase in the interferon α/β signalling pathway. Cell 70, 313–322.

Velasquez, L., Mogensen, K., Barbieri, G., Fellous, M., Uzé, G. and Pellegriru, S. (1995). Distinct domains of the protein tyrosine kinase tyk2 required for binding of interferon-α/β and for signal transduction. J. Biol. Chem. 270, 3327–3334.

Virelizier, J.L. and Gresser, I. (1978). Role of interferon in the pathogenesis of viral diseases of mice as demonstrated by the use of anti-interferon serum. V. Protective role in mouse hepatitis virus type 3 infection of susceptible and resistant strains of mice. J. Immunol. 120, 1616–1619.

Visvanathan, K.V. and Goodbourn, S. (1989). Double-stranded RNA activates binding of NF-κB to an inducible element in human β-interferon promoter. EMBO J. 8, 1129–1138.

Vogel, S.N., Finbloom, D.S., English, K.E., Rosenstreich, D.L. and Langreth, S.G. (1983). Interferon-induced enhancement of macrophage Fc receptor expression: β-interferon treatment of C3H/HeJ macrophages results in increased numbers and density of Fc receptors. J. Immunol. 130, 1210–1214.

von Wussow, P., Pralle, H., Hochkeppel, H.-K., *et al.* (1991). Effective natural interferon-α therapy in recombinant interferon-α-resistant patients with hairy cell leukaemia. Blood 78, 38–43.

Wadler, S. and Schwartz, E.L. (1990). Antineoplastic activity of the combination of interferon and cytotoxic agents against experimental and human malignancies: a review. Cancer Res. 50, 3473–3486.

Wang, C., Constantinescu, S.N., MacEwan, D.J., *et al.* (1993). Interferon α induces a protein kinase C-ε (PKC-ε) gene expression and a 4.7kb PKC-ε related transcript. Proc. Natl Acad. Sci. USA 90, 6944–6948.

Wang, E., Pfeffer, L.M. and Tamm, I. (1981). Interferon increases the abundance of submembranous microfilaments in HeLa-S3 cells in suspension culture. Proc. Natl Acad. Sci. USA 78, 6281–6285.

Watanabe, N., Sakakibara, J., Hovanessian, A.G., Taniguchi, T. and Fujita, T. (1991). Activation of IFN-β element by IRF-1 requires a translational event in addition to IRF-1 synthesis. Nucleic Acids Res. 19, 4421–4428.

Wathelet, M.G., Clauss, I.M., Content, J. and Huez, G.A. (1988). Regulation of two interferon-inducible human genes by interferon, poly(rI):poly(rC) and viruses. Eur. J. Biochem. 174, 323–329.

Wathelet, M.G., Berr, P.M. and Huez, G.A. (1992). Regulation of gene expression by cytokines and virus in human cells lacking the type-I interferon locus. Eur. J. Biochem. 206, 901–910.

Weck, P.K., Apperson, S., Stebbing, N., *et al.* (1981). Antiviral activities of hybrids of two major human leukocyte interferons. Nucleic Acids Res. 9, 6153–6166.

Weidle, U. and Weissmann, C. (1983). The 5′-flanking region of a human IFN-α gene mediates viral induction of transcription. Nature 303, 442–446.

Weisz, A., Marx, P., Sharf, R., *et al.* (1992). Human interferon consensus sequence binding protein is a negative regulator of enhancer elements common to interferon-inducible genes. J. Biol. Chem. 267, 25589–25596.

Whitaker-Dowling, P. and Youngner, J.S. (1983). Vaccinia rescue of VSV from interferon-induced resistance: reversal of translation block and inhibition of protein kinase activity. Virology 131, 128–136.

Whiteside, S.T., Visvanathan, K.V. and Goodbourn, S. (1992). Identification of novel factors that bind to the PRD1 region of the human β-interferon promoter. Nucleic Acids Res. 20, 1531–1538.

Whittemore, L.-A. and Maniatis, T. (1990). Postinduction repression of the β-interferon gene is mediated through two positive regulatory domains. Proc. Natl Acad. Sci. USA 87, 7799–7803.

Wide, L. and Wilander, E. (1989). Treatment of malignant carcinoid tumors with recombinant interferon alfa-2b: development of neutralizing interferon antibodies and possible loss of antitumour activity. J. Natl Cancer Inst. 81, 531–535.

Williams, B.R.G. (1991). Transcriptional regulation of interferon-stimulated genes. Eur. J. Biochem. 200, 1–11.

Williams, B.R.G. and Kerr, I.M. (1978). Inhibition of protein synthesis by 2′-5′ linked adenine oligonucleotides in intact cells. Nature 276, 88–89.

Wilson, V., Jeffreys, A.J., Barrie, P.A., *et al.* (1983). A comparison of vertebrate interferon gene families detected by hybridisation with human interferon DNA. J. Mol. Biol. 166, 457–475.

Wreschner, D.H., James, T.C., Silverman, R.H. and Kerr, I.M. (1981). Ribosomal RNA cleavage, nuclease activation and 2-5A (ppp(A2′p)nA) in interferon treated cells. Nucleic Acids Res. 9, 1571–1581.

Xanthoudakis, S., Cohen, L. and Hiscott, J. (1989). Multiple protein-DNA interactions with the human interferon-β regulatory element. J. Biol. Chem. 264, 1139–1145.

Yan, H., Krishnan, K., Greenlund, A., *et al.* (1996). Phosphorylated interferon-α receptor 1 subunit (IFNaR1) acts as a docking site for the latent form of the 113 kDa STAT 2 protein. EMBO J. 15, 1064–1074.

Yonehara, S., Yonehara-Takahashi, M., Ishii, A. and Nagata, S. (1983). Different binding of human interferon-α1 and -α2 to common receptors on human and bovine cells. Studies with recombinant interferon produced in *Escherichia coli*. J. Biol. Chem. 258, 9046–9049.

Zhang, K. and Kumar, R. (1994). Interferon-α inhibits cyclin E- and cyclin D1-dependent CDK-2 kinase activity associated with RB protein and E2F in Daudi cells. Biochem. Biophys. Res. Commun. 200, 522–528.

Zhou, A., Hassel, B.A. and Silverman, R.H. (1993). Expressing cloning of 2-5A-dependent RNAase: a uniquely regulated mediator of interferon action. Cell 72, 753–765.

Zoon, K.C. and Wetzel, R. (1984). Comparative structures of mammalian interferons. In "Handbook of Experimental Pharmacology" (eds. P.E. Crane and W.A. Carter), pp. 79–100. Springer-Verlag, Berlin.

Zoon, K.C., Smith, M.E., Bridgen, P.J., Anfinsen, C.B., Hunkapillar, M.W. and Hood, L.E. (1980). Amino-terminal sequence of the major component of human lymphoblastoid interferon. Science 207, 527–528.

Zoon, K.C., Miller, D., Bekisz, J., *et al.* (1992). Purification and characterization of multiple components of human lymphoblastoid interferon-α. J. Biol. Chem. 267, 15210–15216.

Zwarthoff, E., Mooren, A. and Trapman, J. (1985). Organization, structure and expression of murine interferon α genes. Nucleic Acids Res. 13, 791–804.

26. Interferon-Gamma

Edward De Maeyer *and* Jaqueline De Maeyer-Guignard

1. Introduction

In 1964, Gresser and Naficy described the presence of viral inhibitory activity in cerebrospinal fluids derived from patients with infectious and noninfectious diseases. They used the term "interferon-like" substance because, by the criteria then in use to characterize antiviral substances, the antiviral activity differed in one important aspect from the other interferons: it was acid-labile. One year later, Wheelock (1965), who was looking for the possible presence of viral-induced interferon in peripheral blood lymphocytes (PBLs) from patients with various virus infections, found an antiviral factor in uninfected human leukocyte control cultures that had been stimulated by phytohemagglutinin (PHA). This "interferon-like substance", as he also called it, again differed from the other interferons in that it was acid-

labile. Four years later, Green and colleagues (1969) demonstrated the appearance of a similar activity in antigen-stimulated leukocyte suspensions. For many years IFN-γ was referred to first as "acid-labile interferon" and subsequently as "type II interferon", as opposed to the acid-stable "type I" interferons, which later turned out to be interferons α and β (De Maeyer and De Maeyer-Guignard, 1988). Since interferons belonged to the virologist's domain, it is probably not too surprising that the first indications for an immunoregulatory role of type II interferon were obtained by virologists. It is rather ironic that for many years immunologists simply were not interested in this substance, whereas presently one cannot open an issue of a journal dealing with the subject of immunology without coming across at least one paper dealing with IFN-γ directly or indirectly. But that is the way science often works!

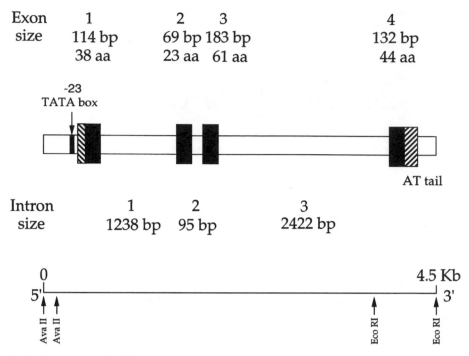

Figure 26.1   Structure of the IFN-γ gene (aa = amino acid). (The figure is derived from Taya *et al.* (1982).)

2. The Cytokine Gene

The single-copy hIFN-γ gene, situated on chromosome 12 in the p12.05-qter region, contains three introns. Intron one is 1238 bp long, intron two 95 bp and intron three 2422 bp (Figure 26.1). The four exons code for 38, 23, 61, and 44 amino acids, respectively, resulting in a 166-amino-acid polypeptide, 23 residues of which constitute the signal peptide. The deduced amino acid sequence reveals two potential *N*-glycosylation sites at positions 25–27 and 97–99, which explains the existence of two species of different molecular mass, 20 and 25 kDa, respectively, glycosylated on either one or both sites (Yip *et al.*, 1982; Rinderknecht *et al.*, 1984) (Table 26.1). The protein is furthermore characterized by a very high content of basic residues, particularly two clusters of four residues each, Lys-Lys-Lys-Arg at positions 86–90 and Lys-Arg-Lys-Arg at positions 128–132. The basic nature of the mature protein probably explains its acid-lability, a characteristic used to distinguish it from the type I interferons before its molecular structure was known. Deletion mutants of hIFN-γ truncated at the COOH terminus show loss of activity when residues in this region are missing (Seelig *et al.*, 1988).

A comparison of the sequences of three different IFN-γ isolated from three individuals reveals moderate allelism, consisting, for example, of a lysine-to-glutamine change at position 6.

3. The Protein

Unglycosylated, *E. Coli*-derived hIFN-γ crystallizes in a dimeric structure, with identical α subunits related by a noncrystallographic twofold axis in the asymmetric unit. The protein has no β-sheet and is principally α-helical, with each subunit consisting of six α-helices, ranging in length from 9 to 21 residues. The 12 helices that make up the dimer are parallel to the dimer twofold axis, with no clear antiparallel four-helix domains present in the molecule. The dimer is stabilized by the intertwining of helices across the subunit interface. The very intimate linkage of the subunits that make up the dimeric structure of hIFN-γ appears to be unusual among globular proteins. Of particular interest is the observation that a comparison of the three-dimensional structures of IFN-γ and IFN-β shows a significant similarity in folding topology (Figure 26.2). Of the 12 helices in the IFN-γ dimer, five form a structural domain that corresponds to the five helices of the IFN-β molecule, and the overall geometrical arrangement and connectivity are conserved. Ealick and colleagues (1991) therefore propose the possibility that IFN-γ and IFN-β may be derived from a common ancestral gene, in spite of the lack of homology between the amino acid sequences and the existence of a separate cell surface receptor for these two cytokines. Similarities in tertiary structure between proteins sharing the same evolutionary origin are frequently more conserved than primary structure,

Table 26.1   Deduced amino acid sequence of human and murine IFN-γ

```
       S1      S10        S20      1      10           20     *    30
Hu  MKYTSYILAFQLCIVLGSLGCYC QDPYVKEAENLKKYFNAGHSDVADNGTLFLGI
Mu  .NA.HC...L..FLMAV.-.....HGTVIESL.S.NN...SSGI..-EEKS...D.

           40         50          60          70          80
Hu  LKNWKEESDRKIMQSQIVSFYFKLFKNFKDDQSIQKSVETIKEDMNVKFFNSNKKK
Mu  WR..QKDG.M..L....I...LR..EVL..N.A.SNNISV.ESHLITT..SNS.A.

       90      *100        110         120        130         140
Hu  RDDFEKLTNYSVTDLNVQRKAIHELIQVMAELSPAAKTGKRKRSQMLFRGRRASQ
Mu  K.A.MSIAKFE.NNPQ...Q.FN...R.VHQ.L.ESSLR.....RC
```

The two ASN glycosylation sites are indicated by an asterisk. (From Taya *et al.*, 1982, with modifications.)

and, in spite of their pronounced differences in primary amino acid sequence, type I and type II interferons could conceivably be derived from the same ancestral gene. An analysis of the interferon genes of more primitive vertebrates may be able to provide some insight into the possible evolutionary relationship of the IFN-γ gene and the type I interferon genes; so far, the existence IFN-γ has been documented in mammals only.

4.   *Cellular Sources and Production*

In contrast to the synthesis of type I interferons, which can take place in any cell, the production of IFN-γ is limited to T cells and NK cells. T cells are stimulated to produce IFN-γ either in a polyclonal manner, via mitogens or antibodies, or in a clonally restricted, antigen-specific manner.

4.1   CELLS OF THE CYTOTOXIC-SUPPRESSOR PHENOTYPE

Human CD8 and murine (m) Ly-2, isolated from individuals after viral infection or vaccination, release IFN-γ upon exposure to the corresponding viral antigens *in vitro* (Celis *et al.*, 1986; Yamada *et al.*, 1986). Similarly, CD8 cytotoxic T cell clones, specific for the MHC class I HLA-B27 antigen, produce large amounts of IFN-γ (Salgame *et al.*, 1991).

4.2   CELLS OF THE T HELPER PHENOTYPE (HUMAN CD4 AND MURINE L3T4)

On the basis of the array of cytokines secreted, the existence of two different subsets of murine T helper (T_H) cells has been described. T_H1 cells secrete mainly IL-2, IL-3 ,TNF-β, and IFN-γ, and T_H2 cells produce mainly IL-3, IL-4, IL-5, and IL-10, and little or no IFN-γ (Mosmann *et al.*, 1986). T_H1 cells mediate cellular immunity such as delayed hypersensitivity responses, whereas T_H2 cells are associated with antibody production and with allergy. T_H1 and T_H2 cells probably derive from a

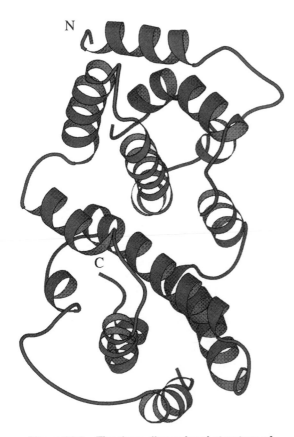

Figure 26.2   The three-dimensional structure of recombinant human IFN-γ. IFN-γ is a homodimer. The protein is primarily α-helical, with six helices in each subunit. There is no β-sheet. (Based on Ealick *et al.* (1991).)

common precursor, T_H0 cells, which make most or all cytokines made by both T_H1 and T_H2 cells before becoming specialized, but it is not known whether the T_H0 population contains already committed precursors of T_H1 and T_H2 cells. IFN-γ preferentially inhibits the proliferation of T_H2 cells, indicating that the presence of IFN-γ will result in the preferential expansion of T_H1 cells during an immune response.

In humans there is also evidence for the existence of a T_H1 and T_H2 type cytokine pattern in CD4 cells, but there are also many examples of overlapping patterns (Parronchi et al., 1991; Salgame et al., 1991).

5.   Biological Activity

5.1   Modulation of MHC Antigen Expression

A critical step in immune responses is the recognition by cells belonging to the immune system of peptide fragments of foreign antigens. Short peptide fragments from proteins degraded in the cytosol are bound to MHC class I molecules and the complex recognized by $CD8^+$ T cells. Peptides from proteins degraded in endosomal cellular vesicles bind to MHC class II molecules and the complex then migrates to the cell surface. The complex of peptide and MHC class II molecules is then recognized by $CD4^+$ T helper cells (as reviewed by Germain, 1994). This critical step in the immune response is stimulated by IFN-γ, which induces or enhances the expression of MHC class II antigens on macrophages and T cells. The stimulation of MHC class II expression by IFN-γ is not limited to these two classes of cells, but is also observed on B cells, and on many different tumor cells. The genes encoding MHC class II molecules belong to the class of genes that need several hours for activation by IFN-γ, as opposed to the "early" genes that are activated in a matter of minutes, via the signal transduction pathway discussed in Section 7 (Blanar et al., 1988; Amaldi et al., 1989; Lew et al., 1991).

5.2   Macrophage Activation

By inducing the production of reactive oxygen intermediates and the secretion of hydrogen peroxide, IFN-γ stimulates the intracellular killing of parasites by macrophages. For example, in both human and in murine macrophages, the IFN-γ induced activation of hydrogen peroxide release results in enhanced killing of the intracellular parasite Toxoplasma gondii (Nathan et al., 1985; Nathan and Tsunawaki, 1986). Interestingly, this activation is antagonized by IFN-α and IFN-β; this is one of the few examples of antagonism between different interferons, which more frequently act synergistically

(Garotta et al., 1986). Similarly, the crucial involvement of IFN-γ-induced reactive nitrogen intermediates in mycobacterial growth inhibition by macrophages has been documented (Flesch and Kaufmann, 1991).

Another important activity of IFN-γ on macrophages consists in the activation of the tumoricidal capacity of these cells. This is an important function of IFN-γ, since tumor-cell lysis by activated macrophages in all likelihood contributes to the mechanism of natural anti-tumor resistance. Although IFN-γ is undoubtedly an important macrophage activating agent, it shares this function with other lymphokines such as TNF-α (Urban et al., 1986).

5.3   Induction of Indoleamine 2,3-Dioxygenase (IDO)

IFN-γ induces IDO, an enzyme of tryptophan catabolism, which is responsible for the conversion of tryptophan to kynurenine. The IDO activity has been implicated in the killing of intracellular parasites such as Toxoplasma gondii or Chlamydia trachomatis and Chlamydia psitacci. It is believed that the inhibition results, at least in part, from tryptophan starvation of the parasites (as reviewed in Taylor and Feng, 1991) and, indeed, in mutant cell lines lacking the capacity to synthetize IDO the inhibition by IFN-γ of intracellular parasites is reduced (Thomas et al., 1993). In patients with advanced HIV-1 infections, the enhanced production of IFN-γ is probably responsible for the significant reduction of tryptophan levels in the serum, and the proposal has been made that this could contribute to the neurological symptoms frequently associated with HIV infection (Werner et al., 1988).

5.4   Activation of NK Cells

Natural killer cells are naturally occurring cytolytic effector cells whose cytotoxicity is not restricted by the MHC complex. Since they do not need previous sensitization, they are in the first line of defense against tumor cells and some infectious agents. They are producers of IFN-γ, and their cytolytic activity is also stimulated by IFN-γ, as well as by IFN-α and IFN-β.

5.5   Regulation of IgG Isotype Production

IFN-γ acts as a regulatory agent in the determination of Ig isotype responses in that it stimulates the expression of immunoglobulin of the IgG2a isotype, while inhibiting the production of IgG3, IgG1, IgG2b, and IgE (Snapper and Paul, 1987). This has led to the implication of IFN-γ in immunoglobulin class switching, but this point is somewhat controversial, and it has been proposed that the increased IgG2a synthesis by IFN-γ treated B cells is merely the result of an enhancement of secretion by cells that were already committed to IgG2a (Bossie and Vitetta, 1991).

5.6   ANTIVIRAL ACTIVITY

Although immunomodulation is undoubtedly the primary role of IFN-γ, it also contributes to the antiviral defense mechanism by its direct antiviral action, via the stimulation of specific genes that are also involved in the antiviral activity of IFN-α and IFN-β. One of the major enzymes responsible for the antiviral action of IFNs is oligoadenylate synthetase. This enzyme, also called (2'-5')A$_n$ synthetase is constitutively present in many cells at very low levels. Its concentration increases at least tenfold after IFN-γ treatment. When activated by double-stranded RNA, the enzyme polymerizes ATP into a series of 2'-5' linked oligomers (ppp(A2'p)n) of which the trimer is the most abundant. These 2'-5'A oligomers then activate a latent endoribonuclease, designated RNase L, which is responsible for the antiviral activity through degradation of viral RNA.

Frequently, the antiviral activity of IFN-γ occurs in synergy with other cytokines, such as, among others, IFN-α, IFN-β, and TNF-α (Feduchi and Carasco, 1991; Thomis and Samuel, 1992).

Significant levels of IFN-γ are present in the lymph nodes, plasma, and cerebrospinal fluid of HIV-infected individuals, and there is evidence that under certain conditions IFN-γ, rather than inhibiting, is able to stimulate the replication of HIV. Indeed, in chronically infected promonocytic U1 cells, IFN-γ activates HIV expression and causes the production of infectious HIV particles that then accumulate in intracellular vacuoles. This suggests that IFN-γ can play a role as an inducer of HIV expression in mononuclear phagocytes, and provides, to the best of our knowledge, a unique example of stimulation of virus production by an interferon (Biswas et al., 1992). The phenomenon is probably limited to chronically infected mononuclear cells, perhaps via the induction of other cytokines, since the usual effect of IFN-γ on HIV replication, even in mononuclear cells, is inhibition (see, for example, Hammer et al., 1986).

Several viruses have developed strategies to block or decrease the antiviral activity of IFNs, including IFN-γ. The adenoviral E1a protein inhibits the action of all types of IFNs, and the hepatitis B terminal protein also impairs the activation of gene expression by IFN-α and IFN-γ (as reviewed by Sen and Lengyel, 1992).

5.7   INDUCTION OF NITRIC-OXIDE SYNTHASE

Nitric-oxide synthase (NOS) is an enzyme that converts 1-arginine to 1-citrulline, resulting in the production of nitric oxide (NO), a highly reactive gaseous molecule.

The enzyme has a constitutively expressed form, cNOS, found in a variety of cell types, that generates low levels of NO and is involved in a host of physiological processes, among which is vasodilatation, and also has an inducible isoform, iNOS, that generates high levels of NO. This latter form is induced in macrophages by a variety of agents, among which are lipopolysaccharide (LPS) and IFN-γ. The induced enzyme generates high levels of NO that play a role in the antiviral, antimicrobial, antiparasitic, and antitumoral activity of IFN-γ. Induction of NO in macrophages by IFN-γ is considered by some as one of the key mechanisms by which IFN-γ inhibits the replication of viruses such as ectromelia, vaccina, and herpes simplex virus (Karupiah et al., 1993; Campbell et al., 1994). Cytokine-induced iNOS can also be produced by microglia and astrocytes, and one wonders whether some of the neurological symptoms that characterize AIDS might not result from IFN-γ-induced iNOS activity, since circulating IFN-γ is frequently present in late-stage AIDS patients. Nitric oxide has indeed been shown to have the potential of being either neuroprotective or neurodestructive, depending on the presence of other agents, for example, superoxide anion (Lipton et al., 1993).

6.   *The Receptor*

The IFN-γ alpha receptor chain   IFN-γ binds to a species-specific cell surface receptor with a molecular mass of 90 kDa, now referred to as the IFN-γ alpha receptor chain. One IFN-γ homodimer binds two receptor alpha chains, which do not interact with one another, but remain separated (Walter et al., 1995). Both the human and the murine receptor alpha chains have an equally large extracellular domain consisting of 288 amino acids, a single transmembrane domain, and an intracellular domain of 222 and 220 amino acids for the human and murine chains, respectively. The structural gene for the receptor alpha chain is on chromosome 6 in humans, and on chromosome 10 in mice.

The IFN-γ beta receptor chains   To be functionally active, the hIFN-γ alpha receptor chain also requires the presence of at least one other component, the beta chain, whose structural gene is on human chromosome 21 (Farrar et al., 1991; Soh et al., 1994). In addition to the alpha chain, the presence of the beta chain is required (and sufficient) for the induction of MHC class II antigens by IFN-γ; it is a transmembrane protein that has an overall structural homology to the class 2 cytokine receptor family. Interestingly, there is evidence that, also on chromosome 21, there are genes encoding other accessory factors involved in the antiviral action of IFN-γ directed against encephalomyocarditis and vesicular stomatitis virus, respectively (Pestka, 1994).

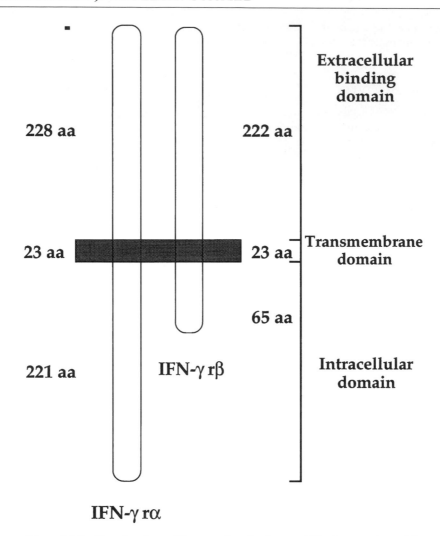

Figure 26.3   The structure of the receptors for human IFN-γ (aa = amino acid).

The murine accessory cofactor, encoded by a gene on chromosome 16, has been called the Mu IFN-γ beta chain, and consists of a large extracellular domain, and a relatively short cytoplasmic domain, estimated at 66 amino acids (Hemmi *et al.*, 1994).

7. *Signal Transduction*

As a result of the binding of IFN-γ to its cell surface receptor, the transcription of many previously quiescent genes is activated. We will briefly summarize the events leading from receptor binding to transcriptional activation of IFN-γ-inducible genes. IFN-γ-induced transcriptional activation can be either late, requiring first the synthesis of specific proteins that are then responsible for gene activation, or immediate, and take place via activation of specific signal-transducing proteins that are already present in the cytoplasm and (or) the nucleus in an inactive form. Mainly as a result of a series of enlightening experiments from the laboratories of J. Darnell, I. Kerr, and G. Stark, based in part on the judicious use of mutant cell lines, we have reached an understanding of the mechanisms of signal transduction involved in early activation of interferon-induced genes (McKendry *et al.*, 1991; Velazquez *et al.*, 1992; Watling *et al.*, 1993; Darnell *et al.*, 1994; Shuai, 1994).

Receptor binding of IFN-γ immediately, in a matter of minutes, activates two tyrosine kinases (PTK) within the JAK (Janus kinase) family: JAK 1 and JAK 2. Subsequently, JAK 1 and JAK 2 then catalyze the tyrosine phosphorylation of the transcription factors involved in signal transduction (Harpur *et al.*, 1992; Muller *et al.*, 1993a). How exactly JAK kinases interact with the IFN-γ receptor has not been established, though it seems likely that they are physically associated with the intracellular part of the receptor, since it has recently been shown that binding of IFN-γ to its receptor results in rapid phosphorylation of the latter (Shuai, 1994). A latent cytoplasmic factor, the gamma activation factor or GAF, is then activated through tyrosine phosphorylation. GAF is a dimer of a 91 kDa protein, called STAT 91 (or STAT 84) protein (STAT is an acronym for signal transducer and activator of transcription) (Imam et al., 1990; Decker et al., 1991; Shuai et al., 1992; Müller et al., 1993b). The STAT 91 and STAT 84 proteins are encoded by the same gene, but STAT 91 contains 38 carboxyl-terminal amino acids that are lacking in STAT 84 (Schindler et al., 1992; Zhong et al., 1994). It is interesting that STAT 91 is also one of the proteins that make up the ISGF-3 (IFN-stimulated gene factor) complex which is involved in the immediate transcriptional activation of IFN-α-inducible genes and is made up of four different subunits, consisting of 113, 91, 84, and 48 kDa proteins (Khan et al., 1993; Muller et al., 1993b). Inactive STAT 91 in the cytoplasm of untreated cells is a monomer, but upon IFN-γ-induced phosphorylation it forms a stable dimer, which then migrates to the nucleus. Only the dimer is capable of binding to the gamma activation site GAS (Shuai et al., 1994).

GAS is a consensus immediate response element of nine nucleotides (TTNCNNNAA) which is present in those genes that are activated immediately after IFN-γ receptor-ligand binding. Although STAT 91 is also present in ISGF3, the transcriptional activator of the IFN-α response, the major DNA-binding component of ISGF3 appears to be the 48 kDa protein, which is normally already present in the nucleus, and which, after activation and inclusion into the ISGF3 complex, binds to the ISRE (interferon-stimulated response element, a consensus sequence present in most genes that respond to IFN-α). Several of the STAT proteins characterized in the IFN-α and IFN-γ signal transduction pathways will undoubtedly turn out to be also involved in the expression control of genes activated by other cytokines, and the knowledge gained from the study of IFN signal transduction has important implications for the understanding of gene expression in general. It has indeed become evident that several cytokines other than interferons also use tyrosine phosphorylation to activate putative transcription factors, and, for example, STAT 91 is phosphorylated after the interaction of epidermal growth factor and its receptor, and then mediates activation of the c-fos gene promoter (Fu and Zhong, 1993; Larner et al., 1993).

8.   The Murine Cytokine

Containing four exons and three introns, the structure of the murine gene is comparable to that of the human gene. The coding part displays an overall nucleotide homology with the human gene of about 65%, and an overall protein homology of 40%. Mature mIFN-γ has 134 amino acids, which is four residues shorter than the human form. Like the latter, mIFN-γ has two N-glycosylation sites and an excess of basic residues. The encoded size of the mature protein, based on the amino acid sequence, is 16 kDa, which is smaller than the size of natural IFN-γ (38 kDa); the difference is due to glycosylation and to dimerization (Dijkmans et al., 1985; Nagata et al., 1986). The gene is located on murine chromosome 10.

Most experiments on the role and biological activity of IFN-γ have been performed by treating cells or animals either with the cytokine, or with antibodies that neutralize its activity. These studies show the various possible activities of IFN-γ, but do not really demonstrate the role played by the cytokine in the normal functioning of the organism. The recent development of two different lines of knockout mice has now provided us with some answers to this question. Dalton and colleagues (1993) have developed mice lacking a functional IFN-γ structural gene, whereas Huang and colleagues (1993) have obtained animals that lack a functional IFN-γ receptor. An analysis of these two mutant mouse strains has been most revealing, and we will summarize the findings of both groups by combining the results obtained in each strain. Mutant mice appear normal, healthy, and fertile; obviously, one should realize that these animals are kept in a mouse colony, in the absence of specific murine pathogens. No gross histological abnormalities are present in the lymphoid organs, and there is no significant difference in the number of cells in the spleens and thymuses of these mice. There are, furthermore, no differences in the expression of CD3, CD4, CD8, cell surface immunoglobulin (IgM), and MHC class I and II on splenic and thymic cell populations or on peripheral blood cells. However, MHC class II antigen expression on macrophages is reduced. Thus, IFN-γ is not essential for the development of the immune system, and is not required for survival under specific-pathogen-free conditions. The importance of IFN-γ in the resistance to some virus infections is confirmed when mutant mice are infected with vaccinia virus. The early defense against vaccinia virus is severely defective: within the first few days after infection, virus replication is about three orders of magnitude above control values, resulting in death. With another virus, vesicular stomatitis virus, the course of infection is identical in wild-type and mutant animals, and the titers of neutralizing antibody that appear as a result of the infection are no different from those of control mice. As discussed in Section 5.5, IFN-γ has been shown to be implicated in immunoglobulin isotype regulation, and it is therefore interesting that mutant mice have decreased total serum IgG2a concentrations, and, after immunization with the appropriate antigens, show decreased titers of hapten-specific IgG2a and IgG3 antibodies. Significantly reduced IgG2a titers are also observed after immunization with pseudorabies virus. We have seen in Section 5.2 that IFN-γ plays a crucial role in the early defense against intracellular parasites. When infected with a dose of *Listeria monocytogenes* that does not kill wild-type animals, mutant mice succumb to the infection and the bacterial titers found in the livers and spleens of these animals are up to a 100-fold higher than those observed in the wild-type mice. Similarly, infection of mutant mice with a normally sublethal dose of *Mycobacterium tuberculosis* (BCG) results in a significantly enhanced mortality.

As was discussed in Section 5.8, mononuclear phagocytes exert an important part of their antimicrobial activity via the synthesis of nitric oxide and possibly other reactive nitrogen intermediates, resulting from the induction of nitric-oxide synthase by IFN-γ, and the generation of nitric oxide is severely decreased in mutant mice, as a result of which the infectivity of intracellular parasites is greatly enhanced in these animals (Kamijo *et al.*, 1993).

9.   *Interferon-γ in Disease and Therapy*

In mice, IFN-γ-mediated necrosis of small intestine has been proposed as a mechanism that underlies the genetic susceptibility of certain mouse strains to peroral infection with *Toxoplasma gondii* (Liesenfeld *et al.*, 1996).

Cachexia is a complex syndrome characterized by weight loss, anorexia, and anemia that is frequently associated with cancer and with some infectious diseases. Cytokines such as TNF-α and IL-1 are involved in the pathogenic mechanism, but, as shown in a mouse model, IFN-γ can also be involved in the pathogenesis of cancer-associated cachexia (Matthys *et al.*, 1991a). Treatment of tumor-bearing mice with potent anti-IFN-γ serum results in a significant inhibition of the wasting syndrome that accompanies tumor development, whereas, in contrast, the administration of IFN-γ accelerates the weight loss of the animals (Matthys *et al.*, 1991b).

IFN-γ has been implicated as a possible component of autoimmune disease. When administered to patients with multiple sclerosis, it accelerates the evolution of the disease. The presence of IFN-γ producing lymphocytes in the pancreas of patients with recent-onset type 1 diabetes has been suggested to contribute to the pathogenesis of insulitis in type 1 diabetes in man (Foulis *et al.*, 1991). IFN-γ can also act as a potent inducer of HIV expression in monocytes, and may thus contribute to the pathogenesis of HIV-infection (Biswas *et al.*, 1992).

In man, two different mutations in the coding sequence of the IFN-γ-receptor alpha chain have been reported, each one responsible for the synthesis of a truncated alpha receptor chain, resulting in the absence of a functional receptor. Individuals homozygous for such a mutation display a greatly enhanced susceptibility to mycobacterial infection (Jouanguy *et al.*, 1996; Newport *et al.*, 1996).

Clinical trials with hIFN-γ have shown its potential in the treatment of cutaneous and also visceral leishmaniasis.

10.   *Acknowledgments*

Work in the authors' laboratory is supported by the Agence National de Recherche sur le SIDA, and the Association de Recherches sur le Cancer.

11.   *References*

Amaldi, I., Reith, W., Berte, C. and Mach, B. (1989). Induction of HLA class II gene by IFN-γ is transcriptional and requires a trans-acting protein. J. Immunol. 142, 999–1004.

Biswas, P., Poli, G., Kinter, A.L., *et al.* (1992). Interferon gamma induces the expression of human immunodeficiency virus in persistently infected promonocytic cells (U1) and redirects the production of virions to intracytoplasmic vacuoles in phorbol myristate acetate-differentiated U1 cells. J. Exp. Med. 176, 739–750.

Blanar, M.A., Boettger, E.C. and Flavell, R.A. (1988). Transcriptional activation of *HLA-DRα* by interferon γ requires trans-acting protein. Proc. Natl Acad. Sci. USA 85, 4672–4676.

Bossie, A. and Vitetta, E.S. (1991). IFN-γ enhances secretion of IgG2a-committed LPS-stimulated murine B cells: implication for the role of IFN-γ in class switching. Cell Immunol. 135, 95–104.

Campbell, I.L., Samimi, A. and Chiang, C.S. (1994). Expression of the inducible nitric oxide synthase. Correlation with neuropathology and clinical features in mice with lymphocytic choriomeningitis. J. Immunol. 153, 3622–3629.

Celis, E., Miller, R.M., Wiktor, T.J., Dietzschold, B. and Koprowski, H. (1986). Isolation and characterization of human T cell lines and clones reactive to rabies virus: antigen specificity and production of interferon-gamma. J. Immunol. 136, 692–697.

Dalton, D.K., Pitts-Meek, S., Keshav, S., Figari, I.S., Bradley, A. and Stewart, T.A. (1993). Multiple defects of immune cell function in mice with disrupted interferon-γ genes. Science 259, 1739–1742.

Darnell, J.E., Jr, Kerr, I.M. and Stark, G.R. (1994). Jak-STAT pathways and transcriptional activation in response to IFNs and other extracellular signaling proteins. Science 264, 1415–1421.

Decker, T., Lew, D.J., Mirkovitch, J. and Darnell, J.E., Jr (1991). Cytoplasmic activation of GAF, and IFN-γ regulated DNA-binding factor. EMBO J. 10, 927–932.

De Maeyer, E. and De Maeyer-Guignard, J. (1988). "Interferons and Other Regulatory Cytokines". Wiley-Interscience, New York.

Dijkmans, R., Volckaert, G., van Damme, J., De Ley, M., Billiau, A. and De Somer, P. (1985). Molecular cloning of murine interferon γ (Mu IFN-γ) cDNA and its expression in heterologous mammalian cells. J. Interferon Res. 5, 511–520.

Ealick, S.E., Cook, W.J., Vijay-Kumar, S., *et al.* (1991).Three-dimensional structure of recombinant human interferon-γ. Science 252, 698–702.

Farrar, M.A., Fernandez-Luna, J. and Schreiber, R.D. (1991). Identification of two regions within the cytoplasmic domain of the human interferon-γ receptor required for function. J. Biol. Chem. 266, 19626–19635.

Feduchi E. and Carrasco, L. (1991). Mechanism of inhibition of HSV-1 replication by tumor necrosis factor and interferon γ. Virology 180, 822–825.

Flesch, I.E. and Kaufmann, H.E. (1991). Mechanisms involved in mycobacterial growth inhibition by gamma interferon-activated bone marrow macrophages: role of reactive nitrogen intermediates. Infect. Immun. 59, 3213–3218.

Foulis, A.K., McGill, M. and Farquharson, M.A. (1991). Insulitis in type 1 (insulin-dependent) diabetes mellitus in man-macrophages, lymphocytes, and interferon-γ containing cells. J. Pathol. 165, 97–103.

Fu, X.-Y. and Zhang, J.-J. (1993). Transcription factor p91 interacts with the epidermal growth factor receptor and mediates activation of the c-*fos* gene promoter. Cell 74, 1135–1145.

Garotta, G., Talmadge, K.W., Pink, J.R.L., Dewald, B. and Aggiolini, M. (1986). Functional antagonism between type I and type II interferon on human macrophages. Biochem. Biophys. Res. Commun. 140, 948–954.

Garotta, G., Ozmen, L., Fountoulakis, M., Dembic, Z., van Loon, A.G.M. and Stüber, D. (1990). Human interferon-γ receptor. Mapping of epitropes recognized by neutralizing antibodies using native and recombinant receptor proteins. J. Biol. Chem. 265, 6908–6915.

Germain, R.N. (1994). MHC-dependent antigen processing and peptide presentation: providing ligands for T lymphocyte activation. Cell 76, 287–299.

Gray, P.W., Leong, S., Fennie, E.H., *et al.* (1989). Cloning and expression of the cDNA for the murine interferon γ receptor. Proc. Natl Acad. Sci. USA 86, 8497–8501.

Green, J.A., Cooperband, S.R. and Kibrick, S. (1969). Immune specific induction of interferon production in cultures of human blood lymphocytes. Science 164, 1415–1417.

Gresser, I. and Naficy, K. (1964). Recovery of an interferon-like substance from cerebrospinal fluid. Proc. Soc. Exp. Biol. Med. 117, 285–289.

Hammer, S.M., Gillis, J.M., Groopman, J.E. and Rose, R.M. (1986). In vitro modification of human immunodeficiency virus infection by granulocyte-macrophage colony-stimulating factor and gamma interferon. Proc. Natl Acad. Sci. USA 83, 8734–8738.

Harpur, A.G., Andres, A.-C., Ziemiecki, A., Aston, R.R. and Wilks, A.F. (1992). JAK2, a third member of the JAK family of protein tyrosine kinases. Oncogene 7, 1347–1353.

Hemmi, S., Böhni, R., Stark, G., Di Marco, F. and Aguet M. (1994). A novel member of the interferon receptor family complements functionality of the murine interferon γ receptor in human cells. Cell 76, 803–810.

Huang, S., Hendriks, W., Althage, A., *et al.* (1993). Immune response in mice that lack the interferon-γ receptor. Science 259, 1742–1745.

Imam, A.M.A., Ackrill, A.M., Dale, T.C., Kerr, I.M. and Stark, G.R. (1990). Transcription factors induced by interferons α and γ. Nucleic Acids Res. 18, 6573–6580.

Jouanguy, E., Altare, F., Lamhamedi, S., *et al.* (1996). Interferon-γ-receptor deficiency in an infant with fatal bacille

Calmette–Guérin infection. N. Engl. J. Med. 335, 1956–1961.

Kamijo, R., Shapiro, D., Le, J., Huang, S., Aguet, M. and Vilcek, J. (1993). Generation of nitric oxide and induction of major histocompatibility complex class II antigen in macrophages from mice lacking the interferon γ receptor. Proc. Natl Acad. Sci. USA 90, 6626–6630.

Karupiah, G., Xie, Q.-W., Buller, R.M.L., Nathan, C., Duarte, C. and MacMicking, J.D. (1993). Inhibition of viral replication by interferon-γ-induced nitric oxide synthase. Science 261, 1445–1448.

Khan, K.D., Shuai, K., Lindwall, G., Maher, S.E., Darnell, J.E., Jr and Bothwell, A.L.M. (1993). Induction of the Ly-6A/E gene by interferon α-β and γ requires a DNA element to which a tyrosine-phosphorylated 91-kDa protein binds. Proc. Natl Acad. Sci. USA 90, 6806–6810.

Larner, A.C., David, M., Feldman, G.M., *et al.* (1993). Tyrosine phosphorylation of DNA binding proteins by multiple cytokines. Science 261, 1730–1733.

Lew, D.J., Decker, T., Strehlow, I. and Darnell, J.E. (1991). Overlapping elements in the guanylate-binding protein gene promoter mediate transcriptional induction by alpha and gamma interferons. Mol. Cell. Biol. 11, 182–191.

Liesenfeld, O., Kosek, J., Remington, J.S. and Suzuki, Y. (1996). Association of CD4[+] T cell-dependent, interferon-γ-mediated necrosis of the small intestine with genetic susceptibility of mice to peroral infection with *Toxoplasma gondii*. J. Exp. Med. 184, 597–607.

Lipton, S.A., Chol, Y.B., Pan, Z.-H., *et al.* (1993). A redox-based mechanism for the neuroprotective and neurodestructive effects of nitric oxide and related nitroso-compounds. Nature 364, 626–632.

Matthys, P., Dijkmans, R., Proost, P., *et al.* (1991a). Severe cachexia in mice inoculated with interferon-γ-producing tumor cells. Int. J. Cancer 49, 77–82.

Matthys, P., Heremans, H., Opdenakker, G. and Billiau, A. (1991b). Anti-interferon-γ antibody treatment, growth of Lewis lung tumours in mice and tumour-associated cachexia. Eur. J. Cancer 27, 182–187.

McKendry, R., John, J., Flavell, D., Müller, M., Kerr, I.M. and Stark, G.R. (1991). High-frequency mutagenesis of human cells and characterization of a mutant unresponsive to both α and γ interferons. Proc. Natl Acad. Sci. USA 88, 11455–11459.

Mossman, T.R., Cherwinski, H., Bond, M.W., Giedlin, M.A. and Coffman, R.L. (1986). Two types of murine helper T cell clone. I. Definition according to profiles of lymphokine activities and secreted proteins. J. Immunol. 136, 2348–2357.

Müller, M., Briscoe, J., Laxton, C., *et al.* (1993a). The protein tyrosine kinase JAK1 complements defects in interferon-α/β and -γ signal transduction. Nature 366, 129–135.

Müller, M., Laxton, C., Briscoe, J., *et al.* (1993b). Complementation of a mutant cell line: central role of the 91 kDa polypeptide of ISGF3 in the interferon-α and -γ signal transduction pathways. EMBO J. 12, 4221–4228.

Nagata, K., Kikuchi, N., Ohara, O., Teraoka, H., Yoshida, N. and Kawade, Y. (1986). Purification and characterization of recombinant murine immune interferon. FEBS Lett. 205, 200–204.

Nathan, C.F. and Tsunawaki, S. (1986). Secretion of toxic oxygen products by macrophages: regulatory cytokines and their effects on the oxidase. In "Biochemistry of

Macrophages", (eds. D. Evered, J. Nugent and M. O'Connor). Pitman, London: Ciba Foundation Symposium 118, pp. 211–230.

Nathan, C.F., Horowitz, C.R., De La Harpe, J., et al. (1985). Administration of recombinant interferon γ to cancer patients enhances monocyte secretion of hydrogen peroxide. Proc. Natl Acad. Sci. USA 82, 8686–8690.

Newport, M.J., Huxley, C.M., Huston, S., et al. (1996). A mutation in the interferon-γ-receptor gene and susceptibility to mycobacterial infection. N. Engl. J. Med. 335, 1941–1949.

Parronchi, P., Macchia, D., Piccinni, M.P., et al. (1991). Allergen- and bacterial antigen-specific T-cell clones established from atopic donors show a different profile of cytokine production. Proc. Natl Acad. Sci. USA 88, 4538–4542.

Pestka, S. (1994). Functional type I and type II interferons receptors : the Odyssey. J. Interferon Res. 14, S87.

Rinderknecht, E., O'Connor, B.H. and Rodriguez, H. (1984). Natural human interferon-γ complete amino acid sequence and determination of sites of glycosylation. J. Biol. Chem. 259, 6790–6797.

Salgame, P., Abrams, J.S., Clayberger, C., et al. (1991). Differing lymphokine profiles of functional subsets of human CD4 and CD8 T cell clones. Science 254, 279–282.

Schindler, C., Fu, X.-Y., Improta, T., Aebersold, R. and Darnell, J.E., Jr (1992). Proteins of transcription factor ISGF-3: one gene encodes the 91- and 84-kDa ISGF-3 proteins that are activated by interferon α. Proc. Natl Acad. Sci. USA 89, 7836–7839.

Seelig, G.F., Wijdenes, J., Nagabhushan, T.L. and Trotta, P.P. (1988). Evidence for a polypeptide segment at the carboxyl terminus of recombinant human gamma interferon involved in expression of biological activity. Biochemistry 27, 1981–1986.

Sen, G.S. and Lengyel, P. (1992). The interferon system. A bird's eye view of its biochemistry. J. Biol. Chem. 267, 5017–5020.

Shuai, K. (1994). Interferon-activated signal transduction to the nucleus. Curr. Opin. Cell Biol. 6, 253–259.

Shuai, K., Schindler, C., Prezioso, V.R. and Darnell J.E., Jr (1992). Activation of transcription by IFN-γ: tyrosine phosphorylation of a 91-kD DNA binding protein. Science 258, 1808–1812.

Shuai, K., Horvath, C.M., Tsai Huang, L.H., Qureshi, S.A., Cowburn, D. and Darnell J.E., Jr (1994). Interferon activation of the transcription factor Stat91 involves dimerization through SH2-phosphotyrosyl peptide interactions. Cell 76, 821–828.

Snapper, C.M. and Paul, W.E. (1987). Interferon-gamma and B cell stimulatory factor-1 reciprocally regulate Ig isotype production. Science 236, 944–947.

Soh, J., Donnely, R.J., Kotenko, S., et al. (1994). Identification and sequence of an accessory factor required for activation of the human interferon γ receptor. Cell 76, 793–802.

Taya, Y., Devos, R., Tavernier, J., Cheroutre, H., Engler, G. and Fiers, W. (1982). Cloning and structure of the human immune interferon-gamma chromosomal gene. EMBO J. 1, 953–958.

Taylor, M.W. and Feng, G. (1991). Relationship between interferon-γ, indoleamine 2,3-dioxygenase, and tryptophan catabolism. FASEB J. 5, 2516–2522.

Thomas, S.M., Garrity, L.F., Brandt, C.R., et al. (1993). IFN-γ-mediated antimicrobial response. J. Immunol. 150, 5529–5534.

Thomis, D.C. and Samuel, C.E. (1992). Mechanism of interferon action: alpha and gamma interferons differentially affect mRNA levels of the catalytic subunit of protein kinase A and protein Mx in human cells. J. Virol. 66, 2519–2522.

Trotta, P.P., Seeling, G.F., Le, T.L., Nagabhushan, L. (1986). Structure–function relations in recombinant gamma interferons. UCLA Symp. Mol. Cell. Biol. New Ser. 50, 497–507.

Urban, J.L., Shepard, H.M., Rothstein, J.L., Sugarman, B.J. and Schreiber, H. (1986). Tumor necrosis factor: a potent effect or effector molecule for tumor cell killing by activated macrophages. Proc. Natl Acad. Sci. USA 83, 5233–5237.

Velazquez, L., Fellous, M., Stark, G.S. and Pellegrini, S. (1992). A protein tyrosine kinase in the interferon α/β signaling pathway. Cell 70, 313–322.

Vijay-Kumar, S, Senadhi, S.E., Ealick, S.E., et al. (1987). Crystallization and preliminary X-ray investigation of a recombinant form of human gamma-interferon. J. Biol. Chem. 262, 4804–4805.

Walter, M.R., Windsor, W.T., Nagabhushan, T.L., et al. (1995). Crystal structure of a complex between interferon-γ and its soluble high-affinity receptor. Nature 376, 230–235.

Watling, D., Guschin, D., Müller, M., et al. (1993). Complementation by the protein tyrosine kinase JAK2 of a mutant cell line defective in the interferon-γ signal transduction pathway. Nature 366, 166–170.

Werner, E.R., Fuchs, D., Hausen, A., et al. (1988). Tryptophan degradation in patients infected by human immunodeficiency virus. Biol. Chem. Hoppe Seyler 369, 337–340.

Wheelock, E.F. (1965). Interferon-like virus-inhibitor induced in human leukocytes by photohemagglutinin. Science 149, 310–311.

Yip, Y.K., Barrowclough, B.S., Urban, C. and Vilcek, J. (1982). Purification of two species of human γ (immune) interferon. Proc. Natl Acad. Sci. USA 79, 1820–1824.

Yamada, Y.K., Meager, A., Yamada, A. and Ennis, F.A. (1986). Human interferon-alpha and gamma production by lymphocytes during the generation of influenza virus-specific cytotoxic T lymphocytes. J. Gen. Virol. 67, 2325–2334.

Zhong, Z., Wen, Z. and Darnell, J.E., Jr (1994). Stat3 and Stat4: members of the family of signal transducers and activators of transcription. Proc. Natl Acad. Sci. USA 91, 4806–4810.

27. Oncostatin M

Mohammed Shoyab, Najma Malik *and* Philip M. Wallace

1. Introduction

The complex processes of development and homeostasis are mediated by a multiplicity of positive and negative regulators of cellular proliferation and differentiation. In 1983 our laboratory focused on the systematic discovery of novel molecules that modulate the growth and differentiation of mammalian cells. By screening conditioned medium from cells and tissue extracts for growth modulatory activities, we succeeded in isolating and characterizing several new cytokines including human oncostatin M (OM) (Zarling et al., 1986).

Human OM (hOM), a single-chain polypeptide, was originally identified for its growth inhibitory effects on human tumor cell lines and contrary growth stimulatory effects on several normal fibroblast lines. hOM was isolated from media conditioned by phorbol 12-myristate 13-acetate (PMA)-treated human histiocytic lymphoma cells (Zarling et al., 1986) and from phytohemagglutinin (PHA)-activated human T lymphocytes (Brown et al., 1987). Recently, a protein was purified from the conditioned medium of a HTLV II-infected human T cell line on the basis of its ability to support the growth of cells derived from AIDS-associated Kaposi's sarcoma (KS) and was found to be hOM (Nair et al., 1992).

hOM is structurally and functionally related to a family of hematopoietic and neurotropic cytokines whose members include leukemia inhibitory factor (LIF), interleukin-6 (IL-6), granulocyte colony-stimulating factor (G-CSF), ciliary neurotropic factor (CNTF), myelomonocytic growth factor (MGF), interleukin-11 (IL-11), and cardiotropin (Bazan, 1991; Rose and Bruce, 1991; Pennica et al., 1994). Likewise, the cellular receptors for each of these factors share structural similarities, and the IL-6 signal transduction subunit (gp130) as a common component(s) of the receptor complexes (Ip et al., 1992; Liu et al., 1992b; Taga et al., 1992; Davis et al., 1993; Stahl et al., 1993; Fann and Patterson, 1994; Gearing et al., 1994; Hirano et al., 1990, 1994).

In the following sections we describe the structure and regulation of expression of the gene for hOM; the structure and biological responses associated with this protein; and the properties of the cellular receptors for hOM, including the current understanding of the signal transduction utilized by the receptors. Mouse oncostatin M has been cloned more recently (Yoshimura *et al.*, 1996) and is therefore less well studied and will be described separately. The final section discusses the basic biology of OM within the context of its role in homeostasis and pathological conditions and highlights potential therapeutic applications.

2. The Cytokine Gene

The human, simian, and bovine genes for OM are approximately 5 kb in length (Malik *et al.*, 1989, 1992a,b). The human gene consists of three exons and two introns and the exon/intron boundary sequences conform to the AT–GT rule for nucleotides flanking eukaryotic exon boundaries. Within the first exon are the 5′ noncoding DNA, the translational initiation codon, and 10 amino acids of the hydrophobic signal sequence. The remaining 14 carboxy-terminal amino acids of the signal sequence and the 34 amino-terminal residues of the mature protein are encoded by the second exon. The third exon encodes the remaining 193 residues of the OM precursor and greater than 1 kb of the 3′ noncoding sequences present in the cDNA clone (Figures 27.1 and 27.2). The structural organization of the OM gene shares homology with genes for LIF, IL-6, G-CSF, IL-11, CNTF, and MGF (Bazan, 1991; Bruce *et al.*, 1992b; Rose and Bruce, 1991).

hOM is a single-copy gene located on chromosome 22q12 within approximately 20 kb of the related LIF gene, suggesting that these two cytokines resulted from the duplication of a common ancestral gene (Giovannini *et al.*, 1993a,b; Jeffery *et al.*, 1993; Rose *et al.*, 1993). hOM cDNA clones obtained from PMA-treated U937 human histiocytic lymphoma cells have an open reading frame that contains 784 nucleotides and encodes a predicted protein of 252 amino acids of which 25 amino acids comprise the leader sequence (Malik *et al.*, 1989). As with many other cytokines, the 3′ untranslated sequence contains five ATTTA repeats and a TTATTTAT octamer motif that are important in the regulation of mRNA stability (Shaw and Kamen, 1986).

hOM mRNA has been detected in PMA-treated U937 histiocytic lymphoma cells, PHA-treated T lymphocytes, activated macrophages and monocytes, and cells derived from AIDS-KS (Malik *et al.*, 1989; Grove *et al.*, 1991a; Miles *et al.*, 1992). The major species of hOM mRNA is ~2 kb in length; however, in PMA-treated U937 cells a minor truncated transcript of ~1 kb is also seen (Malik *et al.*, 1989). In monocytes and T cells, hOM transcripts are first detectable 4 h subsequent to exposure to phorbols, antigens, or mitogens and persists for about 36 h, before declining to basal levels (Malik and Shoyab, unpublished results).

3. The Protein

Natural hOM, isolated from the conditioned medium of PMA-treated U937 lymphoma cells, and recombinant

```
                                                                                    -25 ...GCGGGCCGGAGCACGGGCACCCAGC
                                                                                                                  ***
ATG GGG GTA CTG CTC ACA CAG AGG ACG CTG CTC Agtaagt...ccttag GT CTG GTC CTT GCA CTC CTG TTT CCA AGC ATG GCG AGC ATG GCG     78
GCT ATA GGC AGC TGC TCG AAA GAG TAC CGC GTG CTC CTT GGC CAG CTC CAG AAG CAG ACA GAT CTC ATG CAG GAC ACC AGC AGA CTC CTG    168
GAC CCC TAT gtaagc...cgccag ATA CGT ATC CAA GGC CTG GAT GTT CCT AAA CTG AGA GAG CAC TGC AGG GAG CGC CCC GGG GCC TTC CCC    246
AGT GAG GAG ACC CTG AGG GGG CTG GGC AGG CGG GGC TTC CTG CAG ACC CTC AAT GCC ACA CTG GGC TGC GTC CTG CAC AGA CTG GCC GAC    336
TTA GAG CAG CGC CTC CCC AAG GCC CAG GAT TTG GAG AGG TCT GGG CTG AAC ATC GAG GAC TTG GAG AAG CTG CAG ATG GCG AGG CCG AAC    426
ATC CTC GGG CTC AGG AAC AAC ATC TAC TGC ATG GCC CAG CTG CTG GAC AAC TCA GAC ACG GCT GAG CCC ACG AAG GCT GGC CGG GGG GCC    516
TCT CAG CCG CCC ACC CCC ACC CCT GCC TCG GAT GCT TTT CAG CGC AAG CTG GAG GGC TGC AGG TTC CTG CAT GGC TAC CAT CGC TTC ATG    606
CAC TCA GTG GGG CGG GTC TTC AGC AAG TGG GGG GAG AGC CCG AAC CGG AGC CGG AGA CAC AGC CCC CAC CAG GCC CTG AGG AAG GGG GTG    696
CGC AGG ACC AGA CCC TCC AGG AAA GGC AAG AGA CTC ATG ACC AGG GGA CAG CTG CCC CGG TAG CCTCGAGAGCACCCCTTGCCGGTGAAGGATGCGGCA    795
GGTGCTCTGTGGATGAGAGGAACCATCGCAGGATGACAGCTCCCGGGTCCCCAAACCTGTTCCCCTCTGCTACTAGCCACTGAGAAGTGCACTTTAAGAGGTGGGAGCTGGGCAGACCCC    915
TCTACCTCCTCCAGGCTGGGAGACAGAGTCAGGCTGTTGCGCTCCCACCTCAGCCCCAAGTTCCCCAGGCCCAGTGGGGTGGCCGGGCGGGCCACGCGGGACCGACTTTCCATTGATTCA   1035
GGGGTCTGATGACACAGGCTGACTCATGGCCGGGCTGACTGCCCCCCTGCCTTGCTCCCCGAGGCCTGCCGGTCCTTCCCTCTCATTGACTTGCAGGGCCGTTGCCCCCAGACTTCCTCC   1155
TTTCCGTGTTTCTGAAGGGGAGGTCACAGCCTGAGCTGGCCTCCTATGCCTCATCATGTCCCAAACCAGACACCTGGATGTCTGGGTGACCTCACTTTAGGCAGCTGTAACAGCGGCAGG   1275
GTGTCCCAGGAGCCCTGATCCGGGGGTCCAGGGAATGGAGCTCAGGTCCCAGGCCAGCCCCGAAGTCGCCACGTGGCCTGGGGCAGGTCACTTTACCTCTGTGGACCTGTTTTCTCTTTG   1395
TGAAGCTAGGGAGTTAGAGGCTGTACAAGGCCCCCACTGCCTGTCGGTTGCTTGGATTCCCTGACGTAAGGTGGATATTAAAAATCTGTAAATCAGGACAGGTGGTGCAAATGGCGCTGG   1515
GAGGTGTACACGGAGGTCTCTGTAAAAGCAGACCCACCTCCCAGCGCCGGGAAGCCCGTCTTGGGTCCTCGCTGCTGGCTGCTCCCCCTGGTGGTGGATCCTGGAATTTTCTCACGCAGG   1635
AGCCATTGCTCTCCTAGAGGGGGTCTCAGAAACTGCGAGGCCAGTTCCTTGGAGGGACATGACTAATTTATCGATTTTTATCAATTTTTATCAGTTTTATATTTATAAGCCTTATTTATG   1755
ATGTATATTTAATGTTAATATTGTGCAAACTTATATTTAAAACTTGCCTGGTTTCTAAA...                                                            1814
```

Figure 27.1 Consensus nucleotide sequence of oncostatin M cDNA (capitals), intron junction sequences (lower-case), and deduced protein amino acid sequence. Indicated are potential *N*-linked glycosylation sites (underlined), the signal peptide (black bars), the N-terminus of mature oncostatin M (*), and sequence motifs commonly found in 3′ noncoding regions of cytokine and lymphokine mRNAs (└─┘) and inflammatory mediator RNAs (┌─┐).

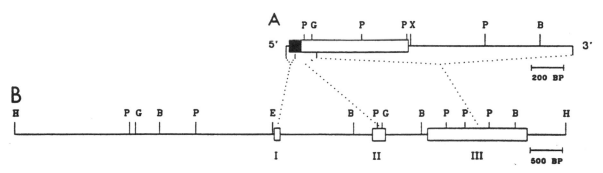

Figure 27.2   Molecular maps of human oncostatin M cDNA and gene. (a) Restriction map of the 1839 bp consensus cDNA, presented 5′ to 3′. The coding region is boxed; the position of the signal sequence is shaded. (b) Map of the 9 kbp *Hind*III genomic fragment with the corresponding exons. The three known exons (I, II, III) are boxed. The extents of the 5′ and 3′ exons have not been determined completely. Restriction enzyme sites: B, *Bam*HI; G, *Bgl*II; E, *Eco*RI; P, *Pvu*II; X, *Xho*I.

hOM, produced by Chinese hamster ovary cells, are single-chain glycoproteins (Zarling *et al.*, 1986; Malik *et al.*, 1989, 1992a). Biochemical analyses of the natural and recombinant proteins have revealed that the mature molecule contains 196 amino acids (Figure 27.3). Molecular cloning and sequence analysis indicate that hOM is synthesized as a 252 amino acid precursor. A hydrophobic signal sequence of 25 amino acids is cleaved during post-translation processing, resulting in a pro-hOM protein of 227 amino acids (Figure 27.4). The mature hOM is generated from pro-hOM by cleavage of 31 C-terminal residues of the pro-hOM at a trypsin-like protease cleavage site (Linsley *et al.*, 1990). Pro-hOM has been reported to be as active in radioreceptor assays as the mature protein but shows 5–60-fold less activity in growth inhibition assays (Linsley *et al.*, 1990). hOM has two potential *N*-linked glycosylation sites and several potential *O*-linked glycosylation sites. The role of oligosaccharides in hOM is unknown at present, but hOM molecules containing no oligosaccharides or only *O*-linked oligosaccharides are indistinguishable in growth inhibitory assays *in vitro* and show biological activity *in vivo* (our unpublished results).

Human OM contains five cysteine residues, four of which form intramolecular disulfide bonds. The disulfide

pair assignments are C^6–C^{127} and C^{49}–C^{167} (Kallestad *et al.*, 1991) and these cysteine residues are conserved in both the bovine and simian proteins. The fifth cysteine, Cys-80, is not involved in disulfide bond formation and is conserved only in the simian protein (Figures 27.3 and 27.4) (Kallestad *et al.*, 1991). The overall sequence homology of hOM with bovine and simian OM at the amino acid level is 52% and 97%, respectively (Rose and Bruce, 1991; Malik *et al.*, 1992a). Site-directed mutagenesis of the human protein has shown that the C^{49}–C^{167} disulfide bond is essential for the expression of biologically active protein, while the C^6-C^{127} linkage is not required for functional activity (Kallestad *et al.*, 1991). These experiments also established that the hOM tertiary structure is influenced by residues on both sides of disulfide bond C^{49}–C^{167}, and that residues in a putative amphiphilic helix extending from Cys-167 to Ser-185 are essential for biological activity of hOM (Kallestad *et al.*, 1991; Radka *et al.*, 1994).

The tertiary and quarternary structures of hOM have yet to be experimentally determined, but circular dichroism studies have shown that 50% of hOM residues are part of α-helical structures (Sporeno *et al.*, 1994). However, it has been predicted by sequence and structure analysis that hOM exhibits a topology formed by a bundle of four antiparallel α-helices linked by three loops of variable length (Bazan, 1991; Rose and Bruce, 1991; Sprang and Bazan, 1993). This is consistent with experimentally elucidated structures of growth hormone and LIF (De Vos *et al.*, 1992; Robinson *et al.*, 1994).

The molecular mass of hOM, as determined by SDS–polyacrylamide gel electrophoresis under reducing or nonreducing conditions, is approximately 28 kDa. In contrast, gel permeation chromatography predicts a size of ~20 kDa (Zarling *et al.*, 1986), suggesting that hOM is a compact single-chain protein. hOM protein is resistant to inactivation. It is stable in 1 M acetic acid, 1 M ammonium hydroxide, 6 M urea, 0.01 M sodium metaperiodate, and at elevated temperatures (56°C for 1 h). However, the biological activity is not preserved

OM Biosynthesis

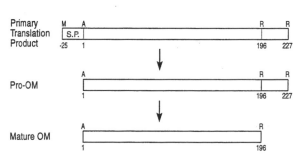

Figure 27.3   OM biosynthesis.

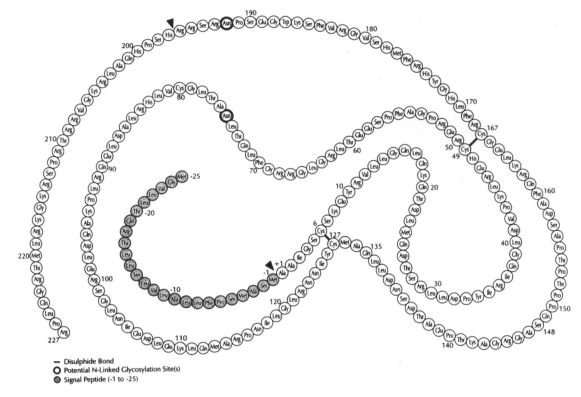

Figure 27.4   Amino acid sequence of OM.

after exposure to trypsin for 90°C for 1 h. hOM, at a concentration of 1 mg/ml, is stable for at least 2 years in 40% acetonitrile, 0.1% trifluoroacetic acid in water either at 4°C or –20°C. hOM retains its functional activity for at least 3 months in aqueous solution (~5 mg/ml in phosphate-buffered saline) and, as a lyophilizate, is stable for a minimum of 6 months at –20°C.

4.   Cellular Sources and Production

Very few cell types have been shown to synthesize and secrete hOM. U-937 histiocytic lymphoma cells do not constituitively express hOM, but upon treatment with PMA these cells acquire many of the characteristics of macrophages and produce hOM (Zarling *et al.*, 1986). Human macrophages, following stimulation with endotoxin also secrete hOM. Human T cells, either PHA-stimulated or chronically infected with one of three human retroviruses, including human immunodeficiency virus-1 (HIV), have hOM present in their conditioned media (Brown *et al.*, 1987; Radka *et al.*, 1993). The expression of hOM is not limited to hematopoietic lineages, as cells derived from AIDS-KS both make, and respond to, hOM (Nair *et al.*, 1992).

5.   Biological Activity

Human OM acts on many cell types, and elicits a variety of diverse responses. *In vivo*, the nature and magnitude of these responses is likely to be modulated by the cellular environment and the presence of other cytokines.

5.1   Growth Modulation and Cancer

The biological activity first associated with hOM was its ability to modulate the proliferation of mammalian cells (Zarling *et al.*, 1986; Brown *et al.*, 1987). Growth inhibition and morphological alteration are observed with many, but not all, human tumor cell lines derived from lung carcinomas, melanomas, breast carcinomas, ovarian carcinomas, and neuroblastomas (Horn *et al.*, 1990). However, hOM does not modulate the proliferation of cell lines derived from colon or prostate carcinomas *in vitro*, even though receptors for hOM are expressed on these cell lines (Horn *et al.*, 1990). Recently, more detailed studies have shown that hOM is capable of preventing the responsiveness of breast cancer cells to a variety of mitogens (including epidermal and fibroblast growth factors) by interfering with transcription of the c-*myc* gene (Liu *et al.*, 1996; Spence *et al.*,

1997). hOM is also capable of acting in concert with other factors in tumor cell growth inhibition. hOM and tumor necrosis factor (TNF) each alone inhibit the proliferation of MCF-7 human breast carcinoma cells but, when combined, have synergistic antiproliferative effects (Todaro *et al.*, 1989). A similar synergy is seen in the inhibition of growth of human melanoma cells when hOM is combined with transforming growth factor-β (TGF-β) (Brown *et al.*, 1987). However, hOM and interferon-γ (IFN-γ) produce only an additive level of growth inhibition in the same cell type. The human melanoma cell line A375 is extremely sensitive to growth inhibitory effects of hOM and IL-6. An A375 variant, selected for its resistance to hOM, was resistant to the growth inhibitory effects of IL-6, suggesting a common pathway in the action of these two cytokines (McDonald *et al.*, 1993). Recently, hOM has been implicated in the progression of malignant disease. It has been reported that human melanoma sensitivity to growth inhibition by hOM and IL-6 is diminished or lost with disease progression (Lu *et al.*, 1993). Also, studies of the cytokine levels present in breast cyst fluids revealed a correlation between the hOM levels and the risk of developing breast cancer (Lai *et al.*, 1994).

Kaposi's sarcoma, a rare malignancy of mesenchymal origin, is the most common neoplasm associated with HIV-infected patients (Rutherford *et al.*, 1989; Ensoli *et al.*, 1991; Miles, 1992). hOM acts as a potent mitogen and morphogen for AIDS-KS cells, and is required for long-term maintenance of KS cells in culture. Furthermore, KS cells in response to hOM also produce IL-6, which is mitogenic for these cells (Miles *et al.*, 1990). However, biological effects of hOM are not solely mediated through IL-6 since antibodies or oligonucleotides that inhibit IL-6 do not inhibit hOM-evoked responses. The majority of KS cell lines produce hOM. Therefore, hOM may play a role in the growth and progression of KS in an autocrine and/or paracrine fashion.

hOM stimulates the growth of several other cell types including human fibroblasts (Zarling *et al.*, 1986), B9 hybridoma cells (Barton *et al.*, 1994), human plasmacytoma cells freshly isolated from a patient with multiple myeloma (Nishimoto *et al.*, 1994), and 2 of 4 human multiple myeloma cell lines (Zhang *et al.*, 1994). Also, it has been reported that hOM can supplant LIF as a maintenance and renewal factor for embryonic stem cells (Rose *et al.*, 1994; Yoshida *et al.*, 1994). Embryonic stem cells cultured in the presence of hOM express stem cell-specific antigen, retain their characteristic morphology and proliferative capacity, and, most importantly, maintain their totipotent potential. The indirect effects of hOM on cell growth may not be limited to KS cells since IL-6, produced by hOM-treated cells, is capable of inhibiting the growth of carcinoma cells (Brown *et al.*, 1991) and is mitogenic for myeloma cells (Zhang *et al.*, 1994). Recently, hOM has been

demonstrated to influence both the growth and function of bone-derived cells (Jay *et al.*, 1996).

5.2 HEMATOPOIESIS

Administration of hOM to mice, dogs, or nonhuman primates results in marked stimulation of thrombopoiesis (Wallace *et al.*, 1995a,b). hOM treatment increases peripheral platelet levels in a dose- and time-dependent manner without changing peripheral erythrocyte or leukocyte counts (Wallace *et al.*, 1995a,b). *In vitro* studies of murine bone marrow cells revealed that hOM, in combination with IL-3, enhances murine megakaryocyte colony formation, though hOM alone has no effect on colony formation. In liquid cultures hOM increases the size and maturation of megakaryocytes both alone and when combined with IL-3. These results indicate that hOM acts as a maturation factor, potentiating the effects of IL-3, but lacks intrinsic colony-forming capacity. The related cytokines as IL-6, LIF, and IL-11 each promote *in vivo* thrombocytopoiesis but with varying effects on cells from other circulating blood lineages (Gordon and Hoffman, 1992). hOM treatment results in accelerated platelet recovery of mice rendered thrombocytopenic by irradiation or cytotoxic drug treatment and prevents the decline in peripheral erythrocyte counts seen in control animals (Wallace *et al.*, 1995a).

Transgenic mice expressing bovine OM under the control of a T cell specific promoter were used to implicate OM as a factor regulating T cell development (Clegg *et al.*, 1996). These and other studies demonstrated a novel thymus-independent pathway by which immature T cells can be produced by OM (Clegg *et al.*, 1996). *In vitro* hOM has been found to act on human cells to induce the synthesis of growth factors which act upon hematopoietic progenitor cells. G-CSF, and GM-CSF (granulocyte-macrophage colony-stimulating factor) and IL-6 are produced by endothelial cells treated with hOM. In addition, IL-6 is secreted by other cell types in response to hOM, either alone or when combined with other factors (Richards and Agro, 1994). Tissue inhibitor of metalloproteinases (TIMP-1) has been found to have erythroid potentiating activity (Murate *et al.*, 1993) and can be upregulated in human cells by hOM (Richards and Agro, 1994). Certain human and murine cells of hematopoietic lineages are induced to differentiate by hOM (Bruce *et al.*, 1992a).

5.3 INFLAMMATION/IMMUNE FUNCTION

The expression of hOM by activated T cells and macrophages suggests that this protein would be present and function physiologically at sites of inflammation. A variety of *in vitro* and *in vivo* properties supports the notion that OM is a component of a feedback mechanism that orchestrates the return to homeostasis necessary following the induction and effector phases of injury/inflammation. The regulation of IL-6 protein production by hOM, discussed earlier, may be important in regulating more than hematopoiesis. IL-6 is a multifunctional cytokine which influences both inflammation, by inducing acute-phase proteins, and immune response, by acting as a potent B cell mitogen (Hirano et al., 1990). In addition, other inflammatory cytokines can influence the production of IL-6 by hOM. Tumor necrosis factor-α (TNF-α), but not IL-1α, synergizes the IL-6 induction in endothelial cells with hOM (Brown et al., 1991); this is despite the fact that IL-1α and TNF-α alone are each capable of causing IL-6 secretion. Synergy is also noted for IL-6 production by fibroblasts treated with hOM and either IL-1α or prostaglandin E2 (Richards and Agro, 1994). Furthermore, in vivo injection of mice with hOM results in a dose-dependent increase of circulating bioactive IL-6. LPS-induced TNF production can also be blocked by hOM treatment of mice (P.M. Wallace et al., manuscript in preparation). TNF and hOM also act in concert to upregulate IL-6 production. Thus hOM is capable of modulating cytokine expression in vivo. The function of IL-1 is also modulated in vitro by OM, resulting in the attenuation of a pro-inflammatory cascade cytokine initiated by this molecule. The expression of IL-8 and GM-CSF following the treatment of synovial and lung fibroblasts with IL-1 is attenuated by hOM (Richards et al., 1996).

Plasminogen activator (PA) activity, which influences cell migration and the activation of latent collagenases, has been shown to be increased on synovial fibroblasts by hOM (Hamilton et al., 1991). Likewise, hOM treatment of vascular endothelial cells results in a significant increase in PA activity (Brown et al., 1990). Conjointly, hOM inhibits upregulation on endothelial cells of class I molecules by TNF-α and class II molecules by IFN-γ (Brown et al., 1994). The constitutive expression of histocompatability antigens is unaltered by hOM.

Human OM selectively regulates the expression of TIMP-1 by fibroblasts in cultures, as previously discussed. In addition to effects on erythropoiesis, TIMP-1 is a potent antagonist of matrix metalloproteinase (MMP) and is a factor controlling matrix integrity in chronic inflammation (Docherty et al., 1992). This increase in TIMP-1 is in the absence of any changes in the MMP level, a finding differing from the results for other cytokines. Stimulation of TIMP-1 production by hOM is also found in human articular cartilage (Nemoto et al., 1996).

In response to injury, infection, and inflammation the serum level of acute-phase proteins (APPs) is increased. These proteins function to inhibit the production and function of cytokines and proteases at sites of inflammation. hOM can induce the production of several APPs including haptoglobulin, α$_1$-antichymotrypsin, α$_1$-acid glycoproteins, α$_1$-protease inhibitor, and ceruloplasmin in HepG2 cells and α$_2$-macroglobulin, thiostatin, and α$_1$-acid glycoprotein in rat hepatocytes (Richards et al., 1992; Baumann et al., 1993; Richards and Shoyab, 1993). hOM is more potent in APP induction than either IL-6 or LIF (Richards et al., 1992). Moreover, IL-1 or glucocorticoids synergize with hOM in APP induction, but hOM actions are only additive with IL-6 and LIF (Richards and Agro, 1994). Several factors regulate the biosynthesis of APPs by hepatocytes (Koj, 1985; Gauldie et al., 1987; Baumann et al., 1989; Heinrich et al., 1990; Baumann and Schendel, 1991). Epithelial cells of various tissues including bronchial, lung, colon, and breast cells have been found to produce acute-phase proteins (Cichy et al., 1995a,b). These cell types are acted upon to produce APPs by a more limited array of cytokines than hepatic cells, the most prominent to date being hOM. The local production of APPs by epithelial cells may play a more critical role than previously appreciated in limiting tissue damage and restoring homeostasis at sites of inflammation. Again, the effects of OM can be amplified by various pro-inflammatory cytokines and glucocorticoids. In vivo APP levels are increased in the blood of mice, dogs, and nonhuman primates shortly following the injection of hOM (Wallace et al., 1995b).

Many cytokines, most notably TNF, and various protease have been implicated in most if not all inflammatory pathologies. Therefore, ability of OM to reduce the production and function of pro-inflammatory cytokines and to induce several protease inhibitors would be predicted to be beneficial at sites of chronic inflammation. Indeed, hOM is effective in inhibiting several animal models of human diseases including experimental autoimmune encephalomyelitis, murine colitis, and antibody-induced arthritis (Wallace et al., manuscript in preparation). These findings indicate that hOM may regulate immune and inflammatory responses either directly or as one component of a cytokine network.

5.4   Cholesterol Metabolism and Atherosclerosis

The low-density lipoprotein (LDL) receptors present on hepatic cells are central in cholesterol homeostasis and metabolism (Goldstein and Brown, 1985; Brown and Goldstein, 1986; Kovanen, 1987) and therefore in atherosclerosis. hOM acts upon the hepatoma cell line HepG2 to increase LDL uptake by upregulating the number of LDL receptors by a novel cholesterol-independent mechanism (Grove et al., 1991a,b). In conjunction with this, hOM significantly inhibits the synthesis of cholesterol in these cells (R. Grove, unpublished observations). Other OM-related cytokines such as LIF, IL-6, CNTF, and G-CSF do not induce this response. Furthermore, hOM stimulates proliferation and dramatically alters the morphology of rabbit aortic smooth-muscle cells (SMC) *in vitro*, in a dose-dependent manner (Grove *et al.*, 1993). Again, this response is limited to hOM, as other related cytokines do not affect the growth of these cells. These data suggest OM may be an important, naturally occurring mitogen for vascular SMC. It may also be an important mitogen in the SMC proliferation responsible for the intimal thickening associated with atherosclerosis, especially in light of the presence of macrophages (a potential source of OM) within such lesions. Taken together, these findings indicate that OM may be involved in cholesterol homeostasis *in vivo* and in atherosclerosis.

5.5   NEUROTROPIC EFFECTS

Human OM has been shown to effect a measurable and reproducible increase in choline acetyltransferase (ChAT) activity and vasoactive intestinal peptide (VIP) levels in cultured rat sympathetic neuronal cells. However, CNTF or LIF produce more potent responses in this system (Rao *et al.*, 1992). Additionally it has been reported that hOM, CNTF, and LIF each upregulate the expression of the neuropeptides and neurotransmitter synthesizing enzymes cholecystokinin, ChAT, enkephalin, somato-statin, substance P, and VIP in rat sympathetic neurons in culture. Again the action of hOM is less pronounced in comparison to that seen with CNTF or LIF (Fann and Patterson, 1994). IL-6 and IL-11 each weakly induce expression of substance P, but not of the other neuropeptides (Fann and Patterson, 1994). Exposure of the human neuroblastoma cell line NBFL to hOM can elevate the expression of VIP and c-*fos* gene. This transcriptional activation is equivalent to that seen with either LIF or CNTF (Rao *et al.*, 1992). Other data on the neurotropic effect of hOM were recently reviewed (Vos *et al.*, 1996).

6.   *OM Receptors*

Oncostatin M receptors are present on many normal cell types including fibroblasts, hepatocytes, neuronal cells, vascular smooth-muscle cells, embryonic stem cells, and endothelial cells (Linsley *et al.*, 1989; Grove *et al.*, 1991b, 1993; Gearing and Bruce, 1992). hOM has also been found to bind to a variety of carcinomas (Linsley *et al.*, 1989), myelomas (Zhang *et al.*, 1994), and AIDS-KS cells (Soldi *et al.*, 1994). These studies have revealed that hOM can bind to at least three distinct cell surface receptors. Two of the receptors, one of high and one of low affinity, cannot be competed by related cytokines. Therefore, these receptors are thought to be bound only by hOM, and will be referred to as the high- and low-affinity OM-specific receptors. A further, better defined,

cellular receptor for hOM is also present on some cells. This receptor, first identified for its binding to LIF, in addition binds hOM. In cell types expressing this OM/LIF shared receptor, OM can reproduce the biological effects exerted by LIF. Human cells have been found to express OM-specific and/or shared receptors (Thoma *et al.*, 1994). In contrast, to date, all binding to murine cells appears to be mediated by receptors of the shared type. The question as to the existence of a murine OM-specific receptor and whether human OM binds to murine OM-specific receptors is still unanswered.

6.1   OM-SPECIFIC RECEPTOR (TYPE II RECEPTOR)

The existence of an OM-specific receptor is supported by the findings that (1) hOM is able to exert its effects on a number of cell lines that do not express functional LIF receptor (Bruce *et al.*, 1992a) or respond to LIF (Liu *et al.*, 1993; Piquet-Pellorce *et al.*, 1994; Soldi *et al.*, 1994) and (2) hOM binding cannot be competed by LIF (Bruce *et al.*, 1992a; Thoma *et al.*, 1994) or related cytokines (Bruce *et al.*, 1992a). Of these receptors, 2–4% belong to the high-affinity class and the remainder of the receptors are of low affinity (Linsley *et al.*, 1989). High-affinity binding sites have K_d values in the range 1–50 pM while those of low affinity are between 0.4–1 nM (Linsley *et al.*, 1989; Brown *et al.*, 1991). Generally, cells exhibit ~5–10 000 receptors, with the highest numbers having been measured on endothelial cells (100 000 sites/cell), although biological responses occur on cells that have as few as 100 receptors/cell (Bruce *et al.*, 1992a).

Results obtained from binding, cross-linking, immunoprecipitation with anti-gp130 antibodies, and direct transfection experiments have revealed that hOM directly binds gp130 (Gearing *et al.*, 1992; Liu *et al.*, 1992b). These findings are in contrast to the related cytokines IL-6, LIF, CNTF, and IL-11. However, anti-gp130 antibodies can block biological responses of all these cytokines (Liu *et al.*, 1992b; Taga *et al.*, 1992). The transfection of gp130 into BAF cells (a murine pro-B cell line that neither expresses gp130 nor responds to hOM) results in only a single class of low-affinity binding sites (Liu *et al.*, 1994), and chemical cross-linking experiments with radiolabeled hOM have revealed a 180 kDa complex from BAF-m130 cells. However, when hOM was cross-linked to the responsive H2981 lung carcinoma cell line, known to express both low- and high-affinity OM receptors, two complexes of ~180 kDa and ~300 kDa were observed (Liu *et al.*, 1994). Taken together, these results support the premise that gp130 is the low-affinity OM receptor and forms a component of the high-affinity receptor. The structure of gp130 has been well characterized. A type I membrane glycoprotein of ~130 kDa, it is a member of a receptor superfamily that is characterized by the conservation of four cysteine residues and a WSXWS motif. The protein is almost ubiquitously expressed in cells (Saito *et al.*, 1992; Taga and Kishimoto, 1992) and is highly conserved between species with a homology between human and murine gp130 of ~77% at both the DNA and protein levels (Hibi *et al.*, 1990; Saito *et al.*, 1992). The cellular receptor complexes for each of the OM-related cytokines IL-6, CNTF, and LIF have gp130 as a component.

The molecular composition of the high-affinity receptor is unknown but studies to date suggest that one or more additional nonbinding affinity-enhancing subunits (AES) are required to interact with an OM/gp130 complex to generate the functional high-affinity OM-specific receptors. This would be consistent with the receptor structure of other members of the IL-6 family of cytokines (Figure 27.5). Recent findings indicate that the generation of high-affinity functional receptors for IL-6, LIF, CNTF, OM, and IL-11 all seem to involve either homo- or heterodimerization of gp130 following interaction of each cytokine with a specific ligand-binding subunit (Liu *et al.*, 1992b, 1994; Davis *et al.*, 1993; Murakami *et al.*, 1993; Stahl *et al.*, 1993). In the case of IL-6 the complex is thought to be composed of a monomeric IL-6R α subunit and a homodimer of gp130. For CNTF the complex apparently consists of a CNTFR, a gp130, and a LIF receptor subunit (LIFR), while for LIF-holoreceptors, a gp130, a LIFR, and an affinity-enhancing subunit are components. The precise composition of the functional ternary receptor complex for OM is not yet known. However, we propose that the functional OM receptor complex is composed of a homodimer of gp130 in association with a putative affinity-enhancing subunit that apparently facilitates the OM-induced dimerization of gp130 and maintains receptors in a high-affinity conformation (Figure 27.5). A preliminary report of the cloning of an affinity conversion subunit for the OMR, which is structurally related to LIFR, is consistent with this model (Mosley *et al.*, 1994).

It has been shown that hOM can antagonize the effects of IL-6. As gp130 is a necessary part of a functional IL-6 receptor, these results suggest that hOM binding to gp130 allows it to sequestrate this protein and thus act as an antagonist for IL-6. This may also be true on cells that express other gp130-containing receptors. Such interactions would establish a form of cytokine hierarchy. However, owing to the high concentrations of cytokine required in these studies, the physiological significance of the putative cross-talk between receptors is unclear.

6.2 OM/LIF Receptor (Type I Receptor)

Binding and competition experiments have demonstrated that hOM binds to the high-affinity LIF receptor complex, although with 5–10-fold lower affinity than LIF (Gearing and Bruce, 1992). hOM does not directly bind to the ~190 kDa glycoprotein that constitutes the low-affinity LIF receptor subunit (LIFR) (Gearing *et al.*, 1991, 1994; Gearing and Bruce, 1992). However, gp190 expression in association with gp130 generates very high-affinity sites for LIF and moderately high-affinity sites for hOM (Baumann *et al.*, 1993; Gearing *et al.*, 1994). The transfection of LIFR into cells expressing only gp130 also confers biological responsiveness to hOM (Baumann *et al.*, 1993). The K_d of this interaction is ~1 nM, making it intermediate in affinity between the OM-specific receptors. In cell types that contain this receptor, hOM can reproduce the biological effects of LIF. As expected from the relative affinities, higher concentrations of hOM, compared to LIF, are required to produce the same effects.

7. *Signal Transduction*

Significant progress has recently been made in understanding the signal transduction events following the binding of OM to its receptors, although the complete details of the mechanism are not yet apparent (Stahl *et al.*, 1993; Kishimoto *et al.*, 1994). This is due, in part, to the signal transduction pathways used by OM being complicated by the existence of the different cellular receptors that may interact in a form of "cross-talk." Beyond this, each receptor uses more than one pathway.

Human OM, in common with other related cytokines that signal through gp130, causes rapid tyrosine

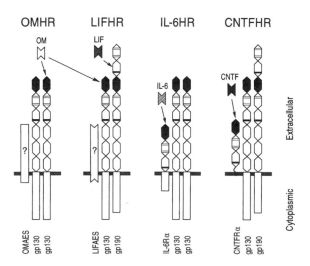

Figure 27.5 Schematic models of OM/LIF/IL-6/CNTF holoreceptor (HR); ◯ indicates a fibronection III/contaction domain; ⬤ indicates an immunoglobin domain; a heavy line indicates the WSXWS sequence and thin lines represent the conserved cysteines; ⸨ indicates GPI linkage: AES, affinity-enhancing subunit.

phosphorylation of several cellular proteins (Grove *et al.*, 1991b, 1993; Ip *et al.*, 1992; Schieven *et al.*, 1992; Taga *et al.*, 1992; Yin *et al.*, 1992; Stahl *et al.*, 1993; Akira *et al.*, 1994; Boulton *et al.*, 1994; Feldman *et al.*, 1994; Lütticken *et al.*, 1994; Narazaki *et al.*, 1994; Thoma *et al.*, 1994; Yin and Yang, 1994). This tyrosine phosphorylation is critical to the biological responses evoked by hOM (Grove *et al.*, 1991b, 1993; Murakami *et al.*, 1991; Ip *et al.*, 1992; Schieven *et al.*, 1992; Yin and Yang, 1994). More detailed analysis of these phosphorylated proteins allowed the preliminary characterization of at least two distinct signal transduction pathways used by hOM, both of which are tyrosine phosphorylation dependent.

The pattern of protein tyrosine phosphorylation has been compared for LIF and hOM using cell lines of human origin that express OM/LIF shared receptors, OM-specific receptors, or both. Analysis of the events following ligand binding revealed that both the OM-specific and OM/LIF receptors each activate mitogen-activated protein kinases (MAPKs). MAPKs are important components of an intrasignaling cascade that links agonists-induced activation of intrinsic-receptor tyrosine kinases or receptor-associated intracellular tyrosine kinases to alterations in gene expression (Thomas, 1992; Blenis, 1993). Comparisons of the patterns of tyrosine phosphorylation and activation of MAPK identified qualitative differences between the responses elicited by the different receptors, even on the same cell types (Thoma *et al.*, 1994). The two types OM receptor complexes employ some common elements, but their signal transduction pathways are distinct and unique. This is despite gp130 being a common component of both known OM receptors; thus the presence of other subunits modulates the gp130 mediated signaling. In some, but not all, cases, the biological response correlated with the activation of MAPK and was inhibited by a MAPK inhibitor.

The nature of the signals initiated by the human OM-specific receptor present on AIDS-derived KS cell lines has also been studied. hOM treatment causes the tyrosine phosphorylation and/or activation of a variety of substrates including phosphatidylinositol 3-kinase and *src* family kinases (Soldi *et al.*, 1994), besides MAPK (Amaral *et al.*, 1993). hOM binding to human umbilical vein cells, which also possess only OM-specific receptors, stimulates tyrosine phosphorylation in association with activation of the *src*-like tyrosine protein kinases p62[yes] and to a lesser extent p59[fyn] (Schieven *et al.*, 1992). In common with human cells, hOM treatment of murine 3T3-L1 preadipocytes results in the phosphorylation of a variety of proteins and triggered the activation of MAPKs, as well as ribosomal S6 protein kinases (Yin and Yang, 1994). It was further demonstrated that the responses elicited by hOM or LIF were indistinguishable. When either IL-6 or IL-11 were used to treat these cells the results were identical, and overlapping with but

distinct from the LIF and hOM results, despite the fact that each factor elicits the same biological response and has gp130 as a component of the cellular receptor. It is likely that the signals described for hOM in these studies are through the shared OM/LIF receptor, as no OM-specific receptor binding hOM has been identified in murine cells.

The absence of a consensus kinase domain from the components of the holoreceptors implicates the involvement of nonreceptor tyrosine kinases in the signal-transduction pathways used by hOM. Moreover, members of the JAK family of tyrosine kinases associate with gp130 homodimers and/or gp130–gp190 heterodimers and undergo rapid activation following stimulation with OM and related cytokines. Also complexed with the receptor and nonreceptor tyrosine kinase are latent transcription factors. These factors are phosphorylated upon receptor ligation. This phosphorylation causes the proteins to form complexes and become active. They are then translocated to the nucleus (Akira *et al.*, 1994; Boulton *et al.*, 1994; Feldman *et al.*, 1994; Lütticken *et al.*, 1994) and, once in the nucleus, bind regulatory sequences within specific cellular genes modulated by OM. These genes are responsible for the biological responses ultimately manifested. The family of latent transcription factors of proteins are termed signal transducers and activators of transcription (STAT) and were initially characterized as molecules important in the regulation of IFN-induced responses (Hunter, 1993; Kishimoto *et al.*, 1994; Mui and Miyajima, 1994; Shuai, 1994). In the IFN system the JAK family has proved to be responsible for the phosphorylation and activation of members of the STAT protein family (Darnell *et al.*, 1994). It is therefore reasonable to assume the JAKs play a similar role in OM signaling. Though the JAK–STAT pathway is utilized for signal transduction by many cytokines, most notably in this instance IL-6, LIF, CNTF, and IL-11, it is clearly capable of generating a diverse array of biological changes. It is probable that this diversity is a result of differing patterns of JAK and STAT utilization.

Little is known of the consequences of activation of OM receptors at the nucleus of responsive cells. *EGR-1*, c-*jun* and c-*myc* early response genes are rapidly and transiently elevated in response to hOM treatment of both murine leukemic M1 cells and human fibroblasts but not in OM-responsive melanoma cells (Liu *et al.*, 1992a). In these studies it is unclear whether the different nuclear events reflect the different biological responses that occur following OM treatment. In studies of murine 3T3-L1 a common set of primary genes (*EGR-1*, *tis*11 and *Jun*B) were activated by hOM and the related cytokines LIF, IL-11, and IL-6 (Yin and Yang, 1994). Interestingly, activation of these nuclear events was shown to be dependent on both serine/threonine and tyrosine kinases, in contrast to the activation of MAPK, which was only dependent on tyrosine phosphorylation. Thus,

MAPK or other serine/threonine kinases are involved in the induction of these genes.

Upregulation of LDL receptors by hOM on HepG2 cells was also accompanied by upregulation of EGR-1. In this instance elevation of EGR-1 expression correlated with the biological response and required tyrosine phosphorylation, but not protein kinase C activation (Liu et al., 1993). EGR-1 may act as the nuclear signal transducer utilized by hOM to induce transcription of the LDL receptor gene as the LDL receptor gene contains an EGR-1 consensus sequence upstream of the transcription initiation site. These changes could not be reproduced by either LIF or IL-6.

In summary, these data show that hOM utilizes at least two signal transduction pathways shared with other cytokines. One involving the activation of MAPKs and other triggered by IFNs and other cytokines. Further studies will be required to detail more carefully the events triggered by the two different receptors and correlate each signaling pathway with particular biological consequences.

8.   Mouse OM

Murine OM (mOM) was recently cloned as a gene produced by myeloid and lymphoid cell lines in response to various cytokines including IL-2, IL-3, and erythropoietin (Yoshimura et al., 1996). mOM cDNA encodes a 263-amino-acid protein which shares 48% identity with hOM and 42% identity with bovine OM. The mouse and human proteins are structurally similar, having in common the four cysteine residues which form disulfides and a signal peptide for secretion. Both human and mouse proteins are expressed in a less biologically active pro-form which requires removal of C-terminal basic sequence, though the mouse has four repeating regulatory units in contrast to the human which contains one.

The mOM gene is located on mouse chromosome 11 proximal to the LIF gene, a location homologous to the human chromosomal location of 22q12. Gene expression is seemingly regulated by a STAT5 recognition site 100 bases upstream of the transcription initiation site. mOM message has a limited tissue expression, being detected by northern blotting in bone marrow, and at a much lower level in the spleen. The cloning of the mouse gene will provide new insights into the physiological role of OM, following the generation of animals lacking the normal expression of this gene.

9.   OM in Disease and Therapy

The ability of hOM to stimulate hematopoiesis in animals and to upregulate the expression of IL-6, G-CSF, GM-

CSF, and TIMP by mesenchymal cells suggests that it may have clinical applications. Various cytopenias associated with bone marrow transplantation, chemotherapy, antiviral agents, and HIV infection remain major clinical problems. Acceleration of the recovery of platelets and T cells has proved the most intractable problem clinically to date. Therefore hOM, alone or in combination with other factors, may be beneficial in treating platelet and/or T cell deficiencies. The effects of hOM on both platelet and thymus-independent T cell production make it unique among the multitude of factors influencing hematopoiesis.

The most promising clinical opportunity and the one which may best exploit the normal physiological function of this protein is the treatment of chronic inflammatory diseases. The use of a variety of cytokine inhibitors including antibodies, soluble receptors, and regulatory cytokines has identified a variety of diseases in which the properties of hOM might be beneficial. These include inflammatory bowel disease, rheumatoid arthritis, and multiple sclerosis, for each of which hOM has been effective in animal models. Other potential uses for hOM may include the treatment of malignancy, infection, septic shock, and cholesterolemia, and to promote wound healing.

Human OM may contribute to a variety of pathological disorders. hOM is a potent mitogen for KS-derived cells and thus has been implicated to be important in KS pathogenesis. Accordingly, hOM antagonists are potential therapeutics for the treatment of KS (Miles et al., 1992). Since IL-6 is also a mitogen for these cells and gp130 is a shared and essential component of OM, IL-6, and LIF signaling pathways, the interruption of gp130-mediated signals may provide the most attractive therapeutic target. For example anti-gp130-neutralizing mAbs, antisense gp130 oligonucleotides, or soluble gp130 may be useful inhibitors of the growth of KS cells. Such antagonists may also be useful in the management of multiple myeloma as gp130 signaling cytokines are present in this malignancy. In addition, it is likely that ectopic or excessive production of hOM is detrimental. Transgenic mice expressing bovine OM under the control of a variety of promoters have provided some insight into potential pathologies in which OM may be a contributory factor (Malik et al., 1995). The use of a promoter that directed OM expression to multiple tissues (metallothionein promoter), or stratified epithelium (keratin-14 promoter), resulted in significantly reduced numbers of offspring (0.8% and none, respectively) than expected statistically (25%). Directed by the insulin promoter, expression of OM in the pancreas resulted in significant pancreatic fibrosis, acinar atrophy, and adema, though presumably without compromise of islet function as animals were not hyperglycemic. Neuron-specific expression was also detrimental, resulting in the runting of animals, severe tremors, ataxia, and early mortality not seen in control littermates. hOM may also be one of many factors that is involved diseases with excessive deposition of connective tissue, as seen in scleroderma. Further clinical studies are needed to define more clearly the influence of hOM in disease. Pleiotropic cytokines invariably present a wide array of potential therapeutic opportunities. However the clinical potential can only be realized in light of the toxicities and side-effects of hOM.

10. Notes Added in Proof

The recent cloning of the affinity enhancing subunit of the hOM-specific receptor (Mosley *et al.*, 1996) has lead to significant advances in understanding the function and signaling of the OM-specific and OM/LIF receptors (Auguste *et al.*, 1997; Ichihara *et al.*, 1997; Korzuset *et al.*, 1997; Kuropatwinski *et al.*, 1997). The OM-specific receptor is now referred to as the Type II OM receptor and the OM/LIF receptor as the Type I OM receptor (Mosley *et al.*, 1996). In addition OM is expressed at high levels in the rat testis during several stages of development and increases survival of Sertoli cells, supporting a role for this protein in spermatogenesis (de Miguel *et al.*, 1997). Analysis by N.M.R. of a mutated hOM protein (in which potential glycosylation sites and cysteine 80 were modified to alanine) is consistent with a four-helix bundle structure (Hoffman *et al.*, 1996).

11. References

Akira, S., Nishio, Y., Inoke, M., *et al.* (1994). Molecular cloning of APRF, a novel IFN-stimulated gene factor 3 P91-related transcription factor involved in gp130-mediated signaling pathway. Cell 77, 63–71.

Amaral, M.C., Miles, S., Kumar, G. and Nel, A.E. (1993). Oncostatin-M stimulates tyrosine protein phosphorylation in parallel with the activation of p42MAPK/ERK-2 in Kaposi's cells. Evidence that this pathway is important in Kaposi cell growth. J. Clin. Invest. 92, 848–857.

Auguste, P., Guillet, C., Fourcin, M., *et al.* (1997). Signaling of type II oncostatin M receptor. J. Biol. Chem. 272, 15760–15764.

Barton, B.E., Jackson, J.V., Lee, F. and Wagner, J. (1994). Oncostatin M stimulates proliferation in B9 hybridoma cells: potential role of oncostatin M in plasmacytoma development. Cytokine 6, 147–153.

Baumann, H. and Schendel, P. (1991). Interleukin-11 regulates the hepatic expression of the same plasma protein genes as interleukin-6. J. Biol. Chem. 266, 20424–20427.

Baumann, H., Won, K.-A. and Jahreis, G.P. (1989). Human hepatocyte-stimulating factor-III and interleukin-6 are structurally and immunologically distinct but regulate the production of the same acute phase plasma proteins. J. Biol. Chem. 264, 8046–8051.

Baumann, H., Ziegler, S.F., Mosley, B., Morella, K.K., Pajovic, S. and Gearing, D.P. (1993). Reconstitution of the response

to leukemia inhibitory factor, oncostatin M, and ciliary neurotrophic factor in hepatoma cells. J. Biol. Chem. 268, 8414–8417.

Bazan, F.J. (1991). Neuropoietic cytokines in the hematopoietic fold. Neuron 7, 197–208.

Blenis, J. (1993). Signal transduction via the MAP kinases: proceed at your own RSK. Proc. Natl Acad. Sci. USA 90, 5889–5892.

Boulton, T.G., Stahl, N. and Yancopoulos, G.D. (1994). Ciliary neurotrophic factor/leukemia inhibitory factor/interleukin 6/oncostatin M family of cytokines induce tyrosine phosphorylation of a common set of proteins overlapping those induced by other cytokines and growth factors. J. Biol. Chem. 269, 11648–11653.

Brown, M.S. and Goldstein, J.L. (1986). A receptor-mediated pathway for cholesterol homeostasis. Science 232, 34–47.

Brown, T.J., Lioubin, M.N. and Marquardt, H. (1987). Purification and characterization of cytostatic lymphokines produced by activated human T lymphocytes. J. Immunol. 139, 2977–2983.

Brown, T.J., Rowe, J.M., Shoyab, M. and Gladstone, P. (1990). Oncostatin M: a novel regulator of endothelial cell properties. In "Molecular Biology of Cardiovascular System", UCLA Symposium on Molecular and Cellular Biology (new series) (ed. R.R. Schneider), pp. 195–206. Wiley–Liss, New York.

Brown, T.J., Rowe, J.M., Liu, J. and Shoyab, M. (1991). Regulation of IL-6 expression by oncostatin M. J. Immunol. 147, 2175–2180.

Brown, T.J., Liu, J., Brashem-Stein, C. and Shoyab, M. (1993). Regulation of granulocyte colony-stimulating factor and granulocyte-macrophage colony-stimulating factor expression by oncostatin M. Blood 82, 33–37.

Brown, T.J., Wallace, P.M. and Gladstone, P. (1994). Regulation of MHC expression by oncostatin M. J. Cell. Biochem. 18B, 316. [Abstract]

Bruce, A.G., Hoggatt, I.H. and Rose, T.M. (1992a). Oncostatin M is a differentiation factor for myeloid leukemia cells. J. Immunol. 149, 1271–1275.

Bruce, A.G., Linsley, P.S. and Rose, T.M. (1992b). Oncostatin M. Prog. Growth Factor Res. 4, 157–170.

Cichy, J., Potempa, J., Chawla, R.K. and Travis, J. (1995a). Regulation of α_1-antichymotrypsin synthesis in cells of epithelial origin. FEBS Lett. 359, 262–266.

Cichy, J., Potempa, J., Chawla, R.K. and Travis, J. (1995b). Stimulatory effect of inflammatory cytokines on α_1-antichymotrypsin expression in human lung-derived epithelial cells. J. Clin. Invest. 95, 2729–2733.

Clegg, C.H., Ruffles, J.T., Wallace, P.M. and Haugen, H.S. (1996). Regulation of an extrathymic T-cell development pathway by oncostatin M. Nature 384, 261–263.

Darnell, J.E., Jr, Kerr, I.M. and Stark, G.R. (1994). Jak-STAT pathways and transcriptional activation in response to IFNs and other extracellular signaling proteins. Science 264, 1415–1421.

Davis, S., Aldrich, T.H., Stahl, N., *et al.* (1993). LIFRβ and gp130 as heterodimerizing signal transducers of the tripartite CNTF receptor. Science 260, 1805–1808.

de Miguel, M.P., de Boer-Brouwer, M., de, Rooij, D.G., Paniagua, R. and van Dissel Emiliani, F.M. (1997). Ontogeny and localization of an oncostatin M-like protein in the rat testis: its possible role at the start of spermatogenesis. Cell Growth and Diff. 8, 611–618.

De Vos, A.M., Ultsch, M. and Kossiakoff, A.A. (1992). Human growth hormone and extracellular domain of its receptor: crystal structure of the complex. Science 225, 306–312.

Docherty, A.J.P., O'Connell, J., Crabbe, T., Angal, S. and Murphy, G. (1992). The matrix metalloproteinases and their natural inhibitors: prospects for treating degenerative tissue diseases. TIBTECH 10, 200–207.

Ensoli, B., Barillari, G. and Gallo, R.C. (1991). Pathogenesis of AIDS-associated Kaposi's sarcoma. Hematol. Oncol. Clin. North. Am. 5, 281–295.

Fann, M.J. and Patterson, P.H. (1994). Neuropoietic cytokines and activin A differentially regulate the phenotype of cultured sympathetic neurons. Proc. Natl Acad. Sci. USA 91, 43–47.

Feldman, G.M., Petricoin, E.F., III, David, M., Larner, A.C. and Finbloom, D.S. (1994). Cytokines that associate with the signal transducer gp130 activate the interferon-induced transcription factor p91 by tyrosine phosphorylation. J. Biol. Chem. 269, 10747–10752.

Gauldie, J., Richards, C., Harnish, D., Lansdorp, P. and Baumannn, H. (1987). Interferon-beta2 B-cell stimulatory factor type 2 shares identity with monocyte hepatocyte-stimulating factor and regulates the major acute phase protein response in liver cells. Proc. Natl Acad. Sci. USA 84, 7251–7255.

Gearing, D.P. and Bruce, A.G. (1992). Oncostatin M binds the high-affinity leukemia inhibitory factor receptor. New Biol. 4, 61–65.

Gearing, D.P., Thut, C.J., Vande, B.T., et al. (1991). Leukemia inhibitory factor receptor is structurally related to the IL-6 signal transducer, gp130. EMBO J. 10, 2839–2848.

Gearing, D.P., Comeau, M.R., Friend, D.J., et al. (1992). The IL-6 signal transducer, gp130: an oncostatin M receptor and affinity converter for the LIF receptor. Science 255, 1434–1437.

Gearing, D.P., Ziegler, S.F., Comeau, M.R., et al. (1994). Proliferative responses and binding properties of hematopoietic cells transfected with low-affinity receptors for leukemia inhibitory factor, oncostatin M, and ciliary neurotrophic factor. Proc. Natl Acad. Sci. USA 91, 1119–1123.

Giovannini, M., Djabali, M., McElligott, D., Selleri, L. and Evans, G.A. (1993a). Tandem linkage of genes coding for leukemia inhibitory factor (LIF) and oncostatin M (OSM) on human chromosome 22. Cytogenet. Cell Genet. 64, 240–244.

Giovannini, M., Selleri, L., Hermanson, G.G. and Evans, G.A. (1993b). Localization of the human oncostatin M gene (OSM) to chromosome 22q12, distal to the Ewing's sarcoma breakpoint. Cytogenet. Cell Genet. 62, 32–34.

Goldstein, J.L. and Brown, M.S. (1985). The LDL receptor and the regulation of cellular cholesterol metabolism. J. Cell Sci. 3, 131–137.

Gordon, M.S. and Hoffman, R. (1992). Growth factors affecting human thrombocytopoiesis: potential agents for treatment of thrombocytopenia. Blood 80, 302–307.

Grove, R.I., Mazzucco, C., Allegretto, N., et al. (1991a). Macrophage-derived factors increase low density lipoprotein uptake and receptor number in cultured human liver cells. J. Lipid Res. 32, 1889–1897.

Grove, R.I., Mazzucco, C.E., Radka, S.F., Shoyab, M. and Kiener, P.A. (1991b). Oncostatin M up-regulates low density lipoprotein receptors in HepG2 cells by a novel mechanism. J. Biol. Chem. 27, 18194–18199.

Grove, R.I., Eberhardt, C., Abid, S., et al. (1993). Oncostatin M is a mitogen for rabbit vascular smooth muscle cells. Proc. Natl Acad. Sci. USA 90, 823–827.

Hamilton, J.A., Leizer, T., Piccoli, D.S., et al. (1991). Oncostatin M stimulates urokinase-type plasminogen activator activity in human synovial fibroblasts. Biochem. Biophys. Res. Commun. 180, 652–659.

Heinrich, P.C., Castell, J.V. and Andus, T. (1990). Interleukin-6 and the acute phase protein. Biochem. J. 265, 621–636.

Hibi, M., Murakani, S., Saito, M., Hirano, T., Taga, T. and Kishimoto, T. (1990). Molecular cloning and expression of an IL-6 signal transducer, gp130. Cell 63, 1149–1157.

Hirano, T., Akira, S., Taga, T. and Kishimoto, T. (1990). Biological and clinical aspects of interleukin 6. Immunol. Today 11, 443–449.

Hirano, T., Matsuda, T. and Nakajima, K. (1994). Signal transduction through gp130 that is shared among the receptors for the interleukin 6 related cytokine subfamily. Stem Cells 12, 262–277.

Hoffman, R.C., Moy, F.J., Price, V., et al. (1996). Resonance assignments for Oncostatin M, a 24-kDa alpha-helical protein. J. Biomol. NMR. 7, 273–282.

Horn, D., Fitzpatrick, W.C., Gompper, P.T., et al. (1990). Regulation of cell growth by recombinant oncostatin M. Growth Factors 2, 157–165.

Hunter, T. (1993). Cytokine connections. Nature 366, 114–116.

Ichihara, M., Hara, T., Kim, H., Murate, T. and Miyajima, A. (1997). Oncostatin M and leukemia inhibitory factor do not use the same functional receptor in mice. Blood 90, 165–173.

Ip, N.Y., Nye, S.H., Boulton, T.G., et al. (1992). CNTF and LIF act on neuronal cells via shared signaling pathways that involve the IL-6 signal transducing receptor component gp130. Cell 69, 1121–1132.

Jay, P.R., Centrella, M., Lorenzo, J., Bruce, A.G. and Horowitz, M.C. (1996). Oncostatin-M: a new bone active cytokine that activates osteoblasts and inhibits bone resorption. Endocrinology 137(4), 1151–1158.

Jeffery, E., Price, V. and Gearing, D.P. (1993). Close proximity of the genes for leukemia inhibitory factor and oncostatin M. Cytokine 5, 107–111.

Kallestad, J.C., Shoyab, M. and Linsley, P.S. (1991). Disulfide bond assignment and identification of regions required for functional activity on oncostatin M. J. Biol. Chem. 266, 8940–8945.

Kishimoto, T., Taga, T. and Akira, S. (1994). Cytokine signal transduction. Cell 76, 253–262.

Koj, A. (1985). Definition and classification of acute phase proteins in the acute phase response to injury and infection. In "The Acute Phase Response to Injury and Infection" (eds. A.H. Gordon and A. Koj), pp. 139–232, Elsevier, Amsterdam.

Kopf, M., Baumann, H., Freer, G., et al. (1994). Impaired immune and acute-phase responses in interleukin-6-deficient mice. Nature 368, 339–342.

Korzus, E., Nagase, H., Rydell, R. and Travis, J. (1997). The mitogen-activated protein kinase and JAK-STAT signaling pathways are required for an oncostatin M-responsive element-mediated activation of matrix metalloproteinase 1 gene expression. J. Biol. Chem. 272, 1188–1196.

Kovanen, P.T. (1987). Regulation of plasma cholesterol by hepatic low density lipoprotein receptors. Am. Heart J. 113, 464–469.

Kuropatwinski, K.K., De Imus, C., Gearing, D., Baumann, H. and Mosley, B. (1997). Influence of subunit combinations on signaling by receptors for oncostatin M, leukemia inhibitory factor, and interleukin-6. J. Biol. Chem. 272, 15135–15144.

Lai, L.C., Kadory, S., Siraj, A.K. and Lennard, T.W.J. (1994). Oncostatin M, interleukin 2, interleukin 6 and interleukin 8 in breast cyst fluid. Int. J. Cancer 59, 369–372.

Linsley, P.S., Hanson, M.B., Horn, D., et al. (1989). Identification and characterization of cellular receptors for the growth regulator, oncostatin M. J. Biol. Chem. 264, 4282–4289.

Linsley, P.S., Kallestad, J., Ochs, V. and Neubauer, M. (1990). Cleavage of a hydrophilic C-terminal domain increases growth-inhibitory activity of oncostatin M. Mol. Cell. Biol. 10, 1882–1890.

Liu, J., Clegg, C.H. and Shoyab, M. (1992a). Regulation of EGR-1, c-jun, and c-myc gene expression by oncostatin M. Cell Growth Differ. 3, 307–313.

Liu, J., Modrell, B., Aruffo, A., et al. (1992b). Interleukin-6 signal transducer gp130 mediates oncostatin M signaling. J. Biol. Chem. 267, 16763–16766.

Liu, J., Shoyab, M. and Grove, R.I. (1993). Induction of Egr-1 by oncostatin M precedes up-regulation of low density lipoprotein receptors in HepG2 cells. Cell Growth Differ. 4, 611–616.

Liu, J., Modrell, B., Aruffo, A., Scharnowske, S. and Shoyab, M. (1994). Interactions between oncostatin M and the IL-6 signal transducer, gp130. Cytokine 6, 272–278.

Liu, J., Spence, M.J., Wallace, P.M., Forcier, K., Hellstrom, I. and Vestal, R.E. (1997). Oncostatin M-specific receptor mediates inhibition of breast cancer cell growth and down-regulation of the c-myc proto-oncogene. Cell Growth and Diff. 8, 667–676.

Lu, C., Rak, J.W., Kobayashi, H. and Kerbel, R.S. (1993). Increased resistance to oncostatin M-induced growth inhibition of human melanoma cell lines derived from advanced-stage lesions. Cancer Res. 53, 2708–2711.

Lütticken, C., Wegenka, U.M., Yuan, J., et al. (1994). Association of transcription factor APRF and protein kinase Jak1 with the interleukin-6 signal transducer gp130. Science 263, 89–92.

Malik, N., Kallestad, J.C., Gunderson, N.L., et al. (1989). Molecular cloning, sequence analysis, and functional expression of a novel growth regulator, oncostatin M. Mol. Cell. Biol. 91, 2847–2853.

Malik, N., Clegg, C. and Shoyab, M. (1992a). Cloning and expression of the Bovine homologue of the cytokine oncostatin M. FASEB J. 6, A1671.

Malik, N., Graves, D., Shoyab, M. and Purchio, A.F. (1992b). Amplification and expression of heterologous oncostatin M in Chinese hamster ovary cells. DNA Cell Biol. 11, 453–459.

Malik, N., Haugen, H.S., Modrell, B., Shoyab, M. and Clegg, C.H. (1995). Developmental abnormalities in mice transgenic for bovine oncostatin M. Mol. Cell Biol. 15(5), 2349–2358.

McDonald, V.L., Dick, K.O., Malik, N. and Shoyab, M. (1993). Selection and characterization of a variant of human melanoma cell line, A375 resistant to growth inhibitory effects of oncostatin M (OM): coresistant to interleukin 6 (IL-6). Growth Factors 9, 167–175.

Miles, S.A. (1992). Pathogenesis of human immunodeficiency virus-related Kaposi's sarcoma. Curr. Opin. Oncol. 4, 875–882.

Miles, S.A., Rezai, A.R., Salazar-Gonzalez, J.F., et al. (1990). AIDS Kaposi sarcoma-derived cells produce and respond to interleukin 6. Proc. Natl Acad. Sci. USA 87, 4068–4072.

Miles, S.A., Martinez-Maza, O., Rezai, A., et al. (1992). Oncostatin M as a potent mitogen for AIDS-Kaposi's sarcoma-derived cells. Science 255, 1432–1434.

Mosley, B., Delmus, C., Friend, D., Thoma, B. and Cosman, D. (1994). The oncostatin-M specific receptor: cloning of a novel subunit related to the LIF receptor. Cytokine 6, 554.

Mosley, B., De Imus, C., Friend, D., et al. (1996). Dual oncostatin M (OSM) receptors. Cloning and characterization of an alternative signaling subunit conferring OSM-specific receptor activation. J. Biol. Chem. 271, 32635–32643.

Mui, A.L.-F. and Miyajima, A. (1994). Cytokine receptors and signal transduction. Prog. Growth Factor Res. 5, 15–35.

Murakami, M., Narazaki, M., Hibi, M., et al. (1991). Critical cytoplasmic region of the IL-6 signal transducer, gp130, is conserved in the cytokine receptor family. Proc. Natl Acad. Sci. USA 88, 11349–11353.

Murakami, M., Hibi, M., Nakagawa, N., et al. (1993). IL-6-induced homodimerization of gp130 and associated activation of a tyrosine kinase. Science 260, 1808–1810.

Murate, T., Yamashita, K., Ohashi, H., et al. (1993). Erythroid potentiating activity of tissue inhibitor of metalloproteinases on the differentiation of erythropoietin-responsive mouse erythroleukemia cell line, ELM-I-1-3, is closely related to its cell growth potentiating activity. Exp. Hematol. 21, 169–176.

Nair, B.F.C., DeVico, A.L., Nakamura, S., et al. (1992). Identification of a major growth factor for AIDS-Kaposi's sarcoma cells as oncostatin M. Science 255, 1430–1432.

Narazaki, M., Witthun, B.A., Yoshida, K., et al. (1994). Activation of JAK2 kinase mediated by the interleukin-6 signal transducer gp130. Proc. Natl Acad. Sci. USA 91, 2285–2289.

Nemoto, O., Yamada, H., Mukaida, M. and Shimmei, M. (1996). Stimulation of TIMP-1 production by oncostatin M in human articular cartilage. Arthritis Rheum. 39, 560–566.

Nishimoto, N., Ogata, A., Shima, Y., et al. (1994). Oncostatin M, leukemia inhibitory factor, interleukin 6 induce the proliferation of human cells via the common signal transducer, GP130. J. Exp. Med. 179, 1343–1347.

Pennica, D., King, K.L., Chien, K.R., Baker, J.B. and Wood, W.I. (1994). Cardiotrophin-1, a novel cytokine that induces cardiac myocyte hypertrophy. Cytokine 6, 577.

Piquet-Pellorce, C., Grey, L., Mereau, A. and Heath, J.K. (1994). Are LIF and related cytokines functionally equivalent. Exp. Cell Res. 213, 340–347.

Radka, S.F., Kallestad, J.C., Linsley, P.S. and Shoyab, M. (1994). The binding pattern of a neutralizing monoclonal antibody to mutant oncostatin M molecules is correlated with functional activity. Cytokine 6, 48–54.

Radka, S.R., Nakamura, S., Sakurada, S. and Salahuddin, S.Z. (1993). Correlation of oncostatin M secretion by human retrovirus-infected cells with potent growth stimulation of cultured spindle cells from AIDS-Kaposi's sarcoma. J. Immunol. 150, 5195–5201.

Rao, M.S., Symes, A., Malik, N., Shoyab, M., Fink, G.S. and Landis, S.C. (1992). Oncostatin M regulates VIP expression in a human neuroblastoma cell line. Neuro Report 3, 865–868.

Richards, C.D. and Agro, A. (1994). Interaction between oncostatin M, interleukin 1 and prostaglandin E2 in induction of IL-6 expression in human fibroblasts. Cytokine 6, 40–47.

Richards, C.D. and Shoyab, M. (1993). The role of oncostatin M in the acute phase response. In "Acute Phase Proteins: Molecular Biology, Biochemistry and Clinical Applications" (eds. A. Mackiewicz, I. Kushner, and H. Baumann), pp. 321–327. CRC Press, Boca Raton, FL.

Richards, C.D., Brown, T.J., Shoyab, M., Baumann, H. and Gauldie, J. (1992). Recombinant oncostatin M stimulates the production of acute phase proteins in HepG2 cells and rat primary hepatocytes in vitro. J. Immunol. 148, 1731–1736.

Richards, C.D., Shoyab, M., Brown, T.J. and Gauldie, J. (1993). Selective regulation of metalloproteinase inhibitor (TIMP-1) by oncostatin M in fibroblasts in culture. J. Immunol. 150, 5596–5603.

Richards, C.D., Langdon, C., Botelho, F., Brown, T.J. and Agro, A. (1996). Oncostatin M inhibits IL-1 induced expression of IL-8 and granulocyte-macrophage colony-stimulating factor by synovial and lung fibroblasts. J. Immunol. 156, 343–349.

Robinson, R.C., Grey, L.M., Staunton, D., et al. (1994). The crystal structure and biological function of leukemia inhibitory factor: implications for receptor binding. Cell 77, 1101–1116.

Rose, T.M. and Bruce, A.G. (1991). Oncostatin M is a member of a cytokine family that includes leukemia-inhibitory factor, granulocyte colony-stimulating factor, and interleukin 6. Proc. Natl Acad. Sci. USA 88, 8641–8645.

Rose, T.M., Lagrou, M.J., Fransson, I., et al. (1993). The genes for oncostatin M (OSM) and leukemia inhibitory factor (LIF) are tightly linked on human chromosome 22. Genomics 17, 136–140.

Rose, T.M., Welford, D.M., Gunderson, N.L. and Bruce, A.G. (1994). Oncostatin M (OSM) inhibits the differentiation of pluripotent embryonic cells in vitro. Cytokine 6, 48–54.

Rutherford, G.W., Schwarcz, S., Lemp, G.F., et al. (1989). The epidemiology of AIDS-related Kaposi's sarcoma in San Francisco. J. Infect. Dis. 159, 569–572.

Saito, M., Yoshida, K., Hibi, M., Taga, T. and Kishinoto, T. (1992). Molecular cloning of a murine IL-6 receptor-associated signal transducer, gp130, and its regulated expression in vivo. J. Immunol. 148, 4066–4071.

Schieven, G.L., Kallestad, J.C., Brown, J.T., Ledbetter, J.A. and Linsley, P.S. (1992). Oncostatin M induces tyrosine phosphorylation in endothelial cells and activation of p62 (yes) tyrosine kinase. J. Immunol. 149, 1676–1682.

Shaw, G. and Kamen, R. (1986). A conserved AU sequence from the 3′ untranslated region of GM-CSF mRNA mediates mRNA degradation. Cell 46, 659–667.

Shuai, K. (1994). Interferon-activated signal transduction to the nucleus. Curr. Opin. Cell Biol. 6, 253–259.

Soldi, R., Graziani, A., Benelli, R., et al. (1994). Oncostatin M activates phosphatidylinositol-3-kinase in Kaposi's sarcoma cells. Oncogene 9, 2253–2260.

Spence, M.J., Vestal, R.E. and Liu, J. (1997). Oncostatin M-mediated transcriptional suppression of the c-myc gene in breast cancer cells. Can. Res. 57, 2223–2228.

Sporeno, E., Barbato, G., Graziani, R., Pucci, P., Nitti, G. and Paonessa, G. (1994). Production and structural characterisation of amino terminally histidine tagged human oncostatin M in E. coli. Cytokine 6, 255–264.

Sprang, S.R. and Bazan, J.F. (1993). Cytokine structural toxomony and mechanisms of receptor engagement. Curr. Opin. Struct. Biol. 3, 815–827.

Stahl, N., Davis, S., Wong, V., et al. (1993). Cross-linking identifies leukemia inhibitory factor-binding protein as a ciliary neurotrophic factor receptor component. J. Biol. Chem. 268, 7628–7631.

Taga, T. and Kishimoto, T. (1992). Cytokine receptors and signal transduction. FASEB J. 6, 3387–3396.

Taga, T., Narazaki, M., Yasukawa, K., et al. (1992). Functional inhibition of hematopoietic and neurotrophic cytokines by blocking the interleukin 6 signal transducer gp130. Proc. Natl Acad. Sci. USA 89, 10998–11001.

Thoma, B., Bird, T.A., Friend, D.J., Gearing, P. and Dower, S.K. (1994). Oncostatin M and leukemia inhibitory factor trigger overlapping and different signal through partially shared receptor complexes. J. Biol. Chem. 269, 6215–6222.

Thomas, G. (1992). MAP kinase by any other name smells just as sweet. Cell 68, 3–6.

Todaro, G.J., Shoyab, M., Plowman, G.D., Marquardt, H. and Linsley, P.S. (1989). Two new growth inhibitory factors: oncostatin M and amphiregulin. In "International Conference Growth Inhibitors Pre-clinical and Clinical Evaluation in Cancer", Wadham College, Oxford (ed. P. Cozens).

Vos, J.P., Gard, A.L. and Pfeiffer, S.E. (1996). Regulation of oligodendrocyte cell survival and differentiation by ciliary neurotrophic factor, leukemia inhibitory factor, oncostatin M, and interleukin-6. Perspect. Dev. Neurobiol. 4, 39–52.

Wallace, P.M., MacMaster, J.F., Rillema, J.R., Burstein, S.A. and Shoyab, M. (1995a). Thrombocytopoietic properties of oncostatin M. Blood 86,1310–1315.

Wallace, P.M., MacMaster, J.F., Rillema, J.R., et al. (1995b). In vivo properties of oncostatin M. Ann. N.Y. Acad. Sci. 762, 42–54.

Yin, T. and Yang, Y.-C. (1994). Mitogen-activated protein kinases and ribosomal S6 protein kinases are involved in signaling pathways shared by interleukin-11, interleukin-6, leukemia inhibitory factor, and oncostatin M in mouse 3T3-L1 cells. J. Biol. Chem. 269, 3731–3738.

Yin, T., Miyazawa, K. and Yang, Y.-C. (1992). Characterization of interleukin-11 receptor and protein tyrosine phosphorylation induced by interleukin-11 in mouse 3T3-L1 cells. J. Biol. Chem. 151, 2555–2561.

Yoshida, K., Chambers, I., Nichols, J., et al. (1994). Maintenance of the pluripotential phenotype of embryonic stem cells by a combination of interleukin-6 and a soluble form of interleukin-6 receptor. Mech. Dev. 45, 163–171.

Yoshimura, A., Ichihara, M., Kinjyo, I., et al. (1996). Mouse oncostatin M: an immediate early gene induced by multiple cytokines through the JAK-STAT5 pathway. EMBO J. 15, 1056–1063.

Zarling, J.M., Shoyab, M., Marquardt, H., Hanson, M.B., Lioubin, M.N. and Todaro, G.J. (1986). Oncostatin M: a growth regulator produced by differentiated histiocytic lymphoma cells. Proc. Natl Acad. Sci. USA 83, 9739–9743.

Zhang, X.G., Gu, J.-J., Lu, Z.-Y., et al. (1994). Ciliary neurotropic factor, interleukin 11, leukemia inhibitory factor, and oncostatin M are growth factors for human myeloma cell lines using the IL-6 signal transducer GP-130. J. Exp. Med. 177, 1337–1342.

28. *Transforming Growth Factor* β_1

Francis W. Ruscetti, Maria C. Birchenall-Roberts, John M. McPherson *and* Robert H. Wiltrout

1. Introduction

Although transforming growth factor-β_1 (TGF-β_1) was originally named for its ability to reversibly cause a phenotypic transformation of rat fibroblasts (Moses *et al.*, 1981; Roberts *et al.*, 1981), numerous bioassays describing cartilage inducing factor (Seyedin *et al.*, 1985), glioblastoma immunosuppressive factor (de Martin *et al.*, 1987), myoblast differentiation inhibition factor (Florini *et al.*, 1986; Massague *et al.*, 1986), and epithelial growth inhibitor (Holley *et al.*, 1980) were found to be measuring the same TGF-β protein. Despite the fact that oncogenic transformation is one of the few biological activities that TGF-β does not stimulate, the misleading name has stuck.

In addition to being multifunctional, TGF-β is a member of a large family of factors with widely diverse activities. The TGF-β superfamily was first recognized in the late 1980s (Massague, 1987; Sporn *et al.*, 1987). Since then, the superfamily has grown to more than 25 members, with some of the molecules being grouped into four distinct subfamilies owing to highly related structures (for recent reviews, see Kingsley 1994; Massague, 1990; Roberts and Sporn, 1990). Information concerning the structure, expression, function, receptors, and mechanism(s) of action of the family members is accumulating at an explosive pace. Herein, our remarks will be limited to the current information pertaining to TGF-β_1 and, where needed, its mammalian counterparts, TGF-β_2 and TGF-β_3.

Cytokines
ISBN 0–12–498340–5

2. The Cytokine Gene Family

Multiple species of TGF-β which are encoded by different genes have been identified. TGF-βs are disulfide-linked polypeptide dimers of molecular mass 25 000 kDa. Each mature chain is 112 amino acids in length and is derived from a larger precursor. At least three genes encode TGF-β precursors in mammalian genomes (TGF-β_1, β_2, and β_3). TGF-β_1, β_2, and β_3 each have a single gene located on a separate chromosome in humans (19q13.1 (Fujii *et al.*, 1986), 1q41, and 14q23 (Barton *et al.*, 1988), respectively).

2.1 GENOMIC ORGANIZATION OF THE TGF-β_1 GENE

The human TGF-β_1 precursor is encoded by seven exons (Derynck *et al.*, 1987; see Figures 28.1 and 28.2). The similarity of their seven-exon structures and the high degree of conservation of their exon/intron borders (6/7 are identical between TGF-β_1 and β_3 (Derynck *et al.*, 1988)) indicate that these genes are derived from a common ancestor. The very large introns have not been completely sequenced and the exact gene size remains unknown. TGF-βs have unique 5′ long untranslated regions (UTR). For TGF-β_1, there are regions of 841 untranslated base pairs and 50 untranslated base pairs in exon 7 (Figure 28.1). In contrast to TGF-β_2 and β_3 promoters, β_1 does not have a traditional "TATA" sequence. Instead, just upstream of the first transcriptional start site, the TGF-β_1 promoter has a very GC-rich region with several Sp1 binding sites.

2.2 TRANSCRIPTIONAL CONTROL

TGF-β_1, β_2, and β_3 each express a predominant mRNA of distinct size (2.4 kb, 4.1 kb, and 3.5 kb, respectively). While there is evidence for alternative splicing with each isoform (i.e., activated T cells produce a 1.9 kb mRNA), the significance of such splicing is not known.

The promoter region of each mammalian TGF-β gene displays distinct regulatory elements, suggesting that expression of each isoform can be independently regulated (Lafyatis *et al.*, 1990). Indeed, the production of each isoform can be stimulated by a specific set of agonists both *in vivo* and *in vitro*. For example, phorbol myristate (PMA) stimulates TGF-β_1 production and retinoic acid stimulates TGF-β_2. Also, TGF-β_1 upregulates its own expression (Van Obberghen-Schilling *et al.*, 1988), while other isoforms have more complex patterns of regulation (Bascom *et al.*, 1989). Both TGF-β_1 and PMA increase c-*jun* and c-*fos* expression which activates TGF-β_1 transcription by binding to AP-1 sites in the TGF-β_1 gene promoter (Kim *et al.*, 1989, 1990). This autoinduction mechanism is important in the accumulation of TGF-βs at sites of injury but can lead to the pathobiology of TGF-βs

(Border and Noble, 1994; Wahl, 1994). In contrast, TGF-β_2 and β_3 promoters are regulated by cAMP-responsive (CRE) and AP-2 binding sites. This diversity increases the specificity and regulatory flexibility of the TGF-β system (Li *et al.*, 1990). In addition, post-transcriptional control of TGF-β *in vitro* has been shown. T cells can be induced to express TGF-β mRNA shortly (2 h) after activation. However, secreted TGF-β is not detected in the culture supernatants until 3–4 days after activation, suggesting that TGF-β production is post-transcriptionally regulated (Kehrl *et al.*, 1986).

3. The Protein

The TGF-β_1 precursor is 390 amino acids (aa). The organization of the mammalian isoforms is outlined in Figure 28.3(a). All the potential *N*-linked glycosylation sites lie outside the bioactive domain. The TGF-β_1 precursor contains an amino-terminal hydrophobic signal sequence (aa 1–23) for translocation to the endoplasmic reticulum to undergo exocytosis. The proregion (aa 24–278) guides the correct folding and disulfide bonding of the bioactive domain during TGF-β synthesis (Gray and Mason, 1990). The bioactive domain (aa 279–390) is the carboxy-terminus of the precursor, preceded by a sequence of four basic amino acids which form the cleavage site. Intracellular proteolytic cleavage probably occurs by a furin peptidase after dimerization of the precursors (Dubois *et al.*, 1995). Active TGF-β is usually a homodimer but naturally occurring TGF-β_1,β_2 and β_2,β_3 heterodimers can also be found.

The amino acid sequences of all three TGF-β isoforms are strictly conserved; there is 98–100% identity between chicken and human. The evolutionary pressure to retain such a degree of conservation suggests that the functions of each isoform are critical. The degree of homology of the bioactive domain between the three human isoforms ranges from 70% to 80%, with much greater sequence divergence in the precursor region (Figure 28.3(b)).

A monomer of active TGF-β contains nine cysteines with conserved spacing which are required for biological activity. They form four interior intrachain disulfide bonds (C^7–C^{16}, C^{15}–C^{78}, C^{44}–C^{109} and C^{48}–C^{111}) and an interchain disulfide bond (C^{77}; Figure 28.4).

The crystal structure of TGF-β_2 has been solved and reveals that six of the cysteines form a rigid central structure known as a cysteine knot (Daopin *et al.*, 1992; Schlunegger and Grutter, 1992) (Figure 28.5). This singular motif is an 8-amino-acid ring enclosed by two cysteine disulfide bonds with a third bond threaded through the ring. This ring locks four segments of the twisted antiparallel β-sheets, which extend from the knot like two fingers, and an α-helix that extends opposite to the fingers, like the palm of a hand. The remaining cysteine forms a disulfide bond with the corresponding

5'UTR.
ACCTCCCTCCGCGGAGCAGCCAGACAGCGAGGGCCCCGGCCGGGGGCAGGGGGGACGCCCCGTCCGGGGGCACCCCCC
CGGCTCTGAGCCGCCCGCGGGGCCCGGCCTCGGCCCCGGAGCGGAGGAAGGAGTCGCCGAGGAGCAGCCTGAGGCCCCA
GAGTCTGAGACGAGCCGCCGCCGCCCCCGCCACTGCGGGGAGGAGGGGGAGGAGGAGCGGGAGGAGGGGACGAGCTGG
TCGGGAGAAGAGGAAAAAAACTTTTGAGACTTTTCCGTTGCCGCTGGGAGCCGGAGGCGCGGGGACCTCTTGGCGC
GACGCTGCCCCGCGAGGAGGCAGGACTTGGGGACCCCAGACCGCCTCCCTTTGCCGCCGGGGACGCTTGCTCCCTCCC
TGCCCCCTACACGGCGTCCCTCAGGCGCCCCCATTCCGGACCAGCCCTCGGGAGTCGCCGACCCGGCCTCCCGCAAA
GACTTTTCCCCAGACCTCGGGCGCACCCCCTGCACGCCGCCTTCATCCCCGGCCTGTCTCCTGAGCCCCCGCGCATCC
TAGACCCTTTCTCCTCCAGGAGACGGATCTCTCTCCGACCTGCCACAGATCCCCTATTCAAGACCACCCACCTTCT
GGTACCAGATCGCGCCCATCTAGGTTATTTCCGTGGGATACTGAGACACCCCCGGTCCAAGCCTCCCCTCCACCACT
GCGCCCTTCTCCCTGAGGACCTCAGCTTTCCCTCGAGGCCCTCCTACCTTTTGCCGGGAGACCCCCAGCCCCCTGCAGG
GGCGGGGCCTCCCCACCACACCAGCCCTGTTCGCGCTCTCGGCAGTGCCGGGGGGCGCCGCCTCCCCC

1. exon1.
ATGCCGCCCTCCGGGCTGCGGCTGCTGCTGCTGCTGCTACCGCTGCTGTGGCTACTGGTGCTGACG
CCTGGCCGGCCGGCCGCGGGACTATCCACCTGCAAGACTATCGACATGGAGCTGGTGAAGCGGAA
GCGCATCGAGGCCATCCGCGGCCAGATCCTGTCCAAGCTGCGGCTCGCCAGCCCCCCGAGCCAGGG
GGAGGTGCCGCCCGGCCCGCTGCCCGAGGCCGTGCTGCCCCTGTACAACAGCACCCGCGACGCGGGT
GGCCGGGGAGAGTGCAGAACCGGAGCCCGAGCCTGAGGCCGACTACTACGCCAAGGAGGTCACCC
GCGTGCTAATGGTGGAAACCCACAACGgtgagctcggaggggcaggggagccgggaggggggcccccagggggcgc
exon 2
gatcgcctcccttcatttctccctgctag***AAATCTATGACAAGTTCAAGCAGAGTACACACAGCATATATATGTT***
CTTCAACACATCAGAGCTCCGAGAAGCGGTACCTGAACCCGTGTTGCTCTCCCGGGCAGAGCTGCG
TCTGCTGAGGCTCAAGTTAAAAGTGGAGCAGCACGTGGAGCTGTACCAGgtgaggacatgagccagaaggaagg
tcagggcatgggctggagagggtga
exon 3
tcgtgacaccatctacccactgtctctctcctgccttcatcatcctctag***AAATACAGCAACAATTCCTGGCGATACCTCAGC***
AACCGGCTGCTGGCACCCAGCGACTCGCCAGAGTGGTTATCTTTTGATGTCACCGGAGTTGTGCGGC
AGTGGTTGAGCCGTGGAGgtgaggattacttgtgtgtcccacccctgttttctccctggggtccaccc
exon4-5
gctgggtgagctgcactctcagactggcttccctctcgccactcctacag***GGGAAATTGAGGGCTTTCGCCTTAGCGCCCACT***
GCTCCTGTGACAGCAGGGATAACACACTGCAAGTGGACATCAACGgtgaggcctgcttcccggccatgcccagtt
gtgacgtgtgtgcgtgtgtgtgtcccatctgccccacgccccacttatctatccctctgagagtgtgtgtgtatgtccctatccctgactcccacacaa
agcag***GGTTCACTACCGGCCGCCGAGGTGACCTGGCCACCATTCATGGCATGAACCGGCCTTTCCTGC***
TTCTCATGGCCACCCCGCTGGAGAGGGCCCAGCATCTGCAAAGTCCCGGCACCGCCGA<u>**GCCCTGGA**</u>
<u>**CACCAACTATTGCTTCAG**</u>gtgagccttgtagcctggatggaggccttccaggctgggggcatgactgc
Exon 6
tctatggtggtagccctccctgccctgatgcgtctctcctgcctgcag**CTCCACGGAGAAGAACTGCTGCGTGCGGCAGCTG**
TACATTGACTTCCGCAAGGACCTCGGCTGGAAGTGGATCCACGAGCCCAAGGGCTACCATGCCAAC
TTCTGCCTCGGGCCCTGCCCCTACATTTGGAGCCTGGACACGCAGTACAGCAAGgtacgtctggccaccgggc
tacgagatgcgcttggggggagccaggacg
exon 7 and 3'-flanking region
TTTTCCTCCCTCCACGAGCCCTGAGCCCTGACCCCGCCCGCCGCCGCCCCGCAG<u>**GTCCTGGCCCTGTACAACCAGCAT**</u>
<u>**AACCCGGGCGCCTCGGCCGCGCCGTGCTGCGTGCCGCAGGCGCTGGAGCCGCTGCCCATHGTGTAC**</u>T
<u>**ACGTGGGCCGCAAGCCCAAGGTGGAGCAGCTGTCCAACATGATCGTGCGCTCCTGCAAGTGCAGC**</u>TG
AGGTCCCGCCCCGCCCCGCCCCGCCCGGCAGGCCCGGCCCCACCCCGCCCCGCCCCCGCTGCCTTGCCCATGGGGGCT
GTATTTAAGGACACCCGTGCCCCAAGCCCACCTGGGGCCCCATTAAAGATGGAGAGAGGGACTGCGGATCTCTGTGTC
ATTGGGCGCCTGCCTGGGGTCTCCATCCCTGACGTTCCCCCACTCCCACTCCCTCTCTCTCCCTCTCTGCCTCCTCCTG
CCTGTCTGCACTATTCCTTTGCCCGGCATCAAGGCACAGGGGACCAGTGGGGAACACTACTGTAGTTAGATCTATTT
ATTGAGCACCTTGGGCACTGTTGAAGTGCCTTACATTAATGAACTCATTCAGTCACCATAGCAACACTCTGAGATG
CAGGGACTCTGATAACACCCATTTTAAAGGTGAGGAAACAAGCCCAGAGAGGTTAAGGGAGGAGTTCCTGCCCACC
AGGAACCTGCTTTAGTGGGGGATAGTGAAGAAGAC<u>aataaa</u>AGATAGTAGTTCAGGCCAGGCGGGGTGGCTCACGCCT
GTAATCCTAGCACTTTTGGGAGGCAGAGATGGGAGGATTACTTGAATCCAGGCATTTGAGACCAGCCTGGGTAACA
TAGTG<u>A</u>GACCCTATCTCTACAAAACACTTTTAAAAAATGTACACCTGTGGTCCCAGCTACTCTGGAGGCTAAGGTG
GGAGGGATCACTTGATCCTGGGAGGTCAAGGCTGCAG

Figure 28.1 Gene sequence of the human TGF-β₁ gene. The seven exons of the human TGF-β₁ gene are indicated in the figure in capital letters. The untranslated region is in capitals while the coding region is in bold capitals with the coding region of the active mononer in underlined bold capitals. The flanking equences of the six introns (Derynck *et al.*, 1987) are in lowercase letters. The polyadenylation signal is in underlined small letters and the adenylated A is an underlined capital letter.

cysteine in the other monomer. The two monomers in an active complex are oriented flat antiparallel to each other, with the inner surface of the fingers of one monomer forming hydrophobic contacts with the palm of the other monomer.

4. Cellular Sources and Production

While most if not all mature cells produce at least one of the TGF-β isoforms, enhanced TGF-β expression in the adult occurs during processes such as tissue repair, bone

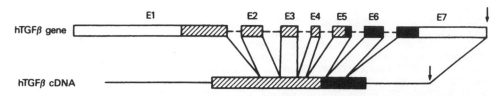

Figure 28.2   Intron–extron structure of the human TGF-β₁ gene. The solid region corresponds to the bioactive domain of the precursor and striped region corresponds to the pro-region. It is divided into seven introns and six introns (Derynck *et al.*, 1987). Arrows indicate the polyadenylation signal. (From Roberts *et al.*, 1988.)

A. TGF-β SUPERFAMILY: GENERAL STRUCTURE

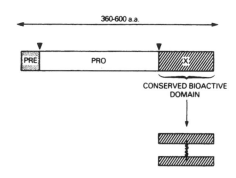

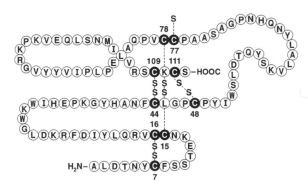

Figure 28.4   The primary and secondary structure of the TGF-β₁ protein. Cysteine residues are indicated in black along with the four intrachain and one interchain disulfide bonds.

B. MAMMALIAN TGF-β ISOFORMS: REGIONS OF HOMOLOGY

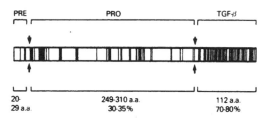

Figure 28.3   The structure of the TGF-β precursor. (a) The general structure of the TGF-β precursor molecule. (b) Regions of homology between the mammalian TGF-β isoforms. PRE = leader signal (aa 1–23); PRO precursor/latency associated peptide (aa 24–278); active monomer (aa 279–390). The black bars indicate sequence homology between TGF-β₁, β₂ and β₃. (From Wakefield *et al.*, 1991.)

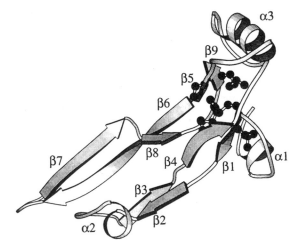

Figure 28.5   The three-dimensional structure of TGF-β₂. A schematic diagram of a TGF-β₂ subunit showing the helix bundle formation (α = α-helices; β = β-sheets) (Daopin *et al.*, 1992). (Figure kindly provided by Peter Sun (aka Sun Daopin).)

remodeling, and inflammation (Wahl, 1992). TGF-β is also stored at high levels in the alpha granules of blood platelets (Assoian and Sporn, 1986), which, along with inflammatory cells (Assoian *et al.*, 1987), can deliver it to sites of tissue injury where it can bind to fibronectin (Mooradian *et al.*, 1989).

4.1   TGF-β: LATENCY AND ACTIVATION

Uniquely among cytokines, TGF-β is predominantly secreted from cells or released from platelets as an

inactive or latent complex that is unable to bind to its receptor (Pircher *et al.*, 1986). The latent complex consists of a mature dimer plus two pro-region peptides (called latency-associated peptides, LAP) noncovently linked to each other (Gentry *et al.*, 1987). An additional complexity is that the LAPs in the latent complex can be disulfide-linked to a glycoprotein of

125–190 kDa (Miyazono and Heldin, 1989). This protein, known as "latent TGF-β binding protein" (LTBP) contains multiple epidermal growth factor-like repeats in tandem as its distinctive feature (Kanzaki *et al.*, 1990; Tsuji *et al.*, 1990). LTBP does not bind TGF-β directly, but it may enhance release and activate latent forms (Miyazono *et al.*, 1991; Flaumenhaft *et al.*, 1993).

The mechanism(s) that activates TGF-β *in vivo* is not known, but may be enzyme mediated. *In vitro*, there is evidence that the proteolysis by plasmin or cathepsin D, heat treatment, extremes of pH, or chaotropic agents can activate TGF-β (Lyons *et al.*, 1990). Endothelial cell cultures can activate latent TGF-β, but only in contact with vascular pericytes (Antonelli-Orlidge *et al.*, 1989). Similarly, primitive thymocytes activate latent TGF-β, but only in contact with CD8⁺ T cells (Mossalayi *et al.*, 1995). This interaction may be mediated by mannose 6-phosphate (man-6-P) residues in the pro-region, which tether the complex to the cell surface man-6-P receptors as a requisite for activation *in vivo* (Kovacina *et al.*, 1989; Dennis and Rifkin, 1991). This, however, is not sufficient, because many cells that express man-6-P receptors fail to activate latent TGF-β. Urokinase-type II plasminogen activator must also bind to its receptor. Transglutaminase and LTBP are needed for optimal activation, suggesting that several molecules must be present on the cell surface of endothelial cells and macrophages for activation (Nunes *et al.*, 1995).

Activated TGF-β can bind to various extracellular matrix components such as β-glycan (Andres *et al.*, 1989), endoglin (McAllister *et al.*, 1994), and decorin (Yamaguchi *et al.*, 1990). Clearance of TGF-β by binding to α_2-macroglobulin is extremely rapid, (<3 min) (O'Conner-McCourt and Wakefield 1987; LaMarre *et al.*, 1991). Thus, the availability of active TGF-β is likely very transient and probably restricted to the local environment.

5. Biological Activity: Multifunctional Regulation

The extraordinary ability of TGF-β family to affect diverse functions in cells from virtually every lineage is unparalleled. These cellular responses mediated by TGF-β fall into four main categories: proliferative responses; effects on cell differentiation; effects on differentiated functions; and responses involving extracellular matrices. TGF-β can also regulate apoptosis (Havrilesky *et al.*, 1995; Zhange *et al.*, 1995). Furthermore, effects of TGF-β *in vitro* can be positive or negative depending on the cell type, state of differentiation, and culture conditions. The biological functions of TGF-β on various cell types have been the subject of several recent

comprehensive reviews (Massague, 1990; Roberts and Sporn, 1990) and will be covered here only in general terms.

5.1 Cellular Proliferation

The effect of TGF-β on cell growth is inhibitory, in most cell types, particularly epithelial (Silberstein and Daniel, 1987), endothelial (Heimark *et al.*, 1986), lymphoid (Ruscetti and Palladino, 1991), and hematopoietic (Keller and Ruscetti, 1992). Cases in which this factor acts as a growth promoter are relatively few, and in most of them the effect of TGF-β may be indirect (Leof *et al.*, 1986; Battegay *et al.*, 1990), by stimulating the production of mitogens or their receptors. It is the growth inhibitory effects of TGF-β that have attracted the most research because effects like morphogenesis, immunosuppression, and tumor suppression are largely a result of its antigrowth function. The mechanism(s) by which TGF-β regulates growth inhibition are not completley known but clearly involve cell cycle regulation. Recent evidence suggests that TGF-β and immunosuppressor drugs like FK506 and rapamycin share the same signal transduction pathways (Wang *et al.*, 1994). Similar positive and negative effects of TGF-β on cellular function have been shown. In-depth reviews of this and the response of individual cell and tissue types to TGF-β have been published (Roberts and Sporn, 1990; Ignotz and Massague, 1990).

5.2 Cell–Environmental Interactions

Many cellular processes such as migration, chemotaxis, homing, tissue formation, repair, remodeling, and wound healing depend upon complex interactions between cells and extracellar matrices. TGF-β₁ regulates the adhesiveness of cells at many levels. These include: increased synthesis of extracellular matrix protein expression; control of matrix-degrading proteases and protease inhibitors; and increased expression of cell surface receptors (integrins) for cell adhesion proteins (Ignotz and Massague, 1987).

Moreover, cell surface expression of growth factor receptors can be downregulated by TGF-β. In fibroblasts EGF receptors (Takehara *et al.*, 1987), in lymphoid cells the γ chain of the IL-2 receptor (Espinosa-Delgado *et al.*, 1994), and in hematopoietic cells several receptors, including SCF, IL-3, and GM-CSF receptors (Jacobsen *et al.*, 1993), are downregulated by TGF-β. This regulation can be transcriptional as in case of the IL-2 receptor (Bosco *et al.*, 1994) or post-transcriptional as in case of the SCF receptor (Dubois *et al.*, 1994).

6. *The Receptors*

Despite the fact that there are many proteins which bind TGF-β, its biological activities are signaled by binding two transmembrane proteins designated TGF-β receptor types I and II (TβR-I and TβR-II) (for recent reviews see Derynck, 1994; Kingsley, 1994; Massague *et al.*, 1994). These are transmembrane serine/threonine kinases distinct from the transmembrane tyrosine kinases used by many of the mitogens, whose action TGF-β opposes. TβR-I and TβR-II have a cysteine-rich extracellular domain, a single transmembrane region, and a cytoplasmic kinase domain (Figure 28.6). The large numbers of cysteines in the extracellular domain with a cluster of three just upstream of the transmembrane are characteristic of all signaling receptors of the TGF-β superfamily. The type III receptor, a transmembrane proteoglycan, has a short, highly conserved cytoplasmic domain without an apparent signaling motif (Lopez-Casillas *et al.*, 1993; Sankar *et al.*, 1995).

The cloned type II receptor (Lin *et al.*, 1992) has a high affinity for TGF-β₁ and TGF-β₃ but not TGF-β₂. High-affinity binding of TGF-β₂ type II receptor occurs in the presence of the type III (β-glycan) receptor (Lopez-Casillas *et al.*, 1993). The presence of another type II receptor which binds TGF-β₂ with high affinity remains a possibility. The kinase domain of TβR-II is flanked by a spacer between the domain and the transmembrane region and has a carboxy-terminal tail (Figure 28.6). The kinase domains of the associated type II receptors of the superfamily show <40% amino acid sequence identity, suggesting that this divergence might be involved in conferring some signaling specificity. Several type I receptors have been cloned and their sequences form a distinct subfamily (Attisano *et al.*, 1993; Bassing *et al.*, 1994; Ebner *et al.*, 1993; Franzen *et al.*, 1993). Distinct from type II receptors, their kinase domains are similar, with shorter cytoplasmic domains, and the spacer region contains a conserved SGSGGLP (GS) domain absent in the type II receptors. Additionally, type I receptors have a much shorter carboxy-terminal tail and the extracellular domain has a distinct clustering of five cysteines.

Surprisingly, type II receptors can bind ligand directly from the medium, while type I receptors can not (Massague *et al.*, 1994). Type I receptors recognize ligand that is bound to type II receptors, with which they form a complex, probably a heterotetramer. Several type I receptors can bind TGF-β in this way. If distinct heterodimers have different functional specificities, this could provide another way of expanding the regulatory flexibility of TGF-β family.

7. *Signal Transduction*

Both TβR-I and TβR-II are needed to generate an activated TGF-β receptor complex (Laiho *et al.*,

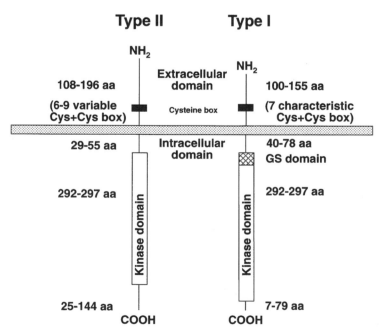

Figure 28.6   The structure of the receptors for TGF-β. Black bars represent characteristic three-cysteine clusters for both receptors (type I receptor has a unique five-cysteine cluster in the extracellular domain); the dotted region represents the transmembrane; and the cross-hatched region represents a characteristic type I receptor GS domain. The number of amino acids for each domain represents the human type I and type II TGF-β receptor. Vertical boxes indicate the Ser/Thr kinase domain.

1990b). TβR-II is phosphorylated on multiple serines and threonines partly through autophosphorylation. Its constitutively active kinase activity is not elevated by ligand binding. However, ligand bound to TβR-II is recognized by TβR-I, which is then recruited into the complex (Wrana et al., 1995). Once in the complex, TβR-I is phosphorylated by TβR-II at serine and threonine residues in the GS domain, located just upstream of the kinase domain of all type I receptors. Mutations that interfere with this phosphorylation block signal propagation (Wrana et al., 1992. TβR-I does not phosphorylate itself or the associated TβR-II, but its kinase activity is essential for signaling (Bassing et al., 1994), suggesting that TβR-I signals by phosphorylating downstream substrates. A constitutively active TβR-I generates inhibitory responses when expressed in the absence of TGF-β and TβR-II (Wrana et al., 1995).

In cells growth-inhibited by TGF-β, early mitogenic events are usually not affected by TGF-β (Chambard and Pouysseur, 1988; Like and Massague, 1986). The exceptions include (1) TGF-β activation of Ras and downstream components in epithelial cells (Mulder et al., 1988); (2) TGF-β downregulation of c-*myc* mRNA in keratinocytes (Coffey et al., 1980) and T cell lines (Ruegemer et al., 1990); and (3) induction of phosphorylation of the transcription factor of cAMP response element binding protein (CREBP) (Kramer et al., 1991). Downregulation of c-*myc* may not be required for growth inhibition (Birchenall-Roberts et al., 1996).

7.1   SMAD PROTEINS AND TGF-β SIGNALING

The Mad (mothers against decapentaplegic) gene in *Drosophila* and related Sma genes in *Caenorhabolitis elegans* are involved in signal by TGF-β family members (Sekelsky et al., 1995; Savage et al., 1996). Cloning of human homologs of Smad1 has shown that these molecules are downstream effectors of TGF-β (Liu et al., 1996; Lechleider et al., 1996). Human Smad1, 2, 3 and 5 are ligand specific docking proteins while Smad4 is a common signaling component binding to other Smads (Zhang et al., 1996). Similarly to the STAT proteins involved in the cytokine mitogenic signaling, Smad proteins exist in the cytoplasm in inactive forms and are rapidly phosphorylated, leading to nuclear translocation after TGF-β binding. The homo- and heterodimerization of Smad proteins necessary for this activation remains to be determined. A new transcription factor with site-specific DNA binding activity has been found to form a complex with the *Xenopus* Smad2 TGF-β sigmal transducer (Chen et al., 1996). There is a direct pathway from the cytoplasm to specific DNA binding in TGF-β-mediated transcription.

Recently, Imamura et al. (1997) and Nakao et al. (1997) reported that Smad6 and Smad7 inhibit Smad based TGF-β signaling. Smad6 and 7 like Smad4 lack the carboxy-terminal phosphorylation site but unlike Smad4 can bind to type I receptors. Since these Smads cannot be activated by phosphorylation and released from the receptors, these Smads probably function by competitive inhihition.

7.2   CELL CYCLE ARREST

When added to exponentially growing cells, the inhibitory effects of TGF-β are marked by an inability of cells to enter the S phase of the cell cycle and the cells arrest at the G_1/S boundary. As cells near the end of G_1 phase, a restriction point (R) is reached beyond which the cell cycle proceeds autonomously, indifferent to mitogens and TGF-β until the cells exit from mitosis (Campisi et al., 1982). Thus, there is a period before late G_1 phase during which cells must perform functions essential for reaching R that are susceptible to TGF-β inhibition. Among such functions, the phosphorylation of the retinoblastoma tumor suppressor gene product (Rb) has been studied extensively (Buchkovich et al., 1989; Sherr, 1994). As cells exit from mitosis, Rb is in a hypophosphorylated state, binding and sequestering E2F and other transcription factors needed for S phase transition. Rb phosphorylation in mid-G_1 induces the release of these factors, which activate a defined set of genes. In TGF-β-treated cells, Rb phosphorylation and cell cycle progression are blocked (Laiho et al., 1990a; Pietenpol et al., 1990a,b). Moreover, cells expressing the SV40 T antigen which binds underphosphorylated Rb (Ludlow et al., 1989) are no longer growth-inhibited by TGF-β even though it can inhibit Rb phosphorylation (Pietenpol et al., 1990b). Similarly, overpression of E2F, which is regulated by Rb, overcomes TGF-β-mediated growth inhibition in fibroblasts (Schwarz et al., 1995). In myeloid cells, the Rb-related proteins p130 and p107 are targets of TGF-β signaling (Bang et al., 1995).

7.3   CYCLIN AND CYCLIN-DEPENDENT KINASES AS TGF-β TARGETS

Rb phosphorylation in G_1 phase results from the action of cyclin D/Cdk4 and cyclin E/Cdk2 (Sherr, 1994). Activation of these kinases requires accumulation of an activating subunit (cyclin) and a catalytic subunit (cyclin-dependent kinase (Cdk)), assembly of cyclin/Cdk complexes, and Cdk phosphorylation by the Cdk-activating kinases (CAK). Active cyclin D/Cdk4 (mid-G_1) and cyclin E/Cdk2 (late G_1) complexes are targets for TGF-β regulation.

Indeed, depending on the cell type, TGF-β can downregulate the expression and/or action of numerous G_1 cyclins and their kinases (Figure 28.7). In mid-G_1,

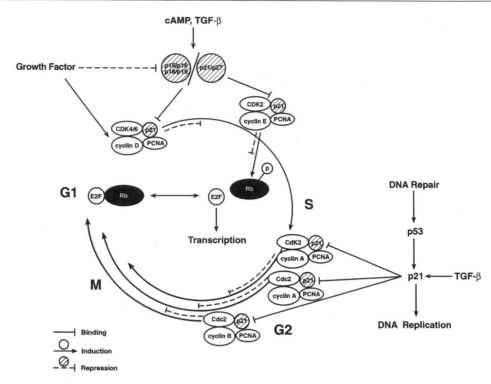

Figure 28.7   The effects of TGF-β on cell cycle progression. Solid lines indicate induction; broken lines indicate repression. Open circles indicate positive regulators of progression, and hatching represents negative regulators of progression.

TGF-β does not affect cyclin Ds but can inhibit the induction of their partner, Cdk4 (Geng and Weinberg, 1993; Ewen *et al.*, 1993). In late G₁, induction of Cdk2 and cyclin E is abolished by TGF-β in keratinocytes but not in epithelial cells (Slingerland *et al.*, 1994), where Cdk2 is inactive because it fails to assemble with cyclin E. In hematopoietic cells, TGF-β decreases both the phosphorylation of Cdk2 and the kinase actvity of Cdk2–cyclin E complexes in myeloid cells (Bang *et al.*, 1995). Finally, the induction and action of another Cdk2 partner, cyclin A, whose expression normally occurs late in G₁ can also be regulated by TGF-β. These effects on expression are generally exerted at the transcriptional level, but TGF-β causes a marked decrease in Cdk4 protein levels in epithelial cells without affecting Cdk4 transcription (Ewen *et al.*, 1993), and Cdk4–cyclin E complexes fail to assemble in TGF-β-treated fibroblasts (Koff *et al.*, 1993).

7.4   CYCLIN-DEPENDENT KINASE INHIBITORS AND TGF-β ACTION

The interest in cyclins and Cdks as targets of negative regulation was further heightened by the discovery of Cdk inhibitors (CDIs), a new class of cell cycle regulators. Two CDI families have been identified to date. The p27^{kip} (Polyak *et al.*, 1994a,b) and p21$^{wafl/cip1}$ (El-Deiry *et al.*, 1993; Harper *et al.*, 1993) proteins are related. The p21 CDI whose expression is a downstream target of the p53 tumor suppressor gene product can inhibit all Cdks (Xiong *et al.*, 1993). Also, p21 blocks PCNA-dependent DNA replication (PCNA: proliferative cells nuclear antigen), suggesting that it is responsible for coordinating cell cycle progression and DNA replication and repair (Waga *et al.*, 1995). Biochemical analysis of extracts from quiescent cells led to the discovery of p27 CDI, which binds G₁ cyclin–Cdk complexes stoichiometrically, inhibiting their activation by CAK (Polyak *et al.*, 1994a). The homology between p21 and p27 is limited to the region that binds and inhibits Cdks (Polyak *et al.*, 1994b; Toyoshima and Hunter, 1994). The second CDI family includes p15 and p16 (Makela *et al.*, 1994; Hannon and Beach, 1994), which were identified as members of the multiple tumor suppressor genes 1 and 2 and are small proteins with multiple ankyrin repeats and no sequence similarity to p21 or p27. Functionally, p15 and p16 target isolated Cdk subunits, not cyclin–Cdk complexes, and are specific for Cdk4 and its close homolog, Cdk6 (Serrano *et al.*, 1993).

TGF-β stimulates p21 production in growth-arrested cells by a p53-independent mechanism (Datto *et al.*, 1995); in fibroblasts, in which growth is stimulated by TGF-β, p21 production is turned off (Raynal and

Lawrence, 1995). Similarly, in cells that are growth-arrested by TGF-β, free p27 is in excess, facilitating its association with Cdk2–cyclin E complexes and preventing their activation (Polyak *et al.*, 1994a,b). However, in cells that are growth-stimulated by TGF-β, p27 levels decrease, thus reducing the amount of p27 complexing with Cdk2 (Ravitz *et al.*, 1995). In human keratinocytes growth-arrested by TGF-β, a down-regulation of Cdk levels and an elevation of p15 mRNA levels leads to an increased association of p15 with Cdk4 and Cdk6 (Hannon and Beach, 1994). These effects are complex and involve multiple pathways (Figure 28.7). The rate of G_1 progression is probably determined by a balance between Cdks and CDIs. This balance can be altered by the interplay between mitogens and antimitogens. Also, recent evidence suggests there are transcription-independent and transcription-dependent effects of TGF-β on cell cycle inhibition (Alexandrow and Moses, 1995). Thus, the effects of TGF-β in limiting cell cycle progression are pleiotropic.

8. *Murine TGF-β*

The genes for murine TGF-β_1, β_2, and β_3 are located on chromosomes 7, 1, and 12, respectively. The murine *TGF-B1* gene encodes seven exons ranging in size from 78 to 357 bp with all the intron–exon junctions conserved except exon 6/7 junction (Guron *et al.*, 1995). The mRNA expression of at least one of the TGF-β isoforms in adult murine tissues has been observed in spleen, testis, kidney, liver, lung, brain, heart, adipose tissue, submaxillary gland, and placenta (Bascom *et al.*, 1989). Several consensus sequences identified in the promoter region of the murine TGF-β1 gene are five Pu boxes, two NF-E1s, one NFκB, two AP2s, eight SP-1s, and one Epo-B2.

Among mammals the TGF-β1 protein is highly conserved: it is identical in man, pig, and cow, and differs by only one amino acid in mice. The primary and secondary structure of murine TGF-β_1 is virtually identical to that of the human molecule. TGF-β is produced and secreted by many cell types including activated lymphocytes, macrophages, and platelets. Most neoplastic and normal fibroblastic and epithelial cells maintained in culture secrete TGF-β in the latent form (Barnard *et al.*, 1990).

In general, the structure, activation, and signaling of murine TGF-β receptors are similar to those of the human receptors.

8.1 ROLE IN MURINE EMBRYOGENESIS

TGF-βs have been implicated in murine embryogenesis by the pattern of mRNA expression. Mouse embryos 9–16 days old express TGF-β_1 mRNAs in their bone, liver, and hematopoietic cells, with high levels seen in megakaryocytes, osteoclasts, and mesenchymal and epithelial tissues (Ellingsworth *et al.*, 1986; Wilcox and Derynck, 1988). TGF-β protein was detected by immunoreactivity in 11- to 18-day-old embryonic tissue-derived mesoderm including palate, larynx, teeth, cardiac valves, hair follicles, bones, and cartilage (Heine *et al.*, 1987). Murine embryonic undifferentiated stem cells secrete latent TGF-β_1 protein and possess few TGF-β receptors (Slager *et al.*, 1993). Upon retinoic acid-induced differentiation the cells expressed TGF-β_2 in latent and active forms and greatly upregulated TGF-β receptors (Rizzino, 1987). Also, TGF-β_3 protein was highly secreted and activated depending on the direction of differentiation. *In vivo*, the TGF-β isoforms are differentially expressed in discrete locations in virtually all tissues during embryonic development, with distinctive spatial and temporal patterns (for reviews, see Akhurst *et al.*, 1990; Kingsley, 1994). For instance, TGF-β_2 plays a major role in the induction of mesoderm (Rosa *et al.*, 1988).

Highly restricted expression of TGF-β receptors during embryonic development is an additional method for increasing specificity. Mouse fetuses during mid-gestation express the type-II TGF-β receptor primarily in mesenchymal and epidermal developing tissues (Lawler *et al.*, 1994; Wang *et al.*, 1995). The expression of this receptor and of the TGF-βs suggests that developmental regulation occurs in an autocrine or paracrine manner (Lehnert and Akhurst, 1988).

8.2 LESSONS FROM MURINE KNOCKOUT AND TRANSGENIC MICE

Information on the homeostatic regulatory effects of TGF-β has been obtained by selectively deleting the genes for various TGF-β isoforms. Targeted disruption of all (Kuilkarni *et al.*, 1993) or part (Shull *et al.*, 1992) of the TGF-β_1 gene results in a rapid wasting syndrome followed by death at 3–5 weeks of age. The major pathology was an extensive multifocal inflammatory response associated with serum autoantibodies and immune complex deposition (Dang *et al.*, 1995), and increased expression of both class I and II MHC antigens (Geiser *et al.*, 1993). Mice homologous for both the TGF-β_1 null gene and the class II null allele are without any of the inflammatory or immune pathology seen in the TGF-β_1 null mouse (Letterio *et al.*, 1996). Instead, these mice have extensive myeloproliferative expansion in marrow, spleen, and liver, suggesting that class II antigens cooperate with TGF-β_1 in regulating hematopoiesis. Thus, TGF-β_1 appears to be critical for physiologically limiting inflammatory, immune (Christ *et al.*, 1994), and hematological responses. Also, its functions are not redundant with other TGF-βs.

Further, Letterio and colleagues (1994) have shown in mice that maternal transfer of TGF-β_1 from *TGFB1*-heterozygous mothers prevents the development of cardiac abnormalities in TGF-β_1 null embryos, implying a critical role of TGF-β_1 in fetal development. Additional studies have revealed that administration of fibronectin peptides that bind to β_1-integrins and other cell surface adhesion structures blocked the inflammatory response that occurred in TGF-β_1 knockout mice (Hines *et al.*, 1994), indicating a role for TGF-β_1 in the regulation of expression and/or activation of adhesion molecules and their ligands. In addition, the TGF-β_1 null mouse has elevated levels of nitric oxide (NO) in the serum and inducible NO synthase (iNOS) in the heart and kidney (Vodovotz *et al.*, 1996).

The overexpression of TGF-β_1 transgene though specific promoters leads to tissue-specific effects. For instance, using an albumin promoter target overpression to hepatocytes, hepatocyte apoptosis and hepatic fibrosis occurred. Extrahepatic lesions including glomerulo-nephritis, renal failure, arthritis, myocarditis, and pancreatic atrophty occurred (Sanderson *et al.*, 1995). However, the general overexpression of TGF-β using a metallothionein promoter resulted in perinatal lethality suggesting that regulated expression of TGF-β_1 is essential for life (McCartney-Francis and Wahl, 1994).

9.   TGF-β *in Disease and Therapy*

With the numerous pathological consequences found as a result of the overexpression or deficiency of TGF-β_1, the TGF-β_1 axis must play a role in numerous disease processes. Furthermore, despite its ubiquitous presence, agonists or antagonists of its production, activation, or function could be clinically useful.

9.1   ONCOGENESIS

Unrestricted growth control owing to defects in the negative regulatory control of TGF-β and its role in oncogenesis has intrigued many investigators. Indeed, resistance to the antiproliferative effects of TGF-β is often observed in human cancer cells (Fyman and Reiss, 1993). Among the CDIs associated with TGF-β growth inhibition, p16 heterozygous deletions accompanied by mutations in the other allele have been described in a significant fraction of cases in various human malignancies, and the deletions sometimes affect the adjacent p15 gene as well (Kamb *et al.*, 1994; Nobori *et al.*, 1994). Recently, a novel candidate tumor suppressor gene, DPC4 (deleted in pancreatic cancer) was found to be identical to Smad4 (Hahn *et al.*, 1996), a downstream effector of TGF-β signaling.

In addition, the absence of TGF-β receptors in human RB tumor cells (Kimchi *et al.*, 1988) suggests a relationship between TGF-β regulation and tumor suppressor genes. Defects in TβR-II expression and TβR-II gene rearrangements have been described in tumor-derived cell lines of gastric and lymphoid cell origin (Capocasale *et al.*, 1995; Park *et al.*, 1994). The *in vivo* development of a dominant negative type II receptor was associated with malignant progression of a T cell lymphoma (Knaus *et al.*, 1996). Whether mutations in these genes are specifically involved in human cancer remains to be determined.

More compelling evidence was found in an inherited form of colon cancer (hereditary nonpolyposis colorectal cancer, HNPCC), where the cells have a genetic instability called "microsatellite instability" due to DNA repair defects (Markowitz *et al.*, 1995). In these cells, the TβR-II routinely has inactivating mutations within a sequence of a "mini-microsatellite" (Markowitz *et al.*, 1995). When a good copy of the receptor gene was introduced into a tumor cell line lacking it, the cell line could no longer form tumors in athymic mice, suggesting a biological consequence of the mutation and linking a DNA repair defect with a pathway of tumor progression.

9.2   FIBROSIS

Perhaps the most widely studied biological functions of TGF-β have been in the areas of fibrosis and wound healing (for reviews, see Border and Noble, 1994; Wahl, 1994). Since these activities have been extensively and repeatedly reviewed, they will not be discussed here. The nature of the ultimate biological outcome, beneficial or deleterious, is determined by a complex network of regulatory events that remains incompletely understood. Elevated levels of TGF-β_1 are often accompanied by fibrosis and/or inflammation in the heart, liver, and kidneys. Similarly, lung and liver fibrosis sometimes occurs as a life-threatening complication following bone marrow transplantation. High levels of plasma TGF-β_1 pretransplant may be predictive for the development of these complications, suggesting that the subsequent fibrotic effects may be associated with overproduction of this molecule (Zugonaier *et al.*, 1991; Anscher *et al.*, 1993). In summary, these findings suggest that the continued presence of high systemic levels of TGF-β result in uncontrolled stimulation of normal biological processes that ultimately result in disease pathology.

9.3   IMMUNITY

The pleiotropism of TGF-βs is also manifested in its ability to upregulate or downregulate cell processes such as immune activation. The downregulating effects can be either beneficial or deleterious depending on the circumstances. Such a downregulating effect is likely to be beneficial in situations where a successful immune response needs to become quiescent. However, the inappropriate production of TGF-β during crucial phases of immune response generation may also blunt the development of beneficial responses and lead to disease progression. For example, the transfection of a highly immunogenic UV-induced tumor with the cDNA for TGF-β_1 prevented T cell-mediated rejection of the tumor, demonstrating that under some circumstances TGF-β can promote escape from T cell-mediated immune surveillance (Torre-Amione *et al.*, 1990). Similarly, the administration of a TGF-β-neutralizing monoclonal antibody prevented the progressive growth of the MDA-231 human breast cancer cell line in athymic mice by an NK cell-dependent mechanism (Arteaga *et al.*, 1993) as well as experimental nephritis (Border *et al.*, 1990). TGF-β_1 can also be an autocrine-negative growth regulator of human colon carcinoma cells *in vivo* as revealed by transfection of an antisense expression vector which blocks tumorigenicity (Wu *et al.*, 1992).

Other studies have shown a potentially critical role for TGF-β in the response to infectious agents such as *Toxoplasma gondii* (Hunter *et al.*, 1995), *Candida albicans* (Spaccapelo *et al.*, 1995), and *Leishmania* (Barral *et al.*, 1993; Barral-Neto *et al.*, 1992) as well as HIV-1 (Kekow *et al.*, 1990). Thus, TGF-β can play a role in disease progression by interfering with productive antitumor or antimicrobial immune responses.

9.4   PROTECTION OF HEMATOPOIETIC AND OTHER TISSUE

A variety of hematopoietic effects of TGF-β *in vivo* have been described. Studies from our laboratory and others have shown that both TGF-β_1 and TGF-β_2 have potent, but reversible, inhibitory effects on myeloid progenitor cells *in vivo*. Goey and colleagues (1989) demonstrated that locoregional administration, via intrafemoral arterial injection of TGF-β_1, to the bone marrow inhibited endogenous and factor-driven proliferation of bone marrow cells, and impaired the colony-forming ability of multipotential and single-lineage progenitor cells. Also, Migdalska and colleagues (1991) noted that intra-peritoneal injection of TGF-β_1 for 5 days inhibited the proliferation of colony-forming units-spleen (CFU-S), but that growth of these progenitors began recovering within 24 h after treatment stopped.

These antiproliferative effects of TGF-βs suggested that TGF-βs might be useful for protecting stem and progenitor cells from the lethal effects of cell cycle-active chemotherapeutic drugs. Accelerated hematological recovery in mice pretreated with TGF-βs was noted by Grzegorzewski and colleagues (1994) in mice subsequently treated with 5-fluorouracil (5FU). It was also found that the administration of TGF-β_1 or TGF-β_2 during the hyperproliferative phase of bone marrow recovery after administration of a sublethal dose of 5FU

protected most mice from a supralethal dose of 5FU. Thus, a negative regulator of hematopoiesis can be used successfully to protect mice from otherwise lethal doses of chemotherapy.

The potential benefits of negative regulation by TGF-βs may extend beyond the hematopoietic system. Migdalska and colleagues (1991) noted that cells of the intestinal epithelium were also transiently inhibited following repeated systemic administration of TGF-β_1. Also, Grzegorzewski and colleagues (1994) noted that while TGF-βs could protect mice from the lethal toxicity of both 5-fluorouracil (5FU) and doxorubicin hydrochloride, the protective effects for 5FU were hematological, while the mechanism of protection for doxorubicin was nonhematological. The protective effects of TGF-βs for doxorubicin may occur by inhibiting proliferation of intestinal epithelial cells, thereby sparing them from death. A similar rationale was recently used by Sonis and colleagues (1994) for investigating other possible settings of chemoprotection by TGF-βs. They showed that the topical application of TGF-β_3 to the oral mucosa of Syrian hamsters prior to systemic administration of 5FU decreased the level of mucositis induced by the 5FU. This protective effect correlated with decreased growth of epithelial cells by TGF-β_3. Alternatively, TGF-β_1 was reported to protect mice from TNF-α-dependent myocardial ischemia–reperfusion injury (Lefer *et al.*, 1990) by some mechanism that is not dependent on inhibiting cell proliferation (Pinsky *et al.*, 1995). It seems likely that this type of inflammatory toxicity as well as that seen in TGF-β-deficiency is due to nitric oxide toxicity, which allows for development of new therapies for cardiac pathology.

9.5   CLINICAL DEVELOPMENT OF TGF-β

The slow rate of clinical development for TGF-β can be attributed to several unrelated factors. These include technical problems in producing recombinant TGF-β, concerns regarding its pleiotropic effects, and general concerns regarding the safety of such a potent growth factor on cell and tissue function.

The first clinical evaluation of TGF was in patients with psoriasis performed by Celtrix Pharmaceuticals using TGF-β_2. The rationale for this study was based on the previously mentioned potent immunosuppressive and antiinflammatory activities of TGF-β *in vitro* and *in vivo* (Ruscetti and Palladino, 1991; Wahl, 1992). Injection of TGF-β_2 into psoriatic lesions did not significantly reduce the severity of the lesion, but instead induced a fibrotic response at the injection site (Higley *et al.*, 1992), consistent with the stimulation by TGF-β of extracellular matrix production.

Based on the initial human clinical experience with TGF-β_2, as well as published preclinical studies which showed that exogenous TGF-β_2 could significantly enhance or stimulate the wound healing response, Celtrix Pharmaceuticals pursued clinical evaluation of TGF-β_2 in both ophthalmological and cutaneous wound healing indications.

A phase II ophthalmological study was performed to evaluate the utility of TGF-β_2 in stimulating the healing of macular holes. The results of this study were encouraging, indicating that injection of TGF-β_2 at the site of injury stimulated healing of the lesion as judged by visual examination and improved visual acuity of the patients (Glaser *et al.*, 1992). Based on these results, a multicenter phase III pivotal study was performed to evaluate TGF-β_2 in treatment of macular holes. Celtrix Pharmaceuticals announced in November 1994 that a planned interim analysis at 3 months post-treatment of patients revealed no significant difference between treated and control groups in terms of improvement in visual acuity.

In contrast, clinical results with TGF-β_2 in the treatment of cutaneous ulcers appeared more promising. Delivery of TGF-β_2 in a collagen–heparin sponge has been used to treat venous ulcers in two small clinical studies. The results provided evidence that TGF-β_2 can enhance the rate of healing in these kinds of obdurate wounds (Robson *et al.*, 1995). As a result, Genzyme Tissue Repair initiated a phase II clinical study evaluating the utility of TGF-β_2 in the treatment of diabetic foot ulcers.

Genzyme Tissue Repair is continuing a phase I clinical study with TGF-β_2 in patients with multiple sclerosis. The rationale was based on two separate observations. The first was the potential role of endogenous TGF-β in controlling the exacerbations and remissions of the disease in humans (Beck *et al.*, 1991; Link *et al.*, 1994; Mokhtarian *et al.*, 1994). The second was based on results in animal models of multiple sclerosis which indicated that treatment with TGF-β_2 could significantly reduce the frequency and intensity of disease exacerbations (Johns *et al.*, 1991; Kuruvilla *et al.*, 1991; Racke *et al.*, 1991).

Oncogene Science and Ciba-Geigy successfully completed a phase I clinical trial evaluating the safety of TGF-β_3 in the treatment of oral mucositis. The ultimate objective of this therapy is to reduce oral ulceration during chemotherapeutic treatment by reducing the rate of epithelial cell proliferation in the oral mucosa.

There are a number of additional clinical indications for which TGF-β may have utility. These include use of the cytokine to enhance wound strength development in surgical wounds that are prone to dehiscence or whose normal healing time requires extended convalescence for the patient. Studies in animal models of wound healing provide compelling evidence that TGF-β may be effective in this type of clinical indication (Mustoe *et al.*, 1987; Ksander *et al.*, 1990).

In addition to soft-tissue wound repair, TGF-β has a stimulatory role in the healing of hard tissues. Application of TGF-β to bony defects in a variety of animal models has been reported to significantly enhance the healing of these wounds (Beck *et al.*, 1993; Kibblewhite *et al.*, 1993; Lind *et al.*, 1993; Nielsen *et al.*, 1994). If similar stimulatory effects can be achieved in humans, TGF-β may have therapeutic roles in clinical indications such as orthopedic implant fixation, stimulation of healing of hip fractures in the elderly, and craniofacial repair. TGF-β may also have utility in other areas of orthopedics. It has been reported that, in certain tissue sites, TGF-β can induce chondrogenesis *in vivo* (Joyce *et al.*, 1990). Thus, it is possible that TGF-β may be efficacious in stimulating the healing of cartilaginous tissue following traumatic injury.

Given the central role of TGF-β in cellular regulation, it is likely that agonists or antagonists will eventually have therapeutic utility. The major hurdle will be to avoid systemic toxicity through transient or inducible delivery systems targeted to tissue- or cell-specific sites.

10. References

Akhurst, R., Fitzpatrick, D., Gatherer, D., Lehnert, S. and Millan, F. (1990). Transforming growth factor βs in mammalian embryogenesis. Prog. Growth Factor Res. 2, 153–168.

Alexandrow, M. and Moses, H. (1995). Transforming growth factor β1 inhibits mouse keratinocytes late in G_1 independent of effects on gene transcription. Cancer Res. 55, 3928–3932.

Andres, J., Stanley, K., Cheifetz, S. and Massague, J. (1989). Membrane-anchored and soluble forms of beta-glycan, a polymorphic proteoglycan that binds transforming growth factor-β. J. Cell Biol. 109, 3137–3145.

Anscher, M.S., Peters, W.P., Reisenbichler, H., Petros, W.P. and Jirtle, R.L. (1993). Transforming growth factor β as a predictor of liver and lung fibrosis after autologous bone marrow transplantation for advanced breast cancer. N. Engl. J. Med. 328, 1592–1598.

Antonelli-Orlidge, A., Saunders, K., Smith, S. and D'Amore, P. (1989). An activated form of transforming growth factor β is produced by cocultures of endothelial cells and pericytes. Proc. Natl Acad. Sci. USA 86, 4544–4548.

Arteaga, C.L., Hurd, S.D., Winnier, A.R., Johnson, M.D., Fendly, B.M. and Forbes, J.T. (1993). Anti-transforming growth factor (TGF)-β antibodies inhibit breast cancer cell tumorigenicity and increase mouse spleen natural killer cell activity. J. Clin. Invest. 92, 2569–2576.

Assoian, R.K. and Sporn, M.B. (1986). Type-beta transforming growth factor in human platelets: release during platelet deregulation and action on vascular smooth muscle cells. J. Cell Biol. 102, 1712–1733.

Assoian, R.K., Fleurdelys, B.E., Stevenson, H.C., Roberts, A. and Sporn, M. (1987). Expression and secretion of type beta transforming growth factor by activated human macrophages. Proc. Natl Acad. Sci. USA 84, 6020–6024.

Attisano, L., Carcamo, J., Ventura, F., Weis, F.M.B., Massague, J. and Wrana, J.L. (1993). Identification of human activin and TGF-β type I receptors that form heteromeric kinase complexes with type II receptors. Cell 75, 671–680.

Bang, O., Ruscetti, F., Lee, M.H., Kim, S.-J. and Birchenall-Roberts, M. (1995). Transforming growth factor β1 (TGF-β1) inhibits myeloid cell cycle proliferation: correlation with p107 regulation. J. Biol. Chem., 271, 7811–7819.

Barral, A., Barral-Netto, M., Yong, E.C., Brownell, C.E., Twardzik, D.R. and Reed, S.G. (1993). Transforming growth factor β as a virulence mechanism for Leishmania braziliensis. Proc. Natl Acad. Sci. USA 90, 3442–3446.

Barral-Netto, M., Barral, A., Brownell, C.E., et al. (1992). Transforming growth factor-β in Leishmanial infection: a parasite escape mechanism. Science 257, 545–548.

Barton, D.E., Foellmer, B.E., Du, J., Tamm, J., Derynck, R. and Franke, U. (1988). Chromosomal locations of TGF-βs 2 and 3 in the mouse and human. Oncogene Res. 3, 3230–3231.

Bascom, C., Wolfshohl, J. and Coffey, R. (1989). Complex regulation of transforming growth factor β1, β2 and β3 mRNA expression in mouse fibroblasts and keratinocytes by transforming growth factors β1 and β2. Mol. Cell Biol. 9, 5508–5515.

Bassing, C.H., Yingling, J.M., Howe, D.J., et al. (1994). A transforming growth factor β type I receptor that signals to activate gene expression. Science 263, 87–89.

Battegay, E.J., Raines, E.W., Seifert, R.A., Bowen-Pope, D.F. and Ross, R. (1990). TGF-β induces bimodal proliferation of connective tissue cells via complex control of an autocrine PDGF loop. Cell 63, 3039–3045.

Beck, J., Rondot, P., Jullien, P., Wietzerbin, J. and Lawrence, D.A. (1991). TGF-beta-like activity produced during regression of exacerbations in multiple sclerosis. Acta Neurol. Scand. 84, 452–455.

Beck, S., Amento, E., Xu, Y., et al. (1993). TGF-β induces bone closure of skull defects: temporal dynamics of bone formation in defects exposed to TGF-β1. J. Bone Miner. Res. 8, 753–761.

Border, W.A. and Noble, N.A. (1994). Transforming growth factor β in tissue fibrosis. N. Engl. J. Med. 331, 1286–1292.

Border, W.A., Okuda, S., Languino, L.R., Sport, M.B. and Ruoslahti, E. (1990). Suppression of experimental glyomerulonephritis by antiserum against transforming growth factor B1. Nature 346, 371–374.

Bosco, M.C., Espinoza-Delgado, I., Schwabe, M., et al. (1994). The γ subunit of the interleukin-2 receptor is expressed in human monocytes and modulated by interluekin-2, interferon-γ, and transforming growth factor β1. Blood 83, 3462–3467.

Buchkovich, K., Duffy, L.A. and Harlow, E. (1989). The retinoblastoma protein is phosphorylated during specific phases of the cell cycle. Cell 58, 1097–1105.

Campisi, J., Medaqno, E., Morreo, G. and Pardee, A. (1982). Restriction point control of cell growth by a labile protein: evidence for increased stability in transformed cells. Proc. Natl Acad. Sci. USA 79, 436–440.

Capocasale, R.J., Lamb, R.J., Vonderheid, E.C., et al. (1995). Reduced surface expression of transforming growth factor β receptor type II in mitogen-activated T cells from Sezary patients. Proc. Natl Acad. Sci. USA 92, 5501–5505.

Chambard, J.C. and Pouyssegur, J. (1988). TGF-0 inhibits growth factor-induced DNA synthesis in hamster fibroblasts without affecting the early mitogenic events. J. Cell Physiol. 135, 101–107.

Chen, X., Rubock, M.J. and Whitman, M. (1996). A transcriptional partner for MAD proteins in TGF-β signalling. Nature 383, 691–696.

Christ, M., McCartney-Francis, N.L., Kulkarni, A.B., et al. (1994). Immune dysregulation in TGF-β1-deficient mice. J. Immunol. 153, 1936–1946.

Coffey, R.J., Jr, Bascom, C., Sipes, N., Graves-Deal, R., Weissman, B. and Moses, H. (1988). Selective inhibition of growth-related gene expression in murine keratinocytes by transforming growth factor β. Mol. Cell. Biol. 8, 3088–3093.

Dang, H., Geiser, A.G., Letterio, J.J., et al. (1995). SLE-like autoantibodies and Sjogren's syndrome-like lymphoproliferation in TGF-β1 knockout mice. J. Immunol. 115, 3205–3212.

Daopin, S., Piez, K.A., Ogawa, Y. and Davies, D.R. (1992). Crystal structure of transforming growth factor-β2: an unusual fold for the superfamily. Science 257, 369–372.

Datto, M.B., Li, Y., Panus, J.P., Howe, D.J., Xiong, Y. and Wang, X.-F. (1995). Transforming growth factor β induces the cyclin-dependent kinase inhibitor p21 through a p53-independent mechanism. Proc. Natl Acad. Sci. USA 92, 5545–5549.

de Martin, R., Haendler, B., Hofer-Warbinek, R., et al. (1987). Complementary DNA for human glioblastoma-derived T cell suppressor factor, a novel member of the transforming growth factor-β gene family. EMBO J. 6, 3673–3677.

Dennis, P.A. and Rifkin, D.B. (1991). Cellular activation of latent transforming growth factor-β requires binding to the cation-independent mannose 6-phosphate/insulin-like growth factor type II receptor. Proc. Natl Acad. Sci. USA 88, 580–584.

Derynck, R. (1994). TGF-β-receptor-mediated signaling. TIBS 19, 548–553.

Derynck, R., Rhee, L., Chen, E.Y. and Van Tiburg, A. (1987). Intron–exon structure of human transforming growth factor-β precursor gene. Nucleic Acids Res. 15, 3188–3189.

Derynck, R., Lindquist, P.B., Lee, A., et al. (1988). A new type of transforming growth factor-β, TGF-β3. EMBO J. 7, 3737–3743.

Diebold, R., Eis, M., Yin, M., et al. (1995). Early onset multifocal inflammation in the transforming growth factor-β1 null mouse is lymphocyte-mediated. Proc. Natl Acad. Sci. USA 92, 12215–12219.

Dubois, C., Ruscetti, F., Stankova, J. and Keller, J. (1994). Transforming growth factor β regulates c-kit message stability and cell-surface expression in hematopoietic progenitors. Blood 83, 3138–3145.

Dubois, C.M., Laprise, M.H., Blanchette, F., Gentry, L.E. and Leduc, R. (1995). Processing of transforming growth factor beta 1 precursor by human furin convertase. J. Biol. Chem. 270, 10618–10624.

Ebner, R., Chen, R.-H., Lawler, S., Zioncheck, T. and Derynck, R. (1993). Determination of type I receptor specificity by the type II receptors for TGF-β and activin. Science 262, 900–902.

El-Deiry, W.S., Tokino, T., Velculescu, V. E., et al. (1993). WAF1, a potential mediator of p53 tumor suppression. Cell 75, 817–825.

Ellingsworth, L., Brennan, J., Fok, K., et al. (1986). Antibodies to the N-terminal portion of cartilage-inducing factor A and transforming growth factor beta. J. Biol. Chem. 261, 12362–12367.

Espinoza-Delgado, I., Bosco, M.C., Musso, T., et al. (1994). Inhibitory cytokine circuits involving transforming growth factor-β, interferon-γ, and interleukin-2 in human monocyte activation. Blood 83, 3332–3338.

Ewen, M.E., Sluss, H.K., Whitehouse, L.L. and Livingston, D.M. (1993). TGFβ inhibition of Cdk4 synthesis is linked to cell cycle arrest. Cell 74, 1009–1020.

Flaumenhaft, R., Abe, M., Sato, Y., *et al.* (1993). Role of latent TGF-β binding protein in the activation of latent TGF-β by co-culture of endothelial and smooth muscle cells. J. Cell Biol. 120, 995–1002.

Florini, J., Roberts, A., Eaton, D., Falen, S., Flanders, K. and Sporn, M. (1986). Transforming growth factor-β, a very potent inhibitor of myoblast differentiation, identical to differentiation inhibitor secreted by Buffalo rat cells. J. Biol. Chem. 261, 16509–16513.

Franzen, P., ten Dijke, P., Ichijo, H., *et al.* (1993). Cloning of a TGF-β type I receptor that forms a heteromeric complex with TGF-β type II receptor. Cell 75, 681–692.

Fujii, D., Brissendem, J., Derynck, R. and Franke, U. (1986). Transforming growth factor-β maps human chromosome 19 long arm and mouse chromosome 7. Somat. Cell Mol. Genet. 12, 281–288.

Fyman, T.M. and Reiss, M (1993). Resistance to inhibition of cell growth by transforming growth factor-β and its role in oncogenesis. Crit. Rev. Oncol. 4, 493–540.

Geiser, A., Letterio, J., Kulkarni, A., Karlsson, S., Roberts, A. and Sporn, M. (1993). Transforming growth factor β_1 (TGF-β_1) controls expression of major histocompatibility genes in the postnatal mouse: aberrant histocompatibility antigen expression in the pathogenesis of the TGF-β_1 null mouse phenotype. Proc. Natl Acad. Sci. USA 90, 9944–9948.

Geng, Y. and Weinberg, R.A. (1993). Transforming growth factor β effects on expression of G1 cyclins and cyclin-dependent protein kinases. Proc. Natl Acad. Sci. USA 90, 10315–10319.

Gentry, L., Webb, N., Lim, G., *et al.* (1987). Type 1 transforming growth factor-β: amplified expression and secretion of mature and precursor polypeptides in Chinese hamster ovary cells. Mol. Cell. Biol. 7, 3418–3426.

Glaser, B., Michels, R., Kupperman, B., Sjaarda, R. and Pena, R. (1992). Transforming growth factor beta 2 for treatment of full-thickness macular holes. Ophthalmology 99, 1162–1163.

Goey, H., Keller, J., Back, T., Longo, D.L., Ruscetti, F. and Wiltrout, R.H. (1989). Inhibition of early murine hematopoietic progenitor cell proliferation after in vivo locoregional administration of transforming growth factor-β1. J. Immunol. 143, 877–880.

Gray, A.M. and Mason, A.V. (1990). Requirement for activin A and transforming growth factor β1 proregions for homodimer assembly. Science 247, 1328–1330.

Grzegorzewski, K., Ruscetti, F.W., Usui, N., *et al.* (1994). Recombinant transforming growth factor β_1 and β_2 protect mice from acutely lethal doses of 5-fluorouracil and doxorubicin. J. Exp. Med. 180, 1047–1057.

Guron, C., Sudarshan, C. and Raghow, R. (1995). Molecular organization of the gene encoding murine transforming growth factor β1. Gene 165, 325–326.

Hahn, S.A., Schutte, M., Shamsul Hoque, A.T.M., *et al.* (1996). DPC4, A candidate tumor suppressor gene at human chromosome 18q21.1. Science 271, 350–353.

Hannon, G.J. and Beach, D. (1994). p15[INK4B] is a potential effector of TGF-β-induced cell cycle arrest. Nature 371, 257–261.

Harper, J.W., Adami, G.R., Wei, N., Keyomarsi, K. and Elledge, S.J. (1993). The p21 cdk-interacting protein Cip1 is a potent inhibitor of G1 cyclin-dependent kinases. Cell 75, 805–816.

Havrilesky, L.J., Hurteau, J.A., Whitaker, R.S., *et al.* (1995). Regulation of apoptosis in normal and malignant ovarian epithelial cells by transforming growth factor β^1. Cancer Res. 55, 944–948.

Heimark, R., Twardzik, D. and Schwarz, S. (1986). Inhibition of endothelial cell regneration by type-beta transforming growth factor from platelets. Science 233, 1078–1080.

Heine, U. Flanders, K., Roberts, A., Munoz, E. and Sporn, M. (1987). Role of transforming growth factor-β in the development of the mouse embryo. J. Cell Biol. 105, 2861–2876.

Higley, H., Chu, S., Persichitte, K., Ellingsworth, L., Ellis, C. and Voorhees, J. (1992). Intralesional injection TGF-b2 increases collagen, elastin, proteoglycans and fibronectin in lesions of psoriasis. J. Invest. Dermatol. 98, 610A.

Hines, K., Kuilkarni, A., McCarthy, J., *et al.* (1994). Synthetic fibronectin peptides interrupt inflammatory cell infiltration in transforming growth factor β, knockout mice. Proc. Natl Acad. Sci. USA 91, 5187–5191.

Holley, R.W., Bohlen, P., Fava, H., Baldwin, J.H., Kleeman, G. and Armour, (1980). Purification of kidney epithelial cell growth inhibitors. Proc. Natl Acad. Sci. USA 77, 5989–5992.

Howe, P.H., Draetta, G. and Leof, E.B. (1991). Transforming growth factor β1 inhibition of p34[cdc2] phosphorylation and histone H1 kinase activity is associated with G1/S-phase growth arrest. Mol. Cell. Biol. 11, 1185–1194.

Hunter, C.A., Bermudez, L., Beernink, H., Waegell, W. and Remington, J.S. (1995). Transforming growth factor-β inhibits interleukin-12-induced production of interferon-γ by natural killer cells: a role for transforming growth factor-β in the regulation of T cell-independent resistance to *Toxoplasma gondii*. Eur. J. Immunol. 25, 994–1000.

Hunter, T. and Pines, J. (1994). Cyclins and cancer II: cyclin D and CDK inhibitors come of age. Cell 79, 573–582.

Ignotz, R.N. and Massague, J. (1986). Transforming growth factor-β stimulates the expression of fibronectin and collagen and their incorporation into the extracellular matrix. J. Biol. Chem. 261, 4337–4345.

Ignotz, R.A. and Massague, J. (1987). Cell adhesion receptors as targets for transforming growth factor-β action. Cell 51, 189–197.

Imamura, T., Takase, M., Nishihara, A., *et al.* (1997). Smad6 inhibits signalling by the TGF-β superfamily. Nature 389, 622–626.

Jacobsen, S.E.W., Ruscetti, F.W. and Keller, J.R. (1993). Transforming growth factor-β is a bidirectional modulator of colony stimulating factor receptor expression on murine bone marrow progenitor cells. J. Immunol. 151, 4534–4544.

Johns, L.D., Flanders, K.C., Ranges, G.E. and Siram, S. (1991). Successful treatment of experimental allergic encephalo-myelitis by transforming growth factor beta 1. J. Immunol. 147, 1792–1797.

Joyce, M.E., Roberts, A.B., Sporn, M.B. and Bolander, M.E. (1990). Transforming growth factor-beta and the initiation of chondrogenesis and osteogenesis in the rat femur. J. Cell Biol. 110, 2195–2207.

Kanzaki, T., Olofsson, A., Moren, A., *et al.* (1990). TGF-β1 binding protein: a component of the large latent complex of TGF-β1 with multiple repeat sequences. Cell 61, 1051–1061.

Kamb, A., Gruis, N., Weaver-Feldhaus, J., *et al.* (1994). A cell cycle regulator potentially involved in the genesis of many tumor types. Science 264, 436–440.

Kehrl, J.H., Wakefield, L.M. and Roberts, A.B. (1986). Production of transforming growth factor beta by human T lymphocytes and its potential role in the regulation of T cell growth. J. Exp. Med. 163, 1037–1050.

Kekow, J., Wachsman, W., McCutchan, J.A., Cronin, M., Carson, D.A. and Lotz, M. (1990). Transforming growth factor β and noncytopathic mechanisms of immunodeficiency in human immunodeficiency virus infection. Proc. Natl Acad. Sci. USA 87, 8321–8325.

Keller, J.R. and Ruscetti, F.W. (1992). Transforming growth factor β (TGF β) and its role in hematopoiesis. Int. J. Cell Cloning 10, 2–11.

Kibblewhite, D.J., Bruce, A.G., Strong, D.M., Ott, S.M., Purchio, A.F. and Larrabee, W.F. (1993). Transforming growth factor-beta accelerates osteoinduction in a craniofacial onlay model. Growth Factors 9, 185–193.

Kim, S., Jeang, K., Glick, A., Sporn, M. and Roberts, A. (1989). Promoter sequences of the human transforming growth factor-β1 gene responsive to transforming growth factor-α1 autoinduction. J. Biol. Chem. 264, 7041–7045.

Kim, S.J., Andel, P. Lafyatis, R., Sporn, M. and Roberts, A. (1990). Autoinduction of TGF-β1 is mediated by the AP-1 complex. Mol. Cell. Biol. 10, 1492–1497.

Kimchi, A., Wang, X., Weinberg, R., Cheifetz, S. and Massague, J. (1988). Absence of TGF-β receptors and growth inhibitory responses in retinoblastoma cells. Science 240, 196–199.

Kingsley, D. (1994). The TGF-β superfamily: new members, new receptors and new genetic tests of function in different organisms. Genes Dev. 8, 133–146.

Koff, A., Ohtsuki, M., Polyak, K., Roberts, J.M. and Massague, J. (1993). Negative regulation of G1 in mammalian cells: inhibition of cyclin E-dependent kinase by TGF-β. Science 260, 536–539.

Kovacina, K.S., Steele-Perkins, G., Purchio, A.F., et al. (1989). Interactions of recombinant and platelet transforming growth factor β1 precursor with the insulin-like growth factor/mannose 6-phosphate receptor. Biochem. Biophys. Res. Commun. 160, 393–403.

Kramer, S., Koornneef, I., de Laat, S. and van den Eijnden-van Raaij, A. (1991). TGF-βb induces phosphorylation of the cyclic AMP responsive element binding protein in ML-CC164 cells. EMBO J. 10, 1083–1089.

Ksander, G., Ogawa, Y., Chu, G., McMullin, H., Rosenblatt, J. and McPherson, J. (1990). Exogenous transforming growth factor-beta 2 enhances connective tissue formation and wound strength in guinea pig dermal wound healing by secondary intent. Ann. Surg. 211, 288–294.

Kuilkarni, A.B., Huh, C.-H., Becker, D., et al. (1993). Transforming growth factor-β1 null mutation in mice causes excessive inflammatory response and early death. Proc. Natl Acad. Sci. USA 90, 770–774.

Kuruvilla, A., Shah, R., Hockwald, G., Liggitt, H., Palladino, M. and Thorbecke, G. (1991). Protective effect of transforming growth factor β1 on experimental autoimmune diseases in mice. Proc. Natl Acad. Sci. USA 88, 2918–2921.

Lafyatis, R., Lechleider, R., Kim, S.J., Jakowlew, S., Roberts, A.B. and Sporn, M.B. (1990). Structural and functional characterization of the transforming growth factor β3 promoter: a cAMP responsive element regulates basal and induced transcription. J. Biol. Chem. 265, 19128–19136.

Laiho, M., DeCaprio, J.A., Ludlow, J.W., Livingston, D.M. and Massague, J. (1990a). Growth inhibition by TGF-β linked to suppression of retinoblastoma protein phosphorylation. Cell 62, 175–185.

Laiho, M., Weis, F.M.B. and Massague, J. (1990b). Concomitant loss of transforming growth factor (TGF)-receptor types I and II in TGF-β-resistant cell mutants implicates both receptor types in signal transduction. J. Biol. Chem. 265, 18518–18524.

LaMarre, J., Hayes, M.A., Wollenberg, G.K., Hussaini, I., Hall, S.W. and Gonias, S.L. (1991). An α_2-macroglobulin receptor-dependent mechanism for the plasma clearance of transforming growth factor-β1 in mice. J. Clin. Invest. 87, 39–44.

Lawler, S., Candia, A.F., Ebner, R., et al. (1994). The murine type II TGF-beta receptor has a coincident embryonic expression and binding preference for TGF-beta 1. Development 120, 165–175.

Lechleider, R., de Caesteckert, M., Debejia, A. Polymeropoulos, M. and Roberts, A. (1996). Serine phosphorylation, chromosomal localization and transforming growth factor-β signal transduction. J. Biol. Chem. 271, 17617–17620.

Lefer, A.M., Tsao, P., Aoki, N. and Palladino, M.A., Jr (1990). Mediation of cardioprotection by transforming growth factor-β. Science 249, 61–64.

Lehnert, S.A. and Akhurst, R.J. (1988). Embryonic expression pattern of TGF-beta type 1 RNA suggests both paracrine and autocrine mechanisms of action. Development 104, 263–273.

Leof, E.B., Proper, J.A., Goustin, A.S., Shipley, G.D., DiCorleto, P.E. and Moses, H.L. (1986). Induction of c-sis mRNA and activity similar to platelet-derived growth factor by transforming growth factor β: a proposed model for indirect mitogenesis involving autocrine activity. Proc. Natl Acad. Sci. USA 83, 2453–2457.

Letterio, J. Geiser, A., Kulkarni, A., Roche, N., Sporn, M. and Roberts, A.B. (1994). Maternal rescue of transforming growth factor-β1 null mice. Science 264, 1936–1938.

Letterio, J., Geiser, A., Kulkarni, A., et al. (1996). Autoimmunity associated with TGF-β1-deficiency in mice is dependent on MHC class II antigen expression. J. Clin. Invest. 48, 2109–2119.

Li, L., Hu, J. and Olson, E. (1990). Different members of the jun proto-oncogene family exhibit distinct patterns of expression in response to type β transforming growth factor. J. Biol. Chem. 265, 1556–1562.

Lin, H.Y., Wang, X.-F., Ng-Eatron, E., Weinger, R. and Lodish, H. (1992). Expression cloning of the TGF-β type II receptor, a functional transmembrane serine/threonine kinase. Cell 67, 797–805.

Lind, M., Schumacker, B., Soballe, K., Keller, J., Melsen, F. and Bunger, C. (1993). Transforming growth factor beta enhances fracture healing in rabbit tibiae. Acta Orthop. Scand. 64, 553–556.

Link, J., Fredrikson, S., Soderstrom, M., et al. (1994). Organ-specific autoantigens induce transforming growth factor-beta mRNA expression in mononuclear cells in multiple sclerosis and myasthenia gravis. Ann. Neurol. 32, 197–203.

Liu, F., Hata, A., Baker, J., et al. (1996). A human Mad protein acting as a BMP-regulated transcriptional activator. Nature 381, 620–623.

Lopez-Casillas, F., Wrana, J.L. and Massague, J. (1993). Betaglycan presents ligand to the TGF-β signaling receptor. Cell 73, 1435–1444.

Ludlow, J.W., DeCaprio, J.A., Huang, C.M., Lee, W.H.,

Paucha, E. and Livingston, D.M. (1989). SV40 large T antigen binds preferentially to an underphosphorylated member of the retinoblastoma susceptibility ene product family. Cell 56, 57–65.

Lyons, R.M., Gentry, L.E., Purchil, A.F. and Moses, H.L. (1990). Mechanism of activation of latent recombinant transforming growth factor β1 by plasmin. J. Cell Biol. 110, 1361–1367.

Makela, T.P., Tassan, J.-P., Nigg, E.A., Frutiger, S., Hughes, G.J. and Weinberg, R.A. (1994). A cyclin associated with the CDK-activating kinase MO15. Nature 371, 254–257.

Markowitz, S. and Roberts, A. (1996). Tumor suppressor activity of the TGF-β pathway in human cancers. Cytokine Growth Factor Rev. 1, 1–10.

Markowitz, S., Wang, J., Myeroff, L., *et al.* (1995). Inactivation of the type II TGF-β receptor in colon cancer cells with microsatellite instability. Science 268, 1336–1338.

Massague, J. (1987). The TGF-β family of growth and differentiation factors. Cell 49, 437–438.

Massague, J. (1990). The transforming growth factor-β family. Annu. Rev. Cell Biol. 6, 597–641.

Massague, J., Chiefetz, S., Endo, T. and Nadal-Ginard, B. (1986). Type β transforming growth factor is an inhibitor of myogenic differentiation. Proc. Natl Acad. Sci. USA 83, 8206–8210.

Massague, J., Boyd, F.T., Andres, J.L. and Chiefetz, S. (1992). Mediators of TGF-β receptors and TGF-β-binding proteoglycans. Ann. N.Y. Acad. Sci. 593, 59–72.

Massague, J., Attisano, L. and Wrana, J. (1994). TGF-β family and its composite receptors. Trends Cell Biol. 4, 172–178.

McAllister, K., Grogg, K., Johnson, D., *et al.* (1994). Endoglin, a TGF-β binding protein of endothelial cells, is the gene for hereditary haemorrhagic telangiectasia type 1. Nature Genetics 8, 345–351.

McCartney-Francis, N.L. and Wahl, S.M. (1994). Transforming growth factor β: a matter of life and death. J. Leukocyte Biol. 55, 401–409.

Migdalska, A., Molineux, G., Demuynck, H., Evans, G.S., Ruscetti, F. and Dexter, T.M. (1991). Growth inhibitory effects of transforming growth factor-β1 in vivo. Growth Factors 4, 239–245.

Miyazono, K. and Heldin, C.H. (1989). Interaction between TGF-β1 and carbohydrate structures in its precursor renders TGF-β1 latent. Nature 388, 158–160.

Miyazono, K., Olofsson, A., Colosetti, P. and Heldin, C.H. (1991). A role of the latent TGF-β binding protein in the assembly and secretion of TGF-β1. EMBO J. 10, 1091–1102.

Mokhtarian, F., Shi, Y., Shirazian, D., Morgante, L., Miller, A. and Grob, D. (1994). Defective production of anti-inflammatory cytokine, TGF-beta by T cell lines of patients with active multiple sclerosis. J. Immunol. 152, 6003–6010.

Mooradian, D.L., Lucas, R.C., Weatherbee, J.A. and Furcht, L.T. (1989). Transforming growth factor-β1 binds to immobilized fibronectin. J. Cell. Biochem. 41, 189–200.

Moses, H.L., Branum, E.L., Proper, J.A. and Robinson, R.A. (1981). Transforming growth factor production by chemically transformed cells. Cancer Res. 41, 2842–2848.

Mossalayi, M., Mentz, F., Ouaaz, F., *et al.* (1995). Early human thymocyte proliferation is regulated by an externally controlled autocrine transforming growth factor-β1 mechanism. Blood 85, 3594–3601.

Mulder, K.M., Levine, A.E., Hernandez, X., McKnight, M.K., Brattain, D.E. and Brattain, M.G. (1988). Modulation of c-myc by transforming growth factor-β in human colon carcinoma cells. Biochem. Biophys. Res. Commun. 150, 711–716.

Mustoe, T.A., Pierce, G.F., Thomason, A., Gramates, P., Sporn, M.B. and Deuel, T.F. (1987). Transforming growth factor beta induces accelerated healing of incisional wounds in rats. Science 237, 1333–1336.

Nakao, A., Afrakhta, M., Moren, A., *et al.* (1997). Identification of Smad 7, a TGF-β-inducible antagonist of TGF-β signalling. Nature 389, 631–635.

Nielsen, H.M., Andreassen, T.T., Ledet, T. and Oxland, H. (1994). Local injection of TGF-beta increases the strength of tibial fractures in the rat. Acta Orthop. Scand. 65, 37–41.

Nobori, T., Miura, K., Wu, D., Lois, A., Takabayshi, K. and Carson, D. (1994). Deletions of the cyclin dependent cyclin kinase 4 inhibitor gene in multiple human cancers. Nature 368, 753–756.

Nunes, I., Shapiro, R.L. and Rifkin, D.B. (1995). Characterization of latent TGF-β activation by murine peritoneal macrophages. J. Immunol. 1450–1459.

O'Conner-McCourt, M.D. and Wakefield, L.M. (1987). Latent transforming growth factor β in serum. J. Biol. Chem. 262, 14090–14099.

Park, K., Kim, S.-J., Bang, Y.-J., *et al.* (1994). Genetic changes in the transforming growth factor β (TGF-β) type II receptor gene in human gastric cancer cells: correlation with sensitivity to growth inhibition by TGF-β. Proc. Natl Acad. Sci. USA 91, 8772–8776.

Pientenpol, J.A., Holt, J.T., Stein, R.W. and Moses, H.L. (1990a). Transforming growth factor β1 suppression of c-myc gene transcription: role in inhibition of keratinocyte proliferation. Proc. Natl Acad. Sci. USA 87, 3758–3762.

Pientenpol, J.A., Stein, R.W., Moran, E., *et al.* (1990b). TGF-β1 inhibition of c-myc transcription and growth in keratinocytes is abrogated by viral transforming protein with pRB binding domains. Cell 61, 777–785.

Pinsky, D.J., Cai, B., Yang, X., Rodriguez, C., Sciacca, R.R. and Cannon, P.J. (1995). The lethal effects of cytokine-induced nitric oxide on cardiac myocytes are blocked by nitric oxide synthase antagonism or transforming growth factor β. J. Clin. Invest. 95, 677–685.

Pircher, R., Jullein, P. and Lawrence, D.A. (1986). β-Transforming growth factor is stored in human blood platelets as a latent high molecular weight complex. Biochem. Biophys. Res. Commun. 130, 30–37.

Polyak, K., Kato, J.-Y., Solomon, M.J., *et al.* (1994a). p27Kip1, a cyclin-Cdk inhibitor, links transforming growth factor-β and contact inhibition to cell cycle arrest. Genes Dev. 8, 9–22.

Polyak, K., Lee, M.-H., Erdjument-Bromage, H., *et al.* (1994b). Cloning of p27Kip1, a cyclin-CDk inhibitor and a potential mediator of extracellular antimitogenic signals. Cell 78, 59–66.

Racke, M., Dhib-Jalbut, S., Cannella, B., Albert, P., Raine, C. and McFarlin, D. (1991). Prevention and treatment of chronic relapsing experimental allergic encephalomyelitis by transforming growth factor-beta 1. J. Immunol. 89, 7375–7382.

Ravitz, M.J., Shaochun, Y., Herr, K.D. and Wenner, C.E. (1995). Transforming growth factor β-induced activation of cyclin E-cdk2 kinase and down-regulation of p27[Kip1] in C3H 10T 1/2 mouse fibroblasts. Cancer Res. 55, 1413–1416.

Raynal, S. and Lawrence, D.A. (1995). Differential effects of transforming growth factor-β1 on protein levels of p21 WAF on cdk-2 and on cdk-2 kinase activity in human RD and CCL64 mink lung cells. Int. J. Oncol. 7, 337–341.

Rizzino, A. (1987). Appearance of high affinity receptors for type β transforming growth factor during differentiation of murine embryonal carcinoma cells. Cancer Res. 47, 4386–4390.

Roberts, A.B. and Sporn, M.B. (1990). The transforming growth factor-betas. In "Peptide Growth Factors and Their Receptors" (eds. M.B. Sporn and A.B. Roberts), pp. 419–472. Springer-Verlag, Heidelberg.

Roberts, A.B., Anzano, M.A., Lamb, L.C., Smith, J.M. and Sporn, M.B. (1981). New class of transforming growth factors potentiated by epidermal growth factor. Proc. Natl Acad. Sci. USA 78, 5339–5343.

Roberts, A.B., Flanders, K.C., Kondaiah, P., et al. (1988). Transforming growth factor β: biochemistry and roles in embryogenesis, tissue repair and remodeling and carcinogenesis. Recent Prog. Hormone Res. 44, 157–197.

Robson, M., Philip, L., Cooper, D., et al. (1995). Safety and effect of transforming growth factor-β2 for treatment of venous stasis ulcers. Wound Repair Regen. 3, 157–167.

Rosa, F., Roberts, A.B., Danielpour, D., Dart, L.L., Sporn, M.B. and David, I.B. (1988). Mesoderm induction in amphibians: the role of TGF-β2-like factors. Science 236, 783–786.

Ruegemer, J.J., Ho, S.N., Augustine, J.A., et al. (1990). Regulatory effects of transforming growth factor-β on IL-2- and IL-4-dependent T cell-cycle progression. J. Immunol. 144, 1767–1776.

Ruscetti, F.W. and Palladino, M.A. (1991). Transforming growth factor-B and the immune response. Prog. Growth Factor Res. 3, 159–175.

Sanderson, N., Factor, V., Nagy, P., et al. (1995). Hepatic expression of mature transforming growth factor β1 in transgenic mice results in multiple tissue lesions. Proc. Natl Acad. Sci. USA 92, 2572–2576.

Sankar, S., Mahooti-Brooks, N., Centrella, M., McCarthy, T.L. and Madri, J.A. (1995). Expression of transforming growth factor type III receptor in vascular endothelia cells increases their responsiveness to transforming growth factor β2. J. Biol. Chem. 270, 13567–13572.

Savage, C., Das, P., Finelli, A., et al. (1996). Caenorhabditis elegans genes sma2, sma3 and sma4 define a conserved family of transforming growth factor β pathway components. Proc. Natl Acad. Sci. USA 93, 790–794.

Schlunegger, M.P. and Grutter, M.G. (1992). An usual feature revealed by the crystal structure at 2.2 Å resolution of human transforming growth factor β2. Nature 358, 430–434.

Schwarz, J.K., Bassing, C.G., Kovesdi, I., et al. (1995). Expression of the E2F1 transcription factor overcomes type β transforming growth factor-mediated growth suppression. Proc. Natl Acad. Sci. USA 92, 483–487.

Sekelsky, J., Newfeld, S., Raftery, L., Chartoff, E. and Gelbart, W. (1995). Genetic characteristics and cloning of mothers against dpp, a gene required for decapentaplegic function in Drosophilia melanogaster. Genetics 139, 1347–1358.

Serrano, M., Hannon, G. and Beach, D. (1993). A new regulatory motif in cell cycle control causing specific inhibition of cyclin D/Ckd4. Nature 366, 304–307.

Seyedin, S., Thomas, T., Thompson, A., Rosen, D. and Piez, K. (1985). Purification and characterization of two cartilage-inducing factors from bovine demineralized bone. Proc. Natl Acad. Sci. USA 82, 2267–2271.

Sherr, C. (1994). The ins and outs of RB: coupling gene expression to the cell cycle clock. Trends Cell Biol. 4, 15–18.

Shull, M.M., Ormsby, I., Kier, A.B., et al. (1992). Targeted disruption of the mouse transforming growth factor-B1 gene results in multifocal inflammatory disease. Nature 359, 693–699.

Silberstien, G.B. and Daniel, C.W. (1987). Reversible inhibition of mammary gland growth by transforming growth factor-β. Science 237, 291–293.

Slager, H.G., Freund, E., Buiting, A.M.J., Feijen, A. and Mummery, C.L. (1993). Secretion of transforming growth factor-β isoforms by embryonic stem cells: isoform and latency are dependent on direction of differentiation. J. Cell. Physiol. 156, 247–256.

Slingerland, J.M., Hengst, L., Pan, C., Alexander, D., Stampfer, M. and Reed, S.I. (1994). A novel inhibitor of cyclin-cdk activity detected in transforming growth factor β-arrested epithelial cells. Mol. Cell. Biol. 14, 3683–3694.

Sonis, S., Lindquist, L., VanVugt, A., et al. (1994). Prevention of chemotherapy-induced ulcerative mucositis by transforming growth factor β3. Cancer Res. 54, 1135–1138.

Spaccapelo, R., Romani, L., Tonnetti, L., et al. (1995). TGF-β is important in determining the in vivo patterns of susceptibility or resistance in mice infected with Candida albicans. J. Immunol. 155, 1349–1350.

Sporn, M., Roberts, A., Wakefield, L. and de Crombrugghe, B. (1987). Some recent advances in the chemistry and biology of transforming growth factor-β. J. Cell Biol. 105, 1039–1045.

Takehara, K., LeRoy, E.C. and Grotendorst, G.R. (1987). TGF-β inhibition of endothelial cell proliferation: alteration of EGF binding and EGF-induced growth-regulatory (Competence) gene expression. Cell 49, 415–422.

Torre-Amione, G., Beauchamp, R.D., Koeppen, H., et al. (1990). A highly immunogenic tumor transfected with a murine transforming growth factor type β, cDNA escapes immune surveillance. Proc. Natl Acad. Sci. USA 87, 1486–1490.

Toyoshima, H. and Hunter, T. (1994). p27, a novel inhibitor of G1 cyclin-Cdk protein kinase activity, is related to p21. Cell 78, 67–74.

Tsuji, T., Okada, F., Yamaguchi, K. and Nakamura, T. (1990). Molecular cloning of the large subunit of transforming growth factor type βb masking protein and expression of the mRNA in various rat tissues. Proc. Natl Acad. Sci. USA 87, 8835–8839.

Van Obberghen-Schilling, E., Roche, N.S., Flanders, K.C., Sporn, M.S. and Roberts, A.B. (1988). Transforming growth factor-β1 positively regulates its own expression in normal and transformed cells. J. Biol. Chem. 263, 7741–7746.

Vodovotz, Y., Geiser, A.G., Chesler, L., et al. (1996). Spontaneously increased production of nitric oxide and aberrant expression of the inducible nitric oxide synthase in vivo in the transforming growth factor β1 null mouse. J. Exp. Med. 183, 2337–2342.

Waga, S., Hannon, G.J., Beach, D. and Stillman, B. (1994). The p21 inhibitor of cyclin-dependent kinase controls DNA replication by interaction with PCNA. Nature 369, 574–578.

Wahl, S.M. (1992). Transforming growth factor-beta in inflammation: a cause and a cure. J. Clin. Immunol. 12, 61–74.

Wahl, S.M. (1994). Transforming growth factor β: the good, the bad, and the ugly. J. Exp. Med. 180, 1587–1590.

Wakefield, L.M., Colletta, A.A., McCune, B.K. and Sportn, M.B. (1991). Roles for transforming growth factors-β in the genesis, prevention and treatment of breast cancer. In "Genes, Oncogenes, and Hormones: Advances in Cellular and Molecular Biology of Breast Cancer" (eds. R.B. Dickson and M.E. Lippman), pp. 97–136. Kluwer Academic Publishers, Dordrecht.

Wang, T., Donahoe, P.K. and Zervos, A.S. (1994). Specific interaction of type I receptors of the TGF-β family with the immunophilin FKBP-12. Science 265, 674–676.

Wang, Y.Q., Sizeland, A., Wang, X.F. and Sassoon, D. (1995). Restricted expression of type-II TGF beta receptor in murine embryonic development suggests a central role in tissue modeling and CNS patterning. Mech. Dev. 52, 275–289.

Wilcox, J.N. and Derynck, R. (1988). Developmental expression of transforming growth factors alpha and beta in mouse fetus. Mol. Cell. Biol. 8, 3415–3422.

Wrana, J.L., Attisano, L., Wieser, R., Ventura, F. and Massague, J. (1995). Mechanism of activation of the TGF-β receptor. Nature 370, 341–347.

Wrana, L., Attisano, L., Carcamo, J., *et al.* (1992). TGF-β signals through a heteromeric protein kinase receptor complex. Cell 71, 1003–1014.

Wu, S., Theodorescu, D., Kersel, R.S., *et al.* (1992). TGF-β1 is an autocrine-negative growth regulator of human colon carcinoma FET cells in vivo as revealed by transfection of an antisense expression vector. J. Cell Biol. 116, 187–196.

Xiong, Y., Hannon, G., Zhang, H., Casso, D., Kobayashi, R. and Beach, D. (1993). p21 is a universal inhibitor of cyclin kinases. Nature 366, 701–704.

Yamaguchi, Y., Mann, D.M. and Ruoslahti, E. (1990). Negative regulation of transforming growth factor-β by the proteoglycan decorin. Nature 346, 281–283.

Yang, E.Y. and Moses, H.L. (1990). Transforming growth factor β1-induced changes in cell migration, proliferation, and angiogenesis in the chicken chorioallantoic membrane. J. Cell Biol. 111, 731–741.

Zhang, Y., Feng, X., We, R. and Derynck, R. (1996). Receptor-associated Mad homologues synergize as effectors of the TGF-beta response. Nature 383, 168–172.

Zhange, X., Giangreco, L., Broome, E., Dargan, C.M. and Swain, S.L. (1995). Control of CD4 effector fate: transforming growth factor β1 and interleukin 2 synergize to prevent apoptosis and promote effector expansion. J. Exp. Med. 182, 699–709.

Zugonaier, G., Paik, S., Wilding, G., *et al.* (1991). Transforming growth factor β1 induces cachexia and systemic fibrosis without an antitumor effect in nude mice. Cancer Res. 51, 3590–3594.

29. RANTES

Peter J. Nelson, James M. Pattison *and* Alan M. Krensky

1.   Introduction

The C-C chemokine RANTES was originally identified as a cDNA during a general screen for genes selectively expressed by "functionally mature" cytotoxic T lymphocytes and not by B cells (Schall *et al.*, 1988). The subtractive hybridization approach used in this screen yielded several novel cDNA clones, one of which was found to encode for a new member of the family of *chemo*tactic cyto*kines*, now known as chemokines (Schall *et al.*, 1988). The name RANTES is an acronym derived from the original observed and predicted characteristics of the gene and the protein it encodes: *r*egulated upon *a*ctivation *n*ormal *T* cell *e*xpressed and *s*ecreted.

Subsequently, it was shown that RANTES protein could function as a potent chemoattractant for monocytes, specific subsets of T lymphocytes (Schall *et al.*, 1990; Taub *et al.*, 1993), eosinophils (Kameyoshi *et al.*, 1992; Rot *et al.*, 1992), basophils (Dahinden *et al.*, 1994), and natural killer cells (Maghazachi *et al.*, 1994; Taub *et al.*, 1995; Loetscher *et al.*, 1996). RANTES can also cause degranulation of basophils, respiratory burst in eosinophils, and the stimulation of T cell proliferation (Kuna *et al.*, 1992; Alam *et al.*, 1993a; Bischoff *et al.*, 1993; Bacon *et al.*, 1995). RANTES may also have antiviral properties, as it has been shown to suppress replication of HIV *in vitro* (Cocchi *et al.*, 1995). *In vivo*, RANTES is expressed in diseases characterized by a mononuclear cell infiltration such as renal allograft rejection (Pattison *et al.*, 1994), delayed-type hypersensitivity (Devergne *et al.*, 1994), and inflammatory lung disease (Lukacs *et al.*, 1996). These multiple activities suggest a role for the RANTES chemokine in both acute and chronic phases of inflammation.

Expression of RANTES can be induced in a wide variety of cell types including T cells, monocytes, basophils, mesangial cells, fibroblasts, epithelial cells, and endothelial cells (Schall, 1991). Megakaryocytes, which form platelets, appear to make RANTES constitutively (von Luettichau *et al.*, 1996). The rapid expression of RANTES by fibroblasts, endothelial cells, and epithelial cells may be an early response to induced stress by injured tissue. It is thought that this expression results in the attraction and infiltration of a variety of inflammatory cell types, including monocytes and memory T cells, into the stressed tissue. The expression of RANTES, accompanying T cell and monocyte effector function, may represent a means to amplify and propagate the inflammatory response (Pattison *et al.*, 1994; Wiedermann *et al.*, 1993).

2.   The RANTES Gene

Human RANTES is encoded by a single mRNA transcript approximately 1.2 kb in length. The mRNA is composed of a short 5' untranslated region, a coding region of 275 bases, and a long (~870 base) 3' untranslated region which contains a series of Alu repeats (Figure 29.1).

2.1   Genomic Organization of the RANTES Gene

The gene encoding RANTES displays the three-exon/two-intron organization common to the C-C chemokine family (Nelson *et al.*, 1993). In humans, the first exon is 133 bases in length and contains the 5' untranslated region and coding bases for the leader peptide; the second exon, spanning 112 bases, encodes the amino-terminal half of the mature protein; and the third exon, consisting of approximately 1 kb, codes for the carboxyl terminal and 3' untranslated regions (Figures 29.1 and 29.2).

2.2   Chromosomal Localization

The various members of the C-C chemokine family demonstrate a high degree of conservation of both gene structure and chromosomal localization (Oppenheim *et al.*, 1991; Schall, 1991; Sherry and Cerami, 1991; Miller and Krangel, 1992). The C-C chemokine genes are generally found as a gene cluster on human chromosome 17. The gene encoding human RANTES was localized to 17q11.2-q12 by in-situ hybridization (Donlon *et al.*, 1990). The high degree of conserved protein sequence and genomic organization between the chemokine family members suggests that these genes arose by gene duplication and subsequent divergence to their present forms (Oppenheim *et al.*, 1991; Schall, 1991).

2.3   Transcriptional Regulation

Approximately 1 kb of the immediate upstream region of human RANTES has been analyzed for functional regulatory elements. The preliminary results suggest that the transcriptional control of RANTES may vary among the cell types expressing RANTES (Nelson *et al.*, 1993, 1996; Ortiz *et al.*, 1996). This is discussed below.

2.3.1   T Lymphocytes

RANTES expression by T lymphocytes may be, in part, a developmentally controlled event. Early RANTES expression is seen following activation of resting peripheral blood T cells (Turner *et al.*, 1995). This is followed by a strong upregulation occurring 3–5 days later as the T cells become functionally mature (Schall *et al.*, 1988; Turner *et al.*, 1995). A high level of RANTES expression is maintained indefinitely *in vitro* by activated T cells, such as cytotoxic and helper T lymphocytes (Nelson *et al.*, 1993). It has been suggested that the strong expression of RANTES by activated T cells may play a role in amplifying

```
                                                      CCATGGATGAGGGAA
AGGAGGTAAGATCTGTAATGAATAAGCAGGAACTTTGAAGACTCAGTGACTCAGTGAGTAATAAAGACTC
AGTGACTTCTGATCCTGTCCTAACTGCCACTCCTTGTTGTCCCCAAGAAAGCGGCTTCCTGCTCTCTGAG
GAGGACCCCTTCCCTGGAAGGTAAAACTAAGGATGTCAGCAGAGAAATTTTTCCACCATTGGTGCTTGGT
CAAAGAGGAAACTGATGAGCTCACTCTAGATGAGAGAGCAGTGAGGGAGAGACAGAGACTCGAATTTCCG
GAGGCTATTTCAGTTTTCTTTTCCGTTTTGTGCAATTCACTTATGATACCGGCCAATGCTTGGTTGCTA
TTTTGGAAACTCCCCTTAGGGGATGCCCCTCAACTGCCCTATAAAGGGCCGCCTGAGCTGCAGAGGATTC
CTGCAGAGGATCAAGACAGCACGTGGACCTCGCACAGCCTCTCCCACAGGTACC    ATG AAG      Exon 1
GTC TCC GCG GCA GCC CTC GCT GTC ATC CTC ATT GCT ACT GCC CTC TGC
GCT CCT GCA TCT GCC TCC CCA T*
```

                                           1.4 kb Intron

```
                      *AT TCC TCG GAC ACC ACA CCC TGC TGC
TTT GCC TAC ATT GCC CGC CCA CTG CCC CGT GCC CAC ATC AAG GAG TAT      Exon 2
TTC TAC ACC AGT GGC AAG TGC TCC AAC CCA GCA GTC G*
```

                                           4 kb Intron

```
                      *TC TTT GTC ACC
CGA AAG AAC CGC CAA GTG TGT GCC AAC CCA GAG AAG AAA TGG GTT CGG      Exon 3
GAG TAC ATC AAC TCT TTG GAG ATG AGC TAG GATGGAGAGTCCTTGAACCTGA
ACTTACACAAATTTGCCTGTTTCTGCTTGCTCTTGTCCTAGCTTGGGAGGCTTCCCCTCACT
ATCCTACCCCACCCGCGCCTGAAGGGCCCAGATTCTGACCACGACGAGCAGCAGTTACAAAA
ACCTTCCCCAGGCTGGACGTGGTGGCTCAGCCTTGTAATCCCAGCACTTTGGGAGGCCAAGG
TGGGTGGATCACTTGAGGTCAGGAGTTCGAGACAGCCTGGCCAACATGATGAAACCCCATGT
GTACTAAAAATACAAAAAATTAGCCGGGCGTGGTAGCGGGCGCCTGTAGTCCCAGCTACTCG
GGAGGCTGAGGCAGGAGAATGGCGTGAACCCGGGAGCGGAGCTTGCAGTGAGCCGAGATCGC
GCCACTGCACTCCAGCCTGGGCGACAGAGCGAGACTCCGTCTCAAAAAAAAAAAAAAAAAAAA
AAAAAAATACAAAAATTAGCCGCGTGGTGGCCCACGCCTGTAATCCCAGCTACTCGGGAGGC
TAAGGCAGGAAATTGTTTCAACCCAGGAGGTGGAGGCTGCAGTGAGCTGAGATTGTGCCAC
TTCACTCCAGCCTGGGTGACAAAGTGAGACTCCGTCACAACAACAACAACAAAAAGCTTCCC
CAACTAAAGCCTAGAAGAGCTTCTGAGGCGCTGCTTTGTCAAAAGGAAGTCTCTAGGTTCTG
AGCTCTGGCTTTGCCTTGGCTTTGCAAGGGCTCTGTGACAAGGAATGAAGTCAGCATGCCTC
TAGAGGCAAGGAAGGGAGAACACTGCACTCTTAAGCTTCCGCCGTCTCAACCCCTCACAGGA
GCTTACTGGCAAACATGAAAAATCGGGCTTACCAATAAAGTTCTCAATGCAACCAAAAAAAA
AAAAAAAA
```

Figure 29.1   Sequence of the RANTES cDNA and immediate upstream region of the RANTES gene. The three exons of the RANTES gene are indicated. The cDNA sequence is given in bold and the coding sequence is shown with the individual codons. Asterisks show the splice junctions. The double-underlined codon denotes the coding sequence for the first amino acid in the mature protein. The TATA box and CAAT box are underlined (Schall *et al.*, 1988; Nelson *et al.*, 1993).

the inflammatory response (Wiedermann *et al.*, 1993; Nelson *et al.*, 1996; Ortiz *et al.*, 1996).

The T cell tumor line Hut78 (ATCC TIB 161) expresses RANTES constitutively (Nelson *et al.*, 1993). Reporter gene experiments using the immediate upstream region of the RANTES promoter to drive luciferase reporter gene have identified the minimal region of the RANTES promoter needed for reporter

gene expression in Hut78 and in mitogen-stimulated peripheral blood lymphocytes (Nelson *et al.*, 1993, 1996; Ortiz *et al.*, 1996). This stretch of DNA represents the immediate −195 bases 5′ from the site of transcriptional initiation. Four regulatory sites important for RANTES expression in T cells have been identified within this region of the RANTES promoter (Nelson *et al.*, 1996; Ortiz *et al.*, 1996).

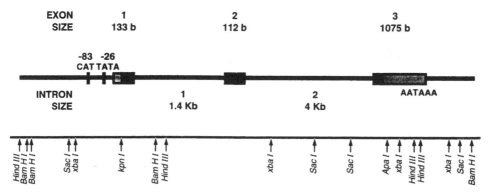

Figure 29.2   Genomic organization of the RANTES locus. The RANTES gene spans approximately 7.0 kb and shows the three-exon two-intron gene organization characteristic of the C-C chemokine family members (Nelson *et al.*, 1993).

The region designated R(E) found at –91 shows a strong homology to a C/EBP-like consensus site (Nelson *et al.*, 1993; Ortiz *et al.*, 1996). The R(E) region binds the transcription factor C/EBP-beta/NFIL6-alpha early in T cell activation (peak expression seen by day 2), but by day 3 a developmental switch takes place and a new as yet unidentified factor related to the C/EBP family of transcription factors replaces C/EBP-beta and binds to this region (Ortiz *et al.*, 1996).

The control region referred to as site R(A/B) is a 32-nucleotide element located between the CCAAT and TATAAA boxes of the RANTES promoter (Nelson *et al.*, 1996). R(A/B) contains two consensus sites for NFκB binding sites; R(A) and R(B). Site R(B) efficiently binds Rel p50-p50 homodimers but will also bind Rel p65-p50 heterodimers at a much lower efficiency. In contrast, site R(A) binds Rel p65-p50 heterodimers preferentially but will also bind p50-p50 homodimers. In activated T lymphocytes the R(A/B) region also binds unknown non-Rel nuclear factors expressed 3–5 days after T cell activation (Nelson *et al.*, 1996).

A site termed R(C) found at –185 is a purine-rich region important for RANTES expression by the T cell line Hut78 (Ortiz *et al.*, 1996). The factor which binds this site is called R(C)FLAT (RANTES site C factor of late-activated T cells) and is transiently induced in activated peripheral blood T cells, where peak expression is found 5 days after activation (Ortiz *et al.*, 1996). R(C)FLAT is comprised of at least two DNA binding subunits of approximately 65 and 45 kDa (Ortiz *et al.*, 1996). R(C)FLAT is not present in cytotoxic T cell lines and does not appear to be related to known transcription factor families (Ortiz *et al.*, 1996).

Study of the RANTES promoter and the nuclear factors regulating RANTES transcription in T cells has provided some explanation for the immediate early expression of RANTES in many cell types and it has allowed the identification of several factors contributing to the late upregulation of RANTES transcription during T cell functional maturation.

2.3.2 Fibroblasts

RANTES mRNA is expressed as an "immediate early gene" (6–20 h) by fibroblasts (Rathanaswami *et al.*, 1993). Cultured human synovial fibroblasts isolated from rheumatoid arthritis patients produce RANTES in both a time- and dose-dependent manner following stimulation with TNF-α and IL-1β. Preincubation with cycloheximide will "superinduce" levels of RANTES mRNA upregulated by IL-1β, but appear to decrease the expression of RANTES mRNA in response to TNF-α. In addition, IL-4 downregulates and IFN-γ enhances the TNF-α and IL-1β-induced increase in RANTES mRNA (Rathanaswami *et al.*, 1993). Nelson and colleagues (1996) demonstrated Rel p65-p50 binding to both κB-like sites R(A) and R(B) following stimulation of human

skin fibroblasts with TNF-α. Unpublished results (P.J. Nelson) show that NF-IL6α (C/EBP-β) is also induced in TNF-α-stimulated dermal fibroblasts and binds to the site designated R(E). These results may explain much of the RANTES promoter activity induced by TNF-α.

2.3.3 Endothelial Cells, Epithelial Cells, and Monocytes

Endothelial cells produce RANTES following stimulation with TNF-α and IFN-γ, while IL-4 and IL-13 appear to inhibit this induction (Marfaing-Koka *et al.*, 1995). Stellato and colleagues (1995) demonstrated that a human bronchial epithelial cell line will upregulate RANTES mRNA and protein by 16 h after stimulation with TNF-α and IFN-γ. This induction was inhibited by treatment with glucocorticoids (Stellato *et al.*, 1995; Wang *et al.*, 1996). Human monocytes and macrophages induce RANTES mRNA expression within hours after activation with lipopolysaccharide (Devergne *et al.*, 1994). Much has been learned about the control of RANTES transcription in murine monocytes. This will be covered in more detail in Section 8.2 of this chapter.

3. The RANTES Protein

3.1 PROTEIN SEQUENCE

The RANTES protein is composed of 91 amino acids, including a 23-amino-acid leader sequence cleaved in the endoplasmic reticulum (Figure 29.3). The "mature" protein is a highly basic polypeptide (pI~9.5) with a predicted molecular mass of 7847 Da. RANTES is probably a dimer at physiological pH and concentrations (Skelton *et al.*, 1995; Fairbrother and Skelton, 1996).

The various branches of the chemokine family have been classified in part, according to the placement of highly conserved cysteine residues (Schall *et al.*, 1992). In the C-C chemokine family, the first two cysteines are adjacent to one another, while in the C-X-C family they are separated by a single amino acid residue (Oppenheim *et al.*, 1991; Schall, 1991). Lymphotactin, the prototypic C chemokine, lacks the first and third cysteine residues (Kennedy *et al.*, 1995). These cysteine residues are important for the secondary structure of the chemokine proteins. Disulfide bridges link the first cysteine to the third (amino acids 10 and 34 in the RANTES protein) and the second to the fourth in the C-C and C-X-C families (RANTES amino acids 11 and 50), giving rise to a secondary structure made up of two peptide loops (Figure 29.3).

3.1.1 Recombinant Protein

Recombinant RANTES protein has been expressed in a variety of eukaryotic and prokaryotic expression systems (Schall *et al.*, 1990; Rot *et al.*, 1992; von Luettichau *et al.*, 1996). As is found with the other chemokine family

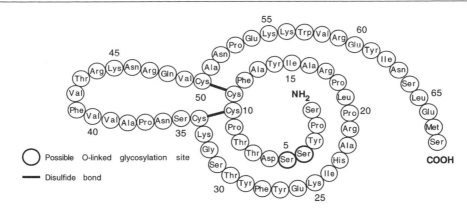

Figure 29.3   Amino acid sequence and secondary structure of the RANTES chemokine. Primary and secondary structure of the RANTES chemokine following cleavage of the signal sequence. The fourth and fifth serine residues (in bold circles) have been shown to undergo O-linked glycosylation. The first cysteine residue (amino acid 10) forms a bridge with the third residue (amino acid 34), while the second cysteine residue (amino acid 11) links to the fourth residue (amino acid 50).

members, recombinant RANTES protein shows a strong tendency to aggregate at high protein concentrations (greater than 1 mg/ml) (Lodi *et al.*, 1994; Skelton *et al.*, 1995; von Luettichau *et al.*, 1996). This effect can be minimized by using high concentrations of salt (1.5 M sodium chloride) or low pH (pH 3.5). Recombinant protein can be stored in HEPES buffer containing 1 mg/ml BSA in small aliquots at –70°C (Rot *et al.*, 1992).

3.2   GLYCOSYLATION

Kameyoshi isolated natural RANTES protein from human platelets (Kameyoshi *et al.*, 1992). Upon peptide analysis it was determined that most of the RANTES protein may have undergone O-linked glycosylation of the serine residues at positions 4 and 5 of the primary amino acid sequence (Figure 29.3). The natural forms of RANTES (glycosylated and unglycosylated) showed strong eosinophil-chemotactic activity, with optimal activity found at protein concentrations near 10 nM, similar to the values found for recombinant sources of RANTES (Rot *et al.*, 1992; Noso *et al.*, 1994; von Luettichau *et al.*, 1996). While O-linked glycosylation does not appear to result in a functional difference in activity with respect to eosinophil chemotaxis, glycosylation may be important in binding to endothelium, control of RANTES protein turnover, solubility, aggregation, or in the interaction with receptors other than those found on eosinophils.

3.3   QUATERNARY STRUCTURE

The structure of RANTES has been determined using nuclear magnetic resonance (Skelton *et al.*, 1995; Fairbrother and Skelton, 1996). The results show that RANTES is extensively aggregated in solution above pH

4.0. At pH 3.7 the protein is primarily dimeric. A set of NMR structures demonstrate that each monomer consists of a carboxyl-terminal α-helix packing against a three-stranded antiparallel β-sheet and two short N-terminal β-strands. Dimerization occurs between the N-terminal region of each monomer and the carboxyl-terminal β-strand of the second monomer. This quaternary structure is very different from that of the C-X-C chemokines such as interleukin-8, where the N-terminal strands of each monomer interact with each other to form the dimer (Clore *et al.*, 1992). The overall structure is very similar to that described for the C-C chemokine MIP-1β (Lodi *et al.*, 1994). RANTES appears to form a slightly more compact dimer relative to that seen for MIP-1β (Skelton *et al.*, 1995; Fairbrother and Skelton, 1996).

4.   Cellular Sources of RANTES

RANTES is expressed by many cell types in response to specific stimuli. Histological staining of a panel of tissues and in-situ hybridization for RANTES mRNA indicate that normal adult tissues contain few, if any, RANTES-positive cells, but that RANTES expression dramatically increases in inflammatory sites. In addition, megakaryocytes, some tumor types, and select fetal tissues express high levels of RANTES message and protein, portending other as yet unappreciated functions (von Luettichau *et al.*, 1996).

4.1   RANTES EXPRESSION IN NONTRANSFORMED TISSUE

RANTES is rapidly induced *in vitro* by inflammatory cytokines such as TNF-α, IL-1β, and IFN-γ, or mitogenic agents which may mimic conditions in the various stages of inflammatory processes *in vivo*. The expression of RANTES in specific disease states will be discussed in more detail in Section 9 of this chapter.

Von Luettichau and colleagues (1996) stained inflamed tonsil with RANTES-specific monoclonal antibodies and showed RANTES expression by a minor population of small scattered cells, probably T cells, found predominantly in the mantle zone and T-zone of lymphoid follicles. In addition, RANTES protein was found in subepithelial mononuclear inflammatory cells (monocytes and T cells) and in rare scattered cells in the germinal centers. Normal adult spleen contained a few scattered positive cells by antibody staining and by in-situ hybridization. Extramedullary hematopoiesis, blood cell maturation within the spleen, however, was associated with very strong RANTES expression by mega-karyocytes. As previously noted, platelets, which are produced by megakaryocytes, are a rich source of RANTES protein (Kameyoshi *et al.*, 1992).

RANTES expression has been studied by in-situ hybridization in human lymph nodes with delayed-type hypersensitivity lesions (DTH) related to either sarcoidosis or tuberculosis. The morphological characteristics of the tissue and distribution of positive cells indicate that both macrophages and endothelial cells are contributors to RANTES production in DTH reactions. RANTES expression by these cells has been confirmed by *in vitro* studies using alveolar macrophages and umbilical vein-derived endothelial cells (Devergne *et al.*, 1994).

Rheumatoid synovial fibroblasts produce RANTES mRNA and protein in response to TNF-α, IL1-β, and IFN-γ *in vitro* (Rathanswami *et al.*, 1993). Stellato and colleagues (1995) have demonstrated *in vivo* RANTES expression by pulmonary epithelial cells.

4.2 RANTES EXPRESSION BY TRANSFORMED CELLS

RANTES mRNA is expressed by many tumor lines such as the rhabdomyosarcoma RD and the osteosarcoma MG63 (Schall, 1991). It is expressed by the myeloid precursor cell lines HL60 and HEL, and by the T cell tumor line, Hut78 (Schall, 1991; Nelson *et al.*, 1993). It is difficult to draw definitive conclusions concerning the tumor-specific expression of a gene from the study of an *in vitro* immortalized cell line since the results may not always faithfully reflect the biology of the gene's expression by the tumor *in vivo*.

Tumor expression of RANTES *in vivo* appears heterogeneous within tumor types (von Luettichau *et al.*, 1996). Ten out of eleven lymphomas showed no expression of RANTES by the tumor, while rare scattered positive cells (dendritic cells or infiltrating lymphocytes) associated with the tumors showed strong

expression of RANTES. One lymphoma, a γδ T cell cutaneous lymphoma, showed staining for RANTES protein. The 10 negative lymphomas consisted of three follicular mixed lymphomas, three follicular small cleaved lymphomas, two small lymphocytic lymphomas, one mantle lymphoma, and one T cell diffuse mixed lymphoma (von Luettichau *et al.*, 1996). A series of solid renal tumors were also examined for RANTES protein expression. All Wilms' tumors tested showed strong RANTES expression by both tumor cells and perivascular macrophages. The "developing" renal tubules, characteristic of this tumor, were weakly positive for RANTES protein, while the subcapsular nephrogenic blastoma showed strong RANTES protein expression (von Luettichau *et al.*, 1996).

In contrast to the relatively "undifferentiated" Wilms' tumors, only one of five "well-differentiated" renal cell carcinomas expressed RANTES, although RANTES again was present in tumor-infiltrating mononuclear cells. The one embryomal rhabdomyosarcoma examined did not express RANTES protein (von Luettichau *et al.*, 1996).

4.3 RANTES EXPRESSION DURING HUMAN DEVELOPMENT

The expression of RANTES by some tumors led to the question whether RANTES is expressed during development (von Luettichau *et al.*, 1996). In a general survey of fetal tissues derived from a 22-week therapeutic abortion, only rare scattered RANTES-positive cells were identified in fetal thymus and lung, and no expression was found in developing brain, heart, or skin by in-situ hybridization and antibody staining. Fetal megakaryocytes, present in extramedullary hematopoiesis in spleen, were an exception in that they showed very strong expression of both RANTES messenger RNA and protein (von Luettichau *et al.*, 1996).

Interestingly, developing human kidney was found to express high levels of both RANTES protein and mRNA (von Luettichau *et al.*, 1996). Fetal kidney tissues stained concurrently with anti-RANTES monoclonal antibody and RANTES mRNA-specific antisense oligomers showed differential expression of message and protein in the developing kidney. The least differentiated cells (outer region of the kidney) showed very strong staining for RANTES message and little or no expression of RANTES protein. The center of the developing organ, which demonstrates a more mature morphology, showed RANTES protein and mRNA in the glomerular mesangium and in the proximal tubules. The authors suggest a developmental-translational control of RANTES expression in fetal kidney (von Luettichau *et al.*, 1996).

Although the function of RANTES in the

development of human kidney remains unknown, it is tempting to speculate that RANTES may play a role in organogenesis quite separate from its function as a pro-inflammatory cytokine. A role for C-C chemokines in development is not without precedent, since MIP-1α has been implicated in early stages of hematopoiesis (Graham and Pragnell, 1992; Broxmeyer et al., 1993).

5. Biological Activities

RANTES is a potent chemoattractant for T lymphocytes, natural killer cells, monocytes, eosinophils, and basophils. It has also been shown to activate these cell types and has been implicated as an important factor in the suppression of HIV replication. These phenomena will be discussed in more detail below.

5.1 ACTIONS ON T LYMPHOCYTES, MONOCYTES, AND NATURAL KILLER CELLS

5.1.1 Chemotaxis

During chemotaxis, cells move in the direction of an increasing concentration of a chemoattractant. Leukocytes sense a concentration gradient of as little as 1% across their diameter. Several groups have shown that RANTES is a potent chemotactic agent for T lymphocytes and monocytes. Schall showed that RANTES induced a significant migration of CD4[+] and CD45RO[+] (memory cells) but not of CD8[+] and CD45RA[+] (naive) cells (Schall et al., 1990). In contrast, Taub demonstrated that both naive and memory T cells migrated equally in response to RANTES although it was most effective on activated T cells (Taub et al., 1993). RANTES was 10 times more potent than MIP-1α or MIP-1β; chemotaxis was maximal at 1 ng/ml. RANTES was also chemotactic for CD4[+] T cell clones. RANTES is also a potent chemotactic agent for monocytes (Schall et al., 1990; Wiedermann et al., 1993; von Luettichau et al., 1996) with maximum effects at 10^{-10} to 10^{-8} M. Chemotaxis can be completely abolished by addition of anti-RANTES antibody (Wiedermann et al., 1993; von Luettichau et al., 1996). Rot and Hub (1995) have suggested that the chemotactic effect of RANTES on monocytes is highly dependent upon the monocyte concentration in Boyden chamber assays, arguing that RANTES may be a chemotactogenic rather than a chemotactic agent for monocytes; that is, it may act to cause the release of another, yet to be identified, chemoattractant from monocytes.

The RANTES protein also induces the chemotaxis and chemokinesis of natural killer cells at concentrations that range between 10 and 100 pg/ml. RANTES induces chemotaxis but not chemokinesis of IL-2-activated natural killer cells. MIP-1α, and MCP-1 share similar actions (Maghazachi et al., 1994; Taub et al., 1995).

5.1.2 Haptotaxis

Haptotaxis is cell migration induced by surface-bound gradients of chemoattractants. In blood vessels, chemotaxis along soluble gradients is theoretically unlikely since such a gradient would be rapidly dispersed by the flow of blood (Rot, 1993). Thus, haptotaxis along a gradient of mediator bound to endothelium or extracellular matrix may be more important in vivo. In modified Boyden chamber assays, Wiedermann (1994) showed that RANTES is a potent haptotactic agent for monocytes, with a maximal response at nanomolar concentrations. Since RANTES protein has been localized to the endothelium and extracellular matrix during inflammation (Wiedermann et al., 1993; Pattison et al., 1994; von Luettichau et al., 1996), it is ideally located to act as a haptotactic agent.

5.1.3 Increased Adhesion

Taub demonstrated that RANTES augmented the adhesion of activated CD4[+] cells (but not CD8[+] cells) to IL-1α-treated human umbilical vein endothelial cells. No change in T cell adhesion was present using resting T lymphocytes or endothelium (Taub et al., 1993). Blocking studies with monoclonal antibodies suggest that adhesion was mediated by T cell VLA-4 binding to endothelial VCAM-1 (Tanaka et al., 1993). Lloyd and colleagues (1996) showed that RANTES and other chemokines will induce T cell adhesion to recombinant ICAM-1 and VCAM-1. The adhesion process was found to occur rapidly and to be dose-dependent, and appeared to be mediated via the β_1- and β_2-integrins. The results suggest that RANTES stimulates the development of a high-affinity state in the integrin molecules.

RANTES also increases monocyte adhesion to activated endothelial cells (Vaddi and Newton, 1994). RANTES upregulates the expression of the α chains of CD11b and CD11c (β_2-integrins) and of CD18 (their common β chain). The upregulation of CD11b and CD11c peaked at 4 h and then declined to basal levels after 24 h. The monocyte binding to activated endothelial cells followed similar dose–response kinetics to the upregulation of the integrins and was blocked by monoclonal antibodies to CD11b and CD11c, suggesting that modulation of integrin expression by RANTES might facilitate tissue trafficking of monocytes during inflammation (Vaddi and Newton, 1994).

5.1.4 Activation of T Cells

In addition to inducing chemotaxis, RANTES acts as an activator of T cells in vitro. Bacon showed that RANTES could induce a biphasic mobilization of Ca^{2+} in a T cell clone (Bacon et al., 1995). The initial peak, a transient increase in cytosolic Ca^{2+}, was mediated by a G protein-coupled pathway, and was associated predominantly with chemotaxis. The second Ca^{2+} peak was associated with a spectrum of cellular responses, including Ca^{2+} channel opening, interleukin-2 receptor expression, cytokine release, and T cell proliferation. All of these events are characteristic of T cell receptor activation.

Slightly different results were reported by Turner and collegues (1995). The authors found that anti-CD3 mAb, but not RANTES, could elicit elevation of intracellular calcium levels in human T lymphocytes. RANTES stimulation did produce a bell-shaped chemotactic response and an increase in polarization of the T lymphocytes. T lymphocytes stimulated with RANTES, showed increased phosphatidylinositol (PI) 3-kinase activity with maximal activity seen at 10–100 ng/ml. Wortmannin, a potent PI 3-kinase inhibitor, inhibited the RANTES-induced T lymphocyte migration. These results suggest that RANTES activation of T lymphocytes is independent of detectable elevation of cytosolic free calcium, and that RANTES-induced chemotaxis involves the putative PI 3-kinase signal transduction pathway. The reason for the contrary results regarding Ca^{2+} flux is, at present, not clear.

RANTES also enhances effector function of cytotoxic T lymphocytes and natural killer cells. Taub found that RANTES is capable of augmenting CTL and NK but not lymphokine-activated killer cell or antibody-dependent cell cytotoxicity-specific cytolytic responses (Taub et al., 1995). RANTES was also found to modulate antigen-driven T cell proliferative responses as well as effects on lymphokine production. The costimulation with RANTES was interleukin-2-dependent and required the presence of free extracellular calcium. Neutralization of endogenously produced RANTES with specific antibody during antigen-specific T cell response blocked cellular proliferation, suggesting that RANTES may have an autocrine role in antigen-induced T cell proliferative responses. Together, these results suggest that chemokines such as RANTES play a role in the activation of polyclonal as well as antigen-specific helper and cytotoxic T cells during the development of an immune response.

5.1.5 T Cell and Monocyte Transendothelial Migration

RANTES was found to elicit dose-dependent transendothelial migration of unstimulated peripheral blood lymphocytes in vitro. Phenotype analysis showed monocytes and $CD4^+$, $CD8^+$, and $CD45RO^+$ T lymphocyte subsets were recruited. No migration of natural killer cells or significant numbers of B cells was seen (Roth et al., 1995). RANTES has also been demonstrated to upregulate the secretion and activity of matrix metalloproteinases 2 and 9 by infiltrating T cells (Xia et al., 1996). The activation of these enzymes is thought to facilitate leukocyte transmigration through the basement membrane and extracellular matrix and thus may be important to the transendothelial migration of leukocytes.

5.2 EOSINOPHIL CHEMOTAXIS, TRANSENDOTHELIAL MIGRATION, AND ACTIVATION

Kameyoshi purified an eosinophil chemotactic factor from thrombin-stimulated platelets, and by amino terminal sequencing identified this protein as RANTES (Kameyoshi et al., 1992, 1994). Alam showed that RANTES induced chemotaxis of eosinophils in a dose-dependent manner at 10^{-11} to 10^{-7} M (Alam et al., 1993a). Comparison with other eosinophil chemoattractants indicates that RANTES is as potent as C5a (Rot et al., 1992), MCP-3 (Baggiolini and Dahinden, 1994), and eotaxin (Jose et al., 1994) and is 2-fold to 3-fold more potent than MIP-1α (Rot et al., 1992). Recently, much attention has been paid to the interaction among the cytokines IL-3, IL-5, and GM-CSF, potent differentiation and proliferation factors for eosinophil precursors, and the chemokines. Schweitzer and colleagues (1994) showed that eosinophils migrated toward RANTES at concentrations of 10^{-9} to 10^{-7} M, only after priming with IL-5 (10 pM). Unprimed eosinophils only showed a significant migratory response at higher concentrations of RANTES (10^{-6} M). This difference between the optimal concentration of RANTES found by different groups is probably related to the fact that, in the earlier studies, eosinophils were obtained from donors with a peripheral blood eosinophilia (Kameyoshi et al., 1992) or allergy (Alam et al., 1993b) in whom in vivo priming took place.

In an in vitro transendothelial migration assay, RANTES induced eosinophil transendothelial migration in a dose-dependent manner (Ebisawa et al., 1994). The other C-C chemokines MCP-1, MIP-1α, MIP-1β, and I-309 showed no effect. The effect of RANTES on transmigration was greatly potentiated by priming the eosinophils with IL-5.

In addition to promoting eosinophil chemotaxis, RANTES activates eosinophils, causing release of eosinophil cationic protein in a dose-dependent manner. RANTES is a relatively weak stimulator of exocytosis and leukotriene formation, even when the eosinophils are primed (Rot et al., 1992). In addition, RANTES is a potent activator of the respiratory burst in eosinophils, leading to the release of highly toxic oxygen radicals (Chihara et al., 1994; Kapp et al., 1994). By scanning electron microscopy, RANTES induces morphological changes in eosinophils, including decreased numbers of specific granules and increased numbers of small dense granules, probably indicating intracellular translocation of toxic granular proteins. This localized "piecemeal degranulation" may explain the changes of eosinophil density observed upon stimulation with RANTES (Alam et al., 1993a).

5.3 BASOPHIL CHEMOTAXIS AND ACTIVATION

RANTES and MCP-3 are some of the most potent basophil chemoattractants known (Kuna *et al.*, 1992; Bischoff *et al.*, 1993; Baggiolini and Dahinden, 1994). In contrast, MCP-1 and MCP-3 are more effective inducers of histamine release than RANTES (Bischoff *et al.*, 1992, 1993; Dahinden *et al.*, 1994). RANTES also enhanced the adherence of basophils to activated human umbilical vein endothelial cells. The RANTES-induced adhesion was dose-dependent and increased following priming with IL-5. Adhesion was blocked by monoclonal antibodies to the β_2-integrins. All C-C chemokine effects on basophils are blocked by pertussis toxin, indicating that they activate basophils via G-protein-coupled receptors (Bacon *et al.*, 1994).

5.4 RANTES SUPPRESSION OF HIV REPLICATION

The phenomenon of noncytolytic CD8[+] T cell suppression of HIV replication is a well documented effect in which soluble factors derived from CD8[+] T cells limit the ability of HIV to copy itself (Levy, 1993). This HIV-suppressing activity of CD8[+] T cells may in part be responsible for the phenomenon of the "slow progressor" phenotypes seen in some HIV-infected individuals. The chemokines RANTES, MIP-1α, and MIP-1β produced by CD8[+] T cell will inhibit the replication *in vitro* of monocyte tropic strains of HIV-1 (Cocchi *et al.*, 1995). In addition, the CCR2b, CCR3, and CCR5 chemokine receptors (discussed in Section 6 of this chapter) have been demonstrated to act as coreceptors for HIV-1 entry into cells (Alkhatib *et al.*, 1996; Choe *et al.*, 1996; Deng *et al.*, 1996; Doranz *et al.*, 1996; Dragic *et al.*, 1996). Understanding the biology of these chemokines and their receptors will open new avenues of research into pathogenesis of AIDS and possibly the treatment of HIV-infected individuals.

5.5 *IN VIVO* EXPERIMENTS

In an attempt to study the *in vivo* effects of RANTES, Murphy (1994) used a human/severe combined immune deficient (SCID) mouse model. SCID mice received human peripheral blood lymphocytes, followed by three injections of RANTES over 3 days. RANTES induced mononuclear cell infiltration 72 h after injection. The majority of cells recruited were human CD3[+] T cells with equal numbers of CD4[+] and CD8[+] cells.

Although, the injection of RANTES intradermally into rats or rabbits induced no infiltrate, it did cause an eosinophil- and macrophage-rich infiltrate within 4 h in beagle dogs (Meurer *et al.*, 1993). The cell infiltration peaked at 16–24 h after injection. There was histological evidence of intravascular activation of eosinophils at 4 h, with adherence of eosinophils to the endothelium and homotypic aggregation of eosinophils, but there was no evidence of eosinophil degranulation. The lack of response in other species tested may reflect species differences in the RANTES receptor. The abundance of peripheral blood eosinophils (and possibly eosinophil priming in dogs) may explain the florid eosinophilic infiltrate seen in the beagle.

Although our knowledge of the actions of RANTES and the other chemokines has expanded greatly in recent years, most of this information has been obtained from *in vitro* assays and many questions remain about the *in vivo* significance of these findings. The development of overexpressing and knockout mice for RANTES, as well as specific antagonists, will be instructive in this respect.

6. *RANTES Receptors*

As described earlier, the chemokine receptors have multiple functions. Not only do they mediate chemotaxis but they also mediate the upregulation of integrins, actin polymerization, respiratory burst, degranulation, and cell proliferation. Most of the biological effects of the chemokines appear to be mediated primarily through interactions with different classes of G-protein-coupled receptors that span the membrane seven times (so-called serpentine receptors) (Gerard and Gerard, 1994; Murphy, 1994) (Figure 29.4). The chemokine receptor genes are expressed in a cell type-specific manner and this differential expression seems to be the basis for the specificity of chemokines for subsets of leukocytes.

The known chemokine receptors can be grouped into four different classes: the "specific", "shared", "promiscuous", and "virally encoded" receptors (Gerard and Gerard, 1994; Murphy, 1994). The "specific" chemokine receptors bind only one chemokine. To date, a specific RANTES chemokine receptor has not been described. The "shared" chemokine receptors bind more than one chemokine within either the C-X-C or the C-C subfamily. The chemokine receptors CCR1, CCR3, and CCR5 all bind RANTES, and other C-C chemokines, and thus fall into the shared category (Gao *et al.*, 1993; Neote *et al.*, 1993a; Daugherty *et al.*, 1996; Kitaura *et al.*, 1996; Raport *et al.*, 1996; Samson *et al.*, 1996). The Duffy antigen receptor for chemokines is a promiscuous chemokine receptor that binds both C-X-C and C-C chemokines including RANTES (Horuk *et al.*, 1993; Neote *et al.*, 1994; Horuk and Peiper, 1996). The fourth type of chemokine receptor is encoded within viral genomes. The US28 gene found in the cytomegalovirus (CMV) genome binds the RANTES protein (Neote *et al.*, 1993; Murphy *et al.*, 1996).

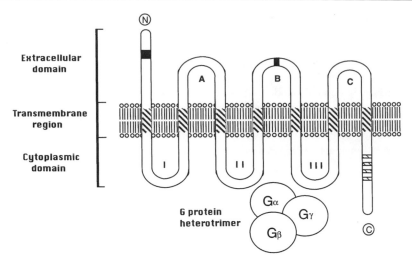

Figure 29.4   The RANTES receptors span the membrane seven times (so-called serpentine receptors). The chemokine receptors identified to date that bind the RANTES protein include; the "shared" receptors CCR1, CCR3, CCR4, and CCR5; the "promiscuous" chemokine receptor DARC, and the "virally encoded" US28 receptor. The transmembrane regions are detailed with dark hatched bars. The amino terminus and second extracellular domain are thought to be involved in ligand specificity (black bars). The G-protein heterotrimer probably interacts with the II and III intracellular domains (of the shared and virally encoded receptors). The carboxyl terminus contains serine and threonine residues (shown as light hatched bars) involved in phosphorylation of the receptor (Murphy, 1994; Gerard and Gerard, 1994; Horuk and Peiper, 1996; Murphy et al., 1996).

In addition to specific high-affinity receptors on leukocytes and erythrocytes, RANTES also binds to proteoglycans on the endothelium that may present the chemokine to circulating leukocytes. The biochemical nature of this proteoglycan–chemokine interaction is poorly understood (Witt *et al.*, 1993).

6.1   General Features of Chemokine Receptors

The second and third intracellular loops of serpentine receptors interact with a G-protein heterotrimer and, upon ligand binding, exchange GDP for GTP, resulting in activation of the G-protein subunits (Neote and McColl, 1996). In turn, the activated G-proteins signal effector enzymes, such as phospholipase $C\beta_2$. GTP is hydrolyzed to GDP, and the GDP form of the G-protein completes the cycle by complexing with unoccupied receptors. Pertussis toxin causes ADP-ribosylation of the $G\alpha_1$ subunits and thus irreversibly inactivates their action. Most known biological effects of the chemokines appear to be inhibited by pertussis toxin (Wu *et al.*, 1993; Neote and McColl, 1996).

Serine and threonine amino acid residues found at the carboxyl terminus of the serpentine receptors act as substrates for phosphorylation. Phosphorylation of the carboxyl terminus leads to binding of arrestin, which prevents binding of G-proteins to the receptor and hence prevents signaling. This may be the mechanism of desensitization by which prior exposure to a ligand

blocks the subsequent response to the same ligand. Desensitization causes the rapid cessation of the ligand-induced responses critical to the cellular response to a concentration gradient, and thus inhibits chemotaxis (Murphy, 1994).

6.2   Ligand Binding Studies and Receptor Expression

Ligand binding to chemokine receptors is complex, with multiple chemokines binding to a single receptor and multiple receptors binding to a specific ligand (Murphy, 1994; Kelvin *et al.*, 1993; Gerard and Gerard, 1994; Neote and McColl, 1996). Such complexity presumably increases the potential for sophisticated intercellular communication. The CCR1 chemokine receptor found on neutrophils, eosinophils, monocytes, T cells, natural killer cells, and B cells binds MIP-1α, RANTES, and MCP-3 (Neote *et al.*, 1993a; Gao *et al.*, 1993; Combadiere *et al.*, 1995b,c; Ben-Baruch *et al.*, 1995; Neote and McColl, 1996). The CCR3 receptor binds RANTES, eotaxin, and MCP-3 and is expressed by monocytes, eosinophils, and neutrophils (Combadiere *et al.*, 1995a; Daugherty *et al.*, 1996; Kitaura *et al.*, 1996; Neote and McColl, 1996). The CCR5 receptor binds MIP-1α, RANTES, and MIP-1β and is macrophage and T cell expressed (Raport *et al.*, 1996; Samson *et al.*, 1996; Neote and McColl, 1996).

Neote and colleagues (1993b, 1994) demonstrated that the receptor expressed by human erythrocytes has a broad specificity for chemokine binding. This receptor is the Duffy blood group antigen, which is also the receptor for the human malarial parasite *Plasmodium vivax* and is now referred to as DARC (Duffy antigen receptor for chemokines) (Horuk *et al.*, 1993). Its mRNA is detectable in brain, kidney, spleen, lung, and thymus (Chaudhuri *et al.*, 1994; Neote *et al.*, 1994; Horuk and Peiper, 1996). In human embryonic kidney cells transfected with Duffy antigen cDNA, there was no evidence of signal transduction as measured by transient increases in intracellular calcium ion concentration. The physiological role of the erythrocyte DARC is uncertain, although it may function in the clearance of chemokines from the systemic circulation (Chaudhuri *et al.*, 1994; Horuk and Peiper, 1996).

The virally encoded chemokine receptors appear to exhibit superior binding and signaling properties relative to the human receptor analogs. The role of these receptors in viral pathogenesis is at present unclear. The open reading frame designated US28 found in human cytomegalovirus encodes a protein that is similar in sequence to CCR1 (33% homology), with 56% homology in the amino-terminal domain, and binds the RANTES chemokine (Gao *et al.*, 1993; Neote *et al.*, 1993a). US28 may have been acquired by the virus to impart some selective advantage. Expression of US28 may alter the responsiveness of infected cells to C-C chemokines (Ahuja *et al.*, 1994; Murphy *et al.*, 1996).

7. Signal Transduction

Many details of the signal transduction pathways are still unclear, but it seems that both G-protein-dependent and -independent activation by chemokines can occur. Receptor–ligand interaction leads to activation of phospholipase $C\beta_1$ and β_2, production of inositol 1,4,5-trisphosphate (IP_3) and diacylglycerol, a rapid and transient increase in intracellular calcium, and the activation of protein kinase C (Howard *et al.*, 1996; McColl and Neote, 1996). Different kinases may be involved in signal transduction, including serine/threonine and tyrosine protein kinases as well as the MAP kinase cascade (Howard *et al.*, 1996; McColl and Neote, 1996). Bacon used antisense oligonucleotides against $G\alpha16$ and $G\alpha11$ to block chemotaxis, calcium flux, and IP_3 generation in T cells in response to RANTES (Bacon *et al.*, 1994) (see also Section 5.1.4 of this chapter).

The next few years will see further dissection of the signal transduction pathway downstream from the RANTES receptor. Special focus will be placed on the role of specific G-protein subunits, and the various second messengers involved in the various cellular responses found to be elicited by specific chemokines.

8. Murine RANTES

Mouse and rat express a single RANTES mRNA transcript of about 560 bases. These transcripts display a much shorter 3′ untranslated region relative to that found in human cells (Schall *et al.*, 1988).

8.1 THE MURINE RANTES GENE

The gene for murine RANTES (small inducible cytokine A5 (Scya5)) was isolated by Danoff and found to contain a 4.5 kb transcriptional unit. The first exon is 138 bases long and contains the 5′ untranslated region and 76 bases of coding region; the second exon contains 112 bases of coding sequence, while the third exon contains the remaining 85 bases of the coding region and the complete 3′ untranslated region. The murine RANTES gene was localized to chromosome 11 using Southern blot analysis of somatic cell hybrids containing known complements of mouse chromosomes on hamster or rat backgrounds (Danoff *et al.*, 1994).

DNA sequences important for regulation may be conserved through evolution. A comparison of the immediate upstream region found in the mouse (Danoff *et al.*, 1994) and human genes (Nelson *et al.*, 1993) demonstrate a high level of conservation within the first 200 bases 5′ to the site of transcriptional initiation (Ortiz *et al.*, 1995). This homology extends through the TATAAA box and includes the CAP site and the complete 5′ untranslated region to the start of translation, suggesting that these sequences may be important in regulation of RANTES expression.

8.2 EXPRESSION IN MOUSE EPITHELIAL AND MESANGIAL CELLS

RANTES is expressed as an immediate early gene (2–20 h) by stimulated murine renal tubular epithelial and mesangial cells. Wolf reported that a mouse mesangial cell line upregulated murine RANTES within 2 h in response to stimulation with TNF-α (Wolf *et al.*, 1993). The level of RANTES mRNA remained elevated in these cells for 24–48 h after stimulation. IL-1β, TNF-α, and lipopolysaccharide were also found to increase expression of RANTES mRNA these cultured murine mesangial cells. Heeger demonstrated a significant elevation in mRNA transcripts encoding RANTES following the stimulation of mouse tubular epithelium with TNF-α and IL-1α in culture, but not with TGF-β, IFN-γ, or IL-6 (Heeger *et al.*, 1992).

8.3 TRANSCRIPTIONAL CONTROL IN MOUSE MONOCYTES

The expression of RANTES by primary murine macrophages was described by Orlofsky (1994), who showed that RANTES mRNA was induced by LPS treatment of bone marrow-derived macrophages and resident peritoneal macrophages. Shin (1994) used a murine monocyte cell line and promoter deletion and mutational analysis to define LPS-responsive elements within the murine RANTES promoter. The region termed LRE (lipopolysaccharide-responsive element) consists of two motifs of TCAYR at −169 and −172 which contain half of an AP-1 site with two flanking bases, and a site (A/T)(G/C)NTTYC(A/T)NTTY found at −154. The binding of nuclear protein to this region was observed after stimulation with LPS. Interestingly, this site, which resembles in part the interferon-stimulated responsive element (ISRE), is also highly conserved between the mouse and human promoters (Ortiz et al., 1995).

8.4 SPECIES SPECIFICITY OF THE RANTES PROTEIN

Schall found that recombinant murine RANTES can attract *human* monocytes in a dose-dependent fashion (Schall et al., 1992). A reciprocal phenomenon is seen in the rat (R.A. Stahl, unpublished observation), where human RANTES was found to attract rat monocytes *in vitro*. Monoclonal antibodies to human RANTES were also found to cross-react with inflamed rat kidney tissue, suggesting that anti-human RANTES antibodies recognize rat RANTES *in vivo*. This cross-reactivity may be attributable to the high degree of amino acid homology between human, murine, and rat RANTES, which share approximately 85% amino acid identity (Schall et al., 1992; von Luettichau et al., 1996).

8.5 SEPSIS SYNDROME

On a tissue level, sepsis is characterized by endothelial cell injury, edema formation, and the influx and activation of neutrophils and monocytes. VanOtteren (1995) demonstrated expression of both RANTES mRNA and protein in lung homogenates of mice following intraperitoneal endotoxin administration. Cellular sources of the RANTES antigen, as determined by immunohistochemical techniques, were localized to the epithelial cells lining the alveoli. Neutralization experiments with anti-RANTES antibodies indicated that RANTES contributed to the accumulation of macrophages but not neutrophils or lymphocytes in the lung after intraperitoneal administration of endotoxin. Pretreatment of animals with soluble TNF receptor:immunoglobulin constructs attenuated the LPS-induced RANTES expression, suggesting a role for TNF in the induction of RANTES in this murine model of RANTES.

9. RANTES in Disease and Therapy

The role of the RANTES chemokine in disease should be considered in the larger context of leukocyte–endothelial interactions and the generation of an inflammatory response. The conventional model envisages a three-step process (Butcher, 1991; Springer, 1994; Adams and Shaw, 1994). First, selectins allow the attachment of flowing leukocytes to the vessel wall through labile adhesions that permit leukocytes to roll in the direction of flow and allows them to sample the signals presented by the endothelium. Second, chemokines direct the migration of leukocytes by chemotaxis and haptotaxis. Chemokines also upregulate integrins on leukocytes and may allow the third step, which is the tight adhesion between integrins and their ligands on the endothelium. Chemokines are bound to the endothelium by noncovalent interactions with the proteoglycans on the vessel wall. Chemokines provide molecular signals for specific cells to migrate into the interstitium (Shaw et al., 1993), perhaps explaining the characteristic cellular infiltrates seen in various diseases. Structural variation of proteoglycans expressed on specific endothelial beds may add additional specificity by regulating which chemokines are retained. Chemokines also act within the interstitium to direct migration of cells.

RANTES mRNA and protein are highly expressed during acute cellular rejection of human renal allografts (Pattison et al., 1994). Cellular rejection is characterized by a mononuclear cell infiltrate of T cells, macrophages, and occasional eosinophils. Given the activities of RANTES *in vitro*, it is a prime candidate for attraction of this infiltrate. RANTES mRNA and protein were barely detectable in 12 biopsies taken 1 h after formation of the vascular anastomosis during transplantation surgery or in three native renal biopsies from cardiac transplant recipients with cyclosporin nephrotoxicity. In contrast, RANTES mRNA and protein were present in 17 out of 20 biopsies showing cellular rejection. RANTES mRNA and protein were detected in infiltrating mononuclear cells and in renal tubular cells. On peritubular capillaries only RANTES protein was found, with very low corresponding mRNA signal, suggesting that the RANTES protein was not produced by the endothelial cells but rather was made by other cell types and deposited on the vessel wall.

The *in vivo* demonstration of RANTES protein on the endothelium is particularly important because it is ideally situated to mediate haptotaxis. Shaw (1993) showed that MIP-1β binds selectively to the high endothelial venules of lymph nodes. Rot (1993) demonstrated that the glycosaminoglycan binding site and the neutrophil receptor binding site of IL-8 are positioned on opposite sides of the IL-8 molecule, allowing IL-8 to bind simultaneously to its neutrophil receptor and to the endothelial proteoglycan.

It is tempting to speculate that RANTES may be important in the recruitment of monocytes and lymphocytes in other renal diseases such as interstitial nephritis and certain types of glomerulonephritis. One of the characteristic features of HIV nephropathy is an interstitial mononuclear cell infiltrate. Kimmel assayed tissue levels of RANTES and MCP-1 in renal biopsies from patients with HIV nephropathy (Kimmel et al., 1993). They isolated the chemokines from the biopsy by high-performance capillary electrophoresis and quantified them using biotinylated antibodies with chemiluminescence enhancement. RANTES and MCP-1 levels were significantly elevated (11.5 and 11.3 pg/mg, respectively) in patients with HIV nephropathy compared to biopsies without a cellular infiltrate (<2.0 pg/mg).

Enhanced recruitment of monocytes from the bloodstream to the arterial intima is also a feature of atherosclerosis. MCP-1 has been detected in human carotid endarterectomy specimens, primarily in macrophage-rich regions underlying the core of the plaque. RANTES mRNA and protein expression by infiltrating monocytes and lymphocytes, and by myofibroblasts and endothelial cells in coronary arteries undergoing the accelerated atherosclerosis, has been described in cardiac transplants (Pattison et al., 1996). Thus RANTES and the other chemokines may play an important role in the pathogenesis of atherosclerosis.

Rheumatoid arthritis is another chronic inflammatory state characterized by the accumulation of mononuclear cells. Rathanaswami and colleagues (1993) found very high levels of RANTES mRNA and protein induced by IL-1 and TNF in rheumatoid synovial fibroblasts in culture. Snowden used reverse transcriptase-polymerase chain reaction to detect RANTES mRNA and mRNA for the CCR1 chemokine receptor in synovial tissue from patients with rheumatoid arthritis (Snowden et al., 1994). RANTES mRNA was detected in 4 of 7 rheumatoid synovial tissue samples, and the receptor was detected in all 7 biopsies.

Lymph nodes from patients with tuberculosis or sarcoidosis contain granulomas consisting mainly of memory T lymphocytes and macrophages. Devergne and colleagues demonstrated RANTES mRNA and protein expression by macrophages and endothelial cells in 15 such lymph nodes, but not in lymph nodes with reactive hyperplasia (demonstrating a predominant B lymphocyte response).

Endometriosis is associated with increased numbers of peritoneal macrophages. Khorram, using a RANTES ELISA, showed that RANTES levels were elevated in pelvic fluid collected from patients with moderate-severe endometriosis (average 29.1 ng/ml) and that the levels correlated with the severity of the disease (Khorram et al., 1993).

Since RANTES is an important eosinophil and basophil chemoattractant and causes mediator release, it is a good candidate as an intermediary in allergic disease. It is likely that many of the previously described histamine-releasing factors and eosinophil chemotactic factors of anaphylaxis are actually C-C chemokines. RANTES is found in the bronchoalveolar fluid of asthmatic individuals at levels higher than that of normals (Alam et al., 1993b). RANTES protein has also been detected in nasal polyp tissue (Beck et al., 1994).

Thus, although our knowledge of the expression of RANTES is still fragmentary, it seems that RANTES is produced in a wide variety of inflammatory and neoplastic conditions. It will be important to determine RANTES expression in other diseases, and to simultaneously correlate it with the expression of the other C-C chemokines and their receptors. Established anti-inflammatory treatments, such as glucocorticoids, may work, in part, by suppressing chemokine secretion (Mukaida et al., 1994). Novel agents such as anti-RANTES monoclonal antibodies, or RANTES receptor antagonists may be applicable to a variety of autoimmune and inflammatory disorders. Immunotherapy of certain forms of cancer might be improved by engineering tumor cells to produce RANTES. Finally, the control of RANTES expression may play an important role in new therapies for the treatment of AIDS.

10. References

Adams, D.H. and Shaw, S. (1994). Leukocyte–endothelial interactions and regulation of leukocyte migration. Lancet 343, 831.

Ahuja, S.K., Gao, J.L. and Murphy, P.M. (1994). Chemokine receptors and molecular mimicry. Immunol. Today 15, 281.

Alam, R., Stafford, S., Forsythe, P., et al. (1993a). RANTES is a chemotactic and activating factor for human eosinophils. J. Immunol. 150, 442.

Alam, R., York, J., Boyars, J., et al. (1993b). Detection and quantitation of RANTES and MIP-1a in bronchoalveolar lavage (BAL) fluid and their mRNA in lavage cells. J. Allergy Clin. Immunol. 93, 183. [Abstract]

Alkhatib, G., Combadiere, C., Broder, C.C., et al. (1996). CC CKR5: A RANTES, MIP-1alpha, MIP-1beta receptor as a fusion cofactor for macrophage-tropic HIV-1. Science 272, 1955.

Bacon, K.B., Flores-Romo, L., Aubry, J.P., Wells, T.N. and Power, C.A. (1994). Interleukin-8 and RANTES induce the adhesion of the human basophilic cell line KU-812 to human endothelial cell monolayers. Immunol. 82, 473.

Bacon, K.B., Premack, B.A., Gardner, P. and Schall, T.J. (1995). Activation of dual T cell signaling pathways by the chemokine RANTES. Science 269, 1727.

Baggiolini, M. and Dahinden, C.A. (1994). C-C chemokines in allergic inflammation. Immunol. Today 15, 27.

Beck, L.A., Schall, T.J., Beall, L.D., et al. (1994). Detection of the chemokine RANTES and activation of vascular endothelium in nasal polyps. J. Allergy Clin. Immunol. 93, 234. [Abstract]

Ben-Baruch, A., Xu, L., Young, P.R., Bengali, K., Oppenheim, J.J. and Wang, J.M. (1995). Monocyte chemotactic protein-3 (MCP3) interacts with multiple leukocyte receptors. C-C CKR1, a receptor for macrophage inflammatory protein-1 alpha/Rantes, is also a functional receptor for MCP3. J. Biol. Chem. 270, 22123.

Bischoff, S.C., Krieger, M., Brunner, T. and Dahinden, C.A. (1992). Monocyte chemotactic protein 1 is a potent activator of human basophils. J. Exp. Med. 175, 1271.

Bischoff, S.C., Krieger, M., Brunner, T., et al. (1993). RANTES and related chemokines activate human basophil granulocytes through different G protein-coupled receptors. Eur. J. Immunol. 23, 761.

Broxmeyer, H.E., Sherry, B., Cooper, S., et al. (1993). Comparative analysis of the human macrophage inflammatory protein family of cytokines (chemokines) on proliferation of human myeloid progenitor cells. Interacting effects involving suppression, synergistic suppression, and blocking of suppression. J. Immunol. 150, 3448.

Butcher, E.C. (1991). Leukocyte–endothelial cell recognition. Three (or more) steps to specificity and diversity. Cell 67, 1033.

Chaudhuri, A., Zbrzezna, V., Polyakova, J., Pogo, A.O., Hesselgesser, J. and Horuk, R. (1994). Expression of the Duffy antigen in K562 cells. Evidence that it is the human erythrocyte chemokine receptor. J. Biol. Chem. 269, 7835.

Chihara, J., Hayashi, N., Kakazu, T., Yamamoto, T., Kurachi, D. and Nakajima, S. (1994). RANTES augments radical oxygen products from eosinophils. Int. Arch. Allergy Immunol. 104, 52.

Clore, G.M. and Gronenborn, A.M. (1992). NMR and X-ray analysis of the three-dimensional structure of interleukin-8. Cytokines 4, 18.

Choe, H., Farzan, M., Sun, Y., et al. (1996). The β-chemokine receptors CCR3 and CCR5 facilitate infection by primary HIV-1 isolates. Cell 85, 1135.

Cocchi, F., DeVico, A.L., Garzino-Demo, A., Arya, S.K., Gallo, R.C., and Luzzo, P. (1995). Identification of RANTES, MIP-1alpha and MIP-1beta as the major HIV-suppressive factors produced by CD8+ T Cells. Science 270, 1560.

Combadiere, C., Ahuja, S.K. and Murphy, P.M. (1995a). Cloning and functional expression of a human eosinophil CC chemokine receptor. J. Biol. Chem. 270, 16491.

Combadiere, C., Ahuja, S.K. and Murphy, P.M. (1995b). Cloning, chromosomal localization, and RNA expression of a human beta chemokine receptor-like gene. DNA Cell. Biol. 14, 673.

Combadiere, C., Ahuja, S.K., Vandamme, J., Tiffany, H.L., Gao, J.L. and Murphy, P.M. (1995c). Monocyte chemoattractant protein-3 is a functional ligand for CC chemokine receptors 1 and 2B. J. Biol. Chem. 270, 29671.

Dahinden, C.A., Geiser, T., Brunner, T., et al. (1994). Monocyte chemotactic protein 3 is a most effective basophil- and eosinophil-activating chemokine. J. Exp. Med. 179, 51.

Danoff, T.M., Lalley, P.A., Chang, Y.S., Heeger, P.S. and Neilson, E.G. (1994). Cloning, genomic organization, and chromosomal localization of the Scya5 gene encoding the murine chemokine RANTES. J. Immunol. 152, 1182.

Daugherty, B.L., Siciliano, S.J., DeMartino, J.A., Malkowitz, L.,

Sirotina, A. and Springer, M.S. (1996). Cloning, expression, and characterization of the humam eosinophil eotaxin receptor. J. Exp. Med. 183, 2349.

Deng, H.K., Liu, R., Ellmeier, W., et al. (1996). Identification of a major co-receptor for primary isolates of HIV-1. Nature 381, 661.

Devergne, O., Marfaing-Koka, A., Schall, T.J., et al. (1994). Production of the RANTES chemokine in delayed-type hypersensitivity reactions: involvement of macrophages and endothelial cells. J. Exp. Med. 179, 1689.

Donlon, T.A., Krensky, A.M., Wallace, M.R., Collins, F.S., Lovett, M. and Clayberger, C. (1990). Localization of a human T-cell-specific gene, RANTES (D17S136E), to chromosome 17q11.2-q12. Genomics 6, 548.

Doranz, B.J., Rucker, J., Yi, Y., et al. (1996). A dual-tropic primary HIV-1 isolate that uses fusin and the β-chemokine receptors CKR-5, CKR-3, and CKR-2b as fusion cofactors. Cell 85, 1149.

Dragic, T., Litwin, V., Allaway, G.P., et al. (1996). HIV-1 entry into CD4(+) cells is mediated by the chemokine receptor CC-CKR-5. Nature 381, 667.

Ebisawa, M., Yamada, T., Bickel, C., Klunk, D. and Schleimer, R.P. (1994). Eosinophil transendothelial migration induced by cytokines. III. Effect of the chemokine RANTES. J. Immunol. 153, 2153.

Fairbrother, W.J. and Skelton, N.J. (1996). Three-dimensional structures of the chemokine family. In "Chemoattractant Ligands and Their Receptors" (ed. R. Horuk). CRC Press, Boca Raton, FL.

Gao, J.L., Kuhns, D.B., Tiffany, H.L., et al. (1993). Structure and functional expression of the human macrophage inflammatory protein 1 a/RANTES receptor. J. Exp. Med. 177, 421.

Gerard, C. and Gerard, N.P. (1994). The pro-inflammatory seven-transmembrane segment receptors of the leukocyte. Curr. Opin. Immunol. 6, 140.

Graham, G.J., and Pragnell, I.B. (1992). SCI/MIP-1 alpha: a potent stem cell inhibitor with potential roles in development. Dev. Biol. 151, 377.

Heeger, P., Wolf, G., Meyers, C., et al. (1992). Isolation and characterization of cDNA from renal tubular epithelium encoding murine Rantes. Kidney Int. 41, 220.

Horuk, R. and Peiper, S.C. (1996). The duffy antigen receptor for chemokines. In "Chemoattractant Ligands and Their Receptors" (ed. R. Horuk). CRC Press, Boca Raton, FL.

Horuk, R., Chitnis, C.F., Darbonne, W.C., et al. (1993). A receptor for the malarial parasite P.vivax. The erythrocyte chemokine receptor. Science 261, 1182.

Howard, O.M.Z., Benbaruch, A. and Oppenheim, J.J., (1996). Chemokines: progress toward identifying molecular targets for therapeutic agents. Trends Biotech. 14, 46.

Jose, P.J., Griffiths-Johnson, D.A., Collins, P.D., et al. (1994). Eotaxin: a potent eosinophil chemoattractant cytokine detected in a guinea pig model of allergic airways inflammation. J. Exp. Med. 179, 881.

Kameyoshi, Y., Dorschner, A., Mallet, A.I., Christophers, E. and Schroder, J.M. (1992). Cytokine RANTES released by thrombin-stimulated platelets is a potent attractant for human eosinophils. J. Exp. Med. 176, 587.

Kameyoshi, Y., Schroder, J.M., Christophers, E. and Yamamoto, S. (1994). Identification of the cytokine RANTES released from platelets as an eosinophil chemotactic factor. Int. Arch. Allergy and Immunol. 104, 49.

Kapp, A., Zeck-Kapp, G., Czech, W. and Schopf, E. (1994). The chemokine RANTES is more than a chemoattractant: characterization of its effect on human eosinophil oxidative metabolism and morphology in comparison with IL-5 and GM-CSF. J. Invest. Dermatol. 102, 906.

Kelvin, D.J., Michiel, D.F., Johnston, J.A., et al. (1993). Chemokines and serpentines: the molecular biology of chemokine receptors. J. Leukocyte Biol. 54, 604.

Kennedy, J., Kelner, G.S., Kleyensteuber, S., et al. (1995). Molecular cloning and functional characterization of human lymphotactin. J. Immunol. 155, 203.

Khorram, O., Taylor, R.N., Ryan, I.P., Schall, T.J. and Landers, D.V. (1993). Peritoneal fluid concentrations of the cytokine RANTES correlate with the severity of endometriosis. Am. J. Obstet. Gynecol. 169, 1545.

Kimmel, P.L., Bodi, I., Abraham, A. and Phillips, T.M. (1993). Increased renal tissue cytokines in human HIV nephropathy. J. Am. Soci. Nephrol. 4, 279. [Abstract]

Kitaura, M., Nakajima, T., Imai, T., et al. (1996). Molecular cloning of human eotaxin, an eosinophil-selective CC chemokine, and identification of a specific eosinophil eotaxin receptor, CC chemokine receptor 3. J. Biol. Chem. 271, 7725.

Kuna, P., Reddigari, S.R., Schall, T.J., Rucinski, D., Viksman, M.Y. and Kaplan, A.P. (1992). RANTES, a monocyte and T lymphocyte chemotactic cytokine releases histamine from human basophils. J. Immunol. 149, 636.

Levy, J. (1993). HIV pathogenesis and long-term survival. AIDS 7, 1401.

Lodi, P.J., Garrett, D.S., Kuszewski, J., et al. (1994). High-resolution solution structure of the b chemokine hMIP-1β by multidimensional NMR. Science 263, 1762.

Loetscher, P., Seitz, M., Clark-Lewis, I., Baggiolini, M. and Moser, B. (1996). Activation of NK cells by CC chemokines. Chemotaxis, Ca^{2+} mobilization, and enzyme release. J. Immunol. 156, 322.

Lukacs, N.W., Strieter, R.M., Chensue, S.W. and Kunkel, S.L. (1996). Activation and regulation of chemokines in allergic airway inflammation. J. Leukocyte Biol. 59, 13.

Maghazachi, A.A., al-Aoukaty, A. and Schall, T.J. (1994). C-C chemokines induce the chemotaxis of NK and IL-2-activated NK cells. Role for G proteins. J. Immunol. 153, 4969.

Marfaing-Koka, A., Devergne, O., Gorgone, G., et al. (1995). Regulation of the production of the RANTES chemokine by endothelial cells. Synergistic induction by IFN-gamma plus TNF-alpha and inhibition by IL-4 and IL-13. J. Immunol. 154, 1870.

McColl, S.R. and Neote, K. (1996). Chemotactic factor receptors and signal transduction. In "Chemoattractant Ligands and Their Receptors" (ed. R. Horuk). CRC Press, Boca Raton, FL.

Meurer, R., Van Riper, G., Feeney, W., et al. (1993). Formation of eosinophilic and monocytic intradermal inflammatory sites in the dog by injection of human RANTES but not human monocyte chemoattractant protein 1, human macrophage inflammatory protein 1 alpha, or human interleukin 8. J. Exp. Med. 178, 1913.

Miller, M.D. and Krangel, M.S. (1992). Biology and biochemistry of the chemokines: a family of chemotactic and inflammatory cytokines. Crit. Rev. Immunol. 12, 17.

Mukaida, N., Morito, M. and Ishikara, Y. (1994). Novel mechanism of glucocorticoid-mediated gene repression. J. Biol. Chem. 269, 3289.

Murphy, P.M. (1994). The molecular biology of leukocyte chemoattractant receptors. Annu. Rev. Immunol. 12, 593.

Murphy, P.M., Ahuja, S.K. and Goa, J.L. (1966). Viral chemokine receptors. In "Chemoattractant Ligands and Their Receptors" (ed. R. Horuk). CRC Press, Boca Raton, FL.

Nelson, P.J., Kim, H.T., Manning, W.C., Goralski, T.J. and Krensky, A.M. (1993). Genomic organization and transcriptional regulation of the RANTES chemokine gene. J. Immunol. 151, 2601.

Nelson, P.J., Ortiz, B.D., Pattison, J.M. and Krensky, A.M. (1996). Identification of a novel regulatory sequence which is critical for expression of the RANTES chemokine in T lymphocytes. J. Immunol. 157, 1139.

Neote, K. and McColl, S.R. (1996). C-C chemokine receptors. In "Chemoattractant Ligands and Their Receptors" (ed. R. Horuk). CRC Press, Boca Raton, FL.

Neote, K., DiGregorio, D., Mak, J.Y., Horuk, R. and Schall, T.J. (1993a). Molecular cloning, functional expression, and signaling characteristics of a C-C chemokine receptor. Cell 72, 415.

Neote, K., Darbonne, W., Ogez, J., Horuk, R. and Schall, T.J. (1993b). Identification of a promiscuous inflammatory peptide receptor on the surface of red blood cells. J. Biol. Chem. 268, 12247.

Neote, K., Mak, J.Y., Kolakowski, L.F., Jr and Schall, T.J. (1994). Functional and biochemical analysis of the cloned Duffy antigen: identity with the red blood cell chemokine receptor. Blood 84, 44.

Noso, N., Proost, P., Van Damme, J. and Schroder, J.M. (1994). Human monocyte chemotactic proteins-2 and 3 (MCP-2 and MCP-3) attract human eosinophils and desensitize the chemotactic responses towards RANTES. Biochem. Biophys. Res. Commun. 200, 1470.

Oppenheim, J.J., Zachariae, C.O., Mukaida, N. and Matsushima, K. (1991). Properties of the novel proinflammatory supergene "intercrine" cytokine family. Annu. Rev. Immunol. 9, 617.

Ortiz, B.D., Nelson, P.J. and Krensky, A.M. (1995). The RANTES gene and the regulation of its expression. In "Biology of the Chemokine RANTES" (ed. A.M. Krensky) (Molecular Biology Intelligence Unit). R.G. Landes and Springer-Verlag.

Ortiz, B.D., Krensky, A.M. and Nelson, P.J. (1996). Kinetics of transcription factors regulating the RANTES chemokine gene reval a developmental switch in nuclear events during T lymphocyte maturation. Mol. Cell. Biol. 16, 202.

Pattison, J., Nelson, P.J., Huie, P., et al. (1994). RANTES chemokine expression in cell-mediated transplant rejection of the kidney. Lancet 343, 209.

Pattison, J.M., Nelson, P.J., Huie, P., Sibley, R.K. and Krensky, A.M. (1996). RANTES chemokine expression in accelerated atherosclerosis following cardiac transplantation. J. Heart Lung Transplant. 15, 1194.

Power, C.A., Meyer, A., Nemeth, K., et al. (1995). Molecular cloning and functional expression of a novel CC chemokine receptor cDNA from a human basophilic cell line. J. Biol. Chem. 270, 19495.

Raport, C.J., Gosling, J., Schweickart, V.L., Gray, P.W. and Charo, I.F. (1996). Molecular cloning and functional characterization of a novel human C-C chemokine receptor (CCR5) for RANTES, MIP-1β and MIP-1α. J. Biol. Chem. 271, 17161.

Rathanaswami, P., Hachicha, M., Sadick, M., Schall, T.J. and McColl, S.R. (1993). Expression of the cytokine RANTES in human rheumatoid synovial fibroblasts. Differential regulation of RANTES and interleukin-8 genes by inflammatory cytokines. J. Biol. Chem. 268, 5834.

Rot, A. (1993). Neutrophil attractant/activation protein-1 (interleukin-8) induces *in vitro* neutrophil migration by a haptotactic mechanism. Eur. J. Immunol. 23, 303.

Rot, A. and Hub, E. (1995). Leukocyte migration to RANTES. In "Biology of the Chemokine RANTES" (ed. A.M. Krensky) (Molecular Biology Intelligence Unit). R.G. Landes, Springer-Verlag.

Rot, A., Krieger, M., Brunner, T., Bischoff, S.C., Schall, T.J. and Dahinden, C.A. (1992). RANTES and macrophage inflammatory protein 1 alpha induce the migration and activation of normal human eosinophil granulocytes. J. Exp. Med. 176, 489.

Roth, S.J., Carr, M.W. and Springer, T.A. (1995). C-C chemokines, but not the C-X-C chemokines interleukin-8 and interferon-gamma inducible protein-10, stimulate transendothelial chemotaxis of T lymphocytes. Eur. J. Immunol. 25, 3482.

Samson, M., Labbe, O., Mollereau, C., Vassart, G. and Parmentier, M. (1996). Molecular cloning and functional expression of a new human C-C chemokine receptor gene Biochemistry 35, 3362.

Schall, T.J. (1991). Biology of the RANTES/SIS cytokine family. Cytokine 3, 165.

Schall, T.J., Jongstra, J., Bradley, B.J., *et al.* (1988). A human T cell-specific molecule is a member of a new gene family. J. Immunol. 141, 1018.

Schall, T.J., Bacon, K., Toy, K.J. and Goeddel, D.V. (1990). Selective attraction of monocytes and T lymphocytes of the memory phenotype by cytokine RANTES. Nature 347, 669.

Schall, T.J., Simpson, N.J. and Mak, J.Y. (1992). Molecular cloning and expression of the murine RANTES cytokine: structural and functional conservation between mouse and man. Eur. J. Immunol. 22, 1477.

Schall, T.J., Bacon, K., Camp, R.D., Kaspari, J.W. and Goeddel, D.V. (1993). Human macrophage inflammatory protein alpha (MIP-1 alpha) and MIP-1 beta chemokines attract distinct populations of lymphocytes. J. Exp. Med. 177, 1821.

Schweitzer, R.C., Welmers, B.A., Raaijmakers, J.A., Zanen, P., Lammers, J.W. and Koenderman, L. (1994). RANTES- and interleukin-8-induced responses in normal human eosinophils: effects of priming with interleukin-5. Blood 83, 3697.

Sherry, B. and Cerami, A. (1991). Small cytokine superfamily. Curr. Opin. Immunol. 3, 56.

Skelton, N.J., Aspiras, F., Ogez, J. and Schall, T.J. (1995). Proton NMR assignments and solution conformation of RANTES, a chemokine of the C-C type. Biochemistry 34, 5329.

Snowden, N., Hajeer, A. and Thomson, W. (1994). RANTES role in rheumatoid arthritis. Lancet 343, 547.

Sozzani, S., Molino, M., Locati, M., *et al.* (1993). Receptor-activated calcium influx in human monocytes exposed to monocyte chemotactic protein-1 and related cytokines. J. Immunol. 150, 1544.

Sozzani, S., Rieppi, M., Locati, M., *et al.* (1994). Synergism between platelet activating factor and C-C chemokines for arachidonate release in human monocytes. Biochem. Biophys. Res. Commun. 199, 761.

Springer, T.A., (1994). Traffic signals for lymphocyte recirculation and leukocyte emigration: the multistep paradigm. Cell 76, 301.

Stellato, C., Beck, L.A., Gorgone, G.A., *et al.* (1995). Expression of the chemokine RANTES by a human bronchial epithelial cell line. Modulation by cytokines and glucocorticoids. J. Immunol. 155, 410.

Tanaka, Y., Adams, D.H., Hubschen, S., Herano, H., Siebenlist, U. and Shaw, S. (1993). T cell-adhesion induced by proteoglycan-immobilized cytokine MIP-1b. Nature 361, 79.

Taub, D.D., Conlon, K., Lloyd, A.R., Oppenheim, J.J. and Kelvin, D.J. (1993). Preferential migration of activated CD4[+] and CD8[+] T cells in response to MIP-1 alpha and MIP-1 beta. Science 260, 355.

Taub, D.D., Sayers, T.J., Carter, C.R. and Ortaldo, J.R. (1995). Alpha and beta chemokines induce NK cell migration and enhance NK-mediated cytolysis. J. Immunol. 155, 3877.

Taub, D.D., Ortaldo, J.R., Turcovski-Corrales, S.M., Key, M.L., Longo, D.L. and Murphy, W.J. (1996). Beta chemokines costimulate lymphocyte cytolysis, proliferation, and lymphokine production. J. Leukocyte Biol. 59, 81.

Turner, L., Ward, S.G. and Westwick, J. (1995). RANTES-activated human T lymphocytes. A role for phosphoinositide 3-kinase. J. Immunol. 155, 2437.

Vaddi, K. and Newton, R.C. (1994). Regulation of monocyte integrin expression by beta-family chemokines. J. Immunol. 153, 4721.

van Riper, G., Siciliano, S., Fischer, P.A., Meurer, R., Springer, M.S. and Rosen, H. (1993). Characterization and species distribution of high affinity GTP-coupled receptors for human rantes and monocyte chemoattractant protein 1. J. Exp. Med. 177, 851.

van Riper, G., Nicholson, D.W., Scheid, M.P., Fischer, P.A., Springer, M.S. and Rosen, H. (1994). Induction, characterization, and functional coupling of the high affinity chemokine receptor for RANTES and macrophage inflammatory protein-1 alpha upon differentiation of an eosinophilic HL-60 cell line. J. Immunol. 152, 1055.

von Luettichau, I., Nelson, P.J., Vandereijin, M., *et al.* (1996). RANTES chemokine expression in normal and diseased tissue. Cytokine 8, 89.

Wang, J.H., Devalia, J.L., Xia, C., Sapsford, R.J. and Davies, R.J. (1996). Expression of RANTES by human bronchial epithelial cells *in vitro* and *in vivo* and the effect of corticosteroids. Am. J. Respir. Cell. Mol. Biol. 14, 27.

Wang, J.M., McVicar, D.W., Oppenheim, J.J. and Kelvin, D.J. (1993). Identification of RANTES receptors on human monocytic cells: competition for binding and desensitization by homologous chemotactic cytokines. J. Exp. Med. 177, 699.

Wiedermann, C.J., Kowald., E.N., Reinish, N., *et al.* (1993). Monocyte haptotaxis induced by the RANTES chemokine. Curr. Biol. 3, 735.

Witt, D.P. and Lander, A.D. (1994). Differential binding of chemokines to glycosaminoglycan subpopulations. Curr. Biol. 4, 394.

Wolf, G., Aberle, S., Thaiss, F., *et al.* (1993). TNF alpha induces expression of the chemoattractant cytokine RANTES in cultured mouse mesangial cells. Kidney Int. 44, 795.

Wu, D., La Rosa, G.J. and Simon, M. (1993). G-Protein coupled signal transduction pathways for interleukin-8. Science 261, 101.

Xia, M., Leppart, D., Hauser, S.L., *et al.* (1996). Stimulus-specificity of matrix metalloproteinase-dependence of human T cell migration through a model basement membrane. J. Immunol. 156, 160.

30. NAP-2/ENA-78

Alfred Walz

1. Introduction

NAP-2 and ENA-78 are members of a new class of chemotactic cytokines sharing structural features and biological activities with interleukin-8 (IL-8). This class of cytokines, called C-X-C chemokines, is composed of platelet factor 4 (PF4) (Poncz et al., 1987); platelet basic protein (PBP) (Paul et al., 1980; Holt et al., 1986) and its cleavage products connective-tissue activating peptide-III (CTAP-III) (Niewiarowski et al., 1980), β-thromboglobulin (β-TG) (Begg et al., 1978) and neutrophil-activating peptide 2 (NAP-2) (Walz and Baggiolini, 1989, 1990); interleukin-8 (IL-8) (Schröder et al., 1987; Walz et al., 1987; Yoshimura et al., 1987); GROα,β,γ (Anisowicz et al., 1987; Haskill et al., 1990); ENA-78 (Walz et al., 1991a); GCP-2 (Proost et al.,

1993); SDF-1 (Nagasawa, 1994); and the two interferon (IFN)-γ induced proteins γIP-10 (Luster et al., 1985) and MIG (Farber, 1993). Some of these proteins were discovered on the basis of their neutrophil-stimulating activity; IL-8, GROα,β,γ, NAP-2, GCP-2, and ENA-78 have similar profiles of biological activities on human neutrophils (Peveri et al., 1988; Moser et al., 1990; Walz et al., 1991a). The other members of the family were shown to have no effects on neutrophils (Walz et al., 1989; Dewald et al., 1992). An essential structural element for neutrophil activation is a Glu-Leu-Arg (ELR) motif in the 5′-structure of the protein (Clark-Lewis et al., 1991b; Hébert et al., 1991). PF4, γIP-10, and MIG, which are devoid of the ELR sequence, do not bind to the IL-8 receptors on neutrophils and do not induce neutrophil responses such as chemotaxis, enzyme release, respiratory burst and $[Ca^{2+}]_i$ changes. Introduction of the ELR motif into PF4 converted this factor into a neutrophil-activating protein, thus illustrating this motif's importance (Clark-Lewis *et al.*, 1993). An exception to the ELR rule is SDF-1 which has an SYR instead of the ELR sequence and was shown to be active on neutrophils and to induce $[Ca^{2+}]_i$ changes and chemotaxis (Oberlin *et al.*, 1996). Whereas neutrophil chemotaxis and activation functions have been clearly established for the C-X-C chemokines, possible effects on other types of cells are still unclear and will have to be investigated.

NAP-2 and ENA-78 have very similar biological functions and an extensive structural identity (53%). However, they differ markedly in their cellular origin as well as in their mode of synthesis and release. NAP-2 is produced by proteolytic cleavage from a precursor protein (PBP or CTAP-III) contained in the α-granules of platelets. ENA-78, on the other hand, is induced and secreted preferentially from epithelial cells. This suggests that the two molecules, in spite of their similar neutrophil-activating functions, play different roles in inflammatory pathophysiology due to compartmentalization of their production. Therefore, NAP-2 might play a role in processes related to the vascular system, whereas ENA-78 might be more involved in inflammatory lung and gut diseases.

2. The Cytokine Gene

2.1 NAP-2

NAP-2 is a proteolytic cleavage product of platelet basic protein (PBP) (Walz and Baggiolini, 1990). The cDNA for PBP has been cloned from a λgt11 expression library which was prepared from human platelet mRNA (Wenger *et al.*, 1989). The 690 base pairs (bp) long cDNA codes for a peptide with 128 amino acids; 94 amino acids are identical to the protein sequence reported for PBP (Holt *et al.*, 1986). Both the 34-amino-acid leader sequence as well as the coding region for PBP show significant homology to the sequence of platelet factor 4 (PF4) (Poncz *et al.*, 1987), another protein contained in the α-granules of platelets. Interestingly, there is no homology between the leader sequences of PBP and IL-8. In contrast, the leader sequence of ENA-78 (Chang *et al.*, 1994; Corbett *et al.*, 1994) shares a leucine-rich stretch with PBP, PF4, and GROα in which up to 7 out of 9 adjacent amino acids are identical (Wenger *et al.*, 1989). The genomic structure of the human PBP has been determined using a clone from a DAMI cDNA library (Majumdar *et al.*, 1991) (Figure 30.1). The overall structure closely resembles that of the PF4 gene (Eisman *et al.*, 1990); it is 1139 bp long and contains two introns and three exons (Figure 30.2). The PBP gene differs from the genes of the other C-X-C chemokines such as IL-8, GROα, ENA-78, and γIP-10 by the fact that it lacks the intron which is located 4 or 5 amino acids from the 3′ end of the coding sequence. The gene for PBP was mapped to the long arm of chromosome 4 (q12-21) (Wenger *et al.*, 1991) together with the genes of the other C-X-C chemokines.

2.2 ENA-78

A partial cDNA for ENA-78 has been isolated by PCR using degenerate oligonucleotide primers and mRNA obtained from stimulated alveolar epithelial cells A549 (Walz *et al.*, 1991a). The structure of the 5′ end of the approximately 2.5 kb long mRNA has been determined using a 5′ RACE protocol (Chang *et al.*, 1994; Corbett *et al.*, 1994). The ENA-78 cDNA codes for a protein of 114 amino acids (aa) with a putative 31-aa leader sequence (cleavage at proline 31-glycine 32). This suggests that the 78-aa protein isolated from human alveolar epithelial type-II cells is proteolytically processed by 5 amino acids upon secretion (Corbett *et al.*, 1994). A full-length cDNA encoding ENA-78 has also been obtained from human platelet mRNA (Power *et al.*, 1994). The 105 bp 5′ flanking region of this cDNA is slightly shorter and has a different nucleotide sequence within the first 24 residues than the A549 cDNA. The genomic structure of the ENA-78 gene has been determined using clones obtained from a human chromosome 4 flow-sorted cosmid library (Chang *et al.*, 1994; Corbett *et al.*, 1994) (Figure 30.3). The ENA-78 gene resembles the genes for IL-8, GRO, and γIP-10: they all contain four exons and three introns (Figure 30.4). The 5′ flanking region comprises potential binding sites for several nuclear factors such as AP-2, NFκB, and interferon regulatory factor-1. It was suggested that the AP-2 site might play a role in cell-type specific expression of ENA-78 (Corbett *et al.*, 1994). Upon induction with IL-1 or TNF-α, ENA-78 is highly expressed in epithelial cells and to a lesser degree in other cell types such as monocytes/macrophages (Walz *et al.*, 1993b). IL-8, which does not have AP-2-like binding sites in the same upstream region, is produced by almost every cell type in the body. A fusion gene containing the first 125 bp of the 5′ flanking region transfected into human embryonic kidney cells was inducible with TNF-α, IL-1β, and phorbol myristate acetate, suggesting that the NFκB element is sufficient for ENA-78 induction in these cells (Chang *et al.*, 1994). IFN-α and IFN-γ have been demonstrated to effectively downregulate ENA-78 mRNA and protein release from human monocytes (Schnyder-Candrian *et al.*, 1995). Recent results demonstrate that LPS causes a delayed, but long-lasting upregulation of ENA-78 in human monocytes (Schnyder-Candrian and Walz, 1997).

```
TTTCATGTATACGTATTCATTGATTTATTAATTATTTAAGACCTACTTTATATTGAAGTA      60
TCCTACATGTTATAGGATACATGAGACATTCAAACAAACACAGGACACGCCACTCCCTAC     120
AAAAGTCATGGTTCAATTGAGGAAGGAAGTCTTACAGATGGGACAGAGAACCAGAAAATA     180
ACATAAAGCCAACATGTAATGTGTAGGTCCACCTCTCTATCTTTATATAGCATTTCAGAC     240
CAAAACCTTTAATTTGGGTCTGAGAATTCAGGAACAAATATTTTCCTTGGTAAGGACATT     300
TCTTGCACAATTTCTAATAGTGCAAAATTGGAAACACCGTGAAGGGGCACCAACAAGGCA     360
AGGTTGAGTAACTATTGTTCTTATCACAGGACGTCACATGGTCATTAAGAAGAATGGGGG     420
AAAACTACATTAAGATGACCTGAAAAGATAAGGAACACATATTTTTTGAAGGATAAAAGT     480
CAGCTGTATAGTGATCCATATCAATACTTCTATGTATATGTGTACCTCTATGTCTTAAAC     540
GTGTCTCATTGTATACATATACATGTGTGCATATAAGTGCAGACAGAATTGAAAACTAAT     600
GCAAAATTGTTGACCAGTGTTTACCACTGAATATGAGGAATGTGGAGAAATGATGAAGAG     660
AAATTTGTGTTTTATTTCAAAGACATCTGCAGCATACGAATCTTTTGCAATAAATTATAT     720
AATTTTTCTGTTAAATTAAAAATAAAAAGAAAAATAAATATGCTGGGTCCTACTTTTTAG     780
GTATTCTTAGGTGGTAGAAACAAGTAGCTTCTTTTGTAATGTAAAGGAGGATGAGTTTCA     840
TTTGTAGTTTCTAGATGAAATTACAAAAGATAATATAGATTGAAGGCAGAAAGAATCCA     900
ATAAGCTAAAATCCAAAGACAACCTTTGACGGTAATTGGCTCTTATTTTACTTACATGCT     960
CCAAACCAATCCCAATATTTATCAACATTCATGAAAAGAAGATTTGCTTTCTTTGGCAAG    1020
CAAAGATAACTTTCTCTTTCAAGGGCCACATGTTTGTGCCTTCAGGTGCTCCTAAAGTTT    1080
CTATTAAGCCTAAAAACTGCAGTATAGAAAAGGCAGGGCGTATTGTTATAAATCATACAA    1140
AAGAAAAATGCAATGTAGTATTTCAGTCTAGTTCTTACCTTCCTGAACGGAGTTCTTACA    1200
CAGGTGTAAGGAAGATAAGTATTGAGAAAGGGAGAGTGGGAATGTGAAGTGATGCACATT    1260
AAGCAAGTTAGTAGGAATTTGACCTGTCTGGTCTTTCTCTGGGTTGGGCACAGCTTCAAA    1320
TGCTTATGTGTGTATCACCACATACCCTCACTTCCTCCTTTCCTACCTCTTCCTTCTTAC    1380
TGGCTTTGAGAAACAGCATATAAATGACATCTTCAGGGCATGAGAAGCCACTTATCTGCA    1440
GACTTGTAGGCAGCAACTCACCCTCACTCAGAGGTCTTCTGGTTCTGGAAACAACTCTAG    1500
CTCAGCCTTCTCCACCATGAGCCTCAGACTTGATACCACCCCTTCCTGTAACAGTGCGAG    1560
ACCACTTCATGCCTTGCAGGTGCTGCTGCTTCTGTCATTGCTGCTGACTGCTCTGGCTTC    1620
CTCCACCAAAGGACAAACTAAGAGAAACTTGGCGAAAGGCAAAGGTAGAGGCCCTGCTTC    1680
TCTGCACTTGTTGCTGCTTCTGCTACACCTGTCTCGGGGTAAATAGCATGCTTGGTGCCT    1740
TTGGGGCTGGAGAGGGCCATTATACCAATAACTCCAATTGGAGGAGACACACAGGGGGGT    1800
CACTTCTCACTTCTTGTGTGCTGGGCAATCTTCTGGGCACTTTACTAAAGCGTTACAGAT    1860
CATATTCACAATGGCTTTATGAGAGAGGTCAATTGCCCTCAATCTGCAAATAAGAGACCT    1920
GAGGAAAATATTCATGACCACCAATAGGTCACATTTTCTACCCTAGAGGAAAGTCTAGAC    1980
AGTGACTTGTATGCTGAACTCCGCTGCATGTGTATAAAGACAACCTCTGGAATTCATCCC    2040
AAAAACATCCAAAGTTTGCAAGTGATCGGGAAAGGAACCCATTGCAACCAAGTCGAAGTG    2100
ATGTAAGTTGCTGTTTCTGTGCTATTGCCTTATCAGAGAAACCCTCTACCTCCATCCACA    2160
TATGCACTCGTTTTCCTCCAGTCTCATGGATGGATTAGTTCTGATATTCAGATCAGGACA    2220
CCCACAGATAACCCTGTTCTCTTTTGCAGAGCCACACTGAAGGATGGGAGGAAAATCTGC    2280
CTGGACCCAGATGCTCCCAGAATCAAGAAAATTGTACAGAAAAAATTGGCAGGTGATGAA    2340
TCTGCTGATTAATTTGTTCTGTTTCTGCCAAACTTCTTTAACTCCCAGGAAGGGTAGAAT    2400
TTTGAAATTGATTTTCTAGAGTTCTCATTTATTCAGGATACCTATTCTTCTGTATTAAAT    2460
TTGGATATGTGTTTCTTCTGTCTCAAAAATCACATTTTATTCTGAGAAGGTTGGTTAAAA    2520
GATGGCAGAAAGAAGATGAAAATAAAATAAGCCTGGTTTCAACCCTCTAATTCTTGCCTAA    2580
ACATTGGACTGTACTTTGCATTTTTTTTCTTTAAAAATTTCTATTCTAACCACAACTTGGT    2640
TGATTTTTCCTGGTCTACTTTATGGTTATTAGACATACTCATGGGTATTATTAGATTTCA    2700
TAATGGTCAATGATAATAGGAATTACATGGAGCCCAACAGAGAATATTTGCTCAATACAT    2760
TTTTGTTAATATATTTAGGAACCTTAATGGAGTCTCTCAGTGTCTTAGCCTAGGATGTCT    2820
TATTTTAAAA                                                    2830
```

Figure 30.1 Gene sequence of the PBP/NAP-2 gene. The "TATA" sequence and the poly(A) signal sequence are underlined. The position of the amino terminal of the mature PBP protein is boxed. The cleavage point of the signal peptide is identical with the amino terminus of PBP. Putative sites of initiation of transcription and poly(A) are asterisked.

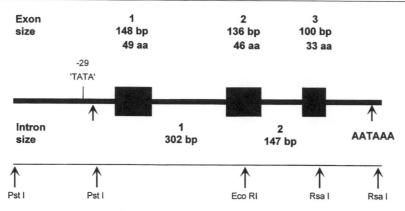

Figure 30.2 Gene structure of the PBP/NAP-2 gene. The three exons are indicated in the figure by black boxes. Amino acid residues encoded by each individual exon are indicated above the exon. The sizes of the introns are indicated below the scheme. Arrows indicate transcription initiation and poly(A) signal sequences. Relative positions of restriction sites are indicated.

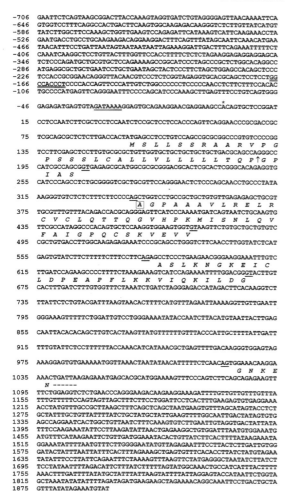

Figure 30.3 Gene sequence of the ENA-78 gene. The "TATA" and the "CAT" sequence are underlined. The position of the amino terminal of the mature ENA-78 protein is boxed. The cleavage point of the signal peptide is indicated by an arrow. Transcription initiation site determined by a 5′-RACE protocol is asterisked.

3. The Protein

3.1 NAP-2

NAP-2 was discovered and isolated on the basis of its neutrophil-activating properties (Walz and Baggiolini, 1989). Amino- and carboxy-terminal sequencing identified NAP-2 as a cleavage product of PBP (Paul *et al.*, 1980; Holt *et al.*, 1986) and its derivatives connective-tissue activating peptide III (CTAP-III; also called low-affinity platelet-activating factor 4) (Niewiarowski *et al.*, 1980), and β-thromboglobulin (β-TG) (Begg *et al.*, 1978) (Figure 30.5). They all share an identical carboxy terminus, but differ at the amino-terminus by +24 (PBP), +15 (CTAP-III), and +11 residues (β-TG) in respect to NAP-2. While PBP, CTAP-III, and β-TG may have other biological functions, only NAP-2 has neutrophil-stimulating activities (Walz *et al.*, 1989). NAP-2 consists of 70 amino acids and has a calculated molecular mass of 7628 Da (Figure 30.6). Owing to its basic isoelectric point of about 8.7 it can be readily purified by cation exchange and reversed-phase chromatography. Similar to IL-8, NAP-2 is a fairly stable molecule, resisting harsh conditions such as pH 2.0, elevated temperature, and treatment with a number of proteases such as chymotrypsin and cathepsin G (Peveri *et al.*, 1988; Padrines *et al.*, 1994). Furthermore, NAP-2 was shown to be stable in fresh human serum for more than 2 h at 37°C (Walz *et al.*, 1991c). Under the same conditions the classical receptor agonist C5a is rapidly inactivated. Similarly, other chemoattractants such as fMet-Leu-Phe and the lipids leukotriene B$_4$ and platelet-activating factor are rapidly inactivated by oxidation or hydrolysis (Colditz *et al.*, 1990). In contrast to IL-8, NAP-2 is significantly more susceptible to degradation with trypsin and proteinase-3 (Padrines *et al.*, 1994). In common with all the other members of the C-X-C chemokine family, NAP-2 contains four conserved

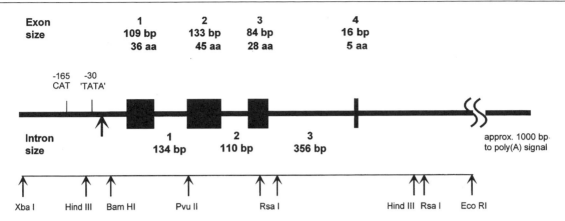

Figure 30.4   Gene structure of the ENA-78 gene. Boxes represent exons. Amino acid residues encoded by each individual exon are indicated above the exon. The sizes of the introns are indicated below the scheme. Arrow indicates transcription initiation site. Relative positions of restriction sites are indicated.

PBP	SSTKGQTKRNLAKGKEESLDSDLYAELRCMCIKTTSGIHPKNIQSLEVIGKGTHCNQVEVIATLKDGRKICLDPDAPRIKKIVQKKLAGDESAD
CTAP-III	NLAKGKEESLDSDLYAELRCMCIKTTSGIHPKNIQSLEVIGKGTHCNQVEVIATLKDGRKICLDPDAPRIKKIVQKKLAGDESAD
β-TG	GKEESLDSDLYAELRCMCIKTTSGIHPKNIQSLEVIGKGTHCNQVEVIATLKDGRKICLDPDAPRIKKIVQKKLAGDESAD
NAP-2	AELRCMCIKTTSGIHPKNIQSLEVIGKGTHCNQVEVIATLKDGRKICLDPDAPRIKKIVQKKLAGDESAD
ENA-78	AGPAAAVLRELRCVCLQTTQGVHPKMISNLQVFAIGPQCSKVEVVASLKNGKEICLDPEAPFLKKVIQKILDGGNKEN
boENA	VVRELRCVCLTTTPGIHPKTVSDLQVIAAGPQCSKVEVIATLKNGREVCLDPEAPLIKKVIQKILDSGKN

Figure 30.5   Amino acid sequences of PBP and its cleavage products CTAP-III, β-TG, NAP-2; ENA-78, and boENA. Sequences were aligned according to their conserved cysteine residues (bold). Abbreviations: PBP, platelet basic protein; CTAP-III, connective-tissue activating protein III; β-TG, β-thromboglobulin; NAP-2, neutrophil-activating protein 2; ENA-78, epithelial-cell derived neutrophil-activating protein 78; boENA, bovine ENA.

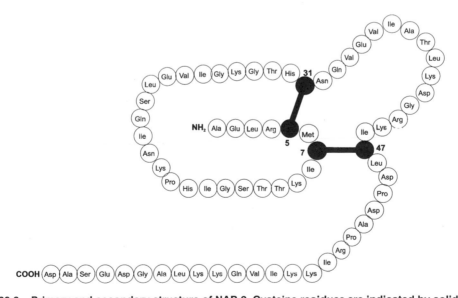

Figure 30.6   Primary and secondary structure of NAP-2. Cysteine residues are indicated by solid circles.

cysteine residues, the first two spaced by one amino acid. These residues form two characteristic intramolecular disulfide bonds linking Cys^5–Cys^{31} and Cys^7–Cys^{47} (Begg *et al.*, 1978). The activity is lost upon reduction of the disulfide bridges. NAP-2 contains potential sites for protein kinase C, casein kinase-II phosphorylation (Thr^{39}), N-myristoylation (Gly^{28}) and amidation (Asp^{42}), but no sites for N-linked glycosylation (Walz and Baggiolini, 1989). The fact that synthetically produced NAP-2 is fully active suggests that such modifications are not required for neutrophil-stimulating activity (Clark-Lewis *et al.*, 1991a). Whereas PF4 was shown to preferentially form tetramers (Moore *et al.*, 1975; Mayo and Chen, 1989), CTAP-III monomers were shown to be favored under physiological conditions (Mayo, 1991), suggesting that the active form of NAP-2 is a monomer also. The discrepancy in molecular nature of PF4 and CTAP-III may account for the difference in binding strength to heparin, CTAP-III eluting from heparin–agarose at about 0.35 M NaCl and PF4 eluting at 1.4 M NaCl (Holt *et al.*, 1986). Recombinant NAP-2 has been crystallized and the crystal structure has been resolved at 1.9 Å resolution (Kungl *et al.*, 1994; Malkowski *et al.*, 1995). The overall tertiary structure was found to be similar to PF4 and IL-8. It includes an extended amino-terminal loop, three strands of antiparallel β-sheets arranged in a Greek key fold, and one α-helix at the carboxy-terminus. A similar structure for monomeric NAP-2 was also determined in solution using ^{1}H nuclear magnetic resonance spectroscopy (Mayo *et al.*, 1994).

3.2   ENA-78

ENA-78 was isolated from human alveolar type-II epithelial cell line A549 (Walz *et al.*, 1991a) and consists of 78 amino acids with a molecular mass of 8357 Da (Figure 30.7). No *N*-linked glycosylation sites were detected, but ENA-78 contains a putative phosphorylation site for protein kinase C and casein kinase-II. ENA-78 is a typical C-X-C chemokine and has its highest identity in amino acid sequence to GCP-2 (77%) (Proost *et al.*, 1993), NAP-2 (53%) (Walz and Baggiolini, 1989), and GROα (52%) (Haskill *et al.*, 1990). However, its identity to IL-8 is rather low (22%) (Walz *et al.*, 1987). Incubation of recombinant ENA-78 with cathepsin G and chymotrypsin *in vitro* yielded a transient increase in biological activity predicting amino-terminal variants with higher potency (Walz *et al.*, 1993b). A similar proteolysis might occur *in vivo* in the presence of

neutrophils or monocytes. Bovine ENA isolated from alveolar macrophages and blood monocytes contains only 5 amino acids preceding the first cysteine residue and thus represents such a proteolytically processed form of ENA-78 in a different species (Allmann-Iselin *et al.*, 1994) (Figure 30.5).

4.   Cellular Sources and Production

4.1   NAP-2

NAP-2 was originally isolated from supernatants of stimulated mononuclear cells (Walz and Baggiolini, 1989). Subsequently it was demonstrated that NAP-2 was exclusively formed by proteolytic cleavage from PBP or CTAP-III, which are stored in the α-granules of platelets and are released into the bloodstream upon platelet activation (Holt *et al.*, 1986; Walz and Baggiolini, 1990; Car *et al.*, 1991). The discovery of NAP-2 was therefore based on platelet contamination in mononuclear cell preparations. Indeed, platelets tend to adhere to monocytes when buffy coats are prepared (Pawlowski *et al.*, 1983). Highly purified monocytes, platelets, or thrombin-induced platelet release supernatant (PRS) by itself did not yield any significant amounts of NAP-2 (Walz and Baggiolini, 1990; Walz *et al.*, 1990). Upon mixing of unstimulated monocytes and platelets or PRS, NAP-2 (70-amino-acid form) was produced. Lipopolysaccharide (LPS)-stimulated monocytes and platelets or PRS yielded, in addition to NAP-2 (70 amino acids), three other variants of NAP-2 of length 75, 74, and 73 amino acids, possibly owing to induction of additional proteases by monocytes (Walz and Baggiolini, 1990; Walz *et al.*, 1990). *In vitro* proteolysis of CTAP-III or β-thromboglobulin with chymotrypsin and cathepsin G yielded active NAP-2 (Car *et al.*, 1991; Cohen *et al.*, 1992) (Figure 30.8). Upon incubation with highly purified CTAP-III, cathepsin G, which is known to

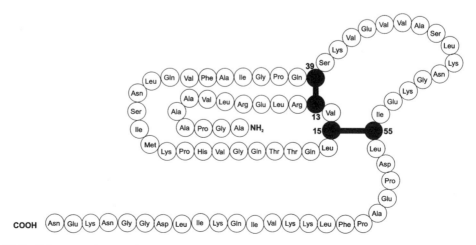

Figure 30.7   Primary and secondary structure of ENA-78. Cysteine residues are indicated by solid circles.

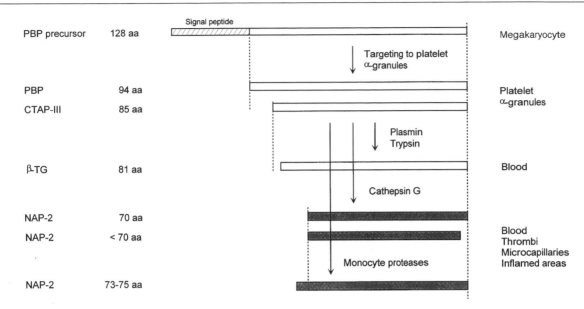

Figure 30.8   Processing of PBP precursor to β-TG and NAP-2 isoforms. Empty bars indicate inactive forms and shaded bars indicate proteins with neutrophil-stimulating activity.

be released from monocytes and degranulating neutrophils, yielded two active peaks with markedly different HPLC mobility but identical NH$_2$-terminal sequence (Car et al., 1991). This suggested that in addition to the formation of NAP-2 (70 amino acids) an additional variant with processed C-terminal end was produced. Subsequently, such a C-terminal truncated form of NAP-2 was detected and characterized in supernatants of mononuclear cells (Brandt et al., 1993). C-terminal truncation of NAP-2 has revealed that the 66-aa molecule has a 3-fold higher potency than the native 70-aa protein (Ehlert et al., 1995). Cathepsin G, which binds to platelets and activates the release of granule content, is capable of producing NAP-2 (De Gaetano et al., 1990; Selak and Smith, 1990; Cohen et al., 1992). The interaction of neutrophils or monocytes with platelets appears to be an essential prerequisite for NAP-2 production, whereas activated lymphocytes were apparently unable to convert CTAP-III to neutrophil-activating peptide 2 (Walz and Baggiolini, 1990).

In contrast to human platelets, porcine platelets contain two active forms of NAP-2, a 73-aa form and a variant with seven additional amino acids at the NH$_2$-terminus (Yan et al., 1993). The short form was shown to have comparable activity to human NAP-2, whereas the longer form had markedly reduced activity. Porcine NAP-2 is highly homologous to human NAP-2 (68% identity).

4.2   ENA-78

ENA-78 was purified from supernatants of IL-1β or TNFα-stimulated human alveolar type-II epithelial cell line, A549 (Walz et al., 1991a). Using a synthetic 43-mer

oligonucleotide probe, the production of ENA-78-specific mRNA has been detected in IL-1β-stimulated alveolar type-II epithelial cells, monocytes, endothelial and mesothelial cells as well as in monocytes stimulated with lipopolysaccharide (Walz et al., 1993b). ENA-78 mRNA levels were low or not detectable in keratinocytes and in some fibroblast lines. Using a specific ELISA for ENA-78, protein release has been observed from IL-1β-stimulated alveolar type-II epithelial cells, from blood monocytes, from normal pulmonary fibroblasts but not from embryonic lung fibroblasts and keratinocytes, and from blood monocytes stimulated with IL-1, TNF, and LPS (Strieter et al., 1992a; Schnyder-Candrian et al., 1995; Walz et al., 1997). Primary cultures of renal cortal epithelial cells expressed significantly increased levels of ENA-78 mRNA and ENA-78 antigen in response to IL-1β stimulation. However, stimulation of these cells with TNF-α, IFN-γ, and LPS failed to induce ENA-78 or its mRNA (Schmouder et al., 1994). Release of low levels of ENA-78 was also observed from IL-1- and TNF-stimulated pulmonary smooth muscle and arterial endothelial cells (Lukacs et al., 1995). Overall, the production of ENA-78 differs markedly from IL-8 or GROα, which are readily induced with IL-1 or TNF-α in various fibroblast lines, alveolar macrophages, and keratinocytes (Schröder, 1992; Baggiolini et al., 1994).

N^G-monomethyl-L-arginine (NMA), an inhibitor of the nitric-oxide synthetase pathway, enhanced the production of ENA-78 and IL-8 from human peripheral blood monocytes in a mixed lymphocyte reaction by 5-fold and 2-fold, respectively (Orens et al., 1994). In contrast, the production of MCP-1 and MIP-1α was not significantly altered in the presence of NMA, suggesting that products of the L-arginine oxidative metabolism may negatively effect the synthesis of inducible C-X-C chemokines. At present, it is unclear whether this inhibitory effect is caused by nitric oxide, by other intermediate metabolites of L-arginine, or by second messengers.

Bovine ENA (boENA) was identified and purified from endotoxin-stimulated monocytes and alveolar macrophages (Allmann-Iselin et al., 1994). boENA is closely related to the human homolog and is identical to bovine granulocyte chemotactic protein-2 (GCP-2) isolated from bovine kidney tumor cells (Proost et al., 1993). Naming this factor boENA instead of bovine GCP-2 is more appropriate owing to the higher identity of boENA to human ENA-78 (73%) than to human GCP-2 (67%) and to its pattern of expression and biological activity, which closely resembles that of ENA-78. The immunohistochemical identification of boENA in the hyperplastic type-II alveolar epithelial cells and in pulmonary alveolar leukocytes of pneumonic bovine lungs strongly supports a role for ENA-78 in the genesis of pulmonary inflammation (Allmann-Iselin et al., 1994). Human ENA-78 was shown to have a stretch of 12 amino acids upstream from the first cysteine residue; cathepsin G is able to truncate this NH_2-terminus in vitro by 8 residues yielding a 70-aa peptide with about 5-fold higher potency (Walz et al., 1993b). boENA had a similarly truncated NH_2-terminus and only a single species with 70 amino acids was observed. Potency differences with truncated NH_2-terminal variants have been observed earlier with IL-8 produced by tissue cells such as epithelial cells (77 amino acids), and monocytes (72 amino acids) (Lindley et al., 1988; Walz et al., 1991a).

5. Biological Activities

5.1 ACTIVITIES ON NEUTROPHILS

NAP-2 was first observed as a minor peak with elastase-releasing activity on cytochalasin B-treated human neutrophils during HPLC purification of IL-8 (Walz and Baggiolini, 1989). In the elastase-release assay, highly purified as well as recombinant NAP-2 was active at concentrations ranging from 1 to 100 nM, whereas the NAP-2 precursors PBP and CTAP-III were practically inactive. IL-8 was consistently more effective than NAP-2 especially at concentrations >3 nM (Walz and Baggiolini, 1989; Walz et al., 1989). Similar concentration-dependent effects of NAP-2 were also observed for exocytosis of lactoferrin or elastase as markers for the activation of either specific or azurophilic granules (Van Damme et al., 1990; Petersen et al., 1991; Brandt et al., 1992; Holt et al., 1992). Pretreatment of neutrophils with TNF-α enhanced enzyme release by

NAP-2 and IL-8 from both types of granules (Brandt et al., 1992). NAP-2 induced neutrophil chemotaxis and changes of cytosolic free calcium in a concentration-dependent fashion (Walz et al., 1989; Clark-Lewis et al., 1991a). At equimolar concentrations, both NAP-2 and ENA-78 induced similar responses, with significant effects observed down to 10^{-10} M (Walz et al., 1991a). Again, PBP and CTAP-III were about 1000-fold less active, giving rise to a weak signal at the highest concentration tested (100 nM) (Walz et al., 1989). Similarly, PF4, which was reported to be chemotactic for neutrophils (Deuel et al., 1981; Bebawy et al., 1986) was found to be inactive at concentrations up to 10^{-7} M (Walz et al., 1989). NAP-2 is an inducer of the respiratory burst in human neutrophils and its activity is comparable to the activity of GROα (Moser et al., 1990; Walz et al., 1991b). Their efficacy, however, is low compared to that of IL-8 or other agonists such as fMet-Leu-Phe (FMLP) or C5a (Baggiolini and Dewald, 1984; Moser et al., 1990). IL-8, NAP-2, and GROα increased surface expression of CD11b/CD18 and CR1 on neutrophils, indicating that they cause the exocytosis of specific granules and change the adhesive interactions of neutrophils with their surroundings (Detmers et al., 1991). In agreement with these results, binding of CD11b/CD18 to its ligand C3bi was strongly enhanced by IL-8, GROα, and NAP-2, and changes in F-actin content were observed (Detmers et al., 1991). While CTAP-III does not bind specifically to the IL-8 receptors and does not activate neutrophils to release elastase, it was shown to decrease the activating effects of NAP-2 and IL-8 in a concentration-dependent manner (Haerter et al., 1994). Desensitization of neutrophils by CTAP-III was due to NAP-2 formed by proteolytic processing of CTAP-III. Neutrophils pretreated with 2 nM of NAP-2 also significantly downregulated a subsequent effect with 100 nM NAP-2, suggesting a modulatory mechanism that might be important under conditions of platelet activation to prevent an overshoot of neutrophil function.

ENA-78 was purified on the basis of its elastase-releasing activity on cytochalasin B-treated human neutrophils (Walz et al., 1991a). Highly purified natural as well as recombinant ENA-78 was active at concentrations ranging between 1 and 100 nM, as was the case for NAP-2, whereas IL-8 was more active. ENA-78 induced a concentration-dependent chemotaxis between 0.1 and 100 nM in vitro. This response was identical to chemotaxis induced by NAP-2, but somewhat weaker than observed with IL-8. The lower potency of ENA-78 with respect to IL-8 was also apparent by the measurements of stimulus-dependent changes in cytosolic free calcium (Walz et al., 1991a). Incubation of ENA-78 with cathepsin G yielded a NH_2-terminal truncated variant with about 5-fold higher activity (Walz et al., 1993b). ENA-78 upregulated cell surface expression of integrin Mac-1 on human neutrophils with an EC_{50} of 22 nM, while IL-8 was more potent (EC_{50} = 0.6 nM) (Bozic et al., 1996).

5.2   ACTIVITIES ON BASOPHILS AND EOSINOPHILS

Basophil leukocytes express functional IL-8 receptors as shown by binding studies and changes in $[Ca^{2+}]_i$ after IL-8 stimulation (Krieger et al., 1992). $[Ca^{2+}]_i$ changes were also observed with NAP-2 but not with CTAP-III and PF4 (Krieger et al., 1992). In unprimed basophils, IL-8 induced histamine release which was enhanced severalfold by pretreating the cells with IL-3. In contrast, NAP-2 only induced histamine release in IL-3 primed basophils but to a significantly smaller extent than IL-8 (Krieger et al., 1992). CTAP-III and PF4 were inactive. Other authors observed histamine release with NAP-2 as well as CTAP-III without IL-3 pretreatment (Baeza et al., 1990; Nourshargh et al., 1992; Kuna et al., 1993).

Contradictory results exist on the effects of IL-8 on $[Ca^{2+}]_i$ changes in eosinophils. Weak responses were observed in a study using eosinophils from patients with hypereosinophilic syndrome (Kernen et al., 1991). In other studies using cells from healthy volunteers, no or only an insignificant $[Ca^{2+}]_i$ change was detected with IL-8, NAP-2, or PF4 (Schweizer et al., 1994; Simon et al., 1994) and ENA-78 (Walz, unpublished results), suggesting that low neutrophil contamination in eosinophil preparations may have caused $[Ca^{2+}]_i$ fluxes. Eosinophils from normal individuals were shown to be chemotactically responsive to C5a, PAF, LTB_4, and some CC-chemokines but not to IL-8, PF4, NAP-2, and β-thromboglobulin. However, eosinophil chemotaxis with IL-8 was observed when cells were pretreated with IL-3, GM-CSF, or IL-5 (Warringa et al., 1991; Ebisawa et al., 1994; Schweizer et al., 1994). Eosinophil accumulation has also been observed in vivo, in response to intradermal injection of human IL-8 into guinea-pigs or dogs (Collins et al., 1993; Meurer et al., 1993).

5.3   ACTIVITIES ON ENDOTHELIAL CELLS

C-X-C chemokines containing the ELR motif, such as ENA-78, NAP-2, and IL-8, were demonstrated to promote endothelial cell proliferation and chemotaxis in vitro, and to cause neovascularization of rat and rabbit cornea in vivo (Koch et al., 1992b; Strieter et al., 1992b,c). In contrast, non-ELR C-X-C chemokines such as γIP-10, MIG, or PF4, behaved as angiostatic factors and inhibited in a concentration-dependent manner endothelial cell proliferation and chemotaxis induced by ENA-78 and IL-8 (Strieter et al., 1995c). Similar results were also obtained in a rat model of cornea neovascularization. Interestingly, γIP-10, MIG, and PF4

also inhibited the response of the unrelated angiogenic molecule basic fibroblast growth factor (Angiolillo et al., 1995; Strieter et al., 1995c; Luster et al., 1995).

IL-8 molecules having the ELR motif replaced by TVR (corresponding motif in γIP-10) or DLQ (corresponding motif in PF4) were no longer active in inducing endothelial cell chemotaxis; instead, these IL-8 variants inhibited the response of wild-type IL-8 by more than 80% (Strieter et al., 1995c). If, on the other hand, the ELR motif was introduced into MIG, it transformed from an angiostatic to an angiogenic protein. These results point to the importance of the ELR motif as an essential requisite for angiogenic activity of C-X-C chemokines. While the angiogenic activity may possibly be transmitted via an IL-8 receptor, the transduction of the inhibitory effect may involve heparan sulfate proteoglycans on the cell surface of endothelial cells (Ryback et al., 1989; Luster et al., 1995).

5.4   ACTIVITIES ON OTHER CELL TYPES

In contrast to IL-8 and GROα, which induce a transient $[Ca^{2+}]_i$ signal in human monocytes, no such effect was observed with NAP-2 or ENA-78 (Walz et al., 1991b).

CTAP-III and NAP-2 isoforms (at 10^{-6} M and higher) were shown to stimulate $[^{14}C]$glycosaminoglycan synthesis in human synovial cell cultures (Castor et al., 1989) and to induce $[^3H]$thymidine incorporation into DNA of human fibroblasts cultures (Castor et al., 1990). These responses were demonstrated at agonist concentrations 10- to 500-fold higher than used for neutrophil activation. At similar concentrations human CTAP-III and NAP-2 (but not IL-8) stimulated 2-deoxyglucose uptake into murine 3T3-F442A fibroblasts and increased the number of GLUT-1 glucose transporters (Ku Tai et al., 1992).

CTAP-III, as well as PF4 have been shown to inhibit megakaryocyte colony formation in vitro (Gewirtz et al., 1989; Han et al., 1990). Using concentrations of 10 μg/ml, a 100% inhibition was obtained with PF4, whereas CTAP-III yielded a 60% inhibition of megakaryocyte colony formation (Han et al., 1990). NAP-2 was recently shown to be about 1000-times more potent than PF4 (Gewirtz et al., 1995). The presence of mRNA for both receptors IL-8R1 and IL-8R2 in megakaryocytes suggests that they mediate the inhibitory effects of C-X-C chemokines on megakaryocyte colony formation in vitro, and that they may be involved in the regulation of megakaryocytopoiesis in vivo. ENA-78 and other C-X-C chemokines were reported to synergistically suppress the proliferation of multipotential myeloid progenitor cells when applied in combination with macrophage-stimulating protein (Broxmeyer et al., 1996).

5.5 *IN VIVO* EFFECTS

Intradermal injection of NAP-2 into the rabbit skin caused plasma leakage and ^{111}In-labeled neutrophil accumulation at doses of 2×10^{-11} mol/site and higher, while virtually no effect was observed with CTAP-III (Van Osselaer *et al.*, 1991). The effects of NAP-2 and IL-8 have also been compared histologically in the skin of rabbits and rats (Walz *et al.*, 1991c, 1993a; Zwahlen *et al.*, 1993). In rabbits, human NAP-2 was clearly less active than IL-8 at 0.01 nmol/site, whereas at 0.1 and 1 nmol/site a similar massive neutrophil migration from venules into the surrounding tissue was observed with NAP-2 and IL-8. In the rat skin, IL-8 and GROα induced a similar dose-dependent neutrophil infiltration, whereas NAP-2 was significantly less potent (Walz *et al.*, 1993a; Zwahlen *et al.*, 1993). The reason for this interspecies difference is not clear. Data on systemic application of C-X-C chemokines are only available for IL-8 that was applied to Chincilla rabbits (Zwahlen *et al.*, 1993). A single injection of IL-8 caused a marked neutrophilia that peaked at 1–2 h and was still present 7 h later.

6. *The Receptors*

Evidence that NAP-2 acts on neutrophils via a receptor shared with IL-8 was obtained with functional studies by measuring intracellular calcium changes and respiratory burst in response to sequential stimulation (Walz *et al.*, 1991b). When cells were stimulated twice with the same agonist at a given concentration, the response of the second challenge was abolished or markedly diminished, presumably as a consequence of receptor desensitization, a mechanism likely to be associated with receptor internalization (Besemer *et al.*, 1989; Samanta *et al.*, 1990). Sequential stimulations with NAP-2 and IL-8 or vice versa demonstrated that the IL-8 response was not markedly affected when given after a first NAP-2 stimulation. In contrast, $[Ca^{2+}]_i$ rise and respiratory burst response were almost completely abolished when an equimolar concentration of NAP-2 was applied after a preceding stimulation with IL-8 (Walz *et al.*, 1991b). Similarly, binding of fluorescinated IL-8 to neutrophils was only partially competed by unlabeled NAP-2 (Leonard *et al.*, 1991). Data analysis of competition studies using radiolabeled IL-8 and unlabeled NAP-2 suggested the existence of two IL-8 receptors: one with high affinity for IL-8 and NAP-2 ($K_d = 0.1$–0.4 nM) and one with high affinity for IL-8 and low affinity for NAP-2 ($K_d > 100$ nM) (Moser *et al.*, 1991; Schnitzel *et al.*, 1991). Binding studies using synthetic NAP-2 with C-terminal tyrosine substitutions for iodination confirmed earlier studies and indicated that 30–45% out of a total of 60 000–90 000 receptors bind NAP-2 with high affinity

and 55–70% bind NAP-2 with low affinity (Schumacher *et al.*, 1992). Using citrated blood instead of heparinized blood for neutrophil isolation yielded a similar total number of IL-8 receptors, but significantly lower receptor numbers for NAP-2 (~2200 high-affinity and 11 500 low-affinity sites) (Petersen *et al.*, 1994). Using the rabbit IL-8 receptor as a probe, two classes of cDNAs, termed IL-8R1 and IL-8R2, were isolated from a human library (Holmes *et al.*, 1991; Murphy and Tiffany, 1991). Both IL-8 receptors are seven-transmembrane-domain neutrophil chemokine receptors with 78% amino acid identity. The two human IL-8 receptor subtypes expressed in monkey kidney cells (COS-7) or chinese hamster ovary cells bound iodinated IL-8 with similar high affinity (LaRosa *et al.*, 1992; Cerretti *et al.*, 1993). Interestingly, only one of them (IL-8R2) bound GROα and NAP-2 with high activity, whereas the other (IL-8R1) did not bind NAP-2 and GROα, confirming the functional characterization of the two IL-8 receptors in neutrophils. Jurkat cells which normally fail to express IL-8 receptors (Moser *et al.*, 1993) were stably transfected with either IL-8R1 or IL-8R2 (Loetscher *et al.*, 1994). IL-8R1 bound IL-8 with high affinity and NAP-2 and GROα with low affinity, whereas IL-8R2 bound all three ligands with high affinity, confirming earlier results with COS-7 cells. In agreement with the binding affinities, IL-8, NAP-2, and GROα were equally potent in inducing chemotaxis in Jurkat cells expressing IL-8R2, but differed by 300- to 1000-fold in cells expressing IL-8R1 (Loetscher *et al.*, 1994). COS-7 cells and hamster lung fibroblasts (CCL-39) stably transfected with the rabbit neutrophil IL-8 receptor F3R cDNA bound labeled human IL-8 but did not bind related peptides such as NAP-2, GROα, or PF4 (Thomas *et al.*, 1994). The rabbit 5B1a receptor binds human IL-8, NAP-2, and GROα with apparent K_i values of 4, 120, and 320 nM, respectively (Prado *et al.*, 1994). On the basis of its binding profile the 5B1a receptor resembles the human IL-8R2 receptor; this might explain the fact that NAP-2, but not GROα induces a $[Ca^{2+}]_i$ rise in rabbit neutrophils. However, intracellular calcium mobilization in 5B1a-transfected Chinese hamster ovary cells has only been demonstrated with IL-8 (Prado *et al.*, 1994). Studies with chimeric human IL-8 receptors have indicated that high-affinity binding of NAP-2 to IL-8R2 is not only determined by the N-terminal segment but involves other structural elements probably located in an area containing the second extracellular domain, while exchange of the C-terminal end was ineffective (Ahuja *et al.*, 1996). These studies also indicated that determinants in the third extracellular loop were essential for receptor activation, suggesting that high-affinity binding may not be critical for receptor activation. Cross-linking experiments revealed that, at physiologically relevant concentrations, IL-8 bound to IL-8R2 as a dimer or oligomer, while NAP-2 did not (Schnitzel *et al.*, 1994).

As expected from the functional properties on neutrophils, NAP-2 and ENA-78 showed a similar binding profile to IL-8 receptors. While ^{125}I-labeled ENA-78 did not bind to IL-8R1, it specifically bound to IL-8R2 with a binding affinity of K_d = 2.2 nM (Bozic et al., 1996). Cold ENA-78 competed for high-affinity ^{125}I-labeled IL-8 binding to IL-8R2 in stable transfected HEK 293 cells similarly to IL-8 or GROα. In cells transfected with IL-8R1, ENA-78 was a weak agonist and induced a calcium transient at threshold concentrations 25- to 100-fold greater than for IL-8 (Ahuja et al., 1996).

A promiscuous IL-8 receptor has been found on human erythrocytes. About 1000–9000 receptor sites/cell were detected which bind the C-X-C chemokines IL-8, GROα, and NAP-2 and the C-C chemokines MCP-1 and RANTES reversibly with a K_d of approximately 5 nM (Neote et al., 1993). This receptor, which appears not to be regulated by G-proteins is believed to act as a sink for chemokines in circulation.

As demonstrated by functional studies, IL-8 receptors are likely to be present on human monocytes (Walz et al., 1991b). $[Ca^{2+}]_i$ signaling was observed with IL-8 and GROα, which elicited similar responses and induced mutual cross-desensitization. However, NAP-2 and ENA-78 failed to induce $[Ca^{2+}]_i$ fluxes. NAP-2 does not induce chemotaxis and also failed to desensitize monocytes for either IL-8 or GROα. The discrepancies observed for the effects of GROα and NAP-2 on monocytes were unexpected, particularly in view of the fact that monocytes contain mRNA for both types of receptors (IL-8R1 and IL-8R2) (Colombo et al., 1992). Cell-specific accessory proteins might possibly alter receptor specificity and thereby prevent NAP-2 and ENA-78 from triggering monocyte signal transduction pathways.

7. Signal Transduction

Phospholipase D (PLD) catalyzes the production of second messenger molecules such as phosphatidylethanol and phosphatidic acid (PA), which in turn can be converted to diglyceride (DG) or hydrolyzed to lyso-PA. Both PA and DG have been implicated in the activation of neutrophils, including mobilization of intracellular calcium, activation of phosphokinase C, and the oxidative burst. IL-8, NAP-2, and GROα were shown to have similar levels of protein tyrosine phosphorylation, but only IL-8 enhanced formation of phosphatidyl-ethanol and induced a significant respiratory burst (L'Heureux, 1995). Using monoclonal antibodies for IL-8R1 and IL-8R2 it has been elucidated that only IL-8R1 mediates activation of phospholipase D and the generation of superoxide, explaining the negative results obtained with NAP-2 and GROα (Jones et al., 1996). In

contrast, the release of elastase from neutrophils in response to IL-8, NAP-2, GROα, and ENA-78 could not be blocked by either of the two receptor antibodies, suggesting that both receptors can signal independently for granule release. The capacity of neutrophils to generate superoxide by FMLP can be enhanced by prior exposure to IL-8, GROα, and ENA-78. Monoclonal antibodies blocking IL-8R1 were able to block solely priming effects of IL-8, while blocking IL-8R2 eliminated priming effects exhibited by GROα and ENA-78 (Green et al., 1996). All these results predict differences in the coupling of the two IL-8 receptors to the signal transduction cascade.

Dibutyryl-cAMP inhibited IL-8- and NAP-2-induced elastase release from neutrophils in a concentration-dependent manner, suggesting the involvement of cAMP in the signal transduction pathway (Petersen et al., 1991). In contrast to FMLP and TPA, neutrophil activation with IL-8 and NAP-2 did not stimulate membrane-bound phosphokinase C (PKC) enzyme activity, suggesting that PKC does not participate in granule release (Petersen et al., 1991).

8. Cytokine in Disease and Therapy

8.1 ARTHRITIS

A number of independent studies have documented the presence of high levels of IL-8 in synovial fluid of inflamed joints of patients with rheumatoid arthritis (RA) (Brennan et al., 1990; Koch et al., 1991; Seitz et al., 1991, 1992). The amount of IL-8 present in synovial fluid accounted for only less than half of the neutrophil chemotactic activity, suggesting the presence of other chemoattractants. In contrast to patients with osteoarthritis, strongly elevated levels of ENA-78 were observed in the synovial fluid of patients with RA (Koch et al., 1994). ENA-78 levels were about 15 times higher than the mean levels observed for IL-8 and about 9 times higher than mean levels of the monocyte chemo-attractant MCP-1 (Koch et al., 1991, 1992a). In addition, anti ENA-78 antibodies were able to suppress neutrophil chemotactic activity contained in synovial fluid by up to 60%, confirming the presence of ENA-78 in a biologically active form. Thus, ENA-78 represents one of the most abundant chemokines present in RA synovial fluid. Mononuclear cells were identified as ENA-78 producers in synovial fluid by ELISA, and synovial tissue macrophages and to a lesser degree endothelial cells and fibroblasts were shown to be positive for ENA-78 by immunohistochemistry (Koch et al., 1994). ENA-78 blood levels in normal individuals are below 0.2 ng/ml, but levels in patients with RA reached mean levels of 70 ng/ml. This can be attributed, at least in part, to the fact that blood monocytes from RA patients constitutively release ENA-78. Similar observations were made for IL-8 production by blood monocytes (Seitz *et al.*, 1991).

8.2   ADULT RESPIRATORY DISTRESS SYNDROME (ARDS)

Neutrophils appear to be the important effector cells mediating the development of multiple organ failure syndrome, adult respiratory distress syndrome (ARDS), and idiopathic pulmonary fibrosis (IPF). A role has been suggested for IL-8 in the attraction of neutrophils to the lung in patients with IPF and pulmonary sarcoidosis (Carré *et al.*, 1991; Kunkel *et al.*, 1991; Rolfe *et al.*, 1991; Car *et al.*, 1994) and in patients at risk to develop ARDS (Donnelly *et al.*, 1993), but there is increasing evidence that mediators such as NAP-2 and ENA-78 may be of similar importance. Angiographic study of patients with ARDS have revealed a high incidence of artery filling defects (Greene *et al.*, 1981). Pathological analysis of lung specimen of ARDS patients has demonstrated that enmeshed platelets, neutrophils, and fibrin account for such occlusions in the small pulmonary arteries (Pratt *et al.*, 1979; Pietra *et al.*, 1981). This suggests that platelets become activated in the lungs of ARDS patients and degranulate. Using a radioimmunoassay (RA) for β-thromboglobulin-like proteins, high levels were detected in the bronchoalveolar lavage (BAL) fluid of patients with ARDS or in pulmonary edema fluids of patients with ARDS or congestive heart failure (Idell *et al.*, 1989; Cohen *et al.*, 1993). Since the RIA used is based on an antibody made against β-thromboglobulin, the test measures all CTAP-III-related proteins (CTAP-III, β-TG, as well as all NAP-2 isoforms) and therefore cannot distinguish between inactive and neutrophil-activating forms of the protein. HPLC separations of NAP-2 from CTAP-III and β-TG in pulmonary edema fluids indicated that only a small portion of the β-thromboglobulin-like proteins eluted in the NAP-2 elution volume (Cohen *et al.*, 1993). This study cannot clarify the contribution of NAP-2 to ARDS and congestive heart failure in relation to other chemokines; however, it shows that under certain pathological conditions high levels of NAP-2 precursors (up to 800 ng/ml in pulmonary edema fluid) accumulate in the lung and represent a potential source for NAP-2 production. There is now evidence that under pathophysiological conditions the immune system may respond to the presence of high chemokine levels by the production of autoantibodies. Free as well as complexed IL-8 autoantibodies have been detected in BAL fluid of ARDS patients (Kurdowska *et al.*, 1996). Recent studies with BAL fluid of patients with ARDS demonstrated that the elevated levels of IL-8 and ENA-78 observed correlated with neutrophil concentrations in the lung fluids but not with the outcome of the disease (Goodman *et al.*, 1996).

8.3   ISCHEMIA–REPERFUSION INJURY

Ischemia–reperfusion injury contributes to the pathophysiology of many clinical disorders including infarction, stroke, organ transplantation, and circulatory shock. ENA-78 which is formed at high level by pulmonary type II epithelial cells *in vitro* might be expected to contribute significantly to the recruitment of neutrophils into the lung *in vivo*. Two lines of evidence support the importance of ENA-78 or ENA-78-like factors in lung pathology. One of the striking consequences of liver injury is the associated pulmonary dysfunction, which may be mediated by the production of hepatic cytokines. In a rat model of lobar hepatic ischemia–reperfusion injury, TNF-α produced by the liver is believed to mediate the microvascular injury, neutrophil sequestration, and the induction of ENA-78 observed in the lung (Colletti *et al.*, 1990a,b). Passive immunization with TNF-α antiserum significantly reduced the production of pulmonary-derived ENA-78, neutrophil sequestration, and lung damage. Similar effects were observed when ENA-78 immune serum was applied to the animals, strongly suggesting that ENA-78 is an important mediator of lung injury (Colletti *et al.*, 1994). Similarly, neutralization of TNF-α attenuated the production of hepatic-derived ENA-78, and neutralization of ENA-78 with antibodies significantly decreased hepatic neutrophil sequestration and liver injury (Colletti *et al.*, 1995, 1996). These findings suggest that ENA-78 can be produced in the lung or liver in the context of ischemia–reperfusion and may thus contribute to neutrophil-dependent tissue injury.

Renal tubule cells appear to participate in acute inflammation of the kidney by interacting with the cytokine network. The immune functions of these cells may be of particular importance during the early, acute phase of the rejection of allografts owing to the release of effective chemoattractants for monocytes and neutrophils (Schmouder *et al.*, 1992, 1993). It was recently demonstrated that nontransformed primary human renal tubule cells release ENA-78 upon stimulation with IL-1β (Schmouder *et al.*, 1994). Whereas IL-8 expression has not been found to be a reliable marker for acutely rejected human allografts (Lipman *et al.*, 1992), ENA-78 expression was shown to be strongly elevated in rejected kidneys compared to tolerated renal allografts. ENA-78 may thus be a more reliable marker of ongoing rejection.

8.4   PNEUMONIC PASTEURELLOSIS

The bovine homolog of ENA-78 (boENA) has been isolated and characterized (Allmann-Iselin *et al.*, 1994). The immunoreactivity of ENA-78 was tested in cases of bovine pneumonic pasteurellosis. A striking positive signal was observed in the hypertrophic and hyperplastic type II alveolar epithelial cells, in mesothelial cells, and in fine capillary endothelia, whereas fibroblasts were negative (Allmann-Iselin *et al.*, 1994). However, beside boENA, which likely is one of the most prominent C-X-C chemokines produced by bovine blood monocytes and alveolar macrophages, other factors such as boGRO may be of similar importance (Rogivue *et al.*, 1995).

8.5   Heparin-associated Thrombocytopenia

Autoantibodies to a complex of heparin and PF4 (H-PF4) have been associated with platelet activation and possible thrombosis in about 85% of patients with heparin-associated thrombocytopenia (HAT). In HAT devoid of H-PF4, autoantibodies to IL-8 or CTAP-III/NAP-2 have been demonstrated (Amiral *et al.*, 1996). Autoantibodies to C-X-C chemokine-heparin complexes may bind to platelets and endothelial cells, and via the Fc portion lead to platelet activation and aggregation.

8.6   Angiogenesis and Tumorigenesis

The production of chemoattractants for monocytes and granulocytes by tumor cells suggests possible roles of these mediators in tumorigenesis. Monocyte chemoattractant protein-1 (MCP-1) was shown to induce intratumoral infiltration of macrophages which were related to suppressive effects on tumor growth *in vivo* (Yamashiro *et al.*, 1994). The C-X-C chemokine PF4 was reported to inhibit angiogenesis and the growth of melanoma cells (Maione *et al.*, 1990; Sharpe *et al.*, 1990). On the other hand, C-C and C-X-C chemokines were suggested to play a role in mediating tumor cell migration and metastasis by passive countercurrent invasion (Opdenakker and Van Damme, 1992). C-X-C chemokines containing the ELR motif, such as ENA-78 and IL-8 were demonstrated to promote endothelial cell proliferation and chemotaxis *in vitro*, and to cause neovascularization of rat and rabbit cornea *in vivo* (Koch *et al.*, 1992b; Strieter *et al.*, 1992b, 1995c; see also Section 5.3). In comparison to normal lung, tissue of bronchogenic carcinoma contains elevated levels of IL-8 and ENA-78. Activity of tumor tissue extracts has been demonstrated by endothelial cell chemotaxis and neovascularization of rat cornea. Interestingly, neutralizing antiserum to IL-8 was more effective in inhibiting these activities than antiserum to basic fibroblast growth factor (Smith *et al.*, 1994). The finding that ENA-78 is elevated similarly to IL-8 in adenocarcinomas and squamous cell carcinomas suggests a role for both C-X-C chemokines in tumor promotion by supporting neovascularization of tumor tissue (Strieter *et al.*, 1995b).

9.   Acknowledgments

Studies from my laboratory have been supported by the Swiss National Science Foundation, grants 31-36162.92 and 3100-045538.95.

10.   References

Ahuja, S.K. and Murphy, P.M. (1996). The CXC chemokines growth-regulated oncogene (GRO) α, GROβ, GROγ, neutrophil-activating peptide-2, and epithelial cell-derived neutrophil-activating peptide-78 are potent agonists for the type B, but not the type A, human interleukin-8 receptor. J. Biol. Chem. 271, 20545.

Ahuja, S.K., Lee, J.C. and Murphy, P.M. (1996). CXC chemokines bind to unique sets of selectivity determinants that can function independently and are broadly distributed on multiple domains of human interleukin-8 receptor B—determinants of high affinity binding and receptor activitation are distinct. J. Biol. Chem. 271, 225.

Allmann-Iselin, I., Car, B.D., Zwahlen, R.D., *et al.* (1994). Bovine ENA (boENA), a new monocyte-macrophage derived cytokine of the interleukin-8-family: structure, function and expression in acute pulmonary inflammation. Am. J. Pathol. 145, 1382.

Amiral, J., Marfaing-Koka, A., Wolf, M., *et al.* (1996). Presence of autoantibodies to interleukin-8 or neutrophil-activating peptide-2 in patients with heparin-associated thrombocytopenia. Blood 88, 410.

Angiolillo, A.L., Sgadari, C., Taub, D.D., *et al.* (1995). Human interferon-inducible protein 10 is a potent inhibitor of angiogenesis in vivo. J. Exp. Med. 182, 155.

Anisowicz, A., Bardwell, L. and Sager, R. (1987). Constitutive overexpression of a growth-regulated gene in transformed Chinese hamster and human cells. Proc. Natl Acad. Sci. USA 84, 7188.

Baeza, M.L., Reddigari, S.R., Kornfeld, D., *et al.* (1990). Relationship of one form of human histamine-releasing factor to connective tissue activating peptide-III. J. Clin. Invest. 85, 1516.

Baggiolini, M. and Dewald, B. (1984). Exocytosis by neutrophils. Contemp. Top. Immunobiol. 14, 221.

Baggiolini, M., Dewald, B. and Moser, B. (1994). Interleukin-8 and related chemotactic cytokines—CXC and CC chemokines. Adv. Immunol. 55, 97.

Bebawy, S.T., Gorka, J., Hyers, T.M. and Webster, R.O. (1986). In vitro effects of platelet factor 4 on normal human neutrophil functions. J. Leukocyte Biol. 39, 423.

Begg, G.S., Pepper, D.S., Chesterman, C.N. and Morgan, F.J. (1978). Complete covalent structure of human beta-thromboglobulin. Biochemistry 17, 1739.

Besemer, J., Hujber, A. and Kuhn, B. (1989). Specific binding, internalization, and degradation of human neutrophil activating factor by human polymorphonuclear leukocytes. J. Biol. Chem. 264, 17409.

Bozic, C.R., Gerard, N.P. and Gerard, C. (1996). Receptor binding specificity and pulmonary gene expression of the neutrophil-activating peptide ENA-78. Am. J. Respir. Cell Mol. Biol. 14, 302.

Brandt, E., Petersen, F. and Flad, H.-D. (1992). Recombinant tumor necrosis factor-α potentiates neutrophil degranulation in response to host defense cytokines neutrophil-activating peptide 2 and IL-8 by modulating intracellular cyclic AMP levels. J. Immunol. 149, 1356.

Brandt, E., Petersen, F. and Flad, H.-D. (1993). A novel molecular variant of the neutrophil-activating peptide NAP-2 with enhanced biological activity is truncated at the C-terminus: identification by antibodies with defined epitope specificity. Mol. Immunol. 30, 979.

Brennan, F.M., Zachariae, C.O.C., Chantry, D., et al. (1990). Detection of interleukin 8 biological activity in synovial fluids from patients with rheumatoid arthritis and production of interleukin 8 mRNA by isolated synovial cells. Eur. J. Immunol. 20, 2141.

Broxmeyer, H.E., Cooper, S., Li, Z.H., et al. (1996). Macrophage-stimulating protein, a ligand for the RON receptor protein tyrosine kinase, suppresses myeloid progenitor cell proliferation and synergizes with vascular endothelial cell growth factor and members of the chemokine family. Ann. Hematol. 73, 1.

Car, B.D., Baggiolini, M. and Walz, A. (1991). Formation of neutrophil-activating peptide 2 from platelet-derived connective-tissue-activating peptide III by different tissue proteinases. Biochem. J. 275, 581.

Car, B.D., Meloni, F., Luisetti, M., Semenzato, G., Gialdroni-Grassi, G. and Walz, A. (1994). Elevated IL-8 and MCP-1 in the bronchoalveolar lavage fluid of patients with idiopathic pulmonary fibrosis and pulmonary sarcoidosis. Am. J. Respir. Crit. Care Med. 149, 655.

Carré, P.C., Mortenson, R.L., King, T.E., Jr, Noble, P.W., Sable, C.L. and Riches, D.W.H. (1991). Increased expression of the interleukin-8 gene by alveolar macrophages in idiopathic pulmonary fibrosis. A potential mechanism for the recruitment and activation of neutrophils in lung fibrosis. J. Clin. Invest. 88, 1802.

Castor, C.W., Walz, D.A., Ragsdale, C.G., et al. (1989). Connective tissue activation. XXXIII. Biologically active cleavage products of CTAP-III from human platelets. Biochem. Biophys. Res. Commun. 163, 1071.

Castor, C.W., Walz, D.A., Johnson, P.H., et al. (1990). Connective tissue activation XXXIV: effects of proteolytic processing on the biologic activities of CTAP-III. J. Lab. Clin. Med. 116, 516.

Cerretti, D.P., Kozlosky, C.J., Vanden Bos, T., Nelson, N., Gearing, D.P. and Beckmann, M.P. (1993). Molecular characterization of receptors for human interleukin-8, GRO/melanoma growth-stimulatory activity and neutrophil activating peptide-2. Mol. Immunol. 30, 359.

Chang, M., McNinch, J., Basu, R. and Simonet, S. (1994). Cloning and characterization of the human neutrophil-activating peptide (ENA-78) gene. J. Biol. Chem. 269, 25277.

Clark-Lewis, I., Moser, B., Walz, A., Baggiolini, M., Scott, G.J. and Aebersold, R. (1991a). Chemical synthesis, purification, and characterization of two inflammatory proteins, neutrophil activating peptide 1 (interleukin-8) and neutrophil activating peptide 2. Biochemistry 30, 3128.

Clark-Lewis, I., Schumacher, C., Baggiolini, M. and Moser, B. (1991b). Structure–activity relationships of interleukin-8 determined using chemically synthesized analogs. Critical role of NH₂-terminal residues and evidence for uncoupling of neutrophil chemotaxis, exocytosis, and receptor binding activities. J. Biol. Chem. 266, 23128.

Clark-Lewis, I., Dewald, B., Geiser, T., Moser, B. and Baggiolini, M. (1993). Platelet factor 4 binds to interleukin 8 receptors and activates neutrophils when its N terminus is modified with Glu-Leu-Arg. Proc. Natl Acad. Sci. USA 90, 3574.

Cohen, A.B., Stevens, M.D., Miller, E.J., Atkinson, M.A.L. and Mullenbach, G. (1992). Generation of the neutrophil-activating peptide-2 by cathepsin G and cathepsin G-treated human platelets. Am. J. Physiol. Lung Cell. Mol. Physiol. 263, L249.

Cohen, A.B., Stevens, M.D., Miller, E.J., et al. (1993). Neutrophil-activating peptide-2 in patients with pulmonary edema from congestive heart failure or ARDS. Am. J. Physiol. Lung Cell. Mol. Physiol. 264, L490.

Colditz, I.G., Zwahlen, R.D. and Baggiolini, M. (1990). Neutrophil accumulation and plasma leakage induced in vivo by neutrophil-activating peptide-1. J. Leukocyte Biol. 48, 129.

Colletti, L.M., Burtch, G.D., Remick, D.G., et al. (1990a). The production of tumor necrosis factor alpha and the development of a pulmonary capillary injury following hepatic ischemia/reperfusion. Transplantation 49, 268.

Colletti, L.M., Remick, D.G., Burtch, G.D., Kunkel, S.L., Strieter, R.M. and Campbell, D.A., Jr (1990b). Role of tumor necrosis factor-alpha in the pathophysiologic alterations after hepatic ischemia/reperfusion injury in the rat. J. Clin. Invest. 85, 1936.

Colletti, L.M., Kunkel, S.L., Walz, A., et al. (1994). Chemokine expression during hepatic ischemia/reperfusion-induced lung injury in the rat. J. Clin. Invest. 95, 134.

Colletti, L.M., Kunkel, S.L., Walz, A., et al. (1995). Chemokine expression during hepatic ischemia/reperfusion-induced lung injury in the rat. The role of epithelial neutrophil activating protein. J. Clin. Invest. 95, 134.

Colletti, L.M., Kunkel, S.L., Green, M., Burdick, M. and Strieter, R.M. (1996). Post-ischemic shunt following hepatic ischemia/reperfusion does not affect tissue chemokine levels or tissue injury. Shock 5, 371.

Collins, P.D., Weg, V.B., Faccioli, L.H., Watson, M.L., Moqbel, R. and Williams, T.J. (1993). Eosinophil accumulation induced by human interleukin-8 in the guinea-pig in vivo. Immunology 79, 312.

Colombo, M.P., Maccalli, C., Mattei, S., Melani, C., Radrizzani, M. and Parmiani, G. (1992). Expression of cytokine genes, including IL-6, in human malignant melanoma cell lines. Melanoma Res. 2, 181.

Corbett, M.S., Schmitt, I., Riess, O. and Walz, A. (1994). Characterization of the gene for human neutrophil-activating peptide 78 (ENA-78). Biochem. Biophys. Res. Commun. 205, 612.

De Gaetano, G., Evangelista, V., Rajtar, G., Del Maschio, A. and Cerletti, C. (1990). Activated polymorphonuclear leukocytes stimulate platelet function. Thromb. Res. Suppl. 11, 25.

Detmers, P.A., Powell, D.E., Walz, A., Clark-Lewis, I., Baggiolini, M. and Cohn, Z.A. (1991). Differential effects of neutrophil-activating peptide 1/IL-8 and its homologues on leukocyte adhesion and phagocytosis. J. Immunol. 147, 4211.

Deuel, T.F., Senior, R.M., Chang, D., Griffin, G.L., Heinrikson, R.L. and Kaiser, E.T. (1981). Platelet factor 4 is chemotactic for neutrophils and monocytes. Proc. Natl Acad. Sci. USA 78, 4584.

Dewald, B., Moser, B., Barella, L., Schumacher, C., Baggiolini,

M. and Clark-Lewis, I. (1992). IP-10, a gamma-interferon-inducible protein related to interleukin-8, lacks neutrophil activating properties. Immunol. Lett. 32, 81.

Donnelly, S.C., Strieter, R.M., Kunkel, S.L., et al. (1993). Interleukin-8 and development of adult respiratory distress syndrome in at-risk patient groups. Lancet 341, 643.

Ebisawa, M., Yamada, T., Bickel, C., Klunk, D. and Schleimer, R.P. (1994). Eosinophil transendothelial migration induced by cytokines: III. Effect of the chemokine RANTES. J. Immunol. 153, 2153.

Ehlert, J.E., Petersen, F., Kubbutat, M.H.G., Gerdes, J., Flad, H.-D. and Brandt, E. (1995). Limited and defined truncation at the C terminus enhances receptor binding and degranulation activity of the neutrophil-activating peptide 2 (NAP-2). Comparison of native and recombinant NAP-2 variants. J. Biol. Chem. 270, 6338.

Eisman, R., Surrey, S., Ramachandran, B., Schwartz, E. and Poncz, M. (1990). Structural and functional comparison of the genes for human platelet factor 4 and PF4$_{alt}$. Blood 76, 336.

Farber, J.M. (1993). HuMIG: a new human member of the chemokine family of cytokines. Biochem. Biophys. Res. Commun. 192, 223.

Gewirtz, A.M., Calabretta, B., Rucinski, B., Niewiarowski, S. and Xu, W.Y. (1989). Inhibition of human megakaryocytopoiesis in vitro by platelet factor 4 (PF4) and a synthetic COOH-terminal PF4 peptide. J. Clin. Invest. 83, 1477.

Gewirtz, A.M., Zhang, J., Ratajczak, J., et al. (1995). Chemokine regulation of human megakaryocytopoiesis. Blood 86, 2559.

Goodman, R.B., Strieter, R.M., Martin, D.P., et al. (1996). Inflammatory cytokines in patients with persistence of the acute respiratory distress syndrome. Am. J. Respir. Crit. Care Med. 154, 602.

Green, S.P., Chuntharapai, A. and Curnutte, J.T. (1996). Interleukin-8 (IL-8), melanoma growth-stimulatory activity, and neutrophil-activating peptide selectively mediate priming of the neutrophil NADPH oxidase through the type A or type B IL-8 receptor. J. Biol. Chem. 271, 25400.

Greene, R., Zapol, W.M., Snider, M.T., et al. (1981). Early bedside detection of pulmonary vascular occlusion during acute respiratory failure. Am. Rev. Respir. Dis. 124, 593.

Haerter, L., Petersen, F., Flad, H.-D. and Brandt, E. (1994). Connective tissue-activating peptide III desensitizes chemokine receptors on neutrophils: requirement for proteolytic formation of the neutrophil-activating peptide 2. J. Immunol. 153, 5698.

Han, Z.C., Bellucci, S., Walz, A., Baggiolini, M. and Caen, J.P. (1990). Negative regulation of human megakaryocytopoiesis by human platelet factor 4 (PF4) and connective tissue-activating peptide (CTAP-III). Int. J. Cell Cloning 8, 253.

Haskill, S., Peace, A., Morris, J., et al. (1990). Identification of three related human GRO genes encoding cytokine functions. Proc. Natl Acad. Sci. USA 87, 7732.

Hébert, C.A., Vitangcol, R.V. and Baker, J.B. (1991). Scanning mutagenesis of interleukin-8 identifies a cluster of residues required for receptor binding. J. Biol. Chem. 266, 18989.

Holmes, W.E., Lee, J., Kuang, W.-J., Rice, G.C. and Wood, W.I. (1991). Structure and functional expression of a human interleukin-8 receptor. Science 253, 1278.

Holt, J.C., Harris, M.E., Holt, A.M., Lange, E., Henschen, A. and Niewiarowski, S. (1986). Characterization of human

platelet basic protein, a precursor form of low-affinity platelet factor 4 and β-thromboglobulin. Biochemistry 25, 1988.

Holt, J.C., Yan, Z., Lu, W., Stewart, G.J. and Niewiarowski, S. (1992). Isolation, characterization, and immunological detection of neutrophil-activating peptide 2: a proteolytic degradation product of platelet basic protein. Proc. Soc. Exp. Biol. Med. 199, 171.

Idell, S., Maunder, R., Fein, A.M., et al. (1989). Platelet-specific alpha-granule proteins and thrombospondin in bronchoalveolar lavage in the adult respiratory distress syndrome. Chest 96, 1125.

Jones, S.A., Wolf, M., Qin, S.X., Mackay, C.R. and Baggiolini, M. (1996). Different functions for the interleukin 8 receptors (IL-8R) of human neutrophil leukocytes: NADPH oxidase and phospholipase D are activated through IL-8R1 but not IL-8R2. Proc. Natl Acad. Sci. USA 93, 6682.

Kernen, P., Wymann, M.P., von Tscharner, V., et al. (1991). Shape changes, exocytosis, and cytosolic free calcium changes in stimulated human eosinophils. J. Clin. Invest. 87, 2012.

Koch, A.E., Kunkel, S.L., Burrows, J.C., et al. (1991). Synovial tissue macrophage as a source of the chemotactic cytokine IL-8. J. Immunol. 147, 2187.

Koch, A.E., Kunkel, S.L., Harlow, L.A., et al. (1992a). Enhanced production of monocyte chemoattractant protein-1 in rheumatoid arthritis. J. Clin. Invest. 90, 772.

Koch, A.E., Polverini, P.J., Kunkel, S.L., et al. (1992b). Interleukin-8 as a macrophage-derived mediator of angiogenesis. Science 258, 1798.

Koch, A.E., Kunkel, S.L., Harlow, L.A., et al. (1994). Epithelial neutrophil activating peptide-78: a novel chemotactic cytokine for neutrophils in arthritis. J. Clin. Invest. 94, 1012.

Krieger, M., Brunner, T., Bischoff, S.C., et al. (1992). Activation of human basophils through the IL-8 receptor. J. Immunol. 149, 2662.

Kungl, A.J., Machius, M., Huber, R., et al. (1994). Purification, crystallization and preliminary X-ray diffraction analysis of recombinant human neutrophil-activating peptide 2 (rhNAP-2). FEBS Lett. 347, 300.

Ku Tai, P., Liao, J.-F., Hossler, P.A., Castor, C.W. and Carter-Su, C. (1992). Regulation of glucose transporters by connective tissue activating peptide-III isoforms. J. Biol. Chem. 267, 19579.

Kuna, P., Reddigari, S.R., Schall, T.J., Rucinski, D., Sadick, M. and Kaplan, A.P. (1993). Characterization of the human basophil response to cytokines, growth factors, and histamine releasing factors of the intercrine/chemokine family. J. Immunol. 150, 1932.

Kunkel, S.L., Standiford, T., Kasahara, K. and Strieter, R.M. (1991). Interleukin-8 (IL-8): the major neutrophil chemotactic factor in the lung. Exp. Lung Res. 17, 17.

Kurdowska, A., Miller, E.J., Noble, J.M., et al. (1996). Anti-IL-8 autoantibodies in alveolar fluid from patients with the adult respiratory distress syndrome. J. Immunol. 157, 2699.

LaRosa, G.J., Thomas, K.M., Kaufmann, M.E., et al. (1992). Amino terminus of the interleukin-8 receptor is a major determinant of receptor subtype specificity. J. Biol. Chem. 267, 25402.

Leonard, E.J., Yoshimura, T., Rot, A., et al. (1991). Chemotactic activity and receptor binding of neutrophil attractant/activation protein-1 (NAP-1) and structurally related host defense cytokines: interaction of NAP-2 with the NAP-1 receptor. J. Leukocyte Biol. 49, 258.

L'Heureux, G.P., Bourgoin, S., Jean, N., McColl, S.R. and

Naccache, P.H. (1995). Diverging signal transduction pathways activated by interleukin-8 and related chemokines in human neutrophils: interleukin-8, but not NAP-2 or GROα, stimulates phospholipase D activity. Blood 85, 522.

Lindley, I., Aschauer, H., Scifert, J.M., et al. (1988). Synthesis and expression in Escherichia coli of the gene encoding monocyte-derived neutrophil-activating factor: biological equivalence between natural and recombinant neutrophil-activating factor. Proc. Natl Acad. Sci. USA 85, 9199.

Lipman, M.L., Stevens, A.C. and Strom, T.B. (1992). Cytotoxic T lymphocyte (CTL) and immunosuppressive cytokine gene expression in human renal allograft biopsies. Am. Soc. Nephrol. 3, 867.

Loetscher, P., Seitz, M., Clark-Lewis, I., Baggiolini, M. and Moser, B. (1994). Both interleukin-8 receptors independently mediate chemotaxis. Jurkat cells transfected with IL-8R1 or IL-8R2 migrate in response to IL-8, GROα and NAP-2. FEBS Lett. 341, 187.

Lukacs, N.W., Kunkel, S.L., Allen, R., et al. (1995). Stimulus and cell-specific expression of C-X-C and C-C chemokines by pulmonary stromal cell populations. Am. J. Physiol. Lung Cell. Mol. Physiol. 268, L856.

Luster, A.D., Unkeless, J.C. and Ravetch, J.V. (1985). γ-Interferon transcriptionally regulates an early-response gene containing homology to platelet proteins. Nature 315, 672.

Luster, A.D., Greenberg, S.M. and Leder, P. (1995). The IP-10 chemokine binds to a specific cell surface heparan sulfate site shared with platelet factor 4 and inhibits endothelial cell proliferation. J. Exp. Med. 182, 219.

Maione, T.E., Gray, G.S., Petro, J., et al. (1990). Inhibition of angiogenesis by recombinant human platelet factor-4 and related peptides. Science 247, 77.

Majumdar, S., Gonder, D., Koutsis, B. and Poncz, M. (1991). Characterization of the human β-thromboglobulin gene. Comparison with the gene for platelet factor 4. J. Biol. Chem. 266, 5785.

Malkowski, M.G., Wu, J.Y., Lazar, J.B., Johnson, P.H. and Edwards, B.F.P. (1995). The crystal structure of recombinant human neutrophil-activating peptide-2 (M6L) at 1.9-Å resolution. J. Biol. Chem. 270, 7077.

Mayo, K.H. (1991). Low-affinity platelet factor-4 ^{1}H NMR derived aggregate equilibria indicate a physiologic preference for monomers over dimers and tetramers. Biochemistry 30, 925.

Mayo, K.H. and Chen, M.J. (1989). Human platelet factor 4 monomer-dimer-tetramer equilibria investigated by ^{1}H-NMR spectroscopy. Biochemistry 28, 9469.

Mayo, K.H., Yang, Y., Daly, T.J., Barry, J.K. and La Rosa, G.J. (1994). Secondary structure of neutrophil-activating peptide-2 determined by [1]H-nuclear magnetic resonance spectroscopy. Biochem. J. 304, 371.

Meurer, R., Van Riper, G., Feeney, W., et al. (1993). Formation of eosinophilic and monocytic intradermal inflammatory sites in the dog by injection of human RANTES but not human monocyte chemoattractant protein 1, human macrophage inflammatory protein 1α, or human interleukin 8. J. Exp. Med. 178, 1913.

Moore, S., Pepper, D.S. and Cash, J.D. (1975). Platelet antiheparin activity. The isolation and characterisation of platelet factor 4 released from thrombin-aggregated washed human platelets and its dissociation into subunits and the isolation of membrane-bound antiheparin activity. Biochem. Biophys. Acta 379, 370.

Moser, B., Clark-Lewis, I., Zwahlen, R. and Baggiolini, M. (1990). Neutrophil-activating properties of the melanoma growth-stimulatory activity. J. Exp. Med. 171, 1797.

Moser, B., Schumacher, C., von Tscharner, V., Clark-Lewis, I. and Baggiolini, M. (1991). Neutrophil-activating peptide 2 and gro/melanoma growth-stimulatory activity interact with neutrophil-activating peptide 1/interleukin 8 receptors on human neutrophils. J. Biol. Chem. 266, 10666.

Moser, B., Barella, L., Mattei, S., et al. (1993). Expression of transcripts for two interleukin 8 receptors in human phagocytes, lymphocytes and melanoma cells. Biochem. J. 294, 285.

Murphy, P.M. and Tiffany, H.L. (1991). Cloning of complementary DNA encoding a functional human interleukin-8 receptor. Science 253, 1280.

Nagasawa, T., Kikutani, H. and Kishimoto, T. (1994). Molecular cloning and structure of a pre-B-cell growth-stimulating factor. Proc. Natl Acad. Sci. USA 91, 2305.

Neote, K., Darbonne, W., Ogez, J., Horuk, R. and Schall, T.J. (1993). Identification of a promiscuous inflammatory peptide receptor on the surface of red blood cells. J. Biol. Chem. 268, 12247.

Niewiarowski, S., Walz, D.A., James, P., Rucinski, B. and Kueppers, F. (1980). Identification and separation of secreted platelet proteins by isoelectric focusing. Evidence that low-affinity platelet factor 4 is converted to β-thromboglobulin by limited proteolysis. Blood 55, 453.

Nourshargh, S., Perkins, J.A., Showell, H.J., Matsushima, K., Williams, T.J. and Collins, P.D. (1992). A comparative study of the neutrophil stimulatory activity in vitro and pro-inflammatory properties in vivo of 72 amino acid and 77 amino acid IL-8. J. Immunol. 148, 106.

Oberlin, E., Amara, A., Bachelerie, F., et al. (1996). The CXC chemokine SDF-1 is the ligand for LESTR/fusin and prevents infection by T-cell-line-adapted HIV-1. Nature 382, 833.

Opdenakker, G. and Van Damme, J. (1992). Chemotactic factors, passive invasion and metastasis of cancer cells. Immunol. Today 13, 463.

Orens, J.B., Lukacs, N.W., Kunkel, S.L., et al. (1994). Regulation of chemokine production by the oxidative metabolism of L-arginine in a human mixed lymphocyte reaction. Cell. Immunol. 156, 95.

Padrines, M., Wolf, M., Walz, A. and Baggiolini, M. (1994). Interleukin-8 processing by neutrophil elastase, cathepsin G and proteinase-3. FEBS Lett. 352, 231.

Paul, D., Niewiarowski, S., Varma, K.G., Rucinski, B., Rucker, S. and Lange, E. (1980). Human platelet basic protein associated with antiheparin and mitogen activities: purification and partial characterization. Proc. Natl Acad. Sci. USA 77, 5914.

Pawlowski, N.A., Kaplan, G., Hamill, A.L., Cohn, Z.A. and Scott, W.A. (1983). Arachidonic acid metabolism by human monocytes: studies with platelet-depleted cultures. J. Exp. Med. 158, 393.

Petersen, F., Flad, H.-D. and Brandt, E. (1994). Neutrophil-activating peptides NAP-2 and IL-8 bind to the same sites on neutrophils but interact in different ways: discrepancies in binding affinities, receptor densities, and biologic effects. J. Immunol. 152, 2467.

Petersen, F., Van Damme, J., Flad, H.-D. and Brandt, E. (1991). Neutrophil-activating polypeptides IL-8 and NAP-2 induce identical signal transduction pathways in the regulation of lysosomal enzyme release. Lymphokine Res. 10, 35.

Peveri, P., Walz, A., Dewald, B. and Baggiolini, M. (1988). A novel neutrophil-activating factor produced by human mononuclear phagocytes. J. Exp. Med. 167, 1547.

Pietra, G.G., Ruttner, J.R., Wust, W. and Glinz, W. (1981). The lung after trauma and shock—fine structure of the alveolar-capillary barrier in 23 autopsies. J. Trauma 21, 454.

Poncz, M., Surrey, S., LaRocco, P., et al. (1987). Cloning and characterization of platelet factor 4 cDNA derived from a human erythroleukemic cell line. Blood 69, 219.

Power, C.A., Furness, R.B., Brawand, C. and Wells, T.N.C. (1994). Cloning of a full-length cDNA encoding the neutrophil-activating peptide ENA-78 from human platelets. Gene 151, 333.

Prado, G. N., Thomas, K. M., Suzuki, H., et al. (1994). Molecular characterization of a novel rabbit interleukin-8 receptor isotype. J. Biol. Chem. 269, 12391.

Pratt, P.C., Vollmer, R.T., Shelburne, J.D. and Crapo, J.D. (1979). Pulmonary morphology in a multihospital collaborative extracorporeal membrane oxygenation project. I. Light microscopy. Am. J. Pathol. 95, 191.

Proost, P., Wuyts, A., Conings, R., et al. (1993). Human and bovine granulocyte chemotactic protein-2: complete amino acid sequence and functional characterization as chemokines. Biochemistry 32, 10170.

Rogivue, C., Car, B.D., Allmann-Iselin, I., Zwahlen, R.D. and Walz, A. (1995). Bovine GRO (boGRO), a new monocyte-macrophage derived cytokine of the interleukin-8-family: partial structure, function and expression in acute pulmonary inflammation. Lab. Invest. 72, 689.

Rolfe, M.W., Kunkel, S.L., Standiford, T.J., et al. (1991). Pulmonary fibroblast expression of interleukin-8: a model for alveolar macrophage-derived cytokine networking. Am. J. Respir. Cell Mol. Biol. 5, 493.

Rybak, M.E., Gimbrone, M.A., Jr, Davies, P.F. and Handin, R.I. (1989). Interaction of platelet factor four with cultured vascular endothelial cells. Blood 73, 1534.

Samanta, A.K., Oppenheim, J.J. and Matsushima, K. (1990). Interleukin 8 (monocyte-derived neutrophil chemotactic factor) dynamically regulates its own receptor expression on human neutrophils. J. Biol. Chem. 265, 183.

Schmouder, R.L., Strieter, R.M., Wiggins, R.C., Chensue, S.W. and Kunkel, S.L. (1992). In vitro and in vivo interleukin-8 production in human renal cortical epithelia. Kidney Int. 41, 191.

Schmouder, R.L., Strieter, R.M. and Kunkel, S.L. (1993). Interferon-gamma regulation of human renal cortical epithelial cell-derived monocyte chemotactic peptide-1. Kidney Int. 44, 43.

Schmouder, R.L., Strieter, R.M., Walz, A. and Kunkel, S.L. (1995). Epithelial-derived neutrophil activating protein-78 (ENA-78) production in human renal tubule epithelial cells and in renal allograft rejection. Transplantation 59, 118.

Schnitzel, W., Garbeis, B., Monschein, U. and Besemer, J. (1991). Neutrophil activating peptide-2 binds with two affinities to receptor(s) on human neutrophils. Biochem. Biophys. Res. Commun. 180, 301.

Schnitzel, W., Monschein, U. and Besemer, J. (1994). Monomer-dimer equilibria of interleukin-8 and neutrophil-activating peptide 2. Evidence for IL-8 binding as a dimer and oligomer to IL-8 receptor B. J. Leukocyte Biol. 55, 763.

Schnyder-Candrian, S., Strieter, R. M., Kunkel, S. L., and Walz, A. (1995). Interferon-α and interferon-gamma down-regulate the production of interleukin-8 and ENA-78 in human monocytes. J. Leukocyte Biol. 57, 929.

Schnyder-Candrian, S. and Walz, A. (1997). Neutrophil-activating protein ENA-78 and IL-8 exhibit different patterns of expression in LPS- and cytokine-stimulated human monocytes. J. Immunol. 158, 3888–3894.

Schröder, J.-M. (1992). Generation of NAP-1 and related peptides in psoriasis and other inflammatory skin diseases. Cytokines 4, 54.

Schröder, J.-M., Mrowietz, U., Morita, E. and Christophers, E. (1987). Purification and partial biochemical characterization of a human monocyte-derived, neutrophil-activating peptide that lacks interleukin 1 activity. J. Immunol. 139, 3474.

Schumacher, C., Clark-Lewis, I., Baggiolini, M. and Moser, B. (1992). High- and low-affinity binding of GROα and neutrophil-activating peptide 2 to interleukin 8 receptors on human neutrophils. Proc. Natl Acad. Sci. USA 89, 10542.

Schweizer, R.C., Welmers, B.A.C., Raaijmakers, J.A.M., Zanen, P., Lammers, J.-W.J. and Koenderman, L. (1994). RANTES- and interleukin-8-induced responses in normal human eosinophils: effects of priming with interleukin-5. Blood 83, 3697.

Seitz, M., Dewald, B., Gerber, N. and Baggiolini, M. (1991). Enhanced production of neutrophil-activating peptide-1/interleukin-8 in rheumatoid arthritis. J. Clin. Invest. 87, 463.

Seitz, M., Dewald, B., Ceska, M., Gerber, N. and Baggiolini, M. (1992). Interleukin-8 in inflammatory rheumatic diseases: synovial fluid levels, relation to rheumatoid factors, production by mononuclear cells, and effects of gold sodium thiomalate and methotrexate. Rheumatol. Int. 12, 159.

Selak, M.A. and Smith, J.B. (1990). Cathepsin G binding to human platelets. Evidence for a specific receptor. Biochem. J. 266, 55.

Sharpe, R.J., Byers, H.R., Scott, C.F., Bauer, S.I. and Maione, T.E. (1990). Growth inhibition of murine melanoma and human colon carcinoma by recombinant human platelet factor 4. J. Natl Cancer Inst. 82, 848.

Simon, H.-U., Tsao, P.W., Siminovitch, K.A., Mills, G.B. and Blaser, K. (1994). Functional platelet-activating factor receptors are expressed by monocytes and granulocytes but not by resting or activated T and B lymphocytes from normal individuals or patients with asthma. J. Immunol. 153, 364.

Smith, D.R., Polverini, P.J., Kunkel, S.L., et al. (1994). Inhibition of interleukin 8 attenuates angiogenesis in bronchogenic carcinoma. J. Exp. Med. 179, 1409.

Strieter, R.M., Kunkel, S.L., Burdick, M.D., Lincoln, P.M. and Walz, A. (1992a). The detection of a novel neutrophil-activating peptide (ENA-78) using a sensitive ELISA. Immunol. Invest. 21, 589.

Strieter, R.M., Kunkel, S.L., Elner, V.M., et al. (1992b). Interleukin-8: a corneal factor that induces neovascularization. Am. J. Pathol. 141, 1279.

Strieter, R.M., Polverini, P.J., Arenberg, D.A. and Kunkel, S.L. (1995a). The role of CXC chemokines as regulators of angiogenesis. Shock 4, 155.

Strieter, R.M., Polverini, P.J., Arenberg, D.A., et al. (1995b). Role of C-X-C chemokines as regulators of angiogenesis in lung cancer. J. Leukocyte Biol. 57, 752.

Strieter, R.M., Polverini, P.J., Kunkel, S.L., et al. (1995c). The functional role of the ELR motif in CXC chemokine-mediated angiogenesis. J. Biol. Chem. 270, 27348.

Thomas, K.M., Taylor, L., Prado, G., *et al.* (1994). Functional and ligand binding specificity of the rabbit neutrophil IL-8 receptor. J. Immunol. 152, 2496.

Van Damme, J., Rampart, M., Conings, R., *et al.* (1990). The neutrophil-activating proteins interleukin 8 and β-thromboglobulin: in vitro and in vivo comparison of NH_2-terminally processed forms. Eur. J. Immunol. 20, 2113.

Van Osselaer, N., Van Damme, J., Rampart, M. and Herman, A.G. (1991). Increased microvascular permeability in vivo in response to intradermal injection of neutrophil-activating protein (NAP-2) in rabbit skin. Am. J. Pathol. 138, 23.

Walz, A. and Baggiolini, M. (1989). A novel cleavage product of β-thromboglobulin formed in cultures of stimulated mononuclear cells activates human neutrophils. Biochem. Biophys. Res. Commun. 159, 969.

Walz, A. and Baggiolini, M. (1990). Generation of the neutrophil-activating peptide NAP-2 from platelet basic protein or connective tissue-activating peptide III through monocyte proteases. J. Exp. Med. 171, 449.

Walz, A., Peveri, P., Aschauer, H. and Baggiolini, M. (1987). Purification and amino acid sequencing of NAF, a novel neutrophil-activating factor produced by monocytes. Biochem. Biophys. Res. Commun. 149, 755.

Walz, A., Dewald, B., von Tscharner, V. and Baggiolini, M. (1989). Effects of the neutrophil-activating peptide NAP-2, platelet basic protein, connective tissue-activating peptide III and platelet factor 4 on human neutrophils. J. Exp. Med. 170, 1745.

Walz, A., Dewald, B. and Baggiolini, M. (1990). Formation and biological activity of NAP-2, a neutrophil-activating peptide derived from platelet alpha-granule precursors. In "Molecular and Cellular Biology of Cytokines", pp. 363–368. Wiley-Liss, New York.

Walz, A., Burgener, R., Car, B., Baggiolini, M., Kunkel, S.L. and Strieter, R.M. (1991a). Structure and neutrophil-activating properties of a novel inflammatory peptide (ENA-78) with homology to interleukin 8. J. Exp. Med. 174, 1355.

Walz, A., Meloni, F., Clark-Lewis, I., von Tscharner, V. and Baggiolini, M. (1991b). [Ca^{2+}]$_i$ changes and respiratory burst in human neutrophils and monocytes induced by NAP-1/interleukin-8, NAP-2, and gro/MGSA. J. Leukocyte Biol. 50, 279.

Walz, A., Zwahlen, R. and Baggiolini, M. (1991c). Formation and biological properties of neutrophil activating peptide 2 (NAP-2). Adv. Exp. Med. Biol. 305, 39.

Walz, A., Meloni, F., Zwahlen, R., Clark-Lewis, I. and Car, B. (1993a). Generation and properties of neutrophil-activating peptide 2 (NAP-2). In "Host Defense Dysfunction in Trauma, Shock and Sepsis" (eds. Faist, Meakins, and Schildberg), pp. 593–598. Springer Verlag, Berlin.

Walz, A., Schmutz, P., Mueller, C. and Schnyder-Candrian, S. (1997). Regulation and function of the CXC chemokine ENA-78 in monocytes and its role in disease. J. Leukocyte Biol., in press.

Walz, A., Strieter, R.M. and Schnyder, S. (1993b). Neutrophil-activating peptide ENA-78. Adv. Exp. Med. Biol. 351, 129.

Warringa, R.A.J., Koenderman, L., Kok, P.T.M., Kreukniet, J. and Bruijnzeel, P.L.B. (1991). Modulation and induction of eosinophil chemotaxis by granulocyte-macrophage colony-stimulating factor and interleukin-3. Blood 77, 2694.

Wenger, R.H., Wicki, A.N., Walz, A., Kieffer, N. and Clemetson, K.J. (1989). Cloning of cDNA coding for connective tissue activating peptide III from a human platelet-derived lamdagtll expression library. Blood 73, 1498.

Wenger, R.H., Hameister, H. and Clemetson, K.J. (1991). Human platelet basic protein/connective tissue activating peptide-III maps in a gene cluster on chromosome 4q12-q13 along with other genes of the β-thromboglobulin superfamily. Hum. Genet. 87, 367.

Yamashiro, S., Takeya, M., Nishi, T., Kuratsu, J., Yoshimura, T., Ushio, Y. and Takahashi, K. (1994). Tumor-derived monocyte chemoattractant protein-1 induces intratumoral infiltration of monocyte-derived macrophage subpopulation in transplanted rat tumors. Am. J. Pathol. 145, 856.

Yan, Z., Holt, J.C., Stewart, G.J. and Niewiarowski, S. (1993). Neutrophil-activating intercrine secreted by porcine platelets is active without proteolytic processing. Am. J. Physiol. Cell Physiol. 265, C1396.

Yoshimura, T., Matsushima, K., Tanaka, S., *et al.* (1987). Purification of a human monocyte-derived neutrophil chemotactic factor that has peptide sequence similarity to other host defense cytokines. Proc. Natl Acad. Sci. USA 84, 9233.

Zwahlen, R., Walz, A. and Rot, A. (1993). In vitro and in vivo activity and pathophysiology of human interleukin-8 and related peptides. Int. Rev. Exp. Pathol. 34(Pt B), 27.

31. Macrophage Inflammatory Protein 1-α

Robert J.B. Nibbs, Gerard J. Graham *and* Ian B. Pragnell

1. Introduction

Macrophage inflammatory protein-1α (MIP-1α), a member of an ever-expanding chemokine family (at least 30 at present), was first discovered as a complex released as macrophage inflammatory protein (MIP) after lipopolysaccharide (LPS) stimulation of a murine macrophage cell line (Wolpe *et al.*, 1988). Subsequently, MIP was fractionated into MIP-1α and the closely related MIP-1β and cDNA clones were isolated (Sherry *et al.*, 1988). The human homolog was first described by Obaru and colleagues (Obaru *et al.*, 1986) and the nomenclature LD78 was used to denominate human MIP-1α, though more recently the name MIP-1α is used for all homologs. Progress in our knowledge of the activities and potential physiological role of MIP-1α since then has been dramatic and the literature testifies to the interest in this fascinating protein. MIP-1α is primarily an inflammatory chemokine as evidenced by the inducibility in hemopoietic cells and the demonstrated chemotactic action on a range of hemopoietic target cells. Whilst earlier reports suggested a chemotactic action on neutrophils, it is generally accepted nowadays that MIP-1α has little, if any, effect on neutrophils; macrophages, eosinophils, basophils, and certain T cell subsets are accepted as the primary target cells. Thus MIP-1α, in common with other chemokines, acts at the onset of inflammation and can clearly be seen as an important immunoregulatory cytokine. However, MIP-1α has a number of other biological activities, most notably as an inhibitor of the proliferation of particular populations of stem cells, such as transiently engrafting hemopoietic stem cells (CFU-S) and clonogenic keratinocytes. Further, a plethora of more recent observations concerning the complex field of MIP-1α receptor biology have made an exciting connection between chemokines and AIDS. A number of general chemokine reviews have been published (Driscoll, 1994; Friedland, 1995; Kunkel *et al.*, 1995; Petrek *et al.*, 1995, Verfaillie, 1996) and a brief review on aspects of MIP-1α (Cook,

Cytokines
ISBN 0–12–498340–5

1996). This review attempts to bring the reader up to date on the various aspects of the structure, expression, and physiology of this protein and its receptors.

2. The MIP-1α Gene

As a result of the original identification of the protein produced by macrophages (Wolpe *et al.*, 1998), a MIP-1α cDNA clone of 753 base pairs in length was subsequently isolated (Figure 31.1(a)) and shown to contain a short 5′ noncoding region, an open reading frame of 276 nucleotides, and a longer 3′ untranslated region containing a polyadenylation signal (AATAAA, boxed in Figure 31.1(a)) and three repeats of an AT-rich motif implicated in cytokine mRNA stability (underlined in Figure 31.1(a)) (Caput *et al.*, 1986; Davatelis *et al.*, 1988). Protein produced from this cDNA contains a signal sequence (boxed) which is cleaved off at the position before the protein is secreted from the cell. MIP-1α is a single-copy gene containing two introns and three exons (Figure 31.1(c)), and is located on chromosome 11, clustered with other members of the C-C chemokine family (Grove *et al.*, 1990; Oppenheim *et al.*, 1991). The arrows in Figure 31.1(a) and (b) show the position of the splice junction.

LD78, also known as AT464 or GOS19S, is considered to be the human homolog of MIP-1α by virtue of the 75% identity shared over their protein sequence (Figure 31.1(b)) (Obaru *et al.*, 1986; Zipfel *et al.*, 1989). There appear to be three distinct forms of LD78 clustered with other β-chemokines on chromosome 17 in the region q11-q21 (Blum *et al.*, 1990; Irving *et al.*, 1990; Nakao *et al.*, 1990). LD78α and LD78β differ by only three amino acids in their mature protein sequence, and like MIP-1α these genes are organized into three exons split by two introns, a conserved structural arrangement in chemokine genes (Figure 31.1(c)) (Nakao *et al.*, 1990). The 5′ flanking regions of LD78α and β and MIP-1α contain a TATA box and are highly homologous to one another over 2 kb upstream of the transcription start site, suggesting common transcription control mechanisms (LD78β contains a repetitive *Alu* sequence at position –294 relative to the transcription initiation site, but this does not affect transcription) (Blum *et al.*, 1990; Nakao *et al.*, 1990; Nomiyama *et al.*, 1993). The α gene exists as a single copy, but the number of β genes on chromosome 17 varies between individuals (Hirashima *et al.*, 1992). α and β both produce mRNA and comparisons of proteins produced in a recombinant form have shown no functional differences. Finally, LD78γ is an unexpressed

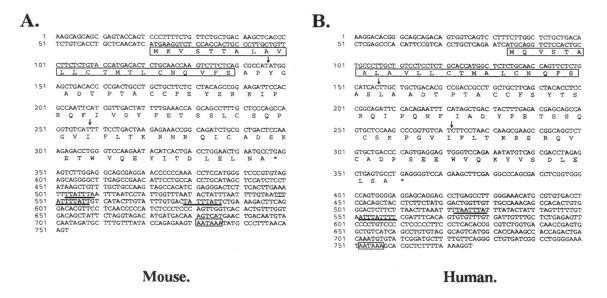

Figure 31.1   The cDNA and genomic structure of MIP-1α genes.

pseudogene lacking the upstream control regions and most of the first intron found in α and β. Like LD78β, the γ form is present in variable amounts at 17q11-q21, and can in fact be absent from some individuals (Hirashima *et al.*, 1992).

2.1 MIP-1α EXPRESSION

Perhaps the most pertinent feature concerning the control of MIP-1α expression, is the large number of diverse stimuli which cause a rapid and transient induction of MIP-1α transcription in a number of different cell types (discussed in greater length in Section 4). In unstimulated cells, it is, with a few exceptions, difficult to detect MIP-1α transcripts and protein detection remains elusive. For instance, MIP-1α mRNA is only detectable in the bone marrow of normal animals using sensitive PCR protocols, although this may be sufficient to create high local concentrations around a small population of target cells, aided perhaps by the proteoglycan binding properties of the protein (see below). The inducibility features of MIP-1α suggest that like other chemokines it falls into the class of rapid-response, or immediate-early, genes. Bearing in mind the inflammatory roles of MIP-1α, it is likely that this is an important step in a cascade of events leading to chemoattraction and activation of leukocytes typical of inflammatory invasion of sites of injury and infection.

How diverse stimuli converge to activate the MIP-1α promoter poses complex and interesting questions. It is known from the inclusion of protein synthesis inhibitors in induction experiments that some cell types (e.g., T cells) require *de novo* protein synthesis before the MIP-1α gene is expressed, while others do not, such as macrophages (Zipfel *et al.*, 1989; Nakao *et al.*, 1990; Grove and Plumb, 1993). As a model system for studying the transcriptional control of MIP-1α, Grove and Plumb (1993) in our laboratory performed a detailed analysis of the murine MIP-1α promoter in a macrophage cell line before and after stimulation with serum and LPS. They demonstrated that proximal promoter sequences (+36 to −220 bp relative to the transcription start site) are sufficient for the low level of basal macrophage-specific transcription that is observed, but that LPS and serum response elements are within −160 to +1. This region harbors functional binding sites for nucleoprotein complexes containing C/EBP, NFκB, and PU.1/SpiB and/or closely related proteins which appear to be responsible for the low level of basal expression observed in macrophages. Alterations in the specific contents of these complexes correlate with the transcriptional activation of the gene. Interestingly, several of these control elements are found in the promoters of other cytokine and chemokine genes, including human MIP-1α, implying the existence of common control mechanisms in their expression (Nakao *et al.*, 1990;

Grove and Plumb, 1993). Indeed, Nomiyama and colleagues (1993) have shown that in T cells one of the C/EBP-like binding sites at position −100 in the LD78α and β genes can be recognized by both positive and negative factors: the positive factors play a key role in phorbol ester- and phytohemaggluttinin-induced transactivation, whilst the negative factors maintain basal T cell LD78 expression at an undetectable level. Thus, it appears that even during low or undetectable MIP-1α expression, sequence-specific DNA binding proteins are bound to the promoter: this may maintain an open chromatin structure to allow the rapid induction of transactivation, with subtle changes in the constituents of the bound nucleoprotein complexes. More extensive analysis of the control of MIP-1α gene expression is required to fully understand how its promoter is being activated in response to varied inflammatory signals. Further, since a number of diseases, such as rheumatoid arthritis (Hosaka *et al.*, 1994), and some virally infected cells (Canque and Gluckman, 1994; Baba *et al.*, 1996; Sprenger *et al.*, 1996) are typified by elevated production of MIP-1α (see below), promoter studies could elucidate the mechanisms responsible for this aberrant expression.

3. The MIP-1α Protein

The chemokine family of proteins is defined on the basis of limited sequence homology and on the presence of four positionally conserved cysteine residues (Schall, 1991; Stoeckle and Barker, 1990). It can be further subdivided on the basis of the specific arrangement of the four cysteines, with the α-chemokines having a −−CXC−−C−−C−− motif and the β-chemokines having a −−CC−−C−−C−− motif. MIP-1α is a member of the β chemokine family, and the relationship in primary sequence between MIP-1α and the other members of the β-chemokine family is summarized in Figure 31.2. The closest relative of MIP-1α is MIP-1β, which was initially identified as a copurifying molecule from macrophages (Wolpe *et al.*, 1988). MIP-1β is 60% identical to MIP-1α and appears to be coordinately expressed. The human homolog of murine MIP-1β is ACT-2. The disulfide bonding pattern has been elucidated for the chemokines and appears to involve disulfide bridge formation between cysteines 1 and 3, and 2 and 4 (Figure 31.3) (Tanaka *et al.*, 1988).The integrity of these disulfide bridges appears to be critical for chemokine function.

Human MIP-1α is synthesized as a 92-amino-acid (aa) precursor from which a 22-aa leader sequence is cleaved during secretion to yield the mature 70-aa peptide which has a calculated molecular mass of 7787.6 kDa and an isoelectric point of 4.6. The peptide lacks *N*-linked glycosylation sites but has putative *O*-linked sites, although the peptide does not appear to be glycosylated as both the mammalian and bacterial recombinant

```
MIP-1α    APYGADTPTA  CC FSY SRKIPRQFIVDYFE TSSL  C SQPGVIFLTKRNRQI  C ADSKETWVQEYITDLELNA
MIP-1β    APMGSDPPTS  CC FSYTSRQLHRSFYMDYYE TSSL  C SKPAVVFLTKRGRQI  C ANPSEPWVTEYMSDLELN
LD 78     ASLAADTPTA  CC FSYTSRQIPQNFIADYFE TSSQ  C SKPGVVFLTKRSRQV  C ADPSEEWVQKYVSDLELSA
ACT-2     APMGSDPPTA  CC FSYTARKLPRNFVVDYYE TSSL  C SQPAVVFQTKRSKQV  C ADPSESWVQEYVYDLELN
MCAF      QPDAINAPVT  CC YNFTNRKISVQRLASYRRITSSK  C PKEAVIFKTIVAKEI  C ADPKQKWVQDSMDHLDKQTQTPKT
RANTES    ASPYSSDTTP  CC FAYIARPLPRAHIKEYFY TSGK  C SNPAVVFVTRKNRQV  C ANPEKKWVREYINSLEMS
MCP2      QPDSVSIPIT  CC FNVINRKIPIQRLESYTRITNIQ  C PKEAVIFKTKRGKEV  C ADPKERWVRDSMKHLDQIFQNLKP
MCP-3     QPVGINTSTT  CC YRFINKKIPKQRLESYRRTTSSH  C PREAVIFKTKLDKEI  C ADPTQKWVQDFMKHLDKKTQTPKL
JE        QPDAVNAPLT  CC YSFTSKMIPMSRLESYKRITSSR  C PKEAVVFVTKLKREV  C ADPKKEWVQTYIKNLDRNQMRSE...
```

Figure 31.2   The amino acid sequence alignment of the C-C family of chemokines. Proteoglycan binding sites are as indicated: underlined and bold.

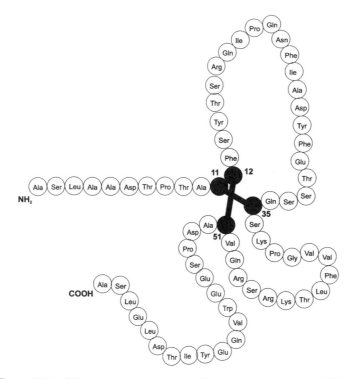

Figure 31.3   Primary structure and disulfide bonding of human MIP-1α.

peptides migrate with identical molecular masses in SDS gels. This is in contrast to the closely related murine MIP-1β peptide which has been suggested to contain a putative N-linked glycosylation site; however, the presence of a proline residue as the X in Asn-X.Ser/Thr would suggest that glycosylation in this area is unlikely to occur in the mature peptide (Shakin-Eshelman *et al.*, 1996). The human homolog of MIP-1β, ACT-2, does not have an N-linked glycosylation site.

Murine and human (h)MIP-1α are both remarkably stable proteins (Graham *et al.*, 1992) which can withstand acidification and heating to elevated temperatures. Full biological activity in stem cell inhibitory assays is retained following heating to 75°C for 1 h and up to 80% activity is maintained following heating to 100°C for 10 min. The protein is inactivated by a range of proteolytic enzymes including trypsin, chymotrypsin, *Staphylococcus aureus* V8 (Graham *et al.*, 1992). MIP-1α is also stable on prolonged storage at 4°C, although we routinely store the protein at −20°C. It

is stable to repeated freeze–thaw cycles. It is recommended that storage of recombinant forms of the peptide is in polypropylene containers (such as Eppendorf tubes) as it tends to stick avidly to polystyrene surfaces. Storage of MIP-1α at very high concentrations (>5–10 mg/ml) is not recommended owing to the formation of insoluble aggregates (see below). Carrier protein is not required for solutions of 0.1 mg/ml and above; however, it is recommended that carrier protein (e.g., bovine serum albumin) be added to more dilute solutions. For use in assay systems that do not contain serum, it is again recommended, where appropriate, that a carrier protein be added.

3.1 PROTEOGLYCAN BINDING

All members of both the α- and β-chemokine families have been characterized as being "heparin binding", a property used to great effect in the purification of many of these peptides. In our experience MIP-1α binds to heparin affinity columns and requires solutions of 0.25–0.5 M NaCl to remove it at physiological pH. Among members of the α-chemokine family, such as interleukin-8 (IL-8) and platelet factor 4 (PF$_4$), the heparin binding site has been mapped to the highly basic carboxy tail region (Loscalzo *et al.*, 1985). However, the site of heparin binding is not as clearly defined for the β-chemokine family, many members of which possess a neutral or even acidic carboxy tail. We have attempted to analyze the site of glycosaminoglycan binding on MIP-1α and have concentrated our studies on the highly conserved tribasic region underlined in Figure 31.2. Synthetic peptides spanning this region of the protein are able to bind to heparin, and mutagenesis of this region to neutralize the basic charges removes the heparin binding ability of MIP-1α (Graham *et al.*, 1996). It is our belief therefore that in the β-chemokine subfamily the proteoglycan binding is dependent on the conserved tribasic region. It is assumed that this may also be a contributor to α-chemokine binding, but a minor one as one can delete the PF$_4$ carboxy-terminus and remove heparin binding capacity (Maione *et al.*, 1991).

The functional significance of chemokine proteoglycan binding is not yet clear. There is, however, evidence from studies on IL-8, an α-chemokine, which suggests that although proteoglycan binding is not required for bioactivity, soluble heparan sulfate was observed to enhance IL-8 bioactivity up to 4-fold in neutrophil chemotaxis assays (Webb *et al.*, 1993). Our own data on MIP-1α (Graham *et al.*, 1996) suggest that proteoglycan binding is not a prerequisite for function of this peptide in either stem cell inhibitory or monocyte chemoattraction assays.

The other area in which a role for glycosaminoglycan binding of chemokines has been proposed is in the function and migration of leukocytes, particularly with respect to presentation of chemokines within sites of active inflammation (Tanaka *et al.*, 1993; Gilat *et al.*, 1994).

3.2 TERTIARY STRUCTURE STUDIES

While the tertiary structure of MIP-1α has not yet been reported, a number of chemokine structures have been resolved by both NMR and x-ray crystallography that are likely to be representative of the MIP-1α structure.

The first member of the chemokine family with a complete tertiary structure elucidated was platelet factor 4, which has been crystallized as a tetramer (StCharles *et al.*, 1989). The secondary structure of the single PF$_4$ peptide from amino- to carboxy-terminus consists of an extended loop, three strands of antiparallel β-sheet arranged in a Greek key, and one carboxy-terminal α-helix. The individual peptides appear to build up the

tetrameric structure by forming dimers by hydrogen-bonding between individual residues within the two component chains. Interaction between the two dimers to form the tetramer involves hydrophilic residues and results in a tetrameric structure with the carboxy-terminal α-helices on the surface of the protein. This may be important for bioactivity.

Following determination of the crystal structure of PF$_4$, the three-dimensional structure of IL-8 was determined by both NMR (Clore *et al.*, 1990) and x-ray crystallography (Baldwin *et al.*, 1991). The structure unambiguously corresponds to a dimer with a similar dimerization surface to that seen for PF$_4$. Again the monomer consists of three antiparallel β-sheets connected with loops and a long carboxy-terminal α-helix. It has been observed that the overall architecture is similar to that seen for the a1/a2 domains of the human class 1 histocompatability antigen (HLA-A2) and the suggestion has been made that the two α-helices which are presented to the outside of the PF$_4$ structure and are external to the IL-8 dimer are involved in receptor binding.

More recently, the first reports of structural analysis of β-chemokine family members have appeared. This initially involved modeling of the structure of the monocyte chemoattractant and activating protein (MCAF/MCP-1) on the basis of the known solution structure of IL-8 (Gronenborn and Clore, 1991), an approach that has subsequently proved to be inappropriate for approximation of higher order β-chemokine structure. More recently, the high-resolution structure of the β-chemokine ACT-2 (human MIP-1β) has been determined (Figure 31.4) by multidimensional NMR (Lodi *et al.*, 1994). This structure, like IL-8, is a symmetric homodimer; however, while the monomer structure is very similar to that observed for IL-8, the quaternary structures of the two proteins are distinct. This distinction is most noticeable in that the dimer interface is formed by a different set of residues and, furthermore, while the IL-8 structure is globular, the structure of the MIP-1β dimer is elongated and cylindrical. The publication of the structure of the β-chemokine RANTES (Skelton *et al.*, 1995) has confirmed these broad structural differences between the α- and β-chemokines. It is argued that the general lack of cross competition between α- and β-chemokines for receptor binding can be explained on the basis of these differences in tertiary structure (see below).

We are in the process (in collaboration with Neil Isaacs and John MacLean, University of Glasgow) of resolving the tertiary structures of dimeric and tetrameric mutants of MIP-1α (Graham *et al.*, 1994). Our preliminary analysis indicates that the structure of MIP-1α is essentially superimposable on those of MIP-1β and RANTES, suggesting a high degree of conservation of tertiary structure in the β-chemokine family.

carboxy terminus

amino terminus

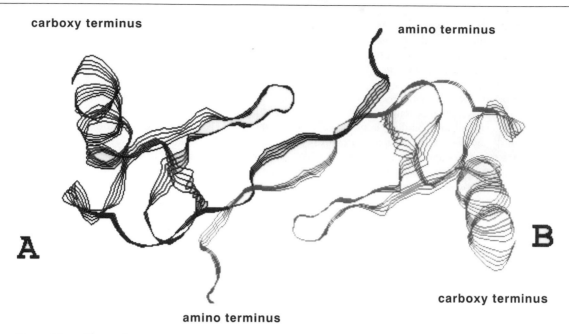

A

B

amino terminus

carboxy terminus

Figure 31.4 **The tertiary structure of human MIP-1β. This structure was defined by NMR analysis. Identical monomers are indicated as A and B.**

3.3 QUATERNARY STRUCTURE: AGGREGATION

As discussed above, it has been assumed on the basis of crystal structure that the stable form of the chemokine proteins is a homodimer. In the case of MIP-1α and MIP-1β and their human homologs LD78 and ACT-2, however, the dimeric forms are able to undergo further self-aggregation to form very large multimeric structures (Graham *et al.*, 1992; Patel *et al.*, 1993). The extent of aggregation appears to be dependent on the concentration of the peptide and on the buffer in which the peptide is dissolved. Under certain circumstances, molecular masses in excess of 10^6 Da have been observed for MIP-1α and the aggregated form of human MIP-1β has been reported to be of sufficient size to be easily visualized by electron microscopy (Lodi *et al.*, 1994). In our experience, MIP-1α has a native molecular mass of around 100 kDa in physiological buffers such as PBS at a concentration of 0.1 mg/ml (Graham *et al.*, 1992). It appears, therefore, that under these circumstances MIP-1α is a dodecamer.

We and others have demonstrated this self aggregation to be noncovalent and probably electrostatic in nature with substantial reversal of aggregation being seen in high salt concentrations (Graham *et al.*, 1992; Patel *et al.*, 1993). We have extended these observations and have demonstrated the ability of dilute solutions of acetic acid (10 mM) to be effective in disrupting the aggregates into their monomeric form. Interestingly, despite distinct mechanisms for dimerization as mentioned above, the IL-8 dimer is also reduced to the monomeric form by addition of acetic acid (Gayle *et al.*, 1993).

We have been able to demonstrate that the self-aggregation seen in MIP-1α, and presumably with the other β-chemokines, is a consequence of the interaction of carboxy-terminal acidic amino acid residues with clusters of internal basic residues (Graham *et al.*, 1994, 1996). Simple neutralization of the carboxy-terminal acidic amino acid residues using PCR-based mutagenesis has allowed generation of nonaggregating mutant variants of MIP-1α. To date, we have generated stable monomeric, dimeric, and tetrameric variants of MIP-1α. Bioactivity analysis of these mutants in both stem cells and inflammatory function assays indicates that they are all bioactive. Interestingly, all three mutants and the wild-type molecule have identical activity profiles in both assay systems. It appears that the only way to argue for identical activity profiles resulting from dodecameric, tetrameric, dimeric, and monomeric variants of MIP-1α is to assume that all forms of the molecule ultimately interact with their receptors as monomers and that under conditions of the assay all forms are monomeric.

We have confirmed this by dilution studies in which we have observed that sequential dilution of wild-type MIP-1α progressively breaks up the aggregate via tetramers and dimers to the monomeric form. It appears therefore that the aggregation of MIP-1α is a dynamic and reversible phenomenon and that the protein is almost exclusively in the monomeric form at concentrations less than 100 ng/ml (Graham *et al.*, 1994).

The observations that MIP-1α appears to interact with its receptor as a monomer are in apparent contrast to the suggestions from structural analysis indicating MIP-1α and other chemokines to be stable dimers. In this respect it is interesting to note that IL-8, which appears to be dimeric by NMR and crystallographic analyses, is able to function as a monomer which is blocked in the dimerization domain (Rajarathnam *et al.*, 1994). It may be, therefore, that all the chemokines can actively function as monomers and that dimerization observed in crystallographic studies results from the high concentrations of peptide required for crystal formation. This suggestion has been confirmed by a study reporting that other chemokines exist in an equilibrium between the monomeric and aggregated states and that the predominant form in bioassays is the monomeric form (Paolini *et al.*, 1994).

4. Cellular Sources and Production

It is now clear that MIP-1α is inducible in virtually all mature hemopoietic cells and many cell lines of hemopoietic origin. As mentioned above, it is entirely conceivable, however, that very low levels of the protein are expressed constitutively; indeed, transcripts for MIP-1α are seen in both peripheral blood and bone marrow cells from healthy individuals, indicating a constitutive level of expression within the hemopoietic system in the absence of overt inflammation (Cluitmans *et al.*, 1995). Detection of such transcripts, however, requires the use of sensitive PCR techniques and thus it is likely that transcript levels are low. In accordance with this suggestion, we have been unable to detect any evidence of the murine MIP-1α protein in extracts from normal murine bone marrow.

Monocytes or macrophages from a range of tissue sites are also able to act as abundant sources of MIP-1α; however, although sensitive PCR-based techniques allow detection of MIP-1α transcripts in resting cells, induction is again required for production of easily detectable amounts of either mRNA or protein. In macrophages, MIP-1α mRNA and protein can be strongly induced by a range of agents such as the bacterial endotoxin, lipopolysaccharide (LPS), phytohemagglutinin, IL-1, IL-2, IL-3, and IL-6, viral infection, and endothelial cell adhesion, reflecting the central role of this chemokine in inflammatory reactions (Davatellis *et al.*, 1988; Martin and Dorf, 1991; Christman *et al.*, 1992; Lukacs *et al.*, 1993, 1994a; Baba *et al.*, 1996; Canque *et al.*, 1996). We have also reported potent induction of MIP-1α transcription and translation following growth factor starvation and re-feeding of M-CSF-dependent murine bone marrow macrophage populations (Maltman *et al.*, 1993). The induction of mRNA peaks at 4 h after growth factor refeeding and remains elevated for up to 24 h. Similar kinetics of induction are seen in macrophages induced with LPS.

Other hemopoietic sources of MIP-1α include polymorphonuclear leukocytes (PMNs) which can be induced to secrete detectable MIP-1α transcripts and protein products following LPS induction. In the context of PMNs, GM-CSF synergizes with LPS to increase the level of induction (Kasama *et al.*, 1993). Eosinophils (Costa *et al.*, 1993), mast cells (Selvan *et al.*, 1994, Ebisawa *et al.*, 1996), peripheral blood basophils (Li *et al.*, 1996), and platelets (Klinger *et al.*, 1995) are also a source of MIP-1α. It should be noted, however, that identification of a source does not implicate the protein alone in any particular inflammatory reaction, since in many cases a range of chemokines and cytokines are inducible in mature blood cells. Indeed, it is likely that there is considerable synergy between individual chemokines in the cascade of events during an inflammatory response after the initial stimulus. The constitutive expression of MIP-1α in Langerhans cells is exceptional and is considered below. There have been no reports of red blood cells being a source of MIP-1α.

MIP-1α expression can also be detected in cells at a number of non-hemopoietic sites. For example, stimulation of either primary human fibroblasts or human glioma cells induces expression of MIP-1α to levels detectable by northern blotting (Nakao *et al.*, 1990). In another tissue, PCR analysis of intact epidermis reveals evidence for the presence of MIP-1α transcripts (Matsue *et al.*, 1992). Upon further analysis it has been demonstrated by us, and others, that the source of these transcripts is the epidermal Langerhans cells (ELCs) and that keratinocytes do not express MIP-1α (Heufler *et al.*, 1992a,b; Matsue *et al.*, 1992; Parkinson *et al.*, 1993). Indeed, MIP-1α appears to be one of the major cytokines produced by resting ELCs. Intriguingly, expression of MIP-1α by ELCs appears to be confined to ELCs in situ or immediately following isolation as the levels of MIP-1α drop to undetectable levels on culture of the cells (Heufler *et al.*, 1992b). Thus, in contrast to the situation that is observed with T cells, macrophages, and PMNs, it appears that MIP-1α is constitutively expressed in ELCs and that it is downregulated rather than upregulated following induction of ELCs. We have postulated that the role for ELC-derived MIP-1α is to act as an endogenous regulator of epidermal cell proliferation (see below). There is also evidence for expression of MIP-1α within dendritic epidermal T cells (Matsue *et al.*, 1993).

A number of studies have demonstrated the ability of a range of factors to suppress the production of MIP-1α by hemopoietic cells. For example, IL-4 appears to be able to suppress production of MIP-1α by both monocytes and alveolar macrophages following stimulation with LPS, PHA, or IL-1, and the mode of action of IL-4 in this respect appears to be to increase the rate of decay of the MIP-1α mRNA (Standiford *et al.*, 1993a). We have also demonstrated the ability of transforming growth factor-β (TGF-β) to potently downregulate MIP-1α expression in murine bone marrow-derived macrophages with substantial suppression of MIP-1α transcription being seen at femtomolar concentrations of TGF-β₁ (Maltman *et al.*, 1993). This property is shared by all three TGF-β isoforms; however, more divergent members of this family such as activin or bone morphogenetic proteins appear to be incapable of exerting similar effects at equivalent concentrations. We have also used the TGF-β latency associated protein to block endogenous TGF-β function in bone marrow macrophage cultures. This results in an upregulated expression of MIP-1α in the macrophages, indicating a role for TGF-β in suppressing MIP-1α expression by macrophages in an autocrine manner. There also appears to exist a reciprocal relationship in which MIP-1α acts to induce TGF-β expression in bone marrow macrophages (Maltman *et al.*, 1996), suggesting that a powerful negative regulatory loop is set up between TGF-β and MIP-1α and that all interactions between these two peptides act to minimize MIP-1α expression. It may be, therefore, given the widespread expression of TGF-β, that this suppression is of physiological relevance. It must, however, be recognized that the relative ease with which MIP-1α can be detected in the peripheral blood and bone marrow of healthy individuals (Cluitmans *et al.*, 1995) suggests that there are mechanisms *in vivo* through which this TGF-β mediated block on MIP-1α expression is alleviated.

Other suppressers of MIP-1α expression include interferon (IFN)-γ (Kasama *et al.*, 1995a) and IL-10 (Kasama *et al.*, 1994). The IL-1 receptor antagonist also blocks MIP-1α production in mixed lymphocyte reactions, indicating the involvement of IL-1α in the cytokine cascade (Lukacs *et al.*, 1993).

5. Biological Activities

A wide-ranging series of studies have revealed significant pro-inflammatory properties attributable to MIP-1α. In many cases these conclusions have been made from *in vitro* studies indicating potential as a potent inflammatory agent *in vivo*, and although there are relatively few of the latter studies published to date, it is now clear that MIP-1α is likely to play a key role in a number of inflammatory processes *in vivo*.

5.1 Pro-inflammatory Activities

The β-chemokine family comprises a number of potent mediators of inflammatory responses which are likely to play a major role in recruiting various inflammatory cells to the sites of infection. This family, which includes MIP-1α, MIP-1β, MCP-1, and RANTES among others, has been studied intensively over the years. An early activity ascribed to MIP-1α was modulation of

macrophage function (Fahey *et al.*, 1992; Wang *et al.*, 1993). The finding that MIP-1α can induce cytokine production in macrophages *in vitro* (Fahey *et al.*, 1992) suggests that it may contribute to potentiation of priming of effector cells in inflammatory reactions. The fact that macrophages are an inducible source of MIP-1α suggests that autocrine mechanisms must feature in the biology of this protein. Thus, a potential key role would be to function as an autocrine mediator for the macrophage production of TNF-α (Shanley *et al.*, 1995). The conclusion that MIP-1α must play an important role in inflammatory reactions is further underlined by the fact that MIP-1α will activate basophils or mast cells (Alam *et al*, 1992; Bischoff *et al.*, 1993) and recruit eosinophils (Rot *et al.*, 1992; Lukacs *et al.*, 1995). These findings suggest a probable role in allergic inflammation (Baggiolini and Dahinden, 1994) and this activity may well be crucial in protection against parasite infections by virtue of the induced histamine release and activation of eosinophil cytotoxic activity. The potential role of chemokines in the specific targeting of lymphocyte subpopulations during immune challenge or inflammatory response is an area of considerable interest and importance in lymphocyte biology and MIP-1α most likely plays a role in the recruitment of lymphocytes. Thus, when analyzed *in vitro* in microchemotaxis experiments, MIP-1α has been shown to be a potent lymphocyte attractant with differing specificities dependent on the concentration of chemokine tested (Schall *et al.*, 1993). MIP-1α is chemotactic for resting and activated CD4⁺, CD8⁺ T cells and B cells, but it appears that it is primarily chemotactic for activated CD8⁺ lymphocytes, whereas the closely related MIP-1β is chemotactic for activated CD4⁺ cells (Schall *et al.*, 1993; Taub *et al.*, 1993). MIP-1α has been shown to induce B cell migration as well as natural killer (NK) cell migration and cytolysis (Schall *et al.*, 1993; Maghazachi *et al.*, 1994; Taub *et al.*, 1995). Many of these activities are likely to involve changes in adherence properties and there are a number of observations, such as enhancement of the ability of CD8⁺ T cells to bind endothelial cells (Taub *et al.*, 1993), induction of T lymphocyte adherence (Lloyd *et al.*, 1996), and induction of monocyte integrins (Vaddi and Newton, 1994a) and the observation that MIP-1α restores the integrin-dependent adhesion of chronic myeloid leukemia (CML) cells (Bhatia *et al.*, 1995), which support this conclusion. All these observations suggest that MIP-1α may provide a means by which monocytes adhere to vessel walls and direct T cells to the site of antigenic challenge (Cook, 1996). In view of the gene knockout studies with MIP-1α (see below), it is likely that the observations on modulation of T cell adherence and recruitment or chemoattraction of T cell subsets point to an important physiological role in this area.

In addition to the range of pro-inflammatory activities demonstrated, activities in other settings are indicated by observations that many different cell types from a range of tissues are responsive to MIP-1α *in vitro*. For example, while MIP-1α will stimulate immature spermatogenic cell proliferation, the division of more differentiated intermediate or type B spermatogonia is inhibited (Hakovirta *et al.*, 1994). Astrocyte proliferation has also been reported to be inhibited by MIP-1α (Khan and Wigley, 1994), but more immature precursors appear to be unaffected (G. Graham, unpublished); Schwann cell proliferation can be stimulated by this chemokine (Khan and Wigley, 1994). Interestingly, it has also been postulated that MIP-1α may play a role in prostaglandin-independent induction of fever (Davatelis *et al.*, 1989; Zawada *et al.*, 1994; Minano *et al.*, 1996). Modulation of adherence is again suggested by the observation that motility of rat osteoclasts is affected by MIP-1α (Fuller *et al.*, 1995). In many cases the biological activities observed are not unique to MIP-1α, as other chemokines such as RANTES have been shown to have similar properties. This is not too surprising in view of the observations that MIP-1α and other chemokine receptors are fairly promiscuous in their range of ligand binding (see below).

5.2 STEM CELL PROLIFERATION

Apart from the pro-inflammatory properties which have been described above, MIP-1α has also been shown to inhibit proliferation of hemopoietic stem cells (Graham *et al.*, 1990) and keratinocytes (Parkinson *et al.*, 1993). In the case of the hemopoietic stem cell, these effects have also been demonstrated *in vivo* (Dunlop *et al.*, 1992; Lord *et al.*, 1992; Maze *et al.*, 1992). There is some evidence that the inhibitory action of MIP-1α is restricted to the CFU-A/CFU-S part of the stem cell compartment known as transient engrafting stem cells (Jacobsen *et al.*, 1994; Keller *et al.*, 1994), whereas cells with a more primitive stem cell phenotype are not inhibited (Quesniaux *et al.*, 1993; Keller *et al.*, 1994). Moreover, it appears that MIP-1α may even stimulate more primitive stem cells under certain conditions (Verfaillie and Miller, 1995) as well as mature progenitors (Broxmeyer *et al.*, 1990). It has also been suggested that MIP-1α has the potential role of inducing self-renewal in the more primitive CFU-S compartment (Verfaillie *et al.*, 1994; Lord, 1995). Thus, potentially important roles for MIP-1α in hemopoietic physiology are suggested by these reports. In addition, we have shown that MIP-1α will also inhibit clonogenic keratinocyte proliferation *in vitro* (Parkinson *et al.*, 1993) and that a likely source of MIP-1α in the skin is the Langerhans cells (see references above for various sources of MIP-1α). Interestingly, MIP-1α produced in bacteria is completely inactive in this assay, and production in COS cells is required to elicit an inhibitory response. We postulated that this may be due to a

requirement for an accessory factor present in COS cells, but the retention of inhibitory activity upon purification to homogeneity of COS cell-produced MIP-1α suggests that some form of post-translational modification of the protein occurs in COS cells that unmasks keratinocyte inhibition potential (Parkinson *et al.*, 1993; G. Graham, unpublished).

Thus, a number of activities have been described for MIP-1α, mostly based on *in vitro* experimentation which may not necessarily indicate an important role *in vivo*. However, more recent experiments using homologous recombination or monoclonal anti-MIP-1α antibodies have provided evidence for a fundamental role for MIP-1α in inflammatory responses, and possibly proliferative inhibition.

5.3 GENETIC APPROACHES TO THE ROLE OF MIP-1α *IN VIVO*

One very useful approach to the elucidation of the role of a cytokine/chemokine in physiology is to inactivate the gene of interest by homologous recombination. Recently, gene disruption has revealed some unexpected roles for MIP-1α *in vivo* (Cook *et al.*, 1995). Mice homozygous for the disruption showed no overt abnormalities in development nor were any changes in peripheral blood and bone marrow indices noted in untreated null mice, suggesting that MIP-1α is not required to maintain normal stem cell quiescence in normal physiology and that other stem cell inhibitory activities must compensate, since the cycling status of the stem cell compartment is low in untreated null mice (B. Pragnell, unpublished). However, MIP-1α null mice exhibited interesting phenotypes upon infection with certain viruses. For example, they had a reduced inflammatory response to influenza virus characterized by a significantly reduced pulmonary edema at necropsy and less mononuclear cell infiltration than is seen in control infected mice. There is also a delay in the viral clearance from the lungs of the null animals, a process which is known to be dependent on T cells. Thus, MIP-1α seems to be essential to mount a proper inflammatory response to this virus. Remarkably, the null mice also demonstrated a complete absence of myocarditis when infected with coxsackie virus. Substantial evidence suggests that in normal mice these cardiac lesions result from cell killing mediated primarily by cytotoxic T cells, although direct virally mediated effects may also occur. This again indicates that MIP-1α may be essential for efficient recruitment of the T cells. A somewhat surprising conclusion from these observations is that, at least during these particular viral infections, there is a lack of functional redundancy. The β-chemokines are a large family of proteins with what appear to be overlapping functions *in vitro* and, as discussed below, they often bind and signal through common receptors. However, these studies provide genetic evidence that MIP-1α has an indispensable physiological role in the inflammatory response to specific viral infections. It remains to be seen whether MIP-1α has such a central role in the pathology of other diseases induced either by viral infection or by other agents. Diseases in which MIP-1α may be involved and thus which are worthy of investigation are discussed below.

As mentioned previously, another biological activity ascribed to MIP-1α *in vitro* is suppression of proliferation of epidermal keratinocytes (Parkinson *et al.*, 1993) and it has been shown that Langerhans cells from skin are a source of MIP-1α *in vitro* (Heufler *et al.*, 1992a,b). It was therefore of interest to examine the skins of null mice and their normal littermates. We have observed a striking hyperproliferation of keratinocytes in the skins of the knockout mice (S. Holmes, J. De Bono, G. Graham, and I.B. Pragnell, unpublished), which is consistent with a novel role for Langerhans cells and MIP-1α in keratinocyte physiology (Parkinson *et al.*, 1993). It is likely that manipulation of this gene knockout model will reveal more about the role of MIP-1α in inflammation and various disease states.

6. Receptors

The mechanism by which MIP-1α exerts its pleiotropic effects on its numerous target cells is poorly understood. However, the recent cloning of receptors for MIP-1α and related β-chemokines is beginning to provide an insight into the molecular mechanisms of MIP-1α function, and has also had some exciting implications for research into the human immunodeficiency virus (HIV) which causes AIDS. Historically, classical receptor binding experiments using radioiodinated chemokine have demonstrated the presence of specific cell surface receptors for MIP-1α on many cell types, and in a number of these studies the generation of a Ca^{2+} flux within the treated cell has been used as an indication of intracellular signaling (Oh *et al.*, 1991; Yamamura *et al.*, 1992; Graham *et al.*, 1993; Sozzani *et al.*, 1993; Wang *et al.*, 1993; Zhou *et al.*, 1993; Avalos *et al.*, 1994; Van Riper *et al.*, 1994). Consistent with the widespread effect of MIP-1α, almost all primitive and mature hemopoietic cells tested possess receptors for this chemokine. Such experiments have also determined (a) the affinity of interaction between the ligand and receptor, which is given as the dissociation constant (K_d)—the concentration of ligand at which 50% of the receptors are occupied; (b) the approximate number of receptors per cell; and (c) which other β-chemokines are able to displace and compete for MIP-1α binding, a property which indicates receptor sharing by related molecules. In many cases, but not all, pretreatment of a cell with a

competing ligand prevents it from responding to subsequent MIP-1α challenge. This phenomenon, termed desensitization, may in part be due to temporary inhibitory covalent modifications of the receptor induced by a signal elicited on binding of the first ligand and is discussed at greater length below.

MIP-1α receptor biology has been complicated by variations observed between cell types with respect to the affinity of binding and the competition with other β-chemokines. First, there appears to be at least two affinity subclasses: one with high affinity, with K_d values in the picomolar range (Graham *et al.*, 1993; Avalos *et al.*, 1994), and a second with lower affinity, K_d being in the nanomolar range (Avalos *et al.*, 1994). This is perhaps analogous to binding of human melanocyte growth-stimulatory activities (MGSA), an α-chemokine, to the two cloned IL-8 receptors: IL-8 receptor type A binds MGSA weakly ($K_d = 450$ nM), while IL-8 receptor type B interacts with higher affinity ($K_d = 2$ nM), a property dependent on the extracellular amino terminus of the receptors (reviewed in Horuk, 1994). Incidentally, both receptors bind IL-8 with equivalent avidity, with K_d values of between 1 and 4 nM. Second, detailed cross-competition studies by Wang and colleagues (1993) suggest that on monocytes alone there are at least three types of β-chemokine receptor: one that binds MIP-1α and MIP-1β; one that is specific for the β-chemokine MCP-1; and a third more promiscuous receptor able to bind MCP-1, MIP-1α, and MIP-1β. Neutrophils, on the other hand, appear to express a low-affinity receptor able to bind to MIP-1α and the related protein RANTES (Gao *et al.*, 1993; McColl *et al.*, 1993). These biochemical observations were hypothesized to be indicative of the existence of a number of different genes encoding MIP-1α receptors, or alternatively that the properties of a single receptor could be influenced by covalent modifications or protein–protein interactions within a specific cell type.

Since the mid 1990s there has been a considerable effort to understand the molecular basis of MIP-1α receptor biology, and the existence of multiple MIP-1α receptor genes – possibly five to date – has been demonstrated. By analogy with the α-chemokine receptors, it was highly likely that MIP-1α and the other β-chemokines utilize serpentine receptors, proteins that contain seven helical membrane-spanning domains and are coupled to heterotrimeric G-proteins (reviewed in Horuk, 1994). Moreover, chemokine-induced Ca^{2+} fluxes into many different target cells could be prevented by pretreatment with cholera or pertussis toxin, which prevent G-protein–receptor coupling. As a consequence, it has been possible to design degenerate oligonucleotide primers from conserved regions within this superfamily for use in PCR cloning protocols. Neote and his colleagues used such an approach to clone the first β-chemokine receptor from HL60 mRNA, a promyelocytic cell line known to express both high- and low-affinity

MIP-1α receptors (Neote *et al.*, 1993; Avalos *et al.*, 1994). This gene was originally called C-C CKR-1, but has recently been renamed CCR-1 to fit with accepted nomenclature. The predicted amino acid sequence and the presumed structure of the protein in the membrane are shown in Figure 31.5. The four conserved cysteines residues are circled: these residues are essential in ligand binding and are thought to maintain the receptor in a cylindrical structure. CCR-1 had ~30% identity to the two IL-8 receptors and ~20% identity to the receptors for C5a and fMet-Leu-Phe (FMLP), peptides also involved in inflammation (Neote *et al.*, 1993). When CCR-1 was expressed in human embryonic kidney cells (which do not have β-chemokine receptors), they became able to bind human and murine MIP-1α with a K_d of ~5 nM and elicit a transient increase in intracellular calcium ions when the transfected cells were challenged with MIP-1α at a concentration of 10 nM. Radioiodinated MIP-1α could be displaced by excess unlabeled MIP-1α and also by excess β-chemokines RANTES, MIP-1β, or MCP-1, although 20–100-fold more of these other chemokines was required: α-chemokines were unable to bind or signal through CCR-1. Furthermore, although MIP-1β and MCP-1 appeared not to signal through CCR-1 except at high concentration (>1 mM), RANTES generated a Ca^{2+} flux at concentrations as low as 1 nM, an effect abolished by MIP-1α pretreatment. However, RANTES was not able to prevent a subsequent MIP-1α-induced signal, a discrepancy which may reflect different interactions between the two ligands and the receptor,

resulting in the generation of signals distinct to each chemokine. Interestingly, the reverse is observed on eosinophils: RANTES pretreatment prevents a subsequent MIP-1α signal, but MIP-1α does not interfere with a RANTES-induced Ca^{2+} flux (Rot *et al.*, 1992). More recently, MCP-3 and MCP-2 have been shown to bind to and signal through CCR-1, further demonstrating the variable ligand specificity that this receptor displays (Combadiere *et al.*, 1995; Ben-Baruch *et al.*, 1995; Gong *et al.*, 1997).

Northern blot analysis of hemopoietic cell mRNA demonstrated the presence of a 3 kb transcript in HL-60, THP-1, U937, bone marrow-derived macrophages, and B and T cell lines, and CCR-1 protein is probably responsible for some, if not all, of the MIP-1α binding exhibited by these cells (Neote *et al.*, 1993; Gao *et al.*, 1993; Nomura *et al.*, 1993). The murine homolog has been identified, but ligand recognition is altered in this protein: RANTES and human MCP-3 appear not to bind as efficiently to the murine protein, suggesting that slight sequence differences may affect ligand–receptor interaction (Gao and Murphy, 1995; R. Nibbs, unpublished observations).

Recently, five other ligand-binding β-chemokine receptors have been identified, plus a number of orphan receptors, which all show strong sequence homology to one another (reviewed in Murphy, 1996; Premack and Schall, 1996). CCR-2 was cloned shortly after CCR-1 and has been shown to interact with MCP-1, -2 and -3 when expressed in heterologous cells (Charo *et al.*, 1994;

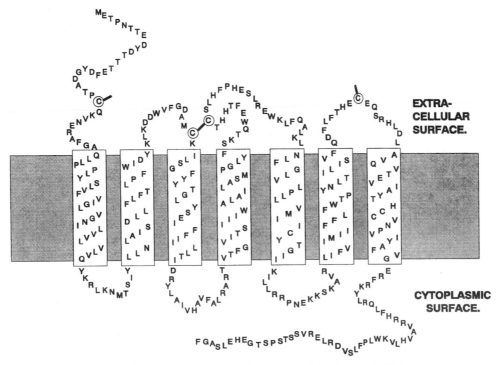

Figure 31.5 Sequence and hypothetical structure of the human CCR-1 receptor.

Combadiere *et al.*, 1995; Gong *et al.*, 1997). This gene is currently unique among chemokine receptors in that it can be produced as two alternatively spliced forms, allowing coupling to distinct subsets of G-proteins, potentially widening the biological responses in which this protein may be involved (Kuang *et al.*, 1996). The inability of CCR-2 to interact with MIP-1α has enabled the domains of CCR-1 involved in MIP-1α binding to be identified through the production of CCR-1/CCR-2 chimeras (Monteclaro and Charo, 1996). While the amino-termini of CCR-2 and the two IL-8 receptors appear to be the main region of ligand interaction, CCR-1 uses the three extracellular loops (in particular the third loop) to interact with MIP-1α (Horuk, 1994; Monteclaro and Charo, 1996). We have shown that a mutant of murine MIP-1α, in which two basic residues between the third and fourth cysteines are neutralized, is unable to bind to murine CCR-1, defining this region of MIP-1α as a potential interface for CCR-1 binding (Graham *et al.*, 1996).

Despite the binding properties of CCR-1, several cell types are responsive to MIP-1α and the other CCR-1 ligands in the absence of detectable CCR-1 expression, suggesting that genes exist which code for other MIP-1α receptors. This has proved to be the case and a steadily increasing family of heptahelical receptors have been cloned which, when expressed in heterologous cells, are able to bind to β-chemokines (reviewed in Murphy, 1996; Premack and Schall, 1996). Human CCR-3, which is expressed on eosinophils, is able to bind to eotaxin, an eosinophil-specific β-chemokine, and RANTES, MCP-3, and MCP-4 with K_d values in the low nanomolar range (Daugherty *et al.*, 1996; Kitaura *et al.*, 1996; Ponath *et al.*, 1996). Human CCR-5 is able to bind to RANTES, MIP-1α and MIP-1β with a somewhat higher affinity than CCR-1 and it is expressed in monocytes, macrophages, and certain T-cells (Boring *et al.*, 1996; Raport *et al.*, 1996; Samson *et al.*, 1996). CCR-5 is very similar in sequence to CCR-2, with only the amino-termini exhibiting extensive amino acid differences. This domain of divergence is probably responsible for the differences in ligand recognition that exist between the two proteins. Similarly, a murine gene, called CCR-3, has been cloned which is highly homologous to human CCR-3, diverging at the amino terminus. Despite the sequence similarities, and the fact that both are expressed in eosinophils, murine CCR-3 may not be a functional homolog of human CCR-3. This is supported by its ability to bind to human and murine MIP-1α and MIP-1β (K_d in the low nanomolar range), lack of binding to RANTES and MCP-3, and its abundant expression in cells of the monocyte/macrophage and T cell lineages (Post *et al.*, 1995; Nibbs *et al.*, 1997). CCR-4, present on T cells, basophils, and platelets, has also been reported to bind to MIP-1α when expressed in *Xenopus* oocytes or HL-60 myeloid cells (Power *et al.*, 1995; Hoogewerf *et al.*, 1996), but this appears to be a fairly weak interaction and we have been unable to confirm these observations when the murine receptor is expressed in hamster CHO cells (R. Nibbs, unpublished observations). We have independently cloned all the murine counterparts of these genes in our laboratory using degenerate oligonucleotide-primed PCR and demonstrated similar affinity and specificity data. Furthermore, we have also isolated a more divergent member of this gene family, called D6, that is expressed in T cells and monocytes and acts as a very high-affinity binding site for MIP-1α ($K_d = 110$ pM): it is also able to bind with high affinity to the related molecules MIP-1β, MCP-1, RANTES, and MCP-3 (Nibbs *et al.*, 1997; Nibbs *et al.*, in press).

In summary, it is apparent that there are at least four, and possibly more, heptahelical receptors that are involved in mediating the biological effects of MIP-1α, and none of these is specific for MIP-1α alone. These receptors are expressed on mature hemopoietic cells and are therefore probably involved in mediating MIP-1α-induced chemotaxis. None of the cloned MIP-1α receptors is expressed on keratinocytes (R. Nibbs, unpublished) so the inhibition of these cells uses another mechanism (Parkinson *et al.*, 1993). However, murine D6 is present in populations of bone marrow cells enriched for hemopoietic stem cells, and may potentially be involved in MIP-1α-induced stem cell inhibition (Nibbs *et al.*, 1997). Other cytokines and growth factors rarely have such a large number of potential receptors, and the reason for this complexity is unclear. However, unlike other receptor types, the open reading frames of these genes are usually contained on a single exon. Thus, with the exception of CCR-2, it appears that rather than evolving alternative splicing to generate groups of related but functionally distinct proteins, gene duplications and subtle sequence changes have occurred. There is a quite dramatic variation in the affinity of MIP-1α for the presently cloned receptors, such that they can be divided into two potential classes: high-affinity (D6 and CCR-5) and low-affinity (CCR-1, -3 and -4). This may be important functionally as it could allow a cell expressing one receptor from both classes the ability to respond differentially to a wide variation in MIP-1α concentrations, and may have relevance when considering cell migration along a chemokine concentration gradient.

Surprisingly, β-chemokine receptors show homology, particularly in the amino-terminus, to the US28 open reading frame (ORF) encoded by human cytomegalovirus (CMV) (Neote *et al.*, 1993). Infection with CMV, a β herpesvirus, is usually asymptomatic, but can cause mononucleosis syndrome in normal individuals and severe gastrointestinal, pulmonary, and retinal inflammation in immunocompromised hosts, in particular patients with AIDS or those undergoing immunosuppressive therapy during organ transplantation. US28 can act as a receptor for the β-chemokines MIP-1α, MIP-1β, RANTES, and MCP-1 ($K_d = 2-6$ nM), and elicit an intracellular Ca^{2+} flux in response to ligand binding; α-chemokines cannot bind (Neote *et al.*, 1993; Gao and Murphy, 1994). Although US28 can therefore be classified as a promiscuous β-chemokine receptor that can transduce a signal, the biological role of this protein in CMV infections is unknown. It is possible that CMV-infected cells may become responsive to these inflammatory mediators and thus US28 may play a role in the induction of viral replication or, conversely, be involved in the establishment of a latent state. Interestingly, CMV is not alone in employing what has been termed "molecular mimicry", and ORFs in pox virus and herpes virus have been identified that show extensive structural and functional homology with a variety of immunoregulatory molecules, such as ECRF3 from herpes virus saimiri which acts as a promiscuous signaling receptor for α-chemokines (reviewed in Ahuja *et al.*, 1994).

6.1 RECEPTORS AND HIV

A dramatic development in chemokine receptor biology was the elucidation of their role in cellular infection by the human immunodeficiency virus (HIV) that causes AIDS (reviewed in Wilkinson, 1996). HIV can be broadly divided into two subtypes, namely macrophage (or M)-tropic, which can infect macrophages and T cells, and T cell line (or T)-tropic strains which show a restriction to T cells. Generally, the M-tropic strains are responsible for the transmission of AIDS, while the T-tropic strains appear late in the disease, possibly evolving from M-tropic subtypes, and may be associated with the onset of the pathogenic symptoms of the disease. The role of chemokine receptors was initiated as a consequence of two observations: first, host factors that were able to prevent macrophage infection with HIV were identified to be identical to RANTES, MIP-1α, and MIP-1β (Cocchi *et al.*, 1995); and second, high levels of endogenous β-chemokines were detectable in uninfected individuals at high risk of HIV infection, potentially accounting for the lack of infection (Paxton *et al.*, 1996). The reason for these phenomena was quickly shown to be the fact that M-tropic viruses use CCR-5 as an obligate coreceptor with CD4 for cell infection, although CCR-2 and CCR-3 could also be used by some viral isolates (Wilkinson, 1996). This appears to occur via a stepwise series of protein–protein interactions: the HIV surface glycoprotein gp120 associates with CD4; gp120 then undergoes a conformational change that unmasks the tropism-determining hypervariable V3 domain, which allows gp120 to interact directly with CCR-5 (Cocchi *et al.*, 1996; Trkola *et al.*, 1996; Wu *et al.*, 1996). The ligands for these receptors are able to abrogate this interaction to prevent cell infection. Once gp120 and CCR-5 interact, it is thought that a second hydrophobic transmembrane HIV envelope protein,

gp41, is able to interact with the membrane of the target cell to catalyze fusion of the viral and cellular membranes (reviewed in Wain-Hobson, 1996). Interestingly, HIV infection of macrophages induces MIP-1α expression (Canque *et al.*, 1996) which may act to dampen subsequent infections of neighboring cells. Moreover, some individuals whose cells are resistant to HIV infection have been shown to be homozygous for a short deletion in CCR-5 that prevents proper presentation of the protein on the cell surface (Dean *et al.*, 1996; Huang *et al.*, 1996; Liu *et al.*, 1996; Samson *et al.*, 1996). In parallel, it was shown that the T-tropic strains use a different but related chemokine receptor CXCR-4 (also called fusin and LESTR) as a coreceptor along with CD4 for entry into T cells via a mechanism similar to that seen with CCR-5 (reviewed in Lapham *et al.*, 1996; Wilkinson, 1996). This protein normally acts as a receptor for stromal cell-derived factor-1 (SDF-1), an α-chemokine, and, like the observation that CCR-5 ligands block M-tropic viral infection, SDF-1 prevents T-tropic HIV infection of T cells (Bleul *et al.*, 1996; Oberlin *et al.*, 1996). These intriguing observations may have useful implications toward the development of AIDS therapies targeted at disrupting HIV binding to chemokine-binding coreceptors (see below).

7. *Signal Transduction*

Although little is understood concerning the precise molecular nature of signaling by MIP-1α–receptor complexes, it is highly likely that the chemotactic properties of MIP-1α are manifested by using G-protein-linked pathways that are employed by other leukocyte chemoattractants which use similar heptahelical receptors. The details of these potential signaling pathways are beyond the scope of this review and are summarized elsewhere (Bokoch, 1995). The use of G-proteins in MIP-1α signaling is supported by a number of observations. First, it has been demonstrated that the effects of MIP-1α on certain responsive cells are prevented by pretreating the cells with *Bordetella pertussis* toxin, which specifically inhibits a subset of G-proteins, namely Ga_i (Bischoff *et al.*, 1993; McColl *et al.*, 1993). Second, MIP-1α induces a rapid but transient increase in the cytosolic concentration of Ca^{2+} ions in responsive cells due to an increase in inositol trisphosphate generated by G-protein-activated phospholipase C (PLC) (Rot *et al.*, 1992; Sozzani *et al.*, 1993; Gao *et al.*, 1993; Van Riper *et al.*, 1994; Vaddi and Newton, 1994b). Third, as mentioned above, MIP-1α is able to desensitize cells to a subsequent challenge with MIP-1α (homologous desensitization) or other related β-chemokines (heterologous desensitization). In the case of the β-adrenoreceptor and the rhodopsin "light receptor", a specific serine–threonine kinase activated by components of the trimeric G-protein complex phosphorylates carboxy-terminal serine residues on the receptor to abrogate signaling (Inglese *et al.*, 1993); β-chemokine receptor desensitization may work through a similar mechanism, although specific kinases have yet to be identified.

The heterotrimeric G-protein complex consists of α, β and γ subunits (reviewed in Neer, 1995; Neer and Smith, 1996). Gα binds to GTP and catalyzes it conversion to GDP; Gβ and Gγ form a highly stable dimer which is considered to function essentially as a monomer. When GDP is bound to Gα it allows this protein to associate with the Gβγ dimer, which in turn associates with the cytoplasmic domains of receptors. Upon ligand binding, a conformational change in the receptor causes GDP to be displaced from Gα and, owing to the high ratio of GTP to GDP in the cytoplasm, GTP binds in its place and causes Gα to lose contact with Gβγ and the receptor. Gα-GTP and Gβγ then act as the second messengers, altering the activity of a variety of enzymes and thus initiating the cascade of events that will eventually result in the phenotypic alteration of the cell. Once the GTP is hydrolyzed, Gα-GDP reassociates with Gβγ, returning to an inactive state attached to the receptor ready for the next stimulus: desensitization of receptors possibly interferes with this final step of the cycle.

The number of potential targets for the activated Gα-GTP and free Gβγ is large and steadily increasing with new discoveries. For example, Gα regulates phospholipase C, membrane cation channels, and the cAMP-generating enzyme adenylyl cyclase (Neer, 1995), while free Gβγ can also have a direct effect on adenylyl cyclase, increase arachidonic acid synthesis via phospholipase A_2 activation (reviewed in Clapham and Neer, 1993), and stimulate the ras/mitogen-activated protein kinase (MAPK) cascade (Crespo *et al.*, 1994; Faure *et al.*, 1994). Monocyte chemotaxis in response MIP-1α can be partially inhibited by antisense oligonucleotides to phospholipase C, and MCP-1 and IL-8 are both able to stimulate the MAPK cascade, so both these pathways are likely to be involved in chemokine function (Druey *et al.*, 1996; Knall *et al.*, 1996; Locati *et al.*, 1996; Wu *et al.*, 1993). However, many isoforms of α, β, and γ exist which vary with respect to (a) their spatial and temporal expression, (b) their receptor specificity, (c) their efficiency of GTP hydrolysis and (d) their extent of inhibition or activation of effector molecules (reviewed in Neer, 1995). For example, pertussis toxin does not inhibit signaling through some serpentine receptors as they use Gα proteins outwith the $G\alpha_i$ subclass. Also, whilst $G\alpha_i$-GTP proteins inhibit adenylyl cyclase to reduce intracellular cAMP levels, the GTP-bound form of the $G\alpha_s$ subclass actually activates the same enzyme to increase the cAMP concentration. Thus, it seems possible that the same ligand–receptor complex could elicit quite different biological effects depending on the type of G-protein complexes available

within the target cell and also that two different receptors may work through different pathways when expressed in different cell types. This hypothesis has obvious implications with respect to MIP-1α, which signals through several receptors to produce a varied range of biological effects, and care must be taken when extrapolating the results generated with one chemokine in a particular cell type to the effect of MIP-1α on a different cell type. The identification of the G-protein complexes involved in specific cell types with each of the receptors discussed above should be one of the first steps in understanding MIP-1α-induced signaling.

The study of the stem cell inhibitory function of MIP-1α has been simplified by the characterization of cell lines whose proliferation is reduced in the presence of MIP-1α. FDCP-Mix, a mouse multipotent hemopoietic cell line, and MO7e cells derived from a patient with acute megakaryocytic leukemia are both inhibited to some extent by MIP-1α (Graham *et al.*, 1993; Mantel *et al.*, 1995), and these can now be used to begin to unravel the complexities of the signal transduction cascade. Indeed, Mantel and colleagues have demonstrated that in MO7e cells treated with MIP-1α there is an increase in phosphatidylcholine (PC) turnover and a rise in the level of cAMP, probably due to activation of phospholipase C (PLC) and adenylyl cyclase, respectively (Mantel *et al.*, 1995). Interestingly, PC breakdown by PLC is known to activate the transcription factor NFκB, which in turn stimulates the production of mRNA for the p53 protein, a potent growth suppressor (Schutze *et al.*, 1992; Wu and Lozanno, 1994). Furthermore, increase in cAMP has for many years been known to inhibit the proliferation of some hemopoietic cells (Kurland *et al.*, 1977; Rock *et al.*, 1992) and, although the exact nature of this suppression remains unknown, it is intriguing that cAMP has been shown to prevent ras–raf association and therefore abrogate the activation of the MAPK cascade, a major mitogenic stimulus (Cook and McCormick, 1993). Indeed, it has recently been demonstrated that the MAPK pathway is inhibited in MO7e cells by MIP-1α treatment (Aronica *et al.*, 1995), which suggests that MIP-1α may inhibit cell proliferation by directly impinging on the signals initiated by proliferation inducers.

Signaling targets for MIP-1α are only beginning to be identified and more extensive studies are required before we can fully understand the molecular basis for the biological effects of MIP-1α.

8. *MIP-1α in Disease and Therapeutic Implications*

The potential for MIP-1α to inhibit normal hemopoietic stem cell proliferation suggested a therapeutic role for MIP-1α as a stem cell protection agent during chemotherapy for cancer some time ago (Graham and Pragnell, 1991), and indeed clinical trials are now under way to test this possibility in a number of centers. MIP-1α may also play a role in the development of leukemia. We have found, along with other investigators, that in contrast to normal hemopoietic stem cells, the proliferation of the most primitive progenitors from bone marrow and peripheral blood from patients with chronic myeloid leukemia (CML) is not prevented by MIP-1α treatment *in vitro* (Eaves *et al.*, 1993; Holyoake *et al.*, 1993; Nirsimloo and Gordon, 1995). In another hemopoietic neoplasm, acute myeloid leukemia, there is a range of responses to MIP-1α treatment, some blast cells being refractory to MIP-1α *in vitro* and others from different patients displaying a moderate response (Ferrajoli *et al.*, 1994; Basara *et al.*, 1996; Owen-Lynch *et al.*, 1996). Thus use of MIP-1α for purging in these diseases may be possible, but more development of this approach will be required. As with CML, the proliferation of cell lines established from various stages during the development of squamous cell carcinoma is unaffected by MIP-1α treatment, despite their being derived from keratinocytes whose proliferation can be inhibited by this chemokine (G. Graham, unpublished). Whether the development of these neoplasms requires loss of MIP-1α responsiveness remains to be determined. It has also been observed that there is marked increase in the expression of MIP-1α mRNA in patients with bone marrow failure, aplastic anemia, myelodysplastic syndrome, and hypereosinophilia, and in several types of leukemia (Yamamura *et al.*, 1989; Maciejewski *et al.*, 1992; Costa *et al.*, 1993), though the significance of these results remains to be explored.

The wide range of the biological activities described above indicates that MIP-1α expression may feature in a range of inflammatory diseases, both acute and chronic, and indeed this is the case. Considerable attention has focused on inflammatory lung disease, which can be induced experimentally in laboratory animals and is often associated with a considerable upregulation of MIP-1α expression (Standiford *et al.*, 1993; and reviewed in Strieter *et al.*, 1996; Driscoll, 1994). Pulmonary alveolar macrophages have been shown to be a potent source of MIP-1α, although it is likely that other cell types may also contribute to chemokine production during inflammatory reactions. Recently, a number of reports have demonstrated that neutralizing antibodies to MIP-1α can significantly abrogate leukocyte infiltration into the lung and reduce associated tissue damage. For example, pulmonary fibrosis and mononuclear cell accumulation induced by treatment with bleomycin are reduced when the animals are inoculated with anti-MIP-1α antibodies (Standiford *et al.*, 1993; Smith *et al.*, 1994). Similar antibody inocula reduce neutrophil numbers in bronchoalveolar lavage from mice with acute lung injury resulting from intratracheal administration of LPS or IgG complexes (Shanley *et al.*, 1995). These two

models are quite distinct: bleomycin stimulates a T cell-dependent inflammatory response, while LPS injury operates mainly through neutrophils. Both, however, appear to use MIP-1α as a chemoattractant at some stage during the development of the pathology. Furthermore, MIP-1α has been implicated in allergic inflammation of the lung, characterized by eosinophil influx and activation, and again in this example anti-MIP-1α antibodies can partially abrogate eosinophil accumulation, although an associated neutrophil influx is unaffected (reviewed in Strieter *et al.*, 1996). Thus, animal models of lung inflammatory diseases individually characterized by infiltration of a broad spectrum of leukocyte populations exhibit a dependency on MIP-1α for full pathological expression. Finally, as discussed above, the reduced inflammation and viral clearance seen in MIP-1α null mice infected with the influenza virus shows that MIP-1α has a role in protecting the lung from certain viruses. In conjunction with these observations is the demonstration that MIP-1α, and other β-chemokines, are induced in human monocytes infected with influenza virus *in vitro* (Sprenger *et al.*, 1996). It is of interest also to note here that HIV-1-infected alveolar macrophages express MIP-1α and this expression may contribute to the CD8+ alveolitis (Denis and Ghadirian, 1994).

MIP-1α has also been implicated in rheumatoid arthritis. In synovial fibroblasts from arthritic patients, LD78 is highly abundant and may play a role in the infiltration of the joint by leukocytes apparent in the disease (Hosaka *et al.*, 1994; Koch *et al.*, 1994; and see review by Kunkel *et al.*, 1995). The murine type II collagen-induced arthritis has been used as a model for the human form of the disease and shows an upregulation of MIP-1α expression paralleling the incidence and magnitude of the arthritis, with macrophages, chondrocytes, and fibroblasts as the likely source of the chemokine. Antibodies to MIP-1α induced a significant reduction in the extent of the disease when inoculated into test animals during the period of arthritic induction (Kasama *et al.*, 1995b).

In another animal model, experimental autoimmune encephalitis (EAE), MIP-1α antibody treatment once more leads to an amelioration of the severity of the disease (Godiska *et al.*, 1995; Karpus *et al.*, 1995; Miyagishi *et al.*, 1995). EAE is a T cell-mediated autoimmune demyelinating disease which serves as a model for the human disease multiple sclerosis. Both acute and relapsing disease was prevented in this study and the accompanying mononuclear infiltration into the CNS was very much reduced. It is possible that astrocytes, microglia, and encephalitogenic T cells serve as sources of MIP-1α in this disease (Karpus *et al.*, 1995; Murphy *et al.*, 1995). Moreover, MIP-1α appears to be partially responsible for PMN and mononuclear influx into the central nervous systems of mice infected with *Listeria monocytogenes*, which is used as a model for bacterial meningitis in humans (Seebach et al., 1995).

The models described above are just a few examples in which a role for MIP-1α in inflammatory responses has been clearly demonstrated. It is likely that this list will increase, and in fact MIP-1α has already been implicated in cutaneous inflammation (Schroder et al., 1996), cutaneous Leishmaniasis (Ritter et al., 1996), and inflammation in the kidney induced with anti-alpha-3(IV) collagen antibodies (Danoff et al., 1995). It is of interest to note here that HIV-1-infected alveolar macrophages express MIP-1α and this expression may contribute to the CD8⁺ alveolitis in the disease (Denis and Ghadirian, 1994). A detailed examination of all of inflammatory disease models in the MIP-1α null mice should provide further evidence to support the implications from the studies using anti-MIP-1α antibodies, and one may predict that these mice may show a less severe pathology as is observed in coxsackie virus infection (Cook et al., 1995).

The therapeutic implications of the association of MIP-1α with these various disease states may be manifold, and one could hypothesize that MIP-1α antagonists may prove to be useful weapons with which to combat certain chronic and acute inflammatory diseases in man, such as rheumatoid arthritis, asthma, and multiple sclerosis. However, it is likely that MIP-1α is just one of the players in a highly complex cascade of events involving many other chemokines and cytokines responsible for the development of the inflammatory response, and it is debatable whether abrogation of MIP-1α alone will be sufficient to avert the pathological manifestations of the disease in question. Indeed, inhibiting the function of other α- and β-chemokines has also been shown to reduce inflammatory cell influx into affected tissue (reviewed in Strieter et al., 1996) and a combinatorial approach to chemokine inhibition may prove to be more successful. On a more cautionary note, accumulating evidence discussed above implies a crucial role for MIP-1α in fighting viral infection and it is possible that regimes aimed at preventing MIP-1α function may compromise protection from certain viruses.

Perhaps the most exciting potential for MIP-1α is suggested by the recent observations that MIP-1α, RANTES, and MIP-1β are major HIV-suppressive factors produced by CD8 T cells (Cocchi et al., 1995). This field has become even more exciting as the initial discoveries that chemokine receptors are cofactors for HIV entry into CD4⁺ lymphocytes are rapidly extended. As discussed at length above, it has now been shown that HIV resistance is seen in individuals who are homozygous for a mutation in the CCR-5 gene, a receptor for MIP-1α, RANTES, and MIP-1β. This mutation appears to result in the improper trafficking of CCR-5 to the cell surface, and one would assume a complete loss of function: individuals with this mutation

are nonetheless phenotypically normal, with the exception of being resistant to HIV (Wilkinson, 1996). Therefore, it is unlikely that small-molecule antagonists to this receptor will have an adverse affect on the health of the individual, and the development of such drugs is being intensively pursued in many laboratories. In fact, mutants of RANTES have been described that are able to antagonize HIV infection without activating the CCR-5 receptor (Arenzana-Seisdedos et al., 1996; Simmons et al., 1997). Such a molecule may have potential use as an anti-HIV drug without being hampered by the chemotactic and activating effects of the wild-type protein; similar MIP-1α-based antagonists could also be developed. A better understanding of receptor–ligand interactions should permit the development of other potential drugs and is awaited with anticipation.

9. Summary

There has been much exciting progress in the development of our knowledge of MIP-1α, particularly concerning protein structure, the complexities of its receptor biology, and the involvement of this chemokine in a range of chronic and acute inflammatory diseases. The specific roles attributed to MIP-1α by in vivo and in vitro studies strongly suggest therapeutic potential for either the protein or antagonists, particularly in important diseases like AIDS, multiple sclerosis, and rheumatoid arthritis, and also as a possible hemoprotective agent during cancer treatment. More detailed studies of the relevant animal models may reveal further activities to be exploited, particularly in the area of virus-induced disease, and the MIP-1α null animals should prove to be a valuable tool. Finally, the availability of genes encoding MIP-1α receptors should now allow molecular dissection of the mechanisms involved in mediating the pleiotropic effects of this fascinating protein.

10. References

Ahuja, S.K., Gao, J.-L. and Murphy, P.M. (1994). Chemokine receptors and molecular mimicry. Immunol. Today 15, 281.

Alam, R., Forsythe, P.A., Stafford, S., Lettbrown, M.A. and Grant, J.A. (1992). Macrophage inflammatory protein-1-alpha activates basophils and mast-cells. J. Exp. Med. 176, 781.

Arenzana-Seisdedos, F., Virelizier, J.-L., Rousset, D., et al. (1996). HIV blocked by chemokine antagonist. Nature 383, 400.

Aronica, S.M., Mantel, C., Gonin, R., et al. (1995). Interferon-inducible protein 10 and macrophage inflammatory protein-1 alpha inhibit growth factor stimulation of Raf-1 kinase activity and protein synthesis in a human growth factor-dependent hematopoietic cell line. J. Biol. Chem. 270, 21998.

Avalos, B.R., Bartynski, K.J., Elder, P.J., Kotur, M.S., Burton, W.G. and Wilkie, N.M. (1994). The active monomeric form of macrophage inflammatory protein-1α interacts with high- and low-affinity classes of receptors on human hematopoietic cells. Blood 84, 1790.

Baba, M., Imai, T., Yoshida, T. and Yoshie, O. (1996). Constitutive expression of various chemokine genes in human T-cell leukemia-virus type-1-role of the viral transactivator tax. Int. J. Cancer 66, 124.

Baggiolini, M. and Dahinden, A.D. (1994). CC chemokines in allergic inflammation. Immunol. Today 15, 127.

Baldwin, E.T., Weber, I.T., StCharles, R., et al. (1991). Crystal structure of Interleukin 8: symbiosis of NMR and crystallography. Proc. Natl Acad. Sci. USA 88, 502.

Basara, N., Stosic-Grujicic, S., Sefer, D., Ivanovic, Z., Antunovic, P. and Milenkovic, P. (1996). The inhibitory effect of human macrophage inflammatory protein-1 alpha (LD78) on acute myeloid leukemia cells in vitro. Stem Cells 14, 445.

Ben-Baruch, A., Xu, L., Young, P.R., Bengali, K., Oppenheim, J.J. and Wang, J.M. (1995). Monocyte chemotactic protein-3 (MCP3) interacts with multiple leukocyte receptors. J. Biol. Chem. 270, 22123.

Bhatia, R., McGlave, P.B. and Verfaillie, C.M. (1995). Treatment of marrow stroma with interferon-alpha restores normal beta-1 integrin-dependent adhesion of chronic myelogenous leukemia hematopoietic progenitors—role of MIP-1-alpha. J. Clin. Invest. 96, 931.

Bischoff, S.C., Krieger, M., Brunner, T., et al. (1993). RANTES and related chemokines activate human basophil granulocytes through different G protein-coupled receptors. Eur. J. Immunol. 23, 761.

Bleul, C.C., Farzan, M., Choe, H., et al. (1996). The lymphocyte chemoattractant SDF-1 is a ligand for LESTR/fusin and blocks entry. Nature 382, 829.

Blum, S., Forsdyke, R.E. and Forsdyke, D.R. (1990). Three human homologues of a murine gene encoding an inhibitor of stem cell proliferation. DNA Cell Biol. 9, 589.

Bokoch, G.M. (1995). Chemoattractant signalling and leukocyte activation. Blood 86, 1649.

Boring, L., Gosling, J., Monteclaro, F.S., Lusis, A.J., Tsou, C.-L. and Charo, I.F. (1996). Molecular cloning and functional expression of murine JE (Monocyte chemoattractant protein 1) and murine macrophage inflammatory protein 1α receptors. J. Biol. Chem. 271, 7551.

Broxmeyer, H.E., Sherry, B., Lu, L., et al. (1990). Enhancing and suppressing effects of recombinant murine macrophage inflammatory proteins on colony formation in vitro by bone marrow myeloid progenitor cells. Blood 76, 1110.

Canque, B. and Gluckman, J.C. (1994). MIP-1-alpha is induced by and it inhibits HIV-infection of blood-derived macrophages. Blood 84, 480.

Canque, B., Rosenzwajg, M., Gey, A., Tartour, E., Fridman, W.H. and Gluckman, J.C. (1996). Macrophage inflammatory protein 1-alpha is induced by human immunodeficiency virus infection of monocyte-derived macrophages. Blood 87, 2011.

Caput, D., Beutler, B., Hartog, K., Thayer, R., Brown-Shimer, S. and Cerami, A. (1986). Identification of a common nucleotide sequence in the 3'-untranslated region of mRNA molecules specifying inflammatory mediators. Proc. Natl Acad, Sci. USA 83, 1670.

Charo, I.F., Myers, S.J., Herman, A., Franci, C., Connolly, A.J. and Coughlin, R. (1994). Molecular cloning and functional expression of two monocyte chemoattractant protein 1 receptors reveals alternative splicing of the carboxyl-terminal tails. Proc. Natl Acad. Sci. USA 91, 2752.

Christman, J.W., Blackwell, T.R., Cowan, H.B., Shepherd, V.L. and Rinaldo, J.E. (1992). Endotoxin induces the expression of macrophage inflammatory protein-1 alpha mRNA by rat alveolar and bone marrow macrophages. Am. J. Resp. Cell Mol. Biol. 7, 455.

Clapham, D.E. and Neer, E.J. (1993). New roles for G-protein βγ-dimers in transmembrane signalling. Nature 365, 403.

Cluitmans, F.H.M., Essendam, B.H.J., Landegent, J.E., Willemze, R. and Falkenberg, J.H.F. (1995). Constitutive in vivo cytokine and haemopoietic growth factor gene expression in the bone marrow and peripheral blood of healthy individuals. Blood 85, 2038.

Cocchi, F., DeVico, A.L., Garzino-Demo, A., Arya, S.K., Gallo, R.C. and Lusso, P. (1995). Identification of RANTES, MIP-1α, and MIP-1β as the major HIV-suppressive factors produced by CD8+ T cells. Science 270, 1811.

Cocchi, F., DeVico, A.L., Garzino-Demo, A., Cara, A., Gallo, R.C. and Lusso, P. (1996). The V3 domain of the HIV-1 gp120 envelope glycoprotein is critical for chemokine-mediated blockade of infection. Nature Med. 2, 1244.

Combadiere, C., Ahuja, S.K., Van Damme, J., Tiffany, H.L., Gao, J.-L. and Murphy, P.M. (1995). Monocyte chemoattractant protein-3 is a functional ligand for CC chemokine receptors 1 and 2B. J. Biol. Chem. 270, 29671.

Cook, D.N. (1996). The role of MIP-1-alpha in inflammation and hematopoiesis. J. Leukocyte Biol. 59, 61.

Cook, D.N., Beck, M.A., Coffman, T.M., et al. (1995). Requirement of MIP-1α for inflammatory response to viral infection. Science 269, 1583.

Cook, S.J. and McCormick, F. (1993). Inhibition by cAMP of Ras-dependent activation of Raf. Science 262, 1069.

Costa, J.J., Matossian, K., Resnick, M.B., et al. (1993). Human eosinophils can express the cytokines tumour necrosis factor-alpha and macrophage inflammatory protein-1 alpha. J. Clin. Invest. 91, 2673.

Crespo, P., Xu, N., Simonds, W.F. and Gutkind, J.S. (1994). Ras-dependent activation of MAP kinase pathway mediated by G-protein βγ subunits. Nature 369, 418.

Danoff, T.M., Cook, D.N., Neilson, E.G. and Kalluri, R. (1995). Murine anti-alpha-3(IV) collagen disease is abrogated in MIP-1-alpha deficient mice. J. Am. Soc. Nephrol. 6, 827.

Daugherty, B.L., Siciliano, S.J., DeMartino, J.A., Malkowitz, J.A., Sirotina, A. and Springer, M.S. (1996). Cloning, expression, and characterisation of the human eosinophil eotaxin receptor. J. Exp. Med. 183, 2349.

Davatelis, G., Tekamp-Olson, P., Wolpe, S.D., et al. (1988). Cloning and characterization of a cDNA for murine macrophage inflammatory protein (MIP), a novel monokine with inflammatory and chemokinetic properties. Exp. Med. 167, 1939.

Davatelis, G., Wolpe, S.D., Sherry, B., Dayer, J.-M., Chicheportiche, R. and Cerami, A. (1989). Macrophage inflammatory protein-1: a prostaglandin-independent endogenous pyrogen. Science 243, 1066–1068.

Dean, M., Carrington, M., Winkler, C., et al. (1996). Genetic restriction of HIV-1 infection and progression to AIDS by a deletion allele of the CKR5 structural gene. Science 273, 1856.

Denis, M. and Ghadirian, E. (1994). Alveolar macrophages from subjects infected with HIV-1 express macrophage inflammatory protein-1-alpha (MIP-1-alpha)—contribution to the CD8(+) alveolitis. Clin. Exp. Immunol. 96, 187.

Driscoll, K.E. (1994). Macrophage inflammatory proteins—biology and role in pulmonary inflammation. Exp. Lung Res. 20, 473.

Druey, K.M., Blumer, K.J., Kang, V.H. and Kehri, J.H. (1996). Inhibition of G-protein-mediated MAP kinase activation by a new mammalian gene family. Nature 379, 742.

Dunlop, D.J., Wright, E.G., Lorimore, S., et al. (1992). Demonstration of stem cell inhibition and myeloprotective effects of SCI/rhMIP1α in vivo. Blood 79, 2221.

Eaves, C.J., Cashman, J.D., Wolpe, S.D. and Eaves, A.C. (1993). Unresponsiveness of primitive chronic myeloid-leukemia cells to macrophage inflammatory protein 1-alpha, an inhibitor of primitive normal hematopoietic cells. Proc. Natl Acad. Sci. USA 90, 12015.

Ebisawa, M., Tachimoto, H., Saito, H. and Iikura, Y. (1996). Cultured human mast cells are capable of producing multiple cytokines and chemokines. J. Allergy Clin. Immunol. 97, 706.

Fahey, T.J., Tracey, K.J., Tekampolson, P., et al. (1992). Macrophage inflammatory protein-1 modulates macrophage function. J. Immunol. 148, 2764.

Faure, M., Vonyo-Yasenetskaya, T.A. and Bourne, H.R. (1994). cAMP and βγ subunits of heterotrimeric G proteins stimulate the mitogen-activated protein kinase pathway in COS-7 cells. J. Biol. Chem. 269, 7851.

Ferrajoli, A., Talpaz, M., Zipf, T.F., et al. (1994). Inhibition of acute myelogenous leukemia progenitor proliferation by macrophage inflammatory protein 1-alpha. Leukemia 8, 798.

Friedland, J.S. (1995). Chemokines and human infection. Clin. Sci. 88, 393.

Fuller, K., Owens, J.M. and Chambers, T.J. (1995). Macrophage inflammatory protein-1-alpha and IL-8 stimulate the motility but suppress the resorption of isolated rat osteoclasts. J. Immunol. 154, 6065.

Gao, J.-L. and Murphy, P.M. (1994). Human cytomegalovirus open reading frame US28 encodes a functional β chemokine receptor. J. Biol. Chem. 269, 28539.

Gao. J.-L. and Murphy, P.M. (1995). Cloning and differential tissue-specific expression of three mouse β chemokine receptor-like genes, including the gene for a functional macrophage inflammatory protein-1α receptor. J. Biol. Chem. 270, 17494.

Gao, J.-L., Kuhns, D.B., Tiffany, H.-L., et al. (1993). Structure and functional expression of the human macrophage inflammatory protein 1α/RANTES receptor. J. Exp. Med. 177, 1421.

Gayle, R.B., III, Sleath, P.R., Srinivason, S., et al. (1993). Importance of the amino terminus of the interleukin-8 receptor in ligand interactions. J. Biol. Chem. 268, 7823.

Gilat, D., Hershkoviz, R., Makori, Y.A., Vlodavsky, I. and Lider, L. (1994). Regulation of adhesion of CD4+ T lymphocytes to intact or heparinase-treated subendothelial extracellular matrix by diffusible or anchored RANTES and MIP-1β. J. Immunol. 153, 4899.

Godiska, R., Chantry, D., Dietsch, G.N. and Gray, P.W. (1995). Chemokine expression in murine experimental allergic encephalomyeletis. J. Neuroimmunol. 58, 167.

Gong, X., Gong, W., Kuhns, D.B., Ben-Baruch, A., Howard, O.M.Z. and Wang, J.M. (1997). Monocyte chemotactic protein-2 uses CCR1 and CCR2B as its functional receptors. J. Biol. Chem. 272, 11682.

Graham, G.J. and Pragnell, I.B. (1991). Treating cancer: the potential role of stem cell inhibitors. Eur. J. Cancer 27, 952.

Graham, G.J., Wright, E.G., Hewick, R., et al. (1990). Identification and characterization of an inhibitor of haemopoietic stem cell proliferation. Nature 344, 442.

Graham, G.J., Freshney, M.F., Donaldson, D. and Pragnell, I.B. (1992). Purification and biochemical characterisation of human and murine stem cell inhibitors (SCI). Growth Factors 7, 151.

Graham, G.J., Zhou, L., Weatherbee, J.A., Tsang, M.L.S., Napolitano, M., Leonard, W.J. and Pragnell, I.B. (1993). Characterization of a receptor for macrophage inflammatory protein-1-alpha and related proteins on human and murine cells. Cell Growth Differ. 4, 137.

Graham, G.J., Mackenzie, J., Lowe, S., et al. (1994). Aggregation of the chemokine MIP-1α is a dynamic and reversible phenomenon: biochemical and biological analyses. J. Biol. Chem. 269, 4974.

Graham, G.J., Wilkinson, P.C., Nibbs, R.J.B., et al. (1996). Uncoupling of stem cell inhibition from monocyte chemoattraction in MIP-1 alpha by mutagenesis of the proteoglycan binding site. EMBO J. 15, 6506–6515.

Gronenborn, A.M and Clore, G.M. (1991). Modeling the three-dimensional structure of the monocyte chemo-attractant and activating protein MCAF/MCP-1 on the basis of the solution structure of Interleukin 8. Protein Eng. 4, 263.

Grove, M. and Plumb, M. (1993). C/EBP, NF-kappa B, and c-Ets family members and transcriptional regulation of the cell-specific and inducible macrophage inflammatory protein 1 alpha immediate-early gene. Mol. Cell. Biol. 13, 5276.

Grove, M., Lowe, S., Graham, G., Pragnell, I. and Plumb, M. (1990). Sequence of the murine haemopoietic stem cell inhibitor/macrophage inflammatory protein 1α gene. Nucleic Acids Res. 18, 5561.

Hakovirta, H., Vierula, M., Wolpe, S.D. and Parvinen, M. (1994). MIP-1-alpha is a regulator of mitotic and meiotic DNA-synthesis during spermatogenesis. Mol. Cell. Endocrinol. 99, 119.

Heufler, C., Parkinson, E.K., Graham, G.J., et al. (1992a). Cytokine gene-expression in murine epidermal Langerhans cells—MIP-1-alpha is a major product and inhibits the proliferation of keratinocyte stem-cells. J. Invest. Dermatol. 98, 515.

Heufler, C., Topar, G., Koch, F., et al. (1992b). Cytokine gene expression in murine epidermal cell suspensions: interleukin 1 beta and macrophage inflammatory protein 1 alpha are selectively expressed in Langerhans cells but are differentially regulated in culture. J. Exp. Med. 176, 1221.

Hirashima, M., Ono, T., Nakao, M., et al. (1992). Nucleotide sequence of the third cytokine LD78 gene and mapping of all three LD78 gene loci to human chromosome 17. DNA Sequence 3, 203.

Holyoake, T.L., Freshney, M.G., Sproul, A.M., et al. (1993). Contrasting effects of rh-MIP1α and TGF-β₁ on chronic myeloid leukemia progenitors in vitro. Stem Cells 11, 122.

Hoogewerf, A., Black, D., Proudfoot, A.E., Wells, T.N. and Power, C.A. (1996). Molecular cloning of murine CC CKR-4 and high affinity binding of chemokines to murine and human CC CKR-4. Biochem. Biophys. Res. Commun. 218, 337.

Horuk, R. (1994). Molecular properties of the chemokine receptor family. TIPS 151, 159.

Hosaka, S., Akahoshi, T., Wada, C. and Kondo, H. (1994). Expression of the chemokine superfamily in rheumatoid-arthritis. Clin. Exp. Immunol. 97, 451.

Huang, Y., Paxton, W.A., Wolinsky, S.M., et al. (1996). The role of a mutant CCR5 allele in HIV-1 transmission and disease progression. Nature Med. 2, 1240.

Inglese, J., Freedman, N.J., Koch, W.J. and Lefkowitz, R.J. (1993). Structure and mechanism of G protein-coupled receptor kinases. J. Biol. Chem. 268, 23735.

Irving, S.G., Zipfel, P.F., Balke, J., et al. (1990). Two inflammatory mediator cytokine genes are closely linked and variably amplified on chromosome 17q. Nucleic Acids. Res. 18, 3261.

Jacobsen, S.E.W., Ruscetti, F.W., Ortiz, M., Gooya, J.M. and Keller, J.R. (1994). The growth-response of LIN(−)THY-1(+) hematopoietic progenitors to cytokines is determined by the balance between synergy of multiple stimulators and negative cooperation of multiple inhibitors. Exp. Hematol. 22, 985.

Karpus, W.J., Lukacs, N.W., McRae, B.L., Strieter, R.M., Kunkel, S.L. and Miller, S.D. (1995). An important role for the chemokine macrophage inflammatory protein-1-alpha in the pathogenesis of the T-cell-mediated autoimmune-disease, experimental autoimmune encephalomyelitis. J. Immunol. 155, 5003.

Kasama, T., Strieter, R.M., Standiford, T.J., Burdick M.D. and Kunkel, S.L. (1993). Expression and regulation of human neutrophil-derived macrophage inflammatory protein 1 alpha. J. Exp. Med. 178, 63.

Kasama, T., Strieter, R.M., Lukacs, N.W., Burdick, M.D. and Kunkel, S.L. (1994). Regulation of neutrophil derived chemokine expression by IL-10. J. Immunol. 152, 3559.

Kasama, T., Strieter, R.M., Lukacs, N.W., Lincoln, P.M. Burdick, M.D. and Kunkel, S.L. (1995a). Interferon-gamma modulates the expression of neutrophil derived chemokines. J. Invest. Med. 43, 58.

Kasama, T., Strieter, R.M., Lukacs, N.W., Lincoln, P.M., Burdick, M.D. and Kunkel, S.L. (1995b). Interleukin 10 expression and chemokine regulation during the evolution of murine type II collagen-induced arthritis. J. Clin. Invest. 95, 2868.

Keller, J.B., Bartelmez, S.H., Sitnicka, E., et al. (1994). Distinct and overlapping direct effects of macrophage inflammatory protein-1-alpha and transforming growth-factor-beta on hematopoietic progenitor stem-cell growth. Blood 84, 2175.

Khan, S. and Wigley, C. (1994). Different effects of a macrophage cytokine on proliferation in astrocytes and Schwann-cells. Neuroreport 5, 1381.

Kitaura, M., Nakajima, T., Imai, T., et al. (1996). Molecular cloning of human eotaxin, an eosinophil-selective CC chemokine, and identification of a specific eosinophil eotaxin receptor, CC chemokine receptor 3. J. Biol. Chem. 271, 7725.

Klinger, M.H., Wilhelm, D., Bubel, S., Sticherling, M., Schröder, J.-M. and Kühnel, W. (1995). Immunocyto-chemical localization of the chemokines RANTES and MIP-1α within human platelets and their release during storage. Int. Arch Allergy Immunol. 107, 541.

Knall, C., Young, S., Nick, J.A., Buhl, A.M., Worthen, G.S. and Johnson, G.L. (1996). Interleukin-8 regulation of the Ras/Raf/Mitogen-activated protein kinase pathway in human neutrophils. J. Biol. Chem. 271, 2832.

Koch, A.E., Kunkel, S.L., Harlow, L.A., et al. (1994). Macrophage inflammatory protein-1-alpha—a novel chemotactic cytokine for macrophages in rheumatoid-arthritis. J. Clin. Invest. 93, 921.

Kuang, Y., Wu, Y., Jiang, H. and Wu, D. (1996). Selective G protein coupling by C-C chemokine receptors. J. Biol. Chem. 271, 3975.

Kunkel, S.L., Lukacs, N. and Strieter, R.M. (1995). Expression and biology of neutrophil and endothelial cell-derived chemokines. Semin. Cell Biol. 6, 327.

Kurland, J.I., Hadden, J.W. and Moore, M.A.S. (1977). Role of cyclic nucleotides in the proliferation of committed granulocyte-macrophage progenitor cells. Cancer Res. 37, 4534.

Lapham, C.K., Ouyang, J., Chandrasekhar, B., Nguyen, N.Y., Dimitrov, D.S. and Golding, H. (1996). Evidence for cell-surface association between fusin and the CD4-gp120 complex in human cell lines. Science 274, 602.

Li, H.M., Sim, T.C, Grant, J.A. and Alam, R. (1996). The production of macrophage inflammatory protein 1 alpha by human basophils. J. Immunol. 157, 1207.

Liu, R., Paxton, W.A., Choe, S., et al. (1996). Homozygous defect in HIV-1 coreceptor accounts for resistance of some multiply-exposed individuals to HIV-1 infection. Cell 86, 367.

Lloyd, A.R., Oppenheim, J.J., Kelvin, D.J. and Taub, D.D. (1996). Chemokines regulate T-cell adherence to recombinant adhesion molecules and extracellular-matrix proteins. J. Immunol. 156, 932.

Locati, M., Lamorte, G., Luini, W., et al. (1996). Inhibition of monocyte chemotaxis to C-C chemokines by antisense oligonucleotide for cytosolic phospholipase A$_2$. J. Biol. Chem. 271, 6010.

Lodi, P.J., Garrett, D.S., Kuszewski, J., et al. (1994). High resolution solution structure of the β-chemokine hMIP-1β by multidimensional NMR. Science 263, 1762.

Lord, B.I. (1995). MIP-1-alpha increases the self-renewal capacity of the hematopoietic spleen-colony-forming cells following hydroxyurea treatment in vivo. Growth Factors 12, 145.

Lord, B.I., Dexter, T.M., Clements, J.M., Hunter, M.A. and Gearing, A.J.H. (1992). Macrophage-inflammatory protein protects multipotent hematopoietic cells from cytotoxic effects of hydroxyurea in vivo. Blood 79, 2605.

Loscalzo, J., Melnick, B. and Handin, R.I. (1985). The interaction of platelet factor four and glycosaminoglycans. Arch. Biochem. Biophys. 240, 446.

Lukacs, N.W., Kunkel, S.L., Burdick, M.D., Lincoln, P.M. and Strieter, R.M. (1993). Interleukin-1 receptor antagonist blocks chemokine production in the mixed lymphocyte reaction. Blood 82, 3668.

Lukacs, N.W., Chensue, S.W., Smith, R.E., et al. (1994a). Production of monocyte chemoattractant protein-1 and macrophage inflammatory protein-1-alpha by inflammatory granuloma fibroblasts. Am. J. Pathol. 144, 711.

Lukacs, N.W., Strieter, R.M., Elner, V.M., Evanoff, H.L., Burdick, M. and Kunkel, S.L. (1994b). Intercellular adhesion molecule-1 mediates the expression of monocyte-derived MIP-1α during monocyte-endothelial cell interactions. Blood 83, 1174.

Lukacs, M.W., Strieter, R.M., Shaklee, C.L., Chensue, S.W. and Kunkel, S.L. (1995). Macrophage inflammatory protein-1-alpha influences eosinophil recruitment in antigen-specific airway inflammation. Eur. J. Immunol. 25, 245.

Maciejewski, J.P., Liu, J.M., Green, S.W., *et al.* (1992). Expression of stem cell inhibitor (SCI) gene in patients with bone marrow failure. Exp. Hematol. 20, 1112.

Maghazachi, A.A., Al-Aoukaty, A. and Schall, T.J. (1994). C-C chemokines induce the chemotaxis of nk and IL-2-activated NK cells. J. Immunol. 153, 4969.

Maione, T.E., Gray, G.S., Hunt, A.J. and Sharpe, R.J. (1991). Inhibition of tumour growth in mice by an analogue of platelet factor four that lacks affinity for heparin and retains potent angiostatic activity. Cancer Res. 51, 2077.

Maltman, J., Pragnell, I.B. and Graham, G.J. (1993). Transforming growth factor β: is it a downregulator of stem cell inhibition by macrophage inflammatory protein-1α? J. Exp. Med. 178, 925.

Maltman, J., Pragnell, I.B. and Graham, G.J. (1996). Specificity and reciprocity in the interactions between TGF-β and macrophage inflammatory protein-1α. J. Immunol. 156, 1566.

Mantel, C., Aronica, S., Luo, Z., *et al.* (1995). Macrophage inflammatory protein-1α enhances growth factor-stimulated phosphatidylcholine metabolism and increases cAMP levels in the human growth factor-dependent cell line MO7e, events associated with growth suppression. J. Immunol. 154, 2342.

Martin, C.A. and Dorf, M.E. (1991). Differential regulation of interleukin-6, macrophage inflammatory protein-1, and JE/MCP-1 cytokine expression in macrophage cell lines. Cell. Immunol. 135, 245.

Matsue, H., Cruz, P.D., Bergestresser, P.R. and Takashima, A. (1992). Langerhans cells are the major source for IL-1β and MIP-1α among unstimulated mouse epidermal cells. J. Invest. Dermatol. 99, 537.

Matsue, H., Cruz, P.D., Bergstresser, P.R. and Takashima, A. (1993). Profiles of cytokine mRNA expressed by dendritic epidermal T cells in mice. J. Invest. Dermatol. 101, 537.

Maze, R., Sherry, B., Kwon, B.S., Cerami, A. and Broxmeyer, H.E. (1992). Myelosuppressive effects *in vivo* of purified recombinant murine macrophage inflammatory protein-1α. J. Immunol. 149, 1004.

McColl, S.R., Hachicha, M., Levasseur, S., Neote, K. and Schall, T.J. (1993). Uncoupling of early signal transduction events from effector function in human peripheral blood neutrophils in response to recombinant macrophage inflammatory proteins-1α and -1β. J. Immunol. 150, 4550.

Minano, F.J., Fernandezalonso, A., Myers, R.D. and Sancibrian, M. (1996). Hypothalamic interaction between macrophage inflammatory protein-1-alpha (MIP-1-alpha) and MIP-1-beta in rats—a new level for fever control. J. Physiol. (London) 491, 209.

Miyagishi, R., Kikuchi, S., Fukazawa, T. and Tashiro, K. (1995). Macrophage inflammatory protein-1-alpha in cerebrospinal-fluid of patients with multiple-sclerosis and other inflammatory neurological diseases. J. Neurol. Sci. 129, 223.

Monteclaro, F.S. and Charo, I.F. (1996). The amino-terminal extracellular domain of the MCP-1 receptor, but not the RANTES/MIP-1α receptor, confers chemokine selectivity. J. Biol. Chem. 271, 19084–19172.

Murphy, G.M., Jia, X.C., Song, Y., *et al.* (1995). Macrophage inflammatory protein-1-alpha messenger-RNA expression in an immortalized microglial cell-line and cortical astrocyte cultures. J. Neurosci. Res. 40, 755.

Murphy, P.M. (1996). Chemokine receptors: structure, function and role in microbial pathogenesis. Cytokine Growth Factor Rev. 7, 47.

Nakao, M., Nomiyama, H. and Shimada, K. (1990). Structures of human genes coding for cytokine LD78 and their expression. Mol. Cell. Biol. 10, 3646.

Neer, E.J. (1995). Heterotrimeric G proteins: organizers of transmembrane signals. Cell 80, 249.

Neer, E.J. and Smith, T.F. (1996). G protein heterodimers: new structures propel new questions. Cell 84, 175.

Neote, K., DiGregorio, D., Mak, J.Y., Horuk, R. and Schall, T.J. (1993). Molecular cloning, functional expression, and signalling characteristics of a C-C chemokine receptor. Cell 72, 415.

Nibbs, R.J.B., Wylie, S.M., Pragnell, I.B. and Graham, G.J. (1997). Cloning and characterization of a novel murine β chemokine receptor, D6. J. Biol. Chem. 272, 12495.

Nirsimloo, N. and Gordon, M.Y. (1995). Progenitor cells in the blood and marrow of patients with chronic phase chronic myeloid-leukemia respond differently to macrophage inflammatory protein-1-alpha. Leukemia Res. 19, 319.

Nomiyama, H., Hieshima, K., Hirokawa, K., Hattori, T., Takatsuki, K. and Miura, R. (1993). Characterization of cytokine LD78 gene promoters: positive and negative transcription factors bind to a negative regulatory element common to LD78, interleukin-3, and granulocyte-macrophage colony-stimulating factor gene promoters. Mol. Cell. Biol. 13, 2787.

Nomura, H., Nielsen, B.W. and Matsushima, K. (1993). Molecular cloning of cDNAs encoding a LD78 receptor and putative leukocyte chemotactic peptide receptors. J. Int. Immunol. 5, 1239.

Obaru, K., Fukuda, M., Maeda, S. and Shimada, K. (1986). A cDNA clone used to study mRNA inducible in human tonsillar lymphocytes by a tumour promoter. J. Biochem. 99, 885.

Oberlin, E., Amara, A., Bachelerie, F., *et al.* (1996). The CXC chemokine SDF-1 is the ligand for LESTR/fusin and prevents infection by T-cell-line-adapted HIV-1. Nature 382, 833.

Oh, K.-O., Zhou, Z., Kim, K.-K., *et al.* (1991). Identification of cell surface receptors for murine macrophage inflammatory protein-1α. J. Immunol. 147, 2978.

Oppenheim, J.J., Zachariae, C.O., Mukaida, N. and Matsushima, K. (1991). Properties of the novel proinflammatory supergene "intercrine" cytokine family. Annu. Rev. Immunol. 9, 617.

Owen-Lynch, P.J., Adams, J.A., Brereton, M.L., Czaplewski, L.G. and Whetton, A.D. (1996). The effect of the chemokine rhMIP-1 alpha and a non-aggregating variant, BB-10010, on blast cells from patients with acute myeloid leukaemia. Br. J. Haematol. 95, 77.

Paolini, J.F., Willard, D., Consler, T., Luther, M. and Krangel, M.S. (1994). The chemokines IL-8, monocyte chemo-attractant protein-1 and I-309 are monomers at physiologically relevant concentrations. J. Immunol. 153, 2704.

Parkinson, E.K., Graham, G.J., Daubersies, P., *et al.* (1993). A haemopoietic stem cell inhibitor (SCI/MIP-1α) also inhibits clonogenic epidermal keratinocyte proliferation. J. Invest. Dermatol. 101, 113.

Patel, S.R., Evans, S., Dunne, K., *et al.* (1993). Characterisation of the quaternary structure and conformational properties of the human stem cell inhibitor protein LD78 in solution. Biochemistry, 32, 5466.

Paxton, W.A., Martin, S.R., Tse, D., *et al.* (1996). Relative resistance to HIV-1 infection of CD4 lymphocytes from persons who remain uninfected despite multiple high risk sexual exposures. Nature Med. 2, 412

Petrek, M., Du Bois, R.M., Sirova, M. and Weigl, E. (1995). Chemotactic cytokines (chemokines) and their role in physiological and immunopathological reactions. Folia Biol. (Praha) 41, 263.

Ponath, P.D., Qin, S., Post, T.W., *et al.* (1996). Molecular cloning and characterisation of a human eotaxin receptor expressed selectively on eosinophils. J. Exp. Med., 183, 2437.

Post, T.W., Bozic, C.R., Rothenberg, M.E., Luster, A.D., Gerard, N. and Gerard, C. (1995). Molecular characterization of two murine eosinophil β chemokine, receptors. J. Immunol. 155, 5299.

Power, C.A., Meyer, A., Nemeth, K., *et al.* (1995). Molecular cloning and functional expression of a novel CC chemokine receptor cDNA from a human basophilic cell line. J. Biol. Chem. 270, 19495.

Premack, B.A. and Schall, T.J. (1996). Chemokine receptors: gateways to inflammation and infection. Nature Med. 2, 1174.

Quesniaux, V.F., Graham, G.J., Pragnell, I.B., *et al.* (1993). Use of 5-fluorouracil to analyse the effect of macrophage inflammatory protein-1 alpha on long-term reconstituting stem cells *in vivo*. Blood 81, 1497.

Rajarathnam, K., Sykes, B.D., Kay, C.M., *et al.* (1994). Neutrophil activation by monomeric interleukin-8. Science 264, 90.

Raport, C., Gosling, J., Schweikart, V.L., Gray, P.W. and Charo, I. (1996). Molecular cloning and functional characterisation of a novel human CC chemokine receptor (CCR5) for RANTES, MIP-1α and MIP-1β. J. Biol. Chem. 271, 17161.

Ritter, U., Moll, H., Laskay, T., *et al.* (1996). Differential expression of chemokines in patients with localized and diffuse cutaneous American leishmaniasis. J. Infect. Dis. 173, 699.

Rock, C.O., Cleveland, J.L. and Jackowski, S. (1992). Macrophage growth arrest by cyclic AMP defines a distinct checkpoint in the mid-G1 stage of the cell cycle and overrides constitutive c-myc expression. Mol. Cell. Biol. 12, 2351.

Rot, A., Krieger, M., Brunner, T., Bischoff, S.C., Schall, T.J. and Dahinden, C.A. (1992). Rantes and macrophage inflammatory protein 1-alpha induce the migration and activation of normal human eosinophil granulocytes. J. Exp. Med. 176, 1489.

Samson, M., Libert, F., Doranz, B., *et al.*, (1996). Resistance to HIV-1 infection in caucasian individuals bearing mutant alleles of the CCR-5 chemokine receptor gene. Nature 382, 722.

Sasseville, V.G., Smith, M.M., Mackay, C., Ringler, D.J. and Lackner, A.A. (1996). Chemokine expression in the brain of Macaque monkeys with SIV-induced aids encephalitis. Lab. Invest. 74, 832.

Schall, T.J. (1991). Biology of the RANTES/SIS cytokine family. Cytokine 3, 165.

Schall, T.J., Bacon, K., Camp, R.D.R., Kaspari, J.W. and Goeddel, D.V. (1993). Human macrophage inflammatory protein-1α (MIP-1α) and MIP-1β chemokines attract distinct populations of lymphocytes. J. Exp. Med. 177, 1821.

Schroder, J.M., Noso, N., Sticherling, M. and Christophers, E. (1996). Role of eosinophil-chemotactic C-C chemokines in cutaneous inflammation. J. Leukocyte Biol. 59, 1.

Schutze, S., Potthoff, K., Machiedt, T., Berkovic, D., Wiegmann, K. and Kronke, M. (1992). TNF activates NF-kB by phosphatidylcholine-specific phospholipase C-induced "acidic" sphingomyelin breakdown. Cell 71, 765.

Seebach, J., Bartholdi, D., Frei, K., *et al.* (1995). Experimental *Listeria* meningoencephalitis: macrophage inflammatory protein-1α and -2 are produced intrathecally and mediate chemotactic activity in cerebrospinal fluid of infected mice. J. Immunol. 155, 4367.

Selvan, R.S., Butterfield, J.H. and Krangel, M.S. (1994). Expression of multiple chemokine genes by a human mast cell leukaemia. J. Biol. Chem. 269, 13893.

Shakin-Eshelman, S.H., Spitalnik, S.L. and Kasturi, L. (1996). The amino acid at the X-position of an Asn-X-Ser sequon is an important determinant of N-linked core-glycosylation efficiency. J. Biol. Chem. 271, 6363.

Shanley, T.P., Schmal, H., Friedl, H.P., Jones, M.L. and Ward, P.A. (1995). Role of macrophage inflammatory protein-1-alpha (MIP-1-alpha) in acute lung injury in rats. J. Immunol. 154, 4793.

Sherry, B.P., Tekamp-Olson, C., Gallegos, D., Davatelis, F., Coit, D. and Cerami, A. (1988). Resolution of the two components of macrophage inflammatory protein 1, and cloning of one of those components, macrophage inflammatory protein 1β. J. Exp. Med. 168, 2251.

Simmons, G., Clapham, P.R., Picard, L., *et al.* (1997). Potent inhibition of HIV-1 infectivity in macrophages and lymphocytes by a novel CCR5 antagonist. Science, 276, 276.

Skelton, N.J., Aspiras, F., Ogez, J. and Schall, T.J. (1995). Proton NMR assignments and solution conformation of RANTES, a chemokine of the C-C type. Biochemistry 34, 5329.

Smith, R.E., Strieter, R.M., Phan, S.H., *et al.* (1994). Production and function of murine macrophage inflammatory protein-1-alpha in bleomycin-induced lung injury. J. Immunol. 153, 4703.

Sozzani, S., Molino, M., Locati, M., *et al.* (1993). Receptor-activated calcium influx in human monocytes exposed to monocyte chemotactic protein-1 and related cytokines. J. Immunol. 150, 1544.

Sprenger, H., Meyer, R.F., Kaufmann, A., Bussfeld, D., Rischkowsky, E. and Gemsa, D. (1996). Selective induction of monocyte and not neutrophil-attracting chemokines after influenza A virus infection. J. Exp. Med. 184, 1191.

Standiford, T.J., Kunkel, S.L., Liebler, J.M., Burdick., M.D., Gilbert, A.R. and Strieter, R.M. (1993a). Gene expression of macrophage inflammatory protein-1 alpha from human blood monocytes and alveolar macrophages is inhibited by interleukin-4. Am. J. Respir. Cell Mol. Biol. 9, 192.

Standiford, T.J., Rolfe, M.W., Kunkel, S.L., *et al.* (1993b). Macrophage inflammatory protein-1 alpha expression in interstitial lung-disease. J. Immunol. 151, 2852.

StCharles, R., Walz, D.A. and Edwards, B.F.P. (1989). The three dimensional structure of bovine platelet factor 4 at 3.0 Å resolution. J. Biol. Chem. 264, 2092.

Stoeckle, M.Y. and Barker, K.A. (1990). Two burgeoning families of platelet factor-4 related proteins: mediators of the inflammatory response. New Biol. 2, 313.

Strieter, R.M., Standiford, T.J., Huffnagle, G.B., Colletti, L.M., Lukacs, N.W. and Kunkel, S.L. (1996). "The good, the bad, and the ugly". The role of chemokines in models of human disease. J. Immunol. 156, 3583.

Tanaka, S., Robinson, E.A., Yoshimura, T., Matsushima, K., Leonard, E.J. and Appella, E. (1988). Synthesis and biological characterisation of monocyte derived neutrophil chemotactic factor. FEBS Lett. 236, 467.

Tanaka, Y., Adams, D.H., Hubscher, S., Hirnao, H., Siebenlist, U. and Shaw, S. (1993). T-cell adhesion induced by proteoglycan-immobilised cytokine MIP-1 beta. Nature 361, 79.

Taub, D.D., Conlon, K., Lloyd, A.R., Oppenheim, J.J. and Kelvin, D.J. (1993). Preferential migration of activated CD4+ and CD8+ T-cells in response to MIP-1-alpha and MIP-1-beta. Science 260, 355.

Taub, D.D., Sayers, T.J., Carter, C.R.D. and Ortaldo, J.R. (1995). Alpha-chemokine and beta-chemokines induce NK cell migration and enhance NK mediated cytolysis. J. Immunol. 155, 3877.

Trkola, A., Dragic, T., Arthos, J., et al. (1996). CD4-dependent, antibody-sensitive interactions between HIV-1 and its co-receptor CCR-5. Nature 384, 184.

Vaddi, K. and Newton, R.C. (1994a). Regulation of monocyte integrin expression by beta-family chemokines. J. Immunol. 153, 4721.

Vaddi, K. and Newton, R.C. (1994b). Comparison of biological responses of human monocytes and THP-1 cells to chemokines of the intercrine-beta family. J. Leukocyte Biol. 55, 756.

Van Riper, G., Nicholson, D.W., Scheid, M.P., Fischer, P.A., Springer, M.S. and Rosen, H. (1994). Induction, characterization, and functional coupling of the high affinity chemokine receptor for RANTES and macrophage inflammatory protein-1α upon differentiation of an eosinophilic HL-60 cell line. J. Immunol. 152, 4055.

Verfaillie, C.M. (1996). Chemokines as inhibitors of hematopoietic progenitors. J. Lab. Clin. Med. 127, 148.

Verfaillie, C.M. and Miller, J.S. (1995). A novel single-cell proliferation assay shows that long-term culture-initiating cell (LTC-IC) maintenance over time results from the extensive proliferation of a small fraction of LTC-I. Blood 86, 2137.

Verfaillie, C.M., Catanzarro, P.M. and Li, W.N. (1994). Macrophage inflammatory protein 1-alpha, interleukin-3 and diffusible marrow stromal factors maintain human hematopoietic stem-cells for at least 8 weeks in vitro. J. Exp. Med. 179, 643.

Wain-Hobson, S. (1996). One on one meets two. Nature 384, 117.

Wang, J.M., Sherry, B., Fivash, M.J. and Kelvin, D.J. (1993). Human recombinant macrophage inflammatory protein-1 alpha and -beta and monocyte chemotactic and activating factor utilize common and unique receptors on human monocytes. J. Immunol. 150, 3022.

Webb, L.M.C., Ehrengruber, M.U., Clark-Lewis, I., Baggiolini, M. and Rot, A. (1993). Binding to heparan sulphate or heparin enhances neutrophil responses to interleukin 8. Proc. Natl Acad. Sci. USA 90, 7158.

Wilkinson, D. (1996). HIV-1: cofactors provide the entry keys. Curr. Biol. 6, 1051.

Wolpe, S.D., Davatelis, G., Sherry, B., et al. (1988). Macrophages secrete a novel heparin-binding protein with inflammatory and neutrophil chemokinetic properties. J. Exp. Med. 167, 570.

Wu, D., LaRosa, G.J. and Simon, M.I. (1993). G protein-coupled signal transduction pathways for interleukin-8. Science 261, 101.

Wu, H. and Lozanno, G. (1994). A potential mechanism for suppressing cell growth in response to stress. J. Biol. Chem. 269, 20067.

Wu, L., Gerard, N.P., Wyatt, R., et al. (1996). CD4-induced interaction of primary HIV-1 gp120 glycoproteins with the chemokine receptor CCR-5. Nature 384, 179.

Yamamura, Y., Hattori, T., Obaru, K., et al. (1989). Synthesis of a novel cytokine and its gene (LD 78) expressions in hemopoietic fresh tumor cells and cell lines. J. Clin. Invest. 84, 1707.

Yamamura, Y., Hattori, T., Ohmoto, Y. and Takatsuki, K. (1992). Identification and characterization of specific receptors for the LD78 cytokine. Int. J. Hematol. 55, 131.

Zawada, W.M., Ruwe, W.D., Clarke, J., Wall, P.T. and Myers, R.D. (1994). Chemokine-induced fever—intracerebroventricular actions of MIP-1 and MIP-1-alpha in rats. Neuroreport 5, 1365.

Zhou, Z., Kim, Y.-Y., Pollok, K., et al. (1993). Macrophage inflammatory protein-1α rapidly modulates its receptors and inhibits anti-CD3 mAb-mediated proliferation of T lymphocytes. J. Immunol. 151, 4333.

Zipfel, P., Balke, J., Irvine, S.G., Kelly, K. and Siebenlist, U. (1989). Mitogenic activation of human T cells induces two closely related genes which share structural similarities with a new family of secreted factors. J. Immunol. 142, 1582.

32. Monocyte Chemotactic Proteins 1, 2 and 3

Paul Proost, Anja Wuyts, Ghislain Opdenakker *and* Jo Van Damme

1. Introduction

In the early 1980s, Bottazzi and colleagues (1983) reported that supernatants of more than 20 human and murine tumors contained chemotactic activity for mononuclear phagocytes. When purified by gel filtration chromatography, the chemotactic activity eluted at an apparent molecular mass of about 12 kDa.

In the late 1980s, several groups isolated and identified from human tumor cell lines related factors with monocyte chemotactic activity: MCF, monocyte chemotactic factor (Matsushima *et al.*, 1989); MCAF, monocyte chemotactic and activating factor (Furutani *et al.*, 1989); GDCF-2, glioma-derived monocyte chemotactic factor (Robinson *et al.*, 1989; Yoshimura *et al.*, 1989a); MCP or MCP-1, monocyte chemotactic or chemoattractant protein-1 (Van Damme *et al.*, 1989; Yoshimura *et al.*, 1989c) and CF, chemotactic factor (Graves *et al.*, 1989). Malignant cells were not the only source of monocyte chemotactic activity, as similar human factors could also be purified from fibroblasts (JE,

Rollins *et al.*, 1989; MCP, Van Damme *et al.*, 1989) and peripheral blood mononuclear leukocytes (LDCF, lymphocyte-derived chemotactic factor, Yoshimura *et al.*, 1989b; MCP, Van Damme *et al.*, 1989; HC11, Chang *et al.*, 1989). The primary structure of the corresponding 76-amino-acid glycoprotein was determined by cDNA sequencing and by protein sequence analysis. All these monocyte chemotactic factors had the same amino acid sequence but appeared as different glycosylation forms. In this chapter, they will be referred to as MCP-1.

Human MCP-1 was homologous to a murine protein encoded by the JE cDNA, a platelet derived growth factor (PDGF)-inducible protein from mouse fibroblasts (Cochran *et al.*, 1983). It belongs to a family of small proteins with four conserved cysteines. This family of chemotactic cytokines, including interleukin-8 (IL-8), was initially referred to as the "intercrines" (Oppenheim *et al.*, 1991). More recently the term chemokines became generally accepted. Chemokines were divided into two subgroups, called C-X-C (or α-) and C-C (or β-) chemokines, depending on whether the first two cysteines were separated by one amino acid or not (for review, see Oppenheim *et al.*, 1991; Miller and Krangel, 1992; Baggiolini *et al.*, 1994; Schall, 1994; Van Damme, 1994). MCP-1 belongs to the C-C branch together with other human chemokines like RANTES, I-309, the macrophage inflammatory proteins (MIP-1α and MIP-1β), and the two related molecules MCP-2 and MCP-3 (Van Damme *et al.*, 1992). The MCP-2 protein sequence has also been deduced from the HC14 cDNA that was co-induced with HC11 or MCP-1 in anti-CD2-stimulated peripheral blood lymphocytes (Chang *et al.*, 1989). However, the cDNA was never used to generate biologically active protein and the protein sequence was not introduced in the data libraries. As a consequence, no biological function could be attributed to this protein until MCP-2 was purified and sequenced from osteosarcoma cells (Van Damme *et al.*, 1992). Its primary structure completely matched the HC14 cDNA-derived sequence. The MCP-3 protein identified from the same cellular source was found to be most closely related to MCP-1 (Van Damme *et al.*, 1992).

These MCPs have attracted much interest with respect to their role in leukocyte migration and activation in inflammatory and neoplastic diseases. MCP-1 can be produced constitutively or upon induction with different cytokines, viruses, endotoxins, plant lectins, or phorbol esters in a wide variety of normal and malignant cell types. mRNA or protein has been demonstrated in several human diseases where monocyte accumulation or activation is an important phenomenon, e.g., cancers, rheumatoid arthritis (RA), atherosclerosis, inflammatory skin diseases, bronchial infections, and liver diseases (for review, see Strieter *et al.*, 1994; Kunkel *et al.*, 1996; Lukacs *et al.*, 1996; Schröder *et al.*, 1996).

The spectrum of cells known to be responsive to MCP-1, MCP-2, MCP-3 and other C-C chemokines has since been broadened considerably. Leukocytes other than monocytes, such as eosinophils, basophils, lymphocytes dendritic cells, and natural killer (NK) cells have been shown to be sensitive (chemotaxis, enzyme release, enhanced intracellular Ca^{2+} concentration) to one or more of these MCPs.

Chemokines, including MCPs, activate their target cells by G-protein-coupled receptors characterized by seven transmembrane regions and belonging to the serpentine receptors (for review, see Kelvin *et al.*, 1993; Ahuja *et al.*, 1994; Murphy, 1994; Wells *et al.*, 1996). The intracellular loops of the receptor interact with G-proteins which, upon activation through ligand–receptor binding, can exchange GTP for GDP. The GTP-bound G-protein can further activate intracellular effector enzymes and the GDP form is restored after GTP hydrolysis.

This chapter will concentrate on the sequences, activities, and producer and effector cells, as well as receptors, of the three human MCPs.

2.   The MCP Gene Sequences and Structure

The MCP-1 and MCP-3 genes (Figures 32.1(a) and 32.2(a)), like all other C-C chemokine genes, have both been allocated to human chromosome 17. Both genes, with locus symbols SCYA2 and SCYA7 (small inducible cytokine genes numbers 2 and 7), respectively, were assigned to the C-C chemokine gene cluster on locus 17q11.2-12 (Mehrabian *et al.*, 1991; Rollins *et al.*, 1991a; Opdenakker *et al.*, 1994). For the MCP-2 protein, the cDNA and genomic DNA sequence have been described recently (Van Coillie *et al.*, 1997a,b).

2.1   MCP-1

The MCP-1 gene (Figure 32.1(b)) consists of three exons (136 bp (base pairs), 118 bp, and 478 bp in size) and two introns (800 bp and 385 bp) (Chang *et al.*, 1989; Shyy *et al.*, 1990). The translated MCP-1 mRNA codes for a 99-amino-acid (aa) protein including a 23-aa, mainly hydrophobic, leader sequence. As well as the normal transcription initiation site (asterisked in Figure 32.1(a)), two other sites which are located farther upstream (6 and 8 bp, respectively) are sometimes used. The promoter region contains the TATA box, 96 bp upstream of the translation initiation site. The polyadenylation site (AATAA) is located 353 bp downstream of the stop codon (Figure 32.1(b)).

In addition to a GC box (at −126 bp), two tetradecanoylphorbol acetate (TPA)-responsive elements (TRE) for the binding of transacting factor AP-1 (TGACTCC and TCACTCA) are found at positions −128 and −156, respectively (Figure 32.1(a)). The GGAAGATCCCT consensus sequence for the κB enhancer element (position −148), which might possibly be involved in lipopolysaccharide (LPS) and tumor necrosis factor (TNF) responses, and the ATTTGCGT consensus sequence for the octamer transcription factor (OTF at position −282) were found farther upstream in the promoter region (Shyy *et al.*, 1990). Several other κB binding sites and TREs were discovered in the enhancer region, between 2 and 3 kbp upstream from the translation initiation site (Ueda *et al.*, 1994). Promoter and enhancer regions thus contain *cis*-elements which are possibly important for the regulation of MCP-1 gene transcription. The two *cis*-elements, essential for the MCP-1 induction and for the maintenance of basal MCP-1 transcriptional activity, were found to be the Sp1 binding GC box at position −126 in the promoter region and the κB binding site at position −2672 in the enhancer region, respectively. All other κB binding sites and TREs are thought to be nonfunctional. The essential nuclear factor κB (NFκB) and Sp1 binding sites however, cannot completely explain the MCP-1 gene transcription since, for example, in lymphocytes virtually no MCP-1

IFN-γ
181 ...atgctttcatctagtttcctccgcttccttccttttcctgcagttttcgcttcagagaaagcagaatccttaaaaataaccctcttagttcacatctgtg

OTF
280 gtcagtctgggcttaatggcaccccatcctccccattttgcgtcatttggtctcagcagtgaatggaaaaaagtgctcgtcctcaccccccctgcttc

                                               TRE   κB   TRE
376 cctttcctacttcctggaaatccacaggatgctgcatttgctcagcagatttaacagcccacttatcactcatggaagatccctcctcctgcttgact
    GC-box   TATA-box   *
474 ccgccctctctccctctgcccgctttcaataagaggcagagacagcagccagaggaaccgagaggctgagactaacccagaaacatcca
563 attctcaaactgaagctcgcactctcgcctccagc ATG AAA GTC TCT GCC GCC CTT CTG TGC CTG
628 CTG CTC ATA GCA GCC AC C TTC ATT CCC CAA GGG CTC GCT CAG CCA Ggtaa...

INTRON

1466 ...ttccag AT GCA ATC AAT GCC CCA GTC ACC TGC TGC TAT AAC TTC ACC AAT A
1517 GG AAG ATC TCA GTG CAG AGG CTC GCG AGC TAT AGA AGA ATC ACC AGC A
1565 GC AAG TGT CCC AAA GAA GCT GTG AT gtgagtt...

INTRON

1972 ...cagC TTC AAG ACC ATT GTG GCC AAG GAG ATC TGT GCT GAC CCC AAG CAG
2021 AAG TGG GTT CAG GAT TCC ATG GAC CAC CTG GAC AAG CAA ACC CAA ACT
2069 CCG AAG ACT tgaacactcactccacaacccaagaatctgcagctaacttattttcccctagcttccccagacaccctgttt
2151 tattttattataatgaattttgtttgttgatgtgaaacattatgccttaagtaatgttaattcttatttaagttattgatgtgtttaagtttatct
2246 ttcatggtactagtgttttttagatacagagacttggggaaattgctttcctcttgaaccacagttctaccctgggatgtttgagggt
2337 ctttgcaagaatcattaatacaaagaattttttttaacattccaatgcattgctaaaatattattgtggaaatgaatattttgtaactatt
                  *
2428 acaccaaataaatatattttttgtacaaaacctgacttccagtgtttttcttgaaggaaatt...

(a)

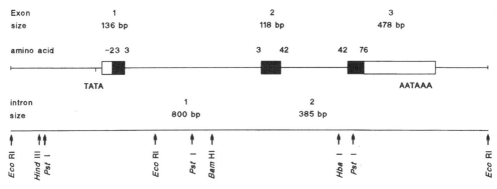

(b)

Figure 32.1   MCP-1 genomic sequence and gene (SCYA2) structure. (a) The MCP-1 (Rollins *et al.*, 1989; Shyy *et al.*, 1990) genomic sequence is shown with the three exons printed in bold. Upper-case letters indicate the sequence which encodes the MCP-1 protein and signal sequence (the box indicates the first amino acid in the mature protein). The 5′ and 3′ ends of the intron sequences are in medium lettering. Both the transcription initiation and poly(A) sites are asterisked. The TATA box, the IFN-γ consensus sequence, the TRE, the κB enhancer element, the GC-box, and the OTF are underlined. (b) The MCP-1 exon sequences are indicated as open and solid boxes for the untranslated and translated DNA sequences, respectively. The intron sequences are indicated as a straight line. The positions of the 23-aa signal peptide and the 76-residue secreted protein are indicated above the exons. Restriction enzyme cleavage sites are marked by arrows.

mRNA is found (Colotta *et al.*, 1992a), although the Sp1 expression is rather high (Ueda *et al.*, 1994).

2.2   MCP-3

As for the MCP-1 gene, three exons (131 bp, 118 bp, and 546 bp long) and two introns (779 bp and 433 bp long) are included in the MCP-3 gene (Figure 32.2(b)). The promoter region of the MCP-3 gene (Figure 32.2(a)) contains two tandem dinucleotide repeats (TDR), $(CA)_{17}$ and $(GA)_7$, a property which is shared with the MIP-1α and I-309 gene sequences. Direct and indirect repeats (DR and IR) as well as palindromic

sequences (Pal), which might enhance DNA recombinatorial events, are also clustered in this region. Other interesting features like the CAAT-box, the TATA-box, a Cap-signal, the transcription initiation site and at the 3′ end a mRNA hairpin loop, an AT-rich mRNA destabilizing region, and the polyadenylation site are indicated on Figure 32.2(a) (Opdenakker *et al.*, 1994).

Similarly to the MCP-1 exons, the MCP-3 exons encode for a protein of 99 amino acids, which includes a 23-aa signal sequence. The codons for aa 3 and 42 are interrupted by the intron sequences (Figure 32.2(b)). Recently, a more detailed analysis of the promoter region of MCP-3 has been described (Murakami *et al.*, 1997).

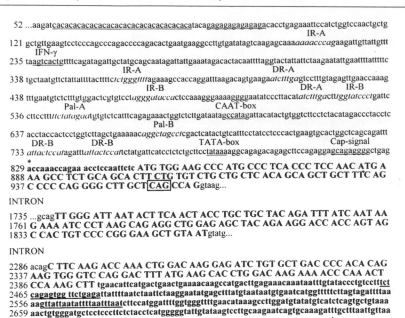

52 ...aagat<u>cacacacacacacacacacacacacacacacacacacaca</u>tac<u>agagagagagagagaga</u>cacctgagaaattccatctggtccaactgctg
                                                                     IR-A

121 gctgttgaagtcctcccagcccagacccagacactgaatgaaggccttgtgatatagtcaagagcaaa*aaaaacccag*aagattgttattgttt
    IFN-γ

235 ta<u>agtcactg</u>ttttcagatagattgctatgcagcaatagattattgaaatagacactacaattttaggtactattattctaagaatattgaatttt*attttttc*
         IR-A   DR-A

338 tgctaatgttctattattttactttt*ctctgggtttt*agaaagccaccaggatttaagacagtgaag*aatctttg*agtcctttgtagagttgaaccaaag
                                 IR-B   DR-A   IR-B

438 tttgaatgtctctttgtggactcgtgtcct*agggataccac*tccaaaggg*aaaaggg*aatatcccttacat*atctttg*actt*tggtatccct*gattc
        Pal-A   CAAT-box

536 cttccttt*ttctataguat*gtgtctcatttcagagaaactggtctctt gataata<u>gccatag</u>attacatactgtggtcttcctctacatagaccctacctc
        Pal-B

637 acctaccactcctggtcttagctgaaaaac*aggctagcct*cgactcatactgtcatttcctatcctcccactgaagtgcactggctcagcagattt
    DR-B   DR-B   TATA-box   Cap-signal

733 *attactccat*agattt*attactccat*tctatgattcatcctctctgcttcc<u>tataaa</u>aggcagagacagagcttccagaggag<u>cagaggggg</u>ctgag
          *

829 **accaaaccagaa acctccaattctc ATG TGG AAG CCC ATG CCC TCA CCC TCC AAC ATG A**

888 **AA GCC TCT GCA GCA CTT** `CTG` **TGT CTG CTG ACA GCA GCT GCT TTC AG**

937 **C CCC CAG GGG CTT GCT** `CAG` **CCA G**gtaag...

INTRON

1735 ...gcag**TT GGG ATT AAT ACT TCA ACT ACC TGC TGC TAC AGA TTT ATC AAT AA**

1761 **G AAA ATC CCT AAG CAG AGG CTG GAG AGC TAC AGA AGG ACC ACC AGT AG**

1833 **C CAC TGT CCC CGG GAA GCT GTA AT**gtatg...

INTRON

2286 acag**C TTC AAG ACC AAA CTG GAC AAG GAG ATC TGT GCT GAC CCC ACA CAG**

2337 **AAG TGG GTC CAG GAC TTT ATG AAG CAC CTG GAC AAG AAA ACC CAA ACT**

2386 **CCA AAG CTT** tgaacattcatgactgaactgaaaacaagccatgacttgagaaacaaataatttgtataccctgtcctt<u>tct</u>

2465 <u>**cagagtgg** ttctgagat</u>tattttaatctaattctaaggaatatgagctttatgtaataatgtgaatcatggttttctttagtagatttaa

2556 aag<u>ttattaatatttt</u>aatttaatcttccatggatttt ggtgggtttgaacataaagccttggatgtatatgtcatctcagtgctgtaaa

2659 aactgtgggatgctcctcccttctctacctcatggggg tattgtataagtccttgcaagaatcagtgcaaagatttgctttaattgttaa
                                                                                          *

2737 gatatgatgtccctatggaagcatattgttattatataattacatatttgcatatgtatgactcccaaatttcacataaaatagatttt

2838 gtataaca gctgc...

(a)

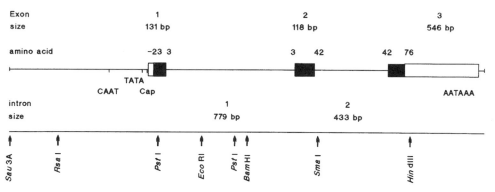

(b)

Figure 32.2 MCP-3 genomic sequence and gene (SCYA7) structure. (a) The MCP-3 genomic sequence (Opdenakker *et al.*, 1994) is shown with the three exons printed in bold. Upper-case letters indicate the sequence which encodes the MCP-3 protein and signal sequence, with the position of the first amino acid in the mature protein boxed. The 5′ and 3′ ends of the intron sequences are in medium lettering. Both the transcription initiation and poly(A) sites are asterisked. The CAAT box, TATA box, Cap signal, hairpin loop TA motif, CA and GA repeats and the IFN-γ recognition element (IFN-γ) are underlined, while the DR and IR and Pal encoded are printed in italic. (b) The MCP-3 gene structure is shown in the same way as the MCP-1 gene structure in Figure 32.1(b).

3.   *The MCP Proteins*

Natural MCP-1, MCP-2 and MCP-3 were copurified from osteosarcoma cell supernatants by their affinity for heparin. Fractionation and purification to homogeneity could be obtained by cation exchange and reversed-phase columns, resulting in pure MCPs identified by protein sequencing (Van Damme *et al.*, 1992; Proost *et al.*, 1996a).

3.1   MCP-1

Human MCP-1 is a glycoprotein of 76 amino acids with four cysteines forming two intramolecular disulfide bridges (Cys^{11}–Cys^{36} and Cys^{12}–Cys^{52}). The protein contains 13 basic Lys and Arg residues, responsible for its high theoretical pI of 10.6. Except for some minor amino-terminal processed forms (Decock *et al.*, 1990; Carr *et al.*, 1994), the amino-terminus of mature MCP-1 is blocked for Edman degradation by a pyroglutamic acid residue, which could not be removed by pyroglutamyl aminopeptidase. Using a combination of sequencing of proteolytic fragments and mass spectrometry, the complete amino acid sequence of human MCP-1 could be determined (Robinson *et al.*, 1989). One *N*-glycosylation site is located close to the amino-terminus (Figure 32.3) and several glycosylated

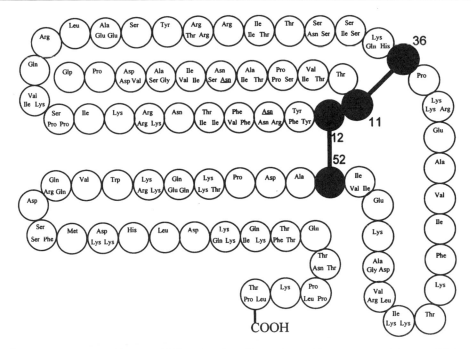

Figure 32.3   Protein sequences of human MCP molecules. The amino acid sequences of human MCP-1, MCP-2 and MCP-3 with their amino terminal pyroglutamic acid (Glp) and *N*-glycosylation sites (underlined) indicated. The MCP-1 protein sequence is indicated uppermost in the circles with the MCP-2 (left) and MCP-3 (right) sequences depicted below. Where a residue is completely conserved in all human MCPs, this amino acid is not repeated for MCP-2 and MCP-3. The cysteine residues are shown as solid circles and both disulfide bridges (Cys[11]–Cys[36] and Cys[12]–Cys[52]) are indicated by bold lines.

forms of MCP-1 have been reported ranging from 9 kDa to 17 kDa on SDS-PAGE (sodium dodecyl sulphate–polyacrylamide gel electrophoresis). So far, the use of such *N*-glycosylation has not been reported. The addition of *O*-linked sugars and sialic acid residues contributes to the different molecular mass forms of MCP-1 (Robinson *et al.*, 1989; Yoshimura *et al.*, 1989a; Decock *et al.*, 1990; Jiang *et al.*, 1990).

Based on the IL-8 solution structure, deduced from NMR data, the three-dimensional structure of MCP-1 was predicted to consist of two MCP-1 peptide molecules linked to form an MCP-1 dimer (Gronenborn and Clore, 1991). NMR studies on the solution structure of MIP-1β, however, revealed a number of structural differences between C-C and C-X-C chemokines (Lodi *et al.*, 1994). IL-8 was found to be globular in shape, whereas for MIP-1β a more cylindrical structure is proposed. In contrast to the IL-8 α-helices, both MIP-1β α-helices were found to be located at opposite sites of the molecule. Thus, the MIP-1β monomer structure is similar to the IL-8 structure but the dimer interface and quaternary structure are completely different. The authors suggest that probably all C-X-C chemokines follow the IL-8 model, whereas all C-C chemokines behave like MIP-1β, indicating that the former modeling of the MCP-1 structure on IL-8 was incorrect (Lodi *et al.*, 1994). Further experiments need to be done to clarify these issues.

Recent studies using size-exclusion HPLC (high-performance liquid chromatography), sedimentation equilibrium, and chemical cross-linking have shown that, at physiological (low nanomolar) concentrations, MCP-1 as well as IL-8 and I-309 occur as monomers instead of dimers (Paolini *et al.*, 1994). MCP-1, as well as IL-8, forms an equilibrium between the monomeric and dimeric forms with a dimer dissociation constant'(K_d) of 33 ± 18 μM. Thus, at concentrations above 100 μM almost all MCP-1 is in the dimeric state, explaining the previously obtained results. At low, but physiologically active concentrations (< 100 nM), however, MCP-1 occurs almost exclusively as a monomer. This is in line with the findings that IL-8 is also active in the monomeric form (Rajarathnam *et al.*, 1994).

3.2   MCP-2 AND MCP-3

Similarly to all other animal MCPs isolated so far, human MCP-2 and MCP-3 also appeared to be blocked at the amino-terminus. Therefore, protein sequence data were obtained by Edman degradation of proteolytic fragments (Van Damme *et al.*, 1992, 1993). The MCP-2 and MCP-3 proteins contained 76 amino acids, including four cysteines which are characteristic for the chemokine family (Figure 32.3). Both peptides display high sequence similarity to MCP-1 (62% and 71% identity, respectively). MCP-2 and MCP-3 are slightly more basic than MCP-1 (pI = 10.6) with theoretical pI values of 10.8 and 10.9.

The MCP-2 sequence does not contain N-glycosylation sites. Since the theoretical molecular mass of MCP-2 is 8893 Da and since natural as well as synthetic MCP-2 both appear as 7.5 kDa proteins on SDS-PAGE under reducing conditions (Van Damme et al., 1992; Proost et al., 1995), no O-glycosylation is to be expected. So far, no studies on the tertiary and quaternary MCP-2 structure are available.

Natural human MCP-3 occurred as an 11 kDa protein on SDS-PAGE (Van Damme et al., 1992). Although the cDNA-derived protein sequence contains one amino-terminally located N-glycosylation site (Figure 32.3; Minty et al., 1993; Opdenakker et al., 1993), natural 11 kDa MCP-3 did not appear to be N-glycosylated (Opdenakker et al., 1995). Moreover, folded synthetic 76-aa long MCP-3 also appeared as an 11 kDa protein on SDS-PAGE, although the theoretical relative molecular mass is only 8935 Da (Proost et al., 1995). In addition to unglycosylated MCP-3, Minty and colleagues (1993) detected four forms (11, 13, 17, and 18 kDa) after expression in COS cells. Here, both N- and O-glycosylation were involved. Electrospray mass spectrometry of the unglycosylated protein confirmed the amino-terminal pyroglutamate and the existence of two disulfide bridges.

4.   Cellular Sources and Production

4.1   MCP-1

MCP-1, together with IL-8, is one of the most extensively studied chemokines. A wide variety of cells have been shown (at the protein or mRNA level) to produce MCP-1. In addition to normal cells like monocytes, fibroblasts, epithelial cells, mesothelial cells, chondrocytes, keratinocytes, melanocytes, mesangial cells, osteoblasts, astrocytes, lipocytes, endothelial cells, and smooth-muscle cells, a number of tumor cells have been identified as sources of MCP-1 (reviewed in Proost et al., 1996b). MCP-1 is expressed only at a very low level by unstimulated normal cells, whereas many tumor cell lines were found to constitutively produce MCP-1 (Graves et al., 1989; Kuratsu et al., 1989; Mantovani et al., 1992).

Monocytes are both target and producer cells of MCP-1. In addition to cytokines (IL-1, interferon-γ (IFN-γ), tumor necrosis factor-α (TNF-α), granulocyte-macrophage colony-stimulating factor (GM-CSF) and macrophage colony-stimulating factor (M-CSF)), mitogens (phytohemagglutinin (PHA) and concanavalin A (Con A)), double-stranded RNA (dsRNA), viruses (e.g., measles), phorbol ester, and α-thrombin were also found to induce MCP-1 in monocytes, mononuclear cells, or monocytic cell lines. Induction with bacterial lipopolysaccharide (LPS) yielded rather controversial results, which might in part be explained by contamination with very low amounts of LPS, resulting in a considerable MCP-1 expression, during monocyte purification. The protein synthesis inhibitor cycloheximide inhibited MCP-1 expression in monocytes, whereas it superinduces MCP-1 expression on endothelial and fibrosarcoma cells. Other leukocytes are rather poor producers of MCP-1. Virtually no MCP-1 could be recovered from lymphocytes or granulocytes (Colotta et al., 1992a; Van Damme et al., 1994).

In general, cytokines such as IL-1, TNF-α, and IFN-γ are effective MCP-1 inducers in most normal cells. In addition, phorbol esters, low-density lipoprotein, thrombin, and shear stress have been identified as MCP-1-inducing factors on endothelial cells (Proost et al., 1996b). IL-4 could amplify the MCP-1 induction by IL-1 in endothelial cells (Colotta et al., 1992b).

Shyy and colleagues (1993) studied extensively the multiple signal transduction pathways of MCP-1 induction in endothelial cells. Dioctanoylglycerol induced a protein kinase C (PKC)-dependent expression of MCP-1. Since staurosporine and genistein could only completely block phorbol ester-induced MCP-1 gene expression when delivered together, both PKC and tyrosine kinase must be involved. In addition, LPS-induced MCP-1 expression could only be partially blocked by PKC and tyrosine kinase inhibitors. Therefore, it was concluded that a third pathway must be involved in MCP-1 gene transcription.

4.2   MCP-2

Natural human MCP-2 was originally identified after stimulation of osteosarcoma cells with human IL-1β or with a semi-purified cytokine mixture. By the use of a specific radioimmunoassay (RIA) for MCP-2, protein induction was detected on human diploid fibroblasts, by dsRNA and IFN-γ (Van Damme et al., 1994). In general, maximal MCP-2 production levels (± 10 ng/ml) were about 10 times lower than those of MCP-1, and IL-1β was a better inducer for MCP-1 than for MCP-2. Measles virus and IFN-γ, but not IL-1β, were good MCP-2 inducers in MG-63 osteosarcoma cells, whereas IL-1β but not IFN-γ significantly enhanced MCP-1 production in these cells. Hep-2 epidermal carcinoma cells stimulated with various MCP-1 inducers were poor producers of MCP-2 (< 1 ng/ml).

Regulation studies of MCP-2 production in peripheral blood mononuclear cells confirmed the lower MCP-2 yields compared to MCP-1. Significant MCP-2 induction was detected after stimulation with IL-1β, IFN-β, IFN-γ, dsRNA, measles virus, Gram-positive bacteria (S. aureus), and Con A, but not with PMA, LPS, or Gram-negative bacteria (*E. coli*) (Van Damme *et al.*, 1994; Bossink *et al.*, 1995). Finally, in contrast to mononuclear cells, neutrophil granulocytes were poor MCP-1 and MCP-2 producers.

In conclusion, MCP-2 is often coproduced with MCP-1 but the production levels are considerably lower. In addition, induction of MCP-1 and MCP-2 could be differently regulated because IFN-γ seemed to be a relatively better MCP-2 (than MCP-1) inducer, whereas IL-1β was a superior inducer of MCP-1 in mononuclear cells, fibroblasts, and osteosarcoma cells. In contrast to MCP-1, MCP-2 was only induced by Gram-positive and not by Gram-negative bacteria.

4.3 MCP-3

At present, little is known about MCP-3 inducers and producer cells. MCP-3 protein was copurified with MCP-1 and MCP-2 from the supernatant of osteosarcoma cells, induced with a cytokine mixture (Van Damme *et al.*, 1992). The amount of protein recovered was comparable for MCP-3 and MCP-2. MCP-3 mRNA was found to be superinduced with cycloheximide in promonocytic U937 cells after pretreatment with phorbol ester (Minty *et al.*, 1993).

Minty and colleagues showed that MCP-1 and MCP-3 mRNAs were coordinately expressed in monocytes (after IFN-γ or PHA stimulation) but not in fibroblasts or astrocytoma cells. After induction of astrocytoma cells or lung and bone marrow fibroblast cell lines with IL-1β and TNF-α, no MCP-3 mRNA expression could be detected, although the expression of MCP-1 mRNA was clearly augmented. IL-13 inhibited both MCP-1 and MCP-3 mRNA expression.

5. *Biological Activities*

Monocyte chemotaxis *in vitro* and *in vivo* were the first activities ascribed to the three MCP molecules isolated so far. *In vitro* chemotaxis could be detected by measuring the migration of purified monocytes through polyvinylpyrrolidone (PVP)-treated micropore filters in Boyden chambers (reviewed in Wuyts *et al.*, 1994). The method can easily be adapted to neutrophil, eosinophil, and basophil chemotaxis provided that PVP-free membranes are used. However, the amount of blood, from normal donors, necessary to purify sufficient numbers of basophils and eosinophils is substantial. In order to study lymphocyte chemotaxis, micropore filters have to be pretreated with endothelial cells or fibronectin to prevent the cells from "falling" through the membrane pores (Carr *et al.*, 1994; Taub *et al.*, 1995a).

A common property of most chemokines is the increase of $[Ca^{2+}]_i$ in their target cells. In order to measure the calcium concentrations, the cells are loaded with a fluorescent indicator (e.g., Fura-2; Grynkiewicz *et al.*, 1985). Other activities of these monocyte chemotactic and activating factors include the release of enzymes (e.g., gelatinase, *N*-acetyl-β-glucosaminidase), arachidonic acid (AA) and leukotrienes, induction of the respiratory burst (H_2O_2 formation), inhibition of tumor growth, and histamine release (Table 32.1 and see below).

5.1 MCP-1

MCP-1 was found to be chemotactic for human monocytes (but not for neutrophils) from 0.1 nM upward and to induce *N*-acetyl β-glucosaminidase (with an EC_{50} of 0.5 nM) and gelatinase B release from monocytes (Table 32.1). Recombinant MCP-1 is also able to induce (from 0.5 nM upward) enhanced $[Ca^{2+}]_i$ and to elicit the respiratory burst (from 16 nM upward) as measured by H_2O_2 formation in human monocytes (Yoshimura *et al.*, 1989a; Zachariae *et al.*, 1990; Rollins *et al.*, 1991b; Sozzani *et al.*, 1991; Van Damme *et al.*, 1992). In addition, thymidine incorporation into tumor cells was inhibited by MCP-1-activated monocytes (Matsushima *et al.*, 1989). After stimulation of human monocytes and monocytic THP-1 leukemic cells with MCP-1 (1–10 nM for 15 s), induction of Ca^{2+}-dependent AA release occurs (Locati *et al.*, 1994).

Expression of the $β_2$-integrins Mac-1 and p150,95 (not LFA-1), but not of $α_4$-integrins, was enhanced by MCP-1. In addition, antibodies to $β_2$- and $α_4$-integrins both inhibited *in vitro* MCP-1-induced monocyte chemotaxis. Both integrins are probably necessary to induce effective *in vivo* chemotaxis (Jiang *et al.*, 1992).

Unlike RANTES and MIP-1α (two other C-C chemokines), MCP-1 is not active on human eosinophils (Rot *et al.*, 1992). However, MCP-1 (10 nM to 1 μM) was able to induce histamine release from human basophils (normal and allergic donors). This effect could be inhibited ± 30% by preincubation with IL-8 or RANTES (Alam *et al.*, 1992a,b). When basophils were pretreated with IL-3, IL-5, or GM-CSF, the histamine-releasing effect of MCP-1 was doubled, and basophils were also activated to release leukotriene C_4 (LTC_4) (Bischoff *et al.*, 1992; Kuna *et al.*, 1992). MCP-1 was found to be a stronger basophil agonist than IL-8, RANTES, MIP-1α, MIP-1β, complement fragment 3a (C3a), or anti-IgE receptor antibodies, but was somewhat weaker than C5a. Triggering of basophils with 30 nM MCP-1 induced a significant increase in $[Ca^{2+}]_i$. MCP-1 also appeared to have chemotactic properties for basophils (Leonard and Yoshimura, 1990). It was active from 3 nM upward with an optimal concentration of 10–30 nM (Table 32.1). RANTES was active with the same optimal concentration but resulted in a higher number of migrating cells, while MIP-1α caused comparable basophil migration to MCP-1 at 3 times lower concentrations (Bischoff *et al.*, 1993).

Table 32.1   Comparison of MCP-1, MCP-2 and MCP-3 activities on leukocytes in vitro

	Minimum effective concentration (nM)		
	MCP-1	MCP-2	MCP-3
Chemotaxis			
Monocytes	0.1	0.1–0.3	0.1–0.3
Neutrophils	> 30	> 100	(1)[a]
Eosinophils	> 300	3	3
Basophils	3	10	3
NK cells[b]	1[c]	ND[d]	ND
NK cells (+ IL-2)	3–6	3	3
T cells[e]	0.01–1	0.01–1	0.01–1
Cloned T cells[f]	0.1	0.1	0.1
Dendritic cells	> 10	> 10	1
Release of intracellular Ca^{2+}[g]			
Monocytes	0.5	> 5	5
Neutrophils	–[h]	ND	ND
Eosinophils	> 300	10	1
Basophils	< 30	< 100	< 50
NK cells (+ IL-2)	10	ND	ND
Cloned T cells[f]	0.05	5	0.5
Dendritic cells	> 10	ND	< 10
Additional activities on monocytes			
Respiratory burst induction	16	ND	ND
Enhanced release of			
AA[i]	1	> 5	5
N-Acetyl β-glucosaminidase	1	3	1
Gelatinase B	+	+	+
Tumor growth inhibition[j]	+	ND	ND
Additional activities on basophils			
Untreated cells: histamine release	10	100	10
IL-3-primed cells			
LTC_4 release[i]	1–3	30	10
Histamine release	1	30	3

[a] The maximal chemotactic response with MCP-3 on neutrophils is weak and at least 10 times lower than the response with IL-8 (Xu et al., 1995).

[b] Chemotaxis through fibronectin-coated membranes (Taub et al., 1995b).

[c] Large differences in the minimal effective dose (0.1–10 nM) for NK cell chemotaxis through fibronectin-coated membranes were found between distinct donors (Taub et al., 1995b).

[d] ND, not determined.

[e] Freshly isolated peripheral blood T lymphocytes (CD4+ or CD8+) (Taub et al., 1995a).

[f] T cell clones (CD4+ or CD8+) (Loetscher et al., 1994).

[g] $[Ca^{2+}]_i$ concentrations were measured on cells preloaded with a fluorescent indicator (e.g. Fura-2; Grynkiewicz et al., 1985).

[h] Negative (–) or positive (+) results were described without indication of concentrations.

[i] AA and LTC_4 release were detected by thin-layer chromatography on cells preloaded with radioactively labeled AA (Bischoff et al., 1992; Sozzani et al., 1994a).

[j] Detected by measuring thymidine incorporation into tumor cell lines (Matsushima et al., 1989).

Using freshly isolated T lymphocytes, MCP-1 acted as a potent chemoattractant for memory T cells (both CD4+ and CD8+) in a lymphocyte transendothelial migration assay (at 1 nM MCP-1) or in Boyden chambers using fibronectin-coated membranes (optimal MCP-1 concentration of 100 pM) (Carr et al., 1994; Taub et al., 1995a). MCP-1 (optimal concentration 10 nM) was also chemotactic for cloned CD4+ and CD8+ T cells. An increase in $[Ca^{2+}]_i$ was detected in these T cell clones from 50 pM upward (Loetscher et al., 1994). Finally, MCP-1 (3–6 nM) was chemotactic for and enhanced the $[Ca^{2+}]_i$ in IL-2-activated natural killer (NK) cells (Allavena et al., 1994) but was inactive on dendritic cells (Sozzani et al., 1995). MCP-1 was as potent as RANTES and MIP-1β in inducing NK cell chemotaxis but less potent than γIP-10 and MIP-1α (Taub et al., 1995b). MCP-1, at low nanomolar concentrations, was able to enhance NK cell cytolysis of tumor cells and induced NK cell degranulation. The sensitivity of NK cells to MCP-1-induced chemotaxis was dependent on the donor, on the coating technique used, and on the NK cell purification method. When NK cells had been incubated at 37°C during the purification, they lost their sensitivity to chemokine-induced chemotaxis.

After intradermal injection of human MCP-1 in Lewis rats, a marked *in vivo* monocyte infiltration could be detected (Zachariae *et al.*, 1990). In rabbits, intradermally injected MCP-1 (10–500 ng/site) also resulted in monocyte recruitment at the injection site (Van Damme *et al.*, 1992). Subcutaneous administration of MCP-1 yielded a significant accumulation of human T lymphocytes in chimeric SCID/hu mice (Taub *et al.*, 1995a).

Surprisingly, in transgenic mice overexpressing murine JE (murine MCP-1) in the basal layer of the epidermis, dendritic cells, but not monocytes, were recruited to the skin (Nakamura *et al.*, 1995). This observation seems to be in contrast with the *in vitro* observation that human MCP-1 is not chemotactic for dendritic cells.

5.2 MCP-2

MCP-2 induced chemotaxis of monocytes *in vitro* (using the Boyden chamber), showing a typical bell-shaped curve. Maximal monocyte chemotactic index was reached at 1 nM, whereas the minimal MCP-2 concentration for monocyte chemotaxis was 0.1 nM (Table 32.1). *In vitro* monocyte chemotaxis could be confirmed *in vivo* by intradermal injection of MCP-2 (10–500 ng/site) into rabbits, resulting in significant infiltration of monocytes into the rabbit skin after 18 h (Van Damme *et al.*, 1992). Natural or synthetic MCP-2 was not able to attract neutrophil granulocytes *in vitro* at concentrations up to 100 nM, whereas IL-8 was active from 0.1 nM upward (Van Damme *et al.*, 1992; Proost *et al.*, 1995). MCP-2 could also activate monocytes but not neutrophils to secrete gelatinase activity as determined by SDS-PAGE zymography (Opdenakker *et al.*, 1993). In contrast to other C-C chemokines (e.g., MCP-1 and MCP-3), MCP-2 was not able to induce an increase in $[Ca^{2+}]_i$ or an AA release on monocytes in suspension (Sozzani *et al.*, 1994a,b). However, when adherent monocytes were studied at the single cell level, 40% of the cells responded to 1 nM of MCP-2 (0.1 pM for MCP-1) with an increased $[Ca^{2+}]_i$ (Bizzarri *et al.*, 1995).

Recently, other leukocytes have also been shown to be responsive to MCP-2. Like other C-C chemokines (e.g., MCP-3, MIP-1α), MCP-2 was found to be chemotactic for basophils (from 10 nM upward). MCP-2 at 100 nM was also able to induce enhanced, Ca^{2+}-dependent histamine secretion from human basophils. No effect could be detected on mouse peritoneal mast cells (Alam *et al.*, 1994). MCP-2 induced *in vitro* chemotaxis of eosinophils from 3 nM upward, but was less potent than RANTES (Noso *et al.*, 1994). Significant chemotaxis of freshly isolated CD4+ and CD8+ T lymphocytes (through fibronectin-precoated membranes) was detected with MCP-2 at 10 pM. This MCP-2-induced T cell chemotaxis could be inhibited with a polyclonal rabbit anti-human MCP-2 antiserum (Taub *et al.*, 1995a).

Chemotactic activity (minimal MCP-2 concentration of 10–100 pM) for T cell clones was also detected (Table 32.1). An MCP-2-induced increase in $[Ca^{2+}]_i$ could only be detected from 5 nM upward, while MCP-1 was already active at 50 pM (Loetscher *et al.*, 1994). MCP-2 was found to be equally chemotactic as MCP-1 for IL-2-activated NK cells (Allavena *et al.*, 1994), but was not active on dendritic cells (Sozzani *et al.*, 1995).

5.3 MCP-3

MCP-3 was first described as a monocyte chemotactic and activating protein without effect on neutrophils (Van Damme *et al.*, 1992; Opdenakker *et al.*, 1993). The minimal (0.1 nM) and optimal (1 nM) concentrations of MCP-3 were comparable to those of MCP-1 and MCP-2 (Table 32.1). In addition, incubation of monocytes with MCP-3 induced an increase in $[Ca^{2+}]_i$, the release of AA and N-acetyl-β-D-glucosaminidase (Sozzani *et al.*, 1994a,b; Uguccioni *et al.*, 1995). *In vitro* monocyte chemotaxis was confirmed *in vivo* by intradermal injection of natural human MCP-3 (50–500 ng/site) into rabbits. After 18 h, monocyte infiltration could be detected into rabbit skin (Van Damme *et al.*, 1992).

Synthetic (Proost *et al.*, 1995) as well as recombinant (Minty *et al.*, 1993) MCP-3 were found to be potent (minimal effective concentration of 3 nM) chemotactic proteins for eosinophil and basophil granulocytes (Table 32.1). MCP-3 also induced an increase in $[Ca^{2+}]_i$ in these cells (Alam *et al.*, 1994; Dahinden *et al.*, 1994; Noso *et al.*, 1994). MCP-3 caused an enhanced histamine release from both unprimed and IL-3-treated basophils and induced the release of LTC_4 from IL-3-treated basophil granulocytes (Alam *et al.*, 1994; Dahinden *et al.*, 1994). Recently, a weak neutrophil chemotactic response (maximal chemotactic index of 2.0) has been detected with 1–10 nM MCP-3 (Xu *et al.*, 1995).

MCP-3 was reported to attract peripheral blood human T lymphocytes (both CD4+ and CD8+) in Boyden chambers using low concentrations (10 pM) of chemoattractants (Taub *et al.*, 1995a). The minimal MCP-3 concentration resulting in chemotactic activity for cloned T cells was 10–100 pM (Table 32.1). Lymphocyte chemotaxis was confirmed in transendothelial assays for both CD4+ and CD8+ T cells (Roth *et al.*, 1995). An increase in $[Ca^{2+}]_i$ could be detected from 500 pM upward (Loetscher *et al.*, 1994). IL-2-activated (but not resting) natural killer (NK) cells were attracted by MCP-3 at concentrations between 3 and 20 nM (Allavena *et al.*, 1994). Like MIP-1α and RANTES, but in contrast to MCP-1, MCP-2, IL-8, and γIP-10, MCP-3 was a potent chemotactic agent for dendritic cells (Sozzani *et al.*, 1995). MCP-3 at 1 nM resulted in significant dendritic cell chemotaxis, and a maximal chemotactic response was obtained at 5 nM. At chemotactic concentrations, MCP-3 elicited an enhanced $[Ca^{2+}]_i$ in dendritic cells.

In conclusion, MCP-1, MCP-2 and MCP-3, originally described as monocyte chemotactic proteins, were also found to be active on other leukocytes. However, mononuclear cells remained the most responsive (lowest minimal concentration) cell types. Although the three proteins are structurally (primary sequence) highly related, in addition to common properties, clear differences in biological activities, active concentrations, and target cells have been observed (Table 32.1). MCP-3 has been characterized as a pluripotent chemokine with high specific activity for most leukocyte cell types.

6. MCP Receptors

Several C-C chemokine receptors (CCR-1 to CCR-5) have been cloned. Their cDNAs encode proteins of about 360 amino acids (aa) with 45–75% amino acid identity. Similarly to both IL-8 receptors, C-C chemokine receptors (CCR) are G protein-linked seven-transmembrane-spanning receptors with an extracellular amino-terminus and an intracellular carboxy-terminal domain. The seven-transmembrane α-helices each contain about 25 amino acids which are interconnected through short (10–20-aa) intra- and extracellular loops. The amino-terminus (40–50 amino acids) together with the extracellular loops and possibly also the transmembrane domains of the receptors determine the ligand specificity. The carboxyterminus (± 50 amino acids) of most CCRs is important for G-protein–receptor interaction and receptor desensitization through phosphorylation (reviewed in Wells et al., 1996). MCP-1, MCP-2, and MCP-3 binding to these receptors as well as the expression of the receptors on different leukocytes has been deciphered only partially (Figure 32.4).

Four receptors were found to be expressed on monocytes. MCP-3, similarly to the earlier-identified ligands MIP-1α and RANTES, was found to be a functional ligand for CCR-1 (Figure 32.4). In contrast, MCP-1 was identified as a weak ligand for this receptor and could not compete for MCP-3 binding (Gao et al., 1993; Neote et al., 1993; Combadiere et al., 1995b; Ben-Baruch et al., 1995b). Both spliced variants of CCR-2 were originally described as MCP-1-specific receptors (Charo et al., 1994). Recently, also MCP-3 and MCP-2 were found to be functional ligands for this receptor (Combadiere et al., 1995b; Franci et al., 1995; Gong et al., 1997). Competition of MCP-3 for ^{125}I-MCP-1 binding to recombinant CCR-2 on transfected HEK-293 cells (human embryonic kidney cells) was 35-fold less effective than with cold MCP-1. MCP-1 was able to block a subsequent MCP-3-induced enhanced $[Ca^{2+}]_i$. However, MCP-3 had no significant effect on MCP-1-induced increases of the $[Ca^{2+}]_i$ through CCR-2. This is

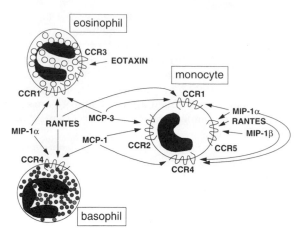

Figure 32.4   C-C chemokine–receptor interactions. Schematic representation of the expression of C-C chemokine receptors (CCRs) on monocytes and eosinophilic and basophilic granulocytes. The interaction of chemokines with the CCRs is indicated by arrows (Horuk, 1994; Murphy 1994; Ben-Baruch et al., 1995a,b; Combadiere et al., 1995a,b; Franci et al., 1995; Power et al., 1995).

in contrast to the results on monocytes, in which MCP-3, but not MCP-2, competed as effectively as cold MCP-1 for ^{125}I-MCP-1 binding to the MCP-1 receptors (Sozzani et al., 1994b). In addition, although MCP-2 bound well to CCR-2 transfected cells, MCP-2 only poorly competed for radioactive MCP-1 binding (Gong et al., 1997; Yamagami et al., 1997). Thus, MCP-2, compared to MCP-1, differently interacts with CCR-2. MCP-1 has also been shown to interact weakly with CCR-4, which is a CCR expressed on monocytes and lymphocytes (Power et al., 1995). CCR-5 signaled through MIP-1α, MIP-1β, and RANTES binding (Combadiere et al., 1995a; Samson et al., 1996).

In basophils, MCP-1 and MCP-3 induced an increase in the $[Ca^{2+}]_i$ and a histamine release through G-protein-coupled receptors (Alam et al., 1992a,b, 1994; Bischoff et al., 1993; Dahinden et al., 1994). Histamine release from basophils could only be obtained in the presence of extracellular Ca^{2+} (Alam et al., 1994). Ca^{2+} seemed to be essential for activation of basophils and therefore it was not surprising that MCP-2 was only a weak chemotactic and activating chemokine for basophils. Basophils might have one common MCP-1- and MCP-3- but not an MCP-2-specific receptor. The MCP-1/MCP-3 receptor(s) might be activated at high MCP-2 concentrations. The first identified CCR on activated basophils was CCR-4 (Power et al., 1995). Since MCP-3 was such a strong chemotactic and activating chemokine for basophils, it will certainly be interesting to investigate whether this receptor signals through MCP-3 in addition to MCP-1, MIP-1α, and RANTES.

In contrast to MCP-2 and MCP-3, MCP-1 was not active on eosinophilic granulocytes. In addition, pretreatment of eosinophils with natural MCP-1 had no effect on subsequent stimulation with MCP-2 or MCP-3. The difference between MCP-1 and MCP-3 with respect to eosinophil chemotaxis and activation might be explained by the different expression patterns of CCR-1 and CCR-2 on eosinophils. CCR-2 mRNA was, in contrast to CCR-1 mRNA, not detected on eosinophils and MCP-1 is only a weak activator of CCR-1. Stimulation of eosinophils with MCP-2 and MCP-3 reduced chemotactic activity in response to RANTES and vice versa. In addition, MCP-3 stimulation reduced the chemotactic response to MCP-2 (Noso *et al.*, 1994). This suggests that, on eosinophils, MCP-2 and MCP-3 share a common receptor and/or signal transduction pathway(s) with RANTES. MCP-1 and MCP-3 desensitization experiments with MIP-1α and RANTES were evaluated by measuring changes in $[Ca^{2+}]_i$ (Dahinden *et al.*, 1994). RANTES-pretreatment inhibited the enhanced $[Ca^{2+}]_i$ induced with MCP-3 and MIP-1α. MCP-3 only partially desensitized for both the MIP-1α and RANTES effects and MIP-1α hardly desensitized for RANTES or MCP-3. A combination of MCP-3 and MIP-1α was able to prevent completely the RANTES-induced increase in $[Ca^{2+}]_i$. Thus, on eosinophils, there is a RANTES receptor activated by RANTES and MCP-3, but not by MIP-1α, and an MIP-1α receptor (probably CCR-1) activated by MIP-1α, RANTES, and more weakly by MCP-3 (Figure 32.4). CCR-3 is an eosinophil-specific receptor, which has been identified as the receptor for a powerful eosinophil chemotactic C-C chemokine, eotaxin (Kitaura *et al.*, 1996). Recently, also RANTES and MCP-3 binding to CCR-3 has been reported (Daugherty *et al.*, 1996).

MCP-1, MCP-3 and, at high concentrations, also MCP-2, induced enhanced $[Ca^{2+}]_i$ in cloned T lymphocytes. An increase in the $[Ca^{2+}]_i$, in response to MCP-2 and MCP-3, was inhibited after prestimulation with MCP-1. Pretreatment with MCP-3 resulted in a complete and partial inhibition of the enhanced $[Ca^{2+}]_i$ induced by MCP-2 and MCP-1, respectively. MCP-2 reduced the response to MCP-3, but had little effect on MCP-1-induced $[Ca^{2+}]_i$ increase. Thus, T cells (both CD4$^+$ and CD8$^+$) seem to bear MCP-1 receptors (probably CCR-2), that are also recognized by MCP-3, and by MCP-2 with lower affinity (Loetscher *et al.*, 1994).

7. Signal Transduction Pathways

The signaling pathway for MCP-1 in monocytes differs from the one used by FMLP (*N*-formyl-methionyl-leucyl-phenylalanine), C5a, PAF (platelet-activating factor), and leukotriene B$_4$ (LTB$_4$). $[Ca^{2+}]_i$ increase and chemotactic activation of mononuclear cells in response to MCP-1 and MCP-3 could be inhibited by pertussis toxin (PT), whereas cholera toxin (CT) had little or no effect. In contrast, MCP-2-induced chemotaxis was only slightly affected by PT treatment but was sensitive to CT. In addition, at chemotactic concentrations, MCP-2 was not able to induce an enhanced $[Ca^{2+}]_i$ in monocytes. About 100 times higher amounts of MCP-2, compared to MCP-1 and MCP-3, were necessary to give a comparable increase in the $[Ca^{2+}]_i$ in monocytes (Sozzani *et al.*, 1991, 1994b; Bizzarri *et al.*, 1995). MCP-1 and MCP-3 activation of CCR-2B-transfected HEK-293 cells inhibited adenylylcyclase activity (cAMP generation) with an IC$_{50}$ of 67 pM. MCP-2 showed a minor effect at 100 times higher concentrations and failed to inhibit 50% of adenylylcyclase activity at concentrations up to 20 nM (Franci *et al.*, 1995). MCP-1 and MCP-3 thus differ from MCP-2 in signaling pathways.

In the absence of extracellular Ca^{2+} and in the presence of EGTA or Ni^{2+}, the $[Ca^{2+}]_i$ increase, induced by MCP-1 or by MCP-3, was prevented, indicating that the rise in $[Ca^{2+}]_i$ is predominantly dependent on the influx of extracellular Ca^{2+} through plasma membrane channels (Sozzani *et al.*, 1991, 1994b). Depolarization with KCl had no effect on $[Ca^{2+}]_i$ in unstimulated monocytes and voltage-operated Ca^{2+}-channels were not used after MCP-1 stimulation. General Ca^{2+}-channel blockers however, diminished the $[Ca^{2+}]_i$ increase induced with MCP-1, suggesting the use of receptor-activated Ca^{2+}-channels (Sozzani *et al.*, 1993). At the single cell level, blocking of the influx of Ca^{2+} into the monocyte (with Ni^{2+} or EGTA) resulted in only a partial inhibition of the MCP-1-induced increase in $[Ca^{2+}]_i$. The induction of an enhanced $[Ca^{2+}]_i$ was only prevented when Ca^{2+} channels were blocked, intracellular Ca^{2+} stores were emptied, and Ca^{2+} reuptake was inhibited (with thapsigargin) before the addition of MCP-1 (Bizzarri *et al.*, 1995).

In monocytes, MCP-1 and MCP-3, but not MCP-2, induced an enhanced release of AA from phosphatidylcholine (PC). Combined treatment with PAF and MCP-1 (or MCP-3) resulted in a synergistic effect. If AA release is involved in the chemotactic response of monocytes, this transduction pathway is important for some (MCP-1 and MCP-3) but not all (MCP-2) MCPs (Locati *et al.*, 1994, Sozzani *et al.*, 1994a,b). The MCP-1-induced release of AA was strongly decreased by PT and inhibitors of phospholipase A$_2$ (PLA$_2$) (mepacrine, manoalide, and dexamethasone). Increase in $[Ca^{2+}]_i$ was required for AA release but a PT-sensitive step, other than activation of Ca^{2+} influx, was also necessary for the activation of PLA$_2$ by MCP-1 (Locati *et al.*, 1994). When the $[Ca^{2+}]_i$ increase or PLA$_2$ activity was inhibited, monocyte chemotaxis in response to MCP-1 decreased. Inhibitors of PLA$_2$ also inhibited monocyte polarization, indicating that PLA$_2$ plays an important role in signal transduction (Sozzani *et al.*, 1993; Locati *et al.*, 1994).

Phospholipase C (PLC)-dependent signal transduction pathways are not expected since the release of Ca^{2+} from intracellular stores was hardly detectable and no inositol trisphosphate (IP_3) production or turnover of phosphatidylinositol biphosphate (PIP_2), was observed after stimulation of monocytes with MCP-1. Neither cAMP-dependent protein kinases nor PKC are thought to be directly involved in the induction of $[Ca^{2+}]_i$ increase by MCP-1. However, PKC may have a role in the control of $[Ca^{2+}]_i$ homeostasis in monocytes by inhibiting the MCP-1-induced $[Ca^{2+}]_i$ increase (Sozzani et al., 1993). PKC and cAMP-dependent kinase inhibitors caused inhibition of MCP-1-induced chemotaxis. Specific inhibitors for cGMP-dependent kinase and for myosin L-chain kinase had no effect on MCP-1-induced migration. This suggests that a serine/threonine kinase, possibly PKC, is involved in signaling by MCP-1 (Sozzani et al., 1991). Serine/threonine or tyrosine kinase inhibitors caused an inhibition of chemotactic activity for MCP-1, MCP-2, and MCP-3. This demonstrates that protein kinase activation is a common pathway in the action of MCP-1, MCP-2 and MCP-3 (Sozzani et al., 1994b).

In basophils, MCP-1 induced $[Ca^{2+}]_i$ increase through G-protein-coupled receptors, as do other chemokines (Bischoff et al., 1992). In CD4$^+$ and CD8$^+$ cloned T lymphocytes, MCP-1, MCP-3 and, at higher concentrations, also MCP-2 induced an increase in $[Ca^{2+}]_i$ that could be inhibited by PT (Loetscher et al., 1994).

8. Murine MCPs

Murine JE mRNA was discovered in 1983 as a PDGF-inducible early competence gene (Cochran et al., 1983). The cDNA encoded a protein of 148 amino acids with a 23-aa long signal sequence. Murine JE is generally considered to be the homologue of human MCP-1. The intron and exon boundaries were located at the same site in murine and human JE/MCP-1 (Rollins et al., 1988) and the gene was found in the C-C chemokine gene cluster on mouse chromosome 11 (Wilson et al., 1990). In the amino-terminal part of murine JE, 55% of the amino acids, including an amino-terminal pyroglutamic acid, were identical in the human MCP-1 sequence and 53% of the amino acids were shared with human MCP-2 and MCP-3. The amino-terminal part of the protein was detected to be responsible for the chemotactic activity (Van Damme et al., 1991; Ernst et al., 1994). In addition, JE contained a long serine- and threonine-rich carboxy-terminal tail of 49 residues which may be heavily O-glycosylated. Natural, biologically active JE had an apparent molecular mass on SDS-PAGE of 7–30 kDa (Van Damme et al., 1991; Luo et al., 1994; Liu et al., 1996). Carboxy-terminal truncation and/or glycosylation explained these large differences in molecular mass.

Using the human MCP-3 cDNA, the homologous murine cDNA has been identified from an LPS-stimulated macrophage cell line. The deduced protein sequence of murine MCP-3 showed 55% and 59% amino acid homology with human MCP-1 and human MCP-3, respectively, and contained 74 amino acids (Thirion et al., 1994). The murine MCP-3 cDNA was almost identical to two other murine cDNA clones, MARC and fic. MARC was isolated from a mast cell line and the transcription of the gene was enhanced after IgE plus antigen challenge of the cells (Kulmburg et al., 1992). fic had been identified as a cDNA of which the transcription was upregulated with serum in a fibroblast cell line (Heinrich et al., 1993).

Murine JE, like human MCP-1, possessed chemotactic activity for monocytes and lymphocytes but not for neutrophils (Van Damme et al., 1991; Luo et al., 1994; Liu et al., 1996). Murine JE and MCP-3 both induced an enhanced $[Ca^{2+}]_i$ in human monocytes (Heinrich et al., 1993). The 7 kDa form of JE had a higher specific activity in in vitro monocyte and lymphocyte chemotaxis experiments than the high molecular mass form (Liu et al., 1996).

JE was induced by a variety of factors including cytokines (IL-1β, TNF α, and PDGF) and bacterial (LPS) and viral (vesicular stomatitis virus, Newcastle disease virus, Sendai virus, dsRNA) products. Upon stimulation, JE was produced in tumor cell lines and in normal cells such as monocytes, fibroblasts, endothelial cells, astrocytes, and mesangial cells (reviewed in Haelens et al., 1996). Estrogen partially inhibited LPS-induced JE mRNA expression in murine macrophages (Frazier-Jessen and Jovacs, 1995). Thus, estrogen may work anti-atherogenically through the prevention of macrophage accumulation in atherosclerotic lesions. In animal models, JE suppressed tumor growth by attracting monocytes which infiltrated the tumor (Bottazzi et al., 1992).

A murine homolog of one of the human MCP-1 receptors, namely CCR-4, has been cloned (Hoogewerf et al., 1996). Murine JE/MCP-1 seemed to use the same signal transduction pathways as human MCP-1. Murine MCP-1 signaled through PT-sensitive G-protein-linked receptors; the $[Ca^{2+}]_i$ was enhanced through the influx of extracellular Ca^{2+} and JE used Tyr-kinases and p42/p44 MAP-kinase for further signal transduction (Dubois et al., 1996).

Transgenic mice that were constitutively over-expressing MCP-1/JE showed high (>10 ng/ml) serum levels of MCP-1, but no accumulation of monocytes was seen in different organs (Rutledge et al., 1995). These mice were more sensitive to bacterial infections. Possible explanations for this phenomenon might be that the high serum MCP-1 levels neutralized local chemotactic gradients or that the MCP-1 receptors were desensitized by the high systemic concentrations of MCP-1.

9. MCPs in Disease

Sylvester and colleagues (1993) have found low MCP-1 levels in serum from normal individuals, whereas significant concentrations of IgG autoantibodies against MCP-1 were also detected. Upon intravenous injection of human volunteers with LPS, MCP-1 serum levels increased to more than 1 µg/ml within 2–4 h and returned to baseline after 5 h. Although free anti-MCP-1 autoantibody levels decreased, no increase in MCP-1–IgG complexes could be detected, suggesting an unknown mechanism for the elimination of the complexes.

Until now, most studies on the appearance of MCPs during diseases, in which recruitment of monocytes or macrophages is important, have concentrated on MCP-1. In 1992, Koch and colleagues had reported enhanced MCP-1 concentrations in synovial fluid and sera of RA patients compared to osteoarthritis or control patients (Koch et al., 1992). Synovial macrophages were found to be a major source of MCP-1 and a positive correlation with the induction of IL-6, IL-8, and IL-1β was reported. MCP-1 was also detected in the synovial fluid of patients with osteoarthritis, gout, and traumatic arthritis. Immunoreactive protein was observed in the lining and sublining cells of the synovium and in the vascular endothelial cells of rheumatoid synovia (Koch et al., 1992; Harigai et al., 1993). Type B (fibroblast-like) synoviocytes of RA patients were found to be inducible for MCP-1 synthesis by TNF-α, IL-1α, IL-1β, and IFN-γ. IL-1β-, but not TNF-α-induced MCP-1 levels, were further enhanced by IL-4 and IFN-γ. After IL-1α injection into the knee joint of rabbits, MCP-1 was produced locally and mononuclear cells were found to infiltrate the joint (Akahoshi et al., 1993; Hachicha et al., 1993; Seitz et al., 1994).

MCP-1 has been detected in the bronchial epithelium of idiopathic pulmonary fibrosis and asthma patients (Antoniades et al., 1992; Sousa et al., 1994). MCP-1 was detected in tuberculosis effusions and it was the predominant chemotactic activity found during pleural infections. The highest MCP-1 levels were measured in malignant pleural effusions (Antony et al., 1993). After phagocytosis of *Mycobacterium tuberculosis* by monocytic cell lines, MCP-1 induction was also detected (Friedland et al., 1993).

Enhanced MCP-1 expression has been detected in a variety of inflammatory skin diseases, e.g., psoriasis, lichenoid dermatitis, and spongiotic dermatitis. In psoriasis, a strong MCP-1 mRNA signal is found above the dermal–epidermal junction in lesional psoriatic skin but not in uninvolved psoriatic or normal skin. The basal keratinocytes, rather than the melanocytes, were the major MCP-1-producing cells (Gillitzer et al., 1993; Yu et al., 1994). Recently, enhanced MCP-3 and RANTES mRNA has been detected in allergen-challenged skin of

human atopic subjects (Ying et al., 1995). MCP-3 (peak after 6 h) and RANTES (peak after 24 h) mRNA enhancement paralleled the kinetics of eosinophil and T cell infiltration, respectively. These results suggest a role for MCP-3 in eosinophil recruitment *in vivo*.

Elevated expression of MCP-1 was also detected during periodontal infections (Hanazawa et al., 1993; Tonetti et al., 1994). In human inflamed gingival tissues, enhanced MCP-1 mRNA and protein levels were found in endothelial cells and macrophages (Yu et al., 1993).

MCP-1 was strongly expressed in cells in the macrophage-rich regions of human atherosclerotic lesions (Ylä-Herttuala et al., 1991). Interestingly, human aortic cell cultures produced MCP-1 after induction with low-density lipoprotein, an effect which was abolished by high-density lipoprotein and antioxidants and almost completely prevented by leumedin (Navab et al., 1991, 1993).

Monocyte infiltration in the central nervous system could also be due to MCP-1 production. MCP-1 has been recovered from glioma cell lines and from surgical specimens of human malignant glioblastomas and ependymomas (Takeshima et al., 1994; Desbaillets et al., 1994) and during experimental autoimmune encephalomyelitis (EAE; Hulkower et al., 1993; Ransohoff et al., 1993). In autoimmune diseases, chemokines can play a pivotal role. MCPs recruit macrophages (antigen-presenting cells) and lymphocytes to the primary focus and stimulate the production of extracellular proteases which contribute to the autoimmune process (Opdenakker and Van Damme, 1994).

In livers from patients with chronic acute hepatitis or fulminant hepatic failure, enhanced MCP-1 levels were measured (Marra et al., 1993; Czaja et al., 1994). Recently, MCP-1 and MCP-2 levels were found to be elevated in serum of sepsis patients (Bossink et al., 1995). Elevated MCP-1 levels were detected during Gram-positive and Gram-negative infection.

Macrophages can be attracted to tumors by their production of MCPs. These tumor-associated macrophages can be activated by MCPs to inhibit tumor growth (Mantovani et al., 1992). The role of these chemotactic factors in invasion and metastasis can also be explained by the countercurrent principle (Opdenakker and Van Damme, 1992). The production of chemotactic cytokines by cancer cells could function as a mechanism which guarantees directional migration of malignant cells towards the blood vessels or lymphatics. Thus it was found that expression of IL-8 correlates with the metastatic potential of melanoma cells (Singh et al., 1994). The production of MCPs by allogeneic or syngeneic cancer cells seems to suppress cancer cell growth and metastatic potential *in vivo*. In these cases, however, antigenic stimuli (allogeneic or from tumor antigens) might contribute to these effects (Matsushima et al., 1989; Huang et al., 1994).

10. Acknowledgments

The authors thank René Conings for editorial help and appreciate the technical support of Jean-Pierre Lenaerts and Willy Put. G.O. is Senior Research Associate and A.W. is a research assistant of the Belgian National Fund for Scientific Research (NFWO). We acknowledge the financial support of the Belgian NFWO and the National Lottery.

11. References

Ahuja, S.K., Gao, J.-L. and Murphy, P. M. (1994). Chemokine receptors and molecular mimicry. Immunol. Today 15, 281–287.

Akahoshi, T., Wada, C., Endo, H., et al. (1993). Expression of monocyte chemotactic and activating factor in rheumatoid arthritis: regulation of its production in synovial cells by interleukin-1 and tumor necrosis factor. Arthritis Rheum. 36, 762–771.

Alam, R., Forsythe, P.A., Lett-Brown, M.A. and Grant, J.A. (1992a). Interleukin-8 and RANTES inhibit basophil histamine release induced with monocyte chemotactic and activating factor/monocyte chemoattractant peptide-1 and histamine releasing factor. Am. J. Respir. Cell Mol. Biol. 7, 427–433.

Alam, R., Lett-Brown, M.A., Forsythe, P.A., et al. (1992b). Monocyte chemotactic and activating factor is a potent histamine-releasing factor for basophils. J. Clin. Invest. 89, 723–728.

Alam, R., Forsythe, P., Stafford, S., et al. (1994). Monocyte chemotactic protein-2, monocyte chemotactic protein-3, and fibroblast-induced cytokine, three new chemokines, induce chemotaxis and activation of basophils. J. Immunol. 153, 3155–3159.

Allavena, P., Bianchi, G., Zhou, D., et al. (1994). Induction of natural killer cell migration by monocyte chemotactic protein-1, -2 and -3. Eur. J. Immunol. 24, 3233–3236.

Antoniades, H.N., Neville-Golden, J., Galanopoulos, T., Kradin, R.L., Valente, A.J. and Graves, D.T. (1992). Expression of monocyte chemoattractant protein 1 mRNA in human idiopathic pulmonary fibrosis. Proc. Natl Acad. Sci. USA 89, 5371–5375.

Antony, V.B., Godbey, S.W., Kunkel, S.L., et al. (1993). Recruitment of inflammatory cells to the pleural space. Chemotactic cytokines, IL-8, and monocyte chemotactic peptide-1 in human pleural fluids. J. Immunol. 151, 7216–7223.

Baggiolini, M., Dewald, B. and Moser, B. (1994). Interleukin-8 and related chemotactic cytokines-CXC and CC chemokines. Adv. Immunol. 55, 97–179.

Ben-Baruch, A., Michiel, D.F. and Oppenheim, J.J. (1995a). Signals and receptors involved in recruitment of inflammatory cells. J. Biol. Chem. 270, 11703–11706.

Ben-Baruch, A., Xu, L., Young, P.R., Bengali, K., Oppenheim, J.J. and Wang, J.M. (1995b). Monocyte chemotactic protein-3 (MCP-3) interacts with multiple leukocyte receptors. C-C CKR-1, a receptor for macrophage inflammatory protein 1α/RANTES is also a functional receptor for MCP-3. J. Biol. Chem. 270, 22123–22128.

Bischoff, S.C., Krieger, M., Brunner, T. and Dahinden, C.A. (1992). Monocyte chemotactic protein 1 is a potent activator of human basophils. J. Exp. Med. 175, 1271–1275.

Bischoff, S.C., Krieger, M., Brunner, T., et al. (1993). RANTES and related chemokines activate human basophil granulocytes through different G protein-coupled receptors. Eur. J. Immunol. 23, 761–767.

Bizzarri, C., Bertini, R., Bossù, P., et al. (1995). Single-cell analysis of macrophage chemotactic protein-1-regulated cytosolic Ca^{2+} increase in human adherent monocytes. Blood 86, 2388–2394.

Bossink, A.W.J., Paemen, L., Jansen, P.M., Hack, C.E., Thijs, L.G. and Van Damme, J. (1995). Plasma levels of the chemokines monocyte chemotactic proteins-1 and -2 are elevated in human sepsis. Blood 86, 3841–3847.

Bottazzi, B., Polentarutti, N., Balsari, A., et al. (1983). Chemotactic activity for mononuclear phagocytes of culture supernatants from murine and human tumor cells: evidence for a role in the regulation of the macrophage content of neoplastic tissues. Int. J. Cancer 31, 55–63.

Bottazzi, B., Walter, S., Govoni, D., Colotta, F. and Mantovani, A. (1992). Monocyte chemotactic cytokine gene transfer modulates macrophage infiltration, growth, and susceptibility to IL-2 therapy of a murine melanoma. J. Immunol. 148, 1280–1285.

Carr, M.W., Roth, S.J., Luther, E., Rose, S.S. and Springer, T.A. (1994). Monocyte chemoattractant protein 1 acts as a T-lymphocyte chemoattractant. Proc. Natl Acad. Sci. USA 91, 3652–3656.

Chang, H.C., Hsu, F., Freeman, G.J., Griffin, J.D. and Reinherz, E.L. (1989). Cloning and expression of a γ-interferon-inducible gene in monocytes: a new member of a cytokine gene family. Int. Immunol. 1, 388–397.

Charo, I.F., Myers, S.J., Herman, A., Franci, C., Connolly, A.J. and Coughlin, S.R. (1994). Molecular cloning and functional expression of two monocyte chemoattractant protein 1 receptors reveals alternative splicing of the carboxyl-terminal tails. Proc. Natl Acad. Sci. USA 91, 2752–2756.

Cochran, B.H., Reffel, A.C. and Stiles, C.D. (1983). Molecular cloning of gene sequences regulated by platelet-derived growth factor. Cell 33, 939–947.

Colotta, F., Borré, A., Wang, J.M., et al. (1992a). Expression of a monocyte chemotactic cytokine by human mononuclear phagocytes. J. Immunol. 148, 760–765.

Colotta, F., Sironi, M., Borré, A., Luini, W., Maddalena, F. and Mantovani, A. (1992b). Interleukin 4 amplifies monocyte chemotactic protein and interleukin 6 production by endothelial cells. Cytokine 4, 24–28.

Combadiere, C., Ahuja, S.K. and Murphy, P.M. (1995a). Cloning and functional expression of a human eosinophil CC chemokine receptor. J. Biol. Chem. 270, 16491–16494 and correction J. Biol. Chem. 270, 30235.

Combadiere, C., Ahuja, S.K., Van Damme, J., Tiffany, H.L., Gao, J.-L. and Murphy, P.M. (1995b). Monocyte chemoattractant protein-3 is a functional ligand for CC chemokine receptors 1 and 2B. J. Biol. Chem. 270, 29671–29675.

Czaja, M.J., Geerts, A., Xu, J., Schmiedeberg, P. and Ju, Y. (1994). Monocyte chemoattractant protein 1 (MCP-1) expression occurs in toxic rat liver injury and human liver disease. J. Leukocyte Biol. 55, 120–126.

Dahinden, C.A., Geiser, T., Brunner, T., et al. (1994). Monocyte chemotactic protein 3 is a most effective basophil-

and eosinophil-activating chemokine. J. Exp. Med. 179, 751–756.

Daugherty, B.L., Siciliano, S.J., DeMartino, J.A., Malkowitz, L., Sirotina, A. and Springer, M.S. (1996). Cloning, expression, and characterization of the human eosinophil eotaxin receptor. J. Exp. Med. 183, 2349–2354.

Decock, B., Conings, R., Lenaerts, J.-P., Billiau, A. and Van Damme, J. (1990). Identification of the monocyte chemotactic protein from human osteosarcoma cells and monocytes: detection of a novel N-terminally processed form. Biochem. Biophys. Res. Commun. 167, 904–909.

Desbaillets, I., Tada, M., de Tribolet, N., Diserens, A.-C., Hamou, M.-F. and Van Meir, E.G. (1994). Human astrocytomas and glioblastomas express monocyte chemoattractant protein-1 (MCP-1) in vivo and in vitro. Int. J. Cancer 58, 240–247.

Dubois, P.M., Palmer, D., Webb, M.L., Ledbetter, J.A. and Shapiro, R.A. (1996). Early signal transduction by the receptor to the chemokine monocyte chemotactic protein-1 in a murine T cell hybrid. J. Immunol. 156, 1356–1361.

Ernst, C.A., Zhang, Y.J., Hancock, P.R., Rutledge, B.J., Corless, C.L. and Rollins, B.J. (1994). Biochemical and biologic characterization of murine monocyte chemoattractant protein-1. Identification of two functional domains. J. Immunol. 152, 3541–3549.

Franci, C., Wong, L.M., Van Damme, J., Proost, P. and Charo, I.F. (1995). Monocyte chemoattractant protein-3, but not monocyte chemoattractant protein-2, is a functional ligand of the human monocyte chemoattractant protein-1 receptor. J. Immunol. 154, 6511–6517.

Frazier-Jessen, M.R. and Kovacs, E.J. (1995). Estrogen modulation of JE/monocyte chemoattractant protein-1 mRNA expression in murine macrophages. J. Immunol. 154, 1838–1845.

Friedland, J.S., Shattock, R.J. and Griffin, G.E. (1993). Phagocytosis of Mycobacterium tuberculosis or particulate stimuli by human monocytic cells induces equivalent monocyte chemotactic protein-1 gene expression. Cytokine 5, 150–156.

Furutani, Y., Nomura, H., Notake, M., et al. (1989). Cloning and sequencing of the cDNA for human monocyte chemotactic and activating factor (MCAF). Biochem. Biophys. Res. Commun. 159, 249–255.

Gao, J.-L., Kuhns, D.B., Tiffany, H.L., McDermott, D., Li, X., Francke, U. and Murphy, P.M. (1993). Structure and functional expression of the human macrophage inflammatory protein 1α/RANTES receptor. J. Exp. Med. 177, 1421–1427.

Gillitzer, R., Wolff, K., Tong, D., et al. (1993). MCP-1 mRNA expression in basal keratinocytes of psoriatic lesions. J. Invest. Dermatol. 101, 127–131.

Gong, X., Gong, W., Kuhns, D.B., Ben-Baruch, A., Howard, O.M.Z. and Wang, J.M. (1997). Monocyte chemotactic protein-2 (MCP-2) uses CCR1 and CCR2B as its functional receptors. J. Biol. Chem. 272, 11682–11685.

Graves, D.T., Jiang, Y.L., Williamson, M.J. and Valente, A.J. (1989). Identification of monocyte chemotactic activity produced by malignant cells. Science 245, 1490–1493.

Gronenborn, A.M. and Clore, G.M. (1991). Modeling the three-dimensional structure of the monocyte chemoattractant and activating protein MCAF/MCP-1 on the basis of the solution structure of interleukin-8. Protein Eng. 4, 263–269.

Grynkiewicz, G., Poenie, M. and Tsien, R.Y. (1985). A new generation of Ca²⁺ indicators with greatly improved fluorescence properties. J. Biol. Chem. 260, 3440–3450.

Hachicha, M., Rathanaswami, P., Schall, T.J. and McColl, S.R. (1993). Production of monocyte chemotactic protein-1 in human type B synoviocytes: synergistic effect of tumor necrosis factor α and interferon-γ. Arthritis Rheum. 36, 26–34.

Haelens, A., Wuyts, A., Proost, P., Struyf, S., Opdenakker, G. and Van Damme, J. (1996). Leukocyte migration and activation by murine chemokines. Immunobiology 195, 499–521.

Hanazawa, S., Kawata, Y., Takeshita, A., et al. (1993). Expression of monocyte chemoattractant protein 1 (MCP-1) in adult periodontal disease: increased monocyte chemotactic activity in crevicular fluids and induction of MCP-1 expression in gingival tissues. Infect. Immun. 61, 5219–5224.

Harigai, M., Hara, M., Yoshimura, T., Leonard, E.J., Inoue, K. and Kashiwazaki, S. (1993). Monocyte chemoattractant protein-1 (MCP-1) in inflammatory joint diseases and its involvement in the cytokine network of rheumatoid synovium. Clin. Immunol. Immunopathol. 69, 83–91.

Heinrich, J.N., Ryseck, R.-P., MacDonald-Bravo, H. and Bravo, R. (1993). The product of a novel growth factor-activated gene, fic, is a biologically active "C-C"-type cytokine. Mol. Cell. Biol. 13, 2020–2030.

Hoogewerf, A., Black, D., Proudfoot, A.E., Wells, T.N. and Power, C.A. (1996). Molecular cloning of murine CC CKR-4 and high affinity binding of chemokines to murine and CC CKR-4. Biochem. Biophys. Res. Commun. 218, 337–343.

Horuk, R. (1994). The interleukin-8-receptor family: from chemokines to malaria. Immunol. Today 15, 169–174.

Huang, S., Singh, R.K., Xie, K., et al. (1994). Expression of the JE/MCP-1 gene suppresses metastatic potential in murine colon carcinoma cells. Cancer Immunol. Immunother. 39, 231–238.

Hulkower, K., Brosnan, C.F., Aquino, D.A., et al. (1993). Expression of CSF-1, c-fms, and MCP-1 in the central nervous system of rats with experimental allergic encephalomyelitis. J. Immunol. 150, 2525–2533.

Jiang, Y., Valente, A.J., Williamson, M.J., Zhang, L. and Graves, D.T. (1990). Post-translational modification of a monocyte-specific chemoattractant synthesized by glioma, osteosarcoma, and vascular smooth muscle cells. J. Biol. Chem. 265, 18318–18321.

Jiang, Y., Beller, D.I., Frendl, G. and Graves, D.T. (1992). Monocyte chemoattractant protein-1 regulates adhesion molecule expression and cytokine production in human monocytes. J. Immunol. 148, 2423–2428.

Kelvin, D.J., Michiel, D.F., Johnston, J.A., Lloyd, A.R., Sprenger, H., Oppenheim, J.J. and Wang, J.-M. (1993). Chemokines and serpentines: the molecular biology of chemokine receptors. J. Leukocyte Biol. 54, 604–612.

Kitaura, M., Nakajima, T., Imai, T., et al. (1996). Molecular cloning of human eotaxin, an eosinophil-selective CC chemokine, and identification of a specific eosinophil eotaxin receptor, CC chemokine receptor 3. J. Biol. Chem. 271, 7725–7730.

Koch, A.E., Kunkel, S.L., Harlow, L.A., et al. (1992). Enhanced production of monocyte chemoattractant protein-1 in rheumatoid arthritis. J. Clin. Invest. 90, 772–779.

Kulmburg, P.A., Huber, N.E., Scheer, B.J., Wrann, M. and Baumruker, T. (1992). Immunoglobulin E plus antigen challenge induces a novel intercrine/chemokine in mouse mast cells. J. Exp. Med. 176, 1773–1778.

Kuna, P., Reddigari, S.R., Rucinski, D., Oppenheim, J.J. and Kaplan, A.P. (1992). Monocyte chemotactic and activating factor is a potent histamine-releasing factor for human basophils. J. Exp. Med. 175, 489–493.

Kunkel, S.L., Lukacs, N., Kasama, T. and Strieter, R.M. (1996). The role of chemokines in inflammatory joint disease. J. Leukocyte Biol. 58, 6–12.

Kuratsu, J.-I., Leonard, E.J. and Yoshimura, T. (1989). Production and characterization of human glioma cell-derived monocyte chemotactic factor. J. Natl Cancer Inst. 81, 347–351.

Leonard, E.J. and Yoshimura, T. (1990). Human monocyte chemoattractant protein-1 (MCP-1). Immunol. Today 11, 97–101.

Liu, Z.-G., Haelens, A., Wuyts, A., et al. (1996). Isolation of a lymphocyte chemotactic factor produced by the murine thymic epithelial cell line MTEC1: identification as a 30 kDa glycosylated form of MCP-1. Eur. Cytokine Netw. 7, 381–388.

Locati, M., Zhou, D., Luini, W., Evangelista, V., Mantovani, A. and Sozzani, S. (1994). Rapid induction of arachidonic acid release by monocyte chemotactic protein-1 and related chemokines. Role of Ca²⁺ influx, synergism with platelet-activating factor and significance for chemotaxis. J. Biol. Chem. 269, 4746–4753.

Lodi, P.J., Garrett, D.S., Kuszewski, J., et al. (1994). High-resolution solution structure of the β-chemokine hMIP-1β by multidimensional NMR. Science 263, 1762–1767.

Loetscher, P., Seitz, M., Clark-Lewis, I., Baggiolini, M. and Moser, B. (1994). Monocyte chemotactic proteins MCP-1, MCP-2 and MCP-3 are major attractants for human CD4⁺ and CD8⁺ T lymphocytes. FASEB J. 8, 1055–1060.

Lukacs, N.W., Strieter, R.M., Chensue, S.W. and Kunkel, S.L. (1996). Activation and regulation of chemokines in allergic airway inflammation. J. Leukocyte Biol. 59, 13–17.

Luo, Y., Laning, J., Hayashi, M., Hancock, P.R., Rollins, B. and Dorf, M.E. (1994). Serologic analysis of the mouse β chemokine JE/monocyte chemoattractant protein-1. J. Immunol. 153, 3708–3716.

Mantovani, A., Bottazzi, B., Colotta, F., Sozzani, S. and Ruco, L. (1992). The origin and function of tumor-associated macrophages. Immunol. Today 13, 265–270.

Marra, F., Valente, A.J., Pinzani, M. and Abboud, H.E. (1993). Cultured human liver fat-storing cells produce monocyte chemotactic protein-1. Regulation by proinflammatory cytokines. J. Clin. Invest. 92, 1674–1680.

Matsushima, K., Larsen, C.G., Dubois, G.C. and Oppenheim, J.J. (1989). Purification and characterization of a novel monocyte chemotactic and activating factor produced by a human myelomonocytic cell line. J. Exp. Med. 169, 1485–1490.

Mehrabian, M., Sparkes, R.S., Mohandas, T., Fogelman, A.M. and Lusis, A.J. (1991). Localization of monocyte chemotactic protein-1 gene (SCYA2) to human chromosome 17q11.2-q21.1. Genomics 9, 200–203.

Miller, M.D. and Krangel, M.S. (1992). Biology and biochemistry of the chemokines: a family of chemotactic and inflammatory cytokines. Crit. Rev. Immunol. 12, 17–46.

Minty, A., Chalon, P., Guillemot, J.C., et al. (1993). Molecular cloning of the MCP-3 chemokine gene and regulation of its expression. Eur. Cytokine Netw. 4, 99–110.

Murakami, K., Nomiyama, H., Miura, R., Follens, A., Fiten, P., Van Coillie, E., Van Damme, J. and Opdenakker, G. (1997). Structural and functional analysis of the promoter region of the human MCP-3 gene: transactivation of expression by novel recognition sequences adjacent to the transcription initiation site. DNA Cell. Biol. 16, 173–183.

Murphy, P.M. (1994). The molecular biology of leukocyte chemoattractant receptors. Annu. Rev. Immunol. 12, 593–633.

Nakamura, K., Williams, I.R. and Kupper, T.S. (1995). Keratinocyte-derived monocyte chemoattractant protein 1 (MCP-1): analysis in a transgenic model demonstrates MCP-1 can recruit dendritic and langerhans cells to skin. J. Invest. Dermatol. 105, 635–643.

Navab, M., Imes, S.S., Hama, S.Y., et al. (1991). Monocyte transmigration induced by modification of low density lipoprotein in cocultures of human aortic wall cells is due to induction of monocyte chemotactic protein 1 synthesis and is abolished by high density lipoprotein. J. Clin. Invest. 88, 2039–2046.

Navab, M., Hama, S.Y., Van Lenten, B.J., Drinkwater, D.C., Laks, H. and Fogelman, A.M. (1993). A new antiinflammatory compound, leumedin, inhibits modification of low density lipoprotein and the resulting monocyte transmigration into the subendothelial space of cocultures of human aortic wall cells. J. Clin. Invest. 91, 1225–1230.

Neote, K., DiGregorio, D., Mak, J.Y., Horuk, R. and Schall, T.J. (1993). Molecular cloning, functional expression, and signaling characteristics of a C-C chemokine receptor. Cell 72, 415–425.

Noso, N., Proost, P., Van Damme, J. and Schröder, J.-M. (1994). Human monocyte chemotactic proteins-2 and 3 (MCP-2 and MCP-3) attract human eosinophils and desensitize the chemotactic responses towards RANTES. Biochem. Biophys. Res. Commun. 200, 1470–1476.

Opdenakker, G. and Van Damme, J. (1992). Chemotactic factors, passive invasion and metastasis of cancer cells. Immunol. Today 13, 463–464.

Opdenakker, G. and Van Damme, J. (1994). Cytokine-regulated proteases in autoimmune diseases. Immunol. Today 15, 103–107.

Opdenakker, G., Froyen, G., Fiten, P., Proost, P. and Van Damme, J. (1993). Human monocyte chemotactic protein-3 (MCP-3): molecular cloning of the cDNA and comparison with other chemokines. Biochem. Biophys. Res. Commun. 191, 535–542.

Opdenakker, G., Fiten, P., Nys, G., et al. (1994). The human MCP-3 gene (SCYA7): cloning, sequence analysis, and assignment to the C-C chemokine gene cluster on chromosome 17q11.2-q12. Genomics 21, 403–408.

Opdenakker, G., Rudd, P., Wormald, M., Dwek, R.A. and Van Damme, J. (1995). Cells regulate the activities of cytokines by glycosylation. FASEB J. 9, 453–457.

Oppenheim, J.J., Zachariae, C.O.C., Mukaida, N. and Matsushima, K. (1991). Properties of the novel proinflammatory supergene "intercrine" cytokine family. Annu. Rev. Immunol. 9, 617–648.

Paolini, J.F., Willard, D., Consler, T., Luther, M. and Krangel, M.S. (1994). The chemokines IL-8, monocyte chemoattractant protein-1 and I-309 are monomers at physiologically relevant concentrations. J. Immunol. 153, 2704–2717.

Power, C.A., Meyer, A., Nemeth, K., *et al.* (1995). Molecular cloning and functional expression of a novel CC chemokine receptor cDNA from a human basophilic cell line. J. Biol. Chem. 270, 19495–19500.

Proost, P., Van Leuven, P., Wuyts, A., Ebberink, R., Opdenakker, G. and Van Damme, J. (1995). Chemical synthesis, purification and folding of the human monocyte chemotactic proteins MCP-2 and MCP-3 into biologically active chemokines. Cytokine 7, 97–104.

Proost, P., Wuyts, A., Conings, R., Lenaerts, J.-P., Put, W. and Van Damme, J. (1996a). Purification and identification of natural chemokines. Im "Methods: A Companion to Methods in Enzymology" 10, 82–92.

Proost, P., Wuyts, A. and Van Damme, J. (1996b). Human monocyte chemotactic proteins-2 and -3: structural and functional comparison with MCP-1. J. Leukocyte Biol. 59, 67–74.

Rajarathnam, K., Sykes, B.D., Kay, C.M., *et al.* (1994). Neutrophil activation by monomeric interleukin-8. Science 264, 90–92.

Ransohoff, R.M., Hamilton, T.A., Tani, M., *et al.* (1993). Astrocyte expression of mRNA encoding cytokines IP-10 and JE/MCP-1 in experimental autoimmune encephalomyelitis. FASEB J. 7, 592–600.

Robinson, E.A., Yoshimura, T., Leonard, E.J., *et al.* (1989). Complete amino acid sequence of a human monocyte chemoattractant, a putative mediator of cellular immune reactions. Proc. Natl Acad. Sci. USA 86, 1850–1854.

Rollins, B.J., Morrison, E.D. and Stiles, C.D. (1988). Cloning and expression of *JE*, a gene inducible by platelet-derived growth factor and whose product has cytokine-like properties. Proc. Natl Acad. Sci. USA 85, 3738–3742.

Rollins, B.J., Stier, P., Ernst, T. and Wong, G.G. (1989). The human homolog of the *JE* gene encodes a monocyte secretory protein. Mol. Cell. Biol. 9, 4687–4695.

Rollins, B.J., Morton, C.C., Ledbetter, D.H., Eddy, R.L., Jr, and Shows, T.B. (1991a). Assignment of the human small inducible cytokine A2 gene, SCYA2 (encoding JE or MCP-1), to 17q11.2-12: evolutionary relatedness of cytokines clustered at the same locus. Genomics 10, 489–492.

Rollins, B.J., Walz, A. and Baggiolini, M. (1991b). Recombinant human MCP-1/JE induces chemotaxis, calcium flux, and the respiratory burst in human monocytes. Blood 78, 1112–1116.

Rot, A., Krieger, M., Brunner, T., Bischoff, S.C., Schall, T.J. and Dahinden, C.A. (1992). RANTES and macrophage inflammatory protein 1α induce the migration and activation of normal human eosinophil granulocytes. J. Exp. Med. 176, 1489–1495.

Roth, S.J., Carr, M.W. and Springer, T.A. (1995). C-C chemokines, but not the C-X-C chemokines interleukin-8 and interferon-γ inducible protein-10, stimulate transendothelial chemotaxis of T lymphocytes. Eur. J. Immunol. 25, 3482–3488.

Rutledge, B.J., Rayburn, H., Rosenberg, R., *et al.* (1995). High level monocyte chemoattractant protein-1 expression in transgenic mice increases their susceptibility to intracellular pathogens. J. Immunol. 155, 4838–4843.

Samson, M., Labbé, O., Mollereau, C., Vassart, G. and Parmentier, M. (1996). Molecular cloning and functional expression of a new human CC-chemokine receptor gene. Biochemistry 35, 3362–3367.

Schall, T.J. (1994). The chemokines in "The Cytokine Handbook" (ed. A. Thomson), pp. 419–460. Academic Press, London.

Schmouder, R.L., Strieter, R.M. and Kunkel, S.L. (1993). Interferon-γ regulation of human renal cortical epithelial cell-derived monocyte chemotactic peptide-1. Kidney Int. 44, 43–49.

Schröder, J.-M., Noso, N., Sticherling, M. and Christophers, E. (1996). Role of eosinophil-chemotactic C-C chemokines in cutaneous inflammation. J. Leukocyte Biol. 59, 1–5.

Seitz, M., Loetscher, P., Dewald, B., Towbin, H., Ceska, M. and Baggiolini, M. (1994). Production of interleukin-1 receptor antagonist, inflammatory chemotactic proteins, and prostaglandin E by rheumatoid and osteoarthritic synoviocytes-regulation by IFN-g and IL-4. J. Immunol. 152, 2060–2065.

Shyy, Y.-J., Li, Y.-S. and Kolattukudy, P.E. (1990). Structure of human monocyte chemotactic protein gene and its regulation by TPA. Biochem. Biophys. Res. Commun. 169, 346–351.

Shyy, Y.-J., Li, Y.-S. and Kolattukudy, P.E. (1993). Activation of MCP-1 gene expression is mediated through multiple signaling pathways. Biochem. Biophys. Res. Commun. 192, 693–699.

Singh, R.K., Gutman, M., Radinsky, R., Bucana, C.D. and Fidler, I.J. (1994). Expression of interleukin 8 with the metastatic potential of human melanoma cells in nude mice. Cancer Res. 54, 3242–3247.

Sousa, A.R., Lane, S.J., Nakhosteen, J.A., Yoshimura, T., Lee, T.H. and Poston, R.N. (1994). Increased expression of the monocyte chemoattractant protein-1 in bronchial tissue from asthmatic subjects. Am. J. Respir. Cell Mol. Biol. 10, 142–147.

Sozzani, S., Luini, W., Molino, M., *et al.* (1991). The signal transduction pathway involved in the migration induced by a monocyte chemotactic cytokine. J. Immunol. 147, 2215–2221.

Sozzani, S., Molino, M., Locati, M., *et al.* (1993). Receptor-activated calcium influx in human monocytes exposed to monocyte chemotactic protein-1 and related cytokines. J. Immunol. 150, 1544–1553.

Sozzani, S., Rieppi, M., Locati, M., *et al.* (1994a). Synergism between platelet activating factor and C-C chemokines for arachidonate release in human monocytes. Biochem. Biophys. Res. Commun. 199, 761–766.

Sozzani, S., Zhou, D., Locati, M., *et al.* (1994b). Receptors and transduction pathways for monocyte chemotactic protein-2 and monocyte chemotactic protein-3. J. Immunol. 152, 3615–3622.

Sozzani, S., Sallusto, F., Luini, W., *et al.* (1995). Migration of dendritic cells in response to formyl peptides, C5a, and a distinct set of chemokines. J. Immunol. 155, 3292–3295.

Strieter, R.M., Koch, A.E., Antony, V.B., Fick, R.B., Standiford, T.J. and Kunkel, S.L. (1994). From the Chicago meetings: The immunopathology of chemotactic cytokines: the role of interleukin-8 and monocyte chemoattractant protein-1. J. Lab. Clin. Med. 123, 183–197.

Sylvester, I., Suffredini, A.F., Boujoukos, A.J., *et al.* (1993). Neutrophil attractant protein-1 and monocyte chemoattractant protein-1 in human serum. J. Immunol. 151, 3292–3298.

Takeshima, H., Kuratsu, J.-I., Takeya, M., Yoshimura, T. and Ushio, Y. (1994). Expression and localization of messenger RNA and protein for monocyte chemoattractant protein-1 in human malignant glioma. J. Neurosurg. 80, 1056–1062.

Taub, D.D., Proost, P., Murphy, W.J., *et al.* (1995a). Monocyte chemotactic protein-1 (MCP-1), -2, and -3 are chemotactic for human T lymphocytes. J. Clin. Invest. 95, 1370–1376.

Taub, D.D., Sayers, T.J., Carter, C.R.D. and Ortaldo, J.R. (1995b). α and β chemokines induce NK cell migration and enhance NK-mediated cytolysis. J. Immunol. 155, 3877–3888.

Thirion, S., Nys, G., Fiten, P., Masure, S., Van Damme, J. and Opdenakker, G. (1994). Mouse macrophage derived monocyte chemotactic protein-3: cDNA cloning and identification as MARC/FIC. Biochem. Biophys. Res. Commun. 201, 493–499.

Tonetti, M.S., Imboden, M.A., Gerber, L., Lang, N.P., Laissue, J. and Mueller, C. (1994). Localized expression of mRNA for phagocyte-specific chemotactic cytokines in human periodontal infections. Infect. Immun. 62, 4005–4014.

Ueda, A., Okuda, K., Ohno, S., *et al.* (1994). NF-kB and Sp1 regulate transcription of the human monocyte chemoattractant protein-1 gene. J. Immunol. 153, 2052–2063.

Uguccioni, M., D'Apuzzo, M., Loetscher, M., Dewald, B. and Baggiolini, M. (1995). Actions of the chemotactic cytokines MCP-1, MCP-2, MCP-3, RANTES, MIP-1α and MIP-1β on human monocytes. Eur. J. Immunol. 25, 64–68.

Van Coillie, E., Fiten, P., Nomiyama, H., Sakaki, Y., Miura, R., Yoshie, O., Van Damme, J. and Opdenakker, G. (1997a). The human MCP-2 gene (SCYA8): cloning, sequence analysis, tissue-expression and assignment to the C-C chemokine gene contig on chromosome 17q11.2. Genomics 40, 323–331.

Van Coillie, E., Froyen, G., Nomiyama, H., Miura, R., Fiten, P., Van Aelst, I., Van Damme, J. and Opdenakker, G. (1997b). Human monocyte chemotactic protein-2 (MCP-2): cDNA cloning and regulated expression of mRNA in mesenchymal cells. Biochem. Biophys. Res. Commun. 231, 726–730.

Van Damme, J. (1994). Interleukin-8 and related chemotactic cytokines. In "The Cytokine Handbook" (ed. A. Thomson), pp. 185–208. Academic Press, London.

Van Damme, J., Decock, B., Lenaerts, J.-P., *et al.* (1989). Identification by sequence analysis of chemotactic factors for monocytes produced by normal and transformed cells stimulated with virus, double-stranded RNA or cytokine. Eur. J. Immunol. 19, 2367–2373.

Van Damme, J., Decock, B., Bertini, R., *et al.* (1991). Production and identification of natural monocyte chemotactic protein from virally infected murine fibroblasts. Relationship with the product of the mouse competence (JE) gene. Eur. J. Immunol. 199, 223–229.

Van Damme, J., Proost, P., Lenaerts, J.-P. and Opdenakker, G. (1992). Structural and functional identification of two human, tumor-derived monocyte chemotactic proteins (MCP-2 and MCP-3) belonging to the chemokine family. J. Exp. Med. 176, 59–65.

Van Damme, J., Proost, P., Lenaerts, J.-P., Conings, R., Opdenakker, G. and Billiau, A. (1993). Monocyte chemotactic proteins related to human MCP-1. In "The Chemokines" (ed. I.J.D. Lindley), pp. 111–118. Plenum Press, New York.

Van Damme, J., Proost, P., Put, W., *et al.* (1994). Induction of monocyte chemotactic proteins MCP-1 and MCP-2 in human fibroblasts and leukocytes by cytokines and cytokine inducers. J. Immunol. 152, 5495–5502.

Wells, T.N.C., Power, C.A., Lusti-Narasimhan, M., *et al.* (1996). Selectivity and antagonism of chemokine receptors. J. Leukocyte Biol. 59, 53–60.

Wilson, S.D., Billings, P.R., D'Eustachio, P., *et al.* (1990). Clustering of cytokine genes on mouse chromosome 11. J. Exp. Med. 171, 1301–1314.

Wuyts, A., Proost, P., Put, W., Lenaerts, J.-P., Paemen, L. and Van Damme, J. (1994). Leukocyte recruitment by monocyte chemotactic proteins (MCPs) secreted by human phagocytes. J. Immunol. Methods 174, 237–247.

Xu, L.L., McVicar, D.W., Ben-Baruch, A., *et al.* (1995). Monocyte chemotactic protein-3 (MCP-3) interacts with multiple leukocyte receptors: binding and signaling of MCP-3 through shared as well as unique receptors on monocytes and neutrophils. Eur. J. Immunol. 25, 2612–2617.

Yamagami, S., Tanaka, H. and Endo, N. (1997). Monocyte chemoattractant protein-2 can exert its effects through the MCP-1 receptor (CC CKR2B). FEBS Letters 400, 329–332.

Ying, S., Taborda-Barata, L., Meng, Q., Humbert, M. and Kay, A.B. (1995). The kinetics of allergen-induced transcription of messenger RNA for monocyte chemotactic protein-3 and RANTES in the skin of human atopic subjects: relationship to eosinophil, T cell, and macrophage recruitment. J. Exp. Med. 181, 2153–2159.

Ylä-Herttuala, S., Lipton, B.A., Rosenfeld, M.E., *et al.* (1991). Expression of monocyte chemoattractant protein 1 in macrophage-rich areas of human and rabbit atherosclerotic lesions. Proc. Natl Acad. Sci. USA 88, 5252–5256.

Yoshimura, T., Robinson, E.A., Tanaka, S., Appella, E., Kuratsu, J.-I. and Leonard, E.J. (1989a). Purification and amino acid analysis of two human glioma-derived monocyte chemoattractants. J. Exp. Med. 169, 1449–1459.

Yoshimura, T., Robinson, E.A., Tanaka, S., Appella, E. and Leonard, E.J. (1989b). Purification and amino acid analysis of two human monocyte chemoattractants produced by phytohemagglutinin-stimulated human blood mononuclear leukocytes. J. Immunol. 142, 1956–1962.

Yoshimura, T., Yuhki, N., Moore, S.K., Appella, E., Lerman, M.I. and Leonard, E.J. (1989c). Human monocyte chemoattractant protein-1 (MCP-1): full-length cDNA cloning, expression in mitogen-stimulated blood mononuclear leukocytes, and sequence similarity to mouse competence gene JE. FEBS Lett. 244, 487–493.

Yu, X., Antoniades, H.N. and Graves, D.T. (1993). Expression of monocyte chemoattractant protein 1 in human inflamed gingival tissues. Infect. Immun. 61, 4622–4628.

Yu, X., Barnhill, R.L. and Graves, D.T. (1994). Expression of monocyte chemoattractant protein-1 in delayed type hypersensitivity reactions in the skin. Lab. Invest. 71, 226–235.

Zachariae, C.O.C., Anderson, A.O., Thompson, H.L., *et al.* (1990). Properties of monocyte chemotactic and activating factor (MCAF) purified from a human fibrosarcoma cell line. J. Exp. Med. 171, 2177–2182.

33. GRO/MGSA

Stephen Haskill *and* Susanne Becker

1. Introduction

The cytokine and growth factor GRO/MGSA is a member of a family of cytokines and growth factors originally associated with either immortal growth of fibroblasts (Anisowicz *et al.*, 1987) or malignant and benign nevus cells expressing chromosomal abnormalities (Richmond *et al.*, 1983). It is interesting that, unlike most of the other members of the C-X-C family of cytokines (Baggiolini *et al.*, 1994), GRO and its synonym melanoma growth-stimulatory activity (MGSA) were the only factors cloned based upon selective expression in malignant cells or through a growth-stimulatory assay, and it was not initially recognized that these apparently distinct factors had pro-inflammatory properties (Sager *et al.*, 1991). Furthermore and unexpectedly, it was only after the cloning of GRO (Anisowicz *et al.*, 1987) and MGSA (Richmond *et al.*, 1988) that the identity of these factors was uncovered.

MGSA was initially characterized as an activity found in the culture supernatant of the Hs294T human malignant melanoma cell line (Richmond *et al.*, 1983). The strong association of this growth stimulatory activity with approximately 70% of primary melanoma cell lines and its absence from benign nevus lines free of overt chromosomal abnormalities suggested that dysregulation of MGSA expression was an important step in autonomous growth of these tumors. cDNA clones of MGSA were obtained from a knowledge of the N-terminal amino acid sequence of highly purified material (Richmond *et al.*, 1988; Thomas and Richmond 1988a,b). Sequence analysis identified a protein with a high degree of sequence conservation with several proteins including β-thromboglobulin (β-TG) and platelet factor 4 (PF₄) (Castor *et al.*, 1983; Holt *et al.*, 1986), which, unlike MGSA, were known to be associated either with connective-tissue regulation or inflammation.

GRO, on the other hand, was cloned by subtractive hybridization techniques (Anisowicz *et al.*, 1987). Sager's group postulated that genes differentially expressed in the CHEF/16 transformed fibroblast line might be associated with the malignant phenotype expressed in these but not the CHEF-18 line, which was nontumorigeneic although derived from the same embryo. Utilizing the Chinese hamster cDNA, a similar human gene was quickly identified. Formal identification that GRO or the hamster equivalent was responsible for the malignant property of the CHEF-16 line has not been reported.

Subsequently, in spite of the early chromosomal identification data which was interpreted as GRO/MGSA representing a single gene (Anisowicz *et al.*, 1987; Richmond *et al.*, 1988; Sakaguchi *et al.*, 1989), several groups succeeded in isolating cDNA clones representing additional genes; GRO/MGSAβ and γ (Haskill *et al.*, 1990; Iida and Grotendorst, 1990; Tekamp-Olson *et al.*, 1990). While the three genes clearly are distinct, showing selective tissue expression, they are 90 and 86% homologous at the protein level. While malignant cell lines were the source of the original isolates of GRO/MGSA, the additional genes were identified from myeloid cells. Direct isolation was accomplished by differential screening of a monocyte adherence library (Haskill *et al.*, 1990). The other groups utilized murine probes of the KC gene (Iida and Grotendorst, 1990) or of the MIP-2 gene (Tekamp-Olson *et al.*, 1990) to isolate additional GRO/MGSA members based on the approximately 60% homology between these distinct murine genes and the GRO/MGSA family. It was thus established that melanoma cells, fibroblasts, various epithelial cells, as well as monocytes and neutrophils either constitutively or immediately following a cytokine stimulus, express high levels of GRO/MGSA transcripts. It will be the purpose of this review to assimilate the data from these diverse arenas to come to see how these closely related gene products are differentially regulated in different cell lineages and what selective roles they play in inflammation, tissue repair, and cancer. To simplify

nomenclature and to accept the more common name, subsequent discussion will refer to GROα, GROβ and GROγ rather than the MGSA names.

The following discussion will review the limited knowledge of GRO expression and function. The paucity of information on GRO is perhaps a result of the profusion of cytokines with somewhat similar activities and the emphasis that has been placed on IL-8 by so much of the immunological community. As we will see during the following discussion, it is clear that while IL-8 and GRO share overlapping activities and indeed receptors, they have evolved to have complementary and distinct activities and are, in part, responsive to distinct signaling events. The interested reader could also consult the review on GRO/MGSA by Horuk (1996).

2. The Cytokine Gene

2.1 Gene Sequence

The exons of the GRO/MGSA gene family are organized in a very similar pattern to those of the other C-X-C members of the chemokine family (Baggiolini *et al.*, 1994), differing from PF₄ and β-TG/NAP-2 in having four rather than three exons. Differences between the three GRO genes are slight (Figure 33.1). All three members have TATA boxes, and the leader sequence is encoded almost solely within the first exon. The most striking feature of the three genes is the high proportion of ATTTA motifs found within the 3′ untranslated regions. As will be discussed later, the presence of varying numbers of these repeats which appear to be associated with mRNA instability is likely to be important in differential regulation of gene expression between the three GROs. GROβ in particular has 11 repeats, six of them within a single stretch. GROγ has six and GROα five ATTTA repeats (Haskill *et al.*, 1990). Only GROα appears to have more than one transcript size; this depends upon the cell and stimulus and appears to be related to aspects of mRNA stability (Anisowicz *et al.*, 1988; Wen *et al.*, 1989; Stoeckle, 1991, 1992; Stoeckle and Guan, 1993; Shattuck *et al.*, 1994) and is discussed below. The close homology between the three GRO forms makes expression studies based upon northern transfer analysis extremely difficult, if not impossible. Although several reports have described studies related to GROβ expression, they often employed the GROβ cDNA as probe, which cannot distinguish between species. Specific primers employed in PCR or as northern probes (Shattuck *et al.*, 1994) are required to specify individual species (Haskill *et al.*, 1990; Cuenca *et al.*, 1992).

2.2 Gene Structure

All of the GRO genes show a similar organization to each other and some but not all members of the C-X-C family

```
α  TCCCACCTCTCAGGTGGT.ATCT.........TCAGCGCAGGCTGCCACTCAGCCCCCCT
B  CGCCTCCTCGCAGGCGGTTATCTcggtatctcTGAGAGC.GGCGGGCTCTCG....C..T
γ  TCCCCCCTCACAGGCTGT.ATCT.........TCAGCGA.GGTGGACTCACTGCCTC..T

α  CCAGGGATCTGG....GGCAGAAGGCGAATATCCCAGAGTCTCAGAGTCCACAGGAGTTA
B  CCCGCTCCAGGGattcGGGGCAGAAAGAG......AACATCCCACAGTTGGCGGGAGTTA
γ  CCAGGAATTTGG....GGCAGAAAATGAATATCCCAAAGTCCCAGAGTGCACGGGGGTTA

α  CTCTGAAGGGCGAGCCGCGGGCTGCATCAGTGGACCCCCACACCCCACCCGCACCCCAAG
β  CGCAAGACAGTCAGACCCGGACGTCACTCGTGAG.TGCC....CCGACCC....CC....
γ  CTCTGGAGGGCGAGGCGTAGGCGTCACCAGTGGGCTCCC....CCTACCCGTATCCGA..

α  CGCTCCACCCTGGGGGCGGGGCCGTCGCCTTCCTTCCGGACTCGGGATCGATCTGGAACT
β  ..CTCCACCCCAGAGGCGGGGCCATCGCCTTCCTTCCGAACTCGGGATCGATCTGGAGCT
γ  ..CTCCACCCCGGGGGCGGG.CCGTCGCCTTC.TTCGGGACTCCGGATCGATCTGGAGCT

α  CCGGGAATTTCCCTGGCCCGGGGGCTCCGCCCTTTCCAGCCCCAACCATGCATAAAAGGG
β  CCGGGAATTTCCCTGGCCCGGGA.CTCCGG.CTTTCCAGCCCCAACCATGCATAAAAGGG
γ  CCGGGAATTTCCCTGGCCCGGCCGCTCCGGGCTTTCCAGTCTCAACCATGCATAAAAAGG

α  GTTCGCGGATCTCGGAGAGCCACAGAGCCCGGGCCGCAGGCACCTCCTCGCCAGCTCTTC
β  GTTCGCCGTTCTCGGAGAGCCACAGAGCCCGGGCCACAGGCAGCTCCTTGCCAGCTCTCC
γ  GTTCGCCGATCTTGGGGAGCCACACAGCCCGGGTCGCAGGCACCTCCC.GCCAGCTCTCC

α  CGCTCCTCTCACAGCCGCCAGACCCGCCTGCTGAGCCCCATGGCCCGCGCTGCTCTCTCC
β  ...TCCTCGCACAGCCGCTCGAACCGCCTGCTGAGCCCCATGGCCCGCGCCACGCTCTCC
γ  CGCTTCTCGCACAGCTTCCCGACGCGTCTGCTGAGCCCCATGGCCCACGCCACGCTCTCC

α  GCCGCCCCCAGCAATCCCCGGCTCCTGCGAGTGGCACTGCTGCTCCTGCTCCTGGTAGCC
β  GCCGCCCCCAGCAATCCCCGGCTCCTGCGGGTGGCGCTGCTGCTCCTGCTCCTGGTGGCC
γ  GCCGCCCCCAGCAATCCCCGGCTCCTGCGGGTGGCGCTGCTGCTCCTGCTCCTGGTGGGC

α  GCTGGCCGGCGCGCAGCAGGAGCGTCCGTGGCCACTGAACTGCGCTGCCAGTGCTTGCAG
β  GCCAGCCGGCGCGCAGCAGGAGCGCCCCTGGCCACTGAACTGCGCTGCCAGTGCTTGCAG
γ  ...AGCCGGCGCGCAGCAGGAGCGTCCGTGGTCACTGAACTGCGCTGCCAGTGCTTGCAG

α  ACCCTGCAGGGAATTCACCCCAAGAACATCCAAAGTGTGAACGTGAAGTCCCCCGGACCC
β  ACCCTGCAGGGAATTCACCTCAAGAACATCCAAAGTGTGAAGGTGAAGTCCCCCGGACCC
γ  ACCCTGCAGGGAATTCACCTCAAGAACATCCAAAGTGTGAATGTAAGGTCCCCCGGACCC

α  CACTGCGCCCAAACCGAAGTCATAGCCACACTCAAGAATGGGCGGAAAGCTTGCCTCAAT
β  CACTGCGCCCAAACCGAAGTCATAGCCACACTCAAGAATGGGCAGAAAGCTTGCCTCAAC
γ  CACTGCGCCCAAACCGAAGTCATAGCCACACTCAAGAATGGGAAGAAAGCTTGTCTCAAC

α  CCTGCATCCCCCATAGTTAAGAAAATCATCGAAAAGATGCTGAACAGTGACAAATCCAAC
β  CCCGCATCGCCCATGGTTAAGAAAATCATCGAAAAGATGCTGAAAAATGGCAAATCCAAC
γ  CCCGCATCCCCCATGGTTCAGAAAATCATCGAAAAGATACTGAACAAGGGGAGCACCAAC
```

Figure 33.1   Nucelotide sequences of the GROα, GROβ, and GROγ genes.

of chemokines (Baggiolini *et al.*, 1994) (see Figure 33.2). For GROα, β, and γ, IL-8, and IP10, this includes four exons with the leader sequence predominantly in the first exon and the C-terminal 3–5 amino acids added from the fourth exon. It is this fourth exon which contains the highest number of amino acid differences between the three GRO gene products. All of the GRO, IL-8, and IP-10 genes are located closely on human chromosome 4q12-21 (Anisowicz *et al.*, 1988; Richmond *et al.*, 1988; Baggiolini *et al.*, 1994). The 80% or greater nucleotide sequence homology between the GRO genes and their close chromosomal location implies that they may have arisen through gene duplication.

2.3   TRANSCRIPTIONAL REGULATION

The role of specific transcription factors and the importance of transcript stabilization in the regulation of gene expression is currently under investigation. Early reports (Anisowicz *et al.*, 1991) suggested a high degree of similarity between the three GRO gene proximal promoters, particularly the −136 bp region. All three genes contain a putative NFκB site, SP-1 and AP-3 motifs (Widmer *et al.*, 1993; Anisowicz *et al.*, 1991). A TATA box is located at the expected location bp −27.

Several groups have used transient transfection reporter assays to examine the NFκB site in GRO. The NFκB site appears to be important in IL-1β and

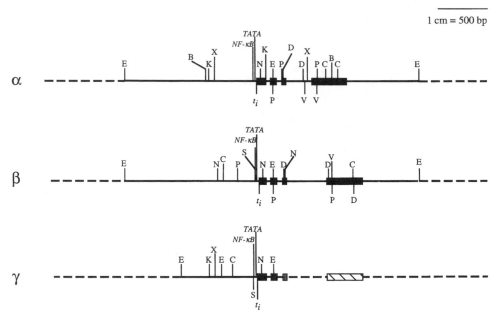

Figure 33.2   Genomic organization of the GROα, GROβ, and GROγ genes.
Sequenced or subcloned, ——————; not available, — — — — — —; Exon ■■■■■■■ ; Exon/Intron Structure not yet
known ▨▨▨▨. Intron sizes (bp): intron 1, α 98, β 95, γ 96; intron 2, α 113, β 118; intron 3, α 531, β 826.
Restriction enzyme sites: E = *Eco*RI; K = *Kpn*I; B = *Bam*HI; C = *Hinc*II/*Hind*II; D = *Hind*III; N = *Nco*I; P = *Pst*I; V = *Pvu*II;
S = *Sac*I; X = *Xba*I.

TNF-2-stimulated fibroblast activation (Anisowicz et al., 1991) and IL-1β stimulation of the A375-C6 melanoma cell line (Joshi-Barve et al., 1993) as well as lipopolysaccharide (LPS) stimulation of the RAW 264.7 murine macrophage cell line (Widmer et al., 1993). Control of gene expression appears to reside only partly within this proximal domain as none of the tissue specificity exhibited by the three GRO genes could be explained by such a high degree of similarity. For example, this is particularly true of monocytes and neutrophils, with neutrophils only producing a GROα transcript (defined by PCR) whereas monocytes expressed all three transcripts when stimulated by adherence to fibronectin but only GROβ and GROγ when exposed to phorbol ester (PMA) stimulation (Haskill et al., 1990). In addition, while IL-1β induces a 10–20-fold increase in transcription in retinal pigmented epithelial cells (RPE), the same stimulus initiates only a 2-fold increase in nuclear runoff activity in Hs294T cells, even though mRNA levels were increased 14-fold following IL-1 stimulation (Shattuck et al., 1994). Using CAT constructs for promoter analysis, these authors also demonstrated that IL-1β stimulated similar differences in activity upon transfection into the respective cells. NFκB was shown to be required in the RPE cells, but NFκB failed to activate κB CAT constructs or translocate in Hs294T cells following IL-1 stimulation. These data indicate that transcription induction dependent upon NFκB sites may not be sufficient in all cell types. More

recent observations in this model suggest that the Hs294T melanoma cells have a high basal level of transcription that does require NFκB elements but C/EPB and an additional structural high-mobility group protein (HMGI(Y)) are necessary in the melanoma cells. The HMGI(Y) protein appears to associate with an AT-rich region located within the NFκB element (Wood and A. Richmond, personal communication).

IL-8 and GRO gene expression are often closely regulated (Baggiolini et al., 1994); however, promoter analysis indicates that there are significant differences between them. While IL-8 is frequently stimulated by PMA, GRO is not (Haskill et al., 1990; Joshi-Barve et al., 1993). This may, in part, be explained by the presence of an NF-IL-6 site in close and interacting proximity with the NFκB proximal site in the IL-8 gene (Kunsch et al., 1994).

2.4   TRANSCRIPT STABILIZATION

Transcript expression is closely regulated via post-transcriptional events. In the case of the IL-1 stimulation of Hs264T cells, the 14-fold increase in mRNA is accounted for almost entirely by stabilization of the high levels of constitutive transcription in these cells (Shattuck et al., 1994). In the case of GROα and GROγ, the increase in half-life was at least 20-fold (15 min to 6 h), while that of GROβ, which is known to contain more

AUUUA repeats, was increased less (1.5 h). Transcript heterogeneity is also a feature of GROα but not GROβ or GROγ expression. FS-2 fibroblasts express 0.9 kb and 1.2 kb transcripts on treatment with cycloheximide (Anisowicz *et al.*, 1987) or TNF but not on IL-1 stimulation (Anisowicz *et al.*, 1991) and a similar observation has been made with endothelial cells (Wen *et al.*, 1989). Neutrophils, however, appear to express a single transcript that is slightly smaller than usual (1.1 vs 1.2 kb) (Iida and Grotendorst, 1990). The cell and the signal appear to play a role in transcript heterogeneity as Hs294T melanoma cells stimulated with IL-1 have both GROα transcripts. The origin of this heterogeneity appears to be degradation intermediates, involving shortening and elimination of the poly(A) tail, a topic that has been examined in some detail (Stoeckle, 1991, 1992; Stoeckle and Guan, 1993). FS-4 fibroblasts show a dramatic increase in mRNA half-life following IL-1 stimulation. In these studies it was concluded that the minor component at 0.9 kb represented a deadenylated species deleted of a 130 bp segment. The deadenylation reaction was sensitive to cycloheximide, suggesting that this was a stable intermediate in the degradation process. In contrast, IL-1 did not stimulate two transcripts in FS-2 fibroblasts while TNF-α did (Anisowicz *et al.*, 1991), indicating that this degradation intermediate is both signal and cell specific.

2.5 TRANSLATIONAL REGULATION

Translational coupling to GRO transcription has not been examined in any detail, thus it is difficult to determine if the enhanced levels of gene transcription actually result in significant extracellular GRO protein. In endothelial cells, the available data suggest that secretion of GROα is correlated with the levels of transcripts. LPS and IL-1, which induced the highest levels of mRNA, also induced the greatest amount of secreted protein. PMA, thrombin, and TNF stimulated low levels of both mRNA and protein (Wen *et al.*, 1989). Translational coupling to transcription remains an area of relative ignorance and one which clearly requires additional investigation.

3. *The Protein*

3.1 AMINO ACID SEQUENCES

The amino acid sequences of the three GRO gene products are amazingly similar, with GROβ being the most similar to GROα (90% identical), and GROγ showing 86% identity with GROα (Haskill *et al.*, 1990) (Figure 33.3). Like all members of the C-X-C family, the protein has two intrachain cysteine bridges. Compared to GROα there are 11 amino acid changes in GROβ and 9 of these are found in the secreted protein while GROγ has 11 changes in the mature protein. While most of the changes are conservative for charge and size, two proline substitutions may be of significance. GROβ has a proline alteration immediately before the ELR motif known to be important for receptor binding of IL-8 and GRO (Geiser *et al.*, 1993).

ASVATELRCQC	GROα
A**P**LATELRCQC	GROβ
ASV**V**TELRCQC	GROγ

This substitution may be responsible for the lower activity of native GROβ as deletion of the N-terminal five

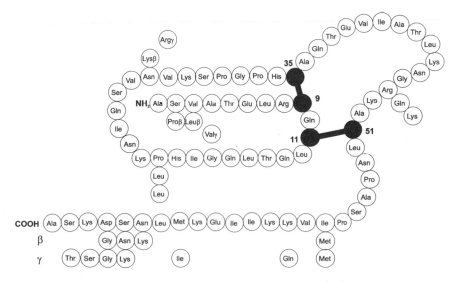

Figure 33.3 Amino acid sequences of the three GRO genes.

amino acids has little effect on GROα or GROγ activity but markedly enhances that of GROβ (Geiser *et al.*, 1993). The importance of the remaining amino acid substitutions is still difficult to determine. Alanine scanning mutagenesis of the charged residues was used to demonstrate that the EGA conversion was sufficient to separate binding by the Duffy antigen receptor and that of the IL-8RB (Hesselgesser 1995). Although there are several potentially interesting changes in the C-terminal five amino acids, their influence on activity has not been discerned. It is possible that alterations within the putative heparin-binding domain characterized by KKIIKKLL in PF$_4$ may alter binding of the different GRO gene products to the extracellular matrices.

KKIIKKLLES	PF$_4$
KKIIK**EH**LES	PF$_4$v
KKIIEKMLNSDKSN	GROα
KKIIEKMLKNGKSN	GROβ
QKIIEKILNKGSTN	GROγ

It has been reported that two amino acid changes within this domain markedly alter heparin binding of a variant of PF$_4$ and CTAP-III (Green *et al.*, 1989), suggesting that the weak heparin binding property of GRO (Richmond and Thomas, 1988) could be differentially altered by the concentration of changes within the C-terminus. Recently, however, it has been reported that IL-8 and GRO preferentially bind to a subfraction of heparin, one that fails to bind PF$_4$, suggesting that matrix binding may be a shared characteristic of IL-8 and GRO (Witt and Lander, 1994). Both the amino- and carboxyl-termini appear to be important for binding to the β-receptor. However, there appears to be little difference in affinity between the three GRO forms (Geiser *et al.*, 1993) when the five N-terminal residues are deleted, suggesting that these differences in the carboxy-termini are of no significance as regards receptor interaction. The absolute requirement of the N-terminal ELR motif for receptor binding has been challenged by the observation that the C-terminal (47–71) GRO protein can bind to melanoma cells and is sufficient to stimulate DNA synthesis (Roby and Page, 1995).

3.2 Secondary Structure

The secondary structure of GROα has been determined by NMR spectroscopy and appears to be similar to that of the other members of the C-X-C family (Fairbrother *et al.*, 1993; Clark-Lewis *et al.*, 1995), which is a consequence of the conservation of cysteine bridges within the different chemokines (Figure 33.4). The monomer is predicted to have a three-stranded antiparallel β-sheet conformation and a C-terminal α-helix from amino acid residue 58–69. GRO exists

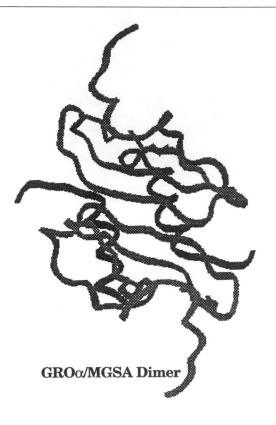

GROα/MGSA Dimer

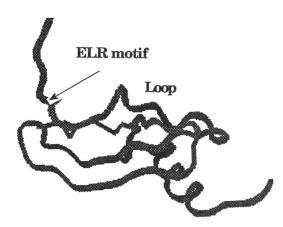

GROα/MGSA Monomer

Figure 33.4   Secondary structure of the GROα/MGSA dimer and a single subunit of the dimer.

apparently as a dimer in solution, with the dimer being stabilized between Gln-24 and Ser-29. As this region is conserved between the three GRO gene products, it is likely that all have a similar dimerization domain in solution. In addition, like IL-8, it appears that the N-terminal amino acids preceding the CQC motif are conformationally disordered but tightly hinged to the central protein via the disulfide bridge (Fairbrother *et al.*,

1993; Baggiolini *et al.*, 1994). In contrast, the C-terminal α-helix of GROα degenerates at Asp-70, unlike in IL-8 where the α-helix continues to the C-terminal residue. It would be interesting to determine whether GROβ and GROγ are more similar to IL-8, as neither has an appropriate Asp residue but each has a Gly which is highly conserved with Ala-74 in IL-8.

3.3   Proteolytic Processing

While considerable interest has been shown in the presence of different processed forms of IL-8 (Baggiolini *et al.*, 1994) much less is known about GROα, GROβ, and GROγ. GROα was identified from natural sources as being slightly smaller than the predicted 73 amino acids (Derynck *et al.*, 1990). It was demonstrated that cleavage of the recombinant protein by dog pancreas microsomes resulted in a protein with the 34-amino-acid leader sequence removed and, in addition, the two final carboxy-terminal residues were also cleaved. There was no evidence of any post-translational modifications. Amino acid determinations on HPLC-purified natural osteosarcoma-derived forms of GROα and GROγ confirmed the initial cleavage site definition from melanoma cells (Derynck *et al.*, 1990; Proost *et al.*, 1993). The size of IL-8 can be dictated by the cellular origin of the protein; monocytes and macrophages release a 72-amino acid form of IL-8, SAKELRCQC, while five additional N-terminal residues **AVLPR**SAKELRCQC are detected in endothelial and fibroblast cultures (reviewed in Baggiolini *et al.*, 1994). As with IL-8, it is to be expected that the different forms of GRO could be further processed. Several N-terminal-truncated forms of GROα differing by deletion of one or two amino acids were detected in osteosarcoma cells. (Proost *et al.*, 1993). While there is no evidence for this, it would be a particularly important process with GROβ, which is apparently less active unless the N-terminal five amino acids are removed (Geiser *et al.*, 1993).

4.   *Cellular Sources of GRO mRNAs and Proteins*

Message and/or protein for GROα has been identified in a large number of cell types, most commonly after an inflammatory stimulus such as can be provided by IL-1β or TNF-α. Information about GROβ and GROγ expression is available from a more restricted number of cell types (see Tables 33.1 and 33.2 for a summary of expression studies). When expression of all three GROs has been investigated in the same cells and in the same study, it appears that while GROα can be induced in most cells tested there is a more selective induction of the other two GRO members. Colonic epithelial and bronchial epithelial cells express GROα and GROγ but

not GROβ (Isaacs *et al.*, 1992; Becker *et al.*, 1994). Synovial fibroblasts express GROα and GROβ but not GROγ (Hogan *et al.*, 1994). Furthermore, the type of signal given to the cells may determine the GRO response. PMA-exposed monocytes express GROβ and GROγ but not GROα, while endotoxin-stimulated monocytes express all three GROs (Haskill *et al.*, 1990). Retinal pigment epithelial cells exposed to IL-1β or TNF-α express all three GROs, while serum-stimulated cells express only GROα and GROγ (Jaffe *et al.*, 1993). It is not clear whether this differential expression is regulated on the transcriptional level or by means of message stabilization.

4.1   *In Vivo* Studies

Constitutive GROα expression has been noted in biopsies of normal colonic (Isaacs *et al.*, 1992) and airway epithelium (Becker *et al.*, 1994), in various cells in normal skin including keratinocytes, melanocytes, hair follicles, and blood vessels (Richmond and Thomas, 1988; Tettelbach *et al.*, 1993). The presence of GRO in these tissues with barrier function may indicate a state of activation possibly induced by ambient environmental sources. In disease states, GRO has been found to be highly expressed in psoriatic scales (Schroder *et al.*, 1992; Kojima *et al.*, 1993) and, as suggested by animal experiments, at sites of tissue injury and inflammation. In psoriatic skin, overexpression of GRO correlated with IL-8 expression and could be downregulated by cyclosporin A *in vivo* (Kojima *et al.*, 1993). Overexpression of GROα was restricted to the suprapapillary layers of psoriatic lesions (Kulke *et al.*, 1996). Biopsies of squamous cell carcinomas and adenocarcinomas showed GRO mRNA expression (Richmond *et al.*, 1988) and GROα was localized immunohistochemically in prostatic carcinoma, sarcoidosis tissue, small-cell lung carcinoma, and nonmalignant nevus tissue (Richmond and Thomas, 1988). GROγ appears to be preferentially expressed in a colonic carcinoma (Cuenca *et al.*, 1992). GROα has been found to be constitutively expressed in cell lines of melanoma, glioblastoma, mammary, bladder, renal, and prostatic carcinoma cells (Richmond *et al.*, 1988; Pichon and Lagarde, 1989; Haskill *et al.*, 1990; Anisowicz *et al.*, 1991; Rangnekar *et al.*, 1992).

4.2   *In Vitro* Studies

The large majority of studies of GRO mRNA or protein expression have been performed with various cell lines: normal cells in primary culture, cultured cells from diseased tissue such as synovial cells and chondrocytes from arthritic lesions, or melanocytes in progressive stages of carcinogenesis. Most of these cell sources in their growing but "unstimulated" state express very low to

Table 33.1   **Expression and stimulation of GRO mRNAs or proteins**

Cell source	Stimulus	Reference[a]
Monocyte/macrophages	Adherence, *in vitro* culture	1–3
	Endotoxin	1, 2, 4
	TNF-α	
	Virus infection	
Neutrophils	Adherence, endotoxin	1, 5
Endothelial cells	TNF-α, IL-1β, thrombin, PMA	1, 6, 7, 18
Epithelial cells		
Eye		
Pigment	TNF-α, IL-1β	23
Lung		
Airway	None, TNF-α	3
Type II	TNF/PMA	8
Mammary	Constitutive, PMA, TNF-α, IL-β, EGF	1, 11, 21
Skin		
Keratinocytes	Constitutive	17, 20
Melanocytes	Culture, serum	15, 19
Fibroblasts		
Gingiva	IL-1β	24
Lung	TNF-α, IL-1β	1
Mammary	TNF-α, IL-1β	1
Skin, foreskin	Serum, PMA, TNF-α, IL-1β	1, 9–11, 31
Synovium	TNF-α, IL-1β	9, 26
Astrocytes	IL-1β	12
Chondrocytes	IL-1β	13
T lymphocytes	PHA/PMA, IL-2	22, 25
Tumor cells		
Melanoma	Constitutive, culture, PDGF, GRO	14–17, 19, 27, 28
Renal carcinoma	Constitutive	10
Colonic carcinoma	Constitutive	1, 30
Bladder carcinoma	Culture	11
Lung carcinoma	Culture	11

[a] 1, Haskill *et al.* (1990); 2, Sporn *et al.*; 3, Becker *et al.* (1994); 4, Schroeder *et al.* (1990); 5, Iida and Grotendorst (1990); 6, Wen *et al.* (1989); 7, 8, Walz *et al.* (1991a,b); 9, Golds *et al.* (1989); 10, Richmond *et al.* (1988); 11, Anisowicz *et al.* (1988, 1991); 12, Legoux *et al.* (1992); 13, Recklies and Golds (1992); 14, Lawson *et al.* (1987); 15, Bordoni *et al.* (1989, 1990); 16, Chenevix-Trench *et al.* (1990); 17, Richmond and Thomas (1988); 18, Introna *et al.* (1993); 19, Mattei *et al.* (1994); 20, Tettlelbach *et al.* (1993); 21, Stampfer *et al.* (1993); 22, Skerka *et al.* (1993); 23, Jaffe *et al.* (1993); 24, Odake *et al.* (1993); 25, Saraya and Balkwill (1993); 26, Bedard and Golds (1993); 27, Rodeck *et al.* (1991); 28, Pichon and Lagarde (1989); 30, Cuenca *et al.* (1992); 31, Stoeckle (1991).

negligible levels of GRO, but the cells are induced to express GRO after stimulation with TNF-α or IL-1β. Induction of expression of one or several GRO genes has been observed in such a variety of cells as endothelium (Wen *et al.*, 1989; Haskill *et al.*, 1990; Introna *et al.*, 1993; Schwartz *et al.*, 1994), retinal pigment epithelium (Jaffe *et al.*, 1993; Shattuck *et al.*, 1994), foreskin fibroblast (Anisowicz *et al.*, 1987, 1988, 1991; Stoeckle, 1991), synovial fibroblasts (Hogan *et al.*, 1994), articular chondrocytes (Recklies and Golds, 1992), gingival fibroblasts (Odake *et al.*, 1993), and airway epithelial cells (Becker *et al.*, 1994). Endotoxin/LPS has been used as the stimulus to demonstrate GRO induction in monocytes, macrophages, and granulocytes (Haskill *et al.*, 1990; Iida and Grotendorst, 1990; Becker *et al.*, 1994).

4.3 SELECTIVE EXPRESSION OF GRO mRNAS

Comparison of expression of the different GRO genes has been performed on psoriatic skin, where GRO is a major neutrophil chemoattractant. GROα was approximately 6-fold more abundant then GROβ and 25-fold more abundant than GROγ (Kojima *et al.*, 1993). In pigment retinal epithelial cells all three GROs were induced in response to IL-1 and TNF; the addition of serum, on the other hand, induced only GROα and GROγ mRNA expression (Jaffe *et al.*, 1993). In primary human nasal and bronchial epithelial cell biopsies from healthy individuals, GROα and GROγ were constitutively expressed. When the cells were put in

Table 33.2   Differential expression of GROα, GROβ and GROγ

Cell type	Stimulus	GRO expression	Reference[a]
Monocyte	Adherence to fibronectin	α>β>γ	1
	Endotoxin	α>β>γ	1
	PMA	β,γ	1
Alveolar macrophage	Culture, endotoxin	α,β,γ	2
Neutrophil	Adherence to fibronectin	α	1
Airway epithelium	Constitutive, TNF-α	α,γ	1
Psoriasis	Constitutive	α>β>γ	7
Fibroblasts	IL-1β	α,β,γ	9
Keratinocytes	Culture, IL-1β, IFN-γ + TNF-α	α,β,γ	7
Synovial fibroblast lines			
Osteoarthritic	Culture, TNFα, IL-1β	α,β	4
Noninflammatory	Culture, TNFα, IL-1β	α,β	4
Rheumatoid	TNFα, IL-1β	α,β	4
Eye epithelium	TNFα, IL-1β	α,β,γ	5, 8
	Serum	α,γ	8
Chondrocytes	IL-1β	α,β,γ	3
Cartilage explants	IL-1β	γ	3
Colonic carcinoma	None	α,γ	1, 6
Melanoma	None	α	5
(Hs294T cells)			

[a] 1, Haskill *et al.* (1990); 2, Becker *et al.* (1992); 3, Recklies and Golds (1992); 4, Hogan *et al.* (1994); 5, Shattuck *et al.* (1994); 6, Cuenca *et al.* (1992); 7, Kojima *et al.* (1993); 8, Jaffe *et al.* (1993); 9, Stoeckle (1991).

primary culture, expression decreased and was then elevated when TNF-α was added to the culture medium (Becker *et al.*, 1994). There was no constitutive or induced message expressed for GROβ. In contrast, alveolar macrophages expressed all three GROs after stimulation with LPS or TNF-α. Similar levels of all the GRO mRNAs were expressed in the stimulated macrophages. Recently, two homologs of GRO have been identified in rabbit alveolar macrophages and bronchial epithelial cells (Johnson *et al.*, 1996). Neutrophils selectively express GROα transcripts in response to several stimuli including LPS, TNF, and opsonized yeast (Gasperini *et al.*, 1995). Recklies and Golds (1992) investigated whether chondrocytes isolated from normal femoral chondrocytes and cartilage produce neutrophil chemoattractant chemokines which may contribute to the destruction of cartilage observed in rheumatoid arthritis. Primary chondrocyte cultures stimulated with IL-1β were found to preferentially express GROγ, although low levels of GROα and GROβ were also detected by PCR. On the other hand, cartilage explants expressed only GROγ after stimulation. There also appears to be preferential expression of GROγ in colonic mucosa and in colonic tumors (Cuenca *et al.*, 1992). Insight into the possible need for an inflammatory signal in GRO expression came from cultures of synovial fibroblasts from rheumatoid, inflammatory (nonrheumatoid), and osteoarthritis (Hogan *et al.*, 1994). Unstimulated cultures from the latter two diseases constitutively expressed GROα and GROβ but not GROγ, while cells from rheumatoid lesions were negative for the GROs. Stimulation with IL-1β or TNF-α increased/induced long-term expression of mRNA for GROα and GROβ in all three synovial cell cultures. No GROγ was induced.

4.4   EXPRESSION OF GRO IN GROWTH-STIMULATED CELLS AND MELANOMA CELL CULTURES

In growth-stimulated hamster, mouse, and human fibroblasts, GRO expression is correlated with growth status of the cells. Within the first hour of serum addition to serum-starved human fibroblasts, GRO mRNA is transiently induced (Anisowicz *et al.*, 1987). Transformed fibroblasts and some tumor cell lines constitutively expressed GRO which led to the suggestion that the GRO gene was involved in growth regulation. It is possible that GRO may function as a growth factor or a cofactor for fibroblasts. Schroder and colleagues (1990) mentioned increased proliferation of fibroblasts in response to purified GRO protein from LPS-stimulated human monocytes, although another study found no effect on growth regulation of rheumatoid synovial fibroblasts despite the expression of a large number of GROα binding sites on the cells (Unemori *et al.*, 1993).

The identification of "melanoma growth stimulatory activity" protein as the gene product of GROα supports the contention that this gene is involved in growth regulation and autonomous growth in certain cell types. The Hs294T melanoma cell line was used for production and isolation of the activity. Generation of antibodies to the protein allowed the identification of GRO expression in a variety of tumor cells in addition to melanomas (Richmond and Thomas, 1988). GROα is produced by approximately 70% of primary cell cultures from melanoma biopsies. In contrast, primary cultures of benign nevus cells are negative for GRO production. GROα levels in the Hs294T which constitutively produce GROα are upregulated by IL-1β, TNF-α, PMA, and LPS, but not by serum; TGF-β downregulates GRO expression in this cell line (Shattuck et al., 1994). The progression of melanocytes into metastatic melanoma is associated with loss of exogenous growth factors and gain of production of autocrine growth factors, among them GROα (Richmond et al., 1983; Graeven and Herlyn, 1992). In an elegant study by Balentien and colleagues (1991) it was shown that deregulation of GROα expression may indeed be associated with events in tumor progression, since overexpression of human recombinant GRO in immortalized rodent melanocytes results in tumor formation in nude mice. This transformation property is shared with GROβ and GROγ.

Exposure of various cell types to IL-1β is associated with both growth stimulatory and inhibitory functions in various cell types. With this stimulus, GRO expression was clearly dissociated from growth regulation. IL-1β is growth inhibitory on the melanoma A375-C6 and induced GRO, while serum which was growth stimulatory did not (Rangnekar et al., 1991, 1992). In human glioblastoma cells, IL-1β was mitogenic as well as inducing GRO, while other mitogens such as serum did not. The action of IL-1 on GRO induction has also been dissociated from growth regulation by other investigators (Anisowicz et al., 1988; Shattuck et al., 1994). Recently, GRO has been identified as one of several cytokines induced by IL-1 in epidermal keratinocytes in situ. The authors employed IL-1R transgenic mice to establish the probable significance of GRO in inflammatory and hyperproliferative skin disease (Groves et al., 1996).

5.   Functions of GRO Proteins

Although GROα protein was originally identified as an autocrine melanoma growth stimulatory activity (Richmond et al., 1983; Richmond and Thomas, 1988), the potential growth regulatory function of the protein was ignored in other tumor systems after the GRO gene products were classified as members of the α-chemokine family. The emphasis of the vast majority of studies has been to demonstrate the neutrophil stimulatory function of the GROs.

5.1   GRO-INDUCED NEUTROPHIL INFLAMMATION

Purified, synthetic, and recombinant GROα proteins have been found to induce a strong neutrophil inflammatory response (Moser et al., 1990; Sager et al., 1992; Zwahlen et al., 1993). GRO was injected into the foot pad or intradermally in mice. Four hours post injection, the site was dominated by an intense intradermal and perivascular inflammatory infiltrate. GROα gave a similar neutrophilic inflammatory response when administered intradermally in the rat.

5.2   GRO AS A STIMULATOR OF LEUKOCYTE CHEMOTAXIS AND DEGRANULATION

There is some controversy about the potency of GRO as a chemoattractant for neutrophils as compared to IL-8. In the early studies of recombinant purified GRO it was found to be active at >10-fold lower concentrations, while being a poor secretagogue for neutrophil primary granules (Sager et al., 1992; Balantien et al., 1990). In subsequent studies, Schroder and colleagues (1990) compared IL-8 and GRO for chemotactic and neutrophil degranulation activity and found that although the ED_{50} was the same for IL-8 and GRO, the number of neutrophils migrating in a chemotaxis assay was only 60% of the cells responsive to IL-8. Furthermore, the response to GRO showed a bell-shaped curve with peak activity at 10 ng/ml and no response at 200 ng/ml. This difference in response between GRO and IL-8 was also observed when triggering neutrophil degranulation. Moser and colleagues (1990) used synthetic GROα to compare potency with IL-8 to induce chemotaxis, production of H_2O_2, and release of elastase. Both chemokines were active at similar molar ranges (10^{-10}–10^{-7} M), but IL-8 was more potent in all the tested functions. The biggest difference in stimulatory capacity was seen in H_2O_2 production, where GRO gave <2% of the IL-8 response. Using chemically synthesized GROs, Geiser and colleagues (1993) compared the potency of the three GRO proteins in activating calcium flux, H_2O_2 production, and granule exocytosis in neutrophils. They found the hierarchy GROα > GROγ > GROβ for all these functions.

GROα also has the ability to influence integrin CD11b/CD18 expression of the surface of neutrophils concomitantly with a decrease in the expression of L-selectin (Detmers et al., 1991). Transiently increased phagocytosis of IgG-coated erythrocytes was also noted in GRO-exposed cells, although Fc receptors were not affected. This phenomenon was apparently mediated by CD11b/CD18, as an antibody to the moiety abolished the increase. GROs may also have some activity on basophils, as these cells mobilized Ca^{2+} in response to exposure (Geiser et al., 1993).

Magazin and colleagues (1992) compared the biological actions of GROβ with IL-8. They found similar levels of chemotactic activity, induction of respiratory burst, and induction of IL-6 in neutrophils for the two chemokines. However, 100 times more GROβ than IL-8 was required for elevation of intracellular free Ca^{2+} and increase in CD11b expression on granulocytes.

5.3 GRO-INDUCED GENE ACTIVATION IN NON-HEMATOPOIETIC CELLS

With the exception of studies on growth competence in melanocytes and melanoma cells, studies on GRO-inducible functions or GRO-induced gene products are scarce in nonhematopoietic cell types. Richmond and colleagues (1988) showed that treatment of Hs294T melanoma cells with GROα increased GRO mRNA expession at least 10-fold. Exposure of endothelial cells to GROα was shown to induce GROα expression in endothelial cells (Wen et al., 1989), but a role for GRO in the regulation of endothelial cell inflammatory or growth characteristics has not been defined. In rheumatoid synovial fibroblasts, Unemori and colleagues (1993) found that procollagen gene expression and collagen production were decreased by exposure of the cells to GROα, while no effect was seen on collagen-degrading metalloproteinases or an inhibitor of metalloproteinases. A role for GRO in angiogenesis of tumors has recently been proposed (Cao et al., 1995). It appears that, while GROα and GROγ may stimulate angiogenesis, GROβ may serve to inhibit this process.

6. GRO Receptors

6.1 EXPRESSION OF GRO-IL-8 RECEPTORS ON NEUTROPHILS

The three GROs share both the high- and low-affinity receptors on neutrophils with IL-8 (see Chapter 8 for details about receptor genes, structure, and signal transduction). Both these receptors, IL-8R2 (GRO high-affinity receptor) and IL-8R1 (GRO low-affinity receptor), have been cloned and localized to chromosome 2q35 (Holmes et al., 1991; Murphy and Tiffany, 1991; Ahuja et al., 1992; Morris et al., 1992; Cerretti et al., 1993). GROα competes effectively with IL-8 for the 67–70 kDa IL-8R2 (K_d = 0.2–2.5 nM), while GROα competes poorly with IL-8 for the other low-affinity 44–59 kDa IL-8R1 (K_d = 200–500 nM) (Moser et al., 1993; Loetscher et al., 1994a). Binding studies with radioiodinated GROα showed that 30–45% of the receptors bound the molecule with high affinity, and 55–70% with low affinity (Schumacher et al., 1992). IL-8R1 and R2 were successfully transfected into Jurkat cells by Loetscher and colleagues (1994b) and assayed for binding of GROα. This experiment elegantly confirmed the high-affinity nature of IL-8R2, both in binding

studies with IL-8 and GROα as well as studies of GROα concentrations required for functional triggering of Ca^{2+} fluxes and chemotaxis. It was also shown that these functions can be triggered by GROα through the IL-8R1 receptors if the cells are exposed to GROα concentrations close to the K_d values.

6.2 GRO-SPECIFIC RECEPTORS

An indication of expression of GRO receptors on other cell types than neutrophils comes from the responsiveness of cells like melanoma cells and synovial fibroblasts to GRO stimulation. The melanoma line Hs294T has recently been shown to express 53 000–68 000 GROα binding sites per cell with a K_d of 3.9–4.3 nM (Horuk et al., 1993). This receptor appears to be specific for GRO and cannot be displaced by IL-8. In the same report, the monocyte-like cell line U937 also expressed receptors for GROα, but this binding was effectively cross-competed with IL-8. The Hs294T cell line also expresses the IL-8R2 receptor as evidenced by PCR (Mueller et al., 1994), as do primary keratinocyte cultures. Two GRO binding proteins were identified in membrane preparations of human placenta (Cheng et al., 1992). These proteins were 50–58 kDa and 70–78 kDa and correspond in size to IL-8R1 and IL-8R2. No further identification of the binding proteins or IL-8 cross-competition or affinity studies were performed on these extracts. Others have found that melanoma cells express transcripts for IL-8R2 but not R1 (Moser et al., 1992). IL-8R2 mRNA was also found in all blood cells and related cell lines, in fibroblasts, melanocytes, and melanoma lines. Synovial fibroblasts were found to express 75 000 binding sites for GROα as determined by Schatchard analysis (Unemori et al., 1993). The result of receptor triggering in these cells was a downregulation of collagen synthesis but no effect on fibroblast cell growth.

6.3 THE PROMISCUOUS CHEMOKINE RECEPTOR

Erythrocytes express a chemokine receptor with the ability to bind both C-X-C and C-C chemokines (Darbonne et al., 1991; Neote et al., 1993). Binding of GROα to this receptor is easily displaced by both IL-8 and C-C chemokines like RANTES and MCP-1, while GROα in turn displaces other chemokines. It has been suggested that these receptors act as scavengers for chemokines in the circulation (Darbonne et al., 1991).

The identity of the promiscuous receptor with the "Duffy antigen" (DARC), the moiety which is used by the malaria parasite *Plasmodium vivax* to invade erythrocytes, has been recognized (Horuk et al., 1993, 1994; Chaudhuri et al., 1994; Hesselgesser et al., 1995). It is possible that GROα and other chemokines are able to inhibit parasite invasion into erythrocytes. The Duffy receptor also appears to belong to the seven-transmembrane-spanning receptor superfamily. However, transfection of the receptor cDNA into an embryonic kidney cell line does not result in a signaling receptor, as measured by a lack of Ca^{2+} mobilization when reacted with either GROα or IL-8 (Neote *et al.*, 1994).

7.   *Signal Transduction by GRO*

Both IL-8/GRO receptors are members of the seven-transmembrane domain, G-protein-coupled receptor family. This characterization provides evidence that the GRO are primarily chemoattractants, as other chemoattractants such as receptors for C5a, fMLP, and platelet-activating factor have all been shown to share this structure (Dohlman *et al.*, 1991). Signaling events in GRO or IL-8 binding to the IL-8R2 receptor involve a GTP binding protein, protein kinase C activity, and rise in free cytoplasmic Ca^{2+} and increase in both GTPase activity and GTPγS binding (Schroeder *et al.*, 1990; Walz *et al.*, 1991b; Kupper *et al.*, 1992). Transfection of receptors into nonexpressing cells has allowed the study of other specific signal-transducing events. Mueller and colleagues (1994) used the transfection approach to study signaling mechanisms of GROα through the IL-8R2/GRO receptor in a placental and a kidney cell line. The transfected protein was found to be basally phosphorylated and GROα markedly increased phosphorylation of its receptor. Phosphoaminoacid analysis indicated that the phosphorylation was on serine residues, suggesting activation of a serine kinase as a signal in GRO binding to its receptor in nonhematopoietic cells. Cheng and colleagues (1992) also investigated phosphorylation events in response to GROα in placental membrane extracts following 1–5

min treatment with GROα. Enhanced phosphorylation of several proteins was observed; one of the phosphorylated proteins corresponded in size to a GRO binding protein in the membrane preparations, suggesting autophosphorylation of the receptor. Although both IL-8 and GRO stimulate similar levels of free calcium, only IL-8 was able to stimulate phospholipase D, demonstrating a divergence of signaling pathways downstream of the two IL-8R members (L'Heureux *et al.*, 1995). Similarly, Jones and colleagues (1996) have reported that both receptors mediate changes in free calcium and granule release, but only the IL-8R1 functions to stimulate the oxidative burst or activation of phospholipase D. Both receptors appear capable of triggering reorganization of the actin cytoskeleton (Norgauer *et al.*, 1996) and both receptors become coupled to the α subunit of the Gi2 protein (Damaj *et al.*, 1996). Several authors report that homologous proteins have been identified in mice (Groves *et al.*, 1996) as well as rabbits (Johnson and Schwartz, 1994), offering additional models amenable to investigation of the role of GRO in disease.

8.   *Animal Models for the Study of GRO Function*

The major neutrophil chemoattractant molecules in the rat have turned out to be homologous to the GROs and not to IL-8. Analysis of these chemokines in various disease models is likely to be important in interpreting information about GRO expression in human disease. We have therefore included information on studies investigating rat CINC/GRO expression and production in disease models and inflammation (Table 33.3). Studies

Table 33.3   Expression of CINC/GRO in rat models of inflammation and disease

mRNA or protein expression	Disease/model	Reference
CINC	LPS-induced inflammation LPS in the air pouch	Watanabe *et al.* (1992, 1993) Iida *et al.* (1992)
CINC	Immune complex glomerulonephritis	Wu *et al.* (1994)
CINC-1 CINC-2α CINC-2β CINC-3	Carragenan-induced inflammation	Nakagawa *et al.* (1994)
CINC	Pituitary gland +/− TNF	Koike *et al.* (1994)
CINC	Focal cerebral ischemia	Liu *et al.* (1993)
CINC	Chronically ethanol-fed rats	Shiratori *et al.* (1993)
MIP-2, KC	Pulmonary inflammation	Huang *et al.* (1992)
MIP-2, KC	Cardiac allografts	Wieder *et al.* (1993)
KC	Wound tissue	Iida and Grotendorst (1990)

exposing various cells to CINC proteins also suggests some new approaches to studies of human GRO function and role in homeostasis.

The rat chemokine "cytokine-induced neutrophil chemoattractant" CINC is believed to be the rat homolog of GROα and of the mouse gene KC (Watanabe *et al.*, 1989). This protein with strong chemoattractant function was originally purified from endotoxin-, TNF-α-, and IL-1β-stimulated kidney epithelioid cells and subsequently from endotoxin-induced inflammatory exudates (Watanabe *et al.*, 1993). Another neutrophil chemoattractant was also present in these exudates, and sequence analysis found it to be highly similar to mouse MIP-2. Two additional highly homologous MIP-2-like neutrophil chemotactic proteins have recently been purified (Nakagawa *et al.*, 1994). The suggested nomenclature for these chemokines is now CINC-1 for the original CINC, CINC-3 for the first isolated MIP-2 protein, and CINC-2α and CINC-2β for the latest MIP-2-like proteins (Nakagawa *et al.*, 1994). At low concentrations (10^{-9} M) the hierarchy of chemotactic activity is CINC-3>CINC-2α >CINC-1>CINC-2β; at 10^{-8} M the CINCs are similar in activity. CINC-2α appears to be the predominant form secreted by LPS-activated macrophages (Nakagawa *et al.*, 1996).

9. Acknowledgments

We would like to acknowledge Ann Richmond and Becky Shattuck for the MGSA genomic information and Ian Clark-Lewis for the ribbon diagrams.

10. References

Ahuja, S.K. and Murphy, P.M. (1993). Molecular piracy of mammalian interleukin-8 receptor type B by herpesvirus saimiri. J. Biol. Chem. 268, 20691–20694.

Ahuja, S.K., Ozcelik, T., Milatovitch, A., Francke, U. and Murphy, P.M. (1992). Molecular evolution of the human interleukin-8 receptor gene cluster. Nat. Genet. 2, 31–36.

Alberta, J.A., Rundell, K. and Stiles, C.D. (1994). Identification of an activity that interacts with the 3'-untranslated region of c-myc mRNA and the role of is target sequence in mediating rapid mRNA degradation. J. Biol. Chem. 269, 4532–4538.

Anisowicz, A., Bardwell, L. and Sager, R. (1987). Constitutive overexpression of a growth-regulated gene in transformed Chinese hamster and human cells. Proc. Natl Acad. Sci. USA 84, 7188–7192.

Anisowicz, A., Zajchowski, D., Stenman, G. and Sager, R. (1988). Functional diversity of gro gene expression in human fibroblasts and mammary epithelial cells. Proc. Natl Acad. Sci. USA 85, 9645–9649.

Anisowicz, A., Messineo, M., Lee, S.W. and Sager, R. (1991). An NF-kappa B-like transcription factor mediates IL-1/TNF-alpha induction of gro in human fibroblasts. J. Immunol. 147, 520–527.

Baggiolini, M., Dewald, B. and Moser, B. (1994). Interleukin-8 and related chemotactic cytokines—exc and cc chemokines. Adv. Immunol. 55, 97–179.

Balentien, E., Han, J.H., Thomas, H.G., *et al.* (1990). Recombinant expression, biochemical characterization, and biological activities of the human MGSA/gro protein. Biochemistry 29, 10225–10233. [Published erratum in Biochemistry (1991) 30(2), 594]

Balentien, E., Mufson, B.E., Shattuck, R.L., Derynck, R. and Richmond, A. (1991). Effects of MGSA/GRO alpha on melanocyte transformation. Oncogene 6, 1115–1124.

Becker, S., Quay, J., Koren, H.S. and Haskill, J.S. (1994). Constitutive and stimulated MCP-1, GRO alpha, beta, and gamma expression in human airway epithelium and broncho-alveolar macrophages. Am. J. Physiol. 266, L278–L286.

Bedard, P.A. and Golds, E.E. (1993). Cytokine-induced expression of mRNAs for chemotactic factors in human synovial cells and fibroblasts. J. Cell. Physiol. 154, 433–441.

Blackwell, T.S., Holden, E.P., Blackwell, T.R., DeLarco, J.E. and Christman, J.W. (1994). Cytokine-induced neutrophil chemoattractant mediates neutrophilic alveolitis in rats: association with nuclear factor kappa B activation. Am. J. Respir. Cell. Mol. Biol. 11, 464–472.

Bordoni, R., Thomas, G. and Richmond, A. (1989). Growth factor modulation of melanoma growth stimulatory activity mRNA expression in human malignant melanoma cells correlates with cell growth. J. Cell. Biochem. 39, 421–428.

Bordoni, R., Fine, R., Murray, D. and Richmond, A. (1990). Characterization of the role of melanoma growth stimulatory activity (MGSA) in the growth of normal melanocytes, nevocytes, and malignant melanocytes. J. Cell. Biochem. 44, 207–219.

Cao, Y., Chen, C., Weatherbee, J.A., Tsang, M. and Folkman, J. (1995). Gro-beta, a C-X-C-chemokine, is an angiogenesis inhibitor that suppresses the growth of Lewis lung carcinoma in mice. J. Exp. Med. 182, 2069–2077.

Castor, C.W., Miller, J.W. and Walz, D.A. (1983). Structural and biological characteristics of connective tissue activating peptide (CTAP-III), a major human platelet-derived growth factor. Proc. Natl Acad. Sci. USA 80, 765–769.

Cerretti, D.P., Kozlosky, C.J., Vanden, B.T., Nelson, N., Gearing, D.P. and Beckmann, M.P. (1993). Molecular characterization of receptors for human interleukin-8, GRO/melanoma growth-stimulatory activity and neutrophil activating peptide-2. Mol. Immunol. 30, 359–367.

Chaudhuri, A., Zbrzezna, V., Polyakova, J., Pogo, A.O., Hesselgesser, J. and Horuk, R. (1994). Expression of the Duffy antigen in K562 cells. Evidence that it is the human erythrocyte chemokine receptor. J. Biol. Chem. 269, 7835–7838.

Chenevix-Trench, G., Martin, N.G. and Ellem, K.A. (1990). Gene expression in melanoma cell lines and cultured melanocytes: correlation between levels of c-src-1, c-myc and p53. Oncogene 5, 1187–1193.

Cheng, Q.C., Han, J.H., Thomas, H.G., Balentien, E. and Richmond, A. (1992). The melanoma growth stimulatory activity receptor consists of two proteins. Ligand binding results in enhanced tyrosine phosphorylation. J. Immunol. 148, 451–456.

Clark-Lewis, I., Kim, K.-S., Rajarathnam, K., *et al.* (1995). Structure–activity relationships of chemokines. J. Leukocyte Biol. 57, 703–711.

Cuenca, R.E., Azizkhan, R.G. and Haskill, S. (1992).

Characterization of GRO alpha, beta and gamma expression in human colonic tumours: potential significance of cytokine involvement. Surg. Oncol. 1, 323–329.

Damaj, B.B., McColl, S.R., Mahana, W., Crouch, M.F. and Naccache, P.H. (1996). Physical association of Gi2alpha with interleukin-8 receptors. J. Biol. Chem. 271, 12783–12789.

Darbonne, W.C., Rice, G.C., Mohler, M.A., et al. (1991). Red blood cells are a sink for interleukin-8, a leukocyte chemotaxin. J. Clin. Invest. 88, 1362–1369.

Derynck, R., Balentien, E., Han, J.H., et al. (1990). Recombinant expression, biochemical characterization and biological activities of the human MGSA/gro protein. Biochemistry 29, 10225–10233.

Dezube, B.J., Pardee, A.B., Beckett, L.A., et al. (1992). Cytokine dysregulation in AIDS: in vivo overexpression of mRNA of tumor necrosis factor-alpha and its correlation with that of the inflammatory cytokine GRO. J. Acquir. Immune Defic. Syndr. 5, 1099–1104.

Dohlaman, H.G., Thorner, J., Caron, M.G. and Lefkowitz, R.J. (1991). Model systems for study of seven-transmembrane-segment receptors. Annu. Rev. Biochem. 60, 653–688.

Fairbrother, W.J., Reilly, D., Colby, T. and Horuk, R. (1993). 1H assignment and secondary structure determination of human melanoma growth stimulating activity (MGSA) by NMR spectroscopy. FEBS Lett. 330, 302–306.

Gasperini, S., Calzetti, F., Russo, M.P., De Gironcoli, M. and Cassatella, M.A. (1995). Regulation of GRO alpha production in human granulocytes. J. Inflamm. 45, 143–151.

Geiser, T., Dewald, B., Ehrengruber, M.U., Clark, L.I. and Baggiolini, M. (1993). The interleukin-8-related chemotactic cytokines GRO alpha, GRO beta, and GRO gamma activate human neutrophil and basophil leukocytes. J. Biol. Chem. 268, 15419–15424.

Golds, E.E., Mason, P. and Nyirkos, P. (1989). Inflammatory cytokines induce synthesis and secretion of gro protein and a neutrophil chemotactic factor but not beta 2-microglobulin in human synovial cells and fibroblasts. Biochem. J. 259, 585–588.

Graeven, U. and Herlyn, M. (1992). In vitro growth patterns of normal human melanocytes and melanocytes from different stages of melanoma progression. J. Immunother. 12, 199–202.

Green, C.J., St. Charles, R., Edwards, B.F.P. and Johnson, P.H. (1989). Identification and characterization of PF4varl, a human gene variant of platelet factor 4. Mol. Cell. Biol. 9, 1445–1451.

Groves, R.W., Rauschmayr, T., Nakamura, K., Sarkar, S., Williams, I.R. and Kupper, T.S. (1996). Inflammatory and hyperproliferative skin disease in mice that express elevated levels of the IL-1 receptor (type I) on epidermal keratinocytes. Evidence that IL-1-inducible secondary cytokines produced by keratinocytes in vivo can cause skin disease. J. Clin. Invest. 98, 336–344.

Haskill, S., Peace, A., Morris, J., et al. (1990). Identification of three related human GRO genes encoding cytokine functions. Proc. Natl Acad. Sci. USA 87, 7732–7736.

Hesselgesser, J., Chitnis, C.E., Miller, L.H., et al. (1995). A mutant of melanoma growth stimulating activity does not activate neutrophils but blocks erythrocyte invasion by malaria. J. Biol. Chem. 270, 11472–11476.

Hogan, M., Sherry, B., Ritchlin, C., et al. (1994). Differential expression of the small inducible cytokines GRO alpha and GRO beta by synovial fibroblasts in chronic arthritis: possible role in growth regulation. Cytokine 6, 61–69.

Holmes, W.E., Lee, J., Kuang, W.-J., Rice, G.C. and Wood, W.I. (1991). Structure and functional expression of a human IL-8 receptor. Science 253, 1278–1280.

Holt, J.C., Harris, M.E., Holt, A.M., Lange, E., Henschen, A. and Niewiarowski, S. (1986). Characterization of human platelet basic protein, a precursor form of low-affinity platelet factor 4 and beta-thromboglobulin. Biochemistry 25, 1988–1996.

Horuk, R. (ed.) (1996). "Chemoattractant Ligands and Their Receptors". CRC Press, Boca Raton, FL.

Horuk, R., Yansura, D.G., Reilly, D., et al. (1993). Purification, receptor binding analysis, and biological characterization of human melanoma growth stimulating activity (MGSA). Evidence for a novel MGSA receptor. J. Biol. Chem. 268, 541–546.

Horuk, R., Wang, Z.X., Peiper, S.C. and Hesselgesser, J. (1994). Identification and characterization of a promiscuous chemokine-binding protein in a human erythroleukemic cell line. J. Biol. Chem. 269, 17730–17733.

Huang, S., Paulauskis, J.D., Godleski, J.J. and Kobzik, L. (1992a). Expression of macrophage inflammatory protein-2 and KC mRNA in pulmonary inflammation. Am. J. Pathol. 141, 981–988.

Huang, S., Paulauskis, J.D. and Kobzik, L. (1992b). Rat KC cDNA cloning and mRNA expression in lung macrophages and fibroblasts. Biochem. Biophys. Res. Commun. 184, 922–929.

Iida, N. and Grotendorst, G.R. (1990). Cloning and sequencing of a new gro transcript from activated human monocytes: expression in leukocytes and wound tissue. Mol. Cell. Biol. 10, 5596–5599. [Published erratum in Mol. Cell. Biol. (1990) 10(12), 6821]

Iida, M., Watanabe, K., Tsurufuji, M., Takaishi, K., Iizuka, Y. and Tsurufuji, S. (1992). Level of neutrophil chemotactic factor CINC/gro, a member of the interleukin-8 family, associated with lipopolysaccharide-induced inflammation in rats. Infect. Immun. 60, 1268–1272.

Introna, M., Breviario, F., dAniello, E.M., Golay, J., Dejana, E. and Mantovani, A. (1993). IL-1 inducible genes in human umbilical vein endothelial cells. Eur. Heart J. 14 (supplement K), 78–81.

Isaacs, K.L., Sartor, R.B. and Haskill, S. (1992). Cytokine messenger RNA profiles in inflammatory bowel disease mucosa detected by polymerase chain reaction amplification. Gastroenterology 103, 1587–1595.

Jaffe, G.J., Richmond, A., Van, L.L., et al. (1993). Expression of three forms of melanoma growth stimulating activity (MGSA)/gro in human retinal pigment epithelial cells. Invest. Ophthalmol. Vis. Sci. 34, 2776–2785.

Johnson, M.C., II, Kajikawa, O., Goodman, R.B., et al. (1996). Molecular expression of the alpha-chemokine rabbit GRO in Escherichia coli and characterization of its production by lung cells in vitro and in vivo. J. Biol. Chem. 271, 10853–10858.

Jones, S.A., Wolf, M., Qin, S., Mackay, C.R. and Baggiolini, M. (1996). Different functions for the interleukin 8 receptors (IL-8R) of human neutrophil leukocytes: NADPH oxidase and phospholipase D are activated through IL-8R1 but not IL-8R2. Proc. Natl Acad. Sci. USA 93, 6682–6686.

Joshi-Barve, S.S., Rangnekar, V.V., Sells, S.F. and Rangnekar, V.M. (1993). Interleukin-1-inducible expression of gro-beta via NF-kappa B activation is dependent upon tyrosine kinase signaling. J. Biol. Chem. 268, 18018–18029.

Koike, K., Sakamoto, Y., Sawada, T., *et al.* (1994). The production of CINC/gro, a member of the interleukin-8 family, in rat anterior pituitary gland. Biochem. Biophys. Res. Commun. 202, 161–167.

Kojima, T., Cromie, M.A., Fisher, G.J., Voorhees, J.J. and Elder, J.T. (1993). GRO-alpha mRNA is selectively overexpressed in psoriatic epidermis and is reduced by cyclosporin A in vivo, but not in cultured keratinocytes. J. Invest. Dermatol. 101, 767–772.

Konishi, K., Takata, Y., Watanabe, K., *et al.* (1993a). Recombinant expression of rat and human Gro proteins in *Escherichia coli*. Cytokine 5, 506–511.

Kulke, R., Todt-Pingel, I., Rademacher, D., Rowert, J., Schroder, J.M. and Christophers, E. (1996). Co-localized overexpression of GRO-alpha and IL-8 mRNA is restricted to the suprapapillary layers of psoriatic lesions. J. Invest. Dermatol. 106, 526–530.

Kunsch, C., Lang, R.K., Rosen, C.A. and Shannon, M.F. (1994). Synergistic transcriptional activation of the IL-8 gene by NF-κB p65 (RelA) and NF-IL-6. J. Immunol. 153, 153–164.

Kupper, R.W., Dewald, B., Jakobs, K.H., Baggiolini, M. and Gierschik, P. (1992). G-protein activation by interleukin 8 and related cytokines in human neutrophil plasma membranes. Biochem. J. 282, 429–434.

L'Heureux, G.P., Bourgoin, S., Jean, N., McColl, S.R. and Naccache, P.H. (1995). Diverging signal transduction pathways activated by interleukin-8 and related chemokines in human neutrophils: interleukin-8, but not NAP-2 or GRO alpha, stimulates phospholipase D activity. Blood 85, 522–531.

Lawson, D.H., Thomas, H.G., Roy, R.G., *et al.* (1987). Preparation of a monoclonal antibody to a melanoma growth-stimulatory activity released into serum-free culture medium by Hs0294 malignant melanoma cells. J. Cell. Biochem. 34, 169–185.

Legoux, P., Minty, C., Delpech, B., Minty, A.J. and Shire, D. (1992). Simultaneous quantitation of cytokine mRNAs in interleukin-1 beta stimulated U373 human astrocytoma cells by a polymerisation chain reaction method involving co-amplification with an internal multi-specific control. Eur. Cytokine Netw. 3, 553–563.

Liu, T., Young, P.R., McDonnell, P.C., White, R.F., Barone, F.C. and Feuerstein, G.Z. (1993). Cytokine-induced neutrophil chemoattractant mRNA expressed in cerebral ischemia. Neurosci. Lett. 164, 125–128.

Loetscher, M., Geiser, T., O'Reilly, T., Zwahlen, R., Baggiolini, M. and Moser, B. (1994a). Cloning of a human seven-transmembrane domain receptor, LESTR, that is highly expressed in leukocytes. J. Biol. Chem. 269, 232–237.

Loetscher, P., Seitz, M., Clark, L.I., Baggiolini, M. and Moser, B. (1994b). Both interleukin-8 receptors independently mediate chemotaxis. Jurkat cells transfected with IL-8R1 or IL-8R2 migrate in response to IL-8, GRO alpha and NAP-2. FEBS Lett. 341, 187–192.

Magazin, M., Vita, N., Cavrois, E., Lefort, S., Guillemot, J.C. and Ferrara, P. (1992). The biological activities of gro beta and IL-8 on human neutrophils are overlapping but not identical. Eur. Cytokine Netw. 3, 461–467.

Mattei, S., Colombo, M.P., Melani, C., Silvani, A., Parmiani, G. and Herlyn, M. (1994). Expression of cytokine/growth factors and their receptors in human melanoma and melanocytes. Int. J. Cancer. 56, 853–857.

Morris, S.W., Nelson, N., Valentine, M.B., *et al.* (1992). Assignment of the genes encoding human interleukin-8 receptor types 1 and 2 and an interleukin-8 receptor pseudogene to chromosome 2q35. Genomics 14, 685–691.

Moser, B., Clark-Lewis, I., Zwahlen, R. and Baggiolini, M. (1990). Neutrophil-activating properties of the melanoma growth-stimulatory activity. J. Exp. Med. 171, 1797–1802.

Moser, B., Dewald, B., Barella, L., Schumacher, C., Baggiolini, M. and Clark, L.I. (1993). Interleukin-8 antagonists generated by N-terminal modification. J. Biol. Chem. 268, 7125–7128.

Mueller, S.G., Schraw, W.P. and Richmond, A. (1994). Melanoma growth stimulatory activity enhances the phosphorylation of the class II interleukin-8 receptor in non-hematopoietic cells. J. Biol. Chem. 269, 1973–1980.

Murphy, P.M. and Tiffany, H.L. (1991). Cloning of complementary DNA encoding a functional human IL-8 receptor. Science 253, 1280–1283.

Nakagawa, H., Komorita, N., Shibata, F., *et al.* (1994). Identification of cytokine-induced neutrophil chemoattractants (CINC), rat GRO/CINC-2 alpha and CINC-2 beta, produced by granulation tissue in culture: purification, complete amino acid sequences and characterization. Biochem. J. 301, 545–550.

Nakagawa, H., Shiota, S., Takano, K., Shibata, F. and Kato, H. (1996). Cytokine-induced neutrophil chemoattractant (CINC)-2 alpha, a novel member of rat GRO/CINCs, is a predominant chemokine produced by lipopolysaccharide-stimulated rat macrophages in culture. Biochem. Biophys. Res. Commun. 220, 945–948.

Neote, K., Darbonne, W., Ogez, J., Horuk, R. and Schall, T.J. (1993). Identification of a promiscuous inflammatory peptide receptor on the surface of red blood cells. J. Biol. Chem. 268, 12247–12249.

Neote, K., Mak, J.Y., Kolakowski, L.F., Jr and Schall, T.J. (1994). Functional and biochemical analysis of the cloned Duffy antigen: identity with the red blood cell chemokine receptor. Blood 84(1), 44–52.

Norgauer, J., Metzner, B., Czech, W. and Schraufstatter, I. (1996). Reconstitution of chemokine-induced actin polymerization in undifferentiated human leukemia cells (HL-60) by heterologous expression of interleukin-8 receptors. Inflamm. Res. 45, 127–131.

Odake, H., Koizumi, F., Hatakeyama, S., Furuta, I. and Nakagawa, H. (1993). Production of cytokines belonging to the interleukin-8 family by human gingival fibroblasts stimulated with interleukin-1 beta in culture. Exp. Mol. Pathol. 58, 14–24.

Pichon, F. and Lagarde, A.E. (1989). Autoregulation of MeWo metastatic melanoma cell growth: characterization of intracellular (FGF, MGSA) and secreted (PDGF) growth factors. J. Cell. Physiol. 140, 344–358.

Pike, M.C., Trask, D. and Sager, R. (1990). Recombinant GRO gene product is chemotactic for human neutrophils. Clin. Res. 38, 479a.

Proost, P., De, W.P.C., Conings, R., Opdenakker, G., Billiau, A. and Van, D.J. (1993). Identification of a novel granulocyte chemotactic protein (GCP-2) from human tumor cells. In vitro and in vivo comparison with natural forms of GRO, IP-10, and IL-8. J. Immunol. 150, 1000–1010.

Rangnekar, V.V., Waheed, S., Davies, T.J., Toback, F.G. and Rangnekar, V.M. (1991). Antimitogenic and mitogenic actions of interleukin-1 in diverse cell types are associated with induction of gro gene expression. J. Biol. Chem. 266, 2415–2422.

Rangnekar, V.V., Waheed, S. and Rangnekar, V.M. (1992). Interleukin-1-inducible tumor growth arrest is characterized by activation of cell type-specific "early" gene expression programs. J. Biol. Chem. 267, 6240–6248.

Recklies, A.D. and Golds, E.E. (1992). Induction of synthesis and release of interleukin-8 from human articular chondrocytes and cartilage explants. Arthritis Rheum. 35, 1510–1519.

Richmond, A. and Thomas, H.G. (1988). Melanoma growth stimulatory activity: isolation from human melanoma tumors and characterization of tissue distribution. J. Cell. Biochem. 36, 185–198.

Richmond, A., Lawson, D.H., Nixon, D.W., Stevens, J.S. and Chawla, R.K. (1983). Extraction of a melanoma growth-stimulatory activity from culture medium conditioned by the Hs0294 human melanoma cell line. Cancer Res. 43, 2106–2112.

Richmond, A., Balentien, E., Thomas, H.G., et al. (1988). Molecular characterization and chromosomal mapping of melanoma growth stimulatory activity, a growth factor structurally related to beta-thromboglobulin. EMBO J. 7, 2025–2033.

Roby, P. and Page, M. (1995). Cell-binding and growth-stimulating activities of the C-terminal part of human MGSA/Gro alpha. Biochem. Biophys. Res. Commun. 206, 792–798.

Rodeck, U., Melber, K., Kath, R., et al. (1991). Constitutive expression of multiple growth factor genes by melanoma cells but not normal melanocytes. J. Invest. Dermatol. 97, 20–26.

Sager, R., Haskill, S., Anisowicz, A., Trask, D. and Pike, M.C. (1991). GRO: a novel chemotactic cytokine. Adv. Exp. Med. Biol. 305, 73–77.

Sager, R., Anisowicz, A., Pike, M.C., Beckmann, P. and Smith, T. (1992). Structural, regulatory, and functional studies of the GRO gene and protein. Cytokines 4, 96–116.

Sakaguchi, A.Y., Lalley, P.A., Choudhury, G.G., et al. (1989). Mouse melanoma growth stimulatory activity gene (Mgsa) is polymorphic and syntenic with the W, patch, rumpwhite, and recessive spotting loci on chromosome 5. Genomics 5, 629–632.

Saraya, K.A. and Balkwill, F.R. (1993). Temporal sequence and cellular origin of interleukin-2 stimulated cytokine gene expression. Br. J. Cancer. 67, 514–521.

Sawada, T., Koike, K., Kanda, Y., et al. (1994a). In vitro effects of CINC/gro, a member of the interleukin-8 family, on hormone secretion by rat anterior pituitary cells. Biochem. Biophys. Res. Commun. 202, 155–160.

Sawada, T., Koike, K., Sakamoto, Y., et al. (1994b). In vitro effects of CINC/gro, a member of the interleukin-8 family, on interleukin-6 secretion by rat posterior pituitary cells. Biochem. Biophys. Res. Commun. 200, 742–748.

Schroder, J.M., Persoon, N.L. and Christophers, E. (1990). Lipopolysaccharide-stimulated human monocytes secrete, apart from neutrophil-activating peptide l/interleukin 8, a second neutrophil-activating protein. NH2-terminal amino acid sequence identity with melanoma growth stimulatory activity. J. Exp. Med. 171, 1091–1100.

Schroder, J.M., Gregory, H., Young, J. and Christophers, E. (1992). Neutrophil-activating proteins in psoriasis. J. Invest. Dermatol. 98, 241–247.

Schumacher, C., Clark, L.I., Baggiolini, M. and Moser, B. (1992). High- and low-affinity binding of GRO alpha and neutrophil-activating peptide 2 to interleukin 8 receptors on human neutrophils. Proc. Natl Acad. Sci. USA 89, 10542–10546.

Schwartz, D., Andalibi, A., Chaverri-Almada, L., et al. (1994). Role of the GRO family of chemokines in monocyte adhesion to MM-LDL-stimulated endothelium. J. Clin. Invest. 94, 1968–1973.

Shattuck, R.L., Wood, L.D., Jaffe, G.J. and Richmond, A. (1994). MGSA/GRO transcription is differentially regulated in normal retinal pigment epithelial and melanoma cells. Mol. Cell. Biol. 14, 791–802.

Shiratori, Y., Takada, H., Hai, K., et al. (1994). Effect of anti-allergic agents on chemotaxis of neutrophils by stimulation of chemotactic factor released from hepatocytes exposed to ethanol. Dig. Dis. Sci. 39, 1569–1575.

Skerka, C., Irving, S.G., Bialonski, A. and Zipfel, P.F. (1993). Cell type specific expression of members of the IL-8/NAP-1 gene family. Cytokine 5, 112–116.

Stoeckle, M.Y. (1991). Post-transcriptional regulation of gro alpha, beta, gamma, and IL-8 mRNAs by IL-1 beta. Nucleic Acids Res. 19, 917–920.

Stoeckle, M.Y. (1992). Removal of a 3′ non-coding sequence is an initial step in degradation of gro alpha mRNA and is regulated by interleukin-1. Nucleic Acids Res. 20, 1123–1127.

Stoeckle, M.Y. and Guan, L. (1993). High-resolution analysis of gro alpha mRNA poly(A) shortening: regulation by interleukin-1 beta. Nucleic Acids Res. 21, 1613–1617.

Tekamp-Olson, P., Gallegos, C., Bauer, D., et al. (1990). Cloning and characterization of cDNAs for murine macrophage inflammatory protein 2 and its human homologues. J. Exp. Med. 172, 911–919.

Tettelbach, W., Nanney, L., Ellis, D., King, L. and Richmond, A. (1993). Localization of MGSA/GRO protein in cutaneous lesions. J. Cutan. Pathol. 20, 259–266.

Thomas, H.G. and Richmond, A. (1988a). High yield purification of melanoma growth stimulatory activity. Mol. Cell. Endocrinol. 57, 69–76.

Thomas, H.G. and Richmond, A. (1988b). Immunoaffinity purification of melanoma growth stimulatory activity. Arch. Biochem. Biophys. 260, 719–724.

Unemori, E.N., Amento, E.P., Bauer, E.A. and Horuk, R. (1993). Melanoma growth-stimulatory activity/GRO decreases collagen expression by human fibroblasts. Regulation by C-X-C but not C-C cytokines. J. Biol. Chem. 268, 1338–1342.

Walz, A. (1992). Generation and properties of neutrophil-activating peptide 2. Cytokines 4, 77–95.

Walz, A., Burgener, R., Car, B., Baggiolini, M., Kunkel, S.L. and Strieter, R.M. (1991a). Structure of neutrophil-activating properties of a novel inflammatory peptide (ENA-78) with homology to interleukin 8. J. Exp. Med. 174, 1355–1362.

Walz, A., Meloni, F., Clark, L.I., von, T.V. and Baggiolini, M. (1991b). [Ca^{2+}]$_i$ changes and respiratory burst in human neutrophils and monocytes induced by NAP-1/interleukin-8, NAP-2, and gro/MGSA. J. Leukocyte Biol. 50, 279–286.

Watanabe, K., Konishi, K., Fujioka, M., Kinoshita, S. and Nakagawa, H. (1989). The neutrophil chemoattractant produced by the rat kidney epithelioid cell line NRK-52E is a protein related to the KC/gro protein. J. Biol. Chem. 264, 19559–19563.

Watanabe, K., Koizumi, F., Kurashige, Y., Tsurufuji, S. and

Nakagawa, H. (1991). Rat CINC, a member of the interleukin-8 family, is a neutrophil-specific chemoattractant in vivo. Exp. Mol. Pathol. 55, 30–37.

Watanabe, K., Suematsu, M., Iida, M., et al. (1992). Effect of rat CINC/gro, a member of the interleukin-8 family, on leukocytes in microcirculation of the rat mesentery. Exp. Mol. Pathol. 56, 60–69.

Watanabe, K., Iida, M., Takaishi, K., et al. (1993). Chemoattractants for neutrophils in lipopolysaccharide-induced inflammatory exudate from rats are not interleukin-8 counterparts but gro-gene-product/melanoma-growth-stimulating-activity-related factors. Eur. J. Biochem. 214, 267–270.

Wen, D.Z., Rowland, A. and Derynck, R. (1989). Expression and secretion of gro/MGSA by stimulated human endothelial cells. EMBO J. 8, 1761–1766.

Widmer, U., Manogue, K.R., Cerami, A. and Sherry, B. (1993). Genomic cloning and promoter analysis of macrophage inflammatory protein (MIP)-2, MIP-1 alpha, and MIP-1 beta, members of the chemokine superfamily of proinflammatory cytokines. J. Immunol. 150, 4996–5012.

Wieder, K.J., Hancock, W.W., Schmidbauer, G., et al. (1993). Rapamycin treatment depresses intragraft expression of KC/MIP-2, granzyme B, and IFN-gamma in rat recipients of cardiac allografts. J. Immunol. 151, 1158–1166.

Witt, D.P. and Lander, A.D. (1994). Differential binding of chemokines to glycosaminoglycan subpopulations. Curr. Biol. 4, 394–400.

Wu, X., Wittwet, A.J., Carr, L.S., Crippes, B.A., DeLarco, J.E. and Lefkowith, J.B. (1994). Cytokine-induced neutrophil chemoattractant mediates neutrophil influx in immune complex glomerulonephritis in rat. J. Clin. Invest. 94, 337–344.

Zwahlen, R., Walz, A. and Rot, A. (1993). In vitro and in vivo activity and pathophysiology of human interleukin-8 and related peptides. Int. Rev. Exp. Pathol. 42.

Summary Tables

Cytokine	IL-1α	IL-1β
Location of gene (chromosome, etc.)	2q12–q21	2q13–q21
Gene size	10 kb	7 kb
Number of exons	7	7
Protein produced mature or proform (number of amino acids in proform)	Proform: 271	Proform: 269
Number of amino acids in mature form	159	153
Molecular mass (kDa)	17.5	17.3
Glycosylated/nonglycosylated (glycosylation sites)	No	No
pI	5	7
Number of S–S bonds and amino acid sites	None	None
Basic structure (monomer/dimer, α-helices, etc.)	Monomer Tetrahedral globular protein	Monomer Tetrahedral globular protein
Cell sources	1. Monocytes 2. Lymphocytes 3. Endothelia 4. Keratinocytes 5. Epithelia 6. Microglia 7. Chondrocytes	1. Monocytes 2. Lymphocytes 3. Endothelia 4. Keratinocytes 5. Epithelia 6. Microglia 7. Chondrocytes
Functions	1. Induction of other cytokines in many cell types 2. Hematopoiesis 3. Costimulates T cells 4. Activates endothelium 5. Neuroendocrine activity 6. Acute-phase response 7. Pyrogenic 8. Fibroblast proliferation	1. Induction of other cytokines in many cell types 2. Hematopoiesis 3. Costimulates T cells 4. Activates endothelium 5. Neuroendocrine activity 6. Acute-phase response 7. Pyrogenic 8. Fibroblast proliferation
Receptor structure (e.g. α-chain 30 kDa, β-chain 130 kD)	Type 1: 80 kDa Type 2: 60 kDa	Type 1: 80 kDa Type 2: 60 kDa
Associated pathological disorders	1. Inflammation 2. Sepsis 3. Diabetes 4. Autoimmune disease 5. CML 6. Ovarian carcinoma 7. Osteoporosis	1. Inflammation 2. Sepsis 3. Diabetes 4. Autoimmune disease 5. CML 6. Ovarian carcinoma 7. Osteoporosis
Therapeutic uses	None: blockage of IL-1 may be beneficial	None: blockage of IL-1 may be beneficial

IL-2	IL-3
4q26–28	5q23–31
5737 bp	~3 kb
4	5
Proform: 153	Proform: 152
133	133
13–17.5 kDa	15.1 kDa
Glycosylated: 1 *N*-linked at aa 3	Two glycosylation sites (aa 15–17, aa 70–72)
6.6–8.2	4–8
One: 58–105	One: 16–84

IL-2	IL-3
Monomer	4 helical bundles (helix A–D) + short
Some dimer through aggregation	helix A′ between A and B
1. Activated T cells	1. Activated T cells
2. Some T cell lines and hybridomas	2. NK cells (stim. A23 187)
	3. Mast cells
	4. WEHI-3 myelomonocytic leukemia cell line
	5. Human thymic epithelial cells

IL-2	IL-3
1. Stimulates proliferation of activated T cells	1. Supports proliferation and differentiation of pluripotent myeloid progenitor cells
2. Stimulates cytolytic activity of activated T cells	2. Prevention of apoptosis
3. Stimulates T cell motility	3. Induces MHC class II and LFA-1 in macrophages
4. Stimulates T cell differentiation	4. Histamine release in basophils
5. Stimulates thymocyte proliferation	5. LTC-4 generation in eosinophils and basophils
6. Stimulates B cell proliferation	
7. Stimulates secretion of immunoglobulin by B cells	
8. Stimulates proliferation of large granular lymphocytes	
9. Stimulates cytolytic activity of large granular lymphocytes	
10. Promotes cytolytic activity and possibly proliferation and differentiation of monocytes	
11. Enhances growth and proliferation of oligodendrocytes	
12. May enhance proliferation of fibroblasts and epithelial cells	

IL-2	IL-3
α chain: 55 kDa; β chain: 70–75 kDa; γ chain: 68 kDa	α-chain: IL-3Rα 41 kDa (expressed: 70 kDa)
	β-chain: β_c 96 kDa (expressed: 120 kDa)
1. X-linked SCID	1. Myelodysplastic syndrome (MDS)
2. Adult T cell leukemia/lymphoma	
3. ? Decline in immune response with age?	

IL-2	IL-3
1. Treatment of cancer	1. Support of hemopoiesis after high-dose chemotherapy
2. Treatment of infectious diseases, e.g., AIDs	2. Patients with bone marrow failure and MDS
3. Use in bone marrow stem cell transplantation	
4. Anti-IL-2 or IL-2R therapy used for immunosuppression to prevent allograft rejection and for treatment of autoimmune conditions, ATL, and other leukemias	

Cytokine	IL-4	IL-5
Location of gene (chromosome, etc.)	5q23.3–31.2	5q23–31
Gene size	~10 kb	6.5 kb
Number of exons	4	4
Protein produced mature or proform (number of amino acids in proform)	Proform: 153	Proform: 134
Number of amino acids in mature form	129	115
Molecular mass (kDa)	15–19	45
Glycosylated/nonglycosylated (glycosylation sites)	Yes. N-glycosylation sites unknown	O- and N-linked Thr-3 Asp-28
pI	10.5	?
Number of S–S bonds and amino acid sites	Three: 3–127; 23–65; 46–99	–
Basic structure (monomer/dimer, α-helices, etc.)	4α-helix bundle	Homodimer (antiparallel) S–S bond
Cell sources		44–86. Two 4-helix bundles
	1. T cells	1. T cells
	2. Mast cells	2. Mast cells
	3. Eosinophils	3. Eosinophils
	4. Basophils	
Functions	1. B cell proliferation	1. Eosinophil growth and differentiation
	2. B cell differentiation	2. Eosinophil chemotaxis
	3. Ig switching	3. Eosinophil activation
	4. T cell proliferation	4. Basophil activation
	5. Activation of monocytes	
	6. Mast cell proliferation	
Receptor structure (e.g. α-chain 30 kDa, β-chain 130 kD)	130 kDa α chain	α chain: 45.5 kD
	IL-2 receptor γ chain	β chain: 100 kDa
Associated pathological disorders	1. Asthma	1. Eosinophilia
	2. Dermatitis	2. Asthma
	3. Multiple sclerosis	3. Allergy
	4. Allergy	4. Dermatitis
	5. Inflammatory bowel disease	5. Biliary cirrhosis
Therapeutic uses	1. Antitumor agents	None: anti-IL-5 strategies in above diseases
	2. Immune stimulator in infection	

IL-6	IL-7
7 p15–p21	8q12–q13
5 kb	>33 kb
5	6
Proform: 212	Proform: 177
186	152
19–28	20–28
Variable	Yes: 3 potential N sites
~5	9
Two: at c_{44}–c_{50} and c_{73}–c_{83}	Six Cys residues but bonding pairs unknown
Monomer, 4 antiparallel α-helices	Unknown

IL-6	IL-7
1. Monocytes/macrophages	1. Bone marrow stromal cells
2. Fibroblasts	2. Fetal liver cells
3. T lymphocytes	3. Thymic stroma
4. Endothelial cells	4. Thymic epithelia
5. Epithelial cells and keratinocytes	5. Keratinocytes
6. Chondrocytes	
7. Astrocytes and microglia	
8. Bone marrow stroma	
9. Mast cells	
1. Stimulates B cell antibody production	1. Pre/pro-B cell proliferation
2. Stimulates T cell growth and CTL differentiation	2. Pre-B cell differentiation
3. Stimulates hematopoiesis (megakaryocytes)	3. T cell proliferation
4. Stimulates liver acute-phase proteins	4. NK cell activity
5. Numerous other *in vitro* actions	5. LAK cell activity
	6. Monocyte cytokine secretion
	7. Myelopoiesis with CSFs
80 kDa IL-6 α chain, 130 kDa gp130 β chain	α chain 130 kDa
	IL-2 γ chain 64 kDa
1. Multiple myeloma	1. ALL
2. Cardiac myxoma	2. Juvenile arthritis
3. Rheumatoid arthritis	3. T cell lymphoma
4. Various inflammatory conditions	4. B cell leukemia
5. SLE	5. Hodgkin disease
6. Bacterial infection	
7. Trauma	
1. Potential antitumor effects	1. Gene transfer in cancer as vaccine and generator of cytotoxic T cells
2. Potential stimulant of low platelet numbers	2. Lymphopoiesis post bone marrow transplantation
3. Antibodies to IL-6 have potential in multiple myeloma	

Cytokine	IL-8	IL-9
Location of gene (chromosome, etc.)	4q12–21	5q31–q35
Gene size	5.25 kb	~3.5 kb
Number of exons	4	5
Protein produced mature or proform (number of amino acids in proform)	Produced as 99 amino acid protein, including 22 amino acid signal sequence	Proform: 144
Number of amino acids in mature form	Most common forms: 72 aas and 77 aas	126
Molecular mass (kDa)	8.3 kDa (72 aa form)	~14 Glycosylated IL-9 ~25 kDa
Glycosylated/nonglycosylated (glycosylation sites)	Nonglycosylated	Four glycosylation sites
pI		~10
Number of S–S bonds and amino acid sites	Two: 7–34; 9–50	Ten cysteines in positions: 3, 27, 29, 36, 38, 46, 50, 84, 89, 93
Basic structure (monomer/dimer, α-helices, etc.)	Homodimeric in solution. Probably active as monomer	Monomer 4 α-helical bundles
Cell sources	1. Monocyte/macrophage 2. Lymphocyte 3. Endothelial cell 4. Epithelial cell 5. Fibroblast 6. Keratinocyte 7. Synovial cell 8. Mesangial cell 9. Smooth-muscle cell 10. Some tumor cells	1. T cells
Functions	1. Neutrophil chemotaxis/activation 2. Triggering of primed basophils 3. T cell chemotaxis 4. Keratinocyte mitogenesis/chemotaxis 5. Angiogenesis 6. Release of hematopoietic progenitors	1. Erythroid and myeloid growth and differentiation 2. T cell activation 3. B cell activation (immunoglobulin production) 4. Mast cell activation and differentiation (in mice) 5. Neuron differentiation (in mice)
Receptor structure (e.g. α-chain 30 kDa, β-chain 130 kD)	7-Transmembrane domain, G-protein coupled	α chain = IL-9R 64 kDa β chain = γ_c 64 kDa
Associated pathological disorders	1. Rheumatoid arthritis 2. Ulcerative colitis 3. Cystic fibrosis 4. Alcoholic hepatitis 5. Ischemia–reperfusion injury 6. Psoriasis	1. Hodgkin disease 2. HTLVI leukemias
Therapeutic uses	None to date	

IL-10	IL-11
1q	19q13.3–q13.4
4.7 kb	7 kb
5	5
178	Proform: 199
160	178
17.0	19
Nonglycosylated	None: no apparent glycosylation
1 potential site	
7.92	>11
Two: 12–108; 62–114	None
Monomer: 6 α-helices	4-helix bundle
1. Monocytes	1. Bone marrow stromal fibroblasts
2. T cells	2. Lung fibroblasts
3. B cells	3. Trophoblasts
4. Epithelial cells	4. Osteosarcoma cells
5. Keratinocytes (mouse)	5. Articular chondrocytes
6. Melanoma	6. Synoviocytes
7. Tumor cells (various origins)	7. Trabecular bone cells
1. Inhibition proinflammatory cytokine production by monocytes, granulocytes endothelial cells and mast cells	1. Synergistic stem and progenitor cell growth factor
2. Inhibition of IL-2 production by T cells	2. Synergistic growth factor for megakaryocyte progenitors
3. Inhibition antigen-specific T cell activation and cytokine production	3. Acute-phase protein inducer
4. Inhibition MHC class II and expression of costimulatory molecules on monocytes	4. Inhibitor of adipocyte differentiation
5. Inhibition of NO production by monocytes/macrophages	5. Inducer of neuronal differentiation
6. B cell costimulator for proliferation and immunoglobulin production	6. Stimulates osteoclast development *in vitro*
7. Mast cell growth factor	
8. Inhibition of osteogenic differentiation	
α chain: 90–120 kDa	(α-chain unknown at present)
β chain: unknown at present	β-chain 130 kDa
1. Systemic lupus erythematosus (SLE)	None known
2. Leishmaniasis	
3. Lymphatic filariasis	
4. Inflammatory bowel disease	
1. Anti-inflammatory and immunosuppressant	1. Acceleration of platelet recovery in myelosuppressed patients
2. Prevention of graft versus host disease	2. Protection against epithelial cell damage in inflammatory bowel disease
3. Induction and maintenance of transplantation tolerance	3. Treatment for septic shock
4. Autoimmune diseases	
5. Infectious diseases	
6. Cancer	

Cytokine	*IL-12*
Location of gene (chromosome, etc.)	The p40 subunit is 5 (5q31–q33) and p35, 3 (3p12–3q13.2)
Gene size	The p40 mRNA is ~2.4 kb, p35 mRNA is ~1.4 kb
Number of exons	Analysis of the human gene structure has not yet been published
Protein produced mature or proform (number of amino acids in proform)	For p40, 328 aa, including a 22-aa signal peptide. For p35, 219 aa, including a 22-aa signal peptide
Number of amino acids in mature form	p40 contains 306 amino acids, p35 contains 197 amino acids
Molecular mass (kDa)	The heterodimer protein backbone has a molecular mass of 57.2 kDa although the glycosylated protein is approximately 75 kDa as determined by SDS-PAGE
Glycosylated/nonglycosylated (glycosylation sites)	Four potential N-linked glycosylation sites on p40 and three putative N-linked sites on p35. At least one of the sites on p40 is glycosylated (Asn^{200}) whereas a second is not (Asn^{103})
pI	4.5–5.3
Number of S–S bonds and amino acid sites	p35 contains 3 intramolecular S–S bonds: Cys^{15}–Cys^{88}, Cys^{42}–Cys^{174}, and Cys^{63}–Cys^{101}. The intermolecular disulfide bond is between p35 Cys^{74} and p40 Cys^{177}. p40 contains 4 intramolecular S–S bonds: Cys^{28}–Cys^{68}, Cys^{109}–Cys^{120}, Cys^{148}–Cys^{171}, and Cys^{278}–Cys^{305}
Basic structure (monomer/dimer, α-helices, etc.)	Heterodimeric protein comprised of two disulfide-linked subunits with the p35 having an α-helix-rich structure
Cell sources	1. Monocytes and myeloid cell lines 2. B lymphocytes and Epstein–Barr virus-transformed B-lymphoma cell lines
Functions	1. Stimulates proliferation of activated T and NK cells 2. Enhances lytic ability of NK/LAK cells 3. Induces cytotoxic T lymphocyte responses to tumor cells 4. Increases IFN-γ production by T and NK cells 5. Promotes the development of T_H1 over T_H2 cells 6. Inhibits production of IgE 7. Synergizes with other hemopoietic factors to enhance hemopoietic stem cell proliferation of both the lymphoid and myeloid lineages 8. Regulates the expression of a number of cell surface markers, including adhesion molecules
Receptor structure (e.g. α-chain 30 kDa, β-chain 130 kD)	Two human IL-12 receptor components have been isolated by expression cloning. One protein is 662 amino acids while the second component is 862 amino acids. Both are glycoproteins and are members of the cytokine receptor superfamily of proteins
Associated pathological disorders	1. Endotoxin-induced shock 2. Granulomatous colitis 3. Autoimmune encephalomyelitis 4. Insulin-dependent diabetes mellitus 5. Rheumatoid arthritis
Therapeutic uses	1. Treatment of infectious diseases and opportunistic infections 2. Vaccine adjuvant 3. Antimetastatic and antitumor agent

IL-13	IL-14
5q23–31	Not known
Genomic 4.6 kb; mRNA 1.3 kb	Not known
4	Not known
Proform: 132	498
112	483
17 (glycosylated form)	60
Four	Three potential *N*-sites
Not known	6.7–7.8
Two	Unknown
Monomer. 3D structure not known	Monomer
1. Activated T cells	1. T cells
2. Mast cells	2. B-lymphoma
3. B cells	
1. B cell growth and differentiation factor	1. Proliferation of activated B cells
2. IgE switch factor	2. Inhibition of immunoglobulin secretion
3. Anti-inflammatory activity on monocytes/ macrophages	
4. Increases ICAM-1 on mast cells	
5. Enhances VCAM-1 expression and induces morphological changes on endochelial cells	
6. Stimulates chemotaxis	
CDw 124 (1L4 Rα 130 kDa)	90 kDa
β-chain; 60–70 kDa	
1. Nephrotic syndrome	1. Non-Hodgkin lymphoma
	2. B cell lymphoma
1. Anti-inflammatory agent	None
2. Antitumor activity	

Cytokine	IL-15	IL-16
Location of gene (chromosome, etc.)	4q31	15q26.1
Gene size	32 kb	Unknown
Number of exons	7	Unknown
Protein produced mature or proform (number of amino acids in proform)	Proform: 162 aa	Putative precursor
Number of amino acids in mature form	114 aa	130?
Molecular mass (kDa)	14–18 kDa	Monomer, 14–17; Homotetramer, 56 kDa?
Glycosylated/nonglycosylated (glycosylation sites)	Two: Asn 79, 112	One potential N-site
pI	?	9.1
Number of S–S bonds and amino acid sites	Two: Cys 35–85; 42–88	Unknown
Basic structure (monomer/dimer, α-helices, etc.)	4 α-helix bundle structure	Homotetramer
Cell sources	1. PBMCs 2. Monocytes 3. Placental tissue 4. Muscle tissue 5. Skeletal tissue	1. CD8$^+$ T cells 2. Eosinophils 3. Epithelial cells
Functions	1. Stimulation of activated T cells, B cells, NK cells 2. Stimulation of CTCC, TDAC 3. Induction of differentiation of CD34$^+$ to NK cells 4. Chemoattraction of T cells	1. Chemotaxis CD4$^+$ cells 2. Competence factor for CD4$^+$ T cells 3. Induces eosinophil adhesion
Receptor structure (e.g. α-chain 30 kDa, β-chain 130 kD)	IL-2 receptor β,γ chain IL-15 receptor α chain (237 aa)	CD4
Associated pathological disorders		1. Asthma 2. Sarcoidosis 3. Inflammation
Therapeutic uses	1. Immunotherapy cancer patients 2. In rheumatoid arthritis as a T cell attractant 3. Antitumor therapy	1. Anti-HIV

IL-17	IL-18
2q31–q35	?
? mRNA = 1.2 kb	? cDNA 1.1 kb
?	?
Proform: 155 amino acids	Proform: 193 amino acids
15 K (non-glycosylated) 20 K (glycosylated)	estimated 18.3 K? (157 amino acids)
1 *N* linked site at Asn 68	No
?	?
? (contains 6 cys residues)	Probably none (4 cys residues)
Homodimer (30–38 K)	Probably monomer
1. Activated CD4⁺ memory T cells	1. Monocyte lineage cells
2. Possibly activated CD8⁺ T cells	2. Macrophages and macrophage-like cells
3. ? others	3. Epithelial cells
1. Activates nFκB	1. Activates NK cells
2. Induces secretion of IL-6 by fibroblasts	2. Enhances IFNγ production by activated T cells
3. Induces secretion of IL-8 by fibroblasts	3. Enhances GMCSF production by activated T cells
4. Induces secretion of GMCSF by fibroblasts	4. Enhances IL-2 production by activated T cells
5. Induces secretion of prostaglandin E2 by fibroblasts	5. Inhibits IL-10 production by activated T cells
6. Induces expression of ICAM-1 by fibroblasts	
7. Costimulates activated T cell proliferation	
Receptor 128–132 kDa glycoprotein	?
?	1. Possibly inflammatory disorders
?	1. Possibly antitumor or anti-microbial

Cytokine	G-CSF
Location of gene (chromosome, etc.)	Murine: chromosome 11D–E2 Human: chromosome 17q11–22
Gene size	Encoding regions of human G-CSF genes ~2500 nucleotides, ~2.5 kb
Number of exons	Five
Protein produced mature or proform (number of amino acids in proform)	Proform
Number of amino acids in mature form	Murine: 178 (19.6 kDa) Human: 174 (18.6 kDa)
Molecular mass (kDa)	
Glycosylated/nonglycosylated (glycosylation sites)	O-glycosylation site Thr 133
pI	Low pH to prevent protein agglutination
Number of S–S bonds and amino acid sites	Two disulfide bridges between positions 36 and 42, and 64 and 74
Basic structure (monomer/dimer, α-helices, etc.)	Abundant helices and a small amount of sheet pleating. G-CSF is known as a long-chain-helical bundle structure. The four-helices are arranged in an antiparallel fashion
Cell sources	1. Stromal cells 2. Macrophages 3. Endothelial cells 4. Fibroblasts 5. Monocytes
Functions	1. Proliferation, differentiation, and activation of committed progenitor cells and functionally active mature neutrophils 2. Production and functional enhancement of neutrophils that occur in response to infection or other causes 3. Terminal differentiation of a murine leukemic cell line (WEHI-3B, and in WEHI-3B[D+]) 4. "Emergency" granulopoiesis during infection
Receptor structure (e.g. α-chain 30 kDa, β-chain 130 kD)	Single 150 kDa chain forming homodimer
Associated pathological disorders	1. Exposure to endotoxin (LPS) results in an increase in circulating G-CSF 2. Gram-negative and fungal infections 3. Neutropenia and correlated fever
Therapeutic uses	1. After bone marrow transplantation (BMT) 2. To treat neutropenias after chemotherapy treatment and from other causes 3. To increase patient's tolerance of high-dose chemotherapy 4. To accelerate neutrophil recovery and function 5. To improve tolerance to gancyclovir, thus allowing delivery of full doses. To help ameliorate the myeloesuppression of antiviral and anti-infective therapies 6. Either alone or in combination with chemotherapy, to help recruit peripheral blood progenitor cells (PBPC), and stem cells 7. Possible use in non-neutropenic infectious diseases

M-CSF	*GM-CSF*
1p13–21	X chromosome (Pseudoautosomal region)
20 kb	756 bp (cDNA); 3043 bp (genomic)
10 Proform: 554, 438, 256	4 Proform: 144
2×222	127
45–90 Two *N*- and several *O*-linked sites	14.5–34 Two *N*-linked sites several *O*-linked sites
3–5 Three: 7–90; 48–139; 102–146	4.2 Two S–S bonds: 54–96; 88–121
Homodimer. Interchain Cys^{31}–Cys^{31}. Four α-helix bundle	4 α-helices in up–up, down–down topology
1. Fibroblasts 2. Muscle cells 3. Endothelial cells 4. Stromal cells 5. T cells 6. Hepatocytes 7. Monocytes 8. Neutrophils	1. T lymphocytes 2. Macrophages 3. Endothelial cells 4. Fibroblasts 5. B lymphocytes
1. Monocyte/macrophage proliferation 2. Monocyte/macrophage differentiation 3. Monocyte/macrophage activation 4. Pregnancy?	1. Survival – inhibits apoptosis of targets 2. Proliferation 3. Differentiation into granulocytic and macrophage lineages 4. Functional activation: wide range of actions
150 kDa single chain	α subunit 84 kDa β subunit 120 kDa
1. Inflammation 2. Ovarian cancer 3. Lymphoid malignancies 4. Atherosclerosis?	1. Inflammatory mediator, e.g., rheumatoid synovium 2. Some eosinophilic states
1. Antitumor 2. Lowers cholesterol 3. Anti-infection 4. Myelosuppression	1. Mobilization of peripheral blood stem cells 2. Stimulation of myelopoiesis after chemotherapy or BMT 3. Possible use in infectious disease – e.g., fungal infection

Cytokine	LIF
Location of gene (chromosome, etc.)	22q12 (human); 11A1 (mouse)
Gene size	~6 kb
Number of exons	3
Protein produced mature or proform (number of amino acids in proform)	203-aa; 23-aa signal sequence
Number of amino acids in mature form	180
Molecular mass (kDa)	32–62
Glycosylated/nonglycosylated (glycosylation sites)	Heavily glycosylated. Potential N-linked glycosylation sites at position 32, 56, 86, 96, 119, 128, 139
pI	>9.0
Number of S–S bonds and amino acid sites	Three: 12–134; 18–131; 60–163
Basic structure (monomer/dimer, α-helices, etc.)	Four α-helical bundle. Monomer
Cell sources	1. Mast cells 2. Endometrium 3. LIF production is stimulated by LPS and pro-inflammatory cytokines
Functions	1. Pregnancy and blastocyst implantation 2. Monocyte/macrophage differentiation 3. Aids survival and proliferation of hemopoietic stem cells 4. Growth factor for embryonic stem cells *in vitro* 5. Enhances survival of neuronal cells and stimulates a switch in neurotransmitter phenotype 6. Enhances survival/proliferation of muscle cells 7. Enhances acute-phase protein synthesis by hepatocytes 8. Affects bone and cartilage resorption
Receptor structure (e.g. α-chain 30 kDa, β-chain 130 kD)	α chain: 200 kDa gp130: 130 kDa
Associated pathological disorders	1. Infection, inflammation 2. Septicemia, septic shock 3. Arthritis
Therapeutic uses	None yet possible; list indicates potential therapeutic uses 1. Elevation of platelet levels 2. Repair of bone, muscle and nerve tissue following injury 3. Gene therapy 4. Hemopoietic reconstitution

Steel Factor	flt3 Ligand
12q22–24	19q13.3
Unknown	6.15 kb
8	8
Proform: 273	Proform: 235 aa
248/220 membrane 164–165 soluble	209 aa
Monomer: 18; Homodimer: 36	30 kDa
Two N-sites: Asp-65; Asp-120	Two: Asn-100, Asn-123
Three O-sites: Ser-142; Thr-143; Thr-155	
5.0	8.17 (murine)
Two: 4–89; 43–138	Three: cysteines 4, 44, 85, 93, 127, 132
4-helical bundle	4-helix bundle type I transmembrane protein
Homodimer	
1. Stromal cells	1. B cell lineage cells
2. Fibroblasts	2. T cell lineage cells
3. Endothelial cells	3. Myeloid cells
4. Sertoli cells	
5. Brain	
6. Granulosa cells	
7. Kidney	
1. Germ-cell proliferation/differentiation	1. Stimulation of proliferation of hematopoietic precursor cells
2. Melanocyte migration/survival	2. Stimulation of proliferation of lymphohematopoietic progenitor cells
3. Hematopoiesis of progenitors	3. Stimulation of proliferation of B cell progenitors to B220+ cells
4. Erythropoiesis	4. Acts in synergy with IL-3, IL-6, IL-7, GM-CSF, Steel factor, TNF-α
5. Mast cell proliferation/activation	
6. Myelopoiesis	
c-kit protooncogene 145 kDa	Type III tyrosine kinase receptor (993 aa)
1. Malignancy	Overexpression may be involved in the maintenance/proliferation of malignant clones in cases of acute leukemia
2. Asthma	
3. Allergy	
4. Inflammation	
1. Immunoreconstitution post BMT	In combination with Steel factor, GM-CSF, and TNF-α for ex vivo generation of dendritic cells after immunotherapies and anticancer therapies

Cytokine	Thrombopoietin	Tumor Necrosis Factor
Location of gene (chromosome, etc.)	3q26–28	6p21
Gene size	6.2 kb	~3 kb
Number of exons	6	4
Protein produced mature or proform (number of amino acids in proform)	Proform: 353	Proform: 233
Number of amino acids in mature form	332	157
Molecular mass (kDa)	35.5 (predicted); 60 (expressed)	174 (mature)
Glycosylated/nonglycosylated (glycosylation sites)	Six: Asn-176, 185, 213, 234, 319, 327	Nonglycosylated. No potential N-glycosylation sites
pI		5.6
Number of S–S bonds and amino acid sites	Two: 7–151; 29–85	One: 145–177
Basic structure (monomer/dimer, α-helices, etc.)	Two domains. Aminoterminal: EPO-domain Carboxyterminus: carbohydrate domain	Homotrimer. Antiparallel β-pleated sheet sandwich
Cell sources	1. Tissue: liver 2. Kidney 3. Smooth muscle 4. Endothelial cells	1. Macrophages 2. T cells 3. Many other cells
Functions	1. Stimulation of proliferation and differentiation of megakaryocyte progenitor cells 2. Production of mature platelets	1. Cytotoxic for tumor cells 2. Antiviral activity; Antibacterial activity; Antiparasitic activity 3. Growth stimulation 4. Immune modulation 5. Pro-inflammatory
Receptor structure (e.g. α-chain 30 kDa, β-chain 130 kD)	Hematopoietin receptor family (68 kDa) contains two extracellular segments each with a cysteine-rich region, a homodimerization domain, and a WSXWS box	55 kDa (TNF-R55; p55) 75 kDa (TNF-R75; p75)
Associated pathological disorders	Thrombocytopenia	1. Cachexia 2. Cerebral malaria 3. Multiple sclerosis 4. Rheumatoid arthritis 5. Crohn's disease
Therapeutic uses	1. Recombinant TPO in thrombocytopenia following bone marrow failures 2. Reduction of platelet recovery time	Cancer

Lymphotoxin α	Lymphotoxin β	IFN-α, -β, -ω
6p21	6p21	9 p21–pter
~3 kb	~2 kb	1.1–1.3 kb
4	4	1
Proform: 205	244	Proform: 189
171	171	166
25 (mature)	33	α, 18.5; β, 23 (glycosylated form)
Glycosylated N96, (T41)	Potential N222	HvIFN-α subtype, mainly nonglycosylated
		HvIFN-β and HuIFN-ω-N-linked glycosylated
5.8	5.25	?
None	None	Two: 1–99; 29–139
Homotrimer. Antiparallel β-pleated sheet sandwich	Heterotrimer LT-α 1β2 01LT-α2β2	Monomer/5 α-helices
1. T cells	1. T cells	1. Leukocytes (α and ω)
2. B cells	2. B cells	2. Fibroblasts (β)
		3. Epithelial cells (β)
		4. Endothelial cells (β)
1. Lymphoid organogenesis	1. Lymphoid organogenesis	1. Antiviral actions
2. Immune response regulation	2. Immune response regulation	2. Antiproliferative action/ Antitumor
3. Cytotoxic for tumor cells	3. Cytotoxic for some tumor lines	3. Immunoregulatory actions e.g. NK cell enhanced cytotoxicity
4. Antiviral		4. MHC antigen, upregulation of expression
		5. Antimicrobial actions
55 kDa (TNF-R55; p55)	61 kDa	α chain: 110–130 kDa
75 kDa (TNF-R75; p75)		β chain: 95 kDa
		1. Various autoimmune disorders, e.g., SLE
		2. AIDS
		3. Graft versus host disease
		1. Certain cancers
		2. Hepatitis B, C
		3. Multiple sclerosis (β only)
		4. Genital warts
		5. AIDS-related Kaposi sarcoma
		6. Hematological disorders, e.g., polycythemia vera

Cytokine	IFN-γ	Oncostatin M
Location of gene (chromosome, etc.)	12p12.05 qter	22q12 (human), 11 (mouse)
Gene size	4.5 kb	5 kb
Number of exons	4	3
Protein produced mature or proform (number of amino acids in proform)	Proform: 166	Proform: 227
Number of amino acids in mature form	143	196
Molecular mass (kDa)	17.147 kDa (monomer)	~28
Glycosylated/nonglycosylated (glycosylation sites)	Two sites: Asn-X-Ser/Thr; residues 25→27 97→99	Two N-linked sites, one glycosylated, plus O linked glycosylation
pI	–	>9.5
Number of S–S bonds and amino acid sites	None	Two S–S bonds; 5 cysteines (human). Two S–S (mouse and bovine)
Basic structure (monomer/dimer, α-helices, etc.)	Dimer	Monomer, 4 α-helical bundles
Cell sources	1. Monocytes-macrophages 2. Dendritic cells 3. T cells 4. NK cells	1. Activated T cells (human) 2. Monocytes, macrophages (human) 3. Kaposi cells (human) 4. Bone marrow (mouse) 5. Spleen (mouse)
Functions	1. MHC class II expression 2. Macrophage activation 3. NK cell activation 4. T cell activation 5. Immunoglobulin isotype regulation 6. Antiviral, antibacterial and antiparasite host defence	1. Modulation of cell growth 2. Regulation of inflammatory proteins 3. Hematopoietic factor 4. T cell development factor 5. Regulation of cholesterol metabolism 6. Modulation of bone metabolism 7. Neurotrophic factor 8. Differentiation factor
Receptor structure (e.g. α-chain 30 kDa, β-chain 130 kD)	α-chain: 54 kDa unglycosylated 90 kDa glycosylated β-chain: 35 kDa unglycosylated	α-chain (not published), β-chain 130 kDa
Associated pathological disorders	1. Autoimmunity	1. Kaposi's sarcoma 2. Myeloma
Therapeutic uses	1. Infection with *Leishmania* and *Toxoplasma*	1. Inhibition of inflammatory diseases 2. Modulation of drug induced damage 3. Treatment of breast cancer 4. Treatment of cytopenias

TGFβ1	*RANTES*
19q13.2	17q11.2–12
Unknown (Genbank accession no. X02812 mRNA)	~7 kb
8	3
Proform: 390	Proform: 91
112 (numbers 279–390)	68
Monomer: 12.5 (active as dimer 25 kDa)	7.847
Glycosylated in propeptide aa numbers 82, 136, 176	O-linked at serines 4 and 5
Mature form not glycosylated	
—	9.5
Five: 7–16, 15–78, 44–109, 48–111, 77 interchain	Two: 10–34, 11–51
Homodimer: each subunit has 2 antiparallel β-sheets and 3 α-helices leading to unusual nonglobular fold and cysteine knot	Probably dimer
1. Most cells (secreted as latent complex)	1. T cells
2. Platelets	2. Monocytes
3. Fibroblasts	3. Natural killer cells
4. Placenta	4. Fibroblasts
5. Olfactory and other epithelia	5. Epithelial cells
6. Hematopoietic and immune system	6. Endothelial cells
7. Glioma, other tumors	7. Megakaryocytes
	8. Basophils
	9. Fetal kidney
1. Tissue remodeling	1. T cell chemotaxis and proliferation
2. Wound repair	2. Monocyte chemotaxis
3. Regulation of cell growth	3. Basophil chemotaxis and activation
4. Embryonic development	4. Eosinophil chemotaxis and activation
5. Differentiation commitment	5. Natural killer cell chemotaxis
6. Regulates extracellular matrix	
7. Immune suppression and inflammation	
8. Tumor suppression	
Heterodimers; type II binds ligand, recruits, and phosphorylates type I, which phosphorylates and activates signaling pathways	Seven-membrane-spanning CCR1, CCR3, CCR4, CCR5
1. Fibrosis	1. Transplant rejection
2. Immunosuppression	2. Rheumatoid arthritis
3. Tumor development	3. Sarcoidosis
4. Autoimmunity	4. Endometriosis
5. Inflammatory diseases	5. Glomerulonephritis
6. Angiogenesis	6. Allergy
7. Myeloproliferation	7. Atherosclerosis
1. Bone marrow protective agent*	1. Suppression of HIV replication
2. Bone remodeling*	
3. Tumor suppression*	
4. Fibrosis**	
5. Block cardiac pathology*	
6. Immune activator**	
*: agonist; **: antagonist	

Cytokine	ENA-78	NAP-2
Location of gene (chromosome, etc.)	4q13–21	Chrom. 4q12–21
Gene size	2.5 kb	1.139 kb
Number of exons	4	3
Protein produced mature or proform (number of amino acids in proform)	Proform: 114	Proform: 128
Number of amino acids in mature form	78	70
Molecular mass (kDa)	8.357	7.628
Glycosylated/nonglycosylated (glycosylation sites)	Nonglycosylated	Nonglycosylated
pI	8.73	8.7
Number of S–S bonds and amino acid sites	Two: 13–39; 15–55	Two: 5–31; 7–47
Basic structure (monomer/dimer, α-helices, etc.)	Similar to IL-8	Monomer
Cell sources	1. Epithelial cells 2. Fibroblasts 3. Endothelial cells 4. Monocytes	1. Platelets
Functions	1. *Neutrophil activation*: chemotaxis, enzyme release, calcium flux, respiratory burst 2. *Endothelial cells*: proliferation, chemotaxis	1. *Neutrophil activation*: chemotaxis, enzyme release, calcium flux, respiratory burst 2. *Endothelial cells*: proliferation, chemotaxis
Receptor structure (e.g. α-chain 30 kDa, β-chain 130 kD)	Acts via IL-8R2	Acts via IL-8R2
Associated pathological disorders	1. Rheumatoid arthritis 2. Ischemia–reperfusion injury 3. ARDS 4. Pneumonic pasteurellosis	1. Heparin-associated thrombocytopenia
Therapeutic uses		

Macrophage Inflammatory Protein 1-α	MCP-1	MCP-2	MCP-3
17q11–21	17q11.2–12		17q11.2–12
Human: 2.057 kb; Mouse: 1.921 kb	2.5 kb	2.5 kb	2.5 kb
Two, 688 and 420 nucleotides	3	?	3
92 amino acids prior to secretion	Proform: 99	Proform: 77	Proform: 99
70	76	76	76
7.787.6	9–17	7.5	11
No N-linked sites; a putative O-linked site, non-glycosylated	No N-sites; O-sites	No N-sites; No O-sites	Unknown
4.6	10.6	10.8	10.9
Two: 11–35; 12–51	2: 11–63; 12–52	2: 11–36; 12–52	2: 11–36; 12–52

Dimer, though probably active as monomer, aggregated states observed; see text

Monomer or dimer depending on conditions

Inducible in macrophages, PMNs, T cells, eosinophils, mast cells, basophils, platelets, constitutive in Langerhans cells

	MCP-1	MCP-2	MCP-3
	1. Monocytes	Osteosarcoma	Osteosarcoma
	2. Fibroblasts	Monocytes	Monocytes
	3. Epithelial cells	Fibroblasts	Astrocytes
	4. Mesothelial cells		
	5. Keratinocytes		
	6. Melanocytes		
	7. Lipocytes		

Modulation of macrophage function, activation of basophils or mast cells, recruitment of eosinophils, recruitment of lymphocytes, modulation of T cell adherence. Inhibition of stem cell proliferation, stimulation of mature progenitor proliferation. Inhibition of keratinocyte proliferation. For other activities, see text

1. Monocyte chemotaxis
2. Monocyte activation
3. Activates basophils
4. T cell chemotaxis
5. NK cell cytolysis

Single chain, seven helical membrane-spanning domains, coupled to heterotrimeric G-protein

Seven spanning receptors

	MCP-1	MCP-2	MCP-3
	CCR-1	CCR-2B	CCR-1
	CCR-2B		CCR-2B
	CCR-4		CCR-3

Inflammatory lung disease, autoimmune encephalitis, rheumatoid arthritis, collagen-induced arthritis, bacterial meningitis, cutaneous inflammation, cutaneous leishmaniasis. For other associations, see text

1. Rheumatoid arthritis
2. Osteoarthritis
3. Asthma
4. Pulmonary fibrosis
5. Malignancy
6. Psoriasis
7. Inflammation

Protein or antagonists in rheumatoid arthritis, multiple sclerosis. Receptor antagonists in AIDS, hemoprotectant in cancer therapy

None

Cytokine	GROα	GROβ	GROγ
Location of gene (chromosome, etc.)	4q12-21		
Gene size	Not fully determined		
Number of exons	4	4	2
Protein produced mature or proform (number of amino acids in proform)	Proform: 107	Proform: 107	Proform: 106
Number of amino acids in mature form	73	73	73
Molecular mass (kDa)	7.894	ND	ND
Glycosylated/nonglycosylated (glycosylation sites)	Not reported		
pI	–	–	–
Number of S–S bonds and amino acid sites	Two	Two	Two
Basic structure (monomer/dimer, α-helices, etc.)	Dimer C terminal; α-helix	ND	ND
Cell sources	1. Monocytes 2. Neutrophils 3. Endothelial cells 4. Epithelial cells 5. Keratinocytes 6. Fibroblasts 7. Astrocytes 8. T cells		
Functions	1. Neutrophil chemotaxis 2. Neutrophil adherence 3. IL-6 secretion by pituitary cells 4. Prolactin secretion		
Receptor structure (e.g. α-chain 30 kDa, β-chain 130 kD)	IL-8R1 IL-8R2 (see Chapter 8) DARC		
Associated pathological disorders	1. Melanoma 2. Renal carcinoma 3. Bladder carcinoma 4. Inflammation 5. Allergy 6. Psoriasis 7. Crohn's Disease		
Therapeutic uses	None as yet		

Glossary

Notes: This glossary is up to date for the current volume only and will be supplemented with each subsequent volume.

α_1, α_2 **receptors** Adrenoceptor subtypes
α_1-**ACT** α_1-Antichymotrypsin
α_1-**AP** α_1-antiproteinase *also known as* α_1-antitrypsin and α_1-proteinase inhibitor
α_1-**AT** α_1-Antitrypsin inhibitor *also known as* α_1-antiproteinase and α_1-proteinase inhibitor
α_1-**PI** α_1-Proteinase inhibitor *also known as* α_1-antitrypsin and α_1-antiproteinase
α_2-**AP** α_2-antiplasmin
α_2-**M** α_2-macroglobulin
A Absorbance
AI, AII Angiotensin I, II
Å Angstrom
AA Arachidonic acid
aa Amino acids
AAb Autoantibody
ABAP 2′,2′-azobis-2-amidino propane
Ab Antibody
Ab1 Idiotype antibody
Ab2 Anti-idiotype antibody
Ab2α Anti-idiotype antibody which binds outside the antigen binding region
Ab2β Anti-idiotype antibody which binds to the antigen binding region
Ab3 Anti-anti-idiotype antibody
Abcc Antibody dependent cellular cytotoxicity
ABA-L-GAT Arsanilic acid conjugated with the synthetic polypeptide L-GAT
AC Adenylate cyclase
ACAT Acyl-co-enzyme-A acyltransferase
ACAID Anterior chamber-associated immune deviation
ACE Angiotensin-converting enzyme
ACh Acetylcholine
ACTH Adrenocorticotrophin hormone
ADCC Antibody-dependent cellular cytotoxicity
ADH Alcohol dehydrogenase
Ado Adenosine
ADP Adenosine diphosphate

ADPRT Adenosine diphosphate ribosyl transferase
AES Anti-eosinophil serum
Ag Antigen
AGE Advanced Glycosylation end-product
AGEPC 1-*O*-alkyl-2-acetyl-*sn*-glyceryl-3-phosphocholine; *also known as* PAF and APRL
AH Acetylhydrolase
AHP after-hyperpolarization
AID Autoimmune disease
AIDS Acquired immune deficiency syndrome
A/J A Jackson inbred mouse strain
ALI Acute lung injury
ALP Anti-leukoprotease
ALS Amyotrophic lateral sclerosis
cAMP Cyclic adenosine monophosphate *also known as* adenosine 3′,5′-phosphate
AM Alveolar macrophage
AML Acute myelogenous leukemia
AMP Adenosine monophosphate
AMVN 2,2′-azobis (2,4-dimethylvaleronitrile)
ANAb Anti-nuclear antibodies
ANCA Anti-neutrophil cytoplasmic auto antibodies
cANCA Cytoplasmic ANCA
pANCA Perinuclear ANCA
AND Anaphylactic degranulation
ANF Atrial natriuretic factor
ANP Atrial natriuretic peptide
Anti-I-A, Anti-I-E Antibody against class II MHC molecule encoded by I-A locus, I-E locus
anti-Ig Antibody against an immunoglobulin
anti-RTE Anti-tubular epithelium
AP-1 Activator protein-1
APA B-azaprostanoic acid
APAS Antiplatelet antiserum
APC Antigen-presenting cell
APD Action potential duration
apo-B Apolipoprotein B
APP Acute-phase protein
APRE Acute-phase response element
APRL Anti-hypertensive polar renal lipid *also known as* PAF

APTT Activated partial thromboplastin times
APUD Amine precursor uptake and decarboxylation
AR Aldose reductase
AR-CGD Autosomal recessive form of chronic granulomatous disease
ARDS Adult respiratory distress syndrome
AS Ankylosing spondylitis
ASA Acetylsalicylic acid *also known as* aspirin
4-ASA, 5-ASA 4-, 5-aminosalicylic acid
ATIII antithrombin III
ATHERO-ELAM A monocyte adhesion molecule
ATL Adult T cell leukaemia
ATP Adenosine triphosphate
ATPase Adenosine triphosphatase
ATP*ys* Adenosine 3′ thiotriphosphate
AITP Autoimmune thrombocytopenic purpura
AUC Area under curve
AVP Arginine vasopressin

β_1, β_2 **receptors** Adrenoceptor subtypes
β_2 **(CD18)** A leukocyte integrin
β_2**M** β_2-Microglobulin
β-TG β-Thromboglobulin
B$_7$/BB$_1$ *Known to be* expressed on B cell blasts and immunostimulatory dendritic cells
BAF Basophil-activating factor
BAL Bronchoalveolar lavage
BALF Bronchoalveolar lavage fluid
BALT Bronchus-associated lymphoid tissue
B cell Bone marrow-derived lymphocyte
BCF Basophil chemotactic factor
B-CFC Basophil colony-forming cell
BCG Bacillus Calmette-Guérin
BCNU 1,3-bis (2-chloroethyl)-1-nitrosourea
bFGF Basic fibroblast growth factor
BFU Burst-forming unit
Bg Birbeck granules
BHR Bronchial hyperresponsiveness

BHT Butylated hydroxytoluene
b.i.d. *Bis in die* (twice a day)
BK Bradykinin
BK$_1$, BK$_2$ receptors Bradykinin receptor subtypes *also known as* B$_1$ and B$_2$ receptors
B1-CFC Blast colony-forming cells
B-lymphocyte Bursa-derived lymphocyte
BM Bone marrow
BMCMC Bone marrow cultured mast cell
BMMC Bone marrow mast cell
BOC-FMLP Butoxycarbonyl-FMLP
BAEC Bovine aortic endothelial cells
bp Base pair
BPAEC Bovine pulmonary artery endothelial cells
BPB Para-bromophenacyl bromide
BPI Bacterial permeability-increasing protein
BSA Bovine serum albumin
BSS Bernard-Soulier Syndrome
B-TCGF B cell-derived T cell growth factor

^{51}**Cr** Chromium^{51}
C1, C2 . . . C9 The 9 main components of complement
C1 inhibitor A serine protease inhibitor which inactivates C1r/C1s
C1q Complement fragment 1q
C1qR Receptor for C1w; facilitates attachment of immune complexes to mononuclear leucocytes and endothelium
C3a Complement fragment 3a (anaphylatoxin)
C3a$_{72-77}$ A synthetic carboxyterminal peptide C3a analog
C3aR Receptor for anaphylatoxins, C3a, C4a, C5a
C3b Complement fragment 3b (anaphylatoxin)
C3bi Inactivated form of C3b fragment of complement
C4b Complement fragment 4b (anaphylatoxin)
C4BP C4 binding protein; plasma protein which acts as co-factor to factor I inactivate C3 convertase
C5a Complement fragment 5a (anaphylatoxin)
C5aR Receptor for anaphylatoxins C3a, C4a and C5a
C5b Complement fragment 5b (anaphylatoxin)
C$_\epsilon$2, C$_\epsilon$3, C$_\epsilon$4 Heavy chain of immunoglobulin E: domains 2, 3 and 4
Ca *The chemical symbol for* calcium
[Ca^{2+}]$_i$ Intracellular free calcium concentration
CAH Chronic active hepatitis
CALLA Common lymphoblastic leukemia antigen

CALT Conjunctival associated lymphoid tissue
CaM Calmodulin
CAM Cell adhesion molecule
cAMP Cyclic adenosine monophosphate *also known as* adenosine 3′,5′-phosphate
CaM-PDE Ca^{2+}/CaM-dependent PDE
CAP57 Cationic protein from neutrophils
CAT Catalase
CatG Cathepsin G
CB Cytochalasin B
CBH Cutaneous basophil hypersensitivity
CBP Cromolyn-binding protein
CCK Cholecystokinin
CCR Creatinine clearance rate
CD Cluster of differentiation (a system of nomenclature for surface molecules on cells of the immune system); cluster determinant
CD1 Cluster of differentiation 1 *also known as* MHC class I-like surface glycoprotein
CD1a Isoform a *also known as* non-classical MHC class I-like surface antigen; present on thymocytes and dendritic cells
CD1b *Known to be* present on thymocytes and dendritic cells
CD1c Isoform c *also known as* non-classical MHC class I-like surface antigen; present on thymocytes
CD2 Defines T cells involved in antigen non-specific cell activation
CD3 *Also known as* T cell receptor-associated surface glycoprotein on T cells
CD4 Defines MHC class II-restricted T cell subsets
CD5 *Known to be* present on T cells and a subset of B cells; *also known as* Lyt 1 in mouse
CD7 Cluster of differentiation 7; present on most T cells and NK cells
CD8 Defines MHC class I-restricted T cell subset; present on NK cells
CD10 *Known to be* common acute leukemia antigen
CD11a *Known to be* an α chain of LFA-1 (leucocyte function antigen-1) present on several types of leukocyte and which mediates adhesion
CD11c *Known to be* a complement receptor 4 α chain
CD13 Aminopeptidase N; present on myeloid cells
CD14 *Known to be* a lipid-anchored glycoprotein; present on monocytes
CD15 *Known to be* Lewis X, fucosyl-*N*-acetyllactosamine
CD16 *Known to be* Fcγ receptor III
CD16-1, CD16-2 Isoforms of CD16

CD19 Recognizes B cells and follicular dendritic cells
CD20 *Known to be* a pan B cell
CD21 C3d receptor
CD23 Low affinity FcεR
CD25 Low affinity receptor for interleukin-2
CD27 Present on T cells and plasma cells
CD28 Present on resting and activated T cells and plasma cells
CD30 Present on activated B and T cells
CD31 *Known to be* on platelets, monocytes, macrophages, granulocytes, B-cells and endothelial cells; *also known as* PECAM
CD32 Fcγ receptor II
CD33$^+$ *Known to be* a monocyte and stem cell marker
CD34 *Known to be* a stem cell marker
CD35 C3b receptor
CD36 *Known to be* a macrophage thrombospondin receptor
CD40 Present on B cells and follicular dendritic cells
CD41 *Known to be* a platelet glycoprotein
CD44 *Known to be* a leukocyte adhesion molecule; *also known as* hyaluronic acid cell adhesion molecule (H-CAM), Hermes antigen, extracellular matrix receptor III (ECMIII); present on polymorphonuclear leukocytes
CD45 *Known to be* a pan leukocyte marker
CD45RO *Known to be* the isoform of leukosialin present on memory T cells
CD46 *Known to be* a membrane cofactor protein
CD49 Cluster of differentiation 49
CD51 *Known to be* vitronectin receptor alpha chain
CD54 *Known to be* Intercellular adhesion molecule-1 *also known as* ICAM-1
CD57 Present on T cells and NK subsets
CD58 A leukocyte function-associated antigen-3, *also known to be* a member of the β-2 integrin family of cell adhesion molecules
CD59 *Known to be* a low molecular weight HRf present to many hematopoietic and non-hematopoietic cells
CD62 *Known to be present on* activated platelets and endothelial cells; *also known as* P-selectin
CD64 *Known to be* Fcγ receptor I
CD65 *Known to be* fucoganglioside
CD68 Present on macrophages
CD69 *Known to be* an activation inducer molecule; present on activated lymphocytes

CD72 Present on B-lineage cells

CD74 An invariant chain of class II B cells

CDC Complement-dependent cytotoxicity

cDNA Complementary DNA

CDP Choline diphosphate

CDR Complementary-determining region

CD$_{xx}$ Common determinant *xx*

CEA Carcinoembryonic antigen

CETAF Corneal epithelial T cell activating factor

CF Cystic fibrosis

Cf Cationized ferritin

CFA Complete Freund's adjuvant

CFC Colony-forming cell

CFU Colony-forming unit

CFU-Mk Megakaryocyte progenitors

CFU-S Colony-forming unit, spleen

CGD Chronic granulomatous disease

CHAPS 3-[(3-cholamidopropyl)-dimethylammonio]-1-propane sulphonate

cGMP Cyclic guanosine monophosphate *also known as* guanosine 3′,5′-phosphate

CGRP Calcitonin gene-related peptide

CH Cycloheximide

CH2 Hinge region of human immunoglobulin

CHO Chinese hamster ovary

CI Chemical ionization

CIBD Chronic inflammatory bowel disease

CK Creatine phosphokinase

CKMB The myocardial-specific isoenzyme of creatine phosphokinase

Cl *The chemical symbol for* chlorine

CL Chemiluminescent

CLA Cutaneous lymphocyte antigen

CL18/6 Anti-ICAM-1 monoclonal antibody

CLC Charcot–Leyden crystal

CMC Critical micellar concentration

CMI Cell mediated immunity

CML Chronic myeloid leukemia

CMV Cytomegalovirus

CNTF Ciliary neurotrophic factor ["h" for British]

CNS Central nervous system

CO Cyclooxygenase

CoA Coenzyme A

CoA-IT Coenzyme A – independent transacylase

Con A Concanavalin A

COPD Chronic obstructive pulmonary disease

COS Fibroblast-like kidney cell line established from simian cells

CoVF Cobra venom

CP Creatine phosphate

Cp Caeruloplasmin

c.p.m. Counts per minute

CPJ Cartilage/pannus junction

Cr *The chemical symbol for* chromium

CR Complement receptor

CR1, CR2 & CR4 Complement receptor types 1, 2 and 4

CR3-α Complement receptor type 3-α

CRE cAMP-responsive element

CRF Corticotropin-releasing factor

CRH Corticotropin-releasing hormone

CRI Cross-reactive idiotype

CRP C-reactive protein

CSA Cyclosporin A

CSF Colony-stimulating factor

CSS Churg–Strauss syndrome

CT Computed tomography

CTAP-III Connective tissue-activating peptide

CTD Connective tissue diseases

C terminus Carboxy terminus of peptide

CThp Cytotoxic T lymphocyte precursors

CTL Cytotoxic T lymphocyte

CTLA-4 *Known to be* co-expressed with CD20 on activated T cells

CTMC Connective tissue mast cell

CVF Cobra venom factor

2D Second derivative

Da Dalton (the unit of relative molecular mass)

DAF Decay-accelerating factor

DAG Diacylglycerol

DAO Diamine oxidase

D-Arg D-Arginine

DArg-[Hyp^3,DPhe^7]-BK A bradykinin B$_2$ receptor antagonist. Peptide derivative of bradykinin

DArg-[Hyp^3,Thi^5,DTic^7,Tic^8]-BK A bradykinin B$_2$ receptor antagonist. Peptide derivative of bradykinin

DBNBS 3,5-dibromo-4-nitroso-benzenesulphonate

DC Dendritic cell

DCF Oxidized DCFH

DCFH 2′,7′-dichlorofluorescin

DEC Diethylcarbamazine

DEM Diethylmaleate

desArg^9-BK Carboxypeptidase N product of bradykinin

desArg^{10}KD Carboxypeptidase N product of kallidin

DETAPAC Diethylenetriaminepentaacetic acid

Dex Dexamethasone

DFMO α1-Difluoromethyl ornithine

DFP Diisopropyl fluorophosphate

DFX Desferrioxamine

DGLA Dihomo-γ-linolenic acid

DH Delayed hypersensitivity

DHA Docosahexaenoic acid

DHBA Dihydroxybenzoic acid

DHR Delayed hypersensitivity reaction

DIC Disseminated intravascular coagulation

DL-CFU Dendritic cell/Langerhans cell colony forming

DLE Discoid lupus erythematosus

DMARD Disease-modifying anti-rheumatic drug

DMF *N,N*-dimethylformamide

DMPO 5,5-dimethyl-l-pyrroline *N*-oxide

DMSO Dimethyl sulfoxide

DNA Deoxyribonucleic acid

D-NAME D-Nitroarginine methyl ester

DNase Deoxyribonuclease

DNCB Dinitrochlorobenzene

DNP Dinitrophenol

Dpt4 *Dermatophagoides pteronyssinus* allergen 4

DGW2, DR3, DR7 HLA phenotypes

DREG-56 (Antigen) L-selectin

DREG-200 A monoclonal antibody against L-selectin

ds Double-stranded

DSCG Disodium cromoglycate

DST Donor-specific transfusion

DTH Delayed-type hypersensitivity

DTPA Diethylenetriamine pentaacetate

DTT Dithiothreitol

dv/dt Rate of change of voltage within time

ε Molar absorption coefficient

EA Egg albumin

EACA Epsilon-amino-caproic acid

EAE Experimental autoimmune encephalomyelitis

EAF Eosinophil-activating factor

EAR Early phase asthmatic reaction

EAT Experimental autoimmune thyroiditis

EBV Epstein–Barr virus

EC Endothelial cell

ECD Electron capture detector

ECE Endothelin-converting enzyme

E-CEF Eosinophil cytotoxicity enhancing factor

ECF-A Eosinophil chemotactic factor of anaphylaxis

ECG Electrocardiogram

ECGF Endothelial cell growth factor

ECGS Endothelial cell growth supplement

E. coli *Escherichia coli*

ECP Eosinophil cationic protein

EC-SOD Extracellular superoxide dismutase

EC-SOD C Extracellular superoxide dismutase C

ED$_{35}$ Effective dose producing 35% maximum response

ED$_{50}$ Effective dose producing 50% maximum response

EDF Eosinophil differentiation factor
EDL Extensor digitorum longus
EDN Eosinophil-derived neurotoxin
EDRF Endothelium-derived relaxing factor
EDTA Ethylenediamine tetraacetic acid *also known as* etidronic acid
EE Eosinophilic eosinophils
EEG Electroencephalogram
EET Epoxyeicosatrienoic acid
EFA Essential fatty acid
EFS Electrical field stimulation
EG1 Monoclonal antibody specific for the cleaved form of eosinophil cationic peptide
EGF Epidermal growth factor
EGTA Ethylene glycol-bis(β-aminoethyl ether) N,N,N',N'-tetraacetic acid
EHNA Erythro-9-(2-hydroxy-3-nonyl)-adenine
EHV2 Equine herpes virus type 2
EI Electron impact
EIB Exercise-induced bronchoconstriction
eIF-2 Subunit of protein synthesis initiation factor
ELAM-1 Endothelial leukocyte adhesion molecule-1
ELF Respiratory epithelium lung fluid
ELISA Enzyme-linked immunoabsorbent assay
EMS Eosinophilia-myalgia syndrome
ENA Epithelial cell-derived neutrophil-activating protein
ENS Enteric nervous system
EO Eosinophil
EO-CFC Eosinophil colony-forming cell
EOR Early onset reaction *also known as* EAR
EPA Eicosapentaenoic acid
EpDIF Epithelial-derived inhibitory factor *also known as* epithelium-derived relaxant factor
EPO Eosinophil peroxidase
EPO Erythropoietin
EPOR Erythropoietin receptor
EPR Effector cell protease
EPX Eosinophil protein X
ER Endoplasmic reticulum
ERCP Endoscopic retrograde cholangiopancreatography
E-selectin Endothelial selectin *formerly known as* endothelial leukocyte adhesion molecule-1 (ELAM-1)
ESP Eosinophil stimulation promoter
ESR Erythrocyte sedimentation rate
e.s.r. Electron spin resonance
ET, ET-1 Endothelin, -1
ETYA Eicosatetraynoic acid

FA Fatty acid
FAB Fast-electron bombardment

Fab Antigen binding fragment
F(ab')$_2$ Fragment of an immunoglobulin produced pepsin treatment
FACS Fluorescence activated cell sorter
factor B Serine protease in the C3 converting enzyme of the alternative pathway
factor D Serine protease which cleaves factor B
factor H Plasma protein which acts as a co-factor to factor I
factor I Hydrolyses C3 converting enzymes with the help of factor H
FAD Flavine adenine dinucleotide
FapyAde 5-formamido-4,6-diamino-pyrimidine
FapyGua 2,6-diamino-4-hydroxy-5-formamidopyrimidine
FBR Fluorescence photobleaching recovery
Fc Crystallizable fraction of immunoglobulin molecule
Fcγ Receptor for Fc portion of IgG
FcγRI Ig Fc receptor I *also known as* CD64
FcγRII Ig Fc receptor II *also known as* CD32
FcγRIII Ig Fc receptor III *also known as* CD16
Fc$_ε$RI High affinity receptor for IgE
Fc$_ε$RII Low affinity receptor for IgE
FcR Receptor for Fc region of antibody
FCS Foetal calf (bovine) serum
FEV$_1$ Forced expiratory volume in 1 second
Fe-TPAA Fe(III)-tris[N-(2-pyridylmethyl)-2-aminoethyl]amine
Fe-TPEN Fe (II)-tetrakis-N,N,N',N'-(2-pyridyl methyl-2-aminoethyl)amine
FFA Free fatty acids
FGF Fibroblast growth factor
FID Flame ionization detector
FITC Fluorescein isothiocyanate
FKBP FK506-binding protein
FLAP 5-lipoxygenase-activating protein
FMLP N-Formyl-methionyl-leucyl-phenylalanine
FNLP Formyl-norleucyl-leucyl-phenylalanine
FOC Follicular dendritic cell
FPLC Fast protein liquid chromatography
FPR Formyl peptide receptor
FS cell Folliculo-stellate cell
FSG Focal sequential glomerulosclerosis
FSH Follicle stimulating hormone
FX Ferrioxamine
5-FU 5-fluorouracil

Ga G-protein
G6PD Glucose 6-phosphate dehydrogenase
GABA γ-Aminobutyric acid
GAG Glycosaminoglycan
GALT Gut-associated lymphoid tissue
GAP GTPase-activating protein
GBM Glomerular basement membrane
GC Guanylate cyclase
GC-MS Gas chromatography mass spectroscopy
G-CSF Granulocyte colony-stimulating factor
GDP Guanosine 5'-diphosphate
GEC Glomerular epithelial cell
GEMSA guanidinoethylmercapto-succinic acid
GF-1 An insulin-like growth factor
GFR Glomerular filtration rate
GH Growth hormone
GH-RF Growth hormone-releasing factor
Gi Family of pertussis toxin sensitive G-proteins
GI Gastrointestinal
GIP Granulocyte inhibitory protein
GlyCam-1 Glycosylation-dependent cell adhesion molecule-1
GMC Gastric mast cell
GM-CFC Granulocyte-macrophage colony-forming cell
GM-CSF Granulocyte-macrophage colony-stimulating factor
GMP Guanosine monophosphate (guanosine 5'-phosphate)
Go Family of pertussis toxin sensitive G-proteins
GP Glycoprotein
gp45–70 Membrane co-factor protein
gp90^{MEL} 90 kDa glycoprotein recognized by monoclonal antibody MEL-14; *also known as* L-selectin
GPIIb-IIIa Glycoprotein IIb-IIIa *known to be* a platelet membrane antigen
GppCH$_2$P Guanyl-methylene diphosphanate *also known as* a stable GTP analog
GppNHp Guanylyl-imidiodiphosphate *also known as* a stable GTP analog
GR Glucocorticoid receptor
GRGDSP Glycine–arginine–glycine–aspartic acid–serine–proline
Gro Growth-related oncogene
GRP Gastrin-related peptide
Gs Stimulatory G protein
GSH Glutathione (reduced)
GSHPx Glutathione perioxidase
GSSG Glutathione (oxidized)
GT Glanzmann Thrombasthenia
GTP Guanosine triphosphate
GTP-γ-S Guanosine5'O-(3-thiotriphosphate)

GTPase Guanosine triphosphatase
GVHD Graft-versus-host-disease
GVHR Graft-versus-host-reaction

H Histamine
H_1, H_2, H_3 Histamine receptor types 1, 2 and 3
H_2O_2 *The chemical symbol for* hydrogen peroxide
HAE Hereditary angiodema
Hag Hemagglutinin
Hag-1, Hag-2 Cleaved hemagglutinin subunits-1, -2
H & E Hematoxylin and eosin
hIL Human interleukin
Hb Hemoglobin
HBBS Hank's balanced salt solution
HCA Hypertonic citrate
H-CAM Hyaluronic acid cell adhesion molecule
HDC Histidine decarboxylase
HDL High-density lipoprotein
HEL Hen egg white lysozyme
HEPE Hydroxyeicosapentanoic acid
HEPES *N*-2-Hydroxylethylpiperazine-*N'*-2-ethane sulphonic acid
HES Hypereosinophilic syndrome
HETE 5,8,9,11,12 and 15 Hydroxyeicosatetraenoic acid
5(S)HETE A stereo isomer of 5-HETE
HETrE Hydroxyeicosatrienoic acid
HEV High endothelial venule
HF Hageman factor
HFN Human fibronectin
HGF Hepatocyte growth factor
HHTrE 12(*S*)-Hydroxy-5,8,10-heptadecatrienoic acid
HIV Human immunodeficiency virus
HL60 Human promyelocytic leukemia cell line
HLA Human leukocyte antigen
HLA-DR2 Human histocompatability antigen class II
HMG CoA Hydroxylmethylglutaryl coenzyme A
HMT Histidine methyltransferase
HMVEC Human microvascular endothelial cell
HMW High molecular weight
HMWK Higher molecular weight kininogen
HNC Human neutrophil collagenase (MMP-8)
HNE Human neutrophil elastase
HNG Human neutrophil gelatinase (MMP-9)
HODE Hydroxyoctadecanoic acid
HO· Hydroxyl radical
HO_2· Perhydroxyl radical
HPETE, 5-HPETE & 15-HPETE 5 and 15 Hydroperoxyeicosatetraenoic acid
HPETrE Hydroperoxytrienoic acid

HPODE Hydroperoxyoctadecanoic acid
HPLC High-performance liquid chromatography
HPS Hantavirus pulmonary syndrome
HRA Histamine-releasing activity
HRAN Neutrophil-derived histamine-releasing activity
HRf Homologous-restriction factor
HRF Histamine-releasing factor
HRP Horseradish peroxidase
HSA Human serum albumin
HSP Heat-shock protein
HS-PG Heparan sulfate proteoglycan
HSV, HSV-1 Herpes simplex virus, -1
5-HT 5-Hydroxytryptamine *also known as* Serotonin
HTLV-1 Human T-cell leukaemia virus-1
HUVEC Human umbilical vein endothelial cell
[Hyp^3]-BK Hydroxyproline derivative of bradykinin
[Hyp^4]-KD Hydroxyproline derivative of kallidin

^{111}In Indium^{111}
i.a. intra-arterial
Ia immune reaction-associated antigen
Ia+ Murine class II major histocompatibility complex antigen
IAP Intracisternal A particle
IB4 Anti-CD18 monoclonal antibody
IBD Inflammatory bowel disease
IBMX 3-isobutyl-1-methylxanthine
IBS Inflammatory bowel syndrome
iC3 Inactivated C3
iC4 Inactivated C4
IC_{50} Concentration producing 50% inhibition
I_{Ca} Calcium current
ICAM Intercellular adhesion molecules
ICAM-1, ICAM-2, ICAM-3 Intercellular adhesion molecules-1, -2, -3
cICAM-1 Circulating form of ICAM-1
ICE IL-1β-converting enzyme
i.d. Intradermal
ID_{50} Dose of drug required to inhibit response by 50%
IDC Interdigitating cell
IDD Insulin-dependent (type 1) diabetes
IEL Intraepithelial leukocyte
IELym Intraepithelial lymphocytes
IFA Incomplete Freund's adjuvant
IFN Interferon
IFNα, IFNβ, IFNγ Interferons α, β, γ
Ig Immunoglobulin

IgA, IgE, IgG, IgM Immunoglobulins A, E, G, M
IgG1 Immunoglobulin G class 1
IgG2a Immunoglobulin G class 2a
IGF-1 Insulin-like growth factor
Ig-SF Immunoglobulin supergene family
IGSS Immuno-gold silver stain
IHC Immunohistochemistry
IHES Idiopathic hypereosinophilic syndrome
IκB NFκB inhibitor protein
IL Interleukin
IL-1, IL-2 . . . IL-8 Interleukins-1, 2 . . . -8
IL-1α, IL-1β Interleukin-1α, -1β
ILR Interleukin receptor
IL-1R, IL-2R; IL-3R-IL-6R Interleukin 1–6 receptors
IL-1Ra Interleukin-1 receptor antagonist
IL-2Rβ Interleukin-2 receptor β
IMF Integrin modulating factor
IMMC Intestinal mucosal mast cell
iNOS inducible NOS
i.p. Intraperitoneally
IP_1 Inositol monophosphate
IP_2 Inositol biphosphate
IP_3 Inositol 1,4,5-trisphosphate
IP_4 Inositol tetrakisphosphate
IPF Idiopathic pulmonary fibrosis
IPO Intestinal peroxidase
IpOCOCq Isopropylidene OCOCq
I/R Ischaemia-reperfusion
IRAP IL-1 receptor antagonist protein
IRF-1 Interferon regulatory factor 1
I_{sc} Short-circuit current
ISCOM Immune-stimulating complexes
ISGF3 Interferon-stimulated gene Factor 3
ISGF3α, ISGFγ α, γ subunits of ISGF3
IT Immunotherapy
ITP Idiopathic thrombocytopenic purpura
i.v. Intravenous

K *The chemical symbol for* potassium
K_a Association constant
kb Kilobase
20KDHRF A homologous restriction factor; binds to C8
65KDHRF A homologous restriction factor, also known as C8 binding protein; interferes with cell membrane pore-formation by C5b-C8 complex
Kcat Catalytic constant; a measure of the catalytic potential of an enzyme
K_d dissociation constant
kDa Kilodalton
KD Kallidin
K_i Antagonist binding affinity
K*i*67 Nuclear membrane antigen

KLH Keyhole limpet hemocyanin
K_m Michaelis constant
KOS KOS strain of herpes simplex virus

λ_{max} Wavelength of maximum absorbance
LAD Leukocyte adhesion deficiency
LAF Lymphocyte-activating factor
LAK Lymphocyte-activated killer (cell)
LAM, LAM-1 Leukocyte adhesion molecule, -1
LAR Late-phase asthmatic reaction
L-**Arg** L-Arginine
LBP LPS binding protein
LC Langerhans cell
LCF Lymphocyte chemoattractant factor
LCMV Lymphocytic choriomeningitis virus
LCR Locus control region
LDH Lactate dehydrogenase
LDL Low-density lipoprotein
LDV Laser Doppler velocimetry
Le^x (Lewis X) Leukocyte ligand for selectin
LFA Leukocyte function-associated antigen
LFA-1 Leukocyte function-associated antigen-1; *also known to be* a member of the β-2 integrin family of cell adhesion molecules
LG β-Lactoglobulin
LGL Large granular lymphocyte
LH Luteinizing hormone
LHRH Luteinizing hormone-releasing hormone
LI Labelling index
LIF Leukemia inhibitory factor
LIFRE LIF response element
LIS Lateral intercellular spaces
LMP Low molecular mass polypeptide
LMW Low molecular weight
LMWK Low molecular weight kininogen
L-**NOARG** L-Nitroarginine
LO Lipoxygenase
5-LO, 12-LO, 15-LO 5-, 12-, 15-Lipoxygenases
LP(a) Lipoprotein (a)
LPS Lipopolysaccharide
L-selectin Leukocyte selectin, *formerly known as* monoclonal antibody that recognizes murine L-selectin (MEL-14 antigen), leukocyte cell adhesion molecule-1 (LeuCAM-1), lectin cell adhesion molecule-1 (LeCAM-1 or LecCAM-1), leukocyte adhesion molecule-1 (LAM-1)
LT Leukotriene
LT lymphotoxin
LTA$_4$, LTB$_4$, LTC$_4$, LTD$_4$, LTE$_4$ Leukotrienes A$_4$, B$_4$, C$_4$, D$_4$, and E$_4$

L$_y$-1$^+$ (Cell line)
LX Lipoxin
LXA$_4$, LXB$_4$, LXC$_4$, LXD$_4$, LXE$_4$ Lipoxins A$_4$, B$_4$, C$_4$, D$_4$ and E$_4$
Lys-BK Kallidin

M Monocyte
M3 Receptor Muscarinic receptor subtype 3
M-540 Merocyanine-540
mAb Monoclonal antibody
mAb IB4, mAb PB1.3, mAb R 3.1, mAb R 3.3, mAb 6.5, mAb 60.3 Monoclonal antibodies IB4, PB1.3, R 3.1, R 3.3, 6.5, 60.3
MABP Mean arterial blood pressure
MAC Membrane attack molecule
Mac Macrophage (also abbreviated to MΦ)
Mac- Macrophage-1 antigen; a member of the β-2 integrin family of cell adhesion molecules (also abbreviated to MΦ1), *also known as* monocyte antigen-1 (M-1), complement receptor-3 (CR3), CD11b/CD18
MAF Macrophage-activating factor
MAO Monoamine oxidase
MAP Mitogen-activated protein
MAP Monophasic action potential
MAPK Mitogen-activated protein kinase
MAPTAM An intracellular Ca^{2+} chelator
MARCKS Myristolated, alanine-rich C kinase substrate; specific protein kinase C substrate
MBP Major basic protein
MBSA Methylated bovine serum albumin
MC Mesangial cells
MCAO Middle cerebral artery occlusion
M cell Microfold or membranous cell of Peyer's patch epithelium
MCGF Mast cell growth factor
MCP Membrane co-factor protein
MCP-1, 2, 3 Monocyte chemotactic protein-1, 2, 3
M-CSF Monocyte/macrophage colony-stimulating factor
MC$_T$ Tryptase-containing mast cell
MC$_{TC}$ Tryptase- and chymase-containing mast cell
MDA Malondialdehyde
MDCK Madin Darby Canine kidney
MDGF Macrophage-derived growth factor
MDP Muramyl dipeptide
MEA Mast cell growth-enhancing activity
MEL Metabolic equivalent level
MEM Minimal essential medium
MG Myasthenia gravis
MGI-2 Macrophage granulocyte inducing factor-2
MGSA Melanocyte-growth-stimulatory activity

MGTA DL2-mercaptomethyl-3-guanidinoethylthio-propanoic acid
MHC Major histocompatibility complex
MI Myocardial ischaemia
MIF Migration inhibition factor
mIL Mouse interleukin
MIP-1α Macrophage inflammatory protein 1α
MI/R Myocardial ischaemia/reperfusion
MIRL Membrane inhibitor of reactive lysis
mix-CFC Colony-forming cell mix
Mk Megakaryocyte
MLC Mixed lymphocyte culture
MLymR Mixed lymphocyte reaction
MLR Mixed leukocyte reaction
mmLDL Minimally modified low-density lipoprotein
MMC Mucosal mast cell
MMCP Mouse mast cell protease
MMP, MMP1 Matrix metalloproteinase, -1
MNA 6-Methoxy-2-napthylacetic acid
MNC Mononuclear cells
MΦ Macrophage (also abbreviated to Mac)
MPG N-(2-mercaptopropionyl)-glycine
MPLV Myeloproliferative leukemia virus
MPO Myeloperoxidase
MPSS Methyl prednisolone
MPTP N-methyl-4-phenyl-1,2,3,6-tetrahydropyridine
MRI Magnetic resonance imaging
mRNA Messenger ribonucleic acid
MS Mass spectrometry
MSAP Mean systemic arterial pressure
MSS Methylprednisolone sodium succinate
MT Malignant tumor
MW Molecular weight

Na *The chemical symbol for* sodium
NA Noradrenaline *also known as* norepinephrine
NAAb Natural autoantibody
NAb Natural antibody
NAC N-acetylcysteine
NADH Reduced nicotinamide adenine dinucleotide
NADP Nicotinamide adenine diphosphate
NADPH Reduced nicotinamide adenine dinucleotide phosphate
NAF Neutrophil activating factor
L-**NAME** L-Nitroarginine methyl ester
NANC Non-adrenergic, non-cholinergic
NAP Neutrophil-activating peptide
NAPQI N-acetyl-p-benzoquinone imine

NAP-1, NAP-2 Neutrophil-activating peptides -1 and -2
NBT Nitro-blue tetrazolium
NC1 Non collagen 1
N-CAM Neural cell adhesion molecule
NCEH Neutral cholesteryl ester hydrolase
NCF Neutrophil chemotactic factor
NDGA Nordihydroguaretic acid
NDP Nucleoside diphosphate
Neca $5'$-(N-ethyl carboxamido)-adenosine
NED Nedocromil sodium
NEP Neutral endopeptidase (EC 3.4.24.11)
NF-AT Nuclear factor of activated T lymphocytes
NFκB Nuclear factor κB
NgCAM Neural-glial cell adhesion molecule
NGF Nerve growth factor
NGPS Normal guinea-pig serum
NIH 3T3 (fibroblasts) National Institute of Health 3T3-Swiss albino mouse fibroblast
NIMA Non-inherited maternal antigens
NIRS Near infrared spectroscopy
Nk Neurokinin
NK Natural killer
Nk-1, Nk-2, Nk-3 Neurokinin receptor subtypes 1, 2 and 3
NkA Neurokinin A
NkB Neurokinin B
NLS Nuclear location sequence
NMA N-monomethyl-L-arginine [IF ≠ L-NMMA]
NMDA N-methyl-D-aspartic acid
L-NMMA L-Nitromonomethyl arginine
NMR Nuclear magnetic resonance
NNA Nω-nitro-L-arginine
1,N^2-NET β-(2-Naphthyl-1,N^2-etheno
1,N^2-PET β-Phenyl-1,N^2-etheno
NO *The chemical symbol for* nitric oxide
NOD Non-obese diabetic
NOS Nitric oxide synthase
c-NOS Ca^{2+}-dependent constitutive form of NOS
i-NOS Inducible form of NOS
NPK Neuropeptide K
NPY Neuropeptide Y
NRS Normal rabbit serum
NSAID Non-steroidal anti-inflammatory drug
NSE Nerve-specific enolase
NT Neurotensin
N terminus Amino terminus of peptide

$^1\Delta O_2$ Singlet Oxygen (Delta form)
$^1\Sigma O_2$ Singlet Oxygen (Sigma form)
O_2^{-} *The chemical symbol for* the superoxide anion radical

OA Osteoarthritis
OAG Oleoyl acetyl glycerol
OD Optical density
ODC Ornithine decarboxylase
ODFR Oxygen-derived free radical
ODS Octadecylsilyl
OH$^-$ *The chemical symbol for* hydroxyl ion
·OH *The chemical symbol for* hydroxyl radical
8-OH-Ade 8-hydroxyadenine
6-OHDA 6-hydroxyguanine
8-OH-dG 8-hydroxydeoxyguanosine *also known as* 7,8-dihydro-8-oxo-2′-deoxyguanosine
8-OH-Gua 8-hydroxyguanine
OHNE Hydroxynonenal
4-OHNE 4-hydroxynonenal
OM Oncostatin M
OT Oxytocin
OVA Ovalbumin
ox-LDL Oxidized low-density lipoprotein
OZ Opsonized zymosan

Ψa Apical membrane potential
P Probability
P Phosphate
P_aO_2 Arterial oxygen pressure
P_i Inorganic phosphate
p150,95 A member of the β-2-integrin family of cell adhesion molecules; *also known as* CD11c
PA Phosphatidic acid
pA_2 Negative logarithm of the antagonist dissociation constant
PAEC Pulmonary artery endothelial cells
PAF Platelet-activating factor *also known as* APRL and AGEPC
PAGE Polyacrylamide gel electrophoresis
PAI Plasminogen activator inhibitor
PA-IgG Platelet associated immunoglobulin G
PAM Pulmonary alveolar macrophages
PAS Periodic acid-Schiff reagent
PBA Polyclonal B cell activators
PBC Primary biliary cirrhosis
PBL Peripheral blood lymphocytes
PBMC Peripheral blood mononuclear cells
PBN N-*tert*-butyl-α-phenylnitrone
PBP Platelet basic protein
PBS Phosphate-buffered saline
PC Phosphatidylcholine
PCA Passive cutaneous anaphylaxis
pCDM8 Eukaryotic expression vector
PCNA Proliferating cell nuclear antigen
PCR Polymerase chain reaction
PCT Porphyria cutanea tarda
p.d. Potential difference

PDBu 4α-phorbol 12,13-dibutyrate
PDE Phosphodiesterase
PDGF Platelet-derived growth factor
PDGFR Platelet-derived growth factor receptor
PE Phosphatidylethanolamine
PECAM-1 Platelet endothelial cell adhesion molecule-1; *also known as* CD31
PEG Polyethylene glycol
PET Positron emission tomography
PEt Phosphatidylethanolamine
PF_4 Platelet factor 4
PG Prostaglandin
PGAS Polyglandular autoimmune syndrome
PGD_2 Prostaglandin D_2
PGE_1, PGE_2, PGF_2, $PGF_{2\alpha}$, PGG_2, PGH_2 Prostaglandins E_1, E_2, F_2, $F_{2\alpha}$, G_2, H_2
PGF, PGH Prostaglandins F and H
PGI_2 Prostaglandin I_2 *also known as* prostacyclin
P_aO_2 Arterial oxygen pressure
PGP Protein gene-related peptide
Ph^1 Philadelphia (chromosome)
PHA Phytohemagglutinin
PHD PHD[8(1-hydroxy-3-oxo-propyl)-9,12-dihydroxy-5,10 heptadecadienic acid]
PHI Peptide histidine isoleucine
PHM Peptide histidine methionine
P_i Inorganic phosphate
pI Isoelectric point
PI Phosphatidylinositol
PI-3,4-P2 Phosphatidylinositol 3, 4-biphosphate
PI-3,4,5-P3 Phosphatidylinositol 3, 4, 5-trisphosphate
PI-3-kinase Phosphatidylinositol-3-kinase
PI-4-kinase Phosphatidylinositol-4-kinase
PI-3-P Phosphatidylinositol-3-phosphate
PI-4-P Phosphatidylinositol-4-phosphate
PI-4,5-P2 Phosphatidylinositol 4,5-biphosphate
PIP Phosphatidylinositol monophosphate
PIP_2 Phosphatidylinositol biphosphate
PIPES piperazine-N, N'-bis(2-ethanesulfonic acid)
PK Protein kinase
PKA, PKC, PKG Protein kinases A, C and G
PL Phospholipase
PLA, PLA_2, PLC, PLD Phospholipases A, A_2, C and D
PLN Peripheral lymph node
PLNHEV Peripheral lymph node HEV
PLP Proteolipid protein
PLT Primed lymphocyte typing

PMA Phorbol myristate acetate
PMC Peritoneal mast cell
PMN Polymorphonuclear neutrophil
PMSF Phenylmethylsulphonyl fluoride
PNAd Peripheral lymph node vascular addressin
PNH Paroxysmal nocturnal hemoglobinuria
PNU Protein nitrogen unit
p.o. *Per os* (by mouth)
POBN α-4-pyridyl-oxide-*N*-*t*-butyl nitrone
PPD Purified protein derivative
PPME Polymeric polysaccharide rich in mannose-6-phosphate moieties
PQ Phenylquinone
PRA Percentage reactive activity
PRD, PRDII Positive regulatory domain, -II
PR3 Proteinase-3
PRBC Parasitized red blood cell
proET-1 Proendothelin-1
PRL Prolactin
PRP Platelet-rich plasma
PS Phosphatidylserine
P-selectin Platelet selectin *formerly known as* platelet activation-dependent granule external membrane protein (PADGEM), granule membrane protein of MW 140 kD (GMP-140)
PT Pertussis toxin
PTCA Percutaneous transluminal coronary angioplasty
PTCR Percutaneous transluminal coronary recanalization
Pte-H$_4$ Tetrahydropteridine
PTH Parathyroid hormone
PTT Partial thromboplastin times
PUFA Polyunsaturated fatty acid
PUMP-1 Punctuated metalloproteinase *also known as* matrilysin
PWM Pokeweed mitogen
Pyran Divinylether maleic acid

q.i.d. *Quater in die* (four times a day)
QRS Segment of electrocardiogram

·R Free radical
R15.7 Anti-CD18 monoclonal antibody
RA Rheumatoid athritis
RANTES A member of the IL8 supergene family (*R*egulated on *a*ctivation, *n*ormal *T* *e*xpressed and *s*ecreted)
RAST Radioallergosorbent test
R$_{aw}$ Airways resistance
RBC Red blood cell
RBF Renal blood flow
RBL Rat basophilic leukemia
RC Respiratory chain
RE RE strain of herpes simplex virus type 1

REA Reactive arthritis
REM Relative electrophoretic mobility
RER Rough endoplasmic reticulum
RF Rheumatoid factor
RFL-6 Rat foetal lung-6
RFLP Restriction fragment length polymorphism
RGD Arginine–glycine–asparagine
rh- Recombinant human – (prefix usually referring to peptides)
RIA Radioimmunoassay
RMCP, RMCPII Rat mast cell protease, -II
RNA Ribonucleic acid
RNase Ribonuclease
RNHCl *N*-Chloramine
RNL Regional lymph nodes
ROM Reactive oxygen metabolite
RO· *The chemical symbol for* alkoxyl radical
ROO· *The chemical symbol for* peroxy radical
ROP Retinopathy of prematurity
ROS Reactive oxygen species
R-PIA (*R*)-(1-methyl-1-phenyltheyl)-adenosine
RPMI 1640 Roswell Park Memorial Institute 1640 medium
RS Reiter's syndrome
RSV Rous sarcoma virus
RTE Rabbit tubular epithelium
RTE-a-5 Rat tubular epithelium antigen a-5
r-tPA Recombinant tissue-type plasminogen activator
RT-PCR Reverse transcriptase/polymerase chain reaction
RW Ragweed

S Svedberg (unit of sedimentation density)
SAC *Staphylococcus aureus* Cowan I
SALT Skin-associated lymphoid tissue
SaR Sacroplasmic reticulum
SAZ Sulphasalazine
SC Secretory component
SCF Stem cell factor
SCFA Short-chain fatty acid
SCG Sodium cromoglycate *also known as* DSCG
SCID Severe combined immunodeficiency syndrome
sCR1 Soluble type-1 complement receptors
SCW Streptococcal cell wall
SD Stranded deviation
SDS Sodium dodecyl sulphate
SDS-PAGE Sodium dodecyl sulphate-polyacrylamide gel electrophoresis
SEM Standard error of the mean
SERPIN Serine protease inhibitor
SF Steel factor
SGAW Specific airway conductance
SHR Spontaneously hypertensive rat

SIM Selected ion monitoring
SIN-1 3-Morpholinosydnonimine
SIRS Systemic inflammatory response syndrome
SIV Simian immunodeficiency virus
SK Streptokinase
SLE Systemic lupus erythematosus
SLe^x Sialyl Lewis X antigen
SLO Streptolysin-O
SLPI Secretory leukocyte protease inhibitor
SM Sphingomyelin
SNAP *S*-Nitroso-*N*-acetylpenicillamine
SNP Sodium nitroprusside
SOD Superoxide dismutase
SOM Somatostatin *also known as* somatotrophin release-inhibiting factor
SOZ Serum-opsonized zymosan
SP Sulphapyridine
SR Systemic reaction
sr Sarcoplasmic reticulum
sR$_{aw}$ Specific airways resistance
SRBC Sheep red blood cells
SRS Slow-reacting substance
SRS-A Slow-reacting substance of anaphylaxis
STAT Signal transducer and activator of transcription
STZ Streptozotocin
Sub P Substance P

T Thymus-derived
α-TOC α-Tocopherol
$t_{1/2}$ Half-life
T84 Human intestinal epithelial cell line
TAME Tosyl-L-arginine methyl ester
TauNHCl Taurine monochloramine
TBA Thiobarbituric acid
TBAR Thiobarbituric acid-reactive product
TBM Tubular basement membrane
TBN di-*tert*-Butyl nitroxide
tBOOH *tert*-Butylhydroperoxide
TCA Trichloroacetic acid
T cell Thymus-derived lymphocyte
TCR T cell receptor α/β or γ/δ heterodimeric forms
TDI Toluene diisocyanate
TEC Tubular epithelial cell
TF Tissue factor
Tg Thyroglobulin
β-TG β-Thromboglobulin
TGF Transforming growth factor
[^{3}]TdR Tritiated thymidine
TFG-α, TGF-β, TGF-β_1 Transforming growth factors α, β, and β_1
T$_H$ T helper cell
T$_H$o T helper o
T$_H$p T helper precursor
T$_H$0, T$_H$1, T$_H$2 Subsets of helper T cells

THP-1 Human monocytic leukemia
Thy 1 + Murine T cell antigen
t.i.d. *Ter in die* (three times a day)
TIL Tumor-infiltrating lymphocytes
TIMP Tissue inhibitors of metalloproteinase
TIMP-1, TIMP-2 Tissue inhibitors of metalloproteinases 1 and 2
Tla Thymus leukemia antigen
TLC Thin-layer chromatography
TLCK Tosyl-lysyl-CH_2Cl
TLP Tumor-like proliferation
Tm T memory
TNF, TNF-α Tumor necrosis factor, -α
tPA Tissue-type plasminogen activator
TPA 12-*O*-tetradecanoylphorbol-13-acetate
TPCK Tosyl-phenyl-CH_2Cl
TPK Tyrosine protein kinases
TPO Thrombopoietin
TPP Transpulmonary pressure
TRAP Thrombospondin related anomalous protein
Tris Tris (hydroxymethyl)-aminomethane
TSH Thyroid-stimulating hormone
TSP Thrombospondin
TTX Tetrodotoxin
TX Thromboxane
TXA_2, TXB_2 Thromboxane A_2, B_2
Tyk2 Tyrosine kinase

U937 (cells) Histiocytic lymphoma, human
UC Ulcerative colitis
UCR Upstream conserved region
UDP Uridine diphosphate
UPA Urokinase-type plasminogen activator
UTP Uridine triphosphate
UV Ultraviolet
UVA Ultraviolet A
UVB Ultraviolet B
UVR Ultraviolet irradiation
UW University of Wisconsin (preserving solution)

VAP Viral attachment protein
VC Veiled cells
VCAM, VCAM-1 Vascular cell adhesion molecule, -1, *also known as* inducible cell adhesion molecule MW 110 kD (INCAM-110)
VF Ventricular fibrillation
V/GSH Vanadate/glutathione complex
VIP Vasoactive intestinal peptide
VLA Very late activation antigen beta chain; *also known as* CD29
VLA α2 Very late activation antigen alpha 2 chain; *also known as* CD49b
VLA α4 Very late activation antigen alpha 4 chain; *also known as* CD49d
VLA α6 Very late activation antigen alpha 6 chain; *also known as* CD49f

VLDL Very low-density lipoprotein
V_{max} Maximal velocity
V_{min} Minimal velocity
VN Vitronectrin
VO_4^- *The chemical symbol for* vanadate
vp Viral protein
VP Vasopressin
VPB Ventricular premature beat
VT Ventricular tachycardia
vWF von Willebrand factor

W Murine dominant white spotting mutation
WBC White blood cell
WGA Wheat germ agglutinin
WI Warm ischaemia

XD Xanthine dehydrogenase
XO Xanthine oxidase

Y1/82A A monoclonal antibody detecting a cytoplasmic antigen in human macrophages

ZA Zonulae adherens
ZAP Zymosan-activated plasma
ZAS Zymosan-activated serum
zLYCK Carboxybenzyl-Leu-Tyr-CH_2,Cl
ZO Zonulae occludentes

Key to Illustrations

 Helper
lymphocyte

 Suppressor
lymphocyte

 Killer
lymphocyte

 Plasma cell

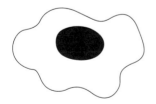 Bacterial or
Tumour cell

 Blood vessel
lumen

 Eosinophil
passing through
vessel wall

 Neutrophil
passing through
vessel wall

Resting
neutrophil

Activated
neutrophil

Resting
eosinophil

Activated
eosinophil

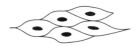

Smooth
muscle

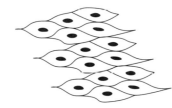

Smooth muscle
thickening

Smooth muscle
contraction

Normal blood
vessel

Endothelial cell
permeability

Resting
macrophage

Activated
macrophage

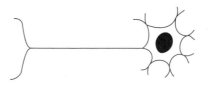

Nerve

Intact epithelium

Damaged epithelium

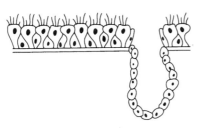

Intact epithelium with submucosal gland

Normal submucosal gland

Hypersecreting submucosal gland

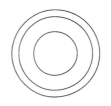

Normal airway

Oedema

Bronchospasm

Resting platelet

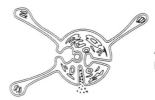

Activated platelet

Airway hypersecreting mucus

Resting
basophil

Activated
basophil

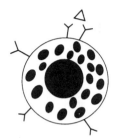

Resting
mast cell

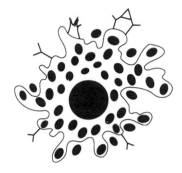

Activated
mast cell

Resting
chondrocyte

Activated
chondrocyte

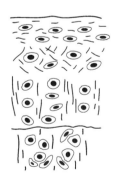

Cartilage

Fibroblast

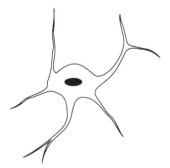

Dendritic cell/
Langerhans cell

Arteriole

Venule

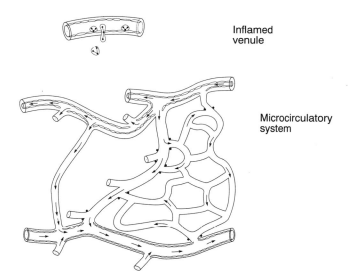

Inflamed
venule

Microcirculatory
system

Index